3rd International Conference on Industrial Engineering (ICIE-2017)

SHS Web of Conferences Volume 35 (2017)

St. Petersburg, Russia
16 - 19 May 2017

Editors:

Andrey A. Radionov
Andrey V. Shmidt
Igor A. Bayev
Tatyana A. Khudyakova
Alevtina V. Keller
Yevgeniy B. Kolbachev
Aleksandr V. Babkin
Viktor V. Savaley

ISBN: 978-1-5108-4329-5

Printed from e-media with permission by:

Curran Associates, Inc.
57 Morehouse Lane
Red Hook, NY 12571

Some format issues inherent in the e-media version may also appear in this print version.

Printed by Curran Associates, Inc. (2017)

For additional information, please contact EDP Sciences – Web of Conferences
at the address below.

EDP Sciences – Web of Conferences
17, Avenue du Hoggar
Parc d'Activité de Courtabœuf
BP 112
F-91944 Les Ulis Cedex A
France

Phone: +33 (0) 1 69 18 75 75
Fax: +33 (0) 1 69 28 84 91

contact-edps@webofconferences.org

Additional copies of this publication are available from:

Curran Associates, Inc.
57 Morehouse Lane
Red Hook, NY 12571 USA
Phone: 845-758-0400
Fax: 845-758-2633
Email: curran@proceedings.com
Web: www.proceedings.com

TABLE OF CONTENTS

ISSN (electronic edition): 2261-2424

PREFACE

International Conference on Industrial Engineering took place on May 16-19, 2017 in Saint-Petersburg, Russian Federation. The conference was organized by 4 universities – Peter the Great Saint-Petersburg Polytechnic University, South Ural State University (national research university), Platov South-Russian State Polytechnic University and Far Eastern Federal University.

The conference was carried out under financial support of the South Ural State University (national research university), Peter the Great Saint-Petersburg Polytechnic University as well as at informational support of the Institute of Electrical and Electronics Engineers (South Ural IEEE Chapter).

The conference was really large-scaled and international. The international program committee has selected more than 1000 reports. The conferees represented 66 Russian cities and towns and 17 cities of 11 states such as Hungary, Egypt, India, Kazakhstan, China (including Taiwan), Iran, Kyrgyzstan, Poland, Tajikistan, Ukraine, Switzerland.

The conference participants submitted papers reflecting recent advances in the field of Industrial Engineering, in Russian and English. The conference was organized in 15 sections, including:

Part 1. Research and development of machines and mechanisms (Dynamics of machines and working processes; Friction, wear, and lubrication in machines; Design and manufacturing engineering of industrial facilities; Surface transport and technological machines);

Part 2. Materials engineering and technologies for production and processing (New functional materials and technologies; Innovation and cost-effective use of resources of metallurgy industry);

Part 3. Control and automation systems for manufacturing in the industrial production areas (Control systems and their industrial application; Industrial mechatronics, automation and robotics; Electric power systems and renewable energy sources; Power electronics, electrical machines and drives; Signal processing and real time embedded control; Modeling and computer technologies; Theory and applications of dynamical measurements);

Part 4. Sustainable development of industrial enterprises;

Part 5. Industrial and civil construction.

The international program committee has selected totally 156 papers for publishing in the "SHS Web of Conferences" journal.

The organizing committee would like to express our sincere appreciation to everybody who has contributed to the conference. Heartfelt thanks are due to authors, reviewers, participants and to all the team of organizers for their support and enthusiasm which granted success to the conference.

Conference Chair, prof. Andrey A. Radionov.

edp sciences

SHS Web of Conferences

Statement of Peer review

In submitting conference proceedings to SHS Web of Conferences, I certify to the Publisher that I adhere to the **Policy on Publishing Integrity** of the journal in order to safeguard good scientific practice in publishing.

1. All articles have been subjected to peer review administered by the proceedings editors.
2. Reviews have been conducted by expert referees, who have been requested to provide unbiased and constructive comments aimed, whenever possible, at improving the work.
3. Proceedings editors have taken all reasonable steps to ensure the quality of the materials they publish and their decision to accept or reject a paper for publication has been based only on the merits of the work and the relevance to the journal.

Title, date and place of the conference

2017 3rd International Conference on Industrial Engineering
(ICIE-2017)
St. Petersburg, Russia
16 - 19 May 2017

Proceedings editor(s):

Editor-in-chief
Dr. Andrey A. Radionov

Editors
Dr. Andrey V. Shmidt
Dr. Igor A. Bayev
Ph.D. Tatyana A. Khudyakova
Dr. Alevtina V. Keller
Dr. Yevgeniy B. Kolbachev
Dr. Aleksandr V. Babkin
Dr. Viktor V. Savaley

Date and editor's signature

31 May 2017

Assessment of investment appeal of industrial enterprises

Mikhail Kuvshinov[1,*], *Anna* Kalacheva[1], and *Andrey* Butrin[1]

[1]South Ural State University (National Research University), the Department of Finance, Money Circulation and Credit, 454080 Chelyabinsk, Russia

Abstract. Expansion of the completeness of examination of the investment appeal of the enterprises considering all set of interests of the specific investor in an integral indicator, and providing sufficiency of reliability represents an actual task. On the basis of refining of the concept "investment appeal of the enterprise" the economic-mathematical model of an assessment of investment appeal of industrial enterprises providing possibility of step-by-step selection of options of forming of its investment image at decrease in the risks decision making corresponding to interests of the specific investor is developed. Methodical approach, the implementing its algorithm and the software product at the choice of options of forming of investment image or a choice of investees with the subsequent assessment of attractive options and receipts of the aggregate result which is most corresponding to preferences of the specific investor is developed. The practical recommendations about a choice of acceptable ranges of change of critical parameters of an integral indicator for the recommended assessment model allowing to prove decisions on the accepted options on management of investment appeal of the enterprises or the chosen investees are developed.

1 Introduction

The real investments in industrial enterprises is one of the defining factors of development of the enterprises, regions and national economy in general. At the present stage industrial enterprises create 24% of GDP of Russia. Their further development in an essential measure depends on the size of the attracted real investments. In the current conditions of economy of Russia limitation of financial resources is observed and each enterprise interested in attraction of external sources for financing of the development needs to possess the sufficient level of investment appeal in view of the high competition of objects of the real investments in the market of investment resources Therefore analytically defined assessment of the current level of investment appeal at risk minimization of adoption of the wrong decision is required to two parties of investment process. First, to management of the enterprise for reasonable determination of a vector of further development of the enterprise for the purpose of improvement of its investment image. Secondly, investors for making decision on a choice of the enterprise recipient need to estimate a condition of the applicant enterprises, perspectives of their development taking into account all concurrent factors, essential to them.

2 Relevance and scientific importance

Different approaches to an assessment of investment appeal of industrial enterprises, known at the moment in economic practice, or have restrictions by quantity of the considered factors and indicators characterizing them, or demand availability of big arrays of initial information which not always is available, or do not consider interests of specific investors concerning the importance of separate factors and indicators, or are very labour-consuming [1, 2, 3, 4, 5, 6, 7, 8, etc.].

In a similar situation there is a problem of expansion of completeness of examination of investment appeal of the enterprises on the basis of the complex assessment reflecting and considering all set of the corresponding preferences and interests of the specific investor in an integral indicator, and providing sufficiency of reliability of such examination.

As investment objects foreign and domestic economists at different times were engaged in research of questions of the theory of the investment analysis, attractiveness of the enterprises [2, 4, 7, 9, 10, 11, etc.]. At the same time, despite considerable achievements in the field, did not find due reflection and the questions connected with an integral assessment of the current level of investment appeal at a choice of potential options of development of the enterprise and forming of its attractive investment image demand further research.

* Corresponding author: msk1954@mail.ru

Similar questions arise and at investors recipients at a choice of potential objects of the real investments.

3 Problem definition

The created assessment of investment appeal of industrial enterprises in view of large volumes and complexity of accounting of interrelations of the used indicators has to exclude processing of redundant information and be carried out in full accordance with interests of the specific investor on the importance of separate factors or indicators. Therefore the methodical tools allowing on a limited set of indicators criteria, critical for the specific investor are required to choose to select from a large number of potential options of development of the enterprise and forming of its investment image or from potential investees significantly smaller group for which then to make more detailed analysis of investment appeal, excepting the options which are obviously unattractive for the investor. Increase of reliability of examination reduces risks at a variable choice of option of forming of investment image of the enterprise or a choice of an investee. The specified problem of a reasonable choice exists for the majority of the enterprises of real production sector, and its decision allows to create attractive investment image and to avoid probable essential economic losses for investors.

The analysis of different approaches to determination of the concept "investment appeal of the enterprise" showed absence of unity in the matter. Thus the existing approaches do not consider or a risk component of investment appeal, or influence of diverse factors, significant for the investor (internal and external, quantitative and qualitative), or only the economic result of investment is taken into account.

4 Theoretical part

Authors offer the specified determination of concept *investment appeal of the enterprise* is the complex characteristic of the enterprise forming under the influence of all factors, significant for potential external investors, internal and external environment and providing information on capability of the enterprise to provide as a result of the activity the indicators demanded on structure and size (economic, social, etc.) at acceptable risk level. This determination allows to take into account variety of internal and external factors of forming of the assessment performed always from a position of requirements of specific investors taking into account an effect ratio, including non-economic, and risk of investment.

The assessment of investment appeal of the enterprise has to consider in a complex all differentiation of requirements of investors to effect and risks of investment. For an assessment of required effect it is necessary to enter the quantitative criterion including an assessment of the economic, social and/or other effect essential to specific users of this information. Rather investment risk it is similarly necessary to define a complex quantitative assessment of the financial, material and/or other risks essential to specific users. Use of separate quantitative indices of effect and risk allows to consider information, most essential to any investor, at an assessment of investment appeal of the enterprise: about return of option of capital investments and about the level of neigh boring risks (for capital repayment, achievement of expected return). In model the result answering to preferences of the specific investor allows to have an opportunity of regulation of the importance of these indicators.

On the basis of the analysis of merits and demerits of known models of an assessment of investment appeal of the enterprises the author's model which represents set of the connected dependences allowing to estimate investment appeal on the basis of the integral $_{Y0\ indicator}$ created based on two group indicators is offered: investment potential of X_{IP} enterprise and factors of its investment risks of X_{IR} by means of linear factor model with weight coefficients of C_{1k}:

$$Y_0 = C_{1IP}X_{IP} + C_{1IR}X_{IR} \tag{1}$$

Investment potential is understood as ensuring the required effect answering the investor's purpose. Authors suggest an indicator of investment potential of the enterprise to create on the basis of the complex estimates of separate structural components of economic capacity of the enterprise and separate estimates reached effect and efficiency of activity for accounting of an enterprise scale. On the basis of the analysis and synthesis of the existing approaches to an assessment [1, 12, 13, 14, 15, 18, 20, 21] the choice of significant factors of economic potential and risks of activity of the enterprise, and also a method of their group in the offered model is made. Investment potential is recommended to be estimated by means of indicators of share, labor, managerial, financial, marketing, innovation potentials, effect and efficiency of activity of the enterprise. Investment risks - on the basis of indicators of the investment climate, financial, production commercial risks of the enterprise and its goodwill.

The model of an assessment is open - allows addition of indicators of other factors of investment appeal and the parameters essential to specific users characterizing them.

It is offered to calculate level 1 indicators in a formula (1) on the basis of linear factor models L of factors with weight coefficients C_{2k}:

$$X_1 = \sum_{k=1}^{L}(C_{2k}X_{2k}), \tag{2}$$

where as the defining factors of X_2 the separate listed components of investment potential of the enterprise and its investment risks representing level 2 considered models act. Indicators of X_2 are estimated by means of linear factor model on the basis of *the M* intermediate indicators of x_{2j} (level 2) characterizing them with weight coefficients of c_{2j}. Intermediate x_{2j} values also represent linear model on the basis of the given values of the corresponding private indicators of x_{3ji} (level 3) quantity of N_j with weight coefficients of c_{3ji}. As a result equality is carried out:

$$X_2 = \sum_{j=1}^{M}\left(c_{2j}x_{2j}\right) = \sum_{j=1}^{M}\left(c_{2j}\sum_{i=1}^{N_j}\left(c_{3ji}x_{3ji}\right)\right) \quad (3)$$

In all specified cases weight coefficients are offered to be determined by the rule of Fishburne [19] providing adoption of the best estimated decisions in conditions when it is not known of the value of N factors anything, except an order of decrease of their importance, and then the importance of i of a factor is defined:

$$r_i = \frac{2(N-i+1)}{(N+1)N} \quad (4)$$

Use of the coefficients reflecting expert evaluations of the importance of factors for the specific enterprise or the investor is in that specific case allowed. Lack of the fixed weight coefficients allows the end user to add independently to model indicators of level 2 and private indicators of level 3, essential to it (to delete insignificant).

Authors on the basis of the analysis [1, 6, 12, 16, 17, etc.] the set more than 300 private indicators allowing to carry out an assessment of investment appeal taking into account a set of various factors is selected and reasonable. All private indicators of model (level 3) are given to the comparable range of values [0; 1]. For quantitative indices reduction formulas are used (for the indicators having direct and retroactive effect on investment appeal of the enterprise, respectively):

$$x_{3i} = \frac{a_i - a_{i\min}}{a_{i\max} - a_{i\min}}, \quad x_{3i} = \frac{(a_{i\max} - a_i)}{a_{i\max} - a_{i\min}}, \quad (5)$$

where a_i - i-value of a private indicator for the considered enterprise; $a_{ima}x$ and a_{imin} - basic i-values of a private indicator. Thus for indicators like "risk" of a_i the return is defined as size, their probabilistic assessment.

As basic values a_{imax}, a_{imin} the maximum and minimum measure values among group of the compared enterprises, or the acceptable levels of an indicator established by the investor can be applied. The second option allows to estimate investment appeal of one enterprise out of group, in the ratio with the certain reference object having the characteristics, the best in the industry demanded by the investor, etc. Value (point) in the range [0 is appropriated to qualitative private indexes; 1] according to the criteria established individually for each similar indicator. As a result of value of all indicators of the model x_{3i}, x_{2i}, X_{2i}, X_{IP}, X_{IR}, Y_0 vary from zero (the worst state) to unit (the best state). Value of integral total of investment appeal Y forms using a correction multiplier of K considering quantity of the private indicators of x_{3i} used for the specific enterprise:

$$Y = KY_0 \quad (6)$$

The correction multiplier of K is calculated on the basis of the same model which will be applied to an assessment of an integral indicator $_{of\ Y0}$, with preserving of a set of the defining indicators and the chosen values of all weight coefficients. But at calculation of a correction multiplier instead of values of private indicators of x_{3i} it is substituted 1 if for the considered enterprise value of the corresponding indicator is known and it is used at an assessment; and 0 if it is not used, no. Then the correction multiplier accepts value from range [0; 1] where 1 means the most complete use of information - at an assessment of the enterprise all indicators provided by model are considered. Use of weight coefficients of model at calculation of K allows to consider the importance of absence of specific primary data at an assessment of degree of reliability of aggregate result. Value of integral total of Y allows to estimate investment appeal of the enterprise from 0 (attractiveness is absent) to 1 (maximum level).

The developed economic-mathematical model can be applied as to a choice of the most attractive options of development of the enterprise with forming of the corresponding investment image, and to a choice of potential investees. The organization of the step-by-step choice allowing on a limited set of indicators criteria, critical for the specific investor is for this purpose provided to select from a large number of options of development of the enterprise or according to the applicant enterprises them significantly smaller group for which then to make more detailed analysis of investment appeal. It allows to reduce volumes of the processed primary information, quantity labor-consuming and not always reasonable calculations.

Authors suggest to carry out primary selection of objects on compliance of values of the private indicators of model of an assessment, most important for the specific investor. It will allow to limit the list of objects of the subsequent analysis significantly. At the following stage to be made the main selection of objects as which criteria values of group indicators of X_{2i} of separate types of investment potential and investment risks of the enterprise act. To each stage there is an increase in volume of the considered information about reasonably the chosen indicators that reduces risk of adoption of the wrong decision by the investor. Thus it should be taken into account situations, first, when applicant object, having some indicators is slightly worse than admissible level, as a result can yield acceptable aggregate integral result. Secondly, when all objects have values of a certain indicator worse than admissible level, then at mutual interest it makes sense to recommend to the investor reduction of requirements. In connection therewith it is offered to create in selection process two lists with the different level of selection: the main, containing objects which all indicators have the values entering the admissible range, and the additional, containing objects which some indicators deviate admissible level for the worse no more than for 15%. Practice of the market shows that accounting of a similar deviation is considered acceptable at accomplishment of procedure of an assessment, but also size can be adjusted in interests of the investor.

The technique implementing author's methodical approach has the following idea.

Step 1. Forming of model of an assessment. The offered model of an assessment has no static structure. Before its use it is necessary to perform a choice of specific characteristics of the varied parameters. It is for this purpose carried out:

1) selection of essential group X_{2i}, x_{2ij} and private x_{3ij} of indicators of model and task of weight

coefficients of model. At the insignificant importance of factors equal coefficients are applied. If only the order of decrease of the importance of the factors characterizing one indicator is taken into account weight coefficients of Fishburne - a formula (4) are applied.

2) in the presence of the reasonable differentiated preferences concerning the importance of the separate factors reflecting interests of the specific enterprise or the investor they are considered in the corresponding weight coefficients;

3)choice of base of reduction of private quantitative indices: the maximum a_{imax} and/or the minimum a_{imin} of the basic values used in formulas 5.

4)a choice of criteria of an assessment of the private qualitative indexes providing their reduction to the comparable range of values [0; 1].

Step 2. Collection of information, the activity of the enterprises and the assessment reflected in model *characterizing internal and external factors* in the form of private indicators.

The quantity of the considered private indicators influences aggregate result through the correction multiplier used in a formula of calculations (6) and accepting values in the range [0; 1]. The higher it value, the is less uncertainty of a situation and risk of receipt of wrong result.

Step 3. Carrying out primary selection (in three stages). At the first stage the assessment of the acceptability of objects of the analysis on 3-4 most critical for the investor to private indicators criteria is performed. At the second stage the acceptability on 7-8 significant private indicators is in addition estimated. At the third stage - in addition 15-18 following on the importance of indicators. The specified quantity is recommended and can be changed according to interests of the investor. If at any stage criteria are not set, pass to the main selection (a step 4). For each stage of primary selection private indicators criteria which acceptability of values is critical for the investor get out. On the basis of the analysis of frequency of use of private indicators of author's model which is carried out in research in different evaluation methods of investment appeal of the enterprise the 10 most used indicators recommended as criteria of primary selection are allocated: coefficients of current liquidity and autonomy and indicator of return on equity for primary selection, and in addition indicators of return on assets, profitability of sales, profitability of the invested capital, absolute liquidity, receivables turnover ratio, a market share and investment appeal of the region for the following selection. This structure of indicators allows to characterize the enterprises for the main results of activity, but it can be and is changed according to requirements of the specific investor.

The choice of admissible levels for indicators criteria of primary selection assumes a task of the lower bound of admissible range of values a_i^{min} for indicators of direct action and the upper bound a_i^{max} for indicators of retroactive effect as which normative values, averages on the industry, or critical for the investor, etc. can be chosen.

Step 4. Carrying out the main selection at which each stage as eligibility criterion of objects one group indicator of the X_{2i} level 2 acts (a component of investment potential or factors of investment risks). The quantity of stages corresponds to quantity of indicators of X_{2i} in model. For each group indicator criterion of X_{2i} admissible level - the lower bound of admissible range of values is set [X_{2i}^{min}; 1] as which value, critical for the investor, gets out.

The sequence of carrying out stages of the main selection is defined as reduction of the importance of factors of X_{2i} for the specific investor, i.e. their weight coefficients in an integral indicator of the Y_0 model. At an assessment of a group indicator criterion of X_{2i} of the current stage of selection for the remained objects only private x_{3i} and intermediate x_{2i} the indicators relating to the current X_{2i} are used. It is for this purpose carried out:

1) determination of basic levels of a_{imax}, a_{imin} of private quantitative indices. In the absence of the acceptable levels set by the investor as basic levels the maximum and minimum measure values among the actual can be used. If use of the minimum basic level of a_{imin} is not provided in model, it is accepted equal to zero. If the actual value of a private indicator of direct action of a_i exceeds the maximum a_{imax} acceptable level established by the investor as the last a_i value is accepted. For indicators of retroactive effect if the actual a_i value is lower than the minimum acceptable a_{imin}, it becomes the minimum basic level. Thus basic levels of a_{imax} and a_{imin} i of an indicator cannot be equal, differently the denominator is equal in a formula (5) to zero;

2) reduction of the actual a_i values of private quantitative indices to a comparable form of x_{3i} for each object of an assessment on the basis of expressions (5). If the actual measure value of direct action is lower than a basic level of a_{imin}, or the actual measure value of retroactive effect is higher than a_{imax}, the given x_{3i} value is considered equal to zero;

3) calculation of a group indicator of X_{2i} of a stage (making potential or reliability) for each object of an assessment according to expression (3).

Step 5. Measure calculation of Y for the selected objects which remained in lists includes calculation of measure values of potential and risks (X_{IP}, X_{IR}) - a formula (2), a correction multiplier *(K)*, a final integral indicator of Y - a formula (1) and (6).

Step 6. Ranging of the selected objects as increase in an indicator of Y as a result of which of group of the remained enterprises the object with the maximum complex result of Y can be chosen.

Step 7. The decision concerning the acceptability of result of selection. As a result of accomplishment of all actions will be received or the ranged sequence of the objects meeting requirements of the investor then is carried out transition to a step 8, or the empty list if any of applicants does not conform to requirements, then reduction of requirements of the investor, and repeated carrying out procedure of selection, or an exception of consideration of all current objects and transition to a step 8 is possible.

Step 8. Decision-making. In the ranged sequence the object with the highest value of total of Y possesses the best characteristics and it should be recommended as the best object for investment from the considered group. In case of the empty list the conclusion about unacceptability of all applicants is drawn and the issue of selection of the new is resolved.

Steps 1 and 3 of the described technique can be executed by the group of experts number of 8-10 people, as a part of heads of divisions of the enterprise in which interests the assessment is made, and also the involved third-party experts - specialists in the field. Steps 4 … 6 can be executed by specialists of the corresponding qualification, for example, the staff of the enterprise in which interests the choice and an assessment, or by means of the software product developed by authors with ensuring input of basic data with the operator is made.

5 Practical importance and results of research

Unlike known methods of an integral assessment the offered author's methodical approach provides: 1) the organization of step-by-step two-level selection with forming of the main and additional lists at which each stage the list of objects for the subsequent analysis is reasonably limited and the number of labour-consuming or unreasonable calculations decreases; 2) the additional list with other level of criteria for selection of separate indicators allows to consider the selection options attractive by an integral assessment; 3) receipt of the aggregate result corresponding to the interests of the specific user finding reflection in sequence of the analysis of selection criteria, the list of conditions of compliance for each stage of selection, structure of indicators of model and their weight coefficients.

The integral result of an assessment of investment appeal of Y can be used by the investor as a sufficient indicator at a choice of the enterprise investment object or as the additional indicator at an assessment of attractiveness of the investment project characterizing reliability of the enterprise as bases of implementation of the considered project.

On the basis of the offered uniform methodical approach to a choice and an assessment of options by authors the algorithm [22] providing the organization of their step-by-step two-level selection with forming of the main and additional lists in full accordance with the given technique and calculation of a final integral indicator of their investment appeal is developed. Based on algorithm the software product [23] in the form of a web application which application provides automation of the provided calculations of economic-mathematical model, and also input and use only of necessary information on the enterprises that allows to process different options of basic data in acceptable time limits is created.

Practical approbation of the developed methodical approach to a choice and an assessment of industrial enterprises on an indicator of investment appeal is carried out on the example of group of companies of a type of economic activity "Production of radio - and the television transferring equipment" on the basis of the option of implementation of model of an assessment proved by authors, including a choice of weight coefficients and basic levels for reduction of private indicators to a comparable form. In case of availability at the specific investor of other preferences concerning the importance of factors and a choice of basic levels, they can be considered in model that will provide forming of the ranged list of the enterprises corresponding to interests of this investor.

The result of primary selection among group of the analyzed enterprises on the basis of the corresponding forms of the financial reporting for 2015 allowed to carry in the main list Joint-stock company "Yaroslavsky Radiozavod" (YaRZ) and Omskoye Open joint stock company "Radio plant of A.S. Popov" (ORZ), and in the additional list to carry the Public joint stock company "Radiozavod" (PRZ) and Joint-stock company "Ryazansky Radiozavod" (RRZ).

For comparison of options 3 types of investment policy for which weight coefficients of investment potential of X_{IP} and factors of investment risks of X_{IR} had ratios were defined: aggressive type as 0,6667:0,3333, conservative type as 0,3333:0,6667, moderate type as 0,5:0,5.

In the table results of a complex assessment of the enterprises which passed primary selection on the basis of data of their reporting for 2015 are provided. The assessment is executed for three cases: a priority of the importance of investment potential and aggressive type of the investor ($P > R$), a priority of the importance of investment reliability and conservative type of the investor ($P < R$), the equal importance of the specified factors and moderate type of the investor ($P = R$).

Table 1. Result of an assessment of investment appeal for group of companies.

Enterprise	**Performance level 1**		**Type of investor**		
	X_{IP}	X_{IR}	Aggressive ($P > R$)	Conservative ($P < R$)	Moderate ($P = R$)
YaRZ	0.254	0.409	Y_0= 0.305 K = 0.648 Y = 0.198	Y_0= 0.357 K = 0.693 Y = 0.247	Y_0= 0.331 K = 0.670 Y = 0.222
PRZ	0.410	0.231	Y_0= 0.350 K = 0.652 Y = 0.229	Y_0= 0.291 K = 0.691 Y = 0.201	Y_0= 0.321 K = 0.671 Y = 0.215
ORZ	0.245	0.235	Y_0= 0.242 K = 0.621 Y = 0.150	Y_0= 0.900 K = 0.675 Y = 0.161	Y_0= 0.240 K = 0.648 Y = 0.156
RRZ	0.161	0.217	Y_0= 0.179 K = 0.639 Y = 0.115	Y_0= 0.198 K = 0.684 Y = 0.136	Y_0= 0.189 K = 0.662 Y = 0.125

According to the maximum values of total of Y for the investor of aggressive type it is most attractive to investment of Public joint stock company "Radiozavod", to the investor of moderate or conservative type - Joint-stock company "Yaroslavsky radiozavod". Dispersion of values of a correction multiplier of K for all objects and options of an assessment less than 5% that testifies to use of approximately identical volume of information for each object. It is almost confirmed that the objects which are a part of the additional list of selection can yield high integral result of an assessment (Public joint stock company "Radiozavod").

The offered author's model of an assessment allows to reveal factors which parameter values are critical, and to select from them the most suitable for development of the necessary adjusting managerial actions for the specific enterprise. So the greatest lag of Public joint stock company "Radiozavod" on value of total of Y is observed in case of conservative type of the investor. In author's implementation of model of an assessment the most essential factor of X2i is the indicator of financial risks of the enterprise entering model XIR. Decrease in this indicator for Public joint stock company "Radiozavod" will provide decrease in investment risks of the enterprise and it will become the most attractive to investment in the considered group.

6 Conclusion

For the purpose of improvement of methodical tools of an assessment of investment appeal of industrial enterprises the economic-mathematical model and methodical approach to a choice of the industrial enterprises conforming to requirements of the investor, and their complex assessment for an integral indicator of investment appeal are offered and proved. It allows to organize the process of a step-by-step two-level choice of investment objects excluding processing of redundant information and which is carried out in full accordance with interests of the specific investor that provides the solution of an actual problem of increase of accuracy of an assessment and allows investors to avoid probable essential economic losses.

The work was supported by Act 211 Government of the Russian Federation, contract № 02.A03.21.0011.

References

1. V.I. Belotserkovsky, I.V. Streshinsky, and R.A. Rostislavov, News of the Tula state university. Economic and jurisprudence, **3-1**, 236-247 (2013)
2. A.V. Bolotin, *Positioning of industrial enterprises in the investment market* (Chelyabinsk, 2009)
3. O.I. Gavrilyuk, *Instrumenty and evaluation methods of investment appeal of industrial enterprises* (Belgorod, 2010)
4. V.A. Nikitina, *Score of investment appeal of the large Russian companies* (Moscow, 2005)
5. M.L. Polumiskov, *Development of economic-mathematical model and tools of an assessment of investment appeal of the gold mining enterprise* (Moscow, 2010)
6. R.A. Rostislavov, *Score of investment appeal of industrial enterprise for the strategic investor* (Tula, 2011)
7. D.S. Sizyck, *Model and methods of an express assessment of investment appeal of the enterprises* (Moscow, 2010)
8. E.N. Staroverova, Quality. Innovations. Education, **7**, 30-36 (2007)
9. R.A. Brealey, S.C. Myers, and F. Allen., *Principles of corporate finance* (McGraw-Hill/Irwin, New York, 2010)
10. B.D. Jordan, and T.W. Miller, *Fundamentals of investments: valuation and management* (McGraw-Hill/Irwin, New York, 2009)
11. U. Sharp, G. Alexander, and D. Bailey. *Investments* (Moscow, INFRA-M, 2001)
12. E.V. Bartova, *Impact of potential production on increase of productivity of activity of industrial enterprise* (Izhevsk, 2011)
13. G.E. Bazhenov, and O.A. Kislitsyna, The Bulletin of the Siberian state space university of the academician M.F. Reshetnev, **3**, 176-181 (2010).
14. I.A. Blank, *Financial risk management* (Nick Center, Kiev, 2005)
15. A.G. Butrin, Z.S. Gelmanova, and D.L. Yaruchin, Metallurgist, **59**, 99-103 (2015)
16. V.I. Denisenko, and A.P. Dyachenko, Risk analysis Problems **3**, 22-28 (2008)
17. E.M. Korotkov, Bulletin of the Southern Russian state technical university (Novocherkassk polytechnical institute). Social and Economic Sciences series **2**, 31-38 (2012)
18. A.A. Tafeeva, *Score of investment appeal of the enterprise in the conditions of the innovation development* (St. Petersburg, 2011)
19. P. Fishburne, *The theory of usefulness for decision making* (Science, Moscow, 1978)
20. A.S. Shapkin and V.A. Shapkin, *Economic and financial risks. Assessment, management, portfolio of investments* (Dashkov and To, Moscow, 2012)
21. A.A. Shaposhnikov, *Development of the complex mechanism of the analysis and assessment of investment appeal of the enterprise* (N. Novgorod, 2010)
22. M.S. Kuvshinov, A.G. Kalacheva, *Algorithm of selection of investment attractive enterprises: Algorithm* (South Ural State University (national research University), 50201650127, 2016)
23. M.S. Kuvshinov, A.G. Kalacheva, *Calculation of an integral indicator of investment appeal of the enterprises: Program* (South Ural State University (national research University), 50201650126, 2016)

Statistical analysis of the integration activity of industrial enterprises as a basis for development of Russian regions

Mariia Karelina[1,*], *Tatiana* Ivanova[1], and *Violetta* Trofimova[1]

[1]Nosov Magnitogorsk State Technical University, Magnitogorsk, Russia

Abstract. The article investigates the integration activity of industrial structures in the regions of Russia. The monitoring of the current legal system of the Russian Federation carried out in the area of statistics and compiling statistical databases on mergers and acquisitions showed the absence of formal executive authority dealing with the compiling collections of information on the integration activity at the regional level. In this connection, the data of Russian information and analytical agencies made the information research base. As research tools we used methods of analysis of structural changes, methods of analysis of economic differentiation and concentration, methods of non-parametric statistics. The article shows the close relationship between social and economic development of the subjects of Russia and integrated business structures functioning on its territory. An investigation of the integration activity structure and dynamics in the subjects of the Russian Federation based on statistical data revealed the increasing heterogeneity of the integration activity of business structures in the regions of Russia. The research results are of practical importance, since they can be used to improve existing federal programs aimed at evening-out disproportions in the social and economic development of Russian regions.

Introduction

Issues relating to the problems and prospects of economic integration in the contemporary world can be attributed to debatable issues. At the same time, the world practice shows that the large companies form the aggregate supply and demand, determine the most important prerequisites for winning in the competition [1]. The integration of production on global scales at the level of modern business structures is inherently a market phenomenon. It acts as a competitive strategy and at once is a manifestation of a self-organizing system in the market competitive environment. At the same time, as M. Lubatkin points out, there is a bilateral interaction. On the one hand, integration processes may affect the market formation; on the other hand, the integration is a reaction to the market [2].

It is effectually to start the study of the integration activity of business structures in the regions of Russia with the specification describing its fundamental categories. A further empirical analysis is impossible without this methodological step. The term "economic integration" appeared in the works of German and Swedish economists in the 1930s, but today the general integration theory has not yet been created. Theoretical schools, focusing attention on specific aspects of the integration process, give different definitions to the integration as an economic phenomenon. Therefore, traditionally there are different approaches to the interpretation of the key categories characterizing the integration process. According to one of the most capacious definitions that are presented in Great Dictionary of Economics of A. B. Borisov, integration (from Latin integer - whole) is the association of economic subjects, their cooperative extension, and the development of relationships between them. At the same time, economic integration takes place both at the level of national economies of the whole countries and between business structures.

Thus, according to authors V. L. Bersenov, S. G. Vazhenin, and A. I. Tatarkin, despite the multiplicity of integration types they result in two options, depending on the subject composition of participants:

territorial (interstate, interregional) integration initiated by central and local administrative authorities;

industrial integration initiated by enterprises and organizations - the subjects of business (economic) activity [3].

In this connection, economic integration in this article refers to industrial integration, which is realized by business structures through the mechanism for executing integration transactions.

According to the synergetic paradigm describing the development of complex systems, evolutionary processes are based on the property of these systems to regulate the internal structure by strengthening the relationship between the structural elements. The application of this attitude to the investigation of the industrial integration process allows us to define the concept of integration activity as economic activity of

* Corresponding author: marjyshka@mail.ru

business entities aimed at providing extension and enhancement of interaction, relationships, and cooperation towards the fuller use by economic entities of their competitive advantages and obtain a synergistic effect from the merger.

Historically, the investigation of specific features in the development of Russia's corporate integrated business structures, problems of social and economic development of Russian regions are carried out largely independently from each other; different scientists, experts and various research and analytical structures are engaged in it. Therefore, despite the fact that a number of research studies of the integration activity of economic entities at the level of the Russian economy have been recently published (I. G. Vladimirova, YU. V. Ignatishin, S. V. Gvardin, M. M. Musatova, N. B. Rudyk and others), the analysis of the integration activity of business structures in Russian regions largely disregards. The integration activity of business structures in Russian regions deserves study that is more intent. In addition, integration transactions at the level of regions are of particular interest to the analysis, since the largest vertically integrated structures are donors of the budget of roughly a quarter of all regions of Russia.

Integration transactions have a significant influence on the economic entity's internal corporate environment. This, in turn, begins to affect the external institutional environment, consisting of economic, social, and political constituents of a particular region where integration projects are implemented [4]. For this reason, business structures affect social and economic development of the region transforming the structure and system of the regional budget, as well as indicators of its strategic development, in general, through the mechanism of the implementation of integration transactions.

Due to the specific features of historical and economic formation of the Russian Federation, economic globalization leads to disproportions in the development of its territories resulting in the concentration of financial control levers and an excessive accumulation of capital assets in a limited number of regions and its lack in others [5]. In terms of territorial aspect, the problem of increased separation of regions appears and, consequently, a conflict of interests between "leading" and "lagging" subjects of the Russian Federation emerges.

Thus, the problem of investigating the differentiation of regional differences in social and economic indicators, including the integration activity of economic entities, becomes of special importance for Russia. Moreover, investigation in this area should be based on comparable statistics, since only in this case continuity research can be carried out.

1 Review of sources of statistical information relating to the integration activity of business structures

When monitoring existing Russian legislation in the field of statistics and compilation of statistical databases it should be noted that, there is currently no federal law that gives a single concept of statistics on mergers and acquisitions, order and principles of formation of statistical databases on the integration activity. The rules governing integration transactions are distributed in various normative legal acts of the Russian Federation.

The fact that there is no formal executive body that is entrusted with a comprehensive compilation of information on integration activities in Russia confirms also and the lack of a single state policy in the field of official statistical accounting for mergers and acquisitions. At the same time, the author thinks that the statistical information system of the integration activity of business structures in the regions of Russia generally seems to be a necessary element of the information system of the state. This provides the government, ministries, organizations, and the public with official data on transactions related to transferring ownership of assets and capital from one owner to another.

It is necessary to solve organizational tasks related to the collection and processing of data on mergers and acquisitions in the regional context in order to improve statistics in the field of the integration activity of business structures:

- the improvement of normative and legal acts of the Russian Federation governing the collection and processing of statistical data on integration transactions of economic entities;
- the enhancement of international cooperation within the framework of statistics on mergers and acquisitions due to greater integration of Russia into the process of data exchange on M&A transactions;
- the improvement of data sources by using surveys of head offices of integrated business structures as a source of information on the integration activity.

The implementation of the recommendations reviewed would significantly improve the quality of Russian statistics on integration activities of economic entities both in the regional and federal context due to full coverage of accurate real-time data, and actual level of detail.

In this situation, various information and analytical agencies are engaged in data collection on integration activities of economic entities in the Russian Federation. Information agencies take into account transactions in which the purchaser and (or) the company acquired are Russian, that is the companies doing main business in the regions of the Russian Federation [6,7]. They publish data that indicate the lower bound of cost and quantitative volume of integration transactions and in many cases they are contradictory since their collection are based on various guidelines; disclosure of share package value (shares) in the majority of transactions is not acceptable, so the use of experts may be necessary to estimate it.

The expert and analytical group M&A-Intelligence (agency Mergers and Acquisitions from the second half of 2013) and ReDeal Group are the most authoritative analytical agencies in Russia engaged in the monitoring of Russian market for corporate control.

Agency Mergers and Acquisitions (Russian M&A-agency) in addition to collection and collation of information on the integration activity of business structures provides in-depth research of the market for corporate control in Russia. The database is updating with daily frequency. Moreover, the journal of Mergers and Acquisitions agency is currently the only Russian analytical edition that is entirely devoted to mergers and acquisitions, business and capital centralization, corporate control; it highlights the problems of legal and illegal market of M&A, the problems associated with protection from corporate takeovers.

The data of analytical group ReDeal Group, established in 2004, may serve the information base for practical analysis of mergers and acquisitions. In 2011, there was the merger of ReDeal Group and Cbonds Information Agency. The database of analytical group ReDeal Group is updated daily and contains information on more than 10400 of integration transactions amounting to more than 620 billion dollars USA.

A database of integration activities of economic entities in the regions of the Russian Federation for the period from 2003 to 2014 was compiled based on statistical data of the analytical group M&A-Intelligence, and allows you to receive the following information:

- the date of settlement of the integration transaction;
- the object of merger and acquisition transaction;
- the amount of merger and acquisition transaction;
- sectoral affiliation of the integration transaction;
- territorial location of the object of the M&A transaction;
- the purchaser;
- territorial location of the purchaser in the M&A transaction.

2 The results of statistical investigations of the integration activity of business structures in the regional context

If we analyse the contribution of each federal district to a national amount of merger and acquisition transactions concluded in 2014 that the Central and Volga Federal District are leading and Southern and North Caucasus Federal District are outsiders. The main share in the number of integration transactions, concluded by the Central federal district comprising 18 subjects, belongs to Moscow (93.08% of the total M&A transactions in this district). Moscow is leading in the number of merger and acquisition transactions concluded and planned in financial, trade, metallurgy, construction, and transport sectors. A significant proportion of transactions is with foreign participation [8].

The maximum share in the number of M&A transactions concluded by the Northwest Federal District comprising 11 regions belongs to St. Petersburg (77.85%), Novgorod region (6.95%), and the Pskov region (6.78%). St. Petersburg is characterized by the integration activity primarily in the financial and food sectors. There is a little volume of large transactions concluded, small and medium businesses are the most popular among investors.

In order to study the mobility of structural changes of the integration activity of regions in the federal context we will analyse structural changes according to the terms "the number of merger and acquisition transactions concluded." An analysis of the relative linear structural change with a permanent comparison base shows that the structural changes in all federal districts of Russia for the period from 2003 to 2014 are characterized as large structural changes. The distribution of federal districts in 2014 according to the value of the linear structural change with a permanent comparison base (2003) is shown in Figure 1.

An analysis of Figure 1 reveals that the Southern and Far Eastern Federal Districts are characterized by the largest structural changes in the quantitative volume of the market for corporate control as in the case of cost volume of the regional M&A market.

This may be explained by the fact that in the federal districts there are many production assets both being stuck in bankruptcy and subject to sale and dynamic and attractive to investors.

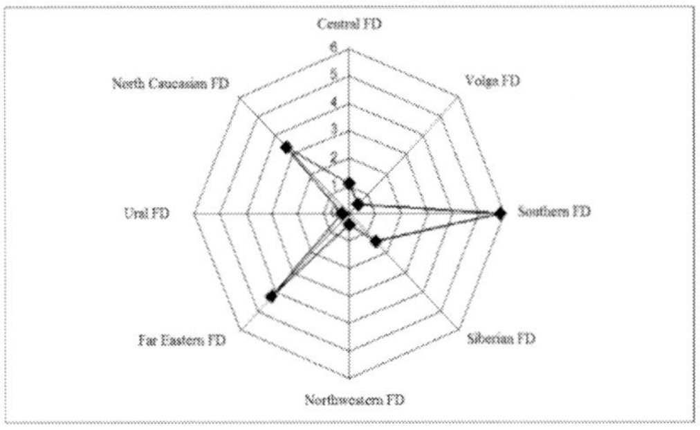

Fig. 1. Linear relative structural changes (basic) in quantitative volume of the M&A market in the context of federal districts.

The smallest structural changes in the quantitative volume of the merger and acquisition market describe the Ural and Northwest federal districts. The dynamic changes in the linear coefficient of absolute structural changes in the Ural Federal District for the period from 2003 to 2014 allowed us to reveal the sharp increase in this rate accounting for 2009, when 28 integration transactions were concluded of 3.955 billion dollars of the total sum. Chelyabinsk region made the main contribution concluded 13 transactions amounting 2837.70 million dollars, most of which falls on metallurgy industry.

The dynamic changes in the linear coefficient of absolute structural changes in the North-West Federal District for the period from 2003 to 2014, calculated by the number of merger and acquisition transactions, allowed us to draw conclusion that the sharp increase in this rate accounts for 2010. It was the year, when 51 integration transactions were concluded with a total value of 3.305 billion dollars. Vologda region and St. Petersburg made the main contribution. The transactions of metallurgical holding Severstal OJSC were the most significant in 2010 in Vologda region.

To avoid cancellation of different specific weight changes of individual elements by sign in the total [9,10] quadratic relative structural changes have been calculated in the structure of the number of M&A transactions and cost value of M&A market in the federal section in 2014 to the crisis of 2009. An analysis of the data shows that all of the federal districts of Russia reduced their integration activity because of the financial crisis. At the same time, the structure of the merger and acquisition market in terms of "cost volume of the M&A market" has undergone more significant changes than the structure of the market in terms of "the number of M&A transactions" in 6 of 8 federal districts.

Many foreign researchers noted the impact of macroeconomic factors on mergers and acquisitions. For example, Wu Changqi, Xie Ningling [11] argued that merger and acquisition transactions are dependent on the external environment that is on such factors as economic growth (crisis), the degree of competition, political and economic changes. Some researchers emphasize the importance of non-economic factors namely political, legal, and others [12].

Thus, in 2008-2009 in implementing large-scale one-off strategic mega transactions (worth over 1000 million dollars) there was a significant amount of might-have-been transactions costing less than 100 million dollars USA, which were agreed before the financial crisis:

- the purchase of a controlling interest of Kyivstar. Es. Em. CJSC by Vimpel-Communications OJSC for 9999 million dollars;
- the buy-out by AFC System OJSC of 8.3% of the shares of Bashkirnefteprodukt OJSC, 51.5% of the shares of Bashneft, 59.1% of the shares of Novoil OJSC, 43.4% of the shares of Ufaneftekhim OJSC, 52.97% of the shares of Ufaorgsintez and 48.15% of the shares of Ufimsky Refinery Plant OJSC for 2500 million dollars;
- the buy-out by the Deposit Insurance Agency of 30% of the shares of Rostelecom from the bank KIT Finance for 230 roubles per 1 share. The total amount (package consisted of 218,600,000 ordinary shares) DIA spent was 1580 million dollars;
- the purchase by VTB Bank OJSC of the territory of the Moscow Stud Farm for 2400 million dollars;
- the purchase by Rostelecom OJSC of 25% + 1 share of Svyazinvest OJSC and 100% of shares of Sky Link CJSC for 1560 million dollars.

During the crisis, Russian business structures have significantly reduced the scope of export expansion and focused on the domestic market in order to solve the problems of debts and restructuring of its assets. Residents sat on the fence, and determined if the gap between asking price expectations and sales price expectations reduced or eliminated, and, in connection with those, they were in no hurry to get access to the regional markets [13].

Less significant changes for the period from 2009 to 2014 in the structure of the M&A market took place in the North-West Federal District. Maximum decrease in terms of "cost volume of the M&A market" was observed in the Far East Federal District, and in terms of "the number of M&A transactions" was in the Siberian Federal District.

To confirm the hypothesis of a significant divergence of mergers and acquisitions of corporate structures in the Russian regions we consider dynamics of changes in the coefficients of the inequality of the integration activity, calculated in terms of "cost volume of merger and acquisition market" for quintile (20 per cent) groups of regions for 2003-2014. The fifth quintile group (leading regions) includes such subjects as Moscow, Moscow region, St. Petersburg, Yamalo-Nenets Autonomous District, Republic of Tatarstan, Chelyabinsk region, and others. In general, the data in Table 4 provide the vision of the concentration process of the integration activity of business structures in the subjects of the Russian Federation.

In the period from 2003 to 2014, the share of the fourth quintile group remained relatively constant. At the same time, there was a decrease in the share of the first, second and third quintile groups - from 7.39% to 5.76%; that is to say, it declined by 1.28 times for 10 years under study. The most significant decrease was in the share of the least integration activity of the first quintile group of regions - from 1.27% to 0.60%, that is it decreased by 2.12 times. At the same time, the share of the most active integration of the fifth quintile group increased by 1.03 times.

Along with the terms the " share of quintile group in the total value of cost value of the M&A market", it is advisable to use special factors, which comprehensively characterize the phenomenon under consideration, in order to analyse the concentration by regional groups. The index of concentration (Gini coefficient) is among these special parameters.

The calculation of the Gini coefficient is based on building the Lorenz curve that characterizes the cumulative characteristic under study depending on cumulative elements. The more the Lorenz curve deviates from a diagonal line, the greater is the degree of inequality in distribution of the integration activities together with analysable subjects of the Russian Federation. This degree of inequality in distribution is expressed as the ratio between the area of the figure, enclosed by the diagonal of the square and the Lorenz curve, to the area of the triangle equal to 0.5. This ratio is defined as the Gini coefficient (factor) [14].

According to the approach of V. A. Litvinov, the Gini coefficient, determining the deviation degree in actual distribution of the existing integration activity from the line of their possible equal distribution and comprehensively characterizing the integration process of activity concentration in groups of subjects of the Russian Federation only indirectly reflects the actual concentration in the narrow sense of the word, that is an intention to "pull" the entire integration activity of economic entities in one region [15].

With all that said, they suggested to use the Herfindahl index when analysing the concentration in the study of regional integration activity of business structures:

$$K=\sum_{i=1}^{5} d_i^2 \quad (1)$$

where d_i is the proportion of each group of regions in the total value of the M&A market.

This index varies from 0 to 1. In contrast to the Gini coefficient, the Herfindahl index is an indicator of "direct action" and is indifferent to the line of theoretically possible equal distribution [16]. In other words, the Herfindahl index takes the unequal distribution of the integration activity as an axiom, and its changes reflect the changes in the existing proportions between groups, that is, the ratio of the proportion of individual groups of regions in the total value of cost value of the merger and acquisition market. At a given (constant) number of groups, the increased Herfindahl index in the current period compared with that for the base period directly indicates the increased concentration; that is the increased concentration of the integration activities of business structures in any single group of regions.

In the analysed period from 2003 to 2014, the Gini coefficient showed high values, exhibiting unequal distribution of the integration activity of the subjects of the Russian Federation. The highest value of the Gini coefficient accounts for 2007 ($G_{2007}=0.92$), the lowest value accounts for 2011 ($G_{2011}=0.74$). In the period from 2003 to 2014 the Herfindahl index did not fall below a value equal to $K_{2005}=0.76$ that indicates high concentrated integration activity (in 2014, there has been an increased concentration of activity on the integration of 7.58% compared to 2013).

The Gini coefficient and the Herfindahl index when characterizing the concentrated integration activity of economic entities by the regions of the Russian Federation indirectly reflect the overall differentiation measure of M&A-activity in Russian regions. However, we should not confuse concentration as the centralization of something with a differentiation as the distinction (difference) of parts of the whole. We define differentiation, above all, as different varying characteristic (the volume of total value of the M&A regional market) as a weighted value in the outermost groups. Consequently, the Gini coefficient and the Herfindahl index cannot be directly used to characterize the different integration activity, and therefore, it is necessary to apply special indicators such as quantile differentiation coefficients when studying it.

The decile differentiation coefficient is the ratio of the value of the regional volume of merger and acquisition market, higher and lower of which 10%, respectively, of the most and least integration active Russian regions are located [17]. The resulting decile differentiation coefficient equal to the ratio of the 9th and 1st deciles made $D_{2014}=168$ in 2014, that is, the minimum cost volume of the market for corporate control of 10 % of the most integration active subjects exceeds the maximum cost volume of M&A market of the less integration active regions of the Russian Federation over 168 times.

Thus, the Russian economy is characterized by deformed spatial structure of the integration activity of business structures in the regions. Both state and regional development policy in the field of the integration activity does not have enough influence to even out imbalances in the distribution of the ways for implementing integration projects. All this shows the need of activation policy in the field of the integration activity of the Russian business structures in the regions of the Russian Federation. One of the tools to achieve this goal is the adoption of approved documents at the regional level, which are relevant to international best practices (in particular, the Code of Good Practices on Fiscal Transparency, the Code of Good Practices on Transparency in Financial Policy, Regional Resources Management Guidelines, and others).

Business structures are an inherent part of economic development in modern conditions. They work in virtually all sectors of the economy and implement their projects in a significant number of regions [18]. At the same time, interregional differentiation manifests itself in the distribution of the number of enterprises and organizations (according to the state registration), which directly affect the integration activity.

The Central Federal District (40.76% of the total Russian enterprises and organizations) is the main leader, where Moscow (64.35%) and the Moscow Region (11.84%) make the great contribution. In 2014, they accounted 76.19% of all enterprises and organizations in the region. This is followed by the Volga and North-West Federal District (15.08% and 12.32% of the total Russian number, respectively), where the Republic of Tatarstan (14.66%), Samara region (13.95%) and St. Petersburg (57.9%) made the largest contribution.

Consolidated budget revenues consist of corporate profit tax, personal income taxes, property tax, and grant revenues. Tax payments of Russian holdings contribute significantly to the formation of the main economic indicators of the national economy and the individual territorial units. Let us analyse the statistical relationship between the amount of mergers and acquisitions of Russian business structures and tax contribution by the Federal Districts based on nonparametric statistics.

The use of the most appropriate rank correlation methods, in particular Spearman's rank correlation coefficient, is effectual in studying the relationship of quantitative characteristics by nonparametric statistic methods [19, 20]:

$$\rho = 1-\frac{6\sum_{i=1}^{n}(d_w-d_c)^2}{n^3-n} \quad (2)$$

where n is a number of observations; dw is the Federal District rank abreast of the amount of M&A transactions by the Federal Districts; dc is the Federal District rank among the tax contribution by the Federal Districts.

The power of Spearman's rank correlation coefficient is inferior to the power of parametric correlation coefficient; however, taking into account the relatively small quantity of observations, this method gives results that are more accurate. This coefficient varies from -1 to 1. The closer its absolute value to one, the closer is relationship between characteristics. According to available data for 2014, Spearman's rank correlation coefficient is equal to 0.7823 that indicates the presence of a well-established and direct relationship between the amount M&A transactions and tax contribution (a significant p-value <0,05).

Thus, we can conclude that the integration activity of the Russian business structures is one of the key factors that influence the formation of the vector, trends, and strategies for economic development of the regions of the Russian Federation. Reduced differences in economic development create favourable conditions for the development of an internal market and optimization of social and economic transformations while differential aggravation hinders carrying out a unified policy in institutional reforms and forming common national market and increases the threat of regional crises and interregional conflicts.

Conclusions

Thus, on the one hand, integration processes of business structures reach significant proportions; on the other hand, the role of the state (the region) as an active subject of the economic impact is often reduced to the monitoring of the dynamics of macroeconomic indicators. The imperfection and divergence of the legal system in Russia are of particular importance in this process, resulting in the state inability to form the necessary basis for property relations as a fundamental element in the search for sustainable development at all levels, including regional.

The work is performed under the grant of the President of the Russian Federation for state support of young Russian scientists — PhDs (MK-5339.2016).

References

1. D.A. Yendovitskiy, D.E. Soboleva, *Economic analysis of mergers and acquisitions of enterprises* (KNORUS, Moscow, 2008)
2. M.H. Lubatkin, Academy of Management Review, **8(2)**, 218-225 (1983)
3. V.L. Bersenev, S.G. Vazhenin, A.I. Tatarkin, *Institutions of economic integration* (Institute of Economics of Ural Branch of Russian Academy of Sciences, Yekaterinburg, 2003)
4. I.S. Glebova, D.V. Rodnyanskiy, Modern problems of science and education, **1**, 284-289 (2013)
5. O.V. Tyutyk, Problems of Contemporary Science and Practice. Vernadsky University, **3(13)**, 241-245 (2008)
6. M.G. Karelina, Economy of Region, **4**, 54-68 (2015)
7. M.G. Polikarpova, Bulletin of USTU, **4**, 62-72 (2010)
8. M.G. Polikarpova, Vestnik of NMSTU, **3**, 73-77 (2010)
9. O.V. Kuznetsova, A.V. Kuznetsov, R.F. Turovskiy, A.S. Chetverikova *Investment strategies of big business and economy of the regions* (Book House Librokom, Moscow, 2013)
10. E.V. Efimova, G.G. Valyayeva, Mathematics application in economic and technical studies, **4**, 90-98 (2014)
11. W. Changqi, X. Ningling, Procedia Social and Behavioral Sciences, **2**, 6896-6905 (2010)
12. L. Yan, L. Ming, Energy Procedia, **5**, 69-73 (2011)
13. M.M. Musatova, *The main trends in the development of integration processes in the contemporary world* (Novosibirsk State University, Novosibirsk, 2013)
14. R. Dorfman, The Review of Economics and Statistics, **1**, 146-149 (1979)
15. V.A. Litvinov, HSE Economic Journal, **2**, 226-237 (1999)
16. S. Calkinst, California Law Review, **71**, 402-429 (1983)
17. I.I. Yeliseyeva, I. I. Yegorova, *Statistic* (TC Velbi, Publishing House Prospect, Moscow, 2004)
18. A.A. Chernykh, Entrepreneurship, **6**, 58-62 (2011)
19. J. Haan, J.E. Sturn, European Journal of Political Economy, **16**, 215-241 (2002)
20. T.A. Ivanova, Economics and Politics, **6**, 31-37 (2013)

Urban regeneration of plants territories in industrial cities of Russia: a case study in Volgograd

Alexey Antyufeev[1,*], and *Galina* Ptichnikova[2]

[1]Volgograd State Technical University, Department of Urban Studies and Theory of Architecture, 400074 Volgograd, Russia
[2]Research Institute of Theory and History of Architecture and Urban Planning, Theory of Architecture Department, 111024 Moscow, Russia

Abstract. Economic transformations and reorganizations of production led to the crisis of many industrial cities in Russia in the first decade of the 21st century. This paper concerns the renovation of the former industrial zones located in coastal areas, which are very important for urban development of cities. In terms of socio-economic changes, this topic is relevant for many industrial cities of Volga region. The study is an attempt to find new roles in post-Soviet cities which face the problems of adapting to economic and social changes. The authors reveal the basic provisions of the project by the example of the renovation of the territory of the Metallurgical Plant "Red October" in Volgograd to alter the function to create a multifunctional cluster.

1 Introduction

At the beginning of the 21st century, post-industrial cities are faced with the necessity of reorientation to diversified specialization, creation of human and social capital, improvement urban fabrics, and improvement interurban transport connections [1, 4, 13].

Contemporary innovation economy requires a livable and prosperous urban environment for attracting and keeping high-skilled human capital [5, 14, 15]. The added value to cities and towns in Russian economy could be an initial impulse for launching drivers of the innovation economy [17].

Today many large cities in Russia began redeveloping their abandoned industrial areas [7, 9, 18]. This process could be called regeneration or renovation. The ongoing process de-industrialization which is freeing up areas in strategic urban places can provide with opportunity to re-assess the role, function and form of important part of the city [8, 19]. The gradual restructuring of some Post-Soviet industrial cities in recent years seem to point toward the relevance of new urban approaches. Different paths of development may only be explained from the distinctiveness of each territory, specific local decisions and successive strategies to tackle decay. These processes of revitalization seem to reveal the importance of local action and the possibility of guiding and easing decline [20]. Even more, they point out to the crucial role of urban planning [2].

The aim of this study is an attempt to reveal new practice, the new approach to regeneration of post-industrial areas in great Post-Soviet cities faced with the problems of adapting to economic and social changes. The case study is in Volgograd, great industrial city with population of 1 million people. Volgograd is part of this urban trend, as it using many of its abandoned industrial zones.

2 Post-Soviet industrial cities: Urban problems

2.1 Decay and abandoned industrial areas

In Post-Soviet industrial cities many plants have been shut down; many industrial areas have been abandoned. Often these areas are located in strategic urban positions. They are areas which were once on the outskirts, but as cities spread they became more central and usually supplied with major urban infrastructure and connected by roads and railways. From their very outset the industrial areas were formed as enclosed enclaves and were seen as autonomous specialized areas distinct from rest the city

2.2 Maintaining the Integrity of the Specifications

The problem is especially acute in large cities of the Volga region. Historically feature of these cities is to place plants in coastal areas. The occupation of the coastal areas of industrial buildings was caused by the rapid industrial development of cities in the Volga to the 19th and 20th centuries, as well as the extensive industrialization in the Soviet period [3]. New factories were located in coastal areas along the river, around a new large enterprise working village and serving

* Corresponding author: antyufeevav@vgasu.ru

infrastructure grew [10]. Then the industrial and residential area joined the city. For example, in Volgograd (former Tsaritsyn, then Stalingrad) during the industrial development of the largest giant plants - Steel ("Red October"), gun ("Barricades"), the Stalingrad Tractor (VPP), the chemical complex "Khimprom" are located along the Volga River (Fig.1).

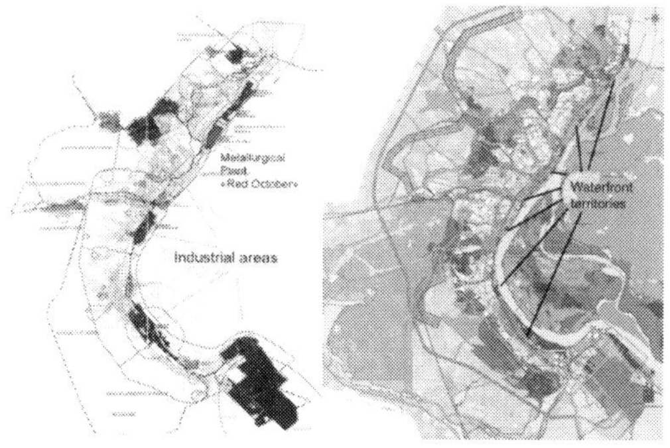

Fig. 1. Plan of waterfronts territories of Volgograd and Plan of locating the existing industrial zones situated on the coastal territories of the city, the most part of which are presently degrading post-industrial areas.

Currently Volgograd ceased to grow only at the expense of industrial giants. But at the same time, the city continues by inertia to grow further to the south and to the north, between its extreme points is almost a hundred kilometers.

Vast stretches of coastal area are still in the grip of production use or turned into a wasteland because of the decline in industrial production. The length of manufactured front in large industrial cities of the Volga region is 30% to 50% of the total coastline [6]. In Volgograd industrial areas occupy more than half of all coastal areas. The problem of creating social and recreational areas of coastal territories in Russian industrial cities situated along major rivers is presently one of the principal tasks which directly affect urban planning development of these cities [10]. Typical for this time of de-industrialization processes significantly influenced the image of many industrial cities of Volga region (Samara, Sizran, Saratov, Volzhsky).

Volgograd has known as an important industrial center and transport hub for more than a century. Today, Volgograd remains a bustling urban center and an industrial powerhouse of the Southern Federal District. So, for historical reasons, the Metallurgical Plant was situated along the Volga in city's north shore (Fig.2).

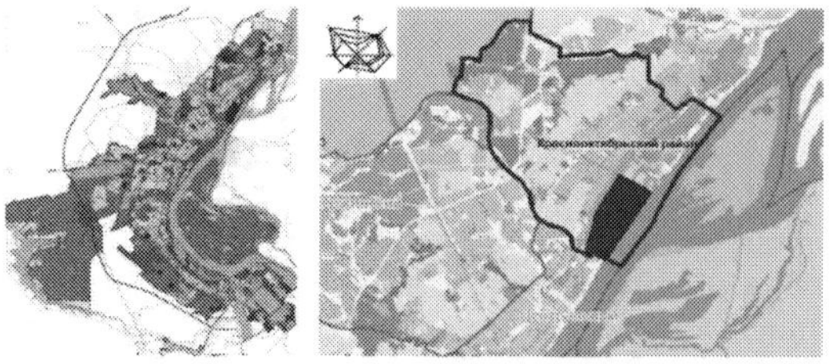

Fig. 2. Location of the Metallurgical Plant and selected study area.

The basis of Volgograd coastal industrial enterprises system is out-of-date. The planning structure of the existing coastal territory is characterized by the presence of randomly located territories of different use and purpose, low density of industrial development, and vast disrupted areas not using coastal territories as active public areas [16]. Today Volgograd is suffering from problem of decay tied to the lost its industrial base, but in the city we can find out of the beginning of a places of regeneration.

3 Project of regeneration of the territory of the Metallurgical Plant "Red October"

The project of urban regeneration of industrial territories of the Plant "Red October" was developed in Volgograd (Project Manager is professor Alexey Antyufeev).

The Metallurgical Plant was established in 1897 by the French joint stock "Ural-Volga metallurgical company". The plant was one of Russia's largest industrial facilities at the South. After the October Revolution the factory became known as "Krasny Oktyabr" ("Red October"). The factory was completely destroyed in the battle of Stalingrad, but was restored shortly after.

Currently, the plant is located in the central part of the city, directly adjacent to the Mamayev Kurgan, the stadium for 45 000 spectators, projected to the FIFA World Cup 2018, and the coastal area along the Volga. The length along the river of plant facade is more than 3 kilometers. Industrial buildings, warehouses, office buildings, engineering infrastructure facilities occupy most of the territory of the plant. Currently, due to reduced steel production factory most of the territory, an area of 320 hectares, is in decay. Large areas are unused, production halls are empty buildings. This area with good transport accessibility and near the city center could not continue to be empty.

By order of the plant management the authors developed a renovation project with the formation of new functional areas, the relevant provisions of the existing Master Plan of Volgograd. The Red October regeneration plan was approved in 2015. The plan will be implemented in stages. In 2015, the project of sport complex was approved. The main goal now is to improve the transport infrastructure. Reorganization tasks and the creation of new connections have been solved for the optimal integration of the new public areas at the city planning structure. (Fig.3).

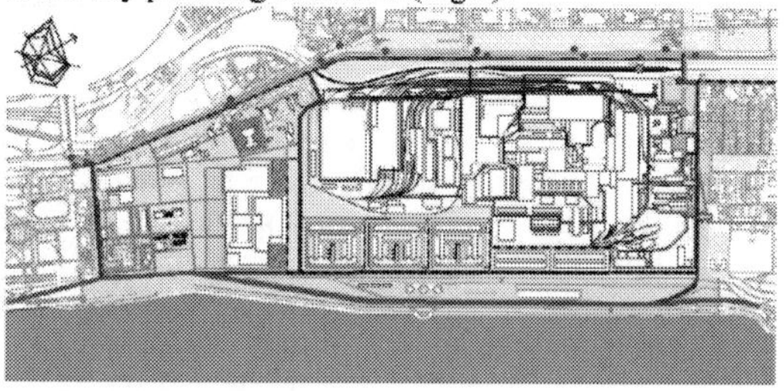

Fig. 3. Plan of modernization of transport infrastructure.of the Metallurgical Plant.

DOI: 10.1051/shsconf/20173501003

The following problems were solved:

- functional reorganization of industrial areas;
- selection of the most effective option from an environmental point of view, the use of coastal areas and the organization of their accessibility to the public;
- the establishment of the planned development of the parameters of the elements of the newly formed planning structure.

It was decided to establish a public area with a focus on the creation of a sports cluster on liberated from the industrial production area of 160 ha [14]. Reconstruction of the repair center with span structures allowed creating space for a universal sports complex citywide importance, including basketball club and other facilities for training children and youth sports. This complex will join the sports zone that is generated around the stadium for the 2018 World Cup, of which the matches will be held in Volgograd. As additional functions exposition and exhibition facilities, shopping mall and two industrial parks were designed.

The transport infrastructure is being developed at a dynamic pace. In 2017-2018 there will be 15 km of new roads and The Rockada road along the river. The authors proposed the formation of a new section of the pedestrian promenade where it is supposed to create a center of extreme sports with the use and adaptation of existing cooling towers smelter.

In this way this practice shows new urban model of regeneration of post-industrial areas for community use. In was determined of some design approaches (Fig.4). First of all, it is the restoration of important industrial building for public uses. Secondly, project has been developed on brown fields to retrieve the sites to the community through its reuse as pedestrian promenade and recreational areas.

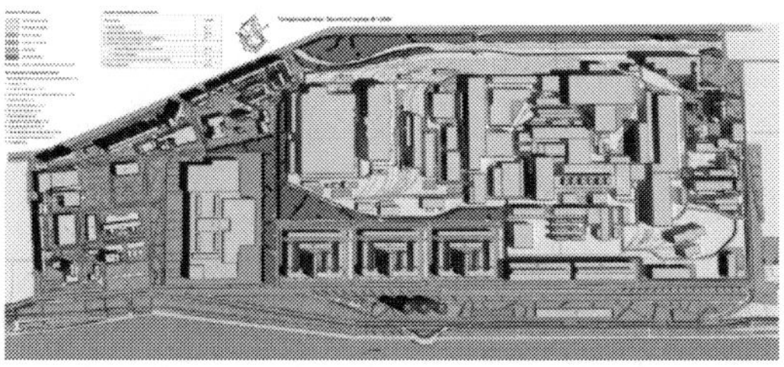

Fig.4. Master Plan of regeneration of post-industrial territories.

4 Conclusions

In Post-Soviet period industrial cities have seen fundamental economic restructuring. The weaknesses of these cities are their strong dependence from industries which have been created under another ideological, economic and planning rule without concern of transportation cost. Recently, various urban oriented initiatives have emerged in different cities. New approaches to solving the problems are revealed by the example of the project of regeneration of industrial areas of Volgograd plant "Red October".

The project has been an opportunity to link landscape architecture, environment restoration of polluted industrial areas and increase of parks and open spaces. Redeveloping industrial area will help improve the investment climate in the city.

To sum up, it seems to conclude that there are certain urban approaches and techniques that can help to face urban and economical decay. They would be:

1. Projects related to different aspects of the regeneration process. They are linked to Regional economic revitalization plans. These projects try to create a new regional economic base as diversified as possible. There are some kinds of development of new public activities.
2. Industrial Culture and Museum Programs. They linked to the restoration and reuse of important industrial heritage [11, 12]. Most important industrial buildings are reconstructed for tourism, cultural uses and renovation for new public services.

Therefore urban planning has some basic techniques to cope with decline of post-industrial cities. An urban model can help these cities become complex and more sustainable. The emerging practice shows that urban discourse is getting the force and become an actual topic for discussion on the problem of economic regeneration of industrial cities.

References

1. B. Aguenda, *The Archivo Digital UPM* (2009). http://oa.upm.es/5976/1/FernandezAgueda_ponencia_2009.pdf
2. B. Aguenda, Territoire en Mouvement, **23-24** (2014) https://tem.revues.org/2527
3. A.V. Antyufeev, Biosphere compatibility: people, region, technology, **1**, 100-105 (2015)
4. D. Brady, M. Wallace, Sociological Forum, **16(2)**, 321-358 (2001)
5. J. Friedrichs, Urban Studies, **30(6)**, 907-917 (1993)
6. S.S. Frolov, *Urban planning reconstruction of coastal industrial territories of the largest cities (by the example of Volgograd)* (SPbGASU, St. Petersburg, 2005)
7. C. Horn, Urban planet. http://urbanplanet.info/urbanism/potential-moscows-industrial-zones-russia/
8. G. Garofoli, *Ristrutturazione industriale e territorio* (Franco Angeli Editore, Milano, 1978)
9. D.V. Knyazev, The economy of the region, **18** (2007). http: //journal.vlsu.ru/index.php id = 1732
10. E.E. Krasilnikova, A.V. Antyufeev, *Strategic decision making on spatial development* (ROAD/SPECTRA Centrum Excellencies, STU, Bratislava, 2014)
11. Y.H. Lou, X.Q. Zhu, F. Zhou, *2010 International Conference on Mechanic Automation and Control Engineering*, 829-832 (2010)
12. O. Mamleyev, Architecture. Building. Design, **1(23)**, 21-27 (2001)
13. H. Molotch, American Journal of Sociology, **82(2)**, 309-332 (1976)

DOI: 10.1051/shsconf/20173501003

14. M.E. Porter, *Clusters and the new economics of competition* (Harvard Business Review, Watertown, 1998)
15. G.A. Ptichnikova, Modern architecture world, **3**, 177-190 (2013)
16. G.A. Ptichnikova, A.V. Antyufeev, Sociology of the city, **2**, 5-19 (2014)
17. E. Rullani, *Complexity and industrial clusters* (Physica-Verlag, New York, 2002)
18. N.N. Sinitsyna, *Architecton:* Math. Universities, **1** (2004). http://archvuz.ru/numbers/2004_1/k23
19. J.W.R. Whitehand, N.J. Morton, Environ. Planning B, Planning Design, **30(6),** 819-839 (2003)
20. Th. Wiechmann, K. Pallagst, International Journal of Urban and Regional Research, **32(2),** 261-280 (2012)

Theoretical and practical aspects of the synthesis of optimal systems of energy complex enterprises for preparation and processing of hydrocarbons

Igor Dolotovskii[1,*], and *Evgeniy* Larin[1]

[1]Yuri Gagarin State Technical University of Saratov, 410054 Saratov, Russia

Abstract. We have developed a technique for structural-parametric synthesis of optimal energy complex enterprises for preparation and processing of hydrocarbons, which employs the decomposition-search principle to create technological and constructive solutions; it is also based on mathematical apparatus of generalized quantitative quality evaluation. Developed the structure and composition of the equipment of the energy complex with the system of electric-, warmly- and water supply. The complex is integrated energy–technological systems, and provide maximum utilization of recycled energy resources and waste. The developed information-program software allows performing an evaluation of the a multi-criterion of the efficiency of systems of energy complex.

1 Introduction

Energy complex (EC) enterprises preparation and processing of hydrocarbons (EPPHC) includes several systems (fuel, heat and electricity, water supply and cold supply, etc.), forming together with technological systems (TS) a complex set of manufacturing equipment. Machines of EC and TS are consumers and generators of various energy resources (ER) derived from external or own sources, and interconnected with production cycle by material and energy flows. Development and improvement of such complicated structured object, taking into account all factors is possible only on the basis of the system of evidence-based methodology and its software, allowing the synthesis and design of EC with any topology TS at every stage of their life cycle. In the present study we offer methodological position of structural-parametric synthesis – the development of the methodology of system studies in the energy and chemical-technological enterprises [1–3], and examples of solutions of the structure of systems and design hardware EC [4].

2 Problem statement

For the synthesis of the EC, optimal in the whole space of changing factors, created the mathematical apparatus of universal quantification of the quality of its functioning within the EPPHC. Need to create a mathematical models of system that include a description of the interrelated processes of generation and consumption of ER, water consumption and water disposal, heat-technological and chemical-technological processes of preparation and processing of raw hydrocarbon. Need to take into account the dynamic parameters of investigated object and hierarchical sets of private indicators and criteria of varying degrees of significance.

The basis of the developed method is decomposition-search principle of synthesis of energy-technology systems [5], which provides the solution of problems of two hierarchy levels: the first is formed of the technological structure of EC, on the second – it's hardware design and design of information and software.

3 Theory

3.1 Synthesis of technological solutions

At the stage of conceptual modeling of synthesis technology solutions from many possible principles Π ($\pi \in \Pi$) chosen: π_1 – maximum insularity with the use of its own of recyclable energy resources (RER), combustible waste and wastewater; π_2 – ecological safety; π_3 – reliability of energy supply throughout the life cycle period.

The task of synthesis of technological solutions illustrates a generalized block diagram of EC (Fig. 1). In scheme realized principles Π, as well multi-functional systems of energy complex and its integration with the systems: technology processing, water supply and wastewater. Electric and thermal energy is generated in own source through the maximum utilization of secondary energy resources – in a combined steam turbine and gas turbine plant (STP and GTP). Process

* Corresponding author: dolotowsky@mail.ru

heat consumer may 2 include evaporator's absorbents drying and gas purification units, boilers distillation columns, heaters process flows, and other high-temperature equipment.

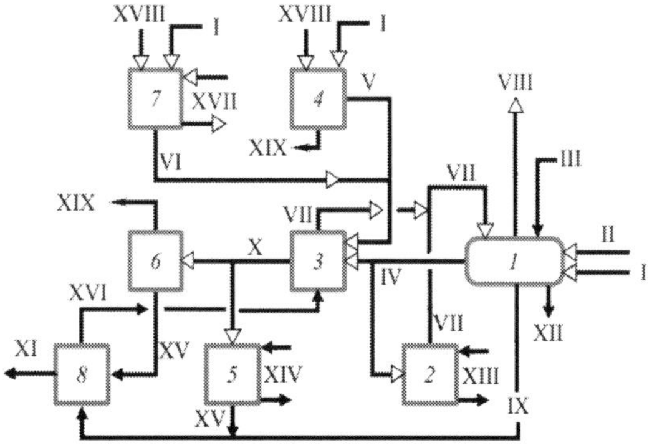

Fig. 1. Scheme of EC with a source of heat-, electric energy, and water supply. The figure: 1 – 8 is the plant: 1 is the combustible waste and wastewater utilization; 2, 5 is the technological heat-consumer; 3 is the steam generation; 4, 6 is the GTP and STP; 7 is the compressing gas; 8 is the water treatment; I – XIX is the technological and energy flows: I, II is the gas fuel and recyclable; III is the industrial wastewater; IV, V, VI is the high-temperature flue gas; VII, VIII is the gases in the drying and gas dehumidified; IX is the water in the treatment system; X – steam; XI –water system for household consumption; XII is the dry waste; XIII, XIV is the flow technology; XV is the condensed water; XVI is the demineralized water; XVII is the gas to compress; XVIII is the air; XIX is the electrical energy.

For auxiliary units of the technological system, for example, for hydrocarbon storage, flue gases neutralizing wastes and gas installation 1 can be used by the consumer 2 in boilers for heating high-temperature intermediate heat carrier used for heating tanks in the winter. Consumer technology 5 depending on technological topology can be formed from equipment: boilers columns; absorption or compression chillers (compressors driven by steam turbines or rotary steam engines); equipment heating and hot water systems and other heat consumers.

3.2 Synthesis of structurally-hardware design

Development of options for hardware design with optimized of constructive and operational parameters was carried out by the algorithm is implemented as a network model (Fig. 2).

Developed solutions of options of EC [5] assessed on energy, techno-economic and the additive-multiplicative factors: exergy efficiency; factors rationalization of fuel and energy and water balance; the integral effect, payback period, the complex criteria [4, 6, 7]. The last group of criteria has the form of functional

$$\Phi(\mathbf{u}(x)) = \sum_{i=1}^{k} \gamma_i u_i(x) + \sum_{i=1}^{k}\sum_{j=1}^{k} \gamma_{ij} u_i(x) u_j(x) + \ldots, \quad (1)$$

where $\mathbf{u}(x)$ is the vector of indicators of efficiency options for EC x; γ_i is the indicator of importance; γ_{ij} is the coefficient taking into account the linkages indicators.

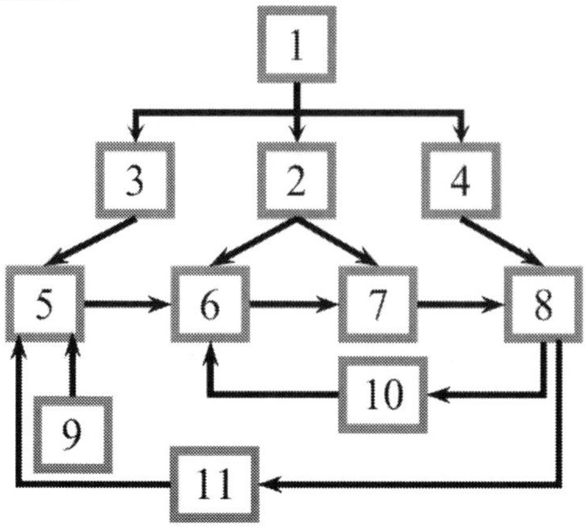

Fig. 2. Scheme of the synthesis of the optimal EC.
The figure: 1 is the terms of reference; 2 is the independent input data; 3 is the demand for the operation; 4 is the effectiveness criteria; 5 is the formation structure; 6 is the formation model; 7 is the description of the model; 9 is the database; 10 is the optimization parameters; 11 is the change of structure.

The sum (1) contains components $\gamma_{ij} u_i(x) u_j(x)$ that enable to assess the mutual influence of the partial indicators of the effectiveness on an integrated criterion. The functional $\Phi(\mathbf{u}(x))$ shows that the impact of the private indicator u_i (e.g., Integral effect) on the evaluation of alternative energy of the complex depends not only on its value for the variant $u_i(x)$, but also from the values $u_j(x)$ of indicator u_j for this variant (e.g., a of fuel economy or water resources).

The cumulative index of efficiency (CIE) of EC of multi-purpose EPPHCs formed from the partial criteria of technological [7], energetic, economic and of ecological efficiency. The criteria are grouped into two sets: the criteria of production and economic activities to ensure the profitability and efficiency of the enterprise as an economic asset (effectiveness) $r_p(t)$; efficiency criteria $\lambda_i(t)$ which characterize the degree of realization of the basic principles $\pi_{1\ldots3} \in \Pi$ during the accounting period.

Elements are normalized relative to the «basic» version of the object: a draft version of the operating enterprise; object with a similar factor of structural complexities and of topologies technology (with weighted average values of indicators).

Set of effectiveness elements $r_p(t)$ includes 6 indicators ($p = 1\ldots6$): $R(t)$ is the value of goods sold; $C_1(t)\ldots C_5(t)$ is the operational costs of the main production (costs of raw and auxiliary materials, energy and water from an external source; maintenance, repair and depreciation, fees for the disposal of waste, industrial emissions and waste water).

The second set $\lambda_i(t)$ containing the elements of effectiveness include 5 vector criteria (combining 17 indicators) and of productivity to hydrocarbon raw materials V_{HRM}. In total, this set of 6 indicators, $l = p = 6$.

Vectors of efficiency criteria $U_1(t)\ldots U_5(t)$ are formed from of the following characteristics and metrics calculated for time t:

- efficiency of use of raw materials and processing products

$$U_1 = 0{,}417\overline{G}_R + 0{,}263\gamma_c + 0{,}160\beta_x + \\ + 0{,}097\overline{\Delta}_{RC} + 0{,}062\eta_{EC}, \quad (2)$$

where $\overline{G}_R$ is the change in the volumes of products sold (relative to the «base» variant); γ_c, β_x is the coefficients of the utilization of raw materials and production sold of by-products; $\overline{\Delta}_{RC}$ is the difference between the cost of products sold and raw material costs; η_{EC} is the rationalization factor energy costs;

- potential of increasing use of ER, water and waste recycling

$$U_2 = 0{,}381\overline{E} + 0{,}252\eta_0 + 0{,}16\eta_{FEB} + \\ + 0{,}101\eta_W + 0{,}064\eta_L + 0{,}043\overline{P}_E, \quad (3)$$

where $\overline{E}$ is the useful used energy of external fuel-energy resources; η_0 is the coefficient of relative energy savings; η_{FEB} is the rationalization factor of the fuel-energy balance; η_W and η_L are factor of rational water consumption from an external source and of water disposal; P_Σ is the technically realizable potential by the use of all types of resources;

- use of ERs and economy fuel in system

$$U_3 = 0{,}417\eta_\Sigma + 0{,}263\eta_{FQ} + 0{,}16\eta_u^{RER} + \\ + 0{,}097\eta_{ex} + 0{,}062\overline{\Delta F}, \quad (4)$$

where η_Σ is the total of energy efficiency; η_{FQ} is the utilization factor of heat fuel for heat generation; η_u^{RER} is the utilization factor of RERs; η_{ex} is the exergy efficiency; $\overline{\Delta F}$ is the system fuel economy;

- costs for assurance the reliability of ERs supply and water supply and for of reservation

$$U_4 = 0{,}5\overline{C}_{HS} + 0{,}5\overline{K}_1(\Omega, Z), \quad (5)$$

where $\overline{C}_{HS}$ is the costs for assurance the reliability of ERs supply and water supply; $\overline{K}_1(\Omega, Z)$ is the capital cost to reserve of technology equipment, as a function of process (Ω) and design (Z) factors;

- compensation damage environmental

$$U_5 = 0{,}540\eta_W + 0{,}297\eta_L + 0{,}163\overline{K}_2(\Omega, Z), \quad (6)$$

where $\overline{K}_2(\Omega, Z)$ is the capital cost to reserve of equipment for waste disposal systems and drains.

The coefficients in (2) – (6) are the rank coefficients of the partial efficiency criteria calculated by the hierarchy analysis method [8].

When deciding about the priority of sets we adopted the concept of equivalency, since when the structural-parametric optimization of EC and the implementation of the principles of efficiency should be retained (or be increased) of profitability of the enterprise.

The formula CIE merges the generated values of fractional deviation of each indicator with consideration of the coefficients of potential.

$$CIE = 4{,}117\frac{R(t)/R^0(t)}{\sum_{p=1}^{6} r_p(t)/r_p^0(t)} - \\ -34{,}72\sum_{i=1}^{5}\frac{C_i(t)/C_i^0(t)}{\sum_{p=1}^{6} r_p(t)/r_p^0(t)} + \\ +25{,}0\frac{\sum_{i=1}^{3} U_i(t)/U_i^0(t)}{\sum_{l=1}^{6}\lambda_l(t)/\lambda_l^0(t)} - \\ -16{,}67\frac{\sum_{i=4}^{5} U_i(t)/U_i^0(t)}{\sum_{l=1}^{6}\lambda_l(t)/\lambda_l^0(t)} + \\ +8{,}33\frac{V_{HRM}(t)/V_{HRM}^0(t)}{\sum_{l=1}^{6}\lambda_l(t)/\lambda_l^0(t)}, \quad (7)$$

where the subscript «0» denotes indicators of the basic variant.

The coefficients of potentials (7) generated from the three factors, allowing to establish the direction of potential (+1 or −1), to ensure a balance between effectiveness $r_p(t)$ and efficiency $\lambda_l(t)$, to determine the importance (weight) of each indicator.

The balance factor of the coefficients of the potentials is 1/6, if the indicator is the only one in the set $r_p(t)$ or $\lambda_l(t)$, and is multiplied by the corresponding number of repeated calculation for each set. Thus, the coefficient of potential of the first indicator (7) is defined as follows:

$$4{,}117 = [+1] \times \left[\frac{1}{6} \times 0{,}5\right] \times \left[\frac{\sum_{p=1}^{6} R_p(t)}{\sum_{p=1}^{6}\sum_{l=1}^{6} R_{pl}(t)}\right] \times 100 \cdot \quad (8)$$

It should be emphasized that in the general case, all indicators have a weighting factor as part of the coefficients of the potentials. However, for performance measurement of the set $\lambda_l(t)$ it is equal to 1.

The values of other coefficients for CIE obtained similarly the (8). The higher the CIE, the more effective option ECs EPPHCs.

3.2 Information and software synthesis systems

The information-analytical system (IAS) [9] consists of software and computational and information-methodical units and modules (Fig. 3).

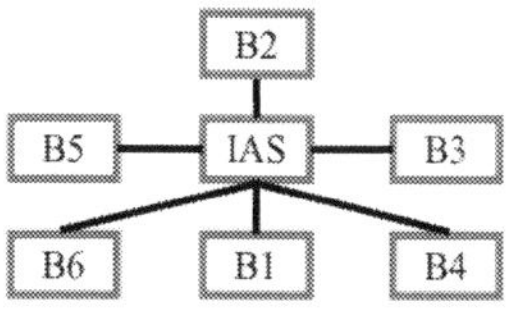

Fig. 3. Scheme IAS. The figure: B1 – B6 – blocs: B1 is the structured data of equipment; B2 is the modeling program; B3 is the methodical documentation; B4 is the normative documents; B5 is the experimental information; B6 is the instructions to the user.

With the help of the IAS implemented numerical method of a computer experiments with mathematical models that describe the behavior of individual components and the whole EC for a specified period or the configured period of time.

Information in single blocks of IAS is structured by the elements of EC and TS EPPHC, which in turn commute to each other via entry and exit signals. Each output signal (Y_{sj}^{n}) from one or component is input to the other component of the system (X_{sj}^{k}), or included in the overall system output signal (the subscripts refer to the subsystems, the top - to complex signals). Generally, $k \neq n$; k and n are different for each system.

The mathematical scheme by which describes the real processes in the components of block B2 (RF computer programs numbers: 2010615353, 2011616340, 2011616684, 2012612727, 2012613267, 2012613266, 2012613269, 2014613737, 2014619488, 2014660962, 2014660407, 2016610685), presented in IAS in the form of aggregate (AG).

At any time the AG can be in one of possible states. The transition of the AG from one state to another is described by a transition operator, which allows you to define by a previous state of AG next state of the AG. The AGs have inputs, which receive inputs from the other AGs, and outputs on which are generated output signals. In addition, the AGs have additional inputs, which receive the control signals. The outputs signals are generated by operator of output from the input and the control signals, because of its interaction with the transition operator. The values of the operators of transition and exit are defined at the approximation of the execution of AGs of the functional action of the real system.

4 The results of the experiments

Developed theoretical principles of synthesis of optimal systems ECs EPPHCs implemented in the project of modernization of heat- and electric power systems for the enterprise of processing sour gas and gas-condensate.

With the using software of IAS have been conducted numerical experiments with models of ECs different structures, composition equipment and modes of operation.

Variants of EC were assessed both by private criteria of efficiency, and by CIE and fuel economy.

Key performance indicators for the investment project of creation and functioning of the EC with an independent source a electricity-, heat-, water- supple system in the structure of the multidisciplinary EPPHC for different variants of equipment are presented in the table.

Table. Efficiency characteristics of EC EPPHC variants.

Index	**Gas turbine plants (GTP)**		
	3×GT8C2 ALSTOM	**2×V-64.3A SIEMENS**	**4×GTX-100 ALSTOM**
	Heat-recovery steam generators (KU)		
	3×KU-93	**2×KU-120**	**4×KU-60**
	Steam turbines		
	3×K-6-2.4 KTZ	**2×K-6-2.4 KTZ**	**2×K-6-2.4 KTZ**
Power (MW):			
electrical	192	158	190
thermal	163	163	163
Fuel rate to generate power:			
electrical (tons of reference fuel/MW·h)	0.161	0.158	0.164
thermal (tons of reference fuel/GJ)	0.045	0.042	0.046
Rationalization factor:			
of energy consumption, η_{FEB}	0.255	0.316	0.217
of water consumption, η_W	0.914	0,912	0.912
of water disposal, η_L	0.974	0,974	0.974
Integral effect, %[a]	137.18	117.47	135.49
Annual fuel savings, 10^3 tons equivalent fuel; in electricity generation:			
is the GTP and STP condensed type	54.1	69.6	39.2
is the separate method with a weighted average efficiency	119.7	123.2	104.1
CIE	25.3	24.1	22.5

a. relative to the base variant, to which effect is accepted 100%

5 The discussion of the results

Developed technological solutions for the structure of the EC have a number of advantages:

- efficiencies of energy at the expense of utilization useful heat from the flue gases in heat recovery boilers to generate steam;
- of use as a fuel gas liquid and gas wastes in the neutralizer waste; this reduces the specific fuel consumption for the joint production (cogeneration) of electricity and heat, cold supply and water supply;
- increase of ecological safety of EPPHCs due to the thermal neutralization of industrial and drainage effluents, the water of separating and water flushing equipment;
- reduced costs for water supply through the use condensate of steam as the working substance in steam-power plant and in systems water for the system of economic-drinking water supply;
- in hardware design of EC its installations as much as possible integrated with TS, so not in need of replacement by decommissioning, or in in the conservation and used at all stages of the life cycle EPPHC, from beginning construction of the facility before its decommissioning.

Evaluation of the effectiveness of technical solutions made out on the basis of a complex of energy-economic-and-environmental indicators.

A balanced approach of technique choice of CIE allows you to exclude bias or subjectivity in the synthesis of the optimal EC, which is especially important in the processes of hydrocarbon processing. Use single criteria for these enterprises can lead to incorrect conclusions, to the detriment of such components of efficiency as the reliability of ERs, ecological safety, system energy efficiencies and other factors.

The developed software has allowed to define, on the basis of simulation, performance indicators EPPHCs with different versions of ECs. For example, for shown in table three of variants heat-, electricity-, and water-supply systems, is the most effective first variant, is characterized by a maximum value of CIE and the integral effect in comparison with a variant of EC of the acting EPPHC.

6 Conclusions

A developed method of structural-parametric synthesis of optimal variant of an independent source of power supply for energy complex enterprises for preparation and processing of hydrocarbons.

Presents the factors that determine the techno-economic, energy, ecology effectiveness and efficiency of enterprises of preparation and processing of hydrocarbons. Developed alternative structure energy complex with a source of electrical and thermal energy and water supply, integrated with the technological system and ensuring maximum utilization of recovery energy resources and waste disposal.

With using the information and software made evaluation of the efficiency of alternative variants of the system of electricity-, heat-, water supply for multidisciplinary enterprises on processing of gas and gas condensate.

This work was supported by the Ministry of Education and Science of the Russian Federation, within the framework of a state assignment.

References

1. G.M. Ostrovsky, T.A. Berezinsky, A.R. Belyaev, *Algorithms for optimization of chemical processes* (Chemistry, Moscow, 1978)
2. V.V. Kafarov, V.P. Meshalkin, *Analysis and synthesis of chemical-technological systems* (Chemistry, Moscow, 1991)
3. D.A. Marca, C.L. McGowan, *SADT: Structured Analysis and Design Techniques* (McGraw Hill Software Engineering Series, 1987)
4. I.V. Dolotovskii, E.A. Larin, N.V. Dolotovskaya, *Design and optimization of plants and systems and energy complex of the industrial enterprises* (Amirit, Saratov, 2015)
5. I.V. Dolotovskii, *A power, heat, and water supply installation for hydrocarbon raw material production, transportation, and processing enterprises* (RU Patent 118360, 20, 2012)
6. N. Korovkin, M. Odintsov, N. Belyaev, O. Frolov, H. Masashi, IEEE Conference Publications, 12-14 (2014)
7. A. Tonchev, C. Tonchev, Oil & Gas Journal, **10** (2005)
8. T. Saaty, L. Thomas, *Multicriteria Decision Making – The Analytic Hierarchy Process* (RWS Publications, Pittsburg, 1992)
9. I.V. Dolotovskij, E.A. Larin, N.V. Dolotovskaya, *The information-analytical system of standardization and optimization of production and consumption of fuel and energy resources in the enterprise* (RU Patent 2465639, 30, 2012)

Issues associated with energy efficiency programs implementation at the housing and utility enterprises

Natalia Borisova[1,*], *Alexander* Borisov[2], and *Natalia* Fedonuk[1]

[1]The department of management and urban development and construction, Volgograd State Technical University, Volgograd, Russia
[2]The department of economics and management of construction projects, Volgograd State Technical University, Volgograd, Russia

Abstract. Energy saving potential is quite huge in the most sectors of the national economy, particularly in housing and utilities and industry. Due to this, energy efficiency increase at the enterprises of housing and utilities and industry through the energy efficiency programs implementation, is one of the priorities in the modern economy of Russia and its regions, and requires radical measures to improve the effectiveness of its implementation. The purpose of the authors is the scientific and practical study of the main problems of energy efficiency programs implementation at the enterprises of housing and utilities services and industry in modern conditions. To achieve this purpose the authors solved the following problems: the current state of the housing and utilities sector and industry, the relevance and the need for energy saving policy at the enterprises of housing and utilities services and industry are studied; the main problems impeding to implement the energy-saving program effectively at the enterprises of housing and utilities services and industry are determined; the possible ways of solving the problems identified in the energy efficiency programs implementation at the enterprises of housing and industry are offered. The team of authors focuses in this study on the problems of the energy audit using in practice as a basic tool for the energy saving programs development at the enterprises of housing and utilities services, industry and their subsequent implementation. The subject of author's researches is the factors that determine the energy efficiency programs implementation at the enterprises of housing and utilities services and industry at the level of individual region and the whole country, and the object is the enterprises of housing and utilities services and industry. Methodologically the scientific and practical research is based on the complex approach using the methods of comparative, statistical and logical analysis.

1 Introduction

The main motive is undoubtedly a factor of energy resources exhaustion for saving energy. The idea of energy saving appeared in the late 20th century during the global energy crisis, when most of the advanced countries of the world has sought to improve the efficiency of existing energy sources and to use new alternative and renewable sources [8; 11; 13; 14; 21; 23; 25; 26].

In the new economic conditions, and for many years the energy saving is a fundamental factor in the effective functioning and perspective development of the most enterprises in the housing and utilities services, the industry of many leading countries, especially Russia [7; 9; 10; 15;19; 20; 24]. The practical realization of energy-saving potential is the high priority of the federal and regional economic policy in Russia, as in the new economic conditions, each percentage of fuel-energy resources saving used in production and economic activity of the national economy enterprises can provide the national income increase more than 0.35% [1; 4]. The main purpose of energy saving policy in the housing and utilities sector and industry is to reduce the cost price of services and goods manufactured in these spheres.

2 Relevance and research of the problem situation

Currently, Russia is the energy surplus country, but the situation is getting worse every year, and therefore, the energy crisis can come, if the measures are not applied in the using of the existing energy saving potential in our country. In 2014, energy intensity has undergone a decline of 5.61% compared to 2007. Despite the positive trend, without the implementation of a competent energy saving policy it is difficult to achieve the result set by the President of the Russian Federation and in the "country's

* Corresponding author: borisovani06@mail.ru

energy strategy until 2020" (reduction of the GDP energy intensity by 40%) [3; 6]. Priority ways of energy-saving policy are the industrial sector of the economy, agriculture, construction industry and housing and utilities services. Thus the greatest losses of fuel-energy resources are observed in the industrial (47%), construction industry (22%), as well as housing and utilities services and agriculture (28%) in Russia and its regions in modern conditions [4; 6].

These sectors are interrelated, but especially the housing and utilities sphere unites each of them. For example, any industry can be either the supplier or the customer or the end service user. It was found that only the residential and industrial buildings consume more than 70% of electricity, 49.8% of natural gas, 40% of fossil energy resources. Therefore, the rational use of energy in housing and utilities sector and in the industry is one of the priority tasks in the sphere of energy saving [5; 19; 26; 27].

The energy saving reserves are most huge in heating in large and small Russian towns. The main reasons for this are, firstly, the huge scale of this scope (for heat generation in the northern regions of Russia more than 150 mln. standard fuel is spent per year), and, secondly, the absence of starting base for the implementation of energy saving arrangements. As a result, it is impossible to perform effectively the accounting and control produced, transmitted and consumed by thermal energy. It should also be noted that in the majority of Russia regions the morally and physically obsolete technical equipment is used in heating systems, especially in the transport of heat, which have been developed over 50 years ago.

Energy losses in distribution (often working with flooding) heating networks and in local heat systems of small objects in utilities sphere are characterized by its scale. Water treatment systems are not effective enough in small local boiler rooms, this leads to the formation of scum in boilers and heat exchangers and to a consequence of this reduction in their efficiency. When the scum overgrow in boilers working in hot water systems, boiler efficiency decreases to 60% after 2-3 months exploitation in conditions of bad water using [1; 2].

Analysis of water and sewer networks characteristics, realized by the Russian Association of Water Supply and Sanitation, showed that the average percentage of water supply systems depreciation (in 70 Water Channels) is 75%, sewerage - 70%, newly constructed, reconstructed and shifted water networks in recent years, is near 0.72%, sewerage - 0.44% of the total length of all networks. If you take annual standardized percentage of about 5% to restore the water supply network, underdone repair of water networks in the past few years will exceed more than 21%, and sewerage networks - more than 22% [2; 3].

Thus, water and sewage networks operating costs increase and unaccounted costs of water resources reach up to 30% in the large and small cities and towns of our country. The number of accidents is also increasing. Water losses of our country are more than 23% per year as a result of the enormous physical and moral depreciation of fixed assets in water supply and sewage enterprises, in some cities the losses of water reached 40%. In addition, the accident rate of water supply and sewerage networks reaches until 6 accidents per 1 km of network per year. The physical and moral deterioration of networks and facilities of water supply and sanitation sector have reached more than 80% in some regions (according to the Ministry of Regional Development). As a result, more than 42% of water supply networks shall be changed, sewerage networks - more than 35% of the total stretch. However, it should be noted that less than 1.5% water and 0.5% sewerage networks is chanced each year [3; 27].

Also, it is important to note that a large number of various types of energy losses in the housing and utilities sector and industry is due to the negligent attitude to the consumption of energy resources by consumers. Through the windows and doors of residential and industrial buildings more than 70% of heat is lost. Very often in our homes and industrial buildings, a heating equipment works at full power, but because of the lack of modern heat regulation systems, consumers need to air their residential and industrial rooms, thereby creating a direct energy losses. As a result, Russia's extravagance, the neglect of the energy, heat, water consumption, energy efficiency of housing and utilities services sector and industry in Russia is 5 times lower than in Europe [2; 3].

Undoubtedly, we are confident that the energy efficiency of each residential and industrial object depends primarily from the consumer. If we, the consumers, want to pay bills according to real indexes of energy, heat consumption, but not the losses inheriting in tariffs, it is necessary to demand installation of energy saving systems with suppliers, intermediaries of utilities services (e.g. installation of water gas, electricity counters, etc., as well as regulators of heat supply).

Such huge savings potential and the annual increase of energy, heat and water cost, made a very topical problem of energy saving for the enterprises of housing and utilities sector and industry in the majority of Russia regions.

3 The research and statement of the realisation problem of energy saving program

One of the ways to solve the above problems in 2000 was the creation of the subprogram "Energy supply saving in housing and utilities services", "Energy efficiency in energy-intensive industries," of the federal program "Energy saving in Russia" for practical implementation. Every town should have developed the energy supply saving programs in housing and utilities sector and energy efficiency in the industry, including specific arrangements to save energy, water and heat resources with the technical and economic assessment of the required costs and expected economic effect. The main tool for the development of energy-saving programs, and their subsequent implementation in practice is an energy audit of different objects which belong to the enterprises of housing and utilities services and industry. Energy audit includes energy research of these objects, as well as an analysis of possible savings reserves and technical and economic efficiency account of the developed energy-saving arrangements.

Typically, energy audit is a comprehensive analysis of all residential and non-residential objects, belonging to the legal entity. Also the current state of the engineering infrastructure of each customer's object is investigated. In other words, all the factors are investigated and identified that contribute to energy, heat and water loss, which in turn leads to overruns of above mentioned resources and to overpayment for their use.

According to the results of surveys g energy certificate is formed, containing all the parameters of the investigated companies, and some recommendations can also be given to use the specific energy-saving technologies, taking into account the specifics of the production activities of the studied enterprise [6].

Since the release and adoption of subprograms "Energy supply saving in housing and utilities services", "Energy saving in energy-intensive industries" more than 10 years have passed, but massive development did not happen at the regional and municipal level of these subprograms.

As the leading energy audit experts, not all heads of regional enterprises of housing and utilities sector and industry aware of the need for energy survey and energy audit of their businesses. Most of the heads of regional enterprises of housing and utilities sector and industry do not think, ignore the real possibility of saving its heat, energy resources. It should be noted that this is extra cost, extra actions for many managers. In addition, energy losses are profitable for some regional enterprises of housing and utilities complex, as their value is on the maxim laid in the tariffs of utilities services

As a result of inactivity of the majority enterprises of utilities sector and industry, the Federal Law № 261-FL has been developed and enacted at regional and municipal level in the sphere of energy saving "About energy saving and energy efficiency increase and about changes in certain legislative acts of the Russian Federation". In accordance with this Federal Law the list of enterprises was approved, which necessarily had to undergo energy audits until 31 December 2012. Also this list includes the organizations such as: public authorities and local self-government, with rights of legal entities; organization with the participation of the state or municipality; organizations, the overall cost of which are more than 10 mln. rub. per year for all kinds of fuel, fuel oil, heat, coal, electric power consumption [3; 18].

However, the enactment of the law has caused a lot of questions from leading economists and market experts of energy audit and energy services. Most of the leading energy audit experts believe that the Federal Law does not work in practice, and has many problems in implementation which are still unresolved. In this regard, we consider the main problems which are created in the work of the law in practice.

Authors agree with the opinion of the majority of the leading energy audit experts, who state that the information which is written in legal document, does not work in real market conditions. For example, until this day, the form and purpose of the energy passport are unclear. In addition, there are no clear requirements for the formalization of energy passport. No one knows what should be in the energy performance certificate, what institution must approve it. Today the Department of Energy has received a quarter of all formed energy certificates [4; 6; 27].

In the current market conditions in our country there are more than two thousand energy auditing organizations that appeared at the time of the popularization of this theme. However, only hundreds of existing energy auditing organizations match to the self-regulatory organization. The law does not prescribe clear rules of the audit, but only puts certain requirements and terms under which managers are required to make energy certificate. As a result, due to the lack of clear guidelines for energy audit, the experts study the potential of energy efficiency companies, according to the "old" forms that were developed in 2000 with the enactment of subprograms "Energy supply saving in housing and utilities services", "Energy efficiency in energy-intensive industries ". It is only in rare cases, the analysis of the energy efficiency of the investigated companies is conducted competently and economically correct, and only because of giving the attention to the process [4; 5]. Thus, there is nothing about the development and implementation of specific measures for energy saving. In turn, the state does not finance the implementation of this law in practice.

Today, only a few regions and cities of our country can boast of a full-scale energy survey of the majority of housing and utilities services, industry enterprises and summarize its arrangement. Novosibirsk, Kazan, Tyumen, Kaliningrad, Tula, Perm, Penza - these are a few of the city, where there is the large-scale movement of the energy audit.

For example, today the Tyumen region is the leading region for realization of energy audits in the enterprises of housing and utilities sector and industry. In the Tyumen region a state contract is signed to conduct the energy surveys of state property objects. The contractor of the energy surveys is the company "Tyumenenergo". Currently the energy audit is conducted in the majority of state property objects, the results of which is developed set of measures for energy efficiency. In general, in the Tyumen region energy survey was conducted in 221 enterprises, so energy passports have been issued and registered in the Ministry of Energy of Russia. This is a rather high number, considering that there is a little more than 300 energy passports are registered in all the country. [4; 5; 18].

It should also be noted that a few years before the enactment of law of energy saving, Tyumen region governance decided to use only modern technologies in carrying out all kinds of work in the construction of residential, industrial and other projects financed from the regional budget. As a result, since 2005 the number of accidents, malfunctions of network engineering through the use of modern energy-efficient technologies decreased by 40%. Most of the buildings used by state and local authorities, are equipped with modern energy metering devices. Also, more than 50% of the housing stock of the Tyumen region is equipped with modern energy metering devices. [4; 6] Tyumen region is a prime example of the Russian region, understanding the significance and the importance of practical implementation of energy saving measures in housing and utilities sector and industry in order to further effective social and economic development of the region.

Another problem, according to experienced professionals' opinion in the sphere of energy audit, no less important is the fact that it is practically impossible to choose a competent auditor. Now more than two hundred energy auditing companies, which are relevant to SRO, function in the Russian market of energy auditing services. However, not all can have competent professionals. The problem is that the duration of the standard training courses of the energy audits basics is only seventy-two hours. But it is impossible to teach energy audit to specialist during this time. As a result, in Russia and its regions, less than one hundred energy auditing companies can offer professional services.

The next problem, which is noted by experts in the sphere of energy audit, is that there is no uniform methodology in Russia for energy audit in the enterprises of housing and utilities sector and industry, in which the clear specified algorithm of the energy enterprise survey would be noted. There is no methodology, which is a benchmark for comparing the energy efficiency results of the analysis of the investigated companies. However, we must remember that the energy survey will not give anything if there is no investment in energy-saving technologies for housing and utilities sector and industry [4; 6]. It is necessary to borrow the experience from foreign companies which offer their modern energy-saving technologies and a clear methodology for conducting energy audits of the enterprises of all economic sectors, including enterprises of housing and utilities sector and industry [7; 8; 10; 16; 20; 22; 24; 25; 26].

We agree with the opinion of leading experts of energy audit that the subprogram "Energy supply saving in housing and utilities services", "Energy saving in energy-intensive industries," of the federal target program "Energy saving of Russia" has no relation with energy efficiency, because energy passports do not show the energy efficiency of the investigated companies and are only a simple inventory of energy equipment, buildings and structures.

It is important to mention that the experience of foreign countries proves that an energy audit within the energy-saving of housing and utilities services, industry and other sectors of the economy as a complex process needs the support of the state in our country. For example, in Finland the general direction of work in energy audit come from the Ministry of employment and economy of Finland. State center of economy, environment and transport offers subsidies to enterprises at the end of the work when the auditor has made the report and checked the quality of work with customer. There are different models of the audit for each sphere of the client's company . The energy audit placed in the hands of business and SRO in Russia, but in Finland the state is interested in the results of the energy audit, because it allocates the funds to the project expertise. In addition, the Finnish auditors submit reports of energy audit, but in Russia – energy passports [4; 12; 16; 27].

The experience of Finland proves that the energy audit is a comprehensive approach to the analysis of a company, it is not only a solution of a technical problems, but also the economic problems, and problems of production development based on the implementation of specialized technologies, on the saving of the environment - the reducing of emissions and the negative impact on it. For Russia, the Finnish experience can be valuable because they have comparing data of the energy saving potential, substantiated theoretically by energy auditors in the technical report, and the actual energy saving potential, realized over the years in their country. The accuracy of theoretical predictions of the Finnish "energy auditors" is about 70 percent [4; 12; 26].

Currently, the energy audit is provided by law, and formally its official status is confirmed. In practice, the term "energy audit" exists only in the law, its real functions are in formalities and uncertainties. Experts say that there is the need for new ideas, new rules, new decision-makers, and all this is not possible to approve by only one law.

Thus, there are problems in the implementation of the Federal Law №261-FZ ""About energy saving and energy efficiency increase and about changes in certain legislative acts of the Russian Federation" and the subprogram "Energy supply saving in housing and utilities services", "Energy efficiency in energy-intensive industries" on practice:

- the lack of clear rules and the absence of any budget for the implementation of the directions envisaged by the law;
- the lack of a single uniform of energy passport, the lack of clear requirements for its registration;
- the lack of competent professionals in the market of energy auditing services who are able to identify the reserves for possible savings of energy resources;
- there is no single clear methodology for energy audit or energy surveys of the enterprises in the utilities and industrial complex in the federal and regional level;
- the lack of investment projects to develop energy-saving technologies for the enterprises of the utilities sphere and also for the enterprises of housing and utilities services and industry;
- the lack of regional services, centers for the organization and implementation of an energy audit in the enterprises of housing and utilities services and industry of various management levels;

However, as we have already noted above there are a lot of difficulties, related with the release of the 261st Federal Law "About energy efficiency and energy saving", but not all of them may be provided by law.

4 The main ways to solve these problems

In our country, the attitude to the use of energy resources should be called wasteful, and it is rather a tradition than a temporary situation. The energy saving, unfortunately, is presented for us as a requirement of regulatory authorities. While the bases of careful attitude to electricity are laid in childhood abroad.

We agree with the Alexander Kuschinski, the head of the Centre of energy saving and energy efficiency increase of the Leningrad Region that we have to live ten years else to change the consciousness of citizens and the attitude to the concept of "economical consumption of resources." In the meantime, our mentality forces us to act very differently, even if the law is traditionally called upon to limit the rights.

According to General Director of NP "Soyuzpetrostroy-Energy Audit", the Candidate of technical science, Jakov Abugov, it is advisable to create a federal and regional (at the level of governors) coordination centers for improving the system of organization of the energy audit and energy saving. They should include the representatives of executive and legislative power, energy-governmental organizations, SROs, the various political parties. It is also necessary to strengthen the explanatory work among the population of energy saving through radio and television.

Thus, having considered many opinions, recommendations of leading economists and experts of energy audit, we have formulated the main directions to improve the effectiveness of the program realization of energy efficiency in the regional enterprises of housing and utilities sector and industry in practice:

- the creation of the federal and regional coordinate centers, organizing the work of complex energy survey of enterprises in utilities sector and industry, the results of which help to create priority courses of energy saving and their implementation;
- the development of a single methodology for energy audit at the federal level, as well as clear methodology of energy audits, energy-saving technologies, different models of the audit;
- the development of a single form of energy passport at the federal level (Ministry of Energy), as well as the establishment of clear requirements for its registration;
- the financial support from the state to conduct an energy audit in the regional enterprises of housing and utilities services and industry;
- the re-certification of the experts of energy auditing companies in according to their competence for the services provided by the energy audits of enterprises in the utilities
- the modification of the licensing criteria (more rigid) of the energy auditing companies providing services for energy survey of the regional enterprises of housing and utilities sector and industry;
- the public, private projects investment to develop energy-saving technologies in the enterprises in the utilities sector and in the enterprises of housing and utilities services and industry;

the change of the attitude in the need for energy-saving measures of regional managers of public utilities sector, industry and the general public, by strengthening advocacy on energy efficiency by means of radio, television and the media.

5 Conclusion

Thus, in order to achieve good results, it is necessary to involve the professionals of all sectors of the economy, all producers and consumers of various fuel-energy resources. For successful implementation it is the need for concernment and involvement of all the participants of the events, among which there should be the executive authorities and local self-government of each city and region of Russia. Only when the maximum mobilization of used resources and the optimal return of energy saving policy subjects the thermal characteristics may increase in housing and utilities sector and industry, the equipment of accounting means, automatic control of production processes and the use of fuel-energy resources may improve and also control of energy consumption may be realized more efficiently as possible. However, as ordinary citizens, as well as all services and authorities, should remember that natural energy resources are finite and non-renewable.

References

1. State program "Energy saving and energy efficiency increase until 2020", approved by the Government RU from 27 December 2010 vol. 2446-p. http://www.consultant.ru.
2. N. Borisova, A. Borisov, Economy and business, **49**, 728 (2014)
3. N. Borisova, A. Borisov, and E.Viprickaya, Economy and business, **12-3,** 412 (2015)
4. V. Bushuev, Energy policy **1**, 3 (2014)
5. N. Borisova., A. Borisov, and A. Taranova Construction Economics **39**, 66 (2016)
6. N Borisova, A. Borisov, *Actual problems of the introduction of energy efficient technologies in the construction and engineering of urban farming systems: Proceedings of the II International Scientific and Practical Conference*, **13** (2015)
7. R. Simanjuntak, P. Dupuis, L. Canale, N. Sinisuka, and G. Zissis, *IEEE Industry Application Society Annual Meeting*, (2014)
8. X. Zhao, H. Li, L. Wu, & Y. Qi, Energy Policy, **66**, 170 (2014)
9. Y. Kim, A. Aravkin, H. Fei, A. Zondervan, and M. Wolf, IBM Journal of Research and Development, **60(1)** (2016)
10. V. Blass, C. Corbett, M. Delmas, and S. Muthulingam, Energy, **65**, 560 (2014)

11. A. Sclafani, *ASME 2010 4th International Conference on Energy Sustainability*, **1** (2010).
12. L. Okkonen, and N. Suhonen, Energy Policy, **38**, 3443 (2010)
13. T. Fleiter, J. Schleich, and P. Ravivanpong, Energy Policy, **51**, 863 (2012)
14. S. G.Al-Ghamdi, and M.M. Bilec, Environmental Science & Technology, **49**, 4048 (2015)
15. P. Sun, and P. Nie, Renewable Energy, **74**, 255 (2015)
16. M. Marques, and R. Neves-Silva, Journal of Cleaner Production, **88**, 105 (2015)
17. Y.P. Gusev, Thermal Engineering, **62(3)**, 160 (2015)
18. V.U. Baldin, N.I. Danilov, and G.I. Khudyakova, *Energy Production and Management in the 21st Century*, (2014).
19. W.Y. Saman, L. Mudge, D. Whaley, and E. Halawa, *Sustainability in Energy and Buildings Smart Innovation, Systems and Technologies*, 247 (2012).
20. A. Al-Mofleh, S. Taib, and H. Al-Gulman, *2008 5th International Symposium on Mechatronics and Its Applications*, (2008)
21. W. Zhang, S. Liu, N. Li, H. Xie, and X. Li, Renewable and Sustainable Energy Reviews, **49**, 395 (2015)
22. A. G. Engelen, & J. S. Collins, *2010 43rd Hawaii International Conference on System Sciences*, (2010).
23. M. Abdel-Wahab, and D. Ali, International Journal of Green Energy, **10(2),** 137 (2013).
24. A. Mansikkasalo, and P. Soderholm, *2012 9th International Conference on the European Energy Market*, (2012)
25. M.O. Melo, L.B. Silva, A.S. Coutinho, V. Sousa, and N. Perazzo, Energy and Buildings, **47,** 35 (2012)
26. O. Maksimchuk, and T. Pershina, Sociology of City, **2**, 104 (2016)
27. O. Maksimchuk, T. Pershina, G. Golikova, N. Borisova, and S. Ivashova, *The concept of energy saving management in housing and communal services: a systematic approach*, **284** (2015)

Development of corporate associations of construction industrial enterprises and subjects of financial and credit environment by financial infrastructure merging

Ramil Mavlioutov[1,*], *Ekaterina* Butenko[1], and *Regina* Menyailova[1]

[1]Volgograd State Technical University, 400067 Volgograd, Russian Federation

Abstract. The article is devoted to solving important problems of the modern economy, which is the formation, development, and improvement of the efficiency of the combined corporate structures, including industrial enterprises of the construction industry, development directions and mechanisms of regulation of the process of interaction with the financial and credit environment, as well as the formation of corporations, financial infrastructure. According to the authors, it is the basis for the successful development of the industry.

An active role in the financial and credit support of industrial enterprises play a variety of financial institutions, especially commercial banks. Industrial enterprises and commercial banks are able to independently generate and conduct their own financial policies. Development of ways and forms of interaction economical industrial enterprise with the bank is among the most important tasks of strategic planning at the enterprise.

Economic relations is first of all, the enterprise the process of interaction with the various economic actors, including with financial and credit environment, in terms of location and construction of the most effective economic interactions [1-3]. This process is quite complex because of the different approach of economic actors in the process of formation and profit on the basis of its production activities.

Building a system of effective interaction with the elements of the environment is an important area of activity for the industrial enterprises, which allows not only effectively respond to changes in environmental conditions, but also to actively use the resources of the environment to enhance and improve its efficiency [4, 5].

The joint economic planning process should be defined [6]:

a system of interrelated objectives, taking into account their priority;

tasks that will have to decide on the various stages of preparation and implementation of banking products;

the mechanisms and methods of achieving these goals.

You must have at their disposal a reliable and sufficient information on the state of the financial and economic environment, the situation in the servicing banks. Primarily need information about the bank's management, its financial position and reputation.

The shares in the stock market and a loan at the bank are the main type of financial activity of the industrial enterprises of the construction industry.

Imperfection of the existing system and the mechanism of interaction of "industrial undertaking - the bank" in the conditions of transitive economy dictates the more radical forms of organizational and economic interaction of industrial enterprises with the bank in the direction of creating integrated (combined) corporate type structures to work on the final result in a single financial and technological chain [7-10]. The main advantages of large integrated corporate structures include: the lack of restrictions on the areas of activity, reducing business risks, the ability to attract an unlimited number of investors and the accumulation of significant amounts of capital, high capacity handling.

Development of modern integrated corporate structures are characterized by the increasing complexity of their internal financial and organizational structure, the expansion of economic ties, increasing the need for additional financial resources, the continuing concentration of capital, the increasing complexity of financial relations in the globalization process [11]. As noted earlier, a major role in the structure of modern integrated corporate structures play financial institutions: banks, insurance and investment companies, the purpose of the formation and activity of which is to ensure financial stability and development corporations.

The significance of combined (integrated) corporate structures in industrial production, their active development associated with the accumulation and concentration of capital and the increasing complexity of structures require adequate scientific analysis of these processes, the timely identification and resolution of problems of formation, regulation and improve the effectiveness of their activities.

* Corresponding author: ramil-2002@mail.ru

In the scientific literature [12-14] to determine the integrated corporate structures (associations) uses the term "corporation", "financial-industrial group", "holding", "complex economic systems", "Industrial and Financial Systems," "complex integrated structures" and others. The most commonly to refer to the combined corporate entities, the term "financial-industrial group". In general, financial and industrial group is the integration of banking and industrial capital and is treated as an association of legally independent parties fully or partially pooling their assets with a view to economic integration. Another term often used to determine the association, is the term "corporation", which, in this case, does not accurately describe the name of the association of legal entities, as it has a wider meaning and is used to characterize the company. The unique characteristics of the term "corporation" in the Russian economic and legal theory and practice is not, so when you use it is necessary to clarify some of the corporations in question is a corporation - a joint-stock company or corporation, which is a group of companies [15].

Formation and development of financial and industrial corporate associations determines the prevailing trends and prospects of the current process of corporate development:

strengthening the process of concentration of capital;

widespread associations conglomerate type;

the development of processes to optimize the asset structure in conglomerates;

globalization of economic growth;

increasing competition between corporations in different sectors of the market;

the emergence and development of new forms of integration;

growth in the number and volume of mergers and acquisitions;

an increase in the role of banks and other financial institutions in the creation and management of corporate entities.

As the external environment in relation to the combined corporate structures, the financial and credit system, the main elements of which are banks, insurance companies, pension and investment funds, is a financial infrastructure that enables corporations to attract the necessary financial resources and allocate temporarily free funds, which makes it possible effectively manage the size and structure of the association.

It was revealed that the considered approaches to the analysis of relations between the combined corporate structures and financial systems, are valid only if the structure of the corporate association lacking the financial and credit institutions. The situation changes when a part of the corporate association incorporated financial institutions. In this case, the two blocks are formed within the corporation: production (production and sales) and financial. Production unit, whose main task is the management of production and marketing of products consisting of industrial, commercial and other non-financial institutions [16, 17]. Financial unit includes a variety of financial and credit institutions: banks, insurance companies, investment and pension funds and other financial institutions, its task is to manage the financial corporation engaging and accumulation of financial assets.

With significant amounts of financial resources and implementing large-scale industrial activity, the corporation ceases to be a recipient of financial services, its demand for basic financial services can be satisfied with the help of its member financial institutions. In this case the financial and credit institutions of corporate associations are both part of the corporate structure and part of the financial system. This approach allows us to allocate the financial infrastructure of the composition of the external and internal components, which can be defined as: foreign financial infrastructure and domestic financial infrastructure. At the same time as the domestic financial infrastructure, understood as a set of financial and credit institutions that are part of the corporate associations, and acting in the interests of the association [18].

Financial institutions that make up the external financial infrastructure, act in their own interests, while the financial institutions that are part of the domestic financial infrastructure, acting in the interests of corporate associations. At the same time corporations are interested in the inclusion of financial institutions in its structure, as a result, they gain access to additional resources and opportunities, and provide their priority position in the competition. In the course of its activities and financial-credit organizations representing domestic financial infrastructure, have the opportunity to combine various financial functions, the main of which are: the centralization and redistribution of resources, internal corporate lending, management of internal corporate cash flows, joint activity in the stock market in the interests of association , co-insurance of risks and other functions.

At the same time, coming into a body corporate, industrial and financial organizations have different goals. Manufacturing enterprises in the merger is expected to receive additional financial resources needed for the modernization and development of production. Finance companies require to be able to more efficiently and with less risk to use available capital, to achieve a higher level of control over the invested capital to be able to monitor and control the corporation [19].

The complexity and diversity of the financial and economic relations in the economy contributed to a large number of different types of integrated corporate structures. Depending on the purpose of cooperation, degree of independence and the nature of economic relations between the parties, evolved a variety of options combining industry with financial and credit environment. Depending on the prevailing conditions and opportunities for association members to realize their objectives, the various forms of relationships established between them [19]:

financial and credit institutions provide financial services to isolated and do not have a significant impact on corporate activities of the association;

financial and credit institutions, providing comprehensive financial services and participate in the activities of corporate association, not being members of the association;

financial and credit institutions, are part of the corporate association, but their activities are limited and dependent nature of the provision of financial services;

financial and credit organizations are part of the corporate associations, and monitor its activities;

financial and credit organizations, manage the activities of the corporate association.

The first two forms of relationships correspond to relationships with external financial corporations infrastructure. The other three forms relate to the domestic financial infrastructure and relationships defined within the corporation [20]. Financial institutions that have these types of relationships are part of the corporate association, and can become a financial and management center under certain conditions.

The basis of the domestic financial infrastructure such corporate associations make up the banks, which often act as initiators of the establishment and other financial institutions: insurance companies, investment and pension funds, venture capital funds, brokerage firms.

The inclusion of financial companies in the corporation leads to the formation of the domestic financial infrastructure, which allows, due to the concentration of financial resources to carry out the structural development of even more rapidly.

The core of the domestic financial infrastructure includes banks that have the ability to concentrate large volumes of financial resources and ensure effective management. Other financial institutions that are part of the corporation, acting together with the bank and usually under its control. The Bank created a body corporate or incorporated into its structure, aims, and after a certain period of time becomes the center of the corporation. In the same way and the banks are not included in the corporation. While external financial infrastructure, they seek to capture or form their own body corporate, becoming thus its center (the parent organization) [21].

Consider a situation where a bank included in the corporate association, but its activity is dependent and limited to the provision of financial services. Educated commercial banks and industrial enterprises complex system aims of its existence, as well as any commercial entity maximizing the economic effect of their activities. The economic effect of the complex functioning of the system (EE_{sc}) is the amount of the economic benefit of the enterprise operation (EE_{ie}) and economic impact of the bank operation (EE_b):

$$EE_{sc} = EE_{ie} + EE_b. \quad (1)$$

The economic effect of the enterprise from product sales is maximal ($EE_{ie\ max}$) for a certain amount of sales (Q_{max}). Its increase can be achieved through the use of modern financial instruments: promissory note loans reduces costs and factoring leads to an increase in sales by accelerating the turnover of handling funds.

Consider the mechanism of the economic effect of the bank - the second part of the reporting system of the complex. The economic effect in general terms can be calculated as the difference between revenues and costs.

The economic effects of the company and the bank, has a maximum point, which correspond to certain parameters now attracting / bank providing financial resources, allowing to obtain the greatest economic benefit of the complex functioning of the system. The dependence of the income and expenses of a particular industrial plant construction industry LSC Sebryakovcement. LSC Sebryakovcement is a Soviet and subsequently Russian company the cement industry. It is located in the city of Mikhailovka of the Volgograd region. Optimum (in this case a maximum of) its economic effect from realization of production is reached at 14 627 t and is $1,305*10^8$ rubles.

However, to achieve it, the company needs to attract credit resources of the bank and shall be paid for the use of borrowed funds is the reward of his income.

To go from an absolute return performance relative to the manufacturing facility will take the duration of operating cycle of 90 days.

The economic effect of the production activity of the enterprise has an extreme (in this case the maximum), which corresponds to 16.9% of annual profitability.

By combining in one figure charts of income and expenses of the bank, get a graphical interpretation. Highs economic benefit obtained by the bank from interacting with businesses, are achieved for different curves for different sizes of bank fees levied:

Maximum income of the bank reached at a rate of 25.3% per annum and is $2.17*10^9$ rubles;

the maximum cost of the bank when the rate reached 26.4% and is $1.28*10^9$ rubles;

maximum economic effect is achieved by the bank at a rate of 22% per annum and is $9.05*10^8$ rubles.

The extremum is achieved when the yield of 18% per annum, while the value of the economic benefit of the complex (added value in the supply chain) is 2.2 bln. rubles.

Another example of inter-system interaction can be a bunch of Real Estate Investment Trust (REIT) with the industrial construction industry. REIT operate in the legal system of the United States since the 1960s [22],

The implementation of the construction project for the construction with the use of REIT can be represented as follows (fig. 1):

a step of forming REIT. Accumulation of assets (cash, land or rights of their lease, design and estimate documentation);

finally, the management company with the company-developer of the investment agreement, the subject of which is the construction of facilities on land that is part of a structure of the REIT assets;

transfer of property rights under the treaty obligations of the REIT assets;

trust transfer of the developer of design and estimate documentation,

company-builder obtaining a building permit;

implementation of construction and installation work for the company-developer of the funds received as funding under the investment agreement.

obtain company-builder permit for commissioning of constructed object;

the transfer of the object of the REIT assets.

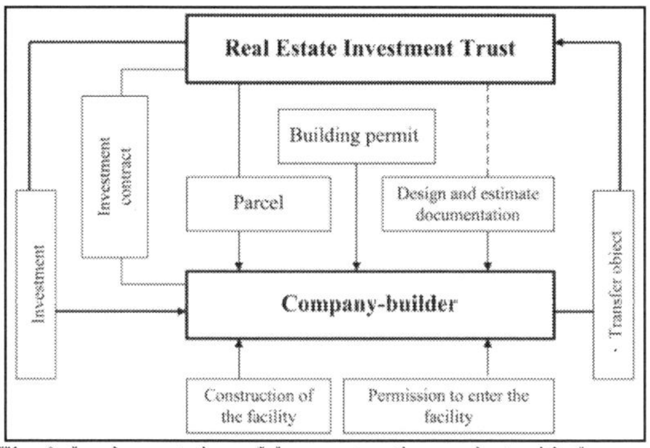

Fig. 1. Implementation of the construction project with the participation of Real Estate Investment Trust.

Industrial construction industry serves as a company-builder. REIT has the potential active participant in the construction market. With the possible direct investment trust for real estate construction projects.

In the Russian legal system analogous to the REIT is a unit investment fund of real estate. This kind of financial structure have been operating for 15 years. As of 2016, the number of unit investment funds of real estate exceeding 500.

The decision of important problems of the modern economy - the formation, development and improvement of the efficiency of the combined corporate structures, including industrial enterprises with a long production cycle, development directions and control the process of interaction mechanisms in the financial and credit environment, as well as the formation of corporations, financial infrastructure, in our opinion, is the basis for successful development of industrial production at the present stage of development.

Intersystem interactions between industrial enterprises and banks have both integrative and disintegrative character. This is due on the one hand the impossibility of the existence of each system separately from the other, and on the other, a conflict of interest between the cooperating parties.

In the system the interaction of industrial enterprise with a bank to obtain additional economic benefit one side automatically leads to the formation of a negative economic effect on the opposite side. This kind of operation is not aimed at the optimization of the interaction between the real and financial sectors, and in the solution of internal problems. For example, reduce the cost of the passage of payments, increased control over transactions produced, obtaining the mechanism of tax optimization and additional sources of financing.

In conclusion, we note that the individual is very important from a practical point of view, are conflict-free distribution of tasks resulting economic effect between the participants of the complex and the construction of intersystem organizational cash flow management services

References

1. L. Cai, J. Cui, H. Jo, Journal of Business Ethics, **139**, **3**, 563-594 (2016)
2. B.N. Cline, C.R. Williamson, Journal of Corporate Finance, **41**, 572-590 (2016)
3. O.K. Tosun, Financial Management, **45**, **4**, 953-979 (2016)
4. N. Vafeas, A. Vlittis, Financial Review, **51**, **4**, 481-506 (2016)
5. K.-W. Lee, G.H.-H. Yeo, Review of Quantitative Finance and Accounting, **47**, **4**, 1221-1250 (2016)
6. B.W. Goh, J. Lee, C.Y. Lim, T. Shevlin, Accounting Review, **91**, **6**, 1647-1670 (2016)
7. G. Jacoby, J. Li, M. Liu, European Journal of Finance, 1-20 (2016)
8. H. Rim, S.-U. Yang, J. Lee, Journal of Business Research, **69**, **9**, 3213-3219 (2016)
9. D. Cairns, Accounting in Europe, **12**, **2**, 187-196 (2015)
10. H. Deangelo, R. Roll, Journal of Finance, **70**, **1**, 373-418 (2015)
11. X. Chang, S. Dasgupta, Journal of Finance, **64**, 1767-1796 (2009)
12. H. DeAngelo, L. DeAngelo, T.M. Whited, Journal of Financial Economics, **99**, 235-261 (2009)
13. E. Fischer, R. Heinkel, J. Zechner, Journal of Finance, **44**, 19-40 (1989)
14. C.A. Hennessy, T.M. Whited, Journal of Finance, 60, 1129-1165 (2005)
15. B.St. John III, R. Arnett, Journal of Public Relations Research, **26**, **2**, 103-116 (2014)
16. M. Alzahrani, R.P. Rao, Financial Review, **49**, **1**, 89-116 (2014)
17. M. Baker, J. Wurgler, Journal of Finance, **57**, 1-32 (2002)
18. G. Bittlingmayer, S.M. Moser, Financial Review, **49**, **1**, 1-19 (2014)
19. A. Pérez, M.M. García de los Salmones, I. Rodríguez del Bosque, European Journal of Marketing, **47**, **1**, 218-238 (2013)
20. Mignon, A. Rüdinger, Renewable and Sustainable Energy Reviews, **65**, 478-488 (2016)
21. Z. Banhalmi-Zakar, Australian Planner, **53**, **3**, 221-231 (2016)
22. Real Estate Investment Trust Act of 1960

Evaluating the efficiency of functioning of integrated structures in industry

Larisa Suvorova[1,*], *Tatyana* Baibakova[1], and *Irina* Pestova[1]

[1]Economics Department, Vyatka State University, 610000 Kirov, Russia

Abstract. In the article the authors review approaches to evaluating the efficiency of the integrated structures in industry in order to improve the methodology of this evaluation. In contrast to existing indicators for evaluating the efficiency, which characterize the performance of some aspects of the integrated structures activities, the authors propose the methodology based on calculating and comparing general and individual efficiency indicators of current (operating), financial and investing activities on the integrated structure as a whole. These indicators are presented in the form of the coefficients. By means of mathematical analysis, the authors have found out the strong and direct correlation among the given indicators, which confirms the objectivity of choosing the indicators. The received results of evaluating the efficiency of functioning of the integrated structures are proposed for use in the developed algorithm of making managerial decisions. The validity of the developed methodology is shown on the example of the integrated structure of the joint-stock company "Kirov Dairy Plant".

Keywords: integrated structures, methods of evaluation, industry, efficiency of functioning

1 Introduction

In the field of economics, the organizational manifestation of the integration is forming the integrated structures. It is made with the help of integrating resources and capital in order to build the integrity of the production and technological, financial and economic chains in the company activities [1, 2].

The integrated structure is viewed as the union of industrial enterprises and companies on the basis of the technological principle in order to improve the quality parameters of the finished product, to optimize its production and marketing costs, to develop mutually beneficial economic relations, to consolidate individual functional structures [3]. The purpose of merging enterprises and companies into the integrated structure is to strengthen its competitive potential, to increase output through applying innovative technologies and increasing the competitiveness of their products.

One of the most important tasks of researching the activities of the integrated structure is to evaluate the efficiency of its functioning.

Nowadays there are three groups of methods for evaluating the efficiency of company functioning.

The first group of methods is based on using the quantitative approach, which allows to evaluate the cost indicators of the company's performance and their dynamics [2, 4].

Using the methods of the second group gives the possibility to make the calculation of indicators of quality of production, the level of qualification of the personnel, the level of technology and of diversification of production [5].

The methods of the third group rely on using the integrated approach, which helps us to evaluate the efficiency indicators of companies functioning [6, 7, 8, 9].

In spite of the existing extensive system of indicators used to evaluate and analyze the effectiveness of the company's activities, there is a need to develop the methodology for evaluating the efficiency of the integrated structure as the whole entity.

2 Defining the task of the research

Today there is a large number of publications concerning the evaluating the efficiency of functioning of the integrated structures [3, 5, 6, 10-17]. The analysis of scientific literature led us to the conclusion that evaluating the efficiency of functioning of the integrated structures is carried out with the use of quantitative, qualitative and complex approaches. The disadvantage of the existing approaches is that there are no causal relationships between the used efficiency indicators in the existing methods of calculation, which does not

* Corresponding author: larsuvorova@mail.ru

result in obtaining the fair evaluation of the efficiency of the integrated structure as a whole.

The aim of this study is to develop the methodology, based on calculating and comparing general and individual indicators of current (operating), financial and investing activities of the integrated structure as an economic system.

3 Developing the approaches to evaluating the efficiency of functioning of integrated structures

Modern scientists` approaches to evaluating the efficiency of functioning of the integrated structures are very diverse and they allow to determine the performance of their activities according to a number of positions. Let us consider some of the main approaches to evaluating the efficiency of functioning of the integrated structures (Table 1).

Table 1. Approaches to evaluating the efficiency of functioning of integrated structures [1].

The Name of the Approach	The Essence of the Approach
The principle of "Triple Helix"	The performance of the integrated structures is evaluated according to the performance of its three components: science, business, government
Contextual approach	The integrated structure has the contextual characteristic of useful results that can be obtained from the system, the evaluation of which shows the efficiency of its functioning
Program and target-oriented approach	The efficiency of functioning of the integrated structures is determined on the basis of evaluating the degree of target achievement. In this case, the evaluation is the ratio of the level of the target implementation with the costs for its achievement
Managerial approach	The efficiency of functioning of the integrated structures is determined at different levels of government: state, industries, region and enterprises
Synergetic approach	Evaluating the performance of the integrated structures is made on the basis of calculating the net synergetic effect
Structural approach	Evaluating the performance of the integrated structures can be carried out on the basis of the structural shift method, which relies on the assumption that the effectiveness is determined by the combined efforts of three factors: the growth of the national economy, the growth of an industry in the country, the growth of the regional economy
Economic approach	Applying different methods of functioning evaluation of the integrated structures, including income, market and costs methods
Aggregative approach	Making up the integral indicators of functioning evaluating of the integrated structures on the basis of its main characteristics, parameters, features and properties: competitiveness, exports, employment, innovation, manufacturing, relationship of participants, science and education, investments, gross value added, area, productivity
Adapting the idea of cross-efficiency	The approach allows to connect different levels of the integrated structures. A level is a standard, and the level of an individual item cannot be higher than the level of the cluster as a system
Adaptive approach	Adapting the idea of a dynamic normal and the order of indicators to evaluating the efficiency of functioning of the integrated structures

For evaluating the efficiency of functioning of the integrated structures one can use different groups of indicators (Table 2).

Table 2. Groups of indicators for evaluating the efficiency of functioning of the integrated structures.

Group of Indicators	Indicators	Authors
The main indicators of functioning	The increase in volume of output. The increase in profit growth. Improving parameters of product quality. The increase in labour productivity. The increase in productivity from the use of labor and capital investments. The decrease in costs in the production process.	Gordeev A. V. Donskova S. V. [2]
Indicators summarizing the results of economic activity	Indicative industry profits. Indicative profitability. Indicative expenditures for a unit of output.	Kosholkina L., Borhunov N., Rodionova O. A. [4]
Indicators of the financial aspect of activities	The increase in the value of the enterprise. The decrease in transaction costs. Savings on tax payments. The decrease in overheads.	Shmidt U. D., Romanova I. M., Mihina I. S. [5]
Indicators of the production aspect of activities	Reduced costs of productive work. Improved product quality. The expansion of product range. The growth of returns from the use of means and objects of labour.	

 DOI: 10.1051/shsconf/20173501007

Indicators of the investment aspect of activities	The increased investments in long-range assets. The increased investments in the development of new technologies in the production of traditional and new products. The increased investments in production equipment.	
Indicators of evaluating the integrated organization potential	Evaluating the peculiarity of funds. Evaluating the market concentration. Evaluating the growth potential of the market power. Evaluating the effect of logistics decisions in the field of supplies.	Sitnikova T. V. [6]
Indicators for selecting potential organizations in order to build the integrated organization and their evaluation	Evaluating the potential of strength and sustainability of the cooperative. Evaluating the market potential. Evaluating the effect of logistics decisions. Evaluating the merger of other organizations into the integrated structure.	
Indicators for monitoring the integrated company	Excluding the organizations, that have exhausted resource efficiency, from the structure. The formation of new integration connections among the entered organizations or the inclusion of new companies.	
Individual indicators	As for federal and state organizations it is the proceeds from selling the products of the companies that are members of the holding. As for regional and municipal organizations it is the number of employees.	Shubin S. A. [7]
Indicators of the efficiency of using raw materials	The level of food consumption by population. The size of the gross product. The yield of marketable products. Gross income. Net profit.	
Indicators revealing the degree of integration	The availability of raw materials for the Production Capability. The distance of transporting raw materials	

A large number of existing indicators allow to make diverse evaluation of activities of companies that are members of the integrated structure [8-17], but there are no indicators to measure the efficiency of functioning of the integrated structure as a whole.

4 Developing the methodology of evaluating the efficiency of functioning of integrated structures in industry

Multidimensional nature of integrated structures activities requires using a certain number of characteristics in the form of appropriate efficiency indicators while evaluating the efficiency of the integrated structures. The appropriate efficiency indicators form the system, which gives them unity in their mutual relations.

The authors propose the methodology for evaluating the efficiency of the integrated structures [18-22]. This methodology consists of three stages and it is based on calculating and comparing general and individual efficiency indicators of current (operating), financial and investing activities (Fig. 1).

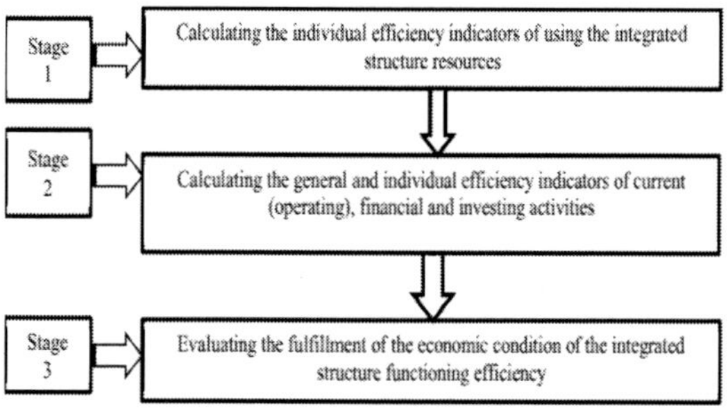

Fig. 1. Stages of evaluating the efficiency of functioning of the integrated structure in industry.

General and individual indicators for evaluating the efficiency of functioning of the integrated structures are presented in the form of dimensionless coefficients (Table 3).

The correlation coefficients were calculated by means of the program Excel using the "Data Analysis" package. These coefficients let us make the conclusion that there is a strong direct correlation among figures presented in Table 3.

After calculating the general indicators (E_{ca}, E_{fa} and E_{ia}) we matched their values in order to make future managerial decisions. The logical arrangement of the general indicators of the efficiency of functioning is determined by the technological expediency of functioning of the integrated structure.

Table 3. General and individual indicators for evaluating the efficiency of functioning of the integrated structures in industry.

A type of activity	Individual indicators of the efficiency of using resources	General indicators of the efficiency of functioning	
		Calculation formulas	Correlation coefficient
Current (operating) activity	R_f – fixed assets turnover ratio; R_t – circulating assets turnover ratio, turnover per year	$E_{ca} = R_f \times R_t$ E_{ca} – efficiency ratio of current (operating) activity	0.94
Financial activity	R_d – debt ratio; P_{bc} – profitability of borrowed capital, coefficient	$E_{fa} = R_d \times P_{bc}$ E_{fa} – efficiency ratio of financial activity	0.63
Investing activity	C_c – capital coefficient; R_c – circulating assets capacity ratio; I_c – capital intensity	$E_{ia} = C_c \times R_c \times I_c$ E_{ia} – efficiency ratio of financial activity	0.94

The developed algorithm of decision-making supposes the choice of actions according to the results of comparing the proposed efficiency indicators of the current (operating), financial and investing activities (Fig. 2).

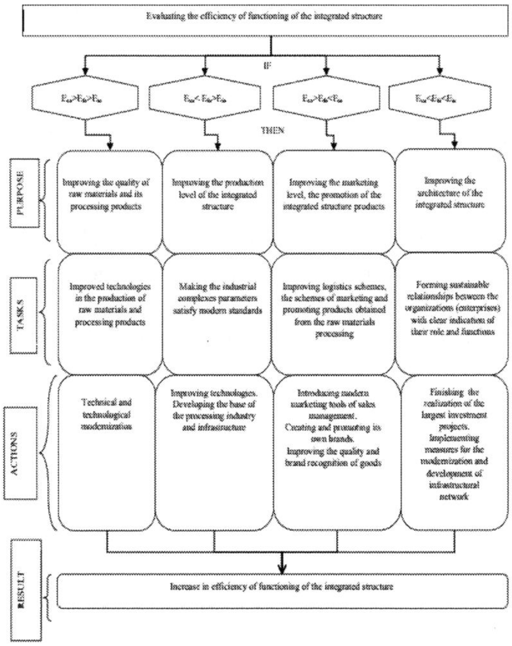

Fig. 2. The algorithm of making managerial decisions of the integrated structure in industry.

Under the condition when $E_{ca}>E_{fa}>E_{ia}$, the integrated structure has the task to improve production technologies through its modernization.

Receiving the condition when $E_{ca}< E_{fa}>E_{ia}$, the integrated structure needs to aim at developing the material-technical base and its infrastructure. It should be done in order to make economic security parameters of industrial complexes satisfy modern standards.

Under the condition when $E_{ca}>E_{fa}<E_{ia}$, the integrated structure needs to improve its logistics schemes of promoting finished products, to create and promote its own brands, to raise its brand name recognition.

In the case of the condition when $E_{ca}<E_{fa}<E_{ia}$, the integrated structure needs to focus on forming sustainable relationships between its constituent organizations with clear indication of their role and functions. This is solved through realizing major investment projects, implementing measures for modernization and development of the infrastructure network.

5 The practical importance of the research

The proposed methodology was tested on the data about functioning of the integrated structure of the joint-stock company (JSC) “Kirov Dairy Plant”, which is the leader of raw milk processing in the Kirov region nowadays.

According to the results of the evaluation it is found out that the condition of the functioning efficiency of the integrated structure of the JSC “Kirov Dairy Plant” has not been fulfilled (Table 4).

Table 4. Indicators of the efficiency of functioning of the JSC “Kirov Dairy Plant” before and after the integration.

A group of indicators	The name of an indicator	Before integration	After integration
Individual indicators of the efficiency of using resources	Fixed assets turnover ratio	1.28	1.87
	Circulating assets turnover ratio, turnover per year	3.34	4.50
	Debt ratio	0.27	0.51
	Profitability of borrowed capital, coefficient	0.41	0.87
	Capital coefficient	0.78	0.54
	Circulating assets capacity ratio	0.30	0.22
	Capital intensity	0.54	0.83
General indicators	Current (operating)	4.28	8.39

	activity, coefficient		
	Financial activity, Coefficient	0.11	0.45
	Investing activity, Coefficient	0.13	0.10
The actual fulfillment of the condition of the efficiency of functioning		$E_{ca} > E_{fa} < E_{ia}$	$E_{ca} > E_{fa} < E_{ia}$

The main reason for failing the condition is the inefficiency of the investing activity of the integrated structure of the JSC "Kirov Dairy Plant", which results in income deficiency.

The positive aspect in the integrated structure functioning is that after the integration all parameters, except for the general indicator of the investing activity efficiency, have significantly increased.

Having analyzed the problem, we offer the following management decisions in order to increase the efficiency of functioning of the integrated structure of the JSC "Kirov Dairy Plant" on the basis of the developed algorithm:

• The increase in the size of production and expansion of the product range of milk products as a result of modernizing production facilities,

• The increase in the number of participants of the integrated structure,

• The development of its own distribution system of produced dairy products on regional and interregional markets,

• The formation of customer loyalty by enhancing the brand of manufactured dairy products.

The realization of the proposed managerial decisions will enable the integrated structure of the JSC "Kirov Dairy Plant" to increase the efficiency of its functioning.

6 Conclusions

1. Our review of existing approaches to evaluating the efficiency of functioning of the integrated structures has revealed the absence of systematization in the indicators for evaluating their activities according to different groups of parameters.
2. We have offered the methodical approach based on introducing the related general and individual indicators of efficiency of current (operating), financial and investing activities of the integrated structure in industry as one entity.
3. The algorithm for making managerial decisions of the integrated structure has been offered. The essence of this algorithm is choosing actions according to the results of comparing the proposed efficiency indicators of current (operating), financial and investing activities.
4. Due to the practical application of the proposed methodology we have found out the causes of ineffective functioning of the integrated structure of the JSC "Kirov Dairy Plant". So, a number of recommendations for realizing certain managerial decisions in the field of investing activities have been given.

References

1. U.N. Androsuk, Economics and Management. Evaluating the performance of the cluster functioning and the sustainable development of the enterprise, **1**, 65-70 (2015)
2. A.V. Gordeev, O.A. Maslennikova, S.V. Donskova, N.K. Dolgushin, A.H. Zaveryuha, E.V. Ulyanov, *Economics of food industry enterprises* (Agrokonsult, Moscow, 2003)
3. T.V. Baibakova, *The system for evaluating the financial condition of the integrated unities of dairy of agro-industrial subcomplex in the Kirov region. Monograph* (Vyatka state university, Kirov, 2015)
4. L. Kosholkina, N. Borhunov, O.A. Rodionova, AIC: Economy, management. Methodology for determining the parity relations in the agricultural sector, **1**, 43-48 (2009)
5. Yu.D. Schmidt, I.M. Romanova, I.S. Mihin, The economic analysis: theory and practice. Complex technique of evaluating the efficiency of the integration processes in bakeries, **30**, 13-19 (2009)
6. T.V. Sitnikova, Possibilities of forming the vertically integrated industrial enterprise. http//uecs.ru/uecs54-542013/item/2204-2013-06-20-08-16-53
7. S.A. Shubin, *Development of the mechanism of forming and evaluating the performance of the vertically-integrated structures* (Russian state library, Moscow, 2007)
8. Kin Lo, and Z. Thomas Lys, *The Ohlson Model: Contribution to Valuation Theory, Limitations, and Empirical Applications: Sauder School of Business Working Paper* (2000).
9. The model of Edwards-Bell-Olson (EBO). http//www.valnet.ru
10. J. Spengler, Journal of political economy. Vertical Integration and Antitrust Policy, **58**, 347-352 (1950)
11. O. Willson, *New developments in the analysis of market structure. Cambridge (Mass). Vertical Integration and Related Variation on a Transaction-Cost Economics Theme* (1986).
12. A.E. Kozhin, *Economic efficiency of the vertical integration of timber enterprises (in the example of the Arkhangelsk region)* (Russian state library, Moscow, 2003)
13. P.Yu. Poluhin, *Evaluating the efficiency of enterprise integration in the sphere of industrial production* (Russian state library, Moscow, 2007).

14. A.A. Popov, *The integration of economic entities in the market economy* (Russian state library, Moscow, 2005).
15. O.A. Rodionova, N.A. Borhunov, O.A. Grebenkova, *Large and small unities: analysis, trends of development and mechanisms of interaction* (NIPKTS Voshod-A, Moscow, 2013)
16. M.F. Tazetdinov, *Evaluating the economic efficiency of enterprise integration* (Russian state library, Kazan, 2006)
17. Yu.V. Yakutin, Russian Economic Journal. Conceptual approaches to evaluating the efficiency of corporate integration, **5**, 71-81 (1998)
18. T.V. Baibakova, Economy, labor, management in agriculture. The sectoral approach to evaluating the efficiency of the integrated agrounities, **1**, 61-63 (2016)
19. T.V. Baibakova, Economics and Management: problems and solutions. Evaluating the efficiency of the integration processes in the agro-industrial complex, **2(56), 8**, 119-123 (2016)
20. L.A. Suvorova, I.V. Pestova, Economics and Management: problems and solutions. The analysis of approaches to evaluating the synergetic effect of functioning of the biotechnology cluster, **2(56), 8**, 59-64 (2016)
21. L.A. Suvorova, *The synergetic effect of the industry clustering: the analysis, evaluation, forecast. Monograph* (the Vyatka State University, Kirov, 2015)
22. L.A. Suvorova, *III International Scientific and Practical Conference. Sustainable Development of Society and Economy. Integration, clustering and synergy as the constituent elements of sustainable economic growth* 397-398 (2016)

Innovative means of energy efficiency management for regional industrial enterprises

Olga Maksimchuk[1], and *Tatyana* Pershina[1*]

[1]The department of management and urban development and construction, Volgograd State Technical University, Volgograd, Russia

Abstract. Currently in the Russian Federation one of the most important strategic goals set by the President of the Russian Federation, is to reduce by 2020 energy intensity of domestic economy by 40%. On the way to achieving this goal, all organizations (institutions) of business, government, society and households should be energy efficient. The paper dwells upon the problem of rational use of energy by regional enterprises. It discusses the theoretical and methodological aspects of developing software that combines the functionality of a training program for beginners and that of universal assistant for energy audit and management practitioners. «sto: Energoauditor.1» has been tested by budget-funded companies and enterprises in Volgograd Oblast to prove the feasibility of using it for making energy audits or filling energy declarations.

1 Introduction

Currently, one of the most important strategic purposes set by the President of the Russian Federation, is to reduce the energy intensity of the national economy by 2020 by 40%. Towards achieving this purpose, all companies (institutions) of business, government, the community and the household must be energy efficient [1]. Energy auditing survey, as one of the main tools of the movement to energy efficiency, does not allow to organize continuous monitoring and energy analysis, which are not free of errors, mistakes, subjective assessment and corruption schemes in implementation so the alternative methods and energy saving management tools are necessary at the local level [2].

The appeal to this theme is also due to problems and trends in the development of modern society, taking into account the priorities of energy saving and energy efficiency of vital activity. According to the assessments of International Energy Agency, the sizes of the effect from energy saving are equivalent to almost 2/3 of the export of natural gas in Russia [3].

The results analysis of implementation of the Federal Law of November 23, 2009 no. 261 "About energy saving and energy efficiency increase and about changes in certain legislative acts of the Russian Federation" has shown the failure of some of the requirements.

In particular it concerns the realization of energy audit surveys. According to the LLC "Center of Energy Efficiency INTER RAO EES", the proportion of companies is no more than one-third which is obliged by the law to carry out mandatory energy audits and actually carried it out.

In the Energy Ministry only 10% of the total received energy passports are registered without comment. All the rest are returned on completion. Often the cause of low audit quality is the choice of unqualified executor that, in turn, happens because of a lack of skilled customer approach to the selection of executor and to control its operation [4].

But it's not the only problem faced to the executive authorities of the R.F. which are responsible for the law implementation. Companies and enterprises in the region are engaged in energy saving by exploiting the old equipment, technologies and they do not seek to develop innovative products and services [5].

To increase the activity and the quality of decisions in the sphere of energy-saving, Energy Ministry of Russia proposes to introduce the rating system of the energy efficiency levels in the regions, based on the off-budget financing of energy-saving programs. 4.94 billion rubles were allocated in 2014 for the development of the state program "Energy efficiency and energy development" in 25 regions [6].

There is necessary to be the leader of the rating to be included in the number of funded regions. But how to achieve this? What actions are necessary to do for this purpose? And what tools to use? In this article, the authors offer answers to these questions.

The need for increasing the effectiveness of the law implementation at all levels of the government has demanded the increase of the specialists' qualification level in the sphere of energy saving and energy management and the development of appropriate instruments to improve energy efficiency of enterprises

* Corresponding author: tatpersh@yandex.ru

and companies of the region, that determined the relevance of this article.

2 The theoretical rationale of tool development expediency of energy efficiency increase in the industrial enterprises

From the standpoint of a systemic and integrated approach the authors fully explore the problem of energy management development at the level of enterprises and companies, the private moments, such as energy auditing survey and the development of energy efficiency programs, which are paid more attention to the practice [7, 8,9,10,11,12,13].

In earlier publications, the authors present the results of the development of energy-economic survey techniques, allowing to realize the survey by the enterprise experts. The survey covers the full cycle of energy expenses management and it is the important management tool which does not require capital expenditures and outsourcing specialists (as in the case of energy audit survey) [14,15,16].

The methodology of energy-economic survey complements the current existing energy audit methodology and software to it in terms of the detail of all the procedures of: 1) energy survey; 2) preparing to the report of energy auditor; 3) development of a comprehensive energy efficiency program.

The first two sets of procedures are sufficiently formalized, but in practice there are non-standard situations, requiring adaptation to the traditional approach to the conditions, such as lack of comparable retrospective measurement base for the company, which rents the room and change location every year. Also the industry aspects are important, which requires skill and qualification level of energy auditors [17].

On average, according to the subjective opinion of the authors and calculations made on the basis of observations in the course of energy auditing and survey work on its own, the loss of time related with the rectification, recalculations, rechecking, arising due to the above circumstances, are 40-50% of the total time spent on energy audits and preparation for the report.

With regard to energy efficiency programs, the content analysis of available energy audit reports in Internet resources, suggests replicating of the typical energy saving measures in the most cases (for example, with the phrases "energy saving lamps", "installation of energy-saving / plastic window frames", "facade insulation " 95% of passports are mentioned).

Typical activities do not always and necessarily lead to the realization of energy saving effect and, in some cases, directly affect the growth of the energy losses [18,19,20].

The author's approach to the development of energy efficiency programs is fundamentally different from the widely practiced: Energy Efficiency Program - a set of investment and innovative solutions and measures to ensure their implementation in the short, medium, long-term periods, taking into account the actual conditions and opportunities for the implementation of funds from the cumulative effect of fast payback measures.

The subject of researches are the conditions and prerequisites for the optimization of energy saving management, the object of the study are the technical-technological and economic aspects of the organization of energy management at the level of companies and enterprises of the region.

Thus, the author's approach allows to resolve the conflict between the regulatory requirements and the real conditions, to harmonize technical requirements of energy audit and economic requirements of energy efficient program development. This tool is management software of local energy efficiency. A particular advantage is the ability to comply with the requirements of confidentiality. Currently, energy-economic survey methodology has the status of "know-how" and the software "sto: Energoauditor.1" - the Certificate of the state registration of the computer program number 2013614783, " sto: Energoauditor.1".

It would be logical to use those software products that are being already developed, but their analysis showed that, unfortunately, none of the existing software matches in full the demands made on them. For example, ARM SRO "Energy Passport" allows only to fill the energy passport and has no additional functions automation calculations, conversion to TOE etc. More successful software product is an automatic working place «E-Pass», which is intended for automation of energy auditors' work for the formation of energy audits results, but it is not suitable for the realization of energy expenses control function in the enterprise.

There are foreign software products in the Russian market, such as AMR (Automatic Meter Reading), PLC (Power Line Communication), ENSI (Energy Saving International), Smart Grid and their localized Russian versions are created [21,22,23].

However, the work with these products is available only to trained specialists- energy auditors, and the authors initially set a goal - to develop software as a tool of local energy efficiency management, that is, directly at the level of companies (institutions) and business enterprises, government, the society and households.

3 The results of software development

In the basis of the software product "sto: Energoauditor.1" is the DBMS Firebird, which is free for the distribution and use, and is ideal for the combination of factors: ease of maintenance / free / speed (much higher Access). The DBMS Firebird offers concurrency, high efficiency, and language support for stored procedures and triggers. This is a commercially independent project of C and C ++ programmers, technical advisors and supporters of the formation and strengthening of multi-platform management systems of relational databases, based on the source code released by Inrise Corporation (now known as BORLAND SOFTWARE.

FireBird – developed industrial software, guaranteeing transactional integrity of the data when using it with multiple users, corresponding to the ACID rules:

1.Atomicity — atomicity of transactions;

2.Consistency — data integrity;

3.Isolation — isolation (user access control);

4.Durability — long duration of data storage [24].

The software designed by the authors can work with the use of a DBMS FireBird:

1) locally, independently of any installed programs, with its database (Firebird Embedded Server mode);

2) in a network with a single base passport data (requires server installation and configuration of client connections);

3) when it is necessary to organize the work via the Internet (with a base located on an accessible server). Customers can work with a single database of passports from anywhere.

The software allows to prevent the likelihood of technical and computational errors and reduces the time required for data processing and preparation of the conclusion.

The advantages can also include:

1) it is an effective tool for training the specialists of energy audit and energy management at the expense of function simulation of practical situation with specified parameters, that is an effective training for audience;

2) it allows to optimize the cost of the survey, compared with the involvement of external energy auditor;

3) it increases the qualification level of employees in the learning process.

The software product "sto: Energoauditor.1" has several beneficial features that distinguish it from other types of software. The first important feature is the function of filling the energy declarations by companies of budget sphere.

Energy declaration is document containing information about energy consumption in the organization of budget sphere during the year, taking into account program obligations improving energy efficiency. Declaration is filled every year after the end of the calendar year by the person who is responsible for energy saving and energy efficiency increasing of company, separately for each building and structures. They are much easier to fill than the energy audit passports, and it is intended that the companies financed from the budget have to do it by themselves, it will bring the cost savings.

However, in practice, the institutions face with difficulties, which helps to solve the software product "sto: Energoauditor.1". It operates the online help system, which contains the legal and background information, legal acts and standards, which help specialist / user to orientate freely with warning about error probability in the vast amount of regulatory information. In addition, the "one hundred: Energy auditor.1" gives a choice probability of a universal energy-saving program precisely those activities that are suitable for the enterprise, as well as filling the energy passport / energy declaration in accordance with the requirements of the Russian Ministry of Energy.

The using of software allows to make the registration process of the survey results / the filling of energy declarations more convenient and simpler, that allows to reduce expenses and to improve the efficiency of enterprises and companies of different ownership forms.

Today using the software " sto: Energy auditor.1" in accordance with the requirements of the Federal Law of November 23, 2009 N 261-FZ "About energy saving and energy efficiency increase and about changes in certain legislative acts of the Russian Federation", the authors developed and implemented the training programs for young specialists (bachelors, masters) and practitioners. Program sets of higher education and qualification increase are shown in Fig. 1 according to the energy management profiles.

DEVELOPED COURSES

BACHELOR'S AND MASTER'S	ENERGY EXPERTS
TRAINING + SOFTWARE PRODUCT + IMITATION MODELING	TRAINING + SOFTWARE PRODUCT + FREE TECHNICAL SUPPORT + CONSULTING SUPPORT

Fig. 1. The program sets of higher education and qualification increase according to the energy management profiles.

In addition, the simulation modeling function of practical situations is built-in the software product (e.g. there is the task of the realization of energy audit survey according to the set parameters). This situation allows to combine the functions of the simulator for beginners and the universal assistant for practitioners in energy audit and energy management.

4 The results of software product approbation «sto: Energoauditor.1»

As a result of the partial approbation of the proposed software product for improvement the energy efficiency in the various companies and enterprises in the region (on materials of the Volgograd area) the audience of the training programs has implemented the energy savings potential assessments and has developed the energy efficiency Programs (Table. 1).

Table 1. The exapmle of the energy saving potential assessment of the enterprises of different ownership.

Companies	Dimension	Energy saving potential			
		Total	Heat energy	Electric energy	Fuel
Consumers					
«Volgograd Building Technical College»	TOE	28.11	24.49	3.62	-
	Rub.	269.69	206.99	62.704	-

«Uryupinsky agro-industrial Technical College»	TOE	42.44	21.26	21.18	-
	Rub.	439.84	179.31	260.53	-
LLC «Electrostroy snab»	Rub.	2708.9	1251.5	1457.4	-
Distributors					
Management company LLC «Zhilcomservice»	TOE	38.39	29.07	9.32	-
	Rub.	387.6	259.64	127.96	-
Homeowners «Zhilservice»	TOE	26.67	19.01	7.66	-
	Rub.	265.90	202.02	63.884	-
Management company LLC «Our House»	TOE.	68.49	68.49	-	-
	Rub.	473.31	473.31	-	-
Suppliers of communal services					
CBM «Volgograd utilities»	TOE	4185.3	4086.4	357.68	258.8
	Rub.	41643.9	34530.5	4399.5	2713

(*Source:* compiled on the basis of authors' calculations)

Currently, the developed energy saving programs implemented at listed enterprises and companies in the Table 1. The example of energy saving measures Program of "Elektrostroysnab" is presented in Table 2.

Table 2. The example of the energy saving measures program of the LLC "Electrostroysnab"

Description of measures	**The annual economic effect, thous.rub.**	**Payback period, years**	**The cost of measures, thous.rub.**
Average payback measures			
Cogeneration installation Viessmann Vitobloc 200	1400	2.32	3250
Thermorenovation of administrative building	210.1	2.1	441.755
Installation of photovoltaic roof tiles TEGOSOLAR	392.4	2.9	1150.6
Installation radiator thermostats RA-N20	114.9	2.5	287.6
Long-term measures			
Installation turbines	152	3,6	548
The introduction of commercial accounting of electric power automated system	439.59	3.25	1430
IN TOTAL	2708.99		7107.955

(*Source:* compiled on the basis of authors' calculations)

The main measures of the energy saving program (Table 2) concern the electricity energy savings, as these measures will be used as exhibition samples to potential buyers. Energy saving measures program is developed for the long term. It is planned to make a profit already in the first year in the amount of 2708.99 thousand rub. when the total expenses are 7107.955 thousand rub. In the future it is also planned to make a profit depending on the payback period of energy-saving measures.

It should be noted that some measures are not for only saving energy, but also for the good environmental situation, which, if possible, should be reflected in the total indexes.

The use of the software product, for example, during the energy audit survey in the institution "Volgograd Building Technical College" has allowed to reduce the time required to prepare a report by 36% and to save 185 thousand rubles due to the fact that the survey was conducted by the training specialists. Commercial offers from energy auditing companies ranged from 200 thousand to 275 thousand rubles, and the training of specialists according to the qualification increase program was 15 thousand rubles. The implementation of the energy saving measures Program in the companies has allowed to reduce energy expenses by 15.8% in the second year. It carried out both by own funds and through debt.

In addition, the use of software product has allowed experts of the Regional Center for Pricing Formation in Construction of Volgograd region as soon as possible to fill the energy declaration of their companies and avoid the penalties.

The software covers the full cycle of energy expenses management and is an important management tool, which received the approval from not only specialists of housing and communal services, but from enterprises of the building and the industry, that is evidenced by the conclusion of contracts and approbation applications.

But it is necessary two powers to change the problem situation. The first power - from the top, i.e. not just calls, but the real actions from the President, the government to the whole administrative system. Fortunately, there are so actions [25]. A second power is the team of trained professionals who are capable in the conditions of total energy savings to direct their companies and enterprises to a new level of energy efficiency with the help of the software product " sto: Energoauditor.1".

5 Conclusion

Received theoretical and methodological results have a bright practical focus and relevance, can be applied in the development process and implementation of energy savings policies aimed at the formation and effective implementation of the energy saving potential of enterprises and business companies, government, society and households in the region. The energy efficiency purposes of the national economy can be achieved as a whole only by joint efforts of energy efficiency management at their local levels.

References

1. M. Belyaev, O. Maksimchuk, and T. Pershina, *Management of energy consumption in the enterprises of housing and communal economy* (VolgGASU, Volgograd, 2009)
2. S. Gheorghe, N. Golovanov, H. Albert, C. Stanescu and D. Ilisiu, *Proceedings of International Conference on Harmonics and Quality of Power, ICHQP*, **430** (2014)
3. V. Dobrodey, Regional Economy, **2**, 220 (2015)
4. I. Bashmakov, Energy saving, **1**, 29 (2011)
5. A. Semenov, *Development of the economic mechanism of national innovation system functioning* (Saratov, 2014)
6. N. Borisova, A. Borisov, *Proceedings of the II International Scientific and Practical Conference*, **13** (2015)
7. C.A. Jenne, R.K. Cattell, Energy Economics, **5**, 114 (1983)
8. M. Marques, R. Neves-Silva, Journal of Cleaner Production, **88**, 115 (2015)
9. Y. Kim, A. Aravkin, H. Fei, A. Zondervan, and M. Wolf, IBM Journal of Research and Development, **1** (2016)
10. V. Blass, C.J. Corbett, M.A. Delmas, and S. Muthulingam, Energy, **65**, 560 (2014)
11. V.U. Baldin, N.I. Danilov, G.I. Khudyakova, WIT Transactions on Ecology and the Environment, **2**, 1208 (2014)
12. G.J. Tsekouras, M.A. Tsaroucha, C.D. Tsirekis, A.D. Salis, E.N. Dialynas, and N.D. Hatziargyriou, International Journal of Electrical Power and Energy Systems, **6**, 1222 (2011)
13. S. Rajabi, S. Behairy, Procedia Engineering, 1088 (2016)
14. O. Maksimchuk, T. Pershina, G. Golikova, N. Borisova, and S. Ivashova, *The concept of energy saving management in housing and communal services: a systematic approach* (Crouton, Volgograd, 2015)
15. T.A. Pershina, *The system development of the energy expenses management in the enterprises of housing and communal services* (Volgograd, 2006)
16. T.A. Pershina, Journal of Volgograd State University, **7**, 133 (2008)
17. A. Thumann, Eric A. Woodroof, *Handbook of financing energy projects* (The Fairmont Press, Lilburn, 2005)
18. V.F. Hershkowitz, *Energy-saving systems of residential buildings. Manual for designing* (2008)
19. C. Diepering, I. Brand, W. Vermeulen, Energy policy, **6**, 773 (2004)
20. O. Seppänen, G. Goeders, *Benchmarking Regulations on Energy Efficiency of Buildings. Executive summary*, **5** (2010)
21. A.G. Van Engelen, J.S. Collins, *Proceedings of the Annual Hawaii International Conference on System Sciences* (2010)
22. A. Al-Mofleh, S. Taib, H. Al-Gulman, *Proceeding of the 5th International Symposium on Mechatronics and its Applications* (2008)
23. R. Hackl, E. Andersson, S. Harvey, *Chemical Engineering Transactions*, **21**, 301 (2010)
24. H. Borry, FireBird. *Guide Database Developer* (BHV-Petersburg, St. Petersburg, 2006)
25. V.V. Zubarev, V.A. Irkov, N.A. Korgin, Management problems, **1**, 27 (2012)

Modeling the dynamic mechanism of the innovative systems self-development

Leonid Baev[1], *Oleg* Golikov[1], and *Natalya* Pravdina[1,*]

[1]High School of Economics and Management, South Ural State University (National Research University), Chelyabinsk, Russia

Abstract. The article is devoted to the creation of efficient economic mechanisms of innovative systems self-development regulation. In the current social and economic conditions stimulating innovative development remains one of the most important tasks of the state economic policy. One of the instruments used to activate the innovative processes is the territories of advancing social and economic development. A special legal regime in such territories provides various preferences for their residents, including tax benefits. The criteria of the preferences provided to the entities are formal, and the management mechanisms of the territory are "indistinct" and subjective. In this regard the development and research of the mechanisms stimulating positive dynamics of economic development according to the actual growth of production efficiency are presented.

1 Introduction

The problem of achieving sustainable increase in productivity and production efficiency is always relevant for the socio-economic system of any country regardless of the management level. Many economists associate solving this problem with the intensification of investments to innovative technology, often reducing development constraints to the insufficiency of financial resources. While recognizing the importance of investments in real assets as well as the feasibility of increasing such investments in the budgets of developing countries, it is very important to pay attention to the managerial component of the development processes. What proves the importance of management quality is the low developmental dynamics our country had in 2004 through 2014. Over this period, the annual average oil prices rose from $28.5 per barrel in 2003 to $99.5 per barrel in 2014, i.e. by 249%. GDP growth amounted to 48.7%, which is below the world average of 49.2% [1].

The economic crisis of 2014–2016 only worsened the situation. The objective now is to reach the world average economic growth figures by 2019 [2].

It is important to note that it is not possible without activating the stimuli that would intensify economic growth and promote better economic management.

The lack of economic mechanisms for regulating hi-tech industrial development has been noted before [3, 4]. At the same time, targeted economic benefits of various kinds that are available to special economic zones, tech parks, industrial sites and innovative groups, often do not have any positive effects because they are mostly implemented on formal grounds [5, 6, 7, 8].

We believe that such formal approach is still visible for instance in the concept of establishing the so-called advancing socio-economic development territories.

Territories of advancing socio-economic development, or TASED, are offered as a tool for intensifying innovation processes 'on-site'. In accordance with the Federal Law [9], TASED is a part of a Russian Federation's entity, which features a special legal regime for entrepreneurship, aimed at creating favourable environment to attract investments, accelerate socio-economic development and create comfortable conditions for the population.

A special legal regime includes a wide range of various preferences to the residents of such TASED, with an emphasis on tax reliefs and exemptions from contributions to state non-budgetary funds.

To date, open information sources contain no details about in-depth studies of how efficient and effective the actually existing TASEDs are. When analysing the legislation and management methodology association with TASED, one can conclude these documents describe no clear mechanisms, tools and criteria for the management of TASED. This casts doubt on the possibility to guarantee the expected outcomes.

For instance, [10] suggests that the Presidential Decree on Assessing the Performance of Executive Bodies in the Entities of the Russian Federation [11] will suffice to compensate the lack of methodological recommendations on how the state could assess the performance of TASED.

Analysing the criteria laid down in that Decree as well as the author's proposals the paper[10] reveals the 'indistinct' and subjective nature of such criteria which are aimed at the ease of measurement while offering no

* Corresponding author: natalyf75@mail.ru

option to objectively converge multiple heterogeneous indicators into an integral criterion.

In this regard, it seems appropriate to develop and research the mechanisms that promote positive economic development; such research should be based on production performance rather than formal grounds. It is the dynamic increase in production performance which we believe is the main sign and objective of innovative development [12, 13].

It seems appropriate to search for the solution to this problem by studying enterprise self-development mechanisms while implementing development-intensifying regulators within the same framework.

The goal of this research is to disclose the essence of the enterprise self-development mechanisms and to substantiate the opportunities for controlling this process in terms of the primary objective of innovative development, which is increase in production performance.

The object of this study is a commercial enterprise. The subject is the self-development mechanism and the impact the intensifying feedback could have on it.

2 Economic mechanism of enterprise self-development

Let us analyse how an enterprise could regulate its self-development; the analysis is to be based on the simplest extended reproduction model. In the absence of credits and where only the corporate tax is taken into account, the extended reproduction model looks as presented in Figure 1.

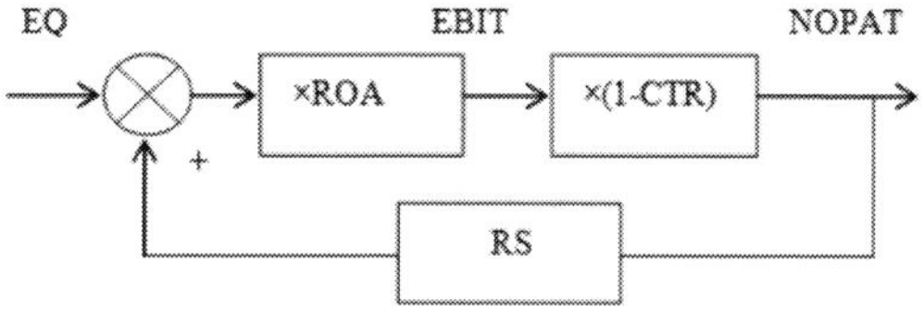

Fig. 1. Extended reproducing model.

This model uses the following abbreviations: EQ for Equity, ROA for Return on Assets, EBIT for Earnings before Interest and Taxes, CTR for Corporation Tax Rate; NOPAT for Net Operation Profit after Tax, and RS for Rate of Savings.

Speaking of self-development of a commercial enterprise, one should note there is a number of positive feedbacks in the extended reproduction mechanism. Given EQ with a certain ROA value, one can obtain an EBIT value. By calculating the CTR-based income tax, we obtain the NOPAT value. Re-investing a portion of NOPAT with a certain RS value results in extended reproduction and enables further innovative self-development and performance increase. This, in turn, further increases NOPAT and enables further development.

In accordance with this model, the rate of return on invested capital is written as

$$ROI = \frac{NOPAT}{I_0} = \frac{ROA(1-CTR)}{1-ROA(1-CTR)RS} \quad (1)$$

Where ROI – Return on Investment;
$NOPAT$ – Net Operation Profit after Tax;
I_0 – Investments;
ROA – Return on Assets;
CTR – Corporation Tax Rate;
RS – Rate of Saving.

Speaking of innovative development, it should be borne in mind that maxing out the innovative potential of a project in particular and of an innovative technology in general is an S-curve rather than linear [14, 15, 16, 17, 18]. To refine the model, let us describe this curve with Pearl's formula [15, 18], by varying the parameters of which we could handle the specifics of innovative development.

$$y = \frac{L}{1+ae^{-b(t-t_0)}}, \quad (2)$$

Where L is the upper limit of variable y;
a and b are the positive parameters affecting the curve;
t is time.

Using the S-shaped function of Pearl's formula as a mathematical description allows one to specify how the ROA shall alter change so as to handle to a large extent the specifics of innovative business development.

Figure 2 presents an S-shaped curve showing how the ROA alters as innovations are implemented. The curve is constructed on the basis of the equation (2). The maximum value of return L equals 25%, whereas the a and b values are chosen so as to obtain a curve with a maximum correspondence to the patterns of innovative development.

Figures 3 and 4 present the dynamics of ROI and ROI growth rate in view of the S-shaped alteration of economic returns.

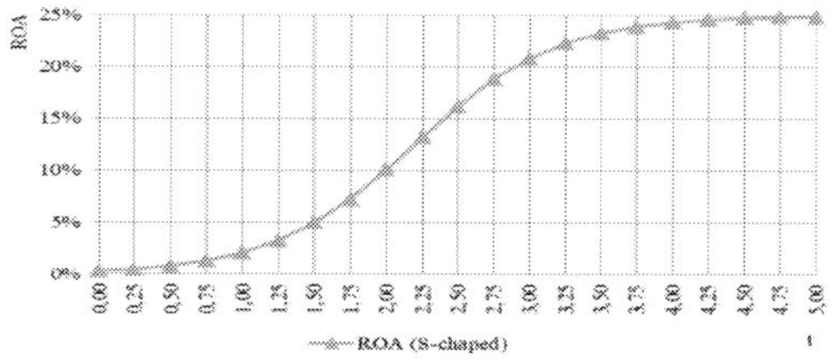

Fig. 2. S-shaped dynamics or ROA.

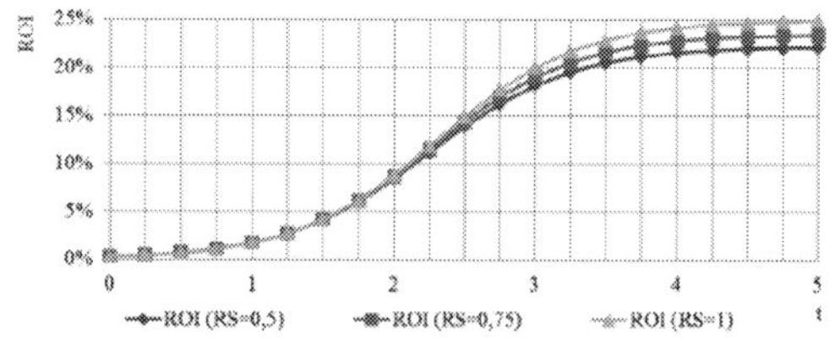

Fig. 3. How ROI alters relative to ROA with a CTR of 20%.

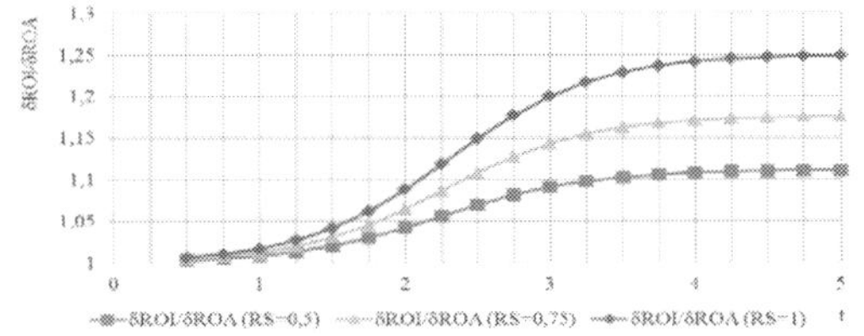

Fig. 4. How ROI increase rate alters relative to the ROA increase rat.

2 Boosting the innovative development by implementing a dynamic tax regulator

Simulation results show that under the 'natural' conditions of enterprise self-development, the stimuli and options to boost the innovative self-development are slim. It is, however, possible to control the intensification of innovative development by adding a dynamic corporate tax rate in this model [12].

The functional dependency of the dynamic corporate tax rate (CTRD) is determined on the following grounds. Where the production performance increases at a regular rate, the dynamic regulator should have the strongest impact.

Tax rate is adjustable from 40% for negative performance increase to 0% for an increase rate that exceeds the regular values by a margin.

In case the performance increase has a regular value, the corporate tax rate has a rated value of 20%. The curvature of the curve, and thus the impact of the regulator depends on the parameters a and b.

Parameter v describes the innovation performance increase rate as specified by the S-shaped curve, relative to the regular efficiency increase rates. In this model, the regular value of return on assets is 5% per annum, which roughly corresponds to the world average innovative business development rates.

The equation for the dynamic corporate tax rate curve corresponds to the Pearl equation (2) subject to adjustments necessary to render the economic intension of CTR_D:

$$CTR_D = CTR_{D,MAX} - \frac{CTR_{D,MAX}}{1 + ae^{b(1-v)}} \quad (3)$$

Where CTR_D is the dynamic tax rate;

a and b are the positive parameters affecting the curve;

v is the innovation performance increase rate relative to the regular performance increase rates.

Figure 5 presents dynamic tax rate values for various equation parameters (3).

From the Figure, one can see that where the efficiency increase rate has a regular value, the tax rate has the standard value of 20%. Where the efficiency increases relative to the regular values, the corporate tax rate falls to zero; or it rises to 40% where the efficiency increase rate is lower.

Integrating the dynamic corporate tax rate in the extended reproduction mechanism adds dynamic positive feedback on the innovative business performance increase rates, which alters the quality of self-development, see Figures 6 and 7.

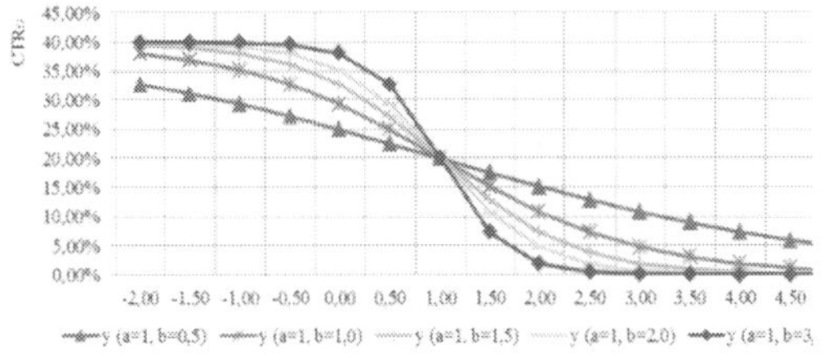

Fig. 5. Dynamic income tax rate at $CTR_{D,MAX}$=40%.

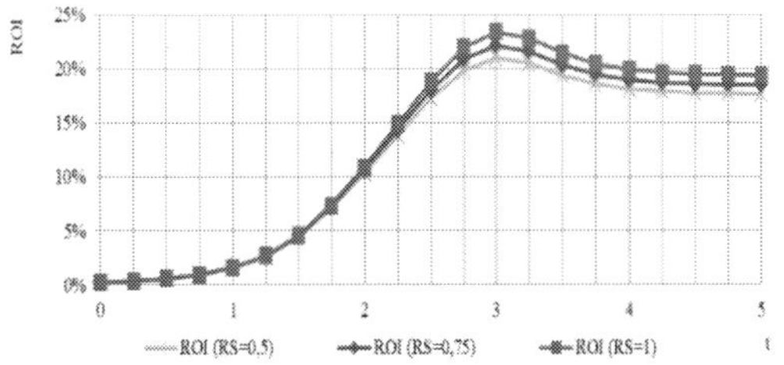

Fig. 6. ROI with the dynamic taxation (a=1, b=2).

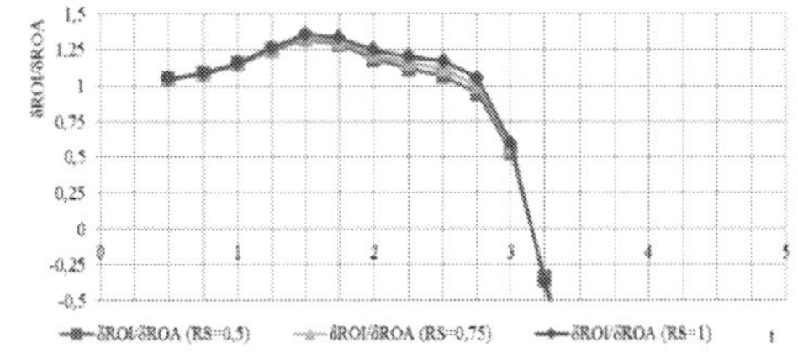

Fig. 7. Rates of growth of the ROI with the dynamic taxation (a=1, b=2).

Comparison of Figures 3, 4 and 6, 7 demonstrates the desired effect of the dynamic regulator. Under dynamic regulation, the ROI increase rate relative to the ROA increase rate is greater until middle of the third year, which creates additional opportunities for self-development. Then this value drops, which should become an additional stimulus to further promote increases in performance and to incentivize search for ways of innovative development.

It should be noted that the relatively small proportion of corporate tax in the total tax deductions and operating profits of an enterprise makes the regulator less impactful, which is why it is necessary to find a way to increase that proportion.

3 More impactful tax regulation

The simplest and the most natural way to make dynamic regulation of intensive self-development more impactful is to add other tax channels of an enterprise to that regulator; the TASED concept actually allows that. To assess the feasibility of dynamic regulation of enterprise self-development, let us analyse the overall tax burden of Russian companies.

According to the calculations made by PricewaterhouseCoopers Audit, the total tax burden for Russian companies is about 50% of their commercial profits [19, 20]. Commercial profits is income before any taxes [21]. According to PricewaterhouseCoopers Audit, commercial profits of Russian companies equals 175% of their operating profits, on average [20].

In view of the foregoing, the self-development regulation model looks like in Figure 8.

This model is different from that in Figure 1 because it has a coefficient of 1.75 in the straight chain that converts operating profits into commercial profits; the updated model also has a wider dynamic regulation range with TR values of $TR_{D,MAX}$=100%, $TR_{D,N}$=50%, $TR_{D,MIN}$=0%.

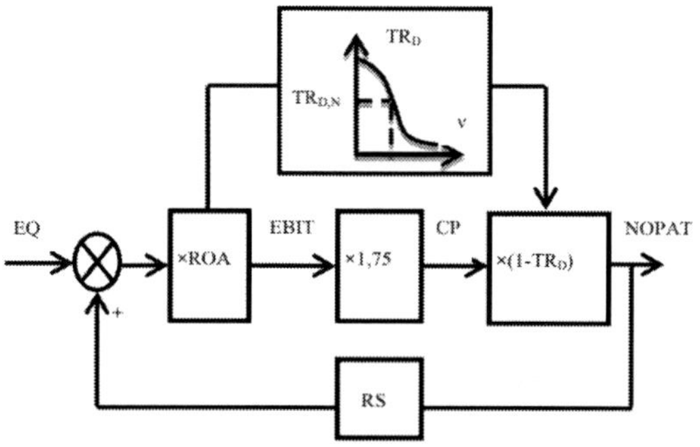

Fig. 8. Model of complete dynamic tax regulation of self-development. Functional diagram.

Here: $TR_{D,N}$, TR_D (Tax Rate) is the dynamic tax rate; CP is Commercial Profit.

In accordance with Figure 8, return on investments, or ROI, could be written as:

$$ROI = \frac{1{,}75ROA(1-TR_D)}{1-1{,}75ROA(1-TR_D)RS}, \qquad (4)$$

Thereby the dynamic tax rate will be as follows (Figure 9).

For the comparative assessment of the impact of the regulator, let us analyse three models. The first model employs a constant corporate tax rate of 20%. The second model is a dynamic regulator with a maximum rate of 40%. The third model is the more impactful tax regulator with a maximum rate of 100%. For all three models, let us specify the same S-shaped dynamics of economic return with a maximum value of 25%. The regular return on assets increase is 5%.

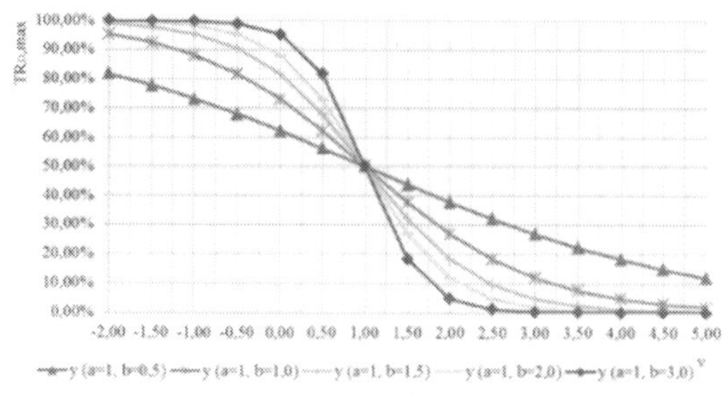

Fig. 9. Dynamic income tax rate at $TR_{D,MAX}$=100%.

Figures 10 and 11 present the dynamics of ROA and ROI for all the three models with the rate of savings of 0.5 and 1.

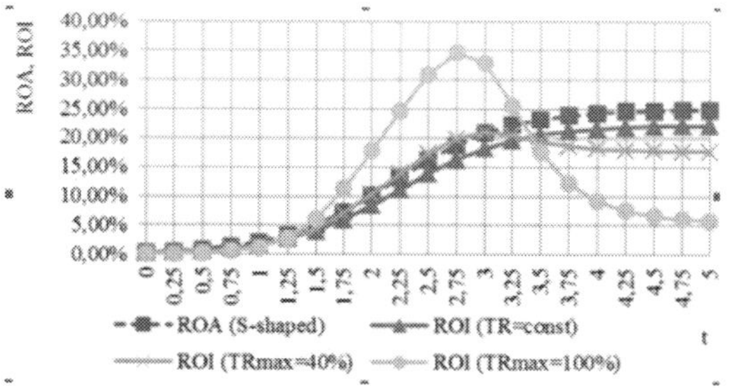

Fig. 10. Alteration of ROA and ROI at RS=0.5.

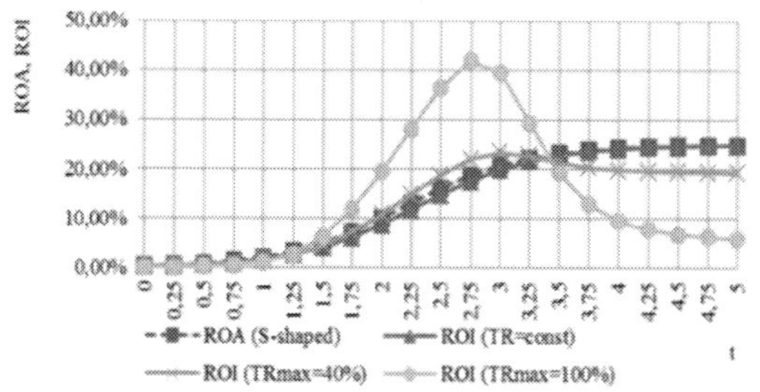

Fig. 11. Alteration of ROA and ROI at RS=1.

As one can see from Figure 10, more impactful tax regulation of the self-development mechanism makes the latter substantially stronger. At a time that corresponds to the middle of the third year, ROI amounts to 16% in the first model, 20% in the second model, and almost 35% in the third model. Setting the rate of savings to 1, see Figure 11, allows rising the ROI in the third model to approximately 42% per annum.

The curves in Figures 10 and 11 are constructed for a situation where the regular (or expected) increase in returns is 5% per annum. Obviously, if we make that value smaller so that it is closer to Russia's average innovative development dynamics, the mechanism under analysis becomes even more impactful, see Figures 12 and 13.

DOI: 10.1051/shsconf/20173501009

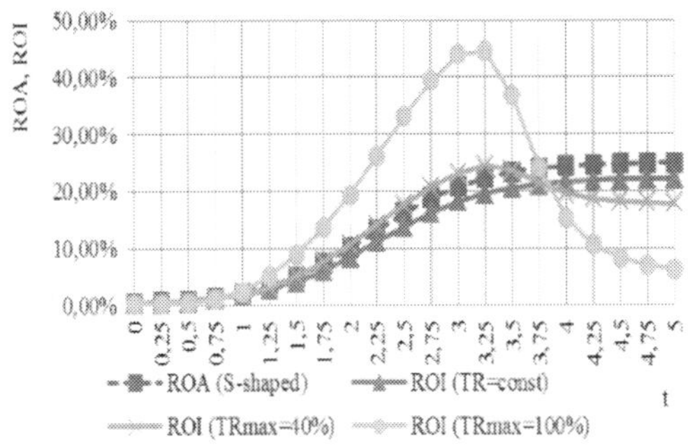

Fig. 12. Alteration of ROA and ROI at RS=0.5, with a regular ROA increase value of 2.5% per annum

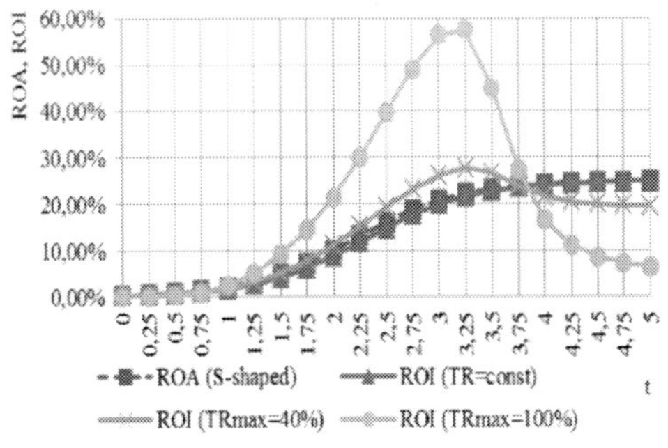

Fig. 13. Alteration of ROA and ROI at RS=1, with a regular ROA increase value of 2.5% per annum

In conclusion, it should be noted that the regulators proposed not just promote, and create opportunities for dynamic self-development at the stage of innovation implementation; they also determine the appropriateness of switching the enterprise to normal (non-preferential) operation as soon as its innovation capacity is maxed out. We have analysed how the innovation capacity could be maxed out over 3.5 to 4 years; it is at the end of this period that switching to regular taxation becomes appropriate.

4 Conclusion

This research has led to the following conclusions. Implementing a dynamic tax regulator of self-development promotes, and creates additional opportunities for dynamic self-development and accelerated exploitation of the innovative capacities of innovations.

Dynamic regulation of the self-development of innovative systems makes the focused on achieving the primary goal of innovative development, i.e. increasing the production efficiency.

The regulator can be made impactful enough by adjusting the entirety of taxes, which is in line with the conceptual fundamentals of the territories of advancing socio-economic development of the advancing social and economic development.

The work was supported by Act 211 Government of the Russian Federation, contract № 02.A03.21.0011.

References

1. World countries rates. http://www.ereport.ru/stat.php?razdel=country/.
2. Message of the president for the Federal Assembly.http://www.kremlin.ru/events/president/news/47173.
3. M.A. Omarov, *Conversion of the military industrial complex in system of state policy of modern Russia* (Moscow, 2010)
4. R.A. Faramazian, *Transformation of military economy: XX - the beginning of the 21st century* (Nauka, Moscow, 2006)
5. In 10 years the SEZ didn't become the efficient instrument of of the economy supportment. http://audit.gov.ru/press_center/news/26369.
6. Bulletin of Audit Chamber of the Russian Federation.http://audit.gov.ru/upload/uf/fac/fac89025a3b0198bcddedfb9d1962bec.pdf.
7. Bulletin of Audit Chamber of the Russian Federation.http://www.ach.gov.ru/upload/uf/f33/f335f242ebb55bf00f87282ee902ab74.pdf.
8. The Audit Chamber has revealed numerous violations during creation of science and technology parks. http://www.cnews.ru/news/top/schetnaya_palata_vyyavila_mnogochislennye.
9. The federal law of December 29, 2014 No. 473-FZ About the territories of the advancing social and economic development in the Russian Federation (2014)
10. E.A. Myakisheva, *Methodological support of the territories of advances social and economic development management* (Perm, 2015)
11. About an efficiency evaluation of activities of executive authoroties of subjects of the Russian Federation, Legislation of the Russian Federation (2012)
12. L.A. Baev, *Intensive self-organization of economic systems: concept, theory, models* (ChGTU, Chelyabinsk, 1992)
13. L.A. Baev, V.E. Shugurov, *System approach to definition of an innovation* (ChGTU, Chelyabinsk, 1995)
14. U.N. Startsev, *S-shaped models of development and technological gaps*. http://www.lib.csu.ru/vch/128/010.pdf.
15. Dg. Martino, *Technologocal forecasting* (Progress, Moscow, 1977)
16. D. Sahal, *Technical progress: concepts, models, assessments* (Finance and Statistics, Moscow, 1985)
17. S.G. Selivanov, S.N. Poezzhalova, Innovative regularities of development of aviation

engineering and technologies. http://innov.etu.ru/innov/archive.nsf/0d592545e5d69ff3c32568fe00319ec1/82425210c23f9c68c3257801003189eb?OpenDocument.

18. U.A. Salikov, I.N. Bulgakova, Improvement of the model of social and economic systems development.http://econ.vsu.ru/downloads/pub/seconomic/2/salikov.pdf.
19. Paying taxes 2016. http://www.doingbusiness.org/~/media/GIAWB/Doing%20Business/Documents/Special-Reports/Paying-Taxes-2016.pdf.
20. Paying taxes 2014. http://www.pwc.ru/ru/events/2013/paying-taxes/2-pwc-uk-packman-ru.pdf.
21. Paying taxes 2016.http://www.pwc.ru/en/tax-consulting-services/assets/paying-taxes-2016-appendix-1.pdf.

Automation of control and analysis of execution of official duties and instructions in the hierarchical organization

A.I. Demchenko[1,*], *L.N.* Palamarchuk[1], and *E.M.* Sartasov[1]

[1]South Ural State University (national research university) Chelyabinsk, Russia

Abstract. The article considers the problem of monitoring over execution of official duties of employees. This problem is characteristic of the enterprises having a hierarchical management structure. The functions and the modes of monitoring are defined, the types of analysis of the staff activities are provided. The description of the program complex allowing distributing functions and instructions for between the employees is given. The developed computer program allows tracking the performance, creating reports. The computer program has a demarcation of access rights and provides the can be operated in both local, and a large-scale network.

1 Introduction

When the bosses are beginning to think about automation of their business or organization, they are usually interested in complete solutions. Automation of the production, financial and accounting aspects are fundamental to the work of the enterprise. But it is equally important for successful business to be automated personnel and managerial aspects, which could be the basis for the optimization of management processes in an enterprise or organization.

If you look at the problem of optimizing of the management process from the outside, we see interpersonal interaction as the structure "the boss - subordinate". The feature of this interaction is that the same person can be the chief and a subordinate. At the same time it is the structure of interpersonal interaction 'boss - subordinate "in general, is a multi-level hierarchy.

In any company bosses give orders to their subordinates. In small companies or in structural units orders are usually given orally. In this case, the boss or manager remembers orders, to monitor compliance. In the medium and large companies or structural units the number of orders is large enough. Orders made in writing in form of orders, resolutions on the various types of documents. There are different special services office. But even in these circumstances, the bosses can have problems controlling their orders. This creates an opportunity for staff of any changes in the terms or execution of orders in general. As a result of this fall performing discipline of employees may lead to a decrease in effectiveness of the company or organization as a whole.

Automation of control over the execution of orders would be true to the company's development direction.

2 Methods of control of execution of orders

There are two types of control: on the essence and timing. Monitoring essentially involves assessing the correctness and completeness of the performance of the assignment. Time-control must ensure the timely execution of orders [1].

There are two basic ways to monitor the performance of official duties and assignments: manual and automatic. The manual method is labor-intensive and highly dependent on the human factor. Currently, with the development of information technologies on the market there are several ways to control automated embodiments, from implementation to function in Excel HRM systems [2-5]. Implementation in Excel is simple enough, but do not have a sufficient number of functions, such as there is no connection to the network. Modules HRM systems have full functionality, but they are quite complex, expensive, and are inextricably linked with other modules, which are not always needed. Therefore, it was decided to set up its program implementation

3 Program description

The Software of accounting performance of official duties and assignments was developed. The software enables to set official duties and assignments for subordinates and to control fulfillment and generate

* Corresponding author: DemchenkoAI@susu.ru

reports. Duties are periodic (daily, weekly, monthly, quarterly, annually, etc.). Orders are ad hoc and are not periodic. The Software can be used in a variety companies which have multi-level hierarchical structure "the boss - subordinate". Duties and assignments are accompanied by the documents. Documents - a regular files, most of them are pdf or word, but there may be other types.

The employee logs in with his or her username and a password. All employee will be divisions bosses and subordinates (bosses only or subordinates only), depending on the position. Bosses can assign duties and orders for subordinates and monitor their implementation. To official duties and assignment files can be attached, such as the implementation of the instructions. Typical duties will be entered by administrators or staff, endowed with the necessary powers.

For example, for the university, the main boss is the rector. He may assign duties and assignments to all employees. He usually appoints duties for closest subordinates, pro-rectors, directors of institutes and higher schools, heads of departments. Vice-rectors may assign duties and assignments to all employees, except the rector and vice-rectors.

Institutes, high schools are located on the second level of the hierarchy. In institutes and high schools the main boss is director. He is subordinated to the rector and vice-rectors and boss for all their divisions. Deputy directors are also the heads of all employees of the division, except for the director and the other deputy directors.

At the third level of the hierarchy there are department. The head of the Department is subordinate to the rector, vice-rector, the director and his deputies and the head of the department for all employees. Deputy head of the department is usually the academic secretary, who is also head of the department for all employees except the head of the department. Other employees are generally not superiors and subordinates only. In exceptional cases, the department may include laboratory. The Chief of the Laboratory is a subordinate head of the department and the boss for all laboratory staff. At the head of the laboratory may be substituent which is subordinate chief of the laboratory, and the chiefs of all other lab employees.

For the plant a similar hierarchy can be built. On the first level is the management (director, deputy), on the second level are departments (head of department, deputy), at the next level are plots (master, foreman, deputy), and at the next level are the other subordinates.

Subordinates receive official duties and assignments and execute them. After work, they make a mark in the program about a date. Bosses perform controlling execution and make organizational decisions.

Administrator performs a special role in this system. He performs the initial filling of the database, introduces a hierarchy of departments and positions of employees indicating the position and the department in which the employee works. Instead of entering the administrator can do the import data from the human resources department, or another database. Each employee is assigned as user name (login) and password, which he can change at their discretion.

After starting, the program opens a window with the following tabs:

1. Duties.
2. Orders
3. Subordinates.
4. Messages.
5. Divisions.
6. Personal data.

On the tab "Duties" employees can:

- assign duties itself, and bosses assign duties for subordinates;
- put a mark on the execution of the assignment, indicating the date of implementation responsibilities;
- download and view documents accompanying duty;
- view the history performance of duties;
- bosses can view as subordinates acted;
- prepare and publish reports on the performance of duties.

On tab "Order" employees can perform the same action as in the tab "Duties", only with the orders, except for watching stories, because orders and one-off stories are not.

On the "Subordinates" tab bosses see the table of their subordinates. For each subordinate boss sees the course of performance of official duties and assignments. He can assign new duties and assignments, and may cancel them, except those appointed a senior chief. On this tab, boss can also see what is not made to date. Staff can prepare and print reports on the execution of duties and assignments subordinates. On the tab "Messages" users can write messages of other users, and send it. Staff can also view sent messages. The message can be attached to the files. Unwanted messages can be deleted.

On the tab "Divisions" the user sees a division in which it operates, as well as divisions for which he is the boss. It is possible to generate and print reports on the execution of duties and assignments the staff of subordinate divisions.

On the tab "personal data", the user can see and update details, such as name, surname, date of birth, home address, passport data, the user name (login) and password.

For example, the head of the department instructs all employees to write an article and abstracts for the conference. The form must be filled in the program shown in fig. 1. In this form should indicate the order text, execution date, department, position and attach the necessary files. After this, all the subordinates see this order in its program in the window shown in Fig. 2. After execution of the order the subordinate makes a note on the implementation and indicates the due date, as shown in Fig. 3. The boss can see information about employees who have not completed at the moment the order. An example is shown on Fig. 4.

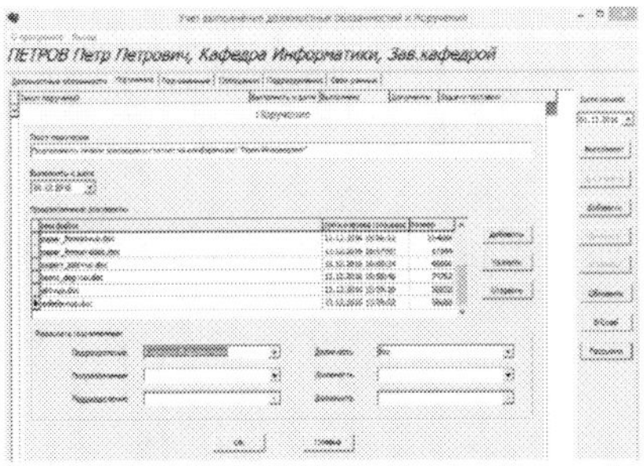

Fig. 1. Enquiry Form subordinate distribution.

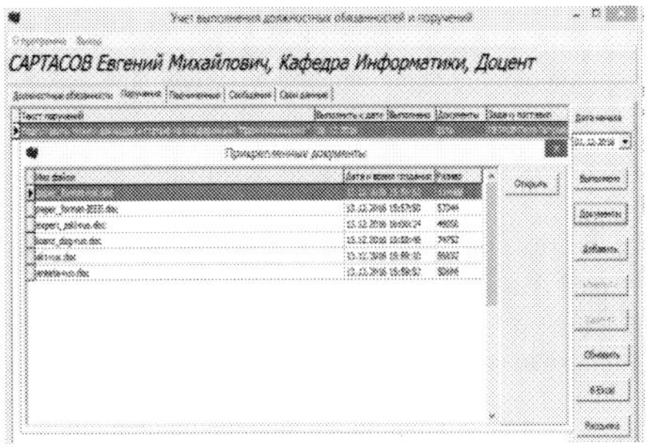

Fig 2. Subordinates browsing orders.

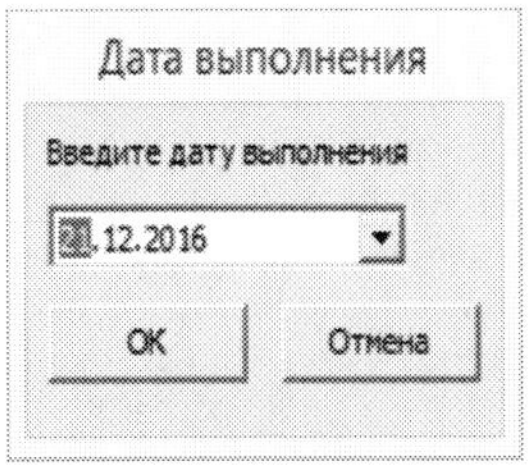

Fig. 3. The mark on the implementation of the order.

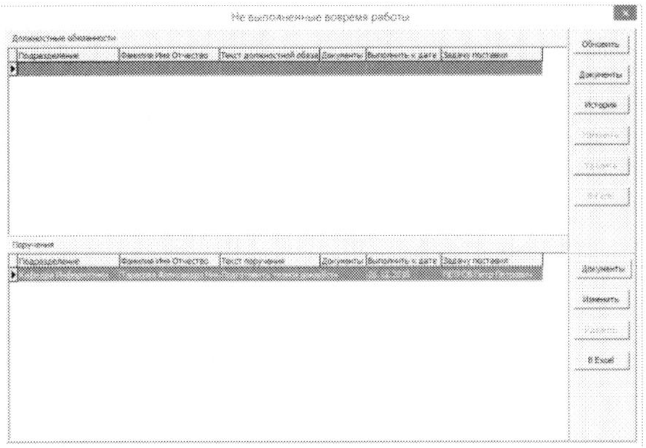

Fig. 4. Window shows not performed official duties and assignments.

Conclusion

As a result, the authors offer a complete software product that allows you to automate routine operations managers of different levels. The software is easy to use, has an intuitive interface. The program supports setting permissions. The program is implemented in the network version, and can operate in either a LAN or via the Internet

The work was supported by Act 211 Government of the Russian Federation, contract № 02.A03.21.0011.

References

1. E.M. Kameneva, Organization of control over the execution of orders. https://www.eos.ru/upload/analitica/kontrol.pdf.
2. Control of execution of instructions in MS Excel. http://delo-press.ru/articles.php?n=11147.
3. LiderTask.http://www.leadercommand.ru/content/view/354.
4. Control orders in IT Audit. http://поручения.рф/.
5. Control of job - ASKIZ. http://softbis.narod.ru/index/kontrol_ispolnenija_zadanij/0-14?yclid=7506766566633509235.
6.

Integral upper-limit estimation method for assessment of project design changes feasibility in industrial engineering

Vyacheslav Pikalov[1], *Aleksandr* Shumov[2,*], and *Julia* Shumova[2]

[1]STC-Geotechnologies LLC, Chelyabinsk, Russian Federation
[2]South Ural State University (NRU), Chelyabinsk, Russian Federation

Abstract. This paper presents the problem of project design risk managements. The empiric evidence is based on the data registered in the analytical GeoTechRemark system integrated in the STC-Geotechnology LLC. It is established that the number of expert comments is loosely connected with the feasibility of design changes. The weighted estimate method, long used as a solution for such tasks, also has a number of drawbacks, particularly the necessity for input data uniformity, which makes this method applicable for a restricted number of design fields. A proprietary methodology has been deduced to involve integral upper-limit estimation in feasibility assessment for project design changes, which is largely correlated with the actual feasibility of changes; the methodology is also available for automation, which makes it eligible for evaluating risks connected with compliance with Russian Labor law and laws regulating industrial engineering.

1 Introduction

According to the Urban Planning Code of the Russian Federation, design documentation is a set of documents containing materials in the form of texts and maps (diagrams) and defining architectural, functional and technological, constructional, and engineering and technical solutions for implementation of construction and reconstruction of capital facilities [1].

Running an enterprise specializing in developing design documentation is a difficult and complex task, which is preconditioned by the dynamic contemporary market. The successfulness of a design organization depends on a number of factors, one of them being the efficiency of design-related risk management.

This issue of risk evaluation the field of project management has been elaborated to a rather high degree. These problems have been studied in papers by such researches as Ansoff H.I. [2], P.A Berdyshev [3-5], V.I. Voropayev [6], Z.M. Galperin [7], P.F. Drucker [8–9], E.M. Korotkov [10], I.I. Mazur [11], S. Mantel [12], J. Meredith [12], N.A. Nesterova [13], N.G. Olderogge [11], M.L. Razu [14], M.M. Romnova [15], A.V. Serov [13], V.M. Serov [13], J. Turner [16], V.V. Trachenko [17–21], J. Frame [22], V.D. Shapiro [11], A.A. Shefov [23].

However, until now, "there have been a lot of problems in project management that require new methodology. 87 per cent of organizations that apply the project management methodology have developed their own software tools to manage specific stages of designing process; the other 13% of organizations have introduced a unified software system to manage entire sets of projects" [17]. These numbers show the overall poor level of project design activity management, including project design risk management.

2 Main risks in construction project design

The main risks in capital construction project design are:

1. Financial responsibility to the contractor for delay in correction of deficiencies identified in the State expert review and making changes to the design documentation as envisaged in Article 44 of Provision of March 5, 2005, No 145 "On conduction of the State expert review of design documentation and results of engineering surveying" [24];
2. Financial responsibility to the personnel and (or) supervision authorities (in case of non-compliance with the requirements of Articles 149, 152 of the Labor Code of the Russian Federation) with engagement of personnel in overtime work due to strong time restrictions for changes to design documentation [25].

Obviously, the aforementioned risks are opposite to each other as the reduced likelihood of events connected with the first risk increases the likelihood of events connected with the second risk, and vice versa. In this context, the upper-limit estimation method for assessment of design changes feasibility is crucial for managing the risks in question.

* Corresponding author: a_shumov@mail.ru

3 Feasibility of changes to design documentation

Feasibility of making changes to design documentation is a total of financial, material, and labor resources the organization needs to spend to make changes to the design documentation. Henceforward, the article studies only labor resources, although a hypothesis is made that the researches in the paper are also true for the financial and material resources of the organization. The main requirement for the deduced method is the simplicity of calculation and possibility of its automation.

4 Number of comments in the State expert review and the feasibility of changes to design documentation

The first logical step to assessment of feasibility of changes to design documentation was accounting the number of comments to the project design [26] (see fig. 1 and 2). The data is retrieved from the analytical GeoTechRemark system applied by the STC-Geotechnology LLC. Here, the linear correlation coefficient between the feasibility of changes and the number of comments was 0.44.

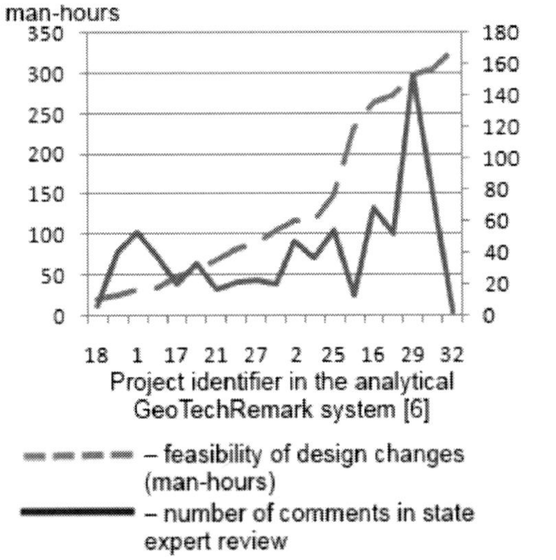

Fig. 1. Dependence of feasibility of design changes on the number of comments in the State expert review (anonymized data).

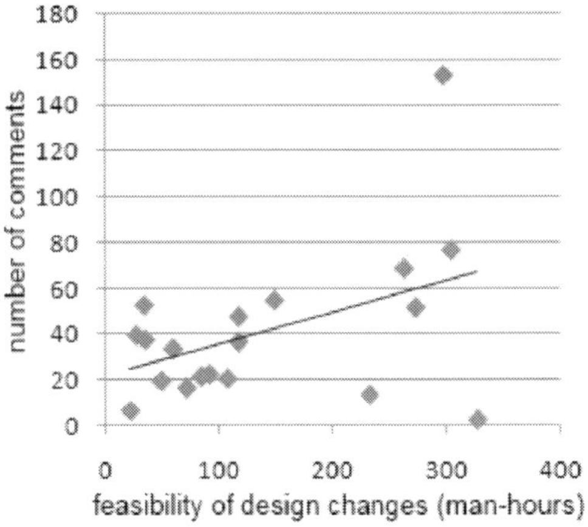

Fig. 2. Scattergram: feasibility of design changes, number of comments in State expert review.

5 Weighted estimate of feasibility of expert comment response

As shown in figures 1 and 2, the feasibility of design changes is loosely connected with the number of comments, leading to the necessity to use weighted estimates and transfer from quantitative to qualitative characteristics, which is the second logical step to the solution of the stated problem. The principle is dividing all comments in categories and assigning a weight to each of them. Then, the total of all weights is calculated. Initially, the weights were calculated with the traditional for this task expert method of optimization with the objective function of highest linear correlation coefficient between the feasibility of design changes and the received estimates [27]. However, the method proved less effective when adding a new project design to the statistics or adding new comment categories to the estimate system, which makes it unacceptable for evaluation or forecasting. Note that the highest correlation coefficient between the feasibility of design changes and the weighted estimate of the feasibility of expert comment response is 0.53 (see fig. 3 and 4).

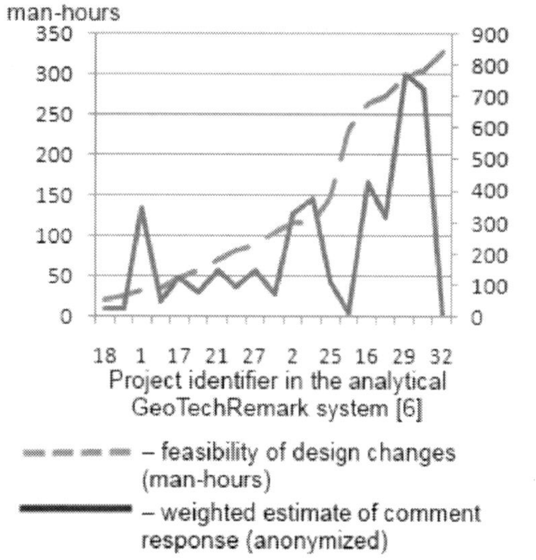

Fig. 3. Dependence of feasibility of design changes on weighted estimate of feasibility of comment reponse (anonymized data).

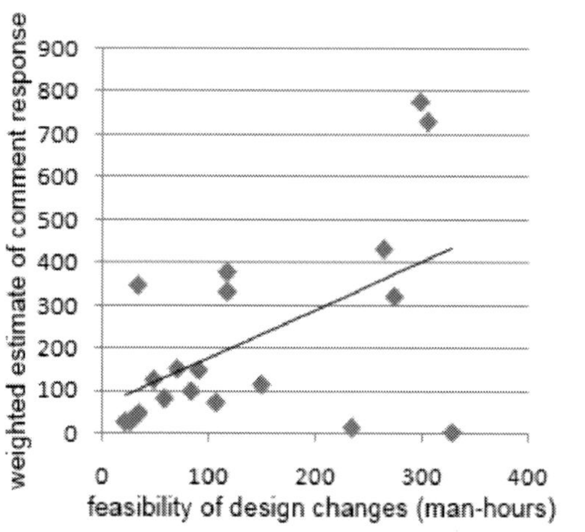

Fig. 4. Scattergram: feasibility of design changes, weighted estimate of feasibility of comment response.

Therefore, the weighted estimate method for definition of feasibility of design changes is, for a number of reasons, unstable for addition of new data such as: comments on new project design and new comment categories; i.e. it is possible to achieve certain results with this method, but it is efficient only for a restricted number of projects, which makes it acceptable only for a narrow field of application.

6 Integral upper-limit estimation method for assessment of feasibility of design changes

The next step in solving the specified problem was separation of feasibility of expert comment response from feasibility of design changes. As part of this separation, it is proposed to assess feasibility of comment response with the traditional expert method, and to assess feasibility of design changes with account of comment structure, design structure, and organization structure.

The upper level the design is defined by Decree 87 of the Government of the Russian Federation "On structure of design documents and requirements to their content" [28].

The next level of the design structure is formed by the structure of the organization: responsibility distribution among the departments, services and individual specialists.

The next level is conditioned by engagement of staff from outside the organization through subcontracts, fixed-term employment, and staff leasing.

Basically, this structure is not solid and can considerably vary depending on a particular organization. The best example illustrating this approach is Wolf number [29-30]. As with the Wolf number formula, the following formula (1) is proposed as solving for upper limit of integral to assess feasibility of design changes:

$$S=\sum\nolimits_{i=1}^{n}(z_i\cdot k_i+5\cdot(T_i-1)+10\cdot(D_i-1)+20\cdot(F_i-1)), \quad (1)$$

where:

S – Integral upper-limit estimate of feasibility of changes to design documentation;

n – number of comments;

z_i – weight of i comment;

k_i – number of corrections (number of areas) caused by compliance with i comment;

T_i – number of volumes "affected" by compliance with i comment;

D_i – number of departments involved in compliance with i comment;

F_i – number of organizations involved in compliance with i comment.

The findings are presented in fig. 5 and 6.

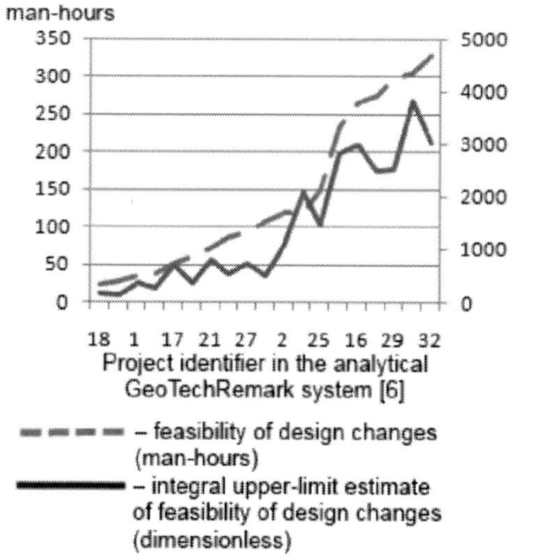

Fig. 5. Dependece of feasibility of design changes on integral upper limit estimate of design changes (anonymized data).

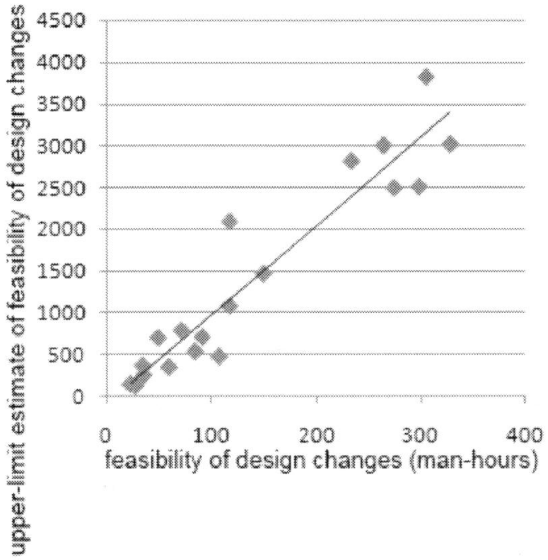

Fig. 6. Scattergram: feasibility of design changes, integral upper-limit estimate of feasibility of design changes.

7 Conclusion (Summary)

The deduced method of integral upper-limit of estimate of feasibility of design changes correlates well with the actual feasibility of design changes, linear correlation coefficient being equal to 0.94 (see fig. 5 and 6). Besides, the proposed method is available for automation, which makes it an acceptable method for estimation of risks connected with compliance of Russian labor law and laws regulating industrial construction design.

The work was supported by Act 211 Government of the Russian Federation, contract № 02.A03.21.0011.

References

1. The Urban Planning Code of the Russian Federation of 2004, No.190-FZ. http://www.consultant.ru
2. H.I. Ansoff, R.P. Declerk, R.L. Hayes, *From Strategic Planning to Strategic Management* (John Willey & Sons, London, 1976)
3. P.A. Berdyshev, *Communication management in design companies* (State University of Saint Petersburg, 2009)
4. P.A. Berdyshev, News of the Russian State Pedagogical University of A.I. Herzen, **114**, 327-335 (2009)
5. P.A. Berdyshev, Recipes for success in project-based management ed. By M.Martinsuo, 5-26 (2009)
6. V.I. Voropayev, *Project management in Russia* (Alans, Moscow, 1995)
7. Z.M. Galperina, Transportation business in Russia: research journal, **12**, 108-112 (2011)
8. P.F. Drucker, *Management Challenges for 21st Century* (Williams, Moscow, 2007)
9. P.F. Drucker, *Management for the future* (Williams, Moscow, 2007)
10. E.M. Korotkov, *Study of management systems* (Urite publishing, Moscow, 2015)
11. I.I. Mazur, V.D. Shapiro, N.G. Oldegorre, *Project Management* (OMEGA-L, Moscow, 2004)
12. J. Meredith, S. Mantel, *Project management* (Piter SPb, Saint Petersburg, 2014)
13. V.M. Serov, N.A. Nesterova, A.V. Serov, *Organization and management in construction* (Academy, Moscow, 2008)
14. M.L. Razu, Project management. Basic project management (KNORUS, Moscow, 2010)
15. M.M. Romnova, Omsk University Herald. Economics, **2**, 30-34 (2011)
16. J. Rodney Turner, Handbook of project-based management (Grebennikov Publishing House, 2007)
17. V.V. Trachenko, Management of design works in a group of companies (The State University of Management, Moscow, 2008)
18. V.V. Trachenko, University Herald, **4(26)** (2007)
19. V.V. Trachenko, Economics. Management. Culture, **13**, **2** (2006)
20. V.V. Trachenko, *Key issues of management – 2005: Materials for an international research and practice conference* (2005)
21. V.V. Trachenko, *Key issues of management – 2006: Materials for an international research and practice conference* (2006)
22. J. Frame, *Project management 2.0*
23. A.A. Shefov, *Multi-project management in design organizations of Russia: summary, traditions, tendencies*
24. Provision of March 5, 2005, No 145 On conduction of the State expert review of design documentation and results of engineering surveying. http://www.consultant.ru
25. The Labor Code of the Russian Federation of 2001 No. 197-FZ. http://www.consultant.ru
26. S.A. Ustinova, A.V. Shumov, Efficient exploitation of mineral resources. NTC Geotechnology LLC, 10 years of experience
27. N.V. Brazovskaya, *Optimization methods* (PASTU, Barnaul, 2000)
28. Decree of the Government of the Russian Federation of 16.02.2008 No. 87, On structure of design documents and requirements to their content. http://www.consultant.ru
29. M. Waldmeier, *Ergebnisse und Probleme der Sonnenforschung* (Leipzig, 1955)
30. Y.I. Vitinsky, M. Kopetsky, G.V. Kuklin, *Statistics of sunspot activity* (Science, Moscow, 1986)

Methodical bases of perceptual mapping of printing industry companies

Pavel Kalinin[1,*], and *Tatiana* Burtseva[1]

[1]Vyatka State University, Kirov, Russia

Abstract. This is to study the methodological foundations of perceptual mapping in printing industry enterprises. This research has a practice focus which affects the choice of its methodological framework. The authors use such scientific research as analysis of cause-effect relationships, synthesis, problem analysis, expert evaluation and image visualization methods. In this paper, the authors present their assessment of the competitive environment of major printing industry companies in Kirov oblast; their assessment employs perceptual mapping enables by Minitab 14. This technique can be used by experts in the field of marketing and branding to assess the competitive environment in any market. The object of research is printing industry in Kirov oblast. The most important conclusion of this study is that in perceptual mapping, all the parameters are integrated in a single system and provide a more objective view of the company's market situation.

1 Introduction

Modern feature of the business is the cut-throat competition. The increased competition, associated with changes in social and economic development of Russia's economy, leads to the increased interest of entrepreneurs to the competitiveness of the enterprise. Techniques of competitiveness assessment are the most popular in the evaluation and analysis of strategic business development prospects.

In the course of time and due to changes in the environment, any successful brand model may stop meet the changing needs of the market and consumer needs. That is why the company needs to continually evaluate its competitiveness in order to create a unique image, which develop a positive association among consumers and is associated with a particular printing service.

Consumers of enterprises of the printing industry measure brands not only by material benefits, but also by intangible (emotional) benefits. Therefore, the manufacturer must evaluate both of these parameters [1.2] in the measurement of the degree of consumers' perception in the market of services. Methodical and methodological bases of perceptual mapping are discussed in the scientific literature in more details [3-9]. It should also be noted that the printing industry enterprises of the Kirov region are often examined by scientists [10-13].

The literature on the subject proves the relevance of using perceptual mapping in order to assess the competitiveness of the printing industry companies.

The main objective of the work is to analyze the competitiveness of the printing industry with the help of the methodology of perceptual mapping.

2 Material and methods

Visualization of the target audience understanding is possible in perceptual mapping, which is constructed by the statistical method of non-metric multidimensional scaling based on the opinions of consumers of printing services.

The authors have developed a perceptual map for 17 printing houses of the Kirov region using the procedure of the factor analysis of questionnaire data of consumers of printing services of the Kirov Region in the program Minitab 14. This map describes the subjective opinion of consumers of printing services of the Kirov region.

In order to define the problem of multidimensional scaling, which includes a specification of objectives, the choice of competing printing houses, we have analyzed 18 brands and 12 brand making elements that affect the perception of the printing house's image. The results are presented in Table. 1.

Table 1. Key elements for determining the brand position.

№	Attributes of perception	№	Attributes of perception
X_1	Staff qualification	X_7	Visual level of printing product quality
X_2	Management quality	X_8	The range of printing services
X_3	Equipment	X_9	The price factor
X_4	Size of constant and loyal customer base	X_{10}	Innovative technologies
X_5	Reputation (goodwill) of enterprise	X_{11}	Deadlines of printing order
X_6	Staff motivation to improve the quality of goods	X_{12}	Relations with client

* Corresponding author: kalinin_p_a@mail.ru

3 Results

According to the results of consumers' survey of the printing industry enterprises on the basis of assessment of 12 elements on a 10-point scale, we obtain expert opinions, which will allow to build a perceptual map in the future (Table 2).

Table 2. Expert opinions for perceptual mapping of printing industry companies.

№	Printing house	Elements of perception of printing houses										
		X_1	X_2	X_3	X_4	X_5	X_6	X_7	X_8	X_9	X_{10}	X_{11}
1	First printing house	9.6	8.9	9.7	9.9	9.9	7.9	9.4	10	3.7	9.7	5.3
2	Kirov regional printing house	8.5	5.6	7.1	8.6	2.1	3.8	4.6	8.7	2.4	5.1	2.9
3	Lona	5.6	6.8	5.2	5.9	6.3	6.3	5.7	6.3	6.8	3.5	8.7
4	Foreks	4.9	4.2	4.7	6.2	4.6	2.9	6.7	4.9	7.1	3.7	7.6
5	Servis Print	9.4	7.4	8.6	9.9	3.4	8.6	9.6	5.8	8.9	9.5	9.1
6	Orma	3.9	2.1	3.1	2.3	2.1	2.5	1.8	4.6	8.7	5.2	8.9
7	Staraya Vyatka	4.7	4.6	2.4	9.8	5.6	3.6	5.7	2.4	9.5	2.3	6.7
8	Raduga	6.7	4.9	7.6	9.9	8.7	7.4	8.4	3.7	6.8	3.1	2.5
9	Printing house of medical statistics	2.9	3.1	4.3	5.6	3.1	2.1	2.5	1.8	5.6	3.4	3.4
10	Funtikov S.I.	6.1	4.6	3.8	6.3	8.7	6.8	3.7	2.8	6.3	1.9	6.7
11	Vyatka-Info	5.7	5.5	6.1	9.9	3.4	5.6	4.9	6.4	8.1	5.4	8.6
12	Printing house of Slobodskoy	4.6	4.8	4.7	6.3	2.5	6.7	4.1	5.4	5.9	2.1	6.7
13	Printing house of Kirovo-Chepetsk	5.3	4.6	3.1	6.7	3.6	4.6	5.1	5.7	5.1	2.3	6.1
14	Printing house of Kumyonsky District	5.2	3.8	2.5	5.1	2.7	7.3	3.7	3.1	3.6	1.4	4.9
15	Vympel - press	6.2	5.7	3.1	9.9	8.7	4.9	7.1	5.9	8.9	8.1	7.1
16	Chauzova E.K.	4.9	4.1	6.4	9.9	4.9	7.1	8.3	4.2	4.7	8.6	6.4
17	Polushkin S.G.	3.8	2.9	2.7	5.6	6.4	5.9	4.1	2.6	8.2	2.5	4.9
18	Perfect printing house	10	10	10	10	10	10	10	10	10	10	10

We have introduced the term "perfect printing house" in order to present the methodological analysis.

"Perfect printing house" is an enterprise of the printing industry, which the consumer would choose among the rest enterprises in the market. This assumption is necessary for the maximum objectification of the research. However, we understand that "perfect printing house" does not exist in reality. "Perfect printing house" is a combination of levels of all elements of the study preferred by the consumer. The introduction of the term "perfect printing house" is necessary due to the requirement to examine the question what kind of expectations of printing industry products consumers have. A study was carried out through the identification of the needs of consumers with the perception of the "perfect printing house".

The next step of perceptual mapping method is to identify the names of.

Identifying correlation estimates on the characteristics of the objects of research (printing houses) with the elements allows to call the axis of the most correlated factors with the elements. Let's represent the data using factor loadings (Table 3).

The model of factors built in the study underlines the importance of the management quality of printing houses, working in the market, as well as the quality of printed products. That is why let's call the axis in accordance with these elements: quality of management and staff qualification (axis 1) and the quality of printed products (axis 2).

Interpretation of the map begins with a study of the location and the company's position relative to the ideal point and with respect to the axes of the perceptual map and each other.

Table 3. Names of axes of perceptual map.

№	The main components of the brand	Axis 1: "Quality management and staff qualification"	Axis 2: "The quality of printed materials"
X_1	Staff qualification	**0.923**	0.057
X_2	Quality management	**0.912**	0.232
X_3	Equipment	**0.905**	-0.078
X_4	Size of constant and loyal customer base	0.772	-0.044
X_5	Reputation (goodwill) of enterprise	0.562	0.177
X_6	Staff motivation to improve the quality of goods	0.313	**0.832**
X_7	Visual level of printing product quality	0.127	**0.925**
X_8	The range of printing services	0.769	0.256
X_9	The price factor	-0.042	**0.814**
X_{10}	Innovative technologies	0.779	0.120
X_{11}	Deadlines of printing order	**0.881**	0.150
X_{12}	Relations with client	0.716	0.188

The map presents visually that only one enterprise is the closest to the "perfect printing house" - "Service Print". However, the distance from the point of "Service Print" to the point of "perfect printing house" is big enough in the perception of customers. It means that there is the need to change the working strategy of "Service Print" in the market. Its leading position in the market is due to the worst competitive position among its competitors, rather than its effective operation in the market, an effective strategy and efficient positioning of the printing press.

"First printing house", LLC "Orma", LLC "Kirov regional printing house", Printing house "Medical Statistics", LLC "Raduga" and Chauzova have unique images and brands. The uniqueness of the image is characterized by the special situation on the map - "individual" location, isolation from other companies.

On the map, there are two groups of companies, which are located nearby. These printing houses have cut-throat competition. As you can see from the map the first group consists of the following printing houses: "Forex", Funtikov, LLC "Staraya Vyatka", Papyrin (Kirovo-Chepetsk), Polushkin. The second group is represented by the following enterprises: "Lona", LLC "Vyatka-Info", "Vimpel-Press".

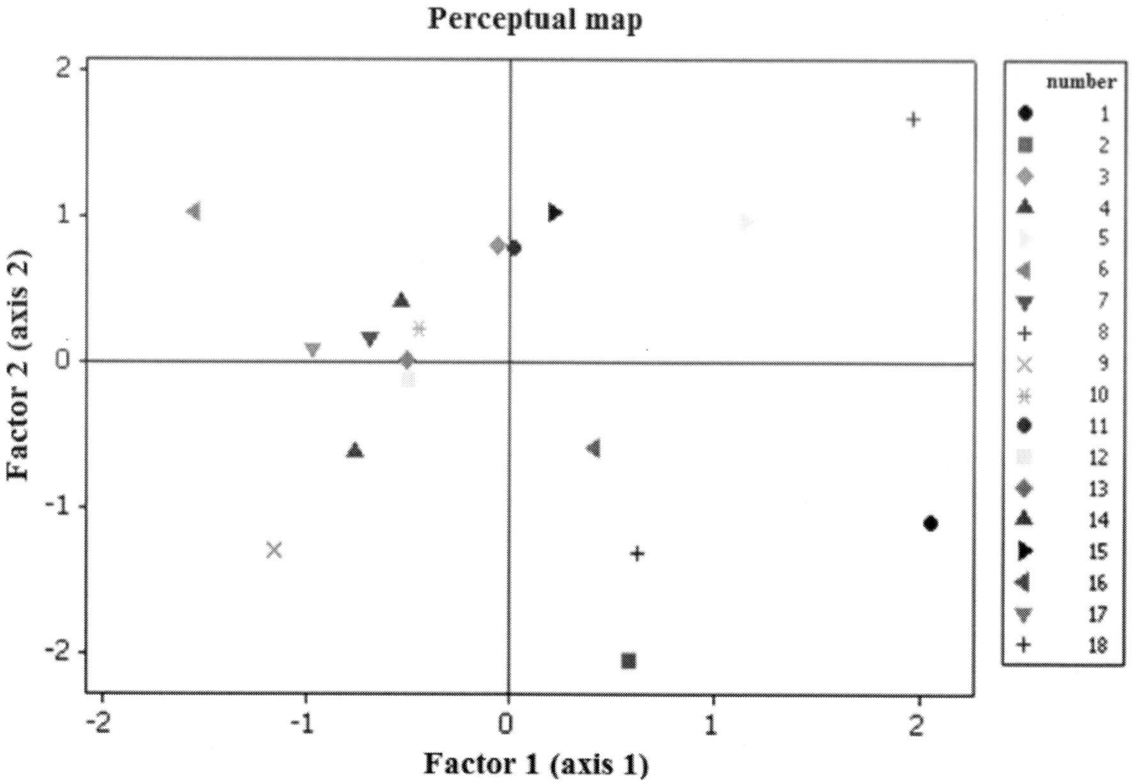

Fig. 1. Perceptual map of competitiveness of printing houses operating in the market of printing services, Kirov.

The outsider in the market for consumers' opinion is the printing house "Medical Statistics" – the state organization, its the target audience is all hospitals, pharmacies and other health care institutions of the Kirov region. That is why orders of other customers of printing services are not fixed for the enterprise and are made according to "residual" principle.

White space on the map indicates the potential for the introduction of new brands on the market or changes in existing businesses positioning.

Printing houses which are far from the beginning of the axes are strong according to the characteristics of printing houses. Thus, LLC "First printing house" is the most powerful on the characteristics of "the quality of management and qualified personnel". It is followed by "Service Print", LLC "Raduga", LLC "Kirov regional printing house" and then comes Chauzova.

However, "First printing house" has weak competitive parameters on the characteristics of the "quality of printed materials".

The following printing houses: "Service Print", "Vimpel Press» and «Orma» have a strong position on the characteristic of "the quality of printed products".

"Service Print" has the strongest position in the framework of the perception of the brand, taking into account its strong position in both directions of the axes.

4 Discussions

The results of the present study show a visualization of business on the perception of printing houses brands. It should be emphasized that the brand of enterprises of the printing industry is a complex and dynamic multi-structural element of the printing industry enterprises. This is due to a large share of the emotional component in the perception of the printing house brand [4,6,7].

It should be noted that each individual enterprise of printing industry needs to evaluate its brand, as well as to take into account the current market conditions, its market situation. Copying someone's experience is not desirable because it can lead to unpredictable consequences and negative effects.

5 Conclusion

This study contributes to the development of brand management of enterprises of the printing industry. The practical application of the research is in the possibility and necessity to use methodical bases of perceptual mapping of the enterprises of the printing industry and to ensure their sustainable development in difficult conditions of the modern market.

The authors suggest that further studies should be carried out within the framework of the formation of the development strategy of the enterprises of the printing industry on the basis of perceptual mapping.

References

1. T.A. Burtseva, M.L. Khalyavina, Bulletin of the Altai State Agrarian university, **12(122)**, 173-178 (2014)
2. T.A. Burtseva, A.Yu. Marinina, Consumer loyalty on the market of consulting services. Marketing, **1(116)**, 79-84 (2011)
3. O.N. Sinitsyna, I.V. Kultysheva, *Modern advertising technologies: theory and practice. Materials of the II Int. scientific-practical conference (part-time)*, 79-85 (2011)
4. V.S. Matyushkin, Perceptual map as a basis of development of brand communication campaign. Advertising: Theory and Practice, **1**, 30-36 (2016)
5. E.V. Lagodina, Europen Social Science Journal, **7-2(46)**, 48-54 (2014)
6. L.M. Kapustina, E.A. Zhad'ko, Application of perceptual maps-schemes for the comparative ranking of educational institutions. Proceedings of the Ural State Economic University, **6(50)**, 113-117 (2013)
7. I. Skorobogatykh, Technique of perceptual mapping. Marketing and Market Research, **5**, 2-7 (2004)
8. V.N. Naumov, Estimates of positioning brands on perceptual maps. Marketing and Market Research, **6**, 530-540 (2006)
9. D.V. Dmitriev, Yu.Yu. Dmitrieva, Application of three-dimensional effects to improve pervceptual maps. Interexpo Kye-Siberia, **1(1-2)**, 121-125 (2012)
10. A.A. Sozinova, P.A. Kalinin, Marketing competence as a success factor for printing companies. In the world of scientific discovery, **5.8(65)**, 3084-3097 (2015). DOI: 10.12731/wsd-2015-5.8-21.
11. A.A. Sozinova, Marketing laws of market development of advertising services. In the world of scientific discovery, **5.2(53)**, 830-835 (2014)
12. A.A. Sozinova, O.V. Fokina, Mediterranean Journal of Social Sciences MCSER Publishing, **6**, **4 S1**, 139-145 (2015). DOI:10.5901/mjss.2015.v6n4s1p.
13. A.A. Sozinova, O.V. Fokina, Life Science Journal, **11(7s)**, 318-321 (2014). http://www.lifesciencesite.com.67.
14. A.A. Sozinova, I.V. Androsova, V.L. Abramov, E.A. Tikhomirov, D.V. Redin, T.L. Oganesyan, International Review Of Management And Marketing, **6(S1)**, 250-254 (2016)
15. E.V. Karanina, I.V. Bykova, N.N. Ilysheva, American Journal of Scientific Research, **1-14**, 112-122 (2014)
16. N. Ilysheva, E. Karanina, I. Bykova, Special Issue Mediterranean Journal of Social Sciences, **6**, **5(S4)**, 348-355 (2015). DOI: 10.5901/mjss.2015.v6n5s4p348
17. A.A. Sozinova, E.V. Zhelnina, V.V. Prokhorova, M.V. Zelinskaya, I.N. Putilina, International Journal of Economics and Financial Issues, **6(S1)**, 52-56 (2016)
18. V.S. Novikov, A.I. Pykhtin, M.A. Gureva, A.A. Sozinova, M.P. Prokhorova, International Journal of Economics and Financial Issues, **6(S1)**, 126-132 (2016)
19. N.R. Saenko, A.A. Sozinova, I.S. Karabulatova, I.V. Akhmetov, O.V. Mamatelashvili, E.E. Pismennaya, International Review Of Management And Marketing, **6(S1)**, 267-272 (2016)
20. N.N. Kataeva, O.V. Fokina, Economics and Management: Problems and Solutions, **2(8)**, 186-188 (2016)

About the new industrial production management concept as the company strategy in the fourth industrial revolution

Julia Kovalchuk[1,*], and *Igor* Stepnov[2]

[1]Market Economy Institute of Russian Academy of Sciences, Moscow, Russia
[2]Ryazan State Radio Engineering University, Ryazan, Russia

Abstract. The new industrial production management requires a review of the third industrial revolution results and accounting for mass adoption of information and communication technologies to create the organizational basis of the fourth industrial revolution. The future changes will affect all components of the organization and management components of industrial enterprises, forming the potential of new competitive advantages in a global economy. The research included the identification of key factors of formation, development and destruction (absorption related branches of knowledge) the industrial production management as the theory and practical activities, given the critical approach to its nature and processes. Revealed common signs of the industrial production management need as a field of knowledge in the framework of previous and current industrial revolutions. It is shown that the industrial production management effectively solves the problem of subsistence economy, and substantiates that the modern digital economy also has the characteristics of subsistence economy. It is important the necessity of formulation of a new organizational thinking, the implementation of which is possible in the modern interpretation of the project office. The article represents the theoretical basis for developing practical recommendations for the formation of the new concept of industrial production management to take advantage of the impact of engineering component on the economic results and the creation of project offices for the development of traditional and created markets in the organization of a new production mode (based on the digital economy).

1 Introduction

The phenomenon of the industrial production management that prevailed in management science, education and practice in the last century in Russia resulted in almost complete disappearance as such of the educational program subject (the study discipline) that once required for all engineers and organizers of production, the absence of a recharge new scientific research, and, in general, lack of popularity among practitioners.

The importance of the industrial production management for the domestic industry was due to the fact that it was the locomotive to unite and lead the industry towards efficiency of technical and technological renewal of industry.

What has caused the role of the industrial production management in organizational and managerial sphere? Why today, the industrial production management is fragmented into a number of components and, in many ways, absorbed by the other branches of knowledge? What are the future prospects of this direction? Here are the key questions which must be answered to researchers in this field of knowledge, putting the fundamental question about the nature of organizational and managerial solutions at the present time.

2 Relevance

A new round of technological progress, which has given a significant number of hypotheses about postcapitalism [1], largely changed the economic model of society. Therefore, in our opinion, the strategic principles and approaches laid down several decades ago to the creation of an area of knowledge as the industrial production management, should be used for the formation of the new industrial production management. In fact, we must recognize that new technology:

i. change the relationship between labor and wages (just as a century ago mass production made the same discrepancy);
ii. destroy the action of the "invisible hand of the market", disrupting the market pricing (just as mass production has fundamentally changed the pricing of commodities);
iii. allow to create a new integration in joint production (similar to the formation of corporations a century ago).

* Corresponding author: fm-science@inbox.ru

Despite the fact that the concept of the third industrial revolution (based on the introduction of new energy sources in conjunction with information and communication technologies [2]) is not fully formed (and the revolution has not ended because of the speed and the magnitude of the current changes), there are some concepts of the next (the fourth) industrial revolution:

- Industrie 4.0 [3] is the creation of digital enterprises based on the digitalization of all systems (physical assets) and their integration into the digital ecosystem with the partners involved in the value chain;

- the new industrial revolution of the "makers" by K. Anderson [4] is the creation of customized mass production with the possibility of interactive exchange of ideas and developments based on the development of 3D design and 3D printing and use the additive technologies;

- the fourth industrial revolution by K. Schwab [5] is the transition from the simple digitalization (as the third industrial revolution) to the innovations, based on combinations of technologies (the fourth revolution), to information platforms, combining the supply and demand and in violation of existing production structures, the creation of new organizational forms and business models like "sharing" economy and "on-demand" economy;

- the new (the fifth) industrial revolution by P. Marsh [6] is a new industrialization of the economy in developed rich countries, which will affect the worldwide industry.

Despite the significant number of opinion and of the terms that generated by the revolutionary transformations in an increasingly technological society, economy and business management should distance themselves from the trappings of change and must focus on finding a compatible principles, approaches and concepts, instead of declarative discussing areas.

The industrial production management in the classical sense is essentially "subsistence economy" [7], and sees "problems" as "reserves" for future development [7], that substantially distinguishes the industrial production management, for example, from the controlling concept [8]. As evidence, it should be noted that many of the actions in the information world have become a kind of subsistence economy, but only informational – the same video or book piracy in its essence is a natural exchange of information when creating a new (libraries, Wikipedia, etc.).

3 Discussion and theoretical justification

Formation, development and destruction of the industrial production management key factors

Later in this article, we examine the factors that influence the formation, development and destruction of the industrial production management.

3.1 Placement and time

At first back to basics and define what the industrial production management solved the problems of interaction in placement and time as the key scarce resource, which had a maximum limit (placement due to high cost of the production area, and time as factor productivity).

Point of view on the concept of " the industrial production management" was formed from mid 30's to the 70's of the last century. Among the most well-known scientists, whose works have formed this base of knowledge should include the following scientists: B.Ya. Kachenbogen, L.Ya. Shukhalter, S.A. Heinman, E.A. Satel, S.E. Kamenitzer, A.I. Neimark, I.M. Razumov, N.A.Slodkevich, V.A. Letenko etc. [7, 9-20].

The industrial production management is presented as the combination and distribution in placement and time personal and physical factors of production with the specific technical conditions of the production. However, later, depending on the prevailing economic conditions, environmental effects, or lack thereof influence, the interpretation of the notion " industrial production management" looked different, refined and specified from the perspective of systems theory and applied aspect for almost all sectors of the economy. The most significant difference between the developed in those years methods is their simplicity, obviousness, the possibility of a recurrence and guarantee a stable result.

A special place in the industrial production management took operational production scheduling, which subsequently spawned a new branch of knowledge, but that operational production scheduling has largely determined the growth of labor productivity in the 60's, reduce lost production time [16].

3.2 Scientific and technical progress

Without a doubt scientific and technical progress had a huge impact on the production processes, in particular it requires a comprehensive solution of the issues of products modernization and the creation of new and development of their production technology and production management. It should be noted that most of the organizers of production, solving the enterprise's problem of organizing placement and time for a short time interval, when considering the issue of productivity growth went to the mandatory consideration of the influence of scientific and technical progress [7, 13]. The development of market economy in Russia caused a change in requirements for industrial production management, which came to be regarded as a particular environment in which the conditions for the best use of people and technology in the production process, thereby increasing its effectiveness [21].

On the forefront there are the problems of competitive products production in the shortest possible time, reducing the serial and batch size, improve the quality in accordance with world standards, skills development and creative potential of employees in terms of improving equipment and technology, improving the flexibility and mobility of production.

In accordance with the changed approaches to determining the industrial production management. So, in [17] define the essence of industrial production management in the integration and the interoperability of personal and physical elements of production, the establishment of necessary links and coordinated actions of participants of production process, creation of organizational conditions for realization of economic interests and social needs of workers in the industrial enterprises.

Thus, in addition to orientation to key resources (which were scarce and crucial not only in the market economy, but also in the planned economy), industrial production management became the integrator of all production processes, flexibly responding not only on industry characteristics but also on the challenges of our time such as technological progress and automation. And for many generations of Industrialists, the industrial production management became a symbol of renewal, a symbol of progress (especially in terms of mass production). Modern development of production infrastructure and a significant increase in the productivity of the equipment (not labor!) led to the reduction of the role of these factors in production.

3.3. Modularity and the possibility of a recurrence

The big advantage of the production was the possibility of organizational solutions recurrence, which guaranteed a stable result. Such a recurrence was possible not only in one industry, but is applicable for various industries (i.e. the production type was more dominant symptom than industry characteristics). One of the directions of development of organizational science was the search for modular integrated solutions. Among them should include the organizational platform [22], when the decisions on the industrial production management was added to the institutional aspect. But, as shown by the development of the economy over the last fifteen years, the organizational platform confirm their objective existence at a level higher than the enterprise, and practically are not in demand at the enterprise level. The organizational platform is inherently designed to limit the market variety of forms of interaction (similarly to the emergence of neo-institutional theory, when the institutional theory was not able to fully replace the monetary theory, then there is an institutional attachment to the main theory). And thus, the organizational platform as a modular design that reinforces the relationship in the market environment, were in demand only at the level of the enterprises and above.

A. The Productivity and Mass Production

The universality of the industrial production management in its golden age was confirmed by the fact that regardless of industry and size of the enterprise, its principles assumed productivity growth and has become an essential element of development.

Mass production was dictated by the demand (market or plan), and it should be noted in the period a large number of research devoted to the increase of serial (from order to small-scale, from small-scale to large-scale and until the mass, as the ideal is mass production). The main impact on this concept's component was the principles of flexibility as a response to market needs and manufacturing automatization [23].

The victory of the new industrial revolution will be achieved only if the production in the digital economy will provide a commensurate in the early stages the level of costs and productivity in mass production. Still, such a commensurability could be observed at the level of replacement of structures and materials (e.g., replacement of steel hull to aluminum polycarbonate, etc.). Development of 3D printing technology can form a distributed network as open to the consumers and does not require skilled workers. This approach to the fourth industrial revolution, more recognized in Germany, could lead to mass de-industrialization large-scale production.

3.4 System approach

Traditionally a strong point of the industrial production management is the proclamation and adherence of a system approach. But, paradoxically, a system approach also made a significant contribution to the reduction of the role of the industrial production management in its attempt to spread the industrial production management in all spheres of industrial knowledge. It is an attempt of generalization and universality led to erosion and separate the industrial production management's thesis to the other fields of knowledge and destroyed the science by the reason of formalization and generalization of knowledge instead of protecting the target segment and increase knowledge about the object of the industrial production management.

Instead of development of the production engineering methods and interaction within the enterprise the industrial production management began to explore all aspects of the enterprise (from industrial engineering to consumer services) that as a result led to the organizational principles began to be developed and enhanced, but the generalization has led to the emergence of the laws of the industrial production management, but did not add new meaning.

3.5 Engineering solution and the cost

The following strong component of the industrial production management was the connection of engineering and efficiency, in which engineers developed the solutions, the justification for which was based on the product cost. This result was achieved without depending on financial and economic departments because the cost was the engineering option[9, 12-14].

This component of the industrial production management was also reformed in the time of arrival of financial managers in companies management. This was achieved due to the replacement of the cost by the budgets and production budgets in the basis of management. Then production managers instead of cost reduction, productivity growth and technical updates have become the busy budgeting. The result is another group of costs, which have led to different results, because the manipulation of figures is more effective for obtaining the financial result, but less useful from the point of view of productivity growth.

Understanding the importance of budget planning and the need to value thinking for every managing at any level, should still be noted that financial science is very well appreciated experience of the industrial production management knowledge and actively replaced it with business processes organization.

On the one hand, this process has been objective in mind of reducing the requirement for mass production and the growth performance of the equipment (i.e. reduce the relevance of issues), but on the other hand, it was initiated by financial department of the companies to strengthen their position in the hierarchy of the company's management. It led to significant contradictions between the financial and operational management with a victory of financial management [24].

3.6 Advance

An important stage of development of the industrial production management was an attempt to make advance nature of production, i.e. attempts to create the industrial production management to meet the challenges of the time: a mechanism of ensuring competitive advantages should be based not only on responding to a dynamic and every time a qualitatively new changes in the external environment, and the pre-emption by changing their own behavior. As a result of changes in the market and offset customers needs has transformed the production concept: focus on efficiency it has passed to focus on quality, flexibility, innovation and, ultimately, to reduce development time for new products.

Therefore, with respect to the industrial production management necessitated orientation of its proactive as a property caused by the dynamic external environment and the requirement of competitiveness of an industrial enterprise and its products. Moreover, the advance is treated in precisely temporally, because, considering the industrial production management as a combination of elements of the production process in time and in placement, ahead in placement, it is objectively impossible [25].

The advance of the industrial production management can be regarded as strategic and tactical. Strategically advance of the industrial production management aimed at accelerating the achievement of the competitiveness of the production system. The advance is determined by the ability to make rapid changes in development and production of new products, anticipating changes in the external environment, in particular, emerging needs of consumers and the corresponding market conditions.

While tactically advance can be seen as directly based on the adaptation of the industrial production management to changing conditions of functioning of the production system based on innovation of industrial production management. The advance and the adaptation are not opposed, but complementary.

Therefore, the proactive industrial production management is seen as such the industrial production management that has the ability to provide rapid changes in the processes of new products setting to the production and to establish its accelerated release by the deadline in accordance with the market requirements.

3.7 Criteria of management efficiency

Another important characteristic of the industrial production management was a rejection of the techniques and methods of management at the production processes level (or rather, their inclusion in the industrial production management, thus ensuring automatic compliance). Proper industrial production management ensures the automatic generation of activity indicators depending on the production method, and many of the issues addressed by the labor regulation, not a set of balanced indicators.

This advantage was also deformed initially by management as a versatile tool (instead of the balance between technical and managerial solutions), and then financial departments. Financial management here won the industrial production management, introducing a system of indicators which must be monitored with a high degree of abstraction, again forgetting the principle that well-organized production system contained specific indicators within the system and not demanded the introduction of a new one.

As an example should bring the cost of information and communication technologies, showing that manufacturers have long understood the importance and value of the cost information, however, financial experts are still not bothered about the introduction of a new element to the cost of information and communication technologies [24, 26].

3.8 Project approach

The industrial production management always had the project approach and was the part of preproduction. The ideas and methods of the industrial production management began to be actively used in project management, but project management methodology did not consider it necessary to delve into the detail of the projects, leaving this part of the work of technical professionals, and focused on planning and reporting.

In particular, it should be noted that activity network and system PERT contained (including the content and duration) and the outcomes of actions. However, modern technologies operate on information reports about the completion of actions, which in turn also alienated the projects management from the industrial production management.

It should be noted that for project management the subject of exchange is information, but for production the subject of the exchange is product. Thus, it is possible to consider project management a kind of derivatives products, which leads to relevant risks despite the fact that brings natural economy to commodity production.

3.9 The Communication protocol

The industrial production management have always created a communication protocol, which is independent from the contractor (and the industrial production management even has introduced the forced nature of such interaction) and the most ignored cognitive distortions. These decisions have resulted in subsequent generations organizational solutions were already implemented in the framework of the new project and the need for the industrial production management decreased. Changing organizational reality leads to the fact that many of the standard methods no longer provide the required quality, and in the new organizational reality there was the possibility of multiplicity of solutions. Currently, it should be noted that the increasing role of cognitive distortions and the relevant risk became prevalent in the production planning, and addressing them is one of directions of development of modern industrial production management [27].

In addition, should talk about the cognitive distortions that can appear at the information and robotic systems not only in their interaction with man, but also in their interaction between themselves because the protocols, written by different programmers, can allow such opportunities, which were not originally planned. Also, if we assume that these devices are also studying the world and while not dangerous in its activities, but in the near future, the protocols of exchange between them will be a significant problem, since we can't even imagine what kinds of interaction and organization calculates by the machine intelligence. Therefore, the new industrial production management should consider communication not only workers, but also information and robotic systems.

4 The decisions and practical significance

The Prospects of a new organizational reality

So, the strength of the emerging industrial production management was engineering and the production cost, which gave the maximum result. The need for further productivity growth has led to the fact that the industrial production management began to distribute their solutions in the design stage and the effects of scientific and technological progress. But in the countries with market economy (unlike Russia) the expansion of operational management was aimed at marketing sector, with the aim of reducing costs.

Today engineering is largely starting to be replaced by startup, which create a wide range of solutions with the possibility of selection. This phenomenon is due to the fact that the majority of engineering solutions transformed from the only allowable in the possible, because the task of consumption or provide supply in a particular segment have been solved, and the vast majority of engineering decisions are in the improvements nature, not revolutionary changes. On the one hand, it is indicates that is no urgent need for these decisions to society, and the industrial production management in the old sense not in demand. On the other hand, preserved the industries, for which engineering has maintained its significance (e.g., energy).

On this basis, the new industrial production management may be presented in the two strategies form:

i. the strategy for those markets that are turning scarce resources and which are mandatory for human life - for example, food, energy, etc. In this area it is maximum necessary to develop standard processes (including, in addition to the foregoing, and quality assurance). It is very common for these industries is a preference for the formation of a system of indicators. But the system of indicators there is one danger - it needs to constantly innovate and prove the usefulness of these indicators. Hence, for modern managers is important to understand that the system of indicators, like the budget system, may lead to manipulation of data, not efficiency. In addition, the system of indicators, like any accounting system, works well with already a fait fact, did not react to expected changes, and may become a factor of destabilization. The industrial production management reduces the future risks (contributing to the dilemma of risk-return), just reducing the risk at the expense of repeatability, stability and higher productivity, giving opportunities for risk taking. A key task of the industrial production management is its upgrading to ensure readiness for the adoption of the fourth industrial revolution;

ii. strategy for unformed markets, on which startups start their activity, and thus future reality design is key for them. For those markets in which demand is determined by marketing, fashion, desire updates, etc., the digital economy can begin to destroy the market pricing, being focused on the market design with the specified parameters. It is therefore very important to focus on organizational training to the new market, not preproduction.

As for marketing [28], here one should set an analogy with the cost. If early mandatory for the success was the unification of engineering and cost control, today for a startup case, in addition to the cost should be included the marketing results. Of course, the industrial production management should not be totally absorb marketing. Marketing should be the cornerstone of market design, i.e. to precisely define the object of exchange in the designed system, in order to proceed with the market design. It should be noted that the organizers of the production have missed there chance in Silicon valley, which the experience was to give a new impulse to the organizers of the production. However, focusing at the moment the other segment (for mass production), they are not considered signals of new time a significant challenge.

5 Conclusion

As paradoxical as it sounds, for the revival of the industrial production management need a cognitive approach, information support and marketing included in the knowledge conductor of the organizational decisions. The Western world has chosen today startup as a conductor instead of engineer, creating a whole sphere of the knowledge economy and including startup to the large corporations activities. But this does not mean that the general orientation to the project approach will create new industries, skipping the stage of the industrial production management. The effectiveness of the projects will not increase by speculation the word "project", and the design thinking can lead to the departure from idea to technology implementation ideas - this approach has never been progressive at a strategically important time. Therefore, we propose to expand the functional framework of the project office, re-establishing the principles of the industrial production management in the formation of a new mode of production (based on the digital economy) and ensuring effective interaction design approach with cognitive technologies, information and design market.

These technologies should be a source of renewal of the industrial production management, but not budgeting, not the Balanced Scorecard, not project management, which are not productive in the early stages of creating a new market. For such tasks, it is necessary to create a project office based on the new organizational reality. The project office, which is based on the real product of the exchange, should ensure the connection all major elements of the new digital economy.

The authors would like to thank the Russian Science Foundation for the financial support in the realization of the research project No. 16-18-10149. The authors are also grateful to the Market Economy Institute of Russian Academy of Sciences for making scientific work.

References

1. P. Mason *PostCapitalism: A Guide to Our Future* (AdMarginem, Moscow, 2016)
2. J. Rifkin, *The Third Industrial Revolution: How Lateral Power Is Transforming Energy, the Economy, and the World Hardcover* (Alpina non-fiction, Moscow, 2014)
3. Industry 4.0: Building the digital enterprise. http://www.pwc.ru/en/technology/publications/industry-4.html.
4. C. Anderson *Makers. New industrial revolution* (Gardners Books, 2011)
5. P. Marsh, *The New Industrial Revolution: Consumers, Globalization and the End of Mass Production* (Yale University Press, 2013)
6. K. Schwab, *The Fourth Industrial Revolution* (The publishing house E, Moscow, 2017)
7. S.A. Heinman, *Industrial production management at the USA enterprices* (The foreign literature publishing house, Moscow, 1961)
8. J.K. Liker, *The Toyota Way: 14 Management Principles From the World's Greatest Manufacturer* (cGraw-Hill, Moscow, 2003)
9. B.Ya. Katsenbogen, *The Industrial Production Management in Machine Building* (The Central scientific-research Institute of industrial organization and management, Moscow, 1937)
10. L.Ya. Shukhalter, *The Industrial Production Management in Machine Building* (USSR Industrial Academy, Moscow, 1938)
11. L.Ya. Shuhgalter, Yu.S. Veniaminov, E.K. Smirnitsky, *Economics, Formation, Planing, Management* (Economika, Moscow, 1985)
12. E.A. Satel, *Formation, Planing and Economics of the Producing Department Machine-Building Plants* (Mashgiz, Moscow, 1962)
13. S.E. Kamenitzer, *The Industrial Production Planning and Management* (Politizdat, Moscow, 1967)
14. A.G. Berman, A.I. Neimark, *Production Line Methods in the Serial Machine Building and Instrument Making* (Mashgiz, Moscow, Leningrad, 1958)
15. I.M. Razumov, L.A. Glagoleva, M.I. Ipatov, V.E. Ermilov, *Formation, Planing and Management in Machine Building Plants* (Machine Building, Moscow, 1982)
16. N.I. Slodkevich, N.A. Salomatin, N.V. Andreeva, *Operative Accounting and scheduling at the Machine-Building plant* (MIYu, Moscow, 1980)
17. V.A. Letenko, O.G. Turovetch, *The Machine-Building Industrial Production Management: Theory and practice* (Machine Building, Moscow, 1982)
18. V.I. Tikhomirov *The Production Management* (MAI, Moscow, 1986)
19. I.M. Razumov, *Production Line and Serial-Scale Production Formation in Machine Building* (The state Committee of the USSR Council of Ministers for the coordination of scientific research, Academy of Sciences of the USSR, Moscow, 1961)
20. V.F. Ershov, N.P. Smirnof, O.A. Yumina, *Industrial The Production Management at the Department of the Machine Building Plant in the market conditions* (University of engineering and Economics, Saint-Petersburg, 2004)

21. R.L. Satanovsky, *Organizational Flexibility of the Machine-Building Production* (Machine Building, Leningrad, 1987)
22. I. Stepnov, *The Theory and methodology of the Innovative Potential Usage in the Regional Industry* (Saint-Petersburg, 2001)
23. I.M. Stepnov, T.G. Sadovskaya, Mashinostroitel, **8** (1990)
24. I. Stepnov, Economics and management in mechanical engineering, **6** (2014)
25. J. Kovalchuk, Organizer of production, **1**, **24** (2005)
26. J. Kovalchuk, I. Stepnov, *Proceedings of The International Conference, Management and controlling in the conditions of instability of markets and external threats* (2015)
27. A. Kazelskaya, I. Stepnov, Modern economics: problems and solutions, **5** (2016)
28. V.V. Glukhov *The Industrial Production Management and Marketing* (Polytechnical University, St. Petersburg, 2012)

Problems of cooperation of food industry enterprises and their consumer market counterparties in Kirov oblast

E.V. Karanina[1,*], and *E.Y.* Selezneva[1]

[1]Department of Finance and Economic Security, Vyatka State University, Kirov, Russia

Abstract. The paper investigates threats to the relations between local industrial enterprises and consumer market retailers, as well as the place of chain structures in this sector of the economy. To achieve the objectives of the study, food industry and consumer market were analyzed statistically, and the activities of chain structures were researched in-field. It discovers the main trends of the regional consumer market, the impact of retail chains on the changes, and especially the effect of federal and international chains on the structure and development of the consumer market and on the way local manufacturers operate. Based hereon, the main focus areas of coordination and progress in regional-level enterprise-to-retail interaction have been identified as to ensure the economic security of the food industry and the region as a whole.

1 Introduction

The food industry of the region is aimed at meeting the needs of its residents in everyday food products. Not only the economic security of an enterprise but also the food and economic security of the region as a whole depends on the rational organization of interaction between the enterprises belonging to this industry and their consumer market counterparties. The threats and challenges for food industry enterprises in the consumer market consist in the monopolization of the market by federal and foreign retail chains. The monopolization of the market and the retailers' focus on profit maximization exacerbates the problem of product quality, resulting in the reduction of the social safety of the region's inhabitants, especially in the period of declining real incomes. It is because the consumer market is a sphere of the direct economic impact on people and a social stability factor of the region and the country as a whole.[11] Its balance of commodity flows, the quantity and quality of goods, and prices is a necessary component for the assessment of the life quality of the population [2].

The food supply for the population is one of the most important social and economic issues, as a person's health and well-being depend on the degree of their satisfaction with food. Development of the food market depends on the situation in the agricultural sector as the main source of the resource base formation, on the trade sector, on the customers' solvency, and on the consumer demand.[15]

The food market experiences an adverse effect from the absence of a developed market infrastructure, high levels of costs in production, and the failure of the output to meet the required standards. Positive dynamics of development have been generally observed in the food market of the Kirov region in recent years, its condition has been stable for nearly all the main types of products. The food market of the region has been characterized by saturation, lack of physical shortages, high competition in the food market, an increase in consumer demand, and the growth of retail trade turnover [16].

The concept of "food security" in relation to regions has not been sufficiently established yet, unlike the definition used at the national level and disclosed in the Food Security Doctrine of the Russian Federation.[17] According to the Doctrine,[12] "Food security of the Russian Federation is the state of the country's economy that ensures food independence of the Russian Federation, physical and economic availability of foodstuffs to every citizen that meet the requirements of the laws of the Russian Federation on technical regulation, in quantities not less than the balanced food consumption rates required to support active and healthy lifestyle".[1]

Therefore, resulting from the above issues, the purpose of this article is to analyze the food industry and consumer market development, to identify the problems of interaction between these two sectors, to propose the ways of coordination and further interaction between retail chains and manufacturers at the regional level in order to ensure the economic security of the food sector, as well as the region in general [3].

* Corresponding author: karanina@vyatsu.ru

2 Methods for the analysis of the operation of food industry enterprises and of the consumer market of the kirov region

The food and food-processing industry of the Kirov region comprises over 400 organizations, including 110 referred to as large and medium ones. The enterprises employ almost 14 thousand people.

This branch of agribusiness is currently one of the most dynamic sectors of the regional economy. The tendency for increasing the output of major foodstuffs continues. In 2015, the turnover of the organizations amounted to 29.4 billion rubles (115.3% as compared to 2014), while the index of production was 103.9%.

Table 1. Industrial output of major foodstuffs in the Kirov region, %[19].

	2010	2011	2012	2013	2014	2015
Industrial output of foodstuffs, including drinks	105,8	99,6	103,9	97,7	99,2	98,2
Meat and edible offal of slaughtered animals	134,8	102,3	106	113	109,5	104,5
Meat and edible offal of poultry	70,3	41,5	139	62,6	180,7	61,2
Sausages	98,7	99,6	104,4	92,9	94,4	93,3
Canned fruit and vegetables	96,4	99,2	97,6	97,5	103,5	96,6
Canned vegetables and mushrooms	112,6	65	100,6	101.3	104,6	107,6
Whole milk products (calculated for milk)	97,8	97,3	104,6	104	97,4	100,7
Butter and butter spreads	68,8	109,6	95,4	104	114	115,5
Cheeses and cheese products	103,2	85,2	101,2	77,2	95,3	134,9
Fermented milk products, other than sour cream and cottage cheese	101,2	101,6	102,4	103,4	95,1	101,3
Flour from cereals, vegetables and other crops	110,1	105	104,2	97,3	99,4	105,4
Flour from grains and legumes	99,3	98,5	97,7	97,3	99,4	105,4
Bread and bakery products	100,1	99,4	98,3	98,5	99,7	101,8
Confectionery	95,4	97	105,2	93,3	99,2	101,4
Pasta					93,4	96,5

Food industry is an integral part of the region's agribusiness. Many of its branches are developing on the basis of local agricultural products – primarily, meat, milk, butter and cheese. Milling and baking, production of canned fruit and vegetables, distilleries, confectioneries and a number of other industries operate on local and, partially, on imported raw materials.

The distribution of food industry enterprises (Figure 1) over the territory of the region is more even as compared to other industries. They are available in all districts, yet the highest concentration is in the regional center. Kirov is a large food industry hub. The region's biggest dairy works, meat-processing works, confectionery factory, pasta factory, margarine plant, flour mill, fruit and vegetables canning plant and other enterprises are located here. Other significant centers of food industry are Yaransk, Malmyzh, Sovetsk, Slobodskoy, Kotelnich, and Urzhum.

The largest enterprises are: OAO (= plc) "Kirov Meat- Processing Works", OAO "Slobodskoy Meat-Processing Plant", OAO "Kirov Margarine Plant", OAO "Kirov Dairy Works", OAO "Kirov Confectionery and Macaroni Works", OAO "Urzhum distillery", OAO "Slobodskoy Distillery", and OAO " Vyatich ".[18]

Fig. 1. Territorial distribution of industrial enterprises in the Kirov region.

A significant advantage of the output of food processing enterprises of the Kirov region is its high quality due to the use of natural raw materials. The following table shows the production capacity of food industry organizations of the Kirov region that is required for our analysis.

Table 2. Utilization level of the average annual production capacity of organizations, %, Kirov region[19].

	2010	2011	2012	2013	2014	2015
Meat and edible offal of slaughtered animals	37.1	42.7	50.9	58	59	64.6
Meat and edible offal of poultry	56.4	20.2	31.7	18.7	25.9	55.1
Sausages	92.5	88.5	87.4	86.8	78	72.2

Whole milk products (calculated for milk)	75.2	75.9	79.6	76.7	72.4	77.3
Butter and butter spreads	37.2	48.7	45.2	48.9	47.4	55.8
Cheeses and cheese products	52.8	41.5	45.5	38.1	36.9	51.2
Flour from cereals, vegetables and other crops	5.9	23.9	22.8	67.8	45.8	22.7
Bread and bakery products	44.3	44	24	23.9	25.8	26.2
Confectionery	78.2	67.5	69.8	60.5	74.3	72.6
Pasta	80.9	75.3	34.3	52.9	52.2	36.1

According to these data, the average capacity utilization level of food enterprises is about 50%, the capacity being reduced in most areas.

The food enterprises could work more efficiently if they had free access to the consumer market and rational interaction with trade organizations.

Next, let us have a closer look at the consumer market of the Kirov region and its transformations. 12 thousand enterprises currently operate in the fields of the wholesale trade, retail trade and catering. This amounts to 30.7% of the total number of business entities in the region. 53.400 people are employed in the consumer market sphere.[7]

The turnover of retail trade amounted to RUR 174.9 billion in 2015, which is 7.7% lower than in the same period of the previous year in comparable prices. The share of food products was 48.3%, that of non-food products - 51.7%.

97.3% of the retail trade turnover in January-December 2015 was formed by trading organizations and individual entrepreneurs selling goods outside retail markets and fairs, the share of sales in retail markets and fairs being 2.7%.

Table 3. Retail trade turnover in the Kirov region, mln rur[19].

Commoditi es	2011	2012	2013	2014	2015
Foodstuffs, including drinks and tobacco products	58130.5	64688.9	71819.8	77855.8	84224. 8
Non-food products (without tobacco)	58486.9	66589.3	77409.6	87208.9	90651. 8

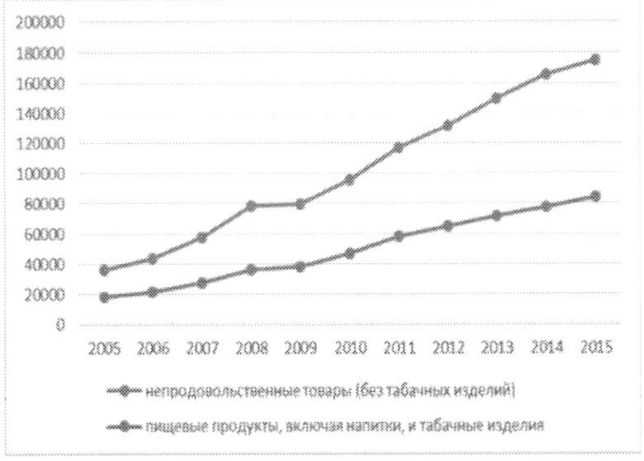

Fig. 2. Retail trade turnover in the Kirov region.

Table 4. Share of retail chains in the retail trade turnover of the subjects of the Russian Federation [19].

Territory	Share of the retail trade turnover of retail chains in the total retail trade turnover					% of change
	2011	2012	2013	2014	2015	
Russian Federation	18.4	20.5	21.9	22.9	25.0	6.6
Volga Federal District	16.8	18.8	20.5	22.3	23.4	6.6
Kirov region	19.3	21.0	23.6	25.6	28.2	8.9

The table gives grounds for the conclusion that there is a trend for an increase of the retail chains turnover, the figure for the Kirov region being above the average for both Russia and the Volga Federal District.

According to the data of the Russian Federal State Statistics Service (Rosstat), the share of retail chains has been steadily increasing, and this trend will continue. The expansion of retail chains is rather heterogeneous, both over federal districts and within the regions of the Volga Federal District. This is due to both the market opportunities and the local authorities' choice of policies in relation to retail chains [6].

National and international retail chains continue to actively involve the leading regional manufacturers in their expanding logistics schemes, as well as to develop production lines within their own companies. This is evidenced, in particular, by the emergence of agricultural holdings within one commercial brand. E.g., one of the largest suppliers of goods under the "Magnit" chain brands is OOO (= pvt. ltd. co.) "TD-Holding." Yet it should be noted that unfair competition is also present in the market: a new manufacturer may be involved in the chain and then deliberately made bankrupt or cut off from the sources of raw materials. All this enables a monopoly on a particular regional market [13].

Based on the Rosstat data, we should study the percentage of the presence of retailers with the participation of foreign capital in some regional markets.

Table 4. Share of retail chains with the participation of foreign capital in regional markets, %[19].

Territory	2010	2011	2012	2013	2014	2015	Rate, +-
Russian Federation, bn RUR	39.6	52.2	46.9	52.4	59.2	74.5	51.3
Volga Federal District	42.8	55.4	50.1	55.6	62.4	77.7	50.3
Kirov region	40.8	53.4	48.1	53.6	60.4	75.7	50.6

On the basis of the data presented, it can be concluded that the presence of non-resident retail chains has increased more than 3 times, which can become a threat to the economic and food security, especially during the implementation of the imports substitution concept.

The major retail chains operating in the market of the Kirov region have been analyzed in our research. The information was taken from official web sites of the chains and from professional portals.

The following retail chains operate in the 39 municipal districts and 6 boroughs of the Kirov region: the hypermarket "Vremya prostora", "Carousel", "Metro Cash & Carry", ZAO "Tander" ("Magnit" stores), OOO "Tamerlan" ("Pyatyorochka" stores); OOO "Sever", "Sever-2", etc. ("Globus" and "Vsyo na svete" stores); OOO "Kirovsnabservis" ("Produktovaya lavka" stores); OOO "Aitek" ("Oasis" stores), etc.[14]

The results are shown in Table VI.

Table 5. Major retail chains operating in the market of the Kirov region [20].

Stores, their format	Number of stores	Year arrived in Kirov	Type of chain	Owners' residence
"Magnit" self-service, next-door store	107	2010	Federal	Russia, Krasnodar, with foreign capital
"Bristol" next-door store, alcomarket	78	2013	Federal	Russia, Nizhny Novgorod, with foreign capital
X5 Retail Group ("Carousel", "Pyatyorochka") supermarket, self-service, next-door store	75	2011.	International	Netherlands
"Globus" supermarket	38	2000	Regional	Russia, Kirov
"Krasnoye I beloye" next-door store, alcomarket	31	2013	Federal	Russia, Chelyabinsk, with foreign capital
"Oasis" next-door store, alcomarket	28	1998	Regional	Russia, Kirov
"Produktovaya lavka" self-service	22	1992	Regional	Russia, Kirov
"Krasnogorsky" self-service	15	2001	Regional	Russia, Kirov
"Nikolsky" (Poley) self-service	15	1996	Regional	Russia, Kirov
"Samobranka" supermarket	8	2004	Regional	Russia, Kirov
"Svetofor" Warehouse-store	4	2015	Regional	Russia, Kirov, with foreign capital
"Zelyony ugol" eco-store	4	2015	Regional	Russia, Kirov
METRO Cash & Carry International GmbH hypermarket	1	2009	International	Germany
"Vremya prostora" hypermarket	1	2006	Regional	Russia, Kirov

In relation to the total number of trade enterprises (including small businesses) operating in the regional consumer market, their share may be small, but in terms of other indicators, primarily turnover, their share is more than 50%, and it tends to grow rapidly.

3 Results of the analysis of the impact of consumer market organizations on the operation of regional manufacturers

Expansion of the world chain companies on regional markets has both positive and negative consequences. Government regulation at both the national and regional level should play a significant role in smoothing the negative consequences of the expansion of chain companies in regional markets. It should be primarily aimed at protecting and supporting small trade enterprises and the implementation of anti-monopoly measures - in particular, combatting unfair competition and monopoly, controlling the product quality and prices, regulating relations with suppliers. All these matters were to be solved by the federal law "On the bases of the government regulation of commercial activities in the Russian Federation", yet unfortunately, it did not solve any in essence.[8]

On June 24, 2016, the State Duma adopted, at its last plenary meeting, in the second and third final reading, amendments to the law on trade, according to which the maximum amount of remuneration received by retail chains from the supplier for the purchase of a certain quantity of goods (retro bonuses) is reduced from 10% to 5%. In addition, it is proposed to significantly decrease the time period within which chains currently pay for the delivery of goods. The law increases the role of the Federal Antimonopoly Service so that this agency monitored the observance of the norms of the new edition of the law on trade, as well as regulated relations between suppliers, manufacturers and retailers. The next year will show to what extent the amendments can solve the problems of the retail market.[10]

The share of federal (international) chains is increasing from year to year, resulting in a decrease in the share of regional and local chains. The problem of the relationship between these chains and local manufacturers should be noted. In fact, the chains force their suppliers to agree to all the proposed terms, and the suppliers do it so that not to lose the market, to be able to sell their products, often perishable ones.

For example, the Director for corporate relations of the "X5 Retail Group N.V.", M.V. Suslov openly stated that in order to survive, suppliers should agree to the conditions of his chain. We quote: "... the proportion of retailers abroad is around 80-90%, while in Russia large retail chains account for only one third of the commercial market. However, the market is moving steadily towards consolidation, and the Russian manufacturers should be prepared for the fact that in the domestic segment, the share of retail chains will continue to grow. Therefore the manufacturers who manage today to build relations with chains, manage to understand how to work with them will secure further development for themselves ...". M.V. Suslov identified the following policy options for the owners of small shops or small chains, as well as for small manufacturers for the time being: either changing to a narrow specialization, in which chains are not able to compete in the assortment of a certain product group, or selling the business to a chain company. In addition, manufacturers can try to take their products to the federal level, that is, to become a federal brand, or begin to make products under a large chain's own brand.[18]

The Kirov manufacturers, in their turn, draw the attention of the retailer's top managers to insufficient flexibility in considering their requests for cooperation. Representatives of the Russian Federal Service for Surveillance on Consumer Rights Protection and Human Wellbeing in the Kirov region noted that federal chains have more problems with the quality of products sold.

There are also many complaints about the federal retail chain "Magnit". Goods manufacturers are dissatisfied with the cooperation with this federal chain, which, in their opinion, washes away Kirov goods from the market. And the figure of 28% that the directors of the branch give as an indicator of the amount of goods from Kirov manufacturers on the shelves of "Magnit" is untrue. Public bodies are also concerned about the fact that apart from other problems, the monthly turnover of "Magnit" is more than RUR 900 million, while the company is not registered in the Kirov region – and the money goes outside the region. This company supplants Kirov manufacturers.

Some entrepreneurs expressed complaints about the quality of the products that citizens acquire in the "Magnit" chain. Among other things, top managers of the "Magnit" chain plan to open a distribution center in the Kirov region in 2017 which should make life easier for our suppliers. The management of the "Magnit" chain promises almost the moon for the region with the opening of this center: major investments (more than RUR 1 billion), about 1 thousand jobs, an annual replenishment of the regional budget in the amount of about RUR 100 million, opportunities for suppliers, numerous social guarantees for employees (pensionable service, discounted tickets to sanatoria, etc.).

Kirov manufacturers view this prospect with suspicion: some of them have already dealt with distribution centers in other regions, where it happens that goods are returned almost without explanation, and it is very difficult to achieve justice at the top of the "Magnit" management.

There is also a problem that federal chains have assumed the right to determine what manufacturers to work with and with what ones not to work, what bonuses and discounts to apply, often exhibiting immoderate appetites. For example, they ask for a 30% discount, while the food industry of the Kirov region for its most part has only 5-10% profitability. Amendments to the federal law "On Trade" might solve this set of problems.

The federal retail chains, in their turn, speak about the non-competitive prices given by the local suppliers. But a low price does not guarantee the quality of the goods. To increase the shelf life, preservatives are used, as well as substitutes, flavor and colour enhancers. Although permitted for the use in production they definitely deteriorate characteristics of the goods.

4 Conclusion

In general, the solution of the acute problems concerning interaction between the manufacturers of the Kirov region and large retail chains includes the following components:

- prevention of excessive concentration in the retail trade in foodstuffs;
- maintaining a balance of large, medium and small forms of retail trade;
- ensuring the preservation of the diversity of goods distribution and of the retail trade organization formats;
- support for small businesses in the retail sector;
- reduction of administrative barriers and administrative regulation in the retail sector, which would contribute to the development of a competitive environment and the improvement of business efficiency;
- coordination of the interaction between manufacturers and chains by modelling the norms of the presence of food industry enterprises in the turnover of retail chains, a program-oriented and regulatory approach that would take into account the interests of local manufacturers;
- use of the instruments for a moderate regional protectionism.[4]

One can identify the following instruments that are necessary to preserve the region's food security:

• additional voluntary certification and labeling;
• establishing specific bans (temporal, territorial, assortment, sales channels of distribution networks);
• directions to distributors requiring to increase sales of locally produced goods;
• price control in conjunction with the Federal Anti-monopoly Service;
• establishment of quotas for imported products.

In order to support the agro-industrial complex, the following measures can be used:

- pressure on municipalities for monitoring the sales of locally produced goods through retail chains;
- preferences to local manufacturers.[9]

The conceptual approach to the development of the food industry should be based on the preservation of the food security of the country and the region, and on the scientific approach of modelling the placement of a trade outlet in the territory.[5] The concept should include instruments for regional protectionism aimed at increasing the efficiency of local industrial enterprises.

References

1. G.S. Bondareva, Bulletin of Kemerovo State University, **3**, **4**, 24-29 (2013)
2. N.A. Vasilyeva, Russian Entrepreneurship, **4**, 103-107 (2012)
3. O.A. Inevatova, E.P. Prokofieva, Bulletin of Orenburg State University, **13**, 210-214 (2011)
4. E.V. Karanina, *Financial and economic security: the regional aspect* (Scientific Publishing Center: Discovery North Charleston, SC, USA, 2015)

5. E.V. Karanina, D.A. Loginov, *Theoretical and methodological aspects of the analysis of the risk system of a regional economy* (Scientific Publishing Center: Discovery, North Charleston, SC, USA, 2015)
6. E.V. Karanina, *The economic security of the state, region and enterprise: formation and provision taking into account the risk factors* (Staraya Vyatka, Kirov, 2015)
7. E.V. Karanina, A.V. Evstratova, Economy and Management: Problems and Solutions, **12**, 146-153 (2015)
8. V.V. Kurchenkov, O.V. Fetisov, T.V. Chigareva, Bulletin of Astrakhan State Technical University, Series: Economy, **1**, 36-41 (2013)
9. N.V. Muratov, *Proceedings of the International scientific and practical conference with elements of scientific schools*, **2**, 16-23 (2013)
10. N.V. Muratov, Business in Law, **6**, 34-39 (2015)
11. N.V. Muratov, Scientific Review. Series: Economics and Law, **6**, 56-63 (2015)
12. On the action plan for the implementation of the provisions of the food safety doctrine of the Russian Federaton: the decree of the Government of the Russian Federation, of 2010 376-p (2010)
13. E.Y. Selezneva, Bulletin of Russian State University for the Humanities. Series: Economics. Management. Law, **12**, 218-222 (2012)
14. E.Y. Selezneva, Concept, **11**, 27-30 (2013)
15. A.E. Chusova, *The consumer market of Russia: the diagnostics of the industry and regional issues* (Institute of Economics, the Ural Branch of the Russian Academy of Sciences, Yekaterinburg, 2012)
16. A.E. Chusova, Scientific and technical bulletin of St. Petersburg Polytechnic University. Series: Economics, **2**, 103-108 (2012)
17. A.E. Chusova, *Proceedings of the I All-Russian symposium on regional economics, Institutions of regional innovation development. Institutions of the self-development of territories of different levels. Current regional policy of the government*, 164-166 (2011)
18. Vyatka Chamber of Commerce and Industry: official web site. http: //www.vcci.ruyu
19. The territorial body of the Federal State Statistics Service for the Kirov region: official web site. http://www.kirovreg.ru/econom/industry/import.php
20. Federal Tax Service: official web site. http://www.nalog.ru/nal_statistik/analit/

Management of logistics chains of industrial enterprises in high-tech industries

Alexander Demchenko[1,*], and *Yulia* Astashova[1]

[1]South Ural State University (National Research University), Department of Marketing, Chelyabinsk, Russia

Abstract. The article analyses the trends of high-tech industries development in the market globalization conditions. The economic growth of high-tech industries largely depends on technological innovations, products and equipment based on usage of the results of applied and fundamental scientific researches. It is shown that characteristics of high-tech productions influence the logistics chain structure. It justifies the importance to include partnering enterprises taking part in creation of intellectual and tangible assets in the logistic chain of the industrial enterprise. It is noted that development of production technologies contributes to formation of innovative logistic chains, which participants are enterprises cooperating for achievement of competitive power. The article offers a model of an industry logistic chain comprising the conventional logistic chain and innovative logistic chains corresponding to the industry market segments. It is proposed to use the customer value of the manufactured products as a criterion of the industrial enterprise's logistic chains management efficiency. Implementation of the offered model of cooperation between the logistic chain participants is aimed at creation of a continuous innovations flow, reduction of technological development costs and uniting of the participants' experience.

1 Introduction

In economic globalization conditions more and more Russian enterprises face the necessity to review their key economic management principles. In the Russian economic structure there appear industries, in which business operations carried out by independent participants are associated with several challenges or almost impossible. We can note a tendency of a transition from the competition between market players to the competition between associations of partners or holdings controlling the key stages of product development, production and market promotion.

Formation of a stable partner cooperation system has acquired a discrete role in high-tech industries, where the main success factor is product development competence.

High scientific and technical development rates, ever-increasing share of R&D costs, uncertainty of the demand and terms of production of high-tech goods precondition the need for cooperation of the industry market players to unite the available resources and improve the competitive power.

2 Topicality of the problem. Literature review

The issues of an efficient management of industrial enterprises' logistics chains have been considered in different enterprise management fields, such as general and strategic management, logistics and innovation management.

One of the up-to-date approaches to business operations is integrated supply chain or logistics chain management. The Supply Chain Management (SCM) concept, which was initially formed in the disciplinary sphere of logistics, currently represents a separate field of knowledge. The SCM concept is based on the fact that the manufacturing enterprise is not a formation independent of the environment but a part of a system chaining "sources of raw materials – production – shipment to the end user".

Term "Supply Chain Management" was introduced for scientific researches in 1983 and became widespread in 1999 after publication of the work of Robert B. Handfield and Ernest L. Nichols [1]. Later the SCM concept was further developed in the works of Donald J. Bowersox [2, 3], Martin Christopher [4], Robert B. Handfield [5], Douglas M. Lambert [6, 7], James R. Stock [8]. It is shown in the works of these authors that the SCM concept is based on the problems of business process management, integration of partners, introduction of information control systems, which is particularly important in the economic globalization conditions.

The SCM concept has been dynamically developed by the Russian schools of science. The works of Russian scientists systematize basic theoretical provisions of the SCM [9-11], develop applied elements of the theory [12, 13], study aspects of functional management within the

* Corresponding author: aidchel@mail.ru

framework of the SCM [14–16] and issues of integration in the chains [17–19].

In spite of the in-depth development of a wide range of issues, it should be noted that logistics chain management in the industries connected with R&D and innovative activities has been studied rather fragmentarily yet. The most important task for organization of innovative production structures is building of value chains with an organized cooperation of developers, suppliers and consumers aimed at creation of innovations.

High-tech enterprises are enterprises, which operations result in products (goods, works, services) with a considerable added value obtained at the cost of application of scientific, technical and technological achievements, characterized by a high share of R&D costs.

Peculiar features of high-tech products are also reflected in the logistic chain structure, which must permanently integrate participants implementing R&D. We believe that presence of such participants modifies the logistic chain structure. Alongside with the main production cycle of the participants' cooperation, the second cycle is formed, which is connected with development and diffusion of innovations and R&D results and which covers one or several links of the production cycle.

Thus, in the conditions of the targeted transition of Russia to the high-tech economic structure, the approaches to the logistics chain formation and management in high-tech industries are subject to in-depth development.

3 Statement of research objectives

The research objective is justification of the structure and development of approaches to industrial enterprises' logistics chain management in high-tech industries.

The industrial enterprise's logistics chain structure in a high-tech industry must consider cooperation with partnering enterprises for joint development and use of intellectual and tangible assets, manufacture and sales of high-tech products in the market conditions.

4 Key research findings

One of the key factors determining adaptability of industrial enterprises is the innovative potential – ability to manufacture innovative products with application of up-to-date production processes, which will allow to ensure competitiveness on the international market.

The noted tendencies of economic openness and globalization are particularly typical of high-tech product markets. These industries are characterized by integration of separate enterprises in order to increase their performance efficiency.

Subject to the international classification, high-tech industries include industries, in which R&D costs comprise at least 10% of the added value and the labor costs for scientists, engineers and technicians exceed 10% of the total labor costs.

The complex of high-tech economic sectors includes: aerospace industry, radioelectronics, production of computers, office equipment and software, as well as communication means, medical equipment, precision and optical-electronic equipment, chemical and pharmaceutical industry. The defense industry complex, which has a vast scientific-technical and industrial potential, occupies a special place.

Subject to the classification of the Organization for Economic Cooperation and Development (OECD), two approaches are used to define high-tech industries.

- Classification by the high-tech sectors. In this case the criterion is intensity of application of up-to-date technologies in the course of production.
- Classification by the product. In this case the criterion of affiliation with a high-tech industry is the end product and its research intensity.

Goods of high-tech industries are characterized by the following features: uncertainty of demand; creative approach to application; developed copying practice; blurred competitive environment. The life cycle of products is short, whereas appearance of such products at the competitors liquidates the enterprise's competitive advantage.

The economic growth of high-tech industries largely depends on technological innovations, products and equipment, which contains progressive knowledge [20,21].

Thus, the key success factor in these industries is product development competence. Consequently, the manufacturing enterprise has to form its own product development and design base or to search for efficient partners in this sphere. The efficiency of partnership in the conditions of a short product life cycle is determined by the development rates, duration of the product launch cycle and flexibility of processes.

It is also necessary to consider that researches and developments at high technology change rates are associated with a high investment risk. They include: indefiniteness of the terms of researches and associated costs, necessity to solve scientific and technical problems, use of unique resources, absence of a forecasting base.

In this connection, the variant of cooperation between the participants is deemed more expedient. Within the framework of the logistic chain formation procedure special attention must be paid to risk distribution between the partners.

In order to mitigate the chain's risks in this sphere cooperation of the participants is characterized by an average integration degree or presence of the chain link of a developer ensuring a high product updating rate. The ability to assume the chain's risks defines the enterprise's position in the logistics chain and its ability to influence other participants of the chain and to make the chain management decisions.

At working out of the manufacturing enterprise development plan in high-tech industries it is necessary to draw attention not only to product development but also to technological development and updating, which is impossible without major investments.

Therefore, we can conclude that for a manufacturing enterprise of a high-tech industry, which lacks a developed base for independent developments, the top-priority task is formation of partner links with enterprises able to conduct R&D or having developments but lacking their own production assets. The manufacturing enterprises' need for cooperation with enterprises developing innovative products and technologies is confirmed by the following data: the volumes of R&D outsourcing have shown a stable growth since the 80s of the XXth century, by the beginning of the XXI century they rose beyond the mark of 10% and currently comprise 15% [22].

The success in the sales link of the supply chain largely depends on the level of marketing project preparation at the R&D elaboration and commercialization stage. Meeting of the customers' needs and expectations, as well as competitive power of the manufactured products allow to form a structure of the product sales participants, which ensures availability and accessibility of the products.

Thus, analysis of the market needs and evaluation of the ability to gain profit from the sales of high-tech products is one of the key technological development factors.

Analysis of the key success factors in the manufacturing enterprise structure assumes identification of common industry parameters, which implementation allows to improve the competitive position on the market. The key success factors in the production sphere include high quality of the manufactured products, low production costs and its planning, strong links with suppliers, qualified work force, advantageous location, etc.

Then, for understanding of the own competitive advantages it is necessary to analyze companies working in this sphere. It is necessary for formation of mutually beneficial cooperation offers to the potential partner. In the conditions of technological depreciation the search of potential partners must by anticipatory, for which purpose the enterprise must have a clear idea of the partners to cooperate with, the advantages, which can attract potential partners to cooperation, and the forms of cooperation, which are most acceptable.

An efficient form of cooperation is inter-company cooperation, when the enterprises' resources, competences and basic technologies are united to achieve a better result. The competition and a high share of R&D costs force the companies to carry out parallel innovative activities with regard to products, implemented processes, organization and management. According to the researches, business-to-business cooperation increases the companies' innovative performance [23].

It is noted that the outlined high-tech industries have a higher level of innovative activities. They, in particular, have higher values of the following indicators: intensity of innovation costs; specific weight of employees engaged in researches and developments; share of new goods in the total volume of the shipped products; specific weight of organizations taking part in joint projects; share of the federal budgetary funds in the sources of innovation cost financing.

Analysis of the conventional supply chain structure within the framework of the industry high-tech product market allows to outline the following interacting links: R&D works; design engineering; procurement; production; sales and distribution.

At the junction of separate links the following industry markets are formed: market of R&D results, market of design engineering works, production equipment market, high-tech product market. The above markets differ by the degree of innovative activities, duration of the life cycle of the manufactured products, copying capabilities, competition level, etc.

Thus, the activity of manufacturing enterprises of high-tech industries leads to formation of an industry logistics chain consisting of the conventional logistic chain and innovative logistic chains corresponding to the segments of the industry market, namely:

- Scientific researches.
- Design engineering.
- Development of industrial technologies (applied researches).
- Manufacture of high-tech products.
- Promotion and sales of high-tech products.

Cooperation of the conventional and the forming innovative logistic chains is shown in Fig.1.

It should be noted that development of industrial technologies includes implementation of applied researches, as well as investment activities ensuring manufacture of products.

The degree of the customers' evaluation of the manufactured products is determined by taking of actions needed for development, production, supply and support of the product. Thus, the value of the products manufactured by the enterprise is characterized by the quality of the supplied good, its conformity with the customers' needs and expectations, as well as the quality of the accompanying service.

The objective of the logistics chain management is creation of the industrial enterprise's value chain, which is more efficient as compared to the competitors. The contribution in adding value of individual processes, structures of the enterprise, as well as its partners generally differs significantly.

Let us consider the value of manufactured products V. Then, it follows from Fig. 1 that

$$V=f(v_i, w_{ij}) \tag{1}$$

where v_i – value added by activity i within the framework of the conventional logistic chain;
w_{ij} – value added at the j-th section of the formed innovative logistic chain corresponding to activity i.

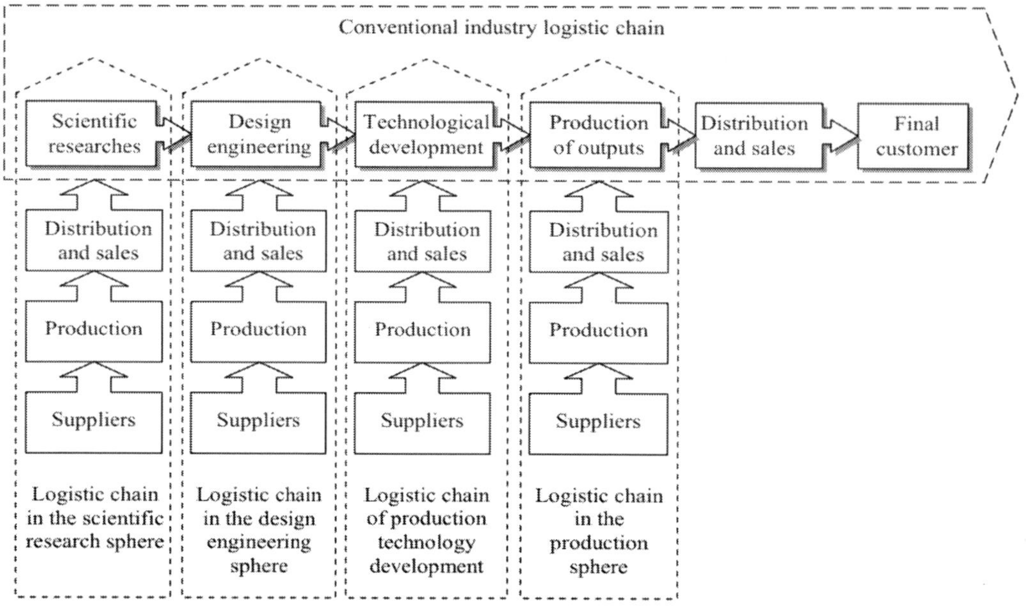

Fig. 1. Logistics chain model of high-tech industries.

Implementation of a specific organizational form of economic entities' integration within the framework of the logistics chain depends on the integration degree, as well as the associated risks. In this connection, the most appropriate way is integration based on business processes. Its applicability is determined by implementation of the process approach to enterprise management. Organization of activities based thereon allows to solve several most important tasks, including optimal resource management, minimization of nonproduction costs, as well as compliance with the needs of all groups of customers.

Use of information systems at modeling and control of the key process parameters allows to integrate the enterprise in the process logistics chain system by reconciliation of the own processes with the partners' abilities and demands, which gives advantages in the joint resource allocation, response to internal and external environmental changes and mutually beneficial coordination of activities on adding customer value of high-tech products.

5 Conclusions

The offered structure of the industrial enterprise's logistics chain management considers cooperation of the high-tech product market players aimed at increase of their competitive power.

The strong points of formation of enterprises' logistic chains comprising both conventional and innovative chains include:

- Creation of a continuous flow of innovative concepts and developments.
- Reduction of technological development costs (currently only one fourth of the patented technologies created inside enterprises is implemented).
- Uniting of the experience of the supply chain participants at development and commercialization of high-tech products.

The work was supported by Act 211 Government of the Russian Federation, contract № 02.A03.21.0011.

References

1. R.B. Handfield, E.L. Nichols, *Introduction to Supply Chain Management* (Prentice Hall, New Jersey, 1999)
2. D.J. Bowersox, D.J. Closs, *Logistical Management: The Integrated Supply Chain Process* (McGraw-Hill Companies, 1996)
3. D.J. Bowersox, D.J. Closs, M.B. Cooper, *Supply Chain Supply Management* (McGraw-Hill, 2013)
4. M. Christopher, *Supply and Supply Chain Management* (Financial Times/Pearson Education, 2016)
5. R.B. Handfield, E.L. Nichols, *Supply Chain Redesign: Transforming Supply Chains Into Integrated Value Systems* (Prentice Hall, New Jersey, 2002)
6. D.M. Lambert, *Supply Chain Management: Processes, Partnerships, Performance* (SCMI, Sarasota, 2008)

7. D.M. Lambert, *The Supply Chain Management and Logistics Controversy* (Pergamon Press, Oxford, 2001)
8. J.R. Stock, D.M. Lambert, *Strategic Logistics Management* (McGraw-Hill, New York, 2001)
9. D.A. Ivanov, *Supply Chain Management* (Polytechnic University, Saint Petersburg, 2009)
10. V.I. Sergeev, Logistics and Supply Chain Management, **5(16)**, **6** (2006)
11. Y.V. Fedotov, K.V. Krotov, Russian Management Journal, **9**, **49** (2011)
12. V.S. Lukinsky, O.V. Badokin, T.A. Blazhenkova, A.A. Bochkarev, E.I. Zaitsev, V.V. Lukinsky, *Problems of Formation of the Applied Theory of Logistics and Supply Chain Management* (SPSUEE, Saint Petersburg, 2011)
13. O.D. Protsenko, Logistics, **4**, **30** (2013)
14. A.G. Butrin, E.I. Rogozhnikov, V.I. Tsaplin, Economic Analysis: Theory and Practice, **15**, **30** (2010)
15. V.V. Lukinsky, T.A. Kulagovskaya, Y.V. Malevich, Bulletin of Saint Petersburg State University of Engineering and Economics. Series: Economics, **1**, **217** (2009)
16. A.N. Sterligova, Logistics and Supply Chain Management, **4(9)**, **36** (2005)
17. V.V. Dybskaya, E.I. Zaitsev, V.I. Sergeev, A.N. Sterligova, *Logistics. Integration and Optimization of Logistical Business Processes in Supply Chains: MBA manual* (Exmo, Moscow, 2008)
18. V.S. Lukinsky, V.V. Lukinsky, T.G. Shulzhenko, Logistics and Supply Chain Management, **4(57)** (2013)
19. L.B. Mirotin, A.G. Nekrasov, *Logistics of the Integrated Supply Chains* (Ekzamen, Moscow, 2003)
20. S.Y. Glazjev, *Nanotechnologies as a Key Factor of the New Technological Structure in the Economy* (Trovant, Moscow, 2009)
21. E.N. Kablov, Science and Life, **4** (2010)
22. E.N. Gorlacheva, I.N. Omelchenko, Science and Education, **12**, **1** (2010)
23. V.S. Verba, E.N. Gorlacheva, A.G. Gudkov, Innovations, **3(149)**, **56** (2011)

Economic-mathematical modelling of placement of gold recovery factory at the development of the group of territorially converged gold deposits

S.I. Yevdokimov[1,*], *E.I.* Meshkov[1], and *T.E.* Gerasimenko[1]

North Caucasian Mining and Metallurgical Institute (The State Technological University) Vladikavkaz, Russia

Abstract. A small volume of production, combined with the high capital costs and operating costs, is an objective reason of the reduction of investment attractiveness of deposits with reserves of gold <25 tons in the link "supply and demand". The object of the research is a group of geographically contiguous succinct gold deposits. Economic result of the development of low-volume gold deposits will be fundamentally different if the production capacity of the company would be increased by combining geographically contiguous fields in small-group. The increase in reserves when the field is developed in a group should outpace the reduction of recoverable reserves value. It is shown that the level of production of gold margin depending on the location within the factory gold industrial area varies by no more than 4-14% when changing the transport arm of the deposit base by 20-40 km, it correlated with specific transportation cost k from 1.5 to 3.0 rubles/t · km profitability 30-40%, and the total cost of an ounce of gold sold U$ 1000-1200. To determine the "center of gravity" of the group for each type of inventory the zone breakeven transport extracted from the feedstock crude concentrate is counted: stocks have in the center of a circle with a radius equal to the maximum possible distance for the transport of crude concentrate to a stationary (bush) gold recovery plant without economic damage to the developer that is recoverable ore value. The place of intersection obtained "break-even" circular zones is the "center of gravity" of the district, where the factory is located.

1 Introduction

Gold mining companies have no competitors in the field of sales, since gold is a standardized product with almost absolute liquidity. According to the Thomson Reuters GFMS, total global demand for gold is more than 4100 tons. Gold is widely used in electronics and pharmacology, but the traditional and the largest consumer (with a share in aggregate demand consistently above 50 %) is the jewelry industry. The ratio between its demand and supply side of the gold mining industry - the steadiest segments of the global gold market - is a key determinant of the price of gold. The biggest consumers of gold are India and China, a strong lead among countries-producers of gold and annually increasing the volume of work. Gold is a common investment tool. Against the background of high risks of inflation in the largest developed economies, Central banks [1, 2, 3, and 4] are an increasing proportion (11-13 %) in the structure of demand.

2 Relevance

In Russia in the undistributed Fund is 30 % of proven gold (with total reserves of over 39 million ounces of metal), centered about 100 small sources of gold (with gold reserves of up to 800 thousand ounces) [5]. The reason is that individual exploitation of geo-resources such unprofitable [1, 6, 7, 8, 9], because of the small volume of production arising from compliance with the normal depreciation period of existence of the mining enterprise, insufficient to recoup the costs of production. Considering the problem of increasing production of gold in Russia as the current state-significant program, it can be argued that the solution to the problem of increasing investment demand for low-volume mineral commodity sources of gold due to cutting costs and reducing the investment payback period will increase the amount deposited to the state Depository metal in the next 10-20 years 650-970 thousand ounces annually.

3 Formulation of the problem

In accordance with Taylor's formula, linking the rational term development of T (years) deposits with balance reserves of M (tons)

$$T = \frac{1}{5}\sqrt[4]{M} \qquad (1)$$

* Corresponding author: eva-ser@mail.ru

the specific value of costs (capital, operational and organizational) decrease with increase in the value of inventories by ΔM. Therefore, the method of increasing the efficiency of a commercial exploitation of small fields is the Association (including accounting for licensing and the preparation of group auction) them to the group for the joint operation of [10, 11, 12, 13].

The mixing of ores with the same gold content, but different on concentration may be disadvantageous in the separation [14, 15, 16]. The content of gold α_Σ (g/t) total reserves at the group deposits M_Σ is calculated based on the gold content α_i (i = 1 ... n) and the number of reserves in the individual fields of the group of $M_i(t)$

$$\alpha_\Sigma = \frac{M_1\alpha_1 + M_2\alpha_2 + ... + M_n\alpha_n}{M_1 + M_2 + ... + M_n} \quad (2)$$

The field may be included in the group to the joint operation, provided that when you combine its inventory with the inventory group to run the criterion, defined by the relations (3) and (4)

$$\frac{M_\Sigma}{M} \times \frac{\alpha_\Sigma}{\alpha} = \frac{Y_\Sigma}{Y} \quad (3)$$

where Y the economic criterion of efficiency of separation of the mixture of ores

$$Y = \left[\frac{Q(A_1 + A_2 + A_3)}{\gamma} + A_4 + A_5 + \sum A + A(\beta)\right]\varepsilon(\beta) \quad (4)$$

where M is the available inventory underlying the Deposit with a gold grade of α; Q - the amount of recycled ore; A_1, A_2, A_3 - respectively, the cost of production, mediation, and enrichment 1 ton of ore; γ - out of the concentrate; A_4, A_5 - the cost of transporting 1 ton of ore to the mill and 1 ton of concentrate to transport hub; ΣA - other costs per 1 ton of ore; $A(\beta)$ - cost metallurgical processing 1 ton of concentrate, depending on the content β of gold; $\varepsilon(\beta)$ - recovery of gold in the metallurgical process, depending on the quality of the concentrate. In economic-mathematical modeling geological and technological parameters of the gold Deposit "Pavlik" (Magadan region) are accepted as basic.

4 Theoretical part

The profitability of gold production (R) depends on the amount of costs for mining and milling (P), gold grade (α_Σ), the extraction of gold (ε_Σ) to the processing plant, transport costs ($\kappa\chi$), the price of 1 g of gold (Z) and is mathematically related formulas in models (5) - (7):

$$R(\alpha_\Sigma, \chi, \varepsilon_\Sigma) = f\left[(Z \times \alpha_\Sigma \times \varepsilon_\Sigma) \times (P + k\chi)^{-1}\right] \quad (5)$$

$$\varepsilon(\alpha_\Sigma, R, \varepsilon_\Sigma) = f\left[(R \times (Z \times \alpha_\Sigma))^{-1} \times (P + k\chi)\right] \quad (6)$$

$$\alpha(\varepsilon_\Sigma, R, \chi) = f\left[(R \times (Z \times \varepsilon_\Sigma))^{-1} \times (P + k\chi)\right] \quad (7)$$

The practical value of the model (5) - (7) consists in the possibility of determining the optimal positioning of GCP within the group of fields on the basis of the calculation of the maximum distance of transportation of ore with an acceptable economic damage to the subsoil user (given profitability).

5 Practical significance

As follows from the equation (5) of model, gold production profitability (R) depends on the amount of production costs and ore processing (P), as well as the variables - the gold content in the ore (α_Σ), gold extraction (ε_Σ), transportation costs ($k\chi$) and the selling price of gold (Z).

Expression (6) of model lets determine the level of gold recovery in mining ε_Σ, if the variable parameters are distance transport χ, quality of ores α_Σ and metal price Z, and the expected (or given) - profitability of gold production R. Expression allows you to determine the quality of the charge, which must be obtained by mixing the ore of grouped fields in order to provide the required profitability under current vehicle operating costs and the level of gold recovery.

Equations (3) - (6) of model allow us to determine the optimal location of the gold recovery plant with respect to gold deposits, included in group development, corresponding to the minimum vehicle operating costs. The relationship of selected economic and technological parameters of the model (3) - (6) is graphically displayed in Fig.1-3.

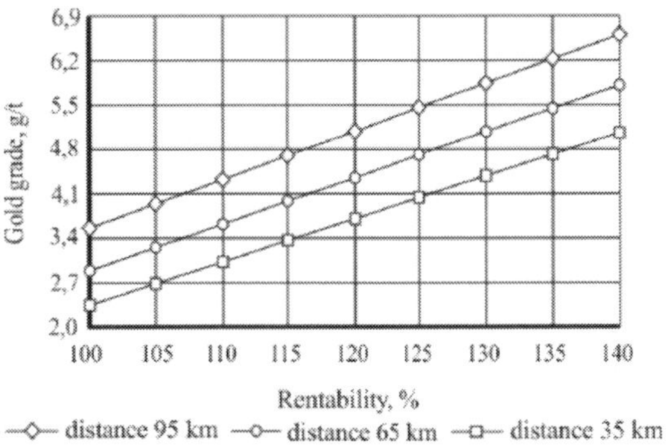

Fig.1. The gold content in the processed ore charge as a function of the profitability of gold production at a different distance transport charge to the gold recovery plant.

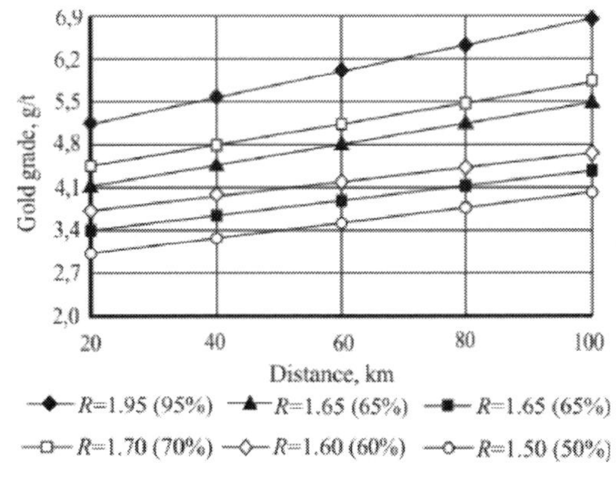

Fig.2. The gold content in the processed ore charge as a function of the distance transportation to the gold recovery plant at different profitability of gold production.

The curves shown in Fig.4 provide a graphical representation of the sensitivity of the project profitability to a change (as a percentage of the base case) of the model parameters (5) - (7).

Get the expected result: $R \to \max$ if $S \to \min$ at various α and ε. Therefore, the model (5) - (7) can be used to solve the problem of determining the optimal placements industrial complex group of fields.

From the results of simulations on the model (5) - (6) follows that The level of profitability of gold production, depending on the placement of the mill in industrial area varies by no more than 4-14% while transport arm is changed from the basic deposit of 20-40 km, specific transportation cost k correlated with it from 1.5 to 3.0 rub/t·km profitability 30-40%, and the total cost (AISC) ounces sold US 1000-1200.

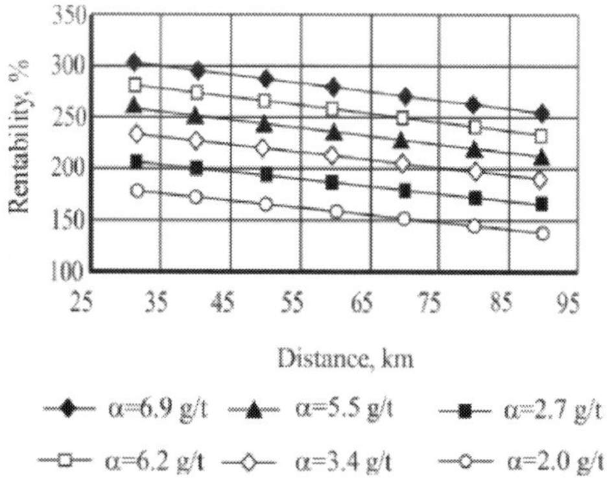

Fig.3. The profitability of gold production as a function of the distance of transportation of processed ore charge to the gold recovery plant at different gold content in the ore charge.

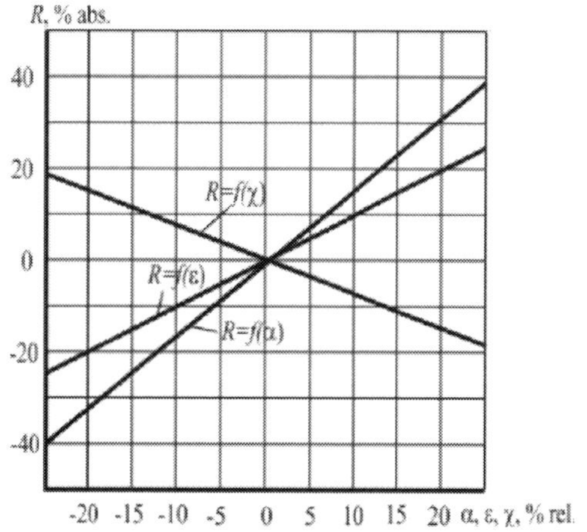

Fig.4. The sensitivity of the profitability of the project to changes in ore transportation distance, its content of gold and extraction at a gold plant.

However, in remote conditions of the succinct fields from basic and auxiliary production facilities, as well as housing and domestic infrastructure of existing mining facilities, the level of profitability of gold production cannot be the only criterion for evaluating the availability of stocks of small-fields If the recoverable value of one of the deposits in a group is significantly higher than the others, the location of this object can be regarded as the "center of gravity" of the group in which it is advisable to place the stationary gold factory (assuming a suitable terrain, availability of energy facilities, based infrastructure and etc.). Otherwise, to determine the "center of gravity" of the group for each type of inventory the zone breakeven transport extracted from the feedstock crude concentrate is counted: stocks have in the center of a circle with a radius equal to the maximum possible distance for the transport of crude concentrate to a stationary (bush) gold recovery plant without economic damage to the developer that is recoverable ore value. The place of intersection obtained "break-even" circular zones is the "center of gravity" of the district, where the factory is located.

Variable parameters for solving the problem of placement processing plant, designed for low-volume processing ore deposits are as follows:

1) the distance of transportation of ores «χ», determined by the position of the gold plant processing capacity based on mobile dressing complexes of modular type;

2) gold content «α_Σ» in charge, formed from ores specially grouped geographically contiguous gold deposits in accordance with the requirements of (5) - (7);

3) recovering gold «ε_Σ», calculated according to the equation type of regression;

4) profitability «R», defined as the ratio of sales price of 1 g of gold to the operational costs of opening deposits, extraction of the rock mass, transport and processing of ores associated with obtaining 1 g of gold (the cost of production of 1 g of gold).

Selected economic and technological indicators of mineral processing are mathematically interrelated with formulas of the model (5) - (7):

The estimated period of operation of gold extraction plant equipment is advisable to take 10 years, the estimated operation - 120 days a year, 20 hours a day. The effectiveness of the seasonal operation mode is based on the fact that the construction of gold processing plant, under these conditions of Siberia and the Russian Far East, the share of the equipment does not exceed 20%, while the share of construction and installation works - 80%. Thus, the value of the gold recovery plant equipment during the transition to a seasonal operation is tripled, and the corresponding decrease in construction and installation works (which for seasonal mobile collapsible gold modular plant will not exceed 25% of equipment cost) reduces the total amount of capital investments 25%. An important part is the factor of terms of factory buildings, which in a preferred embodiment are reduced to 4-5 years to 1 year.

Conclusion

The results allow us to make the important conclusion that for the group developing territorially contiguous fields the character of the terrain in terms of the possibility of rational distribution of industrial site plant and tailings, housing and household products, the availability of water resources and proximity to the grid, the existing road infrastructure, local settlements mainly influence the choice of industrial site location of gold extraction plant having.

In contrast to economic and landscape criterion, transportation costs associated with the transport of ore from the place of production to the gold recovery plant, do not have a decisive influence (are not critical for the commercial effectiveness of the development of stocks grouped fields as a whole) on the economic viability of the development of unconventional resources in the group, construction gold recovery plant efficiently implement near field, the reserves of which have a maximum recoverable value.

The result (a weak dependence of the level of profitability of gold production from the value of the operating costs of transport) is due to the fact that in the structure of production costs of gold ore transportation costs from the quarry to the gold recovery plant is 11 to 23%. With the growth of the selling price of gold compared to the cost of its production the level of profitability increases, while the share of transport costs in the profitability structure decreases significantly (in the actual preservation of the relative share in the total cost of production of gold).

References

1. N.Yu. Samsonov, M.A. Yagolnitser, *Group development of small gold deposits* (IEIE SB RAS, Novosibirsk, 2012)
2. A.A. Petrosov, V.A. Ivanov, Mining information and analytical Bulletin, **12** (2012)
3. B.K. Mikhailov, A.I. Nekrasov, Mineral resources of Russia: Economics and management, **3** (2012)
4. I.M. Petrov, Mineral resources of Russia: Economics and management, **3** (2011)
5. N.Yu. Samsonov, Study areas for cost-effective development of small and medium-sized gold deposits (Institute of Economics and industrial engineering of Siberian branch of RAS, Novosibirsk, 2010)
6. N.Yu. Samsonov, Economy of region, **2** (2010)
7. N.Yu. Samsonov, Regional Economics and sociology, **1** (2010)
8. A.A. Sobolev, Mountain information-analytical Bulletin, **3** (2013)
9. V.V. Kharitonov, U.N. Kurilchuk, Mining magazine, **9** (2014)
10. V.A. Ivanov, *Organizational and economic justification of public-private partnership in the development of small-scale gold deposits* (Moscow state mining University, Moscow, 2013)
11. V.A. Kabirov, *Assessment of economic efficiency of development of group of geographically contiguous ore (metal) deposits* (Saint Petersburg "National mineral resources University "Gorniy", Saint Petersburg, 2014)
12. A.B. Anisimova, *The Economic rationale for involvement in the development of contiguous iron-ore deposits of the Polar Urals* (Russian state geological prospecting University. S. Ordzhonikidze, Moscow, 2011)
13. A.B. Anisimova, The proceedings of the mining University, **191** (2011)
14. M.A. Peshkov, *Substantiation of requirements to quality of ore and technogenic raw materials for the integrated development of copper-pyrite deposits of the Urals* (Institute of comprehensive exploitation of mineral resources Russian Academy of Sciences, Moscow, 2014)
15. Y. Roslavtseva, *Feasibility of mining operations during incremental development of small deposits in the open way* (Irkutsk state technical University, Moscow, 2014)
16. L.I. Boguslavskaya, *Increase of economic efficiency of the mining company on the basis of control of its capitalization* (St. Petersburg state mining University, St. Petersburg, 2012)
17. M.L. Polumiskova, *Development of economic-mathematical models and tools for assessing the investment attractiveness of gold mining companies* (VNIIPVTI, Moscow, 2010)
18. M.L. Polumiskova, Economic science, **10** (2008)
19. M.L. Polumiskova, Economic science, **3** (2010)
20. V.N. Kalmykov, A.A. Gogolin, A.N. Ivashov Mining magazine, **12** (2015)

 DOI: 10.1051/shsconf/20173501017

Specifics of formation the portfolio of orders on the basis of break even analysis for Iron and Steel Works

Evgenia Zambrzhitskaia[1,*], *Tatyana* Kozlova[1], and *Anastasia* Mamaeva[1]

[1]Nosov Magnitogorsk State Technical University, 455000 Magnitogorsk, Russian Federation

Abstract. The issue of orders portfolio formation becomes more and more actual in terms of evolving economy crisis, consequences of which turned into the growing competition both in domestic and foreign markets. Goals of research – formalization of order portfolio formation procedure. The purpose of our academic research – to develop the algorithm for order portfolio formation suitable for iron and steel works. The article suggests the possibility of using break-even analysis indices, in particular, the research introduces a «break-even point» concept in terms of efficient order portfolio. Usage of break-even point indices is more preferable to others (marginal revenue, cost-effectiveness and etc.) the main advantages of the suggested analysis are: 1) great information capacity from the point of management decision making; 2) the suggested analysis is less governed by price factor than marginal revenue.

1 Introduction

In terms of intense competition, innovative technology and periodical crisis events, iron and steel companies have to combine flexibility in working process with target customers and supreme economic viability achievement in the industrial activity. The solution of this task is directly connected to the effective methods searching of order portfolio formation.

It is necessary to mention that today there is no common approach in order portfolio formation (in theoretical or practical aspect) [20, 21]. Thus every business entity develops its own method and approach , which corresponds to industry specific standards and promotes the adoption of effective management decisions

Problem Statement

The task of formation of effective order portfolio usually lies on sales department. [21]. Nowadays the main criteria of order portfolio formation for enterprises of metallurgical complex are:

- Optimal payload for the current production capacities
- Maximization of order effectiveness (generally is achieved by means of benchmarking study applying in the sphere of industry orders).

The current framework concentrates on the valuation of price components, because the optimum work load productive capacity has a satisfactory solution and regularly used in practice. [4, 5, 6].

2 Main Part

A. Production order classification

While solving the problem of effective order portfolio forming there is a necessity to figure out the order classification: to determine classification features and main classification elements.

Roughly all the orders, preceded at a factory, can be divided into three groups (from economic point) (fig. 1) [1].

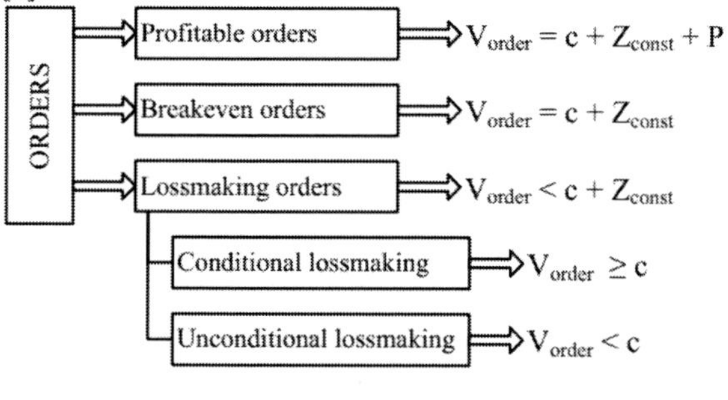

Fig. 1. Order classification (for a factory unit).

On the fig. 1 the following indications are taken:

V_{order} – order value (company commercial proposal);

c – variable costs, according to order (out-of-pocket unit costs);

Z_{const} – constant expenses part moved to an order;

P – profit (in any industry, including smelting sector, it is possible to figure out current profit rate).

Many metallurgical plants in terms of slump in demand, are ready to orders at cost.

All the orders with formula $V_{order} < c + Z_{const}$ are traditionally considered to be unprofitable and not useful for any industry. This group can be roughly devided into two categories:

* Corresponding author: jenia-v@yandex.ru

1 category: conditionally unprofitable – orders, covering all direct and a part of appointed indirect costs.

2 category: unconditionally unprofitable – orders are totally unacceptable for any business and are only taken for the purpose of market gain of for any other strategic reasons (image, technology development and etc.) [1].

B. Analysis of the main order portfolio formation options

C. The following stage is determination of possible order portfolio formation options in terms of iron and steel works, basic of which are in figure 2.

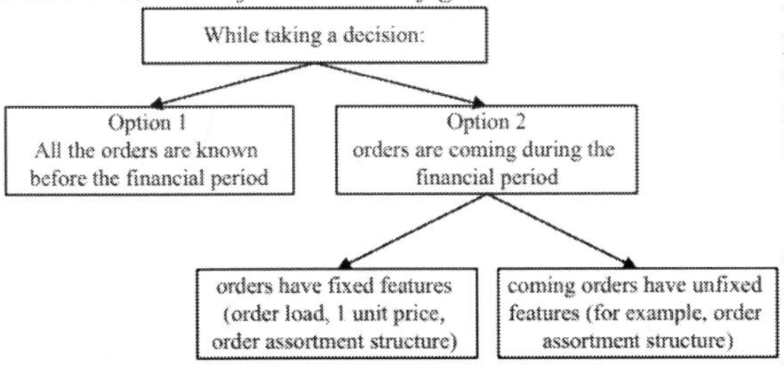

Fig.2. Possible order portfolio formation options in terms of iron and steel works.

The first option is the most preferable for business, because in this case the task of order portfolio formation is quite simple. To achieve this aim special kinds of programs are used. These programs are focused on checking of general order quantities to form references such as «accept – exclude».

As a rule, the general checking criteria are shown in fig 3.

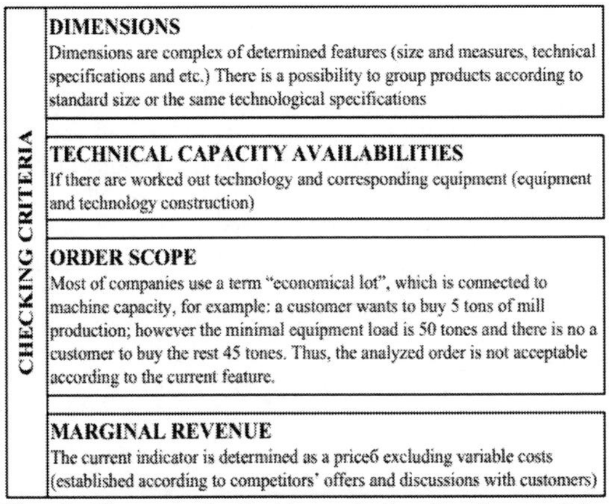

Fig. 3. The general checking criteria of order portfolio construction (version 1).

It is important to mention, that some companies use other indicators instead of "marginal revenue", in particular – efficiency, profit and etc. the most preferable indicator is «break-even-point» (BEP). This factor calculation is recommended to do according to the following formula (in terms of multiproductional industry, like iron and steel works companies) [3, 7, 10, 11, 12]:

$$V_0 = \frac{Z_{const}}{\sum_{i=1}^{n}(p_i - c_i) \times v_i} \quad (1)$$

where

V_0 – overall production account in BEP in physical terms;

Z_{const} – total constant expenses

p_i – price of i-type product;

c_i – out-of-pocket unit costs of i-type product.

v_i – part of i-type product, calculated by formula.

$$v_i = \frac{V_i}{\sum_{i=1}^{n} V_i} \quad (2)$$

Indicator BEP, calculated b formula (1), must be analyzed in view of product range [15]. In fact, a special product measuring unit is used. This unit can be called conditional selection unit. Conditional selection unit – is an assortment (package), including every kind of output in a set lot the sum of all product lots is equal to a unit or 100%.

It is important to mention that while determine BEP one cannot use conventional tons instead of conventional set units, because conventional tons show difficulties in output processing (in iron industry) or other characteristic (for example, cement strength in cement sector), what is not connected to a product range, constant and out-of-pocket unit costs correlation.

The scheme of order portfolio formation for the case, mentioned above, is represented in fig. 4.

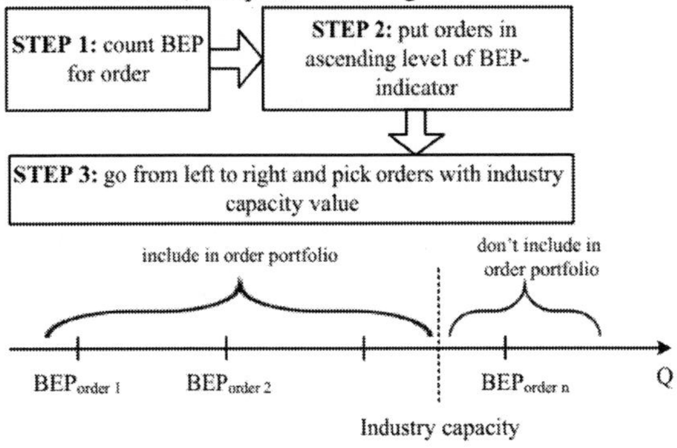

Fig.4. Offered sequence of order portfolio formation in the first option.

In the second option of order portfolio formation is supposed that manufacturing orders are taken in different time and the decision about the order must be made immediately: accept or refuse.

In terms of the second option there can be 2 situations:

- Incoming orders with fixed characteristics (order size, one unit price, assortment order structure);
- Incoming orders with unfixed («float») characteristic (for example, assortment output);

Let's have a closer look at each option, mentioned above. Case, when income orders have fixed characteristics (order size, one unit price, and assortment order structure) is possible in terms of working with big industry companies, which are able to organize further rolled metal products processing.

* Corresponding author: jenia-v@yandex.ru

 DOI: 10.1051/shsconf/20173501017

These customers are usually called «key customers/purchaser». As a rule, industries, where this rule is working are:

- Car industry;
- Industries specialized on different sized pipes and tubes production;
- Car production facilities;
- Defense enterprises;
- Hardware production industry.

On average the mentioned customers' category can consume from 50% till 65% of overall iron and steel work production. To work with this customer group demands order portfolio formation using BEP. In this case order must correspond to certain conditions:

1. BEP of order must not be more than BEP critical, which is describes further;
2. Scope of all income orders will be compared to productive capacity.

Business process of making management decisions is shown in fig.5.

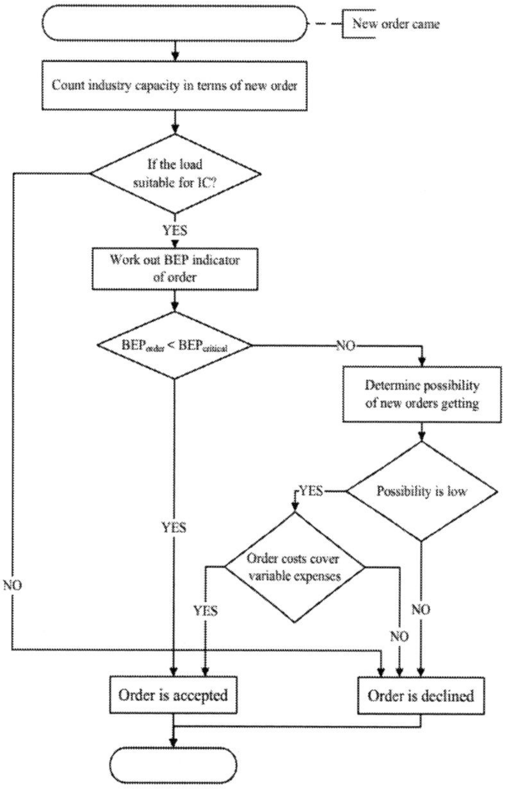

Fig. 5. Business-process flow chart of order portfolio formation.

As one can see in fig.5, when a new order is got, firstly the possibility to add it to the load of current productive capacity (PC). If it is possible, than BEP calculations should be made. Than the mentioned indicator must be compared to some critical meaning by means of calculations. The point of calculations is demonstrated on the example below.

D. Conditional example, explaining the BEP order calculation mechanism

Let's suppose there is a regular order from a car industry company. The order specification is fixed and corresponds to: cold-rolled sheet, cold rolled coil, hot-rolled sheet, hot-rolled band, bar section, hot-rolled round steel bar.

BEP must be worked out for every kind of product i.e. we suppose that the order contains only mentioned products. Worked out BEP scheme for product "A" (cold rolled sheet) is illustrated in fig. 6 [13, 14, 16, 17, 19].

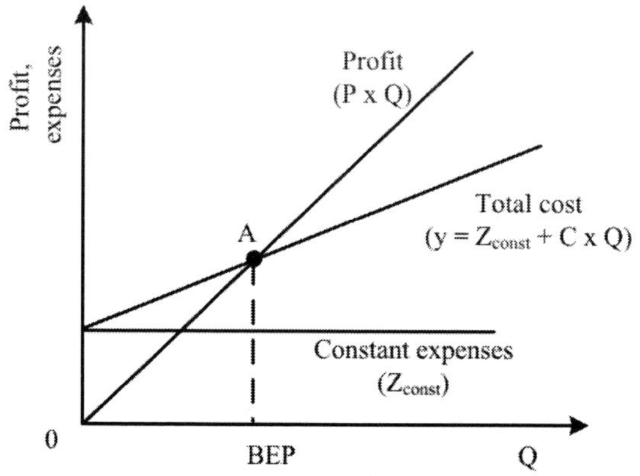

Fig. 6. BEP scheme for product "A" (cold rolled sheet)

While releasing n-type production, the crossing point of profit and expenses has a form of convex, which projection on the strain levels are indicated on the x-coordinate axis gives a possible position oh BEP in terms of selection structure different options. The composition scheme of the hexagon from the example is shown in fig. 7.

Thus, critical BEP meaning is a minimal meaning of the given indicator, i.e. $BEP_{critical} = BEP_{min}$.

It is important to mention that today iron and steel works use marginal revenue as main economic criterion instead of suggested BEP indicator. This indicator, as was mentioned above, has disadvantages, main of which are heavy reliance on price and low information value from the point of management accounting.

* Corresponding author: jenia-v@yandex.ru

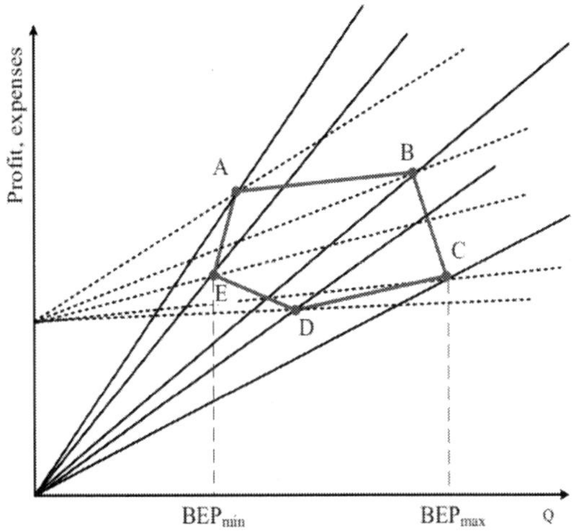

Fig.7. The composition scheme of the hexagon in terms of n-type production releasing.

If the order does not fit to BEP criterion there should be determined a possibility of next order. Statistical techniques are useful on this stage. If the possibility of the next order is not high (less than 70%), than it should be accepted only in case of absolute variable costs covering. Otherwise, the order is better to decline, because it will be lossmaking and provoke losses within the analyzing industry.

Now let's analyze the situation, when orders have unfixed («float») assortment structure. Such case can happen at industries, which resale iron and steel products. The mentioned customers' category is called «metal tradespeople». There total make share can be from 40% till 55%. Typical type of work with this customer's category is signing of "blanket agreements, where assortment structure is not fixed and depends on market condition. Expediency consideration of getting into the order portfolio of mentioned agreements must be done according to the following algorithm:

1. Sphere of possible BEP changes in terms of defferent assortmnet structures [8, 9], mentioned in the "blanket" agreements for the production delivery must be identified (similar to fig. 7);

2. The zone of the analyzing order must be determined. The following zones are possible:

a) Zone of absolute loss ratio (in this case order is declined);

b) Zone of absolute profit (in this case the order is accepted);

c) Ambiguity area or a zone of risk, where financial result depends on assortment structure (in this case production range is to be defind).

Graphically the suggested mechanism is shown in fig. 8.

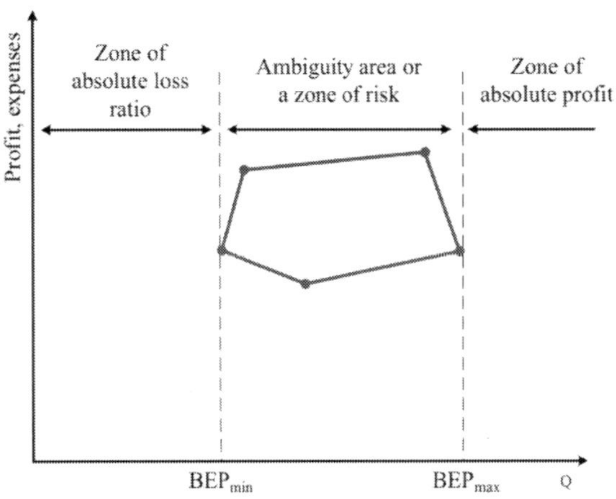

Fig.8. Possible zones of industry work according to "profit" criterion.

3 Conclusion

It is important to mention that practical usage of the suggested method of order portfolio formation can be released as a business process in enterprise information system (EIS) (for example, ORACLE an etc.). Minimal claim to the EIS software is support of the basic business processes.

In conclusion must be said that the suggested method can improve order portfolio formation procedure efficiency in terms of modern iron and steel works. This method also makes the work effectiveness more valid from the economical point of view, what can help to avoid possible losses, caused by ineffective approach to order portfolio formation.

References

1. I.S. Berezovski, Economic systems management: electronic scientific magazine, **31** (2011)
2. M.A. Vahrushina, *Accounting management: Student book for universities* (Омega-L; High School., Moscow, 2003)
3. E.S. Voinova, G.V. Danilov, I.G. Rijova, Scientific and technical reports of St/-Petersburg State technic University. Economic Science, **3-1(58)**, 225-230 (2008)
4. G.V. Danilov I.G., Rijova, E.S. Voinova, Scientific and technical reports of St/-Petersburg State technic University. Economic Science, **4(102)**, 87-90 (2010)
5. G.V. Danilov, I.G. Rijova, E.S. Voinova, Nosov Magnitogorsk State Technic University Massenger, **1**, 79-82 (2012)
6. G.V. Danilov, E.S. Zambrzhitskaia, I.G. Rijova, Analysis of industry capasity smoothness, break-even point and financial safety margin, **43**, 18-23 (2012)

* Corresponding author: jenia-v@yandex.ru

DOI: 10.1051/shsconf/20173501017

7. G.V. Danilov, E.S. Voinova, I.G. Rijova, Modeling of assortment infuence on main industry indicators, **15**, 40-46 (2012)
8. G.V. Danilov, E.S. Voinova, I.G. Rijova, International axxounting management, **21**, 22-28 (2012)
9. G.V. Danilov, I.G. Rijova, E.S. Voinova, Scientific and technical reports of St. Petersburg State technic University. Economic Science, **6-1(90)**, 163-157 (2009)
10. G.V. Danilov, I.G. Rijova, E.S. Voinova, Economic analysis: theory and practice, 26, 35-39 (2009)
11. E.S. Zambrzhitskaia, Joint fund currnt events Science and education, **11(78)**, 95 (2015)
12. R. Zeman, M. Vohoxka, E.S. Zambrzhitskaia, N.I. Ivanova, R. Pihova, M. Popilkova, K. Kabourkova, *Management accounting:teaching guide* (Nosov Magnitogorsk State Technic University, 2015)
13. O.N. Volkova, *Accounting management* (Moscow, TK Welby, 2010)
14. Collin Drury, *Management and cost accounting. Introductive course: Educational complex for university students* (Unity-DANA, Moscow, 2012)
15. V.B. Ivashkevich, *Management accounting: teaching guide* (Master, Infra-M, Moscow, 2013)
16. T.P. Karpova, *Accounting Management: student book for universities* (Unity, Moscow, 2002)
17. N.P. Kondrakova, M.A. Ivanova, *Management accounting: teaching guide* (Infra-M, Moscow, 2013)
18. I.G. Kukunina, *Accounting Management. Losses management Management analysis*
19. Ch. Horngern, G. Fosters, Sh. Datar, *Management accounting* (Piter, St. Petersburg, 2008)
20. I.B. Chachina, C.V. Gorodnichev, Tulskiy State University news. Economic and legal science, **1-1**, 199-203 (2013)
21. N.I. Trishkina, *Science XXI century. Proceedings of materials the international scientific conference*, 321-328 (2015)

* Corresponding author: jenia-v@yandex.ru

Organizational networks with self-standing interests

Vasily Say[1,*], *Sergey* Sizyi[2], and *Igor* Gromov[1]

[1]Ural State University of Railway Transport, Department of railway track and railway construction, 620034 Yekaterinburg, Russian Federation
[2]Ural Federal University, Department of Algebra and Fundamental Informatics, 620083 Yekaterinburg, Russian Federation

Abstract. With the development of world economy gradual replacing of traditional organizational structures by organizational networks, take place. New organizational complexes set difficult tasks for researchers considering optimum building and functioning of organizational networks. In these conditions, a key focus of organization network research is simulating the interaction process as well as network's integrated quantitative assessment based on this simulation. A distinctive feature of the applied simulating approach is formalization of interaction processes between the integrator and network entities taking into account their contradictory interests. The integrated quantitative assessment of a network is offered proceeding from understanding that the general purpose of functioning is satisfying society needs with products and services of a network.

1 Introduction

Nowadays, the task of creating organizational networks is at the peak of its relevance for Russian economy. Such structures, based on associational or holding method, are being built now with consideration of the world's best experience. However, the organizational and economic field, which has developed after the country's changeover to a market economy, was not created in the course of evolution, but by means of strict administrative actions. Universally there was the process of restructuring in the form of downsizing. Thus, a process of radial-flow separation, which undoubtedly influenced the structure of organizational networks, had preceded the process of networking.

The authors of this article apply to the world–renowned examples of companies' network creation and develop the theoretical aspects of the organizational network formation with consideration of regional economy mentality, since a well-organized process can be a key to a network's prosperity.

2 Related literatures

Formation, adjustment and operation of an organizational net-work depend on its type. Manuel Castells divided all the net-works into two groups, the first – a multi-directional network model, introduced into life by small and medium-sized enterprises, the second – licensing and subcontracting production model under an «umbrella» corporation [1–3].

However, it is common to define a network organization as one that is fast and flexible in adapting to changes in the underlying environment [4-7]. Umbrella Corporation is that element of control of this stakeholder group holding the executive power should be a central topic for stakeholder theory [8–9].

In further studies, a network is considered as a formation around a large company, that is, «umbrella» corporation.

Under a network with self-standing interests, we understand the presence in an umbrella corporation and all networking companies their own, in most cases, conflicting interests. Thus, decentralization and increased flexibility of the structure may cause so called «articulation errors», defined by Guy Benveniste as «...the partial or total lack of fit between what is wanted and what is available» (cited from Castells 1996-1998, 178).

The problem of elimination of «articulation errors» in highly decentralized structures is complicated and often not solvable issue, and network adjustment in a constantly changing internal and external environment is a difficult continuous process.

M. Castells confirms our assumptions by considering it as a specific form of an enterprise, which assets are composed by the segments' intersection of autonomous goal systems.

However, further we will consume, that all the companies included in a network are connected by a common aim, and this is one of the key stages of organizational network formation. By intercommunity of the network aim, we suppose satisfying society's needs in products or services or «the production of useful things», as H. Ford claimed.

Our assertions resonate with [10] «Successful partnerships and collaborations in service of sustainable effectiveness will require individual organizations to change their objective function and build new and varied internal and external capabilities», in accordance with

* Corresponding author: vsay@usurt.ru

[11] «Every agent realizes his full potential and has no incentive to deviate from either staying independent or from the endogenously determined partner and payoff. The partnership formation problem contains the widely studied assignment market problem as a special case».

During the network modelling it is also needed to take into account the interests and priorities of its members. J. Tullberg in the work «Stakeholder theory: Some revisionist suggestions» maintains that the priority of stakeholders implies distancing from general altruistic philosophy that argues against, but not for, giving special consideration to the company's stakeholders [8].

The network structure is formed on the basis of integrator properties and the elements included in the network. Thus, the network is formed endogenously: «In equilibrium, the partner-ship formation and the payoff distribution are endogenously determined. Every agent realizes his full potential and has no incentive to deviate from either staying independent or from the endogenously determined partner and payoff» [11].

Another important factor in the formation of networks is that the company gets into tough operating environment. Fernando Vega-Redondo also confirms this fact in «Network organizations»: «The main conclusion obtained is that, as the environmental volatility increases, the optimal functioning mode of the organization sharply switches from being totally flexible to being completely rigid, i.e. no intermediate con-figurations are essentially ever optimal» [4].

3 The model

Formalization of large, sometimes hardly controllable social and economic systems requires a mandatory development of a graphical analytic representation of the processes. The organizational structure of a network, which reflects the main idea of business entities' interaction in the networks with self-standing interests, is presented in the form of the model in Fig. 1.

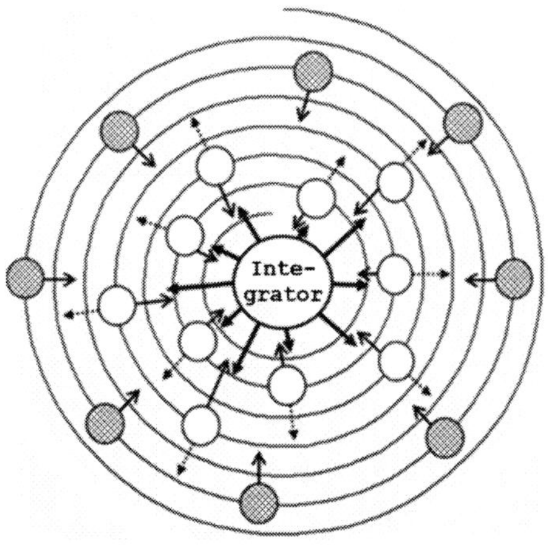

Fig. 1. Graphical analytic model of an organizational network with self-standing interests: ○ – network business entities; ● – market business entities; ——→ – attraction force of entity to a network; ·········→ – attraction force of entity to a market; ——→ – integrator interactive force to network and market entities.

The suggested model is a follow-up of radial-planetary type construction theory for organizational interaction in network structures [12]. A distinctive feature of the proposed modeling approach is formalization of the interaction between the integrator and network elements (business entities) in respect of their conflicting interests [13].

To formalize the relationship between integrator and net-work business entities let us present their interaction by functioning of attraction – repulsion forces [14]. We can draw a successful analogy between interaction of network entities and gravity interaction of material bodies within this approach. Functioning in common economic and legal space subjects inevitably gain certain connections and this phenomenon leads to appearing of attraction and repulsion economic forces.

So, for quantitative evaluation of «attraction forces» non-dimensional indicators accepting values in the interval [0; 1] were introduced. We will consider their content representation [15]. Integrator tending to interact with a business entity (K_{int}) can be calculated by a formula:

$$K_{int} = 1/(1+\alpha_1(R - R_{acc})^2) + \alpha_2 K_{inf} - \alpha_3 P_{fr}, \quad (1)$$

where: R – profitability of a business entity; R_{acc} – acceptable profitability of a business entity; K_{inf} – indicator considering informal relationship between superiors; P_{fr} – average percent of production fraud during a certain period; α_1, α_2, α_3 – weight coefficients.

Attraction of a business entity to a network:

$$K_{att.n} = 1/(1+\alpha_4(R - R_{des})^2) + \alpha_5 K_{inf} - \alpha_6 K_{sup}, \quad (2)$$

where: R_{des} – desirable profitability of a business entity; K_{sup} – indicator characterizing integrator support of business entity; α_4, α_5, α_6 – weight coefficients.

We can calculate K_{sup} indicator by a formula:

$$K_{sup} = \beta K_{adv} + (1 - \beta)(C_{ec}/N_{ec}), \quad (3)$$

where: K_{adv} – indicator accepting value «1», if between the integrator and the subject exists the advance payment system, and «0» – if there is no advance payment system; C_{ec} – economic opportunities regarding the network entity; N_{ec} – economic needs of a business entity for the purpose of maintaining stable and efficient production activity; β – weight coefficient.

Further, we will consider «business entity attraction towards market» indicator – $K_{att.m}$:

$$K_{att.m} = 1/(1+\alpha_7(R_m - R_{des})^2) + \alpha_8(HHI/5000 - 1), \quad (4)$$

where: R_m – business entity profitability in the case of its production capacities realization in the market; HHI – Gerfindal – Girshmana's index for i-business entity branch; α_7, α_8 – weight coefficients.

The entered indicators play an essential role in investigation of organizational networks with self-standing interests. Quantitative evaluation of network attraction forces allows projecting interaction stability of an integrator and a business entity in the conditions of dynamically changing relationship between them [16]. The study of the stability of organizational networks dedicated to a number of works, for example [17–19].

We will note that at research of a concrete organizational network the list of factors needs to be specified depending on research problems and specific nature of network production activity management.

The studied type of a network («umbrella» corporation) is characterized by existence of continuous administrative influence on business entities from the integrator. Administrative influence is understood as organizational, administrative and economic impact of the integrator on economic entities and a network in general.

Administrative influence has an integration character and is directed to a network strengthening. On the one hand, such influence lets the integrator realize its task solving strategy, and on the other hand, it provides for an economic entity effective work in a network.

4 Quantitative network evaluations

The quantitative evaluation of organizational networks with self-standing interests functioning is necessary for giving grounds for the options of their development. Elaboration of quantitative evaluation technique demands carrying out various industrial experiments directed to searching of optimum pro-portions of an organizational network functioning. The lack of opportunity to make the necessary organizational experiments in practice sets for a researcher the task of development a simulating model of organizational network, which allows presenting all the possible options and schemes of business entities interaction in a network. This can also help to test all the administrative decisions by means of numerical experiments, to estimate and choose among them the most acceptable from the point of view, both a separate economic entity, and system in general.

Thus, under organizational network quantitative evaluation we will understand some integrating indicator K_{ind}, which estimates organizational technical and economic level of its development.

We will consider the main idea of K_{ind} indicator formation. As the purpose of organizational network functioning is to satisfy the needs of society in products or services, it will be logical to estimate the activity of a network by the following relation:

$$K_{ind} = (Q_{pr} - R_{ex}) / W_p, \quad (5)$$

where: Q_{pr} – the projected production volume (volume of services), which was defined taking into account technical and economic level of entities' development and the quality of their interaction; R_{ex} – external risks for network functioning; W_{pr} – production capacities of a network.

Q_{pr} is mathematical expectation function of production and economic network activity results, calculated by a set of tests. Testing of an organizational network assumes imitation (generation) of various organizational and production situations for integrator and network entities. Mathematical formalization of Q_{pr} proposed in [20].

Finding R_{ex} is a challenging research task. This can be explained by the fact that casual, hard to project variables, describes the moment of negative external attack on a network, its stability and consequences. The choice of mathematical apparatus for statistical modeling and probability theory can become a rational way of this indicator formalization.

As the generated production situations have various probabilities of happening, the values of Q_{pr}, received during the trials, are not equiprobable. Obviously, various negative consequences from external influence are not equiprobable. Then, the nature of dependences of Q_{pr} and R_{ex} on production situation happening probability P_{sit} can be presented in one system of coordinates (Fig. 2).

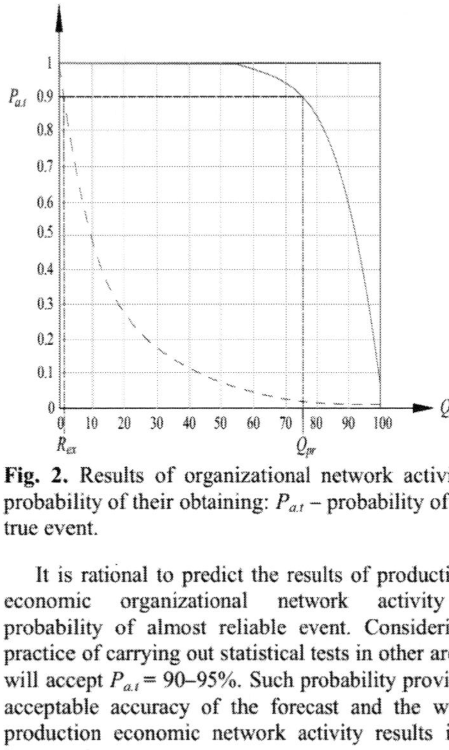

Fig. 2. Results of organizational network activity and probability of their obtaining: $P_{a.t}$ – probability of almost true event.

It is rational to predict the results of production and economic organizational network activity with probability of almost reliable event. Considering the practice of carrying out statistical tests in other areas, we will accept $P_{a.t}$ = 90–95%. Such probability provides the acceptable accuracy of the forecast and the width of production economic network activity results interval [Q_{pr}; W_{pr}].

Practical importance of the presented dependences is that they simplify the process of organizational network adjustment and allow planning its functioning risks.

5 Organizational network adjustments

The obligatory condition of ensuring the effective functioning of network structure in commodity and service market – is its continuous adjusting based on the use of quantitative network evaluation.

We will note that the formalized selection of economic entities from a free market is a guarantee of stability and efficiency of an organizational network. Integrator expects from all the network members positive activity results, increasing a network commercial sufficiency, its sustainability, and functional reliability. It is obvious that business entity network entering is mostly effective if $\Delta K_{ind} \rightarrow$ max. That is, with the choice of the entity from a set of possible options, the integrator aims at boosting the integrated network index.

A possibility of organizational network adjustment using the offered quantitative evaluation is demonstrated through adjustment results of a production organizational network with poor effectiveness. In the studied network, external functioning risks (R_{ex}) are relatively low and were not considered in the course of quantitative evaluation. The main results of adjustment are given in Fig. 3.

It is obvious that logical actions of network adjustment should lead to integrated index increasing. At the initial condition of the network, an integrated index was 0,31, and after four adjustment stages the index reached 0,64. The most popular business administration decisions were tested at different stages of adjustment: change of integrator financial policy towards better satisfaction of network members' interests; increasing network stability due to addition of a new network member; strengthening of network support to the level of network members' needs; reservation of entities' production capacities.

The received results confirm that the quantitative evaluation adequately reflects the change of key network parameters and can be used for further research of organizational networks.

Initial condition of the network: K_{ind} = 0,31.

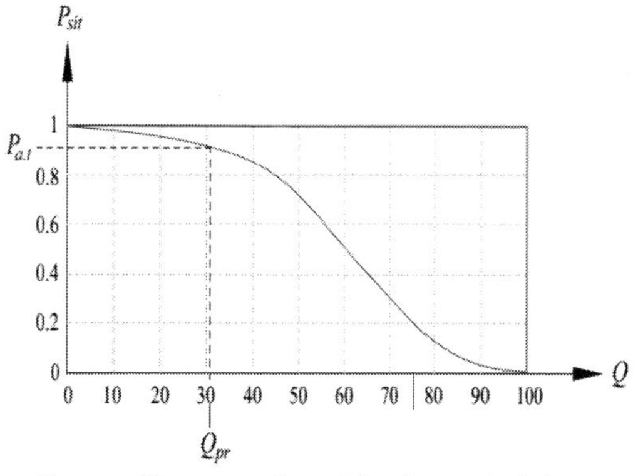

Change of integrator financial policy towards better satisfaction of network members' interests: K_{ind} = 0,46.

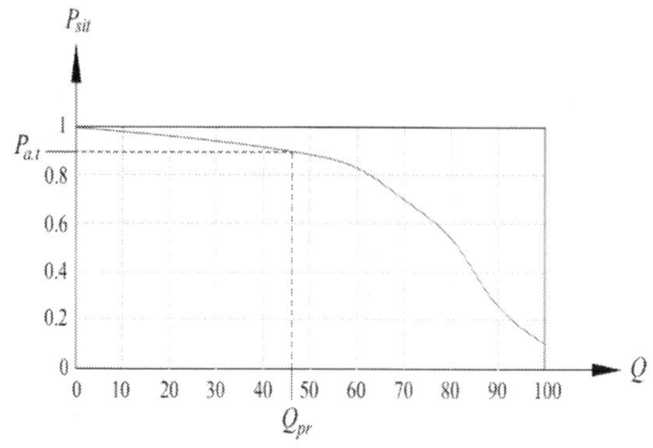

Formation of production capacities reserves due to inclusion in a net-work a new member: K_{ind} = 0,54.

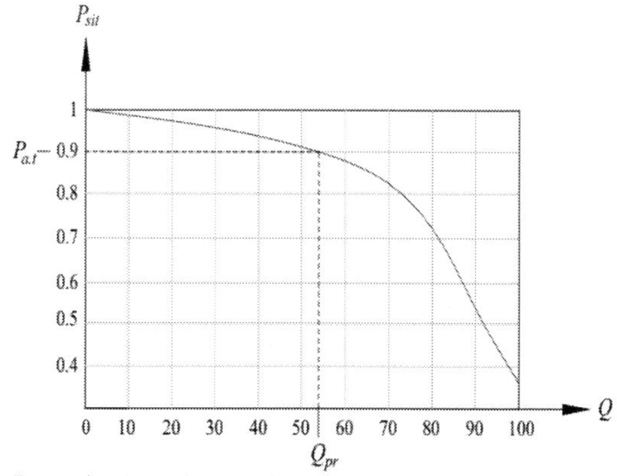

Strengthening of network support to the level of network members' needs: K_{ind} = 0,60.

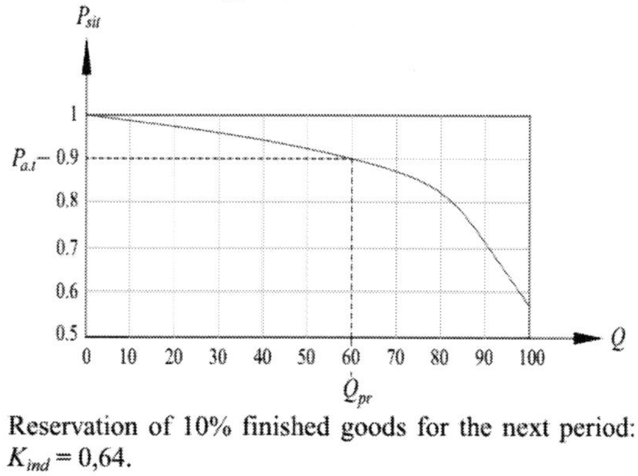

Reservation of 10% finished goods for the next period: K_{ind} = 0,64.

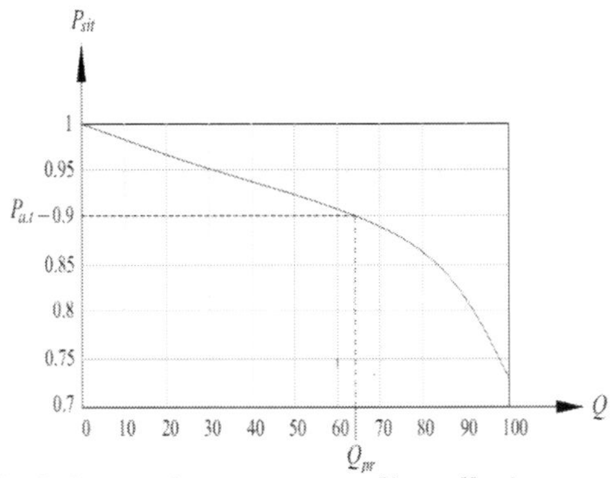

Fig. 3. Consecutive assessments of low effective organizational network's adjustment stages.

5 Summary and conclusion

So, in the offered graphical analytic model the main ideas of the theory of legal entities' interaction simulating in organizational network with self-standing interests are presented. Indicators of integrator and network entities' attraction are formalized in the form of functions. The functions' variables are degrees of satisfaction the corresponding interests. Proceeding from a need of each network, to revise the list of the evaluated interests, linear estimated forms were chosen as the most rational form of convolution.

The suggested approach of a network quantitative evaluation makes it possible to objectively analyze the production and economic activity of the whole network structure, network integrator and included economic entities, to estimate a network interaction, to project the necessary network parameters and to plan a network development.

The main advantages of suggested theoretical ideas and model representations are universality, convenience of computer realization, possibility of stage-by-stage calculation and adjustment of network organizational structures.

Further researches are aimed at development of complex simulating model of organizational network with self-standing interests. This will allow: to generate and quantitatively estimate various production scenarios; to investigate the influence of different load-level and capacity schemes of economic entities on indicators of the whole organizational network; to prove boundary values of production and economic activity of enterprises, which form a network and influence its durability and sustainability; to estimate the risks influence on interaction process of organizational network members.

References

1. M. Castells, *The Rise of the Network Society. The Information Age: Economy, Society and Culture* (Malden, MA, Oxford, 2010)
2. M. Castells, The Power of Identity. The Information Age: Economy, Society and Culture, 145-168 (2010)
3. M. Castells, End of Millennium. The Information Age: Economy, Society and Culture, **3**, 377-378 (2010)
4. Fernando Vega-Redondo, J. of Complex Networks, **1**, 72-82 (2013)
5. M. Van Alstyne, Forthcoming in j. of organizational computing, **7**, 3 (1997)
6. J. Galbraith, J. of Org. Design, **2**, 1-13 (2012)
7. Robert L. Cross, Salvatore Parise, Leigh M. Weiss, McKinsey Quarterly, **39**, 301-323 (2003)
8. J. Tullberg, J. Behavioral and Experimental Ec., **42**, 127-135 (2013)
9. Friedman, L. Andrew, Miles, Samantha, J. of Management Studies, **39**, 1-21 (2002)
10. Philip H. Mirvis, Christopher G. Worley, Building Networks and Partnerships Emerald Group Publishing Ltd., **3**, 1-34 (2013)
11. D. Talman, Zaifu Yang, J. of Math. Economics, **47**, 206-212 (2011)
12. V. Say, S. Sizyi, *Formation, operation and breakup of organizational network* (USURT, Yekaterinburg, 2011)
13. V. Say, I. Gromov, J. modern problems of Russian transport complex, **3**, 199-208 (2013)
14. I. Gromov, Yu. Moskvina, J. herald of USURT, **21**, 34-40 (2014)
15. V. Antropov, I. Gromov, J. herald of USURT, **25**, 29-37 (2015)
16. I. Gromov, J. herald of USURT, **58**, 80-89 (2015)
17. Edward E. Lawler III, Christopher G. Worley, J. of Org. Design, **41**, 265-270 (2012)
18. S. Trajanovski, J. Martin Hernandez, W. Winterbach, J. of Complex Networks, **1**, 44-62 (2013)
19. L. Välikangas, G. Romme, J. of Org. Design, **2**, 44-53 (2013)
20. V. Say, I. Gromov, J. transport of the Urals, **46**, 3-10 (2015

Developing methods of controlling quality costs

A. V. Gorbunova[1,*], *O. N.* Maximova[1], and *V. A.* Ekova[1]

[1]Volzhsky Polytechnical Institute (Branch of) VSTU, Volzhsky, Russian Federation

Abstract. The article examines issues of managing quality costs, problems of applying economic methods of quality control, implementation of progressive methods of quality costs management in enterprises with the view of improving the efficiency of their evaluation and analysis. With the aim of increasing the effectiveness of the cost management mechanism, authors introduce controlling as a tool of deviation analysis from the standpoint of the process approach. A list of processes and corresponding evaluation criteria in the quality management system of enterprises is introduced. Authors also introduce the method of controlling quality costs and propose it for the practical application, which allows them to determine useful and unnecessary costs at the existing operating plant. Implementing the proposed recommendations in the system of cost management at an enterprise will allow to improve productivity of processes operating and reduce wasted expense on the quality of the process on the basis of determining values of useful and useless costs of quality according to criteria of processes functioning in the system of quality management.

1 Introduction

The mechanism of managing quality cost requires scientific understanding and practical solutions to issues related to the introduction of new, effective and consistent with applicable accounting standards, methods of systematization and quantification of quality costs. It is reasonable to develop and implement them due to the need to timely identify and predict adverse situations in the expenditure of resources on quality management, as well as to give the managers of enterprises objective information about the actual level of quality costs.

It is especially important to enhance the effectiveness of assessing and analyzing quality costs in the framework of the widely introduced process approach to quality management. Methods of systematization and evaluation of quality costs used by enterprises should be aimed at timely determination of quality cost within processes, an accurate analysis of their deviations according to the criteria of processes, at identification of the causes of these deviations and at the development of effective corrective actions. This is especially true for companies actively developing a progressive quality management system based on process approach to enhance their competitive positions in domestic and foreign markets. Since enterprises are oriented at achieving managerial progress in the field of quality controlling, it becomes necessary to introduce effective tools for monitoring and evaluating quality costs, that allow optimization of these costs within individual processes, as well as in the context of the entire quality management system.

The necessity to increase the efficiency of quality costs management on the basis of applying the new ways of their systematization and quantification from the standpoint of modern trends in quality management determines the relevance of this article.

The purpose of the article is to give theoretical justification and to develop practical recommendations for improving the mechanism of quality cost management by applying new advanced methods of systematization and evaluation of quality costs on the basis of managerial, scientific and technical progress achievements.

To achieve this goal it is necessary to fulfill particular tasks: to justify theoretically the implications of process approach in enterprises subject to the division of quality costs according to the types of quality management processes; to develop and propose methods of controlling quality costs; to make basic conclusions.

2 Theoretical aspects of the process approach to quality management

The desire to strengthen their competitive position forces economic entities to view quality management as an evolutionary process, based on the constant application of new internationally recognized organization and production management methods based on the achievements of management progress. Among these methods is the process approach to quality management on the basis of existing ISO standards, that ensures regularity of management at the junction of individual processes within their system as well as in case of their combination and interaction [1, 2, 3, 4].

Management of quality costs is vital to companies for the successful application of the process approach to quality management because it provides identification, planning, coordination and steady control of quality costs for the purpose of achieving effectiveness

* Corresponding author: alinavictory@mail.ru

(productivity) of the functioning processes of an enterprise.

All processes of the company according to their functional (target) purpose can be divided into basic, controlling, and maintaining [5]. Therefore, quality costs according to the types of quality management processes can be divided into the following groups: quality costs for basic processes, quality costs for maintaining processes and quality costs for controlling processes. Each cost group according to the types of processes includes a set of useful costs (ensuring compliance with the stated requirements and saving of resources allocated for achieving objectives in the quality field) and unnecessary costs (related inconsistencies and causing the overuse of resources) [6, 7, 8].

Since each process is characterized by a number of criteria allowing us to supervise and analyze the work process, all the cost groups in the proposed classification should be considered in the context of the criteria of the processes taking place in the system of quality management. This gives the opportunity to assess in value terms the resources allocated for achieving the objectives of quality within a particular process, and the totality of the processes in the enterprise [9, 10, 11].

Among the most common and important functions of quality management (planning, organization, coordination, motivation, etc.) control is extremely important because it allows you to develop the strategy and tactics of optimal allocation of resources in the interest of product improvement, to identify the cost of individual errors in the production process and, while ensuring the required level of quality, to evaluate effectiveness and to identify areas for improvement of the quality management system as a whole [12, 13].

Controlling operates as an instrument of quality costs management, allowing to improve the deviation analysis in quality costs management, and the managerial decisions made on its basis [14]. The use of controlling as a tool of deviation analysis according to the criteria of the process from the standpoint of the process approach allows you to increase the effectiveness of control within each individual process, and in the framework of the process approach to the management of quality costs in general.

The process of controlling quality costs is a system of communication, development and implementation of management decisions and their informational support. The system structure of controlling the quality costs means both functional and organizational structure, a set of organizational relations, as well as specific schemes of interaction between management authorities. The technique of controlling quality costs consists of computer and office software, document management system. These subsystems of controlling quality costs are integrated in a single information space, largely defining efficiency of the mechanism for the quality costs management in an enterprise.

The traditional mechanism of managing quality costs involves planning the values of quality costs, expenditure of resources as planned (determination of the actual quality costs values), comparison of planned and actual values of quality costs (identification of deviations) and if there are deviations – the adoption of measures to eliminate them.

The improved mechanism of quality costs management based on controlling contains new unit in its structure – "Controlling quality costs" (Fig. 1), which allows you to analyze deviations of quality costs and find the resources to eliminate them within the ongoing enterprise control.

Regular detection and complete deviation analysis of quality costs within each process provides the basis for preventing the accumulation of negative deviations in the totality of processes in the quality system and, thereby, becomes an integral part of the effective mechanism of quality costs management at an enterprise.

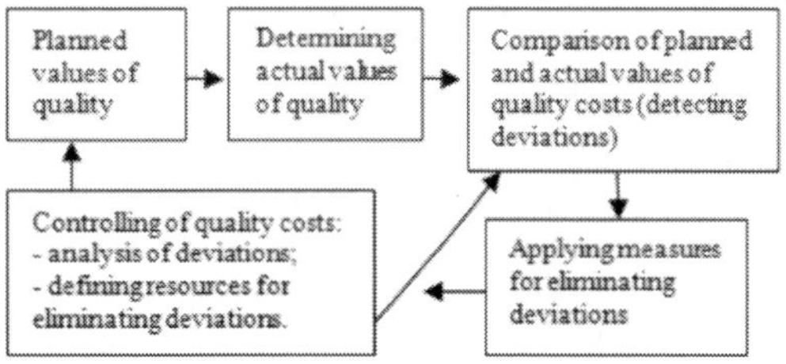

Fig. 1. Improved mechanism of quality cost management.

Applying controlling as a part of the quality costs management mechanism is crucial for businesses. Controlling of quality costs performs as an effective tool of quality cost management and is able to maintain control of these costs from the standpoint of the process approach used at enterprises.

3 Methods of controlling quality costs

The proposed methodology of controlling the quality costs is based on defining the range of allowed values for the processes criteria, on analyzing the corresponding deviations of the actual costs of quality from the standard by the methods of controlling; on identifying the magnitude of useful and unnecessary quality costs according to the criteria of the process and its general effectiveness [10].

The method is based on a common understanding of the process as a set of interacting activities that transforms inputs (input streams) to outputs (output streams) using appropriate resources [15]. Parameters (criteria) of the process represent the characteristics (information) by which the process owner and the senior manager can judge the effectiveness of process execution and to evaluate the degree of achievement of planned results [16].

The methodology of controlling the quality costs allows you to fully and accurately identify the causes of the quality costs deviations and prescribe adequate corrective procedures to return the value and structure of costs in the initial (planned, desired) state. To implement methods of quality costs controlling it is necessary to classify the processes of the enterprise according to their function as basic (B), managing (controlling) (C) and maintaining (M) [17]. Each process is characterized by a number of criteria, allowing you to monitor the work process and to evaluate its quality. The more criteria is involved in assessing the effectiveness of the process, the higher is its reliability and validity [18, 19].

A list of the basic, maintaining and managing processes, functioning as a part of the quality management system, are presented in Table 1. Each company develops its own list of processes that reflect the types and characteristics of production at the enterprise. However, since enterprises are oriented on the common for all economic entities requirements of ISO, a similar approach is used for qualifying managing and maintaining processes as well as to the criteria characterizing them.

Analysis of the processes operating in the system of quality management of an enterprises shows that the substantive part of the criteria of processes (primarily the basic and maintaining) allows you to link process costs directly with the cost of required product quality level, achieved on the basis of the process approach to quality management, which, consequently, justifies the necessity of applying controlling to the management of quality costs. A company has the right to choose its own criteria for the process according to the types of processes occurring in the enterprise, and in accordance with the requirements of the "Quality Manual", applicable instructions and methodological documents of the quality management system.

Table 1. The list of processes and their corresponding criteria in the quality management system of enterprises.

Code of the process	*Name of the process*	*Name of the criterion*
	Basic processes	
B-1	Determining, planning and analyzing the requirements of consumers and production	1. Providing bids for the beginning of the planned period (a month) 2. Providing applications for the end of the planned period (a month) 3. Satisfaction of consumers 4.Fulfillment of a plan on the amount of shipped products, the share of exports in a total shipment volume
B-2	Procurement of basic materials	1. Fulfilling the production needs in basic materials and compliance of the purchased materials with the established procurement requirements 2.Downtimes due to the absence (delays in delivery) of basic materials 3. The volume of inadequate basic material 4. Proficiency level of basic material suppliers
B-3	Procurement of auxiliary materials	1. Fulfilling the production needs in auxiliary materials and compliance of the purchased materials with the established procurement requirements 2. Proficiency level of auxiliary material suppliers
B-4	Production	1. Fulfillment of a production schedule 2. Fulfillment of normatives on nonexceeding the output of inadequate production in the total production volume 3. Technological downtimes
B-5	Storage and shipment of finished products	1. Fulfillment of shipment schedule 2. Existence of rebukes on final product storage conditions
	Managing processes	
C-1	Company planning	1. Fulfillment of expenditure budget 2. Profitability of production shipment 3. Fulfillment of income budget 4. Profit
C-2	Organization and analysis of activities	1.Assessment of the quality management system on the part of senior management 2. Achieving the goals of a company in the sphere of quality 3. Maintaining execution discipline
	Maintaining processes	
M-1	Infrastructure Management	1. Downtimes of mechanical equipment 2. Downtimes of electrical equipment 3. Fulfillment of equipment repair schedule 4. Fulfillment of schedules for examination of measurement and control instrumentation 5. Fulfillment of schedules for buildings and construction inspection
M-2	Personnel management	1. Satisfaction of needs in staff 2. Fulfillment of a staff training plan 3. Level of personnel turnover
M-3	Working environment management	1. Realization of activities on safety arrangement and precautions 2. Level of occupational traumatism

The methodology of controlling the quality costs includes the following main stages:

1. Stating the criteria of a process, including the identification of their type, units of measurement, significance.

2. Calculation of the quality costs according to the criteria of a process, including:

- setting the range (maximum and minimum) of allowed values for process criteria and calculation of the actual values of each criterion; identifying absolute deviations (+/-) of maximum (taken as standard) from the actual criterion value D_i^a;

- calculation of the actual costs value by the criterion of the process, based on resource provision of the actual criterion value C_{ci}^a;

- calculation of regulatory costs by the criterion of the process based on the amount of absolute deviation of the process criterion values:

$$C_{ci}^r = C_{ci}^a \cdot D_i^a \qquad (1)$$

where C_{ci}^r – value of costs on the criterion of the process, accounting for deviations of regulatory criterion value from actual, %;

D_i^a – absolute deviation of process criterion values i, RUB.;

DOI: 10.1051/shsconf/20173501019

C^a_{ci} – actual quality costs by the criterion i, RUB;

i – criterion of a process.

- calculation of the absolute deviation (+/-) of the normative costs value C^r_{ci} by the criterion of a process from actual values of these costs C^a_{ci};

- calculation of the total sum of regulatory quality costs according to the criteria of a process:

$$C^r_{ci} = \sum_{i=1}^{n} C^r_{ci} \quad (2)$$

where C^r_{ci} – regulatory quality costs according to the criteria of a process, rub.;

n – number of criteria.

- calculation of the total sum of actual costs according to the criteria of a process:

$$C^a_{cp} = \sum_{i=1}^{n} C^a_{ci} \quad (3)$$

where C^a_{ci} – actual costs according to the criterion i of a process, rub.

- a comparative analysis of the total actual and regulatory quality costs according to the criteria of a process, aimed at detecting either any cost overruns or saving money according to the criteria of a process, that proves their effective use. Identifying the causes of adverse deviations associated with the overconsumption of resources is carried out by means of the factor multicomponent analysis.

The implementation of factor analysis is carried out in three stages. In the first stage, the adverse deviations in the value of unnecessary quality costs according to the criteria of a processes are associated with various factors, performing as reasons for these deviations. Herewith, the greatest number of factors (causes) that cause deviations was selected. The second stage is the study of the causes of deviations. All the causes are divided depending on the source of problems on the economic (related to prices of supplied raw materials; the value of stocks of finished products, etc.) and production (related to the organization and technology of production and regularity of supply, honesty of suppliers, the presence of complaints, the operation of the equipment and contractors, etc.). By the method of logical reasoning the most important causes of adverse deviations in the unnecessary costs value according to the criteria of processes are defined. The third stage is the development of measures aimed at eliminating the causes of deviations and, respectively, returning the process into a stable state.

3. Calculating the values of useful and unnecessary costs of a process including:

- calculating the useful (allowing you to save money) costs according to the criteria of a process:

$$C_{usf} = \sum_{i=1}^{n} C_{usfi} \quad (4)$$

where C_{usfi} – useful costs of a process according to the criterion i;

- calculating the unnecessary (causing over-expenditure of money) costs according to the criteria of a process:

$$C_{un} = \sum_{i=1}^{n} C_{uni} \quad (5)$$

where C_{uni} – unnecessary costs of a process according to the criterion i;

n – the number of criteria;

4. Calculation of the effectiveness of the process as the relationship of the total amount of useful costs according to the criteria of a process (C_{usf}) and total cost value by the criteria of a process:

$$Ef^c_{pr} = \left(\frac{C_{usf}}{TCP_c}\right) \cdot 100\% \quad (6)$$

where TCP_c – total cost value by the criteria of a process.

Approbation of the methods of quality costs controlling was carried out in the joint-stock company *"V. Makarov Pipe Plant "Profil-Akras"* (Volzhsky) in terms of the basic process (B-2) "Procurement of basic materials (metal)" [20].

Specification on the process requires a number of criteria of the process, "Procurement of basic materials". Their normative value and significance of each criterion are presented in Table 2.

Table 2. Criteria of the process "Procurement of basic materials (metal)" and their normative values.

Criterion	Normative value, %	Significance	Period of control
Fulfillment of the production need in metal	100 (±5)	5	Quarter
Percentage of downtimes in pipe-welding shops (PWS) due to the absence of metal	0 (±0)	3	Quarter
Volume of inappropriate metal	2 (+0;-0,5)	4	Quarter
Proficiency of basic materials suppliers	50 (±25)	5	Quarter

The results of calculating criteria of the process and the quality costs according to the criteria for each month of the reporting period (Q1 of 2016) are presented in tables 3-6.

Table 3. Calculation of the criterion "Fulfillment of the production need in metal".

Criterion	January 2016			February 2016			March 2016		
	A_{pm}, t	Pl_{pm}, t	D_{pm}, %	A_{pm}, t	Pl_{pm}, t	D_{pm}, %	A_{pm}, t	Pl_{pm}, t	D_{pm}, %
Calculating the criterion									
Actual value of the criterion	7990.6	7826.1	102	11116.2	10893.7	102	12778.5	12486.1	102
Calculating process costs by the criterion									
- price of steel in rolls (strip) (P), RUB/t.	19130.0			18832.5			19767.3		
- costs of process by the criterion for a month. thousand RUB	152860.2			209347.7			252596.4		
- total quarterly costs. thousand RUB	614804.3								

Table 4. Calculation of the criterion " Percentage of downtimes in pipe-welding shops due to the absence of metal".

Criterion	January 2016			February 2016			March 2016		
	P_{pr} h.	NP. h.	PD. %	P_{pr} h.	NP. h.	PD. %	P_{pr} h.	NP. h.	PD. %
Calculating the criterion									
Actual value of the criterion	17.3	600	2.9	17.3	648	2.7	17.3	713.6	2.4
Machine 20-76	19.25	552	3.5	19.25	648	2.9	19.25	720	2.7
Machine 10-40	22.1	552	4.0	22.1	648	3.4	22.1	710.4	3.1
Machine 10-32	10.6	696	1.5	10.6	648	1.6	10.6	710.4	1.5
Calculating process costs by the criterion									
- productivity in pipe-welding workshops. t/hour	5.46			5.93			6.23		
- amount of pipes unproduced due to downtimes A_{up}. t	94.5			102.6			107.8		
- actual costs of reproduction C_r. RUB/t.	1931.2			1322.6			1400.2		
– costs of process by the criterion per month. thousand RUB	8890.83			12744.53			16066.83		
– total quarterly costs. thousand RUB	37702.2								

Table 5. Calculation of the criterion " Volume of inappropriate metal"of the process".

Criterion	January 2016			February 2016			March 2016		
	Amount of inaprop. met... ton	*Price. RUB/t.*	*Total. thous. RUB*	*Amount of inaprop. met... ton*	*Price. RUB/t.*	*Total. thous. RUB*	*Amount of inaprop. met... ton*	*Price. RUB/t.*	*Total. thous. RUB*
Calculating the criterion									
Actual value of the criterion	214.9	4505.1	968.1	305.9	4583.0	1401.9	346.9	5342.0	1852.9
V_{im}. %	214.9/7826.1*100%=2.75%			305.9/10893.8%= 2.8%			346.9/12486.1% = 2.78%		
Calculating process costs by the criterion									
- costs of process by the criterion per month. thousand RUB	968.1			1401.9			1852.9		
- total quarterly costs. thousand RUB	4222.9								

Table 6. Calculation of the criterion "Proficiency of basic materials suppliers".

Criterion	January 2016			February 2016			March 2016		
	P_s unit	n. unit	P_{bms}. %	P_s unit	n. unit	P_{bms}. %	P_s unit	n. eð.	P_{bms}. %
Calculating the criterion									
Actual value of the criterion	4	7	57	4	7	57	4	7	57
Calculating process costs by the criterion									
- costs of process by the criterion per month. thousand RUB/.	9.2			9.2			9.2		

* Corresponding author: alinavictory@mail.ru

(informational and consulting services)			
- total quarterly costs. thousand RUB	27.6		

Table 7. Normative and actual criteria values of the process "Procurement of basic materials (metal)".

Criterion	Criterion value		
	normative	*actual*	*deviation. (+/–)*
Fulfilment of the production need in metal	100 (±5) -105	102	-3
Percentage of downtimes in pipe-welding shops (PWS) due to the absence of metal	0 (±0)	2.7	+2.7
Volume of inappropriate metal	2 (+0.-0.5)- 2	2.7	+0.7
Proficiency of basic materials suppliers	50 (±25)- 75	57	-18

Table 8. Improved list of criteria for the process "Procurement of basic materials (metal)".

Criterion	Normative value. %	Significance	Period of control	Criterion
Fulfilment of the production need in metal	%	100 (±5)	5	Quarter
Percentage of downtimes in pipe-welding shops (PWS) due to the absence of metal	%	0 (±0)	3	Quarter
Volume of inappropriate metal	%	2 (+0;-0.5)	4	Quarter
Proficiency of basic materials suppliers	%	50 (±25)	5	Quarter
Costs of the process quality	%	100	5	Quarter

Table 9. Calculation of normative quality costs according to the criteria of the process "Procurement of basic materials (metal)" with regard to deviations.

Criterion	Value of quality costs by the criteria. thousand RUB		
	normative	*actual*	*deviation*
Fulfillment of the production need in coil stock	633248.4	614804.3	-18444.1
Percentage of downtimes in pipe-welding shops (PWS) due to the absence of metal (downtime without explanation)	36711	37702.2	+991.2
Volume of inappropriate metal	4193.3	4222.9	+29.6
Proficiency of basic materials suppliers (informational and consulting services)	32.57	27.6	-4.97
Total quality costs in terms of the process. thousand RUB.	674185.3	656757	17428.3

Comparison of normative and actual values of criteria upon the results of operation of the process. "Procurement of basic materials (metal)" for the 1st quarter of 2016 showed the results presented in Table 7.

In our opinion. to determine the type of corrective measures. and to provide timely warning about the "weaknesses" of the process to the process owner. it is recommended to apply controlling. which will allow you to determine the strengths and weaknesses of the process. to set the normative value of quality costs. to identify favorable or unfavorable deviation of the actual values of the quality costs from the standard and analyze them to find out the causes of deviations and eliminate them. and to develop corrective and preventive actions for the effective functioning of the process.

A new criterion. "Costs of the process quality ". is introduced to the process "Procurement of basic materials (metal)" (see Table 8).

Determination of quality costs by the criteria of a process is presented in Table 9.

The results of approbation showed that in terms of a process useful costs include: the cost of providing the production needs in coil stock; costs of determining the level of proficiency of basic materials suppliers. and the unnecessary costs – costs associated with downtime due to the absence of the metal; the costs associated with the availability of unsuitable metal.

Upon the results of applying factor analysis it was established that the main causes of adverse deviations. i.e. the occurrence of unnecessary quality costs were: 1) a sharp increase in the cost of coil stock supply due to the nonconformity of the metal thickness to the requirements for the procurement that led to the growth of the expenditure coefficients for metal; 2) irregular supply of coil stock production. the reduction in the number of bona fide suppliers due to a low speed of applications rotation within the departments of the company; 3) violation of the pipe production technology. namely. the occurrence of pipe defects due to incomplete penetration of the welds. deformation of the pipes due to improper slinging of packages when loading into railroad cars; 4) low professional level of the production personnel; 5) inadequate supply of finished products in store. which entails an increase in the number of transshipments of production; 6) absence of documented methods for accounting and controlling quality costs. which allows you to fully and accurately identify the composition and structure of useful and unnecessary quality costs of pipe products.

The authors developed a system of corrective actions that will significantly reduce the influence of these factors on the magnitude of quality costs by reducing unnecessary and increasing useful costs.

The results of calculating the value of useful and unnecessary quality costs within the process (Table 10) show that implementation of controlling allowed to reduce unnecessary costs per quarter on 1020.8 thousand RUB.

In fact. the reduction of unnecessary costs means preventing waste of resources on the achievement of required production quality as a part of the process. "Procurement of basic materials (metal)".

The efficiency of the process after implementation of controlling increased by 0.2 p.p.

Table 10. Improved list of criteria for the process "Procurement of basic materials (metal)".

Criterion	Normative value. %	Significance	Period of control	Criterion
Fulfilment of the production need in metal	%	100 (±5)	5	Quarter
Percentage of downtimes in pipe-welding shops (PWS) due to the absence of metal	%	0 (±0)	3	Quarter
Volume of inappropriate metal	%	2 (+0;-0.5)	4	Quarter
Proficiency of basic materials suppliers	%	50 (±25)	5	Quarter
Costs of the process quality	%	100	5	Quarter

Application of controlling in the management of quality costs for all the ten existing enterprise processes over a long period of time will provoke a significant increase of efficiency of enterprise activity in the sphere of resource expenditure for quality management.

Thus saved money can be directed to further development of the enterprise. increasing its competitiveness in internal and foreign markets.

4 Conclusions

The introduction of controlling into the management of quality costs allows based on the detection of the value of useful and unnecessary quality costs according to the criteria of the process to achieve a significant improvement in the functioning of the process. enhance its productivity. reduce unnecessary costs of the process quality.

The proposed method is applied in such a way that in the case of low effectiveness of process it will be possible to identify and analyze its causes.

Since each process is characterized by a number of criteria to supervise and analyze the work process. all the cost groups in the proposed classification should be considered in the context of the criteria of the processes taking place in the system of quality management that provides an opportunity to assess in value terms the resources allocated for achieving the objectives of quality within a particular process. and the totality of the processes in an enterprise.

Regular detection and complete analysis of the deviations in quality cost within each process provides the basis for preventing accumulation of negative deviations in the totality of the processes of the quality system and. thereby. becomes an integral part of the effective management mechanism for quality costs in an enterprise.

References

1. E.M. Karlik. D.S. Demidenko. Economic issues. **9**. 21-29 (1983).
2. P.P. Novichenko. T.D. Popova. *Accounting the costs of product quality improvement* (Finances and statistics. Moscow. 1986)
3. T.D. Popova. Bookkeeping. **11**. 71-72 (2001)
4. P. Freedman. *Audit: controlling costs and financial results while analyzing the quality of products* (UNITY. Moscow. 1997)
5. U.P. Adler. S.E. Schepetova. Standards and quality. **2**. 29-35 (2002).
6. T. Nakamura. Quality. trade and economics. 23-31 (1970)
7. I.H. Ansoff. *Strategic management* (Economics. Moscow. 1989)
8. K.M. Rakhlin. L.E. Skripko. Standards and quality. **3**. 49-51 (1997).
9. D.N. Babenkov. *Management of an industrial enterprise on the basis of process aproach to accounting and analysis of quality costs* (Moscow. 2002)
10. A.N. Kanivets. *Mechanism for evaluating costs efficiency in the quality management system of an enterprise* (Tambov. 2005)
11. Zh.A. Chesnokova. *Improving the organization of quality costs management in the industrial enterprise* (Saratov. 2007)
12. F.I. Palitsyn. V.M. Proleyko. *Economic methods of production quality management* (Mechanical Engineering. Moscow. 1981)
13. N.K. Rozova. *Quality management* (Vector. St. Petersburg. 2005)
14. M.M. Kane. B.V. Ivanov. V.N. Koreshkov. A.G. Shirtladze. *Controlling: tutorial* (Finances and statistics. Moscow. 2006)

15. *British standard BS 6143: Guide to the economics of quality. Part 1. Process cost model* (NTK Trek. Moscow. 2003).
16. I.A. Budischeva. Ya.D. Plotkin. *Regulating costs of ensuring production quality* (Publ. of standards. Moscow. 1989)
17. I.V. Malova. *Practice of implementing the system of accounting and analysis of quality costs* (NTK Trek. Moscow. 2005)
18. S.V. Zakora. *Development of total quality management (TQM) system in the industrial enterprise* (Saratov. 2004)
19. P. Khorvat. Issues of theory and practice of management **4**. 44-56 (2000)
20. A.V. Gorbunova. *Improving quality costs management in the enterprises of pipe and metallurgy industry* (Saratov. 2010)

The application of multiloop diagnostics model to assess and improve the economic security of enterprises

Vladimir Pluzhnikov[1,*], *Sergei* Kukharenko[1], and *Svetlana* Shikina[1]

[1]Department of "Information technologies in economics", South Ural State University, 454080 Chelyabinsk, Russia

Abstract. The author's research is dedicated to the enhancement of the level of the enterprise economic safety. This task involves developing the concept of an integrated system for early prevention of dangers and threats of business activity, substantiation of procedures for regulating the activities of the enterprise in accordance with the changing external and internal factors. Multiloop diagnostics model is proposed to identify causal relations of management dysfunction. It allows you to receive an adequate assessment of the basic parameters of activity of the enterprise and accurately identify its status. Researching problems of economic safety of the enterprise such diagnostic methods as economic and logical analysis, statistical monitoring and strategic management were applied. There was made a conclusion that a qualitative assessment is a key tool of the level assessment of the enterprise economic safety, its control, and monitoring. It allows you to get reliable information about the real possibilities of the enterprise at different stages of development, to monitor and evaluate the level of economic security, find effective solutions to transition to a higher level of economic safety of the enterprise.

1 Introduction

An activity of any business entity in a market economy is characterized by a high level of uncertainty. It is connected to the presence of a large number of the environment external and internal factors, creating the security economic threats of the enterprise. The definition "economic security" is firmly included in the economic theory. The level of economic security and company's operating results are considered as interrelated characteristics of its successful functioning and sustainable development.

It is generally accepted the necessity to develop regulatory procedures and adaptation of the enterprise to changing external conditions with the aim to raise the level of economic security. It is necessary to develop a working system, which allows responding to various (internal and external) adverse changes (threats, dangers) of business development and their further neutralization.

This task is solved by using different tools. A number of authors consider that it's required to establish a system of threat diagnostics in the early stages of their manifestations [1, 2, 3]. However, it is a lack of theoretical research of applying this tool in order to enhance the economic security of the enterprise. In this paper methods of diagnosis are studied.

The goal of the research is to develop the concept of an integrated system for early prevention of internal and external dangers and threats to the business activities of the enterprise. The economic security concept of the enterprise based on the results of statistical monitoring allows selecting the functioning enterprise strategy based on the identification of the predicted state and receiving an adequate assessment of the parameters of its work.

2 Factors of economic security

Within the proposed concept, a set of threats impacting on economic security is considered as a system with a dynamic structure. It is necessary to know the structural elements and be able to predict the possible structural changes. In the functioning process of the enterprise, threat occurrence can lead to another threat, to enhancing their impact on the company's activity or to their neutralization.

The system of the enterprise economic security threats can be classified by various bases. The most significant threats to the system are the ones that give the possibility to minimize the riskiness of the activity, to bring in balance the enterprise internal potential to external conditions of the environment, as well as to ensure the process of its sustainable development in the long term period. Our vision for this issue agrees with the opinion of O.N. Gromova [4]. However, the list of dominant factors affecting the process of sustainable development should be supplemented. The proposed list of system factors is in table 1.

Table 1. Factor system of the economic security category.

Factor	Content
Organizational	Organizational structure of the enterprise, the efficiency management system
Financial	Regulating company financial issues, its paying

* Corresponding author: pluzhnikovvg@susu.ru

	capacity and immune to bankruptcy
Technical (technology)	Creating and applying technology (competitive ideas of scientific development) that enhance the firm's competitiveness
Information	The level of information support of business, information security, etc.
Tax	Regulating the optimality of tax burden and tax administration
Personnel	Recruitment and work with qualified personnel
Legal	Legal maintenance of an organization activity
Spatial (geographical)	The nature of the premises, the enterprise territory, communications, the expandability
Socio-environmental	A social security system degree of the personnel
Innovative	The level of an investment activity and the activity on the financial market

This factor system takes into account the influence of the external internal environment at the level of the enterprise economic security. They form enterprise potential cooperating in the functioning process, both within the enterprise and in the external environment. The basis of this interaction is management processes that can cause certain threats and associated risks of the business operation.

3 Diagnostics as type of the economic analysis

There are two types of: with the regular management of the production process and with the implementation of enterprise development strategies (transition from its current state to the desired state). Accordingly, to run a company under these conditions you must know the source of the threat and the expected consequences. Control over processes can be provided by diagnostic tools, the set of which is shown in Figure 1.

In the economic literature, diagnostics is considered as a special type of economic analysis conducted referring to appearing financial problems of the company. A set of main definitions of the concept of diagnostics is given at table 2. As you can see the diagnostics are considered as a tool to get the reliable information about the real possibilities of an economic entity at any stage of its development. It serves to determine the reasons for changing the company state based on the conducted researches. Early threat diagnostics is a condition for ensuring the enterprise economic security.

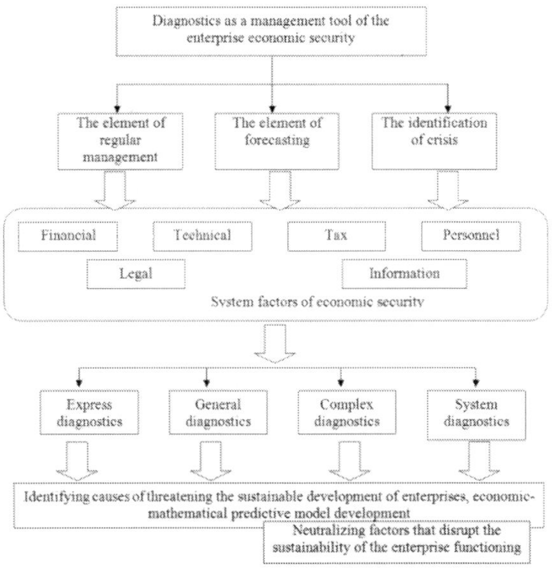

Fig. 1. Set of diagnostics tools.

Diagnostics of the enterprise economic security is a systematic and complex methodology of researching the features of business processes. The main task of diagnostics is to determine and evaluate signs of a multilateral process of deviation identification of parameters at the changes in external situations and internal operating conditions. Comparison of the received information with the normative values allows you to define a secure area of the enterprise [5, 6].

Table 2. The definition of the diagnostics.

Author	The content of definition "diagnostics"
V.Kryzhanovskii, V. Lapenkov, V. Luter, [7]	The estimation of parameters of the organization state to determine the probability of bankruptcy, the period and stage of the crisis
J. Ohlson [8]	Bankruptcy risk evaluation based on logistic regression model of bankruptcy risk prediction
V.V. Kovalev [9]	A special kind of economic analysis for prediction of possible bankruptcy
P.V. Gavrilov, N.F. Shcherbakov [10]	The process of deviation identification in values of economic activity of enterprises from mean value
O.V. Efimova [11]	Evaluation category of the enterprise solvency and liquidity to assess the enterprise bankruptcy risk
V.Yu. Zhdanov [12]	Complex preventive tool for determining the enterprise crisis processes
Yu.A. Loktionova [13]	A special process of abstract-logical operations, inductive and deductive conclusion, which begins with the developing a hypothesis
N. S. Maksakov[14]	The multilateral process of the assessment of company activity, individual and specific features
I.A. Koshkina, E.R. Mubarashkina [6]	Analysis of parameters of external environment and comparison of analytical results with thresholds of indicators (danger zone)

It is important to emphasize that one of the diagnostics tasks is to determine the identities and leveling imbalances business processes development. Diagnostics enables you to identify a causal relationship of management dysfunction and then move to development a predictive and explanatory model of its development and functioning.

Establishment of an effective system of diagnostics of the enterprise economic security requires the following tasks:

• To overcome the information uncertainty of the threat assessment caused by a nonlinearity of the internal and external development of the environment.

• To choose methodological approaches those are most effective in predicting threats.

• To establish an effective diagnostic apparatus for the estimation of the transition state of the production system.

• To account the influence of development factors, the lifecycle of enterprise development, phase portrait of the external environment.

• To find a formalized model predicting system state, boundaries and the limits of structural and phase transitions.

The diagnostic is a process that is a set of logic operations, inductive and deductive conclusion. The diagnostics include a multilevel analysis, which is researching and exploratory and conducted on the basis of a systematic approach.

Applying the systematic approach in diagnostics is explained by the fact that the enterprise as an object cannot be isolated from the influence of the external environment. The enterprise should be considered both as a separate system (a subject, a set of subsystems, a business process) subject, a set of subsystems, business process), and as an element of a large social category (merger, holding, etc.), (fig. 2).

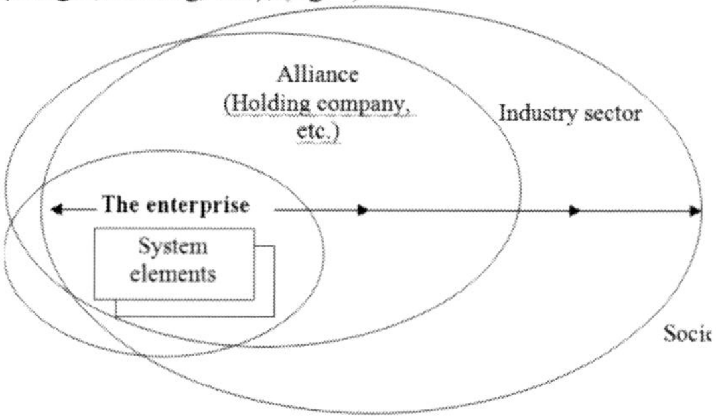

Fig. 2. System approach to the diagnostics.

The main direction of the diagnostic process is from analysis of complex indicators to the analysis of indicators of its constituent elements, from the analysis of the company results to the analysis of the factors of production and resources to forecasting future company activities.

The methodology of developing a system of diagnostics for ensuring the economic security of the enterprise is based on the following approaches: systemic, complex, quantitative, qualitative, behavioral, marketing, reproduction, normative, functional, substantive, operational, integrated, dynamic and situational [15, 16].

4 Multiloop diagnostic model

To solve problems related to the economic security of enterprises we propose to apply the multiloop diagnostic model, which is implemented on the principles presented in the table. 3.

Table 3. The basic principles of diagnostics.

Principle	Description
Complexity	Resource security (human, material, financial and information)
Timeliness	Forecasting of threats, ranging from the initial stage of the crisis
Continuity	Constant monitoring and continuous improvement of the system
Activity	Active monitoring of competitors to perform an effective protection
Legality	Economic security system corresponds to the laws of the Russian Federation
Economic efficiency	Comparing potential damage and the cost of the diagnostics
Differentiation	Security system should be most suitable for the organization
Cooperation and coordination	An interaction of services to ensure optimal functioning of the whole organization
Perfectibility	Optimization of the activity of enterprise services on the basis of the appearance of new mechanisms
Centralization of management	Diagnostics system should operate as an independent control function

Presented principles cover the enterprise completely from all sides (outside and inside), Figure 2. The conceptual basis of the researches of the enterprise is the multiloop diagnostic model, Figure 3.

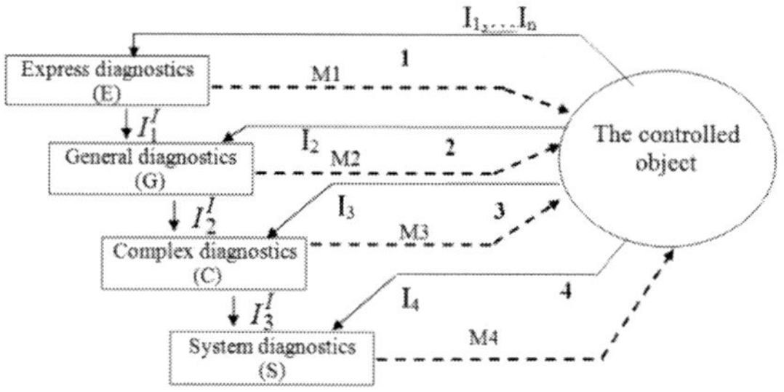

1, 2, 3, 4 – stages of diagnostics; $I_1, \dots I_n$ – information(the parameter type of the state) of automation control object; M1, M2, M3, M4 – management actions by the results of the particular diagnostics stage

Fig. 3. Multiloop diagnostics model in the anti-crisis management.

Multiloop diagnostic model is an implementation of a systematic approach when analyzing the enterprise state and allows you to optimize the resource usage through the multi-step analytical procedures with making the partial decisions. A systematic approach is through consistent, multistage approach to identify the causes of those or other crisis processes and their specifications. The content of the diagnostics stages is presented in table 4.

Table 4. The content of the diagnostics stages.

Name of a stage	Contents
Express diagnostics	Dynamic analysis of financial indicators, identification of factors causing the dynamics of financial indicators

General diagnostics	Identifying the cause factors of production dynamics, assessment of resource effectiveness, identifying the causes of the crisis (the type of crisis situation and factors)
Complex diagnostics	Analysis of the enterprise management system, the parameters of the interrelation and interdependence of production factors
System diagnostics	Developing the economic-mathematical models of business processes, analysis of resource potential, forming set of criteria (institutional rules, corporate standards) adequately reflecting the specifics of the enterprise activity

Multiloop diagnostics model is a reflexive technology of management decisions support at all threats development stages even when the threat causes detection is difficult or impossible.

The diagnostics objectives and content of the analytical procedures are determined by the enterprise specificity [14] and features of the selected analysis type (problem-oriented, forward-looking, operational, etc.).

In the multiloop model of diagnostics, studies begin with the stage of express diagnostics (E) of the enterprise (controlled object) that includes information flow about parameters of the controlled object of (I1), and management action (M1) (fig. 3).

At the first stage of the express diagnostics, the following analysis is conducted: financial performance of enterprises, the structure of assets and sources of their formation, financial performance, a comparison of planned and actual indicators and their dynamics.

At this stage, there are periods that characterized by different behaviors (dynamics) of the initial data, the analysis of initial; data includes:

• periodization (identification of process chronology) and identification of turning points (borders of periods based on dynamic analysis of financial indicators);

• identification of factors causing the dynamics of financial indicators.

At this stage, decisions are made (M1 at Fig. 3) about the rules for the formation of an information database, compares planned (corporate standards or reference) and actual parameters of the enterprise, the specific techniques and research frames for the general diagnostics are identifying. According to the analysis results, there is an evaluation of improvement or deterioration of the enterprise state.

At the second stage there is the general diagnostics (G) of the controlled object that consists of information flow I2 and management action M2 at Figure 3).

At this stage, we identify what group or set the researched object belongs to. Then we evaluate the degree of importance of each group of indicators that is corrected according to the evaluation results of lifecycle stage of the enterprise development [17], the economic sustainability, [18, 19, 20], the identification of business processes [21]. Then factor parameters are identified: technical, production, efficiency and resources usage, etc. [22, 23]. The quantitative assessment of parameters dynamic is conducted is conducted and dynamic characteristics are defined.

Assumptions about the applicability of specific factors based on the assumptions about the properties of identifiable models (like its linear homogeneity). The methods of market analysis, the lifecycle analysis, the investment activity analysis, efficiency analysis are chosen. At this stage, decisions about the choice of factor model are made (ROI, EVA ROA, etc.) for factor analysis of the enterprise.

The third stage includes a complex diagnostics (C) of the controlled object management (information flow I3 and management action M3, Figure 3).

At this stage the following analysis are held: the properties of the production factors (the study of organizational-technical level of enterprises, logistics, resource efficiency, etc.) and the parameters of their interaction; analysis of potential development organization [24, 25, 26]; indicators of statics and dynamics parameters of external and internal environment; efficiency control system (reflexive properties, mechanism of control, elements of corporate culture, the level of institutionalization process of managerial decision-making, etc.); the level of economic security of the enterprise management system.

Based on the results of the analysis of the properties of the enterprise management system, the set of evaluation criteria of what particular class the researched object belongs is corrected.

The fourth stage contains system diagnostics (S) of the controlled object (information flow I4, management action M4, fig. 3).

At this stage there are:

• analysis of the potential of the enterprise development (the potential of labor resources, intangible assets as the competitive advantages of the enterprise are determined by their potential);

• analysis of the quality of corporate standards (institutional rules of the enterprise), the institute of education (development) including subject management system;

• development of economic-mathematical models of business processes, and the sustainability of the development.

The basic economic institutions are actualized in institutional norms, values, philosophy and mission of the enterprise. The diagnostics at this stage involves the identification and classification of new factors that cause the dynamics of the parameters of the controlled object (business processes). The parameters of new factors are included in the monitoring system of the enterprise.

The scheme (fig. 3) is elaborated on the principle “from the general to the special”. It displays the combination of diagnostics methods and the enterprise management processes as a management tool for the economic security. Economic security management involves the consistent standard application which may be corrected depending on the conditions of the enterprise strategy.

Conclusion

Thus, applying the multiloop diagnostics model helps to improve the economic security of the enterprise due to prompt response to the external and internal threats; to find the effective decisions of the problem switching to the next level of economic security with different combinations of resources; to elaborate monitoring system of the enterprise efficiency based on differentiated tools and the implementation measures for the enterprise sustainable development.

It helps to improve a resource effectiveness of the enterprise, a reliability, and accuracy of business process identification within the development strategy of the enterprise and to predict the desired state of the economic security of business operations.

The work was supported by Act 211 Government of the Russian Federation, contract № 02.A03.21.0011

References

1. M. Bendikov, Management in Russia and abroad, **2**, 7 (2000)
2. V. Mamontov, O. Stepicheva, Bulletin of the University of Tambov series: the humanities, **12(104)** (2011)
3. V. Tambovcev, Bulletin of MSU, Series 6, Economics, **3** (1999)
4. O. Gromova, The world of science, culture and education, **3(52)**, 34 (2015)
5. I. Gusev, *The economic security of businesses* (EDP sciences, Moscow, 2006)
6. I. Koshkina, E. Mubarakshina, Contemporary studies of social problems (electronic journal), **8(16)** (2012)
7. V. Kryzhanovskii, V. Lapenkov, V. Luter, *Anti-crisis management: A training manual for technical colleges* (Publishing House "PRIOR", Moscow, 1998)
8. J. Ohlson, J. of Accounting Research, **19** (1980)
9. V. Kovalev, O. Volkova, *Activity analysis* (Prospect, Moscow, 2004)
10. P. Gavrilov, N. Shcherbakova, The economic analysis: theory and practice, **13(412)**, (2015)
11. O.V. Efimova, *How to analyze the financial situation of an enterprise* (Business-shk. INTEL-SINTEZ, Moscow, 1994)
12. V.Yu. Zhdanov, Management of economic systems, **32** (2011)
13. Yu. Loktionova, Social and phenomena and processes, **4(050)**, 88 (2013)
14. N. Maksakov, Social and phenomena and processes, **12(46)**, 202 (2012)
15. G. Kleiner, *"A business strategy," Analytical Handbook* (KONSJeKO, Moscow, 1998)
16. O. Molchanova, T. Aliev, Bulletin of ASTU, **4(45)**, 43 (2008)
17. V. Pluzhnikov, S. Shikina, Finance and credit: economic analysis, **44(443)**, 53 (2015)
18. V. Mokeev, M. Nelyubina, Bulletin of the South Ural State University. Series: Computational mathematics and informatics, **5(1)**, 69 (2016)
19. V. Mokeev, O. Buslaeva, S. Voronina, V. Pluzhnikov, S. Shikina, *Applying the eigenstates for the analysis and forecasting of socio-economic systems* (SUSU, Chelyabinsk, 2015).
20. E. Salimonenko, A. Karpushkina, O. Volozhanina Bulletin of the South Ural State University. Series: Economic and management, **44(303)**, 80 (2012)
21. S. Kukharenko, V. Pluzhnikov, S. Shikina, Bulletin of the South Ural State University. Series: Economics and management, **9(4)**, 57 (2015)
22. V. Pluzhnikov, S. Shikina, Management of economic systems (electronic journal), **10**, (2014).
23. D.Suleymanova, S. Rabadanova, Economics and entrepreneurship, **11(52-2)**, 388 (2014)
24. E. Vyborova, E. Salyahova, The economic analysis: theory and practice, **12(315)**, 23 (2013)
25. L. Kiyashchenko, Yu. Kolesnikova, *Problems of socio-economic development of the regions A collection of articles of the international scientifically-practical* (2015)
26. L. Mikhalina, E. Golovanov, Bulletin of the South Ural State University. Series: Economics and management, **8(3)**, 41 (2014)

Management of mechanical engineering design processes based on product cost estimates

Evgueniy Kolbachev[1,*]

[1]South-Russian State Polytechnical University, Production and Innovation Management dept., Novocherkassk, Russia

Abstract. It is herein conjectured that the cost of a machine can be estimated early in the engineering process. Such estimation is based on an integral index of its informational (parametric) complexity. The index also factors in the properties of materials in use as well as the structural components of machine parts. The author has designed an original approach to the management of design processes, which approach is based on the structural index entropy. The entropy level decreases as the structure is optimized. Using the economic parameters should also help decrease such entropy. The author herein presents the instruments for estimating the parametric complexity of a structure. This index may then be used to calculate the projected costs of the structure.

1 Introduction

Current conditions of the world economy make increasing of the businesses' competitiveness in the market an actual task for the mechanical engineering design and manufacturing enterprises. Main causes of the situation are the re-industrialization processes and development of the knowledge economy.

Reasons for the development of re-industrialization processes that take place in the later years all over the world may differ due to a concrete country [1]. For example, authors of the well-known report [2] contend that the reasons for re-industrialization in the USA are the following: unit labor costs are among the lowest in the industrialized world; the cost of capital is very low given the Fed's ultra-accommodative monetary policy. The similar processes take place in the European countries [3]. Re-industrialization gives the mechanical engineering and manufacturing enterprises a possibility to "come back" to these countries, and also creates incentives for development of the new enterprises. Parameters of these enterprises' product (first of all, the machines and mechanisms) should be very close to the economically optimum.

In the newly industrialized countries (mainly, in the Southeast Asia) modern high-tech industry was developed in the end of 20th century because of the manufacturing move to the region from the industrially developed countries, and the appropriate economic policy of the government. Now the actual task for them is development of the "original" mechanical engineering industry that is independent of the foreign companies' policy of manufacturing development, and switch from copying of the western constructions to the carrying out of the original engineering projects. It's noticeable that the numerous well-known publications in the field of economic management for the engineering design in the last years were presented by authors from the newly industrialized countries [4, 5, 6].

A reason for the re-industrialization of the last years in Russia is understanding by the society and the national government how important the increasing of country's human capital. A catastrophic decreasing of the human capital level in Russia took place in the beginning of the post-Soviet period due to the destructive government policy and resulted in the elimination of many science intensive sectors of Russian economy [7]. The policy of the EU countries and of the USA that restricts the scientific and technical co-operation between Russian and foreign corporations after the coup d'etat in the Ukraine in 2014 finally favours the development of Russian mechanical engineering [8].

So we can say that problem solving in the field of managing the mechanical engineering design processes and working out of the new machines is actual all over the world. Solving of these problems increases the competitiveness level of the engineering products and of the mechanical engineering corporations.

One of the steps in this direction may be a researching of the existing and working out of the new methods and instruments for the valuation of constructs' cost at an early stage of the engineering. Using of the organizational and economic methods for the managing of the engineering design is also appropriate to succeed.

2 Managing the engineering design and entropy of the construct

Valuation of the construct's cost at an early stage of the engineering is an important task because it increases the level of "informational completeness". The higher the

* Corresponding author: kolbachev@yandex.ru

level is the closer construct's parameters will be to the optimum.

The level of construct's approximating to the optimum condition may be presented as a level of the construct's entropy. Decreasing of the entropy level means that the construct's conditions become closer to the optimum. We can use this dependence because during the development and analyzing of the different versions of the construct the number of the possible (potential) versions to choose the optimum construct of them decreases. Each of the versions is a step to the optimum.

We can evaluate the construct's entropy level using the Shannon's formula [9]:

$$H = c\left(p_1 \ln \frac{1}{p_1} + p_2 \ln \frac{1}{p_2} + \ldots + p_n \ln \frac{1}{p_n}\right), \quad (1)$$

The used measurands are: H — entropy of the construct; p_1, ... p_n — state probabilities of the construct's elements; c — the constant.

We have to take into account that during the analysis of the construct's elements and the combinations of them we examine only the material components of the system and structure of these components. The non-material components (such as databases, thesaurus of the certain employees and other informational resources, used in the "classical" Shannon's formula) shouldn't be examined.

As the most applicable in the analyzed situation we are going to use the conditional entropy index, derived by A. Renyi [10]. The essence of the presented approach is the following.

Let B is a randomly chosen version of the construct, and the probability of the choice is positive; ξ is a random variable that possesses the values $x_1, x_2, \ldots, x_n$. We denote by A_k the derivative of B, a version of construct which's $\xi = x_k$ ($k = 1, 2, \ldots, N$). Then basing on the definition, we determine the conditional entropy of ξ under the condition B as an entropy of the random variable ξ, calculated on the base of the conditional probabilities' distribution. It's taken that that the construct version B is created. So we derive the equation.

$$H_B(\xi) = \sum P(A_k|B)\log_2 \frac{1}{P(A_k|B)}, \quad (2)$$

$P(A_k|B)$ is a conditional probability of the version A_k under the condition B (B is taken as the initial version), so the equality is the following.

$$P(A_k|B) = \frac{P(A_k B)}{P(B)} \quad (3)$$

Let η is another version of the construct, which's characteristics possess the values $y_1, y_2, \ldots, y_m$, and B is a version $\eta = y_j$ ($j = 1, 2, \ldots, M$). Then the conditional entropy of ξ when the version η is created is $H_\eta(\xi)$, that is definitely similar to the expectation value of $H_{Bj}(\xi)$.

$$H_\eta(\xi) = \sum P(B_j) H_{B_j}(\xi) = \sum\sum P(A_k B_j)\log_2 \frac{P(B_j)}{P(A_k B_j)}. \quad (4)$$

In the analyzed situation we should take into account the entropy of random variable ξ (the uncertainty of ξ value), and how it decreases during the examining of the randomly chosen version of the construct η. A value of the entropy quantity we designate as $I(\xi, \eta)$. It is a quantity of information about randomly chosen value ξ that we get during the examining of randomly chosen version of construct η.

$$I(\xi,\eta) = H(\xi) - H_\eta(\xi) = \sum P(A_k)\log_2 \frac{1}{P(A_k)} - \sum\sum P(A_k B_j)\log_2 \frac{P(B_j)}{P(A_k B_j)}. \quad (5)$$

Then we use the identity.

$$\sum_j P(A_k B_j) = P(A_k) \quad (6)$$

We can use it because the versions A_kB_1, A_kB_2, ..., A_kB_M are mutually exclusive, and if the construct version A_k, is created, creation of any other version of A_kB_j is possible too

Basing on the above we can label any problem that exists during the engineering design processes as a result of shortage of the information about construct's elements. So, using the information about cost characteristics of the construct's version is a possible way to solve the problem.

3 Methods of valuation the constructs' cost at an early stage of the engineering

Production cost characteristics of the constructs, or their parts include the materials output ratio; labour output ratio (this index also includes the time of using machinery for the processing of the details); total investment costs for the building-up and launching of the production.

The machine's cost is an integral characteristic that includes all types of the production costs that exist during the engineering and production. Exact calculation of the machine's cost may be done only basing on the full set of engineering and design documentation. But at an early stage of the engineering this set of documentation does not exist. To prognosticate the cost value on this stage the numerous rough-and-ready methods may be used. A great variety of methods is a result working out each of them for the certain organizational and technical conditions of engineering. Choosing an appropriate method of cost estimating at an early stage of the engineering is a basis for the right economic justification of the project. Objective cost estimation at an early stage of the engineering stimulates the fastest appearance of the new machine in the market, decreases the costs level and increases the quality. As a result, the cost estimating gives a possibility to make a machine with higher competitiveness indexes in the open market.

The earliest researches in this field were taken out in Germany in the 1950th — 1980th by K. Ehrlenspiel,

A. Kievert, U. Lindemann (1980) [11]; G. Pahl, K. Beelich (1987) [12] et. al. The results of this work were summarized by H. Warnecke, H. Bullinger, R. Hichert, A. Voegle in the well-known work [13].

Some of the researches in the field of economic valuation of the engineering production and working out the economic methods for the engineering were carried out in the USSR in 1980th. The attention is to be paid to the works by A. Kovalev [14], N. Moisseeva [15], et. al. Unfortunately, Russian research projects in this field were broken off, and numerous enterprises that used the results of the researches were eradicated in the 1990th, in the period when management of national economy was ruled by Eltsyn's and Gaidar's cabinet, based on the guidelines, presented by advisers from the USA [7]. The said researches were resumed in Russia in the 2000th and became especially successful after the beginning of Russian industry's rapid development and increasing its competitiveness level in 2014 and after (V. Kovalskiy (2012); D. Zavadskiy (2015); E. Kolbachev and O. Sukharev (2016) [16, 17, 18]). The reason for this process was overcoming the unsuccessful economic blockade of Russia by the USA an EU.

At the same time there was presented a well-known method of cost estimation at an early stage of the engineering by C. Ou-Yang and T. Lin [19]. It was based on the information, used in the CAD systems. Y. Jung [5] worked out an original classification of the machine components to estimate the cost of their processing. The original model of cost estimation at an early stage of the working out the processes for the milling works was presented by E. Nasr and S. Kamrani [4]. Using these models as a basis gives a possibility to work out the universal models of cost estimation. Another interesting research is a known work by E. Maraghy, W. H. Urbanic and R. Jill [20]. These authors worked out an algorithm of estimating the complexity factor for the processes of the machinery details production. The algorithm doesn't deal with the details' costs, but may be improved and used for the cost estimation.

A special attention should be paid to the recently published work by H. Budiono, G. Kiswanto and T. Soemardi (2014). Authors present an original method of cost estimating at an early stage of the engineering design processes. Method is based on the using of correlation between the cost indexes and complexity level of the processing, and takes into account the features of the specific details. The method gives design engineers a possibility to create and analyze the alternative versions of the construct as fast as possible.

In the majority of cases the well-known methods of valuation the constructs' cost at an early stage of the engineering are based on the statistical data about the cost of processing the certain part (or surface) of the detail. This group of methods is worked out by G. Pahl and K. Beelich (1987) [12]; A. Kovalev (1986) [14]; P. Duverli and J. M. Castelain (1999) [21] etc.

Some of the methods were worked out for the details, which can be proceeded by the limited number of ways (for example, milling work), such as methods by H. Budiono, G. Kiswanto, T. Soemardi (2014) [6]. Others (by S. Duglas, T. Gilbert, S. Gilbert (2014)) [22] are worked out for the processing, based on the additive work, or on the special conditions of making the certain product (V. Mallikarjuna (2014)) [22].

In the last years the numerous works in the said field were taken out only to solve the specific task of valuating the costs in CAD and CAM systems (works by P. Karam (2007) [24]; C. Shaw, Y. Eugene (2000) [25]; V. Mallikarjuna (2014) [22] etc.

Besides that, most of the known methods of valuation take into account the processing costs, but not the material costs.

The said features of the known cost estimating methods are extensively reviewed in the Cost Estimating Guide (2011) [26].

The said makes the constructs' cost estimating at an early stage of the engineering design processes much more complicated, and the managing of new machines engineering becomes less efficient.

Our goal is to create a methodology of the constructs' cost estimating at an early stage of the engineering design without these limitations.

4 Construct's informational complexity as a basis for the cost estimating

A basic feature that differs our methodology from the well-known alternatives is that cost estimating of the detail part is based not on the construction parameters (such as size, dimensional accuracy, surface smoothness etc.), but on the integral index of its informational complexity.

Original algorithm of the cost detail parts' cost estimating, based on the said methodology is presented in Fig. 1. Formation of the database "The unit costs" looks like the most tedious part of its realization. The database should contain the information about the detail part's cost characteristics per unit of its informational complexity. The database formation requires realizing of such subgoals as classification of the objects for the empirical study and adaptation of the results for the requirements of the specific enterprise.

It's obvious that realizing of the said goals is impossible without choosing of an index, which index is to be used to indicate a level of the informational complexity. Basing on the results of research we established that the most suitable index for the serial production of the machinery detail parts is the index of parameter complexity per unit of detail's mass. Algorithm of the database "The unit costs" formation is presented in Fig. 2. The index of parameter complexity is being determined here as a code combination of the details' dimensional and other engineering parameters per unit of detail's mass. The index is to be added to the informational valuation of the used material. The valuation is based on the number of material's

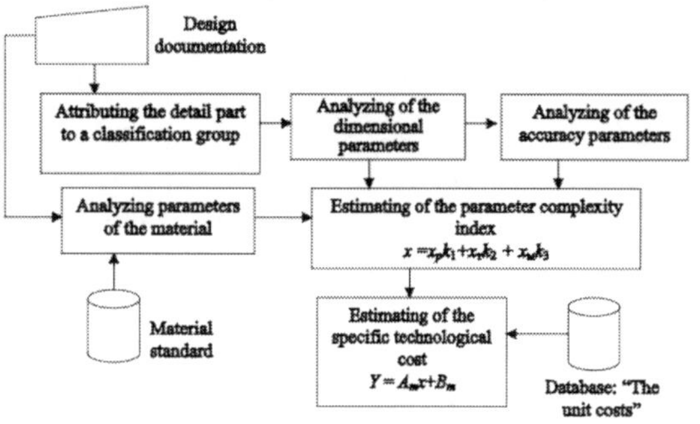

Fig. 1. Schematic diagram of the details' cost estimating based on the integral index of the informational (parameter) complexity.

parameters (these parameters are to be specified in the material standard).

As an example of calculating the indexes of details' unit costs we presented Table 1. It is the empirical dependence of some detail parts categories' costs from their informational complexity. The dependence is ascertained basing on research data we got as a result of the exploring the mechanical engineering enterprises in the South of Russia.

In the presented in Table 1 dependence formula Y is detail parts' specific technological cost (rubles per kg), x is a parametric complexity index. Complexity index was determined for the each detail part according to the algorithm in Fig. 1.

Values of Fisher's criterion and t-statistics ratio mean that the obtained models and regression coefficients are statistically essential. Presented results give us a possibility to suggest a hypothesis that the dependence of the models' cost characteristics (for the widely used in the engineering design detail parts) from their parametric complexity index is a kind of linear dependence.

To our minds, a presented hypothesis may be used as a methodological basis for the future working out of the economic projecting subsystem for the existing engineering design CAD systems.

In the South-Russian State Polytechnical University's department of Production and Innovation Management are being worked out some economic instruments, based on the presented research results. The instruments should be used for the managing of engineering design

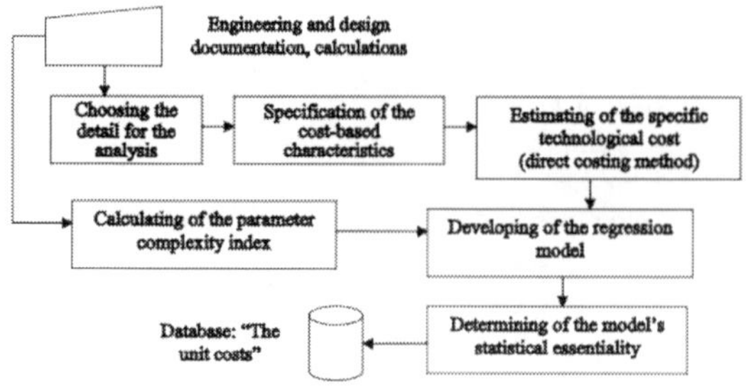

Fig. 2. "The unit costs" database: schematic diagram of the formation.

processes inside the mechanical engineering cluster and its innovation infrastructure [27] (M. Perederiy (2015)).

5 Conclusion

The basic goal of the presented research was determining of the economic (institutional) essence of engineering design. As a basis we used the valuation of construct's cost characteristics. Also on this basis were worked out some instruments of managing the engineering design processes. Using of the instruments should increase the engineering and production enterprises' competitiveness level.

As a result of the research there was established a fact that the ways of increasing engineering and production enterprises' competitiveness level differs greatly in the different countries of the world. The countries, which lead in the technical and economical fields, should choose the ways, based on re-industrialization processes. The newly industrialized countries (first of all, countries of the Southeast Asia) should develop the independent mechanical engineering industries, without western companies' participation. Russia's way is maintenance of the country's political and economic sovereignty and restoration of industry (first of all, it high-tech branches) after the collapse of the 1990th — early 2000th years.

As a result of the analysis of the numerous well-known methods of valuation the constructs' cost at an early stage of the engineering was worked out an original approach to estimating of the construct's prospective cost. The approach is based on using of the

Table 1. Detail parts cost's empirical dependence from their parametric complexity level.

Group of the detail parts	Number of the observations (pairs)	Model	Determination coefficient	Calculated value of the Fisher's criterion	Critical value of the Fisher's criterion, confidence level 0.05	t-statistics ratio of the regression coefficient valuation	Critical value of the t-criterion, confidence level 0.05
Shafts	253	$Y = 0.3965x + 19.63$	0.946	314.05	4.45	17.72	1.7459
Bushes	144	$Y = 0.275x + 30.88$	0.935	131.25	5.32	11.46	1.8595
Gears	185	$Y = 0.904x - 56.51$	0.953	326.61	4.54	18.07	1.7613
Box housings	92	$Y = 0.5529x + 27.749$	0.85	153.88	4.22	12.40	1.7109
Bed-plates	114	$Y = 0.438x + 50.388$	0.825	76.63	4.54	8.75	1.753

integral index of the construct's informational (parametric) complexity. The approach also takes into account the features of used materials and the aggregate of the construct's detail parts.

As an addition for the cost characteristics estimating methods we worked out the approach to managing of engineering design processes, based on the construct's entropy level. The level of the entropy decreases during the construct's characteristics become closer to the optimum.

A presented approach gives a chance to make managing of engineering design processes more efficient, and the worked out machinery — more competitive in the nowadays open market.

The results of the research confirm a theory that using a methodology of the construct's economic valuation at an early stage of engineering is appropriate.

References

1. D. North, B. Goutard, *The Reindustrialization of the U.S. - A 2014 Update* (Euler Hermes, Paris, 2014)
2. *The Reindustrialization of the United States. Spec. Rep. no. 1187* (Euler Hermes, Paris, 2013)
3. *Reindustrialization of Europe – Minerals Industry Contribution. Press release* (European Association of Mining Industries, Metal Ores & Industrial Minerals, Brussels, 2014)
4. E.S.A. Nasr, A.K. Kamrani, Computers & Industr. Engineering, **51**, 389-415 (2006)
5. J.-Y. Jung, J. Intell. Manufact., **4**, 227-238 (2002)
6. D.S. Budiono, G. Kiswanto, T. Soemardi, Intern. J. of Technol., **2**, 183-192 (2014)
7. M. Delyagin, *Russia for the Russians* (Eksmo, Moscow, 2009)
8. I.V. Mukhanova, Econ. J., **2(38)**, 34-41 (2015)
9. C.E. Shannon, Bell System Tech. J., **27**, 423 (1948)
10. A. Renyi, *Dialogues on Mathematics* (Holden-Day, New-York, 1967)
11. K. Ehrlenspiel, A. Kievert, U. Lindemann, VDI J., **17**, 11 (1980)
12. G. Pahl, K. Beelich, Construction, **7**, 10 (1987)
13. H.J. Warnecke, H.J. Bullinger, R. Hichert, A. Voegele, *Cost Estimation and Accaunting for the Engineers* (Hanser-Verlag, Munich, 1996)
14. A.P. Kovalev, *Making the Worked-out Machinery Efficient* (Mechanical engineering, Moscow, 1986)
15. N.K. Moisseeva, *Functional and Costs Analysis for the Machine-building* (Mechanical engineering, Moscow, 1987)
16. D.Ju. Zavadskiy, Fundamental Res., **7**, 387-393 (2015)
17. V.F. Kovalsky, S.V. Kovalsky, M.Ju. Chalova, Railway Track and Track Facilities, **11**, 21-24 (2012)
18. E.B. Kolbachev, O.S. Sukharev, *Economic Methods for the Engineering Design and Machinery Manufacturing* (Akademcenter, Rostov-on-Don, 2016)
19. C. Ou-Yang, T.S. Lin, Int. J. of Advanced Manufact. Tech., **13**, 618-639 (1997)
20. W.H. el Maraghy, R.J. Urbanic, *Modelling of Manufacturing Systems Complexity* (Windsor, Ontario, CA, Intelligent Manufacturing Systems (IMS) Centre, 2003)
21. P. Duverli, J.M. Castelain, Int. J. of Advanced Manufact. Tech., **15**, 895-906 (1999)
22. S. Douglas, D.S. Thomas, S. Gilbert, *Costs and Cost Effectiveness of Additive Manufacturing* (U.S. Department of Commerce, New-York, 2014)
23. V. Mallikarjuna, N. Jashuva, G. Nagaraju, B. Rama Bhupal Reddy, J. of Mech. and Civil Engineer. (IOSR-JMCE), **1**, 53-67 (2014)
24. K.-A. Karam, Engineer. & Tech., **1**, 103-116 (2007)
25. C.F. Shaw, Y.S. Eugene, *Information Modeling of Conceptual Process Planning Integrated with Conceptual Design* (The ASME Design Engineer. Tech. Conf., 2000)
26. *Cost Estimating Guide* (U.S. Department of Energy, Washington, 2011)
27. M.V. Perederiy, Bull. of SRSTU (NPI): Soc. and Econom. Sci., **5**, 30-34 (2015)

A mechanism for encouraging active consumers to optimize the operations of power supply systems

Natalia Aizenberg[1, *], *Nicolai* Voropai[1], and *Elena* Stashkevich[2]

[1]Melentiev Energy Systems Institute of SB RAS, Irkutsk, Russia
[2]National Research Irkutsk State Technical University, Irkutsk, Russia

Abstract. We propose to analyze an interaction of power supply company and consumers as a game in which the retailer's goal is to maximize its profit, and the consumer's goal is to maximize its utility. The utility function of each consumer is described proceeding from the form of the loss function of the consumer, and one of our goals in the submitted paper is to find how to describe this load function mathematically. We want to develop an economic mechanism that help the power retailer to find an optimal strategy on the set of possible loads and tariffs. They should satisfy a set of constraints binding these two variables by natural requirements imposed on the values of the consumer utility functions.

1 Introduction

Consideration is given to the interaction among electricity consumers and an electric power retailer in which each consumer tries to reduce the cost of electricity by redistributing (shifting) its demand (or load) between peak hours and low-demanded ones, whereas the power retailer tries to reduce its expenses associated with storing electricity, especially, at peak hours [1, 2]. We propose to analyze such an interaction as a game in which the retailer's goal is to maximize its profit, and the consumer's goal is to maximize its utility. The utility function of each consumer is described proceeding from the form of the loss function of the consumer, and one of our goals in the submitted paper consists of finding how to describe this load function mathematically. However, the major goal is to develop an economic mechanism that would help the power retailer (considered as a monopoly in the marketplace) maximize its profit based upon the known loads of all the consumers. We consider a finite number of different types of the so-called "active" (load-controlled) consumers based upon their incentives to optimize their loads. Each type of the incentives (and, consequently, consumers) is defined by a utility function of the consumer, which is the function of the hour load volumes within a 24-hour segment. We offer an alternative utility function to describe the behaviour of a consumer who is not an "active" one (since this consumer does not optimize the consumption of electricity by means of shifting its loads from the peak hours to low-cost ones). We formulate the problem of finding an optimal strategy for the power retailer as the one of maximizing the profit that the retailer can make with respect to each type of the "active" consumers on the set of possible loads and tariffs that satisfy a set of constraints binding these two variables by natural requirements imposed on the values of the consumer utility functions.

The use of market mechanisms and development of information and communication technologies encourage active behaviour of another participant of the electricity supply process, i.e. power supply company, as well. The company becomes capable of encouraging consumers to actively manage their electricity consumption [3-8] by establishing price options in real time. The interaction between the power supply company and consumers requires coordination of their operating conditions. Until now the potentialities of the coordination have been studied on the basis of mathematical programming methods, including two-stage stochastic programming [8, 9], a multi-agent [10-14] and a game-theoretic [9, 10, 12-14] approaches.

The behaviour and strategies of load-controlled consumers have been studied by many researchers in the field. The authors of [15] study one such problem in which for any consumer of the power retailer of electricity, a disclosure of any information on its load is not mandatory is considered, and the power retailer tries to offer stimulating tariffs to all the consumers while even determining sources of saving for the consumers. Situations in which the prices for electricity are determined in advance so that each consumer actively changes the load in order to reduce its expenses as much as possible are considered in [12, 13]. Variants of a two-level formulation of the problem in which the interaction in considered for a set of vertically integrated legal entities such as a generating company, a retailer, a network and a consumer, which is represented in an aggregate form, without division into types, are presented in [5]. In some statements of the problem, consumers are allowed to supply any electricity surplus

* Corresponding author: ayzenberg.nata@gmail.com

back to the network [16]. There are statements dealing with an additional generation of electricity, when consumers can only shift load from the peak time via changing their production process, in particular, by using electricity storage systems or by automating the production process [4].

The subject of consideration in the present paper differs from those in the enumerated studies. That is, we consider consumers of several types each having different incentives to optimize their load, and we show how to determine the tariffs harmonized for both fully rational and bounded rational consumers. Here, we consider a customer to be fully rational if it does not vary its load, whereas we consider a customer to be bounded rational if it may change its load curve by shifting part of its load during peak hours to low-demand ones. The latter is offered a tariff stimulating load optimization. Standard solution to the problem of profit maximization of an electricity supply organization under the presence of asymmetric information on the agent type can lead to the formation of conditions for the adverse selection. We form constraints to determine the contracts (tariffs) that make it possible to search for a separating equilibrium where each type of consumers selects "their" tariff and adjusts their consumption as is necessary for the system. The research formulates the incentive rationality constraints and incentive compatibility constraints for the time-of-use and two-rate tariffs.

2 The Model

There are a finite number of different types of the so-called "active" (load-controlled) consumers who have the ability to adjust own load, and consumer who does not optimize the electricity consumption by means of shifting its loads. Θ is the set of N types of consumers $\theta \in \Theta$. There are K consumers in model. Each consumer derives some utility from receiving a certain amount of electricity. Denote this utility through the function $u_k(q_{tk}^{\theta})$ where the load at a certain t-zone of the day is q_{tk}^{θ}, $q_{tk}^{\theta} \geq 0$. R_t^{θ} is a tariff value at hour t, $t \in [1, T]$ for type θ. The problem of each k consumer is the maximizing of gains from electricity buying with respect to q_{tk}^{θ}, $k = \overline{1, K}$:

$$U^{\theta}(\theta, q) = \sum_{t=1}^{T} u_k(q_{kt}^{\theta}) - R^{\theta} \to \max_{q}, \qquad (1)$$
$$q_{kt}^{\theta} \geq 0,\ k \in \overline{1, K},\ \theta \in \Theta.$$

The function of utility usually is monotonically increasing and concave. It should be performed additional property of Spence-Mirrlees:

$$\frac{dU^{\theta}(\theta, q)}{dq} \geq 0,\ \frac{d^2U^{\theta}(\theta, q)}{(dq)^2} \leq 0,\ \frac{d^2U^{\theta}(\theta, q)}{d\theta dq} \geq 0.$$

The number of θ-type consumer is m^{θ} and $\sum_{\theta \in \Theta} m^{\theta} = K$. The power supply company appoints the tariff for θ-type of consumer R_t^{θ} using the total load $\sum_{k=1}^{m^{\theta}} q_{kt}^{\theta}$ and its cost function $C\left(\sum_{k=1}^{m^{\theta}} q_{kt}^{\theta}\right)$. The cost function is convex with respect to $\sum_{k=1}^{m^{\theta}} q_{kt}^{\theta}$. The main purpose of the power supply company is to maximize profits of tariffs R_t^{θ}:

$$\pi = \sum_{\theta \in \Theta} \left(R^{\theta} - \sum_{t=1}^{T} C\left(\sum_{k=1}^{m^{\theta}} q_{kt}^{\theta} \right) \right) \to \max_{R^{\theta}}, \qquad (2)$$
$$R^{\theta} \geq 0.$$

We can decide the problem of select the tariff menu provided optimization the load curve through the coordination of interests of all stakeholders (1) and (2). We will update the form of utility functions and rewrite a problem through the theory of contracts.

The problem (1) is rewritten in terms of the conditions of participation and of individual rationality constraints for consumer types, which mean that the selection of "their" tariff proves more beneficial than the selection of a tariff of another consumer type. Moreover, we will assume that there is always some basic tariff which determines some alternative utility of k-th consumer - $\overline{U}_k$. This alternative utility will be derived by a consumer without participating in the selection of an optimal consumption pattern.

Then the general problem can be formulated as follows.

$$\begin{cases} \pi = \sum_{\theta \in \Theta} \left(R^{\theta} - \sum_{t=1}^{T} C\left(\sum_{k=1}^{m^{\theta}} q_{kt}^{\theta} \right) \right) \to \max_{R^{\theta} q_{tk}^{\theta}}; \\ \sum_{t=1}^{T} u_k(q_{kt}^{\theta}) - R^{\theta} \geq \overline{U}_k, \quad k \in [1, m^{\theta}],\ \theta \in \Theta; \\ \sum_{t=1}^{T} u_k(q_{kt}^{\theta}) - R^{\theta} \geq \sum_{t=1}^{T} u_k^{\theta}(q_{kt}^{o}) - R^{o}, \quad k, j \in [1, K]; \\ \forall \theta, o \in \Theta, q_{kt}^{\theta} \geq 0, q_{kt}^{o} \geq 0; R^{\theta} \geq 0, R^{o} \geq 0. \end{cases} \qquad (3)$$

The variables are the load of the individual consumer q_{tk}^{θ} and the tariffs value R_t^{θ} which is offered by the power supply company. We consider the example of the two types of consumer: with full rationality and with bounded rationality. Let describe utility function of θ-type consumers.

2.1 The determination of optimal tariff for the two types of consumers

I. At first the problem of utility maximization for the participation of the rational consumer $\theta_1 \in \Theta$ is described. The constraints that include the conditions of a positive utility in selecting the tariff are the conditions of participation. In our research we believe that such a consumer can use a two-part tariff. In this tariff is assumed that the electricity unit price corresponds to the marginal cost of a electricity unit, and fixed costs are covered through an entrance fee (or fee peak power). Should take into account that with such scheme due to the absence of constraints the monopolist can confiscate the full consumer surplus, the value of which may exceed the fixed costs of power supply company. The government should regulate the entrance fee.

The utility function of the consumer with the full rationality is written via consumer's loss function or as the tariff $U^{\theta_1} = -R^{\theta_1}$ of θ_1-type offered by the power supply company. R^{θ_1} is two-part tariff.

$$R^{\theta_1} = A_{\theta_1}^{peak} + \sum_{t=1}^{24} T_t^{\theta_1} q_t^{\theta_1}$$

where $A_{\theta_1}^{peak}$ is entrance fee or the price of the maximum power, $T_t^{\theta_1}$ is the tariff value at hour t, $t \in [1,24]$. If the consumer selects the tariff R^{θ_1} and optimize the load, then she wants the expenses less than the costs for her consumption unchanged. Then the condition for the consumer's participation with full rationality is $-R^{\theta_1} \geq \bar{U}$.

The power supply company use the possibility of withdrawal the consumer surplus as much as possible. Therefore, we get that power supply company increases the tariff (under the entrance fee) until the condition of participation allows. Thus the consumer load does not change because of her rationality.

It means that condition of participation for non-active participant fulfils as equality in the problem (3).

II. Now let us write the utility function for consumer with bounded rationality. The costs of this consumer are defined via differentiated tariff. Let $T_j^{\theta_2}$ is the tariff of j-th time zone, $j \in \{I, II, III\}$. Then the differentiated tariff is

$$R^{\theta_2} = T_I \cdot \sum_{j \in I} q_j^{\theta_2} + T_{II} \cdot \sum_{j \in II} q_j^{\theta_2} + T_{III} \cdot \sum_{j \in III} q_j^{\theta_2}.$$

Let's formulate the problem of maximizing the consumer profits when the load shifts from a time zone to another one. The retail price in the spot market is P_t^S, where t defines the purchase zone; P_t^B is the price in the balancing market; x is a potential scope of load optimizing by consumer (here we consider the scope of load shift from the peak time to the night time), $x \geq 0$. Then the payoff of the power supply company from the load shift can be written in the form

$$S(\delta, x) = (P_{III}^B - P_I^S) \cdot x - \delta \cdot x,$$

where δ is step of tariff T_t which are established by the retail company at different hours of the day. For simplicity we can assume that $\delta = T_I - T_{III}$. Here we represent a cost function of power shortage in a simplified form. The function takes into account only the price difference in the balancing and spot markets. There can be additional costs not related to the price difference, to eliminate the shortage.

We assume that the consumer costs related to the load shift is the function of the load $f(x)$, $f(x) \in C^2$. Here, we suppose that some load will be shifted by consumer without any special efforts. For example, this may concern the use of washing and dishwashing machines. At the same time, there is a load whose shift can cause greater inconveniences, for example, a shift of the food cooking time or additional installation of energy saving equipment. The function $f(x)$ reflects an extent to which the consumer is ready to such costs. The total payoff of the consumer will be represented as

$$U^{\theta_2} = -(T_{III} + \delta) \cdot \left(\sum_{j \in I} q_j^{\theta_2} - x \right) - T_{II} \cdot \sum_{j \in II} q_j^{\theta_2} - $$
$$- T_{III} \cdot \left(\sum_{j \in III} q_j^{\theta_2} + x \right) - f(x).$$

Consumer determines the load to be shifted to maximize her payoff U^{θ_2}. For example

$$f(x) = -\beta \cdot \log(1 - x/\alpha). \quad (4)$$

Each parameter can be interpreted in the following way. Parameter α (asymptote) is a maximum possible load that can be shifted by the consumer from the peak time. Parameter β controls the level of costs expressed in some money equivalent. Here this parameter is estimated on the basis of empiric data on the possibilities of load change for the considered categories of consumers. Let us take for an example the function which can describe potential consumer expenses caused by the load shift. The measurement units can be represented by money units, but here we assume that these units are an equivalent of the consumer efforts to change her load curves. The base consumer problem is to maximize the utility:

$$U^{\theta_2} \to \max_x.$$

The solution is $\delta = f'(x)$. Let the function of costs for power supply company be a concave function of x. To find the Nash equilibrium and to match the interests of all the stakeholders, it is necessary to find a point such that the maximum profit for the power supply company and the maximum payoff for the consumer are reached at the same time.

$$S(\delta, x) \to \max_\delta;$$

$$\delta = f'(x), \ x \geq 0, \ \delta \geq 0.$$

The conditions mean that the consumer will shift a certain load at different tariff difference δ:

$$(P_I^B - P_{III}^S) \cdot (f')^{-1}(\delta) - \delta \cdot (f')^{-1}(\delta) \to \max_\delta, \ \delta \geq 0.$$

Using (4) we have that $\delta = \sqrt{\beta\left(P_I^B - P_{III}^S\right)}$.

2.2 Solving the problem on the differentiation of consumer types and determination of an optimal tariff

Let an alternative utility is determined by means a basic linear (i.e., constant in each time of the day) tariff T^L. We want that the consumer of rationality type will prefer two-part tariff, while the consumer of bounded rationality type will prefer differentiated tariff. Let us write the conditions of participation.
For the consumer of full rationality type:

$$U^{\theta_1} \geq -T^L \sum_{t=1}^{24} q_t^{\theta_1} \tag{5}$$

For the consumer of bounded rationality type:

$$U^{\theta_2} \geq -T^L \sum_{t=1}^{24} q_t^{\theta_2} . \tag{6}$$

We already know that restriction (5) in the problem (3) is active for the consumer of full rationality type. The individual rationality constraints for the consumer of this type means that the choice of this tariff is more beneficial for the consumer, than choosing another types of consumer tariff.

$$U^{\theta_1} \geq - \sum_{t \in \{I, II, III\}} T_t \cdot q_t^{\theta_1} , \tag{7}$$

$$U^{\theta_2} \geq -\left(A_{\theta_2}^{peak} + \sum_{t=1}^{24} T_t\, q_t^{\theta_2} \right). \tag{8}$$

It's easy to prove by the methods of the contract theory that the restriction (7) is active. Here, we do not consider the case, when consumers with bounded rationality choose a two-part tariff, and at the same time shift their loads, since the statement of the problem in this situation changes. Our objective is just to encourage consumers with bounded rationality to optimize their load curves. If any consumer can optimize its load curve without an additional action from the power supply company, it refers to the rational type θ_1.
The function of the power supply company profit is concave.

$$R^{\theta_1} + R^{\theta_2} - C\left(q^{\theta_1}, q^{\theta_2}\right) + S(\delta, x) \to \max_{\delta, x, R^{\theta_1}} , \tag{9}$$

where $C\left(q^{\theta_1}, q^{\theta_2}\right)$ is the total cost function of the power supply company for power purchase at the market and its delivery to a consumer. $S(\eta, x)$ is the profitability from the possible shift to another time zone. The variables are the step values in the differentiated tariff and prices in the two-part tariff. Restrictions (5) and (8) are active, (7) run as strict inequalities, (6) is superfluous.
By solving the problem of power supply company profit maximization (8) subject to constraints (4)-(7), we get the equilibrium, i.e., the optimal tariffs. These tariffs match the interests of different types of consumers and suppliers.

2 Example: the search of optimal power supply company strategy in the case of two types of consumers

The several types of consumers is considered in the example of the campus. The set of consumers forms the load curve of a power system. The consumers are hostels, educational buildings, student's catering facilities, municipal health centre and prophylactic sanatorium.

We consider the health centre and prophylactic sanatorium as the unique consumer of rationality type, consumer θ_1, her load is not changed. She should choose a two-rate tariff R^{θ_1} with large payment for connection (power) and non-differentiated small payment per unit of electricity. Another group are consumers with bounded rationality, their loads are changed: θ_2^1 for hostels, educational buildings and students' catering facilities; and (θ_2^2) for cafe). They should choose a differentiated tariff R^{θ_2} per electricity unit at a zero rate for power.

Table 1. The results of the interaction of the power supply company and three customers of power tariffs established by the state.

Consumer spending, rub/day	θ_2^1	θ_2^2	θ_1
R^{θ_1}	**29744.4**	**18847.8**	**13027.5**
R^{θ_2}	32418.9	19202.2	15326.7
Shortage of power, kWt	1065		
Company's income, rub/day	**8767,5**		

Table 1 represents the results of the application of tariffs imposed by the state (see [17], the starting point). These results define the company's income, the consumer spending and the possible shortage of power. All consumers choose identical tariff (indicated in bold), this is a situation of adverse selection with high shortage.

Table 2. The results of the interaction of the power supply company and three customers after optimization of tariffs.

Consumer spending, rub/day	θ_2^1	θ_2^2	θ_1
R^{θ_1}	29744,4	16777,4	**13027,5**
R^{θ_2}	**28476,2**	**16361,2**	14265,7
Shortage of power, kWt	340		
Company's income, rub/day	**10090.3**		

Further, we are determined the type of consumers that can be interested in choosing "not their tariff", then we solve the main problem (8) (see table 2). This way we find the conditions for the mixed equilibrium and for the equilibrium differentiating the types.

2 Example: the search of optimal power supply company strategy in the case of two types of consumers

The proposed methodology of the tariff determination allows both categories of consumers interacting with a power retailer — fully rational (who do not vary their load) and bounded rational (who change their load curve by shifting part of peak load) — to optimize the electricity supply to them under tariffs stimulating their load optimization. The profit maximization problem to be solved by the power retailer under the presence of asymmetric information on the consumer type is formulated in the paper in such a manner that its set of constraints contains the incentive rationality constraints and those of incentive compatibility. This allows consumers of each type to select "their" tariffs and adjust their consumption as it is necessary for the system as a whole.

For an industrial area (the city of Irkutsk), different consumer types have been identified, and utility functions for all the consumer types have been proposed. The choice of these proposed functions has helped implement the system of stimuli for the load optimization (in the sense of shifting the load from peak hours to off-peak ones) by all the consumers with bounded rationality of their preferences. The equilibrium tariffs have been calculated, whereas the tariffs approved by the legislature have been tested. The (tested) inefficiency of the latter follows, in particular, from their unprofitability for the power retailer. Indeed, since as long as part of load is not shifted from peak hours to off-peak ones, the shortage of electricity becomes inevitable, and the retailer expenses associated with shorting electricity increase substantially. In contrast, the tariffs calculated on the basis of the proposed approach can make the above-mentioned system of stimuli for optimizing the consumer loads working, which reduces the probability of emerging shortages in electricity supply.

This work was supported Russian Foundation for Basic Research, № 16-06-00071.

References

1. N.I. Voropai, Z.A. Styczynski, E.V. Kozlova, V.S. Stepanov, K.V. Suslov, Proceedings of the Russian Academy of Sciences. Power engineering, **1**, 84-90 (2014)
2. N. Aizenberg, E. Stashkevich, N. Voropai, Proceedings of the Russian Academy of Sciences. Power engineering, **3**, 15-25 (2016)
3. G.G. Grebenyuk, M.M. Soloviev, Remote Control, **5**, 166-173, (2004)
4. I.O. Volkova, M.V. Gubko, E.A. Salnikova, Control Sciences, **6**, 53-61 (2013)
5. E. Pettersen, A.B. Philpott, S.W. Wallace, Decision support systems, **40(3)**, 427-438 (2005)
6. N. Aizenberg, E. Stashkevich, *Proc. Int. Conf. on Problems of Critical Infrastructures*, 25-27 (2015)
7. K. Sharma, C. Bhattacharya, IEEE Trans. on Smart Grid, **6**, **2**, 795-807 (2015)
8. R. Shigenove, A. Yona, T. Senjyu, *Proc. ICEE* (2015)
9. N. Good, E. Karangelos, A. Navarro-Espinosa, P. Mancarella, IEEE Trans. on Smart Grid, **6(3)**, 2333-2342 (2015)
10. R. Olfati-Saber, J. Fax, Proceedings of the IEEE, **95(1)**, 215-233 (2007)
11. M.R. Sheikhi, Sh. Bahrami, A.M. Ranjbar, IEEE Trans. on Smart Grid, **6(2)**, 675-683 (2015)
12. A.H. Mohsenian-Rad, V.W. Wong, J. Jatskevich, R.Schober, A. Leon-Garcia, IEEE Trans. on Smart Grid, **1(3)**, 320-331 (2010)
13. H. Chen, Y. Li, R.H. Louie, B. Vucetic, Autonomous IEEE Trans. on Smart Grid, **5(4)**, 1744-1754 (2014)
14. J. Chai, Z. Chen, Y. Yang, Zhang, IEEE Trans. on Smart Grid, **5(2)**, 1340-1350 (2014)
15. Z.M. Fadlullah, D.M. Quan, N. Kato, I. Stojmenovic, IEEE Systems journal, **8(2)** (2014)
16. V.S. Stepanov, K.V. Suslov, E.V. Kozlova, Prom. Energetika, **6**, 2-7 (2013)
17. Resolution of the Government of the RU of 2012 no.442.http://www.sbyt.irkutskenergo.ru/qa/2103.html

Introduction of a method for calculating the investment activity of an enterprise (evidence from wood-processing enterprises)

Olga Polyanskaya[1*], *Veronika* Bespalova[1], and *Viktoria* Sokolova[1]

[1]Saint Petersburg State Forest Technical University, Saint Petersburg, Russia

Abstract. In the current economic conditions, the industry is interested in implementing new innovative processes, products and, consequently, it needs investment. This problem also concerns the woodworking industry that suffers from a lack of modern equipment that allows for a more comprehensive use of the available raw materials and proportional quality of the end product. In order to analyze the situation with each industry (particularly woodworking), an innovative production system is proposed that is based on determining the level of an enterprise using design coefficients. Based on this analysis, an appropriate strategy of the company development can be adopted. The calculated index of innovation activity can constitute a reason for the enterprise to strengthen its innovative potential and carry out efficient innovation activity.

1 Introduction

Forests are often called the lungs of the planet, and most of the forest reserves are concentrated in Russia. To date, Russia owns a quarter of the world's wood reserves [13]. However, the country is far from being the best in global woodwork trading. The forest industry's share in the national industry is less than 4% [18]. To date, the woodwork industry in Russia is characterized by low efficiency, which affects the economic performance of the enterprises accordingly and, in general, does not meet the state's economic interests. There are many factors hindering the development of the forest industry [18]. The major ones are:

- high depreciation of fixed assets;
- low investment activity;
- insufficient degree of processing;
- poor quality of all-weather logging roads;
- low competitiveness of the products;
- inefficient use of the working capital.

The development of the woodwork industry is hindered by the lack of innovative technologies for the exploitation of the raw and waste wood, the average share of which in the raw wood is approx. 45% [12]. These technologies do exist and are widely used outside Russia. For instance, in Sweden, wood processing reaches 98%, and in Finland 97%.

In the past century, waste wood was used for producing chip, sawn lumber, split billets, container assemblies, and other types of end products [21]. Today, this kind of material is used in different, more versatile ways mainly due to the innovative development of the chemical industry. In particular, it is proved that chips can be used to produce sulphate cellulose, and charcoal and activated charcoal can be obtained through thermochemical treatment [4, 12]. However, chip is a popular material for decorating kindergartens and playgrounds and for the production of various composite materials widely used in the furniture and construction industries.

Back in 1971, the USSR forest industry had two main objectives:

- provide for the further development of the woodwork industry, taking into account the complex processing of raw wood;
- expand the use of process feedstock in the paper and pulp industry, chipboards and chip.

2 Relevance

The above objectives remain relevant today.

In order to change the current situation in the woodwork industry, we need to implement an innovative approach to its development as the most efficient variant of providing the home market with domestic production and increase the degree of processing.

This problematic has been studied by the following Russian researchers: Z. A. Dikaya, E. D. Shchetinina, M. V. Egorova, E. P. Amosenko, and others.

3 Research objective

In order to solve the formulated problem, we have developed and suggested a method for calculating the innovation activity index for wood-processing enterprises. The result of implementing the method should be the adoption of competent managerial

* Corresponding author: Polyanskaya_78@mail.ru

DOI: 10.1051/shsconf/20173501023

decisions on introducing innovations. The results of their implementation include the satisfaction of the domestic demand (through improving the marketability of domestic products) as well as increasing the industry's contribution to the economic development of the regions.

4 Theoretical part

Innovating is a set of activities aimed at improving the efficiency of the production process and the entire financial and economic activity of the enterprise in general [2].

In modern conditions, this is mainly done by increasing the competitive performance of enterprises through innovative products and processes [1, 3, 5, 7]. As for the innovative products in the woodwork industry, innovations in assortment policies, raw and processed materials, fittings, and tools can be suggested as the most relevant; possible innovative processes include new production technologies and methods of production management [6, 8, 9].

The main trend that is commonly considered to be hindering the innovation activity in the woodwork industry is the lack of investment in this sector. This is mainly due to the high risk and the long payback period of such investments. However, some types of innovating can be carried out by the enterprises routinely without much additional investment: improving the marketing activities, quality of management, and the application and integration of new methods into the management system of the enterprise.

A study conducted by the researchers of the Chair of Economics and Management of Wood Processing Plants at Saint Petersburg Kirov State Forest Technical University on a number of plywood-producing plants showed that the index of innovation activity is rather low: between 0.298 and 0.395. This highlights the low innovative ability of the enterprises. The situation in the wood processing industry is challenging. Therefore, the Innovative Production System (IPS) taking into account both the specifics of the enterprise and the external environmental conditions was suggested for implementation in several enterprises.

Each company is a unique open economic system with its own peculiarities of operation and production. The innovative changes should, therefore, be implemented taking into account the peculiar features of the enterprise and the requirements of the external environment (the market). The main strategic advantage of any enterprise depends on the continuous improvement of its productive capacities through innovation.

Therefore, the following directions of operation have been identified in the introduction of an innovative system:

1. development and implementation of systematic marketing;
2. research and analysis of new growth areas for the enterprise;
3. improvement of the organizational and management structure;
4. comprehensive processing of wood waste and low grade raw materials;
5. providing a modern material and technical base;
6. updating and maximal use of the production capacity;
7. efficient use of the working capital of the enterprise;
8. introduction of innovations in production and management.

Of the many factors that affect innovation systems in production, the main ones in our opinion are: the productive capacity, the human resources capacity and the investment potential [10, 11, 15, 16, 20, 23].

Qualitative and quantitative analysis of business innovation activity measure the ability of the enterprise to innovate [3, 17, 19, 22].

Based on the data collected on the wood-processing industry, a calculation of innovation activity has been made for several enterprises in the industry featuring the following parameters:

1. Low-waste production.

In the woodworking industry, this parameter is assessed through the rate of comprehensive utilization of material (C_u)

$$C_u = \frac{V_0 + V_d}{V_s} \quad (1)$$

where V_o is the volume of principal products produced from raw wood, m³, V_d is the output from the processing of waste from primary products, m³., V_s volume of raw material written off for production, m³.

Thus, it can be assumed that the rate of comprehensive utilization of raw materials should grow and the cost of comprehensive utilization of raw materials should fall. In terms of productive capacities, to be considered innovative, an enterprise needs to comply with the following two conditions:

$$C_u \Rightarrow 1 \quad (2)$$

$$\frac{V_n}{V_c} \Rightarrow max \quad (3)$$

where V_n is the profit before corporate income tax and interest costs,

V_c stands for the costs of production.

2. Availability of the staff engaged in research and development (U_s):

$$U_s = \frac{q_{rd}}{q_{ip}} \quad (4)$$

where q_{ip} is the total number of industrial personnel; q_{rd} is the number of employees engaged in research and development.

There are no restrictions on this parameter. It is the enterprises that decide how many employees should engage in research and development.

3.The share of equipment engaged in innovation in the total equipment (U_{eq}).

In calculating this parameter, it should be noted that, in accordance with accounting rules, the cost of the equipment is reflected with respect to its residual value. If the company acquires new equipment that is going to be directly connected to innovation activity, its cost is going to be relatively high. Therefore, in order to prevent distortion of this indicator, it is recommended to account for all the equipment with respect to its depreciation over the accounting period using the index method.

$$U_{eq} = \frac{F_{inn}}{F_o} \quad (5)$$

Where F_{inn} is the value of the property used for testing and research; F_o is the total cost of machinery and equipment used in the manufacturing of the products.

4.The share of intangible assets connected with implementation of innovation in the total intangible assets of the enterprise U_{ia} .

In calculating this parameter, the comments on the cost and depreciation in p. 3 should be taken into account; the index method should be used for calculation.

$$U_{ia} = \frac{F_{ia}}{F_{onm}} \quad (6)$$

where F_{ia} is the cost of intangible assets employed in innovation; F_{onm} - total value of intangible assets in the enterprise.

5.Availability of funds for the development of innovative activities (U_{inv})

$$U_{inv} = \frac{U_o}{U_{tot}} \quad (7)$$

where U_o is the availability of the enterprise's own funds; U_{tot} is the total amount of investment necessary for the implementation of the innovative project.

The innovation development activity index is the sum of all the five parameters:

$$U_{inn} = \sum (C_u + U_s + U_{eq} + U_{ia} + U_{inv}) \quad (8)$$

The following numerical intervals of the innovation development activity index can be used to describe and further choose the innovation strategy: If the innovation development activity index:

- ranges from 2.41 to 3, it is the advancing strategy;
- ranges from 1.7 to 2.41, it is the defensive strategy;
- ranges from 0 to 1.7, use the imitation strategy.

It should be noted that, for calculating the innovation activity index in the enterprises outside of the woodworking industry, the formula (8) can be written as follows:

$$U_{inn} = \sum (U_s + U_{eq} + U_{ia} + U_{inv}) \quad (9)$$

The numeric intervals of the innovative activity index in this case are going to be somewhat different:

- ranges from 1.41 to 2, the advancing strategy;
- ranges from 0.7 to 1.41, the defensive strategy;
- ranges from 0 to 0.7, the imitation strategy [3, 10, 11].

The formula (9) can include additional parameters depending on the industry of the enterprise. However, its interpretation should be studied and calculated by experts in the respective industry, because the numeric intervals of the index of innovative activity will also be different.

5 Practical significance

Based on the research method suggested above (formula No. 8), the professors of Saint Petersburg Kirov State University analyzed the data of a number of enterprises. All the studies were conducted on the data of existing enterprises, and for information security purposes, the enterprises are going to be presented as A, B, C, D, E, F (A is engaged in furniture production; B in sawing production; c in furniture manufacturing; D in plywood production; E in woodworking (joinery)) [10, 11].

The enterprises that provided the data for the study are typical representatives of the woodworking industry, so let us examine the results of the study as an established pattern.

According to study data and the performed calculations, the integral indices of innovation activity are as follows: Table 1.

Table 1. Determination of the enterprise activity index.

Parameters	Symbol	A	B	C	D	E
1	2	3	4	5	6	7
1. Number of industrial personnel, persons	q_{ip}	88	62	615	471	43

2. Number of employees engaged in research and development, persons	q_{rd}	6	5	31	21	4
3. Cost of fixed assets, thousand rubles	F	24.517	8.901	21.00	17.986	13.27
4. Cost of property used for research and development, thousand rubles.	F_{inn}	213.8	73.6	195.3	180.9	437.8
5. Cost of intangible assets, thousand rubles.	F_{ia}	146.3	91.8	85.7	92.3	117.9
6. Amount of investment necessary, thousand rubles.	U_{tot}	1015.5	916	1200	1315.7	2313
8. Availability of staff engaged in research and development	U_s	0.07	0.08	0.05	0.04	0.09
9. Share of equipment used for innovating	U_{eq}	0.009	0.08	0.09	0.01	0.033
10. Share of intangible assets	U_{ia}	0.06	0.01	0.004	0.005	0.009
11. Availability of own investment sources	U_{inv}	0.31	0.128	0.213	0.261	0.178
12. Rate of comprehensive utilization of material	C_u	0.83	0.71	0.79	0.82	0.76
13. Innovation activity index	U_{inn}	1.225	1.008	1.147	1.136	1.07

On the basis of the proposed simulation model, the following conclusions can be made.

6 Conclusion

The study results indicate that the innovation activity index in the enterprises of the Russian Federation woodworking industry is at a fairly low level and match the simulation strategy for the development of innovation in the industry.

The proposed innovation development activity index for woodworking enterprises includes all the basic parameters that can be used to evaluate the innovation development of the enterprises in the relevant industry. Excluding the rate of the comprehensive utilization of the material from the proposed formula provides us with an integrated index of innovation activity for any enterprises.

Introduction of the innovative production systems in the future will enable the enterprise to move to the defensive and even the advancing strategy, which will promote the efficient development of not only the woodworking industry, but also the country as a whole.

References

1. E.P. Amosenko, V.A. Bazhanov, The Region: Economy and Sociology, **450** (2006)
2. S.D. Arshlanov, Fundamental Studies, **9-1**, 111-115 (2015)
3. V.V. Bespalova, O.A. Polyanskaya, Herald of Belgorod Shukhov State Technological University, **3**, 183-186 (2016)
4. S.K. Goroshko, *All-Soviet Science, Research and Design Institute of Economics, Industrial Management and Information in Forestry, Paper and Pulp and Wood Processing Industry*, **36** (1974)
5. T.N. Danilova, V.A. Grishchenko, Regional Economics: Theory and Practice, **5**, 78-80 (2015)
6. A.G. Devyatkin, Approbation, **9(36)**, 57-59 (2015)
7. Y.A. Doroshenko, Belgorod State Technocal University Publishing, 133 (2012)
8. Y.A. Doroshenko, S.M. Bukhonova, E.I. Tomilina, A.P. Taburchak, Herald of Belgorod Shukhov State Technological University, **1**, 130-133 (2007)
9. M.V. Yegorova, Regional Economics: Theory and Practice, **5**, 80-86 (2007)
10. La Van Thanh, Z.A. Dikaya, The Science and Technical Journal of the Saint Petersburg State Forest Technical University, **6(65)**, 55-56 (2008)
11. La Van Thanh, Z.A. Dikaya, Modern Aspects of Economics, **10(123)**, 15-20 (2007)
12. L.P. Mayorova, *Rational use of timber as the ecological and chemical basis for the operation of the forest sector (evidence from Khabarovsk Region)* (2011)
13. V.I. Patyakin, *The Forest Sector of the XXI Century: Materials of the IVth International Forum*, 130-132 (2002)
14. I.V. Somina, Social and Humanities Insights, **8**, 294-301 (2012)
15. V. Somina, Herald of Belgorod Shukhov State Technological University, **4**, 142-145 (2013)
16. M.S. Starikova, Herald of Belgorod Shukhov State Technological University, **6**, 122 (2014)

17. A.V. Sukhorukov, News of Saratov University. New Series. Series: Economics. Management. Law, **14**, **11-3(52-3)**, 437-443 (2014)
18. N.A. Tyurin, I.V. Griroryev, O.I. Grigoryeva, Relevant Areas of Scientific Research in XXIst Century: Theory and Pracice, **2**, **3(7-3)** 2014. DOI: 10.1273/3266.
19. I.I. Shanin, Issues of Innovative Economy, **4(14)**, 30-39 (2012)
20. E.D. Shchetinina, N.V. Ovcharova, Herald of Belgorod Shukhov State Technological University, **3**, 159-163 (2016)
21. V.I. Yagodin, V.A. Sokolova, News of Saint Petersburg State Forest Technical University, **188**, 246-252 (2009)
22. Ketels C. Sweden's, Globaliserings forum Rapport, **2** (2012)
23. R. Nelson, *National Innovation Systems. A Comparative Analysis* (Oxford Univ. Press, Oxford, 1993)

Salary administration as part of employee incentive system at industrial enterprises

Michail Sagandykov[1,*], and *Galia* Shafikova[1]

[1]South Ural State University (National Research University), Department of Labour and Social Law and Jurisprudence, Chelyabinsk, Russia

Abstract. The article analyses the financial staff incentives at industrial enterprises. The paper concludes that the incentive system should be developed taking into account the developed staff motivation. The only efficient target of influence is the existing staff motivation components. A 3-element employee incentive framework for industrial enterprises is suggested. The article presents the critical analysis of the existing payroll systems of several industrial enterprises, elicits the common flaws of the incentive payments, and develops requirements for such systems. The paper also provides an industrial enterprise needs matrix and an incentive payment types matrix required to build up a rational monetary staff incentive system. These matrices can be adapted for any industrial enterprise with regard to its long-term objectives.

1 Introduction

The present state of the Russian industry can be described as a state of a systemic crisis caused by multiple international, national, political, social, and economic factors. One of them is the minor involvement of the Russian industry in the innovative processes.

Meanwhile, along with R&D and technology the innovations developed at the industrial enterprises should affect the enterprise management system [1], with human resource and staff motivation management systems in particular.

Human resource management should be viewed as a special form of management which deliberately impacts the staff using a specific apparatus of means and methods. The framework of methods is shaped by the fact that the major aim of the staff involved in production is satisfying their personal needs in the long term while being employed at a certain enterprise. By controlling the staff behavior, the management develops a system of motivational drivers, working to boost the productivity and, ultimately, the production performance [2].

2 Analysis of the notions of motivation and staff incentives. Literature review

Quite a lot of theories of motivation were developed over the XX century. Even the shallow analysis of these theories allows for the conclusion that they are based on either one of two groups of factors, i.e. the system of human needs, prompting the human actions (content theories), or the peculiarities of human behavior in certain conditions to achieve some particular aims (process theories) [3].

Quite an arguable [4] theory of human motivation, though the most widely known one and the one applicable to the industrial enterprises [5], Maslow hierarchy of needs, features physiological, safety, love and belonging, esteem and self-actualization needs [6], which are not always in direct correlation with work activities.

Herzberg's Two Factor Theory was widely used in behavioral analysis in industry, both nationally and internationally [7]. According to F. Herzberg the working conditions existing in a company create the basis for the positive perception of labor. The strive for self-actualization, in its turn, provides for the development of an employee's system of motives, determining the performance [8]. In our opinion, such way of influencing the employee motivation is particularly valid for the innovative industrial enterprises, which should strive to boost the creative component in the work of their highly-qualified staff.

At the same time, several experiments showed that enriching labor caused a dramatic increase in the levels of the staff satisfaction with their work, boosting their internal motivation, decreasing absenteeism and personnel turnover, though it did not result in improved performance. This result was particularly due to the most of the employees approaching their work as an instrument to achieve their personal aims. When the employees found out that enriching the labor does not entail the increase in salary and career promotion, the effect of the implemented changes deteriorated considerably [9].

* Corresponding author: mss_lawyer@mail.ru

Employers should understand that the system of employee motivation should be based on their personal needs, most of which are connected with the spheres of life other than work. The external employer's impact on the internal employee motivation, however elaborate, does not always result in achieving the main aim, which is the improved performance.

The most famous theories of staff motivation indicate the direct connection between the management process and the personal mental condition of a member of staff. This, in its turn, underlies the necessity to coordinate the aims of the company management and the motivation of the staff [2]. This coordination should be provided for by the employer's stimulation of the required employee behavior.

L. Mikhailova is right to mention that the mechanisms of staff *stimulation* and motivation are seriously different, while these notions are often unreasonably mixed up or substituted with one another. This results in the lack of efficiency of the motivation and stimulation systems [10].

Some authors either consider these notions as synonymous; or the two types of one and the same process [11]; or address motivation as one of the stimulation methods [12], which is quite difficult to agree with [13]. For instance, A. Kibanov views motivation as a process of the long-term influence on the employees aimed at changing their value systems, aims, and interests to develop their work potential based on that; while he understands stimulation as a tactics used to resolve the issue, i.e. orientation to the existing structure of the employees' values and interests and the efficient implementation of the employees' potential [14].

According to M. H. Mescon "motivation is the process of stimulating oneself and others for the activity aimed at achieving particular and general aims of an organization" [15].

In our opinion, labor motivation and stimulation are obviously interconnected, though different in their nature. Moreover, the fact that motivation describes the attitudes of a person, while stimulation is about the external impact onto a person, is not the only difference.

Labor motivation cannot be viewed as a process or mechanism, it should be perceived as a system of personal attitudes, based on one's needs, expressed in the attitude towards labor.

3 Employee incentive system of a modern industrial enterprise

Equations should be centred and should be numbered with the number on the right-hand side here exist different methods an employer may use to influence the employees' motivation, thus developing the attitude to work required in the short or long term. Such impact is the employee behavior stimulation.

It is worth mentioning again that any employee's with very few exceptions primary concern is their immediate needs, and not boosting the performance. This is a factor the employers have to consider.

So, the employee incentive system should be developed taking into account the existing employee motivation. Only the aspirations already present in the system of motivation can be efficiently impacted. If there is none, any forms of stimulation will fail to produce the necessary impact on the employee behavior in the framework of production

For instance, it is impossible to seek innovative approach to the work from an employee having no such developed motivation. In this case, one might use, for instance, professional re-training to motivate the staff demonstrate the required behavior [16], and only after that stimulate this behavior, by influencing the newly emerged motives.

In other words, the stimulation /incentive/ system should encompass three elements: employee motivation identification, development of the employees' needs, related to the company aims, and, basically, the stimulation of behavior aimed at the realization of these needs in the company interests (Fig. 1).

Furthermore, developing a community of employees with the motivation matching the company aims can be done in two ways: by stimulating the need for the work to emerge in the employees, or by high-quality selection of staff already having the necessary motivational affirmations.

 DOI: 10.1051/shsconf/20173501024

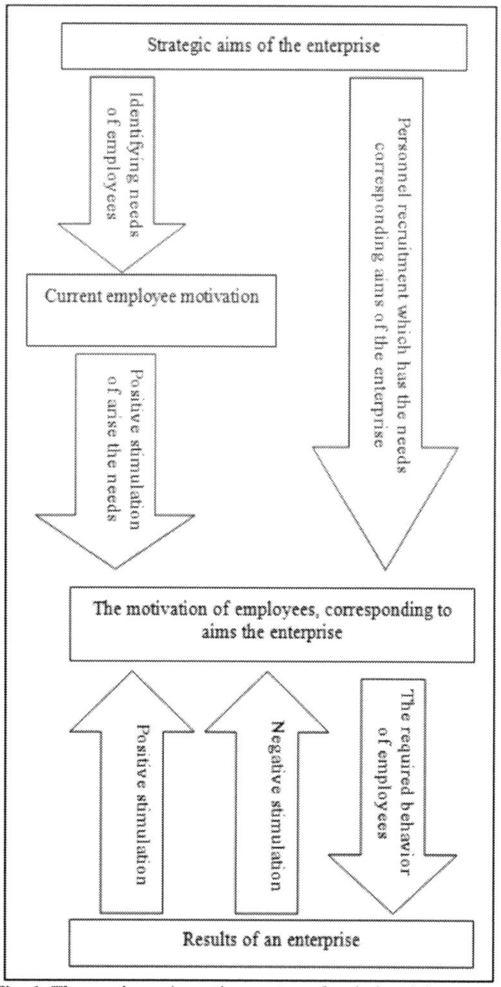

Fig. 1. The employee incentive system of an industrial enterprises.

It was mentioned hereinabove that stimulation is often understood as exclusively positive influence on the employees [2].

It is known that a *stimulus* means *"an incentive, a push"* [17]. We regard the employee liability or other modes of coercion in the cases when the staff fails to fulfill the employer's requirements, as one of such reasons. We suggest calling it *negative stimulation.*

At the same time, S. Ozhegov and N. Shvedova suggest understanding *stimulus* as "an interest in doing something" in the cases, connected with certain "development" [17].

In other words, to develop the needs required for the strategic development of the company in the employees, employers should use positive financial and non-financial motivation. Developing the employee needs related to the company aims with coercion is quite difficult.

At the same time, stimulating the staff behaviors aimed at achieving the performance indices and boosting performance, can be both positive and negative.

4 Flaws of the exisiting financial motivation system and suggestions on their improvement

Online references will be linked to their original source, only if possible. To enable this linking extra care should be taken when preparing reference lists Despite that many companies use non-financial incentives, payroll ones still play the main role in the staff behavior stimulation system.

The analysis of the payroll and other bylaws of several industrial enterprises identified certain features, illustrating the flaws of the financial incentive systems.

Most of the payroll systems are not aimed at the main idea of stimulation, meaning that they lack a correlation between the salary and the achievement of the company strategic goals, while some of the goals are just set incorrectly. The stress is mostly made on the fulfillment of the current performance indices, which does not reflect the connection to the strategic aims.

For instance, the main aims of a major Russian metallurgical enterprise are as follows: "meeting the rising expectations of the consumers in full, balanced accounting for the shareholders' requirements, the interests of the staff and the society in general, bringing down the enterprise-induced environmental load, emergencies and accidents prevention". Some of the quoted aims can hardly be classified as strategic. Besides, the Compensation and Benefits Policy and other bylaws of the company contain no actual stimuli aimed at building up the staff motivation to achieve these aims. The size of compensations largely depends on the current outcomes, particularly, the volume of work, without regard to its quality, innovative development, health and safety.

A matter of the largest and unresolved discussions is balancing the three groups of interests, i.e. societal, collective, and personal, in the employee motivation system framework [3].

The example above shows that the company aims at balancing corporate interests with the interests of the staff and the work conditions; while none of the bylaws explains the way this aim shall be achieved. For instance, no correlation between the salary and the company profit is established.

The salary structure used at almost all of the enterprises does not do anything to stimulate the necessary employee behavior. The salary in general, not some part of it usually called *incentive payments* should perform a stimulating function.

Putting excessive stress on the variable salary component, bonuses, interest, and other incentives can result in pushing the employees to achieve short-term results without relating them to the strategic aims.

The payroll system is mostly aimed at the activation of the motives existing in the employees, and not building up the motivation required for the company to achieve its strategic aims. Such system can be efficient only provided that the staff is thoroughly selected and already has the necessary motivation, which is very complicated to achieve.

The analysis of several compensation and benefits policies indicates that the company employs an irrational approach to the development of certain stimuli, i.e. the bonus payment criteria.

According to A. Kopytova, "the more constituents an incentive system of the company features, the more loyal the staff is going to be loyal to the company, value their positions, and perform the work better" [18]. Along with that the author brings out the direct dependence of the staff behavior stimulation costs on the number of the stimulation criteria.

While we agree with this view in general, it is worth mentioning that this approach is hardly valid for the payroll incentives.

Employers, with only few exceptions, have limited budget for bonuses and other incentive payments. Consequently, the increase in the number of the employee stimulation criteria diminishes the effect and the importance of every one of them.

In many cases the incentive payments system is impractical, as it contains many stimuli having no actual impact on the employee motivation.

The analysis of the needs identification matrix suggested by A. Kopytova, and the possibilities for the implementation of the rational employee implementation system at the enterprise [18] allowed suggesting a matrix of stimuli, featured in an enterprise payroll system.

The first matrix (Table 1) aims at forming the employee behavior stimulation in relation to the long-term tasks of the company. Naturally, each organization has its own list of such tasks and profiles. Moreover, such matrix allows setting the priority short-term tasks and changing them depending on the changes in the long-term tasks and the current situations. For instance when the dramatic innovative development of the enterprise is required, promoting the qualifications of the staff shall become a matter of the primary importance.

Table 1. The matrix of needs of an industrial enterprises.

The long-term task	The direction of the implementation of tasks	The current need
Increased productivity	Increased intensity of labor	Yes
	Improving product quality	Yes
	development of new technologies and equipment	Yes
	Rationalization proposals	No
Human resource development	Stability of the staff	Yes
	Fight against absenteeism	Yes
	Staff development	Yes
Safety	Compliance with safety and health requirements	No

The analysis of the needs identification matrix suggested by A. Kopytova, and the possibilities for the implementation of the rational employee implementation system at the enterprise [18] allowed suggesting a matrix of stimuli, featured in an enterprise payroll system.

The first matrix (Table 1) aims at forming the employee behavior stimulation in relation to the long-term tasks of the company. Naturally, each organization has its own list of such tasks and profiles. Moreover, such matrix allows setting the priority short-term tasks and changing them depending on the changes in the long-term tasks and the current situations. For instance when the dramatic innovative development of the enterprise is required, promoting the qualifications of the staff shall become a matter of the primary importance.

The second matrix (Table 2), in its turn, allows characterizing certain incentive payments, aimed at the implementation of the current tasks of the company. The matrix can serve as a basis to forecast the efficiency of certain incentive payments depending on the target staff and the available budget.

In our opinion, if an company in interested in special stimulation of certain groups of employees, quite limited in terms of their size, for instance, the staff involved in innovative activity, a special incentive system should be developed for them, instead of just introducing additional incentives on top of the existing ones.

The lack of differentiation of the incentive systems depending on the tasks fulfilled by the employees is one of the issues of human resource management at modern industrial enterprises.

An employer will also be armed to make a conclusion that certain payments, though meant for all of the employees, will not me successful, as the budget allocated for them is limited.

Incentive payments, which the employees are not able to earn, or which, on the contrary, require almost no effort to gain and have no considerable material form, can have negative impact on the employee behavior.

We reckon that the payroll system should not feature incentive payments for every possible occasion, otherwise the employees shall think any their actions require additional payment.

Table 2. The matrix of types (criteria) incentive payments.

The direction of the implementation of tasks	Types (criteria) payments	The proportion of staff, for which is intended the payment
Increased intensity of labor	Implementation of the plan	70%
	Increased labor standards	60%
	Execution of works of different qualifications	5%
Improving product quality	Absence of spoilage	50%
	Execution of works requiring high qualification	20%
Development of new	Participation in the implementation process of	1%

The direction of the implementation of tasks	Types (criteria) payments	The proportion of staff, for which is intended the payment
technologies and equipment	equipment	
	Work on the modern equipment	10%
Stability of the staff	Continuous length of service in the enterprise	100%
Fight against absenteeism	Lack of absenteeism and tardiness	100%
	Lack of absence from work for valid reasons	100%
Staff development	Refresher training without interrupting production	25%
	Additional professional training	5%

The efficiency of payroll in terms of stimulation depends on the conditions it is implemented in.

F. Herzberg stated that the flaws of the motivation system may be caused by the unfair or poorly organized working conditions [19].

Companies should use the incentive payment criteria, which allow for the realization of various abilities of the employees, as well as allowing all of the employees to seek these payments based on their basic qualities, i.e. the level of education, work experience, the complexity of the work performed.

The quantitative evaluation of the performance indices, which should be planned and, more importantly, controlled outside the units, is preferable. A system of objective and immediate evaluation of the staff performance should be in place [18].

The duties of the members of staff should be clearly defined based on the position description and other documents. Consequently, the payment incentives rewarding the achievement of individual outcomes should be based on the clear performance of an employee's duties, being independent of the subjective opinion of the head of the unit.

Finally, the compensation system, including the incentive payments, should be as transparent, open, and clear as possible. While working on this article, we found out that at most of the enterprises the policies and bylaws related to compensation are for official use only, being unavailable for the usual members of staff. They are not published on the company web-site, or posted at the information stands. With that in place one can hardly talk about the stimulating function of the salary.

5 Conclusions

All of the existing staff motivation theories indicate the connection between the management process and the psychological state of an individual, or, in other words, the aims of the company and the needs of the employees. Thus, the stimulation system should be developed with regard to the existing employee motivation, as well as the need for balancing three groups of interests, i.e. those of the staff, management, and the company on the whole.

Any employee stimulation, including that by means of salary, should be based on the aims of the company classified into current (short-term) and strategic (long-term).

The payroll system of an industrial enterprise should meet certain requirements.

Firstly, the incentive payment system is defined by the correlation between salary and the achieved strategic aims of the company.

Secondly, the compensation system should balance the fixed and incentive components. The global trend in the development of the compensation system is "the increase in the share of the fixed component of the salary, which results in increased compensation system reliability and the growing assurance of staff in the guaranteed income" [20].

Thirdly, the employee behavior stimulation system should be rational, meaning be based on the direct correlation between the company requirements, the available budget, the corresponding financial stimuli, and, finally, the needs of the employees.

The matrix of the needs of industrial enterprises and the matrix of the incentive payments types (criteria) do not present the exclusive lists of either notions and should be adapted for a certain enterprise with regard to its long-term development aims.

Finally, the prerequisites for the financial stimulation system efficiency are fairness, openness and transparency of the compensation system.

The work was supported by Act 211 Government of the Russian Federation, contract № 02.A03.21.0011.

References

1. A.A. Shchiptsov, Vector of Science of TSU. Series: Economy and Managment, **1**, 141 (2013)
2. N.V. Lobaskova, *Improving the motivation of management personnel in industrial enterprises* (Nizhny Novgorod, 2012)
3. V.I. Gurov, and E.V. Sukmanov, Bulletin of the Kusk State Agricultural Academy, **2**, 72 (2008)
4. A. Neher, JHP, **31**, 3 (1991)
5. M.F. Conforti, Occupational Health Nursing, **20**, 12 (1972)
6. A. Maslow, *Motivation and personality* (St. Petersburg, 2011)
7. F. Herzberg, ILR Review, **18**, 3 (1965)
8. E.P. Kostenko, Terra Economicus, **10**, 115 (2012)
9. C. Orpen, Human Relations, **32**, 3 (1979)
10. L.L. Mikhaylova, Motivation and Salary, **2**, 78 (2011)
11. S.Yu. Shamray, Theory and Practice of Social Development, **4**, 316 (2011)
12. V.V. Rassadin, Labor and Social Relations, **3**, 53 (2008)

DOI: 10.1051/shsconf/20173501024

13. N.A. Sharipova, and D.V. Buss, Siberian Commerce and Economic Journal, **2**, 47 (2016)
14. A.Ya. Kibanov, *Personnel management of the organization: textbook* (Moscow, 2005)
15. M. Meskon, M. Albert, and F. Khedouri, *Fundamentals of management: translated from English* (Moscow, 2008)
16. K. Evans, and E. Waite, Europien Review of Labour and Research, **16**, 2 (2010)
17. S.I. Ozhegov, and N.Yu. Shvedova, *Dictionary of Russian language* (Moscow, 2003)
18. A.V. Kopytova, Modern problems of science and education, **1-1**, 691 (2015)
19. V.A. Krisanov, *Management motivation of personnel in industrial enterprises (by the example of metallurgical enterprises)* (Izhevsk, 2005)
20. S.E. Prokof'ev, A.M. Belyaev, and S.G. Eremin, *Modern technology of staff in government bodies: a monograph* (Moscow, 2015)

The interaction problems between large and small business in modern conditions

Mikhail Belyaev[1], *Svetlana* Sokolova[1*], and *Svetlana* Karpushova[2]

[1]The department of economics and management of projects in the construction industry, Volgograd State Technical University, Volgograd, Russia
[2]The department of economics and finance, Serebryakovsky branch of Volgograd State Technical University, Mikhaylovka, Russia

Abstract. The development of market relations, changes of the conditions in the business environment encourage the enterprises to look for new management methods and to improve forms of interaction. In this regard, the identification of the interaction of large and small businesses, as well as the evaluation of their relation development seems important and urgent problem in modern conditions. The purpose of the survey – the study of the interaction of large and small businesses, as well as the evaluation of the relations between them. The study was conducted on the basis of a comprehensive and systematic approach in which methods of comparative, retrospective, statistical, mathematical analysis are used. In accordance with the purpose there are identified the prerequisites for the development of large and small businesses, features and their functioning problems, and the links between them. The most common form of interaction of large and small enterprises were identified - outsourcing, franchising, leasing, subcontracting, venture financing, the establishment of regional cooperation forms of large and small businesses. However, the cooperative processes of large and small business in Russia developed are not enough today.The authors identified factors that impede the growth of Russian production, offered recommendations for the development of large and small businesses, justified the state's role in this process. In addition, they described the mechanism of state support of small business, including organizational, financial, information and consulting components.

1 Introduction

The transfer of the Russian economy on an innovative way of development and steady growth of large industrial enterprises cannot be without the cooperation with enterprises which are relatively smaller. Such cooperation is beneficial to large and also small businesses. Also in the conditions of market relations the level of development of small business is one of the key indicators characterizing the maturity and stability of the society. In economically developed countries, the share of small enterprises in the structure of the gross domestic product increases the share of large enterprises and is more than 70%. In our country, this index is only 20%. In this regard, stimulating of the small business development, the definition of the relations between large and small businesses, improving their interaction in a particular area is particularly important in the modern conditions.

At various times, these problems of the industry have been studied by R. Nurks, N. Buchanan, A. Lewis, V.V. Ananishnov, M.A. Busygin, N.I. Dorogov, N.N. Koltsov, A.M. Lyalin, V.E. Rokhchin, Y.V. Trifonov and others. To the study of entrepreneurship development problems P. Druker, R. Cantillon, C. Marks, A. Marshall, L. Mises, A. Smith, Z.B. Sei, J. Schumpeter and so on addressed. The interaction of small and large businesses was studued in the works of Russian scientists G.L. Azoev, T.A. Alimova, A.N. Asaul, G.L. Bagiev, A.V. Busygin, E.M. Buchwald, D.N. Zemlyakov, G.B. Kleiner, S.P. Kusch, M.G. Lapusta, M.E. Omarov, B.P. Sapsay, S.A. Sosna, Y.L. Starostin and others. Among foreign scientists V. Anand, J. Brickley, F. Dark, Z. Dilthey, F. Lafontaine, V. Murphy, Y. Murray and others addressed to this problem. However, despite the large number of works dedicated to the problems of the functioning and development of industrial enterprises, including small, the features of large and small businesses interaction are studied insufficiently, a single methodology is absent to assess their link. The research is aimed at studying of the interactions of large and small businesses, as well as evaluating the links between them.

* Corresponding author: svetalekssok@yandex.ru

2 The features of the interaction of the large and small business

The international experience of the industry development confirms the feasibility of the cooperation of large and small businesses. Currently, the most intensive dissemination of this process is in the Eastern countries. Large industrial plants often order the details and components in small enterprises located usually nearby. It promotes sustainability and stability increase of small business, and also allows big business to solve more complex, large-scale and important tasks. As a result, you may experience a synergistic effect. The intensity and nature of the cooperation relations depend, primarily, on the industry sector of the enterprises, on the level of development and structure of the regional economy and the country as a whole.

The need for small businesses to interact with big business is heavily dependent on the sphere of its activities. The companies are more affiliate in providing various services. Today among Russian small businesses repair of motor vehicles, household goods, etc. (39.7%), wholesale trade (24.7%), transactions with real estate, renting and provision of services (20.4%), building (11.0%), retail trade (10.6%), manufacturing (9.5%) are particularly spread.

The analysis showed that at the present time, the amount of small businesses in our country is several times smaller than the subjects of big business, this is contrary to international practice. This trend is typical for the whole territory of Russia. The calculated indexes of small and large enterprises ratio, as well as the industrial production indexes in the Southern Federal District (SFD) are shown in Table 1.

Table 1. The dynamics of small and large enterprises ratio indexes, indexes of the inductrial production in the Southern Federal Disrtict of Russia.

Names of the indexes	Large and small enterprises ration indexes in the SFD of Russia					Sparklines of the enterprises ratio index	Industrial production index in the SFD of Russia					Sparklines of the industrial production indexes
	2012	2013	2014	2015	2016		2012	2013	2014	2015	2016	
Republic of Adygea	0.51	0.51	0.53	0.53	0.54	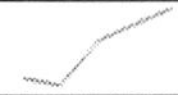	109	103	119	124	129	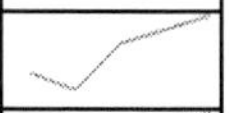
Republic of Kalmykia	0.12	0.18	0.20	0.25	0.33	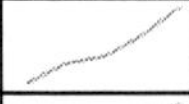	90	108	103	110	116	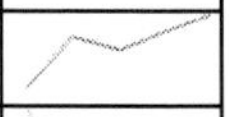
Krasnodar region	0.40	0.44	0.44	0.46	0.48		108	102	105	104	102	
Astrakhan region	0.51	0.50	0.53	0.55	0.56		119	116	103	95	87	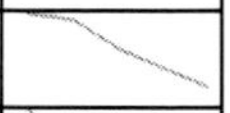
Volgograd region	0.51	0.51	0.52	0.52	0.52	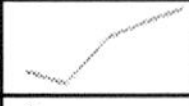	105	102	102	101	99	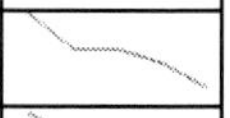
Rostov region	0.64	0.62	0.61	0.59	0.57		108	106	105	104	102	

In Russia, the widespread of cooperative links is characterized for science and scientific services, construction and industry, where the share of cooperating companies is about 71%, 50% and 49%, respectively [1]. In general in our country the processes of interaction of large and small industrial enterprises are not sufficiently developed. However, in conditions of instability and crisis, the interaction of large and small businesses is necessary to significantly improve the use efficiency of workforce, financial, material, information, and time resources. It is possible to flow the time or other resources between entities in both directions.

In the current economic conditions in our country, there are often situations when small businesses can provide the workload of production capacity in large industrial enterprises. With intellectual property, innovative technology the small businesses often have not the necessary equipment and other fixed assets for the commercialization of innovations. This reallocation of resources can improve the financial condition of large enterprises that have received the order. But at the same time, such "reverse" interaction can lead to an increase of expenses and increase in the cost of production due to individual orders and rejection of mass production. If the timely technical upgrade is not implemented in the large industrial enterprises will, it will lose competitive advantages and may even go bankrupt in the long term.

Currently, the main forms of interaction of large and small enterprises in Russia are: outsourcing, franchising, leasing, subcontracting, venture financing, the establishment of regional cooperation forms of large and small businesses. Effective implementation of the above-mentioned forms of interaction of enterprises is possible with the active participation of the state as a regulator that provides the creation of the necessary institutional conditions.

3 The evaluation of large and small bisiness development relations (for example SFD in Russia)

To assess the state of large and small businesses at the regional level may be based on quantitative and qualitative criteria. The authors suggested that there is a link of large and small businesses states. The correlation and regression analysis was used to establish this relation. The state of the industry has been studied in the Southern Federal District of Russia for ten years (2006-2015), and the analysis was implemented according to the following indexes characterizing the small business:

- the number of small businesses per 1,000 people of working-age population in the region;
- the turnover of small businesses per 1,000 people in the region;
- the proportion of employees in the small business objects of total employment in the region, and others.

The following indexes were used to determine the state of a large industrial business:

- the volume of industrial production per 1000 people in the region;
- the number of existing large industrial enterprises per 1,000 people of working-age population in the region;
- the volume of investment in fixed assets per 1000 people in the region;
- the proportion of employees in the large industrial enterprises of total employment in the region, and others.

The correlation link was found between the different indexes as a result of the analysis. It was found that there are more linked indexes such as the volume of industrial production per 1000 people in the region and the number of small business objects per 1,000 people of working-age population in the region. The results of regression analysis of these indexes are presented in Table. 2.

Table 2. The closeness evaluation of small and large indexes link in the South Federal District of Russia.

Names of the indexes	The correlation coefficient	The regression equation	The standard error of the regression
The South Federal Dictrict of Russia	0.97 (very high)	$y = 0.02x + 98095$	2637.52
Republic of Adygea	0.95 (very high)	$y = 0.01x + 2846$	56.98
Republic of Kalmykia	0.63 (remarkable)	$y = -0.03x + 2097$	336.38
Krasnodar region	0.43 (low)	$y = 0.003x + 2097$	2220.14
Astrakhan region	-0.23 (virtually non-existent)	$y = -0.002x + 9494$	460.53
Volgograd region	0.52 (remarkable)	$y = 0.01x + 21872$	1373.35
Rostov region	0.82 (high)	$y = 0.07x - 1364$	6278.31

Thus, there is a connection between above shown indexes of large and small businesses preferably from remarkable to very high. We believe that a very high indexes link in the Republic of Adygea is explained, first of all, by a large number of small businesses in the service sector that provide the needs of large businesses. The reason of low link in certain regions (Krasnodar region, Astrakhan region) is the structure of the regional economy, a large proportion of small enterprises, which operate independently and do not need in active cooperation with big business at the present stage of their development.

Cheddok`s scale was used for qualitative assessment of link closeness. The statistical evaluation of pair correlation coefficient was carried out by comparing it with the absolute value of the critical index when a significance level α corresponds to 0.05, and the degree of freedom f is equal to 8. If $r_{calc} \geq r_{crit}$, then with a given degree of probability (95%) it can be affirmed that there is a close linear link between these number sets. The importance of pair correlation coefficient was also investigated on the basis of Student's t-criteria for small sample size.

4 The fundamentalsof large and small bisiness development

The analysis of scientific literature and materials, presented in the media, allowed to identify factors limiting production growth in Russia and the cooperation development between large and small enterprises, including:

- lack of demand for the products of the domestic market;
- insufficient demand for the products of the foreign market;
- competing imports; the high level of taxation;
- deterioration and lack of equipment; the uncertainty of the economic situation;
- a high percentage of commercial loans;
- lack of financial resources; lack of skilled employees;
- lack or inadequate of legal and regulatory base.

The study showed that the most serious problems in modern conditions are associated with high levels of taxation (54%), lack of demand for the company's products in the domestic market (53%) and the uncertainty of the economic situation (39%). In such conditions, it is important to realize the state support to industrial enterprises, and especially small businesses. When it is inactivity from the government stagnation can create the serious problems of economic development. State support for small enterprises has great importance for the interaction development between large and small businesses. Small business development Program should define the purposes, objects and ways of achieving them, the main indicators, should provide a comprehensive state support of this sector of the economy. The mechanism of state support of small business is schematically shown in Fig. 1.

It should be noted that a large business has become more automated and less labor intensive, that leads to the release of people. These people should be able to become self-employed, income-generating.

Unfortunately, in our country, small business activity is not always perceived by the public as a prestigious activity, there is underdeveloped entrepreneurial culture, increase of which will form civilized connections between large and small businesses, as well as to multiply the resource potential of the country.

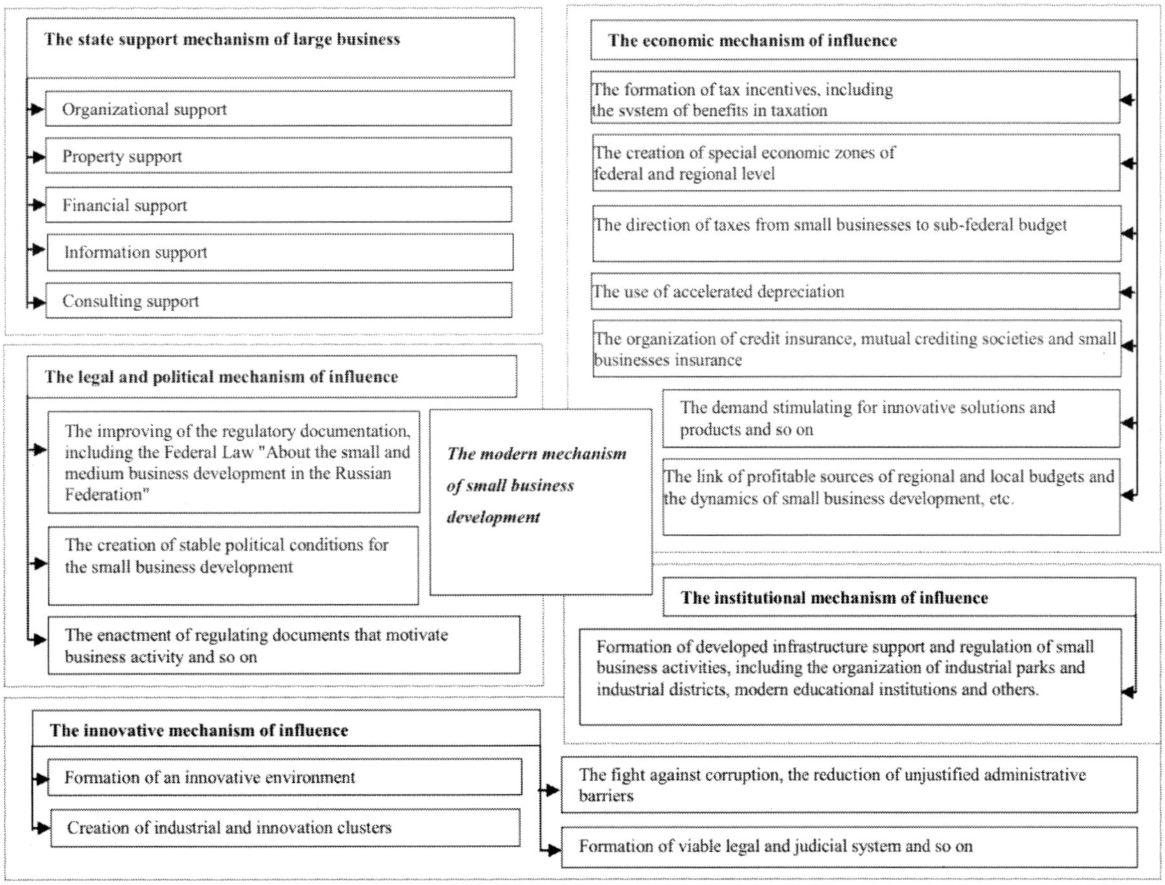

Fig. 1. The mechanism of the small business development.

5 Conclusion

The modern interaction of large and small business has its own characteristics and problems. The most common interaction is carried out in the forms of outsourcing, franchising, leasing, subcontracting, venture capital financing, the creation of regional cooperation forms of large and small businesses. There is a stable link of large and small businesses.

In order to intensify the production of Russia, the state should carry out a competent and effective support for large and small businesses. Particular attention should be paid to the state support mechanism of small business development, including the organizational, financial, information and consulting components.

We believe that the development of cooperation between large and small businesses is particularly important in the current economic conditions, as it provides a whole new level of resources using, based on efficiency and thrift.

References

1. A. Shleifer, R.W. Vishny, Journal of Political Economy, **95**, **461** (1986)
2. A.D. Evans, Studies in Comparative International Development, **51**, **456** (2016)
3. B.S. Min, Journal of Business Ethics, **1** (2016)
4. C. Wickert, Business and Society, **55**, **792** (2016)
5. D. Palia, Journal of Empirical Legal Studies, **13**, **756** (2016)
6. D.H. Kim, M.O. Sambou, M.S. Jung, Empirical evidence from Korea. Sustainability, **8**, **1119** (2016)
7. G. Kaplanoglou, V.T. Rapanos, N. Daskalakis, European Journal of Law and Economics, **1** (2016)
8. G.M. Olevsky, World Economy and International Relations, **60**, **17** (2016)
9. I. Distinguin, C. Rugemintwari, R. Tacneng, World Development, **84**, **18** (2016)
10. K. Ding, T. Gokan, X. Zhu, Annals of Regional Science, **1** (2016)
11. K.G. Palepu, Journal of Accounting and Economics, **8**, **3** (1986)
12. L. Hou, L. Xue, S.N. Bui, W. Kettinger, Information Technology and Management, **17**, **379** (2016)
13. M.C. Jensen, The Journal of Finance, **48**, **831** (1993)
14. M.C. Jensen, W.H. Meckling, Journal of Financial Economics, **3**, **305** (1976)

15. O.E. Williamson, The Yale Law Journal, **93**, **1197** (1984)
16. Q. Guo, C. He, D. Li, Urban Studies, **53**, **2584** (2016)
17. R. Morck, A. Shleifer, R.W. Vishny, Journal of Financial Economics, **20**, **293** (1988)
18. R. Tresca, International Journal of Social Ecology and Sustainable Development, **7**, **35** (2016)
19. R.F. Malaquias, Y. Hwang, Information Development, **32**, **1613** (2016)
20. S. Mayadunne, S. Park, International Journal of Production Economics, **182**, **519** (2016)
21. S.J. Grossman, O.D. Hart, The Bell Journal of Economics, **2**, **42** (1980)

Selection of executors for realization of individual tasks of the project

Askhat Asanov[1,*], and *Irina* Myshkina[2]

[1]Moscow Technological University, 119454 Moscow, Russia
[2]Kazan Federal University, 420008 Kazan, Russia

Abstract. This article deals with the task of selection of executors for the realization of individual tasks of a new project. A short review of approaches to the estimation of personnel based on intellectual methods is provided. Two approaches to the task solution of executors' selection are reviewed. The first approach allows considering estimation of competence of potential executors and data about projects that have already been realized. These data and requirements to the tasks of new projects are presented with the help of cognitive map with correspondent vertices. In the article, a methodologic example of the suggested approach use is introduced. The second approach suggests the use of Hopfield's neural network. Neural networks allow using the already available experience of project activities during realization of new projects.

1 Introduction

Tasks of personnel management of the enterprise/organization in importance rank with the tasks of manufacturing and sales of production. Success of the new project in any sphere of activity directly depends on the competence of the selected executors, on proper distribution of individual tasks of the project between the team members. The majority of works related to the issues of formation of executors' teams, are mostly aimed at acquisition of integral estimations of correspondence of professional, psychologic and personal, as well as other characteristics of applicants to the requirements of individual tasks based on the results of professional skills testing, psychologic testing, expert estimates etc. Further based on the acquired data about the applicants such composition of the team is defined, compliance assessment of which to the requirements for the participation in the project reaches the extreme value of some target function taking into account values of chosen criteria of potential applicants. Similar tasks in optimization theory are related to discrete (combined) optimization and are called assignment tasks. Herewith, the available information is not taken into account if it is present, about the already realized projects, which allows to get a more distinct picture about possibilities of employees of a project organization, their competence in real practical work and about the peculiarities of the project, in which the employees can participate. Value of this information considerably increases in case of absence of universal approaches to the estimation of real competencies of the employees and orientation of the existing methods to the estimation of only "knowledge" component of competences based on application of test jobs.

Data about already realized projects both successful and projects, difficulties in realization of which were conditioned by insufficient competence of the executors, can be used during estimation of new projects and during selection of most suitable executors.

2 General description of the suggested approaches

2.1 Short review of approaches to the estimation of the personnel based on intellectual methods

In conditions of absence of precise mathematic methods of staffing objectives solution of the enterprise (including for the tasks of estimation and selection of personnel) and necessity to handle both quantitative and qualitative indicators during decision making in the context of indefiniteness, it appears that effective solution of the tasks under examination is possible when using intellectual methods of decision making support, popularity of which has recently grown up. Some approaches to the task solution of estimation and selection of personnel based on the intellectual technologies are further considered.

• Feedforward neural network.

Used for the estimation of the perspective of work in the enterprise [1, 2, 3]. The suggested method of estimation of labour potential involves engagement of experts for evaluation, analysis and forecasting of the labour potential of employees. Results of inspection and

* Corresponding author: a.z.asanov@yandex.ru

DOI: 10.1051/shsconf/20173501026

opinions of experts are used for "training" of the neural network.

Trained neural network should hereafter model the work of experts, estimating the labour potential of both separate workers and their groups [1, 2, 3].

For the use of neural networks during prediction of the success of work of the applicants, a quite large training sample is required in conditions of the absence of sufficient and/or slow accumulation of the necessary statistical data. Use of expert evaluations for the formation of statistical sampling requires sufficient consumption of time from the expert.

• Fuzzy logic [4, 5, 6, 7, 8, 9].

Very often, for the evaluation and selection of the personnel fuzzy production and operations with particularly constructed fuzzy sets and numbers are used. This makes it possible to operate with fuzzy initial data that are more natural during evaluation of the applicant according to various criteria. Fuzzy formalization of estimation and comparison criteria becomes possible; it is possible to conduct qualitative assessments. However, when using fuzzy production all knowledge is rigidly prescribed and there is no possibility to add new knowledge.

Use of fuzzy production for the assessment of professional competence of the executor is difficult because of the large amount of input variables of fuzzy rules, while the construction of fuzzy models based on the expert knowledge in practice is applied with a maximum of 2-3 input variables [10]. Setting (training) of fuzzy production system is also difficult.

• Genetic algorithms.

Genetic algorithms are used to solve discrete optimization tasks arising during the formation of the optimal composition of the organizational structure [11]. Moreover, genetic algorithms can be used for the training of the neural networks and setting of fuzzy parameters of models.

• Cognitive maps.

Work [12, 13] describes a method of a test integral assessment of the competence level, based on the application of the cognitive modeling, theory of fuzzy sets and relationship of preferences between different criteria.

Thus, existing approaches tend not to take into account the information about already realized "successful" projects and the executors involved.

2.2 Task setting

Let's assume that the evaluation of competencies of the enterprise's employees is known who are the potential executors of individual tasks of the current project; information about already realized projects: executors, estimations of success of their participation in these projects, demanded competences and their significance. It is necessary based on the requirements to competence, necessary for the realization of individual tasks of the new project to execute ranking of potential executors based on their assessment of compliance with these requirements.

3 General description of possible approaches to the solution of the executors' selection task

3.1 Cognitive model for the evaluation of competences of potential executors

This section examines the possibility of using cognitive maps to obtain assessment of conformity of the executor with the requirements to implement the tasks of the project. This section examines the possibility of using the cognitive maps to obtain estimation of conformity of the executor with the requirements to implement the tasks of the project.

Cognitive model [14, 15, 16] of requirements to the realization of a private task of the project:

$F=\langle G, Y, W\rangle$

где $G=\langle V, E\rangle$ – directed graph,

V – vertex set, where:

$$V = \{V_l\} = \{C_i\} \cup \{RP_j^s\} \cup \{TP_k\} \cup \{I_t\},$$

C_i, i=1..N (N – total number of competences) – vertices–competences,

RP_i^j – vertices–tasks of realized projects, i=1..M, j=1..m_i (M – total number of realized projects, m_i – total number of tasks of the i project).

TP_k – vertices corresponding to the tasks of the current project (k=1..R, R – total number of tasks of the current project).

I_t – vertices corresponding to the potential executors (t=1..L, L – total number of potential executors of tasks of the current project).

Therefore, total number of vertices is equal to:

$$S = N + \sum_{i=1}^{M} m_i + R + L.$$

$E=\{e_{ij}\}$, i,j=1..S – multitude of arcs. Arcs go from vertices–competences C_i to vertices, corresponding to the realized projects RP_j, if at project realization the corresponding competence was engaged. Arcs go from vertices TP_k, corresponding to the tasks of the current project to vertices–competences C_i, if the task demands the possession of these competences. Arcs go from vertices–competences C_i to vertices, corresponding to potential executors I_t, if the executors possess these competences. Arcs go from vertices, corresponding to the realized projects RP_j to vertices I_t, corresponding to the executors, if the executors participated in these projects.

$Y=\{Y^{Vl}\}$, l=1..S – multitude of parameters of vertices V. Values of vertices I_t – estimation of correspondence to the requirements of one task, for which the initial value is set, which corresponds to the significance of the task for the current project in whole. Value of vertices TP_k – significance of the task for the current project in whole.

Weights of influence $W=\{w_{ji}\}$ (i=1..S, j=1..S) between every pair of vertices take values from the interval (0;1]. Weight of connection between the competence and realized project is set as equal to the significance of this competence for this project. Connection weight between the task and competence is set as equal to the significance of this competence for this task. Connection weight between the competence and executor is set as equal to the estimation of this competence for this executor. Connection weight between the realized project and executor is set as equal to the participation successfulness estimation in the already realized project of this executor.

It is required to get an estimate of compliance of the executors with the requirements necessary to perform the task of the project and rank the executors in accordance with these estimates.

Solution of the problem of executors selection for each task of the project includes the following steps:

1st stage. Construction of a cognitive model of requirements to the competences necessary for the realization of individual tasks of the project.

2nd stage. Selection of the project task. Task of the initial value for the corresponding vertex (equal to the significance of the task for the current project in whole). Initial values of the remaining vertices are taken as zero.

3rd stage. Recalculation of the values of all vertices. Model of stages calculation of cognitive maps:

$$Y_i(t+1) = Y_i(t) + \sum_{j=1}^{S} \Delta Y_j(t) \cdot w_{ji},$$

где $\Delta Y_j(t)$ – increment (change) of the value of j vertex on the calculation stage t.

4th stage. Perform ranking of executors on the basis of the resulting values of vertices It. Vertex with the maximum value corresponds to the executor to the fullest extent satisfying requirements of the task.

5th stage. If not all tasks are examined, we pass over to stage 2.

Fig. 1 shows a simple example of a cognitive map of requirements to the competences for the realization of individual tasks of the project.

Fig. 1 describes the following situation:

There are three potential executors (employees of the enterprise). These executors can perform tasks that require possession of three competencies. Executors have already participated in the realization of two projects with one task. The current project demands execution of three tasks. Herewith, the first competence corresponds to the first task, all three competences to the second one and the third competence to the third one.

To get the numerical parameters of the cognitive map (weights of influence, significance of the tasks of the new project) the use of direct expert estimates is suggested with further verification of their coherence [15, 17]. Test methods, including the use of method of a test integral assessment of the level of competence, can be used to determine the competence, also based on the use of cognitive modeling [12, 13].

3.2 Methodic example

Let's consider a simple example, where the cognitive map of requirements to the competences necessary for the realization of individual tasks of the project, is presented in fig. 1.

Assume that weights of influence between the vertices of the map are given in table 1 and significance of tasks for the project in whole are respectively equal to 0.5, 0.8 and 1.

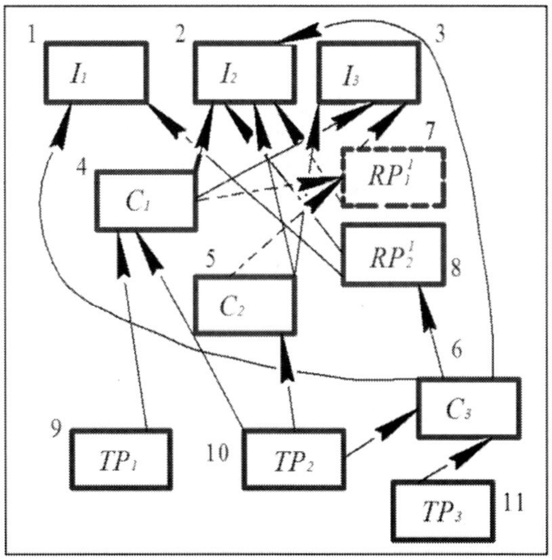

Fig. 1. Example of a cognitive map.

As a result of calculation of the values of all vertices we get the following resulting values for the vertices with the numbers 1, 2 and 3: for the first task – 0, 0.53 and 0.36; for the second task - 0.77, 1.70 and 0.68; for the third task – 1.2, 1.15 and 0.

Thus, the most suitable for the realization of task 1 and task 2 of the project is the second executor and for the task 3 – the first executor.

Table 1. Initial values.

Weights of influence	Values	Weights of influence	Values
w_{61}	0.9	w_{94}	0.9
w_{62}	0.8	$w_{10.4}$	0.5
w_{42}	0.9	$w_{10.5}$	0.6
w_{43}	0.6	$w_{10.6}$	0.8
w_{52}	0.7	$w_{11.6}$	1
w_{53}	0.5	w_{72}	0.4
w_{47}	0.7	w_{73}	0.3
w_{57}	0.8	w_{81}	0.6
w_{68}	0.5	w_{82}	0.7

3.3 Selection of the team based on the history with the help of the Hopfield's neural network

Using a feedforward neural network (for example, a multilayer perceptron) for the problem of team selection requires the presence of sufficient training sample. Problem of absence of sufficient sample scope can be solved through the use of the Hopfield's neural network for the solution of alternative tasks to determine the similarity of the requirements of the new project with the requirements of the already realized "successful" or "unsuccessful" projects.

Hopfield's neural network possesses associative properties. The task solved by this network as associative memory, is stated in the following way. A set of binary signals is known, which is considered to be exemplary. The network should be able, from a random unideal signal supplied to its input, separate (recall based on partial information) a corresponding sample (if any) or give a conclusion on the fact that the input data do not match any of the samples.

A similar problem arises in case of necessity to recognize a potentially unsuccessful project in terms of the insufficient competence of executors. A set of exemplary signals can be formed on the basis of the history of already realized projects. Properly trained network during supply of requirements to its input concerning the new project, should select the appropriate sample (if any), what will mean that the probability of unsuccessful realization of the new project is quite high.

In case of a small number of already realized successful projects Hopfield's neural network can "memorize" them or to be more exact relevant sets of competences and during supply to its input of the constructed neural network of requirements to the new project, it is possible to determine if there are similar projects from the already realized. The probability that the team of executors corresponding to the found sample will successfully implement the new project is quite high.

The number of neurons in the input and output layers of the Hopfield's neural network for these tasks is equal to the total number of competencies possessed by employees. In this case it is more convenient to take initial input values as equal or +1 (if for project realization it is necessary to engage an executor, possessing the relevant competence), or –1 (if there is no such necessity).

On the stage of the network training, weight coefficients are set as follows [18]:

$$w_{ij} = \begin{cases} \sum_{k}^{L} (x_i^k)^T \cdot x_j^k \; i \neq j \\ 0, i = j \end{cases},$$

where x_j^k – j element of k vector-sample, which should "memorize" the network, L – number of memorizable samples. As the activation function of F for network neurons the following function can be used:

$$z_j = F(y_j) = \begin{cases} +1, \textit{if } y_j \geq 1 \\ -1, \text{if } y_j \leq -1 \\ z_j, \text{in the rest cases} \end{cases}.$$

Conditions of neurons are stated using a formula:

$$z_j(t+1) = F(\sum_{i=1}^{n} w_{ij} z_i(t)),$$

$z_i(0)=t_i$, $i=1..n$ (t_i - unknown input signal).

During arbitrary vector supply to the network input the network can "memorize" some sample (by bringing it into a stable condition), as the closest to the presented data or not to find a suitable sample.

Number of memorized samples does not exceed some value L_{max}, which depends on the number n of network neutrons. Precise calculation L_{max} is hindered by the circumstance that samples memorized by the network are usually correlated between each other [19]. For loosely correlated samples an approximate estimation is known [20]:

$$L_{\max} = \frac{n}{2\ln n}.$$

4 Conclusion

Construction of a cognitive map for the task in hand allows to execute the value convolution of partial criteria taking into account their importance for the acquisition of a final estimation; extension of the map is allowed by adding new vertices that characterize new competencies and new realized projects.

Significance of neural networks is concluded in the fact that they allow to use the already available experience of project activities during realization of new projects.

References

1. T.V. Azarnova, V.V. Stepin, and I.N. Schepina, Vestnik VGU. Series: Economics and management, **3**, 121-130 (2014)
2. L.N. Yudina, *System of Estimation and Formation of Labour Potential of the Enterprise* (Tomsk, 2010)
3. L.N. Yudina, Siberian Financial School, **4**, 59-62 (2006)
4. M.V. Guner, and O.I. Pyatkovsky, News of AltGU, **1**(77), 94-99 (2013)
5. E.G. Komarov, *Theoretical Fundamentals of Automated System Construction of Training Management with Consideration of Fuzzy Information* (Moscow, 2011)
6. A.N. Danilenko, *System of Inttelectual Support of Decision Making during Estimation of the Human Factor in the Sphere of Professional Activity* (Samara, 2012)
7. A.N. Danilenko, Electronic magazine Works MAI, **46**, 1-5 (2011)
8. M.A. Sazonov, *Automation of Selection Processes of Employees of Enterprises based on Intellectual methods* (Orel, 2007)

9. S.V. Skorokhod, Investigated in Russia, 1253-1261 (2007)
10. A. Piegat, *Fuzzy Modeling and Control* (Binom. Laboratory of Knowledge, Moscow, 2009)
11. I.E. Egorova, *Intellectual System on the Basis of Genetic Algorithms for Optimization of Content of Organizational Structures in the Sphere of Development and Realization of Innovations* (Volgograd, 2006)
12. I.M. Azhmukhamedov, *Solution of Tasks of Informational Safety Provision based on System Analysis and Fuzzy Cognitive Modelling* (Binom. Laboratory of Knowledge, Astrakhan, 2012)
13. I.M. Azhmukhamedov, Vestnik AGTU, Ser. Management, Computing Technology and Computer Science, **2**, 186-190 (2011)
14. A.Z. Asanov, and I.Yu. Myshkina, News of Computer and Information Technologies, **12**, 29-34 (2012)
15. A.Z. Asanov, I.Yu. Myshkina, and L.Yu. Grudtsyna, International Journal of Soft Computing, **10**, 448-453 (2015)
16. V.V. Borisov, V.V. Kruglov, and A.S. Fedulov, *Fuzzy Models and Networks* (Hot line-Telekom, Moscow, 2007)
17. V.G. Gorsky, A.I. Orlov, and A.A. Gritsenko, Automation and Telemechanics, **3**, 159-167 (2000)
18. V.V. Kruglov, M.I. Dli and R.Yu. Golunov, *Fuzzy Logics and Artificial Neuron Networks* (PHYSMATLIT, Moscow, 2001)
19. Yu.A. Bryukhomitsky, *Neural Network Models for Systems of Information Safety* (Publishing House TRTU, Taganrog, 2005)
20. S. Khaikin, *Neural Networks* (Publishing House: Williams, Moscow, 2006)

SHS Web of Conferences 35, 01027 (2017) DOI: 10.1051/shsconf/20173501027
ICIE-2017

Lack of skilled personnel as threat to HR security of industrial enterprises

Tatiana Ozernikova[1,*], and *Natalia* Kuznetsova[1]
[1]Baikal State University, Department of Labour Economy and Personnel Management, 664003 Irkutsk, Russia

Abstract. The article assesses the needs of Russian industrial enterprises for skilled labor. The authors consider deficiency in qualified personnel as a threat to the human resources security of industrial enterprises, and analyze its causes. The study includes the criteria to distinguish highly skilled employees in the labor force (education level, occupation, correspondence of performed work to education), and analysis of the dynamics of their number. It identifies the internal and external factors that have a significant influence on the number of qualified personnel. The external factors include the size of the salary, population growth due to migration, the regional level of the socio-economic development. The applied methods of analysis are single-factor analysis of variance (F-test), multiple regression and descriptive statistics. The analysis is based on the materials of the Russian Federal State Statistics Service, as well as on the Russian Longitudinal Monitoring Survey - Higher School of Economics.

1 Introduction

The questions of attracting and retaining well-qualified employees are the key task of HR management at industrial enterprises. This fact is proven the data of nation-wide [1, 2] and regional studies [3].

The lack of highly qualified staff has a negative effect on the development of Russian industrial enterprises. According to the Monitoring the Business Climate in Industry that was conducted by the National Research University Higher School of Economics (NRU HSE) in 2016, 18% of Russian companies lacked qualified personnel. The greatest deficiency in skilled workforce was observed in companies that produce leather, shoes and goods made of leather (42%), machinery and equipment (36%) (Fig. 1).

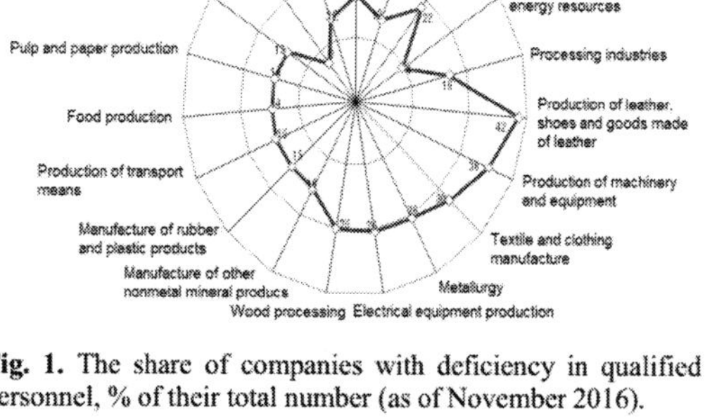

Fig. 1. The share of companies with deficiency in qualified personnel, % of their total number (as of November 2016).

2 Problem statement

Deficiency in qualified staff is one of the major reasons impeding the development of the innovation potential of the Russian industry [4, 5]; it is a threat to its human resources security [6].

The reasons for the qualified personnel deficit are: outflow of scientists and highly skilled professionals from the country since the beginning of the 90s [7], decline in the quality of vocational education, demographic crisis [8], lack of employees' motivation to improve their professional level, a low level of investments in education, training and retraining at enterprises [9], as well as a low level of salary and social protection, which do not allow companies to attract workers with the required qualification [10].

In this paper, we tried to answer the following research questions: What are highly skilled employees? Is there evidence that the industrial enterprises do have a deficit of qualified personnel, which threatens the human resources security? What factors have a significant impact on their quantitative parameters?

The information base of the research was the data of sample surveys of population on employment, the data of the Russian Federal State Statistics Service (Rosstat), the Russian Longitudinal Monitoring Survey - Higher School of Economics (RLMS-HSE).

* Corresponding author: ozernikova@bgu.ru

3 Criteria for determining a high qualification level

It should be noted that in the Russian Federation, the concept of "highly qualified workers" is still not formalized. These obviously include employees whose level of knowledge and skills required for a professional activity is at a "high" level. This approach to defining qualifications was laid down in the International Standard Classification of Education, and the Federal Law #273 "On Education in the Russian Federation" dated December 29, 2012.

It should also be noted that qualification can be considered as a professional activity characteristic not only in terms of its level (formal characteristic), but its content (qualitative characteristic) as well. In terms of content, the concept of "qualification" is often used as a synonym for a profession and its constituent work functions, skills and knowledge [11]. This approach to the estimation of employees' high level of skills was laid down in the All-Russian Classifier of Occupations OK 010-2014 (OKZ), approved by Order of Rosstandart #2020-st on 12.12.2014.

It is common to start the selection of high-skilled workers with a definition of their functions. Employees with a high level of qualification are connected with science and technology, solution of control tasks, management and coordination [12]; they are engaged in provision of analytical (information, financial, commercial, legal and social) services, involved in the production and processing of information that is based on knowledge [13] and capable to enrich and enhance this knowledge for its further practical application and training of others (OKZ).

Scientific studies show that highly skilled employees differ from others in many ways. The have higher learning rates [14], a specific capital that prevents inter-company and occupational mobility [15, 16, 17], higher indicators of job satisfaction [15, 18], and higher levels of salary and better employment terms [19].

The criteria of distinguishing highly skilled employees in the labor force are summarized in Table 1.

Table 1. Overview of approaches to the criteria of qualified workforce evaluation.

Approach	**Evaluation criteria**	**The structure of qualified workforce**
National Research University Higher School of Economics (R. Kapelyushnikov)	Tertiary education	Workers with higher or secondary vocational education
The methodology of calculation of the indicator "The share of the number of highly skilled workers in the total number of skilled workers in the region, per cent" (approved by Order of Rosstat #70 on 21.02.2013)	Higher or post-graduate education, correspondence of the performed work to education, occupation (profession/ position at the basic work is included in Groups 1-3 of the OKZ occupations)	Executives, specialists of high- and middle-level qualification
	Higher or secondary vocational education, correspondence of the performed work to education, occupation (profession/ position at the basic work is included in Group 7 of the OKZ occupations)	Skilled workers of industrial enterprises, construction, transport, geological exploration and prospecting
All-Russian Classifier of Occupations OK 010-2014 (approved by Order of Rosstandart #2020-st on 12.12.2014)	Higher or secondary vocational education (professional training) and experience (length of service) of practical work	Workers with higher or secondary vocational education and experience (length of service) of practical work
Evaluation of the qualification level within the national qualification system (Order #148n of the Ministry of Labor and Social Protection of the Russian Federation dated 12.04.2013)	The qualification level is determined by the skills, knowledge, education in relation to the authority and responsibility of the worker	Employees who meet the requirements of Qualification Levels 5-9

Thus, the main criteria, which may help to determine highly skilled workers, include the level of education, occupation, and correspondence of the performed work to education.

1.1 Changes in the number of highly qualified personnel

Rosstat has been publishing data on the number of highly skilled workers since 2012. The purpose of this statistics is to assess the implementation of the Russian Federation President's Decree #597 "On measures of implementing the state social policy" dated May 7, 2012. According to this Decree, the number of highly skilled workers must be increased by 2020, so that it is not less than one-third the number of skilled workers. In 2015, only 30% of the RF regions managed to come close to this value.

For Russia in general, the proportion of highly skilled workers in the total number of qualified personnel was 32.5% in 2015, having increased by 6.9% over four years (Table 2).

The difference in the index values for the RF regions was more than 3 times (the largest value was registered in Moscow (48.5%), the smallest one - in the Chechen Republic (16.1%)). The Central Federal District is leading among the federal districts, the Siberian Federal District is an outsider.

Table 2. The proportion of highly skilled workers in the total number of qualified personnel.

Territory	2012	2013	2014	2015	Growth rate, 2015 vs. 2012
Russian Federation, including	30.4	31.5	31.9	32.5	106.9
Central Federal District	36.3	37.4	37.0	38.1	105.0
Northwestern Federal District	32.6	32.9	31.9	39.1	119.9
Southern Federal District	27.0	28.4	29.4	29.3	108.5
Volga Federal District	26.7	28.3	28.8	29.6	110.9
Urals Federal District	28.8	30.5	32.2	32.0	111.1
Siberian Federal District	25.7	26.4	27.5	28.5	110.9
Far Eastern Federal District	28.7	30.5	31.5	31.9	111.1
Crimean Federal District	-	-	-	30.9	-

Which groups of employees have the greatest impact on the value of this indicator? In other words, what is the qualitative structure of highly skilled workers?

While analyzing the dynamics of employment in the context of various occupation groups (i.e. the potential availability of highly qualified personnel in the labor market), it may be noted that the number of managers is leading in terms of increase over the past ten years (29.5%), while the general increase in the number of people employed in economy is 4.7%. The number of specialists with high qualification has increased by 25.3%, the number of specialists with midrange qualification – by 11.9%. The number of employees mentioned in Group 7 of the OKZ, on the contrary, decreased by 12.8%, which may be an indirect evidence of deficiency in highly skilled workers.

The proportion of these occupational groups by the end of the indicated period reached 57.3% of the total employment (please, compare it with 53.9% in 2005), which can be regarded as a positive trend.

The distribution of employment by length of service (Table 3) shows that high qualification often not only requires long training, but contributes to the formation of a specific capital, which prevents inter-company mobility.

It is mostly typical for managers (46.6% of them worked for more than 10 years), as well as for specialists with high qualification (41.8% respectively).

Table 3. Distribution of employment by length of service at the basic work.

Category of the employed population	Number of employees by length of service, %					
	less than 1 month	from 1 month to 1 year	from 1 year to 3 years	from 3 to 5 years	from 5 to 10 years	10 years and more
Total employed population	0.9	9.1	15.5	14.8	23.9	35.8
Managers	0.3	4.8	10	12.2	26.1	46.6
Specialists of high qualification	0.4	6.3	13.6	13.8	24.1	41.8
Specialists of midrange qualification	0.5	6.9	14.2	13.7	23.1	41.6
Skilled workers of industrial enterprises, construction, transport, geological exploration and prospecting	1.2	9.9	14.8	14.8	23.7	35.6

1.2 Highly skilled employees' subjective assessment of their professionalism

It is obvious that employers do not regard tertiary education and work experience as an indication of employees' high qualification. A more important characteristic for them is employees' professionalism, which means a high degree of mastery in a certain profession that is characterized by a high skill and competence [20]. For employees, a high level of professionalism guarantees, to a certain extent, a high level of income and employment stability.

Let us consider how Russian employees evaluate their level of professionalism (Fig. 2). This indicator was developed on the RLMS-HSE questionnaire: "Imagine a "professional skill ladder" consisting of 9 steps, where step 1 - is the level of a beginner, a student, and step 9 is the level of a "high-class professional." Which of them would you place yourself?".

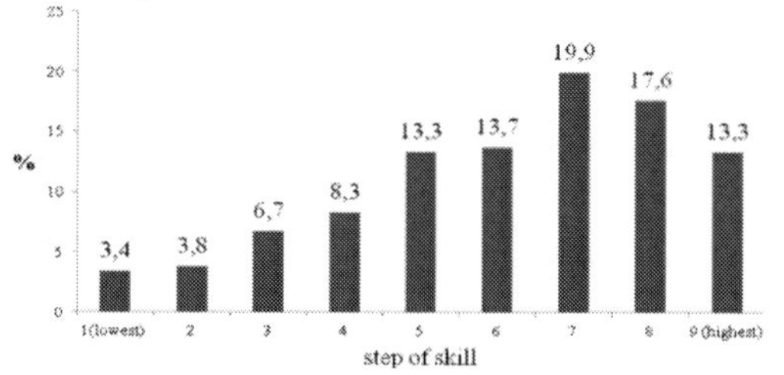

Fig. 2. Distribution of employees by the steps of professional skills, RLMS-HSE.

Fig. 2 shows that the subjective assessment of the professionalism level is quite high. The first three steps of the professional skill ladder are occupied by only 13.9% of respondents. In the second third of the ladder, there is a little more than a third of respondents (35.3% of the total number of the respondents who gave an assessment). The highest steps (7 to 9) of professional skills belong to more than half of the respondents - 50.8%, with only 13.3% of the respondents on step 9 (the level of excellence). Thus, the subjective self-evaluation of professionalism by the employed population generally coincide with the data of the official statistics. It is interesting to compare employees' professional skill assessments in the context of the analyzed occupation groups (Table 4).

Table 2. Assessment of professional skills by occupation groups.

Occupation group	Average rating of the profes sional skill level, points	The proportion of the respondents being on the indicated step of professional skills, %		Absolute deviation from the average value (in the group of the respondents being on the indicated step)	
		Steps 4-6	Steps 7-9	Steps 4-6	Steps 7-9
Senior and middle managers	7.01	24.9	67.3	-8.4	11.3
Specialists of high qualification	7.05	24	69.9	-9.3	13.9
Specialists of midrange qualification	6.51	33.6	54.2	0.3	-1.8
Skilled workers engaged in manual labor	6.54	33.3	54.9	0	-1.1
Skilled workers using equipment	6.67	30.2	57.2	-3.1	1.2
Unqualified workers	5.56	40	32.5	6.7	-23.5
Average value	6.46	33.3	56.0	-	-

Despite the fact that employees from the group of highly skilled workforce gave higher ratings to their professional skills compared with others, a high level of professional skills is characteristic of less than 70% of the respondents. For managers, the highest levels of professional skill ladder are specific only for 67.3% of the respondents, for the specialists with high and midrange qualification - 69.9% and 54.2%, respectively, for skilled workers - 57.2% with the average value of 56%.

4 Identification of factors affecting the number of highly qualified employees

It is important to answer the question of what factors influence the number of highly skilled workforce. It evidently correlates directly with the personnel policy implemented by industrial enterprises in the sphere of labor force recruitment (ensuring competitive salaries, maintaining stable relationships with educational institutions, etc.), as well as its retention (managing the personnel turnover, maintaining a high level of qualification through staff training and motivation, etc.). However, the number of highly skilled workers is largely determined by personnel policy at the level of regions.

Let us introduce the following dependent variables:

y_p – the proportion of the number of highly qualified personnel, y_n – the number of highly qualified personnel, persons. In order to identify the factors that have the greatest influence on one of the dependent variables (y_p or y_n), we introduced the following independent variables:

- x_1 – Nominal salary, thousand rubles;
- x_2 – Migration growth rates per 1000 of population;
- x_3 – The level of the population economic activity;
- x_4 – The average age of the employed population, years;
- x_5 – The innovative activity of enterprises;
- x_6 – The volume of innovative products, works and services, million rubles;
- x_7 – Gross Regional Product per capita, million rubles;
- x_8 – Unemployment rate;
- x_9 – Preparation of qualified personnel per 10000 people of employed population, persons;
- x_{10} – The volume of shipped own produced goods, works performed and services rendered by own forces per capita, thousand rubles;
- x_{11} – Investments in fixed capital per capita, thousand rubles;
- x_{12} – The average subsistent minimum per capita, thousand rubles;
- x_{13} – Actual final consumption of households per capita, thousand rubles;
- $x_{14} = \begin{cases} 2, \text{high level} \\ 1, \text{medium level} \\ 0, \text{low level} \end{cases}$ – Human Development Index [21];

- $x_{15} = \begin{cases} 2, \text{high level} \\ 1, \text{medium level} \\ 0, \text{low level} \end{cases}$ – Quality of life index [22];
- $x_{16} = \begin{cases} 3, \text{high level} \\ 2, \text{above average} \\ 1, \text{below average} \\ 0, \text{low level} \end{cases}$ – The level of socio-economic development of the region [23].

At the beginning of the study, we checked the following hypothesis (assumptions) for each of the dependent variables individually: H_0^1 - the relation of the region to a particular federal district (there are seven groups: Central, North-Western, Volga, Southern, Far Eastern, Siberian, Urals, North Caucasus) has no significant effect on the values of y_p or y_n; H_0^2 - the Human Development Index (there are three groups in the index value: high, medium, low) does not have a substantial impact on the value of y_p or y_n; H_0^3 - quality of life index (there are three groups in the index value: high, medium, low) has no significant effect on the values of y_p or y_n; H_0^4 - the level of socio-economic development of the region (there are four level groups: high, above average, below average, low) has no significant effect on the values of y_p or y_n.

The results are presented in Table 5.

Table 5. The results of single-factor analysis of variance (F-test).

Hypothesis	Dependent variables			
	The proportion of highly skilled employees in the total number of skilled workers (y_p)		The number of highly skilled employees (y_n)	
	The coefficient of determination (R^2), %	Conclusion	The coefficient of determination (R^2), %	Conclusion
H_0^1	4.4	accepted	5.0	accepted
H_0^2	8.0	rejected at a good level of significance	17.0	rejected at a high level of significance
H_0^3	17.0	rejected at a high level of significance	35.5	rejected at a high level of significance
H_0^4	19.63	rejected at a high level of significance	40.0	rejected at a high level of significance

Thus, three of the four above-mentioned factors have a significant impact on the values of the proportion of highly qualified personnel and the number of highly qualified personnel. Here, one factor - relation to a Federal District – is not significant, which can be explained by the strong differentiation of regions in terms of socio-economic development within their Federal District. The strongest factor is the level of socio-economic development of the region.

As a result of regression analysis applied to build models that describe linear relationship between the number of highly qualified personnel and all the independent variables of the form $y = a + b_1x_1 + b_2x_2 + ... + b_{16}x_{16}$, it was found that only 3 variables (out of 16 ones under study) are very significant; they are the variables x_1, x_2, x_{16}.

The final model, which represents the dependency of the number of highly qualified personnel on these three factors, is as follows:

$$y = -214{,}596 + 10{,}15x_1 + 2{,}22x_2 + 153{,}44x_{16} \quad (1)$$
$$(t = -1{,}81) \quad (2{,}63) \quad (3{,}14) \quad (2{,}79)$$

This model is suitable according the Fisher criterion, i.e. the hypothesis about absence of linear functional relationship is rejected at a high level of significance. The coefficient of determination is equal to 40%; the 40% variation in the number of highly qualified personnel can explained by the variation of three indicators: nominal salary (thousand rubles0, migration growth rate per 1000 of population, the level of socio-economic development of the region.

5 Conclusions

We can conclude that the number of highly qualified personnel is directly dependent on the level of socio-economic development of the economy. Concerning the external factors, a positive effect on the attraction of employees with required qualification is provided by improvement of the regional population life quality.

The number of highly qualified personnel at most of industrial enterprises is currently insufficient for their innovative development, and threatens their human resources security.

References

1. A.G. Jefendiev, Ye.S. Balabanova, Ye.O. Jarygina, Economic Sociology, **13**, **4**, 48-81 (2012)
2. Huge deficit in the personnel market. http://www.utro.ru/articles/2011/08/03/990110.shtml
3. N.V. Kuznecova, News of the Irkutsk State Economics Academy, **2(94)**, 80-88 (2014).
4. L.G. Belova, *Attracting highly qualified foreign specialists to Russia: tendencies and ways of improvement of the state policy* (Moscow, 2014)

5. T.Yu. Stuken, T.A. Lapina, Bulletin of Omsk University. Economic Series, **1**, 108-114 (2015)
6. N.V. Kuznecova, A.Yu. Timofeeva, Economy of Region, **12**, **4**, 1123-1134 (2016)
7. S.V. Ryazancev, Ye.Ye. Pis'mennaya, Social Research, **4**, 24-35 (2003)
8. P.M. Kozyreva, Bulletin of the Russian Longitudinal Monitoring Survey of RLMS-HSE (2014)
9. Ye.L. Shershneva, Economic Sociology, **3**, 140-160 (2011)
10. R.I. Kapelyushnikov, The Issues of Economics, **2**, 52-66 (2012)
11. A.N. Lejbovich, I.A. Voloshina, P.N. Novikov, *Development and application of professional standards* (Moscow, Pero Publ., 2014)
12. M. Lavoie, R. Roy, P. Therrien, Research Policy, **32**, **5**, 827-844 (2003)
13. S.A. Korotaev, The Journal of Sociology and Social Anthropology, **XVIII**, **1(78)**, 123-138 (2015)
14. L. Demmou, A. Worgotter, *Boosting Productivity in Russia: Skills, Education and Innovation* (OECD Publ., 2015)
15. R.I. Kapelyushnikov, *How much is human capital of Russia?* (HSE Publ., Moscow, 2012)
16. I.O. Malceva, *Labour mobility and stability: is return from the specific human capital in Russia high?* (HSE Publ., Moscow, 2007)
17. L.I. Smirnyh, *The 5th Int. conf.: Public sector transition* (2002)
18. I.O. Shmerlina, Social Reality, **4**, 46-66 (2007)
19. L.I. Smirnyh, *XI Int. sci. conf. on problems of development of economy and society*, **1** (2011).
20. Dictionary of sociology. http://diclist.ru/slovar/sociologicheskiy/p/professionalizm.html.
21. S.N. Bobylev, *The report on human development in the Russian Federation in 2013* (OOO RA IL'F Publ., Moscow, 2013)
22. The ranking of the regions of the Russian Federation in terms of quality of life. www.riarating.ru/.
23. Economic Research, **2** (2014). http://cyberleninka.ru/article/n/reyting-regionov-rossii-popokazatelyam-sotsialno-ekonomicheskogo-razvitiya-fevral-2013-g.

Analysis of alternative actions for import substitution policy implementation in machine engineering complex

Sergey Chevychelov[1,*], *Igor* Kotov[1], and *Sergey* Klevtsov[2]

[1]Southwest State University, 305040 Kursk, Russia
[2]Kursk Institute of the Cooperation, 305000 Kursk, Russia

Abstract. This paper discusses the considerations for production modernization, replacement of imported spare parts manufactured in-house and redistribution of items manufactured spare parts between managed societies, with a goal to improve the quality and reliability of products, to reduce its cost, and work. Noted that the process of acquiring new foreign equipment and modern materials are complicated. There are examples of restrictive measures of the sanctions impact on industrial products and the dynamics of dependency on imports of industrial products. We used analytic and statistical research methods. Authors conducted a comparative analysis of the engineering assets of «Metalloinvest» holding that showed that all businesses have an impressive fleet of metal-working equipment on all groups of machines on the basis of an analysis of the needs for them to transport spare parts.

1 Introduction

The innovation potential of a machine-building complex is the main competitive advantage of the developed countries, which focused on an integrated approach to the use of tools to support innovation activities of enterprises and enhance public-private partnerships involving research and educational organizations, enterprises and institutions of development.

Production in industrial sectors, in particular of machine-building industry, is carried out using imported equipment from leading manufacturers in the field of metalworking and equipment, and using the equipment procured within the country[1-4].

Among foreign scientists analysis the policy of import substitution in a machine-building complex such as W. Hongyuan, Z. Xunjiang, D. Weifeng [5], M. Jahani, A. Farzanegan, M. Noaparast [7], F. Elskamp, H. Kruggel-Emden [8], G. Wang, X. Tong[9].

2 Results and discussion

However, in connection with the recent complex geopolitical and economic environment and the introduction of a series of sanctions in industrial complex (table 1) is complicated by the acquisition of new foreign equipment and modern materials.

Many countries, not only imports industrial equipment, but also manufactured goods, in table 2 you can see that many countries depend both on imported industrial machinery and products.

Table 1. Examples of restrictive measures of the sanctions impact on the industrial products*.

An example of the impact of sanctions	The country, which sent the impact of sanctions
embargo on dual-use goods and technologies intended for military use or for the military end user	Russia
a ban on the export of certain equipment	Belarus

*http://www.bscn.nl/sanctions-consulting/sanctions-list-countries

Table 2. Dynamics of dependence on imports for industrial production (% of imported industrial products).

Country	2009	2010	2011	2012	2014
Ukraine	52,2	53	53,2	56,3	56,8
Belarus	45,1	47,3	41,7	46,3	47
Russia	76,5	74,5	74,9	83,1	82,2

Therefore, in modern terms, one of the main problems is the problem of import substitution in industry at the national level [10-13].

To date, the process of import substitution is relevant in view of the unstable political and economic situation in the world. On the basis of modern high-quality technologies you should produce more important modern and innovative products. It is advisable to organize their

*Corresponding authors: tschsa@yandex.ru

own production, repairs and servicing, to use of highly qualified specialists, to conclude a cooperation agreement with Russian companies.

The Ministry of trade and industry of the Russian Federation implements state import substitution programme at the national level in the industrial sector, in particular, on the critically important areas. Critically important position, mainly those on the world market are either monopolies or oligopolies, i.e. access is closed or restricted, and the level of dependency on Russian industry from such positions above the critical level.

Financial support measures, such as subsidies to complex investment projects, working capital and the mechanism of the Fund for the development of industry, also focused on import substitution projects. In this sense, import substitution is intended to offset shrinking domestic demand, to ensure technological independence, to ramp up their competence to world market prices[14].

However, import substitution in engineering, from the point of view of some economists, it would take about five to seven years, in the light of the development of scientific, technological and productive capacities. To reach the target level of substitution on the most sought-after items of equipment need additional funding of about 12-15 billion rubles from budget in 2015-2020th, which will be used to develop new technologies and create serial productions.

In the process of import substitution and export products technological level of industry are increased, as well as her willingness to work successfully for long-term development vector[15-17]. The only way to cope with an artificial restriction on the technological level and competitiveness of our products is the implementation of a program to support investment machinery, engineering, contract life cycle. The Government is now considering the possibility to launch a program of support for the development of the production of the means of production, waiting for the next budget cycle, in the framework of anti-crisis plan-2016. On the assessment of the Ministry of industry and trade, in the year 2016 it will need about 20 billion rubles. Another strategic direction is the new point of growth, based on prospective developments that will be converted into new, large-scale global markets in term of 15-20 years [18].

During the project implementation «Import substitution Centre» in close partnership with Russian Engineering Union orientation occurs primarily on the formation of a new industrial paradigm in Russia. Development of industrial infrastructure and establishes a basic funds[19-20].

The key question is the basic funds of enterprises of the machine-building complex. Manufacture of machine tools is an industry that generates funds, which provides technological equipment all strategic spheres of machine-building and the military-industrial complexes[21-23].

Thus, when evaluating the potential of implementing the strategy of import substitution in the framework of a specific industrial sector there is a need to analyse the potential of this complex. The machine-building complex is able to play a vital role in the economy of the region with innovative renewal conditions of radical modernization of facilities engineering, active policies of import substitution, bringing effective investors and strategic partners.

Comparative analysis of engineering assets of «Metalloinvest» holding showed that all businesses have an impressive fleet of metal-working equipment, in relation to all groups of machines.

Let us analyze the needs for imported spare parts for the surveyed enterprises. On request, the management organization for information on requirements for imported spare parts received lists for the year 2016. The results of the analysis checklists are presented in table 3.

Thus, it is possible to recommend a draw in 2016 year engineering and consulting company for the purpose of:
— detailed analysis in managed organization nomenclature import spare parts, to determine whether the development of analogs of own production, in view of the series of products and terms of operation of the equipment;
— development of design documentation for imported spare parts to be procured;
— development of technological documentation for production of spare parts import analogues;
— plot design for manufacture analogues of spare parts are imported;
— development of feasibility study of the construction site for the manufacture of spare parts import analogues.

Table 3. Analysis of the need for import of spare parts.

Managed organization	**Positions on the list, units**			
	critical	***moderately critical***	***not critical***	***total***
JSC «OEMK»	186	40	0	226
Jsc «Mikhailovsky Gok»	119	92	53	264
JSC «Lebedinsky Gok»	1154	1149	12	2315
JSC «Ural steel»	107	43	6	156
TOTAL	1566	1324	71	2961

Authors revealed that preparations for the majority of products are forgings. It was therefore considered the possibility of manufacturing forgings forces managed entities. In parallel, the possibility of replacement purchase its own production of forgings of blanks for already mastered an item.

Cost structure for each of the managed entities is not uniform, each enterprise is allocated their own factors with the greatest influence on the final cost of production and, consequently, the prices of spare parts.

For leveling influence the negative features of the accounting and pricing on each managed organizations in the production of spare parts it is advisable to proceed with the development and approval of unified rules of pricing machine processing managed entities.

Cost reduction redistribution on managed entities download growth workshop at 10%.

As a result of increasing loading machine shops of managed organization at 10% may get cost redistribution at 6% due to the action:

— production scale (increasing the volume of production decreases the magnitude of fixed costs per unit of output spare parts; the result is a decrease in unit cost with the growth of volumes of manufacture);

— synergies between several units of the «Metalloinvest» holding.

We have a comprehensive analysis of alternatives to improve the utilization of engineering assets and their strategic development.

Development of new nomenclature for spare parts purchased from third-party vendors, for the needs of managed societies, divided into 5 options:

Alternative 1: manufacturing of products subject to harness existing unused machinery assets when possible transfer of equipment for 24 hours and increased staffing levels.

Alternative 2: possibility to manufacture products in accordance with the list provided (refinement) of the design documentation (new nomenclature of spare parts of our own production).

Alternative 3: Mastering the products on the list, subject to the acquisition of workpieces or parts, manufacture of which is impossible in conditions of managed organization.

Alternative 4: manufacturing of products subject to the acquisition of equipment and increase in staff (procurement of new equipment).

Alternative 5: castings, which do not require further machining.

Manufacturing of products subject to change operation mode shifts existing engineering assets, with the necessary resources (table 4).

For the implementation phase activities need to be developed for the manufacture of tooling for the year 2016 followed by manufacturing products in quantities 21 names (3122 places) a year weighing 1004.94 tons to the amount of 58388.6 thousand rubles.

Table 4. Alternatives for improving the efficiency of engineering assets when implementing the concept of import substitution.

Alternative 1

Managed organizations	Position claimed managed organizations		Position, planned to manufacture in managed organizations								The necessary resources
	Quantity	*total weight*	*Quantity*	*total weight*	*Positions on the levels of criticality*						*the number of workers*
					critical		*Moderately critical*		*not critical*		
	unit	tons	unit	tons	unit	%	unit	%	unit	%	person
JSC «OEMK»	39	7,43	32	6,06	28	87,5	4	12,5	0	0,0	1
Jsc «Mikhailovsky Gok»	18	243,3	21	220,6	6	28,6	10	47,6	5	23,8	7
JSC «Lebedinsky Gok»	39	494,6	43	518,7	35	81,4	8	18,6	0	0,0	17
JSC «Ural steel»	1	7,7	1	7,7	1	100	0	0,0	0	0,0	1
TOTAL	97	753,1	97	753,1	70	72,2	22	22,7	5	5,1	26

Alternative 2

Managed organizations	Quantity	Positions on the levels of criticality						Necessary resources
		critical		*moderately critical*		*not critical*		
	units	units	%	units	%	units	%	person/day
JSC «OEMK»	53	33	62,3	9	17,0	11	20,8	212
Jsc «Mikhailovsky Gok»	21	4	19,0	14	66,7	3	14,3	84
JSC «Lebedinsky Gok»	107	102	95,3	4	3,7	1	0,9	428
JSC «Ural steel»	0	0	0	0	0	0	0	0
TOTAL	181	139	76,8	27	14,9	15	8,3	724

Alternative 3

Managed organizations	Positions, the administering	Positions planned to manufacture in a managed organizations	Necessary resources

	organizations											
	Quantity	*total weight*	*Quantity*	*total weight*	*Positions on the levels of criticality*						*the number of workers*	*cost of blanks*
					critical		*Moderately critical*		*not critical*			
	units	tons	units	tons	units	%	units	%	units.	%	person	ths. rub.
JSC «OEMK»	4	1,75	4	1,75	1	25,0	0	0	3	75	1	120,5
Jsc «Mikhailovsky Gok»	2	87,4	1	6,4	0	0	1	100	0	0	1	5073,6
JSC «Lebedinsky Gok»	29	227,2	30	308	26	86,7	4	13,3	0	0	10	13951,0
JSC «Ural steel»	0	0	0	0	0	0	0	0	0	0	0	0
TOTAL	35	316,4	35	316	27	77,1	5	14,3	3	8,6	12	19145,1

Alternative 4

Managed organization	Position claimed managed organizations		Position, planned to manufacture in managed organizations								Necessary resources	
	Quantity	*total weight*	*Quantity*	*total weight*	*Positions on the levels of criticality*						*the number of workers*	*Number of equipment*
					critical		*Moderately critical*		*not critical*			
	units	tons	units.	tons	units	%	units	%	units	%	person	units
JSC «OEMK»	74	6,23	74	6,23	57	77	16	21,6	1	1,4	2	12
Jsc «Mikhailovsky Gok»	2	99,69	2	99,6	0	0	2	100	0	0	3	2
JSC «Lebedinsky Gok»	12	122,72	12	122,7	9	75	3	25	0	0	4	8
JSC «Ural steel»	0	0	0	0	0	0	0	0	0	0	0	0
TOTAL	88	228,64	88	228,6	66	75	21	23,8	1	1,14	9	22

Alternative 5

Managed organization	*Quantity*	*total weight*	*Positions on the levels of criticality*					
			critical		*Moderately critical*		*not critical*	
	units	tons	units	%	units	%	units	%
JSC «OEMK»	0	0	0	0	0	0	0	0
Jsc «Mikhailovsky Gok»	5	130,89	1	20	4	80	0	0
JSC «Lebedinsky Gok»	16	874,05	16	100	0	0	0	0
JSC «Ural steel»	0	0	0	0	0	0	0	0
TOTAL	21	1004,9	17	80,9	4	19,1	0	0

We conducted a comparative analysis of the proposed alternatives based on several criteria: 1) financial costs, 2) time lag, 3) the efficiency, 4) product quality, 5) the level of closing critical positions. On the basis of a mark estimation according to the criteria proposed alternatives held the ranking for each of them. Study a fragment is shown in table 5.

However, alternative embodiments of the manufacturing process are not only single phases but interrelated steps that implement the manufacturing process (table 6).

Thus, the level of "close" critical positions in import substitution and the sum of ranks, you can choose the following combinations:

A 2- A 1, A3- A 2- A 5- A 4- A 1, A 3- A 2- A 5- A 1, A 3- A 2- A 5.

Table 5. Comparative analysis of options for implementing enhanced.

Alternatives	Criterias					
	Financial expenses	*Time lag*	*Efficiency*	*Quality of products*	*The level of "close" critical positions*	*Sum*
A1	1	1	2	5	3	12
A2	2	5	1	3	5	16
A3	3	3	3	4	2	15
A4	5	4	5	1	4	15
A5	4	2	4	2	1	13
Characteristics	5 the most expensive	5 the longest	5 the lowest	5 the lowest	5 the worst	

Table 6. Analysis of Alternatives for Activities on Import Substitution.

Possible combinations of alternatives for the implementation of measures for import substitution	**The sum of ranks**	**The abbreviation "critical" level of production (due to import substitution) ***
A3- A 2- A 5- A 4- A 1	71	181
A 3- A 2- A 5- A 1	56	181
A 3- A 2- A 5	44	181
A 2	16	181
A 2- A 1	28	181
A 3- A 4- A 1	42	97
A 3- A 4	30	74
A 2- A 1	28	181
A 2- A 5	29	181
A 4- A 1	27	97

* Index is defined as the maximum possible among the series on the basis of Table 5

The economic effect of the development of the manufacture of spare parts of managed organization, purchased from third party suppliers is presented separately for each alternative.

Table 7. Analysis of The Development of New Nomenclature of Spare Parts in Stages.

Alternatives	**The number of "closed" critical positions**	
	units	*tons*
A1	97	753,11
A2	181	-
A3	35	316,4
A4	74	6,23
A5	21	1004,94

3 Conclusion

The first alternative (97 items) expected annual positive economic effect in the amount of 27719.56 thousand rubles total for all managed organization, including JSC «Lebedinsky Gok»-12755.69 thousand rubles; by JSC «Mikhailovsky Gok» -12808.09 thousand rubles; on JSC «Ural steel» - 365.87 thousand rubles; at OEMK is 1789.92 thousand rubles.

When implementing a third alternative (35 items) expected annual positive economic effect in the amount of 12348.5 thousand rubles, including OEMK-262.02 thousand rubles; by JSC «Lebedinsky Gok»-6312.3 thousand rubles; by JSC «Mikhailovsky Gok» - 5774.1 thousand rubles.

When the fifth alternative (21 position) annual positive economic effect will be 30502.12 thousand rubles on all managed organization, including JSC «Lebedinsky Gok»-9449.16 thousand rubles; by JSC «Mikhailovsky Gok» -21052.96 thousand rubles.

The economic effect is the sum of the amounts of profit and a conditional cost savings in the production of additional amounts of nomenclature and spare parts.

Thus, the machine-building complex is a priority industry innovation policies of import substitution. Developing innovations in machine-building complex of developed countries should be geared to modernization and implementation of technological innovation in the strategic management of the economy.

References

[1] A.V. Maslennikov, S.A. Chevychelov, M.S. Merzhoeva, M.S. Gatiev, V.V. Sidorova, Russian Engineering Research, **34(11)**, 722 (2014)

[2] A.V. Maslennikov, S.A. Chevychelov, I.G. Golubev, Russian Engineering Research, **33(5)**, 295 (2013)

[3] S.G. Emelyanov, S.A. Chevychelov, P.P. Chistyakov, Applied Mechanics and Materials, **698**, 546 (2015)

[4] M.S. Razumov, A.N. Grechukhin, A.O. Gladyshkin, *Proceedings of the Institution of Mechanical Engineers, Part C: Journal of Mechanical Engineering Science*, **230(18)**, 3298 (2016)

[5] W. Hongyuan, Z. Xunjiang, D. Weifeng, Measurement, **69**, 126 (2015)

[6] L. Little, M. Becker, J. Wiese, A.N. Mainza, Minerals Engineering (2015)

[7] M. Jahani, A. Farzanegan, M. Noaparast, Powder Technology, **283**, 32 (2015)

[8] F. Elskamp, H. Kruggel-Emden, Advanced Powder Technology, **26(3)**, 679 (2015)

[9] G. Wang, X. Tong, Mining Science and Technology (China), **21(3)**, 451 (2011)

[10] L.I. Aimin, L.V. Ruiling, L.I.U. Chusheng, Journal of China University of Mining and Technology, **18(2)**, 300 (2008)

[11] T. Dyr, P. Wodzinski, Physicochemical Problems of Mineral Processing, **35**, 147 (2002)

[12] M. Maleki-Moghaddam, M. Yahyaei, S. Banisi, Minerals Engineering, **46-47**, 157 (2013)

[13] B. Axelrod, *TRIZ Future 2005. Graz, Austria, University of Leoben*, **November,** 325 (2005)

[14] B. Axelrod, *ETRIA Conference*, **November**, 131 (2007).

[15] A.I. Barbot'ko, A.V. Maslennikov, M.S. Razumov, Russian Engineering Research, **31(12)**, 1253 (2011)

[16] M.S. Razumov, A.I. Pykhtin, A.V. Maslennikov, Russian Engineering Research, **32(9-10)**, 681 (2012)

[17] A.I. Barbot'ko, M.S. Razumov, A.I. Pykhtin, P.A. Ponkratov, Russian Engineering Research, **35(5)**, 344 (2015)

[18] Yu. Vertakova, M. Klevtsova, T. Babich, Economical chasopis-XXI, **157(3-4)**, 4 (2016)

[19] Y. Vertakova, M. Klevtsova, S. Klevtsov, *Proceedings of the 26th International Business Information Management Association Conference - Innovation Management and Sustainable Economic Competitive Advantage: From Regional Development to Global Growth, IBIMA 2015 26, Innovation Management and Sustainable Economic Competitive Advantage: From Regional Development to Global Growth*, 3230 (2015)

[20] Y. Vertakova, Y. Polozhentseva, M. Klevtsova, Procedia Economics and Finance, **24**, 750 (2015)

[21] R. Nania, D. Regazzoni, Elsevier, Engineering Procedia, 665 (2011)

[22] P. Lepšík, M. Petr, O. Novak, Modern Methods of Construction Design. Proceedings of ICMD, **IV**, 459 (2013)

[23] D. Russo, T. Montecchi, D. Regazzoni, Proceedings of IDMMEVirtualConcept, **May**, 1 (2010)

Game based learning of lean manufacturing: decreasing personnel resistance

Elena Ganebnykh[1*], *Irina* Altsybeeva[1], *Ekaterina* Gurova[1]

[1]Vyatka State University, 610000 Kirov, Russia

Abstract. Developing production systems based on the principles of lean manufacturing is getting more and more popular nowadays. At the initial stage one of the main issues is high personnel resistance. Training helps to overcome the staff resistance. The research aims at proving a hypothesis of the positive effect of game based learning methods with imitating the production process on lessening the personnel resistance at the stage of implementing lean manufacturing in the company. The paper describes the game model worked out by the authors. This model imitates the production process consisting of successive operations. In the model two main factors influencing the productivity are taken into account: defective goods and stand by time. The model was tested at an enterprise producing children toys made of rubber and plastic. As a result, the participants have learnt some basic principles of lean manufacturing, as well as a principle of forming the key index of Overall Equipment Effectiveness (OEE). The task of lessening the personnel resistance was solved, the participants got to understand the necessity of improving the production process. The model is universal and is easily applicable to any productions and non-production processes. The authors believe the research has a broad sphere of applicability.

1 Introduction

A shift to market economy made many Russian enterprises focus their attention on the users meeting any of their requirements, as well as on sales-markets forecasting their depth and dynamics and stimulating the demand. As a result, while in the advanced economies they have been studying and implementing the Japanese philosophy of sustainable development of production processes that was shown in 1986 by the Japanese economist Masaaki Imai in his book «Kaizen: The Key to Japan's Competitive Success» [7], in Russia production aspects actually have not been improved, they are staying on the level of the 90-s of the 20th century. They started to implement the system of lean manufacturing only from 2004, mostly at big, not to say "gigantic" enterprises.

The basic principles, trends, ways and methods of implementing the lean manufacturing system have been widely researched by foreign and Russian scientists and described in scientific and publicistic works [5,7,8,13,14,11]. The value of the system of lean manufacturing is difficult to overestimate, it is the most likely that this philosophy has contributed to the fact that the Japanese economy not just quickly recovered after the WWII, but also has become one of leaders of the world business.

The advantage of developing production systems on the principles of lean manufacturing on a long-term horizon could be demonstrated by the results of the world-famous producers: Toyota, Honda, Nissan, Canon, SIBUR, RUSAL, KAMAZ, TATNEFT and many others which have pushed on to the next level due to the system of lean manufacturing [1,3,15,16,17]. Still a 10-20-years gap in developing and mastering the methods of lean manufacturing between Russia and Japan, the USA, Germany, as well as other mature economies is striking indeed.

Having begun implementing kaizen-principles, entrepreneurs in Russia admit that the process is mainly interfered with by resistance of the employees at their working places [2]. It could be explained from the scientific viewpoint: implementing a new system causes organizational changes connected with re-organizing the organization itself, which is a fundamental change, urged by the necessity of forming a new organizational philosophy of production. Any organizational changes, especially the basic ones, lead to the employees' resistance which is caused by the fear of uncertainty, loss of safety, loss of power, lack of any reliable information, conservatism (habit). Having overcome this resistance at an industrial enterprise, at assembly line production, each employee comes to understand that the result is dependable on the work of each participant and that at considering and implementing the kaizen system in a daily production process (operation) which seems improved one can find some elements which require optimization. On the basis of eliminating the shortcomings the time necessary for an operation shortens, the amount of defective goods decreases, the

* Corresponding author: ganebnykh@mail.ru

quality improves, the stand-by time also shortens, etc. As a result at each working place labour effort is eased, and the wages grow. The personnel who have experienced this effect as a result of implementing the system are shocked, and at the same time they feel grateful [10]. The workers get surer and feel a considerable importance of their part in the whole production process, and, most of all, they understand their responsibility for the work done.

Thus the main issue of implementing the system of lean manufacturing is resistance of the employees of the enterprise. How can this issue be solved? How can one explain to the personnel effectively that the main principles of lean production (little setup time, error protection, preventing defective goods, shortening of stand by time) do good not just to the enterprise but as well to them personally?

In this case one can rely on the instruction of implementing management principles of Toyota: it is necessary to teach the employees to contemplate on what they have learnt, to use their knowledge, and to think of the process and improve it so that it would lead to growth and flourishing of the enterprise at a long-term horizon [12].

So, the main method of overcoming the personnel resistance is training and there are many methods of doing it. A huge amount of modelings and games develops dynamically [18]. We think that the most telling and close to reality way is the method of game based imitating modeling:

- «Imitative modeling» means that such models of production process occur within which one cannot calculate and foresee the result beforehand [4];
- Imitating a production process (cycle), taking into account the impact of negative factors, can in a game form demonstrate efficiency/inefficiency of the production actions.

Thus, the key hypothesis of the research consists in stating a positive role of games based learning which is connected with imitating the production process for the sake of lessening the personnel resistance at the stage of implementing lean manufacturing system at an enterprise.

We think that games based learning not just clearly shows the results of the personnel actions during the production cycle, but also demonstrates the level of the factors' influence on the productivity, thus enhancing the responsibility of each one participating in the production cycle.

2 Methods

The main part of the research was devoted to creating a game model imitating the production process. The model takes into account two factors which influence efficiency ("wastes" in terms of lean manufacturing): defective goods and stand-by time. As a result, the gamers learn how to calculate the key index, i.e. Overall Equipment Effectiveness (OEE) which is a recognized world standard if calculating efficiency [9]. This index is used as an indicator of efficiency increase.

At the stage of first introducing lean manufacturing principles in the work of the company personnel resistance is unavoidable. One of the ways of decreasing this resistance is personnel training. Games based learning in training the personnel helps to introduce lean manufacturing tools to them and to clearly demonstrate the principles and approaches of this system.

In order to make the game as close to real life as possible during production it is necessary to collect the following data on a modelled assembly line: cycle time (TC), average rate of defective goods for a certain line (D), and average standby time due to machinery break down (WB).

The game consists in imitating an assembly line with five stations responsible for successive operations. Each station is represented by one gamer (operator). Each operator, beginning with the first one simulated certain actions with a component (a gambling chip): doing well, producing defective goods, or standing-by. The first operator can stand by only because of machinery breakdown, the next operators can also stand by due to lack of a certain component (for example, in case the previous operator has made a defective component).

A 6-sided die can be used as a random number generator. Thus, there is a range of possible variants from 1 to 6. Knowing the average rate of defective goods on the line it is possible to determine a number of conditioned variants at a die roll meaning "defective goods". Should one die be used the minimum possible variant meaning "defective goods" is one of its planes. The probability of that plane is 1/6 of all the throws, and each gamer is supposed to throw the die. It means that the average percentage of defective goods is to be 15% at each level. This figure does not coincide with the real state of affairs, it is overrated. But exaggeration of defective goods index can demonstrate the effect of lessening the wastes caused by this reason.

A number of stand-byes due to machinery breakdown is calculated in the same way. Thus, we get at minimum that: one of the planes denotes "defective goods", one – machinery break and a stand-by, and four other planes stand for a useful action.

In the beginning, before playing the game, it is necessary either to determine the amount of the product to produce, or restrict the number of gambling chips. In the first case, it is possible to show how much time is spent on producing a certain amount of the product, in the second case it is possible to show the yield at a restricted period of time (a number of moves).

After the run of the first iteration the players get to know about low efficiency of the operating system. The described variant with one 6-sided die and 18 gambling chips the output is usually 1-2 details (gambling chips). The remaining chips belong to in-process inventory.

During the imitation, each operator is offered to fill in the table reflecting their losses at each move (Table 1). Taking into account the results shown in Table 1 each participant counts their own indexes (1, 2, 3):

$$K_{availability} = 1 - \Sigma_{availability\ loss} / n \quad (1)$$

$$K_{performance} = 1 - \Sigma_{performance\ loss} / n \quad (2)$$

$$K_{quality} = 1 - \Sigma_{quality\ loss} / n \quad (3)$$

Table 1. Wastes during the game.

Waste	I iteration						Sum
	1	2	3	4	…	n	.
Lacking detail							
Machinery breakdown							
Defective goods							

On the basis of the calculations the participants themselves estimate their overall equipment efficiency (4):

$$OEE = K_{availability} * K_{performance} * K_{quality} * 100\% \quad (4)$$

After finishing the first iteration and drawing the conclusion the trainer offers to make one improvement. This improvement consists in eliminating one of losses. Then the second iteration takes place, it brings more success. It is necessary to calculate the coefficients of availability, performance, and quality, as well as to estimate the OEE. Comparing the OEE of the first and the second iterations shows improvement as performance coefficient grows due to eliminating one of loss kinds.

During the third rub the loss left is eliminated. It is evident that at eliminating the second loss the output is 100%. Like before, the conclusion is drawn and the coefficients and the OEE are calculated. As there are no losses, the OEE is 100%.

To transfer the skills acquired at the imitation game to production process all the gamers should calculate the real OEE for their production lines and suggest some improvement of the processes ("kaizen").

The game can be improved, made more complex introducing some economical grounding into it. It is possible to set a price of every chip (die), of the ready goods, as well as to set constant expenses (wages imitation, electricity costs, costs, space rental, etc.), as well as variable costs (machinery maintenance for eliminating machinery breakdowns). Calculating economical effect at each iteration shows the results of the production line operating.

The game can be varied, for example, the production input can be done in lots of several chips, or two or more 6-sided dies can be used for a more true-to-life imitation of a definite production.

It is evident that the exact calculation of production indexes allows to imitate production process really true-to-life. Still we think that the simplest variant with one 6-sided die demonstrates the influence of losses on the performance to the gamers and reduces their resistance at the initial stage of implementing lean manufacturing conception at an enterprise

3 Results

The worked-out model was tested at an enterprise producing children toys from rubber and plastic. The rotary line of producing plastisol intermediates (heads) for dolls was chosen as the experimental area. Plastisol is the input, a formed intermediate for assembling the finished goods is the output. The line consists of 12 successive operations:

1. Placing the initial shapes on the electroformed moulds;
2. Filling the first halves of the shapes with plastisol;
3. Closing the first halves of the shapes hermetically;
4. Turning over the electroformed moulds;
5. Filling the second halves of the shapes with plastisol;
6. Closing the second halves of the shapes hermetically;
7. Turning over the electroformed moulds for equidistribution of plastisol in the shape;
8. Putting the disks with electroformed moulds into the rotary oven where plastisol gelatinates. The process is automatized, at the end the shapes go out for making a hollow;
9. Taking out the shapes from the first half of the form;
10. Turning over the electroformed moulds;
11. Taking out the shapes from the second half of the form;
12. Cutting off the edges of plastisol of the shapes.

They produce 10 intermediates during a cycle. The time of a cycle is 425 seconds + baking time (480-960 seconds depending on the model).

Possibility of making defective goods was of interest. It was observed that actually at all the stages when hand work of an operator is used some defective goods are made. About 2.5% defective goods are produced at the seventh stage at turning over the electroformed moulds: insufficient or uneven turning over leads to defective distribution of the material in the form, as a result the shape gets a non-commodity look. About 3% defective goods are formed at the last stage at cutting off the edges of plastisol of the shapes: hand cutting off is vulnerable to human error.

At producing 208-320 items (depending on the model) during an 8-hour shift 11-17 items of goods are with defects. Each month this loss accounts for 242-374 items of goods which is about the amount of the output during one shift. So, in fact one day monthly the enterprise produces just defective goods. And the costs, both constant and variable remain the same.

An important feature of the line is the fact that defective goods are produced even before putting the shapes into the oven. Which means that all the work done with a faulty, defective shape, up to its taking out, cause direct losses to the enterprise.

There are two analogous lines at the enterprise. One of them was made in Italy, it was bought in 2009, it is in a good technical form and is maintained regularly according to the producer's instructions. The second line is much older and is maintained only at current planned repairs. As we did not have an opportunity to see any machinery breakdowns or repairs, the statistical data were asked for which showed that during the last 3 months the line was standing for 8 working days because the oven was repaired (12% working time).

To demonstrate it more clearly, game modelling included two kinds of losses: defective goods (5,5%) and stand-by (12%). Two 6-sided dies were used as random number generators. To denote the losses doublets were used (two same figures on the both dies) (Halprin, R. and Naor, M., 2009), at the same time the doublets 1-1 и 2-2 meant "defective goods", other doublets meant "stand-by". The process imitation was integrated to the following operations:

1. Placing the initial shapes on the electroformed moulds;
2. Filling the forms with plastisol;
3. Closing the shapes hermetically;
4. Turning over the electroformed moulds for equidistribution of plastisol in the shape;
5. Putting the disks in the rotary oven;
6. Taking out the shapes from the forms;
7. Cutting off the edges of plastisol of the shapes.

Seven gamers imitated the production process in accordance with the game model offered to them. The trainer (the role "customer") set a task to make 10 items of goods. As a result, after the first iteration 6 items were produced with OEE = 63%.

For the second iteration, it was settled to refuse from the defective goods, thus for the "run" just the doublets 3-3, 4-4, 5-5 and 6-6 were left. As a results 8 items of goods were produced, OEE was 80%.

So, it is evident that at eliminating machinery stand-byes during 10 runs the required number of goods will be produced at 100% efficiency of the machinery.

To assess the degree of the personnel resistance before game modelling the information was required from the HR-Department. We got to know that in October this year the operators of the shop (12 people) were questioned, one fourth of the employees showed negative attitude to implementing lean manufacturing at their stage of production. A bit more than a half of all the employees showed their concern about the innovation, estimating their attitude to it as "rather negative than positive". The rest operators were neutral having answered that they did not have any definite meaning of the innovation. It is notable that none of the employees showed any wish to try the innovation. Thus, the level of resistance was quite high.

At the end of training the employees were asked again, what their meaning of implementing lean manufacturing tools was. As a result, two thirds gave a positive answer. Only one employee had kept a negative attitude. All the rest answered "rather positive than negative". It is evident that this rather exaggerated demonstration of impact of losses on the ultimate output contributed to decreasing resistance of the personnel of the enterprise.

3 Conclusion

To prove the hypothesis, we worked out a game model imitating assembly line production with successive operations. The theoretical model was probated at an enterprise producing children toys from rubber and plastic. There was a high degree of resistance of the personnel at the enterprise at the initial stage of implementing the system of lean manufacturing.

As a result of the game the participants realized the impact of production losses on the result and they got the skill of calculating the OEE which indicates the degree of performance. At the same time the gamers understood certain basics of lean manufacturing tools. This created a good basis for implementing lean manufacturing system at the enterprise.

The model worked out by the authors adds to a whole set of practical recommendations of developing production systems. The model is universal and is easily adaptable to a production process of any kind. The game is enough illustrated, so it can be used for explaining the processional approach in a non-production sphere. The model is easily applicable, which is its main advantage.

One of the means of checking the efficiency of the model in question is providing a feedback. To better understand how the personnel resistance depends on the level of their training one can use a detailed questionnaire before and after the game. The authors of the paper provided a feedback only after the training and it was done with the help of the forms offered by the HR-Department of the enterprise. The questionnaire before the training was done be the representatives of the enterprise. Incomplete understanding of the issues causing general resistance took place because of the fact that the questionnaires were processed by the HR-Department. The final information was offered only in a summarized form. Still the managers of the enterprise put an aim of lessening the resistance of the personnel and it is to be admitted that the aim was achieved.

We think that the research has a broad sphere of practical applicability due to the fact that production development based on the principles of lean manufacturing has become very popular recently. A compilation of case-studies of different spheres is of great interest from the point of view of statistical data analysis

References

1. D. Bondarenko, The Algorithm of Success, 9-13 (2010)
2. J. Bruce, P. Scott, and M. Roberts, *Investigating Employee Resistance to Lean Transformation* (UK Academy for Information Systems, 2011)
3. N. Davydova, *Lean Production* (Institute of Economics and Management, Izhevsk, 2012)
4. E. Efromeeva, On the Classification of Methods and Models of IT Analysis. Universities for Tourism and Service Association Bulletin, **1** (2008)
5. A. Grinin, *Management in a Kaizen style: How to reduce costs and increase profit* (Alpina Publishing, 2012)
6. R. Halprin, and M. Naor, *Games for Extracting Randomness* (Weizmann Institute of Science, 2009)

7. M. Kaizen Imai, *The Key To Japan's Competitive Success* (Irwin/McGraw-Hill, 1986)
8. M. Imai, *Gemba Kaizen: A Commonsense Low-cost Approach to Management* (McGraw-Hill, 1997)
9. K.E. Chong, *A case study. 2015 IEEE International Conference on Industrial Engineering and Engineering Management (IEEM)*, 1427-1421 (2015)
10. A. Kovacheva, *Challenges in Lean implementation* (Aarhus School of Business, 2010)
11. W. Lareau, *Office Kaizen: Transforming Office Operations Into a Strategic Competitive Advantage* (ASQ Quality Press, 2002)
12. J. Liker, and D. Meier, *The Toyota Way* (McGraw-Hill Education, 2006)
13. R. Maurer, *One Small Step Can Change Your Life: The Kaizen Way* (Workman Publishing, 2004)
14. *Kaizen for the Shop Floor: A Zero-Waste Environment with Process Automation* (Productivity Press Development Team, 2002)
15. P. Rabunets, About the Production System of SIBUR (2011)
16. J. Van Ede, Shoveling waste out of the factory works better than using a bulldozer. Retrieved from Business-improvement.eu: Inspire Businesses to Flow. http://www.business-improvement.eu/lean/Oce_Lean_Transformation.php.
17. J. Womack, and D. Jones, From Lean Production to the Lean Enterprise. Harvard Business Review (1994)
18. J.-M. Yien, The Turkish Online Journal of Educational Technology, **10(2)**, 1-10 (2011)

The financial modeling in industrial building using the method of hierarchies analysis

Olga Maksymchuk[1], *Oksana* Baulina[1], and *Vladislav* Klyushin[1,*]

[1]Volgograd State Technical University, The Department Of Management And Urban Development And Construction, 400074 Volgograd, Russia

Abstract. The processes in the activity of construction companies which are involved in the construction of industrial objects, bridges and roads, and the factors affecting the value of these processes and the organization's financial stability have been investigated. A comparative analysis of the construction companies according to the criteria for stability (financial, economic, organizational, innovation, investment) has been implemented using the analytic hierarchy method; financial models of studied construction companies have been developed, the forecast evaluation of their expediency has been made.

1 Introduction

In the conditions of the stagnant domestic economy, due to the current unfavorable macroeconomic situation, the theme of the construction industry is particularly important. In the conditions of a quick rise of mortgage loans rates and the reduction in real incomes, the building sector has a co-financing problems of activity from the population, on the other hand there is a quick rise of the cost price of building work in connection with the growth rates of the energy, rail and road transport, the rising cost of construction materials. So, according to the Federal Service of state statistics in the Russian Federation, in the 1st quarter 2014 24% of the respondents of the building sector mentioned the high cost of building materials among the factors limiting production activities of construction companies but in the 1st quarter 2015 this index is already 32%. This is due to the use of imported components, this also concerns the production of dry building mixes using imported modifiers and production of aluminum constructions, cable and wire products, and roofing materials. At the same time, according to the study "Business surveys Lab by E.T. Gaidar", 22% of enterprises in the construction industry faced with a lack of domestic analogues of equipment and raw materials, 18% of companies noted the low quality of domestic analogues. In these unfavorable economic conditions, the participants of the construction sector face with problem of its own stability providing, including the use of various financial models in the development of management decisions.

2 The definition of the terminology

Financial modeling is a process of building a real and expected situation in the activity of the enterprise in the market. All the indexes that will be used in the financial model of the enterprise, can be divided into the following groups: property (assets), liabilities, cash flow (receipts and payments funds), income and expenses. That is, all the results of the calculations which were done using the model presented in the usual form of a balance sheet, the report of financial results and the report of cash flow movement according to the arbitrary (selected) date in the future.

The financial model is a tool for monitoring the current situation in the company and the development of financial policy which provides sustainability of the enterprise in the conditions of changing external environment. The financial model is iterative in nature and should be developed according to the principle "from simple" to "complex". During the year, the company's financial model can be corrected, to put down the actual information can be noted in the passed months instead of the planned and the financial results can be controlled so, that the negative trends will be shown and there will be a clear understanding of what they will lead the company.

Comparison of models of different companies allows to choose the most stable and on the basis of integrated indexes to make decisions, such as choosing a potential partner, it also is important to ensure the own stability of the enterprise.

3 Question, Hypotheses, and methods

Among the methods which can be used in the solving the tasks of enterprises stability evaluation there is

* Corresponding author: vladislav-klushin@yandex.ru

regression analysis method, the theory of unclear plurality and multivariate analysis method, known as the analytic hierarchy method (AHM), which is the procedure of system analysis for hierarchy presentation of elements that define the essence of any problems. The method consists in decomposing of the problem into simpler components and its further processing sequence of statement of person, who makes decision by pairwise comparisons. The intensity of elements interaction is expressed by a hierarchy of elements in numerical form. The AHM includes synthesis procedures of multiple statements to obtain the priority criteria and find alternative solutions. The identity and decomposition principles involve problem structuring in the form of a hierarchy or a network, that is the first step in the application of the AHM to solve that problem.

Hierarchy stability model presentation is built from the top (the purpose - in terms of management - assessment of the stability of enterprises, level 1), through the intermediary of the second level (the stability criteria which influence subsequent levels) to the lowest third level (which is a list of alternatives - the analyzed companies in which the level of stability is investigated).

The author's model identified five criteria that determine the stability, namely: 1) Financial stability; 2) Economic stability; 3) Organizational stability; 4) Investment stability; 5) Innovative stability. The main task of our model realization is the ranking of enterprises as well as the compilation of the sample by the criterion of "the most high stability." For this we introduce the mathematical description of this concept. Let a linear combination

$$\sum_{k=1}^{n} W_k r_k = W_c \qquad (1)$$

defines the sought degree of enterprises stability by the criteria adequacy of the chosen linguistic variable. Then the average value of the coefficient of enterprise stability level can be calculated using the formula:

$$K_{acp} = \frac{\sum_{i=1}^{n} K_i}{n} \qquad (2)$$

The excess K_{al} compared with mean value is calculated according to the formula:

$$\delta\%_{K_{al}} = \left| \frac{K_{al} - K_{acp}}{K_{acp}} \right| \cdot 100\% \qquad (3)$$

Where K_{al}, K_{all}, K_{alll} – K_{an} – analysis objects (analyzed enterprises).

For the selection of the most stable enterprises, the number of enterprises can be arbitrary and is established by the user independently, for each of the analyzed companies there is necessary the following information for the financial report: F1-balance sheet, F2, financial results report, F4, cash flow movement report, F5 application to the balance sheet. On the basis of these financial reports the calculation of analytic indexes (coefficients) is carried out, the indexes are grouped by each type of stability: financial, economic, organizational, investment, innovation. For receiving of this system of indexes we have carried out financial and economic analysis of economic activity of the analyzed companies for 2013- 2015 and produced a sample of indexes that form the system of linguistic variables. In the analysis the attention is paid to all expenses profitability assessment, which shows how much profit from main activity falls on 1 monetary unit of general operating expenses. The results of the analysis were used for the automated creation of matrices of binary significances ratio according to the criteria of stability. Integral priorities vectors of financial stability; economic stability; organizational stability; investment stability; innovative stability, calculated for each of the n chosen enterprises, become the basis for matrices formation of binary significances ratio for stability criteria at level of the third hierarchy.

The next step is to apply the synthesis principle that allows to calculate the integral (global) vector of priorities through the final value of the enterprise stability. To identify the global (integral) enterprises priority (according to the previously specified formulas) in the matrix, the local priorities are arranged to each criterion, each vector column is multiplied by the corresponding criteria priority and the result is summarized along each line. It allows to finally rank the investigated companies according to their stability, and that is the solution of the problem.

The analysis of domestic and foreign researchers who pay attention to the stability of the business, allowed the authors to identify a range of indexes which allow to calculate the local stability vectors. When there is the financial component evaluation of the enterprise stability, Russian economists underlined the following factors: solvency, liquidity, creditworthiness and profitability. In assessing these factors of the financial component stability, there is the most spread method of financial coefficients.

In assessing the economic component of the company's stability, there are methods based on the calculation of financial coefficients (intraeconomic financial analysis), as well as indexes of economic activity. Among the factors of the economic component of the enterprise stability, the domestic economists include: success in the market of products, production and technological opportunities of the enterprise, financial opportunities, employment opportunities.

Analysis of the organizational component of the enterprise stability involves the consideration and assessment of such factors as: the organizational structure of the enterprise, the level of corporate culture, the rationality of information flow, development of motivation system.

The indexes for assessing the innovative component of enterprise stability include: indexes of social and technical innovation.

Assessment of the investment stability is implemented on the basis of accounting report information by calculating a range of coefficients.

Thus, we have identified the indexes of financial, economic, institutional, investment and innovation enterprise stability, these indexes are based on the analysis which help to select the most stable companies; integrated assessment is carried out using the analytic hierarchy method, allowing to rank companies on the basis of the calculation of global priorities vector.

4 Result

The model developed by the authors, was the basis of the computer program FAAC (Functional Analysis Of Activity Companies) (certificate of state registration №2016618552 from 2nd August 2016) and it was tested by the example of the five major domestic construction companies. The program is designed for a wide range of users, including: managers and employees of economic departments, state authorities and local self-government, university professors, students and graduates of economic specialties.

The testing results of the model developed by the author, are presented below (Table 1-5). We note that the first two tables ("Financial Stability" and "Economic stability") are filled with an objective method - comparative calculation of financial and economic factors the ratio between the enterprises with each other, and the next three tables ("Organizational stability", "Innovative stability", "Investment stability") - with the use of the subjective method, by expert statement evaluation of person who makes decisions with the use of relative importance scale from 1 to 9.

Table 1. The calculation of integrated priority vectors for financial stability index with yhe local vectors use of financial coefficients.

Local vectors of financial coefficients	**The coefficient of current liquidity**	**The coefficient of autonomy**	**The coefficient of financial leverage**	**The coefficient of capitalization**	**The coefficient of quick liquidity**	**The coefficient of absolute liquidity**	**Integral priority vector «Financial stability»**
JSC «Volgomost»	0.13	0.15	0.17	0.03	0.24	0.52	0.18
JSC «Dalmoststroi»	0.19	0.04	0.61	0.45	0.00	0.01	0.27
JSC «Mostostroi-11»	0.30	0.40	0.04	0.16	0.02	0.04	0.20
JSC «Mostotrest»	0.15	0.23	0.07	0.04	0.68	0.31	0.17
JSC «Sibmost»	0.22	0.17	0.11	0.32	0.05	0.12	0.18

Table 2. The calculation of integrated priority vectors for economic stability index with the local vectors use of economic coefficients.

Local vectors of economic coefficients	**Sale profitability**	**Capital productivity**	**The coefficient of availability**	**The turnover of reserves**	**The turnover of receivables**	**Integral priority vector «Economic stability»**
JSC «Volgomost»	0.30	0.27	0.16	0.36	0.12	0.20
JSC «Dalmoststroi»	0.00	0.15	0.01	0.14	0.14	0.09
JSC «Mostostroi-11»	0.44	0.15	0.31	0.09	0.29	0.31
JSC «Mostotrest»	0.25	0.34	0.35	0.21	0.19	0.25
JSC «Sibmost»	0.01	0.09	0.17	0.19	0.26	0.16

Table 3. The calculation of integrated priority vectors for organizational stability index with the matrix use of significance binary ratio.

Binary ratio of enterprises significance by criteria «Organizational stability »	**JSC «Volgomost»**	**JSC «Dalmost-stroi»**	**JSC «Mostostroi-11»**	**JSC «Mostotrest»**	**JSC «Sibmost»**	**Integral priority vector «Organizational stability »**
JSC «Volgomost»	1.00	1.67	0.83	0.63	1.00	0.19
JSC «Dalmoststroi»	0.60	1.00	0.50	0.38	0.60	0.11
JSC «Mostostroi-11»	1.20	2.00	1.00	0.75	1.20	0.22
JSC «Mostotrest»	1.60	2.67	1.33	1.00	1.00	0.27
JSC «Sibmost»	1.00	1.67	0.83	1.00	1.00	0.21

The integral priority vectors of financial stability; economic stability; organizational stability; investment stability; innovative stability, calculated for each of the five companies, have become the data for the compilation of the of binary ratio matrix of significances by stability criteria at the level of the 3d hierarchy.

Table 4. The calculation of integrated priority vectors for innovative stability index with the matrix use of significance binary ratio.

Binary ratio of enterprises significance by criteria «Innovative stability »	**JSC «Volgomost»**	**JSC «Dalmost-stroi»**	**JSC «Mostroi-11»**	**JSC «Mostotrest»**	**JSC «Sibmost»**	**Integral priority vector «Innovative stability »**
JSC «Volgomost»	1.00	3.00	1.20	0.86	1.00	0.23
JSC «Dalmoststroi»	0.33	1.00	0.40	0.29	0.33	0.08
JSC «Mostostroi-11»	0.83	2.50	1.00	0.71	0.83	0.19
JSC «Mostotrest»	1.17	3.50	1.40	1.00	1.00	0.26
JSC «Sibmost»	1.00	3.00	1.20	1.00	1.00	0.24

Table 5. The calculation of integrated priority vectors for investment stability index with the matrix use of significance binary ratio.

Binary ratio of enterprises significance by criteria «Investment stability »»	**JSC «Volgomost»**	**JSC «Dalmost-stroi»**	**JSC «Mostroi-11»**	**JSC «Mostotrest»**	**JSC «Sibmost»**	**Integral priority vector «Investment stability »**

JSC «Volgomost»	1.00	3.00	1.20	0.86	1.00	0.23
JSC «Dalmoststroi»	0.33	1.00	0.40	0.29	0.33	0.08
JSC «Mostostroi-11»	0.83	2.50	1.00	0.71	0.83	0.19
JSC «Mostotrest»	1.17	3.50	1.40	1.00	1.00	0.26
JSC «Sibmost»	1.00	3.00	1.20	1.00	1.00	0.24

The values of global priorities vector of matrix binary significances ratio by constituting criteria of enterprises stability assessment, are presented in the table below (Table 6).

Table 6. The value of the global priority vector for the index «Enterprise stability».

	Stability components	The value of integral priority vectors	Rank
1	Financial stability	0.477405058	1
2	Economic stability	0.271534589	2
3	Organizational stability	0.03962003	5
4	Investment stability	0.145676217	3
5	Innovative stability	0.065764105	4

Furthermore, in the following table VII authors also calculated deviation of integral priority vector component from their average values (absolute and relative values). This calculation makes it possible to fully and more accurately assess which companies have the high stability in terms of their criteria assessment.

Table 7. The integral index of building enterprises stability.

Enterprises	Ranking companies based on the vector of global priorities			
	Integral priorities	*Rank (rating)*	*Absolute deviation from average value of global priority vector*	*The deviation from average value of global priority vector, %*
JSC «Volgomost»	0.198474408	3	-0.001525592	-1%
JSC «Dalmoststroi»	0.172136855	5	-0.027863145	-14%
JSC «Mostostroi-11»	0.227262582	1	0.027262582	14%
JSC «Mostotrest»	0.217078421	2	0.017078421	9%
JSC «Sibmost»	0.185047732	4	-0.014952268	-7%

How to interpret these priorities? From Table 7 it becomes clear that the presence of high financial support of the company (financial stability, rank 1) is perceived by the administrator of the enterprise as the most important criterion in the assessment of stability. In fact, it is more important than the economic stability (3 grade) (vector of local priorities 0.477405058 against 0.271534589) almost in 2 times and is much more important than organizational stability (rank 5), which has the lowest priority, equal to 0.03962003 (almost in 18 times less).

Thus, implemented stability assessment with the use of AHM, showed that only two companies of 5 companies, (JSC "Mostotrest", JSC "Mostostroi-11") received a positive difference between the average value of global priorities and values of the corresponding linear combinations for each of the studied enterprises (+ 14%; + 9%, respectively). Other enterprises are outsiders according to the values of stability.

5 Conclusion

The stability assessment method using analytic hierarchy procedure is universal and easy to apply in practice. Its application will increase the accuracy of the management decisions (federal, regional, municipal, departmental, and others) with regard to inter-regional and sectoral distribution of resources, and at the level of individual enterprises – to work most effectively with their own and borrowed capital. In addition, application of this method in an expanded form (with an increase in the number and quality of evaluation criteria of financial, economic, organizational, investment, innovation components stability, and involving the most possible number of assessed companies) allows to evaluate stability of the certain sector.

Developed software product FAAC provides: 1. the calculation of analytic indexes (coefficients) which are grouped for each type of stability: financial, economic, institutional, investment and innovation; 2. the calculation of the local vectors of priorities for each type of stability; 3. the calculation of the global priority vector, allowing the company to rank for the selection of the most stable of them; 4. Visualization of calculations with diagrams. With the results of the calculations, it is possible not only to select the most stable enterprise, but also to select the most "narrow places" in the activities of each company. The software product "FEP Analiz.1" was used to develop practical recommendations for optimizing the financial models of the analyzed companies, authoring (certificate of state registration number 2014660103 from 01.10.2014).

Application of optimizational financial model of expenses management allows JSC "Dalmoststroy" to increase the level of the marginal income in 2016 from -250 961 thousand rub. to 17 152 thousand rub. It is also expected the increase of the rate of profit from -4.76% to 0.34%, reflecting the increase of the profitability of the company and out of the loss zone.

Application of optimizational financial model of expenses management allows JSC «Volgomost» " to increase the level of the marginal income in 2017 from 2235165 th. rubles to 2715034 th. rubles. It is also expected the increase of the rate of profit from 11.5% to 16.3%, reflecting the increase of the company's profitability.

Application of optimizational financial model of expenses management allows JSC «Mostostroi-11» " to increase the level of the marginal income in 2017 from 2281077 th. rubles to 2452689 th. rubles. It is also expected the increase of the rate of profit from 14.7% to

17.7%, reflecting the increase of the company's profitability.

Application of optimizational financial model of expenses management allows JSC «Mostostroi» " to increase the level of the marginal income in 2016 from 6531913 th. rubles to 8366770 th. rubles. It is also expected the increase of the rate of profit from 5% to 8.3%, reflecting the increase of the company's profitability.

Application of optimizational financial model of expenses management allows JSC «Mostotrest» " to increase the level of the marginal income in 2016 from 520100 th. rubles to 681297 th. rubles. It is also expected the increase of the rate of profit from 4.9% to 6.5%, reflecting the increase of the company's profitability.

In general, on the basis of the study the authors proposed a universal-optimization approach to financial modeling in the industrial construction with the use of the analytic hierarchy method, which allows to increase the rate of profit for each of the companies. This approach is representative and authentic, has universal nature, and can be applied to the analog organizations and be adapted to the different spheres of economic activity.

References

1. B.D. Baysinger, H.N. Butler, Journal of Law, Economics, and Organization, **1(1)**, 101-124 (1985)
2. T.B. Bell, S. Szykowny, J.J. Willingham, *Assessing the Likelihood of Fraudulent Financial Reporting: A Cascaded Logit Approach* (Working paper, KPMG Peat Marwick, Montvale, NJ, 1991)
3. J.A. Brickley, C.M. James, The Journal of Law and Economics, **30(April)**, 161-180 (1987)
4. R.S. Chaganti, V. Mahajan, S. Sharma, Journal of Management Studies, **22(July)**, 400-417 (1985)
5. H. DeAngelo, L. DeAngelo, The Journal of Finance, **45(December)**, 1415-1431 (1990)
6. H. DeAngelo, L. DeAngelo, D.J. Skinner, Journal of Accounting and Economics, **17(1-2)**, 113-143 (1994)
7. E.F. Fama, Journal of Political Economy, **88**, 288-308.
8. E.H. Feroz, K. Park, V.S. Pastena, Journal of Accounting Research, **29(Suppl.)**, 107-142 (1991)
9. S.C. Gilson, Journal of Financial Economics, **27(2)**, 355-387 (1990)
10. S.J. Grossman, O.D. Hart, The Bell Journal of Economics, **2(Spring)**, 42-64 (1980)
11. B. Hermalin, M.S. Weisbach, The Rand Journal of Economics, **19(Winter)**, 589-606 (1988)
12. M.C. Jensen, Journal of Law and Economics, **26(June)**, 301-325 (1983)
13. M.C. Jensen, The Journal of Finance, **48(July)**, 831-880 (1993)
14. M.C. Jensen, W.H. Meckling, Journal of Financial Economics, **3(4)**, 305-360 (1976).
15. S.N. Kaplan, D. Reishus, Journal of Financial Economics, **27(2)**, 389-410 (1990)
16. R.D. Kosnik, Administrative Science Quarterly, **32**, 163-185 (1987)
17. M.L. Mace, *Directors: Myth and Reality* (Harvard Business School Press, Boston, 1986)
18. G.S. Maddala, The Accounting Review, **66(October)**, 788-807 (1991)
19. R. Morck, A. Shleifer, R.W. Vishny, Journal of Financial Economics, **20(C)**, 293-315 (1988)
20. G. Novikova, A. Chub, J. Gushina, *Globalization and its socio-economic consequences. 16 th International Scientific Conference*, **1**, 1547-1555 (2016)
21. K.G. Palepu, Journal of Accounting and Economics, **8(1)**, 3-35 (1986)
22. Z.V. Palmrose, Auditing: A Journal of Practice & Theory, **6(Spring)**, 90-103 (1987)
23. A. Patton, J. Baker, Harvard Business Review, **65(November)**, 10-12 (1987)
24. K. Pincus, M. Rusbarsky, J. Wong, Journal of Accounting and Public Policy, **8(4)**, 239-265 (1989)
25. S. Rosenstein, J.G. Wyatt, Journal of Financial Economics, **26(2)**, 175-191 (1990)
26. A. Shivdasani, Journal of Accounting and Economics, **16(1-3)**, 167-198(1993)
27. A. Shleifer, R.W. Vishny, Journal of Political Economy, **95**, 461-488 (1986)
28. A.A. Sommer, Accounting Horizons, **(JUNE)**, 91-93 (1991)
29. M. Stone, J. Rasp, The Accounting Review, **66(January)**, 170-187 (1991)
30. Financial Management, **20(Winter)**, 101-112 (1991)
31. R. Vancil, *Passing the Baton: Managing the Process of CEO Succession* (Harvard Business School Press, Boston, 1987)
32. D. Vicknair, K. Hickman, K.C. Carnes, Accounting Horizons, **7(March)**, 53-57 (1993)
33. J.B. Warner, R.L. Watts, K.H. Wruck, Journal of Financial Economics, **20(C)**, 461-492 (1988)
34. M.S. Weisbach, Journal of Financial Economics, **20(C)**, 431-460 (1988)
35. O.E. Williamson, The Yale Law Journal, **93**, 1197-1230 (1984)

Effective services promotion management in Far North (Russia)

Anna Grabar[1,*]

[1]Management and marketing department, Vyatka State University, Kirov, Russia

Abstract. The problem of the effective management of services promotion in the Far North (Russia) is topical due to the severe weather and climatic conditions and the high price barrier in this industry. The objective of the article is the analysis of consumer market characteristics and the formation of specific recommendations on services promotion for the industries. Statistical methods and methods of analysis, synthesis, decomposition and expert assessments used in the study revealed the features of the consumer market. The article highlights the factors and patterns of communication services market functioning and its development dynamics. Specific recommendations on the actions in competitive market are given to the companies providing communications services.

1 Introduction

Unlike other industries: metallurgy, mechanical engineering, mineral extraction, energy, where the main facilities were built in Soviet times, the telecommunications sector has arranged from scratch in new Russia. For this reason, on one hand, our country is experiencing lag from the most economically developed countries in the sphere of information and telecommunications; on the other hand, there is no need to support and maintain outdated technologies in the industry.

One of the priorities of the Innovation Development Strategy of the Russian Federation for the period until 2020 is the formation of modern information and telecommunication infrastructure, ensuring a high level of its accessibility, and provision of high quality services on its basis [1].

Objective of the study was to explore consumer factors that affect the development of the provisions of the strategy to promote services of telecommunication companies.

2 Methodological framework

To achieve the goal the following tasks were set: to give a general description of the behavior of mobile users in the region, to identify factors affecting the penetration of mobile Internet, to reveal consumers' satisfaction level and positive attitude to cellular communication companies, represented in the region, and to identify channels of information used by consumers and which are particularly trusted.

The object of the research is a set of users of communication services among Yamal-Nenets Autonomous District urban residents.

Research method: results study of various market research reports, sampling results and the formation of a common understanding of the state of telecommunications services in the regions of the far north. The analytical study database includes the works of these authors [4-22], who in their studies considered various market aspects of the Yamalo-Nenets Autonomous District

3 Results

3.1 The number of subscribers who use the services of the mobile operator is defined

Thus, it was found that 64% use one provider, while 36% use several providers, which indicates that the vast majority use one operator, however, the number of subscribers seeking to maximize the benefit from using services of several companies can increase. [10]

As a rule, subscribers who need to optimize the cost of long-distance and international communication services use several operating companies. Thus, 16% of subscribers for long-distance and international communication services use the minor operator [10].

Only 4% of respondents believe that competing companies offer better mobile Internet

3.2 Customer loyalty is estimated

According to limitation period of connection to the main telecom operator it was found that 66% of respondents use services more than 5 years, 19% – from three to five years, 12% - from one to three years, 3% – less than a year. This demonstrates the preference of consumers to choose operators sensibly and have long-term relationships with them. [7]

* Corresponding author: grabar@vyatsu.ru

DOI: 10.1051/shsconf/20173501031

3.3 The main factors of operator choice

Among the most important factors in the choice of mobile operator the interviewed emphasize:

- optimum by price tariff considering traffic and discount – 85%;
- quality of services and communication – 75%;
- low prices – 72%;
- high quality of innovative and additional services – 40%;
- reliability and reputation of the operator – 38%;
- image – 14%;
- recommendations of friends and acquaintances – 11%. [8]

Evaluation of the factors shows that subscribers' demand to telecommunication companies at the moment is to create optimal tariffs (or tariff package) for different users, providing quality of services, fair prices and innovation.

3.4 Impact of substitute products on the market

About 20% use the services of IP-telephony (Skype, GoogleVoice, etc.), while 80% of respondents do not use them. A total of 15% use special applications instead of SMS, and 85% - do not use them; it is not common among subscribers to use substitute services of cellular communication [5].

3.5 Prospects for promotion of mobile Internet services on the market

A positive aspect in the mobile communication segment is that in the test group the number of smartphones (71%) exceeded the number of regular phones (29%), which contributes to the implementation of forward-looking strategies of the company in the mobile segment [6].

3.6 Channels for disseminating information on the market

Mass communication plays an important role in the dissemination of information on new services and opportunities of companies, with official website, a means of self-initiation of consumer communication, being the most convenient and informative according to customers (41%). In the presence of home Internet access, 79% of the respondents consider online advertising as the most promising means of information. Also, consumers trust the recommendations of friends and acquaintances (39%), TV advertising (29.6%), consult with specialists from company office or contact center (28.8%), see outdoor advertising, read printed leaflets, brochures (26 with the official website recognized as the most convenient and informative by consumers (41 %), %) and listen to advertising on radio (17%) [12].

The high rate of the recommendations of friends and acquaintances (39%) demonstrates an important role of accounting gain of horizontal linkages in society for marketing communications. Frequent use of social networks and blogs in Russia also confirms the horizontal linkages gain – with the help of mobile Internet 83% of respondents visit these resources. At sufficiently high rate of obtaining information through TV advertising (29%), it was possible to understand consumers' attitude to regional TV advertising:

- 34% of respondents do not watch TV, they watch videos on the Internet;
- 31% easily watch TV with turned-on sound and have their favorite video clips;
- 29% try not to watch ads and immediately switch the channel
- 5% turn the sound off.

One can say that TV advertising has a communications capacity in the regional market. Out of 31% of those who watch commercials easily, 29% acquire some information. However, 68% somehow try to avoid advertising, which under high advertising charge can make it economically inefficient [3].

The importance of interactive and personalized communications in relation to the company and the subscriber is confirmed by consumer behavior in the field of tariff plan change or the connection of a new service. For this respondents usually choose: consultation with a specialist from the office or contact center – 56%, company's official website – 33%, the system of self-service activation / deactivation – 31 % [14].

3.7 Regional consumption of additional mobile services, emphasizing the services of the mobile Internet

The popularity and revenue from SMS and MMS services sale decreases, but in the segment of additional services they still occupy leading positions – 79%. The second most popular additional service is the mobile Internet [9].

Today, the segment mobile data transmission has become one of the fastest growing segments of the additional services in cellular networks in Russia. The basis of the active growth of this segment is actively deploying the third and fourth generation networks, increase of smartphones and tablets' user audience and traffic consumed by users (Fig.1)

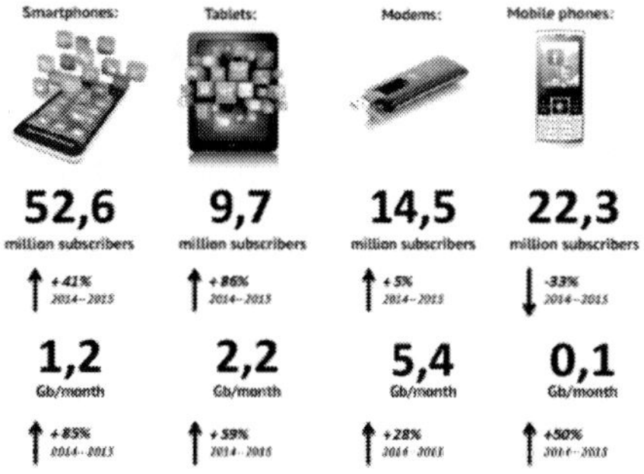

Fig. 1. The number of active subscribers and average traffic of mobile Internet per active user per month in Russia in 2014[19].

From the analysis of the activity of the mobile Internet service use, depending on the age of the subscriber, it is obvious that the service is especially in demand among young people aged 18-24 years, i.e. the student's age. The penetration rate reached more than 55%. In the age group of young people from 25 to 35, the level reduced to 45%, which is also high. For consumers between the ages of 16-17 years, that is, the level of high school students, it was 30%. And for respondents aged 36-55 years, the level of mobile Internet penetration reached 40%.

According to the survey, subscribers often use mobile Internet in places of recreation and leisure activities – 76%, in transport - 66%, at work - 63%, at home - 54%. With the help of mobile internet, 83% of respondents visit social networks or blogs, 72% visit various websites, 66% check emails, 54% download music and books, 30% download/watch videos, 28% use geolocation services, 14% pay for goods and services via the Internet. [10]

On the regional market mobile Internet is developing and being used quite actively. In the next 12 months respondents: think they will spend more time on-line – 17%, think they will spend the same amount of time – 65%, think they will spend less time – 10%, will not use mobile Internet at all – 8%. The perspective of mobile Internet market development is obvious [16].

However, the barriers to the mass distribution of «Mobile Internet» service in the region, according to respondents, are: poor quality of networks, lack of «high speed internet» (56%), too high price for mobile data transmission traffic (55%), high cost of smartphones and tablets (25%), no need for mobile Internet (13%) [17].

At present, with the package offer of cellular service subscribers spend on mobile communication on the average per month: up to 200 rubles – 22% of respondents, 201-499 rubles – 45%, 500-999 rubles – 24%, more than 1000 rubles – 8%. At that, 26% of those who spend up to 200 rubles and 34% of those who spend up to 499 rubles are ready to increase spending on cellular services, provided unlimited mobile internet and introduction of a variety of additional services; however, 57% of this group are strictly not willing to increase their spending; of those who spend no more than 999 rubles 26% are ready, and of those who spend more than 1000 rubles13% are ready. [14]

In general, the subscribers are ready for services based on new technologies and some of active mobile Internet users are willing to spend more on mobile communications.

3.8 Market monopolization level assessment

There are five operators on the communication market of the Yamalo-Nenets Autonomous District:

- Rostelecom;
- telecommunication group MOTIV;
- MegaFon;
- MTS;
- VimpelCom (Beeline).

62% of respondents consider Rostelecom as their primary provider, 14% – Motive, 9% – Megafon and MTS, 6% – Beeline. It is obvious that Rostelecom is the absolute leader of the largest subscriber base in the market [20].

In the process of comparing their major operator with competitors, respondents noted that their operator is: much better (22% of respondents), better in some ways (37%), approximately the same (22%) [10].

According to the study results it can be concluded that the regional market of basic telecommunication services is mature. It is also obvious that the services based on mobile Internet are in demand among the population of working age, especially among young active audience.

For the improvement of strategies to promote telecommunication services on this regional market is necessary to carry out its segmentation to determine the market structure and define the guidelines for strategic actions.

To determine the formation conditions of services promotion strategy, it is also necessary to take into account the main characteristics and problems of the region's population.

3.9 Study of the regional market features

The Russian Federation is quite a unique country, which has a large share of polar and subpolar areas with a harsh climate. They are referred to as the «regions of the Far North and equated localities» (Fig. 2). Out of 17,1 million km2 of Russian territory, 11 million km2 or 2/3 of its area is the territory of the Russian North, which is home to 10 039 065 people or 6,99% of Russian population [13].

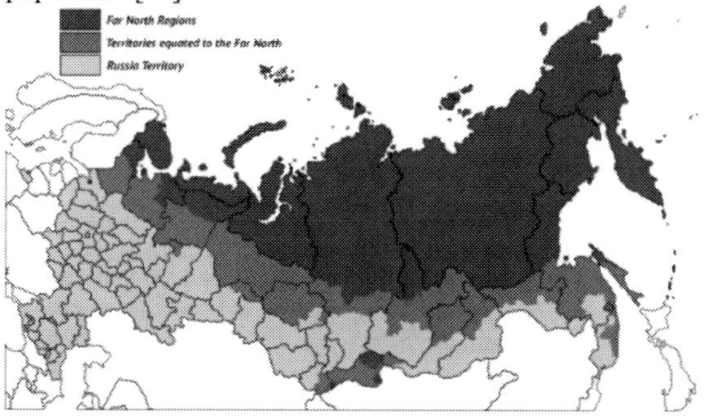

Fig. 2. Regions of the Far North and equated localities.

In general, according to the approach adopted in the Russian Federation, the Far North and equated localities can be characterized by a number of common features from the economic and natural points of view. They include unfavorable climate, permafrost, long distances from the main economic and cultural centers, high transport costs, rise in price of construction and production, high cost of living, and environmental vulnerability of the territory [11].

Because of the severe climatic conditions communications networks service of the Far North is significantly hampered. However, low density of population and its geographical dispersion makes telecommunications services consumers more susceptible to the development of new services, such as broadband access, video conferencing, intelligent network, etc. [13-18].

According to the official statistics website, the total population of the Yamal-Nenets Autonomous District in 2014 amounted to 539,7 thousand people. The share of urban population is 83.1% [13].

Yamalo-Nenets Autonomous District has significant economic value, the basis of the autonomous region's economy is oil and gas extraction. According to the rating of Russian regions in living standards, Yamalo-Nenets Autonomous District is the 45th among 83 subjects of the Russian Federation, but it is the 2nd after Moscow by the level of per capita income and the 6th by the level of economic development [23].

In the ranking of life quality Yamalo-Nenets Autonomous District has maintained the negative trend for the past two years: in 2014 it has lost positions in the eight groups of ten. Despite the fact that the subject of the Russian Federation is among the top ten in socio-economic conditions rating of Russian subjects on the results of 2013 according to the RIA Rating Agency [22], it affects the quality of people's life less than one would expect: this significant advantage nullifies a lot of drawbacks, including poor territory development, severe weather conditions, insufficient small business development, poor housing conditions quality [22].

4 Discussion

One of the traditional problems of the telecommunication market companies in the northern regions is shortage of main canals, which leads to the fact that operators construct optical lines in those areas where population density is more or less acceptable. In most areas the focus is on the use of satellite communication channels, through which most of the traffic is transferred.

The demand for mobile services is growing both among businessmen and among people. Given the rapid turnover of information technology, it is not effective for companies operating in the mobile market, only to sale services or to lease communication channels. Therefore, the percentage of their participation in information and services provision increases every year.

Transport infrastructure communication development, active promotion of broadband Internet access and other innovations are the basis of effective implementation of a business model of any telecommunication company in the Russian market. Thus, of communication services development is one of the most competitive areas of the Russian Far North economy.

The main regularities of communication services spread in the Far North area has developed as a result of historical, cultural and natural development of the region. For the Russian North regions with their vast territories, low population density, and unequal living conditions, the formation of the telecommunications market through the telecommunication and information infrastructure development is of particular importance and it has its competitive features. For example, the short northern summer is the time of construction and repair work, related to communication networks laying, base stations installation and their adjustment and other works. At the same time, summer is the traditional vacation period, and most residents spend it in the south of our country or in foreign countries. During this period, amounting, as a rule, to two-month period, subscribers' activity drops significantly.

It is possible to make recommendations on improving the promotion of communication services, to find and justify solutions to the identified problems through strategic actions in the field of services promotion (Table 1).

Table 1. Problems of communication sector and recommendations for their solution.

Problem	Directions strategic decisions in promotion field
The impossibility of further extensive development of basic services as a result of market saturation	Maintenance of a high degree of customer loyalty and satisfaction of existing customers needs in order to save customer bases, strengthening of relations. Attracting of newly appearing customers, dissatisfied with competitors customer services
The highest growth rates of consumption and the amount of revenue from Mobile Internet service in the central regions, low innovation value, low solvency of the population	Learning of preferences and living conditions of regional consumers, development of relevant special methods of services promotion, together with the stimulation of change of mobile phones to smartphones and tablets, supporting technologies to provide additional services to create potential customers. Information and conclusiveness of advantages and additional mobile Internet opportunities, promotion of high-traffic use of additional services
The high cost of traditional advertising to reduce its effectiveness	To seek the optimum set of marketing mix in the individualization of demand conditions, with an emphasis on personalized interactive marketing communications tools, which will reduce costs and increase communication effectiveness by targeting, consumer interest to information, and absence of negative attitude of consumers. To promote the popularity of self-services and contact centers
With the development of 4G technologies, new communications services imminent appearance on the market is possible from third-party communications companies (OTT)	Formation of long-term relationship of trust with consumers and a positive attitude to the company brand, continuous techniques improvement. Creation of special cooperation terms with the aim of non-appearance of the need for new types of communication or competitors services

Thus, as a result of the work, the authors found general problems, which are faced by telecommunication companies, working in the Far North at the present stage of telecom market development and formulated basic recommendations.
Also it is necessary to highlight key communication tasks of services promotion strategy:

- development of high loyalty level, strengthening of relationships with consumers, attraction of unsatisfied customers of other companies and making loyal customers of them;
- specialized approaches to promotion of new services on the regional markets, increasing the level of customer involvement in telecommunication companies products, as well as the involvement of consumers in the brand.

5 Concluding remarks

As a result of the work, the authors found general problems, which are faced by telecommunication companies, working in the Far North at the present stage of telecom market development and formulated basic recommendations.
1) Building long-term relationships between the company and the consumer in the telecom market is both the factor of company stability in the market and success of the market and competition of its development. Marketing communications of most operators are still based on the traditional approach and on standardization of technologies. However, the consumer behavior model has undergone profound changes in recent years.
2) To achieve long-term cooperation between the company and the consumer on mutually beneficial conditions customers should be considered as heterogeneous audience of different groups of loyalty. In this respect, companies' communication actions should simultaneously adjust risk groups and retain stable group of consumers.
3) According to traditional approaches, the essence of promoting is in the complex of measures aimed at demand formation and sales stimulation, i.e., in the end, to sales increase. Also promoting is interpreted as the concept of marketing communications, which refers to bonds formed by the company with contact audiences through a variety of influence means, including advertising, PR, sales promotion, promotion, personal selling, as well as informal sources of information e.g. rumors and gossip.
4) Specific features of the cellular market in the Far North are found. Among them are: preferential use of satellite communication channels due to natural and climatic peculiarities; subscriber base limitation due to the low population density in the region; dependence of need for mobile communication services on the level of socio-economic development of inhabited locality; available natural resources and the economic feasibility of their development, influencing the structure and volume of information services consumption.

References

1. Russian Federation Government Resolution of 15.04.2014 no. 313 "On approval of the Russian Federation "Information Society state program (2011-2020)".
2. *Marketing research of Russian market of telecommunication systems* (2014). http://marketing.rbc.ru/download/research/demofile_562949992562127
3. *The concept of long-term Russia's social and economic development until 2020.*
4. Y.A. Kuznetsov, S.E. Markova, Structural approach. Nizhny Novgorod State Technical University, **3** (2013)
5. *The future of the telecom market belongs to services.* http://www.iksmedia.ru/news/5021990-Budushhee-telekomrynka-za-servisami.html
6. *IKS-consulting summarized the development of the telecommunications market and gave the forecast for 2014-2018.* http://www.content-review.com/articles/25920/
7. *The main trend of telecom is profitability stagnation.* http://www.cnews.ru/reviews/new/telekom_2014/articles/glavnyj_trend_telekoma__stagnatsiya_rentabelnosti/
8. *Telecom 2015: the icy serpentine road.* Available at: http://www.mskit.ru/news/n176857/
9. *The review and assessment of development prospects of world and Russian markets of information technologies.* http://fs.moex.com/files/9216 (2014)
10. N.M. Rozanova, D.A. Bulichenko, Terra Econ, **9** (2011)
11. E.M. Grishanova, E.A. Krasnoslobodtseva, T-Commm, **7** (2014)
12. *Regulation of the telecommunications industry in Russia: trends and perspectives* (2015). http://json.tv/ict_telecom_analytics_view/regulirovanie-v-telekommunikatsionnoy-otrasli-rossii-tendentsii-i-perspektivy-20150305052453
13. *The economic and social indicators of the Far North and equated localities in 2000-2013* (2014) http://www.gks.ru/bgd/regl/b14_22/Main.htm
14. V.A. Vityazeva, E.S. Kotyrlo, Syktyvkar State University publ. (2007)
15. A.A. Sozinova, *Analytical and methodological support of regional development projects. In the World of Scientific Discoveries*, **11** (2013)
16. V.V. Fauzer, G.N. Fauzer, I.G. Nazarova, G.V. Korshunov, *Trends and prospects of socio-economic development of the northern regions of Russia: demography, labor, migration, resettlement* (Ekon-Inform publ., Moscow, 2012)
17. *The analysis of the Russian market of additional mobile services (VAS),* Advanced Communications & Media (2008). http://www.acm-

consulting.com/data-downloads/doc_download/26-vas-forum-presentation.html

18. A. Rachinsky, The spread of mobile communication in Russia. Litres (2014)
19. *The development of mobile Internet access in Russia and worldwide: Preliminary results of 2014* (2015). http://json.tv/ict_telecom_analytics_view/razvitie-mobilnogo-dostupa-v-internet-v-rossii-i-mire-predvaritelnye-itogi-2014-goda-20150317043818
20. A.V. Kalinin, Economics publ., **3** (2014)
21. *CNews Telecom 2014: The largest telecommunications company in Russia.* http://www.cnews.ru/reviews/new/telekom_2014/
22. A.A. Kobylko, *Virtual operators as a system resource of the telecommunications market in Russia* (Moscow, CEMI RAS publ., 2012)
23. *Ranking of the Russian regions on life quality. Moscow,* (2014). http://vid1.rian.ru/ig/ratings/life_2014.pdf
24. *Ranking of socio-economic situation of the Russian Federation. Results of 2013,* (2014). http://vid1.rian.ru/ig/ratings/rating_regions_2014.pdf

Evaluation of "direct input" effectiveness for industrial enterprises power supply

Tatyana Malinina[1,*], *Serguei* Smolovik[2], and *Mikhail* Ivanov[1]

[1]Peter the Great St. Petersburg Polytechnic University SPbPU, Saint-Petersburg, Russia
[2]Joint Stock Company «Scientific and Technical Center of Unified Power System», Saint-Petersburg, Russia

Abstract. The cost of energy has increased significantly for the industrial enterprises over the recent years, due to an increase in the electricity tariffs and the changes in the rules of the electricity market. Tariffs for electricity sold to consumers, differentiated according to various parameters, one of which is the voltage range. The higher the voltage range, the lower the tariffs, and thus, the consumer pays less for the supplied energy. Currently only for big consumers made energy supply "direct input" through the construction of an overhead line (OL) with a nominal voltage of 110 kV or 220 kV. However, it is advisable to consider the issue of power supply efficiency of big and medium-sized industrial enterprises based on the "direct input". The questions of voltage level study based on empirical formulas are considered in the article, expressing the dependence of the voltage from the transmission distance and the amount of transmitted power. It was proven that these formulas give a large spread, so intermediate voltage classes 35 kV and 110 kV, which compared with the traditional option of power supply on the basis of the cable line 10 kV, have been selected. Based on the technical and economic calculations it has been proved the value of the transmit power and length of the line at which the power supply of industrial enterprise on the basis of " direct input" with a nominal voltage of 110 kV becomes effective.

1 Introduction

One of the major stages in the planning of production start-up is the selection of the optimal power supply scheme, given that energy costs and industrial power systems constitute from 5 to 60% of the cost, depending on the type of product.

There are centralized and decentralized power supply scheme. In the first scheme, the consumer receives the electrical energy from the power grid, with the second one in the company is expected the construction of its own combined heat and power generation or other means (the local power plant, wind turbines, solar panels, etc.). Centralized electricity supply used in the availability of connection to the grid. This process is characterized by high reliability and efficiency, but requires the construction of lines, substations, and much depends on external conditions (grid). Independent electricity supply is used when it is impossible connection, at a considerable distance the consumer from grid.

Consider a centralized power supply of industrial enterprises. In this case, there is the problem of choosing the optimal voltage class, the number of transformation steps, determining the placement of shop, distributive and main step-down substation. In addition, it is necessary to satisfy the requirements for the quality of electricity. The overall objective of the optimization of industrial power supply (other than those mentioned above) systems includes the task of selection of cross-sections of wires and cable wires, automation and control, accounting and other value indicators.

2 Analytical ways of problem solving

It should be noted that overseas devoted considerable attention to the issue of finding a non-standard voltage rational analytical way. In foreign practice offered the following expressions for the non-standard rational voltage.

In American practice for non-standard voltage is applied Still formula:

$$U = 4{,}34 \cdot \sqrt{l + 16P}, \quad (1)$$

where P - is the transmitted power, MW; l - the distance, km.

Formula Still was transformed by S.N.Nikogosov and reduced to a more convenient form:

$$U = 16 \cdot \sqrt[4]{Pl} \quad (2)$$

In the directory of Swedish engineers nonstandard voltage is defined as follows:

* Corresponding author: t_malinina@mail.ru

$$U = 17 \cdot \sqrt{\frac{l}{16} + P}, \quad (3)$$

where P - is the transmitted power, MW; l - the distance, km.

Formula Still and Nikogosov recommended for enterprises that have a design load of more than 5 thousand kV·A [2].

The above empirical formulas expressing the dependence of the voltage on the transmission distance and the amount of transmitted power, give a large spread. As it is difficult to judge which voltage should be given preference. Moreover, these formulas do not take into account many factors, both economic and technical, so they are not applied in practice.

3 Technical and economic calculations for choosing effective option

Based on all the above we can conclude that a rational power diagram of a specific enterprise is selected on the basis of technical and economic calculations, taking into account the perspective of enterprise development and district, indicators power supply systems and industrial enterprises, and other geographic location.

In recent years, along with the traditional way of power supply using a cable line rated voltage of 10 kV power began to be used the power supply way of an industrial enterprise "direct input" through the construction of overhead lines with a nominal voltage of 110 kV.

High voltage class is characterized by the lowest currents in conductors, and thus the least losses, requires one step of transformation 110/10, as well as the construction of the substation near the place of consumption.

This method is characterized by the largest capital investment in a single chain, the most expensive electrical equipment (switches, circuit breakers), however, the capacity of a circuit is very large. This makes it possible to use the network for the transit of electricity to neighboring businesses, that is, to perform the functions of the electricity supply company and earn extra income.

It is also interesting to consider the option of power supply with the intermediate voltage class 35 kV, accordingly with average values of currents and losses in transmission. The cost of capital in line comparable with the investments in 110 kV lines but less bandwidth. The main disadvantage of this option is the need for an intermediate voltage level and, accordingly, several transformers: 110/35 and 35/10, which leads to the need for the construction of two substations and a significant rise in the cost of this option, the annual increase in the costs of maintenance, repairs and depreciation.

Consider the problem of selecting the optimal variant of the centralized power of the industrial enterprise installed electric capacity of 19 MW, located 4 km from the main network. To do this, we define the optimal voltage class by empirical formulas for transmission of electricity, given above recommended for companies with an electrical load of more than 5 thousand kV·A:

1) Still Formula:

$$U = 4{,}34 \cdot \sqrt{l + 16P} = 4{,}34 \cdot \sqrt{4 + 16 \cdot 19} = 76{,}167 kV; \quad (4)$$

2) Formula of S.N.Nikogosov:

$$U = 16 \cdot \sqrt[4]{Pl} = 16 \cdot \sqrt[4]{19 \cdot 4} = 47{,}241 kV; \quad (5)$$

3) "Swedish" formula:

$$U = 17 \cdot \sqrt{\frac{l}{16} + P} = 17 \cdot \sqrt{\frac{4}{16} + 19} = 74{,}59 kV. \quad (6)$$

The results confirm the conclusions previously made, a considerable spread and uncertainty of results. The average value of the voltage is 66 kV, which is intermediate between the voltage of 35 kV and 110 kV classes, so these voltage classes should be considered in the justification option energy industrial plant. As a result, three cases were considered (see Fig. 1):

1) electricity supply by constructing the cable line (CL) of rated voltage 10 kV;

2) "direct input" of electricity supply that is, through the construction of an overhead line (HVL) with a rated voltage of 110 kV;

3) electricity supply with the use of overhead line (HVL) an intermediate voltage class 35 kV.

Initial data for calculations under option are shown in Table 1.

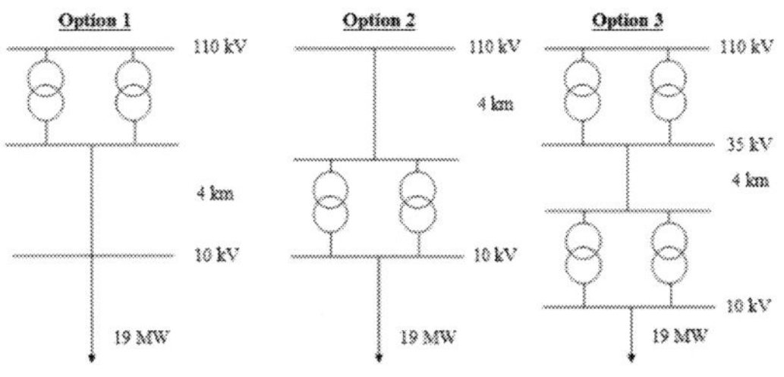

Fig. 1. Scheme of power supply of industrial enterprises.

Table 1. Baseline data on energy supply options.

Index	Option 1	Option 2	Option 3
Rated voltage	***10 kV***	***110 kV***	***35 kV***
The length of the line	4 km		
Using the maximum load	5500 hour		
load factor	cosφ=0,85		
Viewed operation period	15 years		
Location	Leningrad region		

The following assumptions have been made when performing calculations that introduce some uncertainty in the end result:

1) for all considered lifetime electric power consumption is assumed constant;

2) only active losses in the wires were taken into account in the calculation of losses;

3) the construction of substations and overhead lines and putting them into operation is carried out for 1 year;

4) all calculations were carried out based on the integrated value indicators (IVI), given in [3];

5) recalculation of the basic price 2000 year in to price of 2015 year was carried out on the basis of the conversion factors set out in the inter-regional information-analytical bulletin "Price indices in construction", published by KO-INVEST. The conversion factor prices 2000 year in to prices of 2015 year for capital investments in the branch "Electric power industry" is 8.33 [4];

6) zonal multipliers are taken to the European part of Russia;

7) all rates are assumed equal the unit to account complicating conditions.

For more accurate results, it was necessary to specifically consider the situation taking into account the logistics and production capabilities.

Capital expenditures for all of the options under consideration are composed of capital investment in the substation and line (aerial or cable depends on options):

$$I=I_{sub}+I_{line}, \tag{7}$$

where I_{sub} - capital investment in the substation (include the cost of the transformer cell, the cost of electric vehicles, for an accomplishment costs, temporary facilities, design and survey works, supervision, other works and the value of the land);

I_{line} - capital investment in the construction of air / cable line (include the material cost of the work on laying, landscaping, temporary structures, design and survey works, supervision, other works and costs, the cost of land).

To calculate the cost of capital used aggregated cost parameters without VAT according to [3], given the prices of 2000, which were recalculated in the prices of 2015 with using conversion factors [4].

The annual costs were calculated for comparative analysis, which include the cost of the substation and cable line, as well as losses in transmission C_{los} and mathematical expectation of damages from unreliable energy supplies M(d):

$$C=C_{sub}+C_{line}+C_{los}+M(d) \tag{8}$$

The value of M(d) is determined by multiplying the specific damage to the undersupply of electricity, ie $M(d) = d \cdot \Delta W$.

The costs of the substation include the following components:

$$C_{sub}=C_{a\_tr}+C_{m\_tr}+C_{r\_tr}+C_{a\_eq}+C_{m\_eq}+C_{r\_eq} \tag{9}$$

Conventions and formulas for the calculation of the substation costs are shown in Table 2.

Table 2. Annual costs for substation.

Index	Annual costs	The formula for calculating *)
C_{a_tr}	Annual depreciation expense on transformers	$C_{a\_tr}= n_a \cdot FA_{tr}/100$
C_{m_tr}	Costs for maintenance of transformers	$C_{m\_tr}= n_m \cdot FA_{tr}/100$
C_{r_tr}	The cost of repairs of transformers	$C_{r\_tr}= n_r \cdot FA_{tr}/100$
C_{a_eq}	Annual depreciation expense on electrical equipment	$C_{a\_eq}= n_a \cdot FA_{eq}/100$
C_{m_eq}	Costs for maintenance of electrical equipment	$C_{m\_eq}= n_m \cdot FA_{eq}/100$
C_{r_eq}	The cost of repairs of electrical equipment	$C_{r\_eq}= n_r \cdot FA_{eq}/100$

*) FA – fixed assets; n_a, n_m, n_r - norms of depreciations, maintenance and repair according to the data [3].

For the air (cable) line the costs were determined by the formula:

$$C_{line}=C_{a\_line}+C_{m\_line}+C_{r\_line}. \tag{10}$$

Symbols and formulas for the calculation of the substation costs are shown in Table 3.

Table 3. Annual costs of air / cable lines.

Index	Annual costs	The formula for calculating **)
C_{a_line}	Annual depreciation expense on air / cable lines	$C_{a\_line}= n_a \cdot FA_{line}/100$
C_{m_line}	Costs for maintenance of air / cable lines	$C_{m\_line}= n_m \cdot FA_{line}/100$
C_{r_line}	The cost of repairs of air / cable lines	$C_{r\_line}= n_r \cdot FA_{line}/100$

**) n_a, n_m, n_r – norms of depreciations, maintenance and repair according to the data [3].

The costs associated with payment of electricity losses, defined as:

$$C_{los}=p_e \cdot \Delta W_{los}, \tag{11}$$

where p_e – the average price of electricity for 2015 for the tariff group "network organizations that buy electricity for compensation of losses of electric power";

ΔW_{los} – value of transmission loss, kW.

According to the LLC "RKS-Energo" - guaranteeing electricity supplier in the territory of the Leningrad region - unregulated electricity prices to compensate for losses by months in 2015 ranged 1.47-1,93 rub. / kW·h .

On the basis of these data, the average price value was taken into account, which amounted to $p_e = 1.7101$ rub. / kW·h.

The value of the losses in the transmission is determined according to the formula:

$$\Delta W_{los}=3 \cdot I^2_{max} \cdot R_{line} \cdot \tau \cdot n_{line} \tag{12}$$

where I_{max} - maximum normal current of the line;

R_{line} - the line resistance;

n_{line} - the number of lines (lines);

τ - equivalent time of loss, which is determined from the graph according to the data load factor and peak load time using [5].

The main criterion for choosing the best option energy efficiency of industrial enterprise is the criterion of maximum net present value, which is the difference between the discounted results of R and discounted cost C:

$$NPV = R - C \rightarrow max . \quad (13)$$

Income and costs are the same for all three options, so instead of the criterion of a maximum net present value, we can use the criterion of the minimum discounted costs.

$$C = \sum C_t(1+E_{nr})^{-t}, \quad (14)$$

where C_t - the costs for the year t,

E_{nr} - the discount rate (the interest rate of the bank).

Given that the calculation of the efficiency variant carried in basis (constant) prices, we can use the real interest rate. There are nominal and the real rate. According to Fisher's equation:

$$E_{nr}=E_r+r_i+E_r \cdot r_i, \quad (15)$$

where E_{nr}, E_r - nominal and real rates of bank, r_i - the percentage of inflation.

For small values of interest rate E_r and inflation r_i product $E_r \cdot r_i$ can be neglected. Then, instead of the nominal bank rates in the calculations we can use the real rate, which can be taken as equal to 2-3% [6].

The results obtained by the choice of power options of industrial enterprise in the transmission power equal to 19 MW at a length of 4 km line are shown in Figure 2.

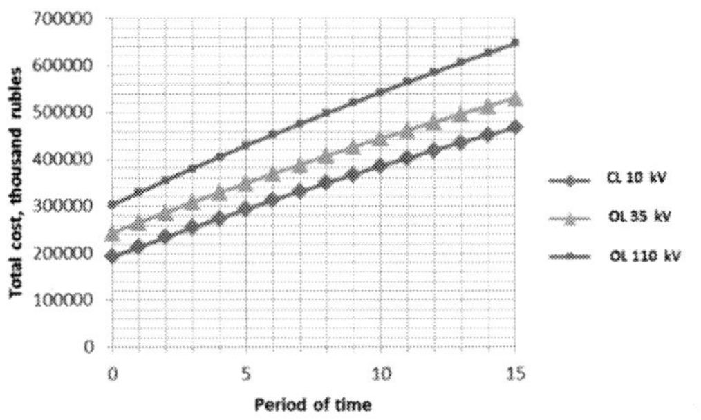

Fig. 2. Comparative characteristics of the three power supply options, depending on the time of operation.

The calculation results show that the option of power supply "direct input" is not effective.

Compare options of energy supply company also carried out when changing the transmitted power from 10 MW to 29 MW with using the minimum of resulted expenses taking into account the security of energy supply:

$$RE = E_{nr} \cdot I+C+M(d) \rightarrow min, \quad (16)$$

where I - capital investment, thousand rubles / year;

C - the annual operating cost, assumed constant during the entire period of operation, rubles / year;

E_{nr} – normative coefficient of comparative effectiveness of capital investments, which in a market economy is accepted at the refinancing rate of the Central Bank;

M(d) – the mathematical expectation of damages from unreliable energy supplies the mathematical expectation of of loss from energy insecurity, thou. / year.

The results are shown in Figure 3.

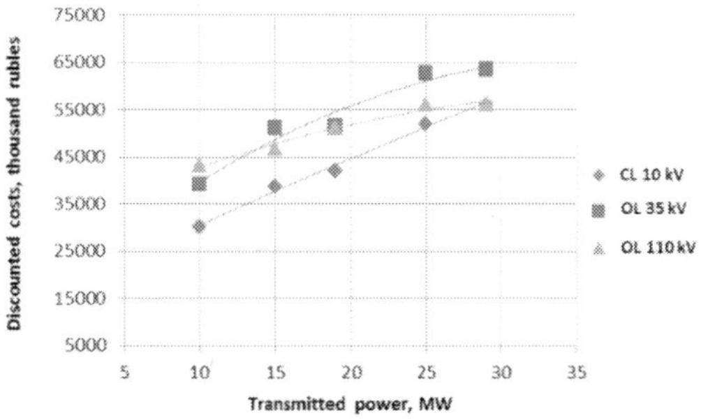

Fig. 3. Comparison of energy options depending on the transmitted power.

On the basis of the calculations we can conclude that the most effective option at the lowest cost given transmission capacity to 29 MW is a variant of the cable line 10 kV. In the transmission more than 29 MW of power supply option overhead line 110 kV becomes preferable the option of the cable line 10 kV. Option 35 kV overhead line is the least effective.

There was also a comparative analysis of options for the industrial supply company, depending on the length of the line (from 1 km to 8 km) using a minimum criterion of reduced costs, taking into account reliability of energy supply (see. Fig. 4).

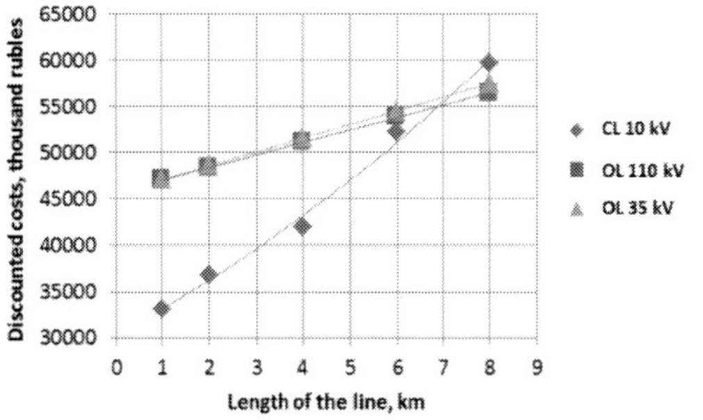

Fig. 4. Dependence of the discounted costs of the line length.

As can be seen from Figure 4 for line lengths of up to 7 km of the highest priority option when transferring 19 MW power is a cable line 10 kV, with a length of more than 7 km of line the best option is the option of using the overhead line 110 kV.

3 Conclusions

1. Existing analytical approaches to find rational voltage for electric power supply of industrial enterprises give a considerable variation in voltage level depending

on the parameters of transmit power and line length. It is expedient to consideration of voltage 35 kV and 110 according to the resulting medium voltage 66 kV.

2. Three energy supply options have been considered In justifying of energy supply of industrial enterprises: energy supply is carried out on the cable line 10 kV overhead line at 110 kV and 35 kV overhead line. On the basis of the calculations It can be concluded that the optimal variant of the power supply of industrial enterprise with capacity of 19 MW, and the 4 km long for connection to the network is the option of using a cable line 10 kV. Other power supply options have as an advantage - a large energy capacity. In this case, the company can act as a supplying company for closely located energy consumers, and thus can extract additional profit.

3. Choosing the best option depends on the installed power capacity of the enterprise and line length. On the basis of calculations, it was determined that the option 110 kV overhead line become preferred over the other options, from the transmit power 29 MW and the line length of 4 km ("direct input" option). Influence of the line length on the choose the optimal variant showed the following. The most effective option is a power option using a cable line 10 kV transmission power 19 MW with a cable length of up to 7 km, with a cable length of more than 7 km the most profitable option is the option of using the overhead line 110 kV.

References

1. A.A. Fedorov, V.V. Kamenev, *Fundamentals of power industry* (Energy, Moscow, 1979)
2. O.P. Balashov, Y.S. Burdochkin, *Power supply systems: Textbook for course designing for students of specialty 140211 all forms of learning* (Rubtsovsk Industrial Institute, Rubtsovsk, 2009)
3. D.L. Faibisovich, *Handbook for the design of electrical networks* (ENAS, Moscow, 2012)
4. CO-Invest. Price indices in construction, **91** (2015)
5. A.A. Glazunov, Electrical networks and systems (Gosenergoizdat, Moscow, 1960)
6. T. Malinina, R. Shulginov, Problems of modern economics-Eurasian Intern. Scientific and analytical magazine, **2**, 322-325 (2012)

Leadership styles and managerial culture at industrial enterprises with different modernization and innovation processes

Sergei Glukhov[1,*], *Petr* Glukhov[2,*]

[1]South Ural State University, Logistics, Economics and Management in Transport, Supply and Distribution Department of Higher School of Economics and Management, Chelyabinsk, Russia
[2]South Ural State University, Laboratory «Psychology and Psychophysiology of Stress Resistance and Creativity» of Higher School of Medicine and Biology, Chelyabinsk, Russia

Abstract. The article presents the results of an empirical research into the transformational leadership and managerial culture at industrial enterprises with different modernization and innovation processes. The authors suggest the criteria for differentiation of options of making changes at the industrial companies: the attitude to their own research base, the focus on introduction of innovations, the investment potential for the introduction of innovations, contacts with research and educational agencies, presence of institutional innovation structures in the structure. It was found out that the managerial profile of an industrial enterprise, focused on innovation activities, involves the index of transformational leadership and the managerial culture with a subject-oriented management, i.e. democratic, which is manifested in the commitment to the group decision making process; humanizing, which is manifested in the attitude towards the employees as being independently valuable regardless the context of achieving organizational results; and innovative, which is manifested in the constant quest for changes.

1 Introduction

Quality and speed of introduction of changes at industrial enterprises determine their adaptability to changing market conditions and, eventually, affect their competitiveness. To a large extent, effectiveness of introduction of changes depends on the characteristics of personnel management. Tools for assessment of human capital of the managers of industrial enterprises become highly sought in addressing the problem of introduction of changes.

Additionally, the importance of the task is related to the cultural chasms and psychological barriers towards introduction of changes, formed as a result of traditionally extensive development of the Russian economy, easier access to resources, and fear of changes [1, 2].

Thus, it would be logical to focus the research on the intersection of such sciences as Economics, Management and Psychology, on the economic and psychological aspects of Management – on assessment of a relationship between the individuality of a manager and the level of managerial culture at an industrial enterprise. It will let us identify and describe a profile of a manager who can effectively manage both the introduction of external innovations (modernization process), and the introduction of innovations created within the enterprise (innovation process). This approach can be an effective tool for human resources management and it can be used in the system of evaluation and training of personnel at industrial enterprises.

2 Topicality. Scholarly Importantce of the Issue

Contemporary study of personality and managerial culture of an industrial organization in the context of modernization and innovation processes is conducted both by foreign and Russian authors.

Some authors studied leadership position of the managers during introduction of changes at an enterprise. For example, Alain Vas studied leadership through management in complex, vague and ambiguous conditions, due to a specific set of such characteristics as commitment to support organizational changes, courage, belief in the staff's ability to take over responsibility [3]. Richard L. Daft analyzed leader's strategy of changes in management through the individual scenarios of managers' activities. He singled out a strategy of "creative self-expression", which implies a non-standard behaviour, creative actions, which approve new norms and values; a strategy of "alliances" which consists in formation of a group of allies of innovations around an innovative leader, and a number of other strategies [4]. B. Bass described and empirically identified a transformational style of leadership, aimed to initiate in the employees an aspiration to achieve higher

* Corresponding author: glukhovsv@susu.ru

DOI: 10.1051/shsconf/20173501033

organizational goals. A manager with such style instigates the staff to move to a higher level of motivation, he raises their ethical expectations. B. Bass singled out a specific character of the influence on the part of a manager with a high level of transformational leadership: the staff gets motivated and gains sight of the set goals; their self-actualization needs are satisfied; their interests expand beyond their personal fulfillment (a turn to organizational actualization) [5, 6].

V.G. Griazeva-Dobshinskaia suggested using a system approach to study innovative leadership of managers. It is the most sophisticated and modern approach, which takes into account manifestations of behaviour of the managers at different levels of managerial communication [7-9]. At the same time, the majority of the above listed studies of leadership in the context of changes is based on the well-developed theoretical concepts and on high expertise of its authors. A comparison of the socio-economic characteristics of industrial enterprises with a distinct particular management attitude, described in the theory, remains central for identification of a profile of a manager who can effectively manage introduction of changes.

Research of the managerial culture during introduction of changes. R. Florida described the specificity of a social environment which favours self-actualization of creative subjects and suggested a three "T" approach (technology, talent and tolerance) [10]. He found out that creative staff, contributing to the innovative development of enterprises, accumulates in spaces characterized by a high degree of tolerance (acceptance of diversity) and technology (acceptance of complexity and novelty).

Arie de Geus described conditions of a managerial environment in which adaptation occurs efficiently. Such conditions include tolerance for alternative behaviour and, as a consequence, decentralization, high cohesion and identity of personnel, enabling synergetic behaviour, sensitivity to the environment, which allows to accurately assess the events taking place inside and outside [11]. Iu.D. Krasovsky suggested paradigms of managerial culture, which can promote and block circulation of information and display of initiative [12-14].

In order to study the personal profile of a manager and the managerial culture of an industrial enterprise under conditions of constant changes, we chose a concept of the transformational style of leadership, suggested by B. Bass in his works [5, 6], proved to be efficient in foreign studies, and a concept by Iu.D. Krasovsky, which is the most easily diagnosed and reflects the specifics of the Russian management culture [12-14]. An explorative framework was developed within these concepts in order to identify individual characteristics of the manager and of the administrative culture of an industrial company.

Accordingly, the aim of our research is to identify a personal profile of a manager who can effectively manage both the introduction of external innovations (modernization processes) and the introduction of innovations created within the enterprise (innovation processes), and it will be formed within the framework of the indicators suggested in these theoretical concepts.

3 Statement of the Aim

We set out the following tasks in order to achieve the aim of this article.

The first task is to differentiate the specifics of the introduction of changes at the industrial enterprises, whose managers participated in the study.

The second task is to study the relationship between leadership styles of the managers and the formed managerial culture at the industrial enterprises.

The third task is to draw up personal profiles of the managers who effectively manage the processes of modernization and innovation at the industrial enterprises.

4 The Oretical Part

B.M. Bass and his co-author B.J. Avolio defined transformational leadership as a new type of leadership and suggested understanding of transactional leadership as a leadership of the old school, based on interaction. They analyzed transformational and transactional leadership in a unified complex, which let them develop a model for their description [5, 15-17].

In our study, we are going to focus only on the description of transformational leadership (TL). There are four elements of transformational leadership.

The first one is idealized influence (II). The leaders with this ability give their subordinates a sense of purpose, they foresee the future, and set up new high standards. The transformational leaders suggest bright and meaningful images of the joint future. Ultimately, the employees feel that they take part in reaching a common goal, they feel concurrence of the objectives and values of organizational behaviour.

The second one is inspirational motivation (IM). The transformational leaders provide inspiration, high expectations, they inspire pride in the employees. They emphasize motivational managerial manifestations as much as possible. The atmosphere of openness and trust is important for those employees who generate new and creative ideas, and therefore the function of control, which is of top-priority during the stabilizing stages of development of the company, becomes irrelevant in times of changes.

The third element is intellectual stimulation (IS). The managers who possess this ability, help their employees to be more creative and innovative, they encourage the creation of new modes of action, and stimulate emergence of new ideas. They create a space for growth through open communication, through support for innovative ideas, and respect for those who suggest them.

The fourth element is individualized consideration (IC). The managers, leaders, who transform reality, communicate on a personal level, they consider each employee as an individual, give them the tasks which would contribute to their personal growth. Such

DOI: 10.1051/shsconf/20173501033

approach allows them to build up a successful communication with the most creative and non-standard employees, taking into account their differences and peculiarities.

J. Burns described the leaders transforming reality as those who provoke the followers to go beyond their own interests and, thus, beyond their self-actualization. This type of leadership is considered as the highest level of leadership, when the leader expands the interests of the subordinates, when s/he generates awareness and acceptance of the goals and objectives of the group in them, when s/he awakens in them a desire to look beyond their own interests for the good of the whole group [18].

Thus, the general formula of the integrative index of transformational leadership of the managers, taking into account the proportion of indicators (determined by the number of the questions in the test), is as follows:

$$TL = (II*8) + (IM*4) + (IS*4) + (IC*4), \quad (1)$$

where TL stands for the integrative index of transformational leadership of the managers; II stands for idealized influence; IM stands for inspirational motivation; IS stands for intellectual stimulation; IC stands for individualized consideration.

Personal fulfillment of a leader is possible if a corresponding managerial culture is present at the enterprise. Iu.D. Krasovsky identifies the following parameters for the analysis of management culture, aimed precisely to perform analysis in the context of introduction of changes at the company: "openness" or "closeness" of the managerial culture, an emphasis on strength or flexibility of authority, a focus on business or personal communications, domination of fragmented individualistic or integrative group processes [12-14].

"Openness" of the organizational culture of a company is characterized, in the first place, by freedom of opinion exchange, an advanced communication system, completeness of information about solvability of management problems at employees' disposal, understanding of interdependence of everyone in the company, organizational clarity and transparency of business relations, solution of the "joint" problems of different departments, freedom of maneuver for employees with authority and rights, their inclusion into direct discussions and solution of the most important management issues.

"Closeness" of the organizational culture of a company is a directly opposite socio-cultural situation when a free exchange of opinions is hindered, communication systems are limited, the channels of information are blocked and narrowed, the interdependence of employees and departments is difficult to understand, when one person solves his or her own problems and does not worry about how this decision will affect the others, when the most important problems are solved privately, and the staff is caught in the need of execution of the managerial decisions taken. If such organizational culture exists and expands in the company, it means that the employees are being increasingly alienated from the business organization, and they conceive and evaluate their position as unsatisfactory.

Iu. D. Krasovsky suggests to conduct the analysis of managerial culture through more detailed categories, including administrative orientations of managers and, therefore, suggests an exploratory technique for managerial culture called "KRAB" [12-14].

The built-in model of this technique consists of two polar groups of managerial orientations: object and subject orientations. Object-oriented management can be converted to one-sided ways of "top-down" influence: a) autocratic, when the employees are subject to the power (the will) of the head; b) technocratic, when the employees are subject to the production process; c) bureaucratic, when the employees obey the organizational order, sometimes to the detriment of the business.

Subject-oriented management can be: a) democratic, when the head constantly takes into account the opinion of the workers during the adjustment of the workflow management; b) humanizing, when the head treats the workers with respect for their human dignity; c) innovative, when the head organizes introduction of initiative suggestions by giving the employees freedom of creativity.

The subject managerial orientations correspond to manifestations of transformational leadership, that is why our empirical research has been focused on them.

5 Results of the Empirical Research

Socio-economic characteristics of the three selected industrial enterprises of the Ural Federal District of the Russian Federation were analyzed in order to complete the first task of the empirical research.

Characteristic of the industrial enterprise No. 1.

The main production assets of the enterprise No. 1 have not been modernized, in large part, for a long period of time and can be included in the category of the off-market equipment. Currently, most competitive businesses use more modern compatibles, which allow them to gain a competitive advantage at the stage of production.

The company produces products of lower and middle-price segment. This situation can be explained by the level of technological development of the enterprise, which does not allow the company to produce products of a higher level.

Technical re-equipment of the production processes of the enterprise is being slowly implemented, without expanding production facilities. At the same time, the company No. 1 proclaims to have a task to conform to the modern requirements for the technical development of certain sections of the production through the introduction of new equipment and technology, mechanization and automation of the production processes. There are virtually no examples of modernization and replacement of obsolete and worn-out equipment with the new, more productive one at the vast majority of the plants of the enterprise No. 1.

The organizational structure of the enterprise includes the units, which are responsible for the introduction of innovations, but the share of the production of these units is relatively small in total volume of production. Practically, at the moment, the company has implemented only one project, claiming to be of high technology. However, the project deadlines have been extended several times, which also gives evidence of the difficulties with the introduction of innovations.

For a long time, the company No. 1 has had a severe shortage of funding due to a high debt load in the form of loans taken out for business support and in order to reinforce operating assets, rather than to modernize production.

There is a negative trend in revenues and operational costs, as well as in financial and production indicators in comparison with the average values. It can be stated that the modernization carried out at the company No. 1 is fragmented.

Characteristic of the industrial enterprise No. 2.

The company is actively developing the use of modern materials and new equipment with varying degrees of novelty. First of all, it is explained by implementation of a strategy of acquisition of new markets (increase of sales in the territory of the Siberian Federal District of Russia), and, secondly, by severization of certification requirements for the company products.

The company No. 2 faces a task to match the severed requirements (in particular, an increase in the warranty period for the products of the company), which create a need to provide an intense introduction of modern technologies aimed at improving the quality of the industrial products.

This task is being solved through a search for new technologies and is associated with the appearance of modern equipment. The activity of the enterprise falls within a number of programs of development of the industrial branch, which include measures of innovative procurement of introduction of scientific research and development activities aimed to improve theoretical foundations and computational methods for enhancing the reliability and durability of the production.

Implementation of the modernization strategy of development of the enterprise is, on the one hand, a requirement of state programs of development of the industrial branch, and on the other hand, it is a choice of the company's management, and awareness of the advantages of the modern technologies.

The company has approved a program of renovation of production, which includes tight time limits of completion of the modernization activities. All this makes it possible to characterize the enterprise No. 2 as a company with an intense level of modernization.

Characteristic of the industrial enterprise No. 3.

The company No. 3 enjoys a status of a research and production association and belongs to knowledge-intensive industries in the Russian Federation. The company's products are diversified. The company produces products both for defense and civil purposes.

Assimilation of new products is one of the priorities of development of the enterprise. When placing orders, great attention is paid to formation of competitive value of the enterprise's products: to a large extent, quality characteristics of the developed products are not inferior to the foreign-made analogues; therefore, the cost for the consumers becomes a key factor in decision-making. Also, the strict terms are set for manufacturing of the first lots of products produced within the tenders won by the enterprise. All this requires high speed and intensification in all structural divisions of the enterprise and the overall high production standards.

The specific requirements for the quality of the products are, in general, related to the implementation of the state defense order of the Russian Federation and to the presence of the system of state military technical reception at the enterprise. The quality management system of the company complies with the specific requirements, and this determines high processability and research intensity of the entire production process.

The enterprise No. 3 has a production base, which is relatively similar to that of the leading Russian and foreign companies. The enterprise cooperates with the Russian Union of Scientific and Engineering Public Organizations and with the Academy of Engineering Sciences of the Russian Federation. The company cooperates with higher educational institutions of the country, which is considered as an important part of the personnel policy of the enterprise, it possesses its own research center, institutionalized in the form of a technology park. The latter is considered as an institutional resource dedicated to the interaction of specialists engaged into innovative activities.

The company No. 3 is following a concept of formation of a team of managers, professionals and executives, which can meet the challenges of innovative development and whose members should have the necessary knowledge and skills allowing them to carry out innovative activities. Therefore, the staff of the enterprise takes part in annual international, sectorial and cross-sectorial exhibitions, conferences, round tables dedicated to new types of products and modern production technology; they systematically attend courses of advanced vocational training in the sphere of their professional expertise and in personnel management.

We can say that this enterprise is in good position for elaboration and development of scientific and technical solutions, it determines directions of innovative business investment, it has a detailed plan to achieve the goals of innovation activities, aimed at strengthening the company's competitiveness, not only nationally, but also internationally.

We can conclude that the level of development of the enterprise No. 3 corresponds to an innovative company.

The summarized results of evaluation of the level of development of the three industrial enterprises of the Ural Federal District based on the characteristics of modernization and innovative development are presented in the Table 1.

This comparative evaluation lets us legitimately characterize the enterprise No. 1 as a company with fragmented modernization, the enterprise No. 2 as a company with intensive modernization, and the enterprise No. 3 as an innovative company.

Table 1. Comparative Evaluation of the Level of Development of the Industrial Enterprises Based on the Characteristics of Modernization and Innovative Development.

Compared characteristic	Enterprise No. 1	Enterprise No. 2	Enterprise No. 3
Attitude to their own research base	Absent and is not forming	Absent but is forming	Present and is being maintained
Orientation towards introduction of innovations	Oriented towards a partial replacement of worn-out equipment	Oriented towards import of technology	Oriented towards elaboration of its own high-technology products
Investment potential of introduction of innovations	Low investment potential	High investment potential	World-class investment potential
Contacts with scientific and educational agencies	Yes, at the level of staff training and quality control	Yes, at the level of staff training	Yes, at the level of staff training, R&D conduction
Presence of innovational structures in the structure of the enterprise	No	No	Yes: a technology park
Conclusion based on the analysis results	With fragmented modernization	With intensive modernization	Innovative enterprise

Now we would like to construe the results of the empirical research aimed at studying the relationship between managers' leadership style and manifestations of managerial culture at the industrial enterprises (Tables 2 and 3).Mann–Whitney U-test was used to investigate the differences [19-20].

Table 2. Differences in the Manifestation of Transformational Lleadership Style of the Managers of the Industrial Enterprises with Different Modernization and Innovation Processes.

Industrial enterprises (Number of people)	Transformational leadership		
	Rank sum	**Average number**	**U emp.**
Enterprise No. 1: fragmented modernization (80 people)	6333,5	56,4	3093,5
Enterprise No. 2: intensive modernization (89 people)	8031,5	58,3	
Enterprise No. 1: fragmented modernization (80 people)	4416,0	56,4	**1176,0** **
Enterprise No. 3: innovative activity (59 people)	5314,0	63,6	
Enterprise No. 2: intensive modernization (89 people)	5548,0	58,3	**1543,0** **
Enterprise No. 3: innovative activity (59 people)	5478,0	63,6	

[a] ** Differences significant at $\rho \leq 0{,}01$

The Table 2 shows that transformational leadership style is more significantly represented by the managers of the enterprises focused on innovative activity, than by the managers of the enterprises oriented towards fragmented modernization, and the managers of the enterprises with intensive modernization level.

The Table 3 shows that the innovative managerial orientation is more evident at the enterprise, focused on innovation activities, and at the enterprise, oriented towards intensive modernization, than at the enterprise with fragmented modernization.

Table 3. Differences in the Manifestation of Managerial Culture at the Industrial Enterprises with Different Modernization and Innovation Processes (Through the Example of Managerial Orientations).

Industrial enterprises (Number of people)	Subject-oriented management					
	Innovative managerial orientation		**Humanizing managerial orientation**		**Democratic managerial orientation**	
	Average number	U emp.	Average number	U emp.	Average number	U emp.
Enterprise No. 1: fragmented modernization (80 people)	39,3	**2897,5** *	40,0	2996,5	42,4	3345,0
Enterprise No. 2: intensive modernization (89 people)	41,5		41,5		42,5	
Enterprise No. 1: fragmented modernization (80 people)	39,3	**1895,5** *	40,0	**1705,5** **	42,4	**1591,0** **
Enterprise No. 3: innovative activity (59 people)	41,7		42,9		46,2	
Enterprise No. 2: intensive modernization (89 people)	41,5	2605,0	41,5	2319,5	42,5	**1993,5** *
Enterprise No. 3: innovative activity (59 people)	41,7		42,9		46,2	

[b] * Differences significant at $\rho \leq 0{,}05$; ** Differences significant at $\rho \leq 0{,}01$

The managerial culture at the industrial enterprise focused on innovation activities is notable for the highest degree of democratic and humanizing managerial orientations.

6 Conclusions

The results of the conducted research show that such criteria as an attitude to their own research base, a focus on innovation, an investment potential for introduction of innovations, contacts with research and educational agencies, presence of institutional innovation structures in the corporation structure, allow us to differentiate industrial enterprises according to how their modernization and innovation processes differ. This is proved by the differences in the managers' leadership styles, aimed at introducing changes into the managerial organizational culture, which would support such type of managers' leadership.

The profile of a manager of the industrial enterprise, focused on innovations, includes an index of transformational leadership, which demonstrates managers' orientation towards changes through creation of inspirational motivation, intellectual stimulation and individualized consideration.

The managerial culture at the innovative industrial enterprise includes subject-oriented management, which can be democratic, manifested in commitment to a group decision making process; humanizing, which is manifested in an attitude towards the employees as being valuable and important, occasionally, even without taking into account their organizational results.

The managerial culture at the innovative industrial enterprise and the enterprise with intensive modernization includes innovation-oriented management, which means that the managers are aware of the importance of the changes and they are ready to initiate their constant introduction.

The data obtained can be used to carry out selection procedure of managers at innovative industrial enterprises and enterprises with intensive modernization.

The work was supported by Act 211 Government of the Russian Federation, contract № 02.A03.21.0011.

References

1. N.M. Lebedeva, E.G. Yacin, Forsyth, **3**, 16-26 (2009)
2. A.P. Prokhorov, *Russian management model* (A. Lebedev studio publishing house, 2014)
3. A. Vas, Journal of General Management, **27**, **1**, 71-89 (2001)
4. R.L. Daft, P.G. Lane, *The Leadership Experience* (Eksmo, 2006)
5. B.M. Bass, B.J. Avolio, *Improving Organizational Effectiveness through Transformational Leadership* (Thousands Oaks, CA, 1994)
6. B.M. Bass, R.E. Riggio, *Transformational Leadership* (2005)
7. V.G. Gryazeva-Dobshinskaya, Social psychology of creativity, 200-207 (2007)
8. V.G. Gryazeva-Dobshinskaya, *Psychology of innovation management of organization* (Publishing house of the SUSU, 2013)
9. V.G. Gryazeva-Dobshinskaya, Y.A. Dmitrieva, *Leadership in innovations: manager's activity resources modeling: monograph* (Publishing house of the SUSU, 2016)
10. R. Florida, *The Rise of the Creative Class. And How It's Transforming Work, Leisure and Everyday Life* (Basic Books, 2002)
11. A. Geus, *The Living Company: Growth, Learning and Longevity in Business* (HBS Press, 1997)
12. Y.D. Krassovsky, *Organizational behaviour* (UNITI-DANA, 2012)
13. Y.D. Krassovsky, *Socio-cultural basis for business-organization management* (UNITI-DANA, 2007)
14. Y.D. Krassovsky, *Behaviour management in a company: effects and paradigms: practical manual* (INFRA-M, 1997)
15. B.M. Bass, Journal of Leadership & Organizational Studies, **7**, **3**, 18-40 (2000)
16. B.J. Avolio, M. Bass, Journal of Occupational and Organizational Psychology, **72**, 441-462 (1999)
17. B.M. Bass, P. Steidlmeier, Leadership Quarterly, **10**, 181-217 (1999)
18. J.MG. Burns, *Transforming Leadership: A New Pursuit of Happiness* (Atlantic Monthly Press, 2003)
19. A. Nasledov, *SPSS. Computer analysis of psychology and social sciences data* (Piter, 2007)
20. E.V. Sidorenko, *Methods of mathematical processing in psychology* (2004)

Program-oriented approach to resource saving issues in construction materials industry

Galina Novikova[1,*], *Margarita* Lukyanitsa[1], and *Larisa* Chizho[1]

[1]The Department of the Economics and Projects Management in the Building, Volgograd State Technical University,Volgograd, Russia

Abstract. The construction as a sector of the economy is one of the largest consumers of energy resources, and the building materials industry is today one of the most energy-intensive construction industry. At the enterprises of the building materials industry the different approaches and methods are used to solve resource and energy problems. Energy saving is considered not as an complex approach in the enterprise activity, but as activity for the implementation of specific energy-saving projects, which have limitations in time and in resources. The authors suggest to use a software-oriented approach to solving the problems of resource and energy saving. For practical application of program-oriented approach we offer to use a structuring method of the decision-making, not previously used to solve problems of resource and energy saving.

1 Introduction

Activation of resource and energy saving is now regarded as one of the major global problems of our time due to mostly non-renewable nature of big part of the energy resources and the negative ecological impact of industrial production. Resource and energy saving is one of the most important factors providing the efficient functioning of the industry and the economy as a whole. This is achieved by implementing a complex of measures that provide the transition to the use of new environmentally friendly materials, energy saving technologies, machinery and equipment, as well as the further development of energy management systems.

Construction as a sector of the economy is one of the largest consumers of energy resources, and the building materials industry is today one of the most energy-intensive construction industry and, therefore, has significant potential for use of complex resource and energy saving technologies, which focus on the development of resource and energy saving program, based on calculations and reasoned statements of managers and specialists who are familiar with the state of affairs and prospects of development in the various practice spheres of the enterprise [1,2].

In this regard, the development of resource and energy programs formation methods at the enterprise has now not only theoretical but also practical importance. Creation of tool for the programs formation of resource and energy saving is a topical scientific and practical task that is aimed at improving the energy efficiency of production, reduce price cost and improve the economic efficiency of the construction company [1]. Formation of resource and energy saving programs based on program-oriented approach will allow even while reducing the amount of financing to achieve high results due to the growth of labor productivity, the decrease of energy and resource capacity, the improvement of capital productivity and management efficiency [2].

Literature review. Various aspects of this problem are shown in the works of foreign and domestic scientists.

Software - oriented approach to the solution of industrial problems has been studied by following scientists: D.A. Nasimov, I.I. Doronina, V.N. Borobov, E.A. Ivanova, E.V. Gorynya, B.M. Zhukov, V.N. Leksin, B.N. Porfiryev, I.V. Osinovskaya, O.V. Lenkova, N.A. Serebryakova, Y.A. Salikov, O.Y.Kolomytseva, T.A. Pakhomova, N.V. Grishchenko [3,4-7].

Problems of resource and energy saving on the scale of the national economy are explored scientists: X. Guo, Q. Zhu, L. Lv, J. Chu, J. Wu, I.J. Faber, J. Sambor, A. Bagley, Y. Niu, R.Q. Si, Z. Li, J.W. Ma, S. Dasgupta, J. Roy, Y. Bian, M. Hu, Y. Wang, H. Xu, and others [8,9-12].

Theoretical foundations of the energy management system, are described in the works of V. Blass, C.J. Corbett, M.A. Delmas, S. Muthulingam and others [13].

The study of various aspects of resource and energy saving in the sphere of investment and construction, the definition of the economic substance of this process have become the subject of scientific interest of authors such as A.A. Volodchenko, V.S. Lesovik, A.N. Volodchenko, E.S. Glagolev, L.H. Zagorodnjuk, Y.V. Pukharenko, K.A. Aleksandrovna, G. Balčiūnas, S.Vėjelis, L. Lekūnaitė, A. Kremensas, M. Garg, A.Pundir, M.P. Devaki, B. Balakrishnan, A.S.M. Abdul Awal and etc [14,15-20].

The problems of energy saving and energy efficiency increase at the enterprise are investigated in the works of

* Corresponding author: volgalan@yandex.ru

P.-H. Huang, S.-S. Sun, W.Liao, A. Gawlak, J. Szkutnik, Y. Fu, Z. Wei, K. Xu, L. Pang, G.J. Tsekouras, M.A. Tsaroucha, C.D. Tsirekis, A.D. Salis, E.N. Dialynas, N.D.Hatziargyriou, Y.-H. Lin, H.-H. Cho, P.-C. Tsou, J.-M. Chang, H.-C. Chao and etc [21,22-25].

Analysis of the works of domestic and foreign scientists has shown that the problems, related to the formation of resource and energy saving programs as a basis for improving the economic efficiency of the enterprise, are not reflected enough, and they need the further study. It shows the need for a separate study of this problem

The purpose of the study is solution of resource problems in the building materials industry with the program-oriented approach.

The subject of the study is a set of organizational and economic relations arising in the process of resource and energy saving programs in order to improve the economic efficiency of the enterprise activity.

The objects of the study is enterprises of building materials industry, with the potential to improve energy efficiency.

Theoretical basis is fundamental and applied researches of foreign and domestic scientists and experts, these researches are dedicated to energy saving and energy efficiency increase in the enterprise.

The methodological basis of the research is the methods of system and synthesis analysis, method of analogy, comparison, expert evaluation, structural, technical-economic, logical analysis, statistical methods, the methodology of energy, strategic and innovation management.

In addition, some of the conclusions and recommendations resulting from the research are used in the practice activity of companies working in the building industry of Volgograd.

2 Program-oriented approach to solving the problems of resource and energy saving in the enterprise

Energy saving is now generally considered as an integrated approach to the activities of the enterprise, and also as an activity for the implementation of individual energy-saving projects that are inherent in time and resource constraints.

Energy saving is considered not as an complex approach in the enterprise activity, but as activity for the implementation of specific energy-saving projects, which have limitations in time and in resources.

For the implementation of resource and energy efficiency programs in the construction company it can be used different approaches to the formation of a complex of measures.

Theoretical part. One of the highly effective approaches, which is recommended in the practice of foreign companies management, is program-oriented approach that focuses on the achievement of clearly defined, sufficiently important purpose of enterprise development. The need to solve the problems with program-oriented method caused the fact that the problems are complex, and they need the coordination of joint efforts and resources.

Program-oriented method of planning as compared to estimate planning, has several advantages and allows to provide:

- resource concentration for solving the priority complex tasks of energy economy;
- long planning horizon, comparable with the terms of the impact and return of investment and it allows to assess the medium and long term socio-economic effects;
- systematic approach to the formation of complex of interrelated according to resources and terms of measures to achieve these purposes;
- the open procedure of the formation and implementation of target programs and their realization;
- more rational use of the allocated financial resources and ensuring the transparency of their developing.

The basis of program-oriented method is the system analysis, the essence of which is to examine the certain problem for rationale of decision-making and choice of development direction by examining its operational purposes, evaluating the effectiveness and risks associated with each alternative.

However, it is necessary to remember that the target complex program is extremely sensitive to the disturbance of any of its elements because of its complexity, connectedness of measures with terms, resources, performers

Considering this information and the experience in implementation of energy efficiency policy at the studied companies, it is proposed the use of a structuring method of the decision-making for practical use within the program-oriented approach; this method was not previously used to solve problems of resource and energy saving

This method allows to realize the structuring of purposes, objects and measures; to evaluate the economic efficiency of the enterprise; to calculate the performance order of measures. As part of this method, it is proposed to combine a targeted multistage interview with «opening» of the problem in time, that is quite feasible in the algorithmization conditions of this procedure and using computer technology.

In accordance with established methodology of program-oriented approach, the general purposes of the program should be disaggregated to the level of program measures to implement the set of program activities to ensure the achievement of program purposes. Development of the program is carried out in stages. At first, as a rule, targeted development is carried out (structuring of the original purpose, the construction of "purpose tree").

Development of the "purposes tree" allows to reveal the hierarchy of purposes in accordance with the hierarchy levels and to determine the place of program activities in achievement of specified purpose. Forming the principles of selection, evaluating the expenses and benefits associated with the implementation of another variant, the most effective options for activities are determined.

As a result, a complex of functionally expressed program activities is shown with the assessments of the relative importance of each event and quantified target specifications, complex options of objectively expressed purposes are developed, while purposes are formed in the form of functional or objective expressed needs, which are met in the case of achieving the main purposes.

The essence of the structuring method is in construction of "purpose tree" by multistage expert interview, using the so-called "delphic procedure" characterized by three main features: the anonymity, feedback and group response. As a result of each stage of the expert survey, the elements of one level of "purposes tree" are formed, comparative assessment is realized, the elements which do not match the accepted selection indexes, are discarded. The remaining elements are necessary to further structuring at a lower level.

This approach makes it possible to more fully explore and evaluate all the proposals of the experts. Every time experts should assess the elements of only one level "tree", that reduces the dimension of the problem and increase the validity of expert statements. This approach accelerates the process of building a "purposes tree ", because indexes of each level are evaluated according to accepted indexes of selection, and the elements, that do not correspond to them, are discarded and do not take part in the further consideration. This method is proposed for the formation of a complex of resource and energy saving measures at the enterprise JSC "Volgograd ceramic factory".

3 The use of the decision-making structuring method for forming resource and energy saving programs at the enterprise of building sphere

There are following stages in the formation of program: selection of problems for the including in program; decision-making of a target program development and its formation; expertise and evaluation of the target program; approval of the target program; managing the implementation of the target program and control of its realization. To form resource and energy-saving program at Implementation of resource and energy saving program will help to improve the organization and management of energy saving and economic efficiency of enterprise of building materials industry and evaluate this company, it is proposed to implement the following stages (Fig. 1).

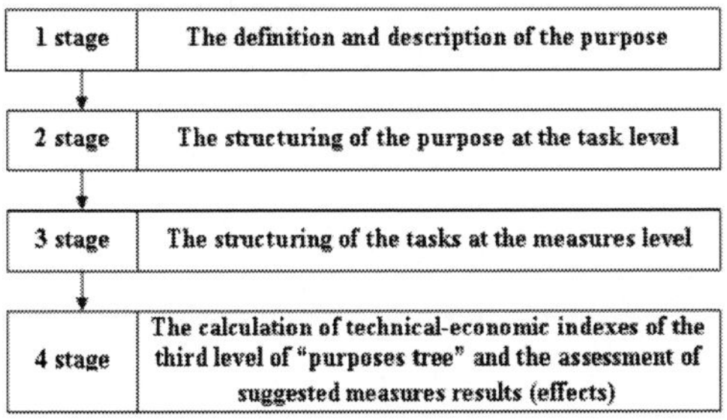

Fig. 1. The stages of resource and energy saving program formation.

At the first stage we define the purpose - the development of an energy management system in the enterprise in order to improve the economic efficiency of the enterprise. Plan or program of actions of the resource and energy efficiency actually "translates" the energy policy of the enterprise in the language of the purposes and tasks and defines the actions required to achieve them, establishes the responsibility of specific employees, and allocates necessary for the implementation of the program (plan) of human and financial resources.

At the same time the definition of the priority purposes has the great importance.

There is purpose structuring at the task level at the second stage.

1) A group of experts is formed. The group consists of 10 people. The external energy consultant was included, who represents the energy service company. The chief engineer is in Head of the group. To compare options of achievement this purpose we use the following two groups of quantitative and qualitative indexes (Table 1).

Table 1. The elements assessment indexes of «purposes tree» at the task level.

Index	Possible values	Marks (points)
Conformity of the purpose (h^1)	Conforms Does not conform	1 0
Purposefulness (h^2)	Solve basic problem of purpose Helps to solve the main purpose along with other tasks Helps to solve the main purpose a little	100 50 10
Novelty (h^3)	The task is a new solution The problem is an improvement of the existing solution	100 50
Involvement (h^4)	The results of the purpose solution can be involved: - in the scale of the sphere; - in the scale of the enterprise; - in the certain subdivision of the enterprise	 100 50 10
Perspectives (h^5)	The results of the task realization will be the basis for the solution of new problems The results are only meaningful to solve this problem	100 50

2) Next, the coefficient of experts' competence is calculated. It can be defined as the arithmetic mean value of self-assessment of the expert group. In this case, the level of group competence is 0.96, so it is a reliability of expertise is over 90%.

3) A common list of tasks is formed, which is presented by all the experts. The generalized expert evaluation of the index is calculated.

4) The expert evaluation of each indexes of the feasibility of the tasks h^1 and h^2 is received. If most experts believe that any problem contributes a little to solve the purpose, it will be excluded from further consideration.

5) The coefficients values of novelty (h^3), involvement (h^4), perspectives (h^5) are arranged in decrease order of their importance to determine whether the expert approach is the same to the parameters evaluation.

6) Each expert assesses the task with the use of indexes of novelty (h^3), involvement (h^4), perspectives (h^5), while these indexes h^3, h^4 can have values 100, 50, 10, and the index h^5 have the value 100, 50.

At the third stage, the structuring of tasks is implemented at the level of events. At this stage the elements of the third-level "purpose tree" are defined, i.e. directly supposed events. It is possible to change the composition of the expert group. It includes professionals primarily of those spheres that match selected events for further consideration.

The results of the experts' survey of purposefulness (h^2) are presented in the Table 2.

Table 2. The experts' assessment of purposefulness of supposed events.

Events	Tasks		
	1	*2*	*3*
1. The introduction of variable frequency drives of electric engine	100	80	70
2. Strategy of organizational changes of energy policy in the enterprise	20	70	100
3. The learning of new methods of financing programs of resource and energy saving	20	80	90
4. Organization of in-house training in the sphere of energy saving	60	100	90
5. Dispatching of heat supply systems	90	18	18
6. Analysis and evaluation of the efficiency functioning of the energy management system	40	100	90
7. Creation of an information system (database) of the regulatory, technical and regulatory documentation in the sphere of energy efficiency	70	80	90
8. Defining and updating of the policy in the sphere of energy management	30	100	90
9. Installation of meters, the introduction of the automated monitoring and the accounting systems of energy resources	80	20	20

The relative indexes of importance are calculated. The results of calculation of the relative importance factors are presented in the Table 3.

Table 3. The table of coefficients of relative importance.

3d level of «purposes tree» (events)	2d level of «purposes tree»			The value of each event for purpose realization
	Task 1	*Task 2*	*Task 3*	
1	0.25	0.13	0.11	0.15
2	0.07	0.09	0.13	0.09
3	0.08	0.11	0.14	0.10
4	0.11	0.16	0.12	0.14
5	0.12	0.01	0.01	0.05
6	0.08	0.15	0.14	0.12
7	0.07	0.08	0.06	0.07
8	0.10	0.24	0.27	0.21
9	0.12	0.03	0.02	0.07

On the basis of the assessments "purposes tree" is built, which will determine the position of program activities to achieve a certain purpose - the development of an energy management system at enterprise of building materials industry (Figure 2).

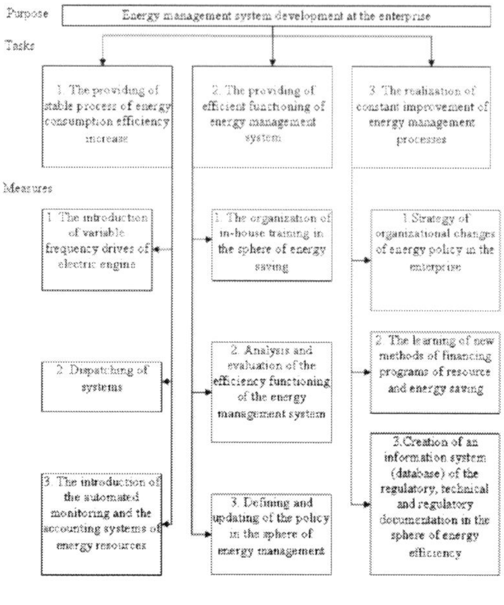

Fig. 2. «Purposes tree» of the energy management development of JSC «Volgograd ceramic factory».

At the fourth stage all the priority ways of enterprise of building materials industry activity are studied in the sphere of resource and energy saving. The measures, implementation of which will require significant financial resources, are defined.

Funding for measures of energy saving and energy efficiency increase will be carried out due to: equity of enterprise of building materials industry, the funds received as a result of reduced consumption of energy resources while the activities realization of energy saving and energy efficiency increase, and other sources.

Also periodic reports of the program activities implementation and the use efficiency of funds are considered. The evaluation provide a proposal to reallocate or change the volume of funds allocated for the implementation of the program activities.

Evaluation of the resource and energy saving program realization at the enterprise of building materials industry is presented in Table 4.

Suggested volumes of investment for the whole period of the program realization will be t 37,708.45 thousand rub. Economic efficiency of the implementation of the program for the period 2017-2019 will be 45,450.66 thousand rub. The average payback period is 2.3 years.

As a result, we can conclude that these activities have a relatively high economic efficiency, their implementation is a priority for the company.

Table 4. The evaluation of economic eficiency of resource and energy saving program realization at the enterprise.

Measures	The volume of reduction of annual consumption			Investments, thous. rub.	The term of payback, years
	The quantitative value	*The price value*			
		Thous. rub.	%		
1. The providing of equipment with devices of variable frequency drives , thous. kW·hours	804.0	2 580.8	12.3	11 355.5	4.4
2. The installation of metering devices, implementation of the automated system for control and accounting of energy resources	-	-	-	18195.4	1.2
- natural gas savings, thous. meters^3	3276.1	12229.1	57.1	-	-
- drinking and technical water saving, thous. meters^3	21.74	739.6	3.4	-	-
- electric energy saving, тыс. kW·hours	1 219.8	3 915.6	18.2	-	-
3. Dispatching of heat supply systems	-	-	-	6432.45	3.3
- heat resources saving, Gcal	2 357.5	1 933.2	9.0	-	-
4. The creation of information system	-	-	-	1725.1	-
In total	-	21398. 3	100	37708.5	2.3

4 Conclusion

Implementation of resource and energy saving program will help to improve the organization and management of energy saving and economic efficiency of enterprise of building materials industry.

Solving the problems of energy saving and increase enterprise efficiency is advantageously carried out on the basis of program-oriented method by implementing the set of measures associated by the purposes, tasks, resources, and terms of realization. As a result of the introduction of program-oriented method, the organizational and financial-economic energy saving mechanisms are realized. The projects are implemented to introduce energy-saving equipment and technologies, to produce energy-efficient equipment and materials, to increase energy efficiency of the company, to reduce emissions of harmful substances and the consumption of non-renewable energy resources.

References

1. O.A. Baulina, V.V. Klyushin, and A.E. Shchukin, *Globalization and its socio-economic consequences*, 2016, **I**, 127-134 (2016)
2. G. Novikova, A. Chub, J. Gushina, *Globalization and its socio-economic consequences*, 2016, **IV**, 1547-1554 (2016)
3. D.A. Nasimov, Indian Journal of Science and Technology, **9**, **16**, 90049 (2016)
4. I.I. Doronina, V.N. Borobov, E.A. Ivanova, E.V. Gorynya, B.M. Zhukov, Int. Journal of Econ. and Fin. Issues, **6**, **1S**, 295-299 (2016)
5. V.N. Leksin, B.N. Porfiryev, Studies on Russian Econ. Develop., **26**, **6**, 555-560 (2015)
6. I.V. Osinovskaya, O.V. Lenkova, Int. Business Man., **9**, **2**, 164-168 (2015)
7. N.A. Serebryakova, Y.A. Salikov, O.Y. Kolomytseva, T.A. Pakhomova, N.V. Grishchenko, Asian Soc. Science, **11**, **20**, 193-205 (2015)
8. X. Guo, Q. Zhu, L. Lv, J. Chu, J. Wu, Jour. of Clean. Prod., **140**, 1313-1321 (2017)
9. I.J. Faber, J. Sambor, A. Bagley, *IIE Annual Conference and Expo 2014*, 3069-3076 (2014)
10. Y. Niu, R.Q. Si, Z. Li, J.W. Ma, Adv. Mater. Research, **1030-1032**, 2573-2577 (2014)
11. S. Dasgupta, J. Roy, Energy Efficiency, 1-19 (2016)
12. Y. Bian, M. Hu, Y. Wang, H. Xu, Renew, Sust. Energy Reviews, **55**, 990-998 (2016)
13. V. Blass, C.J. Corbett, M.A. Delmas, S. Muthulingam, Energy, **65**, 560-571 (2014)
14. A.A. Volodchenko, V.S. Lesovik, A.N. Volodchenko, E.S. Glagolev, L.H. Zagorodnjuk, Y.V. Pukharenko, Int. Jour. of Pharm. and Techn., **8**, **3**, 18856-18867 (2016)
15. A.A. Volodchenko, V.S. Lesovik, A.N. Volodchenko, L.H. Zagorodnjuk, Int. Jour. of Appl. Engin. Research, **10**, **24**, 45142-45149 (2015)
16. A.A. Volodchenko, V.S. Lesovik, L.H. Zagorodnjuk, A.N. Volodchenko, K.A. Aleksandrovna, Research Jour. of Appl. Sciences, **10**, **12**, 931-936 (2016)
17. G. Balčiūnas, S. Vėjelis, L. Lekūnaitė, A. Kremensas, Environmental Engin. and Manag. Jour., **15**, **3**, 699-705 (2016)
18. M. Garg, A. Pundir, Eur. Jour. of Environmental and Civil Engin, 1-17 (2016)
19. M.P. Devaki, Key Engin. Materials, **678**, 135-143 (2016)
20. B. Balakrishnan, A.S.M. Abdul Awal, Key Engin. Materials, **678**, 99-108 (2016)
21. P.-H. Huang, S.-S. Sun, W.Liao, IEEE Transact. on Mob. Comp., **16**, **1**, 143-157 (2017)

22. A. Gawlak, J. Szkutnik, *Proceedings of the 8th Int. Scien. Symposium on Electric. Pow. Engin.*, 208-210 (2015)
23. Y. Fu, Z. Wei, K. Xu, L. Pang, *44th Int. Conf. on Large High Voltage Electric Syst.*, **10** (2012)
24. G.J. Tsekouras, M.A. Tsaroucha, C.D. Tsirekis, A.D. Salis, E.N. Dialynas, N.D. Hatziargyriou, Int. Jour. of Electric. Pow. and Energy Syst., **33(6)**, 1220-1228 (2011)
25. Y.-H. Lin, H.-H. Cho, P.-C. Tsou, J.-M. Chang, H.-C. Chao, *Int. Conf. on Advanc. Communication Technology, ICACT*, 1409-1413 (2011)

The genesis of the Russian mergers and acquisitions market: the role of the industrial sector

Ivanov Aleksey[1,*]

[1]South Ural State University, Department of Accounting, Analysis and Audit, Russian Federation

Abstract. The article presents the Mergers and Acquisitions quantity and aggregate cost dynamics analysis with the involvement of Russian companies as of from 2001 to 2015. The statistical information on the quantity and volume of M&As in Russia and worldwide was used as an initial benchmark data for the research. The article deduces the conclusion that the dynamics of Russian M&A market development resembles the international one in the whole period up to the year 2014. Starting from 2014 Russian and international M&A markets shows polar trends. This fact connected chiefly to the negative trends in Russian economy due to economic sanctions and weak exportable raw products prices. The article also gives the analysis of Russian M&A market structure from several different perspectives: transactions prices range, capital flow, and integration fields. The main specific features of Russian M&A market are outlined in the light of the reviewed aspects: almost a half of M&A market value is formed through major transactions; internal transaction dominance (sellers and buyers are Russian companies);most of the deals are conducted with involvement of industrial enterprises. The results of this study can be used in forecast development of Russian M&A market trends.

Keywords: Merges and acquisitions, Business combinations, Russia, Industry

1 Introduction

Mergers and acquisitions markets of Western countries have been researched for more than one hundred of years. M&A market in the USA is the most thoroughly researched market. The first major review of its trends was conducted by R.L. Nelson in his work [1]. Nelson summarized the segmentary pieces of information contained in multiple works of the first half of the XX century, devoted to research of the USA M&A market dynamics. J.C. Narver's monograph was published ten years later [2]. In this monograph USA M&A market was researched on the grounds of US Federal Trade Commission annual reports. Since 1963 W.T. Grimm & Co, and later its legal successors (now it is Business Valuation Resources), have started to collect statistical data on publicly announced mergers and acquisitions and publish the Mergerstat Review bulletin, which contain the statistical data on M&As with the involvement of USA companies [3]. Starting from 1980s the activity of European M&A researchers [4-11], and from 2000s the activity of the Asian ones [12-17], has increased. Nowadays there are a lot of scientific papers devoted to the research of global and several national M&A markets trends. A classification of such papers is covered in M. Martynova, L. Renneboog [18], and P.A. Gaughan [3] publications.

Russian M&A market is one of the most dynamically developing markets in the world, and it takes up 2-3% of the international market. Several hundreds of major M&As of public companies happen annually and it is extremely difficult to estimate the amount of transactions between lesser non-public companies [19]. However Russian M&A market research works are hardly presented at all in the international academic literature. The author has conducted the analysis of the Scopus peer-reviewed literature database and this analysis showed a really small amount of such publications [20-22]. The aim of this article is to locate the trends of Russian M&A market development and introduce it to the international professional association.

2 Methodology and results

The title is set in bold 16-point Arial, justified. The first letter of the title should be capitalised with the rest in lower case. You should leave 35 mm of space above the title and 6 mm after the title. The author summarized the annual research results of integration transactions of the Institute for Mergers, Acquisitions and Alliances – IMAA (https://imaa-institute.org), of the "Mergers and Acquisitions in Russia" analytical project (http://mergers.ru), and KPMG group (www.kpmg.com). On the basis of the mentioned data author conducted the

* Corresponding author: ivanov.chel@list.ru

analysis of Russian M&A market trends, that were active during the 2001 and 2015. Table 1 contains the data on quantity and aggregate cost of M&As in Russia and worldwide (according to the IMAA annual research data).

Table 1. The volume and quantity of M&As in Russia and worldwide in 2001-2015.

Year	International M&A market volume, bln USD	The number of transactions on international market	Russian M&A market volume, bln USD	The number of transactions on Russian market
2001	1871	31205	3	398
2002	1243	27456	9	400
2003	1413	29773	35	488
2004	2094	33151	25	398
2005	2859	36206	63	469
2006	4070	41668	52	654
2007	4960	47821	159	966
2008	3097	45747	82	1716
2009	2198	41282	36	3285
2010	2745	43901	109	3684
2011	2657	43090	88	3209
2012	2530	40703	115	2532
2013	2523	38816	67	2019
2014	3936	43008	18	1914
2015	4679	46167	31	1760

Volume percent of Russian M&A market from international market varies from 0.16% in 2001 to 4.54% in 2012. Median rate comprises 2.20%. Quantity percent of Russian M&A market from international market varies from 1.28 in 2001 to 8.60% in 2010. Median rate comprises 3.44%.

As a whole, the dynamics of Russian M&A market development had been showing trends similar to the international one up until the 2014. There was a steady growth that reached its peak value in 2007, that turned into a decline, and then the rate reached its minimum value in 2012-2013. The largest mega transaction in the history of Russian M&As influenced the statistical data of 2012 heavily. It is when the "Rosneft" have bought the "TNK-BP" holding company for 53.9 bln USD (according to IMAA). If it wasn't for this mega-transaction, the market in 2012 would continue to decline. Statistical analysis confirms this deduction. Pearson product-moment correlation coefficient between the shift in the quantity of the transactions in the Russian and international markets over the same period comprised 0.56. This means that there is a direct connection between the variations under consideration. The author deduced the Pearson product-moment correlation coefficient between the shift in the quantity of the transactions in the Russian and international markets in 2002-2013 (the sixth wave of M&As in the international market) and it comprised 0.72. This means that there is a strong and direct correlation between the reviewed variations.

The author has built the equations, that show trends in change of the amount of transactions in international and Russian markets in 2002-2013. A polynomial form of the growth curve was chosen, because in the most cases it gave the highest coefficient of determination.

The dynamics of the amount of transactions in the global market is described by the equation

$$y=-389.4x^2+6209.5x+19865 \quad (1)$$

Determination coefficient R^2 amounted to 0.8887.

The dynamics of the amount of transactions in the Russian market is described by the equation

$$y=-16.621x^2+497.89x-684.32 \quad (2)$$

Determination coefficient R^2 amounted to 0.6703.

Adequate trend models of changes in the transactions volume on Russian and global markets appeared to be impossible to compose. The main reason for that is the significant amount of the world-largest mega transactions on Russian M&A market, and those transactions influence its aggregate volume massively.

Fig. 1 shows the fluctuations dynamics of integrative transactions mean prices in Russia and worldwide in 2001-2015.

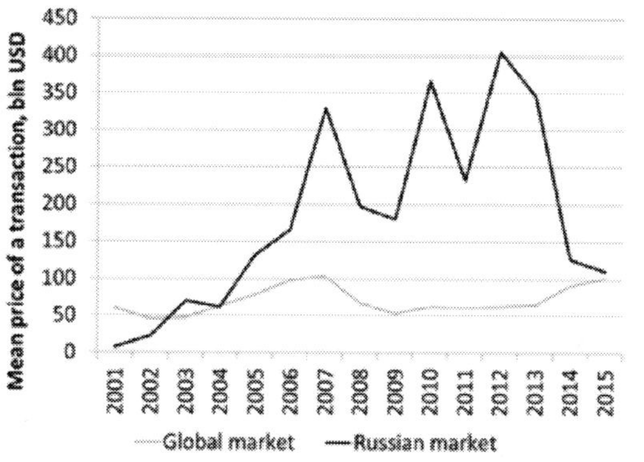

Fig. 1. Fluctuations dynamics of transactions mean prices in Russian and global M&A markets in 2001-2015.

Mean price of the transaction on the global market scale varied from 45.3 mln USD in 2002 to 103.7 mln USD in 2007. In Russia from 7.5 mln USD in 2001 to 405.7 mln USD in 2012. Median rate comprised 63.2 mln USD and 165.7 mln USD respectively. Such substantial difference could be caused by the large number of mega-transactions in 2005-2013, which constituted 43% and 85% of the market volume.

Starting from 2014, Russian and global M&A markets show polar trends. Substantial decline of Russian market performance contrasted with the record-breaking growth of the global market performance in

2014-2015. It is connected to the influence of the two major factors, according to the majority of experts:

– economic sanctions against Russia in light of the events in the Ukraine, that exert the negative effect on the investment environment in the country;

– global raw materials prices decline (especially oil), that leads to absence of free cash in the hands of extractive industry companies, and they are traditionally the most active Russian acquiring companies.

Fig. 2 shows the fluctuation of Russian M&A market volume dynamics in the light of changes in GDP of the country. The author re-estimated the GDP in bln USD, by reference to the official data of the Census Bureau on GDP amount, in roubles and annual average rouble's rate to the dollar according to The Bank of Russia's data.

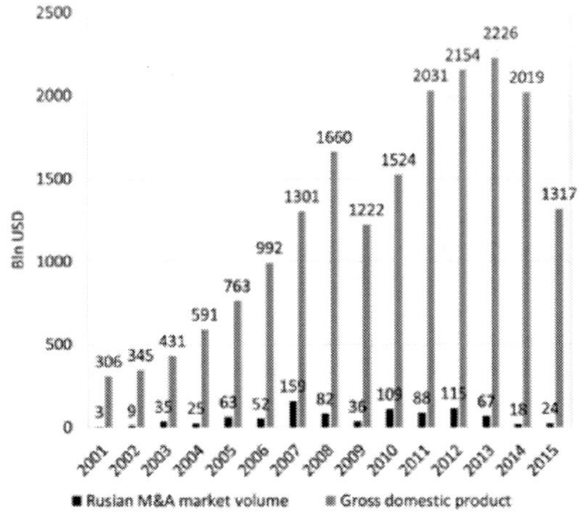

Fig. 2. Fluctuation of Russian M&A market volume dynamics in the light of changes of GDP in Russia.

The conclusion can be drawn from the given data on the fact that the M&A market is a significant constituent part of the Russian economy and it amounts to 5% of the GDP in general (median rate – 4.81%). In the estimation of specialists, that were presented in overviews of the project "Mergers and acquisitions in Russia" such correlation is roughly corresponds with worldwide average. However, in 2014-2015, the volume of M&A market was declining with priority rate to the GDP decline (0.89% in 2014 and 1.82% in 2015).

Table 2 contains the data on quantity structure of M&A in Russia in various price spectrum. Y.V. Ignatishin [23] has thoroughly studied the researches of the Russian M&A market structure in the early XXI century. It is clear from his studies that analytical reviews, published until 2004, never covered all market and was conducted via different approaches. That's why they can't be associated to the data of the subsequent researches. The "Mergers and acquisitions in Russia" project was created in 2004 and it conducted research of the Russian M&A market annually using the unified methodology. The reviews of this project served as the data source for 2004-2012. In 2013 this project ceased to conduct such researches, and therefore, the data source for 2013-2015 became the annual analytical overviews of KPMG group. It tends to segment the price range of transactions, but apart from this fact, its methodology is similar to the "Mergers and acquisitions in Russia" project. So, the author remade the price ranges of transactions into the comparable forms and re-estimated the shares of quantity and transactions amounts in every range.

Table 2. Evaluation of the Russian M&A market quantity structure in 2004-2015.

Market share	Transaction amount				
	More than 500 mln USD	**100 mln USD - 500 mln USD**	**10 mln USD - 100 mln USD**	**Less than 10 mln USD**	**Total:**
2004	37.1%	30.9%	27.4%	4.6%	100.0%
2005	55.6%	25.6%	16.5%	2.3%	100.0%
2006	43.4%	35.4%	18.6%	2.6%	100.0%
2007	69.9%	20.7%	8.1%	1.3%	100.0%
2008	65.1%	24.3%	9.5%	1.1%	100.0%
2009	67.9%	20.8%	9.6%	1.7%	100.0%
2010	56.2%	27.2%	14.9%	1.7%	100.0%
2011	69.7%	18.4%	10.6%	1.3%	100.0%
2012	63.0%	23.4%	12.5%	1.1%	100.0%
2013	85.2%	11.3%	3.5%	0.0%	100.0%
2014	62.9%	25.2%	11.3%	0.6%	100.0%
2015	61.5%	26.3%	11.3%	0.9%	100.0%

According to the given data, the author noted the stable predominance of the transactions from the low-range and mid-range price segment. Almost 80% of all the transactions are conducted in the range up to 100 mln USD (in 2013 this tendency was violated due to the big number of mega-transactions). An increase of the amount of transactions in the upper-range price segment in 2013-2015 is connected to the general increase of the amount of this kind of transactions and also to the various entry thresholds of the minor transactions into the analytical overviews of the "Mergers and acquisitions in Russia" project and KPMG group. In the first case this threshold comprised 1 mln USD, and in the second case – 5 mln USD.

Table 3 contains the data on structure of the aggregate value of M&As in Russia in different price ranges.

A consistent tendency of Russian M&A market became the composition of almost a half of its volume with major transactions (more than 500 mln USD). Median rate of such transactions in 2006-2015 amounted to 62.9%.

Table 3. Evaluation of the Russian M&A market cost structure in 2004-2015.

Market share	Transaction amount				
	More than 500 mln USD	100 mln USD - 500 mln USD	10 mln USD - 100 mln USD	Less than 10 mln USD	Total:
2004	37.1%	30.9%	27.4%	4.6%	100.0%
2005	55.6%	25.6%	16.5%	2.3%	100.0%
2006	43.4%	35.4%	18.6%	2.6%	100.0%
2007	69.9%	20.7%	8.1%	1.3%	100.0%
2008	65.1%	24.3%	9.5%	1.1%	100.0%
2009	67.9%	20.8%	9.6%	1.7%	100.0%
2010	56.2%	27.2%	14.9%	1.7%	100.0%
2011	69.7%	18.4%	10.6%	1.3%	100.0%
2012	63.0%	23.4%	12.5%	1.1%	100.0%
2013	85.2%	11.3%	3.5%	0.0%	100.0%
2014	62.9%	25.2%	11.3%	0.6%	100.0%
2015	61.5%	26.3%	11.3%	0.9%	100.0%

Table 4 shows the quantity structure of Russian M&A market from capital transfers perspective. Internal integration is understood here as a purchase of a Russian company by a Russian buyer, export integration is a purchase of a foreign company by a Russian buyer, and import integration means a purchase of a Russian company by a foreign buyer.

Table 4. Quantity structure of the Russian M&A market in 2004-2015 from capital transfers perspective.

Market share	Integration direction			
	Internal (in Russia)	Export (from Russia)	Import (to Russia)	Total:
2004	85.6%	8.6%	5.8%	100.0%
2005	82.7%	11.2%	6.1%	100.0%
2006	82.0%	9.3%	8.7%	100.0%
2007	79.6%	11.0%	9.4%	100.0%
2008	78.8%	11.7%	9.5%	100.0%
2009	80.9%	10.3%	8.8%	100.0%
2010	83.8%	9.2%	7.0%	100.0%
2011	81.0%	10.5%	8.5%	100.0%
2012	84.0%	8.6%	7.4%	100.0%
2013	69.1%	12.0%	18.9%	100.0%
2014	76.2%	12.4%	11.4%	100.0%
2015	71.4%	14.9%	13.7%	100.0%

Table 5 gives the cost structure of the Russian M&A market from a capital transfer perspective.

Table 5. Cost structure of the Russian M&A market in 2004-2015 from capital transfers perspective.

Market share	Integration direction			
	Internal (in Russia)	Export (from Russia)	Import (to Russia)	Total:
2004	73.6%	10.8%	15.6%	100.0%
2005	72.9%	14.4%	12.7%	100.0%
2006	68.5%	18.3%	13.2%	100.0%
2007	58.9%	20.7%	20.4%	100.0%
2008	61.0%	19.1%	19.9%	100.0%
2009	64.0%	14.3%	21.7%	100.0%
2010	65.6%	15.1%	19.3%	100.0%
2011	57.8%	23.8%	18.4%	100.0%
2012	74.0%	11.0%	15.0%	100.0%
2013	81.4%	4.0%	14.6%	100.0%
2014	72.4%	17.6%	10.0%	100.0%
2015	64.7%	15.4%	19.9%	100.0%

As the tables above show, the strategy of internal integration prevails in Russia and it is hardly surprising. From the one hand, Russian economy shows the high level of corruption and bureaucracy, so the foreign investors aren't sure in the stability of the rules for "the game" in long-range perspective. From the other hand, the rate of return of business abroad is lower that the same rate in Russia.

Table 6 gives the quantity structure of the Russian M&A market from an integration fields perspective.

Table 6. Quantity structure of the Russian M&A market in 2006-2015 from an integration fields perspective.

Integration field	Market share, %									
	2006	2007	2008	2009	2010	2011	2012	2013	2014	2015
Industrial sector	27.8	28.7	24.8	23.6	25.4	26.9	28.9	23.4	17.2	19.0
Agricultural sector	14.2	11.3	10.6	13.0	12.1	10.7	14.4	4.7	5.3	5.2
Property development	7.7	6.7	11.6	12.0	13.0	13.4	10.8	19.9	31.3	17.5
Telecom. and media	16.9	20.5	22.2	20.1	18.4	19.1	16.3	9.2	9.1	10.1
Financial sector	11.2	12.3	11.9	10.5	13.5	10.5	8.4	10.4	9.5	12.3
Trade and services	18.3	15.7	14.0	15.9	13.2	12.0	16.6	5.7	8.0	9.5
Transport sector	3.9	4.8	4.9	4.9	4.4	7.4	4.6	10.4	6.6	9.1
Other economic sectors	0.0	0.0	0.0	0.0	0.0	0.0	0.0	16.1	13.0	17.3
Total:	100	100	100	100	100	100	100	100	100	100

The data on the 2004-2005 time period, contained in "Mergers and acquisitions in Russia" project analytical overviews can't be matched to the subsequent data because of the fundamental differences in organization of the fields into groups, so the analysis of this time period wasn't conducted. The data on 2006-2012 time period ("Mergers and acquisitions in Russia" project) and 2013-2015 (RPMG group) was consolidated into 8 aggregative groups of fields.

Table 7 gives the cost structure of the Russian M&A market from an integration fields perspective.

Table 7. Cost structure of the Russian M&A market in 2006-2015 from an integration fields perspective.

Integration field	**Market share. %**									
	2006	**2007**	**2008**	**2009**	**2010**	**2011**	**2012**	**2013**	**2014**	**2015**
Industrial sector	52.0	75.5	61.1	77.9	40.6	47.0	30.0	50.0	50.7	49.2
Agricultural sector	4.3	3.5	3.2	2.2	8.3	9.0	7.2	0.9	1.3	2.5
Property development	2.8	3.2	6.5	6.8	7.6	8.1	3.0	7.8	15.2	13.4
Telecom. and media	16.2	5.1	11.5	3.9	24.1	16.3	17.7	15.5	6.3	5.2
Financial sector	9.3	5.5	9.1	4.5	7.0	10.5	14.9	5.2	4.8	4.5
Trade and services	12.4	5.6	6.1	3.6	9.6	4.8	9.9	0.9	5.1	3.2
Transport sector	3.0	1.6	2.5	1.1	2.8	4.3	17.3	9.5	9.9	16.5
Other economic sectors	0.0	0.0	0.0	0.0	0.0	0.0	0.0	10.3	6.7	5.5
Total:	100	100	100	100	100	100	100	100	100	100

Analysis of the industry specific characteristics of Russian M&A market showed that quantitative structure of the market is quite stable and is characterized by the low volatility (standard deviation in 2006-2015 hovers from 1.1% for the transport sector to 2.6% for the property development field). However, the cost structure is less stable and is characterized by low volatility (standard deviation in 2006-2015 hovers from 2.3% for the property development field to 17.7% for the industrial sector). It is connected for the most part to the annual conclusion of transactions, which cost is above the mean rate. For example, in 2009 three large mergers were conducted with the total amount of more than 9 bln USD in Fuel-power complex. In the setting of the general market decline it spiked the industrial sector share in the sectoral structure of M&A market. Following on from the data form table 6 and 7 the author makes the conclusion that industrial sector is the key sector for the Russian M&A market: median rate of the cost market share is 50.3%, quantity rate is 25.1%.

3 Practical relevance of research and conclusion

An integration between enterprises is the most important feature of the modern economy. Mergers and acquisitions in XXI century become more and more trans-border and more and more countries get involved in global M&A market. The results of this research prove that the M&A market occupies the major place in the Russian economy and the dynamics of its development is similar to the global one (exception in 2014-1015 time period). It allows to make the following assumption: when Russian economy will finally adapt to the new conditions of economic sanctions and low prices on exported materials, its M&A market will show the steady growth following the global trend. Another important conclusion is in the fact that the development of Russian M&A market largely influences the industrial sector and integrational processes in this sector, in its turn, largely influence the volume and quantity structure of the market. So, the increase of integrational processes effectiveness in industrial sector is a relevant objective, that has high socio-economic and business importance for the whole country.

The work was supported by Act 211 Government of the Russian Federation, contract № 02.A03.21.0011.

References

1. R.L. Nelson, *Merger movements in American industry, 1895-1956* (Princeton university press, 1959)
2. J.C. Narver, *Conglomerate mergers and market competition* (University of California press, 1969)
3. P.A. Gaughan, *Mergers, acquisitions and corporate restructurings* (John Wiley & Sons, Inc., 2011)
4. R. Buhner, International Journal of Industrial Organization. The success of mergers in Germany, **9(4)**, 513-532 (1991)
5. R. Conn, A.D. Cosh, P.M. Guest, A. Hughes, Journal of Business Finance and Accounting. The impact on U.K. acquirers of domestic, cross-Border, public and private acquisitions, **5-6(32)**, 815-870 (2005)
6. A. Dickerson, H. Gibson, E. Tsakalotos, Oxford Economic Papers. The impact of acquisitions on company performance: evidence from a large panel of U.K. firms, **49(3)**, 344-361 (1997)
7. J.R. Franks, R.S. Harris, Journal of Financial Economics. Shareholder wealth effects of corporate takeovers: the U.K. experience 1955-1985, **23**, 225-249 (1989)
8. M. Goergen, L. Renneboog, European Financial Management. Shareholder wealth effects of European domestic and cross-border takeover bids, **10(1)**, 9-45 (2004)
9. K. Jansson, M. Kirk-Smith, S. Wightman, European Business Review. The impact of the single European market on cross border mergers in the UK manufacturing industry, **94(2)**, 8-13 (1994)
10. A. Rodríguez-Pose, H. Zademach, Tijdschrift voor Economische en Sociale Geografie. Industry dynamics in the German merger and acquisitions market, **97(3)**, 296-313 (2006)
11. J.W. Whitelock, M.R. Ree, European Business Review. Trends in mergers, acquisitions and joint ventures in the single European market, **93(4)**, 26-32 (1993)
12. K. Bae, J. Kang, J. Kim, Journal of Finance. Tunneling or value added? Evidence from mergers by Korean business groups, **57(6)**, 2695-2740 (2002)
13. A.K. Gopalaswamy, D. Acharya, J. Malik, Investment Management and Financial Innovations. Stock price

reaction to merger announcements: an empirical note on Indian markets, **5(1)**, 95-103 (2008)

14. J. Kang, A. Shivdasani, T. Yamada, Journal of Finance. The effect of bank relations on investment decisions: an investigation of Japanese takeover bids, **55**, 2197-2218 (2000)
15. J. Ma, J.A. Pagan, Y. Chu, International Journal of Revenue Management. Revenue enhancement and wealth effects of mergers and acquisitions in Asian emerging markets, **4(2)**, 179-194 (2010)
16. C. Nam, D. Yang, M. Park, G. Oh, J. Park, Review of Pacific Basin Financial Markets and Policies. Stock market reaction to mergers and acquisitions in anticipation of a subsequent related significant event: evidence from the Korean telecommunications industry, **8(2)**, 185-200 (2005)
17. B. Xu, *Proceedings of China Private Economy Innovation International Forum 2007*, 300-305 (2007)
18. M. Martynova, L. Renneboog, Journal of Banking and Finance. A Century of corporate takeovers: what have we learned and where do we stand? **32(10)**, 2148-2177 (2008)
19. A.E. Ivanov, Economic Analysis: Theory and Practice. Russian market of merges and absorption: in search of synergetic effect, **41**, 60-70 (2013)
20. S. Avdasheva, D. Tsytsulina, Research in International Business and Finance. The effects of M&As in highly concentrated domestic vis-à-vis export markets: by the example of Russian metal industries, **34**, 368-382 (2015)
21. M.G. Karelina, Economy of Region. Empirical analysis of the integration activity of business structures in the regions of Russia, **4**, 54-68 (2015)
22. H. Lorentz, L. Häkkinen, O. Hilmola, Baltic Journal of Management. Analysis of the Russian retail sector: prospects for cross-border M&A activity, **1(2)**, 148-168 (2006)
23. Y. Ignatishin, *Mergers and acquisitions: strategy, tactics, finances* (Piter, 2005)

Modelling efficient innovative work: integration of economic and social psychological approaches

Yulia Babanova[1,*], *Vera* Gryazeva-Dobshinskaya[2], and *Yulia* Dmitrieva[2]
[1]South Ural State University (national research university), Management of innovation in business Department, Chelyabinsk, Russia

Abstract. The article deals with the relevance of integration of economic and social psychological approaches to the solution of enhancing the efficiency of innovation management. The content, features and specifics of the modelling methods within each of approaches are unfolded and options of integration are considered. The economic approach lies in the generation of the integrated matrix concept of management of innovative development of an enterprise in line with the stages of innovative work and the use of the integrated vector method for the evaluation of the innovative enterprise development level. The social psychological approach lies in the development of a system of psychodiagnostic indexes of activity resources within the scope of psychological innovative audit of enterprise management and development of modelling methods for the balance of activity trends. Modelling the activity resources is based on the system of equations accounting for the interaction type of psychodiagnostic indexes. Integration of two approaches includes a methodological level, a level of empirical studies and modelling methods. There are suggested options of integrating the economic and psychological approaches to analyze available material and non-material resources of the enterprises' innovative work and to forecast an optimal option of development based on the implemented modelling methods.

1 Introduction

In the context of stiffer competition and expedited changes in the social and economic areas, today's managers feel the need "to be the leaders of innovations" that are capable of performing complex tasks of managing innovative work in a project team. Formation of leadership competences requisite for the implementation of innovative projects is a relevant managerial task. Innovation management has certain specifics related to the actual innovative work, namely novelty and speed of changes, maintenance of the traditional long-term functioning, and risks, uncertainty and strain for all its subjects. Efficient management of the subjects' innovative work is determined by a complex of external (objective – technological, institutional and cultural) and internal (subjective – social psychological and psychological) factors.

The study objective is the modelling of efficient innovative work based on the system integration of economic and social psychological technologies used at a number of its subjects' (individuals and organizations) activity levels. Within the scope of the economic approach, the study objects are enterprises that perform innovative work. The object of the social psychological approach is a manager that manifests one innovative activity or another.

Integration of two approaches (economic and social psychological) to the solution of the challenge of enhancing the efficiency of innovation management makes it possible to study available material and non-material resources of the enterprises' innovative work in an integrated manner and to forecast the optimal option of their development.

2 Economic approach

The economic approach is based on the study of interconnection between the enterprises' management efficiency and their innovative activity. Modern enterprises are capable to various degrees to manage their own innovative activity. To enhance the efficiency of the innovative work management, the management concept must be improved in the part of integration of methodological economic and social psychological approaches [1, 2].

The generated integration matrix concept of managing innovative development of an enterprise comprises three fields that are hierarchically dependent structures reflecting the main factors of the enterprise's work [1, 3]. The resource field, as an aggregate of indexes of the enterprise's material assets, represents production factors (labor, land, capital, information and others) requisite to maintain the enterprise functioning.

* Corresponding author: dmitrieva.julia.86@mail.ru

Table 1. Evaluation of the innovative state of an enterprise.

Horizontal level / **Vertical level**	**Novation formation stage** **INPUT**	**Novation transformation into an innovation stage** **MECHANISM**	**Innovation commercialization stage** **OUTPUT**
Mental field of the enterprise MENTAL FIELD	MIn Composite index of the mental field development that determines the quality of input into the innovative process	MM Composite index of the mental field development that determines the quality of mechanism of innovative work	MOut Composite index of the mental field development that determines the quality and number of innovative outputs
Knowledge field of the enterprise KNOWLEDGE FIELD	KIn Composite index of the knowledge field development that determines the quality of input into the innovative process	KM Composite index of the knowledge field development that determines the quality of mechanism of innovative work	KOut Composite index of the knowledge field development that determines the quality and number of innovative outputs
Resource field of the enterprise RESOURCE FIELD	RIn Composite index of the resource field development that determines the quality of input into the innovative process	RM Composite index of the resource field development that determines the quality of mechanism of innovative work	ROut Composite index of the resource field development that determines the quality and number of innovative outputs

The knowledge field, as an aggregate of indexes of non-material assets, represents the system of accumulated knowledge, skills, capacities and competences that are implemented to turn the resource field factors into the final product of the enterprise's work. The mental field, as an aggregate of indexes of the enterprise's intangible assets, represents the system of values, strategic benchmarks and overall corporate culture (table 1).

The stages of innovative work, namely formation of ideas, implementation of an introduced novation and commercialization of an innovation, superimpose on the field structure of the enterprise [4, 5, 6]. The aggregate of indexes at the cross-section of all elements enables to determine the level of the enterprise's innovative development (fig. 1).

The method allows detecting an innovative state of the enterprise and forecasting its future development.

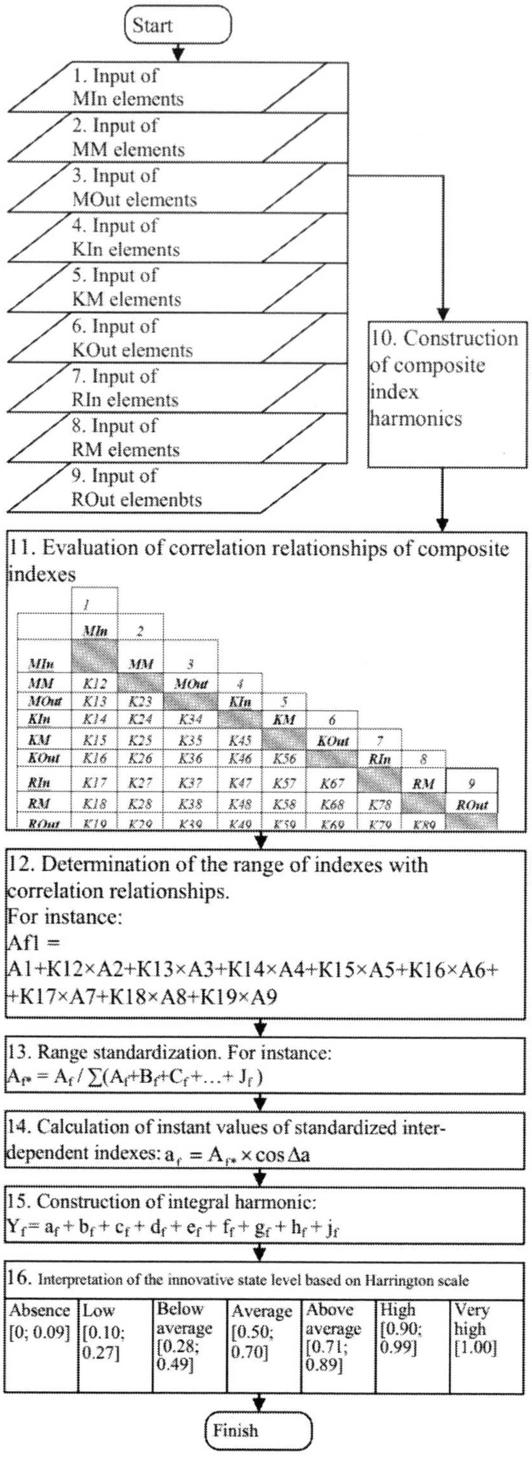

Fig. 1. Integration vector method of evaluation of the innovative state.

 DOI: 10.1051/shsconf/20173501036

3 Social psychological approach

The social psychological approach to the management of innovative work draws on the understanding of general requirements to the psychological resources of managers to unfold all stages of an idea transforming into an innovation. This approach is aimed to detect an individual profile and resource level of managers according to a generated index system. This index system is implemented in mathematical models: to forecast the development activity of managers, to determine an optimal option of an individual leader and his/her team using psychological resources and to form new managerial competences in an efficient manner.

Modern psychological technologies of managing the innovative work are aimed at the controlled stimulation of various stages of the innovative process, starting the author's creative idea to its commercialization by the project team, and at the work efficiency enhancement [7-15]. Within the scope of these technologies, there are used various index systems, including indexes of the leadership styles and management styles, motivation structure, creativity structure, and structure of the organizational cultural attitudes [16-20].

The core of the generated concept of the psychological innovative audit aimed to study the prospects of leadership implementation in the innovation is the notion of innovative leadership [21]. Innovative leadership is perceived as a complex system phenomenon that has two subsystems – focusing the activity on the change and focusing the activity on the retention of the functioning parameters, and various levels of the activity manifestation: intra-, inter- and meta-individual levels (table 2). The intra-individual activity level includes motivational, intellectual-and-creative and existential-and-personal resources. The inter-individual activity level includes team-oriented, management and leadership resources. The meta-individual activity level reflects the sociocultural identity and the organizational culture attitudes [22].

To determine all of the aforementioned indexes, there is a diagnostic toolset available that has been standardized and validated on the selections of enterprise managers with different innovative development levels [2].

See the indexes of the managers' activity resources to determine the resources of innovative leadership in line with the three specified levels in table 2 below.

Based on the factor analytical study of the outcomes of psychological diagnostics of the managers' innovative leadership resources, a two-factor structure for the activity resources (creative, motivational, leadership) was discovered. Detection of this structure confirms an existence of two (search and stabilizing) activities in the managers [23]. By now, based on the factor analytical study of the outcomes of psychological diagnostics of the managers' innovative leadership resources (leadership, creative, motivational and team-and-role), two levels of innovative leadership have been detected: intra- and inter-individual ones [24].

In the practice of the psychological innovative audit, there was revealed a problem of the general evaluation of the managers' activity resources where a great number of psychodiagnostic indexes is present (table 2). For this purpose, there was developed a calculation technology for the integral indexes of activity based on the modelling methods.

Table 2. Indexes of the managers' activity resources.

Activity levels	Resources of the managers' searching activity focused on the change	Resources of the managers' stabilizing activity focused on the retention
Intra-individual	Motivation – "hoping for success"	Motivation – "failure avoidance"
	Personal choice type (creative)	Personal choice type (realistic)
	Creative thinking (flexibility, original mindset)	Intellectual control
	Active non-adaptiveness (intellectual initiative and emotional reactivity)	Adaptiveness (realistic perception; anxiety)
Inter-individual	Transformational leadership	Transactional leadership
	Team-and-role identification (by role: idea generator, resource prospector, motivator, analyst)	Team-and-role identification (by role: implementer, controller, coordinator, harmonizer)
Meta-individual	Tolerance towards creative personnel (to the roles of a creator and a crisis-time rescuer)	Intolerance towards creative personnel (to the roles of a creator and a crisis-time rescuer)
	Values of cognition, freedom	Values of labor, supremacy
	Innovative and humanistic management orientations	Bureaucratic and autocratic management orientation
	Organizational culture attitudes: team-oriented, mobile	Organizational culture attitudes: paternalistic, pyramidal

The generated method to model the managers' activity resources is based on various types of indexes interaction within each of the subsystems (search and stabilizing activity): "cooperation", "competition", "predator vs victim" and "neutralism" [25, 26]. It is determined that the optimal type of index interaction when modelling the structural activity resources is "cooperation" [27].

Table 3. Integral indexes of the managers' activity resources.

Activity levels	Resources of the managers' searching activity focused on the change	Resources of the managers' stabilizing activity focused on the retention
Intra-individual	Ym Integral index for motivational resources	Xm Integral index for motivational resources
	Ycr Integral index for creative resources	Xcr Integral index for creative resources
Inter-individual	Yl Integral index for leadership resources	Xl Integral index for leadership resources
	Yr Integral index for team-and-role resources	Xr Integral index for team-and-role resources

Mathematical modelling of activity resources was performed individually for each manager based on the system of 2 linear equations. The first equation describes resources of the search activity that is focused on the change (y variable) and the second equation describes resources of the stabilizing activity focused on the retention (x variable). Ratios a1, a2, c1 and c2 equal psychodiagnostic indexes of the relevant activity resources.

The overall look of this equation system is set forth below:

$$a2 + x - c2*y = 0 \quad (1)$$

$$a1 + y - c1*x = 0 \quad (2)$$

Solution of the above equation system enables us to calculate Y and X coordinates which represent, in the Cartesian system (on the plane quadrant), the values of competing parameters for each activity trend.

This method allows calculating integral indexes of the managers' activity that accounts not simply for the aggregate of all psychodiagnostic indexes but also for their structure and interaction types for each activity resource (table 3).

Based on the modelling method, the overall evaluation of the managers' activity resources is performed; the shift from a greater number of psychological diagnostics indexes to two integral indexes for each activity resource is ensured; the balance of activity trends for each resource is determined; and the range of the resource optimum for an efficient performance of innovative work is forecasted.

4 Integration of economic and social psychological approaches

Integration of economic and social psychological approaches to the innovative work modelling is performed both at the level of the general academic methodology (system approach to the study of innovative activity of subjects) and at the level of empiric studies (generation of the unified system of the innovative activity indexes) and modelling methods (generation of the overall evaluation model for the level of an enterprise's innovative development).

Integration of the economic and social psychological approaches to the modelling of efficient innovative work is possible in three options.

The first option lies in analyzing the coordination of outcomes of economic and social psychological technologies, including the numerical analysis of compliance of the evaluation results for an innovative development level at various enterprises. One of the methods of such study can be the discriminant analysis where the economic index of non-material resources, i.e. evaluation of the enterprise's innovative development level (knowledge field) acts as a single independent variable and the psychological indexes act as dependent variables that are parts of the structure matrix with various scores.

The second option lies in using the psychological diagnostics results as the managers' activity resource indexes (social psychological approach) in the system of basic data of the integration vector method for evaluation of the innovative state of an enterprise.

The third option lies in using integral indexes as results of the modelling psychological diagnostics data (social psychological approach) in the system of basic data of the integration vector method for evaluation of the innovative state of an enterprise.

5 Conclusion

Thus, the modelling of efficient innovative work based on the system integration of economic and social psychological technologies demonstrated that innovative development of an enterprise is determined by factors formed at the social psychological level of the manager's personality and his/her team, thereby proving the feasibility of integration of these approaches for a further study and implementation in the applied consulting.

Integration of the economic and social psychological approaches to the modelling of innovative activity resources enables to evaluate the activity resource level of both individual managers or management teams and overall enterprises more precisely.

The modelling of the subjects' innovative development level allows constructing more efficient programs of training the personnel in new competences essential for the expansion of the innovative work.

The work was supported by Act 211 Government of the Russian Federation, contract № 02.A03.21.0011.

References

1. Y.V. Babanova, N.V. Kireeva, SUSU newsletter, Economy and Management series, **3**, **7** (2013)
2. V.G. Gryazeva-Dobshinskaya, Y.A. Dmitrieva, *Resources of innovative leadership managers:*

psychological innovative audit (SUSU publishing house, Chelyabinsk, 2016)
3. Y.V. Babanova, V.P. Gorshenin, SUSU newsletter, Economy and Management series, **22** (2012)
4. A.E. Luzin, Y.V. Babanova, Management in Russia and abroad, **6** (2013)
5. Van M. Assen, Van den Berg, G.P. Pitersma, *Key management model. 60 models that every manager needs to know* (Bean. Knowledge Laboratory, Moscow, 2011)
6. P. Ericksen, R. Suri, El-Jawhari, APICS Magazine, **15** (2005)
7. T. Amabile, *How to Kill Creativity Creative thinking in business* (Alpina Business Books, Moscow, 2006)
8. R.L. Daft, *The Leadership Experience* (EKSMO, Moscow, 2007)
9. I.K. Adizes, *Management/Mismanagement Styles* (Alpina Publishers, Moscow, 2011)
10. G. Van Wulfen, *The Innovation Expedition. A visual toolkit to start innovation* (Mann, Ivanov and Ferber, Moscow, 2014)
11. H. Courtney, J. Kirkland, P. Viguerie, *Strategy under Uncertainty, Management under Uncertainty* (Alpina Business, Moscow, 2006)
12. P.F. Drucker, *Innovation and Entrepreneurship* (OOO I.D. Williams, Moscow, 2007).
13. A. Hargadon, *Management of innovations* (OOO I.D. Williams, Moscow, 2007)
14. D. Dile, G. Cangemi, C. Kowalski, Psychology, Higher School of Economics Journal, **2**, **1** (2004)
15. I.K. Adizes, *Leading the Leaders, How to Enrich Your Style of Management and Handle People Whose Style is Different from Yours* (Alpina Business Books, Moscow, 2009)
16. G. Fischer, Psychology, Higher School of Economics Journal, **4**, **2** (2005)
17. B.J. Avolio, B.M. Bass, D.I. Jung, Journal of Occupational and Organizational Psychology (1999)
18. Y.D. Krasovsky, *Sociocultural fundamentals of managing a business organization* (UNITY-DANA, Moscow, 2007)
19. A.A. Tikhomirov, U.D. Spangler, Russian management journal, **2**, **3** (2005)
20. V.G. Gryazeva-Dobshinskaya, *Social psychology of creative work – 2007: collection of treatises* (2007)
21. V.G. Gryazeva-Dobshinskaya, SUSU newsletter, Psychology series, **9** (2010)
22. V.G. Gryazeva-Dobshinskaya, Y.A. Dmitrieva, SUSU newsletter, Psychology series, **1**, **9** (2016)
23. Y.A. Dmitrieva, SUSU newsletter, Psychology series, **3**, **8** (2015)
24. V. Volterra, *Mathematical Theory of the Struggle for Existence* (Institute of computer research, Moscow, 2004)
25. V.G. Gryazeva-Dobshinskaya, Y.A. Dmitrieva, South Ural State University newsletter, Psychology series, **13** (2011)
26. V.G. Gryazeva-Dobshinskaya, Y.A. Dmitrieva, Leadership in innovations: modelling the managers' activity resources: monography (SUSU publishing house, Chelyabinsk, 2016)

New approaches for identification factors of cost management

Anatolii Kholkin[1,*]

[1]Vyatka State University, Department of Accounting, Analysis and audit, 610000 Kirov, Russia

Abstract. This article deals with determination and classification factors, that affect the costs and whose changes are result of management decisions. The modern condition of scientific researches in this field of economic science is analyzed. The classification of impact factors is made for machine building industry. Also is given new costs classification according to impact factors. New cost classification includes three classes and eighteen groups of cost. Every class and group is described. The need of this classification for purposes of cost management is defined. Also the practical significance from the use of new costs classification in management decisions and control is described

1 Introduction

Costs are the most important object of management in each enterprise operating in every branch. Every kind of enterprise attempt to estimate of the optimal level of costs, because it is necessary for survive. The estimation of the optimal level of cost makes by purposeful action on determined factors by made management decisions.

Thus, the amount of costs as the object of management is the result of earlier made decisions on different management levels.

Control of decisions execution is very important for the whole management system. But managers need information that illustrating reflection for decisions affecting to costs factors for making control execution of decisions. For this purpose administration has to understand clearly what kind of decisions or their classes are affecting to costs factors and researches the mechanism of action. Also managers need information about amount of costs variance as result of decisions, that making necessary the new classification or grouping of cost – the classification on impact factors.

For this reason, in this article were discovered the factors, was defined their classification was designed the new grouping of costs. The new grouping of costs is required for identification and evaluation of effects of factors and decisions.

2 Relevance and scientific significance of problem with brief review of the literature

The relevance of research topic is dedicated to the existence of following problem: modern methods and procedures of cost management and accounting do not provide the correct measurement of the effect of factors that determining amount of costs that appear as consequences of taken decisions

Also there is no correct definition which factors effect on amount of specific kind or item of cost. It all makes cost management process more complicated, because it is impossible to define clearly what kind of management decisions caused changes of costs and therefore it is impossible to evaluate quantitatively the achievement of expected result of taken and executed decision.

Moreover, different factors have impact on costs with various natures. Therefore, it is more logical to define what kind of costs are affected on specific factors and create on this base the new grouping of costs, as accounting object. This is necessary for requirements setting for forming of the introduced cost grouping in management accounting.

The scientific significance of this article is the reconsideration of previously existing approaches for cost management in the part of quantitative efficiency evaluation of adopted decisions and control of actual parameters of decisions execution. Also, for control purposes the new grouping of cost has been designed. The introduction or this grouping or classification requires modification in using methods of calculation of costs.

Unfortunately, this problem is not enough described in modern scientific literature. Existing problems of cost management were discovered in books of such authors as L.G. Eldenburg, S.K. Wolcott[1],R.W. Hilton, M.W. Maher, F.H. Selto[2],V.G. Lebedev, T.G. Drozdova, B.P. Kustarev[3],L.E. Chertok[4], E.A. Gomonko, T.F. Tarasova[5],A.N. Asaul, V.K. Sevek, C.S. Manchyk-Sat[6]. Existing problems of cost accounting and analysis as information base of cost management were introduced in books of such authors as C. Drury[7], C.T. Horngren[8], S.M. Bragg[9],V.E. Kerimov[10],M.A. Vahrushina[11],G.V. Savitskaya[12].

In most of these books, authors are operating with definition of relevant cost, but they do not designate approaches of determination the dependence between

* Corresponding author: khav76@mail.ru

specific changed costs and executed decision. Also there are no approaches for comparison of factors impact on costs and made decisions. Authors indicated the volume of production and sales as major factor of cost changes and variations. Thus, modern scientific literature does not contain approaches to the solution to the stated above problem.

3 Target setting

The purpose of this article is identification and determination of factors, which affect to production costs in machine building industry, that are changing under executed decisions, that was needed for costs optimization.

This purpose could be achieved by solving following tasks:

- The identification of factors, which affect to amount of costs, the description of the nature of their action,
- Classification and systematization of factors,
- Drafting of factors group and forming costs grouping based on which this factors affect.

And because factors are not the accounting object and only information about the costs are provided in accounting system, which is reasonable to design cost grouping on the basis of impact factors instead of group of factors.

4 Theoretical aspects of identification of factors and design of costs grouping on the basis of impact factors

Methods that were used for solution of identified problems are the following: analysis and synthesis, deduction and induction, a thought experiment, classification, grouping and summarizing.

The result of using of these methods was determination the composition of factors, which affect to costs, and conclusion about differences between compositions of factors in various branch. Also, the influence of the factors is different in various branch and enterprise. Therefore in this article defined the factors, which have impact to cost, for machine building enterprise.

All factors separated by the possible impact on them of management on internal and external. Also environmental and ecological factors were separated, and here declared that no one could affect to them.

Internal factors are ones, which administration of enterprise could affect, which could change under management decisions.

External factors are ones, which administration of enterprise or company could not affect, because it depend on decisions were taken on higher management level and these decisions were given to administration through laws and other regulations.

On the base of grouping and classification of factors was designed the new grouping of costs. This grouping or classification includes classes of costs as given below:

1. Costs affected by internal factors:

- Costs depending on construction and technical factors or parameters,
- Costs determined by technological factors,
- Costs depending on the level of used technics,
- Costs determined by organizational factors,
- Costs depending on marketing policy of the company,
- Costs stipulated by ecological policy of enterprise,
- Costs defined by the policy of relationship with suppliers,
- Costs depending on logistic or logistic decisions,
- Costs determined by depreciation policy,
- Costs depended on social policy of company,
- Costs stipulated by accounting and tax policy of enterprise,
- Costs depending on personnel policy.

2. Costs affected by external factors:

- Costs determined by state tax policy,
- Costs depended on governmental social policy,
- Costs stipulated by state technical policy,
- Costs determined by governmental policy in labor relations,
- Costs depending on customs policy.

3. Costs depending on environmental and ecological factors

In addition, it was revealed, that a number of costs are indicated in this grouping could belong to several or different classes or groups, because essentially different decisions could affect to the same article or class of costs.

It is reasonable to describe all classes or groups of cost introduced, and its content was determined approximately.

Costs depending on construction and technical factors or parameters. These are the costs, that are depending on product design, product functions and technical parameters of product so that are the factors that determine usefulness and purpose of goods. This group of costs includes: raw materials costs, auxiliary materials, other materials costs, components and semi-finished products, fuel and energy costs, tools and equipment, salary, losses from manufacturing defects, operation costs, package costs, transportation cost.

Product design, product functions and technical parameters set contents of raw materials spent for production of goods. It is because these factors define performance specifications for materials, which spend for production of goods. The effect of this factor for components and semi-finished products is the same.

This factor affects for auxiliary materials costs by determination of ingredients of these materials, which use for given the product certain technical parameters. For example, corrosion resistance.

Also fuel and energy costs depend on this factor, because technical parameters of product and parameters of used material determine requirements for carrying out technological operations, which result is fuel and energy expense. For example, physical parameters of hardening process determined by physical properties of material. Physical parameters of hardening process define fuel and energy expenses.

This factor makes a decisive influence on choice of tools and equipment and its constructive and technical parameters. It explains the application of modes of technological operations, which has been determined by the composition of raw materials and product design.

Also, salary depends on this factor, because it determines the laboriousness of technological operations. For example, laboriousness of assembly depends on the weight of the product, the complexity of the design and the number of assembly units. All of that define requirements for qualification of personnel and finally for salary.

Losses from manufacturing defects are depending on this factor, because the complexity of product construction increases the probability of defected goods production. Besides of it, if requirements to product and materials become higher the probability of defects detention becomes also higher, because the quality control becomes complicated.

This factor affects to operation and maintenance costs, because it determinates the operating time of equipment, spent for execution of technological operations and technical parameters of the equipment

Similarly, package costs depend on this factor, because construction and technical parameters of product define requirements for packaging materials and time consumption per package.

Also, this factor effect to transportation costs, because it defines requirements to the vehicle, which deliver product and requirements to material handling.

Costs determined by technological factors. These are costs, which depend on selected method operations for manufacturing products and the technological process of manufacturing. This group of costs includes: raw materials costs, auxiliary materials, other materials costs, fuel and energy costs, tools and equipment, salary, operation and maintenance costs. It happens because selected method of operations for products manufacturing affects to quantitative consumption of materials, that determines all material costs. Also, this factor effect to laboriousness of technological operations, that defines salary and operation and maintenance costs.

Costs depending on the level of used technics. These are the costs depending on selected equipment for execution of operations to manufacture of product, its operating or maintenance parameters. This group includes: operation and maintenance costs, repairing costs, because technical parameters of using machinery and equipment determine these costs.

Costs determined by organizational factors. These are the costs depending on the level of organization of labor and production, rationing and control. This group includes: components and semi-finished products costs, energy costs, costs of works and services of production character, tools and equipment, intrafactory moving of cargoes, losses from manufacturing defects, administrative costs and losses from shortages.

Costs depending on marketing policy of the company. These are costs depend on taken decisions in the management of marketing activities of the company. It includes: advertising costs, package costs, travel expenses.

Costs stipulated by ecological policy of enterprise. These are costs are determined of made decisions in the sphere of ecological management and environmental protection. This group includes: payments for pollution, expenditures on environmental measures.

Costs stipulated by the policy of relationship with suppliers. These costs depend on executed decisions in the field of relationship management with suppliers. It includes all material and transportation costs, because the policy of relationship with suppliers defines prices of purchased materials, terms of delivery and the choice of supplier.

Costs depending on logistic or logistic decisions. These costs are determined executed decisions in logistic. This group includes: transportation costs, intrafactory moving of cargoes costs.

Costs determined by depreciation policy. This group includes sum of fixed assets depreciation, which depend on selected depreciation method.

Costs depending on social policy of company. These costs are determined by made decisions in social policy of company. This group includes: salary, payments for not worked time, compensations and other similar payments.

Costs stipulated by accounting and tax policy of enterprise. These costs depend on selected accounting methods, tax planning and tax accounting methods. It includes: material costs and other cost, which could be taken into account by different method, tax, which are included in costs.

Costs depending on personnel policy. These costs depend on taken decisions in the field of personnel management. It includes: costs of recruitment, costs of personnel teaching and training

Costs determined by state tax policy. These costs depend on tax regulation. This group involves sum of taxes and fees. Also this group includes expenses for tax accounting, which depend on the complexity of the procedure of calculation of taxes and fees.

Costs depending on governmental social policy. These costs determined by legislation in the sphere of social insurance. It includes: compulsory social insurance payments, social insurance benefits, which are paid the employer.

Costs stipulated by state technical policy. These costs are implemented in the field of technical policy and technical regulation based on the requirements of the legislation. This group includes: certification costs, costs of preparation and mastering of production, research and development costs, other similar costs.

Costs determined by governmental policy in labor relations. These costs prescribed by labor legislation. It involves salary in parts of payments for not worked time, compensations and other similar payments, which provided by labor law.

Costs depending on customs policy. These costs determined by customs legislation. It includes: customs duties and taxes, material costs.

Costs depending on environmental and ecological factors. These costs are depended on climate and other natural factors. It involves costs of heating and lighting of buildings. In present time, nobody and nothing could to affect to these costs.

5 Practical significance of the results

Group of costs and determination of factors were represented in this article allows:

1. Relevant costs could be determined for each managerial decision clearly. Also reflection of costs for specific management action could define.
2. The comprehensive search of reserves of cost reduction could be performed.
3. The effective control of execution decisions in the sphere of cost management could be made.
4. The elimination of costs from effect of external factors could be produced. It is necessary for correct evaluation the contribution of management in cost optimization.

6 Summary

The economic situation in our country and in the world, its trends has result that problems of cost optimization in cost management become more and more important. For this purpose factors, which affect to cost, were determined, defined and described. The nature of effect to different costs of these factors was explored. It was estimated by forming of new cost grouping by costs factor. The grouping of costs was designed in this article has not analog in modern economical literature.

Also, possibilities of using of this grouping in costs management practice were described. The usage of new grouping will improve quality and operability of costs management, and it will increase efficiency of activity of company.

References

1. L.G. Eldenburg, S.K. Wolcott, *Cost Management: Measuring, Monitoring, and Motivating Performance* (John Wiley and Sons, Ltd, Hoboken, NJ, 2004)
2. R.W. Hilton, M.W. Maher, F.H. Selto, *Cost Management: Strategies for Business Decision* (Irwin/McGraw-Hill, New York, 2007)
3. V.G. Lebedev, T.G. Drozdova, B.P. Kustarev, *Cost management in the enterprise. Textbook for high schools* (Piter, St. Petersburg, 2012)
4. L.E. Chertok, *Fundamentals of cost management in the enterprise: textbook.* (Publisher of Sankt-Petersburg University of management and economic, St. Petersburg, 2012)
5. E.A. Gomonko, T.F. Tarasova, *Cost management in the enterprise* (KnoRus, Moscow, 2009)
6. A.N. Asaul, V.K. Sevek, C.S. Manchyk-Sat, *Cost management and controlling: textbook* (RPD TuvSU, Kyzyl, 2014)
7. C. Drury, *Management and Cost Accounting. New Edition (2nd & Subsequent)* (Cengage Learning, Boston, 2012)
8. C.T. Horngren, Srikant M. Datar, Madhav V. Rajan, *Cost Accounting: A Managerial Emphasis* (Pearson Education Limited, Harlow, 2012)
9. S.M. Bragg, *Management Accounting Best Practices: A Guide for the Professional Accountant* (Wiley, Hoboken, 2007)
10. V.E. Kerimov, *Management accounting: textbook* (Dashkov&Co, Moscow, 2014)
11. M.A. Vahrushina, *Management accounting* (Omega-L, Moscow,2007)
12. G.V. Savitskaya, *A comprehensive analysis of the economic activity of the enterprise: textbook* (Infra-M, Moscow, 2013)

The research of the production function of an industrial enterprise

Alexandr Rezepin[1], *Taya* Amirova[1], and *Vera* Mishina[2,*]

[1]South Ural State University (national research university), Department of Economic theory, regional economics, state and municipal management, 454080 Chelyabinsk, Russia

Abstract. The article deals with the use of a production function model for the description of production process and the solution of practical problems, such as choice of technological method of production, rational and effective use of invested funds. Analysis of the production process is carried out by the example of Urals Stampings Plant. The analysis consists of two parts. In the first part with the use of regression analysis the production function is evaluated and the elasticity of revenue is calculated on the cost for main types of resources. In the second part of the analysis with the help of artificial neural networks constructing authors investigate the significance of influence the dynamics in the number of production factors and productivity on the physical volume of the issue. In conclusion, the authors provide recommendations for the implementation to increase the level of productivity of the investment strategy for the Urals Stampings Plant.

1 Introduction

The development of market relations demands much of business entities. They should be able to make independent and effective decisions based on analysis and evaluation of current and future economic situation, clearly formulate the goal of development and work out a strategy to ensure long-term competitive advantages of the company's development. The problem of choosing the technological mode of production and form of reproduction of the basic production assets is largely a problem of rationalizing the volume of invested funds and providing their effectiveness in the long term. Development of the investment strategy of the enterprise is based on the assessment of the production process and the use of economic resources.

2 The Theoretical Model of Production

The development of a theoretical model of the production process is the subject of numerous scientific papers. The central place in the theory of production is taken by the question of using a model that can sufficiently describe the manufacturing process. The basic model is a production function. The production function model can be used as a practical tool for solving a number of planning and analytical tasks, such as planning, forecasting and analysis of the enterprise activity.

Theoretical and methodological basis for the application of production functions in the analysis of the product creation process were laid in the works by: C. Cobb and P. Douglas [1], W. Leontief [2], J. Robinson [3], M. Brown [4], R. Solow [5], F. Fisher [6]. The practical issues of the production functions use for solving the problems of production technology optimization and the improvement of resource efficiency were considered by: G. Williams [7], J. McCombie [8], K. Kim [9], W. Dai [10], M. Machado [11], P. McCarthy [12], Y. Dissou [13], J. Sauer [14]. In the Russian scientific works it has developed a usual practice of Cobb-Douglas multiplicative production function application for the applied problems solving in order to provide sustainable development of the company, effective reproduction of economic resources, optimization of material and financial flows. These studies are presented in the works by: B.V. Revazov [15], Yu.A. Shamara [16], R.M. Nizhegorodtsev [17], S.A. Dobrotin [18].

The development of methods for modeling of economic and productive situations and decision-making on their basis for planning and forecasting of business activity is a necessary condition to ensure the effectiveness entrepreneurship.

The production function is an economic-mathematical relationship between the manufactured products amount and production factors used in their creation. The multiplicative production function is characterized by a partial substitutability of production factors. In practice the most commonly used production function assumes power dependence of production volume on labor and capital. For the first time the power function was applied by C. Cobb and P. Douglas in their work "A Theory of Production" [1] in 1928. They

* Corresponding author: mishinavd@susu.ru

empirically established a connection between the labor input, capital and the amount of US manufacturing industry products in the 1899–1922. The multiplicative production function can be used at different levels of the particular enterprise and the entire industry production and the national economy as a whole. It is defined by:

$$Q = AK^{\alpha}L^{\beta} \qquad (1)$$

where Q – production volume, A – factor of neutral technical progress, K – the amount of production assets, L – the amount of labor used in the production, α and β – coefficients of elasticity for the funds and labor.

3 The production function estimation of an industrial enterprise

The development of investment strategy of an industrial enterprise bases on an analysis of the production process, which allows identifying the direction of change in technology giving the best return [19, 20]. To identify the most effective directions of Urals Stampings Plant (Chebarkul, Chelyabinsk Region, Russian Federation) development authors suggested, first, an assessment of the production function and calculation of the revenue elasticity on the costs of the main types of resources, and secondly, to evaluate the contribution of extensive and intensive factors in the change of shipped products volume.

The multiplicative production function is defined by a time series of issues and resource costs. Since the material and human resources used in the manufacture of Urals Stampings Plant products are varied and differ significantly in their productivity, let's consider a function that assigns the revenue and production costs associated with the use of capital and labor resources: (1) amortization of production assets and (2) the salary fund.

The multiplicative production function in logarithms takes a linear form:

$$\ln Rev_t = \ln A + b_1 \ln Dep_t + b_2 \ln Sal_t, \qquad (2)$$

where $\ln Rev_t$ – the natural logarithm of the enterprise revenue in the period t, $\ln Dep_t$ – the natural logarithm of the fixed capital amortization in the period t, $\ln Sal_t$ – the natural logarithm of the wage fund in the period t, b_1 and b_2 – the model parameters.

Parameters of b_1 and b_2 functions can be determined by least square method of multiple regression using the statistical package in IBM SPSS Statistics.

The regression model has been constructed on the Urals Stampings Plant quarterly financial statements for the 2009–2016 years.[1] The analysis results are reported in tables 1, 2 and 3.

Table 1. Model Summary[b].

Model	R	R Square	Adjusted R Square	Std. Error of the Estimate
1	0.930[a]	0.864	0.855	0.10183

[a] Predictors: (Constant), ln*Sal*, ln*Dep*

[1] The official web-site of Urals Stampings Plant. – URL: http://www.mechel.com/sector/steel/urals_stampings_plant.

Model	R	R Square	Adjusted R Square	Std. Error of the Estimate

[b] Dependent Variable: ln*Rev*

Table 2. ANOVA[a].

Model	Sum of Squares	df	Mean Square	F	Sig.
1 Regression	1.849	2	0.924	89.141	0.000[b]
Residual	0.290	28	0.010		
Total	2.139	30			

[a] Dependent Variable: ln*Rev*
[b] Predictors: (Constant). ln*Sal*. ln*Dep*

Table 3. Coefficients[a].

Model	Unstandardized Coefficients		Standardized Coefficients	t	Sig.
	B	Std. Error	Beta		
1 (Constant)	8.994	0.491		18.322	0.000
LnDep	0.151	0.040	0.344	3.779	0.001
LnSal	0.649	0.053	1.113	12.238	0.000

[a] Dependent Variable: ln*Rev*

R Square equal to 0.864 indicates the significant share of explained variance and high quality of a regression model. Dispersion analysis indicates the significance of the regression equation. The coefficients of the regression equation are significant on the basis of Student's t-test.

According to the analysis of the production function of the company is as follows:

$$Rev = 0.026Dep^{0.151}Sal^{0.649} \qquad (3)$$

where Rev – company revenues. Dep – amortization of fixed capital. Sal – the salary fund.

The elasticity of revenue on the cost of labor is much higher than the elasticity of revenue for fixed capital costs. The increase of labor productivity and salary costs by 1% is accompanied by the increase in revenue up to 0.649%. This indicates the high significance of human resources in the formation production cost.

The sum of exponents in powers of resulting model is less than one. In theoretical issues this means that the production is associated with negative economies of scale. but with regard to this model it is connected with taking into account capital assets only.

4 Analysis of the production factors

To investigate the importance of the changes in the number of production factors impact and their productivity on the physical volume of Urals Stampings Plant. let's analyze the method of artificial neural networks. This method is realized through the defining as the dependent variable the volume of shipped products (Prod). as independent variables: (1) the amount of fixed assets (Capital). (2) Average number of employees (Labour). (3) labor productivity (LabProd) and (4) return on assets (CapProd). The model is constructed on the quarterly data of Urals Stampings

Plant financial statements in 2009–2016. Network Diagram is represented in Fig. 1.

The model is characterized by high quality of Relative Error in Training and Testing samples. They are equal to 0.004 and 0.007 respectively. Predicted and observed values generally coincide (Fig. 2).

Table 4. Model Summary[a].

Model		Sum of Squares
Training	Sum of Squares Error	0.042
	Relative Error	0.004
	Stopping Rule Used	1 consecutive step(s) with no decrease in error[b]
	Training Time	0:00:00.00
Testing	Sum of Squares Error	0.035
	Relative Error	0.007

[a.] Dependent Variable: ln*Rev*

[b.] Error computations are based on the testing sample

According to the results of the analysis of the independent variables importance it may be noted that changes in the number and productivity of the labor force account for 83.7% of the dynamics in shipped products volume. But only 16.3% fell to share fixed assets. The intensity change factors provide 51% of the production volume, and extensive factors provide 49% (Fig. 3).

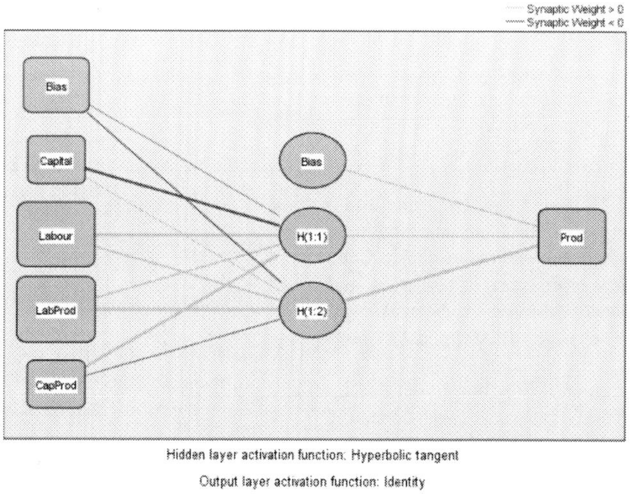

Fig. 1. Network Diagram.

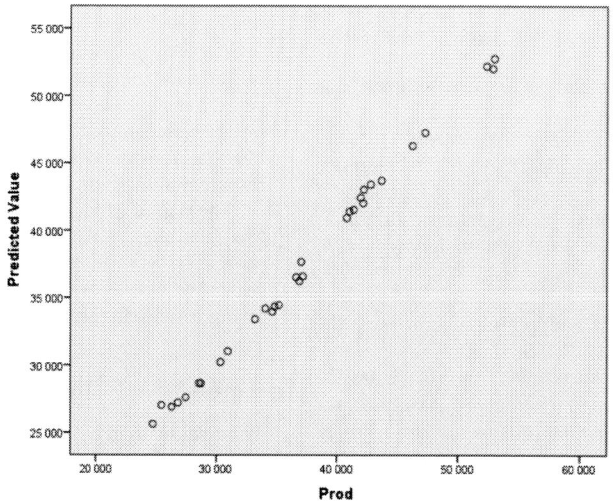

Fig. 2. Predicted by Observed Chart.

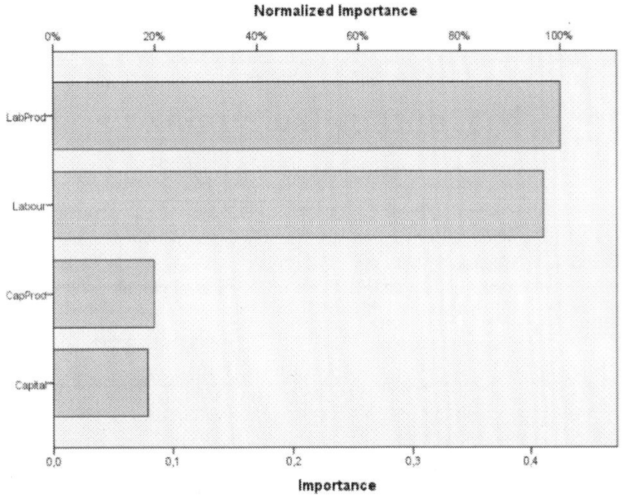

Fig. 3. Independent Variable Importance Chart.

Thus the most significant improvements in physical quantity and value of production can be achieved by changing technologies that promote the productivity of labor in the Urals Stampings Plant. The investment strategy of the enterprise should be changed in accordance with the studied parameters.

According to the result of the research the authors suggest that the point of productivity increase is in the area of human resources. The investment strategy of the enterprise should include the development and use of technologies aimed at the following points.

- Improvement the safety at work. prevention of occupational accidents and diseases.
- Improvement of labor norming system. the implementation of scientifically based set of norms and regulations.
- Establishment of a balanced scorecard of material and moral stimulation and motivation system.
- Provision of training. education and development of human resources within the organization.

The work was supported by Act 211 Government of the Russian Federation. contract № 02.A03.21.0011.
The authors would like to thank the management of Urals Stampings Plant for the full and timely provided information in the field of accounting and management reporting.

References

1. C.W. Cobb. P.H. Douglas. A Theory of Production. **18**. **1**.139 (1928)
2. W.W. Leontief. A Note on the Interrelation of Subsets of Independent Variables of a Continuous Function with Continuous First Derivatives. **3**. 343 (1947)
3. J. Robinson. The production function and the theory of capital. **21**. 81 (1953)
4. M. Brown. W.W. Chang. Capital Aggregation in a General Equilibrium Model of Production. **44**. 1179 (1976)
5. F.M. Fisher. R.M. Solow. J.M. Kearl. Aggregate production functions: Some ces experiments. **44**. 305 (1977)
6. F.M. Fisher. Aggregate Production Functions Revisited: The Mobility of Capital and the Rigidity of Thought. **49**. 615 (1982)
7. G.H. Williams. Use of a production function to estimate the impact of work fragmentation on labor productivity. **11** (2011)
8. J.S.L. McCombie. M.R.M. Spreafico. Cambridge Journal of Economics. **40**. 1117 (2016)
9. K.I. Kim. A. Petrin. S. Song. Journal of Econometrics. **190**. 267 (2016)
10. Wei Dai. Xiaojun Zhou. *International Conference of Information Technology. Computer Engineering and Management Sciences*. **4**. 53 (2011)
11. M.M. Machado. M.C.S. de Sousa. G. Hewings. Energy Economics. **59**. 290 (2016)
12. J. Sauer. C.J. Morrison Paul. Applied Economics. **45**. 1461 (2013)
13. Y. Dissou. L. Karnizova. Q. Sun. Atlantic Economic Journal. **43**. 107 (2015)
14. P. McCarthy. A. Urmanbetova. Applied Economics. **43**. 2883 (2011)
15. B.V. Revazov. Terra economicus. **4**. 310 (2007)
16. Yu.A. Shamara. E.B. Smirnov. Bulletin of Civil Engineers. **6**. 237 (2013)
17. R.M. Nizhegorodtsev. N.P. Gorid'ko. Journal of Economic Theory. **3**. 272 (2013)
18. S.A. Dobrotin. E.A. Suchkov. A.N. Lyabina. Science and world. **3**. 8 (2015)
19. B.O. Blashentsev. M.A. Shatalov. Scientific discussions. **8**. 51 (2015)
20. G.A. Ryskulova. Bulletin of KRSU. **13**. 51 (2013)

Cost-based industrial enterprise human capital management

Sergei Glukhov[1,*], *Evgeniya* Gonchar[2]

[1]South Ural State University, Logistics, Economics and Management in Transport, Supply and Distribution Department of Higher School of Economics and Management, Chelyabinsk, Russia
[2]South Ural State University, Accounting, Analysis and Auditing Department of Higher School of Economics and Management, Chelyabinsk, Russia

Abstract. The paper focuses on importance of human capital in development of industrial enterprises, issues of their management and methods of achieving balance between interests of owners and employees. Difference between such concepts as human and labor potential as well as human and working capital is clarified. The special attention is paid to the fact that an effective system of motivation and incentivation of labor is to serve as means of rapprochement of human and labor capital. The authors point out the limited scope of the traditional forms of labor motivation and incentivation mechanisms and highlight the complexity of their encouragement for collective work results. The authors suggest using the cost-based approach to workforce management, which is based on assessment of market and intrinsic value of human capital in view of investment and quality characteristics. The proposed approach is attended by methodological support and operational calculations. The study states that the suggested human capital management model can be introduced into practice, as well as substantiates the necessity for boosting the performance of industrial enterprises and their cost by increasing production, stimulating the staff for self-realization and self-improvement by bringing the owners' corporate interests closer to the interests of the employees.

1 Introduction

Improving efficiency of human capital management is one of the most common challenges, faced by modern enterprises. Workforce, who fulfils a number of functions and has a closer relation to specific operations, sees weaknesses in a system of enterprise development management much better than top management. An opportunity of getting desired economic benefit by owners of an enterprise depends on work ethics of staff as well as on its readiness to reveal reserves of enterprise economic efficiency increase. Often employees, seeing gaps in activities of close to them sites and knowing, how to resolve them because of specialization in specific operations, do not put rational suggestions forward to company management and unwish to increase amount of their own work. As a result, weak zones remain uneliminated a great while as long as a talented and attentive manager identify them.

This situation is common in many enterprises. It is obvious that the reason for it is weak mechanism of motivation and incentivation of work and inefficient human capital management system as a whole. Economic losses from it are particularly important in the industry, where specific nature of activities is more complex, and management effectiveness depends on attitude of all employees of a large team to work. In order to identify reserves of economic efficiency increasing, it is important to regulate corporate relations for purposes of improvement of efficiency of human capital management through creating command interest in development of an industrial enterprise.

2 Ease of Use

An approach of top management to appraisal of enterprise human capital value is equally important in developing workforce corporate management system. In theory and practice there are known three main approaches to appraisal of human capital value.

According to the first approach, which is called by employers "the cost-based approach", value of human capital is measured through amount of labor costs. It is believed that human capital costs so much, how much a company spends for its maintenance. The utopian nature of such an approach is established by the following circumstances. Firstly, companies offer different wage levels for the same job and the same qualification. Therefore, human capital valuation, made by one enterprise, is not accurate. Secondly, even if a company has a quite functioning system of motivation and incentivation of labor, workers do their best as far as it mainly corresponds with financial incentives. In this case we can not speak about an objective evaluation of all workforce opportunities and consequently about human capital value.

* Corresponding author: glukhovsv@susu.ru

In accordance with the second approach, which is called "the investment approach" [8, 5, 3, 14], human capital value is measured through amount of investment in education, training and health of workforce. It is believed that human capital is worth as much, how much it is spent on education, training and health support of staff. This approach is also questionable for the reason that it is impossible to consider return on investments. Investments in one worker can bring benefit to an enterprise, while investments in other will not. This is due to the fact that not all workers can effectively apply their knowledge in practice, bringing benefit to an enterprise. Not all workers think outside the box and can solve non-standard problems in practical activities. For example, there are even situations, when employees, in whom an enterprise invested nothing, solved practical problems better and acted as initiators of projects, while the other workers, whose education and training were invested by enterprise owners, did not bring real benefit to a company. It follows that investment cost does not match with actual value of human capital.

In concordance to the third approach, which is not so popular yet and which we would call "the qualitative approach", human capital value is determined on the basis of its quality. This approach is more reasonable, because it allows taking into account real benefit, which may be brought to a company by every employee of the work collective.

Nevertheless, in the opinion of the authors, individual application of each approach does not allow to appraise human capital value completely and objectively. The combination of the investment approach and the qualitative one is deemed to be the most correct. Concerning the investment approach, we mean value of investments in education, training and health support of workforce without reference to a specific source of investment. It does not matter, who has invested in education and training of an employee: the company, the employee himself or the state. In this case it is important that it is of some value to the employer. The use of the qualitative approach makes allowances in valuation of human capital, as it enables taking into account individual abilities of employees at acquisition and application of knowledge, skills and experience.

Thus, for appraisal of real human capital value it is advisable to use the integrated investment and qualitative approach, which may be the basis for elaboration of effective incentivation mechanisms in system of industrial enterprise development management. However, issues of methodology for assessing human capital quality and methodology for human capital management based on the integrated investment-qualitative approach are not sufficiently elaborated in science yet. Therefore, the aim of this paper is to develop a comprehensive methodology for assessing human capital value on the basis of the integrated investment-qualitative approach and its application under generation of an effective system of management of industrial enterprise human capital based on the cost approach.

3 Thesis Statement

We set the following objectives for this goal achieving within the framework of this paper.

Firstly, it is necessary to clarify differences between such close concepts as human and labor potential, human and working capital in order to specify nature of human capital quality.

Secondly, it is required to elaborate a methodology for assessing human capital value on the basis of the integrated investment-qualitative approach.

Thirdly, there is need to develop a methodological support of human capital management system based on the cost approach.

Fourthly, it is obligatory to carry out operational calculations in assessment of human capital value in its management system in order to confirm practical relevance of the author's model.

3 Theoretical Part

Such close categories as capital, human capital, working capital, potential, human potential and labor potential are to be considered in order to clarify the nature of the concept of human capital.

The concept of human capital was formed in the process of overcoming the traditional views on capital, which essence was explained for the first time by K. Marx. "...Capital is definite social relation, which is represented in the thing and gives this thing a specific social tone" [9]. At the beginning of the 20th century I. Fisher suggested considering as capital anything and everything, which matches the criteria of earnings generation for specified time, provided that any income is always a product of one of the capital types [7]. Under this approach capital is understood as any stock of goods, which may accumulate and produce a profit.

T. Schulz [16, 17, 18, 19] and his follower G. Becker, who have formulated the components of human capital (knowledge, production skills and motivation) in 1960, are justly the founders of the definition of "human capital" [6]. It should be noted that T. Schultz shared the concepts of "human capital" and "human potential", arguing that "all human resources and powers are either inborn or acquired. Everyone is born with a set of individual genes, determining its inborn human potential. Goodness, acquired by a human, which may be stimulated through corresponding investments, is called human capital" [16]. A number of scientists, including L. Thurow [10], Edwin G. Dolan and David E. Lindsey [3], S. Fischer and R. Dornbusch [15], L. I. Abalkin [11, 13], A. I. Dobrynin and S. A. Diatlov [1], G. N. Tuguskina [4], interprets the essence of the category of "human capital" as total, universal potential of human labor.

According to V. N. Belkin, human capital includes final performance, effectiveness of labor activity of an employee. In other words, when it comes to human capital, we mean human abilities and qualities, which are really involved in the production process, whereas potential productive capacity of a human is called labor potential [20, 21]. This point of view is the closest one to our attitude, but some points are to be clarified. Proceeding from the common concept of potential, which derives from the Latin word "potentia" and means strength, power, opportunity and ability, which exist in the open form and may manifest under certain conditions [2], we clarify the concepts of "human potential" and "labor potential". Human potential is to be understood as a set of inborn and acquired human capabilities, which exist in the open form and are able to appear under certain circumstances in any living environment. Labor potential is not so wide concept, which, in the opinion of the authors, is to be understood as a set of inborn and acquired human capabilities, which exist in the open form and are able to appear under certain conditions in labor activity (a working life).

According to the authors, human capital of a company is a percent of labor potential, already embodied in a company or possible to be embodied. Then, working capital of a company is an integral part of human capital and represents only that part of labor potential, which is actually embodied in labor force of an enterprise. In other words, human capital differs from labor capital: in the first case this is in reference to all explicit and implicit human capacities to act productively, and in the second case – only to those human capabilities, which straight manifest in work. Obviously, an effective system of motivation and incentivation of labor may act as means of rapprochement of human and labor capital [12].

The exact concept of "quality" should follow us to interpretation of the essence of the category of "human capital". It is commonly known that generally the quality is a combination of essential features, properties, characteristics, which distinguish an object or an occurrence from the other and give definition to it. Thereafter the quality of human capital is a set of essential properties and characteristics of human resources. It makes sense to carry out assessment of human capital quality, using a points system, marking investment and qualitative characteristics of employees. Investment characteristics: 1- Education, 2 - Advanced professional training, 3 - Health. Qualitative characteristics: 4 - Total relevant professional experience, 5 - Length of service with a company, 6 – Promptness, 7 – Initiativity, 8 - Decision-making ability, 9 - Operational efficiency, 10 - Self-sufficiency, 11 – Competence, 12 - Innovation activity, 13 - Teamwork skills, 14 - Social skills, 15 – Discipline, 16 - Learning capability, 17 - Self-development capability, 18 - Economic activity, 19 - Labor capacity.

Based on the presented points system, which criteria list may be extended, it is possible to form a ranking of every employee in a company (quality of individual capital of an employee) by the Formula 1:

$$QHC_n = \sum_{i=1}^{19} P_i^n \qquad (1)$$

where QHC_n – quality of individual capital of an n-employee, point;

P_i^n – a number of points within a parameter i of n-employee, point;

i – a parameter;

n – an employee.

Then the total amount of human capital quality may be determined using the Formula 2:

$$QHC = \sum_{n=1}^{m} QHC_n \qquad (2)$$

where QHC – quality of human capital of an enterprise, point.

It is possible to estimate value of human capital of an apprised company based on the data of labor market analysis, which allows determining market wage of corresponding levels of employees. At the same time market wage is to be understood as the maximum for every quality level of individual capital in the country, taking into account that human capital is resilient and is differently estimated by the market in every region that determines the market value of human capital. Furthermore, a certain period is to be considered under evaluating the total human capital cost. For example, it may be an average term of company project implementation, as it is reasoned to believe that valuation of human capital is to be carried out, taking into account a forecast period, which is used under evaluation of provided revenue from projects. An average length of service with a company may be used as an evaluation period.

Based on the foregoing, market value of human capital may be determined by the Formula 3:

$$MVHC = \sum_{n=1}^{m} \frac{QHC_n}{QHC_{R.W.}^n} \cdot MS_{R.W}^n \cdot T_A^n \qquad (3)$$

where $MVHC$ – market value of human capital of a company, rub.;

$QHC_{R.W.}^n$ – quality of human capital of the best employee, compared with a n-employee of an enterprise, point;

$MS_{R.W}^n$ – the maximum market wage of the best employee, compared with a n-employee of an enterprise, rub./month;

T_A^n – an average term of work (length of service) of a n-employee of an enterprise, months;

m – a number of employees in an enterprise.

However we must distinguish market value of human capital and economic benefit of an enterprise from possession of this capital. Economic benefit of an enterprise from possession of human capital is to be understood as economy and income, arising from the use of human labor (Formula 4).

$$EB_{h.c.} = MVHC - WF \cdot T_A + IR + CS \qquad (4)$$

where $EB_{h.c.}$ – economic benefit of an enterprise from possession of human capital, rub.

WF – monthly wages fund, rub.;

T_A – an average term of work (length of service) of employees in an enterprise, months;

IR – increase in revenues of an enterprise for the period, rub.;

CS – cost savings of an enterprise, rub.

The reasoning of evaluation of economic benefit of an enterprise from ownership of human capital predefines possibility to consider it as intrinsic value of human capital.

Comparing intrinsic value of human capital (economic benefit of an enterprise from possession of human capital) and market value (identifying difference between them), we get an amount, which may be considered as increase in cost of human capital (Formula 5).

$$AV_{h.c.} = EB_{h.c.} - MVHC = IR + CS - WF \cdot T_A \quad (5)$$

where $AV_{h.c.}$ – increase in cost of human capital of a company, rub.

Regarding applying the Formula 5 we make the following reservation: values of revenue increment and cost savings are to be analyzed over the same period under allowance of average term of work (length of service) of workforce in a company.

It is possible to form a separate balance sheet statement (let us call it balance sheet statement of human capital), using the figures of the given methodology for assessing value of human capital. The assets show intrinsic value of human capital, the liabilities – its market value and growth.

The proposed system of valuation of human capital may be taken as a basis of motivation and incentivation of labor. In actual practice it is not always possible to draw a parallel between eventual outcomes and individual contribution of employees. Any particular achievement is a result of teamwork, where performed by employees functions are integrated. The cost-based approach in management of human capital can be used in order to distribute overall obtained economic benefit. In this case the mechanism of motivation and incentivation of labor has a bilevel structure.

The traditional forms of motivation and incentivation of labor are realized at the first level through encouragement of individual achievements that is possible, when association between labor input of an employee and obtained enterprise economic or other kind of benefit may be obviously observed. The cost-based approach is offered to be applied at the second level of mechanisms of motivation and incentivation, especially where there is a question of distribution of economic performances, achieved by a team.

We enlarge upon the point of the second level of the mechanism of motivation and incentivation of labor. It is important to bear in mind that under its implementation the final economic results of a company are achieved not only through the use of human capital. At this point there are important financial, organization, innovation and market capital and commercial goodwill, contributing to efficiency of an enterprise and promoting formation of its intrinsic value in addition to human capital.

Financial capital makes contribution by funding current and strategic business needs. Organization capital contributes in the form of corporate standards and rules of behaviour, work of units and staff, organization of business processes, ensuring effective communication within workplace for the benefit of achieving current and strategic objectives and solving operational tasks of an enterprise. Innovation capital, consisted of costs of developed and implemented innovation projects, contributes to strategic development of an industrial enterprise. Market capital is involved in development of an enterprise through established business contacts with buyers, customers, suppliers and contractors.

Commercial goodwill provides additional benefit of a company with a high credit rating, fiscal discipline and trust of partners through simplification of procedures for acquisition of credit resources, empowerment to use investment tax credit and lack of likelihood of capital pulling out of business by its founders. All these admit that it makes sense to distribute final economic results at first between a company, represented by its owners, and human capital and then to distribute economic benefit, providing human capital increment, between employees. Let us refer to developing methodological support of the second level of implementation of mechanism of motivation and incentivation of labor.

A proportion of economic benefit of a company, accounted for human capital, may be determined by means of the Formula 6.

$$S_{h.c.} = \frac{IV_{h.c.}}{IV_{h.c.} + IV_{o.c.} + IV_{i.c.} + IV_{f.c.} + IV_{m.c.} + IV_{r.c.}} \quad (6)$$

where $S_{h.c.}$ – a share of intrinsic value of human capital in intrinsic value of an enterprise, taken as a basis of total economic needs distribution;

$IV_{h.c.}$ – intrinsic value of human capital of an enterprise, rub.;

$IV_{o.c.}$ – intrinsic value of organization capital of an enterprise, rub.;

$IV_{i.c.}$ – intrinsic value of innovation capital of an enterprise, rub.;

$IV_{f.c.}$ – intrinsic value of financial capital of an enterprise, rub.;

$IV_{m.c.}$ – intrinsic value of market capital of an enterprise, rub.;

$IV_{r.c.}$ – intrinsic value of commercial goodwill of an enterprise, rub.

Value of enterprise economic benefit, accounted for human capital (we call it a collective bonus pool), may be determined according to the Formula 7.

$$BF_{c.} = AV_{h.c.} \cdot S_{h.c.} \quad (7)$$

DOI: 10.1051/shsconf/20173501039

where $BF_{C.}$ – a collective bonus pool, rub.

A distributed amount of economic benefit of an enterprise, accounted for a particular employee, may be calculated by virtue of the Formula 8.

$$BF_W^n = BF_{C.} \cdot \frac{QHC_n}{QHC} \qquad (8)$$

where BF_W^n – a distributed amount of economic benefit of an enterprise, accounted for a n-employee, rub.

Summarizing the undertaken study, we draw attention to those results, which contain scientific novelty.

1. Difference between the concepts of "human potential", "labor potential", "human capital" and "labor capital" is clarified. Human potential is a set of inborn and acquired skills and human capabilities, which exist in the open form or can appear under certain conditions in all living spheres.

Labor potential is a set of inborn and acquired skills and human capabilities, which exist in the open form or can appear under certain conditions in the course of labor activity (work life).

Human capital is a share of labor potential, which is already embodied in a company or is possible to be embodied therein.

Labor capital of an enterprise is an integral part of human capital and represents only that portion of labor potential, which is actually embodied in labor activity of an enterprise.

2. The methodology of human capital appraisal on the basis of the integrated investment and qualitative approach is elaborated. Quality of human capital, which is estimated on investment and quality characteristics, and the corresponding market and internal appraisal of value are the foundation of the methodology.

3. The methodological support of human capital management system is developed based on the cost approach. The instrumentarium of labor motivation and incentivation system is its foundation. The methodology of distribution of collectively obtained economic benefit of an enterprise between its owners and its human capital with the further distribution between its employees is included therein.

4 Practical Relevance

In order to justify the practical relevance of the proposed methodological support of human capital management system we carry out operational calculations, using project targets of a small industrial enterprise.

At first, we assess quality of human capital of an enterprise and compare individual performance with performance of the best employees (Table 1).

Table 1. Appraisal of Human Capital Quality of an Enterprise.

Employee	Quality of human capital			Average length of work, months	Maximal market wage, rub.
	Quality of human capital of an company, point	*Quality of the best human capital, point*	*Correlation (2):(3)*		
Employee 1	89	145	0,613793	48	100 000
Employee 2	116	156	0,743590	30	135 000
Employee 3	94	122	0,770492	60	80 000
Employee 4	138	150	0,920000	84	130 000
Employee 5	141	169	0,834320	54	150 000
Total	578	742	–	55,2	–

The market value of human capital, calculated according to the Table 3, is 26 460 494 rub.

Assuming that increase in revenues and skimping on enterprise cost in the whole for a period, equal to the average length of work in a company, is 27 600 000 rub., the payroll for the same period – 12 420 000 rub., then intrinsic value of the human capital is 41 640 494 rub., and its share in the intrinsic value of the company will be 16%, when the intrinsic value of the organization capital is 5 000 000 rub., the intrinsic value of the innovation capital – 80 000 000 rub., the intrinsic value of the financial capital – 90 000 000 rub., the intrinsic value of the market capital – 20 000 000 rub., the intrinsic value of the commercial goodwill – 20 000 000 rub., and the general intrinsic value of the company, taking into account the intrinsic value of the human capital, is 256 640 494 rub.

Then, the overall economic benefit of the enterprise for the average length of work in the company accounts for 15 180 000 rub., where the share of the human capital accounts for 2 462 989 rub., which is 44 619 rub. per month.

Table 2. Distribution of the Collective Monthly Bonus Pool.

Employee	Correlation of individual and the best quality of human capital (Table 1)	Distributed economic benefit of the enterprise per employee, rub.
Employee 1	0.613793	6 870
Employee 2	0.743590	8 955
Employee 3	0.770492	7 256
Employee 4	0.920000	10 653
Employee 5	0.834320	10 885
Total	–	44 619

In conclusion to the practical section, we distribute the collective monthly bonus pool among the employees in proportion to the ratio of their individual qualities and the best quality of human capital of the relevant category (Table 2).

The obtained results indicate that different levels of the individual qualities of employees correspond to different amounts of the distributed collective bonus. The higher quality of individual employee is, the more amount of distributed collective bonus is. Such a system of collective incentivation for team results, combined with the traditional encouragement for individual performance, allow improving efficiency of the mechanism of motivation and, thus, improving effectiveness of human capital management for the benefit of industrial enterprise development.

5 Conclusions

The results of this study confirm that it makes sense to implement human capital management system, based on the cost approach, in practice of industrial enterprises. Human capital as well as financial, organization, innovation, market capital and commercial goodwill contribute to economic benefit gaining and increment of its internal and market value. Therefore, determination of correlation of material incentivation for employees with contribution to intrinsic value of enterprise human capital promotes balance of interests of employees and owners and creates conditions for effective labor motivation. The possibility to provide command interest in improving economic efficiency of an enterprise and its cost, using implementation of the author's model, is stipulated by the following circumstances.

Firstly, distribution of economic benefit between an enterprise, represented by the owners, and its workforce, represented by the human capital, in proportion to a share of cost components in intrinsic value of a company determines strengthening corporate relationships and interest of all employees in company development.

Secondly, application of evaluation of human capital quality as a part of value appraisal system in distribution of economic benefit creates conditions for incentivation of employees to self-improvement, development and self-realization that will improve efficiency of an enterprise in the future.

Embodiment of the results of the above circumstances predetermines a cycle of positive effects in industrial enterprise development system through improving productivity and motivation to self-realization in work and development of workforce.

The work was supported by Act 211 Government of the Russian Federation, contract № 02.A03.21.0011.

References

1. A.I. Dobrynin, S.A. Diatlov, V.A. Konnov, S.A. Kurganskii, *Productive Power of The Human: Structure and Forms of Appearance* (Saint-Petersburg University of Economics and Finances, 1993)
2. A.V. Babkin, A.A. Moshkov, News of the St. Petersburg State Economic University, **6(84)**, 46-47 (2013)
3. E.G. Dolan, D.E. Lindsey, *Market. Microeconomic Model* (Copyright, 1992)
4. G. Tuguskina, Personnel Management, **3**, 205 (2009)
5. G.A. Khmeleva *Human Capital as A Condition of Formation of Innovation Economy of The Region: monograph* (SaSMU, 2012)
6. G.S. Becker, *Economic Analysis and Human Behaviour* (Selected Works on Economic Theory, 2003)
7. I. Fisher, 15-17 (1927)
8. I.V. Soboleva, *Pardoxes of Human Capital Measuring: Scientific Report* (Institute of Economics of RAS, 2009)
9. K. Marx and F. Engels, *Capital* (1961)
10. L. Thurow, *Investment in Human Capital* (Academia, 1970)
11. L.I. Abalkin, *Russian Strategic Response to The Challenges of The New Century* (Ekzamen, Moscow, 2004).
12. M.K. Akhtiamov, E.A. Gonchar, Economics and Entrepreneurship, **8-1(61-1)**, 522-527 (2015)
13. R.D. Galieva, V.A. Mishin, *Collection of Conferences of Research and Development Centre "Soziosfera"*, **11**, 71-72 (2012)
14. R.M. Nureev, *Russia: Characteristics of Capacity Buildings* (Norma, 2009)
15. S. Fisher, R. Dornbusch, R. Schmalensee, *Economics* (Delo, 1994)
16. T.W. Shultz, *Human Capital in the International Encyclopedia of the Social Sciences* (1968)
17. T.W. Schultz, University of California Press, 149-166 (1981)
18. T.W. Schultz, *Investment in Human Capital: the Role of Education and of Research* (Free Press, 1971)
19. V.N. Belkin, V.I. Vinogradova, *Innovation Activity of Human Capital* (Institute of Economics of UD of RAS, 2009)
20. T.W. Schultz, *The Economic Value of Education* (Columbia University Press, 1963)
21. T.W. Schultz, *Investment in Human Capital* (Economic Growth – an American Problem, Englewood Cliffs, 1964)

The cross-industry cluster formation as a reengineering mechanism for the production relations of the old-industrial regions

Dmitriy Kotov[1], *Irina* Burenina[1], and *Evgeniya* Byl[1*]

[1]Ufa State Petroleum Technological University, Economics and Management in the Oil and Gas Industry, 450000 Ufa, Russia

Abstract. The article considers the formation of a strategic regional industrial policy in conditions of the new technological order occurrence on the basis of trends in the global technological development, conducts a strategic analysis of industry of the Republic of Bashkortostan on types of economic activities: mining, production of petroleum products, chemical production, manufacture of rubber and plastic products and construction, identified positive and negative trends and proposed future directions of development taking into account assumed scenarios of regional, national and global markets development. The study proposed options for the sectoral and cross-sectoral clusters establishment, the estimation of the resulting synergies in terms of creating inter-industry construction cluster based on the element wise estimation method of clustering efficiency and performed the benefits of the cross-sectoral cluster formation for all market participants.

1 Introduction

The basis of leadership in the modern world becomes a success in the technology markets. The forecast estimation of the Russian development for the upward stage period of the upcoming sixth "Kondratiev cycle" is calculated for two scenarios – inertial (GDP of Russia for the period is expected to grow by 2.2 times) and the innovative (growth of GDP by more than 9.5 times, and the average annual economic growth will exceed 9%) [1]. Synchronization methods of state innovation and industrial policy with the upward phase of a large cycle of conjuncture is able to give a synergistic effect and lead to significant and rapid increase in the welfare level of the population and the country as a whole.

In connection with the new technological order beginning the experts identify the following trends in global technological development: major technological innovations based on the use of new structural and composite materials created using nanotechnology, the formation of a new core of information technology on the basis of the transition from microelectronics to nanoelectronics, the alternative energy technologies development due to reducing the dependence on hydrocarbon energy resources [2,3].

The Russian development strategy cannot be implemented without synchronization with the strategy of regions. One of the most industrially-important regions is the Republic of Bashkortostan. In this case supporting technological initiatives will be the basis of the transition to a new technological level. The study proposes to assess synergy in the context of sectoral or cross-sectoral clusters based on element-by-element method of cluster creation estimation efficiency.

2 Strategic analysis

Republic of Bashkortostan is an old industrial region with elements of the fourth and fifth technological order. On its territory historically formed the concentration of industrial sectors [4], with its potential to increase production of high technology products through innovative transformation of its production complex in order to improve economic relations.

Strategic analysis of the economy of the Republic of Bashkortostan showed the presence of the following industries:
- mining;
- production of coke and petroleum products, chemical production;
- production of rubber and plastic products;
- construction.

Almost whole list of minerals is extracted on the territory of the Bashkortostan, which is available on the Russian territory. There are 194 deposits of mineral resources (without oil fields) at the state balance [5].

The largest share is oil and natural gas and mining of metal ores.

The analysis of indicators allows to make a conclusion about the ineffectiveness of a purely industry cluster mining, but there is potential for creating cross-sectoral clusters.

Refinery and petrochemical complex is the main source of economic and social development of

* Corresponding author: ms.byl@mail.ru

Bashkortostan, occupies the first place in the Russian Federation by the volume of processed raw materials, produces half of the petrochemical industry products of the Ural economic region and accounts for about 40% of the industry profits. This complex is the largest Russian producer of a number of products, ranking first in production of butyl and isobutyl alcohols (50%); second place – in production of synthetic resins and plastics (15%); third place – in production of caustic soda (20%); sixth place – on polyethylene (11%) and synthetic rubbers (12%). More than seventy types of products are exported [6]. The share of chemicals and petrochemicals in total exports of the Republic is about 20%.

Table 1. Cross-sectoral clusters.

The type of mineral	The variant cluster
Production of bauxite and construction natural resources	Construction cluster
The extraction of copper, iron, manganese ore, precious metals	Automobile cluster, mining and metallurgical cluster
Oil and gas extracting	Fuel-energy cluster, petrochemical cluster
Peat extraction	Agricultural cluster, biochemical cluster

Positive and negative trends in the development of the enterprises for manufacture of coke and refined petroleum and chemical industries in the Republic of Bashkortostan are given in table 2.

Table 2. Positive and negative trends in the development of the enterprises for manufacture of coke and refined petroleum and chemical industries in the Republic of Bashkortostan

Positive trends	Negative trends
1 Labor productivity growth (28.6% for 2005-2013) 2 Exceeding average Russian «labor productivity» indicator in the Republic of Bashkortostan (13%in 2013) 3 The growth of investments in industry (for 23.2% in 2014) 4 Increase in the share of economic activity in the total investment in the Volga Federal district (from 17,1% to 17,9% in 2014) and Russian Federation (from 7,2% to 7,9%in 2014) 5 Leading positions in Russia: production of alcohols, butyl and isobutyl; soda ash; chemical plant protection products; soda; synthetic rubbers; synthetic resins and plastics; of polyethylene. 6 The only manufacturer in Russia of synthetic glycerin, perchloroethylene, ethylene-propylene rubber, baking soda	1 The slowdown in the growth of production volumes in monetary terms 2 The reduction in the volume of production in value terms in Russia (from 9.9% to 8.05% in 2013) 3 The reduction in the workforce (30% for2005-2014 or 14.8 thousand people).

One of the problems of the backwardness of petrochemical industry development in Russia – loss of vertical and horizontal integration. The reason for this is that the petrochemical industry as in the Republic of Bashkortostan and Russian Federation as a whole is structurally fragmented: regions of extraction and major centers of processing removed both geographically and divided by the ownership structure. The solution is to rebuild previously existing linkages between companies in the sector, which can be done by using the following approaches: the conclusion of mutually beneficial long-term agreements between participants, the creation of clusters, joint ventures.

Most popular products of the petrochemical industry at the present time are polymers (polyvinylchloride, polypropylene, polyethylene, etc.), which is primarily due to the mass transfer on plastic pipes, plastic building materials, as well as active usage of polymer packaging [6,7].

The main consumers of chemical and petrochemical products in Russia are the population, the enterprises of chemical and petrochemical industry deeper processing, as well as enterprises of mechanical engineering and the automotive industry, construction and agriculture.

Petrochemical products are in demand due to their high consumer characteristics. While in Russia there is a high untapped potential for growth in domestic consumption of petrochemical products (table 3).

Table 3. The comparison of the share of polymeric materials used in various industries in Russia and abroad, %

Industry	USA	Western Europe	Central Europe	Russia
Highway engineering	65	60	46	6
Pipe	75	70	40	15
Rooting material	70	47	38	3

It is estimated that Russia's current level of GDP per capita should consume 1.5-3 times more plastics than is consumed currently. The country needs to encourage domestic demand for petrochemical products due to changes and introduction of new national and supranational standards on consumers, the introduction of new requirements in the field of energy conservation and sustainability, involving the use of products with high technical characteristics and consumer properties.

The share of economic activity "Manufacture of rubber and plastic products" in the volume of manufacturing in Republic of Bashkortostan is 0.01%, Volga Federal district – 2,33%, Urals Federal district – 0,79%, Russian Federation – 1,83%. Industrial production Index of rubber and plastic products manufacture in 2013 decreased by 77.9% to 94.4 per cent respectively [7]. The decline in the production of rubber and plastic products is due to reduced orders for the automotive industry, rising prices for materials. Annual 30% growth in shipments until 2012changed to fall across Russia, but 2014 it was 7%, which exceeded the 6% growth in Russia and 1% in Volga Federal district.

There is a forecast for the change in the structure of consumption of plastic products in the long term to 2030 [8].All segments of the market related to the processing

DOI: 10.1051/shsconf/20173501040

of polymeric materials in Russia are developing dynamically for several years. Primarily this is due to steadily rising demand and a rapid return on investment. Investment attractiveness of the plastics processing sectors, especially for the enterprises of small and medium business, is due to the national economy development, economic activities consuming products from polymers such as construction, agriculture, production of vehicles and equipment, food production, and improvement of the business environment.

The increase in construction (annually more than 10%) [8] causes an increase in the demand for profiled molded and other products from polymeric materials. The structure of the Russian market capacity profile-molded products include 82-85% of products based on polyvinyl chloride, 10-12% of products from polystyrene, 5-6% of products from polyolefins. Market profile-molded products largely import-dependent, however, is developing towards an active import substitution. The share of domestic production in ensuring the demand of the domestic market has increased from 45% to 80% from 2007 to 2014.

The development of this sector is linked with the development of industries that consume products from polymers such as construction, agriculture, manufacture of transport equipment,food products.

The following main business segments production of plastics has been formed in the Russian market during the period from 2005 to 2014: packaging with a share of 30 – 40%, products and parts for production purposes — 15 – 18%, film – 16-17%, profile-molded products – 5 – 18%, products of cultural and household goods – 10 – 14%, pipes – 4 – 9%, leaves – 2 – 3% [7].

Table 4. Consumption sectors of rubber products.

Market	Automobile industry	Mechanical engineering	Construction
Market capacity, mln rub	881	3081	9676
Type of competition in the market	The average level of competition, a high level of concentration	The average level of competition, a high level of concentration	The high level of competition, a low level of concentration
Market share, %	0,04	0,10	0,13
Growth rate market share	1,05	1,01	1,10
Forecast market share	0,03	0,12	0,20

Industrial rubber products are used in almost all industries (table 4). In most cases, the reliability and durability of structures, machines or mechanisms are limited to the reliability and durability of components of rubber products, the contribution of which to the cost and weight of the machine or mechanism is usually negligible.

The main consumers of rubber are the automotive industry and agricultural machinery. The most significant growth in the production segment for construction is possible by the following products: gaskets, various coatings, isolation material, etc. One of the promising areas is the production of thermoplastic elastomers used in road construction. The use of such technology in the repair and construction of roads allows to increase the service life of pavement from two to three times and save gasoline due to more efficient coupling of tyres with road. In Russia less than 1% of the roads constructed using this technology, while in Scandinavian countries this figure reaches 25%.

According to the Russian Ministry of energy during the period of 2010-2020 reconstruction and construction of new roads (for implementation of the Transport strategy of the Russian Federation until 2020) would require the consumption of plastics of about 192 thousand tons, including 48 thousand tons of polypropylene geogrids and 144 thousand tons of geotextile from polypropylene and polyethylene terephthalate. In this case the existing power in Russia provides only 53 thousand tons of these materials, which requires the commissioning of new capacities [9].

Table 5. Consumption sectors of plastic products.

Market	Packaging	Construction	Automobile industry
Market capacity, mln rub	540	4345	176,2
Type of competition in the market	The average level of competition, a high level of concentration	The high level of competition, a low level of concentration	The average level of competition, a high level of concentration
Market share, %	0,80	21	8
Growth rate market share	1,2	1,03	1,02
Forecast market share	0,90	25	10

The construction sector is one of the interindustry economic complexes, which is a set of material production sectors and design activities to ensure the reproduction of fixed assets.

The result of the activity of the construction organizations is the creation or reproduction of fixed assets residential and nonresidential use.

There were 7 in 2013 and 11 in 2014construction companies of Bashkortostan in the list of 135best construction companies of Russia, that positively characterizes the quality of construction services in the Republic.

The Republic of Bashkortostan by the end of 2014 ranked 5th among the subjects of the Russian Federation for the commissioning of residential homes, 10th place for the amount of work performed by activity "Construction" and 12th place in terms of investment in fixed assets.

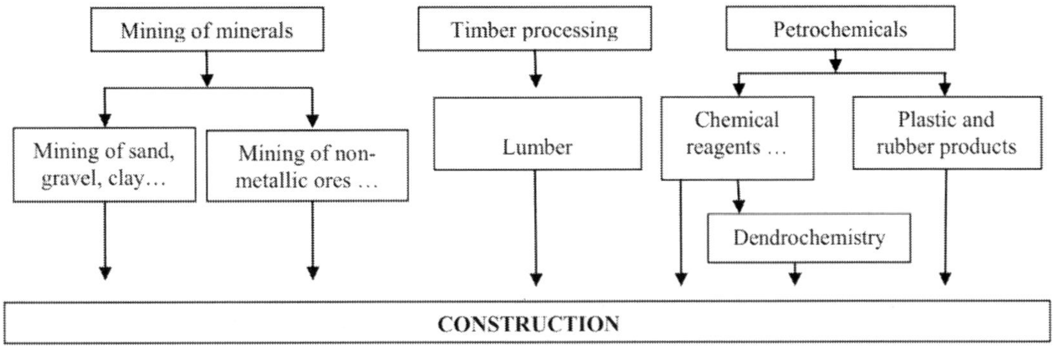

Fig.1. The common scheme of inter-sectoral linkages.

However, despite the high level of performance subject, the main problem of the sector remains a low industrial orders endowment.

The construction industry is one of the main consumers of products of other industries (figure 1).

It is needed to focus on the use of own mineral resource base, which will reduce the cost of new housing by reducing transportation costs of materials for development construction industry in the Republic of Bashkortostan.

3 Cluster formation

The analysis of economic activities in the Republic of Bashkortostan and the assessment of potential and priority directions of development in these industries enables it to identify opportunities for the formation of 2 types of cluster:

1. The petrochemical cluster. There is the Program "Development of petrochemical territorial cluster of the Republic of Bashkortostan" for 2014-2016 (app. the resolution of the Government of the Republic of Bashkortostan, dated December 9, 2014 No. 564) [10]. The effectiveness of this cluster is low. It is required to encourage the active cooperation of enterprises in the cluster and revision of instruments of administrative influence.

2. Cross-sectoral construction cluster (figure 2).

Effectiveness evaluation of the creating inter-industry cluster (construction, production of construction materials, development of new materials, structures and technologies of construction) is given in table 6.

The advantages of the cross-sectoral construction cluster:

- The ability to attract financial resources in the construction industry (Association of General financial capacity of the members of the cluster, attracting investment, joint participation in tenders of projects funded by the grant, cooperation of financial capabilities to provide guarantees for a loan).
- The possibility of reducing the cost of construction products and services of the organizations participating in the cluster.
- Having our own production based on local raw material stimulate reducing the cost of transportation, preservation, trading services, which reduces the cost of living area and repairs. Moreover, there are no problems with the sale: the product is completely used within a cluster.
- Association of organizations of the construction sector of the region allows the cluster members to effectively defend their interests at the level of the local government.

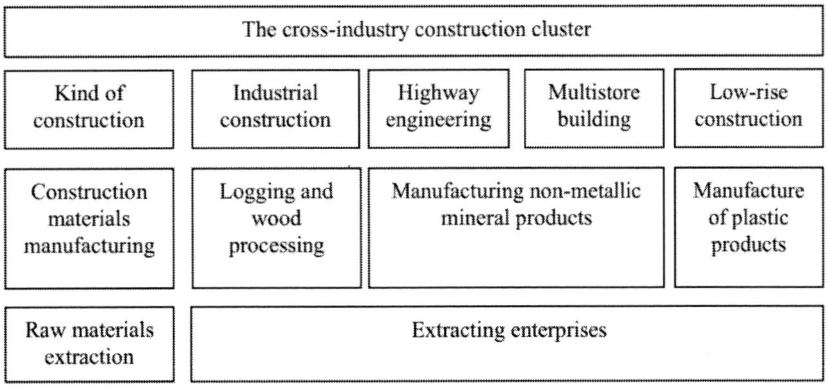

Fig 2. The cross-industry construction cluster.

- Realization the state programs of providing affordable and modern housing by members of the cluster is the provision of a constant workload of the construction industry and related sectors; creating new jobs; contributing to the revival of industry of certain districts with deposits of minerals used in construction.

Table 6. Effectiveness evaluation of the creating inter-industry cluster.

Evaluation criteria characteristics	Coefficient of influence on the characteristic (Ki)	Amount of points (ki)	Evaluation of characteristics (Ni)
1	2	3	4
1. The presence of competitive enterprises (N1)			
k1 – the determination of the current development potential of enterprises	0,4	10	9
k2 – the effect of cluster enterprises on the industry in the region	0,4	10	
k3 – the image of the cluster enterprises in the industry	0,2	5	
2. The implementation of innovative projects (N2)			
k4 – the degree of innovativeness of the industry	0,5	10	10
k5 – the commercialization of new technologies	0,5	10	
3. The growth of economic indicators of the industry (N3)			
k6 – the volume of works (services)	0,4	5	6,5
k7 – the number of people employed in the industry	0,3	5	
k8 – the volume of investments in fixed capital	0,3	10	
4. The linkages and interaction between members of the clusters (N4)			
k9 – the level of cooperation	0,4	10	8,5
k10 – the creation of the infrastructure necessary for cluster development	0,3	10	
k11 – the enter on new level of control	0,3	5	
5. Geographical proximity of cluster members, the right choice of cluster members and their number (N5)			
k12 – the geographical proximity of cluster members	0,5	10	7,5
k13 – the right choice of cluster members and their number	0,5	5	
Performance indicator of cross-sectoral cluster creating			8,3

- Association of organizations at the regional level – a new level of relations (new ways of thinking), which is based on honesty and trust between members of the cluster.
- The increase in innovative activity in the region. In this case, the benefit to the scientific organizations of the region is to expand the possibilities of introducing research results into practice.
- For financial institutions the creation of a cluster is the opportunity to participate in cluster projects in terms of the provision of funding under the large-scale investment projects.

The benefits of creating inter-industry construction cluster is the increase of budget revenues due to the growth in tax revenues, increase competitiveness and investment attractiveness of the region.

4 Conclusion

In the Republic of Bashkortostan there are objective opportunities for the creation of successful intersectoral construction cluster in the framework of strategic planning of sustainable territorial development of the subject. This cluster will allow to realize competitive advantages and development priorities involved in it industries with benefits for all participants, including improving affordability of housing for the population through the innovative materials and technologies using, which is going to reduce the cost of housing and to ensure the further development of the state programs supporting the population in acquiring housing.

References

1. V.A. Sadovnichij, A.A. Akaev, A.V. Korotaev, S.Ju. Malkov, *Modelling and forecasting world dynamics* (ISPI RAN, Moscow, 2012)
2. I. Agamirzjan, D. Belousov, E. Kuznecov, A. Zotov, I. Danilin, D. Holkin, A. Shtajnmjuller, K. Shtajnmjuller, Je. Projdakov, E. Lukin, A. Raevskij, S. Luk'janenko, *Pervushin Challenge 2035* (Olimp–Business, Moscow, 2016)
3. Public report on the results of a sociological study of the behavioral and institutional preconditions for technological development of the Russian regions. http://www.rvc.ru/upload/iblock/d10/attitudes_to_technologies_and_innovations_in_Russia.pdf
4. I. Burenina, E. Evtushenko, D. Kotov, A. Battalova, V.V. Birukova, M. Gayfullina, D. Gamilova, S. Saifullina, A. Tasmukhanova, I. Zaharova, *The formation of fuel and energy complexes strategic priorities of development of the Bashkortostan Republic* (USPTU, Ufa, 2016)
5. I. Burenina, D. Kotov, A. Battalova, A. Tasmuhanova, I. Zaharova, Science, **7**, 6 (2015). DOI: 10.15862/104EVN615
6. I. Burenina, M. Gayfullina, S. Saifullina, E. Evtushenko, A. Zats, Science, **7**, 6 (2015). DOI: 10.15862/105EVN615

7. I. Burenina, D. Kotov, V. Birukova, M. Gayfullina, D. Gamilova, Science, **8**, 1 (2016). DOI: 10.15862/33EVN116
8. Forecast long-term socio-economic development of the Russian Federation for the period up to 2030. http://garant.ru
9. I. Burenina, A. Battalova, V. Birukova, S. Saifullina, Science, **8**, 6 (2016). http://naukovedenie.ru/PDF/36EVN616.pdf
10. Official information portal of the Ministry of economic development of the Russian Federation. http://www.economy.gov.ru/
11. I. Burenina, D. Kotov, V. Birukova, D. Gamilova, A. Tasmukhanova, Science, **8**, 1 (2016). DOI: 10.15862/34EVN116
12. I. Burenina, A. Ovchinilova, Almanac of modern science and education, **12**, 174-176 (2010)
13. D. Kotov, O. Efimova, The scientific review, **4**, 539-549 (2012)
14. D. Kotov, Management problem of innovation development of economy: state, region, cluster. Oil and Gas Business, **1** (2010). http://ogbus.ru/article/problema-upravleniya-innovacionnym-razvitiem-ekonomiki-gosudarstvo-region-klaster/
15. D. Kotov *Management of innovative development of socio-economic systems* (USPTU, Ufa, 2011)
16. National report on innovation in Russia 2016. http://www.rvc.ru/upload/iblock/b6d/RVK_innovation_2016_v.pdf
17. A. Ovchinnikova, I. Burenina, Economy and management: scientific-practical journal, **4**, 23-27 (2011)
18. Official municipal website of the Ministry of economic development and the coat of arms of the Republic. http://www.minecon.bashkortostan.ru
19. E. Tukhvatullina, E. Evtushenko, Oil and Gas Business, **11-3**, 160-164 (2013)

A comprehensive approach to optimization of industrial enterprises management structure

Aleksey Salov[1,*], *Anna* Gavrilova[1], and *Lyaisan* Sagitova[1]

[1]Samara State Technical University, Samara, Russia

Abstract. The conditions of functioning of the power plants management systems are considered in the article. Comprehensive analysis of. The comprehensive analysis of the current and previously used management structures showed a variety of possible options for constructing them. There is proposed a set of partial criteria of the analyzed management structures grouped in local criteria that determine the features of the power plants functioning. Functions of local optimization criteria are identified. The formulated partial and local criteria of the management structures characterize balance, integrity, controllability, and stability of the analyzed control systems and in complex determine the effectiveness of the energy companies. Analysis of the performance of private structures analyzed showed that the assessment of the quality of the particular criteria give different, uncorrelated and conflicting results. It is necessary to develop methods of multi-criteria evaluation to get an unequivocal result of the system effectiveness of the analyzed management structures.

1 Introduction

Effective management of industrial plants, including power plants and systems is not possible if the control systems of the structure are not consistent with the objectives and tasks facing them, particularly nowadays – in the period of intense restructuring and reorganization processes to ensure successful entry into the market of [1-3].

During the transition period, the organization of production management system should be based on the concept of a complex approach, which means that the management of production should be based on holistic properties and patterns of behavior of companies that did not exist in its constituent parts [4-6].

While developing the methods and optimal control systems we should take into account the totality of energy, technological, economic, financial, institutional and other factors characterizing the specifics of the operation of the control object.

With the change of the goals of the companies, there should be changes in their management structure aimed at their effective independent functioning in the reformation conditions, ensuring the competitiveness of the company as a whole [4].

Based on system analysis methods there are studied several structures of power plants management systems to determine their compliance with the objectives and environmental factors in the current operating conditions using the generalized effectiveness criterion [7-11].

The analyzed structures include the one that operated until recently at the generating plants, and is used now with some internal changes.

2 Description of the research object

The object of this research work is the power station with co-generation (heat and power plant). It means the use of power station to generate electricity and useful heat at the same time for the needs of the industry of the region.

The company operates as a part of the energy company, which unites several similar power plants in the industrial region. The management system of power station is a hierarchical multi-level structure.

The CEO, director or chief engineer heads the company. The second level of management includes deputies they are decision makers, as well as the company CEO. First two levels approve the modes of functioning of the equipment, its repair or reservation periods. The second level include the positions listed below:

- Deputy Director for General Affairs (COO),
- Deputy Director of Economics (CFO),
- Deputy Director for Personnel and Social Affairs,
- Chief Engineer,
- Deputy Chief Engineer (Maintenance),
- Deputy Chief Engineer (Repairs),
- Legal Adviser,
- Inspector of orders execution control,
- Chief of administration.

The third level includes heads of various departments: heads of divisions, services and workshops. Such as:

- Human Resources Department,
- Maintenance department,

* Corresponding author: aleksey-salov@rambler.ru

- Planning and economic Department,
- Office Department,
- Department of capital construction,
- Finance Department,
- Accounting Department,
- Shift manager,
- Technical Engineering Department,
- Boiler Department,
- Turbine Department,
- Electric Department,
- Department of thermal automatics and measurements,
- Fuel-transport Department,
- Chemical Department,
- Shift supervisors,
- Repairs Department,
- Metals Laboratory,
- Mechanical repair shop,
- Repair and construction shop,
- Emergencies and Civil Defense Department,
- Labor protection Department.

Three-tier management structure is shown in Fig.1.

Prior to the restructuring of the economy, ie, until 1990 at power generating companies there operated typical management structure, since 2000 energetic enterprises management structure changed dramatically. The changes occurred in the direction of combining the functions of individual deputies, for example: Chief Engineer mixed with the position of director. Furthermore, the departments that did not directly relate to the production function of the enterprise, like repair services, were removed or reduced.

In our opinion, not all of the structural changes that took place at the enterprises provide the necessary reliability and continuity of the heat and electric energy supply in the emergencies during the heating season. Due to that, we created three more management structures, which are considered in the article.

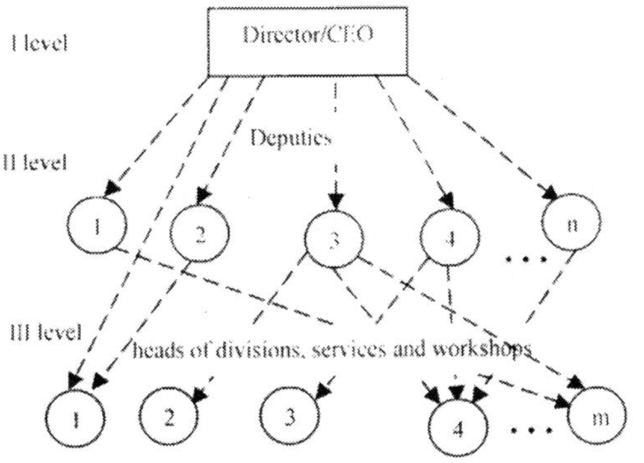

Fig. 1. Scheme of the enlarged management structure of the power plant.

3 Determination of the criteria of assessment of the management structures

System analysis of the structural management schemes is conducted using the methodology of graph theory. These indicators are used as local criteria of effectiveness of the analyzed management structures [1, 2].
- The balance of the structure;
- The integrity of the structure;
- Controllability of the structure;
- The structure stability.

A group of the partial criteria that are formulated further assesses each of the local criteria.

1. The balance of the structure
- Information load at the first level of management.

Information load at the highest level of the hierarchy - from the first head (director) – the value of the balance of the information traffic load among decision makers. When decision-makers have a large number of connections, there is an overload of information and the effectiveness of management decisions falls. In practice, it is widely used that a rational amount of information links for CEO is no more than seven.
- Information load at the second level of management.

The structure is also more balanced, if the load of the deputy heads - decision-makers – is balanced. It is possible in the case when each deputy would have approximately the same number of administrative relations, indicating the similarity of their traffic load.

Equability of the information load distribution among the managers of the first and the second levels

To improve the efficiency of the system it is necessary to seek to ensure that the information load is distributed evenly among all decision makers. It is normal for managerial functions at different levels of the management structure of the hierarchy to be materially different.

The analyzed structures differ in the number of deputy directors, directors of the company, the number of divisions and departments directly under each of the deputies and the Director himself.

2. The integrity of the structure
- The number of the structure centers.

Management decisions are made in the centers of the structure. In the analyzed structures, these centers are the director, chief engineer and deputy director of capital construction. There should be the smallest distance from these centers to the most remote departments and workshops. Based on the index of integrity and compactness of the structure of management, the most effective is the structure with the largest number of centers;
- The diameter of the structure.

The diameter of the management structure characterizes the compactness of the management structure - the shortest path connecting the two elements of the structure – for example the connection path between heads of divisions, services and workshops. The preferred structure should have the smallest diameter from the CEO to the second and third management levels. Compact and integrated structures are the ones with the shortest length between the management positions of the third level;
- The radius of the structure.

Radius of the management structure describes the minimum length of administrative communications to the most distant elements of the organizational structure. The smaller radius the graph has, the higher is the compactness of the management structure.

3. The controllability of the structure

- The number of information channels.

The number of information channels through which management decisions are passed. The more channels, the more opportunities for the management has the appropriate structure. This number of information channels equals the number of lines that connect the various departments of administrative structures among themselves, with the deputy head and his deputies;

- the number of closed feedback loops that occur in the considered management structure. Obviously, the greater the number of closed circular information transfer cycles in the structure, the greater the controllability of the structure.

4. The controllability of the structure

- The stability of the management system is evaluated with the use of the following indicators:

- the structure is less stable, if it has a minimum number of persons taking management decisions, with the direct impact on all participants in the organizational structure. This partial criterion determines the amount of the main persons in the management structure who control all processes and events. The greater the number of such positions, the more stable and more reliable the structure. Otherwise, when management is provided by a small number of decision makers in the loss of control in parts of the processes situations, or in case of non-availability of one of the decision makers loss of control of the whole management structure may occur;

- the structure is internally stable if it has the maximum number of directly unrelated members of the organizational structure. The greater the number of positions with independence in making and implementing management decisions, the greater the stability of the management structure in cases of the loss of working capacity, or failure of individual elements of the structure.

Analysis of partial criteria characterizing the effectiveness of the systems shows that some of them have to take the maximum values, and the other - the minimum. Therefore, the most holistic structure is the one that has the greatest number of centers, but the smallest diameter and radius. Furthermore, it should have the greatest number of information channels, and of feedback loops to become more manageable.

It follows that the quality assessment of different management systems with the use of various partial criteria gives conflicting uncorrelated results.

It is necessary to develop methods of multi-criteria evaluation for unambiguous results of the analysis of the effectiveness of management structures [12-21].

4 Conclusions

1. Analysis of the management structures of power plants has shown a variety of possible options for their construction. It is impossible to make a definite conclusion on the conformity of one of the analyzed structures to the goals and challenges of the enterprises in terms of restructuring and competitiveness.

2. Local criteria that jointly characterize the efficiency of the analyzed management structures are formulated.

3. Partial criteria that convolute to the local criteria and their physical meaning are defined.

4. It was determined that the assessment of the quality of management structures with the use of various partial criteria gives different, uncorrelated and conflicting results.

5. It is necessary to develop methods of multi-criteria evaluation for unambiguous result of the research of the efficiency of the analyzed management structures.

References

1. A.G. Salov, *Proceedings of the higher educational institutions of the North-Caucasian region. Technical science*, **1(143)**, 32-37 (2008)
2. N.V. Diligensky, A.G. Salov, Bulletin of the Samara State Technical University. Tehnical science, **(24)**, 17-25 (2009)
3. M.V. Posashkov, A.G. Salov, V.I. Nemchenko, Problems of Economics and Management of oil and gas complex, **1**, 27-32 (2014)
4. A.A. Gavrilova, A.G. Salov, *Russia, as a transforming society: economics, culture and management*, 24-29 (2011)
5. V.I. Nechiporenko, *Structural analysis systems (efficiency and reliability)* (Soviet Radio, Moscow, 1977)
6. E.A. Matveeva, A.R. Diyazitdinova, O.N. Chernyh, A.A. Gavrilova, *Design of complex business objects based on the system analysis: monograph* (PGUTI, Samara, 2016)
7. N.V. Diligensky, A.A. Gavrilova, M.V. Tsapenko, *Construction and identification of mathematical models of production systems: tutorial* (OOO Etching, Samara, 2005)
8. A.A. Gavrilova, D.S. Kolmykov, A.A. Alpheus, Herald of the Samara State Technical University. A series of Engineering, **40**, 155-162 (2006)
9. A.A. Gavrilova, A.G. Salov, V.K. Gavrilov, Herald of the Samara State Technical University. A series of Engineering, **2(22)**, 186-195 (2008)
10. A.A. Gavrilova, A.G. Salov, A.V. Kyhareva, J.V. Gavrilova, Herald of the Samara State Technical University. A series of Engineering, **3(31)**, 201-207 (2011)
11. A.A. Gavrilova, A.G. Salov, Industrial energetics, **12**, 31-34 (2011)
12. F. Harari, *Graph Theory* (KomKniga, Moscow, 2006)

13. N. Hibiki, Toshiyuki Sueyoshi, Omega, The International Journal of Management Science, **27**, 139-153 (1999)
14. Th. Post, J. Spronk, European Journal of Operational Research, **115**, 472-487 (1999)
15. A.D. Athanassopoulos, N. Lambroukos, L. Seiford, European Journal of Operational Research, **115**, 413-428 (1999)
16. V.L. Arnold, I.R. Bardhan, W.W. Cooper, S.C. Kumbhakar, Annals of Operations Research **66**, 996, 255-277, 9/25 9/27.
17. R.D. Banker, A. Charnes, W.W. Cooper, Omega, The International Journal of Management Science, **30**, **9**, 1078-1092 (1984)
18. A. Charnes, W.W. Cooper, E. Rhodes, European Journal of Operational Research, **2**, 429-444 (1978)
19. A. Charnes, W.W. Cooper, A.Y. Lewin, L.M. Seiford, *The DEA Process, Usages and Interpretations Data Envelopment Analysis: Theory, methodology and applications* (Kluwer Academic Publishers, Boston 1994)
20. A.T. Sik, A. Charnes, W.W. Cooper, Managerial and Decision Economics, **9(3)**, 251-253 (1988)
21. Al–Faraj, N. Taqi, S. Abdulaziz Alidi, Khalid A. Bu–Bshait, International Journal of Operations and Production Management, **13(9)**, 45-52 (1993)

Construction and analysis of the production function for complex structured productions

Mikhail Kharitonov[1]

[1]Mathematics and IT Department, Volgograd State University, 400062 Volgograd, Russia

Abstract. The production function of the complex structured productions is represented as a superposition of Leontief' s production functions corresponding to each of the modules. The optimization problem is reduced to a linear programming problem with a parameter, which describes the operational structure of the core.

1 Introduction

The high efficiency of modern organizations in a competitive environment, within globalization processes, information, technical and technological innovations depends on the constant high adaptability, ie continuous organizational change. The problem of organizational (structural) changes in the past decades has been widely discussed in many works on management and organizational theory. The most notable works are Adizes I. [1], Blake R.R. and Mouton J.S. [2], Champy J.[3], Grant R.M Rami S., Krishnan R. [4], Kauffman S. [5], Mincberg G. [6], Watzlawick P., Weakland J., Fisch R [7], Tran Q. and Tian Y. [8], Weick K.E. и Quinn R.E. [9], Sah R.K. [10], Hart O. [11]. At the same time against the backdrop of the many informal theories and models of adaptation and development of organizations there is a very few relevant conceptual mathematical models (Barnett W.P. and Carroll G.R. [12], Novikov D.A. [13, 14], Voronin A.A. and Mishin S.P. [15, 16], Gubko M.V. [17-18]).

To date, this area doesn't have yet a more or less general conceptual and algorithmic mathematical apparatus, and the available results are fragmentary. Various grounds for the classification of an organization's management problems produce different approaches to the problems of optimization of the organizational structure. One of these grounds is the time interval separating the management of the operation and management of changes. Optimization of structure traditionally is in the last set of administrative problems.

It appears impossible to construct large-horizon control models for an organizational system (OS) based on the system of control mechanisms [13] by disregarding its structural and functional variability. Actually, the variability of internal and external relations, short-, medium- and long-term goals and criteria of OS development stimulates further generalization of the models of the basic mechanisms [13] and the notion of a control mechanism. Such generalization lies in embedding (in addition to decision making methods) a complex of organizational structures, norms and parameters of an OS, that vary on a large control horizon according to the actual and desired dynamics of the OS as the result of implementing a sequence of simple and complex control mechanisms [13]. These generalized control mechanisms represent systems with a complicated structure, whose subsystems and elements are generalized, complex or simple mechanisms. Therefore, the optimality of each mechanism is defined within the framework of the optimization problem of the whole system.

As a matter of fact, this concept coincides with the concept of an organizational system pioneered by H. Minzberg [6]. And so, the notions of an organizational structure and a generalized control mechanism act as synonyms; accordingly, Minzberg's structural configurations can be called optimal generalized control mechanisms. Consequently, the structural and organizational changes in an OS can be treated as transitions between conditionally optimal structural configurations or generalized control mechanisms, with separation of different paths (adaptation, development, etc.) depending on necessary attributes. It seems feasible to relate the development of an OS to the interaction and development of its generalized control mechanisms.

Researchers identify several stages of generalized mechanisms synthesis, i.e., modeling of the structural dynamics and dynamic structural optimization of an OS. These stages include construction of a family of corresponding dynamic equations and an efficiency criterion (in the structure-dependent form), description of the space of structural states, correct formulation of the dynamics "principles," as well as elaboration of efficient optimization algorithms. Among first fruitful steps in this direction, we refer to the publications [15-18], where the authors studied the problems of static and dynamic optimization of control hierarchies in OSs with the efficiency criterion in the form of a structurally-dependent cost function. Generalized control

* Corresponding author: kharitonov@volsu.ru

mechanisms synthesis calls for associating these models described by the structurally-dependent production function (PF) of an OS with game-theoretic models of hierarchical control. The afore-said motivates designing a certain system of multiparametric dynamic models of optimization and control, that form an hierarchical structure of generalized control mechanisms in an OS.

This paper presents an optimization model for the operating core of an organization; following the terminology of Minzberg [6], we comprehend such core as a set of structural units of an OS directly implementing the basic production and auxiliary production (support facilities including personnel and equipment). The structure of the operating core is described by a special oriented graph with a variable number of layers. Here nodes indicate the basic technological module and the modules of support facilities, whereas edges stand for the inputs and outputs of the technological modules. The goal function has the structure-dependent form as the superposition of Leontief production functions corresponding to each of the modules.

Under given inputs (production factors) and structure of the operating core, the value of the production function represents the solution to a linear programming problem.

Increasing the number of support facilities, on the one hand, guarantees stable production output in the case of violating optimal proportions among production factors. On the other hand, such initiative causes higher prime costs. An alternative method to complicate the structure of the operating core consists in purchasing missing production factors on a market (if available). This approach appears most efficient in economies with a highly developed institutional environment. As a result, the structure of an OS gets simplified and special outsourcing organizations come into service. The latter provide support facilities, as well as implement traditional functions of an OS. Economies with a low level of institutional environment demonstrate a totally different tendency—vertical integration by absorption of related and support facilities. However, the model suggested in this paper does not explore the optimal balance problem of these tendencies. Here structural optimization of the operating core lies in seeking for the minimal number of its layers, which ensures the maximal value of the production function under given inputs (production factors). In other words, further increase in the number of such layers would not raise this value.

To tackle the problem, we develop a certain algorithm for automatic construction of the basic equations for each number of layers. Finally, the optimization problem is solved numerically for a wide range of the model variables and parameters.

In [19-20] presented some models of static and dynamic optimization of support structure of OS operating core. In this paper, we present a synthesis of model for control of support structure of OC OS with structure-dependent PF as a superposition of elementary Leontief PF as in [19-20].

2 Problem of support structure optimization

The model of a structure-dependent production function (PF) constructed in this paper bases on the models [21]. As the elementary PF of each structural element of the operating core with invariable technology (a simple transformer—ST), we involve the Leontief PF,

$$f = k\min\left(\frac{f_1}{a_1}, \frac{f_2}{a_2}, \dots, \frac{f_m}{a_m}\right),$$

where f_i indicate the values of arguments (transformation factors of production); $a_1,\dots,a_m$ mean the technological coefficients; and k is the normalizing factor.

Stable interfactor disproportions or fluctuations in transformation factors appreciably reduce the value of the PF of an ST (in comparison with its maximum). Within an OS, the flexibility of the PF with respect to its arguments is guaranteed through constructing a vertical structure of support facilities that produce the missing elements of transformation factors for the maximal efficiency of the basic ST. Therefore, the structure of factor production flows in the operating core of an OS can be expressed as a complex transformer (CT)—a multilevel structure of STs, where outputs of some STs represent inputs of the others. The structure of a CT acquires the form of an oriented graph, whose nodes answer for STs and edges correspond to factor flows. The last node is responsible for the basic production, whereas the rest nodes relate to auxiliary production (support facilities). The original production factors are the inputs of all STs, and the last (basic) ST outputs the product of the OS.

Consider a CT with the production factors N, R^1N, $R^2N,\dots,R^{m-1}N$ (R^1N, $R^2N,\dots$, $R^{m-1}N$ are the elements of the factors R^1, $R^2,\dots$, R^{m-1} respectively, that produce the missing element of the factor N). It is feasible to produce the element of the factor N via n serial STs (see Figure 1).

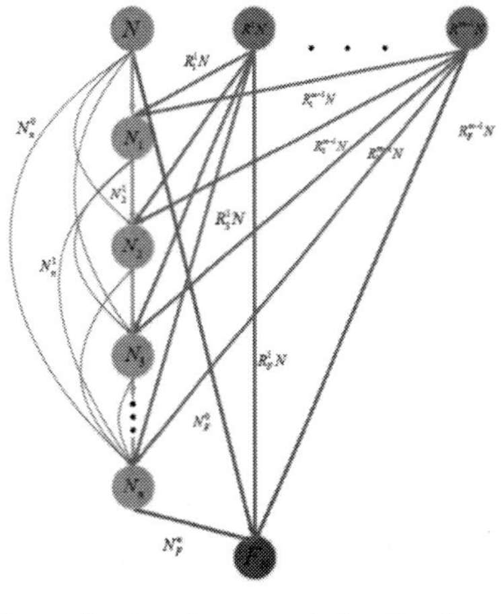

Fig 1. A complex transformer producing the element of the factor N

The PF of this CT satisfies the following system of equations:

$$\sum_{i=1}^{n} N_i^0 + N_F^0 = N, \sum_{i=1}^{n} R_i^j N + R_F^j N = R^j N, j = \overline{1, m-1}.$$

$$f_1 = S_{a_1^i} \min(\frac{N_1^0}{a_1^0}, \frac{R_1^1 N}{a_1^1}, \ldots, \frac{R_1^{m-1} N}{a_1^{m-1}}), S_{a_1^i} = \sum_{i=0}^{m-1} a_1^i,$$

$$f_2 = S_{a_2^i} \min(\frac{f_1 + N_2^0}{a_1^0}, \frac{R_2^1 N}{a_2^1}, \ldots, \frac{R_2^{m-1} N}{a_2^{m-1}}), \quad (1)$$

$$f_k = S_{a_k^i} \min(\frac{1}{a_k^0}(f_{k-1} + \sum_{i=1}^{k-2} N_k^i + N_k^0), \frac{R_k^1 N}{a_k^1}, \ldots, \frac{R_k^{m-1} N}{a_k^{m-1}}),$$

$$k = \overline{3, n}.$$

$$F_N = S_{A^i} \min(\frac{1}{A^0}(f_n + \sum_{i=1}^{n-2} N_n^i + N_F^0), \frac{R_F^1 N}{A^1}, \ldots, \frac{R_F^{m-1} N}{A^{m-1}}). \quad (2)$$

Take the variables of the system (1)-(2) and compile the factor flow vector of the CT:

$$\boldsymbol{\varphi}(n,m) = \left(\{\{N_i^j\}_{i=j+1}^n\}_{j=0}^{n-1}, \{\{R_i^l\}_{i=1}^n\}_{l=1}^{m-1}, \{N_F^i\}_{i=1}^n, \varphi_F \right).$$

Under fixed N, R^1N, R^2N,..., $R^{m-1}N$ the optimization problem for the PF of the CT (1)-(2) with variable structure takes the form

$$F_N = S_{A^i} \min(\frac{f_n + \sum_{i=1}^{n-2} N_n^i + N_F^0}{A^0}, \frac{R_F^1 N}{A^1}, \ldots, \frac{R_F^{m-1} N}{A^{m-1}}) \to \max_{n, \boldsymbol{\varphi}(n,m)}.$$

3 Analytical Estimating

Let's make some statements on a possible estimation of the values of PF (3) for different values of technological coefficients, inter-factors imbalances and complexity of the CT structure.

Statement 1. For m-factor CT with positive technological coefficients a_i^j, A^j, $i = \overline{1, n}$, $j = \overline{0, m-1}$, and proportions between entries

$$\frac{N}{R^i N} = x, \quad \frac{R^j N}{R^i N} = 1, \quad x < 1, \quad i, j = \overline{1, m-1}$$

the solution of the optimization problem (3) has the following upper bound:

$$F_m(n,x) \le \begin{cases} \frac{S_{A^i}}{A^0} \prod_{j=0}^{n} \frac{S_{a_j^i}}{a_0^j} x, n \le n^* \\ x + m - 1, n > n^* \end{cases},$$

$$\text{where } n \in \left(\frac{\ln(\frac{A^0(x+m-1)}{S_{A^i} x})}{\ln(\max\limits_{j=\overline{1,n}} \frac{S_{a_j^i}}{a_j^0})}, \frac{\ln(\frac{A^0(x+m-1)}{S_{A^i} x})}{\ln(\min\limits_{j=\overline{1,n}} \frac{S_{a_j^i}}{a_j^0})} \right), n \in Z.$$

Statement 2. For m-factor CT with technological coefficients $a_i^j = 1$, $A^j = 1$, $i = \overline{1, n}$, $j = \overline{0, m-1}$, and proportions between entries

$$\frac{N}{R^i N} = x, \quad \frac{R^j N}{R^i N} = 1, \quad x < 1, \quad i, j = \overline{1, m-1}$$

the solution of the optimization problem (3) is

$$F_m(n,x) = \begin{cases} m^{n+1} x, n \le n^* \\ x + m - 1, n > n^* \end{cases}, \quad n^* = \left\lfloor \frac{\ln \frac{m-1+x}{x}}{\ln m} \right\rfloor.$$

Statement 3. For m-factor CT with technological coefficients $a_i^j = 1$, $A^j = 1$, $i = \overline{1, n}$, $j = \overline{0, m-1}$, and proportions between entries

$$\frac{R^i N}{N} = x, \quad \frac{R^j N}{R^i N} = 1, \quad x < 1, \quad i, j = \overline{1, m-1}$$

the solution of the optimization problem (3) is $F_m(x,n)=mx$.

4 Analytical Estimating

Note that the problem (3) is equivalent to the following linear programming (LP) problem [21]:

$$F_N \to \max_{n, \varphi_N}, -\frac{S_{A^i}}{A^0}\left(f_n + \sum_{i=1}^{n-2} N_n^i + N_F^0 \right) + F_N \le 0,$$

$$-\frac{S_{A^i}}{A^j} R_F^j N + F_N \le 0, j = \overline{1, m-1}.$$

Similarly, the problem (1)-(3) gets reduced to the parametric LP problem

$$F = \varphi_F \to \max_{\boldsymbol{\varphi}(n,m), n},$$

$$\sum_{j=1}^{p(n,m)} \mathbf{M}(n,m)_{ij} \boldsymbol{\varphi}(n,m)_j \le \boldsymbol{\beta}_i, i = \overline{1, u(n,m)}, \quad (4)$$

$$\boldsymbol{\beta} = \left(N, R^1, R^2, \ldots, R^{m-1}, 0, \ldots, 0 \right)^T.$$

Here $p(n,m)$ means the number of variables and $u(n,m)$ indicates the number of inequalities; these quantities are defined by the formulas

$$p(n,m) = \frac{n(n+2m+3)}{2}, \quad u(n,m) = m(n+2).$$

Denote by $\varphi_F(n)$ the solution to the problem (4). We believe that the optimal structure of the CT described by the system (1)-(2) is the quantity $n^* = \min_n \left(\arg\max_n \varphi_F^* \right)$.

Set $\lambda=F/(N+R^1N+R^2N+\ldots+R^{m-1}N)$. The external parameters of the problem (4) comprise the initial interfactor proportions $N{:}R^1N{:}R^2N{:}\ldots{:}R^{m-1}N$ and the proportions of the technological coefficients. For each CT, the quantities technological coefficients are assumed given. Numerical experiments employ random realizations of these parameters; their values are fixed while varying the number of layers.

Figure 2 illustrates the solution of the problem (4) under certain values of the variables N, R^1N, R^2N,...,$R^{m-1}N$ and the technological coefficients taking random values from the segment $K \subset R^+$, $m=5$.

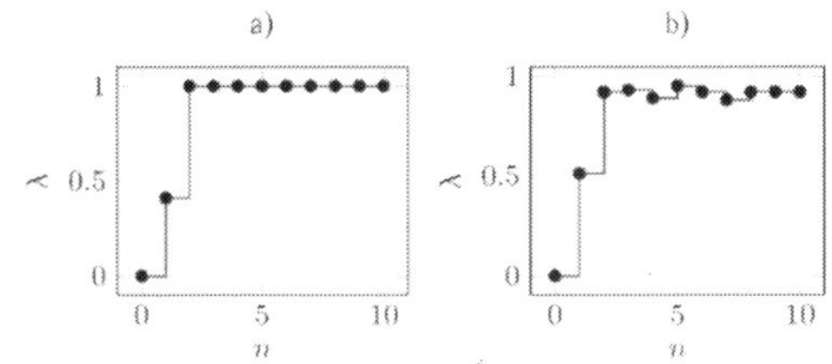

Fig 2. The curve of $\lambda(n)$: (a) $N = 1$, $R^iN = 10$, $i=1,...,4$, $K = \{1\}$, $n^*= 2$ and (b) $N = 1$, $R^iN = 10$, $i=1,...,4$, $K = \{1,2\}$, $n^* = 3$.

Figs. 3 show several solutions of the problem (4), viz., the curves of λ under $R^1N \in [0,1]$, $R^2N \in [0,1]$ with the step $\Delta=0.1$ and the parameter n varying from 1 to 4.

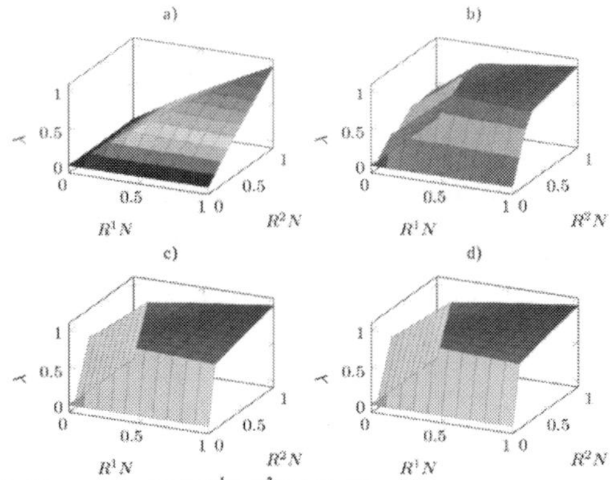

Fig 3. The curve of $\lambda(1,R^1N:R^2N,1,1)$, $K = \{1\}$: (a) $n=1$; (b) $n=2$; (c) $n=3$; (d) $n=4$.

4 Conclusion

The results of numerical simulation of the structural optimization model for the operating core of an OS have testified to the following:

— considerable reserves of the factors in support facilitiesand a sufficiently high number of STs in all lines of the complete CT allow to achieve the stable maximum of the PF under appreciable deviations in the proportions of the basic production and fluctuations in the technological coefficients of auxiliary production;

— under large reserves of the factors, the increasing variation of the technological coefficients of auxiliary production gets compensated by the growing number of layers in an ST;

— reducing the reserves of the factors and the number of STs of auxiliary production impairs stable production output as the proportions of the basic production deviate from their optimal values.

By incorporating parameters that characterize the internal and external environment of an OS into this model, one can construct complex structural optimization models for the OS Another line of further development of the model concerns designing models of structural dynamics and optimal control. Integration of these two lines of investigations yields control models for the adaptation and development of OSs.

Kharitonov Mikhail has been supported by the Ministry of Education and Science of the Russian Federation (government task No.2.852.2017/4.6).

The study was supported by RFBR, project 16-48-340147-r_a.

References

1. I. Adizes, (Adizes Institute Publications, Los Angeles, 1992)
2. R.R. Blake, J.S. Mouton, (Addison-Wesley Publishing Company, Boston, 1969)
3. J. Champy, (Harper Collins Business, London, 1995)
4. R.M. Grant, S. Rami, R. Krishnan, Sloan management review, **35** (1994)
5. S. Kauffman, (Oxford University Press, New York, 1993)
6. G. Mincberg, (Prentice-Hall, NJ, 1983)
7. P. Watzlawick, J. Weakland, R. Fisch, (Norton Books, New York, 1974)
8. Q. Tran, Y. Tian, American Journal of Industrial and Business Management, **3** (2013)
9. K.E. Weick, R.E. Quinn, Annual Review of Psychology, **50** (1999)
10. R.K. Sah, The American Economic Review, **76**, **4** (1986)
11. O. Hart, The Journal of Political Economy, **113** (2005)
12. W.P. Barnett, G.R. Carroll, Annual Review of Sociology, **21** (1995)
13. V.N. Burkov, M.V. Goubko, N.A. Korgin, D.A. Novikov, (Boca Raton: CRC Press, USA, 2015)
14. V.N. Burkov, N.A. Kuznetsov, D.A. Novikov, Automation and Remote Control, **63**, 12 (2002)
15. A.A. Voronin, S.P. Mishin, Automation and Remote Control, **63**, 5 (2002)
16. A.A. Voronin, S.P. Mishin, Automation and Remote Control, **63**, 8 (2002)
17. M.V. Gubko, Automation and Remote Control, **63**, 12 (2002)
18. M.V. Gubko, Automation and Remote Control, **70**, 1 (2009)
19. M. Kharitonov, A. Svetlov, International Journal of Pure and Applied Mathematics, **107**, **4** (2016)
20. A.A. Voronin, M.A. Kharitonov, Automation and Remote Control, **75**, 1 (2014)
21. G.B. Kleiner, Finance and Statistics, (1986)

System performance indicators of regional economic security

Yuliya Lysenko[1], and *Janna* Zelenskaya[1,*]

[1]Higher School of Economics and Management, South Ural State University (NIU), Chelyabinsk, Russia

Abstract. The article presents the scientific analysis of the elements of the developed methods for constructing effective system of the regional economic security, which is important for the national security in the whole. The purposes of economic security management, which determines the ways of applying the proposed methods, are outlined. The notion of "regional system of providing economic security" is introduced and the objects of providing security are identified. The relevance of creating the direct and reversed data streams is clarified with federal executive bodies, especially with fiscal and law enforcement, and the development of the forms of interaction with social groups and economic agents is presented. The socio-economic indicators, which serve as criteria for effectiveness of the economic security system, are stressed. All the elements of the given system integrated into the system of management of socio-economic region development have block representation. The article emphasizes the necessity of identifying specific weight value of each constituent element of security during the complex estimation. Keywords – economic security, effective system, interconnected components of the system, the regional program, the public demands, minimizing internal and external threats.

The economic development under the influence of market mechanisms of the Russian integration into the world economic relations has identified the economic security which is the basis of national and state interests as the most important functions of the state. This is due to the fact that the market economy is not able to perform this function, and the economic interests of economic entities may be contradicted to the national interest, because in their activity they, first of all, pursue income maximizing, maximum exploitation of natural, industrial and labor resources [1, 2].

The purpose of economic security management is to counteract the threats to prevent their transforming and damaging the regional social and economic systems. Achieving this goal is based on the position of the National Security Concept of the Russian Federation and state economic security strategy should be provided by developed strategy for socio-economic development of regions based on the general concept of long-term socio-economic development of the Russian Federation. All above mentioned is aimed at carrying out targeted regional policy.

This is due to the fact that the national interests of the country, its holistic territorial socio-economic development priority are based on the stable operation of the regions. So it is urgent to construct an effective system to ensure the economic security of the region, which involves the creation and implementation of opportunities focused on the impact of internal and external threats for the regional market [3, 4].

In modern conditions of territorial differences in level approach indicators of economic security of the country related to the uneven social development of the RF subjects, peculiarities of their economies functioning and the allocation of productive forces on their territories. The degree of economic security areas is affected by the quality of human potential, the level of development of innovative processes (which, unfortunately, in some regions, are very low), the investment climate dynamics and other factors.

The necessity of an effective system of regional economic security development in the aspect of national security of the country determines the significance of the analysis of the key indicators of regional development, assessment and identification of differences in terms of economic security level between the individual regions. It is necessary to conduct a comprehensive assessment of the regional economic security, allowing to identify the main directions of the regional policy in the aspect of economic security [5].

The aim of the research in this article is to review the constituent elements of the method of constructing an effective system to ensure the regional economic security such as objects, subjects, factors, criteria and indicators and the relationship between them.

Federal authorities provide protection only on the largest and most terrible threats. And the population of the country is subjected to a much wider circle of threats in their daily lives, many of which are territorial in nature. Measures to detect and locate such problems can be most effectively carried out by public authorities of the RF subjects, due to their presence in the territory and its proximity to the population.

Before revealing the economic security indicators, it is necessary to propose such notion as a "regional system of economic security", which is recognised as a set of

* Corresponding author: zhalenz@yandex.ru

DOI: 10.1051/shsconf/20173501043

conditions and factors, institutions and mechanisms for protecting the regional economic interests, improvement of the scope of the population. Economic security on the regional level is provided by the identification and prevention the internal threats of action in various spheres of regional life [6, 7].

Thus the objects of the regional economic security are the territory of the Federation subject, the population and everything related to the economy, located in the territory, including production facilities, infrastructure, natural and other resources.

Under current conditions in Russia the necessity of modern mechanisms and instruments development, priority associated with the economic policies development directed to improving the population's quality of life, is defined as a priority to ensure the economic security of the country regions. Detailed regional economic security programs must have a clear social orientation. The quality of life in this aspect is the main criterion for the safety of social and economic development of the region. In accordance with the selected model of improving the standard of living of the population in the region development policies should be directed to the population growth, the growth of its real income, improve their living conditions.

So, one of the main criteria that characterizes the region interests in the sphere of security and acceptable for the majority of the population quality of life, includes the following:

- the growth of real incomes of the population;
- an increase in natural population growth;
- improvement of housing and infrastructure conditions.

As a basis for the construction of the region economic security a common approach to the formation of a mechanism to ensure the economic security of the country was taken.

As a result, the economic security system, adapted to the specific regional problems, is proposed to include the following blocks:

- the organizational structure as a system of legislative, executive and judicial powers, public or other organizations and associations;
- legal framework regulating relations in the sphere of economic security of the region;
- priorities of ensuring economic security of the region in the context of long-term socio-economic development and national security of Russia in the economic sphere;
- internal threats in the economic field, affecting the standard of living;
- housekeeper condition monitoring system to detect and predict threats to economic security;
- tools for assessing the level of regional security through economic security indicators, their threshold values;
- a combination of organizational, legal and economic measures to prevent threats, ensure economic security.

A method for constructing an effective system to ensure the regional economic security is reflected in the form of the system of the elements and relationships between them (Fig. 1).

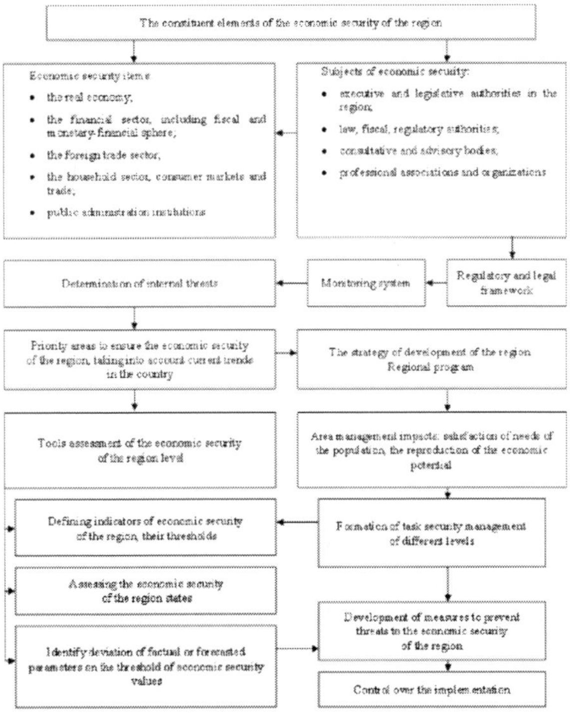

Fig 1. The system of the regional economic security.

The key factor determining the methodological approaches to the construction of an effective system of economic security is the integration of this system into the management of regional socio-economic development, which requires [8]:

- forming permanently stable relations with all the structural units of the legislative and executive authorities of the Federation authorities, professional associations and organizations engaged in control functions;
- attracting experts to solve specific problems of assessment, diagnosis and forecasting of social and economic processes and phenomena;
- creation of the direct and reverse data stream with federal executive bodies, especially fiscal and law enforcement;
- the development of forms of interaction with social groups and economic entities.

Being the subject of economic security, legislative regional authorities form the institutional environment through the development of legal acts, the regional programs and standards, a set of other tools and methods to ensure the economic security of the region. This requires regulatory legal space of the Federation, a mandatory feature which must be coherence and consistency of the provisions of the documents adopted in the particular priority of preserving and improving the level of economic security of the region.

To minimize the possibility of negative impacts from the introduction and not the introduction of norms and regulations, a full role in the structure of the subjects of management is assigned to advisory and advisory bodies, as well as to professional associations and organizations.

The starting point for development of policies and programs to ensure the regional economic security is a monitoring and data analysis system. It covers various aspects of life in the region and is aimed at identifying threats to economic security. Regional authorities should always use a variety of tools to analyze associated with the definition of real and potential threats. At the same time a central place in the monitoring system should be allocated to internal threats, taking into account specific regional focus of policies related to the solution of problems of social protection and raising the standard of living of the population of the region [9].

The assessment criteria is population needs, which vary depending on income, membership of social, national and ethnic groups, and other factors. Therefore, at the regional level it is necessary to fix a reference list-standard services, products and activities, reflecting at least some basic biological and social needs, as well as civil rights, material and spiritual goods, guaranteed by the state. The challenge of meeting these needs is the responsibility of government of the Federation.

Each specific area has its own peculiarities in the aspect of economic security and its maintenance, so there is a necessity of the system development of indicators and parameters that are taken into account by regional specificities. It should be noted that the indicators existed at the moment, are related to federal authorities. At the same time, when taking into account the specific characteristics of the regional problems it is important to work out a clear methodology for containing the regional safety performance, purpose of use which can be summarized as follows:

- assess the crisis and the threat of their occurrence in the socio-economic sphere of the region;
- assess the impact of crisis situations in the region to the national economic security of the subject of the Russian Federation and the country as a whole;
- develop and validate software and targeted activities to ensure economic security of the region;
- to establish an effective economic security of the region management system.

The main requirements to the system of socio-economic development indicators (SEPR) regions must include the following:

- the relationship between general methods of analysis SEPR region system and the indicators used on the federal and branch levels;
- the compatibility of regional security with the existing indicators in the state accounting system, statistics, analysis and forecasting;
- optimal inventory SEPR for the analysis and display of the actual situation in the region;
- simplicity and clarity as to the consistency of the results of the analysis of the status quo;
- compliance SEPR one time period when describing aspects of the socio-economic situation;
- the presence of the possibility of using these indicators for the regular monitoring and forecasting of the factors affecting the level of economic security of the region.

In formulating the purpose of the analysis of the economic security of the region on the basis of certain indicators, the following groups of objects:

- indicators incomes of the population (compared with the average cost of living and minimum wages and pensions, the different levels of income in different sectors of the population, employment dynamics and others.);
- demographic indicators (birth and death rates, life expectancy, incidence of disease, and others.);
- indicators of the state of housing and communal services (number of infrastructure projects for 10 000 people, moral and physical deterioration of assets, technological accidents, etc.).

According to the set of indicators analyzes the state of the region, on the basis of which, with the existing appropriate, covers the development of the crisis exit scenarios, depending on the chosen approach to the assessment of internal threats. This technique is versatile, and can in principle be selected such a set of indicators that would fully satisfy the safety development of specific sectors of the region concept.

Along with the problem of choosing the optimal list of indicators to measure economic security urgent question of determining the specific gravity with a comprehensive assessment of each of the constituent elements of security. If we consider the figures depend on each other in a linear or non-linear dependence, it could mean that the information system contains data that distort the results of the analysis and, as a consequence, the prediction results, which requires multiple analysis of the totality of the set parameters.

The use of modern information technology makes it possible to simulate the control system, taking into account not only the input data, but also the structure and data communications in a single economic process. Also there is the possibility to simulate the control system on the basis of the various options, taking into account the non-linear nature of the relationship, erroneous input. In any case, the introduction of amendments to the original data needs to be able to quickly change further refine ready model management process.

The key issue, the solution of which depends on the effectiveness of the development measures for the prevention of threats and damage assessment of their impact, is to determine the threshold values of economic security that can be called a response to the action of various factors risk. Important condition is that matching components of economic thresholds security parametric list of regions, reflecting the national interests of Russia as a whole. The composition of these thresholds should not be identical with the general list, as the economy of each individual region has its own specific characteristics that should be shown in the list, and the numerical characteristics of the threshold values of economic security.

The basis of the system of thresholds of economic security of the region are the basic provisions contained in the government's economic security strategy, developed and approved by the regions in consultation with the Ministry of Economic Development of the Russian Federation. At the same time finding all set of indicators within acceptable limits and compliance with the conditions for achieving certain performance is not to the detriment of others provides the highest degree of security [10–12].

It is important to emphasize that the most demonstrative and easy-to-use is considered to be a functional analysis of economic security, which aims to identify deficiencies and potential use of a number of measures that ensures each functional component of economic security of the territory as a whole. This type of analysis also helps to use the possibility to carry out an adjustment function to ensure economic security.

The procedure of analysis of the economic state of the region's security in the most optimal form can be represented as follows:

- determination of the composition, structure, adverse effects of each component in terms of the performance of the functions of economic security areas;
- separation of the objective and subjective elements;
- assessment of the likelihood and timing of the certain negative impacts, as well as the estimated amount of damage;
- the formation of a list of measures aimed at assessing the level of economic security and eliminating the impact of adverse impacts;
- assessment of the effectiveness of the measures taken in the aspect of neutralizing the negative effects of each component in terms of the running for the economic security functions;
- determining the causes of lack of effectiveness of the measures taken for the elimination of existing and prevention of possible negative effects, the definition of the persons responsible for the ineffective implementation of the measures taken;
- development of recommendations for the prevention and prevention of adverse effects;
- assessment of the value of each of the measures to rely on the elimination of negative impacts and the definition of responsible persons;
- definition of material, financial and other resources required to remedy it;
- development and adoption of legal documents governing the activities to eliminate prejudice and eliminate its consequences.

Single indicator, which characterizes the economic security risks associated with the fact that it assessed the situation in any area. This allows us to conclude the feasibility of using a system of indicators that allows realistically assessing the level of threat and getting an objective assessment. This objective of this type of assessment can be achieved by comparing the safety performance of a number of neighboring regions.

It is important not simply to evaluate the macro-economic indicators, which are difficult to affect in an expeditious manner, but also to use indicators that are able to influence the authorities in both the strategic and the operational order, using different methods of influence.

If the value of economic security indicators in the individual territories is crucial, it is not always indicative of the crisis situation in the socio-economic area or separate areas. This suggests the need to implement management bodies of measures to surgery in order to change and eliminate dangerous trends, as well as reducing the threat level.

Timely decision-making on minimization of various threats at the federal state level is based on the creation of a list of indicators mandatory for the submission of reporting forms by regional authorities.

Thus, the need to ensure economic security determines the priorities of the regional administration. In this context, the measures taken to increase the gross regional product should be considered as a tool for improving the standard of living. The economic development programs are necessary to correlate target prospects with indicators of regional economic security. In turn, the economic security of the system should be integrated into the socio-economic development of the regional management system.

The work was financed under the Act 211 of the Government of the Russian Federation, contract no. 02.A03.21.0011.

References

1. *State strategy of economic security of the Russian Federation Main provisions,* **608** (1996)
2. V. Vasenkov, *National security: theory, policy, strategy* (NPO Module, 2000)
3. *The concept of long-term socio-economic development of the Russian Federation until 2020,* **1662-R** (2008)
4. *About bases of state regulation of foreign trade activity,* **164-FZ** (2003)
5. *The national security strategy of the Russian Federation until 2020,* **683** (2015)
6. *On security,* **390-FZ** (2010)
7. S.Y. Glazyev, *The Basis of economic security of the country: alternative reformation rate* (1997)
8. N.P. Vashchekin, M.I. Dzliev, A.D. Ursul, *The Security and sustainable development of Russia* (MGUK, 1998)
9. G.S. Alimov, *Formation of the organizational-economic mechanism of management of competitiveness of regions (by example of Central Federal district regions)* (2012)
10. L. Abalkin, QE, **12**, 48-59 (1994)
11. M. Alekperov, BIC, **1** (2000)
12. V.V. Vinogradov, *The Russian Economy* (Yurist, 2007)

Features of forecasting the development of enterprise structures of the regions in the conditions of global economic change

Elvira Taipova[1], and *Maksim* Lysenko[1,*]

[1]Higher School of Economics and Management, South Ural State University (NIU), Chelyabinsk, Russia

Abstract. The article substantiates the role of theory in predicting the development of the economies of all levels, in the development of enterprise structures of different scales. The basis for economic and social analysis is the system parameters, which may vary depending on the conditions of economic development, the goals and objectives of the planning period. Prediction of enterprise structures of regional trends should contribute to solving the problem of increasing the productivity and efficiency of the leading sectors of the economy of the region in a globalized economy. Monitoring and forecasting of structural change of business development in a particular region involves the development and formation of appropriate administrative, institutional and financial structures for the creation of investment attractiveness and optimize resource utilization. Keywords – business structures, region, methods of forecasting, economic globalization, management of economic social development and the cluster approach.

In a globalized economy, forecasting the development of enterprise structures in the regional markets of goods and services is a major milestone for the administrative structures of regions [1]. Improving business activity lies in the implementation of strategies for improving the competitiveness of business entities, taking into account the regional market needs. In this connection it is necessary to improve the methods of development strategies and management decisions, to expand and deepen the integration processes.

Based on the Russian model of development, to provide a favorable investment climate, achieve the goal of increasing the competitiveness of goods and services forecasting of development of enterprise structures in the regional markets should be focused on improving living standards and search options for economic and social development of the region.

Monitoring of scientific literature [2–5] allows to consider in detail the methodological aspects of management at the regional level development forecasts for the development of enterprise structures, taking into account the relationship of external and internal objects and prediction of subjects. Analysis of scientific papers covering issues of development forecasts, development business, helped summarize the existing methods, the use of which is possible in the construction of structural changes in the regional economy scenario (Fig. 1). In developing forecasts of the classical methods used by business organizations.

Prediction of development of enterprise structures in the conditions of globalization of economy believes the passage of the following stages:

- monitoring, the results of the preparation and processing of surveys of economic entities in the region;
- the development of scenarios of formation of state economic forms of development;
- systematic assessment and analysis of trends in the development of the regional market;
- processing of monitoring results and expert assessments;
- construction of conceptual approaches based on expert assessments and monitoring.

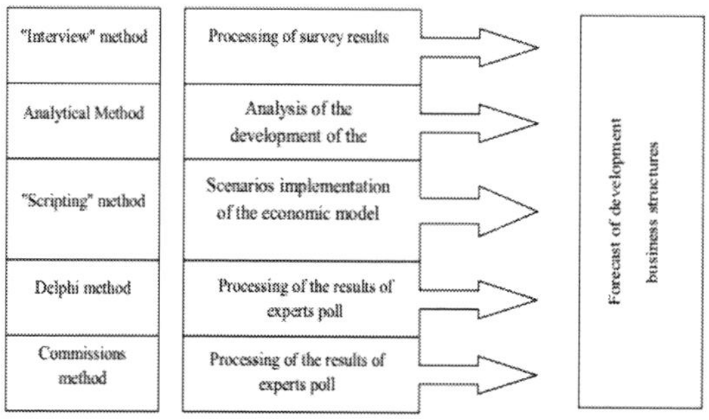

Fig. 1. Methods of forecasting the development of business structures at the regional level of government.

In scientific studies [6–8] all prediction methods for the development of entrepreneurial activity in the region in the scenarios formalizes the intuitive (expert estimates) and formal (Table 1).

* Corresponding author: dec_eib@mail.ru

Table 1. Characteristics of the main methods of forecasting the development of business structures at the regional level.

Method (Group practices)	Summary of specific features
Intuitive methods	They are used when it is impossible to consider all the conditions that affect the forecasting object, there is the difficulty of forecasting object or a large number of exposure conditions
Interview method	Specialist expert asks questions concerning the prediction object. After that, all the results of the interview are processed, and based on the forecast for construction companies or organizations
Script method	The basis of the method - the logic of further development of the economic process
Analytical method	The forecast, based on expert analysis of further developments writes policy briefs for decision by top management of the enterprise solutions
Methods of individual assessments	The accuracy of the result depends on the degree of professional (including educational) training expert
Commissions method	A working group, which is a list of questions in the form of a list or table, select experts, interrogates them on pre-established questions, processes and analyzes the data. An analysis of the expert group produces results, the main ones are the generalized opinion and coherence expert evaluations
"Delphi" method	It consists of multi-stage consecutive individual interviews of experts. The basic principles of the method of "Delfi" are the anonymity of the participants: the survey participants fill out questionnaires and do not interact personally. The range of estimates is reduced and produces a generalized opinion about the prospects of development of forecasting object
System-structural method	Creating a mathematical model predicting the object with the original data. As a result of processing of input data model provides predictable performance
Extrapolation method	It is based on the inertia of economic processes. Forecasts based on the extrapolation method, constructed as follows: according to statistics predicting the object is determined by its dynamics and applied to the current indicators. extrapolation method can be widely used for short-term forecasting of business structures
Economic-mathematical methods	Feature EMM - formalized description of economic phenomena and processes in the form of systems of mathematical equations
Sampling method	It allows you to distinguish between the real and the occasional signs of the research object. You can explore the possible relationship and communication between random variables

In carrying out the prediction of structural changes in the activity of enterprise structures of the regional economy in the conditions of globalization of the world economy need to take a greater role of economic and mathematical methods. A special place in the forecasting tools occupy variational methods and factor analysis, because these methods help to minimize the risks in the development of scenarios for economic and social development of the region [9]. In the study of indicators of business development at the regional level of government must take into account particular state approach to promote growth of the national economy, which is applied taking into account trends in the regional economy [10].

When implementing the forecast at the regional level is necessary to set goals and objectives of business management at all levels, in cooperation with the expert assessments to develop and systematize factors external and internal environment affecting the business development trends (Fig. 2).

To assess the impact of microeconomic factors to the development of business organizations in the region should:

- to predict the expected production volumes and market demand for products or technologies;
- calculate forecasts of production costs and planned sales according to market research.

Based on the situation of the world market globalization of goods and services at the meso level necessary to perform the following management actions:

- develop industry forecasts the development in accordance with the territorial division of labor;
- to forecast price trends in the world market;
- foresee the development of production chains and innovative directions of technological development;
- to predict the development of certain market segments;
- to develop logistics systems in accordance with the volumes of the production chain and product delivery.

At the regional level is necessary to control:

- develop a forward-looking regional development program for the future;
- base scenario of economic and social development of the regions.

At the macro level of government need to implement in reality the following types of integrated assessment:

- forecasts for different productive activities in order to minimize the risks of entrepreneurial activity;
- forecast of financial stability and the possibility of acquisition of investments for business organizations;

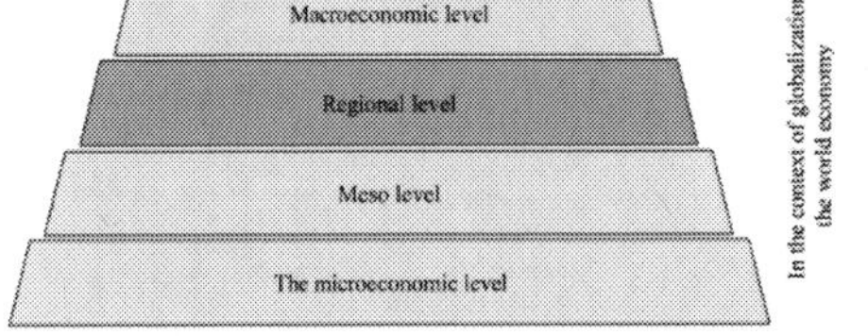

Fig. 2. The levels of implementation forecasts, formation and management of business structures.

- comprehensive forecasts of the state budget on the basis of balance of performance and increase gross domestic product, the degrees of consumption and unemployment, inflation and macroeconomic indicators.

For the prediction in the context of globalization of the world economy is necessary to perform a number of these macroeconomic objectives:

- predict the parameters of business development in the regions and countries;
- an evaluation and prognosis of recurrence of development of the world economy;
- to assess the rate and situations for economic growth (Fig. 3).

As the assessment of researchers [11–14], when considering the small and medium sized business organizations to implement the prediction at the microeconomic level are used mainly intuitive methods of forecasting, since these methods require small investment volumes.

To adjust at the microeconomic level of economic entities in the region using an integrated approach, in particular cluster. Local level cross-industry complexes are considered as part of regional research clusters, which include unification of garment factories, agricultural enterprises, advertising agencies, retail and banking institutions engaged in common activities for profit. The data within business organizations developing small and medium business, the area of activities is limited to the regional level, the aim – the growth of competitiveness in the regional market of goods and services.

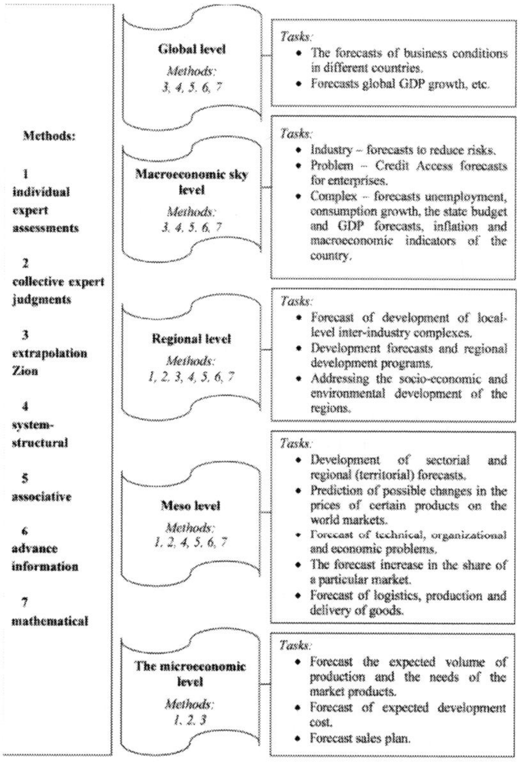

Fig. 3. Relationship management levels and methods of forecasting of development of enterprise structures.

By using the cluster approach identifies priority directions of development of economy of the region and identifies ways to reduce the level of stratification of the population's income. On this basis, translated into reality in a globalized economy, the forecast sectorial parameters of economic and social development of regional areas.

Thus, the regional clusters of business structures that are business entities, provides progressive formation of administrative territorial units or economic zones to meet the challenges of economic and social development of the regions.

Forecast of development of productive activities of business structures, taking into account changes in the global economy in the future is based on a comprehensive assessment of trends in the final of the regional market of goods and services. It is important to pay attention to the forecasting of possible changes in demand and prices for products and raw materials.

At the state level of government forecasting problem mainly affects types of business activities that are in line with the new needs of the regional market of goods and services in the future form the cluster of innovative development of economy.

The work was supported by Act 211 Government of the Russian Federation, contract № 02.A03.21.0011.

References

1. V.M. Voronin, EATP, **18**, 46-49 (2007)
2. N. P. Golubacka, EM, **12(74)**, 45-48 (2011)
3. G.A. Parsadanov, V.V. Egorov, *Forecasting of the national economy* (Higher school, 2002)
4. *Doing Business 2013. Smarter Regulation for small and medium size Enterprises* (2013)
5. M.N. Gregory, D. Romer, D.N. Weil, QJE, **107(2)**, 407-437 (1992)
6. H.K. Anyigba, Scientific notes of Saint-Petersburg University of management and Economics, **1(41)**, 76-84 (2013).
7. J. William, PE, **98(5)**, 893-921 (1990).
8. *International Monetary Fund. World Economic Outlook: World Economic and Financial surveys* (2013)
9. R.G. King, R. Levine, ME, **32(3)**, 513-542 (1993)
10. S. Alvarez, L. Busenitz, M., **27**, 755-775 (2001)
11. I.V. Antonova, *Methods of forecasting socio-economic processes* (ESSTU publisher, 2004)
12. S.V. Arzhenovskiy, I.N. Molchanov, *Statistical forecasting methods* (State economic University, 2001)
13. A.A. Miasnikov, *Synergetic effects in modern economy: an introduction to the issues* (Lenand, 2011)
14. E.E. Tikhonov, *Methods of forecasting in conditions of the market* (2006).

SHS Web of Conferences 35, 01045 (2017) DOI: 10.1051/shsconf/20173501045
ICIE-2017

Optimization of power plants management structure based on the generalized criteria of the efficiency

Aleksey Salov[1,*], *Anna* Gavrilova[1], and *Yulia* Chirkova[1]

[1]Samara State Technical University, Samara, Russia

Abstract. In the article, the analysis of the operation of power plants in the conditions of economic restructuring to ensure successful entry into the market is carried out. The analysis of the five management structures, including current, typical structure and re-designed by the authors is presented. There are developed the partial efficiency criteria of the management structures that characterize the most important properties - the balance, integrity, controllability and stability. Local criteria of the analyzed structures do not allow to make a definite conclusion about the effectiveness of one of the structures analyzed, formulated global efficiency criterion. There is developed the global criterion of the comparative effectiveness of the management systems based on the DEA method (Data envelopment analysis), taking into account the complex of the proposed local criteria. The considered management structures are ranked based on the generalized criterion of efficiency.

1 Introduction

In [1-3] on the basis of system analysis methods there are studied several management structures of enterprises to determine their efficiency in the conditions of market relations development in Russian Federation.

Transition to the market economy and the constantly changing conditions of functioning of domestic enterprises in a changing economic environment imply a periodical restructure of these companies management organization aimed at their effective independent functioning in the reformation conditions, ensuring the competitiveness of the company as a whole [4-6].

These changes should be based on new patterns of behavior of enterprises in the changed circumstances, and include new units, solving the existing problems, restructure in the hierarchy of subordination and nomenclature of decision makers.

In the previous research works [6,7] the analysis of several hierarchical management structures of local and partial criteria of quality of functioning of the management structures was formulated, but it turned out that these criteria sufficiently fully describing the local properties of the system, do not permit to make an unequivocal choice of the optimal management system. Because of the fact that for each of the systems there is a part of the local criteria that testify to its preference and the other part speaks about the optimality of other structures.

To form one optimal generalized criterion of efficiency of the management structures in this paper we propose to use the graph theory. Based on the graph method it is possible to analyze the structures of management in numerical values of the partial criteria that convolute to the generalized criterion. The obtained values of the generalized criterion of efficiency allows to rank the analyzed structures.

2 Description of the research object

The objects of this research work are 5 management structures of the power plant with co-generation (heat and power plant). It means the use of power station to generate electricity and useful heat at the same time for the needs of the industry of the region.

The enterprise operates as a part of the energy company, which unites several similar power plants in the industrial region. The director (CEO) heads the power plant. He controls deputies, heads of various departments: heads of divisions, services and workshops.

Since 2000 energetic enterprises management structure changed dramatically. The changes occurred in the direction of combining the functions of individual deputies, for example: Chief Engineer mixed with the position of director. Furthermore, the departments that did not directly relate to the production function of the enterprise, like repair services, were removed or reduced.

Some structural changes in the power plants management system negatively affect the efficiency, reliability and trouble free operation and power supply of thermal and electric energy during the heating period. Furthermore the structural changes increase the possibility of accidents. A comprehensive analysis and examination of the effectiveness of the management structures (the current and the three alternative proposed re-designed ones) is carried out.

* Corresponding author: aleksey-salov@rambler.ru

3 The solution of the problem

System analysis of the structural management schemes is conducted using the methodology of graph theory [8-10]. This method allows us to consider each position and department in the management structure (director, deputy directors, heads of departments and etc.) as a vertex of the graph, links (management and information channels) as edges of the graph connecting distinct vertices of the management structure.

Analyzed structures as graphs are shown in Figure 1.

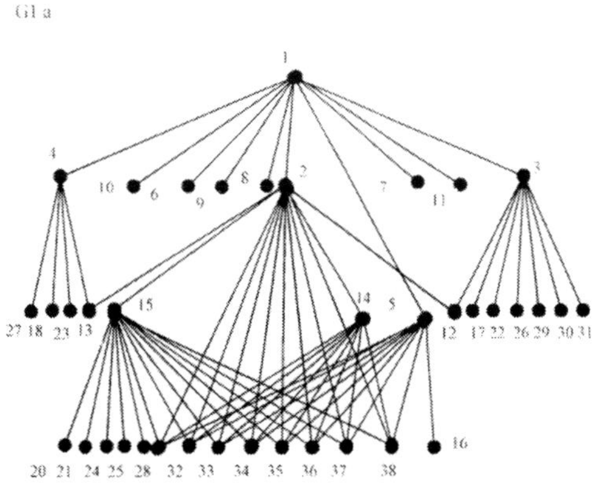

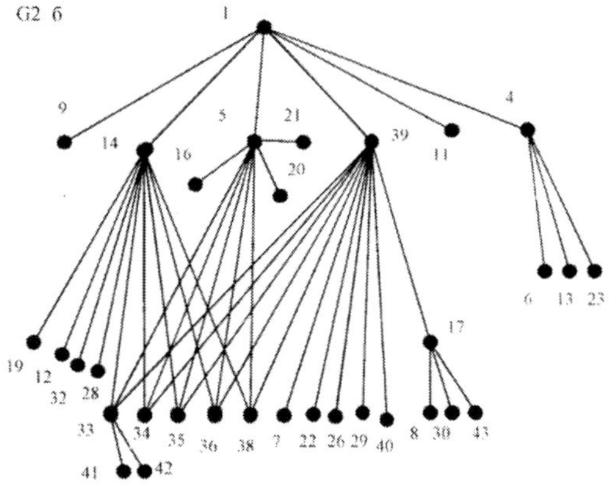

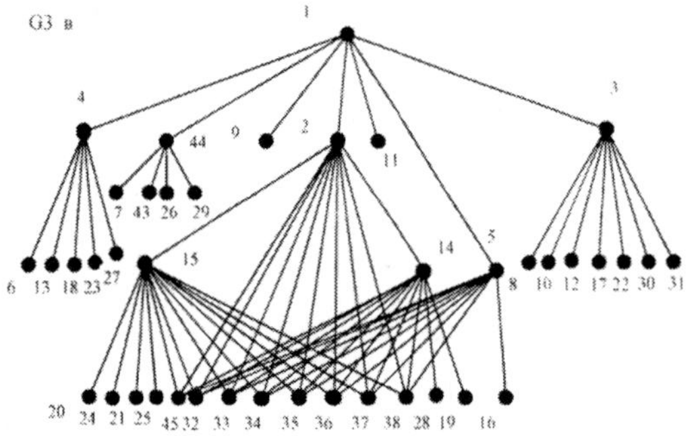

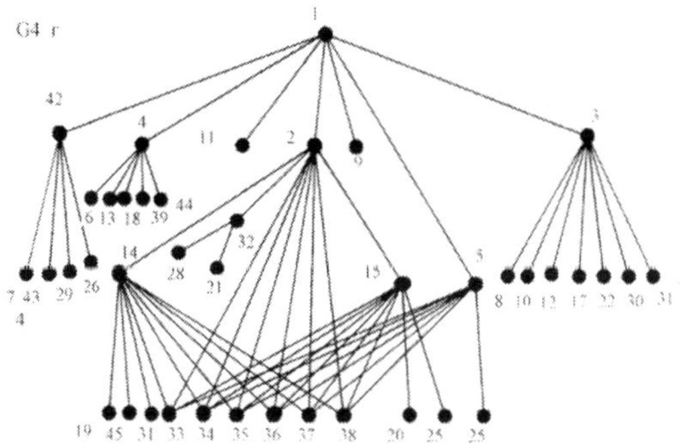

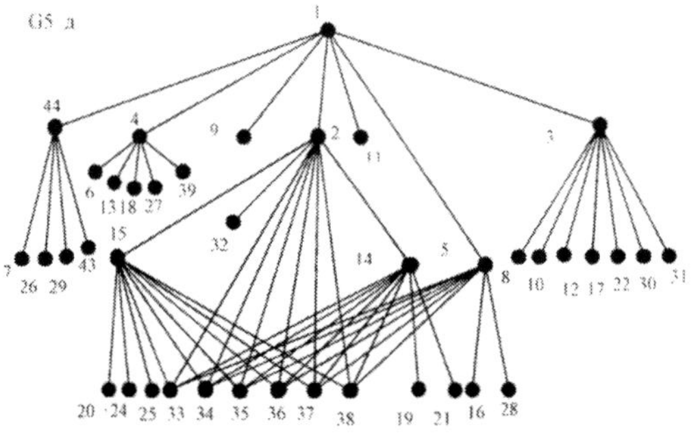

Fig. 1. Graphs of the analysed managemebt structures: a – current management structure, б – standard management structure, в, г, д – modified management structers.

1 - Director (CEO),
2 - Chief Engineer,
3 - Deputy Director for General Affairs (COO),
4 - Deputy Director of Economics (CFO),
5 - Deputy Director,
6 - Accounting Department,
7 - Human Resources Department,
8 - Office Department,
9 - Legal Adviser,
10 - Inspector of orders execution control,
11 - Emergencies and Civil Defense Department,
12 - Department of Logistics,
13 - Planning and economic Department,
14 - Deputy Chief Engineer (Maintenance),
15 - Deputy Chief Egineer (Repairs),
16 - Department of capital construction,
17 - Maintenance department,
18 - Finance Department,
19 - Shift supervisors,
20 - Repairs Department,
21 - Metals Laboratory,
22 - Garage,
23 - Department of real estate registration,
24 - Mechanical repair shop,
25 - Repair and construction shop,
26 - Dining room,
27 - Tenders Department,
28 - Labor protection Department,
29 - Medical Center,
30 - Landscaping Department,

31 - Security Department,
32 - Technical Engineering Department,
33 - Boiler Department,
34 - Turbine Department,
35 - Electric Department,
36 Department of thermal automatics and measurements,
37 - Fuel-transport Department,
38 - Chemical Department,
39 - Chief of administration,
40 - Hostel,
41 - Fuel station,
42 - Boiler room,
43 - Maintenance of buildings Department,
44 - Deputy Director for Personnel and Social Affairs,
45 - Shift manager.

Graphs are tree structures containing loops. Features of the built graphs are assessed by the following criteria:
- the balance of the structure;
- the integrity of the structure;
- controllability of the structure;
- the structure stability.

These indicators are local criteria of effectiveness of the analyzed management structures. A group of the partial criteria that are formulated assesses each of the local criteria.

1. The balance of management structures is estimated by the following local criteria:

1.1. Information load at the highest level of the hierarchy - from the first head (director).

In practice, it is widely used that a rational amount of information links from leaders at all levels of management activities is 7. When decision-makers have a larger number of connections, there is an overload of information and the effectiveness of management decisions falls. Let us take 7 for the optimal value of the number of administrative connections. Then for count "a" this feature will be equal to 1.43. The values of the local criterion for the other graphs are shown in Table 1.

Table 1. Local criterion – information load of the top management positions

Graph «a»	Graph «b»	Graph «c»	Graph «d»	Graph «e»
1.43	1	1	0.86	1
The criterion of optimization - minimum				

Consequently, in the implementation of a standard management structure the Director will have administrative overload of 43%.

1.2. As the next local criterion we will take the value of the balance of the information traffic load among decision makers. These are the deputy directors and chief engineer. The results of the calculation of this index for the structures are shown in Table 2.

Table 2. Local criterion – the value of the balance of the information traffic.

Graph «a»	Graph «b»	Graph «c»	Graph «d»	Graph «e»
1.71	1.71	1.29	1.57	1.29
The criterion of optimization - minimum				

According to this particular criterion, the best management structures of the others are Graphs - "c" and "e".

1.3. To improve the efficiency of the system it is necessary to seek to ensure that the information load is distributed evenly among all decision makers. Therefore, as the next local criterion we will take the index characterizing the degree of heterogeneity of the information load in the management structure that expresses the ratio of the maximum and minimum degrees of heterogeneity. The value of this local criterion is shown in Table 3.

Table 3. Local criterion – the degree of heterogenity of the information load.

Graph «a»	Graph «b»	Graph «c»	Graph «d»	Graph «e»
3	3	4.5	3.67	2.25
The criterion of optimization - minimum				

The above analysis shows, that the perfect management structure for the degree of heterogeneity of the information load criterion must have minimum values of partial criteria - close to 1.

2. The integrity of the management structures will be evaluated by the following local criteria:

2.1. Number of graph centers.

Graph center - is the most important top, it is expedient to form the managerial decisions. Center of the graph minimizes the distance to the most remote peaks. The less centers the graph has, the more effective is the management structure.

Table 4 shows the number of the graph centers of the analyzed management structures.

Table 4. Local criterion – number of centers of the graphs.

Graph «a»	Graph «b»	Graph «c»	Graph «d»	Graph «e»
3	3	2	1	3
The criterion of optimization - minimum				

Thus, in this particular criterion Graph «d» has the optimal management structure.

2.2. Diameter of the graph.

Graph diameter is the maximum distance between its two randomly selected vertices. It characterizes the compactness of the management structure - the shortest path connecting the two vertices. The preferred structure should have the smallest graph diameter.

Graph diameters of the analyzed structures are shown in Table 5.

Table 5. Local criterion – diameter of the graphs.

Graph «a»	Graph «b»	Graph «c»	Graph «d»	Graph «e»
5	5	5	6	5
The criterion of optimization - minimum				

According to this criterion, the best structures are Graphs "a", "b", "c" and "d ".

2.3. Radius of the graph.

Graph radius is the minimum distance from any fixed vertex to all vertices of the graph. It describes the minimum length of administrative communications to the most distant elements of the organizational structure. The smaller radius the graph has, the higher is the compactness of the management structure. Table 6 shows the values of the radius for graphs of the considered management structures.

Table 6. Local criterion – radius of the graphs.

Graph «a»	Graph «b»	Graph «c»	Graph «d»	Graph «e»
3	3	3	3	3
The criterion of optimization - minimum				

According to this criterion, the considered structures are equivalent.

3. Controllability of the structures is evaluated with the use of the following partial criteria:

3.1. The number of information channels through which management decisions are passed.

That structure has more control, when it has more links. The number of communication channels is the number of edges of the graph connecting distinct vertices of the management structure. Table 7 shows the calculated value of the number of information channels for analyzed management structures.

Table 7. Partial criterion – the number of information channels.

Graph «a»	Graph «b»	Graph «c»	Graph «d»	Graph «e»
61	64	58	42	57
The criterion of optimization - maximum				

Based on the criterion of the number of information channels, the preferred structure is the one with graph «b».

3.2. The number of different closed control loops.

The larger the number of the circular, closed information transmission control loops in the structure, the better is the controllability. The number of closed loops of the graph equals the cyclomatic number, which is determined by Euler's formula using the number of vertices and the number of edges by the formula:

$$v = N - n + 2 \qquad (1)$$

where: v - the number of cycles, N - the number of edges, n - the number of vertices.

Table 8 presents data about the number of closed information transfer cycles for the analyzed management structures.

Table 8. Partial criterion – the number of closed information transfer cycles.

Graph «a»	Graph «b»	Graph «c»	Graph «d»	Graph «e»
24	25	19	11	19
The criterion of optimization - maximum				

The perfect management structure for the above criterion should have maximum values – graph «b».

4. The criterion of the sustainability of management is estimated on the following partial criteria:

4.1. The value of the external sustainability.

External stability is determined by the number of decision management solutions and a direct impact on all participants in the organizational structure. This criterion is determined by the amount of the key decision makers directly involved in the management of the enterprise. The higher the value of this criterion, the better are the reliability and stability parameters of the considered management system.

When the value of the sustainability is small, the system is controlled by a small number of decision-makers the leaving of one or two decision-makers leads to loss of the controllability of the system.

Table 9. Partial criterion – the values of external stability.

Graph «a»	Graph «b»	Graph «c»	Graph «d»	Graph «e»
7	8	9	7	8
The criterion of optimization - maximum				

4.2. The value of the internal stability of the system.

It is determined by the maximum amount of members that are not directly linked in the organizational structure. The more decision-makers are not interconnected, independent decision-making, the more stable the system would in the case of the loss of one or more persons who make decisions and do not have the opportunity to participate in management. A Bunches of graph vertices that are not adjacent to each other or independent determine the internal stability of the system.

Table 10 shows the values of the internal stability of the considered management systems.

Table 10. Partial criterion – the values of internal stability.

Graph «a»	Graph «b»	Graph «c»	Graph «d»	Graph «e»
32	33	32	27	32
The criterion of optimization - maximum				

Analysis of the particular criteria of the internal stability shows that the best internal stability has the management structure graph «b».

Analysis of the partial criteria of efficiency of the created systems show that they (the partial criteria) give contradictory, unrelated assess of the effectiveness of a structure under study. This result is inherent to the multicriteria evaluation of the functioning of complex systems. Each of the local criteria evaluates certain aspects of the control system, without revealing its general law. Therefore, according to one local criteria system is the best, and for other worse. It is not possible to make a clear conclusion based only on partial criteria.

To identify the most effective structure for the set of local criteria, we would construct a generalized criterion using the local criteria clotting, which is based on the analytic hierarchy process.

Applying Data Envelope Analysis (DEA) methodology [11-18] we construct the generalized functional of the following form:

$$f_n = \max_{u_{in}, v_{jn} \in U} \frac{u_{1n}Y_{1n} + u_{2n}Y_{2n} + \ldots + u_{kn}Y_{kn}}{v_{1n}X_{1n} + v_{2n}X_{2n} + \ldots + v_{mn}X_{mn}} \quad (2)$$

where n - number of management structure, u_i $(i = 1,2...k)$ and v_j $(j = 1,2...m)$ – positive weights that characterize the relative contribution of each of the minimized and maximized local criteria of structures quality, Y_i - maximized local quality criteria, X_j - minimized local quality criteria.

The weights coefficient for the various structures are arbitrary and unclear.

To find the particular value of the generalized criteria of the efficiency of the management structure we find the maximum of the functional (2) on the set of values of weights $u_{in}, v_{jn} \in U$ for each of the structures under constraints that determine the V - domain of values of weights u_i and v_j. Restrictions are formulated as follows:

$$\frac{u_{1n}Y_{11} + u_{2n}Y_{21} + \ldots + u_{kn}Y_{k1}}{v_{1n}X_{11} + v_{2n}X_{21} + \ldots + v_{mn}X_{m1}} \le 1,$$

$$\frac{u_{1n}Y_{12} + u_{2n}Y_{22} + \ldots + u_{kn}Y_{k2}}{v_{1n}X_{12} + v_{2n}X_{22} + \ldots + v_{mn}X_{m2}} \le 1,$$

...

$$\frac{u_{1n}Y_{1N} + u_{2n}Y_{2N} + \ldots + u_{kn}Y_{kN}}{v_{1n}X_{1N} + v_{2n}X_{2N} + \ldots + v_{mn}X_{mN}} \le 1,$$

where N = 5 - number of the structures; $i = (1,2\ldots5)$ - the number of criteria that must strive to its maximum value, $i = (1,2\ldots5)$ - the number of criteria that must strive to their minimum values.

A system of equations provides a solution in the form of a specific value - generalized criterion of the effectiveness of each of the management structures on the interval [0,1] and weighting values.

Table 11 shows the calculated values of the generalized criterion of efficiency for each of the structures.

Table 11. Generalized criterion of efficiency.

Graph «a»	Graph «b»	Graph «c»	Graph «d»	Graph «e»
0.9868	0.8529	1	0.9049	0.9321

The above analysis of the generalized efficiency criteria shows that the least effective is the existing energy enterprises management structure, and the structure "c" has the best performance of all of the analyzed management structures.

4 Conclusions

1. In the article there is given the technique of estimating the effectiveness of the management structures of industrial complexes, including industrial enterprises.
2. There are developed the local efficiency criteria of the management structures that characterize the most important properties - the balance, integrity, controllability and stability.
3. On the basis of the developed algorithm, using the methodology of the DEA, the analysis of 5 structures of the power plant management systems is made. As a result, the generalized criterion of efficiency that allowed to rank the considered management systems is obtained.
4. The generalized criterion of the comparative effectiveness is obtained. It allowes to rank the examined management structures by effectiveness and to identify the optimal one.

References

1. A.G. Salov, Proceedings of the higher educational institutions of the North-Caucasian region. Technical science, **1(143)**, 32-37 (2008)
2. N.V. Diligensky, A.G. Salov, Bulletin of the Samara State Technical University. Tehnischeskie science, **(24)**, 17-25 (2009)
3. M.V. Posashkov, A.G. Salov, V.I. Nemchenko, Problems of Economics and Management of oil and gas complex, **1**, 27-32 (2014)
4. A.A. Gavrilova, A.G. Salov, Russia, as a transforming society: economics, culture and management, 24-29 (2011)
5. E.A. Matveeva, A.R. Diyazitdinova, O.N. Chernyh, A.A. Gavrilova, *Design of complex business objects based on the system analysis: Monograph* (PGUTI, Samara, 2016)
6. N.V. Diligensky, A.A. Gavrilova, A.G. Salov, V.K. Gavrilov, Proceedings of the higher educational institutions of the North-Caucasian region. Technical science, **5(147)**, 37-40 (2008)
7. A.A. Gavrilova, Bulletin of the Volga University. Tatischev, **1(23)**, 11-16 (2015)
8. F.Harari, *Graph Theory* (KomKniga, Moscow 2006)
9. I.V. Prangishvili, *A systematic approach and system law* (SINEG, Moscow 2000)
10. Th. Post, J. Spronk, European Journal of Operational Research, **115**, 472-487 (1999)
11. A.D. Athanassopoulos, N. Lambroukos, L. Seiford, European Journal of Operational Research, **115**, 413-428 (1999)
12. V.L. Arnold, I.R. Bardhan, W.W. Cooper, S.C. Kumbhakar, Annals of Operations Research **66**, 996, 255-277 9/25 9/27.

13. A. Charnes, W.W. Cooper, E. Rhodes, European Journal of Operational Research, **2**, 429-444 (1978)
14. A. Charnes, W.W. Cooper, A.Y. Lewin, L.M. Seiford, *The DEA Process, Usages and Interpretations Data Envelopment Analysis: Theory, methodology and applications* (Kluwer Academic Publishers, Boston, 1994)
15. A.T. Sik, A. Charnes, W.W. Cooper, Managerial and Decision Economics, **9(3)**, 251-253 (1988)
16. M. Lothgren, M. Tambour, European Journal of Operational Research, **115**, 449-458 (1999)
17. Al–Faraj, Taqi N., Abdulaziz S. Alidi, Khalid A. Bu–Bshait, International Journal of Operations and Production Management, **13(9)**, 45-52 (1993)
18. A.O. Bafail, R.A. Aal, S.A. Karuvat, *International DEA Symposium 2002 – Efficiency and Productivity Analysis in the 21st Century*, 40-52 (2002)

Impact of Platon ETC system on intercity trucking cost

Natalya Pogotovkina[1*], *Ekaterina* Demakhina[1], and *Sergey* Ugay[1]

[1]Far Eastern Federal University, Department of Transport Machinery and Transport-Technological Processes, Vladivostok, Russia

Abstract. In 2015 Platon ETC System, a system of charging trucks with gross vehicle weight exceeding 12 tons, was implemented in Russia. The payment is collected as a compensation fo0 the damage caused to the federal public roads. Platon system is an additional source of financing for the road sector. However, its implementation made the carriers face the increasing costs. This paper presents the first results of the system functioning and the problems, associated with it. We consider the foreign systems of truck charging. The results of calculations, which show the effect of the toll collection on the prime cost of road freight transportation, are also presented.

1 Prerequisites for toll imposition in the Russian Federation

Haulage trucking is an integral part of state infrastructure. They allow making sales at various distances; contribute to the development of economic relations and growth of welfare of the country as a whole. A huge number of companies build their business on providing services for the cargo transportation at different levels of the administrative system from urban and regional down to international.

Road trucking accounts for about half of the volume of cargo traffic in the country. The constant increase in traffic flows and haulage intensifies usage of the road network. Traffic density, total weight, axle load and vehicles speeds are increasing, thus it leads to a rapid deterioration of road surface. Heavy trucks have a greater impact on the destruction of the pavement.

To generate additional financing for the construction, repair and maintenance of roads, "User pays" toll collection systems are implemented all over the world.

On November 15, 2015, the Russian Federation put Platon system into a force. This is a system of charge collection as the compensation for harm caused to the federal public roads by trucks with gross vehicle weight exceeding 12 tons. The name "Platon" derives from abbreviation of Russian phrase "plata za tonny" – PLATON (it means pay for tons). Initially, it was planned to charge 3.73 rubles per 1 kilometer, but due to the mass protests that had swept the country before the system should come into effect the charge was reduced to 1.53 rubles per 1 kilometer [1].

1.1 A brief review of the literature

Having caused a massive public response, Platon system could not fail to catch the interest of researchers. A number of scientific papers have been published in just a year and a half from the system integration. Most of them are devoted to issues related to the performance of system, as well as the impact of a new road toll on the costs of carriers. Thus, the work of D.S. Anisimova and O.V. Psareva "The forecast of state revenues increase after the implementation of road toll for trucks over 12 tons" considers the problems, associated to road toll imposition, and analyze the state income and carriers costs arouse from the Platon system implementation [2]. A number of authors conducted a comparative analysis of the Platon system with its foreign counterparts in their works [3, 4]. A.T. Naniev deals with the legal aspects of the charging system in Russia. He also raises the question of "double" compensation for harm causing to the road surface when transporting heavy cargo [5].

1.2 The size and structure of the motor vehicle fleet in Russia

Even Russian classics called roads one of the main Russian trouble. Throughout the history roads of Russia has been given insufficient attention. Dense forests and swamplands created certain difficulties in the construction of roads. The majority of Russian towns and villages was located along the riverbanks, where there was navigation in summer and sledding in winter.

However, the development of civilization and territorial expansion triggered the necessity of roads construction. Special significance of roads became apparent with the growth of territory in XVI-XVII centuries. The first mention about position of a stagecoach official, who was responsible for the condition of postal stations and roads, one can find in Charter of Ivan III, dated by June 6, 1481. Later in the XVIII century, Catherine II had treated the road building as the matter of national importance.

Industrial development and the occurrence of the first motor vehicles at the turn of XIX-XX centuries contributed to the change of government opinion on the

* Corresponding author: pogotovkina.ns@dvfu.ru

 DOI: 10.1051/shsconf/20173501046

state of the roads. Russian motorization began in the early 30-ies of the last century with the construction of the plant in Nizhny Novgorod (later Gorky Automobile Plant), where the production of license conveyor automobiles had started. Before the 1970s, cars were considered as a luxury, and only the release of the famous car VAZ-2101 gave rise to the mass motorization of the country [6]. Since that time, the level of car ownership has increased more than 50 times and in July 2016 amounted to 287 cars per thousand inhabitants [7, 8]. The motor vehicle fleet of the country has doubled only for the last 15 years (fig. 1).

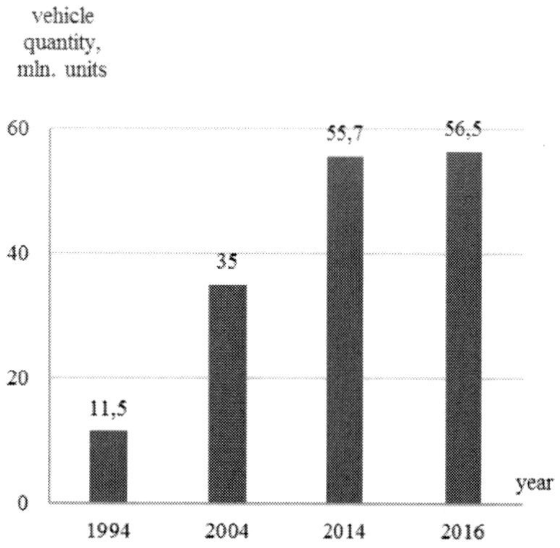

Fig. 1. The number of motor vehicle fleet, mln units.

The entire network of existing roads in Russia has been built since the beginning of the 60th to the 90th years of the last century. Since that time, the number of cars and trucks increased significantly. The influence of cars on the roads is not so much, but the same is not true for trucks.

According to the data of the Russian National Road Traffic Safety Inspectorate in December 2016 there were registered 56.6 million vehicles, including more than 44.2 million cars, and more than 6.2 million trucks (table 1) [9].

Table 1. The number of truck fleet in the Russian Federation.

Type of vehicle	The number of vehicles, un.
Truck	6,229,696
Full trailer	2,328,537
Semi-trailer	675,567

Figure 2 demonstrates the structure of the truck fleet depending on the gross vehicle weight (m).

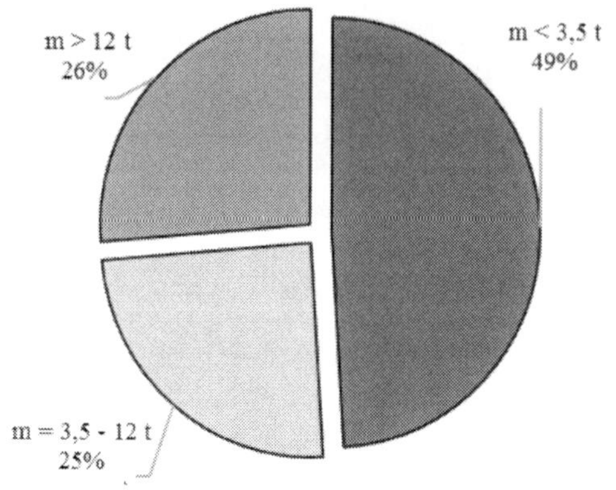

Fig. 2. Structure of the truck fleet in the Russian Federation.

As can be seen from the figure, 26% (which is 1,641,634 units) of the total number of trucks has a maximum mass exceeding 12 tons. In addition, about 400 thousand foreign heavyweight trucks are constantly moving throughout the country. These 2 million trucks cause 56% of all pavement damage [10]. Road freight transport together with cars creates heavy load on the pavement. Cracks appear and turn into the potholes after a short period. It was estimated long ago that the major damage to road is caused by vehicles with gross mass exceeding 12 tons or more. Thus, one pass of a truck with an axial load of 10 tons is equivalent to the damage caused by passing from 10 to 25 thousand cars. Moreover, the trucks cause much greater damage to the environment as compared to cars [10].

1.3 Condition of the road network of the Russian Federation

Vehicle fleet is growing and give rise to the problems of road and transport infrastructure, from snow removal on out to the bridges collapse. The pavement quality leaves much to be desired; constant patching does not give proper results, "patches" do not serve long. There are several reasons for this phenomenon:

- Poor quality design, related to the age of the designers who were graduated from universities long ago and studied an outdated program, as well as their reluctance to advance the skills and adopt foreign countries experience;
- Corruption and the use of substandard construction materials, undersealing of soil and asphalt mixtures and violation of the other regulations, including asphalt laying under the conditions of snowing and sub-zero temperatures [11];
- Outdated regulations and normative documents;
- Lack of funding;
- Geographical and climatic factors;
- The inability of the majority of Russian roads to bear the load to which they are exposed.

Currently, the total length of public road network in the Russian Federation is 1452.2 thous. km, including:

- 51.9 thous. km – federal roads;
- 515.8 thous. km – regional roads;
- 884.5 thous. km – local roads.

8.1% of regional roads and 43.6% of local roads do not have hard surface. 64.4% of regional roads and 29% of local roads have an improved roadway pavement, capable to pass the flows of trucks, regardless of climatic conditions [12]. The density of road in the western (European) part of Russia is much higher than in the eastern part in the Far Eastern Federal District. Thus, in addition to road quality problems in Russia, there is also the problem of the number of roads, that is, the length of roads.

2 Platon ETC system and its influence on the cost of cargo transportation

On November 15, 2015, Platon Electronic Toll Collection (ETC) system began its work to facilitate and process the collection of toll charges from vehicles exceeding 12 tons of gross vehicle weight. Actually, it is a new source of financing for the road sector. The rules and procedure of charging are set by the Russian Government [13].

2.1 Platon ETC system

Trucking in Russia is not the most profitable business; it is mainly due to the geography of our country. Before the Platon system should come into force, the mass protests had swept the country, and it produced results. Initially, the project cost of travel was 3.73 rubles for one km of road, but from the first day, the charge rate was 1.5293 rubles per km (3.73 with a coefficient of 0.41 rubles / km). The reduced rate must have remained until 29 February 2016, but its validity was extended until April 15, 2017.

During the first days after the system starts working the process came with a variety of problems. Carriers had to return the trucks back from haul due to failure of the on-board device, they had to make extra schedules, but the system still gave a failure: the money multiply debited from accounts, reaching five to seven debiting [14]. There have also been cases of debiting the account of the car drive down a country road, parallel to the federal highway, or during the idle period.

The main result of a year's work of Platon system was the provision of 22.9 billion rubles to the Russian regions for road repair, as well as the construction and repair of bridges and highway-over crossings. For example, a third part of 230 city streets in Petrozavodsk was repaired at the expense of Platon system [15].

Most of the countries follows the principle of "user pays" to address the challenges of financing activities related to the road sector. In order to obtain profit, owners of trucks enjoy the useful properties of highways, inflicting excess damage. Charging system is introduced as the mean of compensation for this damage.

The EU countries have different charging systems, and the payers are different groups of vehicle owners. For example, Germany is charging the vehicles with the gross weight more than 7.5 tons [16], while Austria, Switzerland and the Czech Republic are charging the vehicles with gross weight more than 3.5 tons. Eurovignette time-based payment system was commissioned in early 1993, it operates in Belgium, Netherlands, Luxembourg, Denmark and Sweden. Eurovignette is a special sticker on the windscreen. It allows using the motorways in all countries, which have signed the agreement, within a certain period. Differentiation of payments also depends on the number of axles and the vehicle environmental class [17]. The neighboring Belarus introduced a BelToll system of vehicle charging from August 1, 2013 [18]. There are 1,189 kilometers of toll roads in the country. The toll payments apply to vehicles over weight of 3.5 tons. The cost of travel amounts from 0.04 to 0.145 euros for 1 km. Thus, the rates in all countries are different due to the different income levels and living standards, but the tariff in force in Russia is one of the lowest in the world. Despite this, a fee for each kilometer traveled has an impact on the prime cost of transport services.

2.2 Trucking prime cost

A prime cost of trucking is an expense of motor carrier to perform the transportation expressed in monetary terms. The structure of the transportation prime cost is a composition and ratio of expenditures and the cost of elements in the total operating expenses [19]. Figure 3 illustrates expenditures and their ratio in total cost of transportation [20].

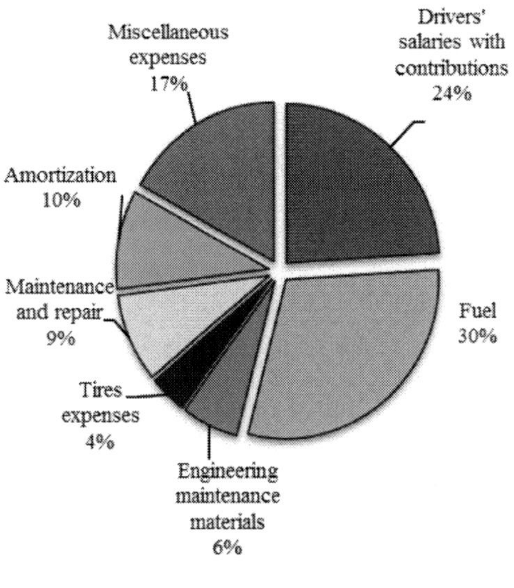

Fig. 3. Trucking expenses structure.

Miscellaneous expenses make up 15-30% from all other expenditures and include:
- Salaries of all employees, except for driver;
- Utility fees;
- Insurance fees;
- Transport tax;

- Safety inspection;
- Toll payments etc.

The introduction of Platon system entails additional costs and an increase in transportation prime costs on the miscellaneous expenses.

2.3 The Impact of Platon ETC System on the Cost of Trucking

The authors of this paper studied the influence of Platon system on the cost of road freight transportation. The largest volume of road freight transportation in Primorsky region accounts for local territory carriage and Far East region carriage. Currently, the largest cargo-points are the seaports of Primorsky region and a warehouse logistics center in the city of Artem, which was built only a few years ago. It is located on the federal highway A 370, 5 km from the "Knevichi" airport and 35 km from the city of Vladivostok. To estimate the impact of Platon traffic toll system on the cost of the trucking we considered four most requested routes in Primorsky region (present numbering of the routes will be used hereafter):

1) Artem – Ussuriysk;
2) Artem – Khabarovsk;
3) Artem – Blagoveshchensk;
4) Artem – Yakutsk.

The starting point of the routes is a logistics warehouse complex in the Artem city.

All the considered routes partially pass on the highway A370. In addition, the track route Artem-Yakutsk runs along the federal highway R297 Chita-Khabarovsk and A360 Never-Yakutsk. Table 2 shows the runs on the considered routes (L), the runs on the federal roads (L_f), the runs on the regional and local roads (L_r), and the ratio of the run on federal roads (R_f) in total run of the route.

Table 2. Routes runs.

Route	L,km	L_f, km	L_r, km	R_f, %
1	82	64	18	78
2	726	700	26	96
3	1,470	1,300	170	88
4	2,998	2,950	48	98

Table 3 shows the costs and fees for one truck ride on each route, provided by the motor carrier company.

Table 3. Prime cost and tariff for a truck ride.

Route	Prime cost, rub/truck ride	Tariff, rub/truck ride
1	12.050	13.500
2	43.256	50.000
3	73.042	90.000
4	222.358	279.000

After the introduction of Platon system the miscellaneous expenses article include tolls on federal roads, which increases the overall cost of transportation. Now the rate for 1 travelled km (P_{km}) equals to 1.53 rubles, when the tax concession period ends and tariff raises it will be 3.06 rubles.

Toll for travel on federal road:

$$P_f = L_f P_{km}. \quad (1)$$

The prime cost of transportation on route is the expenses on the truck ride performance (C_t). Prime cost of the transportation with account of toll for federal road travel

$$C_{t\,pay} = C_t + P_f. \quad (2)$$

Trucking prime cost increase.

$$P_{toll} = P_f \cdot 100 / C_t. \quad (3)$$

Table 4 shows the prime cost of travel on federal road for each route, counted for tax concession period and for normal period.

Table 4. Federal road toll.

Route	C_t, rub/truck ride	Truck ride cost increase, %	
		P_{km} = 1.53 rub/km	P_{km} = 3.06 rub/km
1	12,050	0.81	1.62
2	43,256	2.47	4.95
3	73,042	2.72	5.45
4	222,358	2.03	4.06

The table 4 demonstrates that the implementation of Platon toll collection system causes an increase in trucking cost in the range of 0.8 to 2.7% within the existing (preferential) rate and from 1.6% to 5.4% in the case of an increase rate to 3.06 rubles/km.

Figure 4 illustrates the comparison of prime costs with different tariff rates for federal road travel.

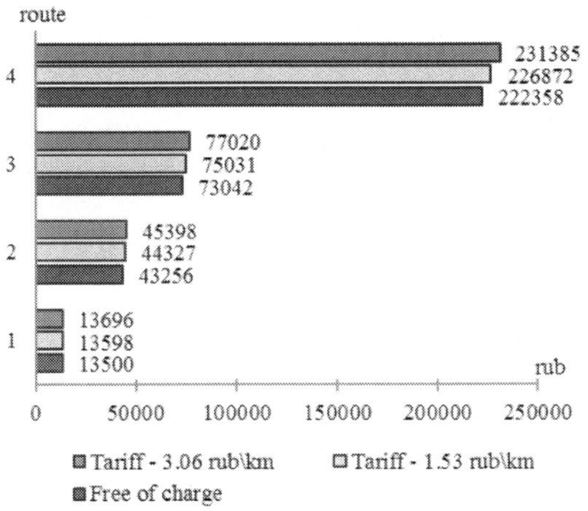

Fig. 4. Trucking prime costs changings caused by Platon ETC system

Thus, the introduction of Platon ETC system leads to an increase in the cost of road freight transportation. The increase in cost results in an increase in tariffs that affects the value of all goods from food and medicine on down to the raw materials, resources and components. As a result, a final consumer of goods and services suffers.

In addition, the increase of the final product cost results in the raise of value added tax, and it turns out that the money goes twice to state budget.

Conclusion

Making a conclusion, we have to notice that the complaints of carriers, caused by road toll imposition, are justified. However, it is an additional source of road sector financing. It is obvious that the road conditions motor vehicles are to operate under have significant effect on the cost of transportation. Poor road conditions increases fuel and lubricants consumption, advances tire wear. It also causes frequent maintenance of motor vehicles; it is necessary to repair trucks more often. Thus, if the funds collected through the Platon system are aimed on improvement of the roads condition, the owners of freight vehicles will save on repairs of automobiles, spare parts and fuel while using the federal roads.

References

1. Electronic Toll Collection System (Platon). Retrieved on January 25, (2017). http://platon.ru/ru/about/
2. D.S. Anisimova, O.V. Psareva, Prospects for development and improvement, **1,** 18-22 (2016)
3. Y.S. Karpova, V.V. Lanskih, *Materials of the PNU 56th Student Scientific and Practical Conference breakout sessions*, **2**, 125-129 (2016)
4. I.V. Biryukova, Actual problems of modern Russian economy, **3**, 200-205 (2016).
5. A.T. Naniev, *Problems of damages recovery in the Russian legal order: collection of articles of the VI annual international research and practical conference*, 465-474 (2016).
6. Page of the history of Russian Roads (JSC "Road Operator Enterprise no. 7"). Retrieved on November 20, (2016). http://m7dep7.ru/stranichka-istorii-dorog-rossii/
7. Cars ownership rating of Russian regions (Autostat Analytic Agency). Retrieved on January 15, (2017). https://www.autostat.ru/press-releases/27115/
8. G.I. Klinkovshtein, M.B. Afanas'ev, *Organization of traffic (*Transport, Moscow, 2001)
9. Information about the road safety indicators. In Road Traffic Police (2017). http://www.gibdd.ru/stat/
10. The goods adjusted for truck (Platon ETC System). Retrieved on January 19, (2017). http://platon.ru/ru/smi/26-10-2015/3640/
11. Asphalt laying during snowfall to be traditional in Russia (Vladivostok city site). Retrieved on January (2017). http://www.newsvl.ru/ggg/2016/11/26/153908/
12. The road infrastructure development. A reference to the teleconference call on measures to improve the state of regional and local roads (The Russian Government site). Retrieved on January 25, (2017). http://government.ru/info/22865/
13. Order of the Government of the Russian Federation of June 14, 2013 No. 504. About collection of the payment on account of the indemnification caused to highways of federal importance by the vehicles having the permitted maximum weight over 12 tons (Garant Database). Retrieved on December 9, (2016). http://base.garant.ru/70397412/
14. Platon-3: system failure, carriers complaints and first millions (PPT.RU). Retrieved on January 20, (2017). http://ppt.ru/news/134727
15. A year of Platon system execution: regional roads get 22.9 bln rubles (Platon ETC System). Retrieved on January 27, (2017). http://platon.ru/ru/front-page/15-11-2016/5981/
16. Truck toll collection in Germany (Tollcollect). Retrieved on January 15, (2017). http://www.vdnk.ru/site/ru/transport-articles/toll-roads
17. Road toll (VDNK Cargo Transportation). Retrieved on January 15, (2017). http://www.vdnk.ru/site/ru/transport-articles/toll-roads
18. Belarus electronic toll collection system (BelToll official site). Retrieved on January 15, (2017). http://www.beltoll.by/
19. O.N. Larin, *Cargo transportation arrangement* (SUSU, Chelyabinsk, 2006)
20. N.S. Pogotovkina, *Road cargo trucking* (FEFU Publishing, Vladivostok, 2012)

Resource recycling as new field for innovative technologies

L.L. Kamenik[1*]

[1]Peter the Great St. Petersburg Polytechnic University, St. Petersburg, Russia

Abstract. This study substantiates the necessity of transition from the natural resource model of socioeconomic development towards an industrially reproducible type of raw material, which is particularly relevant in the context of the global resource crisis. The key role of innovative technologies in the solution to this problem is questioned. Theoretical and methodological principles of the modern economy functioning are examined based on the resource factor. A new concept of "resource recycling", which reflects industrial resource recovery, is introduced. An innovative model of a resource base for economic reproduction is provided, the necessity of transition from the existing linear economic model towards a closed resource cycle model is shown, and three resource cycle models are examined in terms of their objectives, forms, and content. The major problems in the implementation of the innovative model and ways of solving them are defined, which makes it possible to reduce the risk of a resource provision crisis. The conclusion that resource recycling serves as a new sphere of innovative technologies is substantiated. The historical analogy method and the evolutionary systems approach are used.

1 Introduction

Humanity is on the brink of a new period in its history, with nature's capability of sustaining the society decreasing rapidly. Humanity is entering an era of resource crisis that affects every area of life in the society.

"The previous sustainable development has come to an end… However, rejecting the idea of sustainable development is similar to giving permission to commit a murder on a planetary scale." [1, p. 436, 7] The world is in turmoil. The level of uncertainty and anxiety is very high. It is true that we live in a global crisis, and resource depletion is its primary cause. It has spread across all countries, including Russia. People are fighting over what is left of the planet's resources. The world has changed. It will never be the same again.

The question is whether a new civilization model for the development of the society will be found, and whether another long period of economic growth will begin. The author believes that a new lasting period of economic growth can only facilitate the transition from a natural resource model of socioeconomic development towards an industrially reproducible type of raw material. This is a fundamental and difficult task. However, there is no other way to stabilize the situation. The author believes that "resource recycling" serves as a strategic direction for the solution to the above-stated problem. Let us consider the nature of this problem, its origins, concept and prospects.

2 The necessity of changing resource provision models and the role of new technologies

The theory of a resource transition from a natural resource model of socioeconomic development towards an industrially reproducible type of raw material has been conceived by the author (L.L. Kamenik). [2] "It is based on an evolutionary-revolutionary transition towards a new model of resource provision. According to this model, a revolution in the resource use is expected in the upcoming decade". [3, p. 3]

History shows that before the Neolithic revolution, humanity faced an acute crisis: the existing technologies of hunting and gathering were not enough to sustain humanity. According to the estimation of specialists, the population reduced by 90% during the crisis. People had to find a new path to the future, and so they did: that path involved the technologies of grain crop cultivation and domestication of animals, which helped solve the problem of hunger.

The solution to this problem fits into the foundations of the theory of great cycles developed by the Russian economist N.D. Kondratiev, who concluded that the world economy developed in cycles and waves. [4] The theory of great cycles can be applied when looking for a solution to the problem in question.

The current situation can be characterized by three main points. Firstly, according to the estimation of experts, only 2% of substances extracted from the earth are used beneficially, while 98% goes to waste, one way

*Corresponding author: llkamenik@yandex.ru

or another. Secondly, the technologies that are currently used by the society are remarkably wasteful and are hardly aimed at a beneficial use of the resulting waste. Thirdly, at the turn of the 19th century, the great Russian chemist D.I. Mendeleev communicated a message to the future generations, which says, in a nutshell, that there is no waste in chemistry – only unused material. Mendeleev's second commandment was to reuse all mineral resources extracted from the earth. [5] There is another aspect that deserves attention. Academician V.I. Vernadsky [6,7] was far ahead of his time when he developed the foundations of his concept of harmonious interaction between mankind and nature – the noosphere. The ideas of these brilliant scientists should find their way into a new model of resource provision, shifting its focus towards evolution, technological innovations, and systems approach. This imposes new requirements on the science itself and the formation of scientific foundations of the economy's resource base.

What is so special about the present situation is that the leading countries are currently engaged in a transition towards the sixth wave of innovation based on interdisciplinary scientific approaches. It is now being defined which technologies are going to be at the cutting edge. [8] Solving the resource problem requires an interdisciplinary approach, which comprises engineering and technology, chemistry, physics, biology, economics, philosophy, etc.

The stone age ended not because there were no stones left, but because new technologies were introduced.

The planet is running out of natural resources, while the amount of waste resources is enormous. A feature of these reserves is that they are reproducible and the reproduction is permanently ongoing. The transition towards an industrially reproducible type of raw material is inevitable. This task is more complex and difficult than the transition towards an industrially reproducible food economy, which has already been successfully implemented. No need to panic. We need to consider this problem, keeping in mind the gravity of the moment. The time factor suggests there is not much time left – about 13 years. By 2030, the problem of resource transition must be solved. In order to achieve this, new innovative technologies are required.

3 Fundamental contradiction and its solution

Science has made a mistake. The logic behind the analysis of the economic reproduction process indicates that we currently have a linear economic model, which requires continuous involvement of new natural resources in the economic turnover in order to achieve reproduction. The current economic model is a linear natural resource model based on the principle of constant consumption of natural resources, which are non-reproducible. There are two conclusions to be made in connection herewith. Conclusion one. There is a fundamental contradiction: continuous economic reproduction based on non-reproducible resources. Conclusion two. The existing economic model is based on the principles of a closed resource cycle. In the context of increasingly depleting resources, exacerbating resource crisis and the ensuing economic crisis on the one hand, and large volumes of production and consumption waste on the other, there is an urgent need for the return of waste into the economic turnover as well as for a new resource base. This will help resolve the fundamental contradiction of continuous economic reproduction and non-reproducible resource base. Besides, it will help tackle the environmental problem.

It is necessary to close the resource and economic cycles. This can be achieved using a single principle – the principle of industrial reproduction of raw material based on innovative technologies and innovative economic methods. It will be a different kind of model – an innovative model for the resource base of economic reproduction based on innovative technologies, innovative economic methods and active participation of progressive business.

4 Resource recycling. Closed resource cycle

The terminology of industrial reproduction of raw material is yet to be developed. This is essentially a targeted large-scale process of industrial production of raw material from production and consumption waste. Considering the scale of work on creating a new resource base and its importance for characterizing this peculiar process, we believe that a new term should be introduced into the economic turnover – one that would reflect the nature of the process – "resource recycling". [9] The term "resource recycling" shows that the goal is not to extract new resources from nature, but rather to industrially produce (reproduce) them from the resources that are already at hand but in a different form (waste) as a result of the initial consumption of resources. Waste resources are our new resource base as we have no other option... We believe that this process of targeted transformation of one form of resource into another as a result of its industrial reproduction can be termed as resource recycling. The prefix "re" means renewal, repeatability. A cycle (from the Greek "kyklos" – wheel, circle, circuit) is a combination of certain phenomena, work, processes that complete a full circle of development during a specific time period; a production cycle is the complete circle of work that results in a final product or a final range of phenomena within the same subject matter. Now let us trace the process of waste treatment as a whole and determine the potential role of business from the point of the proposed term "resource recycling" and its understanding as a process of industrial resource (raw material) reproduction.

From the standpoint of business interests and organization of its activity in the field of resource recycling, it should be noted that flow cycles can be different. In terms of their purpose, they can have three models depending on their form and content [9]:

1. Product cycle. This is a product flow cycle: product1→ waste1→raw material2→product2. Its objective is to track a product flow and ensure the continuity of production of a new product (product 2). The key word here is "product".

2. Waste flow cycle: waste1→raw material1→ product1→waste2... Its objective is to track a waste flow and generation with due regard to its reproduction (waste2). This is waste recycling. Here, the key word is "waste".

3. Resource (raw material) cycle. This is a resource flow cycle: raw material1→ product1→waste1→raw material2. Its objective is to track a raw material flow and production of a new raw material (raw material 2), i.e. resource reproduction. This is resource recycling. Here, the key word is "resources". The resource cycle is closed: it starts with a raw material and ends in a raw material, but in a new form.

With this procedure that reflects the entire raw material flow cycle we get a closed resource cycle. According to the author's interpretation, the closed resource cycle and recycling are fundamentally different from the similar terms that are being used, which get down to waste use in a narrow sense. [10, 11] Recycling is reprocessing of utility waste. Recycling as an international symbol of reprocessing.

For those involved in business organization it is important to know where the entrepreneurial cycle begins and ends, and what is the key word in his business profile. The truth is that one can engage in any of the three model types: collect data on waste generation, be its "keeper", sell it, categorize it according to the amount, composition, and region of this valuable raw material; take part in the process of resource recycling, i.e. industrial (re)production of resources, be the keeper of "new technologies". The business has already begun working in this direction. However, it can be stated that in terms of the required (necessary) volumes of the recycled waste, particularly with the prospect of transition towards an industrially reproducible resource base, this process is still in its early stages of development. There are enormous opportunities for business activities and very good prospects for business development.

The modern practice lacks clear terminological definition in the field of waste treatment and its branches. Multiple terms are being used: "waste handling" – too abstract, describes the problem as a whole, does not specify; "waste management"[12] – inherently wrong, one cannot manage waste, one can only manage something dynamic; "waste recycling"[13] – long existing, rather narrow but popular, yet does not correspond to the long-term objectives of the formation of a new resource base. Literally, "waste recycling" means "waste reproduction", which is incorrect in terms of semantic meaning: firstly, one cannot reproduce waste on purpose; secondly, it is population that reproduces waste. Thus, it can be stated that the proposed term "resource recycling" complements and expands the concepts in this business area, serves its intended purpose, points directly at the set objective and the ultimate goal – industrial reproduction of resources with further production of raw materials from these resources.

The logic behind the validation for the project of a new resource base and its role in the implementation of innovative technologies is presented in general terms in the diagram (fig. 1).

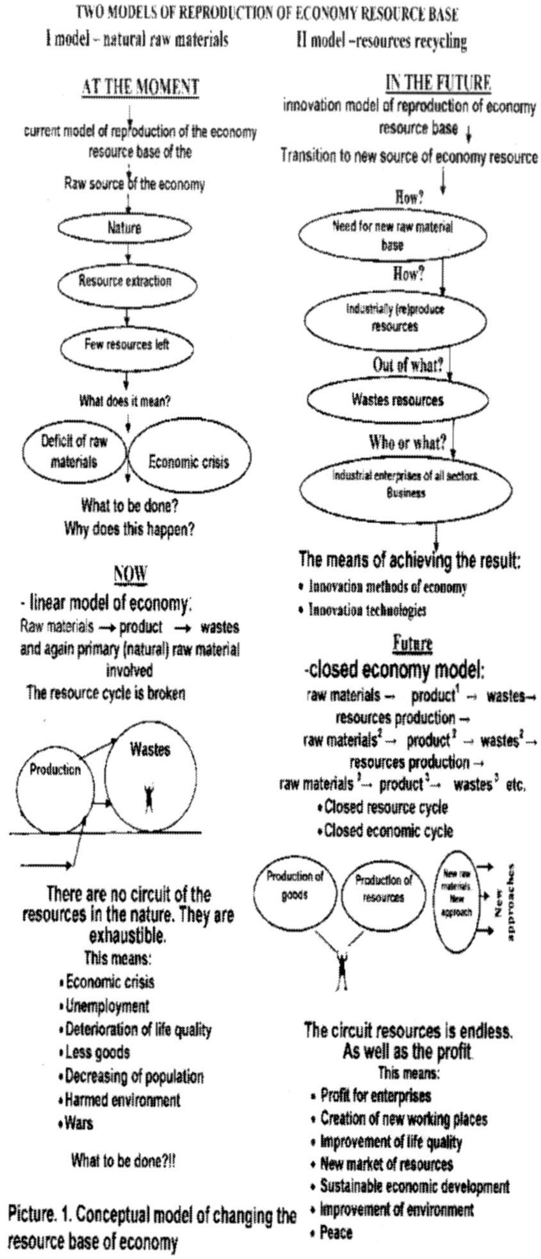

Fig. 1. Conceptual model of changes in the resource base of economic reproduction.

4 Problems of resource recycling implementation

In order to implement resource recycling as a new economic reproduction resource base model, four categories of problems need to be solved:

- **political.** Undertaking such a large-scale project as formation of a new resource base requires political will at the government level. In terms of sheer scale and importance, this project is similar to the development and implementation of the GOELRO plan, plan on the electrification of Russia;

- **technological.** Innovations must have a specific purpose. The development of innovative technologies is very much in demand and is more efficient if it has a clear focus, i.e. it is "attached" to the solution of specific problems. Here, in the field of resource recycling, there is a strong potential for the development of cutting-edge technologies. In Russia, there is already a full-scale bottom-up transition of small and medium enterprises towards non-conventional, original, innovative technologies of resource recycling and environmentally friendly technologies [14, 15, 16, 17, 18];

- **economic.** Economists have overlooked the fundamental contradictions in economic reproduction, which has led the society to a dead end. Nowadays, the resource factor of the economy is a blind spot in science. New economic mechanisms capable of ensuring the functioning of resource recycling need to be developed.

- **organizational**. Our future lies with companies involved in resource recycling and handling the full range of issues: from the development of innovative technologies to their implementation and realization of products.

The future of industrially developed countries will depend on how quickly, cost-efficiently and environmentally friendly they will be able to return production and consumption waste into the economic turnover. The country that will travel this path shall prevail.

5 Conclusions

Resource recycling is a social mandate for the development of innovative technologies. It is a test for the maturity of society as well as its ability to organize themselves and survive. The world has changed. We need to work for the future. It can be stated with confidence that resource recycling will have top priority in the seventh wave of innovation

References

1 Alan Atkisson, *How Sustainability Can Improve Organizational Performance and Transform the World* (BINOM, Laboratory of Knowledge, Moscow, 2015)

2 L.L. Kamenik, Resource-Saving Policy and the Mechanism of Its Realization in the Format of Transformational Growth (SUAI, St. Petersburg, 2012)

3 G.G. Malinetsky, Waste recycling, **6**, 3 (2014)

4 N.D. Kondratiev, D.N. Oparin, *Great Business Cycles, reports and their discussion at the Institute of Economics* (Moscow, 1928)

5 D.I. Mendeleev, *Cherished Thoughts* (Thought, Moscow, 1995)

6 V.I. Vernadsky, *Biosphere and Noosphere: monograph* (Science, Moscow, 1989)

7 G.B. Naumov, *Three Syntheses of Space* (LENAND, Moscow, 2014)

8 A.I. Tatarkin, *Foresight Russia: New Industrial Policy Design, compendium of the St. Petersburg International Economic Congress 2015,* 65-87 (2016)

9 L.L. Kamenik, *Foresight Russia: New Industrial Policy Design, compendium of the St. Petersburg International Economic Congress 2015,* 147-158 (2016)

10 *GOST 30772. Recycling as a Process of Waste and Emission Recovery* (2001)

11 Robin Murray, *Zero Waste* (OMNNO Greenpeace Council, Moscow, 2004)

12 A. Mavropoulos, PDW Solid Domestic Waste, **1(102)** (2015)

13 Waste Recycling (St. Petersburg)

14 S.D. Bodrunov, *Forthcoming. The new industrial society: the reboot: monograph* (Culture of revolution, Moscow, 2016)

15 S.Y. Glazyev, New course: breakthrough strategy. Economic strategy, **16**, **1(117)**, 6-15 (2014)

16 S.Y. Glazyev, *Strategy of advanced development of Russia in conditions of global crisis* (Economy, Moscow, 2010)

17 S.S. Gubanov, *Sovereign breakthrough. Neoindustrialization of Russia and vertical integration* (Book world, Moscow, 2012)

18 A.J. Toynbee, *Civilization on Trial* (Ayris-Press, Moscow, 2006)

Assessment and analysis of energy infrastructural potential of Russian regions

I.A. Baev[1*], *I.A.* Solovieva[1], *A.P.* Dziuba[2]

[1]South Ural state University, Chelyabinsk, Russia
[2]«Gazenergosbyt», Chelyabinsk, Russia

Abstract. The paper is devoted to the questions of assessment and analysis of the energy-infrastructural potential of Russian regions. The characteristics of energy infrastructure in a region play an important role in the evaluation of investment attractiveness of potential places for industrial investment. This paper describes the author's methodical approach to the assessment and analysis of the energy-infrastructural potential of the regions by plotting energy-infrastructural maps and by building a matrix of the regions. It allows distinguishing similar territorial entities based on the indicators of energy infrastructure with the aim of increasing the efficiency of the selection procedure for potential places for industrial investment. The developed method was tested on the example of assessment and analysis of characteristics of Russian region's energy infrastructure. That resulted in the construction of rating of the regions by the level of energy-infrastructural potential and generating of energy-infrastructural maps. The obtained results have theoretical and practical importance and can be used in the development of activities in the field of energy efficiency in the regional context and in the analysis of the energy-infrastructural capacity of potential locations for industrial investments.

Keywords: energy-infrastructural potential, investment attractiveness, the ratio of regions, the coefficient of energy tariffs, the coefficient of technological connection, industrial investment, the region's economy, energy infrastructure, energy tariffs, technological connection.

1 Introduction

A key feature in the evaluation of potential places for industrial investment has the analysis of investment potential of the territories. It determines not only the level of future investment costs but probability and volume of risks of project realization. There are many approaches to the definition of investment potential and its components [1-7].

From our point of view, infrastructure component plays a significant role in the assessment of investment potential of the regions. It includes characteristics of the geographical location, transport and energy infrastructure, the availability and quality of means of communication, etc. Considering the investment attractiveness of investing in the industrial sector it is worth mentioning one crucial characteristic of the energy infrastructure namely the topographic availability of generating capacities, the cost of technological connection to electric networks, the level of electricity tariffs in the region [8-12].

2 Research methodology

Let us define the "energy-infrastructural potential of the region" as an integral feature which reflects the conditions of transportation and selling of energy resources to consumers and influence on the investment appeal of the territory.

Accession conditions to the energy infrastructure determine the investment costs. The existing tariffs for energy resources in the region influence the level of current expenditures of the investment project during its implementation.

Since electrical energy is the main energy resource consumed by humanity, it is reasonable to use two components: the cost of electric energy in the region [13, 14] and the availability of technological connection to electric networks.

The cost of electric energy in the region reflects the level of tariffs for electricity, tariffs for power and transmission services of electricity. The availability of technological connection reflects the tariffs for technological connection operating in the region and

* Corresponding author: baev@susu.ru

availability of generating the capacity to the regional electric power system.
For evaluation of components and an integral value of energy-infrastructural potential, we propose to use a number of indicators.

3 The ratio of energy tariff

This indicator is developed for the analysis of the cost of electric energy in a regional section. It is calculated as the ratio of electricity tariff in a given region to average value on the set of territorial units (formula 1)

$$K_{T,i} = T_{P,i} / \overline{T}_P \quad (1)$$

where $K_{T,i}$ – the coefficient of energy tariffs for the considered territorial unit i;

$T_{P,i}$ – electricity tariff in the considered territorial unit;

$\overline{T}_{P,i}$ – average tariffs in the researched population of territorial units.

The tariffs for electricity $T_{P,i}$ for a territorial e ntity i defined by the formula (2) are the arithmetic sum of the rates directly on electricity, electric power, services in the transfer of electric energy and sales rate of extra charges of retail companies.

$$T_{P,i} = T_{E,i} + T_{M,i} + T_{CH,i} + T_{\pi,i} \quad (2)$$

where $T_{E,i}$ – the tariff for electric energy in the considered territorial unit;

$T_{M,i}$ – the tariff for electric power in the considered territorial unit;

$T_{\pi,i}$ – the tariff for service of electricity transmission in the considered territorial unit;

$T_{CH,i}$ – the rate of sales extra charges of retail companies in the considered territorial unit.

4 The coefficient of technological connection

To evaluate the accessibility of technological connection the indicator "coefficient of connection" is calculated for every investigated region according to the formula (3).

$$K_{T\pi,i} = K_{W,i} \times KT_i \quad (3)$$

where $K_{T\pi} i$ – coefficient of accessibility of technological connection in the reporting territorial unit;

$K_{W,i}$ – the coefficient of dynamics of demand for electricity consumption territorial entities i (defined by the formula 4);

KT_i – the ratio of the cost of technological connection to the territorial entity i (defined by the formula 5);

$$K_{W,i} = W_i^X / W_i^Y \quad (4)$$

where W_i^X, W_i^Y – the volume of demand for electricity consumption of territorial entity in the periods X and Y. X is accepted as later of the studied periods and Y is accepted as earlier of them.

$$KT i = T_{T\pi} i / \overline{T}_{T\pi} \quad (5)$$

where $T_{T\pi} i$ - the tariff for technological connection to power grids in the considered territorial unit;

If $K_{TП} i > 1$, it means that the availability of technological connection for the territorial entity is less than average availability for the set of territorial entities and vice versa.

5 Integral coefficient energy-infrastructural potential

Private indicators of energy-infrastructural potentials, such as the level of energy tariffs and the availability of the technological connection coefficient give the resulting index, called "integral coefficient of energy-infrastructural potential" $K_{I,i}$:

$$K_{I,i} = (K_{T,i} \times 0,7) + (S_{P,i} \times 0,3) \quad (6)$$

The values of integral coefficients of energy-infrastructural potential can be used for comparative assessments in the regions and any other territorial entities with the aim of making more reasonable investment decisions.

For a more detailed analysis and comparison of components of the energy-infrastructural potential of several regions it is advisable to use "the map of the energy-infrastructural potential of the regions", built in the coordinates "coefficient of energy tariffs and the coefficient of technological connection". This representation allows identifying regions with similar characteristics of energy-infrastructural potential and reducing the number of regions for analysis of potential locations for industrial investments.

6 Practical application

The testing of the proposed methodological approach to the assessment and analysis of the energy-infrastructural potential of territorial entities have conducted on the example of the energy characteristics of Russia's regions. Information base of a research was made by the actual values of energy characteristics of regions, acting in 2016: free non-regulated prices for electric energy and for electric power in retail electricity markets [15-17];

uniform regional boiler tariffs for electric power transmission for voltage class HV (110 kV) in one rate [18]; rate of sales extra charges of the basic regional guaranteeing suppliers for consumers with the maximum capacity of power receivers >10 MVA [19]; regional values of rates per unit of maximum capacity for technological connection to electric networks [20, 21], and magnitude of the demand for electricity consumption in the Russian regions for the periods of 1990 and 2014 [22].

"Coefficient of energy tariff" was calculated for each region of Russia (an example of calculation of the coefficient for the regions of Central Federal district is presented in table 1), the "coefficient of technological connection" and "integral coefficient of energy-infrastructural potential " (example of calculation for regions of the Central Federal district is presented in table 2). The rating of regions of Russia by the value of energy-infrastructural potential is shown in Fig. 1.

Table 1. Calculation of the coefficient of energy tariff for regions of the Central Federal district

The region	Amount of the tariff, RUB/kWh					Energy tariff coefficient
	For electricity	For electricity	For transfer	Sales extra charges	Total	
Belgorodskaya oblast	1.18	0.60	1.45	0.16	3.38	1.05
Bryanskaya oblast	1.17	0.55	1.62	0.07	3.41	1.06
Vladimirskaya oblast	1.21	0.55	1.55	0.03	3.33	1.03
Voronezhskaya oblast	1.23	0.55	1.21	0.15	3.14	0.97
Ivanovskaya oblast	1.10	0.55	1.05	0.10	2.80	0.87
Kaluzhskaya oblast	1.23	0.56	1.58	0.04	3.41	1.05
Kostromskaya oblast	1.17	0.55	1.53	0.10	3.34	1.03
Kurskaya oblast	1.10	0.55	1.09	0.10	2.84	0.88
Lipetskaya oblast	1.20	0.57	1.33	0.05	3.16	0.98
Moskovskaya oblast	1.22	0.55	1.14	0.04	2.95	0.91
Orlovskaya oblast	1.24	0.54	1.69	0.07	3.53	1.09
Ryazanskaya oblast	1.23	0.56	1.08	0.07	2.94	0.91
Smolenskaya oblast	1.15	0.57	1.58	0.08	3.37	1.04
Tambovskaya oblast	1.22	0.54	2.13	0.08	3.97	1.23
Tverskaya oblast	1.15	0.55	1.84	0.09	3.62	1.12
Tulskaya oblast	1.28	0.57	1.52	0.15	3.51	1.09
Yaroslavskaya oblast	1.25	0.56	1.11	0.17	3.09	0.96
Moskva	1.22	0.55	0.98	0.11	2.86	0.89

The analysis has shown that the regional values of the integral coefficient energy-infrastructural potential range from 0.43 to 1.47, indicating significant differentiation of regions. The difference between the high and low energy-infrastructural potential is almost 3.5 times. This fact underlines the relevance and importance of evaluation, analysis, and accounting of energy-infrastructural potential of the regions in the process of making investment decisions in terms of selecting the most appropriate locations for industrial investments.

For detailed component analysis of the parts of energy-infrastructural potential of Russian regions and their groupings, the map of the energy-infrastructural potential of the regions shown in Fig. 2 was built. The location of the regions in the given coordinates (the value of the coefficient of energy tariff and the value of the coefficient of technological connection) allows making a separation into groups with similar characteristics of energy-infrastructural potential. This separation will help to improve the validity of management decisions on the choice of locations for the most efficient allocation of industrial investment (table 3).

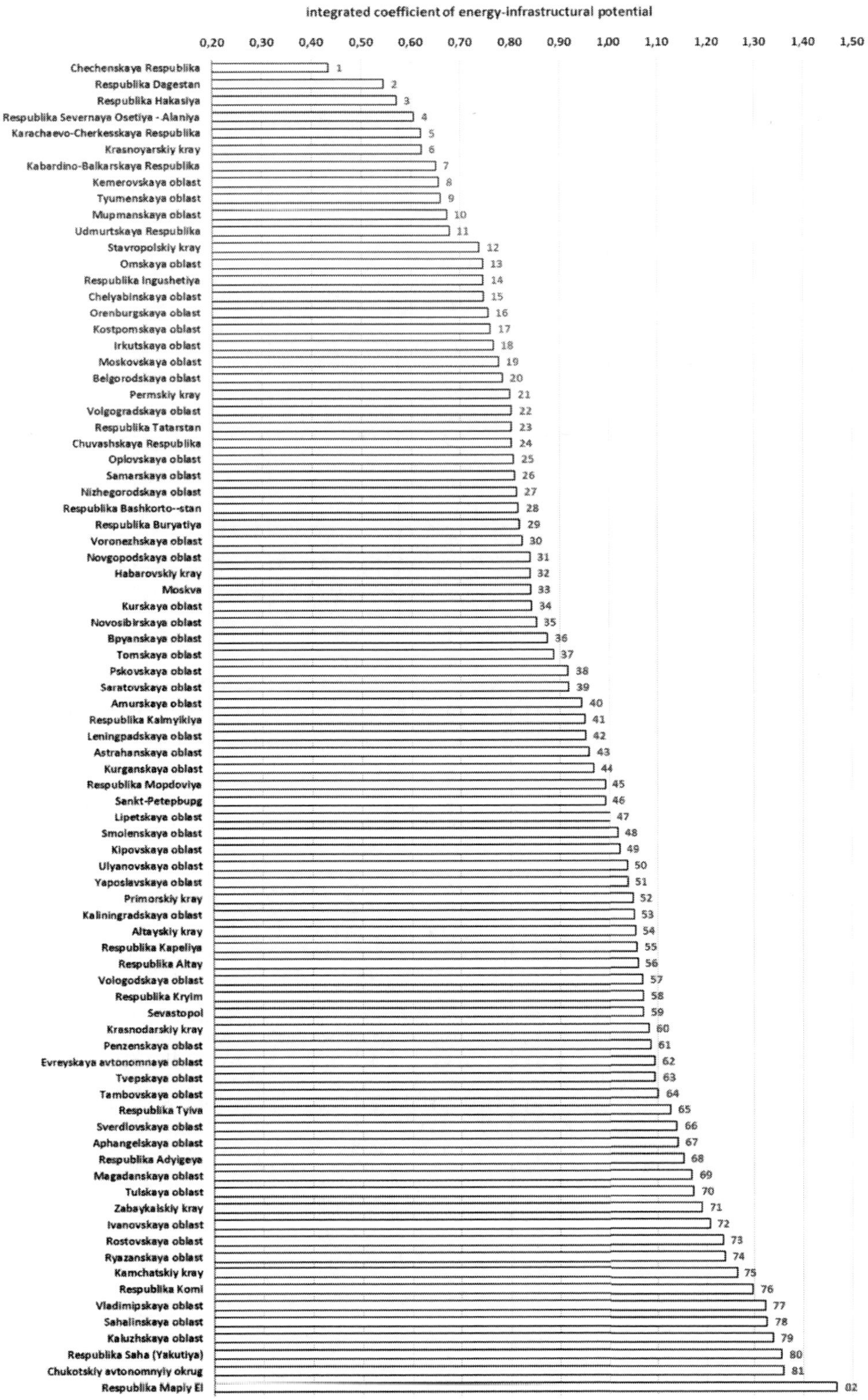

Fig. 1. Rating of Russian regions based on the value of integrated coefficient of energy-infrastructural potential.

Table 2. Calculation of the coefficient of technological connection and integral coefficient of energy-infrastructural potential for regions of the Central Federal district

Region	Demand for a power consumption, million kWh		Coefficient of dynamics of demand for a power consumption	Tariff on technological connection, rub/kW	Coefficient of a tariff for technological connection	Coefficient of technological connection	Integrated coefficient of power infrastructure potential
	1990	2014					
Belgorodskaya oblast	11 987.5	14 932.0	1.25	131.7	0.14	0.18	0.79
Bryanskaya oblast	5 670.0	4 010.0	0.71	601.4	0.64	0.46	0.88
Vladimirskaya oblast	9 042.5	7 276.0	0.80	2 351.6	2.52	2.02	1.33
Voronezhskaya oblast	12 218.1	10 791.0	0.88	509.1	0.54	0.48	0.82
Ivanovskaya oblast	6 696.9	3 431.0	0.51	4 345.9	4.65	2.38	1.32
Kaluzhskaya oblast	4 784.7	6 905.0	1.44	1 686.1	1.80	2.60	1.52
Kostromskaya oblast	4 462.7	3 617.0	0.81	144.7	0.15	0.13	0.76
Kurskaya oblast	9 774.9	8 451.0	0.86	818.5	0.88	0.76	0.84
Lipetskaya oblast	12 553.5	12 251.0	0.98	1 014.9	1.09	1.06	1.00
Moskovskaya oblast	36 692.0	42 424.0	1.16	371.1	0.40	0.46	0.78
Orlovskaya oblast	4 273.3	2 686.0	0.63	207.1	0.22	0.14	0.81
Ryazanskaya oblast	7 900.2	6 522.0	0.83	3 518.7	3.76	3.11	1.57
Smolenskaya oblast	8 501.9	6 095.0	0.72	1 249.6	1.34	0.96	1.02
Tambovskaya oblast	5 950.3	3 420.0	0.57	1 316.6	1.41	0.81	1.10
Tverskaya oblast	9 250.6	8 097.0	0.88	1 108.3	1.19	1.04	1.10
Tulskaya oblast	15 820.1	10 008.0	0.63	2 026.7	2.17	1.37	1.17
Yaroslavskaya oblast	9 034.2	7 579.0	0.84	1 375.0	1.47	1.23	1.04
Moskva	37 433.2	55 114.0	1.47	466.7	0.50	0.74	0.84

Table 3. The matrix of energy-infrastructural potential of Russian regions

Group	Parameter	Region
Group A	High capacity factor electricity tariffs and the coefficient of technological connection	Voronezhskaya oblast. Kurskaya oblast. Moskovskaya oblast. g. Moskva. Leningradskaya oblast. Murmanskaya oblast. Novgorodskaya oblast. Pskovskaya oblast. Volgogradskaya oblast. Respublika Dagestan. Respublika Ingushetiya. Kabardino-Balkarskaya Respublika. Karachaevo-Cherkesskaya Respublika. Respublika Severnaya Osetiya – Alaniya. Chechenskaya Respublika. Respublika Bashkortostan. Respublika Tatarstan. Udmurtskaya Respublika. Chuvashskaya Respublika. Permskiy kray. Orenburgskaya oblast. Samarskaya oblast. Saratovskaya oblast. Tyumenskaya oblast. Chelyabinskaya oblast. Respublika Hakasiya. Krasnoyarskiy kray. Kemerovskaya oblast. Novosibirskaya oblast. Omskaya oblast. Tomskaya oblast. Habarovskiy kray. Amurskaya oblast
Group B1	High capacity factor electricity tariffs and low coefficient of technological connection	Ivanovskaya oblast. Lipetskaya oblast. Ryazanskaya oblast. Yaroslavskaya oblast. Respublika Kapeliya. Vologodskaya oblast. Kaliningradskaya oblast. g. Sankt-Petepbupg. Astrahanskaya oblast. Kipovskaya oblast. Sverdlovskaya oblast. Respublika Altay. Altayskiy kray. Zabaykalskiy kray. Irkutskaya oblast. Evreyskaya avtonomnaya oblast
B2 Group	A low capacity factor electricity tariffs and the high rate of technological connection	Belgorodskaya oblast. Bpyanskaya oblast. Kostpomskaya oblast. Oplovskaya oblast. Smolenskaya oblast. Tambovskaya oblast. Aphangelskaya oblast. Respublika Kalmyikiya. Krasnodarskiy kray. Rostovskaya oblast. Stavropolskiy kray. Respublika Mopdoviya. Nizhegorodskaya oblast. Penzenskaya oblast. Ulyanovskaya oblast. Kurganskaya oblast. Respublika Buryatiya. Kamchatskiy kray. Respublika Kryim. g. Sevastopol. Magadanskaya oblast
Group C	A low capacity factor electricity tariffs and the coefficient of technological connection	Vladimipskaya oblast. Kaluzhskaya oblast. Tvepskaya oblast. Tulskaya oblast. Respublika Komi. Respublika Adyigeya. Respublika Mapiy El. Respublika Tyiva. Respublika Saha (Yakutiya). Primorskiy kray. Sahalinskaya oblast. Chukotskiy avtonomnyiy okrug

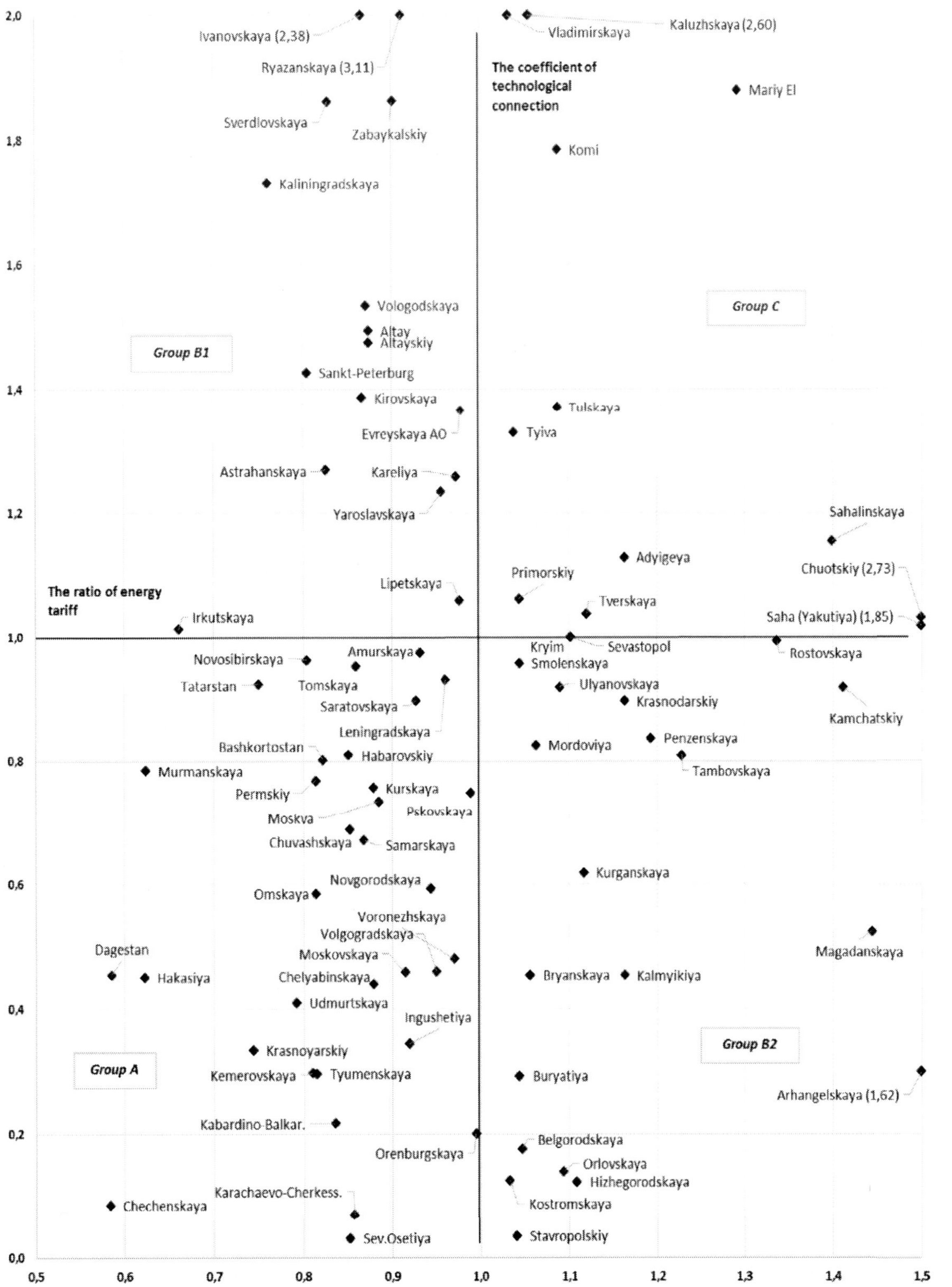

Fig. 2. The map of energy-infrastructural potential of Russian regions.

7 Conclusion

The characteristics of the infrastructure capacity play a significant role in the formation of investment attractiveness of territorial entities. The role of infrastructure potential in case of assessment the allocation of industrial investments is especially high.
The energy-infrastructural potential is defined as an integrated characteristic. which reflects transportation and selling conditions of energy resources to consumers. and influence on an investment appeal of the territory.
The methodical approach developed in the paper to assess the energy-infrastructural potential of any territorial entity. The cost of electricity and the availability of technological connection to electric networks are proposed to use as characteristics of the energy-infrastructural potential of the region.

The proposed methodological approach to the assessment of the energy-infrastructural potential of the regions consists in the calculation the number of coefficients such as the coefficient of energy tariffs. the coefficient of technological connection and integrated coefficient of energy-infrastructural capacity. Next step of this approach is making the rating of regions by the level of energy-infrastructural potential and building maps of energy-infrastructural capacity with the allocation of regional groups with similar energy-infrastructural characteristics.

The practical importance of the assessment method of the energy-infrastructural potential of the regions lies in the possibility of accounting for energy characteristics of the region in the process of making investment decisions in the industrial sector and in development and implementation of management decisions in the field of energy efficiency of territorial entities.

References

1. M. Delgado. M. Porter. S. Stern. *Clusters. Convergence. and Economic Performance* (Institute for Strategy and Competitiveness. 2011)
2. B. Robert. Jr. Stobaugh. Harvard Business Review. **5(47)**. **100-8** (1969)
3. *Competitive Strategy: Techniques for Analyzing Industries and Competitors* (Free Press. New York. 1998)
4. S. Nagaev. A. Woergoetter. Regional Risk Rating in Russia (Bank Austria. Vienna. 1995).
5. V.V. Litvinova. *Investment attractiveness of the region's investment climate: monograph* (Financial University. Moscow. 2013)
6. Rating of investment attractiveness of the regions in 2015 Report of the rating agency RAEX. http://raexpert.ru/ratings/regions/2015/
7. O.A. Denisova. G.P. Byistray. Economy of the region. 1. 221-230 (2012)
8. I.A. Baev. I.A. Soloveva. A.P. Dzyuba. Economy of the region. **3**. 180-189 (2013)
9. A.A. Kuklin. A.L. Myizin. P.A. Pyihov. M.M. Potanin. Bulletin of Transbaikal State University. **10**. 134-149 (2013)
10. I.A. Soloveva. A.P. Dzyuba. Energy Policy. **5**. 70-82 (2016)
11. International Energy Agency. https://www.iea.org
12. Eurostat. Materials of the official site. http://ec.europa.eu/eurostat
13. I.A. Baev. I.A. Soloveva. A.P. Dzyuba. Bulletin of Siberian Science. **4(19)**. 116-129 (2015)
14. Electricity: time to cut costs. Energy Bulletin of the Analytical Center under the Government of the Russian Federation (2016). http://ac.gov.ru/files/publication/a/9764.pdf
15. Resolution of the Government of the Russian Federation "On Approval of the Rules for the Wholesale Electricity and Capacity Market" (2010)
16. Components of the limit levels of unregulated prices for electricity (capacity). http://www.atsenergo.ru
17. Decree of the Government of the Russian Federation "On the functioning of retail electricity by markets. full and (or) private restriction of electricity consumption regulation" (2012). http://www.consultant.ru/document/cons_doc_LAW_130498/
18. Resolution of the Government of the Russian Federation "On approval of the Rules for non-discriminatory access to services for the transmission of electricity and the provision of these services" (2014) http://www.consultant.ru/document/cons_doc_LAW_51030/
19. Order of the Federal Tariff Service of Russia "On approval of the Methodological Guidelines on the expenditure of knock-down surcharges guaranteeing supplies and production volumes of guaranteeing suppliers" (2012).
20. Materials of the official site of PJSC "Rosseti". http://www.rosseti.ru/
21. The Order of the Federal Tariff Service of Russia "On the approval of methodological guidelines for determining the difference in fees for technological connection to electric grids" (2012). https://rg.ru/2012/11/30/elektro-dok.html
22. Electricity consumption in the Russian Federation. Materials of the official site of the Federal Service of State Statistics. http://www.gks.ru

Evaluation of innovative risks when choosing the innovative strategy at the enterprise

Sergey Chernogorskiy[1,*], *Konstantin* Shvetsov[1]

[1]Peter the Great St. Petersburg Polytechnic University, Institute of Industrial Management, Economics and Trade, 195251 St. Petersburg, Russia

Abstract. The tensor model of innovative risk of an enterprise is developed in the paper. It is proposed to introduce an integral innovative enterprise risk index. The result is formation of the selection matrix Innovative strategies.

1 Introduction

Innovation is a prerequisite for economic development in industrialized countries, and therefore it has one of the leading roles in the implementation of innovation policies of states and businesses [1], [2]. Active policy in scientific and technical spheres determines the growth rate of the national economy as a whole, as well as the competitiveness of selected industries and clusters. It also develops, produces and promotes technological innovations and products to the market [3].

Currently, the specificity of the Russian entrepreneurship does not allow full implementation of the existing in the developed countries techniques to assess the innovation potential [4]. Some adaptation is needed, since many of the innovative activity indicators, such as R&D expenditures (both for external and internal research), patent-licensing strategies, cost of the obtained licenses and structure of the license payments, coverage dynamics of the main and auxiliary markets, etc., either generally not tracked by the enterprises, or are of secondary importance compared to the standard indicators of total costs, output, profits, etc [5], [6].

2 Problem statement

In order to make a reasonable selection of criteria for innovation of domestic enterprises, and for working out of the necessary application procedures, one should first formulate a list of problems that are more likely to influence the innovative activity [7], [8], [9].

The major problems that negatively affect the innovative activity of domestic enterprises are the following [10].

Number one in the list of priorities concerning the activities of enterprises is the task of stabilization and optimization of the production process, but not bringing changes into it. As a consequence, the tasks of marketing, perspective technological policies, and market strategy play a relatively small role in organizational and managerial activities within an enterprise [11].

Small and medium-sized enterprises do not have enough successful business experience on a competitive market, the management there is primarily oriented on governmental support or patronage. The senior management is not familiar enough with the existing trends, perspectives and problems of the world economy development and they often don't realize the opportunities to participate in global production processes [12], [13].

The next priority is the staff of the enterprises, primarily the middle-level financial management and engineering specialists that in many respects turned out to be not prepared for working in the market conditions [14]. The specialists are dominated by an active orientation on the commercial prospects of the R&D results, on the diversity and dynamics of funding, on the existing of highly ramified cooperative links and on high rate of renewal of both the activity areas and the products. This very fact stipulates the lack of entrepreneurial activity and initiative for mastering of a true market attitude.

The third problem in the list consists of organizational structures and technological base of many businesses that do not comply with the characteristics of modern production facilities [15]. The production lines of these businesses have become both technologically and physically obsolete. The vast majority of enterprises do not have international quality certificates. The technology advance available in the manufacturing sector (new technologies), in the most cases do not have the properties of a product in the modern sense. The goods are usually not provided with legal protection and do not comply with international industrial standards. The problem can be compounded with the absence of sound cost estimations, lack of standard services, etc.

These circumstances initially put such developments in unfavourable competitive conditions, with reduction of their real value. Moreover it makes the processes of

* Corresponding author: chernog_sa@spbstu.ru

DOI: 10.1051/shsconf/20173501049

bringing the product to the market even more complicated and costly, even if the price advantage is available [16].

The fourth problem is the market activity of domestic enterprises. Mainly it boils down to the attempts to promote traditional products and to restore familiar industrial cooperation links. Along with this companies prefer to compete alone without giving due attention to the development of networks, including the ones in the newly open niches [17]. The lack of investments needed for the technical modernization and for implementing of new technologies is caused not only by objective reasons such as high interest rates, but also by inability to raise funds on favourable terms [18].

The slow pace of renovation of fixed assets and high capital intensity of the industrial upgrade within the domestic enterprises can also be explained with the fact that most major productions were designed and launched in the Soviet times with no regard to frequent product renovation [19]. This marks the fifth problem.

Due to the novelty of the issue of the innovation potential estimation there is a need to develop evaluation criteria, the implementation procedures and innovative strategies for many businesses. All these points should correspond to the development of the Russian economy.

3 Methodology description

Also innovative activity is characterized by high degree of uncertainty of business processes. Hence, some greater attention to the analysis and evaluation of innovative risk of an enterprise should be paid. This necessitates the development of methodological approaches to the assessment of the innovative risk of a company [20].

The tensor model of innovative risk of an enterprise considers innovative business risk as a vector in the space of three groups of factors. The three axes of the model are the innovative projects of a company, the novelty of the underlying innovation and the project environment. The novelty of the innovation characterizes the innovation by three classification features: the type of novelty, the extent of radicalism and the depth of the introduced changes. Herewith it becomes evident that higher novelty, higher radicalism and greater depth of the changes brought in by the innovation means increasing innovative uncertainty and, hence, increasing innovative risk.

The term "environment" means the space where the project is being realized: the in-house (back-office) environment, the national and the international environment.

The environment refers to the level of innovation. The environmental risk grows as the level of complexity increases (from in-house to international).

The tensor model of innovative risk of a company is presented in Fig. 1.

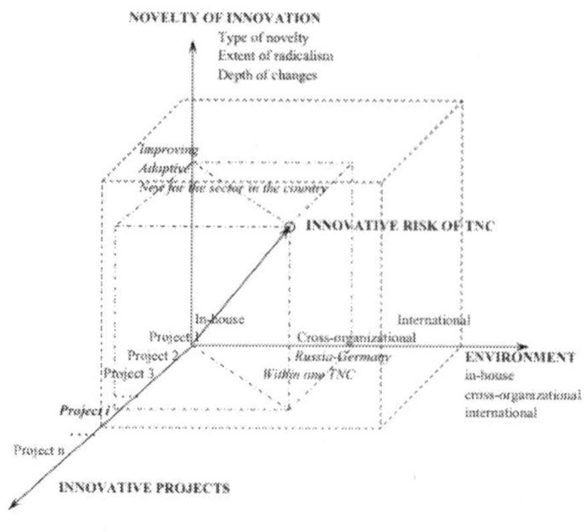

Fig. 1. Tensor model of innovative enterprise risk.

We assume, that on the basis of strategic and SWOT-analysis the innovation matrix is built and a list of innovative projects that are consistent with the business objectives of the company is obtained. The challenge is to determine which of the proposed projects are to be included in the business plan, i.e. a portfolio of innovative projects should be created [21].

The technique consists of the following steps.

1. Determining of the rankings of the innovative projects (by the impact on the business objectives). The "problem factor" of a business objective is being determined. It is calculated as the ratio of the importance of this objective for achieving of the general goal of the company (the numerical coefficient) to reachability of this goal through the implementation of this specific project.

Projects	**Business objectives of a firm**							
	Objective 1 Kpr=a		**Objective 2 Kpr =b**			**Objective m Kpr =x**		**Ra-ting**
	K	HK	K	HK	…	K	HK	
Innovative project 1	K_{11}	HK_{11}	K_{12}	HK_{12}	…	K_{1m}	HK_{1m}	P_1
Innovative project 2	K_{21}	HK_{21}	K_{22}	HK_{22}	…	K_{2m}	HK_{2m}	P_2
…	…	…	…	…	…	…	…	…
Innovative project n	K_{n1}	HK_{n1}	K_{n2}	HK_{n2}	…	K_{nm}	HK_{nm}	P_n

where Kpr - the problem factor;

- K - correlation of the project and the general goal in points. It can be +9, +3, 0, -3, -9, i.e. strong positive, weak positive, zero, weak negative and strong negative. The question that experts are to answer in determining the correlation reads as follows: "How the realization of this project will influence the achievement of the general goal?" The possible answers are: "will strongly improve" (+9 points), will improve "(+3), "in no way" (0), "will worsen" (-3) and "severely worsen" (-9);

HK - coefficient equal to the product of the correlation and the problem factor.

P - rating or the sum of a row of HK-coefficients.

 DOI: 10.1051/shsconf/20173501049

2. Determining of the time needed for realization of each of the innovative projects.

3. Determining of the cost of the project. If it's impossible to calculate the project cost, the expert estimation can be applied.

4. Building of the chart with the projects and objectives. The bubble chart is built with three variables: the rating of the project, the implementation time and cost of the project [22]. An example of the chart is given in Fig. 2.

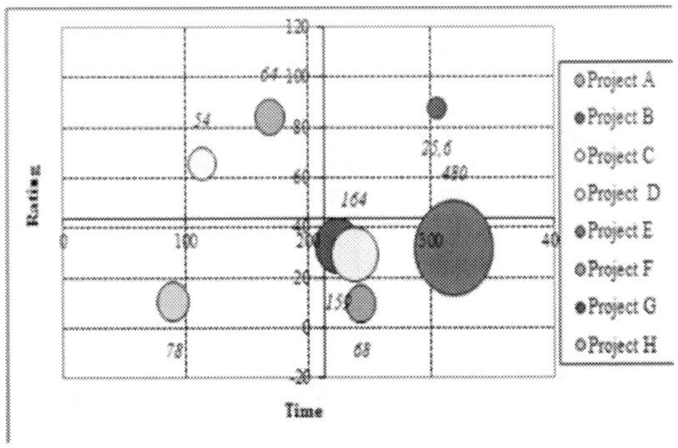

Fig. 2. Project ranking chart.

The chart data visualize the main aspects of the ranking process:

- the higher is a chart element that corresponds to the project, the more the project contributes to the achievement of the objectives;
- the more an element is shifted to the right, the less time it takes to implement the project;
- the larger the element, the greater is the cost of the project.

The intersection points of the axes are the arithmetic averages of the corresponding parameters. Thus, four quadrants are obtained. The projects are distributed among the quadrants.

Risk assessment of innovation activity of an enterprise is proposed to make by calculating the integral innovation risk index. The basis for calculation is the innovative risk model described above.

4 Key results

It is proposed to introduce an integral innovative enterprise risk index. The structure of the index is represented in Table 1, where a_1, a_2, a_3 - coefficients (weights) of importance of risk aspects for the enterprise: the project portfolio risk, the risk of innovation novelty and risk of the environment;

$R_{portfolio}$, $R_{novelty}$, R_{envir} - indicators of the total risk of the project portfolio, the total risk of innovation novelty and overall risk of the environment respectively.

$R_{portfolio} = \prod R_i$, where R_i is the total risk indicator of the innovation project.

$R_i = \prod r_t$, where r_t is partial innovative risk of the project.

$R_{novelty} = \prod n_k$, where n_k is partial innovation novelty risk (estimated for three parameters – the novelty type, the radicalism type and the level of implemented changes).

$R_{envir} = \prod s_z$, where s_z is private risk of the environment (estimated for the in-house, national and international environment).

Table 1. Structure of integral innovative risk index.

Innovative portfolio	Novelty of innovation	Environment (level of innovative activity)
The risks of innovative project: - Negative R&D results - Project planning mistakes - Wrong selection of economic objectives of the project - Project financing - Property rights within the project - Untimely launch of the product to the market - Marketing mistakes of the project - Unforeseen expenditures and exceeding the cost estimates of the project - Marketing mistakes with procurement and current supply within the project - Conflicts with existing legislation and the public - Complementary or delayed problems connected with newly developed technologies and products - Timing problems, violation of terms of the project	- new to the industry in the world, new to the industry in the country, new for this project; - basic, improving, system, accreting and pseudo-innovations; - regenerating the original properties, quantitative changes, adaptive changes, new variant, new generation, new kind, new type	In-house National International
$R_i = \prod r_t$ $R_{portfolio} = \prod R_i$ - the integral risk indicator for the portfolio of innovative projects	$R_{novelty} = \prod n_k$ - the total novelty risk indicator	$R_{envir} = \prod s_z$ - the total environment risk indicator
$R_{integral} = a_1 R_{portfolio} + a_2 R_{novelty} + a_3 R_{envir}$ - the integral indicator of innovative enterprise risk		

The main steps of the calculation of the integral innovation risk indicator of an enterprise are to be mentioned [23].

1. Selection of the modules that constitute the integral indicator.
2. Determining of the partial risks for each module.
3. Expert estimations of the partial risks.
4. Calculation of the total risk for each module.
5. Determining of the coefficients a1, a2, a3 through expert appraisals.
6. Calculation of the integral innovation risk indicator for the enterprise.

The resulting integral indicator may be used in managerial purposes for assessing of the acceptability of the innovative risk level of an enterprise. If the level of innovation risk is acceptable, then the enterprise implements its innovative portfolio [24]. If the level of risk is unacceptable, the innovative portfolio is being reconsidered and the value of the integral innovation risk indicator is calculated anew.

The innovation strategy of an enterprise can be selected according to the proposed procedure presented in Fig. 3.

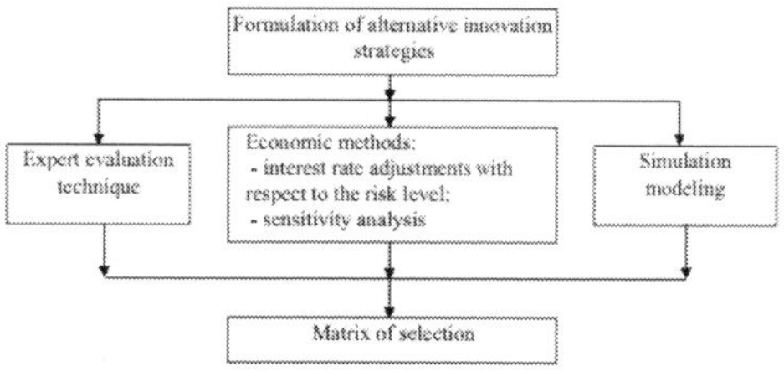

Fig. 3. Procedure of innovation strategy selection.

The result is formation of the selection matrix presented in Table 2. The columns of a matrix are innovative strategies that are applicable to the enterprise, the rows consist the calculated data on a strategy for a specific innovative project or program. The matrix will help to compare the estimates and identify a reliable strategy.

Table 2. Selection matrix.

Innovative strategies / **Evaluation method for innovative project**	**1**	**2**	**3**	**4**
Expert evaluation method				
Economic methods				
Simulation				

Innovative strategies are divided into 3-4 groups according to their compatibility with the competitive strategies:

1. Absolute cost advantage - functional advantage: a group of defensive strategies, namely, traditional, residual and defensive [25].

2. Specialization - a leading position on products: a group of offensive strategies, namely an opportunistic strategy, creating a new market, sharp offensive and medium offensive strategy [26].

3. Concentration - a close relationship with the consumer: a group of simulation strategies, namely simulation, dependent and intermediate strategies.

5 Conclusion

1. The analysis and evaluation of innovation potential and innovation strategy relies on multiple sources of economic information. The reliability of forecasts is much more dependent on the quality of the input data than on the methods of analysis. Even simplified methods of strategic decision making based on accurate input data provide more accurate predictions than complex methods of analysis carried out on the basis of approximate data. The database is formed in accordance with the purposes and objectives of the analysis, combining the data from external and internal sources. However, given the strategic focus of innovations and lack of experience of employing such activities for managerial decision-making, information from external sources turns out to be the predominant both by volume and by value. Managers should give a careful consideration to the formation of the decision-making database, whereas analytics should spend more time and effort for collecting the needed data than on the other stages of analysis.

Thus, the proper formation, rapid updating and control of reliability of the database are the necessary conditions for effective and high-quality analysis and evaluation of innovation potential and innovation strategy as a whole.

2. While selecting a competitive innovation strategy a company, as a rule, follows only one of them, otherwise it will not be a success. When choosing one of these competitive strategies the company is facing a choice of 3-4 innovative strategies. Here the company needs to use the proposed matrix to identify the most profitable innovation strategy within the selected competitive one.

Each of the strategies within the competitive group has its own features. In accordance with these differences the cost estimations, the marketing program and the program of implementation (including the organizational costs) for the selected innovation strategy is elaborated. After the calculations the results are set in a matrix, where they are compared with the proposed procedure. Thereafter the managerial decision is made.

3. Nowadays in Russia only a low level of innovation activity of enterprises can be reported [27]. The problem with this very low resulting indicator of implemented innovations in many respects can be explained with a shortage of qualified personnel, capable to realize innovative projects.

Therefore, management training program should include a subject, revealing completely the process of involvement of new technologies in economic activities. Future managers need to possess a comprehensive knowledge and skills for managing the innovative development.

References

1. J. Tan, J. Bus, Venturing, **16(4)**, 359-376 (2001)
2. A. Madrid-Guijarro, D. Garcia, H. Van Auken, J. Small Bus. Manage, **47(4)**, 465-488 (2009)
3. A.B. Radnejad, H. Vredenburg, IJEIM, **19(1-2)**, 77-98 (2015)
4. R. Agarwal, M. Thiel, IJEIM, **18(5-6)**, 397-408 (2014)
5. P. Herzog, J. Leker, Int. J. Technol. Manage, **52(3-4)**, 322-343 (2010)
6. I.M. Golova, A.F. Sukhovey, Economy of Region, **(1)**, 155-164 (2014)
7. B. Luke, M.-L. Verreynne, K. Kearins, IMPP, **12(2)**, 138-153 (2010)

8. K.-H. Tsai, J.-C. Wang, IJTG, **1(2)**, 239-257 (2005)
9. K.-H. Tsai, J.-C. Wang, Problems and Perspectives in Management, **2(1)**, 62-75 (2004)
10. C. Herstatt, IJEIM, **2(6)**, 473-484 (2002)
11. T. Pihkala, H. Ylinenpää, J. Vesalainen, IJEIM, **2(6)**, 520-536 (2002)
12. O.J. Borch, E.L. Madsen, IJTE, **1(1)**, 109-125 (2007)
13. V. Ramaswamy, Strategy & Leadership, **38(2)**, 22-29 (2010)
14. C. Gray, J. Allan, IJEIM, **2(4-5)**, 373-389 (2002)
15. C. Camelo-Ordaz, A. Pérez-Luño, E. Sousa-Ginel, IJEIM, **10(3-4)**, 243-265 (2009)
16. I.A. Bayev, M.V. Podshivalova, Economy of Region, **(1)**, 189-198 (2013)
17. P. Davis, O. Brady, Innovation, **28(3)**, 324-343 (2015)
18. T.I. Tovt, Actual Probl. of Econ., **152(2)**, 211-218 (2014)
19. A. Jantunen, P. Hurmelinna-Laukkanen, IJLIC, **3(2)**, 153-166 (2006)
20. M. Squicciarini, Small Bus. Econ., **48(2)**, 273-278 (2017)
21. D. Endovitsky, S. Komendenko, Investment in Russia, **3**, 39-46 (2001)
22. P.N. Zavlin, A.V. Vassilyev, *Estimation of the innovation efficiency* (Publishing House "Business press", St. Petersburg, 1998)
23. P.S. Semyanov, *Financial problems of the Russian Federation and their solutions: theory and practice. Proceedings of the IVth international scientific-practical conference* (1998)
24. J. Wonglimpiyarat, Journal of High Technology Management Research, **26(2)**, 186-195 (2015)
25. E. Petuskiene, R. Glinskiene, Engineering Economics, **22(1)**, 69-77 (2011)
26. T. Kuznetsova, V. Rud, Questions of Economics, **11**, 86-108 (2013)
27. N.A. Kravchenko, G.V. Bobylev, O.V. Valieva, A.A. Fyodorov, Studies on russian economic development, **24(5)**, 461-469 (2013)

Probabilistic pattern of risks in company's Quality management system

Vladimir Mager[1], *Tatyana* Leonova[2], and *Liudmila* Chernenkaya[1*]

[1]Peter the Great St. Petersburg Polytechnic University, Highest School of Cyberphysics, 195251 St.Petersburg, Russia
[2]St. Petersburg State University of Economics, Economics and Quality Department, 191023 Sadovaya str., 21, St. Petersburg, Russia

Abstract. One possible approach for calculation of probabilistic rate of risks in Quality management system (QMS), which is prescribed by requirements of International Standard ISO 9001:2015, is proposed in this article. Aspects of the theory of dependability and Markov techniques are used, which are applied for evaluation of probability of failures in complicated technical systems. Representation of QMS processes as a graph with controlled discrete Markov chains is suggested, which allows to evaluate a probability of customer requirements non-fulfillment as a function of an intensity of mistakes that bring to non-conformities in QMS.

1 Introduction

Risk-oriented approach has been announced by the new version of International standards ISO 9000 series (from 2015). Item 6.1 of the standard ISO 9001:2015 [1] prescribes to use this approach during development, implementation and continuous enhancement of the company's Quality Management System (QMS).

Risks could be caused by people (inside and outside of the company), by machines used in the production, by inappropriate management of processes, by the influence of external environment. These risks always accompany any kind of activities and depend on lot of factors. Obviously, the mathematical decision of the task for risks estimation needs attraction of methods of multidimensional statistics. But the task for optimal management of a company with consideration of risks has to be solved by decision-makers, and usually carried on the level of intuition.

The definition of the term "risk" one can find in the item 1.1 of ISO Guide 73:2009 [2]: "effect of uncertainty on objectives", where "an effect is a deviation from the expected – positive and/or negative", which means that estimation of risks could be considered as evaluation of possible scenarios by using of well-known tools from strategic management, like SWOT-analysis. Note 4 of ISO Guide 73:2009 explains that "risk is often expressed in terms of a combination of the consequences of an event (including changes in circumstances) and the associated likelihood", and the word "likelihood" (3.6.1.1) "is used to refer to the chance of something happening … and described using general terms or mathematically (such as a probability)". So, for more accurate, scientific-based estimation of risks, the value of a risk could be reflected in terms of probability for example, in tasks of designing and maintaining of complicated technical systems, which must assure the survival of people.

Studies concerning calculation of the rate of risks in QMS have been done by many authors [3, 4, 5], but procedures of calculations still not introduced. The aim of this report is to provide simple probabilistic procedure for estimation of organizational risks based on the analogy with conceptual approach of the theory of dependability, which is used for calculation of a level of valid level of risk in conditions of partial uncertainty.

2 Rational

The most important notion in the theory of dependability is a fault. Generally, a fault is explained as an event, which brings to the situation of overall or partial loss of capabilities of an object to perform its' specified functions. In organizational activities we can meet initiated faults, which depend on mistakes in decision-making. Let's try to apply basics of the theory of dependability for "constructing" of the simplified procedure, which could be helpful for estimation of organizational risks.

First, we'll consider "company" as a production system, consisting of various processes, interrelating with each other and aimed on performance of customer requirements. These processes should be defined by QMS and monitored in order to assure the quality of production.

Second, we'll adopt the admission that the flow of faults in a system is the simplest, and faults are random and independent events, thus a probability $P_s(t)$ of trouble-free operation in time of a complex system with linear structure of elements will be equal to a product of probabilities of trouble-free operation in time of system's elements [6]:

* Corresponding author: ludmila@qmd.spbstu.ru

$$P_c(t) = p_1(t) \cdot p_2(t) \cdot ... \cdot p_N(t) = \prod_{i=1}^{N} p_i(t), \qquad (1)$$

where p_i – a probability of trouble-free operation of element i, and N – the number of elements in a system.

Third, we'll adopt the condition of stationarity of a flow of faults, which means that $p_i(t) = e^{-\lambda_i t}$, where λ_i is the intensity of faults of an element i.

Hence, in order to calculate a probability of a complex system with linear structure of elements, it is needed to define intensities of faults of each system's element. Evident, that number of faults during selected time interval is a statistical indicator, which should be available from records on system exploitation.

Such task becomes more difficult, if the structure of a system is not linear, i.e. elements of a system are interacting in pairs, in crosses, in feedbacks. In that case the simplest formula (1) will be incorrect, and calculation of probabilities must be done in accordance with the structure of a system.

The important condition in analysis and calculations of probabilities in complex systems is an assumption that at the initial moment, in the beginning of analysis, a system was workable. It allows attracting of Markov techniques into the calculation of dependability. These techniques are widely used in the areas dealing with random processes, which reflects randomized changes of a system's states in time (or in respect to some other argument), described by random functions.

Using of Markov processes in analysis of dependability is recommended by International Standard IEC 60300-3-1:2003 [7]. In particular, the standard tells that introduction of system's behavior through Markov model requires the defining of all possible states of a system, preferably depicted in a scheme (diagram) of states and transitions. Besides, the intensities of transitions from one state to other must be determined, which makes possible to calculate the probability of staying of a system in a present set of states. Moreover, International Standard IEC 61165:1995 [8] specifies the main advantage of application of Markov techniques: "possibility to obtain flexible probabilistic model for analysis of system's behavior…"

The above mentioned assumption about the workable state of a system at the initial moment of analysis could be interpreted as a random process without aftereffects, which is characterized by transitional probabilities depending only of a previous state of a system, regardless to the way of coming (or brining)of a system to those previous state: $P_{i/i+1} = f(S)$. Selection of anchors in behavior of a system, that is determined, for instance, by periods of managerial monitoring or corrective actions, brings to splitting of the random process into discrete states, and this random process is considered as random sequence. If a random sequence has Markov properties, it is named as Markov chain, and if states of a system are divided in time, we come to discrete Markov chain, in which a system can move from one state to the other by leap [9].

Markov chain is characterized by:

(a) matrix of transitional probabilities:

$$\|P_{ij}\| = \begin{pmatrix} P_{11} & P_{12} & \cdots & P_{1n} \\ P_{21} & P_{22} & \cdots & P_{2n} \\ \cdots & \cdots & \cdots & \cdots \\ P_{n1} & P_{n2} & \cdots & P_{nn} \end{pmatrix},$$

(b) vector of initial probabilities, which describes the initial state of a system:

$$P^{(0)}{}_{\langle n \rangle} = \langle P_{01}, P_{02}, ..., P_{0n} \rangle .$$

If we can effect on values of transitional probabilities, we have the controlled Markov process. From this standpoint, the combination of measures aimed on enhancement of quality in frames of QMS could be described as controlled Markov process and introduces by discrete Markov chain.

Some notes from the theory: Markov chains, depending of possibilities to move from one of states to another, could be absorbing or ergodic. The absorbing chain contains an absorb state, where a process, while reaching it, never leaves it, and thus, stops. Ergodic, or returning Markov chain, represents many of states related by matrixes of transitional probabilities, i.e. a process can move from any state to any of others through final number of steps, and do not include absorbing states. Thus, a process, described by ergodic chain, being started in some state, will never stop. From that standpoint, exactly this type of Markov processes is applicable to QMS, taking in consideration the concept of continuous improvement according with Total Quality Management and the PDCA (Deming) cycle.

Each step of ergodic process, i.e. the transition of a system from one specified state to any other state, could be characterized by transitional probabilities. Transitional probability lies on the interval $0 \le P_{ij} \le 1$, it is equal to zero if the transaction from S_i to the other state S_j is impossible by one step, and it is equals one if transaction from S_i is impossible to any states, which means that a system "delays" in S_i, so, $P_{ii} = 1$.

Mathematic model of ergodic Markov chain of a random process could be built if there are determined:

(a) limited set of alternatives K_i, *where* $i \in S$ - number of the state of a system;

(b) matrixes of transitions $\Pi_{[S]}{}^{(k)}$, relevant to changes in a system after impact k, named as partial control;

(c) matrix of result $R_{[S]}{}^{(k)}$, relevant to changes in a system after impact k.

Partial controls in case of QMS are concrete managerial decisions, which should bring to the overall result (improvement of QMS) after limited steps of control due to the chosen management strategy.

Let's use the advice from IEC 60300-3-1:2003 [7] and try to draw a graph of states and transitions. The simplest graph introducing a system with two states S_1 and S_2 looks like following:

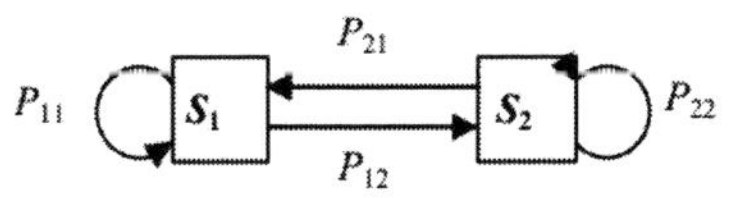

For this graph, matrix of transitional probabilities includes two lines:

$$\left\|P_{ij}\right\| = \begin{pmatrix} P_{11} & P_{12} \\ P_{21} & P_{22} \end{pmatrix}.$$

Here, the condition $\sum_{j=1}^{n} P_{ij} = 1$ is compulsory.

Using this sample, theoretically the graph of any kind of a system could be drawn. For example, let's imagine that the sequence of processes of a production system introduced by a graph Fig. 1:

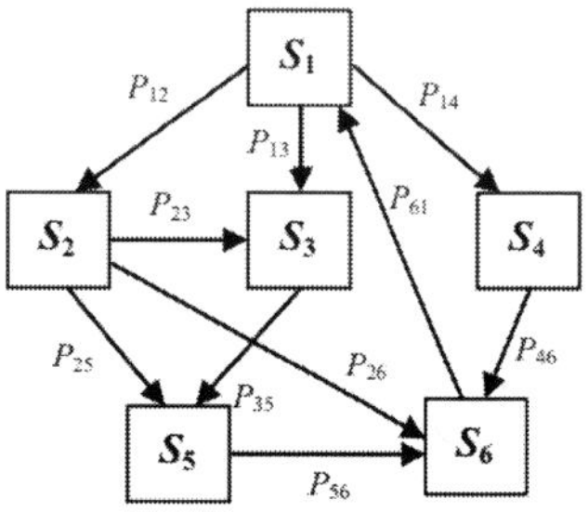

Fig. 1. Graph of states and transitional probabilities of production system:
S_1 – design, S_2 – purchases, S_3 – development of technological documentation, S_4 – planning, S_5 – technological preparation of production, S_6 – production process.

Generally, transitional probabilities for the graph Fig. 1 are determined as:

$$P_{11} = 1 - (P_{12} + P_{13} + P_{14});$$
$$P_{22} = 1 - (P_{23} + P_{25} + P_{26});$$
$$P_{33} = 1 - P_{35};$$
$$P_{44} = 1 - P_{46};$$
$$P_{55} = 1 - P_{56};$$
$$P_{66} = 1 - P_{61}.$$

These equations specify a matrix of transitional probabilities, including probabilities of delays of a system in states S_j (j=1, …, n). If we define the initial state of the system as $P_1(0), P_2(0), \ldots P_n(0)$, it's easy to find probabilities of states $P_1(k), P_2(k), \ldots P_n(k)$ after any of steps k:

$$P_i(k) = \sum_{j=1}^{n} P_j(k-1) P_{ij}, \quad i = 1,\ 2,\ \ldots, n. \qquad (2)$$

This logic could be useful for selecting of optimal strategy, which has to assure maximal effectiveness of QMS. For this, matrix of transitional probabilities may be added by matrix of the same dimension with elements reflecting assumed rates of incomes or expenses, which are linked to transitions of a system between states. As for expenses, they could be related to costs for corrective actions, or implementation of managerial decisions, or loosed benefit due to the delay with transition of a system to the desired state.

But it is evident that in real production all mentioned in Fig. 1 processes are interacting in strict sequence. Initial state is S_1, and final – S_6, resulted in performing of customer order in accordance with his requirements, i.e. the completion of the production of required volumes of goods/services in time. For moving from the state S_1 to the state S_6 four steps are needed:

- step 1: transitions from the state S_1 to states S_2, S_3, S_4;
- step 2: transitions from the state S_2 to states S_3, S_5, and from the state S_4 to S_6;
- step 3: transitions from the state S_3 to S_5 and from S_2 to S_6;
- step 4: transitions from the state S_5 to S_6 and from S_6 to S_1.

Initial probability $P_{11}(0) = 1$. Let's define the transitional probabilities in steps. Taking into consideration only states, from which the direct transition to the following state is possible, we find:

after step 1:
$$P_{22}(1) = P_{11}(0) \cdot P_{12};$$
$$P_{33}(1) = P_{11}(0) \cdot P_{13};$$
$$P_{44}(1) = P_{11}(0) \cdot P_{14};$$

after step 2:
$$P_{33}(2) = P_{33}(1) + P_{22}(1) \cdot P_{23};$$
$$P_{55}(2) = P_{22}(1) \cdot P_{25};$$
$$P_{66}(2) = P_{44}(1) \cdot P_{46};$$

after step 3:
$$P_{55}(3) = P_{55}(2) + P_{33}(2) \cdot P_{35};$$
$$P_{66}(3) = P_{66}(2) + P_{22}(1) \cdot P_{26};$$

after step 4:
$$P_{66}(4) = P_{66}(3) + P_{55}(3) \cdot P_{56};$$
$$P_{11}(4) = P_{66}(3) \cdot P_{61}.$$

As the result, we have the expression for evaluation of the probability of the successful completion of company's work on the customer order:

$$P_{success} = P_{66}(4) = P_{14} \cdot P_{46} + P_{12} \cdot P_{26} + \\ + \left[P_{12} \cdot P_{25} + (P_{13} + P_{12} \cdot P_{23}) \cdot P_{35}\right] \qquad (3)$$

Here, transitional probabilities P_{ij} characterize the reducing of probability of moving from the state i to the state j because of some faults. Now we come to most important question in our proposed analogy: how to treat the term "fault" in Quality management and QMS processes?

Processes of a company are organized for performing of generalized process of production of goods and/or services, which is applicable to any kinds of production and consist of a certain consequence of separate processes [10], and each of them could be introduced as controlled discrete Markov chain. There are tens of processes in companies, and they are interacting – this is the meaning of process approach according with standard ISO 9000 [11]. Technical term "fault" in process-oriented company can be interpreted as the term "nonconformity", defined in the vocabulary of the standard ISO 9001:2015 [1] as "failure to comply with requirement". Prof. Edwards Deming was the first in 1980th, who used this term, and he associated nonconformity with delays, downtimes, mistakes, defects – all causes, which do not allow a company to carry out a production process properly. Thus, nonconformity in QMS is the analog of the fault in technical systems. It means that concepts of the theory of dependability could be applied for evaluation of risks in

QMS, based on probabilistic models. Even though the term "probability" includes some proximity, anyway, it is more reliable to choose a strategy on digital indicators than on some low reasoned hypotheses.

Let's illustrate the simplest method of calculation of risks in QMS, taking into account all assumptions, which we adopted to the considered class of Markov chains. For this, we must define transitional probabilities in (3) and list possible mistakes belonging to arrows of the graph. In general, the probability of a mistake, in case of stationarity, could be calculated as:

$$P_{Mistake.z} = e^{-\lambda_{Mistake.z} \, t_{step\ k}}, \quad (4)$$

where $\lambda_{Mistake.z}$ – intensity of mistake z;

z – number of mistake from the list;

k – number of the step (in our example $1 \le k \le 4$);

$t_{step\ k}$ – normative time for transition on step k.

List of mistakes should be completed by managers, who are in charge of particular processes. Intensities $\lambda_{Mistake.z}$ have to be calculated on the basis of the data from QMS monitoring.

Most part of mistakes in QMS are related to human factor. Let's try to list some of them:

1) mistakes in design, including development of operational manuals;
2) mistakes in planning and organization of production;
3) mistakes in marketing and logistics, which causes inappropriate choose of suppliers;
4) mistakes in production, caused by divergences from requirements of design/technological documentation;
5) breach of requirements of operational manuals on equipment by personnel;
6) inappropriate organization of working places;
7) mistakes in maintenance, which causes inappropriate adjustment and calibration of technological equipment;
8) mistakes in technical inspection, caused by using of incorrect instruments or misconduct of inspection and documenting procedures;
9) misconduct of conditions for storing and shipment of final production;
10) managerial mistakes, including poor motivation of employees.

Obviously, that the whole list of mistakes is much longer, but we'll use this one to finalize our pattern:

$$P_{12} = P_{13} = P_{14} = 1 - e^{-\lambda_{Mistake.1} \, t_{step\ 1}};$$

$$P_{23} = P_{25} = 1 - e^{-\lambda_{Mistake.3} \, t_{step\ 2}};$$

$$P_{26} = 1 - e^{-\lambda_{Mistake.9} \, t_{step\ 2}};$$

$$P_{35} = 1 - e^{-\lambda_{Mistake.7} \, t_{step\ 3}};$$

$$P_{46} = 1 - e^{-\lambda_{Mistake.2} \, t_{step\ 3}};$$

$$P_{56} = 1 - e^{-\lambda_{Mistake.4} \, t_{step\ 3}};$$

$$P_{61} = 1 - e^{-\lambda_{Mistake.8} \, t_{step\ 3}} - e^{-\lambda_{Mistake.10} \, t_{step\ 3}}.$$

This pattern of probabilities, along with (3), allows to calculate the probability of risk of non-fulfillment (or poor quality) of the work, in percentage:

$$P_{Risk} = (1 - P_{success}) \cdot 100\%.$$

3 Conclusion

The proposed approach allows to obtain a quantative rate of risk in QMS. If we are able to evaluate negative consequences of nonconformities (like failures, shutdowns, breaches of technological discipline, delays in deliveries from suppliers, etc.), founded value P_{Risk} could be easily transferred into economic categories.

References

1. ISO 9001:2015. Quality management systems. Requirements
2. ISO Guide 73:2009. Risk management. Vocabulary
3. V.V. Okrepilov, *Quality management* (Science, St. Petersburg, 2007).
4. T.I. Leonova, *Statistical methods in management of processes: handbook* (Printing house of State Economic University, St. Petersburg, 2014).
5. A.V. Chernenkii, International Research Journal, **8(50)**, **1**, 92-96 (2016)
6. V.A. Ostreikovskii, *Theory of dependability: handbook for HEI* (High school, Moscow, 2003)
7. IEC 60300-3-1:2003. Dependability management – Part 3-1: Application guide – Analysis techniques for dependability – Guide on methodology
8. IEC 61165:1995. Application of Markov techniques
9. E.V. Berezhnaia. *Mathematic methods in modeling of economic systems: handbook* (Finances and Statistica, Moscow, 2006)
10. V.E. Mager, *Quality management: handbook* (INFRA-M, Moscow, 2015)
11. ISO 9000:2015. Quality management systems. Fundamentals and vocabulary

Procedure for determining transport and warehousing systems characteristics in industrial enterprises in megapolis

Anton Radaev[1,*], *Valery* Leventsov[1], and *Vladimir* Kobzev[1]

[1]Graduate School of Industrial Management and Economics, Peter the Great Saint-Petersburg Polytechnic University, 195251 Saint-Petersburg, Russia

Abstract. The most common problem among a vast variety of design and performance management problems for the industrial enterprises supply chain networks is determining transport and storage processes' characteristics in the areas of high population density. The divergence of the logistic structure development in the area has a major influence on the choice of location for the warehouses and the characteristics of the corresponding material flows. This article addresses the issues of determining transport and warehousing systems characteristics in a major population center and its suburbs. In particular, this means the warehouse location and the delivery lot sizes. The article suggests a procedure for solving this problem based on a set of mathematical models using linear and non-linear optimization via the standard algorithms of common computer software like Microsoft Excel, Mathcad, MatLab, etc.

1 Introduction

Efficient functioning of the industrial enterprises in modern conditions is impossible without a detailed elaboration on the questions of organization design and performance management. This concerns not only manufacturing systems of said enterprises, but also the corresponding supply chains, which facilitate the material flows with the help of transport and warehousing resources of industrial enterprises and external party contractors. The importance of solving the design and operation problems for industrial enterprises and their supply chains is for the most part determined by the following: the added complexity and growing intensity of the material and assistant flows accompanied by market globalization, and the tightening of the requirements for the products' flow and transformation, which provides competitive ability for the enterprise. Therefore, the parameters of the material flow structure from the supplier to the consumer and the storage buffer locations are considered the most significant characteristics for organization design and performance management of the industrial enterprises' supply chains. It stands to mention that choosing the locations for the storage buffers is especially important for servicing the ultimate consumers in large residential areas, which will be hereafter referred to as megapolises.

This circumstance ascertains the viability of this research. The goal of the research is to develop a procedure for determining transport and warehousing network's characteristics (as a structural unit of a supply chain) for an industrial enterprise in a megapolis environment. The aforementioned procedure should take into account the specific features of logistic processes in megapolises and the surrounding territories. These features include the following:

- products are transferred in delivery lotes in the transport and warehousing network using vehicles of a certain capacity;
- vehicles travel in the complex set of roads, equipped with traffic lights and other means of traffic regulation;
- warehouse upkeep and rent cost (whether rented or owned) is related to its location, which is determined by the specifics of the city infrastructure.

The warehouses' locations and the volumes of transferred goods between the nodes of the supply chain network are considered the most important characteristics of the transport and warehousing network for this research.

It is also worth mentioning that in order to assure feasible results, the proffered procedure should be implemented by the aid of the standard computing algorithms using software like "Microsoft Excel", "Mathcad", "MatLab" etc.

This paper is organized as follows. Section 2 contains the results of the literature review and analysis in the field of determining the supply chain characteristics for enterprises. In section 3 the problem of determining the transport and warehousing network's characteristics in industrial enterprises in a megapolis environment is posed. Section 4 includes the procedure for solving the

* Corresponding author: TW-inc@yandex.ru

problem. Section 5 provides the results and conclusions of the research and also the direction for future work.

2 Literature review

The first stages of the research include the literature review for the field of interest. The results of this review and analysis are presented in the following.

Many of the works are focusing on determining the supply chain characteristics in relation to delivery lot sizes and locations of various supply chain elements. The techniques used are mostly linear and non-linear optimization models. The main objectives are the total operating expenses in the supply chain and some cost indicators derived from those. The suggested solving methods are Lagrangian relaxation-based method [1], genetic algorithms [2, 3, 4, 5], game theory methods [6], heuristic methods [7, 8, 9, 10], decision-making methods [11, 12, 13], fuzzy optimization [14], ant colony algorithm [15], particle swarm optimization [16], modified fruit fly optimization [17], other linear and non-linear techniques [18, 19, 20, 21].

A range of works concerns the development and implementation of techniques and tools based on Economic Order Quantity and Economic Production Quantity models for determining the characteristics of the supply process. The proposed modifications of the aforementioned models account for a large product range [22], the uncertainty of the demand, delivery lot sizes and the corresponding operation cycles [23, 24, 25], lead times [26], backorders [27], vendor-buyer relationships, in regard to consignment stock among other things [28]. The proposed solving methods are numerical methods [25], genetic algorythms [22] and game theory methods [28].

The analysis of the works mentioned above prompted the following conclusions:

- the aforecited works treat transport processes in a simplified, amplified way and they do not account for the specifics of the corresponding process (the arrangement of goods in the cargo area, the traffic in an urban environment), technical features of the vehicle (active storage of the truck body, carrying capacity, mean velocity on the route etc.) and transportation rates (the transportation cost proportional to distance or duration of the traffic route – for intraregional and transregional shipments);
- the suggested mathematical models do not account for the expenses concerning upkeep or rent of the fixed assets in the supply chain structure; for instance, the rent cost per stacking capacity unit of a distribution warehouse for serving the customers in a megapolis will vary considerably depending on the location of a warehouse in said megapolis.

In conclusion, the aforementioned studies do not make allowance for a wide range of specific ways to organize logistic processes in the areas with high population density. Therefore, these studies can only be effective for the approximate, rough design of the supply process of the industrial enterprises. This circumstance serves to prove the high relevance and applicability of the conducted research.

3 Problem definition

This section of the article contains the generalized problem definition for determining the transport and warehousing system's characteristics in industrial enterprises in a megapolis environment. It concerns the main principles of the procedure, the input data and the decision variables.

3.1 Main principles

- The main object of the research is a transport and warehousing network. This network is an element of a supply chain of an industrial enterprise for which the products are transferred in a certain geographical area from a number of suppliers to a number of consumers through an only distribution warehouse where the products are temporarily stored.
- The goods are essentially put in standardized packages assembled on flat pallets.
- The goods are stored in a rented warehouse in a certain geographical area, which includes a megapolis and its suburbs. Given that, there is a functional relation between the warehouse characteristics (storage area, active height and cost per stacking capacity unit of the warehouse) and the warehouse location specifics. For this research, it will be practical to suggest that the location of a warehouse is determined by the shortest distance between the warehouse and the nominal center of the megapolis. The center is a location where the aforementioned characteristics of a warehouse reach extreme values [29, 30]. This point is illustrated by fig. 1.
- The goods are transferred from each supplier to the distribution warehouse and from there to each consumer with the aid of road transport with certain technical features (carrying capacity, velocity etc.) and certain costs. In case of intraregional shipments the costs can be divided into fixed (related to serving the vehicle) and variable costs (determining the monetary equivalent of 1 hour's work of a vehicle).
- The goods are transferred from each supplier to the warehouse in fixed volumes in regular intervals of time equivalent to the preferable supply interval. The goods are transferred from the distribution warehouse to the ultimate consumers daily and in fixed volumes.
- It is required to determine the optimal location of the warehouse and the optimal delivery lot sizes for the shipments from the suppliers to the warehouse. These parameters should ensure the minimal total expenses for warehouse upkeep and rent and for transportation within the confines of the examined transport and warehousing network.

3.2 Input data

z – number of items on stock in the transport and warehousing network;

m, n – accordingly, the number of suppliers and consumers in the examined geographic area;

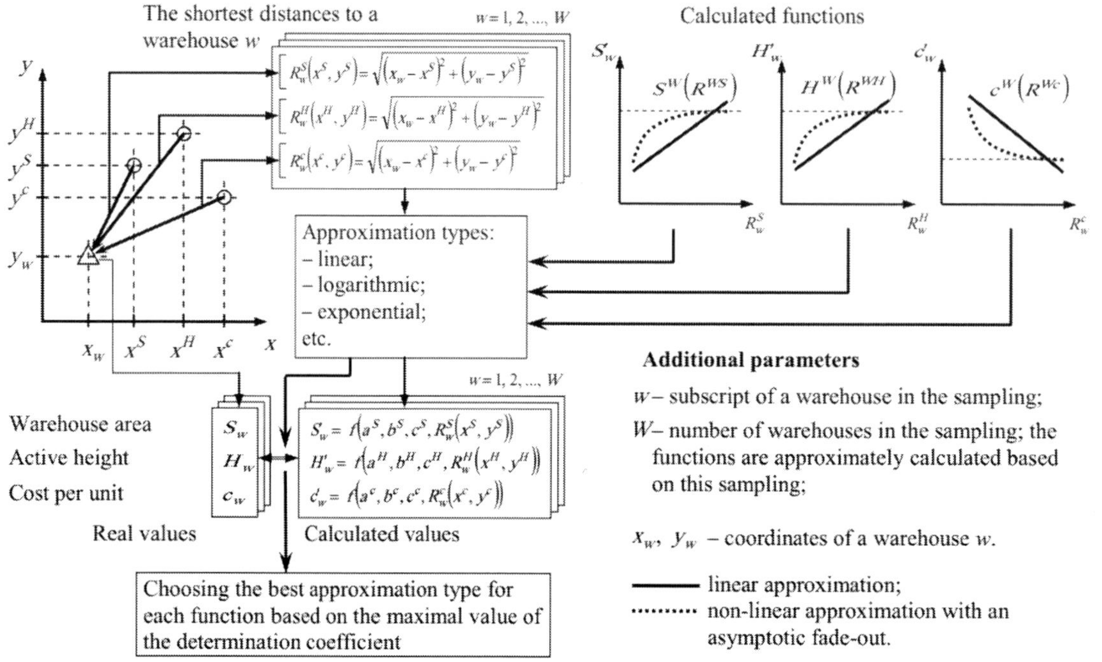

Fig. 1. Description of the calculated functions for the warehouse characteristics in relation to its location, and the procedure for determining the functions.

$k = 1, ..., z$ – subscripts of stock items;

$i = 1, ..., m$ – subscripts of suppliers;

$j = 1, ..., n$ – subscripts of consumers;

L^U, B^U, H^U – accordingly, reference values (m) for the length, width and height of a cargo unit for the store item;

$x_{i(j)}$, $y_{i(j)}$ – location coordinates (km) of supplier i (consumer j) in the Cartesian system, based on a map of the examined geographical area;

$R_{i(j)}$ – preferable supply interval (workdays) for supplier i to the distribution warehouse (from the warehouse to consumer j); $R_j = 1$ workday;

$Q_{ki}^{\max}$ – maximal delivery lot size (pcs/shipment) for stock item k for supplier i;

I_{kj} – expected daily volume of shipments (pcs/shipment) for stock item k from the warehouse to consumer j;

$r_{i(j)}$ – active storage (pcs/unit) of the truck body for the vehicles used by supplier i (consumer j) – calculated based on reference values of mass and dimensional parameters of the truck body and cargo units, which can be evaluated by known formulas [29];

$\upsilon_{i(j)}$ – mean velocity on the route (km/hr) of the vehicles used by supplier i (consumer j) [29];

$c_{i(j)}^{T0}$, $c_{i(j)}^{T1}$ – accordingly, fixed (c.u../run) and variable (c.u./(unit·hr)) costs for the vehicles used by supplier i (consumer j);

$x^{\min(\max)}$, $y^{\min(\max)}$ – minimal (maximal) possible values for the abscess and ordinate coordinates accordingly;

k^{TI} – coefficient accounting for the specifics of the transport infrastructure [29];

$k_{i(j)}^{TL}$ – coefficient accounting for standing time at the traffic lights for the vehicles used by supplier i (consumer j), – it is determined based on the mean velocity on the route $\upsilon_{i(j)}$ using a known formula [29];

Δ_h – vertical distance between the storage units, which is defined by the storage technique, e.g. the design of the rack;

k^S – space occupancy coefficient for the warehouse, which is defined by the storage technique; e.g. for rack storage $k^S = 0.39 \div 0.45$;

T – length (workdays) of the given time period;

$S^W(R^{WS})$, $H^W(R^{WH})$, $c^W(R^{Wc})$ – functions for determining the warehouse area S^W (m^2), active height H^W (m) and rent cost per stacking capacity unit of the warehouse c^W (c.u./(unit·pd)) in relation to the distance from the center $R^{WS(H,c)}$ (km). Each function is defined based on the warehouse sampling in the examined geographical area (the parameters, such as location, area, active height and cost per unit are known). This is accomplished by implementing the following stages (fig. 1): creating the alternative approximations (linear, logarithmic, exponential etc.); optimization of the approximating function coefficients $a^{S(H,c)}$, $b^{S(H,c)}$, $c^{S(H,c)}$ and center coordinates $x^{S(H,c)}$, $y^{S(H,c)}$ using the generalized reduced gradient method; choosing the best approximation based on the maximal value of the determination coefficient. The procedure for determining the approximation coefficients concerning the warehousing system parameters is described in [30];

$x^{S(H,c)}$, $y^{S(H,c)}$ – coordinates (km) of the nominal center used to calculate the area of the warehouse (its height and cost per unit).

3.3 Decision variables

Q_{ki} – delivery lot volume (pcs/shipment) for stock item k for supplier i;

x^W, y^W – coordinates (km) of the distribution warehouse in the Cartesian system, based on a map of the examined geographical area.

4 Solving procedure

The procedure for determining the transport and warehousing system characteristics in industrial enterprises in a megapolis environment is defined by the succession of certain stages. The detailed description will be given further.

The first stage for implementing the procedure is to compose the calculated parameters for the optimization models which will be used in the next stages. The mentioned parameters are defined in Table 1.

Table 1. Calculated parameters, units of measure and equations.

No.	Parameter	Unit	Equation
1	Stock item k, supplied by supplier i		
1.1	Volume of the daily demand	pcs/ workdays	$I_{ki}(Q_{ki}) = \frac{Q_{ki}}{R_i}$
2	Supplier i, consumer j		
2.1	Shortest distance to the warehouse	km	$L^W_{i(j)}\left(x^W, y^W\right) = \sqrt{\left(x_{i(j)} - x^W\right)^2 + \left(y_{i(j)} - y^W\right)^2}$
2.2	Transportation cost for one run to the warehouse (and from the warehouse)	c.u./ run	$C^T_{i(j)}\left(x^W, y^W\right) = c^{T0}_{i(j)} + c^{T1}_{i(j)} \cdot \frac{L^W_{i(j)}\left(x^W, y^W\right) \cdot k^{TI} \cdot k^{TL}_{i(j)}}{\upsilon_{i(j)}}$
2.3	Cargo volume in a shipment	pcs/ shipment	$Q_i(\{Q_{ki}\}) = \sum_{k=1}^{z} Q_{ki}; \quad Q_j = \sum_{k=1}^{z} I_{kj}$
2.4	Mean number of runs in the given time period	ea/ pd	$z^T_i(\{Q_{ki}\}) = \frac{Q_i(\{Q_{ki}\})}{r_i} \cdot \frac{T}{R_i}; \quad z^T_j = \frac{Q_j}{r_j} \cdot \frac{T}{R_j}$
3	Transport and warehousing network		
3.1	Needed warehouse capacity	pcs	$\overline{V}^W(\{Q_{ki}\}) = \sum_{i=1}^{m} Q_i(\{Q_{ki}\})$
3.2	Distance from the nominal center to the warehouse used to calculate the area of the warehouse (its height and cost per unit)	km	$R^{WS(H,c)}\left(x^W, y^W\right) = \sqrt{\left(x^{S(H,c)} - x^W\right)^2 + \left(y^{S(H,c)} - y^W\right)^2}$
3.3	Expected warehouse capacity	pcs	$V^W\left(x^W, y^W\right) = \left\lfloor \frac{S^W\left(R^{WS}\left(x^W, y^W\right)\right) \cdot k^S}{L^U \cdot B^U} \right\rfloor \cdot \left\lfloor \frac{H^W\left(R^{WH}\left(x^W, y^W\right)\right)}{H^U + \Delta_h} \right\rfloor$
3.4	Warehouse rent cost in the given time period	c.u./ pd	$C^W_\Sigma\left(x^W, y^W\right) = c^W\left(R^{Wc}\left(x^W, y^W\right)\right) \cdot V^W\left(x^W, y^W\right)$
3.5	Assumed storage cost for the supply stock in the given time period	c.u./ pd	$\widetilde{C}^W_\Sigma(\{Q_{ki}\}) = c^W\left(R^{Wc}\left(x^W, y^W\right)\right) \cdot \overline{V}^W(\{Q_{ki}\})$
3.6	Total transportation cost from suppliers to the warehouse in the given time period	c.u./ pd	$C^{T in}_\Sigma\left(\{Q_{ki}\}, x^W, y^W\right) = \sum_{i=1}^{m} C^T_i\left(x^W, y^W\right) \cdot z^T_i(\{Q_{ki}\})$
3.7	Total transportation cost from the warehouse to consumers in the given time period	c.u./ pd	$C^{T out}_\Sigma\left(x^W, y^W\right) = \sum_{j=1}^{m} C^T_j\left(x^W, y^W\right) \cdot z^T_j$

3.8	Total transportation cost in the given time period	c.u./ pd	$C_{\Sigma}^{T}\left(\{Q_{ki}\}, x^{W}, y^{W}\right)=C_{\Sigma}^{T\,in}\left(\{Q_{ki}\}, x^{W}, y^{W}\right)+C_{\Sigma}^{T\,out}\left(x^{W}, y^{W}\right)$
3.9	Total rent and transportation cost in the given time period	c.u./ pd	$C_{\Sigma}\left(\{Q_{ki}\}, x^{W}, y^{W}\right)=C_{\Sigma}^{W}\left(x^{W}, y^{W}\right)+C_{\Sigma}^{T}\left(\{Q_{ki}\}, x^{W}, y^{W}\right)$

Note: $\lfloor ... \rfloor$ – means the result of rounding down the value to a nearest whole number.

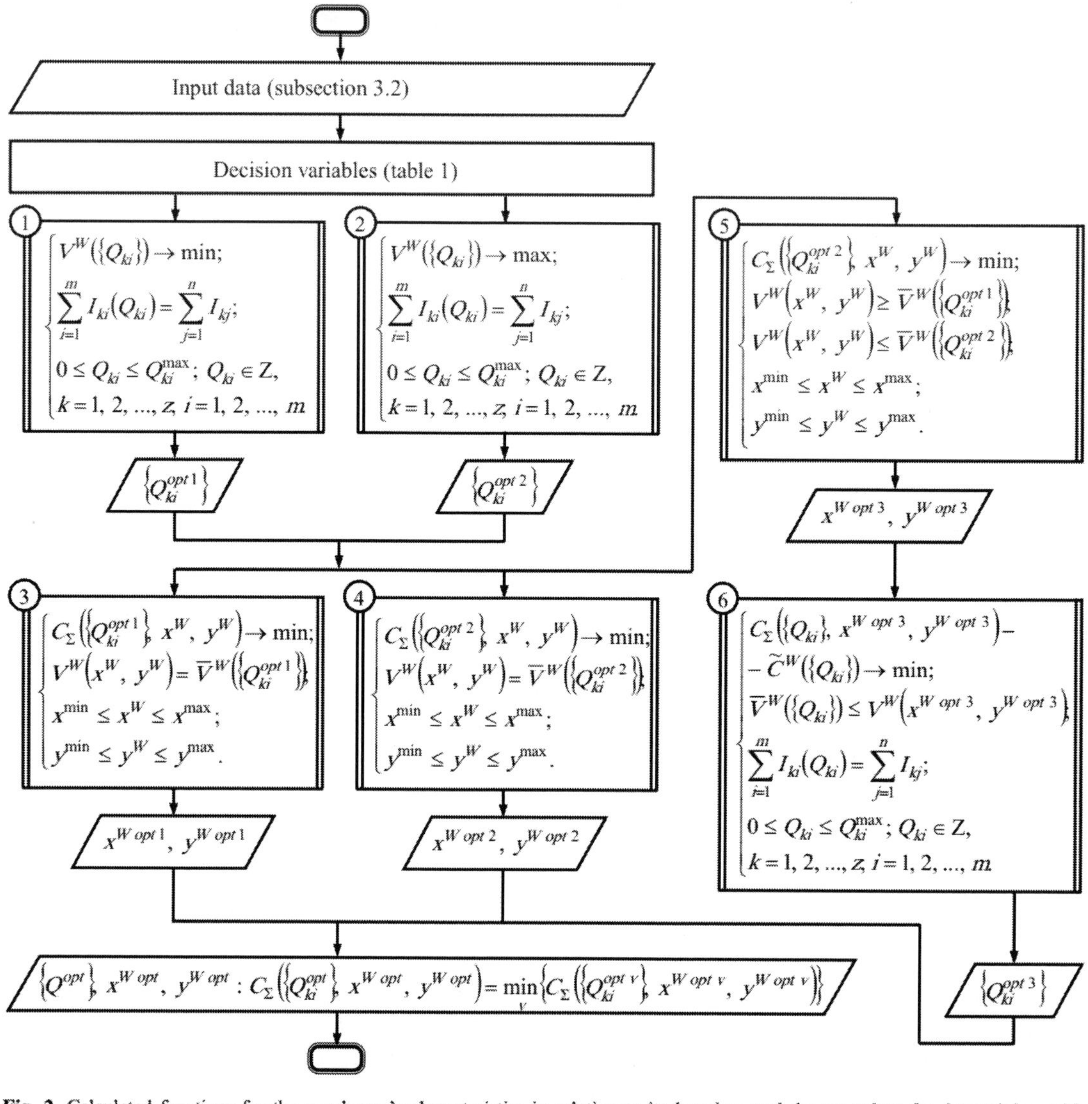

Fig. 2. Calculated functions for the warehouse's characteristics in relation to its location, and the procedure for determining said functions.

The next stages of the procedure are best described by the flow chart in fig. 2. It seems to be impossible to calculate all the decision variables for the given problem by a single optimization model and its implementation via the standard methods, because there are too many miscellaneous variables. Therefore, the procedure suggests a consequent (step-by-step) optimization for groups of variables using certain guidelines.

The third stage of the procedure shows the implementation of two optimization models for determining the delivery lot volumes with minimal (the goods are transferred to the warehouse at a fast rate in small delivery lots) and maximal (the goods are transferred rarely and in large delivery lots) needed warehouse capacity (boxes 1 and 2 in fig. 2).

The fourth stage of the procedure describes the optimization of the location of the warehouse of a minimal and maximal capacity (boxes 3 and 4 in fig. 2). It is based on the optimal (see stage 3) delivery lot volumes. Thus, upon implementing this stage the first two alternate solutions for the problem (sets of optimal values of the decision variables) are derived.

The fifth stage is used to create the last alternative solution for the problem. To begin with, the optimal location of the warehouse is determined using the total cost minimization criteria (box 5 in fig.2). The total cost is related to the delivery lot size, which is in turn determined by maximizing the warehouse capacity (box 2 in fig.2). At that, the expected warehouse capacity is restricted by the threshold values calculated in stage 3 of the procedure (boxes 1 and 2 in fig. 2). After that the delivery lot sizes are optimized using the threshold criteria of minimal difference between the total cost and the assumed storage cost in the given time period. Using this objective function allows for the minimal difference between expected and needed warehouse capacity.

In the last (sixth) stage of the procedure, the choice between all of the alternative solutions is made. The best choice is enabled by using the minimal total rent and transportation cost criteria for the given time period.

Therefore, the suggested procedure supposes the use of linear (boxes 1, 2 and 5 in fig, 2) and non-linear (boxes 3, 4 and 6 in fig, 2) optimization models. The models' structure allows for efficient implementation using the standard computing algorithms – simplex algorithm and evolutionary algorithm (genetic algorithms) – in common software, such as "Microsoft Excel".

5 Conclusions and directions for future work

This article contains a detailed procedure for determining the supply chain characteristics in industrial enterprises in a megapolis environment. As a result of the relevant works analysis it was concluded that the examined problem is highly relevant. The problem was posed and the corresponding main principles, input data and decision variables were determined. The procedure for solving the aforementioned problem was suggested, which is based on implementing a set of mathematical models including linear and non-linear optimization of delivery lot sizes and warehouse coordinates accordingly. These models can be efficiently used with standard computer software; however, they have the following demerits:

- when optimizing integer data of delivery lot sizes, it is possible that the strict correspondence between the real and provided demand volume proves impossible; in that case making an emergency supply or allowing stockouts should be taken into account and included in the threshold criteria and the objective function of the corresponding optimization model;
- no theoretical comparison of the sensitivity of transportation or rent cost to the warehouse coordinates is made in the optimization process; the analysis of said sensitivity would allow for determining the preferable order of optimization.

The follow-up research will allow eliminating the aforementioned demerits. It is planned to refine the procedure and subsequently put it into practice.

References

1. A. Diabat, J.-P. Richard, C.W. Codrington, Ann. Oper. Res., **203**, 55-80 (2013)
2. A. Diabat, T. Aouam, L. Ozsen, IJADS, **2**, 389-405 (2009)
3. S.M. Mousavi, V. Hajipour, S.T.A. Niaki, N. Alikar, Ap. Math. Mod., **37**, 2241-2256 (2013)
4. S.M. Mousavi, S.T.A. Niaki, Appl. Math. Model, **37**, 5109-5119 (2013)
5. S.M. Mousavi, S.T.A. Niaki, E. Mehdizadeh, M.R. Tavarroth, IJSS, **44**, 1897-1912 (2013)
6. O.D. Palsule-Desai, Omega, **41**, 780-796 (2013)
7. Y.-C. Tsao, G.-J. Sheen, Int. J. Prod. Econ., **135**, 106-115 (2012)
8. J. Li, L. Liu, Int. J. Prod. Econ., **101**, 89-98 (2006)
9. T.-Y. Lin, CIE, **66**(1), 53-62 (2013)
10. K.-L. Huang, C.-W. Kuo, M.-L. Lu, Eur. J. Oper. Res., **234**, 77-85 (2014)
11. Y.C. Chou, H.C. Yao, Y.S. Chun, Eur. J. Oper. Res., **189**, 132-145 (2008)
12. T. Ozcan, N. Celebi, S. Esnaf, Expert Sys. Appl., **38**, 9773-9779 (2011)
13. Y.C. Dong, H.J. Zhang, KBS, **58**, 45-57 (2014)
14. M. Ramezani, A.M. Kimiagari, B. Karimi, T.H. Hejazi, KBS, **59**, 108-120 (2014)
15. Nia A. Roozbeh, Far M. Hemmati, S.T.A. Niaki, Int. J. Prod. Econ., **155**, 259-171 (2013)
16. S.M. Mousavi, A. Bahreininejad, S.N. Musa, F. Yusof, J. Intell. Manuf., **1-16**, 10.1007/s10845-014-0970-z (2014)
17. S.M. Mousavi, N. Alikar, S.T.A. Niaki, A. Bahreininejad, CIE, **87**, 543-560 (2015)
18. S. Karakaya, I.S. Bakal, CIE, **64**, 696-707 (2013)
19. M.A. Rodriguez, A.R. Vecchietti, I. Harjunkoski, I.E. Grossmann, Comp. Chem. Eng., **62**, 194-210 (2014)
20. Al-e-hashem S. Mirzapour, A. Baboli, Z. Sazvar, Eur. J. Oper. Res., **230**, 26-41 (2013)
21. K. Subramanian, J.B. Rawlings, C.T. Maravelias, Comp. Chem. Eng., **64**, 71-80 (2014)
22. S. Mondal, M. Maiti, CIE, **44**. 105-117 (2002)
23. M.J. Liberatore, Oper. Res., **27**, 391-396 (1979)
24. A.L. Guiffrida, *Inventory Management: Non-Classical Views* (CRC Press, Taylor and Francis Group, FL, Boca Raton, 2009)
25. S.-C. Chang, J.-S. Yao, H.-M. Lee, Eur. J. of Oper. Res. **109**, 183-202 (1998)
26. L.-Y. Ouyang, J.-S. Yao, Comput. Oper. Res. **29**, 471-487 (2002)
27. N. Kazemi, E. Ehsani, M.Y. Jaber, IJAR, **51**(8), 964-972 (2010)
28. S. Bylka, Int. J. Prod. Econ., **143**, 424-433 (2013)
29. A.E. Radaev, V.V. Kobzev, St. Peters, State Polytech. Univ. J. Econ., **233**(6), 183-195 (2015)
30. A.E. Radaev, V.V. Kobzev, *Cluster Economy and Industrial Policy: Theory and Instruments* (Polytechnical University Publishing House, Saint-Petersburg, 2015)

Sector specific features of innovative development in the Russian economy

I.A. Krasyuk[1*], *V.V.* Bakharev[1], and *Y.Y.* Medvedeva[2]

[1]Saint Petersburg Polytechnic University Peter the Great, 195251 Saint-Petersburg, Russia
[2]Don state technical university, 344000 Rostov-on-Don, Russia

Abstract. The present-day development of the Russian Federation is mostly shaped by the innovative development capacity of the national economy. Industry forms the core of such development. The paper presents the study of the current state of the innovative development in industry being the source of the basic innovations, as well as identifies the limitation factors. The authors have found out that low innovative activity of the industrial enterprises holds back the implementation of innovations in the related sectors, for instance, trade. It has also been found out that the course of the innovative development in the field of trade is also shaped by the sector-specific factors, which should be considered to adjust the innovation policy of a trading enterprise.

1 Introduction

To function successfully in the modern ever-changing environment, an enterprise requires efficient innovations management. Innovative development in multiple fields of the market economy features various peculiarities, impacting its efficiency. The innovative development of the Russian industrial enterprises is presently shaped by a number of external factors, and is largely determined by the structural shifts and imbalances of the Russian economy, and depends on the degree of the internal enterprise potential utilization.

2 Relevance

The activities of the modern industrial enterprises are largely shaped by the modern industrial and commercial policy, the state and the developmental prospects of the Russian retail trade. The development of the latter is mostly considered to be defined by the market influence of the retail chains. The regional expansion of the major retail food store chains usually features purchasing of local structures chains of the required form and scope as well as franchising. Trading enterprises are forced to use innovative methods, which allow for the arrangement of the shopping space planning, modeling the functioning of the stores, setting up the staff schedules. As the customers' incomes increase, and the customers become more discerning, they can be segmented into smaller market groups. With the quickly changing and flexible market most of the retail sellers are forced to introduce new products, enter new international markets, or take some shares of the market over from the competitors. Thus the efficient innovative activity in trade can provide the commercial enterprises with the required competitiveness level.

The study of approaches to the innovation management shows that the peculiar features of the innovative development mechanism formation are quite well-established. J. Schumpeter, an American and Austrian economist considered as the founder of the innovations concept, defined innovations as a new combination of the factors of production motivated by entrepreneurial spirit [1]. 2. Clayton M. Christensen, Michael E. Raynor developed the ideas on the innovative development by specifying basic notions, extending the typology and the array of innovations management methods [2]. P. Druker, D.Stiglic studies the opportunities for business development based on innovation implementation [3,4].

J. Schumpeter defines basic and secondary innovations. Trade, with its peculiarities and features different from production, is the field for secondary innovations. It is impossible to pass the accumulated experience in innovation activity organization from other fields over to trade. Quite few authors study the innovative development, and innovation changes in the field of trade, and their typology. The leader among those is B. Hallier, the founder of EHI Retail Institute, who studies the retail trade modes and formats genesis [5].

3 Studied problem and hypothesis

The core problem of the present research consists in the resolution of the controversy existing between the large number of new developments in the fields of information technology, marketing, and trade, which are potentially applicable in successful commercial companies, and the

* Corresponding author: iri-krasjuk@yandex.ru

inefficient use of these instruments in the enterprise management. Based on this problem, the following hypothesis can be suggested. The organization of efficient innovation activity of a trade enterprise is based on the study and establishing the system of a large number of defining factors to select the most efficient management alternative in the field of innovations policy.

4 Theoretical part

Modern innovation theory understands an innovation as the ultimate result of the innovative activity delivered in the form of a new or modernized technological process, technology, product, or service. The 2016 GII (*The Global Innovation Index)* data, presented in Table 1, show that Russia is consistently improving its rank [6]. The improving of innovative activity efficiency is very much dependent on the effectiveness, employability, and tradeability of innovations.

Table 1. The dynamics of Russia's position in GII in 2014–2016.

Years	GII	Resources of innovation	The results of innovation	The effectiveness of innovation
2016	43	44	47	69
2015	48	52	49	60
2014	49	56	45	-

The innovations implemented in industrial production will allow for a considerably greater pace of modernization processes, and stronger competitiveness of the country. With the import substitution policy in place filling the consumer market in with the goods produced in Russia, both food and non-food, becomes very important. The development of the internal market and meeting the consumer demand allow solving a wide range of tasks, including but not limited to the employment generation, increase in the tax base, lowering social tension, industrial diversification process development, etc. The efficient functioning of commerce provides for the interconnection of trade and production, boosts business activity, stimulates the implementation of innovation.

The innovative development of trade features the industrialization of major commercial processes, establishment of the integrated information and technology product distribution system, implementation of advanced product retail modes, establishment of direct production – consumption links, building up regional electronic trade infrastructure, setting up a bank of electronic product passports [7]. The development of chain retailing brings forward the issues of commercial businesses management modernization. More rigorous competition creates the conditions stimulating further implementation of innovations. The innovative instruments which can be used in the activity of the trading organizations are organizational and managerial in character.

The innovative environment is based on the development of creative thinking, which secures bigger competitive advantage. Creative thinking is the search for novel, original problem solving approaches, creating the conditions stimulating innovative thinking, securing better commercial innovations output; it is performed in different - micro-, meso-, and macro-economic – levels [8].

Innovative activity in trade tends to be continuous, while taking the form of a stage-by-stage process. Regional factors can influence the innovative activity of the trade enterprises, so the discovery of the regional differences facilitates the development of the innovative processes in trade. When conducting the innovative activity in trade, one has to account for the specific peculiarities of the commercial sector. The innovative development in trade unfolds at various stages of the commercial and technological process.

Thus the innovations in trade are usually associated with the improvement of the existing service and commercial and technological novelties, while marketing innovations are also of high importance as the consumers influence the producers indirectly via trade. Moreover, major trade companies, being at the leaders of the sectoral innovative development also have major impact in this respect. Trade and commerce are characterized by the emergence of innovations, which are generated both inside them and in the related sectors, primarily, in production. The implementation of innovative management results in the higher added value of the service for the end consumers, more efficient trade and technological process, increased competitiveness, more precise satisfaction of the consumer demand, he emergence of new trade formats, modes, forms, methods and technologies, improved product quality, and wider product range [11]. To implement such innovations a trade enterprise has to use all of its potential, including talent, finance, technology, information, equipment, and marketing.

5 Research application

The level of innovation development is reflected in the general R&D potential of the industry, the core of innovations and the crucial component of the innovations environment for the trade companies, providing for the goods circulation. According to the State Statistics Service of the Russian Federation the share of the organizations performing innovations in technology in Russia in 2014, 2015, and 2016 equaled 10,1%, 9,9% and 9,3%[a], while the share of innovative products in the total volume of goods and service produced equaled 11,2 in 2013, 12,8% in 2014, and 12,9 in 2015. (Fig. 1.)

Sectoral components characterize the innovative potential and the prospects of the sector depending on its specific peculiarities. As the trade sector is where the commercial services are provided and the goods circulation is executed, the innovative activity in this field is not material in its form, which means that the novelties mostly concern the trade processes and technology, marketing, sales methods [12].

[a] The state statistics Committee of the Russian Federation

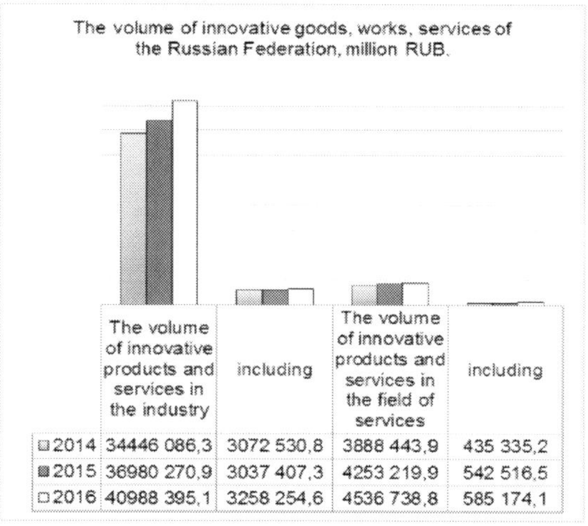

Fig. 1. The volume of innovative goods, works, services in the Russian Federation, million RUB[b].

The analysis of the investment in the development of trading activity by the retail trade companies shows that equipment modernization and renewal expenditure was cut down in 2014, the trend for advertising budgets cut down is also present (Fig. 2).

However, despite the complex economic situation, the share of *no expenditure* has also decreased. The dynamics of the innovative goods in the services sector, where trade belongs in terms of the Russian classification, is still very insufficient.

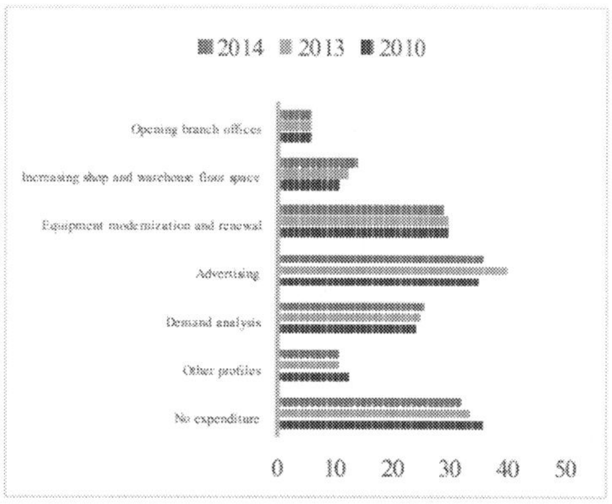

Fig. 2. The designation of development resources of the Russian trade organizations, Percent[c].

The Russian commodity circulation per capita, which is a proxy indicator of both nationwide and regional socio-economic state as of 01.01.2016 equaled 188,000 RUR. However, the regional distribution is extremely uneven. For instance, the Central Federal District has 237,700 RUR per capita, while the North Caucasus Federal District and the Siberian Federal District levels are 155,700 RUR and 141,800 RUR. Consequently, the distribution of the retail commodity turnover according to the territory shows the uneven character of development of this sector of the regional consumer market, thus manifesting the regional irregularity of the supply and demand. Taking into account that the difference in income per capita in different territories of the Russian Federation is negligible, one can conclude that such a lag in product turnover per capita can indicate the lack or insufficiency of retail trade enterprises in several regions of the country [13].

With the low floor space availability and insufficient product turnover the innovative development of the Russian retail trade is stagnating. The factors capable of kick starting the innovative development mechanisms in the sector are those related to the competition and the innovative development in industry, being the source of basic innovations.

On the other hand, the Russian industry sector, impacted by major market turbulence, may not serve as the source of the innovative development of the required level. Low rate of acquisition of technology by the business fails to generate the demand for innovation as the major factor of production and the source of growth and competitiveness. Private investors do not support public investment expenditure, which results in the Russian industry falling back in terms of the number of innovative companies and the share of innovative products and services.

The experts[d], consider the following factors to be the barriers for the innovative development of the industry:
- Low return on public and private investment in innovation;
- The innovative markets are low-competition markets due to low demand for innovation;
- Lack of institutional and legal regulations for the interaction of the innovation owners and investors;
- The inability to attract and keep talent, active in the field of innovation.

There exist both the general parameters, determining the sectoral innovative development, and a range of specific factors, doing so (Table 2).

Table 2. Factors of innovative development of Russian industry and trade.

Specific for the Russian industry the parameters of innovative development	Common factors of innovative development	Specific for Russian trade the parameters of innovative development
Low interest of the major businesses in innovation	Low demand for innovation	High sectoral competition
Innovative markets are low-competition ones	Insufficient level of human capital acquisition and development	The influence of foreign trade companies being the leaders of innovative activity in the market
Low return on	Irregular legal and	High importance

[b] Ibid
[c] Ibid

[d] National report on innovation in Russia-2016

public and private investment	institutional innovations infrastructure	of marketing innovations (consumers influence the producers indirectly via trade)

These groups of parameters should be reviewed separately to improve the efficiency of managerial decisions in the field of innovations policy.

6 Conclusions

The elicited peculiarities of innovation activity in industry and trade allow for the evaluation of the promising innovative development profiles for both sectors. For instance, due to the issues in the industry innovative development, Russian retailers should not expect gaining access to affordable Russian product innovations. However, process, technological, and marketing innovations can presently serve as the alternative instruments for the innovative development of trade.

In the present conditions the retailers try to cover several profiles at once to absolutely dominate the market: they select the sales spot with maximum customer traffic, attract many loyal customers, while cutting down the shop floor and equipment usage costs to minimum levels, increase the profitability of the company along with higher average purchase cost. Though, usually such « rush » pushes the customer satisfaction issue aside, which is a critical error for a business selling goods and services. Standard, traditional marketing instruments are trite for the modern consumer, who is lost among similar products, loses interest to the brand of the shop. This situation calls for the innovations, which would make radical change to the image of the company, while the major such innovations should come from marketing, the dominant aim of which is to plant the ideas capable of meeting the demands of the most discerning customer.

Innovative development can play a major role when such decisions are made. The stock, staff, and the shop floor constitute the most important resources for a retail trader, while the rigorous overall control over them is of crucial importance to maintain one's competitive position.

References

1..Y. Shumpeter, *Theory of economic development: monograph* (Progress, Moscow, 1982)

2. Clayton M. Christensen, Michael E. Raynor, *The Innovator's Solution: Creating and Sustaining Successful Growth* (Alpina Pablisher, Moscow, 2016)

3. P. Druker, *Business and Item innovations Druker* (Williams, Moscow, 2007)

4. D. Stiglic, Steep Dive: America and the New Economic Order after the Global Crisis (Eksmo, Moscow, 2011)

5. B. Hallier, *Modern Stores: A History of Technical Development* (Interekspert, Moscow, 2009)

6. Innovation in Russia. Official website. http://innovation.gov.ru/ru/node/75123.

7. I.A. Krasyuk, Practical Marketing, **6**, 8-18 (2015)

8. I.A. Krasyuk, N.G. Kuznetsov, Yu.Yu. Medvedeva, T.S. Stepchenko, V.P. Fedko, *Marketing, trading* (Rostov-on-don, 2013)

9. National report on innovations in RUSSIA-2016. http://www.rvc.ru

10. Federal State Statistics Service. http://www.gks.ru.

11. Yu.Yu. Medvedeva, Modern science: actual problems of theory and practice. Series: law and Economics, **4**, 68-73 (2016)

12. I.A. Krasyuk, Modern science: actual problems of theory and practice. Series: law and Economics, **12**, 87-91 (2014)

13. Yu.Yu. Medvedeva, *Synthesis of science and society in the solution of global problems of the modern world: collected articles of International scientific practical conference*, 68-75 (2016)

Impact of company capital structure on of its value growth estimation

Oksana Pirogova[1,*], and *Vladimir* Plotnikov[2]

[1]Saint Petersburg State Polytechnic University, Industrial Management Institute, Economy and Trade, Graduate School of Commodity and Service, Saint-Petersburg, Russia
[2]Saint Petersburg State University of Economics, Department of General Economic Theory, Saint-Petersburg, Russia

Abstract. The article analyses the influence of the use of commercial loans on the free cash flow generation based on cost-based management concept. It is shown that the use of commercial credit instead of the usual short-term lending reduces the force of the operational lever. On the other hand, when calculating the economic added value indicators, the use of conventional lending leads to the increase in the economic added value and as a result of the higher cost of growth. This effect is associated with the redistribution of cash outflows and changes in the level of risk for the stake-holders of the enterprise.

1 Introduction

Development of new approaches to the development of enterprises can not manage without a thorough analysis of the relationships that arise between the company and stakeholders in the course of business and the implementation of strategic management plans.

Nowadays more and more widely known gets cost management concept (Value Based Management - VBM). The concept of value-based (in the broader interpretation - the concept of value-based management) requires that an enterprise as an open social and economic system in its development should strive to maximize its value [1 - 4]. Despite the many advantages, the introduction of the concept of value in the work of modern domestic enterprises comes into serious difficulties. The difficulties of implementation of the concept in trade relate primarily to the following aspects [1, 2]:

- Developed nowadays methodological approaches and economic models are primarily intended for large enterprises, however, for small and medium enterprises, the concept is not very applicable;
- Indicators to measure the value of a company or its growth have different shapes and are often not comparable;
- Most of the indicators to measure the growth of the cost of a commercial product, contained in the methodology of the application a number of patented and protected know-how, which are known only to the manufacturer. In addition, the introduction of such products requires significant organizational and financial resources, including the adaptation of the product for a specific company and its licensing support.

Despite these difficulties, the cost-based approach to this stage of economics development is the most progressive approach that allows to manage the planning and evaluation of the enterprise on the basis of an integrated indicator - the cost. It is obvious that the sustainable development of enterprises can only be achieved on the basis of the successful implementation of the basic ideas of cost-based approach.

Development of common views and approaches to the management to ensure the fundamental value growth requires the development of models and methods to comprehensively link parameters such economic activities of commercial enterprise, the cost structure, structure and sources of funding, the level of economic risk. An important aspect of cost-based approach is the recognition of paid sources of capital, which had previously belonged to the free (eg, equity) or shareware (eg accounts payable). It should be noted that the question of taking into account the various sources of funding in the capital structure in assessing the value of the moment is controversial [2, 4, 5]. The main controversy related to the problem of accounting accounts payable.

2 Sources of venture capital

Sources of financing business activities are divided into two main groups: companies' own funds and borrowed funds. Almost all enterprises in varying degrees, use both sources of funding, however, its volume and structure are due to many factors such as enterprise size, growth dynamics, history of development, and others. For many businesses, especially retail feature is the extensive use of commercial credit to finance current operations. Commercial loans used to finance

* Corresponding author: kafedra17@rambler.ru

inventories and other elements of current assets. Many of the successful functioning enterprises have negative indicators of their own working capital, which is defined as the difference between current operating assets and liabilities, excluding loans and credits [5].

Moreover, in the structure of liabilities of modern retail enterprises accounts payable in the form of a commercial loan takes an important place. The following Table. 1 provides information on the capital structure of the largest domestic retail chains [6,7,8,9].

Table. 1 shows that, even in large retail chains, which in size can be compared with the average level of corporations and have access to enough cheap credit resources, the share of accounts payable is an average of 20 - 35%. For smaller commercial enterprises, such as the district consumer society, large stores, etc., the proportion of accounts payable may be even higher. Thus, the accounts payable appears essential element of equity trading businesses. Without the use of commercial credit traders it would be difficult to ensure high rates of economic activity and acceptable growth rate.

Table 1. Capital Structure of retail networks of the Russian Federation for 2015.

Structure capital	LLC Lenta, ths. Rub.	OKey LLC, ths. Rub.	PJSC Magnit, ths. Rub.	X5, ths. Rub.
Equity and liabilities	168485373	63420152	404192973	402115000
Own funds and others, equal to own	34145908	13328975	165140596	104718000
fixed liabilities	74178051	11494273	59481188	101545000
current liabilities	60161414	38596904	166642595	195852000
Accounts payable	49012520	27918597	88372216	195852000
In% of the capital	29.09%	44.02%	21.86%	48.71%

According to many authors [1, 2, 3] accounts payable should not be taken into account when considering the weighted average cost of capital of the enterprise. However, this approach is the situation when a significant part of the enterprise working capital needs to be excluded from the analysis, and therefore should not be considered when determining the cost drivers.

One would agree with the authors of works only to the ex-tent that it is often this type of capital is adopted as shareware and can not count. However, the existence of accounts receivable of this size has a significant impact on the distribution of risks between stakeholders competing for free cash flow enterprise.

Thus, at the present stage of commercial credit being one of the most important sources of finance for enterprises, can have a significant influence on the formation and distribution of free cash flow, as well as the redistribution of financial risks between the stakeholders of the enterprise.

Short-term payables – capital's structure element of the enterprise. Analysis of short-term debt should be carried out in accordance with the structure of accounts payable. In accordance with the accounting rules are divided into short-term liabilities [5]:

- Short-term borrowings (with maturities of less than one year)
- Payables and short-term creditor obligations;
- Payable to participants (founders by income payments)
- Revenue of the future periods;
- Reserves for future expenses;
- Other short-term liabilities.

As can be seen, accounts payable is an element of short-term debt. In terms of value concepts for commercial enterprises significant first and second elements are represented. Short-term borrowings in accordance with the cost-based approach are taken into account in the average cost of capital (WACC) [1,3,4].

The very payables (short-term credit liabilities) is also not uniform in structure and includes the following main elements [5]:

- Payable to suppliers and contractors - the amount owed to suppliers and contractors for delivered and not paid for tangible assets (work performed and services rendered);
- Debt to the staff of the organization - the Organization's debt to workers on wages;
- Debts to the state budget funds - the amount of debt on insurance payments, contributions for compulsory pension insurance, etc .;
- Debt to budget - debt company for all types of payments to the budget;
- Advances received - advance payments received by the organization for the upcoming delivery of goods (works, services);
- Debts to other creditors - debt to insurance companies for property insurance and workers' organizations; debt to the accountable persons; unpaid and not listed penalties (fines, penalties).

For commercial enterprises it is common when the debts to suppliers and contractors amounts to 80-95% of the total ac-counts receivable. Such amounts receivable on settlements with suppliers due to the widespread use of commercial credit in the implementation of operations. Commercial loans in accordance with the Civil Code is in our case in the form of deferred payment or installment [10]. The use of commercial credit due to the difference in the duration of the buyer's and supplier's production and financial cycles.

In the analysis of the trade capital of the enterprise from the point of view of cost-based approach to the fore issues come such elements of capital like cost (ie the cost of raising capital), as well as the risk that generates some form of capital to the enterprise and the creditor.

3 Methods of assessment of payables' attraction cost

It is well known that the cost of individual items in ac-counts payable is different. So payable to personnel, the state budget funds, tax arrears are actually virtually free sources, as this are temporary funds, which were formed as a result of activity of the enterprise, but not paid due to deadline hasn't already come. However, the proportion of these elements in the structure of accounts payable of the trade enterprise is not great.

With debts to suppliers and contractors it is more complicated. Traditionally, financial analysis, it was believed that this type of payables forms shareware borrowed funds. In the future, with the development and introduction of cost-based approach, it became clear that the gratuity of this element is rather conventional.

Many authors suggest to estimate the payment for accounts payable, depending on the size of the potential discounts, which can get a buyer, paying for the goods delivered earlier than proposed in the supply agreement. [5] This means that the supplier in the event of payment delay increases the cost of the delivered goods. The greater the delay - the greater the margin, or vice versa, when paying for goods in a short period of time the buyer receives a discount on goods. As a rule, this discount is tied to the size of bank interest rate on short-term loans.

In [5] proposed the following formula for estimating the value of payables to suppliers:

$$k_{кр} = \frac{d}{100-d} \times \frac{360}{t-k} 100\% \quad , \qquad (1)$$

where d - discount, which can be obtained by paying the goods within days,%;

t - postponement of payment of the goods provided by the supplier, days;

k_{cr} - product payment term in the case of discount days.

The formula (1) consists of two elements, which reflect the relative discount and the term of its provision, which allows to bring the cost of the source of funds to the standard value.

Based on the above, the weighted average cost of capital of commercial enterprise in mind the high homogeneity of accounts payable can be determined as follows:

$$WACC = k_e w_e + (1-n)(k_d w_d + k_{кр} w_{кр}) \quad , \qquad (2)$$

where w_e, w_d, w_{cr} - the cost of equity, debt capital and accounts payable respectively.

k_e, k_d, k_{cr} - share in the total capital structure of equity, debt capital and accounts payable respectively.

n - income tax rate.

However, the account payable in the assessment of the cost of trading company is not limited to the calculation of the weighted average cost of capital. The fact that emerging hidden charges for the use of commercial credit impact on other economic indicators such as profits from the sale, as well as operational and dual levers.

It is important to note that cash flow in the form of a payment for a commercial loan, which is obtained by vendors, there is at the stage of repayment obligations for delivered goods to suppliers. Thus, the fee for the use of this type of capital is inherently (implicitly) included in these payments. [11] The mechanism of the hidden cash outflow as a fee for the use of trade credit is demonstrated in Figure 1. In addition, payment of the commercial value of the loan takes place before payment of the basic interest on trading business' debt, as well as the maturity of the enterprise cost. Therefore, commercial loans, affecting the amount of profit from the sale, also affects the strength of the operating and the conjugate lever.

3 Evaluation of the impact of commercial credit on the strength of the operating and financial leverage

To assess the impact of commercial credit will take a number of assumptions, to assess the impact of commercial credit:

1. When using commercial credit at the price per unit of output the value of the potential discounts is added, which can get a commercial enterprise from the supplier, i.e. increases turnover at purchase prices.
2. If you use a commercial loan commercial enterprise does not increase the sales volume and value of retail turnover due to competition in the market. The increase in the retail price of the goods immediately lead to lower sales volumes.

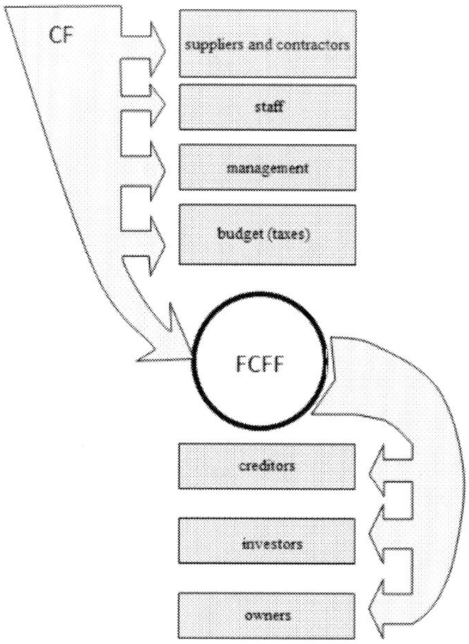

Fig. 1. Scheme of the distribution of company's cash flow.

In [11, 12] have shown that by comparing the dual forces of the lever based on commercial loans and excluding them() can be obtained the following expression:

$$J_{cp} - J'_{cp} = \frac{EBIT+FC}{EBIT-I} - \frac{EBIT'+FC}{EBIT'-I} = \\ = \frac{EBIT+FC}{EBIT-I} - \frac{EBIT - T_{pц} \cdot Y_{cк} + FC}{EBIT - I - T_{pц} \cdot Y_{cк}} \quad , \qquad (3)$$

where $EBIT$ – earnings before interest and taxes (adjusted profit on sales), excluding commercial loans;

$EBIT$ ' – earnings before interest and taxes (adjusted profit from sales) from the register of commercial credit;

T_{rc} – turnover at retail prices;

Y_{sk} - discounts level per one ruble turnover at retail prices;

FC - fixed costs at a given time interval of the enterprise;

I - interest on the loan.
It can be shown that this indicator is positive, the sign of the indicator depends on the value of the numerator of the expression (3), which is calculated by the formula:

$$(FC + I)\times(T_{pu} \cdot Y_{cx}) \qquad (4)$$

This expression indicates that the value of the dual arm, calculated taking into account of a commercial loan should be less than when commercial credit is not included and accepted as payables.

The following Table 2 shows the dual lever calculations performed on data for the year 2015 of LLC "Lenta". The calculations were carried out based on the basic formula for determining the operational and financial leverage, as the ratio of the growth rates of key performance indicators for the reporting year (from sales revenue, gross profit, sales profit, net profit).

Table 2. Calculation of the dual arm with and without taking into account the commercial loan according to the activity.

Parameter	Excluding commercial loans	
	2015	2014
Revenues, ths. Rub.	288201963	220337828
Gross profit, ths. Rub.	69519265	55730544
Selling, general and administrative expenses, ths. Rub.	49696179	39334077
Operating profit, ths. Rub.	19823086	16396467
Interest on loans, ths. rub.	11958615	6782822
Adjusted pre-tax profit, ths. Rub.	8881322	7699325
Adjusted net pr ibyl, ths. Rub.	6291576.8	6159460
revenue growth rate	0,308	
The growth rate of gross profit	0,247	
The growth rate of profit on sales	0,209	
The growth rate of the adjusted net profit	0,0214	
The impact of operating leverage	0,845	
The impact of financial leverage	0,103	
The strength of the impact of the dual arm	0,087	
Revenues, ths. Rub.	288201963	220337828
Gross profit, ths. Rub.	76416207	59810907
Selling, general and administrative expenses, ths. Rub.	49696179	39334077
Operating profit, ths. Rub.	26720028	20476830
Interest on loans, ths. rub.	18855557	10863185
Adjusted pre-tax profit, ths. Rub.	8881322	7699325
Adjusted net pr ibyl, ths. Rub.	6291576.8	6159460
revenue growth rate	0,308	
The growth rate of gross profit	0,278	
The growth rate of profit on sales	0,305	
The growth rate of the adjusted net profit	0,0214	
The impact of operating leverage	1,098	
The impact of financial leverage	0,070	

Thus, it is clear that if, instead of accounts payable, the company would use the normal short-term loan, the strength of the impact of operating leverage has increased from 0,845 to 1,098, but the impact of the financial strength and the dual levers - decreased from 0.103 to 0.070 and 0.087 to 0.077 respectively.

This change is explained by the fact that when using a conventional loan the cash outflow occurs at the stage of payment of the loan, and in the case of accounts payable outflow occurs at the stage of the calculation with suppliers.

This factor can be used in the planning of the company: in the conditions of intensive growth of trade turnover widest use of accounts payable is possible, it will provide profit growth and other indicators of a higher rate. At the same time, in terms of sales fall more appropriate is the use of short-term loans instead of short-term debt. This will provide a slower drop in profits and other indicators of the outcome of the enterprise.

Conclusion

In assessing the activity of the enterprise is an important factor to attract valuation of accounts payable. An adequate account of the cost of accounts payable as a result of the refusal of purchased goods' discounts to evaluate its contribution to the profits of the enterprise and the flexibility to man-age the pace of profit growth depends on the economic situation.
Another important aspect that requires further investigation – in case of changing of credit conditions changes the value of the weighted average cost of capital, the value of the invested capital and adjusted earnings before tax. This in turn leads to changes in the economic added value.

The research is financed by the Government task of Minis-try of education and science of the Russian Federation № 26.3546.2017/PCH "Development fundamentals of analysis and prediction of structural and dynamic parameters of the regional economy are based on the integration of the Russian and world experience of management of territorial development and modern scientific doctrines".

References

1. T.V. Teplova, HSE (2000)
2. M.V. Kudina (FORUM, INFRA-M, 2010)
3. F. Modigliani (Delo, 1999)
4. T. Copeland, T. Kohler, J. Murin (Moscow, 1999)
5. V.V. Kovalev, Prospect (2014)
6. https://www.e-disclosure.ru/portal/files.aspx?id=9679-&type=3
7. https://www.x5.ru/ru
8. http://www.dixygroup.ru
9. http://www.lenta.com
10. RF Civil Code (2014)
11. O.E. Pirogova, Bulletin of St. Petersburg State University of Economics, **6** (2014)
12. O.E. Pirogova, Proceedings of the St. Petersburg State University of Economics, **2** (2014)

DOI: 10.1051/shsconf/20173501054

Industry and environment: modeling the global production impact on CO2

N.I. Didenko[1,*], *D.F.* Skripnuk[1], and *O.V.* Mirolyubova[1]

[1]Peter the Great St. Petersburg Polytechnic University, St. Petersburg, Russia

Abstract. The article analyses the impact of global industrial production on CO2 emissions. The global economy comprises such industrial sectors as the primary economic sector (mining, wood processing industry), various processing industrial sectors, energy sector, housing and communal services and transportation sectors. We present the methodology of modelling global economy industrial impact on CO2 emissions. ADL-model (autoregressive distributed lags model) has been chosen as a theoretical basis. Eight variables affecting CO2 emissions per unit of production were chosen as exogenous: the reduction of forest area; the output of energy industries, utilities and extraction industries; wood processing industry; the volume of goods transported; the volume of transportation and communications sectors; the length of roads; the output of industries producing material resources. The primary statistic information on the endogenous and exogenous variables in many countries was collected. The endogenous and exogenous parameters for global economy based on the initial statistic information were defined as the geometric mean indices in the year *t*. The article presents a model of the global economy industries impact on CO2 emissions. The conclusions about the impact of industrial factors on CO2 emissions are drawn.

1 Introduction

The current state of the environment today is described as critical and there are certain factors that can lead to global environmental disasters. The first is pollution, environment contamination, the depletion of oxygen in the atmosphere, ozone gaps, etc. The second is the existence of radiation hazards, especially nuclear weapons, which is the most dangerous of all existing types of weapons. Its application can cause disastrous and irreversible damage to the natural environment. The third is the excessive use of available natural resources as well as traditional sources of energy and raw materials. The fourth factor is a rapid and virtually unregulated population growth. According to the forecast, the world population will reach 13 billion people by 2040. It is clear that with such a growth of the world population, the negative effects of human activities will sharply increase. Not only the population growth, which contributes to the negative effect on the environment, but also the need for more intensive extraction of nature resources leading to the larger accumulation of industrial waste, result in adverse effects on the biosphere. These four factors that lead to global environmental disasters, depend majorly on the industrial production in the global economy. Industry has an impact on the environment, and one of the influential factors is the growth of industrial production. The industrial development, the constant growth of the world population and consumption of natural resources lead to a continuous flow of various anthropogenic substances and compounds into the environment. Due to the production growth, the volume of emissions into the atmosphere has been steadily increasing. The concentration of carbon dioxide (CO2) in the atmosphere in the last decades is rapidly increasing.

2 Overview of the relevant problem research

According to up-to-date research, the level of carbon dioxide in the atmosphere is the highest in the last 800 thousand years. Because of the global industrial production emissions to the atmosphere have become a global environmental problem. With the accelerated pace of industrial production increases the environmental threat due to the various efforts to satisfy human needs. The only solution is to slow down the growth of industrial production at a level that does not harm the environment, which can be hardly possible. It is impossible to give up industrial growth and development, while maintaining gross national income at a constant level, as this will impact the quality of life.

* Corresponding author: didenko.nikolay@mail.ru

Why is it important to monitor how much CO2 the global industry emits into the atmosphere? The answer is that this gas is largely responsible for the increase of the greenhouse effect on our planet. The global balance of industrial emissions of carbon dioxide was measured for the first time in 1958 by Dave Keeling, the American scientist from the Scripps Institute of Oceanography. The researchers analyzing data on industrial CO2 emissions, the main greenhouse gas, united their efforts in Global carbon budget project. They study the main sources of atmospheric pollution, i.e. fuel and energy complex, processing industry and transport.

Grossman and Krueger [1] conducted an empirical study uncovering an inverted U-shaped relationship, also known as the EKC, between income and local air and water pollutants. Subsequent studies [2], [3], [4] were dedicated to measuring and analyzing the relationship between the environment, economy and industrial policies. Grossman and Krueger [5], Chimeli and Braden [6] examined the models of the effect of industrial production on the environment. Lahiri [7] examines the influence of the factors of income and environmental quality on the growth rates for open economies. The results are different from the standard convergence result for identical economies. This asymmetric outcome is then analyzed in terms of the scale, composition, and intensity effects known to underlie Kuznets' environmental curve. Stokey [8] explored how different production technologies affect the environment, while May, Stahl and Taisch [9] investigated how a firm's choice of alternate energy resources affects the environmental quality. Xu [10] analyzed the factors of competitiveness of environmentally friendly products and made the conclusion that the environmental standards reduce the international competitiveness of environmentally friendly products. Abatement technologies, analysis of the concentration of ground-level ozone and its impact on ecosystems are considered in the works [11], [12]. Botta and Koźluk [13] studied the quantitative evaluation of the environmental policies in industrialized countries. The interaction of the environment, global industrial CO2 emissions and economic development is studied on the basis of different models. Liu and Ulgiati [14] have built a network model of the interaction of economy, energy and the environment. Robison [15], Ederington and Minier [16] developed a model of the influence of environmental factors and industrial emissions on the trade balance. Cole and Elliott [17] study the degree of the influence of capital, labor and industrial emissions on the foreign trade. Arce, López and Guan [18] consider the model of eco-efficient countries. The so-called post-China countries, distinguished by low wages and high economic growth, will replace China as the "world's factory". The aim of the paper is to assess the effect of these changes on global CO2 emissions. Meng, Liu, Guo, and Tao [19] developed the model quantifying the impact of international trade on the concentration of harmful substances in the urban economy. Sakamoto and Managi [20] defined energy efficiency and environmental quality as factors of comparative advantages in the industries, showing the empirical results of the impact of eco-efficiency indicators on industrial production. The reviews of the works of Carson [21], Vyboldina and Fedoseev [22], Didenko and Skripnuk [23], Tcvetkov and Strizhenok [24] indicate the need for further structural theoretical models of the relationship between industrial development and environmental situation.

3 The approach to modeling the influence of industrial development on CO2 emissions

3.1 Theoretical model and its characteristics

Autoregressive distributed lags (ADL-model) was chosen as a theoretical model, in which the current values of the series depend on the past values of the series, and from the current and past values of other time series. The model is generalized in the case of several exogenous variables. In general, we can assume that all exogenous variables are included in the model with the same number of lags, with the possible exception of a lag of some variables. ADL - model is as follows

$$y_t = a_0 + \sum_{i=1}^{n} a_i y_{t-i} + \sum_{j=0}^{q1} b_j x_{t-i}^1 + \ldots + \sum_{j=0}^{qk} b_j x_{t-i}^k + E_t \quad (1)$$

where k is a number of exogenous variables; q is a number of lags in the i-th exogenous variable (i = 1,2, .. k); n is a lag depth for endogenous variable; Et is the residues developing the white noise.

This model suggests that, if at a certain time t, the independent variable x_t changes, then this change will impact the value of the variable y_t at the following time points.

3.2 Data

The initial data was processed following two stages: a) the analysis and selection of endogenous and exogenous variables that correspond to the analyzed process and reflect the essence of the problem; b) the collection of statistical information for the selected variables. The analysis and the choice of endogenous and exogenous variables. The endogenous variable, i.e. the emissions CO2 per unit of production, g/USD (grams per USD in constant prices) is set, per se being a specific target. We denote emissions per product unit at a time t point as y_t^2. Our task is to build the relationship y_t^2 on the past values of the series y_{t-j}^2 and of the past values of other exogenous variables.

Exogenous variables were y_t^2 chosen from the following list: X_{1t}^1 - GDP, World, US $ Per Capita; X_{2t}^1 - Employed Population, World, Unit 000; X_{3t}^1 - Economically Active Population, World, Unit 000; X_{4t}^1 - Exports (fob) by Commodity + Imports (cif) by

Commodity, World, USD million; X_{1t}^{2} - Energy, Utilities and Recycling: Production (turnover) MSP, USD million, USD million; X_{2t}^{2} - Primary Materials. Forestry, Production (turnover) MSP, USD million, Million net ton-kilometres; X_{3t}^{2} - Road Freight Traffic, Million net tonne-kilometres; X_{4t}^{2} -Transport and Communications: Production (turnover) MSP, USD million, USD million; X_{5t}^{2} - Road Network, Kilometres; X_{6t}^{2} - Material Resource Productivity, USD per kg in constant prices; X_{1t}^{3} - Greenhouse Gas Emissions from Energy, 000 tonnes of CO2 equivalent; X_{2t}^{3} - Extraction of Crude Petroleum and Natural Gas, USD million; X_{3t}^{3} - Mining of Coal and Lignite; Extraction of Peat, USD million; X_{4t}^{3} - Energy, Utilities and Recycling: Production (turnover) MSP, USD million; X_{5t}^{3} - Railway Freight Traffic, Million tonne-kilometres; X_{6t}^{3} - Waste Generated by Manufacturing, 000 tonnes; X_{1t}^{4} - Machinery for Food, Beverage and Tobacco Processing: Production (turnover) MSP, USD million; X_{2t}^{4} - Agricultural and Forestry Machinery: Production (turnover) MSP, USD million; X_{3t}^{4} - Animal Husbandry. Industrial: Primary Materials, USD million; X_{4t}^{4} - Arable Land, 000 sq km; X_{5t}^{4} - Animal waste - Production, Terajoules; X_{6t}^{4} - Waste Generated by Agriculture, Forestry and Fishing, 000 tonnes; X_{1t}^{5} - Methane emissions (kt of equivalent); X_{2t}^{5} - Nitrous oxide emissions (thousand metric tons of equivalent); X_{3t}^{5} - Greenhouse Gas Emissions, 000 tonnes of equivalent; X_{4t}^{5} - CO2 Emissions from the Consumption and Flaring of Fossil Fuels, 000 tonnes; X_{5t}^{5} - Waste Generated by Electricity, Gas, Steam and Air Conditioning Supply, 000 tonnes; X_{6t}^{5} - Waste Generated by Households, 000 tonnes; X_{1t}^{6} - Annual fresh water withdrawals, (% of internal resources); X_{2t}^{6} - Agricultural Materials and Live Animals Wholesale: Retail and Wholesale, USD million; X_{3t}^{6} - Farm Animal Feeds: Production (turnover) MSP, USD million; X_{4t}^{6} - Hydrological Disasters, USD million; X_{5t}^{6} - Total population supplied by water supply industry, %; X_{6t}^{6} - Renewable freshwater resources, million cubic metres; X_{7t}^{6} - Net freshwater supplied by water supply industry, million cubic metres; X_{1t}^{7} - Agricultural Land, 000 sq km; X_{2t}^{7} - Total Population, 000 per.; X_{3t}^{7} -Wood and Paper Products: Production (turnover) MSP, USD million; Y_{t}^{7} - Forest Land, 000 sq km.

As a result of the analysis of the list of indicators, the following seven variables impacting y_t^2 - the emissions CO2 per production unit were chosen as exogenous: Y_t^7; $X_{1t}^{2}, X_{2t}^{2}, X_{3t}^{2}, X_{4t}^{2}, X_{5t}^{2}, X_{6t}^{2}$.

Data. The data was collected for the period from 1998 to 2015. It contains the values of endogenous and exogenous variables for the corresponding year in different countries. The data was taken from the following sources: Euromonitor Passport Database, http://www.euromonitor.com/; World Bank Open Data, http://data.worldbank.org/.

3.3 Methodology of the model's empirical verification

The structural form of the model is the following:

$$\left\{y_t^2 = f\left(Y_t^7, X_{1t}^2, X_{2t}^2, X_{3t}^2, X_{4t}^2, X_{5t}^2, X_{6t}^2\right)\right\} \quad (2)$$

The empirical verification of the model consisted of several stages. Below are the main stages.

The first stage of the method was the development of the above form of ADL-model with the given endogenous and chosen exogenous variables.

For the structural form of ADL- model (2) the reduced form is as follows:

$$y_t^2 = a_0 + \sum_{i=1}^{T} a_i y_{t-i}^2 + \sum_{i=1}^{T} b_{t-1-i} Y_{t-1-i}^7 + \sum_{i=1}^{T} \sum_{k=1}^{6} c_{k,t-1-i} X_{k,t-1-i}^2 \quad (3)$$

The second stage of the method was the calculation of endogenous and exogenous variables of the model for the global economy. The initial statistic information on endogenous and exogenous variables was collected for different countries. Endogenous and exogenous parameters for global economy were defined as geometric mean indices for different countries in the year *t*. The calculation was performed in Microsoft Excel. This calculation resulted in developing time-series data for endogenous and exogenous variables for the period from 1997 to 2015.

At the third stage of the method, a test of temporal series of variables for stationarity was carried out using the Dickey–Fuller test. The Dickey–Fuller test is finding a coefficient of the autoregressive equation:

$$y_t = a\, y_{t-1} + \varepsilon_t \quad (4)$$

where y_t is a temporal series, ε_t is an error.

If $|a|<1$ than the series is stationary. If $a=1$ than the process has a unit root and in this case the series is non-stationary and is an integrated temporal series of the first order [25].

The fourth stage included a test of autocorrelation of an endogenous variable for selecting lags which have a strong correlative relationship with the index value in the last period; a test of strength of the endogenous variable's relationship with exogenous variables; a test of the exogenous variables for multicollinearity; a test of significance of the autocorrelation coefficients using the Ljung–Box test; a test of pair correlation coefficients for significance using a Student's t-test for indices.

At the fifth stage, the model coefficients were determined using the regression analysis. The significance and coefficients of the regression equation were tested. The equation was tested for certainty using an F-test and the determination coefficient. The significance of the regression equation's coefficients was tested using a Student's t-test. On the basis of the obtained coefficients, theoretical values of the endogenous variables of the model in a certain year were determined. The conclusions were made.

4 Empirical testing of the model

The quantitative testing of the model is carried out according to the developed method which is given above.

Endogenous and exogenous variables are given in Table 1.

Table 1. Endogenous and exogenous variables of the model.

	y^2_t	y^7_t	x^2_{1t}	x^2_{2t}	x^2_{3t}	x^2_{4t}	x^2_{5t}	x^2_{6t}
2015	755.8	3998. 4	10 876 990.0	270 392.5	13 628 785.7	9 961 604.0	38 934 149.6	00.9
2014	848.8	40 005.4	12 749 580.0	283 001.4	13 325 858.5	10 354 430.0	38 468 770.3	00.9
2013	944.1	40 040.8	12 927 950.0	274 270.9	12 625 047.9	10 032 770.0	37 865 778.3	00.8
2012	1 111.7	40 073.8	12 681 730.0	262 075.0	12 793 170.7	9 656 216.0	37 089 303.3	00.8
2011	1 183.1	40 106.7	12 300 570.0	263 658.3	11 927 459.4	9 594 466.0	36 632 863.4	00.8
2010	1 466.1	39 867.3	10 182 960.0	232 551.5	10 950 618.8	8 507 177.0	36 196 054.8	00.8
2009	1 684.6	39 899.6	8 531 825.0	203 412.3	10 169 687.2	7 738 202.0	35 629 943.2	00.7
2008	1 751.6	39 931.9	10 470 040.0	220 305.4	9 912 228.0	8 361 470.0	34 899 714.6	00.8
2007	1 858.6	39 964.2	8 435 699.0	209 197.7	7 694 557.3	7 538 459.0	34 422 332.7	00.7
2006	2 305.4	39 996.4	7 358 770.0	172 286.4	7 274 432.0	6 651 129.0	33 842 412.1	00.7
2005	2 797.4	39 997.7	6 371 666.0	151 341.1	6 916 817.7	6 068 108.0	33 346 998.2	00.7
2004	3 374.4	40 041.6	5 143 339.0	141 165.5	6 682 466.2	5 522 844.0	32 879 283.3	00.7
2003	4 501.3	40 085.5	4 240 525.0	121 836.7	6 289 926.3	4 879 918.0	32 545 009.2	00.7
2002	5 407.6	40 129.4	3 614 337.0	108 938.5	6 103 291.3	4 295 864.0	32 080 270.7	00.7
2001	7 550.9	40 173.3	3 645 993.0	106 174.2	5 901 409.5	4 159 924.0	31 601 067.8	00.7
2000	8 163.9	40 217.2	3 680 902.0	111 296.8	5 748 609.5	4 130 435.0	30 681 228.0	00.6
1999	12 709.9	40 280.6	3 099 467.0	112 658.6	5 562 868.2	3 927 919.0	30 334 900.1	00.6
1998	18 346.8	40 351.5	2 925 955.0	109 677.0	5 331 136.1	3 762 397.0	29 938 785.4	00.6
1997	21 717.7	40 422.4	3 114 773.0	115 947.4	5 115 996.2	3 726 223.0	29 684 185.1	00.7

The test of temporal series for stationarity was carried out using the Dickey–Fuller test.

The Dickey–Fuller test aimed to find the coefficient of the first-order autoregressive equation in Excel. For a temporal series with an endogenous variable – CO2 emissions per one production unit, gr./USD (grams per USD in constant prices) (Y_{2t}) the coefficient of the autoregressive equation a=0.67829393 (t_T=2.39239415 ≤ t_p=26.015215). Consequently, the temporal series is stationary.

Similarly, for exogenous variables:

For Y_{7t} - Forest Land, 000 sqkm., the coefficient of the autoregressive equation a=0.718665(t_T =2.39239415 ≤ t_p=6.306241). Consequently, the temporal series is stationary.

For X^2_{1t} - the coefficient of the autoregressive equation a=0.59011572 (t_T =2.39239415 ≤ t_p =22.6332371). Consequently, the temporal series is stationary. For X^2_{2t} - the coefficient of the autoregressive equation a=0.47480575 (t_T =1.6746759 ≤ t_p =18.2106505). Consequently, the temporal series is stationary. For X^2_{3t} - the coefficient of the autoregressive equation a=0.54263514 (t_T =1.91391532 ≤ t_p=20.812172). Consequently, the temporal series is stationary. For X^2_{4t} - the coefficient of the autoregressive equation a=0.61046453 (t_T =2.15315473 ≤ t_p =23.4136935). Consequently, the temporal series is stationary. For X^2_{5t} - the coefficient of the autoregressive equation a=0.48158869 (t_T =1.69859985 ≤ t_p =18.4708027). Consequently, the temporal series is stationary. For X^2_{6t} - the coefficient of the autoregressive equation a=0.67151099 (t_T =2.36847021 ≤ t_p=25.7550629). Consequently, the temporal series is stationary.

According to the results of the Dickey–Fuller test, all temporal series are stationary.

The test of exogenous variables for multicollinearity was carried out in Excel. Correlation coefficients are given in table2.

Table 2. Correlation coefficients of exogenous variables.

	Y7	**X2-1**	**X2-2**	**X2-3**	**X2-4**	**X2-5**	**X2-6**
Y7	1						
X2-1	- 0.65167 53	1					
X2-2	- 0.62433 32	0.98825 107	1				
X2-3	- 0.59748 91	0.95950 49	0.97252 26	1			
X2-4	- 0.68759 14	0.99151 663	0.69421 101	0.97276 801	1		
X2-5	- 0.63362 96	0.95221 146	0.66163 634	0.97056 943	0.67884 75	1	
X2-6	- 0.56508 71	0.84833 017	0.56684 673	0.89165 416	0.57399 322	0.59673 6593	1

Table 3. Calculated values of the student's t-test for the indices.

Y7	Y7	X2-2	X2-4	X2-5
X2-2	3.295352871			
X2-4	3.904449189	3.976686		
X2-5	3.376937014	3.638178	3.811846551	
X2-6	2.824031177	2.836979	2.890157728	3.066168975

With the significance level α=0.05, numbers of the degree of freedom are n-2=17, T_{table}=2.11. Since in all cases $T_{calc} > T_{table}$, the coefficients are considered significant.

Lags of the endogenous variable, which have a strong correlational relationship with the index value in the last period, were selected by calculating autocorrelation coefficients. The test of significance of the autocorrelation coefficients was carried out using the Ljung–Box test.

According to both the Box-Pierce test and the Ljung-Box test, if $Q > \chi^2_{1-\alpha,m}$ the coefficients are considered significant, $\chi^2_{1-\alpha,m}$ is determined according to the table. Autocorrelation analysis was done with the Statistica software. The results are given in table 4.

Table 4. Autocorrelation coefficients of the endogenous variable.

Autocorr.	St. error	Ljung-Box Q	p
0.805588	0.240906	11.18228	0.000827
0.565891	0.231455	17.15996	0.000188
0.316687	0.221601	19.20225	0.000249
0.081502	0.211289	19.35104	0.000672
-0.136611	0.200446	19.81553	0.001356
-0.205442	0.188982	20.99731	0.001841
-0.269033	0.176777	23.31343	0.001507
-0.325823	0.163663	27.27677	0.000636

Since the Q-statistic of Ljung-Box is more accurate, it is more preferable for the analysis. Autocorrelation analysis in Statistica, apart from autocorrelation coefficients, also automatically calculates the Q-statistic of Ljung-Box and significance for each coefficient. The significance test by the Q-statistic of Ljung-Box equals to $> \chi^2_{0.95,3}$. The coefficients are significant. The endogenous parameter y^2_t signifies the dependence from one previous period y^2_{t-1}.

The lags of each exogenous variable closely connected with the endogenous variable y^2_t were chosen by calculating correlation coefficients. The significance test of correlation coefficients was carried out using the Ljung-Box Q test.

For endogenous variable y^2_t with only one exogenous variable y^7_{t-1} (Forest Land), with lag *t-1*, the correlation coefficient was equal to 0.798856 and significant (p=0.000914, the Q-statistic of Ljung-Box =10.99617$> \chi^2_{0.95,3}$). Endogenous variable y^2_t signifies the dependence from one previous period of exogenous variable y^2_{t-1}.

The correlation coefficients of endogenous variable y^2_t with other lags of exogenous variables are less than |07| and the coefficients are not significant.

Taking into account previous analyses, the ADL-model has the following form:

$$y^2_t = a_0 + a_1 y^2_{t-1} + a_2 y^7_t + a_3 y^7_{t-1} + a_4 x^2_{2,t} + a_5 x^2_{4,t} + a_6 x^2_{5,t} + a_7 x^2_{6,t} \quad (5)$$

The model's coefficients were found in Excel using the regression analysis. Besides, the certainty check of the regression equation was carried out basing on a Fisher's F-test, the equation coefficients were calculated by the OLS method, the coefficients' certainty was assessed on the basis of the Student's t-test.

For equation (5) F_{table} 9.01 < F_{actual} 62.503378 => the model is statistically significant, the regression equation is reliable for the significance level of α=0.05, R^2 = 0.97765481 and the equation has the form of

$$y^2_t = 40717.9548 + 0.83918843 y^2_{t-1} - 0.1956906 y^7_t - 0.8451041 y^7_{t-1} - 0.0330472 x^2_{2,t} + 0.00100218 x^2_{4,t} - 0.0001067 x^2_{5,t} + 4854.7012 x^2_{6,t} \quad (6)$$

5 Conclusions

CO2 emissions per one production unit depend on the previous period. The dependency is positive, which means the more emissions there are during the current period, the more there will be in the next one. The endogenous parameter is also affected by another parameter – deforestation. It should be noted that the dependency between them is negative, both in current and previous periods, which shows that with deforestation, CO2 emissions per one production unit increases. The indices - the volume of freight traffic and the volume of producing and reprocessing energy were excluded from the analysis due to their strong correlation with other exogenous variables, therefore they are not available in the equation. The rest of the variables signified the following influence on the volume of CO2 emissions. The indices of the transport and communication industries as well as the capacity of the natural resources industry when increasing lead to an increase in CO2 emissions. The indices of the roundwood production as well as expanding the road system, on the contrary, lead to emissions decrease. Previous periods of all exogenous parameters, except for the index of deforestation, do not influence the endogenous variable.

The paper is based on research carried out with the financial support of the grant of the Russian Science Foundation (Project No. 14-38-00009. The program-targeted management of the Russian Arctic zone development). Peter the Great St. Petersburg Polytechnic University.

References

1. G.M. Grossman, A.B. Krueger, *Environmental Impacts of a North American Free Trade Agreement* (NBER Working Paper3914, 1991)
2. R. Lopez, J. Environ. Econ. Manage, **27**, 163 (1994). DOI: http://dx.doi.org/10.1006/jeem.1994.1032
3. K.E. McConnell, Environ. Develop. Econ., **22**, 383 (1997). DOI: https://doi.org/10.1017/S1355770X9700020X
4. J. Andreoni, A. Levinson, J. Pub. Econ., **80**, 269 (2001).http://econweb.ucsd.edu/~jandreon/Publications/JPubEKuznets.pdf
5. G.M. Grossman, A.B. Krueger, Q. J. Econ., **110**, 353 (1995). DOI: https://doi.org/10.2307/2118443
6. A.B. Chimeli, J.B. Braden, Environ. Develop. Econ., **14**, 541 (2009). DOI: https://doi.org/10.1017/S1355770X08004981
7. B. Lahiri, East. Econ. J., **43**, 104 (2017). DOI: 10.1057/eej.2015.3
8. N.L. Stokey, Inter. Econ. Rev., **39**, 1 (1998). DOI: 10.2307/2527228
9. G. May, B. Stahl, M. Taisch, Appl. Ener., **164**, 628 (2016). DOI: http://dx.doi.org/10.1016/j.apenergy.2015.11.044
10. X. Xu, A global perspective, World Dev., **27**, 1215 (1999). DOI: http://dx.doi.org/10.1016/S0305-750X(99)00055-8
11. A. Levinson, M.S. Taylor, Int. Econ. Rev. (Philadelphia), **49**, 223 (2008). DOI: http://dx.doi.org/10.1111/j.1468-2354.2008.00478.x
12. P.E. Karlsson, J. Klingberg, M. Engardt, C. Andersson, J. Langner, G.P. Karlsson, H. Pleijel, Scien. Tot. Environ., **576**, 22 (2017). DOI: 10.1016/j.scitotenv.2016.10.061
13. E. Botta, T. Koźluk, OECD iLib (2014). DOI: http://dx.doi.org/10.1787/18151973
14. G.Y. Liu, Z.F. Yang, B.D. Fath, L. Shi, S. Ulgiati, Appl. Ener., **186**, 96 (2017). DOI: 10.1016/j.apenergy.2016.06.132/
15. H.D. Robison, Can. J. Econ., **21**, 187 (1988)
16. J. Ederington, J. Minier, Can. J. Econ., **36**, 137 (2003). DOI: http://dx.doi.org/10.1111/1540-5982.00007
17. M.A. Cole, R.J.R. Elliott, J. Environ. Econ. Manage, **46**, 363 (2003). DOI: http://dx.doi.org/10.1016/S0095-0696(03)00021-4
18. G. Arce, L.A. López, D. Guan, Appl. Ener., **184**, 1063 (2016). DOI: http://dx.doi.org/10.1016/j.apenergy.2016.05.084
19. J. Meng, J. Liu, S. Guo, Y. Huang, S. Tao, Appl. Ener. **184**, 853 (2016). DOI: http://dx.doi.org/10.1016/j.apenergy.2015.09.082
20. T. Sakamoto, S. Managi, Appl. Ener., **185**, 615 (2016). DOI: 10.1016/j.apenergy.2016.10.126
21. R.T. Carson, Rev. Environ. Econ. Pol., **4**, 3 (2010). DOI: https://doi.org/10.1093/reep/rep021
22. E. Vyboldina, A. Cherepovitsyn, S. Fedoseev, P. Tsvetkov, Ind. J. Scien. Technol., **9** (2016). DOI: 10.17485/ijst/2016/v9i5/87633
23. N. Didenko, D. Rudenko, D. Skripnuk, International *Multidisciplinary Scientific GeoConference Surveying Geology and Mining Ecology Management, SGEM*, **3**, 267 (2015)
24. P. Tcvetkov, A. Strizhenok, J. Ecolog. Engineer., **17**, 56 (2016). DOI: 10.12911/22998993/61190
25. D. Rudenko, N. Didenko, *Proceedings of the 3rd International Conference on European Integration*, 2016

Network technologies as innovative solution in commodity circulation

I.A. Krasyuk[1,*], *T.V.* Kirillova[1], and *N.A.* Kozlova[1]

[1]Saint Petersburg Polytechnic University Peter the Great, 195251 Saint-Petersburg, Russia

Abstract. The paper identifies the specific features of innovations in the trade industry. Innovations, created in other industries, are preferably used in the area of trade. The types of innovations in the field of trade have been specified. The establishment of a networking commercial enterprise has been defined as an innovative solution, widely used in Russian and world practice. The current state of network trade has been investigated and the tendency of retail network expanding in the Russian Federation has been discovered. The concept of a multi-format establishment and the most perspective trade formats have been identified based on the example trading company X5 RetailGroup.

1 Introduction

Retail trade is a promising area for innovations and effective combinations of various internal and external factors. The innovative approach allows to adapt the activities of trade organizations to market demands and compete with other participants. A distinctive feature of the trade industry is the continuous and systematic implementation of innovations in activity of the industry. Innovations have successfully passed the commercialization, are viable and affect all areas of the commercial enterprise. Innovations in the field of trade include not only modern technologies but also innovations related to marketing, organizational and managerial aspects of the enterprise, with the development of its infrastructure, social sphere and ecology. Despite the fact that a significant share of innovations in trade have been created in other industries, they are also widely used particularly in the trading enterprises.

2 Topicality of the research

In the current reality innovations act as a necessary condition for the state economy development as they are aimed at creating new products and new markets. The introduction of innovative solutions contributes to the overall level of economic development, helps to change food processing technologies and improve the means of delivery of goods, provides the intensification of production and technological processes, promotion of goods and services. The sphere of commodity circulation, which is responsible for delivering goods from the production to the final consumer, plays a vital role in modern society. It is rather essential to enhance the innovation potential for further development of trade industry. Innovations improve the efficiency of trade and technological processes, in particular, they have an impact on pricing, help to reduce distribution costs and improve the quality of customer service.

3 The problem and hypothesis of the study

The issue that stands in the centre of this study is to determine the prospects of development of retail trade networks as an innovation typical for the modern state of the sphere of commodity circulation. The hypothesis of the study, formulated according to the advanced research problem consists in the fact that the innovation potential of multi-format retail chain affects a wide range of factors, the identification and registration of which contributes to the optimal size of the sales network, as well as the definition of a rational correlation of enterprises of different retail formats included in particular sales network.

4 The theoretical part

A wide variety of innovations in the field of commodity circulation requires the classification. Innovations in trade can be divided into several types.

Table 1. Types of innovations in trade [complied by the authors].

Type of innovations	Description	Example
Technological	Changes in the sphere of trade and technological processes, trade and technological	The use of the Internet, the application of new refrigerants

* Corresponding author: iri-krasjuk@yandex.ru

	equipment, exchange of information	
Social	Changes in trading technologies in order to preserve and strengthen the buying habits	Trade enterprises with non-stop 24-hours operation; launch of own production
Technical	Define market leadership and reduce distribution costs	Display of goods in a trading hall in partitionig packaging of manufacturers; self-checkouts
Processing	Design of new organizational structures, both within the firm and inter-firm level	Multi-format retail network «X5 RetailGroup» includes hypermarkets, supermarkets, discounters and "convenience stores".
Marketing	Development of private label brands of retail chains	Private labels "Metro", "Dixie", "Red price"

Such process innovation as the creation of a retail trade network is on focus of particular interest. This process intensified in the last few decades due to the concentration of trading capital [1].

The use of network technologies in trade is based on the application of various tools, one of which is the modern information technologies, including the Internet opportunities. Information communication in modern trade includes firstly a database of information between buyers and trading company (for the formation of the product range) ; secondly- communication at the divisional level and management (to increase turnover and operating profit), and thirdly, interacting with suppliers of goods and products. Thus, network ties are formed both in the internal environment of the enterprise and the external contour interaction and, therefore, they become the basis of global activities of the trade industry.

Networking takes various organizational forms, including the forms of vertical inter-firm interactions (cooperation), and the development of horizontal linkages. Network organizational structures have become a natural stage in the evolution of organizational forms and optimally tuned to meet the requirements of modern markets. [2] The economists believe that the network form of business organization while expanding the company's business operate more profitably due to saving internal costs and the small scale of a particular network enterprise allows to adapt to the market demands dynamically. [8]

Achrol R. distinguishes four forms of network interaction. [3]

1. Internal market networks. In such a network the various functional departments of the same organization, endowed with relative autonomy, operate with other internal departments and external agents; satisfy their needs based on market conditions, but act on the basis of the general policy and principles adopted in the organization.

2. Vertical market networks are located around a manufacturing company and represent the channel of supply and distribution.

3. Intermarket networks are grouped around a single major financial structure, or on the basis of intermediary trade firm or manufacturing company.

4. Opportunity networks are the temporary structures created for a particular purpose. In the center of the structure there is, as a rule, the marketing organization. Such networks specialize in gathering market information and provide this information to all the network agents.

In accordance with presented classification, the retail chains are a mixture of all forms of networking. However, a dominant position can be taken by one of the following forms of networks interaction under the particular circumstances.

5 Applied aspect of research

The outspread of network trade in the Russian Federation is confirmed by the data presented in table 2.

Table 2. The share of retail trade networks in formation of retail trade turnover of federal subjcts of the Russian Federation in 2010-2015 years.[a]

Federal subject of RF	The share of retail chains in the total turnover, in current prices, %					
	2010	2011	2012	2013	2014	2015
CFD	20,7	20,8	21,2	23,2	23,8	26,3
Moscow	19,6	18,7	19,1	20,4	20,4	21,8
N-WFD	32,2	34,0	37,6	36,9	39,1	41,5
Saint-Petersburg	48,8	51,1	53,7	49,3	52,1	52,8
SFD	15,7	17,9	20,0	20,9	22,1	23,8
N-CFD	5,0	5,8	6,3	6,2	5,8	5,8
PFD	14,8	16,8	18,8	20,5	22,3	23,4
UFD	12,1	13,6	17,7	19,2	21,3	24,8
SFD	14,7	14,9	20,7	22,4	23,2	25,7
FEFD	7,7	7,6	9,0	9,8	10,8	11,5
CRFD	-	-	-	-	4,2	20,2

The trend of annual increasing of the share of retail chains in total turnover of trade has been found out in the Russian Federation recently. The reasons are the following: the increasing the profitability of network trade for retailers and the attractiveness of the network for consumers. In the future we can predict either the immutability of the current trend, or maintaining the current share of trade network in the regions of the country.

One of the process innovations used by retail chains is multi-format, which appears as the result of diversification of trade formats aimed at a specific group of customers, unified pricing, trading and technological processes. The purposes of multi-format trade at the level of the trade network are the following:

- different formats specialization;
- maximum coverage of a diverse customer audience;
- bet on the ratio of price and quality. [4]

[a] The state statistics Committee of the Russian Federation

Large operators are developing such trade formats like hypermarkets, discounters, supermarkets and "convenience stores". [5]

Table 3. Shares of retail trade formats in the Russian Federation in 2007-2014 years. [6]

Retail trade formats	2006	2007	2008	2011	2014
Hypermarkets	14%	20%	28%	32%	28%
Supermarkets	40%	35%	30%	12%	17%
Discounters	31%	30%	27%	32%	33%
"Convenience stores"	4%	6%	8%	14%	12%
Others	11%	9%	7%	10%	10%

According to Table 3, the most common formats of trade in the Russian Federation are hypermarkets and supermarkets. The country's largest trading company X5 RetailGroup includes commercial properties of different sizes **(Table 4.).**

Table 4. Dynamics of the number of retail properties belonging to X5 RetailGroup trading company for 2007-2016., items.

Retail trade formats	2007	2010	2012	2016	2016. к 2013, %
Discounters	674	1999	3220	8363	259,7
Supermarkets	179	310	370	539	145,7
Hypermarkets	15	71	78	91	116,7
"Convenience stores"	-	45	134	194	144,8
Total	868	2425	3802	9187	241,6

The discounters are developing the most rapidly. It should be noted that in this trading network the format of a hypermarket has not received proper development. Recently, the company has actively used the tool of franchising, including franchise back. [7]

6 Conclusions

Retail trade is a promising field for the introduction of different kinds of innovations. The complexity and uniqueness of the retail chains, as well as the prospects for their development determine the variety of innovations applied. Trade network is the process innovation in the field of commodity circulation and is defined as a trading enterprise with a single point of management, coordination and control, centralized supply system, a unified development strategy (product range, pricing, advertising), operates under a single brand.

The multi-format trade network is the kind of process innovation, which combines a variety of formats in a single trading network. Other innovations in trade are associated with technological, technical, social, marketing and other aspects.

References

1. T.V. Kirillova, Practical Marketing, **2-1(240-1)**, 22-30 (2017)
2. M.A. Beck, N.N. Beck, E.V. Buzulukova, *The methodology of the study of network forms of business organization: monograph* (Publishing house of the Higher school of Economics, Moscow, 2014)
3. R.S. Achrol, Journal of the Academy of Marketing Science, **25**, **1**, 56-71 (1997)
4. Z.V. Kotelnikov, *Development of retail networks as a backbone element of the transformation of the trade (for example, food sector). Organizational forms and competitive strategies in the Russian retail trade (regional aspect)* (Vershina, Moscow, 2008)
5. D.A. Popov, V.A. Sinyukov, *Formats of retail: evolution and modernity: monograph* (RICHIEP, Khabarovsk, 2013)
6. www.planet.retail.net
7. V.P. Cheglov, *Innovative retail companies. Organizational leadership and effective technology: monograph* (Publishing house FORUM, Moscow, 2014)
8. I.A. Dudakova, *Marketing solutions network technologies in the retail trade: monograph* (Publishing house Dashkov and K, Moscow, 2007)

Specifics of multi-project management: interaction and resources constraints

Nadezhda Tsvetkova[1,*], and *Iosif* Tukkel[1]

[1]Peter the Great Saint-Petersburg Polytechnic University, Saint-Petersburg, Russia

Abstract. Multi-project management is fundamentally different from the control of a particular project or a set of slightly interconnected projects in terms of complexity and specifics. In multi-project management of the company production it is important to analyze the innovation interaction and its impact on the commercialization stage. A multiparameter factor of innovations interaction was introduced. The optimization problem which considers this factor was mathematically defined. The solution of this problem produces a schedule of innovations launches. This problem definition allows updating the objective function that corresponds to the aims of a manufacturing company. For example, it can help maximize the number of interdependent innovations with restrictions to current tangible and intangible resources or minimize the number of used tangible resources at a fixed number of innovations implemented. In order to verify the optimization problem an evolutionary approach based on genetic algorithm and local search is used. The verification was performed by the Solver a Microsoft Excel add-in. The readiness for practical use of the proposed solution was proved by the experiment.

1 Introduction

A multi-project management differs fundamentally in problem complexity and specifics from the control of a particular project (mono-project) or a set of slightly interconnected projects both in terms of the rules of resources allocation and possible conflict of local targets.

When constructing a multi-project control system it is important to analyze how the projects interact on each other and how this interaction may have an impact on a product commercialization stage of a manufacturing company production.

A lot of scientific papers of many scientists such as V. Anshin [1], R. Archibald [2], K. Clark [3], R. Cooper [4], S. Edgett [4], A. Kozlovskiy [5], D. Novikov [6], S. Wheelwright [3] et al. were dedicated to the case study of project selection and portfolio construction.

Project interaction issues have been studied by such scientists as M. Dickinson [7], S. Grasman [8], W. Lomax [9], B. Pelegrín [10], D. Pertcev [11], Sh. Srinivasan [8] et al.

It should seem that the above and other researchers have exhausted the problem of determining priority of projects launch in a portfolio. However a specific feature arises in respect to innovations portfolio in the first instance. Innovations interdependence needs extra research related to a specific nature of its interaction.

This paper is devoted to development of such control over sequence of innovations launch that would assume the mentioned specifics as well as limitations of a manufacturing company resources.

2 Necessary definition

Let's introduce two descriptors of potential interaction of innovations in a portfolio:
- positive interaction leading to increase in total effect of a portfolio implementation;
- negative interaction leading to decrease in total effect of a portfolio implementation.

There is another definition of the negative impact called cannibalization [10, 12].

The required resources for implementation of multi-project control are divided into two types as follows: tangible and intangible [13].

1. The tangible resources R mean a set of conditions enabling to succeed in each project's objective. A distinctive feature of the tangible resources of a project is its time dependence t. The basic types of these resources are as follows:
- financial feasibility of a project;
- technological resources;
- human resources;
- information resources.

When analyzing a portfolio of innovations U the tangible resources R are set with two parameters:
- available stock $Q_i(t)$ at the moment t of the resource of type $r_i \in R$;
- demand $D_{ij}(t)$ at the moment t for the resource of type $r_i \in R$ to implement a project $u_j \in U$.

* Corresponding author: nadezhdaat@gmail.com

2. Intangible resources are the ones that initially belong to a project, do not depend on project time t and are not material. Intangible resources include intellectual property assets: essence, innovation idea. Intangible resources separation is required for recording project interaction, its impact on future volume innovative product sales market and on a company competitiveness and so on.

The introduced interaction factor $a_{k,l}$ will be determined as follows.

Innovations portfolio that has been built without consideration of project interaction is defined as $U = \{u_1, u_2, ..., u_n\}$. The groups of dependent projects on intangible resources are found out in a portfolio using an expert approach. A set of these groups is determined as $Z \subseteq U$, where:

- $Z = \{Z_1, Z_2, ..., Z_k, ..., Z_\zeta\}$, $Z_1 \cap ... \cap Z_\zeta = \emptyset$, $Z_k = \{z_{k1}, ..., z_{km}, ..., z_{k\varepsilon}\}$, $Z_k \subseteq Z$;
- k – index number of a group of dependent projects, $k = 1,2, ..., \zeta$;
- ζ – number of groups of dependent projects, $1 \leq \zeta \leq n/2$;
- m – index number of a dependent project, $m = 1,2, ..., \varepsilon$;
- ε – number of dependent projects in a group, $1 < \varepsilon \leq n$.

A project $z_l \in Z_k$ and a subgroup of projects $Z_{kl} = \{z_{km} \in Z_k | Z_{kl} \subseteq Z_k \backslash z_l, m \neq l\}$ will be stand out in each group of dependent projects Z_k, where l is an index number of a group at issue, $l = 1,2, ..., \varepsilon$.

Interaction factor $a_{k,l}$ is calculated for each group of dependent projects Z_k and demonstrates the extent of impact subgroup of projects Z_{kl} on the project $z_l \in Z_k$, $a_{k,l} > 0$.

In order to solve the problems of scheduling it is also important to consider for the extent of the impact on a change of a project z_l completion date with respect to completion dates of the projects Z_{kl}: $f = f_k - f_l$, $f \in \mathbb{Z}$. Considering the above the factor is defined as $a_{k.l.f}$.

If $a_{k.l.f} < 1$, then a concerned subgroup Z_{kl} has a negative impact on innovation project z_l, otherwise, if $a_{k.l.f} > 1$, a subgroup Z_{kl} has a positive impact. No impact at $a_{k.l.f} = 1$.

Intangible resources determination results in introducing interaction factor $a_{k,l} = a_{k,l}(G, C, O, M)$ into a project identification summary, where:

- purpose of innovation G: 1 – similar, 0,5 – inter crossing, 0 – different. If $G = 0$, interaction is not considered anymore;
- innovation category C: at what stage of a commodity life cycle an innovation is implemented (innovation of end product, processes, procedures, cycles);
- kind of innovation O: value of made variations (modifying, improving, breakthrough, integrating);
- innovation scale M: relation of variation value to a sold product value (global, local).

The stated parameters describe an essence of products of innovation projects and must be considered in a project identification summary. The interaction factor is determined by experts for a certain innovation portfolio.

The factor $a_{k.l.f}$ is a function of many parameters. An expert analysis shall be carried out according to iteration algorithm in this case, for example, [14, 15].

In order to assess an efficiency of multi-project management it is necessary to define an objective function of control.

Let X_V be the effect of implementation of independent projects in the portfolio U and X_Z be the effect of implementation of dependent projects. The objective function will be as follows:

$$X = X_V + X_Z \rightarrow max \quad (1)$$

So the defined objective function allows for providing a content of the effect with different meaning depending on current tasks of a manufacturing company. For example:

- maximization of the number of interdependent innovations with restrictions to current tangible and intangible resources.
- minimization of the number of used tangible resources at a fixed number of innovations implemented.

A problem of determining a sequence of launch of the project in the innovation portfolio comes down to defining a problem of scheduling. In general, a problem of scheduling can be laid down as follows. Using some set of resources some fixed multiparameter system of tasks (works, projects) shall be carried out a purpose of which is to optimize the selected objective function.

3 Mathematical formulation of a problem

Let n projects be available in the innovation portfolio $U = \{u_1, u_2, ..., u_n\}$ as before, in which project interaction has not been taken into account yet. All the projects are supposed to be carried out without interruptions.

Let $T = \{1, 2, ..., \tau\}$ be a set of time units admissible for a schedule, within which limitations may be carried out. Then a duration $\tau \in T$ is a desirable and limit time of implementation of all projects in a portfolio, $\tau \in T$ is a time supposed by the algorithm. Completion time $t_i \in T$ of each project $u_i \in U$ can be defined in days, weeks, months depending on a project scale.

Each project is described by such parameters as s – number of a starting event of a project u_i, f_i – number of final event of a project u_i, thereat $s_i, f_i \in T$.

A set of $Z \subseteq U$ is a set of innovation projects determined as dependent according to intangible resources, where $Z = \{Z_1, Z_2, ..., Z_k, ..., Z_\zeta\}$, $Z_1 \cap ... \cap Z_\zeta = \emptyset$, $Z_k = \{z_{k1}, ..., z_{km}, ..., z_{k\varepsilon}\}$, $Z_k \subseteq Z$; k is an index number of a group of dependent projects, $k = 1,2, ..., \zeta$; ζ is a number of groups of dependent projects, $1 \leq \zeta \leq n/2$; m is an index number of a dependent project, $m = 1,2, ..., \varepsilon$; ε is a number of dependent projects in a group, $1 < \varepsilon \leq n$.

The interaction factor $a_{k.l.f}$ is introduced in the given range of endings $f = f_k - f_l$ for each pair consisting of subgroup of projects $Z_{kl} = \{z_{km} \in Z_k | Z_{kl} \subseteq Z_k \backslash z_l, m \neq l\}$ and the project $z_l \in Z_k$.

Let $R = \{r_1, r_2, ..., r_\rho\}$ be a set of types of tangible resources required for project implementation and having direct impact on launch sequence in a portfolio. Then $Q_j(t)$ is a stock of resource type $r_j \in R$ available in a given time, $D_{ij}(t)$ is a demand for the resource $r_j \in R$ to implement the project u_i.

A portfolio U may include independent projects according to intangible resources as well: $Y \subset U$ is a set of independent projects, $Y = \{y_1, ... y_v\}$, $0 \leq v \leq n\text{-}2$.

Then a sum of dependent and independent projects will be equal to the number of all projects in a portfolio n.

Let X_i^Y be an effect of implementation of the independent project y_i and X_i^Z be an effect of implementation of the dependent project z_i.

Vector $S = (s_1, s_2, ..., s_n)$ is called an implementation schedule of the projects in a portfolio. The schedule S is called admissible, if:

- at any specific time $t \in T$ the demand for resources for project implementation will be met for $\forall r_j \in R$ $\sum_{i=1}^{n} D_{ij}(t) \leq Q_j(t)$, $t \in T$ and limitations assume all types of tangible resources at that;
- the project u_i can be started within the period no later than a difference between a number of available periods τ and project duration t_i plus one period, as the counting begins with one: $1 \leq s_i \leq \tau\text{-}t_i + 1$.

The effect of independent project implementation is equal to:

$$X_Y = \sum_{i=1}^{Y} X_i$$

The effect of implementation of interconnected projects depends on the project interaction factor: $X^Z = \sum_{m=1}^{\zeta} X_m^Z$, where: $X_m^Z = \sum_{\alpha=1}^{\varepsilon} a_{k.l.f} * X_{m\alpha}$.

As shown above, the objective function (1) is equal to:

$$X = X_Y + X_Z \rightarrow max$$

An optimum launch time shall be estimated, that means such s_i for each project shall be found that will provide for maximum value of the target function X and meet all limitations.

4 Optimization problem solving

The defined optimization problem is demonstrated as a diagram (fig. 1).

The defined discrete optimization problem in the context of set resource limitations is NP-hard [16–18]. Due to complexity of the problem it makes sense to use algorithms such as approximation methods [19–21] that base on priority rules, Lagrange relaxations, evolution ideas, local search etc.

Approximation methods are frequently used in a scheduling problem solving [22-27].

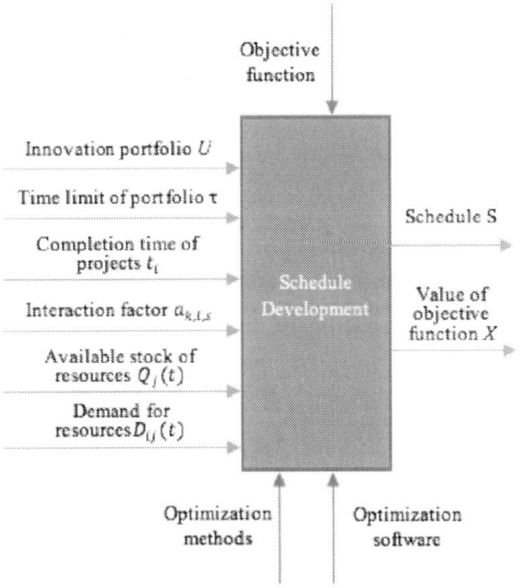

Fig. 1. Problem diagram.

Nowadays there are a number of tools allowing for solving the defined optimized problem in the software market.

For the purposes of the study the widely available Microsoft Excel add-in the Solver was selected for verification of the optimized problem (fig. 2). The evolutionary approach based on genetic algorithm and local search incorporated into the Solver [28] was used. A genetic algorithm is an heuristic algorithm of searching reminding biological evolution. The specific feature of the local search method is that the selection of new decision at each iteration of the process is made not deterministically, but as a result of implementation of some random process [18, 21].

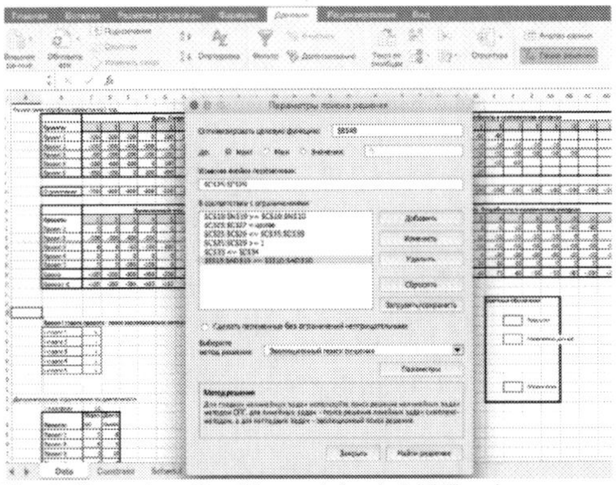

Fig. 1. Examples of problem solving in MS Excel.

The defined optimization problem was verified both on test and real innovation portfolio. Finally the opportunity for improvement of multi-project management efficiency due to taking innovations interaction into account was confirmed.

5 Conclusion

Unlike existing approaches to multi-project control the proposed solution allows for increasing efficiency by virtue of consideration of innovation interaction on intangible resources. A multiparameter factor of innovations interaction was introduced. The optimization problem which considers this factor was defined in terms of mathematics. The solution of this problem produces a schedule of innovations launches. This problem definition allows to update the objective function that corresponds to the aims of a manufacturing company. For example, it can help to maximize the number of interdependent innovations with restrictions to current tangible and intangible resources or minimize the number of used tangible resources at a fixed number of innovations implemented.

In order to verify the optimization problem an evolutionary approach based on genetic algorithm and local search is used. Verification was performed by the Solver a Microsoft Excel add-in. Readiness for practical use of the proposed solution was proved by the experiment.

References

1. V.M. Anshin, I.V. Demkin, I.M. Nikonov, I.N. Tcarkov, *Managing Portfolios of Projects under Uncertainty* (MATI, Moscow, 2008)
2. R.D. Archibald, S. Archibald, *Leading & Managing Innovation. What Every Executive Team Must Know about Project, Program & Portfolio Management* (West Conshohocken, Infinity Publishing, 2013)
3. K.B. Clark, S.C. Wheelwright, *Managing New Product and Process Development: Text and Cases* (Free Press, NY, 1993)
4. R.G. Cooper, S.J. Edgett, E.J. Kleinschmidt, *Portfolio management for new products* (Basic Books, 2001)
5. A.N. Kozlovskiy, A.O. Nedosekin, Z.I. Abdulaeva, *Portfolio management of R&D projects* (SpbSTU, Saint-Petersburg, 2016)
6. D.A. Novikov, I.V. Gontareva, R.M. Nizhegorodcev, *Project management* (Librokom, Moscow, 2009)
7. M.W. Dickinson, A.C. Thornton, S. Graves, IEEE Transactions on Engineering Management, **48(4),** 518-527 (2001)
8. S. Srinivasan, S. Ramakrishnan, S. Grasman, International Journal of Business Innovation and Research, **1,** 92-110 (2006)
9. W. Lomax, K. Hammond, R. East, M. Clemete, Journal of Product & Brand Management, **6(1)**, 27-39 (1997)
10. B. Pelegrín, P. Fernández, M.D.G. Pérez, Annals of Operations Research, 1-19 (2014)
11. D.V. Pertcev, Modern Management: Theory and Practice, 59-69 (2008)
12. J. Heskett, *Marketing* (Macmillan, New York, NY, 1976)
13. I.L. Tukkel, S.A. Golubev, A.V. Surina, N.A. Tsvetkova, *Methods and tools of management of the innovative development of industrial Enterprises* (BHV, Saint Petersburg, 2013)
14. G. Rowe, G. Wright, Int'l Journal of Forecasting, **15,** 353-375 (1999)
15. S.N. Jashin, I.L. Tukkel, E.V. Koschelev, Y.S. Korobova, Y.V. Zaharova, *Development and decision-making in the management of innovation* (N. Novgorod, 2017)
16. J. Blazewicz, J.K. Lenstra, A.H.G. RinnooyKan, Discrete Applied Mathematics, **5,** 11-24 (1983)
17. P. Brucker, A. Drexl, R.H. Mohring, K. Neumann, E. Pesch, European Journal of Operational Research, **112(1)**, 3-41 (1999)
18. Y.A. Zak, *Applications of the theory of scheduling and traffic routing* (Librokom, Moscow, 2012)
19. C. Artigues, S. Demassey, E. Neron, *Resource-constrained project scheduling: models, algorithms, extensions and applications* (Wiley-ISTE, 2007)
20. A.A. Lazarev, E.R. Gafarov, *Scheduling Theory: problems and algorithms* (MSU, Moscow, 2011)
21. M. Uetz, *Algorithms for Deterministic and Stochastic Scheduling* (Cuvillier Verlag, 2002)
22. S. Asta, D. Karapetyan, A. Kheiri, E. Özcan, A.J. Parkes, Information Sciences, **373,** 476-498 (2016)
23. Ph. Baptiste, C. Le Pape, W. Nuijten, *Constraint-based scheduling: applying constraint programming to scheduling problems* (Kluwer Academic Publ., 2001)
24. R.K. Chakrabortty, R.A. Sarker, D.L. Essam, *IES 2016*, 75-86 (2017)
25. R. Kolisch, S. Hartmann, *Manuskripte aus den Instituten fu ̈r Betriebswirtschaftslehre* (Kiel, Germany, 1998)
26. B. Naderi, Journal of Optimization in Industrial Engineering, **13(13),** 65-72 (2013)
27. J. Weglarz, *Project scheduling. Recent models, algorithms and applications* (Kluwer acad. Publ., Boston, 1999)
28. Basic Solver - Algorithms and Methods Used. http://www.solver.com/content/basic-solver-algorithms-and-methods-used

 DOI: 10.1051/shsconf/20173501057

Assessing impact of changes in human resources features on enterprise activities: simulation model

Svetlana Kalmykova[1,*], *Alexander* Krasnoshchekov [1], and *Tatyana* Gileva[2]

[1]Peter the Great St. Petersburg Polytechnic University, 195251 St.Petersburg, Russia
[2]Ufa State Aviation Technical University, 450008 Ufa, Russia

Abstract. The need for creating programs of human resources development is shown; the impact of these programs on organizational effectiveness is taken into account. The stages of development tools and HRD programs on the basis of cognitive modelling are disclosed; these stages will help assess the impact of HR-practices on the key indicators of organization activity at the design stage. The method of HR-practices'pre-selection in professional development of the employees is represented.

1 Introduction

To solve the problem of assessing the impact of changes in the characteristics of human resources in the performance of the enterprise the dynamic simulation modeling environment PowerSim Studio 7 was selected; this environment has the following advantages: there is room for solving optimization problems; advanced tools for presentation of graphs and reports; user-friendly interface. Models built with the help of PowerSim Studio, are dynamic: they make it possible not only to investigate the structure of the system, but also to get the forecast for the development by simulating its behavior over in time. The structure of the model, built according to the investigated enterprise is shown in Fig. 1.

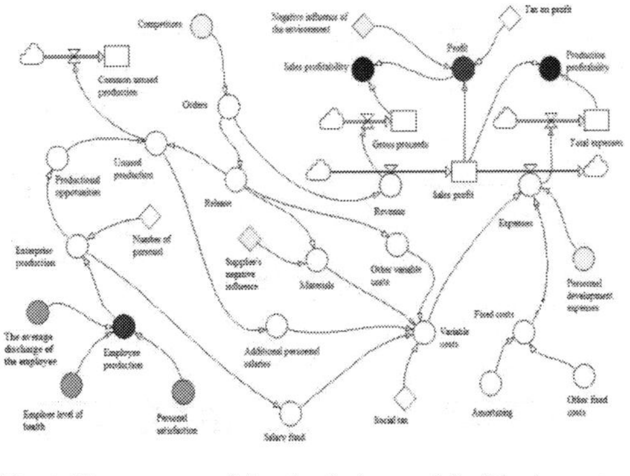

Fig. 1. The structure of the simulation model of the impact human evaluation resources on the economic results of the enterprise.

As the enterprise is an open system, it is a subject of external factors' impact, we introduced constants and variables that reflect the influence of these factors to the simulation model for the realization of the situational approach. For example, the constant "negative influence of the environment" reflects the possibility of negative charges, which affect the value of profit on sales (non-payment of a receivable, penalties and interest on overdue accounts payable, etc.). The constant "suppliers' negative influence" shows the influence of raw material suppliers (increase in raw material prices, the change of the supplier). Variable "competitors" was introduced to take into account, in particular, possible changes in the scope of the enterprise orders, which can be caused by a decrease in the prices of competitors, the entry of new manufacturers of similar products, and others. PowerSim Studio allows alternative calculations using the controls, which are the following human resources indicators for this model: "The average discharge of the employee» (X1), «The employer's level of health» (X2) and "coefficient of staff satisfaction» (X3). Using controls, you can change the values of variables and see how to change the behavior of the model.

2 Selecting HR-practices

We shall dwell on the employee satisfaction in more detail, because the results of the correlation analysis (Table 1) show that this factor have the greatest impact on productivity. The indicators that got the lowest value in the satisfaction questionnaire are: opportunities for professional development (9 persons); social security from the enterprise (7 persons); the level of stress at work (6 people). We can assume that if we do not carry out any changes to improve these indicators, there is a high probability that these people will leave the enterprise in the near future. The dismissal of these workers will be reflected in the coefficient of employee satisfaction. Possible values of this coefficient, calculated on the basis of the total headcount of the

* Corresponding author: kalmykovas@mail.ru

analyzed unit, at different developments are presented in Table 1.

Table 1. Possible values of the employee satisfaction coefficient (X_3).

Case	Values of the employee satisfaction coefficient
No one is dismissed	1
9 people stay, 7+6 people are dismissed	0,87
7 people stay, 9+6 people are dismissed	0,75
6 people stay, 9+7 people are dismissed	0,67
All 22 people are dismissed	0,52

Alternatively when no one is dismissed is possible in case of the implementation of positive changes for all three employees dedicated dissatisfaction factors. Alternatively, when 9 people remain and 13are dismissed is possible when a great opportunity for professional development exists. Alternatively when 7 people remain, and 15 are dismissed is possible if the improvements are implemented only in social security part of the enterprise. Alternatively when 6 people stay and 16 are dismissed is possible if the level of stress at work is reducing and the other two directions stay in constant conditions. Finally, if improvement is realized option, all 22 discontented man will be dismissed.

As the issues related to the increase of professional qualifications on the one hand, are the most significant cause of the enterprise's employees' dissatisfaction, and, on the other, they determine the competitiveness of goods and services of the enterprise it was decided to start a program of employees' professional development. It should be noted that the results of analysis on the private enterprise are consistent with the conclusions of the research [10] in defining the relationship between human resource management and efficiency of Russian companies. This research showed that the most important from the standpoint of organizational efficiency for the companies is the practice of training.

For the formation of a professional development program it has been proposed a technique based on the analysis of employee position competencies profiles and, as well as the systematization of the most common forms and training methods. [1] This method includes two enlarged steps:

- Identification of competence development needs;
- The selection of teaching methods (competency development).

In the first phase:

- By expert evaluation in the framework developed by the enterprise competency models for a 5-point scale specified degree of each competency importance for the task (IL);
- The certification of employees for compliance with competency model their positions is carried out; assessment result is staff level (EL), from 1 to 5 points;
- For the evaluated employee the degree of development problematical character for each competence profile is determined (PC = 5 - EL);
- The highest priority for the development of competence are determined (the sum of IS and EL); competencies ranking matrix is formed on the criteria of "degree of importance" - "the degree of problem".

Explanation and results and of competence assessment for marketing department specialist of the enterprise are shown in Table 2. Priority development requires competence C3, C9 and C13. This situation is largely determined because the employee has recently started to work in the company, where he came from the outside.

Table 2. Evaluation of professional competence of the enterprise marketing department.

Group of competences	Competence	EL	IL	PC	Priority
Functional	1. Education	3	4.3	2	6.3
	2. Working experience	3	3.7	2	5.7
	3. Knowledge of market	2	4.7	3	7.7
	4. Sales technique	4	4.3	1	5.3
	5. Problem solvation	3	3.3	2	5.3
Cognitivee	6. System approach knowledge	2	2.3	3	5.3
	7. Marketing knowledge	4	4.3	2	7.3
	8. Knowledge of economy	3	3.3	2	5.3
	9. Knowledge of technical and design features of enterprise's production	2	3.7	3	6.7
Personal	10. Communicativeness	4	3.3	1	4.3
	11. Team work	3	3.7	2	5.7
	12 Collaboration	4	4.3	1	5.3
	13. Understanding of the company's values	3	5.0	2	7.0
	14. Development orientation	4	3.7	1	4.7

At the second stage, the teaching methods are selected (competency development). Previously, as a result of generalization of experience [5, 15, 16, 19, 27], a list of the 16 HR-practices in the field of professional development of employees was compiled and their systematization in the context of the criteria set forth in Table 3 was held. The list of practices and indication of their criteria respective the values of the selection are presented in Table 4. We should note that the list of HR-practices must be constantly updated and supplemented

Table 3 Criteria for personnel training selection.

Criterion	Assessment		
	1	2	3
A. Price	Low (0 – 10 th.rub.)	Medium (10 – 50 th.rub.)	High (more than 50 th.rub.)
B. Duration	Short-term (1 – 7 days)	Medium-term (1 week – 3 month)	Long-term (more than 3 month)
C. Practicality	Theoretical	Mixed	Practical
D. Professionalism	Inter-individual	Mixed	Professional
E. Complexity	Highly specialized	Number of specializations	Complex
F. Number of students	Individual	Group (2 – 20 people)	Mass (more than 20 people)

Table 4 Characteristics of training methods (fragment).

№	Method	Criteria values
1.	Instructing	A.1; B.1; C.3; D.3; E.1; F.1
2.	Rotation	A.1; B.2; C.3; D.2; E.2; F.1
3.	Mentoring	A.1; B.2; C.2; D.2; E.3; F.1
4.	Method of coping	A.1; B.1; C.3; D.3; E.1; F.1
5.	Delegation method	A.1; B.2; C.2; D.2; E.3; F.1
6.	Education by action	A.1; B.2; C.3; D.2; E.2; F.2
7.	Analysis of practical situations	A.1; B.1; C.2; D.3; E.3; F.2
8.	Training	A.2; B.1; C.3; D.2; E.3; F.2
9.	Group discussion	A.1; B.1; C.2; D.2; E.3; F.2
10.	Business and role games	A.1; B.1; C.2; D.2; E.3; F.2
11.	Thematic seminars	A.2; B.1; C.2; D.3; E.1; F.2
12.	Basket-method	A.1; B.1; C.3; D.3; E.2; F.1
13.	Behavior modeling	A.1; B.1; C.2; D.2; E.3; F.2
14.	Benchmarking	A.1; B.2; C.3; D.3; E.3; F.2
15.	Coaching	A.2; B.2; C.2; D.2; E.3; F.1
16.	E-learning	A.2; B.2; C.2; D.3; E.3; F.2

To select the method of competence we apply the following scheme:

- Preliminary selection of the methods, taking into account their compliance with the content of the identified need for competence development is carried out. This correspondence is determined by the criterion of efficiency, unifying significant method characteristics such as, a practicality, professionalism and complexity (Ri = f {Ci, Di, El});

-Elaboration of the selection is carried by taking into consideration criteria price (Ai) and duration (Bi) of training. The most preferred method is to achieve a minimum amount of points according to the criteria (Table 5). In our case to the competence of the "knowledge market" is a method of analysis of practical situations, for competence "knowledge of technical and design features of enterprise's production "- instruction followed by observation and copying the experience of more experienced colleagues, and improvement of the company's goals and values understanding will contribute to training with and the group discussion.

Table 5 Recommendations for the training methods' selection.

Competences	Effectiveness	Method	Price (A)	Duration (B)	Sum
K3 – Knowledge of market	C.2; D.3; E.3	M7	1	1	2
		M13	1	2	3
		…			
K9 – Knowledge of technical and design features of enterprise's production	C.3; D.3; E.1	M1	1	1	2
		M4	1	1	2
K13 – Understanding of the company's values	C.3; D.2; E.3	M3	1	2	3
		M5	1	2	3
		M8	1	1	2
		M9	1	1	2
		M15	2	1	3

Formed software for the whole enterprise has been focused on the maximum possible retention of employees, but the set of measures in the area of social security after the simulation has been somewhat curtailed due to a disproportionate growth in costs and benefits.

3 Conclusion

Any program of human resource development (HRD program) must meet the following general requirements:

- Be focused on the implementation of the organization's strategy, integrated with all aspects of its activities: research and development, manufacturing, sales and service, as well as be formed with taking into account not only internal but also external stakeholders of the organization;
- Assessed according to the impact of the key performance indicators of the organization, both operational (productivity, product quality) and financial (income, cost of business, market share);
- Integrate HR-practices of different groups: recruitment and selection of personnel, results-oriented assessment of personnel, training and development, the formation of the involvement and commitment of employees as well as their flexibility, readiness for change and the creation of new knowledge.

What is more, while respecting the general principles, it must be differentiated, it must take into account both internal and external features of a particular organization activities.

The development of such a program requires the use of different methods and technologies: benchmarking, expert estimation, regression analysis, etc. It is useful to have a broader application in the field of cognitive and simulation because of the indirect effects of HR-practices on the results of the organization, as well as existing among various human resource management aspects of the relationship. This approach will allow to assess quantitatively the impact of HR-practices on organizational performance at the design stage of the human resource development program of while providing the option to account for a wide range of "situational" factors for a particular organization.

Perspectives of the research

It is most common today to the release three components in the structure of the organization's intellectual capital o: human, organizational and client capital [9]. At the crossroads of these three components the so-called organizational competence, considered today by an increasing number of theoreticians and practitioners as a basis for long-term development strategy of the organization are formed [2, 3, 6, 25]. The most cited research in the field is the study of Hamel G., Prahalad C.K. (1990) [17], according to which the organizational competence is a set of skills and technologies that enables the company to provide its customers with significant benefits. In the course of its development, organizational competencies are converted into the key ones, the main properties of which are [17]:

- Ability to bring significant benefits in terms of the consumer;
- Being difficult to be copied, and what is even better - not to succumb to the simulation;
- Give potential access to a wide range of markets.

It is obvious that the source of organizational competencies are individual competencies of employees, not only professional but also personal. However, in the same way as in the model of the relationship between human resource management and performance (Figure 1) [14], this connection is not obvious, often it is untraceable, and therefore uncontrollable.

References

1. T.A. Gileva, Management in Russia and abroad, **5**, 108-116 (2008)
2. T.A. Gileva, Scientific and technical sheets of the St. Petersburg Polytechnic University, 2, 166-171 (2008)
3. L.A. Ismagilov, T.A. Gileva, L.V. Sitnikova, Problems of the theory and practice of management, **9**, 62-71 (2015)
4. D. Katalevsky, *Fundamentals of simulation modeling and systems analysis in management* (Publishing House Delo RANHiGS, Moscow, 2015)
5. T.N. Klemina, Russian Management Journal, **6**, **3**, 51-58 (2008)
6. B. Slavin, V. Soloviev, Problems of the theory and practice of management, **9**, 72-78 (2015)
7. Y. Fitzents, *Staff investment profitability: measurement of the economic value of personnel* (Publishing house "Peak", Moscow, 2006)
8. R.M. Nizhegorodtseva, S.D. Reznik, *Human capital theory and practice of management in the socio-economic systems* (Moscow, INFRA-M, 2015)
9. L. Edvinsson, Corporate longitude. Navigation in the economy based on knowledge (INFRA-M, Moscow, 2005)
10. A.G. Efendiyev, E.S. Balabanov, A.V. Rebrov, Russian Management Journal, **12**, **1**, 39-68 (2014)
11. M. Alagaraja, Human Resource Development Review, **12(2)**, 117-143 (2012)
12. M. Armstrong, *Strategic Human Resource Management: A Guide to Action* (Kogan Page, London, 2006)
13. B. Becker, M. Huselid, Journal of Management, **32(6)**, 898-925 (2006)
14. P. Boselie, G. Dietz, C. Boon, Human Resource Management Journal, **15(3)**, 67-94 (2005)
15. C.J. Collins, K.D. Clark, The Academy of Management Journal, **46(6)**, 740-751 (2003)
16. T. Garavan, R. Carbery, A Framework. Human Resource Development Review, **6(4)**, 394-418 (2007)
17. G. Hamel, C.K. Prahalad, Harvard Business Review, 5/6 (1990). http://www.enterrasolutions.com/media/docs /2013/08/Prahalad_Hamel_1990.pdf
18. E. Holton, B. Yamkovenko, Human Resource Development Review, **7(3)**, 270-291 (2008)
19. *Human Resources Management and Training. Compilation of Good Practices in Statistical Offices. United Nations* (New York and Geneva, 2013)
20. M. Huselid, Academy of Management Journal, **38(3)**, 635-672 (1995)
21. D. Mankin, *Human Resource Development* (Oxford University Press 2009)
22. J. Paauwe, Journal of Management Studies, **46(1)**, 129-142 (2009).
23. A. Paul, R. Anatharaman, International Journal of Human Resource Management, **14(7)**, 1246-1266 (2003).
24. R. Sanchez, *Knowledge management and Organizational Competence* (Oxford University Press, Oxford, 2003)
25. R. Schuler, S. Jackson, Academy of Management Executive, **1(3)**, 207-219 (1987)
26. R. Schuler, S. Jackson, Journal of Organizational Effectiveness: People and Performance, **1(1)**, 35-55 (2014)
27. J.M. Werner, R.L. DeSimone, *Human Resource Development* (South-Western, Cengage Learning, 2012)

Schemas for the analytical estimation of the operational properties of agile systems

Alexander Geyda[1,*], *Igor* Lysenko[1]

[1]Institute for Informatics and Automation of the Russian Academy of Sciences, 14 Line, 39, St. Petersburg, Russia

Abstract. The article states a brief historical sketch of the research conducted on the effectiveness of targeted processes in the operation of systems performed by Russian authors; an overview of some current problems in operational properties research solved outside Russia is made. A conclusion on the prospects of the analytical estimation of the operational properties is provided. The main directions of the research conducted on the operational properties of systems and processes of their operation, such as system potential, operational properties of application of information technologies, are stated. The main features of the research conducted on the operational properties in the view of the current tasks of their research are provided as well as the schemes of analytical estimation of the operational properties applicable to the new tasks are stated that allows for meeting these research challenges using mathematical models.

1 Introduction

The operational properties of systems and their operation processes (hereinafter referred to as "OP"), as well as the core of the properties defining the quality of systems, processes of their operation and changes, are determined practically (in human activities), as they describe the main aspects of the application of systems in practice. It is better to carry out the improvement of systems and processes of their operation analytically, by solving the mathematical tasks of the OP research. The article describes the main features of the OP study in relation to the current problems being solved by domestic and foreign authors. A conclusion about the prospects of the analytical estimation of OP is made based on the revealed features of the current problems; the schemas of the estimations are also given. These schemas allow for switching to an application of the numerical models of system operation that are necessary for the analytical study of OP.

2 Peculiarities of the study of operational properties by domestic and foreign authors

2.1 Main features of the Russian School of research on system operation effectiveness

Among the proceedings of Russian scientists in the field of "Investigation of the effectiveness of targeted processes", the work of Academician A. N. Kolmogorov [1] is the first to be described; it states a number of methods for estimating the efficiency of firing. The author proposed the estimation of the effectiveness of firing on the basis of a probability measure. Then, these methods were developed by E. S. Wentzel [2], who worked in the Zhukovsky Air Force Engineering Academy for many years. After the war, she sought to solve the task [3] of estimating the effectiveness of different types of weapons and ammunition and ways of firepower arrangement when shooting at flying objects. According to the academician A. N. Kolmogorov, air defense efficiency was assumed as a probability of hitting an air target, calculated on the basis of the distribution of probable points of a bomb explosion neighboring the target and the conditions of the probabilities of target destruction, depending on the point where the explosion could occur. In the 1970 s, these methods were further developed in the works of authors of the Department of combat effectiveness of the A. F. Mozhaisky Military Space Academy, managed by R. M. Yusupov at that time, who now is the director of the Saint Petersburg Institute for Informatics and Automation of the Russian Academy of Sciences [4, 5]. Moreover, there, at the A. F. Mozhaisky Academy, G. B. Petukhov and his students developed the theory of the effectiveness of targeted processes in their work [6, 7]. They assumed the efficiency as a probabilistic measure of compliance of the analytically calculated characteristics of the random effects of a purposeful process to the targeted (directory specified) values of these characteristics. This measure can be described as the probability of a random vector of effects hitting the target area. Around the same time, some research was conducted to improve the efficiency of processes of engineering, production, and application of various types of military equipment, based on the methods proposed

* Corresponding author: geida@iias.spb.su

by A. N. Kolmogorov [8–19]. Thus, for example, the research conducted on the effectiveness of missile weapon systems [8, 9] were carried out by the team of scientists of the Dzerzhinsky Military Academy of Russian Strategic Missile Forces [11–13] as well. The team of scientists from N. G. Kuznetsov Naval Academy [16] under the leadership of I. A. Ryabinin was awarded the USSR State Prize in the field of science and technologies in 1979 for the design and introduction of methods for high reliability and the effectiveness of weapon systems. In the subsequent years, the effectiveness theory was developed in several directions of research that was conducted in the A. F. Mozhaysky Military Space Academy, as well as at the Department of combat effectiveness of Military Space Forces [14] and the Saint Petersburg Institute of Informatics and Automation of the Russian Academy of Sciences [20–23]. Currently, the development of the effectiveness theory includes researching new operational properties [21] of systems and their operation processes, for example the potential of systems [22] and the efficiency of the application of information technologies [23].

2.2 Comparison of the Russian School of effectiveness research with the main trends in effectiveness research abroad

The effectiveness of system operation and other operational properties abroad has been traditionally studied on the basis of economic and econometric methods and models that are closely related to the functioning of the free market, as well as through development of heuristics, i.e. the best practices (according to operational properties), such as PMBOK, CMMI, BABOK, and BPM, especially regarding the application of IT. The description of a number of foreign studies on operational properties and their comparison with domestic results is presented in [21] and [23] (regarding the study of operational properties of IT). It is important to note that due to the differences in models and methods applied, the Russian School of effectiveness study carries out an estimation of operational properties quantitatively, using analytical models of purposeful functioning, based on natural laws. As a result, the prediction on analytical (numerical) models and solving the optimization problems become possible. Meanwhile, foreign scholars have traditionally challenged these tasks with "qualitative", "expert" or "a posteriori" research, while models are rarely analytical. Solutions are gained by the generalization of experience and processing of operational results (a posteriori) using econometric and expert methods. It is obvious that the development of mathematical models for the analytical estimation of operational properties is associated with considerable difficulties. Nevertheless, the need in using analytical models persists. For example, such a need arises when using Decision Management Systems [24], during the control of complex programs [25], application of a new generation IT business analysis [26], and Agile technologies [27], in particular. Analytical estimation could automate the solving of some practical problems that are currently solved with heuristics. The experience accumulated by the domestic school of the effectiveness research can allow for meeting new challenges of current practices based on the analytical estimation of OP, or using both the analytical estimation of OP and the elements of the expert assessment [28]. Let us consider two types of these new, crucial tasks.

3 Tasks of the study of operational properties of agile systems

Basic information about the operational properties as properties characterizing the adaptability of systems required for achieving the possible targets during their operation, peculiarities of the estimation of these properties, and the performance of their estimation were introduced in [22]. This information is used as the basis for the study of OP. Reaching the possible targets during operation means getting the possible effects (major results) of the operation, complying with targets, and while under possible conditions. At the same time, the targets and requirements for effects as well as the possibility of obtaining effects are subject to change depending on the operating conditions. The feature of the earlier introduced concepts and principles of OP estimation regarding the improvement of systems and processes of their operation implies, first of all, that such operational properties, as the potential of a system [23], allow for transition actions implemented with changes in system performance, the results of such transition processes, and the results of subsequent operations. Therefore, it becomes possible to estimate OP, taking into account the effects of transition actions, and to solve the mathematical problems mentioned hereinabove. Secondly, the introduced OP allow for switching to the analytical estimation of OP performance, depending on the selected system characteristics, their functioning and improvement (targeted changing), and capability of purpose updating, which allows solving the tasks arising during improvement, such as mathematical problems of analysis and synthesis as well as optimization problems. Examples of operational properties are efficiency, effectiveness, resource intensity, efficiency of system operation, system potential, operational properties of application of information technologies, controllability, and competitiveness [21]. A study of OP includes an evaluation of the OP performance, analysis of OP, and synthesis of system characteristics and its operation, according to the performance of the operational properties. The result of the OP study is a detailed specification of the actions undertaken, i.e. solving the task (mathematical) of action selection. If digital IT is applied, a solution is obtained in the form of a (digital) action model. This model may be used in practice through the targeted changes in practice (focused improvement of activities). The accuracy of OP estimation influences the results (produced using a particular IT) of solving the problems of the selection of actions (transition or fundamental, targeted) followed by the results of the improvement of operating experience.

The need to investigate the operational properties, solve the problem of choosing actions, and improve practical activities appears from time to time and the processes of its improvement are systematic and can manifest as complex tasks of experience improvement. Thus, for example, the actions planned as a result of solving the problems of OP performance determination can be arranged in the form of a state target program (TP), defined as a set of interrelated activities aimed at addressing the systemic problems of development of the state and society. Such events (actions) can be used to implement changes of different types (control, transition, modernization, innovative). Their planning, preparation, and implementation can cause in their turn (in chain order) the assignment of tasks for the improvement of systems and their components, and improvement of their operation. Such tasks can appear to be newly made, and some additional actions have to be carried out for their addressing. In this regard, a number of following features are to be considered taking into account performance of OP, concepts, and methodology of OP research during improvement of systems and processes of their operation:

Estimation of OP during the improvement of systems requires the predicting of future compliance of effects to the requirements of targeted improvement of systems, their operation and other varying influences. Such improvement of operation and targeted transition actions are associated with possible changes (separate and/or complex) in operation purposes; moreover, the impacts of the environment can lead to changes in the ability to achieve targets (for example, when the given resources are changed). Therefore, it is necessary to investigate the compliance of the effects to the changing requirements of different possible conditions to achieve different targets. For example, this requirement is typical for those tasks of improvement that are nowadays solved with Agile techniques.

All the possibilities of the sequence of transition actions must be kept in mind. Transition actions are purposeful changes from previous to new operation during the operation. Transition actions can cause sequences of repetitive tasks and actions (typical) of improvement and are manifested in the form of the improvement scheme (a loop). It consists of alternating sets of targeted actions of different types, i.e. informative (actions aimed at the assessment of the situation and solving the task of selection of non-informative actions) and non-informative (actions aimed at changing of the system and its operation, as well as targeted actions necessary to obtain final results). Such a loop resembles a control loop with feedback, but it is implemented by a human with the help of engineering means and information technologies. Actions implemented during such sequences are fulfilled in different ways under different conditions because the state of the system changes during the implementation of actions. People are forced to make decisions during the implementation of these actions in order to organize these transition activities and subsequent targeted actions under such changing conditions. For this purpose, certain IT are used (e.g. conventional and/or digital). Each transition action (as one of possible action sequences) can be characterized with separate operational properties. As a result of solving mathematical problems in the research conducted on agile systems and their operation under certain OP performance of certain models of improvement and operation, for example, plans of actions or system projects that are aimed at the achievement of targets under changing environments are obtained.

It should be kept in mind that the need to implement transition actions, characteristics of transition actions, and states after the transition actions that are required for further achievement of targets depends on operational conditions, possible targets, and states in which the functioning system appears before transition actions are implemented. Given the stated conditions, it is required to solve a sequence of tasks of different types (questions that need to be answered) about the ongoing transition and other activities, depending on the observed state and necessary targets. For example, in case it is necessary to obtain an answer to the question, if some actions are required (type of task 1) to reach some other different state (and then, perhaps, an answer to the question of what state and what these actions are), or one can proceed further according to the previous plan of actions, the problem of determining whether the conditions were applied for the continuation of the planned activities arises. If the task of type 1 is solved and shows the result indicating that the actions should be implemented with one of the given alternative methods (for example, according to certain rules), one has to choose a certain method for further actions. Then, a problem (type 2) of the selection of one of the preset modes of actions arises. If the task of type 1 leads to a solution implying the need to move to a new state, requiring the implementation of such actions (from a given set of potential actions), characterized with unknown composition, structure and properties, the following task (type 3) is faced: one should determine the transition actions from a set of allowable actions so that they lead to a desired state. The tasks of the stated types can be combined, and the classification of these tasks can be further continued (for example, for the case when it is necessary to change a set of allowable actions). However, all of these types of tasks are related to the tasks of a choice and need to be solved with any (one or more) IT, irrespective of whether it automated or not.

The loop of changing actions is fulfilled directly or indirectly by someone; this person forms the targets, determines the states and their conformity with targets, predicts (or simulates) the future, updates the tasks of changing operating experience, and then solves the tasks using some IT, and, finally, transforms the practice with or without the help of technical devices.

The study of OP is the basis of implementation of the loop of targeted changes, but such a study can be conducted both on the basis of mathematical methods and models (as in the tasks of the effectiveness theory) as well as heuristically, by the generalization of experiences (as in the above stated examples of the current tasks of improving systems and their operation). The difference between the decisions of the considered problems based on an analytical evaluation of OP from the traditional solutions of the problems, based on the application of the above-mentioned Decision Management Systems, as well as from traditional solutions of tasks of integrated program management, from conventional solutions using IT business analysis implies the automation of planning of improvement processes and increase its relevance through the use of analytical models and methods. Specifically, taking into account transition actions during the improvement, formulation and solution of tasks of estimation, analysis, and synthesis of systems and their operation based on numeric, analytically calculated OP performance, allows for proceeding to the formulation and solving of practical problems of improvement of systems and their operation as mathematical tasks of OP research, application of mathematical models, and digital IT for solution of such problems. A typical OP estimation scheme allowing for an analytical evaluation was determined for the analytical estimation of OP performance.

4 Schemes for operational properties estimation

4.1 Types of schemes for operational properties estimation

Let us introduce the typical schemes for the estimation of operational properties. The first scheme is aimed at the estimation of operation efficiency based on the generalization of the concept of estimating the effectiveness of targeted processes [6, 7]. It is proposed to name the scheme in honor of the author of the theory of effectiveness of targeted processes in system operation, i.e. the Petukhov scheme. According to an estimation with this scheme, it is assumed that the decision on improving actions during their implementation, improving systems and processes of their operation does not affect (or not accepted for) the functioning, while the actions are aimed at achieving one target and are not interrupted.

The second scheme generalizes the first scheme to account for a plurality of possible operations in order to reach different targets under different conditions. According to this scheme, it is assumed that the possible improvements are determined in advance, actions may be interrupted when the target is changed, and then certain transition improvement actions determined before the start of operation.

Finally, the third scheme summarizes the first two schemes in order to account for both possible transition actions selected before their application to achieve different targets and targeted transition actions selected and implemented during operation, depending on the prevailing conditions.

4.2 Petukhov's scheme of operational properties estimation

Let us introduce I_p as a value of measure on the set, p as a function defining this measure, Y-set of vectors of random characteristics of operational effects (characteristics of operational quality), R-set of vectors of the required component-wise relations between random values of effects of characteristics and their desired values, Y_r-set of vectors of characteristics values of the required functioning effects. Then, the estimation of I_p is set by the following scheme:

$$p\colon Y\times R\times Y^r \to [0,1] \tag{1}$$

$$I_p=p(Y\times R\times Y^r) \tag{2}$$

I_p is the measure value of the possibility indicating that the predicate in parentheses takes the value "true" or the value indicating that the random event corresponding to the predicate occurs.

Thus, if $Y=y$ is a random variable defined on the axis of real numbers, R is the ratio "at most", y^r is the point at the real axis (the required limit value of the random variable y), then $p=F_y(x)$, $I_p=F_y(y^r)$ is the value of the function of distribution of the random variable y in the point y^r.

4.3 Scheme of operational properties estimation given that the transition processes are known during target changing

Let us introduce $Y^r(t)$ as a random process modeling any possible accidental changes Y^r in time (which corresponds to the possible changes in targets of system operation), t_0 as the start moment of the system operation, $Y^r=Y^r(t_0)$, T as a target duration of the system operation. Then, the scheme of the estimation of operational properties would be of the following type:

$$p\colon(Y\times R\times Y^r(t),\ t_0,\ T)\to[0,1] \tag{3}$$

$$Iop=p(Y\times R\times Y^r(t)),\ t\in[t_0,T] \tag{4}$$

I_{op} is the measuring value of the possibility that the predicted values of system operation effects (under varying targets) comply with the desired values of effects in the corresponding way. Whereby:

$$I_{op}(t_0)=I_p \tag{5}$$

If the requirements are changed and corresponding changes can be determined before operation, it is necessary to plan a transition process (with the characteristics u) from one operation to another under the stated changes, which makes it possible to estimate the value of I_{op}. However, if characteristics u of transition actions cannot be determined in advance and depend upon the state of the system and the environment during operation, it is necessary to use the third estimation scheme. According to this scheme, all transition actions are a sequence of changing actions depending upon the state s (a state is the set of properties performance at a given time) during the operation of the system and the environment.

4.3 Scheme of the estimation of operational properties given the processes of improvement depend on system states during its operation

The last OP research scheme is used when sequences of transition processes described with the characteristics u are planned and implemented during operation, depending on achieved states of the system and the environment s. Transition actions are a set of actions and relations between them, selected in accordance with different IT applied to change the "target" action and (or) necessary for the better achievement of changed or unchanged targets in view of the state that would result when the target is complete. These actions show the aimed effects because of the actions selected for the "target" process and changing its effects. At the same time, support resources are spent for their implementation. The constraint of $u(s)$ describes the characteristics of transition processes necessary for calculating the effects of transition actions and then targeted actions. Characteristics of the sequence $uc=u_1... u_n$ of such transition actions out from the plurality of possible sequences characteristics Uc depend on the characteristics of the manifested sequences of states $sc=s_1... s_n$ out of the plurality of possible sequences of states Sc. The scheme of estimation of operational properties in this case is as follows:

$$p: Y(Uc(Sc),t) \times R \times Yr(t) \rightarrow [0,1] \quad (6)$$

$$I_{ops}=p(Y(Uc(Sc),t) \times R \times Y^r(t), t \in [t_0, T],$$

$$sc \in Sc, uc \in Uc \quad (7)$$

I_{ops} is the measuring value of the possibility that the predicted values of operation system effects (under varying targets and transition actions) meet the required values of effects in a corresponding way, in accordance with these targets. Note that the possible sequences sc and uc depend on applied IT. At this, the previous estimation scheme complies with $Uc=u$:

$$I_{ops}(Uc(Sc),t)=I_{ops}(u,t)=I_{op}(t) \quad (8)$$

5 Conclusion

According to a brief analysis of the research conducted on the effectiveness of the targeted processes of system operation and a number of other operational properties performed by domestic and foreign authors, it was concluded on the relevance of solving the current problems of OP research based on the analytical estimation of OP performance. The work describes the main features of such important OP research tasks as the tasks of the research of operational properties of agile systems and the tasks of the research of operational properties when using IT. The work states some schemes of analytical estimation of the operational properties and discloses their features. The obtained results should lead to a solution for the contemporary problems of the operational properties research using mathematical models.

Performed under support of the RFBR grant No. 16-08-00953

References

1. A.N. Kolmogorov, *Collection of articles on the theory of firing. Steklov Mathematical Institute*, **12**, 7–25 (1945)
2. E.S. Wentsel, Y.M. Likhterev, U.G. Milgram, I.V. Khudyakov, *Basics of fighting efficiency and operations research* (Zhukovsky Air Force Engineering Academy, Moscow, 1961)
3. G. Averkina, G. Epshtein, I. Grekova, New World, **4** (2008)
4. A.Ya. Ioffe, V.I. Markov, G.B Petukhov, *Probabilistic Methods in Applied Cybernetics* (Leningrad, 1976)
5. L.M. Morozov, G.B. Petukhov, V.N. Sidorov, *Methodological basis of the effectiveness theory* (Leningrad, 1979)
6. G.B. Petukhov, *Basics of the theory of efficiency of targeted processes. Part 1. Methodology, methods and models* (1989)
7. G.B. Petukhov, V.I. Yakunin, *Methodological basis of external designing of targeted processes and purposeful systems* (AST, Moscow, 2006)
8. *Reliability and efficiency in engineering: a guidebook in 10 volumes* (Mechanical Engineering, Moscow, 1988)
9. V. Dubrovsky, M. Kovbich, Independent newspaper, (2013)
10. U.I. Kryuchkov, V.I. Kuznetsov, *Research on effectiveness of weapons* (Dzerzhinsky Military Academy of Russian Strategic Missile Forces, Moscow, 1980)
11. E.B. Volkov, V.Z. Dvorkin, A.I. Prokudin, *Engineering bases of efficiency of missile systems* (Mechanical Engineering, Moscow, 1989)
12. V.D. Roldugin, *Simulation and evaluation of the Strategic Missile Forces fighting* (Dzerzhinsky Military Academy of Russian Strategic Missile Forces, Moscow, 2005)
13. Y.V. Kryuchkov, V.I. Kuznetsov, G.N. Okhotnikov, *Simulation and evaluation of military operations and weapons of Strategic Missile Forces* (Moscow, Dzerzhinsky Military Academy of Russian Strategic Missile Forces, 1986)
14. I.V. Lysenko, *Analysis and synthesis of elaborate engineering systems. Part 1: Analysis and synthesis of systems providing readiness of boost and space vehicles for launching (basic theory)* (Voenizdat, Moscow, 1995)
15. G.P. Zhukov, S.V. Vikulov, *Military and economic analysis and operations research* (Voenizdat, Moscow, 1987)
16. I.A. Ryabinin, Yu.M. Parfenov, *Reliability, capability and safety of ship's power systems* (Naval

Academy named after the Soviet Union Fleet Admiral N. G. Kuznetsov, St. Petersburg, 1997)

17. B.S. Fleishman, *Elements of the theory of potential efficiency of complex systems* (Soviet Radio, Moscow, 1971)
18. V.V. Podinovsky, V.D. Nogin, *Pareto-optimal solutions for multiobjective problems* (Nauka, Moscow, 1982)
19. V.D. Nogin, *Decision-making in multi-objective environment: a quantitative approach* (FIZMATLIT, Moscow, 2005)
20. I.V. Lysenko, *Proceedings of SPIIRAS*, **1**, 49-64 (2002)
21. A.S. Geyda, Z.F. Ismailova, I.V. Klitny, I.V. Lysenko, *Proceedings of SPIIRAS*, **35**, 136-160 (2014)
22. A.S. Geyda, I.V. Lysenko, *Proceedings of SPIIRAS*, **10**, 63-84 (2009)
23. A.S Geyda, I.V Lysenko, R.M. Yusupov, *Proceedings of SPIIRAS*, **42**, 5-36 (2015)
24. J. Taylor, *Decision Management Systems: A Practical Guide to Using Business Rules and Predictive Analytics* (IBM Press, 2011)
25. T. Kendrick, *How to Manage Complex Programs* (Amacom, 2016)
26. Th.W. Dinsmore, *Disruptive Analytics: Charting Your Strategy for Next-Generation Business Analytics* (Apress, 2016)
27. *The Agile Extension to the BABOK® Guide. International Institute of Business Analysis* (2015)
28. I.V. Lysenko, New in public management, **2**, 117-130 (2007)

DOI: 10.1051/shsconf/20173501059

1C software vulnerabilities description

Oleg Ivanov[1,*], and *Dmitry* Silnov[1]

[1]Institute of Cyber Intelligence Systems, National Research Nuclear University MEPhI, 115409 Moscow, Russia

Abstract. This article is devoted to the vulnerability of the application solution based on the "1C: Enterprise 8" platform, which can be used by only built-in tools of the platform. Possible threats and attack algorithm are described.

1 Introduction

Today a large amount of companies from self-employed entrepreneurs on large manufacturers use application solutions based on the "1C: Enterprise 8" platform [1] for financial and regulatory accounting. In the last few years 1C company has made a lot to improve its' software usage ease. Users can now avail of such functions as direct exchange [2] of bank transfer order, electronic document flow, on line receipt of counterparties' folders, direct upload of regulatory reporting to inspection authorities etc. There is little wonder that organizations rely on these app solutions more and more in their work. Databases contain all the information concerning the partners, clients interaction history, all bank details of the company, employees personal data, pricing procedures, information on performed, current and planned marketing actions etc. It is common knowledge that leak or substitution of such information can do extreme damage to the company. At the same time protection actions are usually limited to such standard actions as database access rights differentiation, anti-virus software usage and other common means.

But what if the threat is inside the 1C configuration [3]? This type of situation is possible in case user updates the configuration with an unofficial file.

Thus, eventually an authorized user with full permissions will open the database to update the configuration. Administrator will decompress the update file, and as a result, a new .cfu file with renewed metadata will appear. Launching the update the administrator will see the confirmation (fig. 1) that the update file fits the current configuration and is produced by 1C. He will press the OK button and passively monitor the update procedure.

The intruder only has to create his own update file with the harmful code implemented and to upload it to any illegal software portal in the Internet.

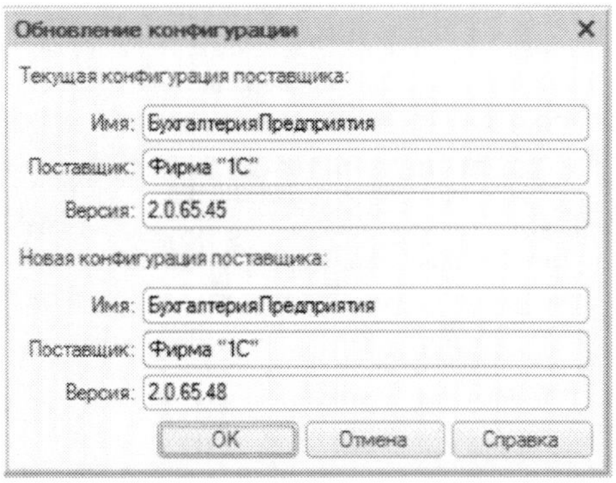

Fig. 1. The confirmation window.

2 Threats

In June 2016 the "Doctor Web" company announced on their website that the first Trojan for 1C has appeared in their virus laboratory [4]. This Trojan was posed as 1C external data processing unit [5] called «ПроверкаАктуальности-КлассификатораБанков.epf» (BankIdentifierCodeReview.epf), distributed via e-mail with the topic "We have our Bank Identifier Code changed" and the following message: "Dear user, Our bank identifier code has been changed. Please update your bank classifier. You can do it in automatic mode if you use 1C: Enterprise 8. File >> Open attached data processing unit for classifiers update >> Yes. The classificatory will be automatically updated in 1-2 minutes if the internet access is on". After that, the dialog window with bank classificatory update proposal appears. Whatever button is pressed at this stage, the processing unit will search for all counterparties e-mail addresses and send them the same e-mail message. Further, Trojan.Endorer.567 [6] is extracted from the processing unit, saved on the HDD and is then launched by the code on fig. 2. Afterwards, Trojan.Encoder.567 works on its own.

* Corresponding author: ivanofoc@gmail.com

DOI: 10.1051/shsconf/20173501059

```
ИмяВременногоФайла = "D:\ОбновлениеБИКБанка.exe";
Попытка
    ДвоичныеДанные.Записать(ИмяВременногоФайла);
Исключение
    ИмяВременногоФайла = "C:\ОбновлениеБИКБанка.exe";
    Попытка
        ДвоичныеДанные.Записать(ИмяВременногоФайла);
    Исключение
        ИмяВременногоФайла = ПолучитьИмяВременногоФайла("exe");
        ДвоичныеДанные.Записать(ИмяВременногоФайла);
    КонецПопытки;
КонецПопытки;
ЗапуститьПриложение(ИмяВременногоФайла);
```

Fig. 2. Trojan.Encoder.567 start up code.

This is a classic scenario of social engineering. To avoid this kind of situation all typical configurations in 1C have an option to turn off the interactive use of external processing units [7]. When on, this function only allows launching the external data processing units approved by administrator. However, if Trojan is not in the form of external data processing unit, but is positioned as a configuration update, external processing interaction launch interdiction will not help.

Obviously, database configuration modification gives numerous opportunities to the intruder. The easiest threat is bank account details substitution and upload of false bank transfer orders. Besides, stock accounting data can be altered: prices increased or decreased addresses of counterparties or clients changed.

Another major concern is that the intruder gets the access to client database. Apart from information pilfering for personal use of for reselling, there is a threat of a forethought reputation damage of the company. For instance, it is possible to configure regular task through the built-in e-mail client to send spam letters to all e-mail addresses found in the system.

A noteworthy detail is that such attack realization is very easy to perform. The update file is easy to create on one's own with standard platform means. All what is needed is a typical configuration file [8] compliant to the update.

3 Update file creation

Let us take a closer look at the process of creation of the update file with the harmful code. For that, we need the 1C: Enterprise 8 platform, the configuration to be updated and the official update file [9] available on 1C website.

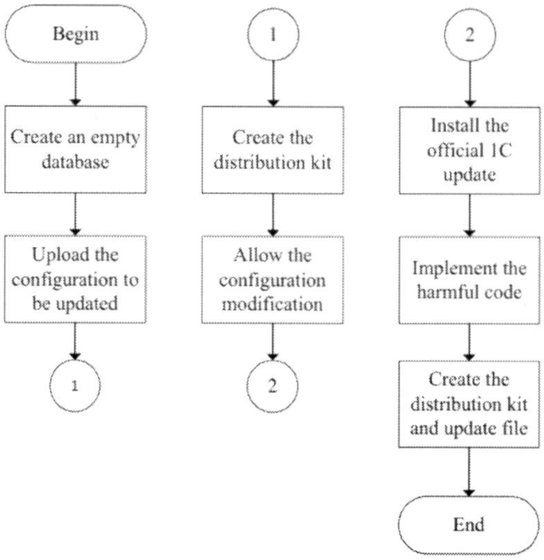

Fig. 3. Upgrade file preparation algorithm.

First, we need to create an empty database [10] and upload the configuration to be update [11]. Then we should create the distribution kit. Then we need to allow the configuration modification [12] and to install [13] the official 1C update.

After the installation, it is possible to implement the harmful code. You can see one simple example of such code below. Here, to the object module [14] of data processing unit "Client bank" we added the procedure «ОткорректироватьЗначениеЭкспорта» ("CorrectExportValue), which changes bank details of the bank transfer beneficiary.

```
Процедура ОткорректироватьЗначениеЭкспорта(СтруктураДляЗаписи)

    Попытка

        СтруктураДляЗаписи.Получатель = "ИНН 000112582645 Иванов Иван Иванович";
        СтруктураДляЗаписи.Получатель1 = "Иванов Иван Иванович";
        СтруктураДляЗаписи.Получатель2 = "";
        СтруктураДляЗаписи.Получатель3 = "";
        СтруктураДляЗаписи.Получатель4 = "";
        СтруктураДляЗаписи.ПолучательБИК = "000000003";
        СтруктураДляЗаписи.ПолучательБанк1 = "АКБ ""АБТ-БАНК""";
        СтруктураДляЗаписи.ПолучательБанк2 = "Г.МОСКВА";
        СтруктураДляЗаписи.ПолучательИНН = "000112582645";
        СтруктураДляЗаписи.ПолучательКПП = "";
        СтруктураДляЗаписи.ПолучательКорсчет = "30101810100000000774";
        СтруктураДляЗаписи.ПолучательРасчСчет = "40802810900001011111";
        СтруктураДляЗаписи.ПолучательСчет = "40802810900001011111";

    Исключение

    КонецПопытки;
КонецПРоцедуры
```

Fig. 4. Harmful code example.

This procedure can only be launched when the payment order is dumped.

```
Функция ПолучитьСекциюДокумент(СтруктураДокумента, КоллекцияРеквизитов, ВерсияФормата)

    Буфер = "";
    Реквизит = "";
    СтруктураДляЗаписи = СформироватьСтруктуруЭкспорта(ВерсияФормата);
    ЗаполнитьЗначениеЭкспорта(СтруктураДляЗаписи,СтруктураДокумента);

    Если СтруктураДокумента.ВидДокумента = "Платежное поручние" Тогда

        ОткорректироватьЗначениеЭкспорта(СтруктураДляЗаписи); //Функция заменяет рекв

    КонецЕсли;

    ДобавитьВСтроку(Буфер, "СекцияДокумент=" + СтруктураДокумента.ВидДокумента);

    Для каждого ВыгружаемыйРеквизит из СтруктураДляЗаписи Цикл
        Значение = ПривестиЗначение(ВыгружаемыйРеквизит.Значение);
        Если НЕ ПустаяСтрока(Значение) Тогда
            ДобавитьВСтроку(Буфер, ВыгружаемыйРеквизит.Ключ + "=" + Значение);
        КонецЕсли;
    КонецЦикла;

    ДобавитьВСтроку(Буфер, "КонецДокумента");

    Возврат Буфер;

КонецФункции
```

Fig. 5. Code inhection point.

Finally, we need to create the distribution kit [15-16] and update file. As a result, we get the .cfu update file. The only thing left is to upload it to any illegal software portal in the Internet.

The system administrator can presumably discover the code: it is very easy to check whether the configuration was developed or not. For this, we need to compare [17] the database configuration [18] or main [19] configuration with the configuration of the supplier [20]. If the database configuration was developed, the difference will be shown in the comparison window. However, if the changes were implemented to the update file, they will be in the supplier's configuration too. Thus, comparison will not discover any implementation done by the intruder. In this case the harmful code can only be discovered through comparison on the database configuration with the external typical configuration file but it is rarely done in practice.

Thus, the harmful code will only be discovered at the next update. Though, it is worth saying that the configuration updated by an unofficial file cannot be further updated by official 1C releases. If we try it, we will see the window in fig. 6 instead of the one in fig.1. It announces that the file has no available updates. This exact message is a sign of altered configuration of the client and of the last update being illegal. Now to resume the right operating of the program we will need to install official 1C configuration.

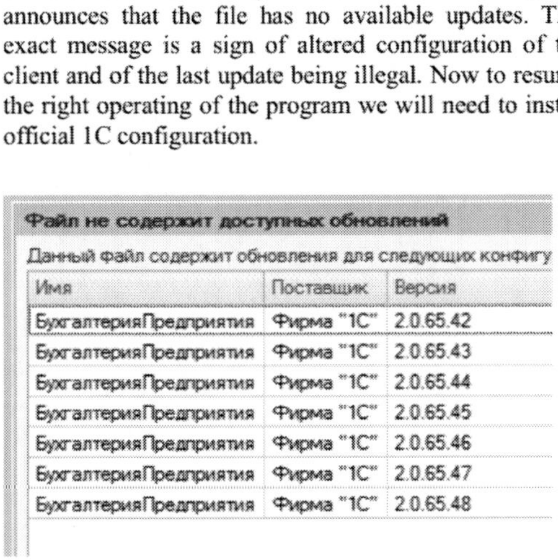
Файл не содержит доступных обновлений

Данный файл содержит обновления для следующих конфигу

Имя	Поставщик	Версия
БухгалтерияПредприятия	Фирма "1С"	2.0.65.42
БухгалтерияПредприятия	Фирма "1С"	2.0.65.43
БухгалтерияПредприятия	Фирма "1С"	2.0.65.44
БухгалтерияПредприятия	Фирма "1С"	2.0.65.45
БухгалтерияПредприятия	Фирма "1С"	2.0.65.46
БухгалтерияПредприятия	Фирма "1С"	2.0.65.47
БухгалтерияПредприятия	Фирма "1С"	2.0.65.48

Fig. 6. The list of suitable configurations.

4 Conclusion

The described algorithm confirms that typical instruments of 1C: Enterprise 8 platform allow the intruder to create an update file with a harmful code implemented that gives him unlimited opportunities to use the databases of careless users. Though the code will only be implemented for the period since illegal update untill the next legal one, this time is enough to do a severe damage to the company. And it was possible to avoid even such short-time harmful code implementation if 1C used digital signature mechanism while creating and installing its update files. For the moment, the only way to avoid this risk is to refuse downloading the updates from untrusted sources.

References

1. 1C: Developer Network. "What is 1C:Enterprise?" http://1c-dn.com/1c_enterprise/what_is_1c_enterprise/
2. Direct bank (direct fraud with the bank) - new technology of the "1C: enterprise 8". system. http://www.v8.1c.ru/edi/edi_app/bank/
3. 1C: Enterprise 8. "Configuration Objects (Metadata)". http://v8.1c.ru/ eng/the-platform/dev_confobjects.html
4. Dr. Web. "1C.Drop.1". http://vms.drweb.ru/virus/?i=8247570.
5. 1C: Enterprise 8. External processing. http://v8.1c.ru/ overview/Term_000000601.htm
6. Dr. Web. "Trojan.Encoder.567". http://vms.drweb-av.pl/virus/?i=3799463
7. Helpme1C.ru "How to open interactive discovery of external reports and downloads?". http://helpme1c.ru/kak-razreshit-interaktivnoe-otkrytie-vneshnix-otchetov-i-obrabotok-dlya-1sbuxgalteriya-8-3-redakciya-3-0
8. 1C:Enterprise 8. "Soxing and configuration configuration". http://v8.1c.ru/ overview/Term_000000602.htm
9. GoodWill. "How to update 1C 8.2 i 8.3 independently through the configurator - a step-by-step instruction". http://programmist1s.ru/obnovit-1s-samostoyatelno/
10. Hello, 1C. "Creating and adding a database of dennics 1C". http://howknow1c.ru/ nastroika-1c/sozdanie-bazy-1s.html
11. Hello, 1C. "Installation of 1C 8.2 - Configuration and Database". http://howknow1c.ru/nastroika-1c/ustanovka-1s-82.html
12. Forum 1C. "The possibility of changing the configuration of 1C". http://forum1s.ru/read/vkliuchit-vozmozhnost-izmeneniia-konfiguratsii-1s/
13. 1S Shop.ru "How to update the typical configuration of 1C 8". http://www.1sshop.ru/index.php3?id=138
14. Professional courses on 1C. "Modules in the 1C platform: enterprise 8.3". http://курсы-по-

1c.рф/articles/модули-в-платформе-1с-предприятие-8-3/

15. Infostart.ru: Information and Analytical Center for Automation of Educational and Management Activities. "As reported on its own update for 1C". http://infostart.ru/public/315534/
16. Portal of Information and Technical Communication 1C. "An Introduction to Post-Conflict Support and Support."https://its.1c.ru/db/metod8dev#content:2294:hdoc
17. 1C: Enterprise 8. "Comparison and integration of configurations, mechanism". http://v8.1c.ru/overview/Term_000000291.htm
18. The life of 1C. "Concepts of configuration and databases. Overview of the three configurations embedded in the love information system 1C". http://life1c.ru/post/400
19. 1C: ITS. Information and technological support of users 1C: Enterprise. "Established configuration and configuration of basic data". http://its.1c.ru/db/pubdevguide83#content:71:hdoc
20. Everything for the 1C programmer. "The general technological information on the support of 1C confuguration". https://1cprogrammistu.ru/index.php?newsid=119&seourl=obschie-tehnologicheskie-svedeniya-o-podderzhke-1s-konfiguracii

Commercial secret as an instrument of company competitive strategy effectiveness increase

Dinara Peskova[1,*], *Yuliya* Vasileva[1], and *Mikhail* Nazarov[2]

[1]Department of Economics, Ufa State Petroleum Technological University, Ufa, Russian Federation
[2]Department of Taxes and Taxation, Samara State University of Economics, Samara, Russian Federation

Abstract. Modern companies are very much diversified in scale, sectoral affiliation, marketing behavior. There are many theoretical and applied studies in effective competitiveness strategies (see Porter, M. (2002, 1998), Kramer, M. (1998), Fatkhutdinov, R. A. (2000), Feigelson, V. M. (1996) and others).They present famous approaches and probably there is no need to repeat them in this article. We would like to feature a different concept (suggested by Yudanov A. and followers) with terminology adopted from natural sciences and show the way the commercial secret can increase effectiveness of competitiveness strategy. We also perform valid methods of commercial secret protection.

In conditions of functioning in highly competitive market companies are looking for innovative instruments to raise the effectiveness of their business.

Information is one of the most significant factors of production in modern economies. Modern entrepreneurship depends as much on the possession of information about potential customers, trade venues, and technical data as it does on personnel, machinery, and other capital goods. The reliance of industry on rapidly developing technologies amplifies the value of information. However, information is also different from other types of economic resources. It can be divided into an infinite amount of parts and, hence, is very complex. Importantly, it is often impossible to guarantee exclusive possession of this resource. The concept of "trade secret" was developed to limit access to such commercially valuable information.

A "commercial secret" is a piece of confidentially kept information that gives an entrepreneur some economic advantage over competitors. This could range from information about specific production techniques or production formulas to marketing strategies or customer data.

There are plenty of ways to classify the companies. According to implemented marketing strategy there can be outlined four types of companies: Explerents, Patients, Violents and Commutants [11]. Using the existent classification we will suggest the most effective pattern of economic behavior for them based on differences of using mechanisms and amounts of commercial secrets.

Explerents are the pioneers of business. The strategies they realize are oriented into risky innovations based on applying the newest highly-efficient technologies leading to maximal profit, venture investments. According to expert's estimations risks of this type of strategy are extremely high. The most positive approximations give three out of ten venture projects reach the break-even point, six loose and only one can become highly profitable.

Patients are pointed into niche specialism and production of small batch of elite goods and overcharge purchasers. Usually their goods are of very elastic demand what commits them to the tough forms of competitive struggle.

Violents as a counter to patients prefer overt price competition. They orientate to the mass-market customer and strife towards cost reduction.

Commutants are small, non-specialized companies. They are very flexible and use any possibilities for business but usually don't possess extended manufacturing nor Research and Advanced Development capacities. Commutants cover the remaining after all others part of market. These companies tend to seize the competitors' commercial secrets rather than create themselves. This is why we do not emphasize their role in our research.

The three basic marketing strategies use commercial secret as an instrument to gain competitive advantage though the content and amount of information classified as a commercial secret differ (Fig. 1).

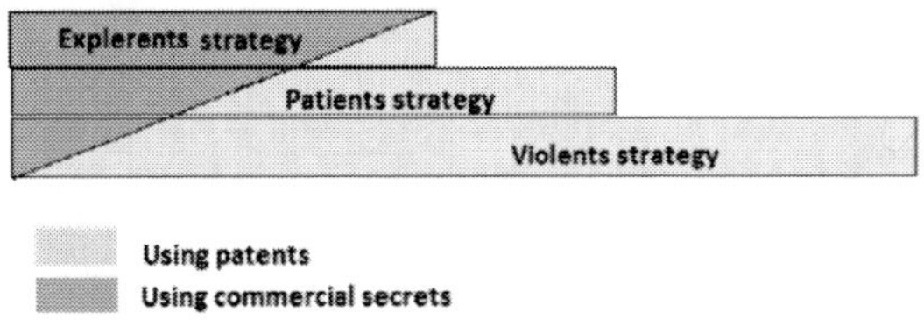

Fig. 1. Basic marketing strategies.

* Corresponding author: peskovad@yandex.ru

Explerent competitive strategy. The companies following this strategy direct most efforts into research and advanced development path-breaker activity. Their best advantage and the condition of high –profit earning is the anticipating implementation of principal innovations. Taking high-level risk explerent companies build up new markets and get the most benefit out of monopoly position. These companies will combine patents and commercial secret instruments. The reasons to use commercial secret versus patent are: 1) the process of filing of an application for patent is time-taking and thus the pioneer position can be disrupted, the competitor can appear and start production first; 2) patenting inherits additional expenditure ; 3) patent provides long-lasting protection (ex. 20 years) which is not a goal for explerents, the supernormal profit is possible to get only in short-run period. These factors explain the propensity of explerents to use commercial secret but some companies will patent technologies aiming to commercialize them to other market participants (for example to Patients). Patent protection facilitates the process of scientific research results circulation. Combing of both systems of information protection allows the companies use commercial secret during the initial stage and the period of patent execution and later in case of favourable decision – use patent. Portfolio models are also possible – some innovations are under the commercial secret protection, others hold a patent. Explerents' commercial secrets consist of engineering solutions, rarely organizational decisions and more rarely client databases.

Patient strategy in Russian economic reality is inclined to production and research programs which are realized by the companies inventing science-intensive technologies.

Patient market strategy is typical for monotechnic companies. It provides production of special, non-ordinal goods for certain narrow circle of customers. Patient-companies draw market strength from their position of manufacturers of irreplaceable goods for some niche of clients. These firms tend to control a small amount of market, addressing their expensive and high-quality articles to those consumers who are not satisfied with standardized goods. Patients shirk direct competition with leading corporation. They will rather consider and gratify the very special consumer requests. Patients vary the markets.

Usage of commercial secrets by companies of patient strategy is determined by mostly the same reasons as for explerents but additionally due to their narrow specialization commercial secret can provide them more advantages. Business of patients companies is already protected by entry barriers such as equipment cost, workers' skills, required technological base. This is why coping of their goods is complicated and unpromising: all possible competitors of patient markets are well-known and appearance of very new participant is seldom. Commercial secrets of patient strategy companies include technological, organizational and commercial information.

Violent competitive strategy is mostly directed to maximal price reduction due to cutback of expenditures. The main point is to orientate into large turnout of standardized goods. This type of production is less-cost in compare with manufacturing small lots of varied goods. Violent-companies aim to dominate in wide market attracting customers with low price and satisfactory quality. Violents acting in a massive demand market are less interested in protection of their technologies with commercial secret because: 1) they are not screened with entry barriers and need to be protected with patent; 2) they are in need of advertisement that leads to necessity to open part of technological advantages to pull custom, 3) they are more susceptible to reverse engineering methods. Though unpatentable information should be classified.

This classification leads to the following conclusions:

1) while selecting the competitiveness strategy the company initially should pay enough attention to commercial secret protection.

2) if companies (especially violents) report of intensive commercial secret usage, it's a sign for regularity authorities about unfair business practice.

The technology of science-intensive products invention and its entry into the goods market includes several stages (Fig. 2):

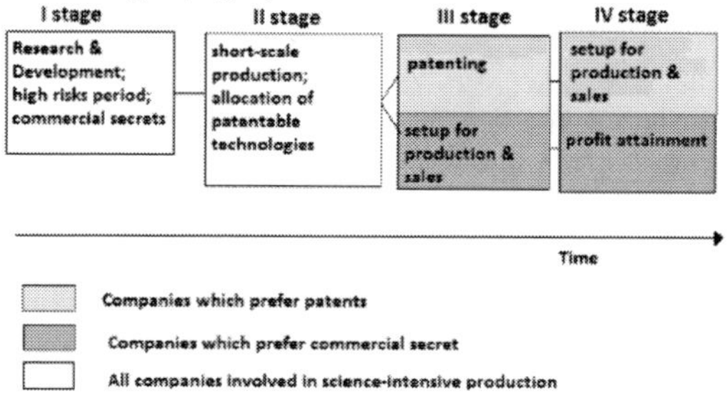

Fig.2. General stages of science-intensive production.

The first stage consists of research and development activities. This is the period of the highest risk of valuable information leakage and the only effective instrument of protection is commercial secret.

The second stage is the start of short-scale test production, demand research and product's determining of patentability. Depending on this stage's results the direction of further performance shall be selected: using patent or commercial secret.

On the third stage companies preferred miscellaneous methods of information protection act differently. Entrepreneurs using commercial secret can start to setup for large-scale production and sales. Patent-focused companies should spend additional time to get through the application for patent procedures and thus can startthe large quantity manufacture only on the next stage.

The presented stages of science-intensive production process, general and universal for companies any competitiveness strategy, show that usage of commercial secret protection significantly increases effectiveness of entrepreneurial activity aiming to innovations.

In compare with patenting the advantage of commercial secret is its ability to protect 1) ideas; 2) non-patentable objects which cannot be treated as discoveries, useful models, utility inventions; 3) delinked information; 4) negative advance (which is information about negative results of explorations); 5) damaging information; 6) information with very short and very long-run usage period.

The downside of commercial secret is low disclosure hardness. It can be subject to simple analysis, expert "data pump" or random seepage.

Provisionally the process of commoditization divided the information kept as commercial secret into two types:

- inalienable commercial secret provides safety and competitive integrity of internal company's activity;
- seizable commercial secret which gained the status of commodity.

In practice the inalienable type of commercial secret is mostly spread while seizable commercial secret is better protected by patents to facilitate the commercialization process.

The complex work on identification of threats to safety should preface the data protection system setup. The sequence of actions will include: detection of threats (probability levelevaluation, systematization); estimation of potential loses; preliminary selection of commercial secret protection; estimation of economic effect and costs. The following classification of threats to the commercial secret owner can be suggested (Table 1):

Table 1. Commercial secret threats classification.

Criteria	Types
Threat source	External (business espionage, competitors' illegal acts, stealing of tangible and intangible assets and valuables); Internal (disclosure of confidential information by the employees, low work motivation, non-effective security service)
Severity level of consequences	Threats with heavy severity level result in all economic indicators breakdown and stop of business activity; threats with mild severity level require high expenses on recovery but not much time; threats with low severity level have potential disruptive affect.
Probability of threat execution	Hardly probable threats; real risk
Stage of firm's functioning	Onstage of initialization; on stage of consistent production. Threats on initializing stage are more dangerous than others.
Object of infringement	Information; finances; tangible assets; good will and others
Subject of infringement	Organized crime; unscrupulous competitors; employees, state authorities.
Disbenefit type	Direct; lost profit.

The existent practical methods of commercial secret protection can be grouped upon functional criteria (Fig.3):

1. Documental methods include rules of circulation of internal papers containing the confidential information. The status "Commercial secret" is considered to be set up after the confidentiality safeguard measures are taken:

a. make a list of confidential information to be kept as commercial secret;

b. take steps to regulate the inside relations upon realization of commercial confidential information by employees according to Labor contracts, by business partners on the ground of civil law treaty;

c. mark the physical media on which confidential data are stored with "Commercial secret' label, pointing the owner of the information;

d. use means and methods of technical protection of information.

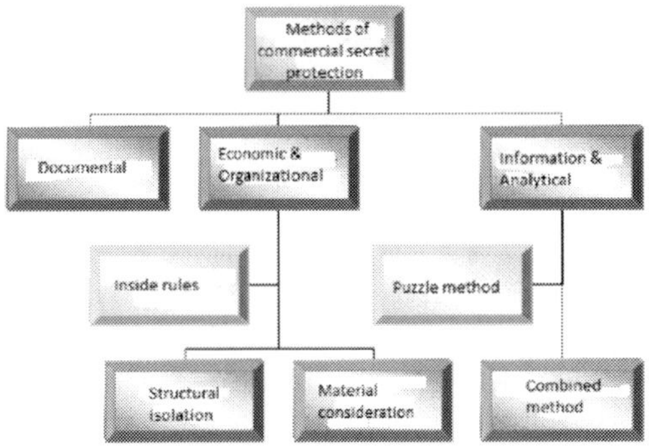

Fig.3. Commercial secret protection methods.

This group of methods consists of formal and legal regulations. Their obvious advantage is apprehensibility and clarity of commercial secret protection policy though in practice can protect mostly from unintended seepage. Execution of named actions will provide an evidentiary foundation in litigation in case of commercial secret loss. However the judicial settlement is usually the measure of last resort which indicates of irrecoverably loss of confidential information. Moreover often austerity measures may result in negative spin-offs: they induce and routinize the infringements.

2. Economic and organizational methods. In the context of institutional economic approach an employer has three basic measures to protect commercial secret for leakage. 1) establish the corporative rules of commercial secret handling; 2) the system of compensational payments to the employees for creation and keeping commercial secrets; 3)the structural isolation of particular departments. Let's take a closer look to each of these methods.

a. Inside corporative rules. Every employer inside the company is authorized to set and request exact obedience of certain rules. Execution of employment agreement implies that the employee agrees to comply with the rules. Institution of exact instructions on commercial secret handling is the most obvious step though constrained with weak points the main of which is ineffective sanctions for rule breakage. The problem is that sanctions don't prevent losses. The inside rules can be of different types: 1) norms which limit social intercommunications of employees. When the confidential information is mostly in viva voce and is not formally arranged, even in a simple personal conversation the secret loss can take place. The standard example of this type of rules prohibits intercorporate communications with competitors existent or potential. And also this includes the rules regulating the right of access to company's documentation, system of passwords and physical entry to certain premises. Elaboration, implementation and enforcement of corporative rules of commercial secret handling build an additional expense item increasing progressively as the rules intensify. Thus The Wall Street Journal provides an example of IBM which has established intracorporate rules of work flow which stipulates a special envelope for each document according to its confidentiality level. There were four different types of envelopes designed in order of secrecy increase: "For internal use only"; "Confidentially"; "Strictly confidentially"; "Particular confidentially". Annual demand for envelopes estimates of 4 million pcs. that makes 400, 000 USD were spent for the envelopes [10]. Here also have to be added the increased transactional costs to obey the safety rules.

b. Inside corporative rules should be backed with financial compensation for creation of marketable information, so called "hush money" It's an analogue to informational rent distribution. Thus personal involving generates what makes this method effective. The difficult side is to calculate the exact amount of the compensation. The upper boundary is the maximal slate of informational rent the employer is able to share, without entrepreneurial motivation decrease. The lower bound is determined by minimal agreeable for the employee amount. Therefore the employer sets the maximum while the employee – the minimum of the compensation level of commercial secret protection. In any case the size of compensation shall relate to the damage the loss of secret can cause. Moral encouragement of workers is worthy of special mention. Moral incentives unit people in a team, raise corporative spirit, give psychological drive to demonstrate loyalty.

c. Structural isolation is the third method of information protection, the main idea of which is access restriction to certain premises for visitors and non-involved employees. The advantage of this method is in narrowing the asset to be protected from the whole company to a certain premise or department. This raises effectiveness of defense measures. Structural isolation might be supplemented by geographical reference.

3.Information and analytical group of methods includes techniques of secure commercial secret usage in production process.

a. Puzzle method suggests to separate commercial secret into parts with different conditions of access and use [5]. All parts together make a mosaic picture of commercial secret content while separate pieces might look like having no commercial value. Some parts of the puzzle can be formally completed as commercial secret, others might not. Only the chief executers understand the puzzle in whole, so this method impedes holistic comprehension of commercial secret content by third party as well as by company's employees. The disadvantage is complexity of puzzle elaboration and unintended loss occurrence (for example providing too much information in scientific conferences, mass media, personal communications, neglectful treating of sensitive documents alleged to be insignificant.).

b. Combing patent and commercial secret. This method suggests that the patentable technologies record a patent while another part of commercially valued information get classified. The described method combines advantages of two system of data protection and compensates for a shortcomings. As a disadvantage of patent experts list the following: limited time of protection, technical narrowness (sometimes patenting of exact technology or invention can give a nudge for competitors of similar but modified model) expensiveness. Free circulation, commercialization, legal legitimacy are undoubtable advantages of patent which are able to improve efficiency to commercial secret if combined.

Trade secrecy is a special economic institution. Commercial secret can be used as an effective instrument for competitiveness raise as it was shown in our paper. Different types of companies use it according to market strategy employed. Methodology of usage the commercial secret is a powerful base for confidential information protection.

References

1. About commercial secret: Federal Law of Russian Federation dated 2004 no. 98-FZ. http://www.referent.ru/1/109717/nav
2. R.A. Fatkhutdinov, *Competitiveness: economics, strategy, management* (Infra-M, Moscow, 2000)
3. V.M. Feigelson, *Basic principles of intellectual property protection and use* (VNIIPI, Moscow, 1996)
4. V.V. Mokryshev, V.M. Aldoshin, *Management of exclusive rights (intellectual property, intangible assets) in competitive struggle* (RosPatent publishing, Moscow, 2002)
5. N. Nyrovova, Industrial property, **11**, 45-54 (2008)
6. D.R. Peskova, Creative economy, **5**, 103-109 (2010)
7. D.R. Peskova, *ESSJ*, **7**, 331-338 (2011)
8. M.E. Porter, *Competitive Strategy: Techniques for Analyzing Industries and Competitors* (Free Press, 1998)
9. M.E. Porter, M.R. Kramer, *The Competitive Advantage of Corporate Philanthropy* (Harvard Business Review, 2002)
10. The Wall Street Journal (1995)
11. A.Yu. Yudanov, Competition: theory and practice (Gnom Press, Moscow, 2001)

Evaluation of alternative institutional arrangements in public utilities

Anna Ermishina[1,*], and *Elena* Mienshenina[2]

[1]Southern Federal University, Economic Faculty, Rostov-on-Don, Russia
[2]Don State Technical University, Road Transport Engineering Faculty, Rostov-on-Don, Russia

Abstract. Since the early 2000s a policy of attracting private operators to public utilities, which should help to increase productivity, reduce costs, and as a result, reduce utility prices, takes place in Russia. The aim of the study is to identify the relationship between institutional arrangements and pricing of water and sanitation services. Statistical and cluster analysis was applied to empirical data on water utilities in the 13 largest cities. There were the differences in the level and dynamics of prices for water supply and sanitation in the group of public utilities and public private water utilities.

1 Introduction

Around the world a search for the alternatives to the institutional arrangements of the municipal economy, which are allowing to provide accessibility and affordability of public utilities services for the majority of the population and the economic efficiency of the utilities, remains relevant. Water supply and sanitation sector is one of the largest of Russia's public utility industries. At the end of XX - the beginning of XXI centuries the reform of the water supply was inspired by the world practices. The experience of the developed European countries in the sphere of production, financing and water provision services was used to form the basic tenets of the reform. The most famous models among Russian scientists and industry professionals are conventionally called "English", "German" and "French" models. They got the titles in accordance with the names of the countries in which these models were widely disseminated.

The "English" model of water supply, developed in the UK, involves the complete privatization of municipal infrastructure in combination with the state regulation of tariffs for water and sanitation services. The "German" model is characterized by a high participation degree of the local authorities as in the price control, so in the production and provision of these services. Municipalities create joint ventures with operators in the form of open joint stock companies (OJSCs). The water supply infrastructure is included in the authorized capital stock as a contribution of the municipality that controls the majority of shares. The "French model" has spread in France. It involves the transfer of the municipal property, owned by the municipality, from the local authorities to the management of a private company on the basis of long-term (30-35 years) concession agreement. Investor's ownership of property and newly created objects are terminated upon the expiration of the concession [1].

The ideologists of the Russia's water supply reform came to the conclusion that the most effective model would be the "French" one in Russia. This water supply organization attracts private business and private investment to the municipal economy and, at the same time, allows controlling the production of socially significant services through the infrastructure, which remains in municipal ownership [2].

Each of the institutional alternative arrangements has its advantages and disadvantages. An indirect measure of economic and social efficiency is the cost of public resources and its dynamics. Many services (including water and wastewater services) are produced in conditions of a natural monopoly in the local economy. The unregulated natural monopoly is a situation of the market inefficiency, which overestimates optimal prices and reduces the socio-effective production. The regulation of natural monopolies aims breaking the power of monopolies to set higher prices. These measures should, in general, improve the economic efficiency of the municipal economy. In addition, the regulated tariff for communal services is the most important indicator, which shows service's affordability for the majority of the population and ensures the effectiveness of the social sector.

In the last decade the regulated tariffs for housing and communal services are growing rapidly, many times ahead of inflation in Russia. At the same time the deterioration of networks remains high, that increases the risk of accidents and reduces the quality of service. To improve service quality and to attract investments a policy of attracting private operators to the industry is held in Russia from the beginning of the 2000s. More efficient operations in terms of public-private partnership should cause an increase in productivity in the industry,

* Corresponding author: AVErmishina@sfedu.ru

costs reduction, and as a consequence, reduction of tariffs for water and wastewater services.

2 Brief literature review and goal of the research

In the foreign literature the operation results of the various institutional alternative arrangements of the municipal economy are actively studied. However, only a small number of empirical studies are devoted to the study of the relationship between institutional options for municipal economy and water tariffs.

S. Garcia et al. examined the influence of technical factors, competition and company's strategy on the water prices in France [3]. Their studies show that the strategy of the utility operator companies has a significant effect on the price level. D. Hall and E. Lobina empirically show that the privatization of the water supply companies sometimes leads to unjustified price increases [4, 5]. E. Chong et al. have also shown the example of the water supply in France, where the choice of any form of public-private partnerships instead of the direct state / municipal government is likely to lead to an increase in the cost of the urban water supply services [6].

A. Carpentier et al. compared the public and the private water utilities organization in France [7]. He found that the prices are higher in the private management, because private operators are faced with a tough environment. In addition, this article concludes that the desire of the local authorities to privatize urban water supply services is driven by a technical complexity of their maintenance. R. Martinez-Espineria et al. also found that privatization lead to an increase in the price of water in the large urban Spanish municipalities [8]. S. Zschille and M. Ruester examined the relationship between the forms of management in the water-supply German companies and prices paid by consumers (controlling economies of scale and technical and structural characteristics of the water utilities) [9]. The main conclusion of this study is the fact that the involvement of the private sector causes an increase in water prices [9].

Studies of V.M. Garcia and others examined the relationship between the various institutional arrangements of municipal economy and the level of the urban water prices for the population [10]. The experience of municipalities in southern Spain (province of Andalusia) is described in the article. The form of the water supply management, which leads to the lowest prices for water, was the provision of services by the municipality. The Urban Water Supply Company, managed by the private operators under contract, took the second place. But municipal enterprises, according to the study, charge higher prices than utility companies with the participation of the private sector [10].

In recent years Russia's state policy in the municipal sector, on the one hand, aims the formation of the business environment through the development of the privatization processes and the use of public-private partnership (PPP). On the other hand, issues in the tariff regulation are fixed at the federal level, while it is one of the major risk factors for attracting investment in the municipal sector. In Russia there are unitary enterprises with a state or municipal property and private operators (OJSC, LLC with a share of private ownership), which exist in the form of the public-private partnership. The bulk of the production and delivery of the water services are provided by municipal unitary enterprises. The share of other forms accounts 26.6% of the total release of water to the population, more than two thirds is provided by private organizations, 1/5 of the water supply infrastructure and sanitation is in lease or concession (19.4% - in lease, 0.6% - in concession).

An analysis of the privatization process in Russia's water and sanitation sector reveals the following distinctive features:

- long-term leases (25-49 years) are close to the classic concession contract;
- private operators work mainly on the long-term lease contracts with investment liabilities;
- the absence of public debate during the competitions and determining the contract conditions;
- trade unions and social organizations of consumers are not represented as a part of the negotiating parties;
- large interregional companies, which are affiliated with the businesses in other sectors, act as tenants;
- tenants are interested, as a rule, in large facilities of cities;
- tenants carry out a significant part of the investment from the state budget [11].
- In addition, according to experts, the PPP in the water supply is characterized by the following facts:
- indicators have improved, but there is no conclusive evidence of the private operators benefits compared to the water supply services as a whole;
- institutional and legal conditions of the organizations are improving, but they are still insufficient for the implementation of concession agreements;
- there are still high barriers to entry into the market of international private operators [12].

The problem needs more qualitative empirical studies of the economic and institutional changes during the reforming process of housing and communal services. Studies on the influence of institutional factors to the results of the tariff policy in the industry are completely absent.

The aim of our research is the development and testing some methodological approaches for the empirical evaluation of the effectiveness of various institutional alternative arrangements. The main purpose is achieved by identifying the relationship between the forms of management and the tariff policy in the water supply sector in Russia.

Table 1. The Russia's largest water utilities.

№	City	Population	Name of the water supply organization
1	Novosibirsk	1 547 910	MUP[a] "Gorvodokanal "[13]
2	Ekaterinburg	1 412 346	MUP "Vodokanal" [14]
3	Nizhny Novgorod	1 263 873	OJSC "Nizhny Novgorod Vodokanal" (since 2006) [15]
4	Kazan	1 190 850	MUP[b] "Vodokanal" [16]
5	Samara	1 172 348	MP "Samaravodokanal" (until 2011); LLC "Samara Municipal Systems" (since 2011) [17]
6	Chelyabinsk	1 169 432	MUP "POVV" [18]
7	Omsk	1 166 092	OJSC "OmskVodokanal" (since 2005) [19]
8	Rostov-on-Don	1 109 835	OJSC "PO Vodokanal" (since 2004) [20]
9	Ufa	1 096 702	MUP "Ufavodokanal" [21]
10	Krasnoyarsk	1 035 528	LLC "KrasKom" (2003) [22]
11	Perm	1 026 477	LLC "Novogor-Kama" (2003) [23]
12	Volgograd	1 017 985	MUP "Gorvodokanal Volgograd" [24]
13	Voronezh	1 014 610	MUP "Vodokanal Voronezh" (2012), LLC "RKS-Voronezh" "(since 2012) [25]

Source: the authors' compilation according to the official pages of companies and the Federal State Statistics Service. The population of the Russian Federation municipalities on January 1, 2016. http://www.gks.ru/free_doc/doc_2014/bul_dr/mun_obr2014.rar

a MUP - Municipal Unitary Enterprise
b MP – Municipal Enterprise

3 Brief literature review and goal of the research

Thirteen major cities of Russia (with a population of over one million people) were chosen to investigate the influence of the management utility on the level and dynamics of tariffs for water supply. The leading manufacturers of the water supply services in the cities are the country's largest water utilities; in fact they are local monopolists in megacities (Table 1).

On 1 January 2014 the municipal water utilities served six of the largest cities of Russia, and the seven largest cities were served by the water utilities organized as a shared ownership with private operators. In all cities at different periods of time private capital was involved. Krasnoyarsk and Perm were ones of the first in Russia, where the private management of water and wastewater networks is carried out since 2003. In Rostov-on-Don, Omsk private operators manage the infrastructure of the water supply services since 2004-2005. Cities with the little experience of the private operating in the water supply are Samara (since 2011) and Voronezh (since 2012). Thus, for more than five years there are private operators in most of the largest cities in Russia.

For the research the tariffs for water supply and sanitation over the past five years in the cities, which are served by the water utilities with a variety of institutional options for utilities is needed to compare.

Table 2 shows the tariffs for the water supply and sanitation in the dynamics in major Russian cities for the last five years.

Table 2. Water and sanitation tariffs in the major cities of Russia in 2010-2014, Rub./m^3.

	2010	2011	2012	2013	2014
MUP "Gorvodokanal" (Novosibirsk)	14.22	21.3	20.09	23.78	23.78
MUP "Vodokanal" (Ekaterinburg)	27.52	31.56	35.52	40.05	40.05
MUP "Vodokanal" (Kazan)	22.78	23.51	26.06	27.13	29.61
MUP "POVV" (Chelyabinsk)	18.87	20.46	21.7	24.3	27.57
MUP "Ufavodokanal"	14.77	17.04	22.51	24.05	24.05
MUP "Gorvodokanal Volgograd"	21.36	20.92	23.18	25.18	29.1
MUP "Vodokanal Voronezh"	21.69	23.34	27.54		
LLC "RKS-Voronezh"				30.25	40.79
MP "Samaravodokanal"	17.56	21.64			
LLC "Samara Municipal Systems"			23.15	27.45	27.45
JSC "Nizhny Novgorod Vodokanal"	16.14	22.38	20.38	23.94	23.94
JSC "OmskVodokanal"	19.51	20.95	25.58	30.23	32.72
OAO "PO Vodokanal" (Rostov-on-Don)	29.35	44.69	48.6	58.12	58.03
Ltd. "KrasKom " (Krasnoyarsk)	16.42	18.14	20.39	22.59	22.59
Ltd. "Novogor-Kama" (Perm)	43	33.77	31.23	33.19	36.12
Correlation	**0.22**	**0.28**	**0.19**	**0.25**	**0.28**

Source: Calculated by the authors according to the official pages of the companies

4 Results

In 2010-2014 in the group of the municipal water utilities Yekaterinburg was a leader in terms of tariffs. And Vodokanal of Rostov-on-Don with the tariffs higher by 41-56% than tariffs in Yekaterinburg took the first place in the group of non-public operators. In the group of private operated water utilities average tariffs were also higher by 11-19%. During the 2010-2014's there was a positive moderate correlation (0.19-0.28) between the forms of the management utilities and tariffs for services. In the group of private management utility prices are higher on average than in the municipal unitary enterprises.The results of the tariff dynamics analysis in the Russia's largest cities are shown in Table 3.

In cities with a municipal form of management there was the highest growth of rates in comparison with the previous year in Novosibirsk in 2011 (it amounted to 49.79%). The following year rates decreased 5.68%, however the overall increase in tariffs for the water supply services for five years was 67.23% in Novosibirsk. And it was also the highest in the group of municipal water utilities.

In the group of the cities where the water and wastewater services are provided by operators with a share of private ownership, the greatest increase in tariffs compared to the previous year took place in Rostov-on-Don: water utility service tariffs were raised by 52.28% compared with the previous 2011. Rostov-on-Don is the leader in rising tariffs for five years (97.72%). In Voronezh the significant growth of tariffs for the water supply services was observed after changing the form of management to a private operator LLC "RKS-Voronezh". In Voronezh overall 5-year growth of rates was more than 88%. However, there is a water utility in Perm in this group, which has reduced tariffs by more than 21% in 2011. And tariff reduction of LLC "Novogor-Kama" has amounted to 16% for five years. It is the only enterprise tariff dynamics which differs from all other water utilities in both groups. Overall, however, during 2011-2014 a faster increase in water tariffs was observed in the non-public water utilities than in the municipal water utilities in the largest cities of Russia.

Table 3. The dynamics of water and sanitation tariffs in the largest cities of Russia in the period from 2011 to 2014, %.

	2011/2010	2012/2011	2013/2012	2014/2013	2014/2010
Municipal Unitary Enterprises					
MUP "Gorvodokanal" (Novosibirsk)	49.79	-5.68	18.37	0	67.23
MUP "Vodokanal" (Ekaterinburg)	14.68	12.55	12.75	0	45.53
MUP "Ufavodokanal"	15.37	32.10	6.84	0	62.83
MUP "Vodokanal" (Kazan)	3.20	10.85	4.11	9.14	29.98
MUP "Gorvodokanal Volgograd "	-2.06	10.80	8.63	15.57	36.24
MUP "Vodokanal Voronezh"	7.61	17.99			
MUP "POVV" (Chelyabinsk)	8.43	6.06	11.98	13.46	46.10
MP "Samaravodokanal"	23.23				
Operators with a share of private property					
LLC "RKS-Voronezh"			9.84	34.84	88.06
OJSC "PO Vodokanal" (Rostov-on-Don)	52.27	8.75	19.59	-0.15	97.72
OJSC "Nizhny Novgorod Vodokanal"	38.66	-8.94	17.47	0	48.33
LLC "Samara Municipal Systems"		6.98	18.57	0	56.32
LLC"paint" (Krasnoyarsk)	10.48	12.40	10.79	0	37.58
LLC"Novogor-Kama" (Perm)	-21.47	-7.52	6.28	8.83	-16.00
OJSC "OmskVodokanal"	7.38	22.10	18.18	8.24	67.71

5 Conclusion and discussion

Thus, the involvement of private capital to the Russian utility sector has not yet led to the expected positive change of the tariff policy. In 2010-2014 the level of tariffs and their dynamics in the group of the water utilities, which are involving private capital was higher than in the water utilities with the municipal form of management in Russia's largest cities. Perhaps tariffs for services of the water utilities managed by the private operators increased due to the development and implementation of investment programs in the largest Russian cities. Other factors should be taken into account, so that creates prospects for the further research. It may be interesting to investigate the results of the private capital, which is attracted to the Russian communal economy, and the effectiveness evaluation of the various institutional arrangements in the municipal economy. However, Perm's example gives a hope for the formation of positive expectations regarding the tariff policy of the water utilities controlled by using a model of public-private partnership.

References

1. V.G. Varnavskiy, A.V. Klimenko, V.A. Korolev, *Public-private partnership. Theory and practice* (Publ. House SU HSE, Moscow, 2010)
2. A.V. Ermishina, Proc. DNTU. Ser.: econ., **38-1**, 193-203 (2010)
3. S. Garcia, L. Guerin-Schneider, G. Fauquert, Water Supply, **5(6)**, 173-181 (2005)
4. D. Hall, E. Lobina, Natural Resources Forum, **28**, 268-277 (2004)
5. E. Lobina, Water Resources Development, **21(1)**, 55-87 (2005)
6. E. Chong, F. Huet, S. Saussier, F. Steiner Rev. of Ind. Org. **29 (1/2)**, 149 – 169 (2006)
7. A. Carpentier, C. Nauges, A. Reynaud, A. Thomas, Economie et Prevision, **174(3)**, 1-20 (2006)
8. R. Martinez-Espineira, M.A. Garcia-Valinas, F. Gonzalez-Gomez, Urb. Stud., **46(4)**, 923-945 (2009)
9. S. Ruester, M. Zschille, Utilities Policy, **18(3)**, 154-162 (2010)
10. M.A. García-Valiñas, F. González-Gómez, A.J. Picazo-Tadeo, Utilities Policy, **24**, 59-69 (2013)
11. A.V. Ermishina, *Innovative Development of Russia's Economy: Regional Diversity: The Sixth International Conference*, 524-532 (2013)
12. A. Rodionov, S. Sivaev, Journal of the head and chief accountant, **10**, 52-53 (2011).
13. MUP Gorvodokanal. http://www.gorvodokanal.com/
14. MUP Vodokanal. http://www.vodokanalekb.rf/
15. OJSC Nizhegorodskiy vodokanal. http://www.vodokanal-nn.ru
16. MUP Vodokanal. http://www.kznvodokanal.ru
17. LLC Samara Utility Systems. http://samcomsys.ru/
18. MUP POVV. http://voda.uu.ru/
19. OJSC Omskvodokanal. http://www.omskvodokanal.ru/
20. OJSC PO Vodokanal. http://vodokanal.rnd.ru/
21. MUP Ufavodokanal. http://www.ufavodokanal.ru/
22. LLC. KrasKom. http://www.kraskom.com/
23. LLC Novogor-Prikame. http://www.novogor.perm.ru/
24. MUP (Municipal Unitary Enterprise). http://www.mupgvv.ru/
25. LLC RKS-Voronezh. http://www.voronezh-rvk.ru

Mathematical modelling as basis for efficient enterprise management

Svetlana Kalmykova[1,*], *Alexander* Krasnoshchekov [1], and *Petr* Pustylnik[2]

[1]Peter the Great St. Petersburg Polytechnic University, 195251, St.Petersburg, Polytechnicheskaya, 29 , Russia
[2]Herzen State Pedagogical University, St. Petersburg, Russia

Abstract. The choice of the most effective HR- management style at the enterprise is based on modeling various socio-economic situations. The article describes the formalization of the managing processes aimed at the interaction between the allocated management subsystems. The mathematical modelling tools are used to determine the time spent on recruiting personnel for key positions in the management hierarchy selection.

1 Introduction

The economic recession revealed explicitly the problem of Russian of industrial enterprises management low efficiency in real time, and that revelation led to the relevance of the theme of this article.
The crisis in the Russian economy started developing since 2008. This situation can be illustrated by the indexes of electricity and steel production in Russia dynamics (see. Fig. 1) [1, p.102]. Line "1" is the dynamics of electricity (MWh); line "2" is the dynamics of steel production (millions of tons).

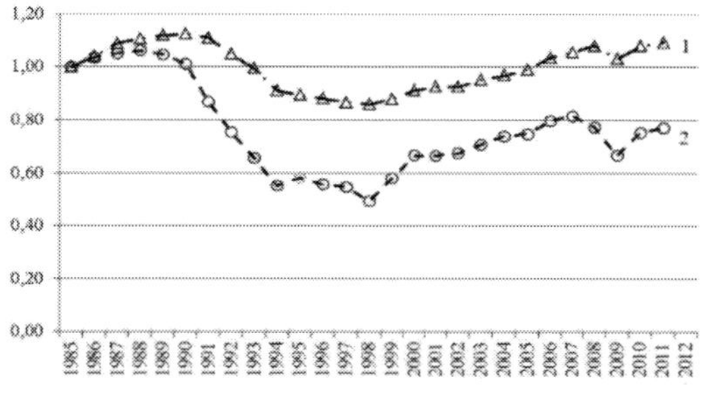

Fig. 1. Dynamics of production indexes (1985 = 1).

In the article [2, p.20] the author attempted to determine the role of the various industrial enterprise management features (see. Fig. 2) based on the assumption that the governance (the management) of the economic system is the top of the management pyramid. With this approach, marketing, controlling and logistics are part of industrial management.

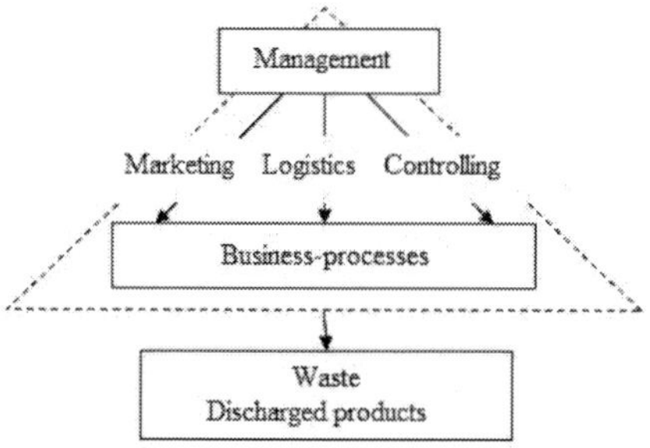

Fig. 2. Enterprise management pyramid.

To describe the process of industrial production with the help of logistics terms, it is necessary to take as a basis the movement of material flow: resource market (0) → supply (1) → production (2) → sale (3) → goods market (4).

2 Formalize the industrial enterprise management process

The chart will help us formalize the industrial enterprise management process (see. Fig. 3)

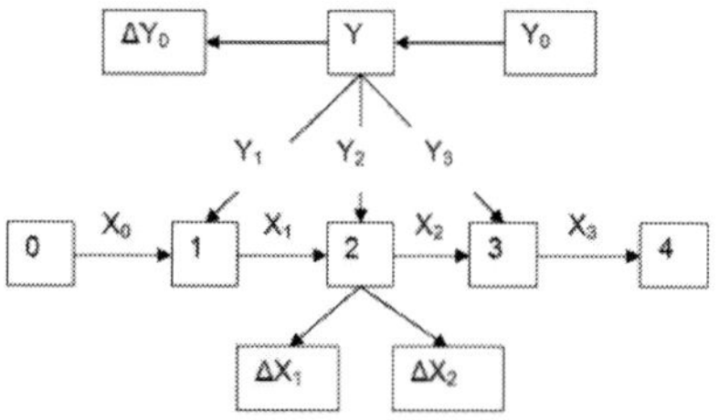

Fig. 3. The formalized enterprise management chart.

* Corresponding author: kalmykovas@mail.ru

Symbols shown on the simplified scheme of industrial enterprise management:

X0 - the movement of materials and semi-finished products from the resource market (from the suppliers) to the warehouse of the enterprise;

X1 - the movement of materials and semi-finished products from the company's warehouse to the production areas;

X2 - the movement of manufactured products to the site storage of finished products;

X3 - the movement of manufactured products from the area of site storage of finished products to the goods market (to the consumer);

ΔX1 - production waste;

ΔX2 - discarded products;

Y0 - the request from a potential customer (the demand for the product on the goods market);

Y - the administrative decision taken at top management level;

Y1 – the command for the supply service to purchase the materials, semi-finished products, etc.

Y2 - the command for the production areas about the number of required products and the timing of production;

Y3 - the command for the sales service to ship the finished products to the customer;

ΔY0 - the reject to manufacture the requested products.

The balance of the material flow is described with the equation:

$$X_0 = X_3 + \Delta X_1 + \Delta X_2 \quad (1)$$

The task of production management can be considered fulfilled if the shown condition is followed:

$$X_3 = Y_0 - \Delta Y_0 \quad (2)$$

The request of the customer (Y0) generates at least three control actions (Y1, Y2, Y3) to perform core business functions: supply, production and sales.

Control actions within production management in aggregated form can be represented by the equation:

$$Y = \sum_{i=1}^{3} Y_i(X_i) \quad (3)$$

If we consider the process of industrial production from the management point of view, depending on the purpose of the research as a basis for reasoning, we choose one of the links in the chain: the forecast → the decision → the production of management commands → the command execution → the comparison the result with the forecast → the grounding of a new forecast → etc.

The processes of globalization change the formalization of development forecasts for the whole industries and for individual companies under the influence of economic crises, which transform the model of industrial enterprises management.

The development of enterprise management model (an approach of Ansoff) was carried out in three stages [3, s.399-413]:

1950-s. - The development of methods to adapt to the changing situation on the market;

1970-s. – The method of management according to the circumstances;

1980-s. - Crisis management.

In 1966, the J. Galbraith (1908-2006) warned: "In the initial stage of development of the corporation, particularly in 1930, the experts concerned that the people who control the company, can make it an instrument of their personal enrichment ... Theft of unprecedented proportions threats the society "[4 s.181-182]. Unfortunately, in Russia this prediction come true at the turn of XX-XXI centuries with the establishment of capitalistic relations and the privatization of industrial enterprises.

Improvement of production management methods involves consideration of certain factors that are uniquely connected with each other and affect the functioning of the enterprise both directly and indirectly.

The production system can identify the main subsystems, each of which can be established as an isolated object of control:

1. Supply management (definition of purchase prices, their minimization in the procurement of quality raw materials).

2. Production management (definition the percentage of discarded products and waste in order to maximize the use of material resources).

3. Sales management (modeling ways to minimize losses during storage of goods (warehouses); model timely goods delivery).

Management of industrial enterprise subsystems is based on the sales forecast in the next period, formed on the basis of statistical data. It can be assumed that the selection an effective personnel management style we should take into account the short economic cycles (Mitchell cycles) which were obtained in the analysis of statistical data for the period 1878-1922 gg. [5]. In Russian Mitchell cycles can be observed in the graph dynamics steel melting production indices (it was shown in Fig. 1).

Modeling the interaction between selected subsystems is possible with the use of game-theoretic models, which provide a basis for decision making in the company's development strategy. In the XXI century strategic development plan of the industrial enterprise should include activities related to the introduction of the concept of "lean production" (LP).

LP concept began to take shape through the works of FW Taylor (1856-1915), who published by the 1911 the principles of scientific management [6], and H. Ford (1863-1947), who introduced the pipeline division of labor in 1913,. [7] The modern version of the LP concept is based on the development of Tahiti It (1948), which in its final form in 1973, in Japan was called «Toyota

 DOI: 10.1051/shsconf/20173501062

Production System» (TPS), and «Lean Production System » LPS in 2000, in the United States [8].

As a result of production management state research in Russia it was concluded, that using the methods of LP "highest estimate had a system of workplace organization 5S », and on down the rated: Kaizen, the TPM, Kanban, J-in-T [9, p. 22].

Currently there is no single answer to the question, how to organize the industrial enterprises management so that the motivation of the staff is not reduced for an extended period.

In a research on human resource management techniques [10, s.155-161], the role of the management team is highlighted, but it should be emphasized that the action of each team is led by the leader who uses a particular style of management.

For various industrial enterprise management styles we will consider:

0 – as the statement of the problem to the subordinates and its perform without interference from the first person;

1 – as the setting an objectives to the subordinates followed by detailed first person control of the all the action.

We form a matrix $A = (a_{ij})$ of control actions possible combinations, where :

$i = 1, 2, 3;\ j = 1, 2, ..., 7$

$$A = \begin{pmatrix} 0 & 1 & 0 & 0 & 1 & 0 & 1 \\ 0 & 0 & 1 & 0 & 1 & 1 & 1 \\ 0 & 0 & 0 & 1 & 0 & 1 & 1 \end{pmatrix} \quad (4)$$

The lines correspond to the three services: supply, production and sales, and the columns corresponds to the combination of management styles.

Each service has its own criteria for evaluating the efficiency of performance^

- Supply - the acquisition of high-quality raw materials and semi-finished products for the least possible cost in the required amount before the start an order in the production;
- Production - manufacturing products according to the working schedule without discharged products from the waste volume not exceeding technological norms;
- Sales - the shipment of finished products, in accordance with a complete set of delivery schedule without disruptions.

Since most understandable for Russian managers of the lean manufacturing techniques is the 5S system, we summarize the experience of its ordinary version application in an industrial enterprise.

The stage of implementation of the principles of 5S system ("sort", "maintenance of order", "keeping clean," "standardization", "improvement") requires management style "1" as the staff perceives the need for an inventory of all items in the workplace as an extra work done without payment.

At the usual use it is better to use the style of "0", as in the team of each business unit an informal leader is detected. This leader initiates the effective application of the 5S system principles.

Possible combinations of control actions on the personnel of the industrial enterprise structural divisions are shown in Figure 4. Letter "D" drown on the chart means "decision" (manager's decision).

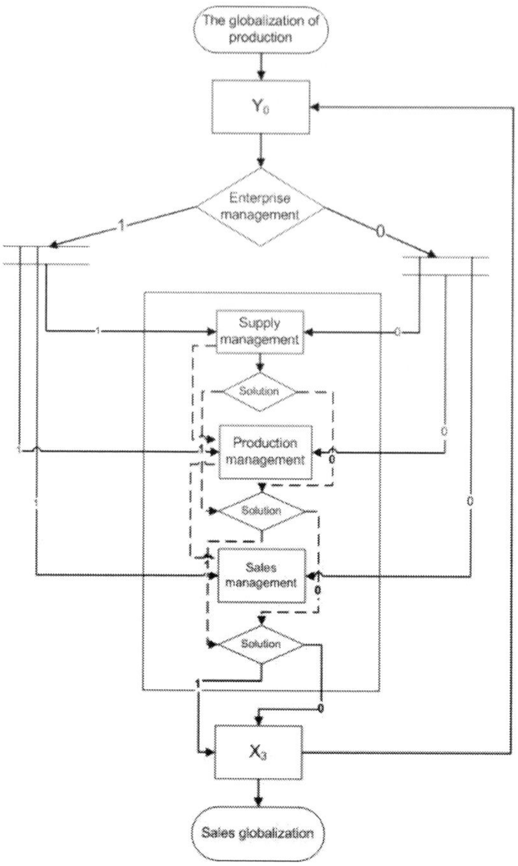

Fig. 4. Possible combinations of control actions.

One of the significant factors affecting the generation of management teams, is the globalization of production and sales, which initiates crises: the crises of underproduction and overproduction crises.

The ideal situation is a complete coincidence of supply and demand, the time when all the products are made entirely for the end-user. This situation is considered in most of mathematical models, which describe the motion of the material flow in the production system.

In general, the objective function for the solving task for an industrial corporation has the following form [11]:

$$OPT(I) = \sum_{i=1}^{n} l(d_i), ..., d_i \in D, ...n = |D| \quad (5)$$

$$D = \{d / d \in D^{s1}, ..., D^{sp}; l(d)^{s1} = ... = l(d)^{sp}; k(d)^{s1} = ... = k(d)^{sp}\} \quad (6)$$

Where:

D^s is a plurality of structural units of the corresponding activities;

s is one of the analyzed activities;

p is one of the industrial enterprises of the corporation;

$k(d)^{si}; l(d)^{si}$ are the parameters of control actions on structural units on each i-th iteration control options;

D is the set of all structural units of the industrial corporation;

I is a task for which the best solution is sought.

The objective function in the absence of constraints determines the maximum possible theoretical estimate of the optimal solution $OPT_{theor}(I)$, representing the amount of control actions on the common structural unit, for which a non-empty intersection of all the group companies exists.

Possible extremes different functions: maximization of the enterprise owner profits; minimization of discharged products amount; minimization of the theft amount in the workplace; maximization of the least paid categories of workers' salaries; minimization of top-management wages, etc.

Going through the possible combinations $k(d)^{si}; l(d)^{si}$ of control parameters effects on the structural divisions of the corporation can get different effects of economic activity (E) for each combination of i:

$$E(Y) = \sum_{i=1}^{3} A_i \cdot k(d)^i \cdot l(d)^i \qquad (7)$$

Detailed top-manager control of the task execution is accompanied by a time-consuming because of continuous action coordination. This control also deprives the subordinate desire to take the initiative. With the tight control theft at the company is only possible with the permission of the head.

Giving carte blanche subordinate reduces the time required to solve tactical problems, frees up time for the formulation of strategic objectives, as well as allows revealing the creative potential. However, in this case, the theft caused by the personal initiative of employees is possible.

While managing microenterprises detailed control is inevitable, as the owner of the company (if it is not firm for one-day) typically performs several functions: recruitment, personnel management, the acquisition of the necessary manufacturing process technology, regulation of labor, accretion of premiums and imposition of penalties, manufactured products quality assessment, customer search, preparation of the contract, search for resources and suppliers etc. The only feature that is rarely performed by managers of micro-enterprises is accounting.

While managing small business owner of the company (if it is not one-day firm), combines different positions: CEO, managing director, chief engineer, director of personnel, and director of quality. This helps to gain a detailed control of all business processes. Some functions are delegated to the hired personnel. Usually it is accounting, process management, legal and patent support.

While managing medium and large enterprises enterprise owner sometimes keeps his position as "CEO", and is always in control financial flows. For key positions recruitment of each management level sorting options combining management styles and selecting managers with the necessary set of qualities are needed.

3 Conclusion

Any industrial enterprise management is the process of interaction of various types of administrative activity, with both dynamic and static manifestation. It is aimed at the adaptation of economic activities to the continuously changing economic conditions.

Changes in the economic situation (crisis or stability) suggest a change of management style that is accompanied by a change in top-management team of the company.

Mathematical modeling of socio-economic situations reduces the time spent on the choice of industrial enterprise and sub-systems management style. This approach simplifies the work of personnel recruitment for key positions in the management hierarchy.

References

1. P.N. Pustyl'nik, Deichite Extraction, **5**, 101-104 (2013)
2. P.N. Pustyl'nik, Scientific and Technical Sheets SPbSPU, **4(52)**, 17-21 (2007)
3. I. Ansoff, *New corporate strategy* (Piter Kom, St. Petersburg, 1999)
4. Dzh. Gjelbrejt, *The New Industrial Society* (Terra Fantastica, St. Petersburg, 2004)
5. C. Mitchell Wesley, *Business Cycles: The Problem and Setting* (National bureau of Economic Research, New York, 1927)
6. Tejlor Frederik Uinslou, *Principles of Scientific Management* (Control, Moscow, 1991)
7. Genri Ford, *My Life* (Finance and economics, Moscow, 1989)
8. Majkl L. Dzhordzh, *Lean Manufacturing + Sigma. Combining the quality of shanti sigma with the speed of lean production* (Al'pina Biznes Buks, Moscow, 2005)
9. *Third Charnov Readings: a collection of abstracts. Materials of the III International Scientific and Practical Conference on the Organization of Production*, 8-22 (2013)
10. M.V. Lopatin, S.V. Tevlina, Scientific and Technical Vedomosti SPbSPU, **6(68)** (2008)
11. S.V. Kalmykova, A.S. Sokolicyn, Scientific and Technical Bulletins of SPbSPU, **6(162)**, 185-190 (2012

Study and assessment of clusters activity effect on regional economy

A. Babkin[1,*], Y. Vertakova[2], V. Plotnikov[3]

[1]Peter the Great St. Petersburg Polytechnic University 195251, Polytechnicheskaya str., 29, Saint Petersburg, Russia
[2]Southwest State University 305040, 50 Let Oktyabrya str., 94, Kursk, Russia
[3]St. Petersburg State University of Economics 191023, Sadovaya str., 21, Saint Petersburg, Russia

Abstract. The cluster approach, i.e., forming basic innovative and industrial clusters is widely applied in modern Russian conditions for the development of the economy. These actions are considered as effective measures for implementing the economic policy stimulating regional development by federal and regional authorities. The analysis we carried out showed that the quantitative approach for assessing the efficiency of cluster creation and performance is still insufficiently used. In this paper we establish and quantitatively estimate the influence cluster have on the regional economy using regression analysis with an example of a number of Russian regional clusters. Expanding the practice of creation and the state support of clusters taking into account the revealed quantitative dependences estimating their efficiency is suggested. We have advanced the hypothesis that clustering has a positive influence on regional economy, and confirmed this influence by means of quantitative methods using representative datasets. Our study of course had a selective character as it is not possible to carry out the calculations for all the existing clusters and cluster initiatives of Russia and discuss the results within a single article. At the same time, following the analysis we performed, we concluded that it is effective to initiate cluster creation in Russian regions. It is shown that cluster activity is capable to have of having a positive impact on GRP growth and the budgetary income in the region. Along with that, we note the dissimilarities in the multiplying influence of clusters on the regional development, its dependence on territorial and branchspecifics that will be the direction for a further in-depth study.

1 Introduction

A cluster as an object of an economic agglomeration of interconnected enterprises in a certain area has been known since the time of the handicraft industry. At that, for the origin of the cluster approach (1830-1890), the economic science is obliged to A. Marshall, who himself called clusters "localized industries" [1]. Later (1890-1950), the term "industrial districts" became widespread in literature. The most popular term today – "cluster" – emerged much later. The notion of "cluster" is considered to be introduced into the economic relations theory by M. Porter (1998) [2]. M. Porter used the term "cluster" to analyze the structure and effectiveness of economic relations in industries and organizations. When analyzing competitiveness of a company, he paid special attention to its economic environment. Now, a cluster is considered as a separate economic system [3 - 9] and an integrated structure [10-12,14,18,20], which significantly affects both: the development of regional and sectoral economies and the national economy in general [13,15-19 and others].

The theoretical foundation of the cluster concept originated at the beginning of the XIX century in the works on economics of agglomeration by J. Von Thunen and his followers W. Launhardt and A. Weber, who studied the enterprise and its location in the economic space in relation to sources of raw materials and sales markets and also factors that influence it [Von Thunen, 1826; Launhardt, 1882; Weber, 1909] (table 1).

Table 1. Characteristics of the cluster theories based on [3-11, 17-20].

Theories of the cluster approach	Authors	Properties of a cluster
Agglomera tion theory	Marshall, Hoover, Ohlin	Geographic proximity of firms and the influence of the triad of external scale economies: labor availability, supporting and

* Corresponding author: babkin@spbstu.ru

		auxiliary industries, specialization of different firms at different stages of the production process, allowance for transport costs.
New economic geography "New industrial districts"	Krugman, Saksenina n	Geographic proximity of firms, the influence of the triad of external scale economies, the influence of information and knowledge spillovers, allowance for transaction costs
Industrial complex	Hoover, Isard	The interaction of firms within one industry in the form of "buyer-seller", "supplier-consumer"
Industrial cluster	Gordon, McCann	The interaction of firms within several interconnected industries both in terms of the production process of goods and services and knowledge flows
Competitiv eness theory by M. Porter	M. Porter	The influence on the cluster's successfulness of interrelations between factors inside the "competitiveness brilliant" region: strategy, structure, competitiveness; demand conditions; factorial conditions; linked industries; state intervention.
Neo-competitiv eness of M. Porter	M. Enright, Solvel, van Grunsven, van Egeraat, Clancy	Taking into account the cluster's lifecycle, the influence of globalization and global markets, transnational companies, participation of foreign investors, specialized cooperative and competitive interrelations
Innovative system theory	Frimen, Lundvall, Edquist, Cook, Morgan	The influence of the surrounding innovative infrastructure (knowledge generation centers, transfers, and commercialization of innovations).
Network theory	Feser, Sviimi, Perucci, Harisson	The flexibility of the cluster's organizational structure, a special role of the social capital (trust, solidarity) and the institutes which ensure stable connections between firms, the processes of mutual and constant learning
Innovative clusters theory	Feldman, Audretsch , Carlson, Johnson	Commitment to creating an innovative product by interaction between the cluster's members (competing firms), external knowledge sources based on formal and informal cooperation processes

However, many researchers [Bathelt, 1998; Ketels, 2003; Krugman, 1991; Scott & Storper, 1992; Audretsch, 1998] think that it was A. Marshall who, in his work "Principles of Economics" [A. Marshall, 1890], was first to empirically prove that productivity of firms and the results of their activity directly depend on their location and geographic proximity of economic agents.

The rise of clusters of our period (starting from the 1970's) is mainly caused by disintegration processes of vertically-integrated companies and structures [10-12, 20], the main reason for which was a radical increase in production complexity – first of all in the machinery-producing industry. The number and range of semi-finished products and components used in assembling final products were greatly increased and expanded. Under these conditions, centralization of all industrial sites within one vertically-integrated structure ceased to be a realistic choice. The alternative to this is cooperation of independent producers on a mutually beneficial basis, i.e. a cluster. As the founder of the cluster approach, M. Porter defines the cluster as "a geographically proximate group of interconnected companies and associated institutions in a particular field, linked by commonalities and externalities" [2]. The given definition fully reflects the cluster's essence as an economic and geographical unit. Besides, it is highlighted that as a rule, a cluster is not something separate but a part of a single system. Here it should be supposed that within a system there is some kind of interconnection and interconditionality determining their successful functioning.

Taking into account distinctive features, the following interpretation is offered for a cluster: it is an industrially limited group of economic agents, innovative formations, state administrative and scientific-educational institutions which combine formal independence and internal competitiveness with cooperation and a shared control center, the purpose of which lies in reprocessing high-tech products and industries and achieving a general synergy effect.

Cluster initiatives are widely adopted in modern Russia. According to the Russian cluster observatory (http://map.cluster.hse.ru/list — the estimates on the quantity of clusters in the Russian Federation, but it isn't concrete numbers, and in qualitative result. Clusters became a noticeable phenomenon, especially at the regional level of the national economy. Because of this, many publications are dedicated to studying them [1, 3, 4, 7, 8, 11—15, etc.] (we should note that the authors of this article also made a certain contribution to studying these issues [3, 7, 8, 11, 12, 17, 20etc.]).

2 Problem statement

While we do not intend to analyze cluster and the publications dedicated to them in detail, we should point out on circumstance that is important in our opinion: cluster organization of economic associations is confirmed to be rather effective both in theory and in practice. This article is dedicated to critical consideration of this statement. We willuse methods of quantitative analysis for assessing the efficiency of cluster organization of economic activity.

3 Results of the study

We used the above-mentioned Map of clusters of Russia as basic data for carrying out calculations. Since it contains numerous objects and our computing and temporal resources were limited, we selected four clusters from the entire set (Tab. 1). We selected the clusters for study based on the premise that if some clusters existed in the same branch, preference was given to the cluster with a higher level of organizational development or functioning during a longer period. We also paid attention to the existence of necessary statistical information on clusters. Besides, when forming our selection, we strived to make it cover various non-related branches to obtain more objective results.

Table 1. Selection of clusters for quantitative analysis.

Cluster, region	Branch	Level of organization al development	Year of establis hment
Textile cluster of the Ryazan Oblast	Production of textile products, clothes, footwear, leather products	Initial	2014
Power cluster Tula region	Electricity generation electric equipment	Initial	2012
Innovative territorial timber industry cluster	Forestry and woodworking; pulp and paper production	Average	2014
Dairy cluster of	Agriculture and fishery	Initial	2015

Source: data of the Russian cluster observatory (http://map.cluster.hse.ru/list).

We considered the following resultant indicators: level of economic development of the region, which is estimated quantitatively through GRP per capita (thousands of rubles), as well as the income of the budget (the average per capitaindicator measured in thousands of rubles per person per year was also considered). We estimated how these indicators changed under the influence of cluster initiatives introduced by theregions under consideration. Since the assessed clusters operate in various branches of specialization, we used similar indicators for quantitative assessment. These included the volume of investment and output, but considered not for the region as a whole but only for branches of cluster specialization (Figure 1). Thus, the average per capita estimates measured in thousands of rubles were also used.

Information from statistical collections of Rosstat was used as basic data for calculations. The period from 2004 to 2014 was covered. The regression analysis method was used which allows to estimate the extent of mutual influence of the considered indicators with a sufficiently highdegree of accuracy.

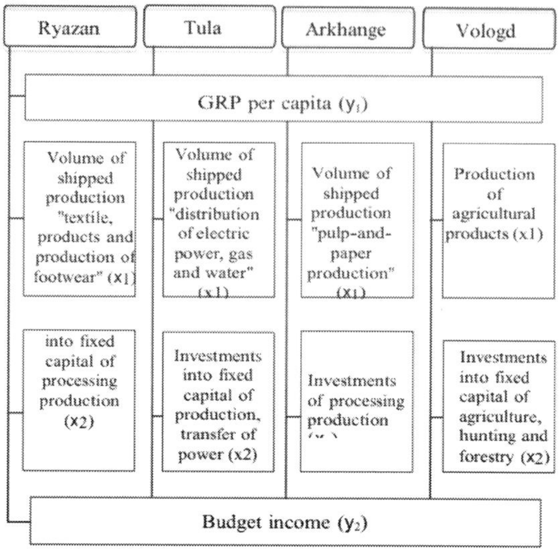

Fig. 1. Resultant and factorial indicators for the analysis of clusters.

Linear one- and two-factorial equations of regression were constructed, given in Tab. 2 (only the dependences which have a high degree of accuracy, estimated by the determination coefficient R2, are presented in the table).

In Tab. 2 indicators : GRP per capita (y1), Budget income (y2)

Table 2. Regression dependences of indicators of analyzed clusters and regions.

Oblast	Regression equation	R^2
Ryazan Oblast	y_1=21.488+13.45209x_2 y_1=17.385+5.776x_1+13.01x_2	0.86 0.88
	y_2=6.03933+2.03368x_2 y_2=5.829+0.296x_1+2.011x_2	0.82 0.82
Tula Oblast	y_1=11.89458+11.858x_1 y_1=85.95291+13.198x2 y_1=15.441+8.267x_1+7.116x_2	0.79 0.67 0.91
	y_2=1.885+2.206x_1 y_2=16.144+2.376x_2 y_2=2.482+1.602x_1+1.197x_2	0.76 0.63 0.89
Arkhangelsk Oblast	y_1=95.97+5.906x_1 y_1=96.077+5.91x_1-0.03x_2	0.62 0.62
	y_2=15562+0.905x_1-0.401x_2	0.62
Vologda Oblast	y_1=-53.685+18.937x_1 y_1=54.295+87.771x_2 y_1=-49.738+16.812x_1+14.602x_2	0.88 0.63 0.88
	y_2=9.864+12.311x_2 y_2=-3.153+2103x_1+3.156x_2	0.70 0.93

4 Analysis of the received results

The analysis of the obtained dependences given in the table allows to draw conclusions as to the extent to which the growth of investments into fixed capital and production volumes influences the amount of the gross regional product and the budget income of the considered regions with clusters:

1. Ryazan Oblast (textile cluster):

- with the investments into the fixed capital of the processing industry increased by 1 thousand rub., the GRP increases by 16565 rub., i.e., the coefficient of a multiplication in this case exceeds 16.5;
- with the production volume for «textile, sewing production, leather products and production of footwear» increased by 1 thousand rub., the GRP will increase by 5776 rub. at the same volume of investment into fixed capital. The increase in the volume of investments into fixed capital of the processing industry by 1 thousand rub. at the same volume of production implies GRP growth by 13010 rub., i.e. the multiplying influence of investments on the GRP exceeds a similar influence of the output increase of the cluster's enterprises by more than two times. At the same time, the production in the cluster is observed to experience a fast growth by almost 5 times with respect to the overall GRP;
- with the volume of investments into fixed capital of the processing industry increased by 1 thousand rub., the income of the budget will increase by 20337 rub., i.e., the coefficient of multiplication in this case exceeds 20.3. It is obvious that granting benefits for the investment activity of textile enterprises is economically justified;
- with the production volume for «textile, sewing production, leather products skin and production of footwear» increased by 1 thousand rub., the income of the budget will increase by 296 rub. at the same volume of investments into fixed capital. The increase in the volume of investments into fixed capital of the processing industry by 1 thousand rub. at the same volume of production leads to the income of the budget increasing by 2011 rub.

Thus, the performance of the enterprises from the «textile, sewing production, leather products and production of footwear» cluster in the Ryazan Oblast has a considerable positive impact on the growth of the GRP and the income of the budget, and investments into fixed capital of the processing industry pay off, having a substantial positive impact on the growth of these indicators. Therefore, the creation of a textile cluster in the Ryazan Oblast was expedient; this cluster has a substantial positive multiplying impact on the regional economy. In this regard, it is possible to conclude that the textile cluster of the Ryazan Oblast is effective.

2. Tula Oblast (power cluster):

- with the production volume for «production and distribution of the electric power, gas and water» increased by 1 thousand rub., the GRP will increase by more than by 11 times, i.e., by 11858 rub.;
- with the investments into fixed capital of production, transfer and distribution of the electric power, gas and water increased by 1 thousand rub., the GRP will increase even more considerably by 13198 rub.;
- with the production volume for «production and distribution of the electric power, gas and water» increased by 1 thousand rub., the GRP will increase by 8267 rub at the same volume of investments into fixed capital. The increase in the volume of investment into the fixed capital of production, transfer and distribution of the electric power, gas and water by 1 thousand rub. at the same production volume leads to an increase of the GRP by 7116 rub;
- with the production volume for «production and distribution of the electric power, gas and water» increased by 1 thousand rub., the income of the budget will increase by 2206 rub;
- with the volume of investments into fixed capital for production, transfer and distribution of the electric power, gas and water increased by 1 thousand rub., the income of the budget will increase by 2376 rub;
- with the production volume for «production and distribution of the electric power, gas and water» increased by 1 thousand rub., the income of the budget will increase by 1602 rub at the same volume of investments into fixed capital. The increase in the volume of investments into the fixed capital for production, transfer and distribution of the electric power, gas and water by 1 thousand rub. at the same production volume leads to an increase of the income of the budget by 1197 rub.

Based on the obtained results, we can draw conclusions for the power cluster of the Tula Oblast similar to the above-formulated ones for the textile cluster of the Ryazan Oblast. Quantitative estimates of the cluster's impact on the indicators of the regional economy in the Tula Oblast are slightly lower than in the Ryazan Oblast. Nevertheless, the economic performance (investments and output) of the cluster's enterprises also has a positive multiplying impact on the regional economy. Quantitative distinctions which we found can be caused by territorial, industrial, technological, institutional and other factors; detailed analysis of these factors is outside the scope of this article.

3. Arkhangelsk Oblast (timber industry cluster):

- with the production volume of the pulp-and- paper industry increased by 1 thousand rub., the GRP will increase by 5906 rub. This means that this type of production is closely tied with the GRP, and the coefficient of multiplication is almost equal to 6 units;
- with the production volume of the pulp-and- paper industry increased by 1 thousand rub., the GRP will increase by 5910 rub. at the same volume of investments into fixed capital; the increase in the volume of investments into the fixed capital of the

processing industry by 1 thousand rub. at the same volume of production will cause a reduction of the GRP by 0.03 thousand rubles. Though this reduction is insignificant (it is possible to explain it by statistical error), but it is still present and thus demands further study;

- with the production volume of the pulp- and-paper industry increased by 1 thousand rub., the income of the budget will increase by 905 rub. (a very significant increase which can indicate, firstly, the considerable multiplying influence of the pulp-and-paper industry on the economic activity in the Arkhangelsk Oblast, and, secondly, the raised tax burden of this branch) at the same volume of investment into fixed capital. With the volume of investment into fixed capital of the processing industry increased by 1 thousand rub. at the same volume of production, the income of the budget will be reduced by 401 rub. It is worth focusing on this last result: the mechanism of its formation is not quite clear. It then follows that the growth of investment activity in the Arkhangelsk Oblast is undesirable from the positions of budget formation. It reduces the income of the budget. This result cannot be explained by calculation errors or inaccuracies of the basic data (as we did in the previous subparagraph): the ratio of investment growth to the decrease in the budgetary income equals 5:2 and is rather significant.

The obtained results testify that the performance of the timber industry cluster in the Arkhangelsk Oblast has a positive impact on the economy of the region, increasing the GRP and the income of the budget. However, investments into the fixed capital of the processing industry in this case are not effective as the increase in their volume leads to a reduction of both the GRP and the income of the budget. Thus, the Pomorinnovales innovative territorial timber industry cluster can be considered effective from positions of development of the regional economy, but a revision of the investment policy of the region, and improvement of its mechanisms are necessary.

4. Vologda Oblast (dairy cluster):

- with the production volume of the agricultural industry increased by 1 thousand rub., the GRP will increase by 18937 rub. In this case we recorded the greatest numerical value of the multiplier, 18.9. This means that the development of the agricultural industry (if we judge by the considered selection of clusters) has the greatest multiplying impact on the regional economy. We connect it with the circumstance that the creation of agroclusters means not only production, but also processing of the agricultural products in the territory of the region where the cluster is located, which in turn generates additional types of economic activity, additional workplaces and added value;

- with the volume of investments into the fixed capital of agriculture, hunting and forestry increased by 1 thousand rub., the GRP will increase by 87771 rub. This means that here we observe an even larger multiplying impact on the development of the regional economy than the one we considered in the previous subparagraph. This confirms the conclusion about the priority of state support of production and processing of agricultural products. The crux of the matter here is not in solving social (support of the rural areas) or political (ensuring food security) tasks, but in purely economic reasons. This support generates a very high economic effect;

- with growth of volume of release of agricultural production increased by 1 thousand rub., GRP will increase for 16812 rub at the same volume of investment into fixed capital. The increase in the volume of investment into fixed capital of agriculture, hunting and forestry increased by 1 thousand rub., at the same volume of production, will cause increase of GRP on 14602 thousand rubles. Thus, and at joint increase in two influencing factors considered by us, the resultant variable shows very high sensitivity;

- with the volume of investments into the fixed capital of agriculture, hunting and forestry increased by 1 thousand rub., the income of the budget will increase by 12311 rub. This circumstance seems very important to us. It is an additional argument for state support (including within the existing institutional regulations and financial) of agrocluster development;

- with the volume of agricultural production increased by 1 thousand rub., the income of the budget will increase by 2013 rub at the same volume of investments into fixed capital. The volume of investments into the fixed capital of agriculture, hunting and forestry increased by 1 thousand rub. at the same volume of production will cause an increase of the income of the budget by 3156 thousand rubles.

Thus, calculations showed that the performance of the agricultural enterprises in the Vologda Oblast and investments into the working capital of agriculture, hunting and forestry have a very significant positive impact on the growth of the GRP and the income of the budget. It follows from this that the dairy cluster created in the region functions in an effective specialization branch, and the full support of its development is necessary and economically expedient.

Conclusions

Summing up the results, we will note that the hypothesis we advanced about the positive influence that clustering has on the regional economy was confirmed quantitatively. This influence was proved by means of quantitative methods using enough representative datasets (covering 11 years). Our study undoubtedly had a selective character: it does not seem possible to carry out calculations for all cluster initiatives existing in Russia and to discuss their results within a single article. At the same time, following the results of the analysis we carried out, it is possible to draw a conclusion on the efficiency of creating clusters in Russian regions. Cluster activity is capable of having a positive impact on the growth of GRP and the budgetary income in the region. Along with it, we noticed differences in the multiplying influence of clusters on regional development, and its dependence on territorial and branch specifics. Apparently, this question requires further in-depth study.

The research is realised by grant № 26.3546.2017 "Development fundamentals of analysis and prediction of structural and dynamic parameters of the regional economy based on the integration of the Russian and world experience of management of territorial development and modern scientific doctrines".

References

1. A. Marshall, *Principles of economics* (Publishing house: Progress, Moscow, 1993)
2. M.E. Porter, Harvard Business Review, **76(6)**, 77-90 (1998)
3. P.A. Arkin, A.G. Golubev, News of the St. Petersburg state economic university, 58-62 (2014)
4. A.V. Babkin, A.O. Novikov, St. Petersburg State Polytechnical University Journal. Economics, **1(235)**, 9-29 (2016). DOI: 10.5862/JE.235.1.
5. D.P. Badgers, V.S. Kudryashov, *Formation of an industrial cluster: theoretical and methodical aspects: monograph* (St. Petersburg, 2014)
6. E.V. Borodkina, I.E. Risin, *Public administration by processes of a clustering of social and economic space of the region: monograph* (Publishing house of VGPU, Voronezh, 2011)
7. Yu.V. Vertakova, V.A. Plotnikov, Economy and management, **3(125)**, 10-18 (2016)
8. Yu.V. Vertakova, Yu.S. Polozhentseva, M.Yu. Hlynin, Technical and technological problems of service, 1(27), 92-99 (2014)
9. A. Babkin, T. Kudryavtceva, S. Utkina, World Applied Sciences Journal, **28(10)**, 1408-1413 (2013)
10. Ya.S. Gorelova, News of the St. Petersburg state economic university, **6**, 103-105 (2015)
11. A.V. Babkin, *Cluster structure of an industrial economy (2014): the monograph* (St. Petersburg state polytechnical university, St. Petersburg, 2014)
12. V.A. Carpenters, P.V. Sergeyev, S.V. Timofeeva, News of Southwest state university. Series: Economy. Sociology. Management, **2**, 92-97 (2011)
13. M.G. Svetovtseva, *Management of formation of an economic kernel of the region* (Voronezh, 2004)
14. A.V. Babkin, S.V. Zdolnikova, *Proceedings of The 28th International Business Information Management Association Conference,* 1-11 (2016)
15. Clusters and Regional Specialisation, *On Geography, Technology and Networks* (Pion, London, 1998)
16. M. Delgado, M. Porter, S. Stern, *Clusters, Convergence, and Economic Performance* (Institute for Strategy and Competitiveness, 2011)
17. A.V. Babkin, T.J. Kudryavtseva, Modern Applied Science, **9**, **1**, 109-118 (2015)
18. A.V. Babkin, *Cluster economy and industrial policy: theory and tools: collective monograph* (Polytechnic University Publishing, St.Petersburg, 2005)
19. G. Ahuja, C.M. Lampert, Strategic Management Journal, **22**, 521-543 (2001).
20. G. Bell, Strategic Management Journal, **26**, 287-295 (2005)

 DOI: 10.1051/shsconf/20173501064

Institute of taxation and its contribution to economic structure development (Russian practice)

Yu. Vertakova[1], *T.* Miroshnikova[2], and *S.* Zdolnikova[3,*]

[1]South-West State University, 305040, 50 years of October str., 94, Kursk, the Russian Federation
[2]Vladivostok State University of Economics and Service, 690014, Gogol str., 41, Vladivostok, the Russian Federation
[3]Peter the Great St.Petersburg Polytechnic University, 195251, Polytechnicheskaya str., 29, St.Petersburg, the Russian Federation

Abstract. Improving the quality of operations and controllability of the Federal Tax Service' (hereinafter – FTS) system is very relevant today. The solution of this problem is complicated by the total amount of regions and their large geographical area, as well as the distances between major cities. One of the necessary tools to increase its efficiency is the creation of a new model, which includes automated data transfer system in the data processing centers (hereinafter – DPC). The composition of the function includes: receiving, registering and digitizing (scanning) of paper-based documents; Massive data input from e-based/paper-based to FTS database; mass printing and mailing tax notifications; archiving paper-based documents. The cost effect is achieved thanks to the elimination of non-core functions, diminution of staff due to centralization, reduction of equipment. The result is saving a share of the national budget.

1 Introduction

The current state of economic climate in Russia shows us that it is very important to analyze a quality of operations of the Federal Tax Service' system. Based on the correlation and regression analysis we can make a decision about a new optimal model of tax system in Russian Federation.

Russia has faced with condition of ineffective operation of the tax system from the point of view of result of the Federal budget. There is a necessity of large-scale changes to create a new economic model which based on the creation of new automated data transfer system in the data processing centers (Ketels K., 2005). According to the results of modeling, cover the entire territory of Russia was defined geographical clusters. Assessment of the economic effects was carried out on the basis of the model developed CMDI activity for 5 years, which includes the production plan, organizational plan and spending plan by major economy CMDI. Modeling activities CMDI based on a combination of formal (mathematical) methods, expert assessments of experts and statistical data. Regardless of the dynamics of volumes of production, the centralization of mass print function will allow you to achieve significant cost savings.

It is a complicated problem in condition of a lack of financial resources in our economy. Thus, a key to the development of the Russian economy is the formation of the new tax service for country budget.

2 Methodological approach

Preliminary approach to the procedure includes four directions of work:

Determination of factors influencing on the placement of DPC

Correlation and regression analysis of the current indicators of DPC' performance

ABC analysis of the current performance of the DPC by main indicators

Positioning of the operational IFTS according different criteria

Correlation and regression analysis performed by consultant with using of statistical modeling of the dynamics of performance indicators of DPC in 7-10 years. Prognosis horizon is set up to 10 years – until 2020.

This analysis is carried out to identify common patterns of change between performance indicators and influencing factors. Predictive model was built on the basis of the factors with high level of correlation.

The task of ABC analysis of DPC' current activities based on main parameters of performance is to determine the individual IFTS or forming regional group for conversion to DPC. The optimal (best) will be to perform an ABC analysis of IFTS in the dynamics of 5-7 years in increments of 2 years. This method is a quantitative and it objectively segregates all IFTS by main indicators on 3 or more segments. The main criteria

* Corresponding author: s.v.zdolnikova@yandex.ru

for the evaluation of all IFTS is segmented to 3 scales (3 – dimensional coordinate system):

On a scale of «A-B-C» - the IFTS segmented by Criterion 1.

On a scale of «X-Y-Z» - the IFTS segmented by Criterion 2.

On a scale of «L-M-N» - the IFTS segmented by Criterion 3.

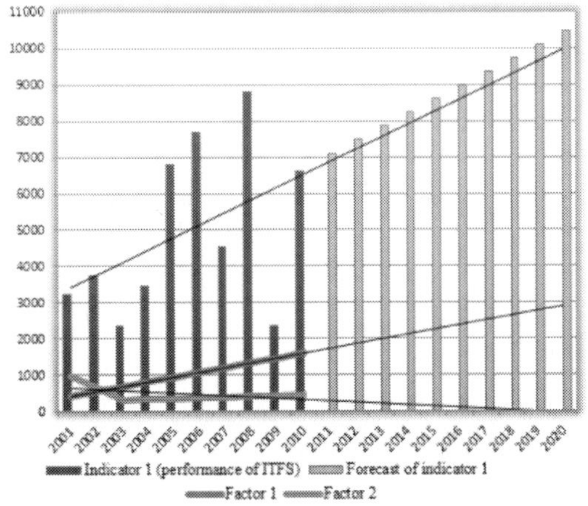

Fig. 1. Correlation and regression analysis according different criteria.

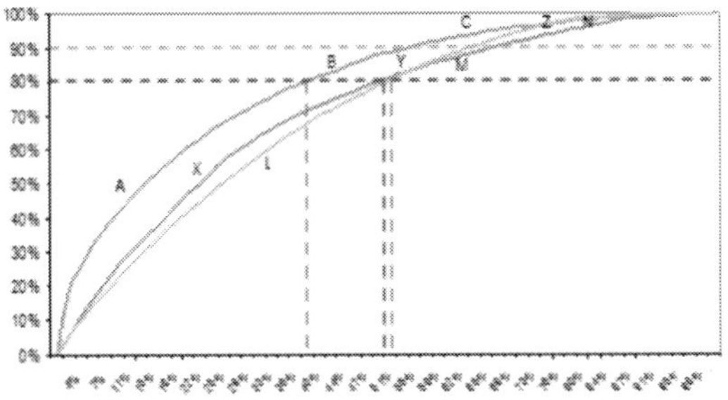

Fig. 2. ABC analysis of DPC' current activities.

After a segmentation of the IFTS analyzed indicators into 3 groups performed cross-analysis, and depending on the number of analyzed indicators and factors of influence can be formed into a targeted IFTS clusters.

The analysis of socio-economic situation of regions was performed by analytical processing on following criteria: the Volume of the GRP (Gross Regional Product); average monthly wages; the amount of companies and organizations; the amount of small businesses; population; population density. As part of the transport infrastructure of the regions were analyzed:

Density of roads

The number of post offices

After analysis of each indicator was compiled overall rating of regions and selected the most relevant of them. Further, based on the analysis of environmental data and performance indicators directly to the IFTS by regions defined target regions for accommodation of the new data centers (mass data input centers).

The next step was trends' forming and IFTS positioning:

In total number of stuff listed factors may be other factors such as: costs (labor, maintenance, etc.); Transport leg; Number of taxpayers; Sectoral specialization of regions of Russia; Real estate' cost; Budget and construction time; Some other factors (etc.)

Practical implementation of the operation is two key phases: Research and Design. In the research stage, the authors developed methodical approach, taking into account experience of their foreign colleagues (USA, Canada, UK), as well as the basic principles of network centralization of the large private and organizations (Plotnikov V.A., Vertakova Yu.V. (2016)).

Based on the methodology of the value chain of M. Porter to optimize management system of the IFTS in its activities processes were grouped into a three groups:

-Main

-Auxiliary

-Service

Then Each group of processes were identified with processes which must be created in centralized DPS. Cabinet part of study also includes a comprehensive quantitative analysis of the statistics of the tax document from all regions of Russia in four types of documents for several years: accounting and tax reporting, required documents, registration documents, printing and mailing. To select the target regions, accommodate the new data center conducted a socio-economic analysis of the regions of the country with the use of correlation and regression analysis, and the ABC analysis. Created a list of key of regional and industrial indictors for benchmarking. The field part of the study included the diagnosis of three of five existing DPS in different cities of the country through personal interviews with key executives, the timing of the operations staff, as well as the personal inspection of manufacturing and warehouse space.

Environmental analysis was performed according to the method of choosing prospective region. To select a promising region for the creation of CMDI) was calculated the correlation coefficient of external factors of influence relatively to indicator "Volume of regulatory documents based on paper (by the number of sheets)."

The correlation coefficient – a measure that determines the extent of connection between motion/values of each represented factor with this exponent indicator. The correlation coefficient may be varied from «1» in case of positive correlation to «-1» at full negative. If the correlation coefficient is equal to «0», variables are completely independent from each other.

Correlation analysis helps determine is it possible to predict the potential value of the indicator "Volume of regulatory documents", knowing the magnitude of external factors of influence. The calculation results are presented in the table below.

These data show that the greatest influence on the performance results of the FTS have 6 factors, including 4 factors of direct influence with relatively high value of the correlation coefficient (the number of companies and

DOI: 10.1051/shsconf/20173501064

organizations, the number of small businesses, the population, the number of inspections) and 2 factors of indirect impact of lower influence (gross regional product, the number of post offices). Other factors can be considered as insignificant and largely derived from the most significant factors.

Table 1. Rating of influence of external factors on the correlation coefficient.

Factors of influence	Correlation coefficient
1. Number of enterprises and organizations	0,92
2. Number of small businesses	0,91
3. Population	0,90
4. Number of IFTS	0,87
5. Gross regional product	0,77
6. Number of post-offices	0,75
7. Population density, thou. pers. on thou. sq. m.	0,24
8. The density of asphalted public roads	0,21
9. The average wage of employees	0,01

The following is an analysis of the existence of the factor for the region and the ranking of regions according to the values of factors.

Analysis of the regions in the number of enterprises and small enterprises.

Allocation of the legal entities in the subject of the Russian Federation is given on basis of applications for the identification of economic entities, codes of Russian classifiers of technical, economic and social information, established in accordance with legislation in the field of technical regulations.

Full rating of Russian Federation' regions was carried out by:

The number of enterprises and organizations; The number of small businesses; The population; The population density; The GRP; The level of wages; The number of post-offices of RPS; The density of asphalted public roads; The natural indicators of IFTS; The performance indicators of IFTS; The external factors:

Analysis of regions by the external factors.

Analysis of regions on the external factors was carried out in the 3 different sections:

The amount of paperwork, mln. sheets (similar to the Analysis of the regions by natural indicators of the tax inspectorate);

The GRP (Gross Regional Product), mln. rubles;

The density of population thou. pers. on thou sq. m.

Analysis of regions by the external factors is conducted to identify the most developed and populous regions with the highest volume of document processing. Borders of groups of regions in terms of volume and the external factors are listed in the table below.

The results of grouping regions by the ABC-analysis method (Chetyrkin E.M., 2008) are listed on the graph below.

ABC-analysis of regions by the external factors.

Table 2. Grouping regions by the ABC-analysis method.

Grouping regions by the ABC-analysis method	**Group**		
	1	**2**	**3**
Shares by the indicator	Maximum	Average	Minimum
Volume of documents to process, sheets	A	B	C
Gross regional product, mln. rub.	F	N	S
Population density, thou. pers. on thou. sq. m.	X	Y	Z
Sum share by regions, %	От 0 до 50%	От 51 до 80%	От 81 до 100%

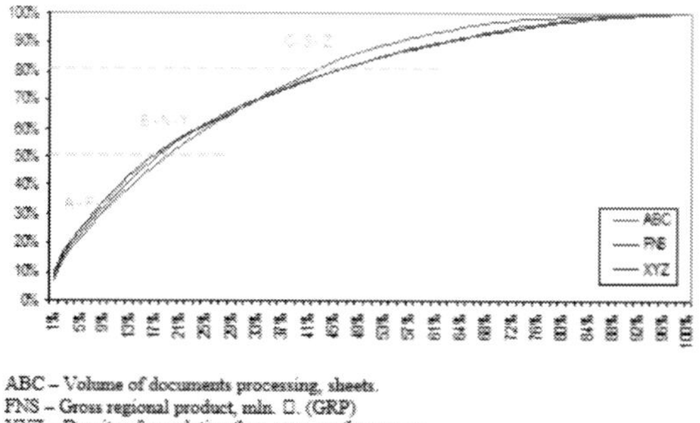

Fig. 3. ABC-analysis of regions by the external factors.

The result of this variant of ABC-analysis was developed a rating with IFTS with lead positions all 3 indicators. Table below.

Table 3. Rating of regions-leaders by the external factors.

Rating of regions-leaders by the external factors	**Position**			
	1	**2**	**3**	**4**
Region	Moskovskaya oblast	Krasnodarsky Krai	Republic of Tatarstan	Samarskaya oblast
Volume of documents to process, sheets	43,3	14,8	14,5	12,4
ABC group	A	A	A	A
Gross regional product, mln. rub.	1685	809	923	707
FNS group	F	F	F	F
Population density, thou. pers. on thou. sq. m.	147	68	56	59
XYZ group	X	X	X	X

From the aggregation of three leading figures of all 4 regions of the country. With total share of the volume of GRP processing not more than a 20% and an average population density of 83 people on 1 sq. km. The low number of represented regions for these indicators. Due to the large proportion of the Moscow region. Particularly, it greatly overstates the average of the leading regions in the density of the population – more than 3 times compare to the average level, and other regions. Positioning of leading regions on the graph (image) below.

This positioning of the regions shows the absence of any clear tendency, however, forms separate segment of the developed regions with large amount of processing. At the same time, vast number of regions have both low population density and small GRP. The main factor affecting the value of these indicators is the uneven distribution of business activity centers (financial, production) in the country (Grechenyuk O.N., Grechenyuk A.V. (2014)). Taking into account the analysis of the factors compiled a list of the leading regions.

In the aggregate made ratings of Russian regions, the resulting analysis of the various factors of the macro environment, external and internal factors the IFTS activity, formed a list of 13 promising regions for the placement of CMDI.

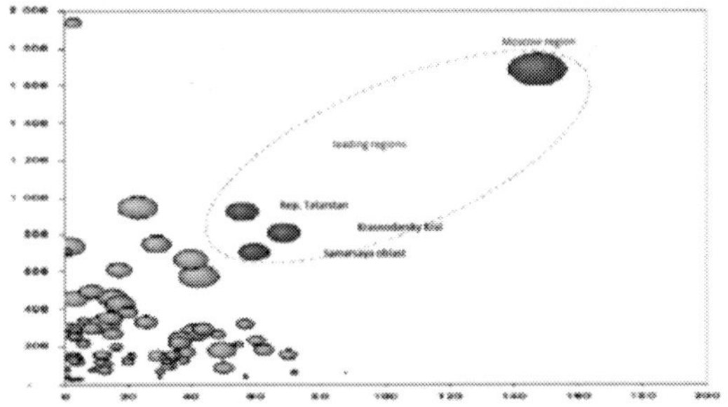

Fig. 4. Positioning of leading regions by the external factors (GRP and population density).

All these regions are leading for most indicators (Altai Krai, Voronezh Oblast, Krasnodar Oblast, Moscow region, Novosibirsk Oblast, Primorsky Krai, the Republic of Bashkortostan, Tatarstan, Rostov Oblast, Samara Oblast, Sverdlovsk Oblast, Stavropol, Chelyabinsk Oblast).

The territorial allocation of the leading regions forms a strongly marked axis of the adjacent densely populated areas of the South and the Black Sea regions through the industrial Ural region to Taimyr. Contact one of the leaders of the northern and Siberian regions due to peculiarity of their industrial development mainly in the mining industries. Also one of the leading positions took by Primorye region due to the high concentration of industry and business and population density.

Process of selection a perspective regions of Russia proved that by geographical location relative to each other and relative to other regions, not all of them can be earmarked for the formation of geographical clusters (William R. Kerr (2010)) and placing CMDI.

As shown at the figure above, regions suitable for dispositioning of CMDI can be divided into 3 groups:

Regions with existing CMDI, Target regions (without alternative), Alternative target regions. The composition of the target regions accommodation of the CMDI and 10 regional clusters is presented in the table below.

Table 4. Regions clusters.

Cluster #1 (center – Moscow)	Cluster #2 (center – S. Petersburg)	Cluster #3 (center – N. Novgorod)
Regions: Moscow	Regions: 1. Arkhangelsk region. and Nenets Autonomous District 2. Vologda. 3. St. Petersburg 4. Kaliningrad oblast 5. Leningrad region. 6. Murmansk region. 7. Novgorod region. 8. Pskov region. 9. Rep. Karelia 10. Rep. Komi	Regions: 1. Kirov region. 2. Nizhny Novgorod region. 3. Penza. 4. Perm kr. 5. Rep. Mari El 6. Rep. Mordovia 7. Ulyanovsk region. 8. Chuvashia Rep.
Cluster #4 (center - Volgograd)	Cluster #5 (center - Kemerovo)	Cluster #6 (center – Dubna or others.)
Regions: 1. Astrakhan region. 2. Volgograd region. 3. Kabardino-Balkar Rep. 4.Karachay-Cherkess Rep. 5. Krasnodar kr. 6. Rep. Adygea 7. Rep. Dagestan 8. Rep. Ingushetia 9. Rep. Kalmykia 10. Rep. North Ossetia Alania 11. Rostov region. 12. Saratov region. 13. Stavropol kr. 14. Chechen Rep.	Regions: 1. Altai kr. 2. Irkutsk region. 3. Kemerovo region. 4. Krasnoyarsk kr. 5. Novosibirsk region. 6. Omsk region. 7. Rep. Altai 8. Rep. Tyva 9. Rep. Khakassia 10. Tomsk region.	Regions: 1. Vladimir region. 2. Ivanovo region. 3. Kostroma region. 4. Moscow region. 5. Smolensk region. 6. Tver region. 7. Yaroslavl region.
Cluster #7 (center - Kaluga)	Cluster #8 (center - Kazan)	Cluster #9 (center - Ekaterinburg)
Regions: 1. Belgorod region. 2. Bryansk region. 3. Voronezh region. 4. Kaluga region. 5. Kursk region. 6. Lipetsk region. 7. Orel. 8. The Ryazan region. 9. Tambov region. 10. Tula region.	Regions: 1. Orenburg region. 2. Rep. Bashkortostan 3. Rep. Tatarstan 4. Samara region. 5. Udmurtia.	Regions: 1. Kurgan region. 2. Sverdlovsk region. 3. Tumen. 4. The Khanty-Mansi Autonomous Area - Yugra 5. Chelyabinsk region. 6. The Yamalo-Nenets Autonomous county
Cluster #10 (center – Khabarovsk)		
Regions: 1. Amura. 2. The Jewish Autonomous Region. 3. Trans-Baikal kr. 4. Kamchatka kr. 5. Magadan region. 6. Seaside kr. 7. Rep. Buryatia 8. Rep. Sakha (Yakutia) 9. Sakhalin region. 10. Khabarovsk kr. 11. Chukotka Autonomous county		

Source: Author's calculations

The number of regions in the cluster may be indicative to the scale of the administrative cooperation perspective, without being a characteristic of the bulk of the future performance of CMDI due to the uneven performance of regions.

As can be seen from the territorial distribution of the target regions of CMDI and regional clusters are not always able to avoid the adjacent location of the target region (see. The figure below). Among the areas of alternative locations of CMDI is recommended to choose: to cluster 8 - Kazan (Republic of Tatarstan); For 10 cluster - Khabarovsk (Khabarovsky Krai), due to a more central location of the city placing CMDI in the entire cluster.

The economic result:

As mentioned above, the centralization of mass data input and printing in CMDI will provide:

Centralization and creating of single database of the tax documents;

Reducing of the number of duplicated functions and data points;

Organization of the required level of quality and technological performance of the functions of mass entering and printing through the use of efficient equipment;

Increase the efficiency of processing incoming documents and notifications;

Redistribution of data-stream load, printing and sending out notices from IFTS employees to CMDI employees as a result liberation of human resources.

In terms of improving economic efficiency CMV creation solves the problem to reduce the unit cost of the notification by the following factors;

Reducing the cost of personnel performing the functions of mass entering and printing, while maintaining the scope of work (Yuya Kajikawa, Yoshiyuki Takeda, Ichiro Sakata, Katsumori Matsushima (2010));

Management of maintenance costs of printing by centralizing this type functions;

Assessment of the economic effects was carried out on the basis of the model developed CMDI activity for 5 years, which includes the production plan, organizational plan and spending plan by major economy CMDI. Modeling activities CMDI based on a combination of formal (mathematical) methods, expert assessments of experts and statistical data.

3 Conclusions

As a result, the preparation of the development strategy of the network of data centers Tax authorities made the calculation and comparison of options for implementing the functions of mass input and print documents. For each function carried out: the construction of an economic model functions, including two scenarios, and if necessary - embodiments of the function; calculation of the economy, which is achieved in the implementation of each of the scenarios and variants of the economic model in the target model management.

In accordance with international methodology of business planning UNIDO for the target model data centers are designed to create new plans for production and logistics, operations, financial and investment plan. These plans are defined reference values of key performance indicators in the data center of the economy, productivity and staff. Built target process model data center operations. The total cost savings over 5 years of the first scenario is 6 871 million. rub. According to the second scenario - 11 315 million. rubles.

This paper is an output of the science project of the government task government task of Ministry of Education and Science of the Russian Federation № 26.3546.2017/PCH "Development fundamentals of analysis and prediction of structural and dynamic parameters of the regional economy are based on the integration of the Russian and world experience of management of territorial development and modern scientific doctrines".

References

1. L.I. Askhatova, A.M. Fatkhiev, L.N. Safiullin, and A.M. Safiullina, *Competitive Strategies* (Sage Publications, London, 2000)
2. Federal State Statistics Service (Rosstat). http://www.gks.ru
3. Weber, Manfred, *Business computing* (Omega-L, Moscow, 2007)
4. E.M. Chetyrkin, *Financial mathematics* (Delo ANCH, Moscow, 2008)
5. William R. Kerr (2010) Journal of Urban Economics, **67**, **1**, 46-60 (2010)
6. V.G. Panskov, Economy Taxes. Right. Finance, 5, 130-138 (2016)
7. V.G. Panskov, Economy. Taxes. Right. Finance, **1**, 114-120 (2015)
8. I.Y. Chugunov, O.M. Zhukevych, Economic Annals-XXI, **1-2(1)**, 61-64 (2014)
9. R.A. Musgrave, P.B. Musgrave, (Public Finance: Theory and Practice, 2009)
10. E.V. Balatsky, he Problems of forecasting, **2**, 88-105 (2003)
11. L.I. Goncharenko, L.M. Arhiptseva, National interests priorities and safety, **6**, 21-26 (2011)
12. I.A. Mayburov, Finance, **8**, 45-49 (2012)
13. Yu. Vertakova, V. Plotnikov, Economical chasopis-XXI, **156(1-2)**, 37-40 (2016)
14. V. Plotnikov, Y. Vertakova, *Proceedings of the 26th International Business Information Management Association Conference - Innovation Management and Sustainable Economic Competitive Advantage: From Regional Development to Global Growth*, 3224-3229 (2015)

15. Y. Vertakova, Y. Polozhentseva, M. Klevtsova, *Conference on Political Sciences Law, Finance Economics & Tourism*, 1, 399-406 (2014)
16. Y. Vertakova, Y. Polozhentseva, M. Klevtsova, E. Leontyev, Innovation management and education excellence vision 2020: from regional development sustainability to global economic growth, **I-VI**, 1134-1144 (2020)
17. Order Federal tax service, 2011 no. MMV-7-2/336 "On changing the composition and structure of the Handbook the Federal tax service of Russia "the Documents should be retained in electronic storage, as well as documents obtained by the tax authorities in the exercise of tax control" and the order of reference, approved by order FTS of Russia from 2009 and no. MM-7-2/517
18. N.E. Egorov, A.V. Babkin, G.S. Kovrov, S.V. Muraveva, Procedia - Social and Behavioral Sciences, **207**, 816-823 (2015)
19. A.V. Babkin, V.S. Lipatnikov, S.V. Muraveva, Procedia - Social and Behavioral Sciences, **207**, 749-758 (2015)
20. A.V. Babkin, S.V. Zdolnikova, *Proceedings of The 28th International Business Information Management Association Conference*, **1-11** (2016)

Research and evaluation of innovative integrated industrial structure potential: theoretical and practical aspects

S.V. Zdolnikova[1,*], *A.V.* Babkin[1], and *N.B.* Smolskaya[1]

[1]Peter the Great St.Petersburg Polytechnic University, 195251, Polytechnicheskaya str., 29, St.Petersburg, the Russian Federation

Abstract. Modern market conditions, as well as the reorientation of the Russian economy from raw materials to the innovative model of development, based, in particular, on strengthening the position of domestic industry being able to compete against the international producers, are the incentives to seek ways to increase efficiency and create additional competitive advantages that at present are innovative products and services. One of the innovative economic entity development factors is the establishment and effective use of its innovative potential. The innovative potential evaluation plays an important role in the development of innovation policy and innovation strategy of market subject activity. This article investigates the innovative potential of the integrated industrial structures (IIS) that presently are one of the most promising forms of economic entities integration. The authors propose an approach to the definition and assessment of innovative IIS potential, as well as the comprehensive and systemic innovative potential estimation procedure. The developed theoretical and methodological provisions were applied to evaluate the innovative potential of the JSC Svetlana, one of the leading Russian electronic instrument engineering holdings.

1 Relevance of the study

The creation of IIS is nowadays one of the current trends of the economic transformations in the Russian economy. Thus, within the period from 2011 to 2015, the average transaction share in the industrial sector amounted to 31%, with the most actively integrated sectors being oil and gas industry, metallurgy and mining. When being integrated, enterprises are able to respond more flexibly to demand changes, and joint production contributes to the various technical product quality increase on the basis of a systemic use of intellectual potential and also allows to save resources and to obtain a synergetic effect [1]. These circumstances contribute not only to the production increase, but also to the development of high-tech innovative products, which leads to increased innovation performance of IIS and industry sectors on the whole.

One of the innovative economic entity development factors is the establishment and effective use of its innovative potential. The innovative potential evaluation plays an important role in the development of innovation policy and innovation strategy of market subject activity. High innovative potential is the primary means of IIS successful adaptation to the uncertain and rapidly changing environmental conditions and it ensures the formation and implementation of such development schemes that will lead to the optimal outcomes in the specific situation.

2 Problem statement

To develop the IIS innovative potential evaluation tool as well as to explore the practical aspect of its application on the basis of the integrated structure essence research and its innovative potential analysis. In order to achieve this goal the following objectives are stated:

to define the "innovative economic entity potential" concept

to develop the approach to the essence presentation and to the innovative IIS potential evaluation

to develop the innovative IIS potential evaluation procedure

to apply the developed evaluation procedure for the innovative potential evaluation of the industrial holding JSC "Svetlana".

3 Results of the study

3.1 The "innovative economic entity potential" concept

The innovative potential is a complex and multifaceted category that has been discussed by the foreign and domestic scholars since the third quarter of the 20th century. The development of this economic category is connected with the concepts of "economic potential", "scientific and technical potential", etc. Conducted by

* Corresponding author: s.v.zdolnikova@yandex.ru

the authors the analysis of approaches to the innovative potential understanding characterizes it as the combination of different kinds of capacities or resources, i.e. the set of certain forces that should guarantee the successful management of innovation to the economic entity [2, 3, 5, 7, 8].

However, the mere existence or absence of such forces does not guarantee effective innovation outcomes, thus the innovative potential should be considered as the economic entity ability with the resources available to achieve maximum innovation outcomes [12, 15, 17, 19]. As the indicator of the innovation activity development level, innovative potential should reflect the effectiveness of innovation development and implementation into the market or into the operating activities of the economic entity, and, therefore, needs to include the resources and the management basis for the creation of innovative goods (construction, services) and to take into account internal and external conditions favorable or unfavorable to their commercialization.

Based on the foregoing, the authors propose to consider the innovative potential as the skills and opportunity of the economic entity to convert the available resources into the innovation outcomes. This approach is based on the evaluation of three main elements (fig. 1):

1) resources as the material basis for the innovation activity implementation;

2) skills, i.e. the know-how to translate available resources into the innovation outcomes (innovation goods, construction, services) that are proposed to be evaluated through the concrete results of innovation activity;

3) opportunities, i.e. the presence of favorable or unfavorable trends in the development of innovation economic entity activity.

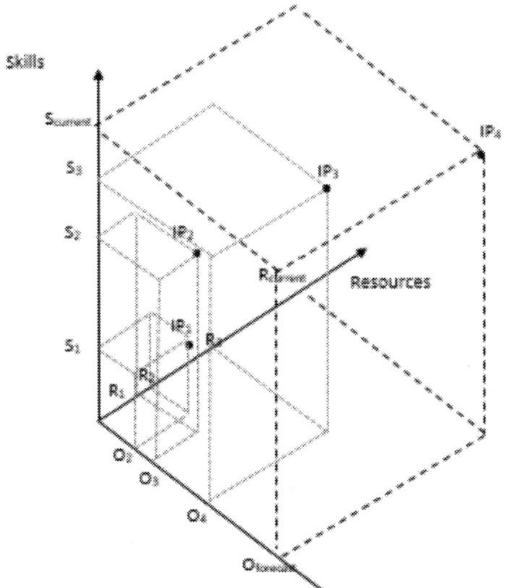

Fig. 1. The innovative economic entity potential in the coordinate system "Resources-Skills-Opportunities".

The difficulty to define the innovative IIS potential is related to the ambiguity of the IIS consideration: on the one hand, it is regarded as a sole structure comparable to the enterprise (the natural integration principle), on the other hand, it is considered as a set of formally independent enterprises involved into the joint operation activity (the quasiintegration principle). From the point of view of the system approach applied by the author to determine the IIS nature, this fact can be explained by two related patterns: progressive systematization and progressive factorization.

3.2. The approach to the essence presentation and to the innovative IIS potential evaluation

The authors considered two variants of innovative IIS potential evaluation:

1) the innovative IIS potential is considered as the IIS ability and opportunity as a whole to convert the available resources into innovation outcomes, in this case IIS is comparable to the enterprise (the natural integration principle), and the integrity coefficient α is measured within the range from 0.94 to 1 ($0.94 < a < 1$);

2) the innovative IIS potential is considered as a set of innovative capacities of the enterprises included in the IIS, i.e. as a set of abilities and opportunities of the organizations included into the IIS, to convert available resources into innovation outcomes, in this case the IIS is formed on the quasiintegration principle, and the integrity coefficient α is measured within the range from 0.5 to 0.94 ($0.5 < a \leq 0.94$).

Thus, the innovative IIS potential is the total potential of the economic entities involved in the IIS that is generated with regard to the structural integrity, as well as to the synergetic effect emerging in it and reflects the IIS ability and opportunity to convert the available resources into innovation outcomes.

The approach to the innovative potential definition and evaluation allows not only to characterize the innovative economic entity development level within a specific period of time, but also to reflect the trend of its change, thereby presenting innovative potential as the integral indicator in the coordinate system "Resources-Abilities-Opportunities" with regard to the structural integrity degree and to the synergetic effect occurring in it. This approach provides a comprehensive view of innovative potential making it possible to fully reveal its nature.

3.3 Innovative IIS potential evaluation procedure

The innovative IIS potential evaluation procedure provides the determination of the IIS innovation activity development level. It includes the developed approach and stages of innovative potential evaluation, the synergetic effect evaluation procedure, the algorithm of IIS innovative potential evaluation and the algorithm for calculating the IIS innovative potential integral index.

The innovative IIS potential evaluation procedure consists of three stages:

1) the preparatory stage including the identification of the IIS integration type, the choice of indicators for evaluating the components of the innovative IIS potential and its normative values determination.

2) the evaluation stage including the calculation of individual and integrated indicators of the innovative potential, determination of the synergetic effect, the calculation of the innovative IIS potential integral index;

3) the final stage including the innovative IIS potential level definition; the development of proposals to improve the innovative IIS potential level.

The preparatory stage begins with the IIS integration type identification on the basis of the integrity coefficient α calculation. Under the proposed approach, depending on the α value the innovative IIS potential is defined either according to the structure as a whole or as the sum of the innovative economic entities capacities included into the IIS.

The choice of indicators for evaluating the innovative IIS potential components is a crucial step of the preparatory stage. The scorecard should [14, 16]:

– provide a comprehensive evaluation of each innovative IIS potential component;

– include universal and easily comparable indicators that will evaluate the innovative potential of each entity included into the IIS as well as other IIS innovative potential;

– include available indicators contained in the IIS accounting, statistical, etc. reporting forms.

The final step at this stage presupposes the normative values identification for each selected innovative IIS potential indicator constituents.

The evaluation stage includes main calculations: separate indicators of the innovative potential constituent parts are identified, integral indicators of each component are calculated, synergetic effect is evaluated, as well as the value of the innovative potential integral indicator is defined.

The synergetic effect describes the IIS innovation activity efficiency increase due to the development of production and organizational relations among entities included into the structure [6, 9, 11]. If entities have partial autonomy and their innovation activities are carried out in cooperation with other entities as well as outside the IIS, a synergistic effect can be evaluated by comparing the effectiveness of the innovation outcomes that have been achieved by the entities independently and jointly [10, 13, 18]. In case the IIS is a sole structure and the entities included in it deprive of their independence, a synergetic effect will be taken into consideration in the IIS innovation activity results as the resources, abilities and opportunities of the entities will be considered not autonomous but jointly as well as their individual results.

Thus, it is reasonable to take a synergistic effect into consideration in those IIS where there is partial independence of the entities included into the IIS, i.e., at $0.5 < a \leq 0.94$.

At the final stage the innovative potential level is determined and the proposals for increasing it are developed. The innovative IIS potential level determination is done using scaling, i.e., assigning numerical values to certain states of the system under study, i.e. innovative potential. Scaling is necessary for bringing the integral index, which combines a large number of diverse indicators, to the only possible interpretation of its value.

The procedure basis is the algorithm of innovative IIS potential evaluation. This algorithm is the combination and the order of entity actions required for obtaining accurate data on the innovative potential state (fig. 2).

The innovative IIS potential integral index is proposed to be evaluated as the arithmetic average of the innovative potential with regard to the integrity coefficient α and the synergetic effect value.

The formulas for calculating the innovative IIS potential integral index are presented in table 1.

Table 1. Formulas for calculating the innovative IIS potential integral index.

The formula for calculating the integral indicator of innovative potential of IIS
1. The coefficient of consistency: 0,5 < *a* ≤ 0,94
short term
1) $IP_{IIS} = \sum_{i=1}^{n} \frac{I_{resources\,i} + I_{resulting\,i} + I_{ab\,i}}{3} \times a \times SE$, if $\sum_{i=1}^{n} \frac{I_{resources\,i} + I_{resulting\,i} + I_{ab\,i}}{3} > 0$; 2) $IP_{IIS} = \sum_{i=1}^{n} \frac{I_{resources\,i} + I_{resulting\,i} + I_{ab\,i}}{3} \times \frac{1}{a} \times \frac{1}{SE}$, if $\sum_{i=1}^{n} \frac{I_{resources\,i} + I_{resulting\,i} + I_{ab\,i}}{3} < 0$;
long term
1) $IP_{IIS} = \sum_{i=1}^{n} \frac{I_{resources\,i} + I_{resulting\,i} + I_{ab\,i}}{3} \times a \times SE^{l}$, if $\sum_{i=1}^{n} \frac{I_{resources\,i} + I_{resulting\,i} + I_{ab\,i}}{3} > 0$; 2) $IP_{ISS} = \sum_{i=1}^{n} \frac{I_{resources\,i} + I_{resulting\,i} + I_{ab\,i}}{3} \times \frac{1}{a} \times \frac{1}{SE^{l}}$, if $\sum_{i=1}^{n} \frac{I_{resources\,i} + I_{resulting\,i} + I_{ab\,i}}{3} < 0$;
2. The coefficient of consistency: 0,94 < *a* < 1
1) $IP_{ISS} = \frac{I_{resources\,i} + I_{resulting\,i} + I_{ab\,i}}{3} \times a$, if $\frac{I_{resources\,i} + I_{resulting\,i} + I_{ab\,i}}{3} > 0$; 2) $IP_{ISS} = \frac{I_{resources\,i} + I_{resulting\,i} + I_{ab\,i}}{3} \times \frac{1}{a}$,

DOI: 10.1051/shsconf/20173501065

$$\text{if } \frac{I_{resources\,i} + I_{resulting\,i} + I_{ab\,i}}{3} < 0;$$

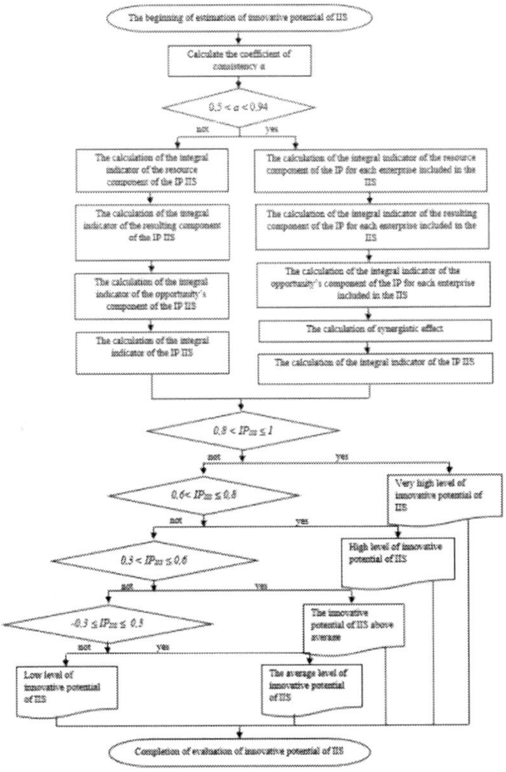

Fig. 2. The algorithm of the innovative IIS potential evaluation.

The innovative IIS potential integral index characterizes the innovative potential value in points, which allows to obtain an objective evaluation based on the use of quantitative indicators covering the main aspects of innovative IIS development.

The developed innovative IIS potential evaluation procedure makes it possible to compare and analyze the level of innovation potential of a single IIS as well as of several IISs through the use of relative indicators. This circumstance facilitates the bottleneck identification, the elimination of which will allow to obtain new competitive advantages and will contribute to stable economic and innovative IIS development.

3.4. Application of the developed evaluation procedure for the innovative potential evaluation (on the example of the industrial holding JSC "Svetlana")

The innovative IIS potential research was carried out on the example of the holding JSC "Svetlana". The holding JSC "Svetlana" (hereinafter – the Holding) has been operating in the market of electronic industry in Russia for more than 125 years. During this time its manufacturing activity has evolved from the incandescent lamps production to the production of unique nanoheterostructures and devices developed on their basis. The analysis shows that innovation activity of the IIS under consideration, on the whole, meets the requirements of the external environment prevailing in the new economy and articulated in the State industrial and innovation policy.

In a number of indicators that characterize the innovative potential, the Holding exceeds the average development level of industrial enterprises operating in the engineering industries, however, the problematic component of the Holding innovative potential is the resulting component (fig. 3).

On the based of the conducted analysis the following proposal for the Holding innovative potential was elaborated, that is the Holding's own assets investment into the innovation activities. As the source for the Holding's own assets the retained earnings were considered [4, 20].

To predict changes in the innovative potential level depending on the amount of the Holding's own assets invested into the innovation activity development four scenarios involving the investment of various company's own assets were considered. The scenario assessment shows that annual investments of 50% of the total cost of research, development and innovation in 2015, i.e. 283141.3 thousand RUR, does not exceed 10% of retained earnings in 2015, will have allowed to raise the level of the integral index of the resulting component, and, with the invariable level of resources component, to increase the opportunity component integral indicator to the level of "very high" by 2020 (table 2). The effectiveness of the proposed practical recommendations on the Holding innovative potential development is presented in table 3.

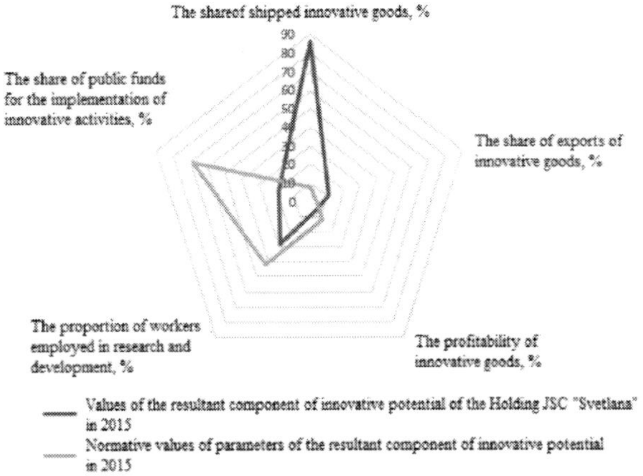

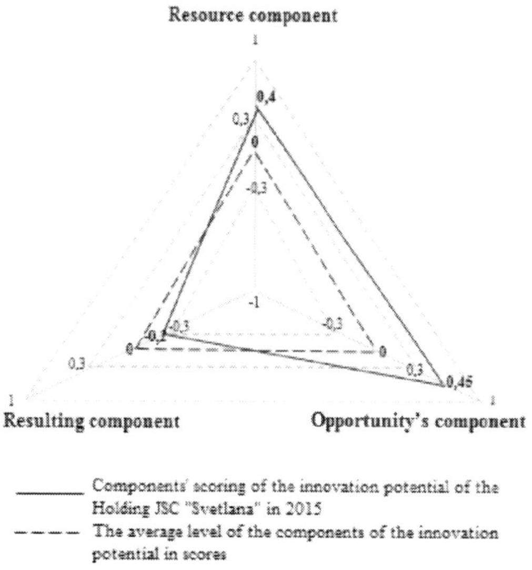

Fig. 3. Analysis of the Holding JSC "Svetlana" development on the basis of the innovative potential.

Table 2. The results of the company's own assets additional investment on research, development and innovation scenarios implementation in 2016-2020.

Indicator	Scenario 1	Scenario 4
2016		
The cost of research, development and innovation, thousand rub.	662959,8	946101,0
The volume of innovative goods (works, services), thousand rub.	4176579	5479028,6
The cost of the produced innovative goods (works, services), thousand rub.	3940168,9	4789592,8
The profitability of innovative goods		
%	6	14,4
in scores	-1	0
The share of public funds for the implementation of innovative activities		
%	37,7	26,4
in scores	-1	-1
The knowledge-intensive innovative goods (works, services), rub./rub.	0,16	0,17
2020		
The cost of research, development and innovation, thousand rub.	1245384	1528525,3
The volume of innovative goods (works, services), thousand rub.	4884805	6187254,5
The cost of the produced innovative goods (works, services), thousand rub.	4608306,6	5457730,5
The profitability of innovative goods		
%	6	13,4
in scores	-1	0
The share of public funds for the implementation of innovative activities		
%	24,4	19,9
in scores	-1	0

Table 3. The effectiveness of the proposed practical recommendations on the Holding JSC "Svetlana" innovative potential development.

Indicator	Result
The resulting component of innovative potential	The increase in 2 times (0,4 scores)
The profitability of innovative goods	An increase from 6% to 13.4 %
The share of public funds for the implementation of innovative activities	The decrease from 24.4% to 19.9 %
Innovative potential	An increase of 33 % (0,13 scores)
The volume of innovative goods (works, services)	An increase of 27 % (1302,4 million rub.)

The practical implementation of the developed proposals will enhance the innovative potential level and ensure the effective development of innovation activities and innovation sector of the Holding.

4 Conclusions

1. Based on the analysis of approaches to the innovative economic entity potential definition and evaluation it was proposed to consider the innovative potential as the sum total of the entity resources, abilities and opportunities to conduct innovative activities.

2. The approach to the understanding and evaluation of the IIS innovative potential representing it as the IIS ability and opportunity to convert the available resources into innovation outcomes as well as taking into account the integrity degree and the synergetic effect occurring in the integrated industrial structures synergistic effect was proposed.

3. The tools were developed for the IIS innovative potential evaluation including stages, algorithm and procedure. The IIS innovative potential evaluation will have three stages: the preparatory stage, which is aimed to determine the IIS integration type, to select the innovative potential components evaluation indicators and to define their normative values; the evaluation stage, which includes the calculation of individual and integrated indicators of the innovative potential components evaluation, the definition of synergetic effect and the calculation of the innovative potential integral indicator; the final stage, which determines the innovative potential level and develops proposals to increase the innovative potential level. The innovative IIS potential evaluation procedure allows to give an objective assessment and to ensure comparability of results as well as to consider such IIS features as the integrity and synergistic effect appearance.

4. On the basis of the developed organizational-economic mechanism the IIS development research was conducted in the course of which the compliance of the state of innovation activity in the IIS under consideration with the requirements of the external environment prevailing in the new economy and articulated in the State industrial and innovation policy was stated, as well as practical recommendations on the Holding JSC "Svetlana" innovative potential development were worked out and their cost-effectiveness was evaluated.

5 Call for further research

The authors associate the future research directions with the modeling of the innovative IIS potential management as the integrated system of measures interacting with other subsystems and processes occurring in the IIS and contributing to the continuous innovative potential development.

References

1. G. Ahuja, C.M. Lampert, Strategic Management Journal, **22**, 521-543 (2001)
2. A. Babkin, T. Kudryavtceva, S. Utkina, World Applied Sciences Journal, **28(10)**, 1408-1413 (2013)
3. A. Babkin, V. Plotnikov, S. Muraveva, *Proceedings of The 25th International Business Information Management Association Conference,* 1286-1294 (2015)
4. A.V. Babkin, S.V. Zdolnikova, *Proceedings of The 28th International Business Information Management Association Conference,* 1-11 (2016)
5. G. Bell, Strategic Management Journal, **26**, 287-295 (2005)
6. G. Bysrtay, Journal of Cybernetics, **2**, 71-88 (2003)
7. Y.T.H. Chiu, Journal of Business and Industrial marketing, 24, 46-55 (2009)
8. R. Cowan, N. Jonard, J. Zimmermann, Management science, 53, 1051-1067 (2007)
9. E.I. Galeeva, News of Orenburg State University, **8(114)**, 118-123 (2010)
10. E.V. Gusev, K.V. Kardapoltcev, Vector of science of Togliatti state University. Series: Economy and management, **2(5)**, 10-16 (2011)
11. G. Haken, *Synergetics* (Mir, Moscow, 1977)
12. M. Hobday, A. Davies, A. Prencipe, *Industrial and Corporate Change*, **14**, 1109-1143 (2005)
13. L. Musaev, Bulletin of SRSTU(NPI), **3**, 132-137 (2011)
14. A. Novikov, A. Babkin, Biosciences Biotechnology Research Asia, **11(30)**, 1563-1571 (2014)
15. M. Schilling, C. Phelps, Management science, **53**, 1113-1126 (2007)
16. L. Serkov, Bulletin of the Chelyabinsk State University. Economy, 9(147), 113-117 (2009)
17. E.A. Shemetov, Modern problems of science and education (2015). http://www.science-education.ru/101-5331
18. E.Yu. Tsyura, Mining information and analytical bulletin (scientific and technical journal), **8**, 175-179 (2006).
19. I. Wilkinson, L. Young, Journal of Business Research, **55**, 123-132 (2002)
20. S.V. Zdolnikova, A.V. Babkin, St. Petersburg State Polytechnical University Journal. Economics, **2(240)**, 78-86 (2016)

Priority directions of innovative development of Russian fuel and energy complex on global investment market

Natalia Sidorova[1], *Irina* Repina[1,*], *Victoria* Timchenko[1]

[1]Far Eastern Federal University (FEFU), Engineering school, 690018 Russky Island, Russia.

Abstract. Energy production is the most important system-forming infrastructure in the Russian Far East. The Asia-Pacific Region retains the position of the most attractive region in terms of investment into new projects, but the investment prospects of the region in terms of development of the fuel and energy sector on the global investment market consists of ten factors. 27 state programs specifically target the Far East, meanwhile small and medium businesses will also be given means for development.

Keywords: infrastructure, energy production, investment, targets, investment prospects, factors, rise for development, Ministry of the Russian Federation for the Development of the Far East priority development area.

The geopolitical risks of relations with Western countries are forcing Russia to seek new markets for its capital, goods and services, as well as new markets for selling its own products. Paying attention to Asian countries (China, India, Singapore, South Korea, Hong Kong, Japan, etc.) as sources of investment to the Russian economy the government focuses on the underdeveloped infrastructure of the country.

Energy production is the most important system-forming infrastructure in the Russian Far East. Attraction of private investments in the Russian economy is possible only on condition of solving problems of the infrastructure of the Far East, which then will attract investments in industrial production in the territorial points of growth.

The Asia-Pacific region (the APR) maintains the position of the most attractive region for investment in new projects: in 2014, the region is ahead of Europe and North America taken together concerning the share of announced foreign direct investment in new projects. When direct foreign investments in new projects were structured over a period of 2012-2014, the statistics were as follows: the APR – 33%, Europe – 37%, North America and other countries – 30%.

The investment prospects of the region for the development of the fuel and energy complex on the global investment market is made up of ten factors, their importance depends on the type of the investor:

1. Access to natural resources (the level of extraction of natural resources, reserves of natural resources);
2. Stability (macroeconomic certainty, stability in a country's politics, legal stability)
3. Corruption
4. Tax advantage
5. Access to home and foreign markets
6. Infrastructure (automobile transport, air transportation, railway transportation)
7. Labour costs
8. Quality of human resourses
9. Suppliers
10. Innovations

In the near future, the APR can become the world's largest investor in new projects: over the past five years, the average dynamics of the region is more favorable than in other regions of the world.

Comparison of the parameters of the investment prospects of the Far Eastern Federal District and the countries of the Asia-Pacific Region.

Economic conditions in the countries of the Asia-Pacific region:

1. Tax on profits (percent of profits / 2016): South Korea – 18.2; China – 11.1; Singapore - 2; USA - 27.4; Canada – 3.9; Japan – 28.9; Russia – 0.
2. Energy prices (US cents per 1 kilowatt of energy / 2016): South Korea – 9.2; China – 11; Singapore – no data available; USA – 9.4; Canada – 7.2; Japan – 28; Russia – 6.3.
3. Connecting infrastructure to power supply source (days / 2016): South Korea – 18; China – 143; Singapore – 31; USA – 90; Canada – 137; Japan – 98; Russia – 160.

At present, the government views the Far East as a Russian super-region. 27 state programs are specifically targeted at the Far East, which is to become a strong outpost of Russia in the Asia-Pacific region. Such a prioritization decision has been made for the first time.

* Corresponding author: repina_ira@mail.ru

So, one of the priorities of making the region a strong outpost is to attract additional investments to it. This became possible due to adopting a legislative proposal aimed at equalizing energy tariffs in the Far Eastern Federal District with the all-Russian level. Energy tariffs for industrial consumers will be reduced since July 1, 2017. The basic level, to which the tariffs will be reduced, is 3.19 rubles per 1 kWh. In this case in Primorsky Krai, the reduction is 21%.

In accordance with the law, tariffs in the Far East will be reduced to the average Russian level due to a premium to the price of capacity in the I and II price zones. The manager of the raised funds will be PJSC RusHydro, which can receive RUB 29.5 billion for these purposes in 2017-2019.

The cheapening of electricity became possible after the Russian President Vladimir Putin approved the amendments to the federal law "On Energy". The corresponding law of December 28, 2016 No. 508-FZ entered into force on January 1, 2017.

According to the data of the Ministry of Regional Development, on average for the Far East tariffs will decrease by 30%. The biggest drop will occur in Chukotka - 65%, in Yakutia - 50%, on Sakhalin - 46%, in the Magadan region - 34%. In Primorsky Krai, the tariff will decrease by 21%, in Kamchatka - by 20%, in the Jewish Autonomous Region by 17%, in the Amur Region and Khabarovsk Territory by 5%.

PJSC RusHydro is supposed to transfer funds intended to reduce tariffs in the Far East to the regions' budgets, where they will go to energy sales companies. The drafts of normative and legal acts prepared by the RF Ministry of Energy say: "The funds received from the introduction of the bonus for the territories of the Far Eastern Federal District are used to provide special-purpose non-repayable contributions to the budgets of the constituent entities of the Russian Federation that represent a part of the Far Eastern Federal District".

RusHydro will act as a premiums gatherer: the money will first be credited to the company's special account, and then within five days it will be transferred to the budgets of the subjects of the Far Eastern Federal District. This financial scheme will allow to reduce tariffs to the all-Russian level.

The development of fuel-energy complex is influenced by many factors. First of all, the scale and pace of productive forces development, especially energy production. At the same time, fuel and energy complex significantly affects the formation and development of all sectors of the national economy and its main components - industry, agriculture, transport and communications, construction and public utilities [14].

The factors shaping it and determining rational structure of fuel and energy complex, affecting its efficiency is divided into classes:

1. Economic - inflation, subsidies, prices of means of production;
2. Organizational - the size of a company, its legal form, system management organization (attraction of innovative solutions), organization of accounting, development of export and import substitution industries in the areas with the most favorable conditions, the formation of special economic zones and technoparks as regional centers for domestic and world science achievements introduction, accelerating economic and social progress.
3. Natural - climatic conditions, territorial disposition of land and water areas for mining and production of mineral resources for the energy sectorReduction in electricity price became possible after amending Federal law №508 "About energetics", which came into effect January 1, 2017.
4. Social – the living conditions of the population consuming the products of fuel and energy complex, working conditions of workers employed in the fuel and energy complex, etc.

Profit is the difference between revenues and production and sales costs [21], therefore, it is important to affect primarily the cost for the manufacturer. He can reduce costs while constant volume of production and sales, increase production at the same cost or increase the volume of production and sales at a faster pace in comparison with rising costs [1].

Factors influencing profit:

1. The cost of production. It depends on the standard costs and consumption rates of labor, material and technical resources, funds, compliance with the technological requirements for the production of fuel resources and the provision with heat and light. The cost per unit of production impacts on unit costs in the unit cost of production [22].
2. The cost of transportation and direct sales of commercial products which is an essential element in the full commercial cost. Their share depends on the type of products, distribution channels, intermediaries and other averaging from 5 to 15 %. This kind of cost is affected by the development of market infrastructure, road network, tariffs for services (transport , information, insurance, communications and so on);

Selling price is a variable opportunistic category, which is influenced by supply and demand, distribution channels, government policy on prices, quality characteristics of product. Prices of fuel and energy complex products are formed under the influence of many factors. Inflation expectations and the interests of producers influence strongly

The degree of company production resources utilization depends on all these factors. Economic efficiency of rationality is measured by profit and its relation to the full cost of products sold and aggregate funds

According to the Ministry of Economic Development of the Russian Federation, the average tariff reduction throughout Far East will reach 30%. The greatest reduction will be on Chukchi Peninsula – 65%, in Yakutia – 50%, on Sakhalin – 46%, in Magadan Region – 34%. In Primorsky Kray it will be 21%, on Kamchatka – 20%, in Jewish Autonomous Region – 17%, in Amur Region and Khabarovsk Territory – 5%.

RusHydro is suspected to transfer money for tariff reduction in Far East to regional budgets, then the money will proceed to energy sales companies.

"The funds raised by the application of premium in the Far Eastern Federal District, are using for non-repayable goal-oriented contributions to the budgets of constituent entities of the Russian Federation, which are members of the Far Eastern Federal District", - stated in the draft legal acts prepared by the Ministry of Energy.

Energy supply companies that guarantee regional supplies will receive money in the form of subsidies directly from the budgets.

"Far Eastern Energy Company" (DEK) - a subsidiary of PJSC RusHydro - now acts as the main guaranteed supply company in the II non-price zone that includes the Far East. DEK also has the status of a single buyer of the II non-price zone. The main shareholders of DEK are PJSC RAO ES Vostok (100% subsidiary RusHydro) with a 51.03% stake and the Donalink Ltd in Cyprus (controlled by Joint-Stock Company SUEK) with a 34% stake.

The Resolution of the Government of the Russian Federation defines the features of the concept "the advanced development of the Far East". This is to show core indicators in all areas above the national average level. Today, the basis for the development of the region is the law on priority development areas (PDA). The Government of the Russian Federation has decided to create PDAs for seventy years. The information on the current benefits and preferences for residents of PDA is presented in Table 1.

All benefits and preferences are based on the proposal of the authorized federal body (Ministry of the Russian Federation for the Development of the Far East). This proposal stipulates:

- a list of types of economic activities performed under a special legal regime for economic activities according to Federal Law;
- the minimum volume of capital investments of residents in the relevant types of economic activity in the priority development areas;
- regulations on the use or non-use of the customs procedure of the free customs zone established by the customs legislation of the Customs Union;
- defining the borders of the priority development areas;
- minimum requirements for the level of technologies, production methods, and equipment used by residents of the PDA realizing various types of economic activities.

Ministry of the Russian Federation for the Development of the Far East executes:

- issuance of construction permits, permits to place objects of construction into use, reconstruction of infrastructure facilities permits, with the exception of the objects and facilities specified in paragraph 5.1 of Part 1 of Article 6 of the Town Planning Code of the Russian Federation, and with the exception of highways of federal importance;

Table 1. Current benefits and preferences for residents of PDA.

Type of tax	Tax rate without benefits and preferences	Tax rate with regard to benefits and preferences	Duration of benefits and preferences	The basis of the benefits and preferences
Income tax federal regional	20%	0% over first 5 years since the date of making profit, then 12% (2% - federal, 10% - regional)	10 years	The second part of the Tax Code of the RF
Corporate property tax	2,2%	0,5% over first 5 years, 1,1% - over second 5 years since the date of resident's activity	10 years	The second part of the Tax Code of the RF
Deductions for social needs	30%	7,6% - Pension Fund of the RF 6%; - Social Insurance Fund of the RF 1,5%; - The RF Compulsory Medical Insurance Fund 0,1%	10 years	Federal Law No. 212-FZ of July 24, 2009 "On Insurance Contributions to the Pension Fund of the Russian Federation, the Social Insurance Fund of the Russian Federation, the Federal Compulsory Medical Insurance Fund"

- coordination of the territorial planning scheme of the subject of the Russian Federation in which the PDA is created;
- harmonization of documents on planning the PDA for the location of capital construction projects of regional importance within the boundaries of the municipality in which the PDA is located, the implementation of state construction supervision in cases provided by the Town Planning Code of the Russian Federation;
- approval of the draft plan of PDA for the purpose of its integrated development;

- approval of the procedure for maintaining a register of residents;
- control over the compliance of the agreement by the resident of the PDA;
- control over the activities of the management company and its subsidiaries;
- harmonization of documents on territorial planning of municipal objects within the boundaries of the PDA and documents on use and development of land;
- provision of land under federal ownership located in the PDA;
- taking decisions on reservation of land and compulsory alienation of land plots (withdrawal of land plots) for state needs for locating infrastructure facilities in the PDA; also creation of a state servitude to land plots for the purpose of locating infrastructure facilities of the PDA;
- consideration of complaints against refusals to make an Agreement on the Implementation of Activities. Minimum growth of producer prices for heat in the pair was awarded in 2011 -by 3.8%. In 2013 compared with the previous year, prices for heat in the pair increased by 12%, heat in hot water by 11.2%, which is close to the price level of 2009 [24, 25].
- In Russian Federation tariff increases in production, transmission and distribution of electricity amounted to 6.2%, heat - 11.5% in 2013. In the production of fuel and energy minerals prices decreased by 5

It should be specially noted that within the frames of this process small and medium businesses would also be given rise to development, thus providing employment for local residents.

References

1. *FEC and the economy of Russian regions. Far Eastern Federal District* (Publishing house ENERGY, Moscow, 2007)
2. A. Bolgov, S. Markov, Energy Market: Portrait of the region, **06(67)** (2009). http://www.e-m.ru/
3. UNCTAD
4. Innovation and entrepreneurship. http://www.innovbusiness.ru/content/document_r_7C5F2FBB-3749-41CF-9706-ED2784B8E677.html
5. I.B. Svetlova, *Fuel and energy complex of the Far East of Russia* (Development and prospects: Report of the director of the ANO "CSI TEK DV", Vladivostok, 2013)
6. Information about the special economic zones of the PRC. Portal Foreign economic information of the Ministry of Economic Development of the Russian Federation. http://www.ved.39gov.ru/export countries/cn/about_cn/laws_ved_cn/special_area_cn/
7. Energy strategy of Russia for the period until 2030. http://minenergo.gov.ru/aboutminen/energostrategy/
8. Ministry of Energy of the Russian Federation. http://minenergo.gov.ru/
9. Far Eastern Federal District. https://en.wikipedia.org/wiki/Дальневосточный_федеральный_округ
10. Strategy of social and economic development of the Far East and the Baikal region for the period until 2025. http: //www.dfo.gov.ru/
11. Federal Service of State Statistics. www.gks.ru
12. zrpress.ru
13. The Law of the Russian Federation of December 28, 2016 No. 508-FZ entered into force on 2017. http://minenergo.gov.ru
14. On the Priority Development Areas in the Russian Federation: Federal Law No. 473-FZ of 2014.
15. Center for Energy Efficiency. http://cenef.ru
16. I.A. Bashmakov, *Development of integrated long-term energy saving and energy efficiency programs: methodology and practice* (Moscow, 2013)
17. World energy model – methodology and assumptions. World Energy Outlook 2007, International Energy Agency, Paris. www.worldenergyoutlook.org/docs/weo2007/WEM_Methodology_07.pdf
18. Innovation strategy. Types of innovation strategies. http: // www.rae.ru/monographs/112-3771
19. M. Filipini, L.K. Hunt, *Measurement of energy efficiency based on economic fundamentals* (Energy Economy, 2015)
20. dv.kapital.ru
21. N.G. Sidorova, Vestnik of the Plekhanov Russian University of Economics, **10(64)**, 110-113 (2013)
22. The state program of the Russian Federation "Socio-economic development of the Far East and the Baikal Region" approved by the decree of the government of Russian Federation of 2013.
23. World Energy Perspective. Cost of Energy Technologies. www.worldenergy.org/wp-content/uploads/2013/09/WEC_J1143_ CostofTECHNOLOGIES_021013_ WEB_Final.pdf;
24. Excerpts from Dmitry Medvedev are opening remarks. government.ru/en/news/15480;
25. To divide the world economy into five. www.gazeta.ru/business/2013/12/25/5819085.shtml

Organizational infrastructure of innovation processes in Russian Russia

Elena Kuznetsova[1,*], *Elena* Cherepanova[1], and *Alexey* Gamberg[1]

[1]Mechanics and Machine Building Institute, Ural Federal University, Russia

Abstract. The study presents the regional and national statistics, allowing drawing the conclusions about the state of the innovation process, its elements, and their relationships. The main result of the study is the identified "gap" between basic science as the object of creating innovations and manufacturing enterprises as the object of the implementation of innovations. According to the authors, there is a necessity for the process control for not only the creation, but the implementation of innovations in production; the formation of an infrastructure that can provide the interaction of science and industry, the transformation of scientific developments into the manufacturing process now. This structure should be a buffer zone between the elements of the innovation process, create an area of mutually beneficial cooperation. The article considers the evolution of scientific development transformation in production. The organizational model of "innovation-to-production" transfer established on the small innovative enterprises affiliated with the leading educational institutions is suggested. The merits and drawbacks of the organizational interaction models in the "higher education institution-the small innovative enterprise-the enterprise" chain are shown.

1 Introduction

The mutual integration of science and business is one of the most pressing problems of the Russian economy associated with the accumulation of the innovative capacity of the country. The vector of the government intervention in this sector was visible most distinctly in the last 3 years. In particular, when the Federal law № 217-FZ "On Amendments to Certain Legislative Acts of the Russian Federation on the Establishment of Budget Scientific and Educational Institutions' Business Entities for the Purposes of Practical Application (Implementation) of Intellectual Activities" was enacted, it directly united the scientific capacity of higher education institutions and business by the allocation of the particular business type – the small innovative enterprises with the special conditions of governmental support, focused on a specific activity kind, which functioning mechanism is described in the article by Lugovtsov, Gamberg [1]. Lately a large number of commercial structures of this type were opened to solve the problem of the diffusion of innovations described by Rogers [2]. Nevertheless, the increase in number of such companies makes it more and more obvious that the existing infrastructure, created to simplify the work process of the subjects in this sphere, does not meet the requirements. The small innovative enterprises, created at higher education institutions, should be granted the commercialization functions. Now various funds support development, but, obviously, the time has come to pay special attention to the implementation financing, which is defined in the Federal law and described in the research by Yershova and.Kopytov [3].

The objective of the research, which results are given in the article, is the identification of the problem elements in the industrial enterprises innovative development process, discussion of innovation degree assessment indicators and the proposition of the ways to increase the innovations introduction process efficiency [4].

2 Assessment of innovate development level

In modern scientific practice a number of indicators are applied to the assessment of the enterprise innovative activity: the "integrated indicator of innovative activity" introduced by Opekun. [5], "Russian index of the innovative level of the enterprises", "functional dependence of the competitiveness level and innovations", etc. used by Mutanov and Esengaliyeva [6]. The international ratings, for example, the Global Innovation Index (in 2014 rating of the Russian Federation was 39,1; 49 place) [7,8] are applied to estimate the innovative activity of the country. These ratings are used for the comparative analysis between the countries based on the level of competitiveness. This method was proposed by Atkinson and. Ezell [9]. The options of estimation based on the level of competitiveness are "competitiveness through cooperation" indicators, suggested by . Teixeira [10]

* Corresponding author: eltcha@mail.ru

DOI: 10.1051/shsconf/20173501067

and "sustainable development" used by Hay and others, Vollebergh, Kemfert, and White in their researches [11-13]. Some researchers, for example Cooke or Buyanova consider the region as a significant unit of consideration in the question of innovation ensuring because the innovative clusters are mostly created at the regional level, the region creates local legislative and infrastructure conditions to provide innovation for the enterprises [14-16], but there are approaches, for example by Hamel . or Kuznetsova which suggest that the development of innovation comes "from the bottom up: the enterprises form competitiveness of the region" [17,18]. As Trifonov and Veretennikova note, that there are no unified index, criteria, or unified approach to the estimation of complex indicators for the innovative activity of regions evaluation. [19]. Just like in case with the country, the assessment of the regional innovative activity is suitable for the comparative assessment. The comparative assessment can be used for the same object in dynamics (for identification of the innovative development trends), and for comparison of the different objects. It is expedient for investors to use the results of such comparative assessment to choose the object of investment.

The known indicators of the innovation level (for the enterprise, region and country) are integrated, and it is difficult to isolate the influencing factors. The integrated indicators do not give the chance to define ways of further development and innovative development efficiency of the research object - the enterprise.

3 Russian practice of innovative development

The Russian practice of the economy innovation assessment level is the following. According to the State Committee of Statistics, expenses on researches and development within the country were about 1% of GDP in 2012. This level is preserved for the last several years (Figure 1). There is a need for the estimation of the innovative activity financing level and science as the source of innovations, from the point of view of its efficiency and sufficiency. Figure 1 shows the expenses on science funding from the GDP in dynamics since 2000.

For example, Aganbegyan suggests that investments in production innovations become innovative result in 3-5 years [21]. Fig. 1 shows dynamics of a number of the used advanced production technologies across the Russian Federation on some mechanical engineering subindustries. In the view of the any economic system development efficiency, the fundamental industries, implementing innovative products and technologies should be manufacturing industries, including mechanical engineering as one of the progress engines because of its position at the beginning of the technological chain of the technological chain of any production.

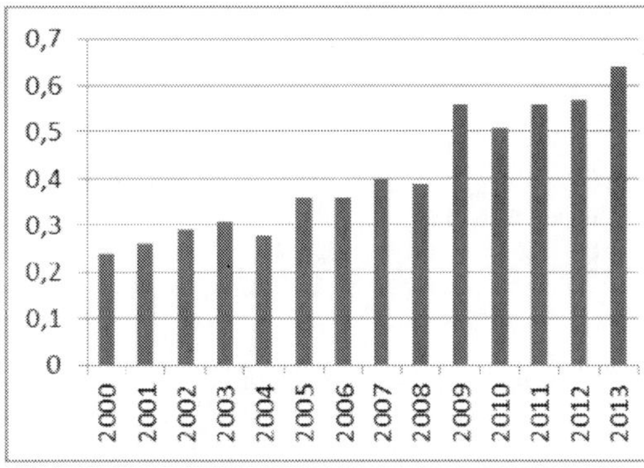

Fig. 1. Financing of the scientific development dynamics (% of GDP) – the State Committee of Statistics [20].

The chart shows that the level of innovative introduction decreases with the growth of the scientific developments financing (Fig. 2) and the need for them increase. This situation can only be estimated as a negative tendency because only the introduced innovative technologies can be considered as successful and as the instrument of the innovative development of the industry in general.

4 Problem of innovative development

Blinkov and Mogilenskikh describe the practical problem of the innovative development in the Russian industry as the "gap" between the efforts made by the government to intensify the innovative development and the reached results. [22] One of the episodes proving the mentioned problem existence is the following. Five years ago the Ural Federal University created innovative structure (a belt of innovative enterprises) to provide the Ural industrial enterprises with innovative products and to stimulate the update of the product line. The analysis of the statistical data showed that in 2012 more than 60 small innovative enterprises were created within the innovative structure of UrFU [23].

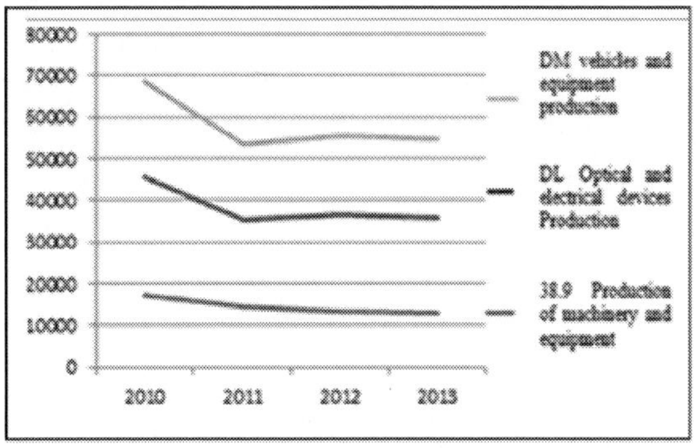

Fig 2. Number of the used advanced production technologies in Russian Federation on some subindustries (units) [20].

At this moment there are more than 80 enterprises, but there is no intensive growth of the innovative industry in Yekaterinburg, or in the Ural region. The created innovative enterprises did not give the multiplicative effect of the innovative production volume growth.

Considering the diverse structure of economy of the Russian Federation, it is possible to assume, that not all regions are equally interested in the introduction of the innovations, for example, some agrarian regions can distort the general statistics by its indicators. Let us consider the main results of Ural Federal District enterprises' activity in 2012 as one of the industrial regions of the country (Table 1) [4,20].

Table 1. The main results of the Ural Federal District enterprises' activity in 2012 [4, 20].

Index	Data on Ural Federal District, rating among sub-federal units of the Russian Federation, value
By the total index of the local production goods shipped volume, by the types of economic activity, "manufacturing activities", place, million rubles.	4th 3211212
By the index of industrial production, place, %	8th 101.3
By the index of industrial production by the types of economic activity "manufacturing activities", place, %	2nd 107.9
The number of the organizations, engaged in the scientific development, place, units	5th 236 (3566-countrywide)
The number of the personnel engaged in the scientific researches, place, number of people	5th 43879 (726318 countrywide)
Self-financing of the basic researches, place, million rubles.	6th 3832.6
Self-financing of the applied researches, place, million rubles.	7th 3084.5
Patent applications, place, units	7th 2326 (42180 countrywide)
Given out patents, place, units	5th 1994 (33633 countrywide)
Level of inventive activity, place	8th
Volume of the innovative goods, works, services, place, million rubles.	5th 148696.2 (2872905.1 countrywide)
Proportion of the innovative goods, works, services in total production, place, %.	8th 2.1 (8 countrywide)

It should be noted that URFO is the smallest industrial region, as on the population (6th place), and on the area. Despite the modest size, rather small amount of the population, the region possesses enormous experience of industrial development and scientific support of production, caused by historical, natural and resource aspects. All this allows the region to reach good results: for example, now, the 2nd place on the level of industrial production. In spite of the rather high level of industrial production output and industrial production index, the level of innovative activity in the region is insufficient, and on the innovative production, works and services implementation volume, URFO is on the last positions in comparison with other regions. Other researchers of the problem also came to the similar conclusions, e.g. Ruban or Baev and Drozin [24,25]. At the same time, URFO industrial complex and areas within, for example, the Sverdlovsk region is one of the largest in the country. But this complex was created a long time ago and initially served for the military purposes. In the modern conditions the survival of such industrial region as URFO demands increase of enterprises and productions competitiveness, which only depends on the success of the existing branches and productions innovative updates. It should be noted that URFO has the smallest difference between submitted patent applications and the given patents that proves the quality of scientists and schools of sciences in the region. At the available industrial development requirements, including the level of the industrial production, share of the industrial output, at the regional level we see lack of created, introduced and expertly confirmed innovations implementation.

Thus, it is necessary to look for the reasons which led to the low level of implementation of innovative development, technologies, and products. Over the past few years the financing level did not change essentially, and the number of enterprises which are engaged in the scientific development did not change either. According to the Federal State Statistics Service, the level of inventive activity (coefficient of inventive activity and the number of the domestic patent applications for inventions submitted in Russia per 10 thousand people of the population) essentially did not change for the last 3 years and makes 2 units. Thus, we cannot speak about decrease in inventive level or number of inventions, but the lack of the worked out system of the scientific researches implementation in production is obvious.

So, the lack of the innovations adaptation mechanism in the existing productions is proved. Obviously, there is a gap between the efforts made by the state government and the regional authorities, and the results in the industry which should be expressed in the realized innovative production volumes.

5 Organizational infrastructure

The innovative activity of the region and state should be shown in the conditions creation for the innovations implementation by the enterprises, i.e. in the creation of policy in the field of innovative development, support, and priorities which, as a result, develop into the regulation of the innovative production market, ensure the stability of demand, and the tax regime allowing the enterprise to allocate the earned funds for the development, in other words the creation of innovative processes organizational infrastructure [4].

Figure 3 shows the logic of the innovation assessment in the enterprise (the left column) and innovation of the region (the right column), and the necessary system of their interrelation (infrastructure for innovations).

We analysed the schematic diagram of interaction between the scientific and innovative enterprises and real production, existing in the Soviet period (Figure 4).

The manufacturing enterprise will not be, and should not be engaged in any form of the basic scientific researches. Industry-specific scientific research institutes in the USSR took on both applied and exploratory research that allowed receiving a finished innovation for the industrial enterprise. After the total closing of the scientific research institutes at the beginning of the 1990s, many enterprises had to stop the innovations implementation completely or independently conduct both applied and exploratory researches. The budget for the research had to be found by the enterprise. Without innovations the product quickly loses the client (decrease in quality, consumer properties, etc.), and researches demand essential financial investments from the enterprise.

Now allocation in the budget of the enterprise for the scientific research and experimental-design development (the Soviet term), or on R&D - researches and development (modern term) is expedient. With these funds the enterprise can buy the innovations in the market and implement them in production; or the enterprise can "order" development for its needs.

In world practice science and technology parks carry out the function of the buffer between the fundamental science and manufacturing enterprises.

The small enterprises working in the science and technology parks are engaged, generally in commercialization of the scientific developments, making a commercial product, and then realizing this product.

There are cases when the major companies buy the innovation together with the company-developer. But the Russian experience of the science and technology parks successful functioning of is low, the majority of the created technology parks remain the formal organization which does not carry out its main function - commercialization, introduction, and development.

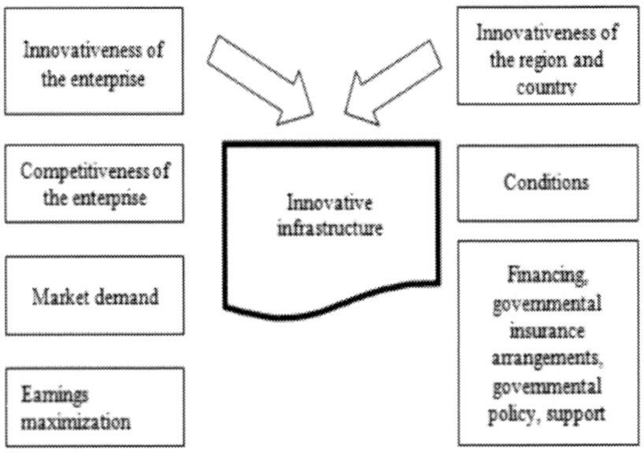

Fig. 3. Infrastructure of the innovative process.

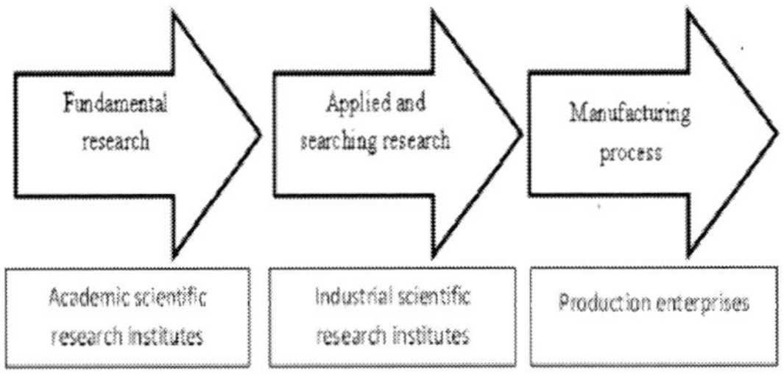

Fig. 4. Research and production realization system of the research results.

The reduced taxation for the science parks land is the reason for this. The results of the science park activity were not controlled, often in the "science and technology park" banal and even handicraft production took place, and the innovative developments were not carried out, as described in the researches by Krivorotov,. Kuznetsova [3,26,27]. In the foreign practice companies have to pass a competition to get into the science park, and time of their presence in the park is limited to the time of the project implementation. At the moment it is necessary to speak about the creation of organizational infrastructure for the transfer of the basic research results to production. Federal law № 217 defines the small innovative enterprise (SIE) activity - practical application (implementation) of the intellectual activity results. Thus, SIEs should replace the disappeared scientific research institutes and ineffective science and technology parks.

We conducted a study of existence and structure of the economic, organizational and scientific and technical barriers in the domestic market of innovations. During the research the heads of the small innovative enterprises created at the leading higher education institutions of Yekaterinburg, Chelyabinsk and Novosibirsk were interviewed. The enterprises founded with the assistance of the Ural Federal University, the Southern Ural State University and Novosibirsk State University were used in the selection [1].

The directors were proposed to list barriers that interfere with the development of their enterprises on the current stage of activity and, which, in their opinion, they may face further. During the interview the respondent was asked to rank the offered types of barriers on extent of their influence on the company activities on a scale from 0 to 3 points, where 0 is no influence, 3 is a very high influence.

The received results were processed by the method of averaging observations, mode and medians among the indicators given by the respondents to each of the barriers. On the basis of these calculations, several diagrams were constructed (Figures 5, 6, 7). The charts show the weight of each estimated barrier during all stages of the small innovative enterprise life cycle [1], according to the director (Figure 6).

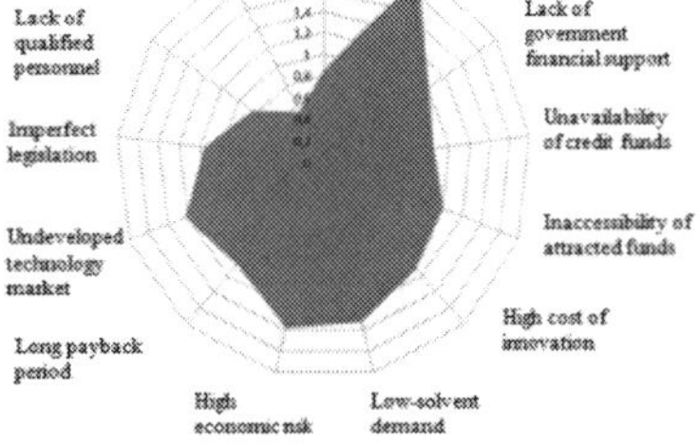

Fig. 5. Average index according to barriers evaluation by the businessmen.

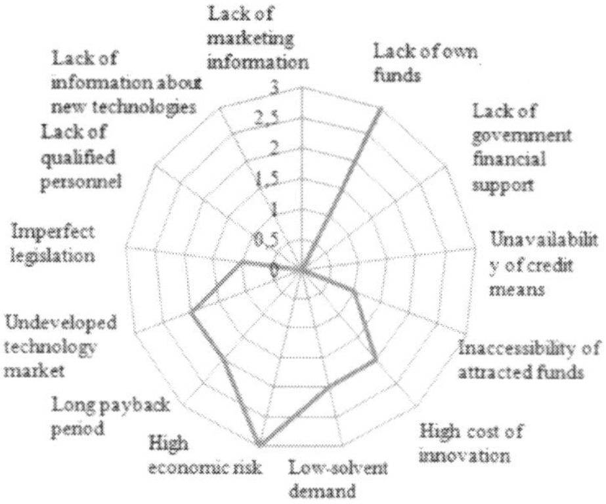

Fig. 6. Size of the mode barriers assessment by the businessmen.

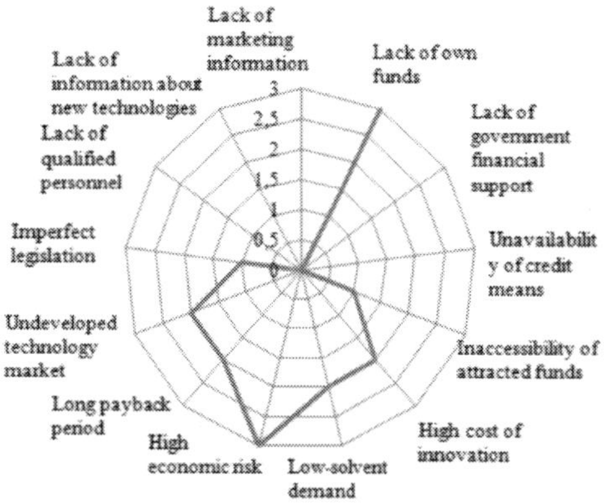

Fig. 7. Size of the median barriers assessment by the businessmen.

At the moment there are several players in the market promoting breaking of the barriers on the way of innovations implementation (Table 2).

Table 2. Classification of the innovative infrastructure elements which promote overcoming of the barriers.

Characteristic	Classification	Example
Size of a property complex	Single organizations	Industrial enterprises, funds, investment companies, business angels, banks, etc.
	Associations	Science and technology parks, scientific clusters, venture funds, etc.
	Special technology development zones	Innovative and implementation centers (Skolkovo)
Purposes and nature of activity	Commercial	Banks, production platforms, etc.
	Non commercial	Support funds, state programs
	Mixed	Data-processing centers, science and technology parks, clusters, etc.
Technology of support	Indirect	Through privileges, subsidies, subintermediary services
	Direct	Through purchase of a share, direct investments, services of production
Activity	Financial	Venture funds; banks, business angels; private investment companies;
	Research and production	Production platforms, research and production clusters, higher education institutions, exchanges of innovations
	Legal	Consulting companies
	Complex	Science and technology parks, data-processing center, business incubators
Used resource	Private	Private investor, venture, commercial enterprises
	State	Funds, unitary enterprises
	Mixed	Government corporations
Capital accessory	National	Foreign venture funds, business incubators and production platforms
	Foreign	
	Mixed	

Table 3. Elements of the innovative infrastructure usage for the barriers orercome.

Barrier	Infrastructure element	Overcoming mechanism
Lack of own funds	Banks, funds of support	Attraction and loan of funds from the state and commercial sources
High economic risk	Business angels, venture funds	Division of implementation risks with strategic and portfolio investors
High cost of innovations	Science and technology park, business incubator	Reduction of the product costs by cooperation with other companies in the science and technology parks, decrease in constant expenses due to the usage of the business incubators resources

Generalizing the above-mentioned and relying on the foreign and domestic experience, the usage of the small innovative enterprises created at higher education institutions and scientific institutions as the engine for innovative process in Russia is possible only on the condition of the necessary infrastructure creation meaning availability of the accurately fulfilled mechanism of overcoming scientific and technical, organizational and economic barriers.

5 Conclusions

We suggest putting into practice such mechanism by forming the organizational infrastructure elements, which are available in the market in a continuous overcoming barriers chain where the innovation moves from its appearance in the form of small innovative enterprise to the issue on a big market.

References

1. A.V. Lugovtsov, A.E. Gamberg, *Proc. of the IVth int. conf.*, 217-225 (2014)
2. E.M. Rogers, *Diffusion of Innovations* (Free Press, New York, 1983)
3. I.V. Yershova, I.V. Kopytov, Questions of management, **11**, 48-53 (1985)

4. E.Y. Kuznetsova, O.O. Podolyak, P.V. Mezentsev, URFU Messenger. Series economy and management, **1**, 96-113 (2015)
5. E.V. Opekun, G.A. Hatskevich, Theoretical and applied questions of economy, **23**, 96-105 (2011)
6. G.M. Mutanov, Zh.S. Esengaliyeva, Basic researches, **3(3)**, 712-717 (2012)
7. The Global Innovation Index — global research and the rating of the countries of the world accompanying it on an indicator of development of innovations according to the international business school (INSEAD). http://gtmarket.ru/ratings/global-innovation-index/info
8. The Global Innovation Index 2014 The Human Factor in Innovation. http://www.globalinnovationindex.org/content.aspx?page=gii-full-report-2014
9. R.D. Atkinson, S.J. Ezell, *Innovation Economics.The Race for Global Advantage* (Yale University Press, New Haven, 2012)
10. A.A. Teixeira, Cambridge Journal of Economics, **38**, 181-214 (2014)
11. L. Hay, A. Duffy, R.I. Whitefield, Journal of Environmental Management, **13(3)**, 232-257 (2014)
12. H.R.J. Vollebergh, C. Kemfert, Ecological Economics, **54**, 133-147 (2005)
13. M.A. White, Ecological Economics, **86**, 213-217 (2013)
14. Ph. Cooke, Geoforum, **23**, 365-382 (1992)
15. Ph. Cooke, Industrial and Corporate Change, **10**, 945-974 (2001)
16. M.E. Buyanova, L.V. Dmitriyev, Bulletin of the Volgograd state university, **2**, 54-62 (2012)
17. G. Hamel, Harvard Business Review, **87**, 91-98 (2009)
18. E.Y. Kuznetsova, E.A. Chopovda, URFU Messenger, Economy and management Series, **2**, 27-35 (2013)
19. Yu.V. Trifonov, A.A. Veretennikova, The Bulletin of the Nizhny Novgorod university of N.I. Lobachevsky, **6(1)**, 244-284 (2013)
20. Federal State Statistics Service. http://www.gks.ru/
21. A.G. Aganbegyan, Forecasting problems, **4**, 3-16 (2014)
22. M.O. Blinkov, O.P. Mogilensky, Problem of economy and management, **3**, 32-34 (2010)
23. Innovative infrastructure of the Ural Federal University. http://www.inno.urfu.ru/
24. D.A. Ruban, Messenger URFU, 78-96 (2015).
25. I.A. Bayev, D.A. Drozin, Messenger URFU, Economy and management series, **2**, 97-110 (2015)
26. E.Y. Kuznetsova, The Organizer of production, **1**, 37-43 (2011)
27. V.V. Krivorotov, A.V. Kalina, E.A. Tikhanov, S.E. Erypalov, Messenger URFU, Economy and management series, **2**, 61-74 (2014)

Structural capacity assessment of machine-building enterprises and associations

Maria Prilutskay[1,*], *Petr* Krilatkov[1], and *Olga* Norkina[1]

[1]Mechanics and Machine Building Institute, Ural Federal University, Russia

Abstract. Multidirectional tendencies of machine-building enterprises integration and disintegration resulted in the emergence of the formal and informal associations. These associations consist of the obviously and/or implicitly affiliated legal entities. Thus, a new element appears in the direct enterprise environment, i.e a management company or a head enterprise. The management company influences the participants even in an informal association. New environment restrictions led to the changes in the management structure. The paper considers the enterprise structures interrelation: organizational, financial, production, resource, and others. The authors draw a conclusion that the structures are hierarchy, and there are the coherence structures assessment criteria. The coordinated structures form the structural capacity of the enterprise. The suggested assessment coherence criteria (for example resource and functional structures) allow estimating the structural potential and defining the directions of the enterprise efficiency increase.

1 Introduction

The theory and practice of machine-building enterprises management convincingly prove that the enterprises possess the integrity property as social and technical systems. The integrity as the immanent organization characteristic can be visualized through the set of the enterprise structures and, also, can be a basis for the assessment of the organization structural capacity.

Scientists have been concerned about the problems of the enterprises structuring since 1880s. During the last decades the conceptual framework of the structural analysis theory was considerably extended [1]. Traditionally the organizational structure of the enterprise was considered the administrative model, and many of the researchers, starting with F. Taylor, studied only the organizational structures, without other approaches to the structural business management.

The essential contribution to the theory and methodology of the organizational structures development was made by R. Ackoff, L. Bertalanffy, S. Beer, V. Gosling D. Mago, M. Mesarovic, H. Mintzberg, U. Parter, I. Takhara, A. Hall, etc. Among the Russian scientists works by L. Gitelman, V. Krasnova, N. Maslennikova, B. Milner, V. Mishin, V. Mukhin, A. Privalov, V. Sai, N. Trenev are known.

Works by V. Hrutsky, V. Gamayunov, T. Sizov, K. Shchiborshch contain approaches to the industrial enterprises financial structures development, but do not offer tools to evaluate the balance between the enterprise structures, i.e. the integrity. Taking into account the current trends in the organization forms of the machine-building business development the question of the balanced structures management formation is relevant.

2 Evolution of the machine-building enterprises organization forms

he main stages of the machine-building complex development in the Russian Federation

Development of the structure definitions is connected to the research object amplification. According to the civil code of the Russian Federation "the enterprise" is the independent property complex used for the business activity implementation [2]. The integration and disintegration processes in the enterprises which have begun in the 1990s led to the emergence of the various associations. The main stages of the modern Russian economic system formation and their influence on the machine-building enterprises development are described in Table 1

Models of the machine-building enterprises associations formation

As a result of the of machine-building enterprises integration and disintegration processes analysis, it is possible to create two models of the associations' formation:

Joining of the independent enterprises in the group controlled by the management company (Fig. 1). One of main goals is the achievement of synergetic effect: providing a full production cycle, formation of a single sales market, diversification for the decrease in risks. The management company is the center of the group, organizes the financial flows and economic relationships

* Corresponding author: mpril74@mail.ru

in the group. Its tasks also include distribution of the investment resources.

Table 1. The main stages of the modern Russian economic system formation and their influence on the machine-building enterprises development [3-5].

Stage	**Characteristic**	**Number of machine-building enterprises**	**Prevailing forms of the business organization**
Until 1985	Planned economy. Prevalence of the closed production structure - "subsistence economy".	Up to 6 thousand	Stateenterprises
1986–1992	Origin of a free market. Formation of the trade houses and marketing departments at the enterprises.	Up to 6 thousand	State enterprises. The first attempts of incorporation through the labor collectives shares.
1992-1994, the first property repartition.	Incorporation and privatization of the state enterprises. Emergence of private property in the industries, split of the assets.	Up to 60 thousand	Emergence of diversified regional associations, transfer of non-core productions to the independent enterprises
1994–1997, the second property repartition.	Monetary privatization, struggle for controlling interest.	Up to 60 thousand	Emergence of private Russian machine-building corporations, acquisition of the Russian enterprises by foreign owners
1998–2000, the first Russian financial crisis	Depreciation of the Russian currency.	Up to 30 thousand	Mass bankruptcy of the enterprises, transition to the improvement of the financial system.
2001–2007, the third property repartition.	Incorporating of the state enterprises.	Up to 30 thousand	Emergence of the Russian state corporations, generally in the military-industrial complex. Mass cutting down of non-core assets.
2008–2013, crisis and post-crisis development	Strengthening of the state support and regulation. Leasing schemes of the equipment updats.	Up to 50 thousand	Bankruptcy of the medium-sized private enterprises, integration and disintegration of the enterprises.
2013 – global crisis, political restrictions	Strengthening of the military enterprises state support.	Up to 50 thousand	Stabilization of the business structure

* The table is made by authors on the basis of data presented in the form of annual reports on the Federal State Statistics Service site

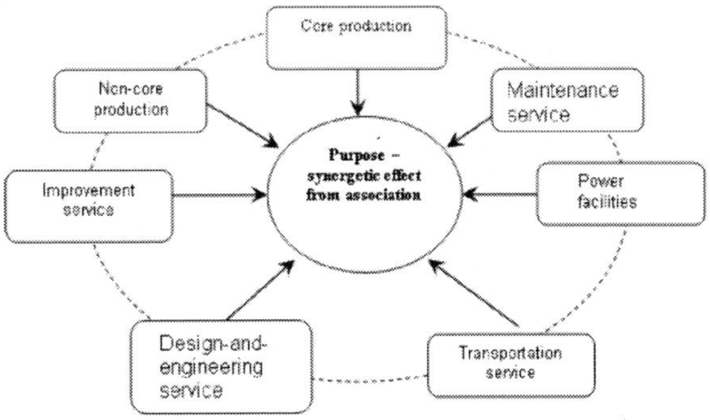

Fig. 1. Group formation on the basis of the independent enterprises association.

The enterprises which become the parts of the group, as a rule, do not lose their independence and work both on internal and external markets.

1. Restructuring of the large enterprise with a full cycle (Fig. 2). The result of this process is the creation of the legally independent companies, usually, these companies are technologically connected. One of the goals is to increase the efficiency by creation of more flexible, operating structure and transition to market relationship.

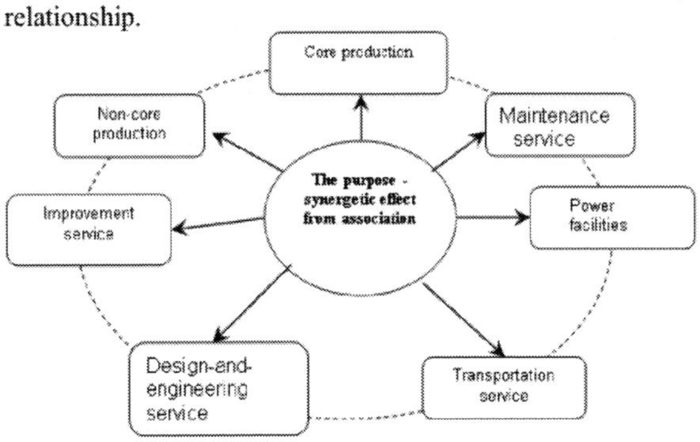

Fig. 2. Group formation on the basis of the large enterprise with a full cycle restructurating.

The initiator of such transformations is the owner. His main goal is the improvement of the finance indexes, which influence potential profitability of the core production and value of the business. The center of the group is the core production transformed into the independent legal entity. The rest of the enterprise in the association can be created as subsidiaries and can operate in the external and internal markets, or only in internal, fulfilling the requirements of the core production.

At the same time, production sites and idle capacities of the core productions are leased. It leads to the creation of a small specialized industrial enterprises buffer zone, which is necessary for the modern economy.

New, more complex organizational forms of business demanded the usage of the better management instruments based on the enterprises (associations) optimization structures and the integrity control. Obviously, modern machine-building enterprise is a polystructural object [6].

3 Polystructural approach to the business management

In addition to the organizational and financial structures mentioned above, the management system should include production, information and legal structures. The management structures correspond to the target and resource restrictions of the organization and should be optimized according to the established criteria. Table 2 contains the short characteristic of the machine-building enterprise key structures.

Table 2. Key types of the machine-building enterprise structures [7-10].

Type of structure	Design goal	Objects (elements)	Solved tasks
Organizational	Administration of the business processes	Divisions created according to the various criteria (functional, matrix, divisional structures)	Distribution of power and responsibility on the management levels
Financial	Control ensuring	Centers of the financial responsibility	Financial planning and management accounting
Production	Organization of the production processes	Core, subsidiary and services productions	Filling of the management accounting and the cost optimization
Informational	Organization of the information flows	Places of the information emergence and consumption, information flows	Definition of the access rights to the information, information flows optimization in the context of the chosen automation method for the company management
Legal	Tax optimization	Legally separated divisions (enterprises)	Assessment of taxes through the systems of transfer pricing and participation in the capital

The structural capacity of the enterprise (association) determines the structures coherence level. The complexity of the potential assessment is attributed to the structural contradiction existence, even if all conditions, the agreed targets, balanced resource and function abilities of the organization, are met. It should be noted that the capacity of the key structures can be initially higher than the goal potential due to SMART restrictions. In modern conditions competitiveness of machine-building enterprises and associations, characterized by the high level of the technological and organizational processes complexity, directly depends on the balance of the structures, i.e. structural potential.

4 Methodical approach to the assessment of the management structures integrity

At the first stage of the enterprise structural capacity diagnostics, the coherence assessment of the functional and resource structures on the basic level is proposed. These structures can be described by the means of the traditional quantitative figures that considerably simplify the assessment procedure [9,10]. Table 3 contains the comparison of three machine-building enterprises activity results. The criterion for comparison is the coherence of the functional and resource structures (integrity). As quantitative indicators of the assessment the following indicators were accepted:

ROA – return on assets shows the generalized index of working efficiency;

Kipm – productive capacity efficiency

VP - output per the worker

Krf - coefficient of claims on functions

Kpmr - coefficient of the material resources proportionality

Kвв – the relation of development one working at the enterprise to the industry average development

Kvio - coefficient of the available equipment involvement

The majority of the indexes given in the table are integrated. For their calculation and analysis the system of private indicators (factors) should be used. When for various reasons calculation of the exact integrated indicator was impossible, private indicators were considered in the final assessment. Influence factors for Kipm: specific weight of the installed equipment, coefficient of the available equipment involvement, coefficient of the equipment disposal, coefficient of the equipment renewal, equipment increase coefficient, specific weight of production made in cooperation. Influence factors for VP: production mechanization coefficient, production automation coefficient, average coefficient of norms performance, efficiency of working hours, coefficient of structural compliance of performers in the specialty, coefficient of structural compliance of performers on qualification.

5 Conclusion

Table 3. Comparison of the three machine-building enterprises activity results on criterion of functional and resource integrity.

.	Year	Indexes					Conclusion
		Generalised	Resource integrity			Functional integrity	
Indexes		*ROA6 %*	*Kipm6 %*	% deterioration	VP (thousand rubles)	*Krf*	Insufficient level of the resource and functional integrity (the enterprise is in a difficult situation)
Standard indicators:		*6 ÷ 7*	*85*	*10÷15*	*1000*	*0*	
Drilling and metallurgical equipment factory	2006	7.1	34.3	97	-	6/12	
	2007	17.55	42.9	85	703.2	6/12	
	2008	28.7	75.2	78	725.3	5/12	
	2009	-2.23	42	76.2	563.8	5/12	
	2010	0.5	42	80.25	434.1	5/12	
	2011	-1.75	42.6	83.8	533.7	5/12	
Indexes:		*ROA*	*Kipm*	deterioration	VP (thousand rubles)	*Krf*	General, structural and resource integrity is at the accepted level for the modern economic situation
Standard indexes:		6 ÷ 7	85	10÷15	0.1÷0.36	1.5 ÷ 2.0	
Telephone factory in Shadrinsk	2006	21.32	46.6	67.2	0.19	1.637	
	2007	30.58	64.3	62.4	0.18	1.610	
	2008	44.67	88.4	57.8	0.18	1.597	
	2009	8.17	46	54.9	0.21	1.763	
	2010	8.83	81.5	56.1	0.25	1.758	
	2011	13.6	80	50.9	0.21	1.813	
Indexes:		*ROA*	*Kipm*	deterioration	*VP* (thousand rubles)	*Krf*	Excess level of the resource integrity
Standard indexes:		6 ÷ 7	85	10÷15	0.8÷0.9	1.5 ÷2.0	
Electroengineering factory in Karpinsk	2006	6.3	55.2	55.4	0.43	1.67	
	2007	8.7	53.6	55.0	0.45	1.65	
	2008	11.1	51.7	56.9	0.456	1.65	
	2009	11.8	52.1	53.4	0.521	1.69	
	2010	5.28	55.3	53.9	0.553	1.61	
	2011	15.35	55.9	56.7	0.559	1.349	

The following conclusions can be made, based on the results of the research:

Modern economic conditions led to the management objects complication: as a result of the machine-building enterprises integration and disintegration processes the polystructural associations were created.

Quality of polystructure demands the formation of the approaches to design the business management (association) key structures: organizational, financial, production, informational, legal.

It is expedient to apply the management structures integrity indicator for the assessment of the difficult objects activity.

The current efficiency and development possibilities of the enterprises (associations) are defined by the structural potential size, characterizing the balance level of the management structures. For the structural potential assessment, indicators of the structures integrity can be used as elements.

By the example of the basic level structures, functional and resource, the application expediency of the integrity assessment approach on the basis of the indicators system of the enterprise overall performance is proved.

References

1. N.Y. Yuryk, Actual Problems of Economics, **117**, 179-183 (2011)
2. Civil code of the Russian Federation. http://base.garant.ru/10164072/
3. G.B. Kleyner, Bulletin of the Samara state economic university, **1(123)**, 26-35 (2015)
4. Russian Federal State Statistics Service. http://www.gks.ru.
5. Trading economics. http://www.tradingeconomics.com/
6. M.A. Prilutskaya, V.G. Ulitin, Scientific and economic magazine, **12**, 16-20 (2010)
7. E.Y. Kuznetsova, The Organizer of production, **1**, 37-43 (2011)
8. O.O. Podolyak, Messenger of the Ural State Technical University, **2(91)**, 58-64 (2008)
9. P.P. Krilatkov, Organizer of production, 30-38 (2012)
10. I.V. Ershova, P.P. Krilatkov, Organizer of production, **3**, 51-60 (2010)

Methods of assessing functioning of organizational and economic mechanism during innovation process implementation

Maksim Blinkov[1,*], and *Oleg* Blinkov[1]

[1]Mechanics and Machine Building Institute, Ural Federal University, Russia

Abstract. This paper proposes methods of assessing the efficiency of organizational and economic mechanism of an industrial enterprise when implementing innovation processes. These methods allow continuous monitoring at all stages of innovation process implementation, lead to reduction in costs of innovation activity and enable maximum use of the creative potential of enterprise personnel. The significance and attractiveness of this method is ensured by the fact that it can be applied by industrial companies in any market fields regardless of the lifecycle stage applicable to the studied goods, company and/or innovative process because the composition and the number of specific indicators can be adjusted by the work group both before the study and in the course of the company's innovative activities (at any stage of their implementation). The multi-sided approach proposed for assessing the efficiency of organizational and economic mechanism of the industrial enterprise when implementing innovation processes ensures full and accurate assessment of the impact of certain factors on the final result.

1 Introduction

In the literature on economics, assessment of functioning of any mechanisms, as a rule, is performed by using the indicators which compare the results of certain activities with input required for their performance, i.e. efficiency [1]. Thus, increased efficiency shall mean better results at reduced or equal input [2]. However, a number of scientific publications express the opinion that when giving assessment, both efficiency and effectiveness should be considered. Thus, for example, in the commonly known work of Mescon, Albert and Khedouri, effectiveness means the capability to "do the right thing", whereas efficiency means "doing the things right" [3]. In terms of assessment of efficiency in implementation of innovation processes, this approach is of special interest, in our opinion. Indeed, when producing a certain innovation (in the form of a new product, technology, management or organization method, etc.), which is a result of innovative process, it is extremely important not only to obtain the innovation with minimum costs but to make the innovation in demand, i.e., it must correspond to certain needs both of the company that initiated its implementation and the market in which the company operates.

2 Methods of assessing the efficiency of organizational and economic mechanism

In addition to the characteristics of innovation process implementation in terms of effectiveness and economic efficiency, the management efficiency should be assessed taking into account another important factor – time [4], i.e., the period of time required to achieve the set goal. If the innovative activities performed are both in demand to the required extent, and economically efficient to the sufficient extent, but the overall process for obtaining these innovations is longer compared with the average duration of innovative processes at similar companies, it is obvious that the efficiency of functioning of organizational and economic mechanisms for implementation of innovation processes at this company or enterprise will be lower that at the companies which achieve similar results within a shorter timeframe [5].

Thus, assessment of functioning of organizational and economic mechanisms (OEM) in implementation of innovation processes shall be performed using the following three key criteria:

- Effectiveness, i.e., capability to produce innovations which contribute to the achievement of set goals as a result of innovation processes;
- Economic efficiency, i.e., capability to obtain necessary innovations at the allowed level of expenses;
- Time or capability to achieve the required results over a certain period of time.

We propose that the functioning of organizational and economic mechanism should be assessed in three stages:

* Corresponding author: agamberg@mail.ru

1. Assess the effectiveness of the innovation process implemented by the company;

2. Assess the practical application of methodological principles of organizational and economic mechanism for implementation of innovation processes;

3. Assess the effects of innovative activities on the basis of integrated indicators obtained at the first and second stages.\

Specific features of information

Additionally, it should be noted that indicators for assessment of OEM functioning were selected considering the following specific features of information:

Objectivity and controllability. Indicators with values comparable with the values for prior periods should be used. The number of cases where the indicators are re-determined over time should be minimized. The method for collection and processing of input data should provide for the possibility to check the accuracy of obtained data in the course of independent monitoring.

Adequacy of reflection of ongoing changes. The indicators shall form a reliable base for assessment of innovation effectiveness. The data must be sufficiently accurate for the purpose of its use and should be obtained on a regular basis (i.e., at strictly defined intervals) and in due time.

Economic efficiency. Capability to obtain necessary data without exceeding the economically feasible cost limits. The indicator should be based on the already available information.

Conformance to the set goal (relevance). Each individual indicator must be aimed at measurement of progress in resolution of a certain task.

Accuracy and unambiguity. The indicators should be selected considering the need to continuously accumulate data and ensure their comparability over certain time periods. The essence of the indicator must be clear to the members of the work group performing the calculations and to the department managers.

Aspects of effectiveness of the innovation process.

In our opinion, effectiveness of the innovation process should be assessed in three aspects: organizational and management, social and production [6].

1. Organizational and management effectiveness reflects the degree of optimization of production system hierarchical pattern, level of correlation between coordination and subordination ties in it, possibilities for implementation of organizational potential and achieving maximum output from management actions. The integrated indicator which determines the organizational effectiveness of the innovative process is represented by (OR), eqn.(1):

$$OR = \sqrt[3]{T_{dep} * T_{sm} * T_m}, \quad (1)$$

where Tdep – coefficient of profitable departments;

Tsm – structuredness of management personnel;

Tm – manageability of management personnel.

Coefficient of profitable departments, i.e., those directly involved in the product lifecycle and making considerable contribution to the economic result achieved by the company, eqn.(2):

$$T_{dep} = K_{department} / Da, \quad (2)$$

where Kdepartment – number of profitable departments;

Da – actual number of departments.

Structuredness of management personnel, eqn.(3):

$$T_{sm} = \sum(Dr * Hr) / \sum(Da * Ha), \quad (3)$$

where Dr – regulatory number of structural divisions;

Hr – regulatory headcount of management personnel;

Ha – actual headcount of management personnel.

Manageability of management personnel, eqn.(4):

$$T_m = \frac{1}{Z} * \sum_i \frac{H_{ai}}{H_{ri}}, \quad (4)$$

where i=1, ..., Z – number of management levels;

Hai and Hri – actual and regulatory number of employees assigned to one manager at the ith management level on the average.

2. Social effectiveness of the innovative process reflects the consumer and social significance for the society and certain individuals observed in the course of consumption of newly created or improved material products. The integrated value which determines the social effectiveness of the innovative process is represented by (SR), eqn.(5):

$$SR = \sqrt[2]{C_{quality} * C_{ecology}}, \quad (5)$$

where Cquality – quality coefficient (determined based on department indicators);

Cecology – coefficient of environmental safety of production activity (determined based on department indicators) [7].

3. Production effectiveness reflects the increase in the level of production and technological activities of the company and quality of work performed by its employees. The integrated indicator which determines the production effectiveness of the innovative process is represented by (PR), eqn.(6):

$$PR = \sqrt[4]{(T_{aut} * T_{tech.prog} * T_{perform}) / (T_{defect} * T_{ritm})}, \quad (6)$$

where Taut – level of mechanization and automation of technological operation performance;

Ttech,prog – level of technological progress;

Tperform – performance;

Tdefect – loss due to defects to the volume of manufactured product;

Tritm – rhythmicity of production.

Level of mechanization and automation of technological operation performance, eqn.(7):

$$T_{aut} = Q_{aut} / Q \quad (7)$$

where Qaut – number of workers performing technological operations using machines and by supervision of automated machinery, people;

Q – number of workers participating in technological processes, people.

Level of technological progress, eqn.(8):

$$T_{tech.prog} = Q_{tech.prog} / Q_{tech} \quad (8)$$

where Qtech.prog – number of new and improved technological processes, pc.;

Q tech – total number of technological processes in accordance with technological documentation of the company, pc.

Performance, eqn.(9):

$$T_{perform} = Q_{volume\_a} / T \quad (9)$$

where Qvolume a – actual volume of issued goods, RUB;

T – time period, hours.

Losses due to defects to the volume of issued product, eqn.(10):

$$T_{defect} = C_{defect} / Q_{volume\_a} \quad (10)$$

where C defect – losses from defects for the period, RUB.

Rhythmicity of production – the actual volume of goods issued in the first ten days period, RUB/hours.

The proposed system of integrated indicators reflects the effectiveness of innovation implementation and its application will contribute to the selection of further organizational modifications associated with this innovation and development of innovative management strategy at large.

After establishment of multi-aspect system of indicators for direct assessment of innovative process effectiveness, we propose that the graphic method should be used. Indices of integrated indicators are marked on the axes (study periods are the same – month, quarter, half year) to reflect the organizational and management, social and production effectiveness of the innovative process; further, the center of gravity of the triangle is identified to reflect the integrated result of studied innovative process and demonstrate graphically which indicators reacted in the greatest degree. Then, assessment of innovative process effectiveness focuses on tracking the dynamics of movement of this active point and analysis of its new position. Thus, the developed methods for assessment of innovation implementation effectiveness enable establishing the parameters for innovation modification and contribute to selection of further organizational modifications associated both with this innovation and with development of innovative management strategy at large; the methods can be applied both in the course of innovation implementation at any stage as the basis control element and for assessment of the final effectiveness of the innovative process.

Therefore, the following integrated indicators are proposed for use in assessment (see the Figure 1) [8].

3 The practical application of methodological principles of organizational and economic mechanism

Further, let us assess in more detail the practical application of methodological principles of organizational and economic mechanism of innovation process implementation at companies. The indicators reflecting the application of methodological principles of organizational and economic mechanism in the business practice in the course of implementation of innovations with various degree of complexity must conform to the requirements which we make to ensure the accuracy of assessment of modifications in conditions institutionalizing the innovative process implemented at the company and stimulating the development of the innovative activities in the company at large.

Namely:

1. Express the essence of organizational and economic mechanism for implementation of the innovative process and give qualitative and quantitative assessment of its functioning;
2. Have complex nature as required to obtain multi-aspect characteristics of conditions impacting the institutionalizing of innovations at companies;
3. Ensure accuracy, completeness and timely manner of information obtaining to ensure flexible and adaptive management of innovative activities at large.

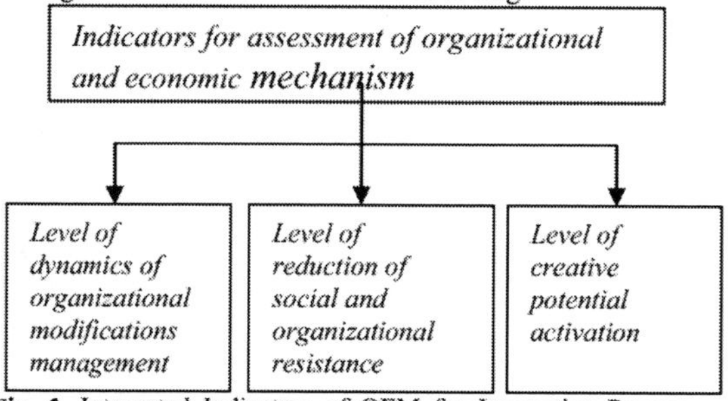

Fig. 1. Integrated Indicators of OEM for Innovative Process Implementation.

1. Take into account the possibility to change the conditions and tasks associated with development of the innovative activities at the company.

Thus, the overall assessment of functioning of organizational and economic mechanism will be performed as follows, eqn.(11):

$$Y = a_1 * Y_D + a_2 * Y_B + a_3 * Y_A \quad (11)$$

where УD, УB, УA – level of dynamics in management of organizational modifications, level of reduction of social and organizational resistance, level of creative potential activation respectively;

a1, a2, a3 - weigh (significance) coefficients of corresponding studied integrated indicators (to be determined by persons responsible for development of innovative activities or experts in the field of innovative solutions).

Additionally, assessment of each integrated indicator shall be performed in accordance with the common methodological basis, eqn.(12).

$$Y = \sum b_i * \frac{T_{i\_act}}{T_{i\_base}}, \tag{12}$$

where Ti act, Ti base – ith indicator (i=1,2,3,….n);

n – number of indicators taken into account;

bi – weigh (significance) coefficient of ith indicator.

The level of dynamics in management of organizational modifications shall be determined as follows, eqn.(13):

$$Y_D = b_1 \frac{T_{durat\_a}}{T_{durat\_b}} + b_2 \frac{T_{qualif\_a}}{T_{qualif\_b}} + b_3 \frac{T_{closed\_a}}{T_{closed\_b}} + b_4 \frac{T_{com\_a}}{T_{com\_b}} + b_5 \frac{T_{progress\_a}}{T_{progress\_b}}, \tag{13}$$

where Tdurat – time for experiment and time for implementation (the ratio tends to minimum);

Tqualif – number of cases where the employee's qualification corresponds to the work performed by the employee (the ratio tends to 100.0 %);

Tclosed – number of vacancies closed as required for implementation of innovative process (the ratio tends to 100.0 %);

Tcom – the indicator in inter-structural communications (the ratio tends to 100.0 %);

Tprogress – number of new technologies and methods of organization and management (the ratio tends to maximum).

The indicator of inter-structural communications shall be calculated as follows, eqn.(14):

$$T_{com} = K_{com\_dep} / K_{dep}, \tag{14}$$

where Kcom dep – number of departments that exchange data in regulated and automated manner;

Kdep – total number of departments in the company.

The level of reduction of social and organizational resistance shall be determined as follows, eqn.(15):

$$Y_B = b_1 \frac{T_{debt\_a}}{T_{debt\_plan}} + b_2 \frac{T_{\inf\_a}}{T_{\inf\_base}} + b_3 \frac{T_{\inf o\_a}}{T_{\inf o\_base}} + b_4 \frac{T_{decent\_a}}{T_{decent\_b}} + b_5 \frac{T_{perf\_a}}{T_{perf\_base}} + b_6 \frac{T_{soc\_a}}{T_{soc\_b}}, \tag{15}$$

where Tdebt – level of current debt owed to the employees in the form of salary (the ratio tends to 1);

Tinf – level of communication automation (the ratio tends to maximum);

Tinfo –indicator reflecting the sources and quality of information available for the employees engaged in the innovative process, falls within the interval (0;1), tends to 1;

Tdecent – indicator reflecting the engagement of personnel into the innovative process (the ratio tends to 100.0 %);

Tperf – performance of the employees engaged into implementation of the innovative process, reflects the quality of work performed, falls within the interval (0;1), tends to 1;

Tsoc –indicator reflecting social tension (the ratio tends to the minimum).

The level of current debt owed to the employees in the form of salary, eqn.(16):

$$T_{debt} = (APS_{accr} - APS_{paid}) / APS_{accr}, \tag{16}$$

where APSaccr – employees' salary accrued for payment (minus income tax and contributions to non-budgetary funds out of salary), RUB.

APSpaid – actually paid salary, RUB.

Level of communication automation, eqn.(17):

$$T_{\inf} = (M_a * K_{\inf\_a}) / (M_r * K_{\inf\_r}), \tag{17}$$

where Ma, Mr – actual and regulatory number of PCs used for making management resolutions in the course of innovative process implementation, pc.;

Kinf a – number of information technologies used for making management resolutions in the course of innovative process implementation, pc.;

Kinf r – number of information technologies required for making management resolutions in the course of innovative process implementation, pc.

Indicator of personnel involvement into the innovative process, eqn.(18):

$$T_{decent} = C_a / C_r, \tag{18}$$

where Ca, Cr – actual and regulated number of employees engaged into the innovative process, people.

Indicator of social tension, eqn.(19):

$$T_{soc} = Q_{conf} / Q, \tag{19}$$

where Qconf – number of conflicts under review in the company departments and at administrative bodies, pc.;

Q – total number of employees, people.

Level of creative potential activation shall be determined as follows, eqn.(20):

$$Y_A = b_1 \frac{T_{patent\_a}}{T_{patent\_plan}} + b_2 \frac{T_{creat\_a}}{T_{creat\_base}} + b_3 \frac{T_{prof\_a}}{T_{prof\_base}} + b_4 \frac{T_{turnover\_a}}{T_{turnover\_base}}, \tag{20}$$

where Tpatent – number of registered rights to intellectual property, pc. (the ratio tends to maximum);

Tcreat – indicator of personnel's creative activity, to be determined based on the number of innovative proposals from employees involved into the innovative process, pc. (the ratio tends to maximum);

Tprof – indicator reflecting the level of professional training and qualification of employees, to be determined on the basis of attestation, falls within the range (0;1), tends to 1;

Tturnover – turnover of personnel (the ratio tends to minimum).

Indicator of personnel turnover, eqn.(21):

$$T_{turnover} = Q_{dismiss} / Q , \qquad (21)$$

where Qdismiss – number of employees who left voluntarily or were dismissed for violation of workplace discipline, people;

Q – total number of employees, people.

To assess the effects of innovative activities in the aspect of implemented innovative process, first of all, we need to identify the effects which reflect the comparison of results of innovative process implementation management with the costs associated with this management. The majority of Russian and foreign authors refer to the list specified in Table 1 [9]:

Table 3. Comparison of the three machine-building enterprises activity results on criterion of functional and resource integrity.

Effect	Result of Impact	Source	Characteristics of selection
Increased performance	Positive	B. Crepon, E. Duguet, J. Mairesse, 1998	4,164 companies in France in 1990
Increased market share	Positive	S. Olav Nes, A. Leolahti, 1997	1,848 Norwegian firms, period from 1990 to 1994
Increased profitability	Unclear (not available)	T. Sandven, 2001	873 Norwegian production firms in 1992 and 640 firms in the period from 1995 to 1999
Temporary monopoly powers	Positive	J. Peters, 1997	410 German industrial firms in 1996
Increased return on investment	Unclear (not available)	Baldwin, Hanel, et al., 2000	5,729 Canadian companies in 1993
Increased firm value	Positive	Loof, Heshmati, et al., 2001	1,062 Finnish companies, 1,315 Norwegian and 746 Swedish companies in the period from 1988 to 1999
Competitive advantages	Positive	Gurkov, 2003	Survey among directors of large and medium industrial companies (in 1998 – 740, in 2000 – 740, in 2002 – 1,140)\

However, it is difficult to identify the exact impact of implemented innovative processes on all the above indicators because in this assessment, the innovative activities of the company should be reviewed as the only events and activities performed by the company, i.e., the current functioning of the company and multiple factors which have impact on it should be disregarded.

5 Conclusion

Thus, integration of results achieved directly through implementation of the innovative process and functioning of OEM enables identification of the key effects. In the aspect of organizational and management effectiveness – effect of reduced time for innovative process implementation. In terms of social effectiveness – effect of competitive advantages of produced product. Production effectiveness shall be interpreted as the effect of increased workforce productivity, reduced losses due to defects, reduced expenses on warranty services.

The, the final economic effect of implementation of a specific innovative process and OEM functioning shall be calculated using the following equation, eqn.(22):

$$E_{IP} = E_{compet.} + E_{loss\_to\_defect} + E_{reduc.time} + E_{warranty} + E_{performance} \qquad (22)$$

where Ecompet. – increase of revenues from sale of company products due to increased demand as compared with sales before implementation of innovative process or analogs existing in the market (including new contracts);

Eloss to defect – economic effect of reduced losses caused by defects (including returns by buyers);

Ereduc.time – savings achieved due to reduced time for implementation of innovative process (measured based on notional value of 1 standard hour);

Ewarranty – savings achieved due to reduced expenses on warranty services (based on reduction in the number of claims);

Eperformance – effect of increased volume of goods over the reviewed period.

Due to timely calculations of indicators of final efficiency, reserves can be identified and strategy of their use can be developed based on concretization of objectives and, finally, a highly efficient mechanism for implementation of scientific and technical achievements can be created in the framework of an industrial company.

Thus, practical application of this method will contribute to production of products which conform to the consumer expectations while minimizing one-time and continued costs for implementation of innovative processes and ensure company profit growth due to increased share and segments of consumers.

References

1. S.N. Yashin, E.V. Koshelev, S.A. Makarov, *Analysis of the effectiveness of innovative activity* (St. Petersburg, BHV, 2012)
2. P. Druker, *Innovation and Entrepreneurship* (Williams, Kiev, 2007)
3. M.N. Mescon, M. Albert, F. Khedouri, *Management* (Delo, Moscow, 1992)
4. A.E. Gerasimov, Innovations, **12(9)** (2001)
5. K.K. Prakhalad, M.S. Krishnan, *The space business innovation. Creating value together with the consumer* (Alpina publisher, Moscow, 2012)
6. Federal State Autonomous Educational Institution of Higher Professional Education "Ural Federal University named after the first President of Russia B.N.Yeltsin" (UrFU). Graduate School of Economics and Management. http://urfu.ru/ru/.
7. *ISO 9000-2008. Quality management systems. Fundamentals and vocabulary* (Standards Publishing House, Moscow, 2009).
8. Ural Branch of the Russian Academy of Sciences. The Institute of Economics. http://www.uiec.ru/.
9. A.A. Chulok, Innovations, **12(5)** (2004)

Mechanism of interaction between large industrial enterprises and small innovative enterprises

Irina Ershova[1,*], *Natalia* Kalinina[1], and *Pavel* Mezentev[1]

[1]Mechanics and Machine Building Institute, Ural Federal University, Ekaterinburg, Russia

Abstract. The innovative development mechanisms were created based on the modern economy need for innovations, but the innovation did not happen. The research is aimed at the identification of the innovative development process problem elements, identification of the small innovative enterprise development problems and prospects, the justification of the need for innovative infrastructure attached to the higher education institution creation. The research analysed the regional and country scale statistical data. That allowed drawing conclusions on the innovative process condition, its elements, and their interrelations. In addition, the research on the small innovative enterprises was conducted. The main result of the research is the revealed "gap" between fundamental science (creating the innovation) and manufacturing enterprise, (implementation of the innovation). According to the research result, there is a need for the innovative infrastructure formation.

1 The need for the small innovative enterprises development

Only the innovative breakthrough will allow our country to overcome the heritage of "collapse of the 90th" and will provide "the state sovereignty" due to the innovative base creation for further successful development of all economic system. Innovations are necessary, first of all, for the enterprises making a difficult industrial output. Despite general understanding of the innovative development need of the economies in general, the separate enterprises, and also active innovative projects public financing, noticeable growth of innovative activity doesn't take place. For innovative development ensuring such structural and organizational forms as science and technology parks, industrial science and technology parks, technopolises, clusters were at different times offered.[1,2,3] The small innovative enterprises (SIE) at the higher education institution creation is one of the instruments for the innovative development crisis overcoming [4]. In 2009 there was a Federal law 02.08.2009 N 217-FZ [5], allowing for the higher education institutions and scientific research institutes to create the small innovative enterprises. As showed the statistical data analysis, there were more than 60 small innovative enterprises (SIE) with Urals Federal University participation in 2012 [6]. Nowadays there are more than 80 small innovative enterprises. But the intensive industry innovation growth in the Ekaterinburg city or in Urals region didn't follow. The created innovative enterprises didn't give multiplicative effect of the innovative production volume gain. The term small innovative enterprises (SIE) isn't a new one, it was used before the Federal law 02.08.2009 N 217-FZ was pass an act [5]. Therefore we consider necessary to specify: in this article we will understand the enterprise created with participation of the scientific organization (the higher education institution or scientific research institute) for the purpose of the innovative product commercialization as the term small innovative enterprises (SIE).

Analysis of the innovative activity level

Rapid growth of the SIE quantity didn't lead to the expected growth of the innovation in industrial production level. In Fig. 1 the share of a new innovative production in the gross national product (GDP) structure is presented.

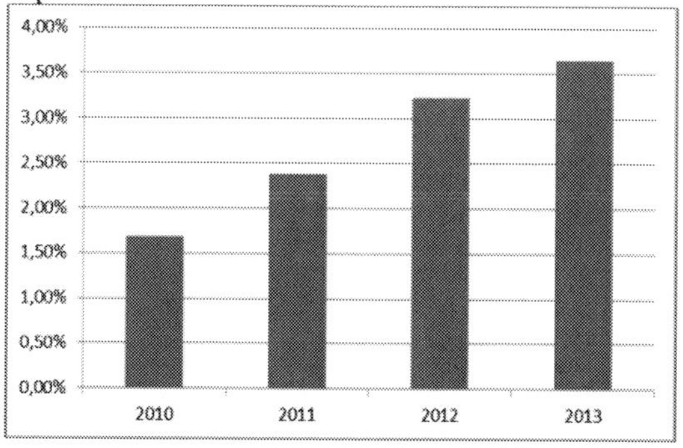

Fig. 1. The share of a new innovative production in the gross national product (GDP) structure [7,8].

The share of innovative goods, services which are introduced for the first time or being exposed to considerable technological changes, in GDP is 3,5% in 2013. So it is possible to draw a conclusion on the

* Corresponding author: eiv.eopm@e1.ru

insufficient level of innovative production introduction. Such situation demands further work in the sphere of ensuring communication between innovative developers and producers of an industrial output.

We will consider the main results of the enterprises activity in the Ural federal district for 2012. Ural federal district is one of the industrial regions of the country (table).

Table 1. The main results of the enterprises activity in the ural federal district for 2012 [7].

Indexes	
On a total indicator of volume of the shipped goods, by types of economic activity - "the processing productions",[a place, million rubles].	4-e 3211212
On an index of industrial production, [place, %]	8-e 101.3
On an index of industrial production, by types of economic activity - "the processing productions", [a place, %]	2-e 107.9
The number of organizations which are engaged in scientific development, [a place, units]	5-e 236 (3566- about the country in general)
The occupied with scientific researches personnel in number, [a place, number of people]	5-e 43879 (726318 about the country in general)
Own financing of basic researches, [place, one million rub].	6-e 3832.6
Own financing of applied researches, [place, million rubles].	7-e 3084.5
Is given patent applications, [place, pieces]	7-e 2326 (42180 about the country in general)
Is given out patents, [place, pieces]	5-e 1994 (33633 about the country in general)
Level of inventive activity, [place]	8-e
The volume of the innovative goods, works, services,[place, million rubles].	5-e 148696.2 (2872905.1 about the country in general
Share of the innovative goods, works, services in total production, [a place, %].	8-e 2.1 (8 about the country in general)

The table is made up by authors on the basis of the data presented in the form of annual reports on the site of Federal State Statistics Service.

It should be noted that Ural federal district is the smallest industrial region, as on population (the 6th place), and on the area. Despite the modest sizes, rather small amount of the population, the region possesses enormous experience of industrial development, scientific support of production that is caused by historical, natural, resource aspects – all this allows the region to reach good results now: for example, the 2nd place on the level of industrial production. Despite rather high the level of the industrial output production, and also the level of the industrial production index, the level of the innovative activity in the region is insufficient, and on the volume of innovative production, works and services introduction Ural federal district is on the last places in comparison with the other regions.

Thus the Ural federal district industrial complex is one of the largest in the country. But this complex was created for a long time ago and initially created under the military purposes. In modern conditions a survival of such industrial region as Ural federal district demands increase of its enterprises and productions competitiveness that depends only on the success of the existing industries and productions innovative updating.

It should be noted that Ural federal district differs in the smallest difference of the submitted patent applications and the taken-out patents, that testifies to the scientists and sciences schools of the region quality.

At the available requirements of the industrial development, including level of industrial production, a share of an industrial output, at the level of the region we again see the lack of due introduction into production created, introduced innovations.

It is necessary to look for the reasons which led to a low introduction of the innovative development, technologies, products. Over the past few years financing level essentially didn't change, and the number of the enterprises which are engaged in scientific development essentially didn't change. According to Federal State Statistics Service, the level of inventive activity (measured by coefficient of inventive activity and equal to number of the domestic patent applications for inventions submitted in Russia per 10 thousand people of the population) essentially didn't change for the last 3 years and makes 2 units. Thus, we can't speak about decrease in the inventive level or the number of inventions, lack of the worked system of bringing results of scientific researches to production is available.

2 Reasons of the low efficiency of the small innovative enterprise

For the identification and assessment of the small innovative enterprise low efficiency reasons the research in the form of questioning was conducted. About 50 small innovative enterprises took part in the research. The results of the questioning are given below.

Following the research results, it is possible to draw a conclusion that the activity of the greatest number of the small innovative enterprises is connect with IT. To some extent their production is based on the innovation creation in the field of the software, collecting, processing, storage and transfer of information. The share of the small innovative

DOI: 10.1051/shsconf/20173501070

enterprises created for commercialization of the technological or grocery innovations in the field of industrial production doesn't exceed 5-7%. Such situation doesn't correspond to the market requirements – according to the data provided in Strategy of Sverdlovsk region development till 2020[9], the enterprises of Sverdlovsk region stated the greatest interest in innovations in the field of production technologies (for the industrial enterprises), power - and resource-saving, and information technologies take only the 3rd place.

Current situation can be characterized as inadmissible: on the one hand, the innovations commercialization mechanism is created and function, there is a requirement in technological innovation, on the other hand, innovations are offered in those spheres which are least interesting to the consumer (in this case to the industrial enterprise). Inconsistency of two subjects of the innovative market actions: small innovative enterprises and the industrial enterprises, doesn't give the chance to effectively rise of the region's innovative level.

In the questionnaire was a question: with what difficulties the company (small innovative enterprise) faces during a first or second year of its existence? Practically all participants noted the need of observance the formal rules of the company working: balance sheets, paperwork, receiving allowing documents, maintaining documentation on emergency, etc. Analyzing the received information, it is possible to speak about two main conclusions:

1. The small innovative enterprises haven't enough support of the higher education institution only as a cofounder, creation of the structure which would provide the small innovative enterprises activity is necessary. In our opinion it can be the structure similar on the functions to science park, technopark, business-incubator, etc.
2. It is necessary to analyse the purposes and problems of the small innovative enterprises creation on the point of view of all the innovative development process participants.

We will consider the need of ensuring the small innovative enterprises activity. What the purposes, tasks didn't face the small innovative enterprise; the similar form of activity assumes formation of the legal entity. The small innovative enterprise as the legal entity has to conduct the activity according to norms and laws of the Russian Federation, i.e. to keep accounting and tax account, to hand over reporting forms, to deal with issues of employment, intellectual property etc. On the other hand, as practice shows, the small innovative enterprise is created by idea carriers, research associates. In the majority they are far from questions of economy and law. As a result, the head of such enterprise has to make a choice between two variants: he understands all subtleties of legal and economic existence of the organization, or forms the team capable to solve all range of questions (Fig. 2).

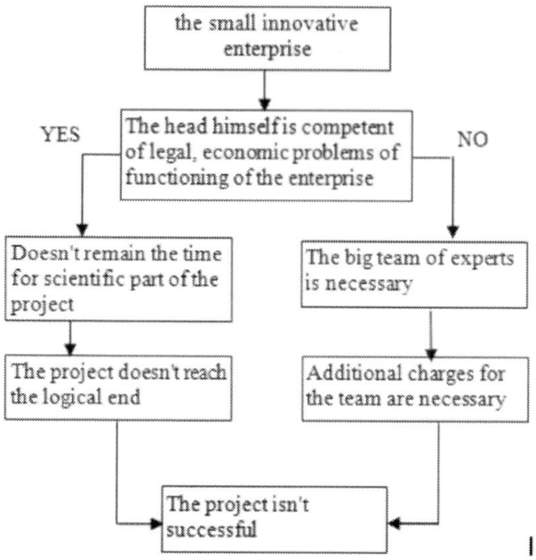

Fig. 2. Forms of the small innovative enterprise functioning.

Apparently the both forms presented at a figure lead to decrease in the project efficiency or its total failure. We consider that for increase of the small innovative enterprise overall performance it is necessary to create a structure which would assume the part of the enterprise ensuring activity tasks. For identification of such concept we suggest to define this system - "innovative infrastructure".

3 The mechanism of the small innovative enterprises interaction with the environment

It should be noted that except the described problem there is a feature in the relationship system between the production workers and scientists. As practice shows, they speak "different languages" what lead to misunderstanding. From our point of view the buffer, the intermediate stage is necessary. This stage will allow to connect the requirements of manufacturing enterprise and the possibilities of the scientific organizations. For example, the enterprise has a problem with the low efficiency of the used technological process or there is a need the import accessories replacement of. At the same time in the higher education institution innovative materials and ways of their processing developments are conducted. If the new material will be directly offered to the enterprise, it will be rather difficult to estimate effect of the innovative material using. The enterprise needs "finished product", for example, the developed technology which can be introduced in production with the minimum completions. In this regard a mechanism allowing "to join" participants of the innovations transfer process from developers to producers is actual. The innovative infrastructure uniting small innovative enterprise can become such mechanism.

The offered interaction mechanism is presented at Fig. 3.

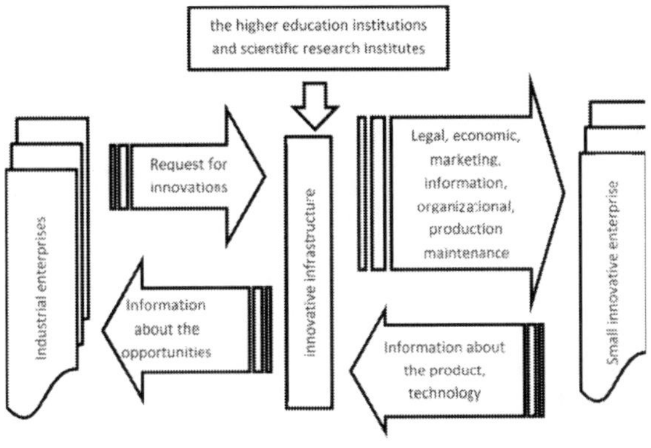

Fig. 3. Interaction of the innovative infrastructure, the small innovative enterprises and the industrial enterprises.

Realization of such mechanism will allow to increase significantly the transfer of innovative development to production system effectiveness.

4 Life cycle of the small innovative enterprise

There is a problem interfering effective operation of the small innovative enterprises like an instrument of innovative development of the country. As showed the conducted survey, most (about 90%) of the small innovative enterprises owners consider it as a long or even the "eternal" project. This situation is not absolutely corresponding the the innovative enterprises development theory. According to the theory of competitive strategy (the author is A.Yu.Yudanov [10]) the small innovative enterprises relate to the explerent ("swallow"). Their possible ways of development are presented at the Fig. 4.

Apparently from the presented scheme, explerent have three ways of the life cikle: bankruptcy, slow development in a patient or in a violent and sale. Most of small innovative enterprises' founders don't consider option of sale in both variants – sale of a product or sale of a technology. The statistics shows that only 5-7% of explerent can become at least patents, the stage of violent only1-2% of explerents can reach. Such "the low survival" is inadmissible in the conditions of sharp need for innovative development of national economy and production, and also is unacceptable for the higher education institution as for the co-owner of the small innovative enterprises.

One more important parameter – it is impossible to create the new enterprise, new plant for each innovation. It is impossible to create the new enterprise for each innovative idea, because the costs of their creation and the contents will exceed effect from their existence.

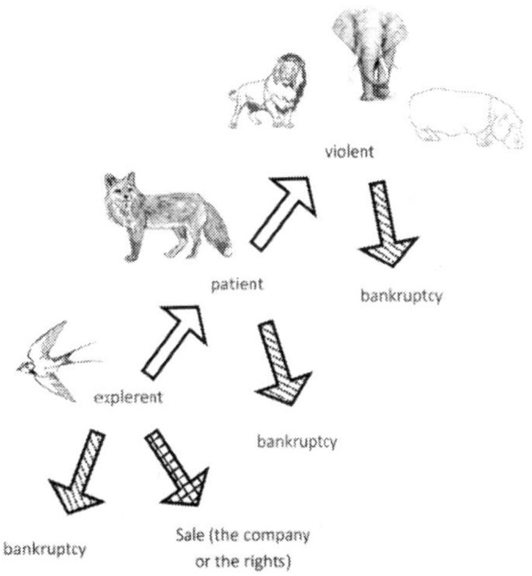

Fig. 4. The type competitive strategy evolution.

On the other hand, the same theory of competitive strategy says that violent in a condition of an elephant and hippopotamus is compelled to look for the ways and the tools for improvement of their products and technologies. In this situation, the innovations developed to a condition of a product within explerent, can be successfully sold to violent that will create mutually advantageous conditions for further development: the violent will have an opportunity to use innovations in the production, and the explerent will have resources for continuation of scientific and innovative activity.

For effective interaction of the small innovative enterprises - explerents and the industrial enterprises – violent (or patient) the innovative infrastructure has to act as the guarantor of safety, honesty and efficiency of transactions on an innovative product and technology sale. This fact once again confirms validity and relevance of the innovative infrastructure creation.

5 The main conclusions following the results of the conducted research

Following the results of the conducted research it is possible to draw the following conclusions:

- The rapid growth of innovative development didn't occur despite the created conditions;
- The small innovative enterprise as the subject of the innovative development was created with the purpose to commercialize scientific results, but the process is going extremely slowly;
- The created small innovative enterprise "wallowed" in the solution of routine problems of the legal entity functioning;
- Founders of the small innovative enterprises consider only one variant the company development– gradual transformation into a patient (or a violent) and that is impossible under objective laws of the economic and production systems development.

On the basis of the conclusions we can speak about the need of the innovative infrastructure creations. The innovative infrastructure provides the small innovative enterprise activity, and the effective interaction the small innovative enterprises with potential customers of innovative development.

References

1. E.Yu. Kuznetsova, The Organizer of production, **1**, 37-43 (2011)
2. I.V. Ershova, I.V. Kopytov, Questions of management, **11**, 48-53 (2010)
3. C. Grimpe, K. Hussinger, Formal and Informal Technology Transfer from Academia to Industry: Complementarity Effects and Innovation Performance. http://ftp.zew.de/pub/zew-docs/dp/dp08080.pdf
4. S.Yu. Alexandrova, *The small innovative enterprises at higher educational institutions of the Russian Federation* (Bauman Moscow State Technical University, Moscow, 2012)
5. The federal law 2009 no. 217-FZ, About modification of separate acts of the Russian Federation concerning creation of economic societies by the budgetary scientific and educational institutions for practical application (introduction) of results of intellectual activity
6. Innovative infrastructure of the Ural Federal University http://www .inno.urfu.ru/
7. Federal State Statistics Service: official site. http://www.gks.ru/
8. The Global Innovation Index 2014. The Human Factor in Innovation. http://www .globalinnovationindex.org/content.aspx? page=gii-full-report-2014
9. Strategy of innovative development of Sverdlovsk region for the period till 2020. The Regional newspaper is approved by the resolution of the government of Sverdlovsk region of 2013 646-PP (2013)
10. A.Yu. Yudanov, *Competitiveness: theory and practice* (GNOM&D, Moscow, 2001)

Model of monopolistic competition with heterogeneous labor

Alexander Filatov[1,2*], and *Yury* Sokolovsky[2]

[1]Far Eastern Federal University, School of Economics and Management, Vladivostok, Russia
[2]Irkutsk State University, Institute of Mathematics, Economics, and Informatics, Irkutsk, Russia

Abstract. The paper presents a tool for modelling monopolistic competition markets, based on Dixit-Stiglitz ideology but taking into account heterogeneity at labor market. We analyse several modifications of a two-sector general equilibrium model. In the basic one with two levels of workers qualification their shares are determined endogenously on the base of comparison between the higher wage of the skilled worker and heterogeneous education costs, also taking into account the labor mobility between the manufacture and agriculture sector. The model is generalized for the case of continuous distribution of labor qualification. The impact of the model parameters (ratio of fixed and variable costs, market size, heterogeneity in productivity, elasticity of substitution, etc.) on the obtained equilibrium prices, quantities, wages, number and size of firms, social welfare is investigated.

1 Introduction

The majority of markets in the modern globalized economy have a structure of the monopolistic competition [1]. They consist of a large number of firms, producing a single variety of a differentiated product, firms don't interact directly, there are no entry or exit barriers, and each firm enjoys market power due to consumers' love for variety.

The classical Dixit-Stiglitz monopolistic competition model [2] and its application to the international trade made by Krugman [3, 4] could explain some effects arisen during last decades in the world economy. These effects, including bilateral flows of very close substitution goods, remained inexplicable within theories based on comparative advantages in labor [5] and in production factors endowments [6]. Nevertheless, this theory based on the assumptions of homogeneous labor productivity and homogeneous firms can't reveal some more observed in reality stylized facts. Particularly, heterogeneity is necessary for majority of the international trade models, though different researches since Bernard and Jensen [7, 8] showed that exporters are usually more productive than non-exporters.

Several economists tried to solve this problem. Melitz [9] used quite complicated constructions based on the initially unknown heterogeneous marginal costs of each firm [10, 11], but it makes almost impossible to analytically investigate the impact of elasticity of substitution, market size, degree of labor heterogeneity, shares of agriculture and manufacture sectors on the equilibrium parameters: prices, quantities, wages, and welfare. There are several other approaches. Manasse and Turrini [12] supposed that entrepreneurs vary in their ability, so they hire homogeneous workers on a perfectly competitive labor market, but the workers work better for a better entrepreneur. Bartel, Lichtenberg, and Sicherman [13, 14] assumed that there exist two technologies – old and more effective new, workers vary in their skills, and the more skilled workers have absolute advantages in both technologies and comparative advantages in the new one. It was shown that at the same time firms with both technologies can co-exist. Bernard, Eaton, Jensen, Kortum [15] and Yeaple [16] also showed that at the same time firms with different productivity can co-exist. They hire different workers and pay them different wages due to their skills.

Our approach also uses this assumption. We don't pay attention on matching mechanisms – they can be different [17], but provide exogenous firm heterogeneity. In general we propose tool for analysing monopolistic competition markets with heterogeneous labor. The worked out models can describe the economical agents' behavior in several situations. They also help estimate consequences of different mechanisms of labor market regulation, possible effects of industrialization and de-industrialization, migration between regions and sectors of economy, the impact of education, taking into account the economic and spatial heterogeneity.

2 The tool for market analysis

Let's consider the economy consists of traditional sector A (usually called "agriculture" in spite of the fact that a wide set of services also have the same structure), and modern sector M (traditionally called "manufacture").

* Corresponding author: alexander.filatov@gmail.com

Agriculture sector produces homogeneous good with constant returns to scale, and manufacture sector – continuum n of horizontally differentiated varieties indexed by $i\in[0;n]$, with increasing returns to scale. L consumers have identical Kobb-Douglas high level utility functions

$$u = M^{\mu} A^{1-\mu} \to \max .$$

It means, that they spend shares $\mu\in(0;1)$ and $(1-\mu)$ of income for manufacture and agriculture good respectively: The low level utility function is CES:

$$M = \left(\int_0^n (q(i))^{(\sigma-1)/\sigma} di \right)^{\sigma/(\sigma-1)} .$$

When elasticity of substitution $\sigma\to1$, we have the low level Kobb-Douglas function also, the modifications of manufacture good are independent; consumers spend the fixed sum of money for each of them. When $\sigma\to+\infty$, we have the linear utility function describing the perfect substitution goods.

Let denote the total income of all identical consumers by I. So under the prices $p(i)$ the i-modification demand $q(i)$ can be found as a solution of a problem

$$\int_0^n (q(i))^{(\sigma-1)/\sigma} di \to \max, \quad \int_0^n p(i) q(i) di = \mu I .$$

Hence we can obtain the very important formula

$$\frac{q(i)}{q(j)} = \left(\frac{p(j)}{p(i)} \right)^{\sigma} .$$

The only production factor is heterogeneous labor. Let the labor demand be

$$l(i) = f + c(i) q(i) .$$

Here value $c(i)$ means the number of workers needed for one unit production at i-firm. If the wage is equal to $w(i)$, the total costs of i-firm will be

$$TC(i) = w(i) l(i) .$$

Each firm maximizes its own profit:

$$\pi(i) = p(i) q(i) - w(i)(f + c(i) q(i)) =$$

$$= q(i)(p(i) - w(i) c(i)) - w(i) f \to \max_{p(i)} .$$

Solving this problem we can find the optimal price for i-modification:

$$p(i) = \frac{\sigma}{\sigma-1} w(i) c(i) .$$

In the long run the profit of all firms should be zero. Let's find the optimal quantity

$$q(i) = \frac{f(\sigma-1)}{c(i)} .$$

Using the ratio of prices and quantities we can find the relation between wages and productivities

$$\frac{w(i)}{w(j)} = \left(\frac{c(j)}{c(i)} \right)^{\frac{\sigma-1}{\sigma}} .$$

The firm size despite heterogeneity is equal for all firms:

$$l(i) = f + c(i) q(i) = f + f(\sigma-1) = f\sigma .$$

3 The basic model

The proposed tool helps us to analyse monopolistic competition markets under some additional assumptions. In the basic model let population L consist of skilled manufacture workers (with share α_1, and productivity $c_1=c$), unskilled manufacture workers (with share α_2, and productivity $c_1=\gamma c$) and farmers (with share $1-\alpha_1-\alpha_2$). Let also agriculture sector be numerator with productivity, wage, and price equal to one, and fixed costs equal to zero: $f_A=0$, $c_A=1$, $p_A=1$, $w_A=1$ Then we can obtain the following formulas:

$$p_1 = \frac{\sigma}{\sigma-1} w_1 c , \quad p_2 = \gamma \frac{\sigma}{\sigma-1} w_2 c ,$$

$$q_1 = f(\sigma-1)/c , \quad q_2 = f(\sigma-1)/\gamma c ,$$

$$w_1 = \gamma^{(\sigma-1)/\sigma} w , \quad w_2 = w , \quad l_1 = l_2 = f\sigma ,$$

$$n_1 = \alpha_1 L / f\sigma , \quad n_2 = \alpha_2 L / f\sigma , \quad \pi_1 = \pi_2 = 0 ,$$

$$TR_1 = TC_1 = \gamma^{(\sigma-1)/\sigma} f\sigma w , \quad TR_2 = TC_2 = f\sigma w ,$$

$$w_1 = kw = \frac{\mu}{1-\mu} \frac{1-\alpha_1-\alpha_2}{\alpha_1 + \alpha_2/k_1} , \quad w_2 = w = \frac{\mu}{1-\mu} \frac{1-\alpha_1-\alpha_2}{\alpha_1 k_1 + \alpha_2} .$$

In the table 1 we show, how the model parameters f, c, L, γ, σ, α_1, α_2, μ impact the equilibrium prices, quantities, revenues, costs, wages, number and size of firms in effective and ineffective sector. Let's denote dependences in the form of primal and inverse proportionality with pluses and minuses, increased and decreased dependences of general form with up and down arrows, independent pairs with points.

Table 1. The impact of model parameters on the equilibrium.

	p_1	p_2	q_1	q_2	TR_1	TR_2	w_1	w_2	n_1	n_2	l
f	•	•	+	+	+	+	•	•	−	−	+
c	+	+	−	−	•	•	•	•	•	•	•
L	•	•	•	•	•	•	•	•	+	+	•
γ	↑	↑	•	−	↑	↓	↑	↓	•	•	•
σ	↓	↓	↑	↑	↑	↓↑	↑	↓	−	−	+
α_1	↓	↓	•	•	↓	↓	↓	↓	+	•	•
α_2	↓	↓	•	•	↓	↓	↓	↓	•	+	•
μ	↑	↑	•	•	↑	↑	↑	↑	•	•	•

The institutional restrictions on labor market are also can be taken into account. Particularly the model with fixed wages has been worked out and explored [18]. The restrictions can be caused, for example, by government policy (uniform scale of wages, etc) or asymmetric information when transactions between principals (firms) and agents (workers) take place. The mechanisms of effective workers stimulation, included firm profit reallocation, were investigated. The obtained equilibrium was compared with equilibrium in the basic model. The possibility for firms to use different production factors was considered by Echazu and Lugovskyy [19], but it was supposed that the production factors endowments are unlimited.

4 The model with education

In both modifications it was supposed that the sector shares are set exogenously. Nevertheless the wage level in industry and agriculture can make workers change the sector. If wages in agriculture are higher than in industry the de-industrialization is quite possible, and in situation of higher wages in industry we can expect the mass migration from countryside to cities. Also we can include into the model the possibility to improve professional skills (at some costs depending on abilities) to get higher wages. Let's formalize these assumptions.

1. The workers can move for free from agriculture to unskilled manufacture sector.

2. The manufacture workers can improve skills under some education costs that depend on $\theta \in [0;1]$ – the uniformly distributed individual "education disability" (see fig.1).

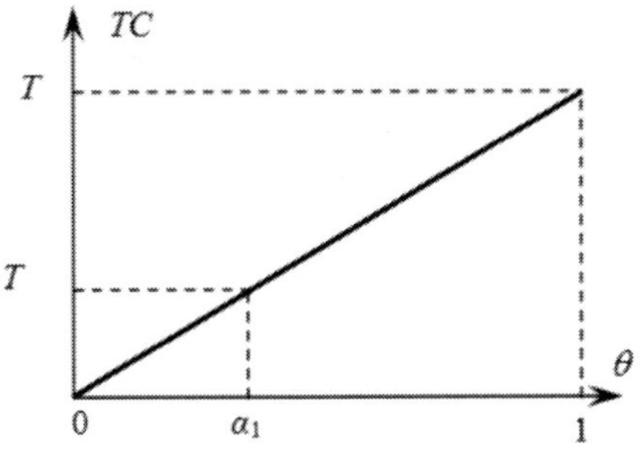

Fig. 1. The education costs.

3. The education costs for the worker with the best ability (θ=0) is equal to zero, and for the most "alternatively gifted" one (θ=1) is equal to T. So share α_1 of the most gifted workers with educational disability $\theta \in [0;\alpha_1]$ will study.

The equilibrium wages in agriculture and unskilled manufacture sector are the same and equal to one. The workers with low education costs (less than the wages gap) move to the effective sector. Their share can be found from the equality $k_1 w_2 - w_2 = \alpha_1 T$. Though the equilibrium wage in unskilled manufacture sector is equal to one, we obtain equality $k_1 - 1 = \alpha_1 T$. So the skilled workers share is equal to

$$\alpha_1 = (k_1 - 1)/T .$$

From the obtained above formulas it follows

$$w_2 = \frac{\mu}{1-\mu}\frac{1-\alpha_1-\alpha_2}{\alpha_1 k_1 + \alpha_2}.$$

We can find also the share α_2 of unskilled workers:

$$\alpha_2 = \mu(1-\alpha_1) - \alpha_1 k_1 (1-\mu).$$

If we substitute α_1 by the obtained above expression we will obtain

$$\alpha_2 = \mu - \frac{(k_1-1)(\mu + k_1(1-\mu))}{T}.$$

The share of "peasants" can be obtained as $\alpha_3 = 1 - \alpha_1 - \alpha_2$:

$$\alpha_3 = (1-\mu)\left(1 + \frac{(k_1-1)^2}{T}\right).$$

Let's estimate how the impact of model parameters has been changed comparing with the basic model. Pluses, minuses, arrows, and points in the table 2 mean the same. Let's also append the line for education costs T and three columns for endogenous shares α_1, α_2, α_3 of each sector. The grey cells – are the ones where we can see the changed form of dependence.

The key new results of the model with education are the following:

1. The increasing heterogeneity of industry leads to the higher wages in the effective sector and to increasing qualification of workers via education.

2. The increasing independence of goods leads to the lower wages in the effective sector and decreasing workers studying.

3. The increasing demand on industrial goods doesn't make higher prices and wages.

Table 2. The impact of the model with education parameters on the equilibrium.

	p_1	p_2	q_1	q_2	TR_1	TR_2	w_1	n_1	n_2	l	α_1	α_2	α_3
f	•	•	+	+	+	+	•	–	–	+	•	•	•
c	+	+	–	–	•	•	•	•	•	•	•	•	•
L	•	•	•	•	•	•	•	+	+	•	•	•	•
γ	↑	+	•	–	↑	•	↑	↑	↓	•	↑	↓	↑
σ	↑↓	↓	↑	↑	↑	+	↑	↑↓	↓	+	↑	↓	↑
T	•	•	•	•	•	•	•	–	↑	•	–	↑	↓
μ	•	•	•	•	•	•	•	•	↑	•	•	↑	↓

5 The generalized model

The constructed above basic model has a lot of modifications and generalizations. Particularly it's possible to investigate the situation with any numbers of productivity classes include continuous distribution of workers qualification.

Let the number of workers essential for the unit production at i-firm depends on the best worker productivity c and expressed by the following formula:

$$c(i) = \gamma(i)c, \;\; \gamma(i) \in [1; \bar{\gamma}].$$

Hence the wage level of *i*-firm worker is connected with the best worker wage *w* and equal to

$$w(i) = (\gamma(i))^{-(\sigma-1)/\sigma} w.$$

In the simplest case of uniform distribution the worker qualification can be calculated as follows:

$$\gamma(i) = 1 + \frac{i}{n}(\bar{\gamma} - 1),$$

and the prices, quantities, firm size, number of firms are obtained using the formulas

$$p(i) = \frac{\sigma}{\sigma - 1} w(i)c(i), \quad q(i) = \frac{f(\sigma - 1)}{c(i)},$$

$$l(i) = f\sigma, \quad n = \alpha L / f\sigma.$$

The total wage of industrial sector workers is expressed by the integral

$$W = \int_0^n l(i)w(i)di = \frac{\alpha L}{n} w \int_0^n \left(1 + \frac{i}{n}(\bar{\gamma} - 1)\right)^{-(\sigma-1)/\sigma} di,$$

$$W = \alpha L w \sigma \frac{\bar{\gamma}^{1/\sigma} - 1}{\bar{\gamma} - 1}.$$

From the supply and demand balance in agriculture we can find the equilibrium wage level:

$$(1 - \alpha)L = (1 - \mu)\left(\alpha L w \sigma \frac{\bar{\gamma}^{1/\sigma} - 1}{\gamma - 1} + (1 - \alpha)L\right),$$

$$w = \frac{\mu}{1 - \mu} \frac{1 - \alpha}{\alpha\sigma} \frac{\bar{\gamma} - 1}{\bar{\gamma}^{1/\sigma} - 1}.$$

6 The Generalized Model with Education

The proposed in the previous section generalized model was used for more generalized analysis of education and migration [20]. Let's make the following assumptions:

1. Education has multistep structure – from initial level to the best universities.

2. The education of high quality increases labor productivity, educational attainment is connected with costs of production by decreasing linear dependence (fig.2).

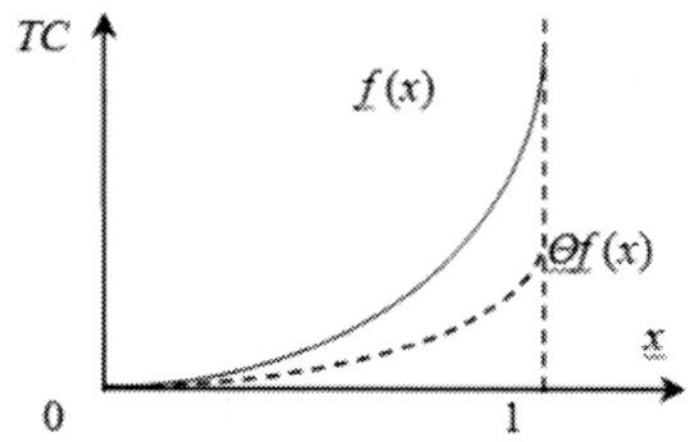

Fig. 2. Educational attainment and costs of education.

3. The education is charged, educational costs are proportional to the "educational skills" θ and they are given by power function $f(x)=Tx^n$, where x is a parameter of educational attainment which is between 0 (with no education) and 1 (the maximum of development) (fig.3).

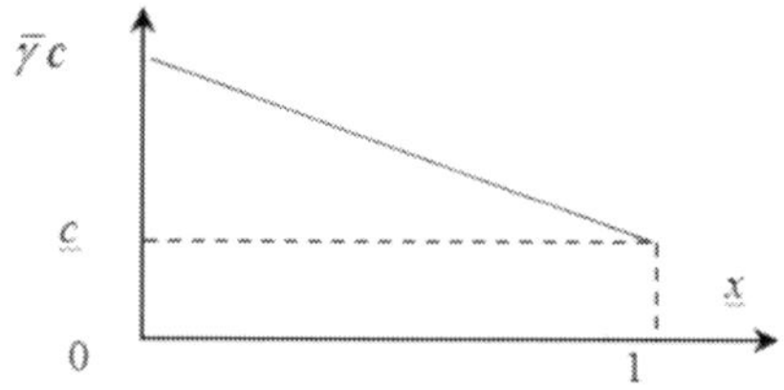

Fig. 2. Educational attainment and productivity of labor.

So, we assume that every additional level of education requires higher increasing of cost incurred and we determine the function of total educational costs as $\theta f(x)$ for taking into account "educational skills".

In this model every person chooses an optimal educational level according to his own skills by focusing on the maximal difference between additional benefits, which are gained after moving to the category of more qualified workers with a higher wage, and costs incurred during the education. Profit can be presented as a financial premium which worker receives having finished study successfully and moved to the necessary sector. Thus, it can be presented in the following way:

$$F = w(c(x)) - w(0) - \theta f(x)$$

which depends on $x(\theta)$.

It has been possible to prove that when costs of education are linear, the results will be equal to the ones received in the basic model. The most gifted students receive elite education and move to the class of high-skilled labor. Untalented ones refuse to obtain education completely and remain in the class on non-skilled workers. Intermediate level of education is not demanded that can be demonstrated by a numerical example with $\rho = 3/4$, $n = 1$, $T = 10$, $\bar{\gamma} = 8$.

Relations between additional incomes which has been gained as a result of training for people with low, below the average, average, and high inclination to education are denoted by letters *a, b, c, d*. It is obvious that the more worker with low ability studies the bigger his losses are. It is explained by high costs of education which cannot cover estimated profit of higher wage. So, he prefers not to enter the educational establishment (fig.4*a*).

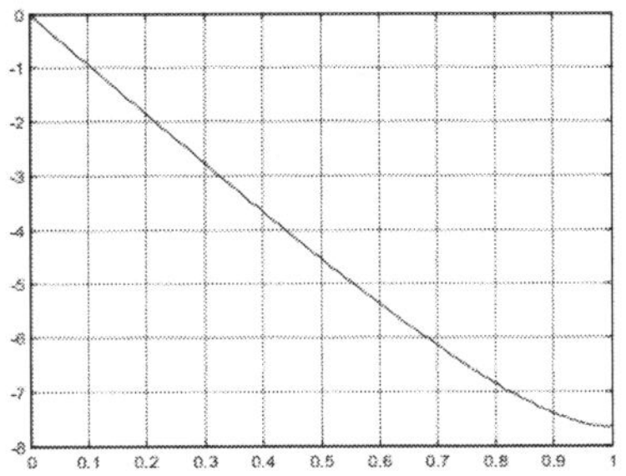

Fig. 4a. Example for "alternatively gifted" person, θ=1, x^*=0.

In case of worker with under-average ability at the beginning of education there are only losses as well but after a critical point profit trend becomes positive but final profit is not enough to cover cost incurred (fig.4*b*). As a result of this process worker makes a decision on refusing to be a student and chooses x^*=0.

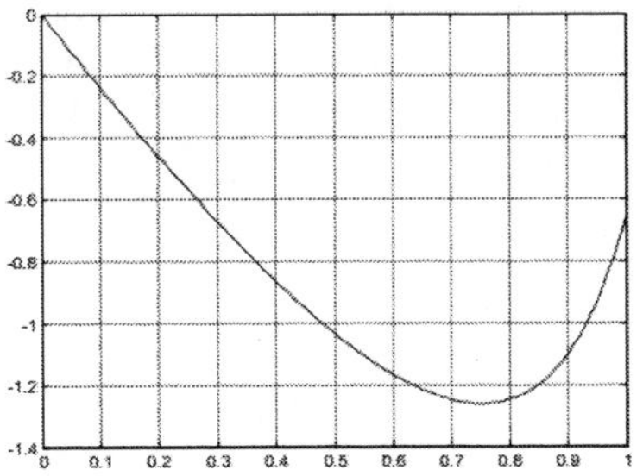

Fig. 4b. Example for medium person, θ=0,3, x^*=0.

If we concern the gifted worker it can be observed that his function of profit increases due to the higher quality of education. So, his benefits are maximal at the point x^*=0 (fig.4*d*). Figure 4*c* illustrates an intermediate case of profound indifference between these extremes.

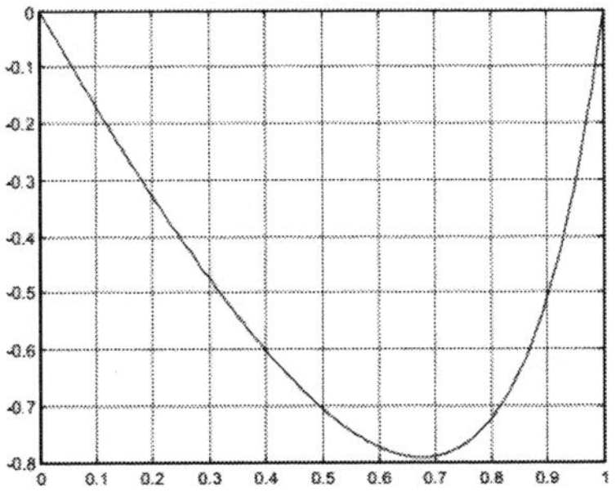

Fig. 4c. Example for talented person, θ=0,234, x^*=0 or x^*=1.

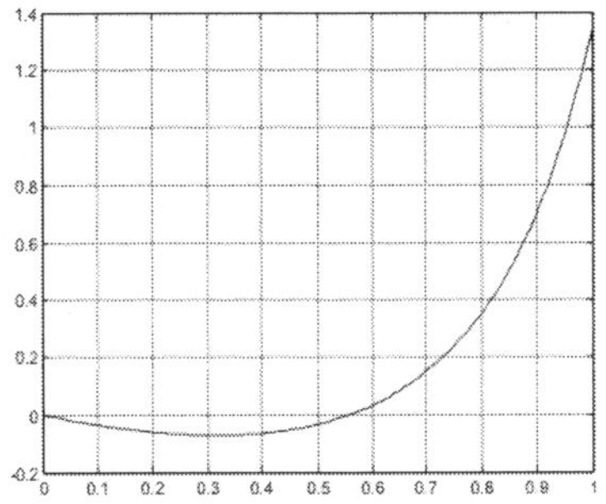

Fig. 4d. Example for genius, θ=0,1, x^*=1.

There is no reason to obtain an education for the first two students with lower skills while the most gifted one should seek for studies at the best expensive universities. As far as we can see, there is no any interim decision in case of linear costs: worker either obtains the best education or continues working without studying.

Other parameter points of educational cost function result in different conclusions. For instance, intermediate level standard of education may be optimal in case of quadratic function. Moreover, some minimal education is needed by everybody. We demonstrate results for numerical example with $\rho = 1/2$, $n = 2$, $T = 10$, $\bar{\gamma} = 8$. Just the same graphics (fig.5) are denoted by letters *a*, *b*, *c*, *d* corresponding with different level of talent for education.

The impact of other model parameters (ratio between fixed and variable costs, economical extent, and elasticity of substitution and so on) on received equilibrium has been also investigated.

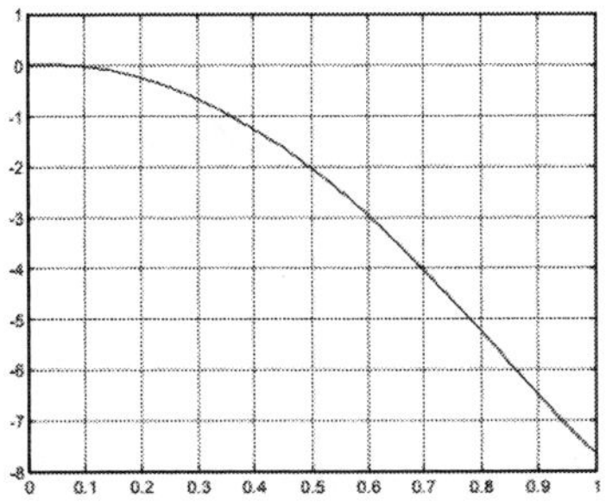

Fig. 5a. Example for "alternatively gifted", θ=1, x^*=0.

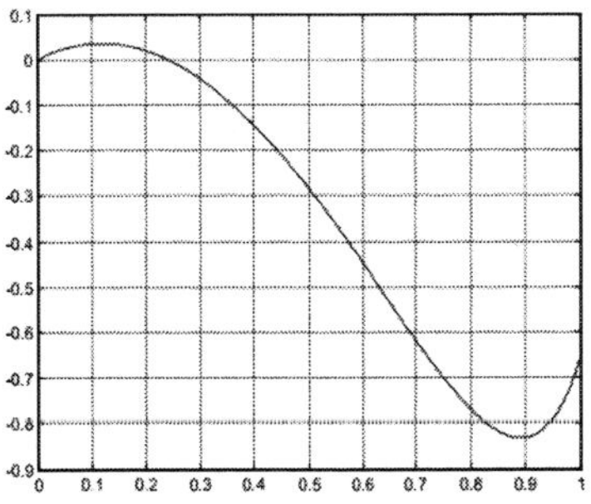

Fig. 5b. Example for medium person, θ=0,3, x^*=0,1.

DOI: 10.1051/shsconf/20173501071

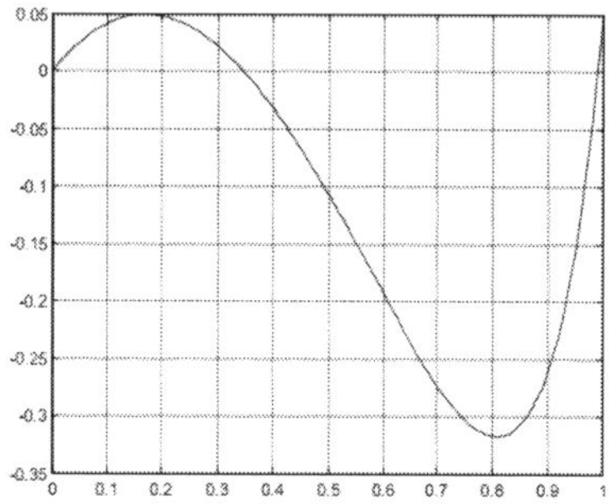

Fig. 5c. Example for talented person, θ=0,229, x^*=0,15 or x^*=1.

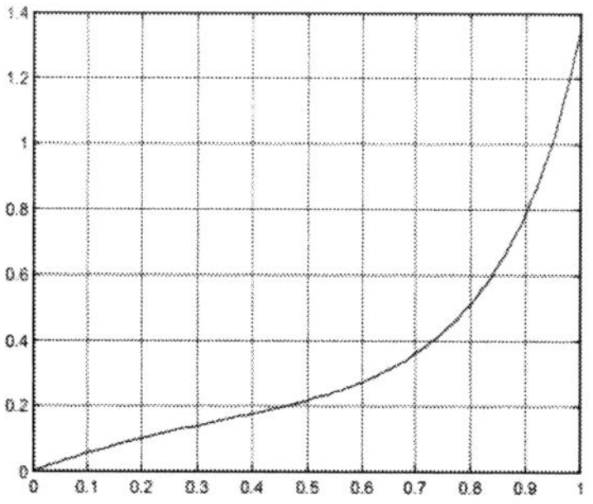

Fig. 5d. Example for genius, θ=0,1, x^*=1.

7 The International Trade Model

The constructed tools for market analysis (especially generalized model from the section 5) can be also applied to the other fields of economics, particularly to the theory of international or interregional trade. For example, let's investigate the simplest situation of two countries. It's interesting to compare autarky (caused by protectionism tariff police or economic sanctions) with the case of free trade, zero tariffs, and zero transportation costs. Let the workers qualification be distributed as it's shown on the fig.6-7. The first picture describes the distribution of workers' qualification before integration, and the second one in the situation of the joint market of two countries – developed with lower marginal costs and developing with higher ones:

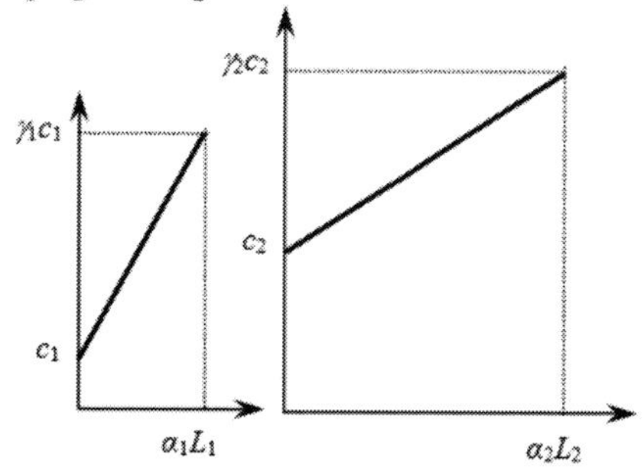

Fig.6. The distribution of workers' qualification before integration.

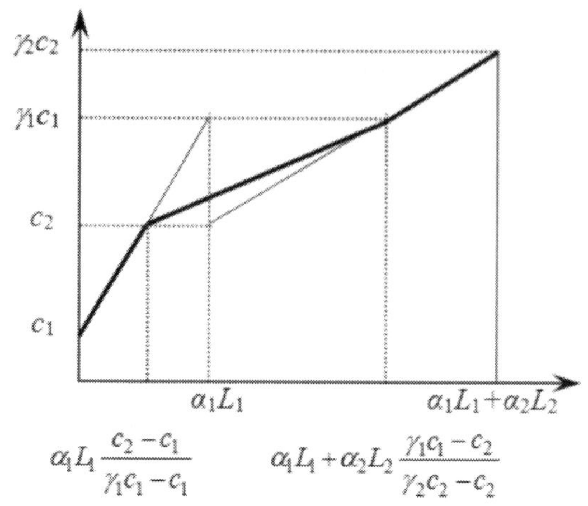

Fig.7. The distribution of workers' qualification after integration.

Using this presented above model it's possible to test some hypotheses. Particularly we showed in [21] that opening borders always expands the variety of goods, increases welfare, converge prices, but sometimes leads to de-industrialization in the country with less effective manufacture sector. It also can be explored the model with transportation cost and tariffs. It's possible to compare equilibriums, obtained in the models with specific and ad valorem tariffs, and give appropriate recommendations.

8 Conclusions

The presented tool due to its simplicity and possibility to obtain non-trivial analytical results can be used for solution different problems and giving policy advices for both firms and market regulator. It's also possible to use it for empirical analysis of the data described monopolistic competition markets of different goods in different regions and countries, making forecasts of such markets dynamics, and comparing different mechanisms of regulation.

This research is supported by the RFBR grant № 16-06-00071, and the European Union's Seventh Framework Program FP7/2007-2013/ under REA grant agreement number 609642.

References

1. E. Chamberlin, *The Theory of Monopolistic Competition* (Harv. Univ. Press, Cambridge, 1933)
2. A. Dixit, J. Stiglitz, AER, **67**, 297-308 (1977)
3. P. Krugman, J. Int. Econ., **9**, 469-479 (1979)
4. P. Krugman, AER, **70**, 950-959 (1980)
5. D. Ricardo, *On the principles of political economy and taxation* (John Murray, London, 1817)
6. B. Ohlin, *Interregional and International Trade* (Harv. Univ. Press, Cambridge, 1968)
7. A. Bernard, B. Jensen, J. Int. Econ., **42**, 3-31 (1997)

8. A. Bernard, B. Jensen, J. Int. Econ., **47**, 1-25 (1999)
9. M. Melitz, Econometrica, **71**, 1695-1725 (2003)
10. H. Hopenhayn, Econometrica, **60**, 1127-1150 (1992)
11. B. Jovanovic, Econometrica, **50**, 649-670 (1982)
12. P. Manasse, A. Turrini, J. Int. Econ., **54**, 97-117 (2001)
13. A. Bartel, F.Lichtenberg, Rev. Econ Stat., **69**, 1-11 (1987)
14. A. Bartel, N. Sicheman, JPE, **107**, 285-325 (1999)
15. A. Bernard, J. Eaton, B. Jensen, S. Kortum, AER, **94**, 1265-1290 (2003)
16. S. Yeaple, J. Int. Econ., **65**, 1-20 (2005)
17. E. Helpman, O. Itskhoki, S. Redding, Econometrica, **78**, 1239-1283 (2010)
18. A. Filatov, Y. Sokolovsky, *Proc. XIV April Int. Acad. Conf. on Econ. and Soc. Development*, **1**, 447-460
19. L. Echazu, V. Lugovskyy (Working Paper, 2007).
20. A. Filatov, Y. Sokolovsky, *Proc. XVI April Int. Acad. Conf. on Econ. and Soc. Development*, **1**, 548-557 (2016)
21. A. Filatov, Y. Sokolovsky, *Scientific Enquiry in the Contemporary World: Theor. Basics and Innov. Appr.*, 42-46 (2012)

Development of algorithm for depreciation costs allocation in dynamic input-output industrial enterprise model

Alevtina Keller[1], and *Tatyana* Vinogradova[1,*]

[1]South Ural State University, 454080 Chelyabinsk, Russia

Abstract. The article considers the issue of allocation of depreciation costs in the dynamic input-output model of an industrial enterprise. Accounting the depreciation costs in such a model improves the policy of fixed assets management. It is particularly relevant to develop the algorithm for the allocation of depreciation costs in the construction of dynamic input-output model of an industrial enterprise, since such enterprises have a significant amount of fixed assets. Implementation of terms of the adequacy of such an algorithm itself allows: evaluating the appropriateness of investments in fixed assets, studying the final financial results of an industrial enterprise, depending on management decisions in the depreciation policy. It is necessary to note that the model in question for the enterprise is always degenerate. It is caused by the presence of zero rows in the matrix of capital expenditures by lines of structural elements unable to generate fixed assets (part of the service units, households, corporate consumers). The paper presents the algorithm for the allocation of depreciation costs for the model. This algorithm was developed by the authors and served as the basis for further development of the flowchart for subsequent implementation with use of software. The construction of such algorithm and its use for dynamic input-output models of industrial enterprises is actualized by international acceptance of the effectiveness of the use of input-output models for national and regional economic systems. This is what allows us to consider that the solutions discussed in the article are of interest to economists of various industrial enterprises.

Introduction

First pointed to the possibility of building an input-output model of the enterprise by the founder of the balance method – W.W. Leontief at work [1]. Building of the input-output model of the enterprise became possible by: a) accumulation of theoretical and practical basis for the study of the models of macro and middle levels, both in Russia and abroad; b) the development of mathematical methods that allow solving degenerate input-output models in Russian schools (including in the Chelyabinsk school: G.A. Sviridyuk [2, 3], S.V. Brychev [4], A.V. Keller [5-7] and etc.)

The common form of the dynamic input-output model of an industrial enterprise is represented by the following equation:

$$x(t) = Ax(t) + B\dot{x}(t) + f(t), \qquad (1)$$

where: A, B are square matrices of size n; A – matrix of specific direct costs; B – matrix of specific capital expenditures; $X(t)$ is the vector of gross output; $f(t)$ is a vector of final consumption [8, 9].

At the enterprise level, the dynamic input-output model will always be degenerate, because in the matrix B there will always be zero rows ($\det B = 0$) on such structural elements of the model as households and corporate consumers.

Using a degenerate dynamic input-output model (DDIOM) for an industrial enterprise will allow, based on the analysis of economic interrelations between the structural units of the company, to plan the dynamics of the growth of gross output of products for the future, including, taking into account investments in fixed assets.

When building the DDIOM for an industrial enterprise the actual issue is the inclusion of depreciation costs in the structural elements of the model. Today is not exist the methodology and algorithm for allocating such costs in the model for an industrial enterprise (the conclusion is made based on the information known to the authors). The purpose of this paper is to construct an algorithm for the allocation of depreciation costs in the DDIOM. Let's note, that an adequate allocation of depreciation costs in the model will allow to implement a competent policy of management of fixed assets. This is relevant for industrial enterprises.

1 Degenerate dynamic input-output model of enterprise and the method of its construction

The study of the construction of balance models for the enterprise is carried out in the works of the following authors: D.V. Ruzakov and P.T. Voronkov [10], P.I. Kuzmin [11], M.I. Letavin and A.A. Banin [12] and others. These researchers combine a unified approach to determining the structural elements of the model, into which they include only the production units, which leads to the construction of static or non-degenerate dynamic input-output models. In addition, in these works, the features of constructing such models are

* Corresponding author: sh.t.a_90@mail.ru

considered only for specific list of companies of a particular type of activity. In our opinion, the use of such an approach to constructing a model with industry specificity limits its application and reduces the degree of adequacy of mathematical modeling of the company's economic activity. The noted limitation allows to eliminate the construction of the DDIOM by the author's method (A.V. Keller and T.A. Vinogradova). In detail this method is presented in [13].

In the construction of the DDIOM of an industrial enterprise to include following structural elements are to be singled out: a) production units; b) service units; c) for the purposes of accounting for the distribution of net profit, as well as changes in the company's own capital, additional conditional units are «allocated equity capital»; d) external consumers of the company's final products: households (in the form of labor of the company's personnel) and corporate consumers (counterparties). Such a combination of structural elements of the model will allow, on the one hand, to take into account the company's economic relations like an input-output model of the country's economy taking into account export and import, and on the other, opens up new possibilities for researching the economic activity of the enterprise.

Note that TA. Vinogradova developed the algorithm of the program, which will automatically build the matrix that is part of the DDIOM on the basis of the accounting data of the industrial enterprise. This algorithm is presented in [14].

In the Russian chart of accounts only such records of depreciation costs are used, such as «02» (fixed assets) and «05» (intangible assets) [15]. At the same time, the purposes of investment of fixed assets or intangible assets may be different, and the economic effect of putting assets into operation in an industrial enterprise may be different. Thus, for the purpose of increasing the effectiveness of managerial decisions made as a result of the use of the DDIOM, it is necessary to take into account the particularity and specificity of the policy of managing of assets. In this connection, the task is to develop methods and, on their basis, algorithms for distributing the depreciation costs of industrial enterprises, which would take into account: the objectives of investment in fixed assets (including their location); the presence of an economic effect from putting into operation (in the form of an increase in X); level of significance of the fixed asset (when making managerial decisions).

2 Methods and algorithm of allocation of depreciation costs in the DDIOM of an industrial enterprise

Depreciation costs are inherent costs of the enterprise, with the help of which it becomes possible to transfer the amount of investments into non-current assets for the purchase of fixed assets and intangible assets in parts by the cost of finished products [16].

The essence of the author's approach to the allocation of such costs in DDIOM is as follows:

a) it is necessary to allocate two groups of fixed assets (NCA) of the enterprise, depending on the presence or absence of growth in gross output when putting it into operation. So, for example, the purchase of a fixed asset to replace of withdrawing asset can only ensure that the output of finished goods is maintained at the same level. Accordingly, such costs in the future should be included in the matrix of direct costs. If the commissioning of the NCA resulted in a growth of a gross output, then the depreciation costs of it are allocated in the capital expenditures matrix.

b) in the matrices A and B themselves, depreciation costs are proposed to allocate on the main diagonal (i = j), depending on the i-th units ($i=\overline{1,n}$), where the NCA will be put into operation.

The algorithm for allocating depreciation costs in the model consists of four stages.

1st stage – Distribution of depreciation costs for the base period – on the first date of the reporting period (for example, 01/01/2016):

1.1. For each NCA it is necessary to determine:

- the presence of an increase in X in the j-th units due to the commissioning of the NCA in the i-th units. Further, for the sake of clarity of the representation of this dependence, we introduce the indicator ER. And for such an algorithm, if there is no increase in gross output, then $ER = 0$, and $ER = 1$ otherwise.

- amount of depreciation costs ($\mathrm{DC}_{\mathrm{p}}, \mathrm{p}_i = \overline{1, \mathrm{b}_i}$) for the base period for each NCA;

- location of the NCA (i-th unit).

1.2. A group of significant NCAs is determined.

The degree of significance (S) of assets is determined by the company independently and is prescribed in its accounting policy. The indicator S is characterized by the degree of influence on the user decisions of owners, founders, shareholders and other interested parties. By analogy with the indicator ER, the value of the indicator $S = 1$ means that the NCA is recognized as significant, and $S = 0$ for otherwise.

1.3. For each NCA, which has the values of $ER = 1$ and $S = 1$, it is also determined:

- indicator G_u (the share of gross output growth for the entire period of its useful life in percentages, where u is the ordinal number of years of useful life of the NCA).

After that, indicator $I_{dc}0$ is calculated (the index of accounting for the distribution of depreciation cost by years of useful life for the base period using the formula:

$$I_{dc}0 = \frac{G_u}{100} \qquad (2)$$

- indicator ERJ (describes the share of the increase in gross output in the j-th units due to the commissioning of the NCA in the i-th unit, at that $i \neq j$).

1.4. Based on the ER indicator, the total depreciation cost values for each NCA are distributed between the types of cost (direct and capital) according to the following scheme: if ER = 0, then such costs are referred to direct, if ER = 1, to capital.

With respect to the NCAs with indicators $S = 1$, $ER = 1$ and a disproportionate coefficient G_u distribution of depreciation costs on them in the accounting period occurs depending on the share of growth in gross output for a given period. For all other NCAs, the actual amount of accrued depreciation is used for calculations.

In the repeated construction of the model, the implementation of point 1.4 is necessary only in case of re-qualification from direct costs to capital type and otherwise.

2nd stage – Correction of depreciation values for NCAs, which are registered with the company on January 1 of the current year.

2.1. Based on the results of the annual revaluation of non-current assets, it is necessary to determine:

- list of NCAs for which the useful life was changed;
- the amount of the residual value for them;
- the amount of increase or decrease in the amount of planned depreciation costs;
- planned values of *ER* for all NCAs, depreciation of which were related to capital expenditures.

2.2. Based on the data from point 2.1., it is necessary to:

- change in depreciation values (for the reporting period) with respect to increasing or decreasing its amounts as of the end of the base period for each type of cost;
- analyze the need to reallocate amounts of depreciation between two types of cost. If for some NCA the economic effect became zero (*ER* = 0), then depreciation on it in the planned period will be a reallocated into direct costs and otherwise to capital expenditures (for *ER* = 1).

3rd stage – Allocation of depreciation costs for new NCAs in the reporting period.

3.1. Based on the planned budget investment in non-current assets for the reporting period is determined: the quantity of the NCAs that are planned to be recorded, their useful life; the future location; the amount of annual depreciation; the value of the ER.

3.2. The group of acquired NCAs is determined, for which ERG = 1.

3.3. For significant NCAs, that having the value of *ER* = 1, the following indicators are additionally determined: G_u and *ERJ* (see 1.3).

3.4. The distribution of depreciation costs for new NCAs between the types of costs is carried out (similar to point 1.4.).

4th stage – Final allocation of depreciation costs in the model for the reporting period.

4.1 At this stage, the total amount of depreciation costs is finally allocated among the matrixes of the model based on the data of the previous stages. Depreciation, referred to direct costs, is allocated in the matrix of direct costs along the main diagonal. Similarly, the costs are allocated in the capital expenditures matrix.

4.2. Exclusion from the rules of point 4.1: if there are significant NCAs in the company for which *ER* = 1 and *ERJ* = 1, changes are made to the order of allocation of costs. For such assets, it is necessary to analyze the presence in the matrix of direct costs of the relationships between these units. In the case when there are no such interrelations, the depreciation costs for them will be taken into account on the *i*-th line depending on the future location of the asset, on *j*-th line, depending on the units in which the growth of gross output is expected.

In the first construction of the DDIOM, the sequence of the stages of matrices compilation to concerning of depreciation costs: 1.1 → 1.2 →1.3 → 4.1 → 4.2. For this process, TA. Vinogradova developed a flowchart (Fig. 1).

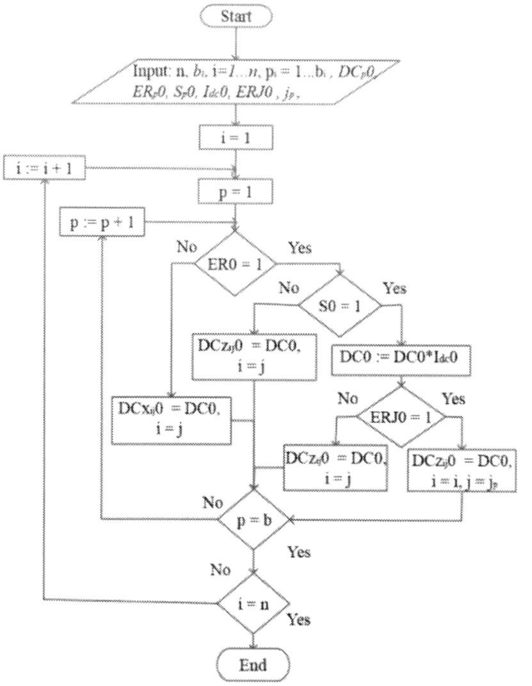

Fig. 1. Flowchart of the allocation of depreciation costs in the construction of the model for the base period.

In repeated the model for subsequent time periods, all the above steps are performed in strict sequence (Fig. 2).

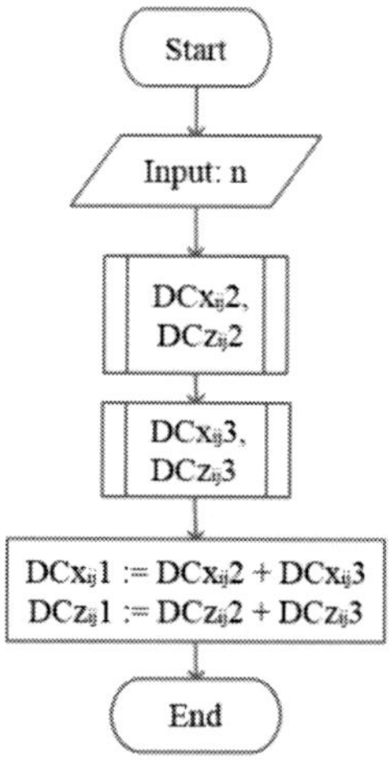

Fig. 2. Flowchart of the allocation of depreciation costs in the construction of the model for the reporting period.

Calculation $DCx_{ij}2$ and $DCz_{ij}2$ is carried out in a similar way to the flowchart (Fig. 1), when the corresponding indices change 0 to 2. The matrices $DCx_{ij}3$ and $DCz_{ij}3$ are compiled to the following flowchart (Fig. 3).

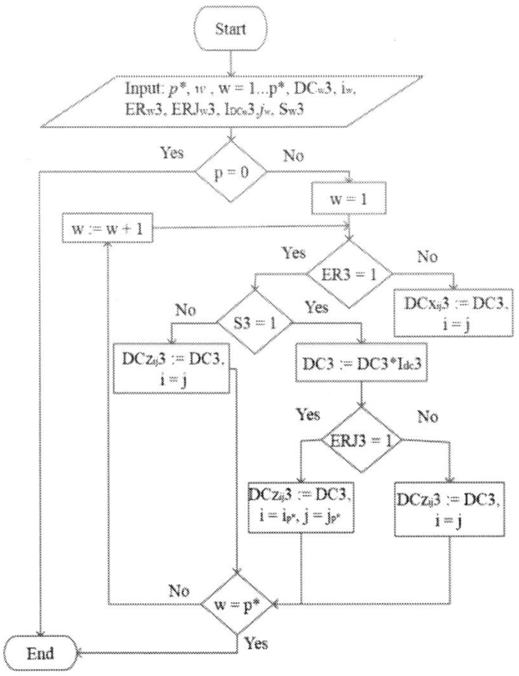

Fig. 3. Allocation of depreciation costs for news NCA.

Let's explain some abbreviations in flowcharts:

- b_i is the NCAs quantity for each i-th structural element of the model, $i = 1 \dots n$;

- j_p is the number of the column in the location of depreciation costs for those NCA for which S and ER = 1;

- p^* is the number of newly acquired NCA that are registered in the reporting period.

3 Advantages of using the proposed method

The use of the author's algorithm creates the possibility of a systematic approach to the management of NCAs of an industrial enterprise.

In the scientific literature, the issue of evaluating the effectiveness of the policy of NCAs management is given to great attention. It is studied in [17, 18, etc.]. In this area, a large number of issues are considered. Describe some of them.

1. Evaluation and validity of the feasibility of investment in the NCAs. This question was considered in [19, 20, etc.].

These authors have similar methods for evaluating the feasibility of investing in the assets. They are based on the assessment of the resulting economic effect by calculating indicators by the payback period of investments, their profitability, the yield index, the calculation of return on assets, and so on. In these methods, the conclusions are based only on generalized parameters, data and representations on the volume of the production capacity of new equipment (or other NCAs), the amount of costs, the estimated income, taking into account the discount rate. At the same time, the whole set of processes occurring in the company and the structural interrelations between the divisions is not taken into account.

The proposed DDIOM for an industrial enterprise will allow to assess the impact of a change in the structure of NCAs on the value of gross output of all structural elements of the model. Thus, this makes it possible to assess the feasibility of investments for the purchase, renovation, etc. NCAs.

Also, within the framework of standard methods for assessing the feasibility of investments in the NCAs and the need for them, it is not possible to determine the need (and volume) of additional investments in other interrelated units of the company. This is especially relevant when assessing the need for additional investments in servicing units. This problem is solved using the model developed by the authors.

2. The effects of the NCAs revaluation, the extension of useful life on the financial result of the enterprise. These issues relate to such a subsection of the management of the NCAs as an amortization policy [21]. The question is presented in detail in [22, 23, 24, etc.]. When using the proposed DDIOM and the methodology for allocating depreciation costs, it becomes possible to assess such influence.

Conclusion

In this paper, the algorithm is proposed for the allocation of depreciation costs in the DDIOM of an industrial enterprise. Briefly list the conclusion:

- the authors present the essence of the methodology of allocation of depreciation costs and the algorithm on its basis in the model of an industrial enterprise;

- the algorithm of allocation of depreciation costs, consisting of four stages, is considered in detail;

- a sequence of actions is indicated at the first construction of the model and its compilation for subsequent time periods;

- for each the sequence are presented flowchart of the algorithm for calculating auxiliary matrices for an accounting of the depreciation cost in the model.

The proposed methodology and algorithm for allocation of depreciation costs in the DDIOM of an industrial enterprise is aimed at developing an input-output method at the micro level. The use of the methodology will make it possible to increase the effectiveness of making managerial decisions in the implementation of the policy of managing NCA in an industrial enterprise.

The work was supported by Act 211 Government of the Russian Federation, contract № 02.A03.21.0011.

References

1. W.W. Leontief, *The Structure of the American Economy* (Oxford Univesity Press, 1941)
2. G.A. Sviridyuk, S.V. Brychev, JVMMF, **11** (2003)
3. G.A. Sviridyuk, S.V. Brychev, News of HEE: Math, **8** (2003)
4. S.V. Brychev, *A study of the mathematical model of the economy of the municipal economy of small towns*, 112 (2002)
5. A.V. Keller, Herald of SUSU, **4** (2011)
6. A.V. Keller, Journal of CEM, **2** (2015)
7. A.V. Keller, *Numerical study of optimal control problems for Leontief type models: the thesis*, 249 (2011)
8. A.G. Granberg, *Dynamic models of the national economy* (Economics, 1985)
9. N.I. Veduta, Socially Efficient Economy, 254 (1999)
10. P.T. Voronkov, D.V. Ruzakov, Herald of MSU, **4** (2001)
11. P.I. Kuzmin, Herald of ASU. Economy, **2** (2002)
12. A.A. Banin, M.I. Letavin, EMM, **4** (2002)
13. A.V. Keller, T.A. Shishkina, Herald of SUSU, **3(7)** (2013)
14. A.V. Keller, T.A. Shishkina, Modern economy: problems and solutions, **9** (2014)
15. D. Volkova, Chart of accounts of financial and economic activity of the enterprise, 122 (2015)
16. A.A. Slobodnyak, I.O. Tsizman, O.A. Prein, Accounting and economic analysis of fixed assets in organizations and institutions, 362 (2014)
17. A.P. Kopteva, RE, **10** (2011)
18. V.A. Fursov, A.A. Solomatina, KANT, **3** (2012)
19. M.E. Grebneva, V.I. Kosik, The role of economic analysis in developing the policy of management of non-current assets, **2** (2016)
20. T.M. Vasilkova, Herald of the MSUM, **5** (2008)
21. E.V. Smolentsev, Depreciation policy as a stage of asset management, **4** (2014)
22. K.V. Gulpenko, PME, **4** (2009)
23. M.L. Pustovalova, Effect of revaluation of fixed assets on return on equity, EB, **1** (2016)
24. I.O. Yursova, I.O. Yakubova, S.M. Baysangurova, IPE, **8** (2011)

The increasing concentration at industrial markets: the social welfare maximization and possible risks

Alexander Filatov[1,2*], and *Yana* Makolskaya[2]

[1]Far Eastern Federal University. School of Economics and Management, Vladivostok, Russia
[2]Irkutsk State University, Institute of Mathematics, Economics, and Informatics, Irkutsk, Russia

Abstract. The paper considers impact of entry barriers on the social welfare. Despite the common opinion that entry barriers are always bad, the excessive number of firms means, all pros aside, duplicated fixed costs. It is shown that the socially effective number of firms is smaller than the equilibrium one for the wide spectre of demand and cost functions, and also for different strategies of companies' behavior. This proposition is satisfied for the homogeneous product markets where output of each company decreases when the number of firms increases, and competition gets stronger. But there is the considerable danger of the increasing probability of collusion in a situation of number of firms limitation. We show that collusion is less dangerous than duplicated fixed costs if the gap between the «choke price» and marginal costs is less than a certain critical value connected with the share of fixed costs. The empirical research on the base of the financial statistics of the biggest world corporations of different industries is carried out.

1 Introduction

Equilibrium states in economy are often socially ineffective. We can refer to different examples of the prisoner's dilemma, the tragedy of commons, insufficient financing of public good, negative selection moral hazard, and other problems of the contract theory. In all of them we need a special market design helps to decrease inefficiency. But we always should compare market failures and inefficiency of regulation taking into account the possible risks of agents' behavior change.

In the paper we will concentrate on the problems of market power. Using theoretical models and empirical data we will answer the following three questions: is there excessive or insufficient number of firms at a market in equilibrium, should we use regulation and what are the possible risks?

Most industries nowadays are high-concentrated markets where every producer maximizes profit taking into account a wide spectre of the competitors' behavior strategies from price wars to collusion. A lot of books and papers are devoted to the different aspects of such oligopoly markets. Among them we would emphasize classical monographs of Tirole [1], Shy [2], Carlton and Perloff [3], and the new standard of studying industrial organization, suggested by Belleflamme and Peitz [4]. We also made a survey [5] of different oligopoly models.

The main feature of oligopoly is a small number of strategic firms interacted at a market. But there is a significant difference between duopoly and oligopoly with, for example, ten firms. So the best strategies, equilibrium price and quantity, social welfare, and deadweight loss are strongly dependent on market concentration, and entry barriers created by the incumbent firms on the base of absolute cost advantages, increased returns to scale, access to resources and technologies limitation and government through the system of licenses and permissions for doing business.

There is a common opinion that entry barriers are always bad for society, because they decrease the number of firms, weaken competition, raise prices and decrease quantities. However we should understand that a lot of firms also mean duplicated fixed costs. Thus, free entry can lead to both situations: excessive and insufficient number of firms in equilibrium. The correct conclusion depends on the market structure, demand and cost functions, and features of strategic interaction between companies at a market.

It was shown [6;7] insufficient number of firms in equilibrium at monopolistic competition markets with differentiated product. It means that entry barriers at these markets are always bad. Particularly it is connected with consumers love for variety. But the opposite situation arises at homogeneous product markets, often having oligopoly structure. Von Weizacker [8] and Perry [9] proposed reasons for the desired limitation of excessive competition. Mankiw and Whinston [10] formulated some conditions (usually, though not always satisfied) under which this reduction would lead to the social welfare maximization.

At the same time even many empirical papers – for radio broadcasting [11], real-estate [12], investment foundations [13] cinema industry [14], etc. didn't convince society that entry barriers and reducing the

* Corresponding author: alexander.filatov@gmail.com

number of firms at a market may be useful. This is partly connected with misconceptions initiated by the basic models of perfect competition where zero fixed costs lead to a definite conclusion «more firms is always better». Partly – with the really existing risks of possible modification of companies' behavior or welfare redistribution within society that can be unfavorable for the majority even when the total welfare increases. Let's consider some of these problems in the paper.

2 The model

Let's consider a homogeneous product market. We assume that there is monotonically declining inverse demand function $p = D^{-1}(Q)$. The cost functions of n identical oligopolists are the same: $TC(q) = f + VC(q)$. Taking into account the number of firms and their interaction strategies we can find quantities $q(n)$ and a market price $p(n) = D^{-1}(nq(n))$.

The social welfare function consists of consumer surplus (integrated spread between the maximum willingness to pay given in the form of inverse demand function, and the current market price), and total profit of firms $SW = CS + n\pi(n)$:

$$SW = \int_0^{nq(n)} D^{-1}(q)dq - n(f + VC(q(n))).$$

If there are no entry barriers at a market, the equilibrium number of firms can be obtained from the zero profit condition

$$\pi(n_1) = p(n_1)q(n_1) - f - VC(q(n_1)) = 0.$$

Let's note that it's not just a theoretical fact. Bresnahan and Reiss on the basis of numerous data of the US food markets with fluctuating demand showed [15] that entry and competition really happens under such conditions.

Now it's possible to estimate how the number of firms impacts social welfare:

$$\frac{\partial(SW)}{\partial n} = \frac{\partial\left(\int_0^{nq(n)} D^{-1}(q)dq\right)}{\partial n} - f - \frac{\partial(nVC(q(n)))}{\partial n} =$$

$$= p(n)\left(q(n) + n\frac{\partial q(n)}{\partial n}\right) - f - VC(q(n)) - nMC(q(n))\frac{\partial q(n)}{\partial n} =$$

$$= \pi(n) + n(p(n) - MC(q(n)))\frac{\partial q(n)}{\partial n}.$$

As noted above, the first term in this equation is equal to zero in equilibrium. Let's also assume that the firm quantities decrease in number of number of firms due to business-stealing effect

$$\frac{\partial q(n)}{\partial n} < 0, \qquad (1)$$

and the price exceeds marginal costs of production

$$p(n) - MC(q(n)) > 0, \qquad (2)$$

Then the second term and the whole derivative of social welfare at the equilibrium number of firms are negative:

$$\frac{\partial(SW)}{\partial n} < 0$$

It means that for social welfare maximization the planner should decrease the number of firms with respect to their equilibrium quantity.

3 Counterexample and discussion

For most demand and cost functions, and strategies of firms' behavior the inequalities (1)-(2) are satisfied. But it's difficult to check the first one even when we know everything about the market. For example assumption (1) is not satisfied for simple case of Cournot model with inverse proportionality demand function and linear costs:

$$p = \frac{a}{Q} = \frac{a}{q_1 + ... + q_n}, \quad TC_i(q_i) = cq_i + f.$$

Let each producer maximize profit

$$\pi_i = pq_i - cq_i - f = \frac{aq_i}{q_1 + ... + q_n} - cq_i - f \to \max.$$

Solving this problem for all symmetric firms we can find equilibrium prices and quantities:

$$q^* = \frac{a(n-1)}{n^2 c}, \quad Q^* = \frac{a(n-1)}{nc}, \quad p^* = \frac{n}{n-1}c.$$

Firms' profits will be the following:

$$\pi^* = \frac{an}{n^2} - \frac{a(n-1)}{n^2} - f$$

The equilibrium number of firms can be found from the zero profit condition:

$$\pi^* = 0, \quad \frac{an}{n^2} - \frac{a(n-1)}{n^2} = f, \quad n_1 = \sqrt{\frac{a}{f}}$$

The social welfare function will be written as follows:

$$SW = \int_0^Q \frac{a}{q} - cQ - nf = a\ln\frac{a(n-1)}{nc} - \frac{a(n-1)}{n} - nf.$$

Let's obtain the derivative of social welfare and estimate its sign at n_1:

$$SW' = \frac{\partial(SW)}{\partial n} = \frac{a}{n^2(n-1)} - f,$$

$$SW'(n_1) = SW'\left(\sqrt{\frac{a}{f}}\right) = \frac{a}{\frac{a}{f}\left(\sqrt{\frac{a}{f}} - 1\right)} - f = f \cdot \frac{2 - n_1}{n_1 - 1}.$$

So when $n_1 \in (1; 2)$ $SW'(n_1) > 0$, and the number of firms growth impacts positively on social welfare. It means insufficient number of firms in equilibrium. We can face this situation if $f > 0.25a$, which happens under a very high level of fixed costs that doesn't allow more than two firms enter the market.

Generally we can say that it's rather the exception than the rule. It's similar to the anti-competitive markets [16] where firms with market power increase prices during demand reduction. They are rare, but can be. And there are also several more questions:

1. We said that it's always necessary to decrease the number of firms from the equilibrium level. But can we still state that the globally best for society number of firms is always smaller than the equilibrium one?

2. What is happening if we face the real situation with heterogeneous firms, for example with firms having different cost functions?

3. How does the possible change of firms' behavior, for example collusion, impact the investigated market situation and social welfare?

4. How are the results connected with shares of fixed and variable costs, and what can we say about real markets on the base of empirical data?

The answer for the first question was given in [10]. Mankiw and Whinston proposed one more assumption

$$\frac{\partial Q(n)}{\partial n} > 0, \quad \lim_{n \to \infty} Q(n) < \infty. \tag{3}$$

The assumption (3) means that equilibrium market output increases with the number of firms entering industry and approaches some finite bound. It provides the monotonic social welfare growth in situation when number of firms falls from the equilibrium to the optimum level, but it's also very difficult to identify if it's satisfied.

The second possible research direction is connected with firms' heterogeneity. In reality producers are very different even within one country. Moreover, contrary to a popular opinion that in each country some industries live well and obtain high profits, and the other ones are on the edge of survival, interindustry distinctions in the firms' productivity are much less essential, than distinctions within industry. Particularly it was shown [17] that the gap in productivity between the best 20% and the worst 20% of Russian enterprises within one industry is equal to 9–24 times. So we should take it into account.

The theoretical models for heterogeneous firms are much more complicated. Instead of number of firms we should use Herfindahl-Hirschman Index as a measure of market concentration. But first preliminary results show the similar conclusions: there are too many firms in equilibrium. Of course we have to assume that when some firms should exit the market it would be the worst ones, the most effective firms would stay on. Otherwise everything is possible. Nevertheless we won't focus on heterogeneity now. Let' consider in detail the last two questions: by-effects and possible risks of the concentration growth, and empirical evidence.

4 Collusion and social welfare

The by-effects of reducing the number of firms at a market can be connected with the possible change of their strategies, particularly with increasing probability of collusion. The principal theoretical and practical problems of the economics of cartels, collusion, and horizontal merger were in detail discussed in [18; 19], but the most important one is undoubtedly deadweight loss caused by reducing quantities under increasing prices. So let's assume that firms maximize their total profit by establishing monopolistic price and then divide profits among them. For linear Cournot model it follows

$$\pi = (a - bnq)q - cq - f \to \max, \quad a - 2bnq - c = 0,$$

$$q^* = \frac{a-c}{2bn}, \quad Q^* = \frac{a-c}{2b}, \quad p^* = \frac{a+c}{2}, \quad \pi^* = \frac{(a-c)^2}{4bn} - f.$$

The social welfare function will be

$$SW = \frac{3}{8}\frac{(a-c)^2}{b} - nf.$$

It should be noted that consumer surplus, total revenue and variable costs are independent from the number of firms, because prices and quantities coincide the monopolistic ones. So increasing the number of firms in a situation of collusion we just duplicate fixed costs and influence negatively on social welfare.

Let's consider the numerical example with demand $p = 55 - Q$ and total cost function $TC(q) = 25+15q$. As it's followed from the obtained above formulas, the equilibrium number of firms will be seven, but at the same time for social welfare maximization it should be three firms.

In the table 1 we will show individual and total market quantities, prices, profits, consumer surplus and social welfare for cases of equilibrium and socially effective number of firms, and also of monopoly. In the table 2 there is a similar information describing the market of cooperating firms

Table 1. Quantities, prices and social welfare in the model without collusion.

n	*Q*	*Q*	*p*	*П*	*CS*	*nπ*	*SW*
7	5	35	20	0	613	0	**613**
3	10	30	25	75	450	225	**675**
1	20	20	35	375	200	375	575

Table 2. Quantities, prices and social welfare in the model with collusion.

n	*Q*	*Q*	*p*	*П*	*CS*	*nπ*	*SW*
7	2,86	20	35	32	200	225	425
3	6,67	20	35	108	200	325	**525**
1	20	20	35	375	200	375	575

We can see from the tables that after the number of firms decrease from seven to three social welfare increases from 613 to 675 (despite rising of prices and reduction of consumer surplus, firms' profits decrease at a greater extent). However, if collusion is possible, the cooperated firms by increasing prices and decreasing quantities to the monopoly level will reduce the social welfare function to 525 which is worse not only than socially effective value 675, but than initial equilibrium state 613. Further let's answer the question, whether the risk of collusion will always dominate the factor of fixed costs reduction.

Let's compare the social welfare in initial situation of equilibrium number of firms with no collusion and in possible under regulation situation of effective number of firms with collusion:

$$SW_{wo/col}(n_1) = \frac{1}{2}\frac{(a-c)^2}{b} + \frac{1}{2}f - \sqrt{\frac{(a-c)^2 f}{b}},$$

$$SW_{col}(n_2) = \frac{3}{8}\frac{(a-c)^2}{b} + f - \sqrt[3]{\frac{(a-c)^2}{bf}}f.$$

Let's estimate under what range of parameters the new situation will be better than initial one, i.e. the inequality $SW_{col}(n_2) - SW_{wo/col}(n_1) > 0$ is satisfied. By defining $x = (a-c)^2/b > 0$ we will get

$$SW_{col}(n_2) - SW_{wo/col}(n_1) = -\frac{1}{8}x + \frac{1}{2}f + \sqrt{xf} - \sqrt[3]{xf^2}.$$

Let's analyze the function

$$g(y) = \frac{SW_{col}(n_2) - SW_{wo/col}(n_1)}{x} = -\frac{1}{8} + \frac{1}{2}y + \sqrt{y} - \sqrt[3]{y^2},$$

where $y = f/x = fb/(a-c)^2 < 1$.

The positivity of function $g(y)$ means that collusion is less dangerous than fixed costs duplication which we can face in equilibrium. Derivative of function $g(y)$

$$g'(y) = \frac{1}{2} + \frac{1}{2\sqrt{y}} - \frac{2}{3\sqrt[3]{y}}$$

is always positive, it means that $g(y)$ is increasing monotonously. Taking into account that

$$g(0) = -\frac{1}{8} < 0, \quad g(1) = \frac{3}{8} > 0,$$

we can state that the only root of function $g(y)$ lies in the interval of [0;1]. By solving numerically the equation $g(y) = 0$ we can find that $y^* \approx 0.064$. Going back to the initial variables we obtain, that $SW_{col}(n_2) - SW_{wo/col}(n_1) > 0$ when the following inequality is satisfied:

$$f > \frac{(a-c)^2}{2b} 2y^* \approx 0.128\frac{(a-c)^2}{2b}. \quad (4)$$

It should be noted that if to forget about fixed costs the expression $(a-c)^2/2b$ means the consumer surplus CS_{PC} in situation of perfect competition, i.e. when goods are on sale at the price equal to marginal costs. Thus, in situation of high fixed costs that exceed 12.8% of CS_{PC} even the inevitable collusion of socially effective number of firms is preferable than competition of equilibrium number of firms.

The obtained boundary level of fixed costs is extremely high and is carried out quite seldom. But if it is known that the collusion is inevitable the lesser of two evils will be the transition from competition of excessive equilibrium number of firms to monopoly.

In case of monopoly social welfare is determined by the following formula:

$$SW_{mon} = \frac{3}{8}\frac{(a-c)^2}{b} - f.$$

We will now estimate the values of model parameters, when the difference

$$SW_{mon} - SW_{wo/col}(n_1) = -\frac{1}{8}\frac{(a-c)^2}{b} - \frac{3}{2}f + \sqrt{\frac{(a-c)^2 f}{b}}$$

is positive. Let's define $z = (a-c)/\sqrt{bf}$ and explore

$$g(z) = \frac{SW_{mon} - SW_{wo/col}(n_1)}{f} = -\frac{1}{8}z^2 - \frac{3}{2} + z.$$

If $g(z)$ is positive, monopoly is the lesser of two evils. By solving the inequality

$$-\frac{1}{8}z^2 + z - \frac{3}{2} > 0$$

we will get that $z \in (2;6)$, which means that the inequality

$$\frac{(a-c)^2}{bf} < 36$$

should be satisfied.

Thus, the transition to monopoly is better than competition of excessive equilibrium number of firms when

$$f > \frac{1}{18} \cdot \frac{(a-c)^2}{2b} = \frac{1}{18} \cdot CS_{PC} \approx 0.056 \cdot CS_{PC},$$

i.e. when fixed costs exceeds 5.6% of consumer surplus in situation of perfect competition.

Thus, with a high share of the fixed component in the cost function we are able not to be afraid of reducing the number of firms to the socially effective level, despite increasing the collusion threat. Let's find out how the danger of collusion related to the share of fixed costs and characteristics of demand function. Let the equilibrium ratio of fixed and variable costs be the following:

$$\alpha = \frac{FC}{VC} = \frac{f}{cq}. \quad (5)$$

Let's suppose also that we know how many times the «choke price» (the maximum price that consumers are ready to pay for the product) exceeds the marginal costs:

$$\gamma = \frac{a}{c}. \quad (6)$$

Under linear demand and costs the optimal quantity of production is calculated by the formula

$$q^* = \frac{1}{n+1} \cdot \frac{a-c}{b},$$

and the equilibrium number of firms is equal to

$$n_1 = \sqrt{\frac{(a-c)^2}{bf}} - 1,$$

So the expression (5) can be rewritten as follows:

$$f = \alpha \cdot cq = \alpha \cdot c \frac{1}{n+1} \cdot \frac{a-c}{b} = \alpha c \frac{\sqrt{bf}}{a-c} \cdot \frac{a-c}{b} = \alpha c \sqrt{\frac{f}{b}}, \quad f = \frac{\alpha^2 c^2}{b}.$$

As it follows from (4), (6), the collusion of socially effective number of firms is the lesser of evil compared to the excessive equilibrium number of firms when

$$f > 0.064\frac{(a-c)^2}{b} = 0.064\frac{(\gamma-1)^2 \cdot c^2}{b}.$$

Using (6), we obtain the inequality

$$\frac{\alpha^2 \cdot c^2}{b} > 0.064\frac{(\gamma-1)^2 \cdot c^2}{b}, \quad \alpha > 0.253(\gamma-1),$$

$$\gamma < \bar{\gamma} = \frac{\alpha}{0.253} + 1 \approx 3.9\alpha + 1.$$

Consequently, the risk of collusion is not critically dangerous if the difference between «choke price» and marginal costs does not exceed a certain critical value associated with the ratio of fixed and variable costs.

5 Empirical research

As it was shown above, the positive or negative correlation between concentration and social efficiency is connected with the share of fixed costs. It's possible to estimate it for the largest world corporations on the base of statistics presented at the Bloomberg terminal. Let's see for example the «Computer Hardware & Services» industry in the 1998-2012 years. Converting original currencies into US dollars and taking into account inflation it's easy to use aggregate corporate data on revenues, costs, and profits to calculate the share of fixed costs by the following formula:

$$\bar{\alpha} = \frac{FC}{TC} = 1 - \frac{VC}{TC} = 1 - \frac{VC}{TR - \pi}.$$

Let's collect the annual shares of fixed costs (in percent) for several biggest corporations from 1998 till 2012, the mean value (M), and the standard deviation (SD) in the table 3. We can see that despite big difference between companies, and substantial fluctuations of sales, costs, and especially profits, the annual shares of fixed costs for all companies are very similar during economic boom and the dotcom crisis, gradual growth and the crisis of 2008-2009.

Table 3. The share of fixed costs in the «Computer Hardware & Services» companies in 1998-2012, %.

	98	**99**	**00**	**01**	**02**	**03**	**04**	**05**	**06**	**07**	**08**	**09**	**10**	**11**	**12**	**M**	**SD**
Lenovo	12	10	12	11	10	11	08	14	12	12	13	10	10	10	10	**11**	1
WD	15	09	15	10	10	11	11	11	07	12	12	12	12	19	24	**13**	4
Dell	16	15	14	14	13	13	13	12	13	15	14	15	15	18	18	**15**	2
HP	24	24	22	26	28	24	20	21	19	19	18	18	18	19	31	**22**	4
Apple	21	20	19	23	27	27	25	22	21	22	23	26	23	22	23	**23**	2
Sony	30	30	31	31	32	32	29	22	21	20	21	23	26	26	21	**26**	5
Fujitsu	28	27	28	31	30	27	26	25	24	25	27	25	27	27	29	**27**	2
Panasonic	30	28	28	29	28	29	29	30	28	27	30	29	26	32	33	**29**	2
IBM	33	30	30	32	34	31	32	34	35	35	37	37	37	38	38	**34**	3
Intel	40	46	45	47	43	47	46	48	43	41	48	49	53	51	52	**47**	4
Cisco	59	58	59	52	59	63	61	57	57	54	55	57	55	55	53	**57**	3
Yahoo	80	81	85	81	82	74	53	38	53	55	55	51	48	60	60	**64**	15
Oracle	59	62	66	68	67	69	70	71	69	71	73	72	69	71	73	**69**	4
Microsoft	76	78	81	75	75	77	77	76	71	73	72	72	67	69	64	**74**	5
Adobe	86	88	91	92	89	91	91	92	86	85	87	88	87	87	86	**88**	2

Let's find the critical ratio between «choke price» and marginal costs for several companies of the «Computer Hardware & Services» industry and estimate for them the risk of possible collusion connected with transition from the equilibrium number of firms to the socially effective one (table 4)

Table 4. Risk of possible collusion depending on the share of fixed costs.

$\bar{\alpha}$	α	Typical Company	$\bar{\gamma}$	Risk
0.1	0.11	Lenovo	1.4	Very high risk
0.25	0.33	Sony, Apple	2.3	High risk
0.5	1	Intel	5	Medium risk
0.75	3	Microsoft	13	Low risk
0.9	9	Adobe	37	No risk

We can see that the costs of collusion for companies engaged in production of computers and digital technologies (such as «Lenovo», «Sony», «Apple») are quite serious in the situation of limited competition. Therefore, the regulation of these markets is impractical. Moreover, the social planner should restrict the possible increasing of market power caused by the companies presented at the market.

On the other hand, for the companies engaged in software development («Microsoft», «Adobe», etc) the high share of R&D fixed costs leads to the fact that the restriction of competition may be socially efficient, despite the increased risk of collusion.

It's possible to do similar empirical investigation for the other industries. Most of them don't have such a big deviation within themselves. Even outliers usually belong to the same group of firms as the whole industry. Moreover the results for most companies are very stable in time, despite economic growth and depressions. But there is a big difference between industries. So we need different mechanisms of their regulation.

There are several industries (include «Diversified Metal & Mining», «Construction», «Oil & Gas Operations», «Auto & Trucks», «Electrical Equipment», «Diversified Chemicals», and «Food Processing») with low share of fixed costs, not exceeding 30%. It means that it's very dangerous to regulate these industries by entry barriers if the regulator can't provide competition at such markets.

Some other industries such as «Personal Care», «Medical Supplies», «Electric Utilities» (we can also add here the sub-industry of software producers unlike hardware companies) due to large investments or R&D costs have a big share of fixed costs. It means that there could be a significant effect of the number of firms decreasing, exceeding the collusion losses even if we can't avoid it.

Let's collect the minimum, maximum, and average shares of fixed costs for the biggest corporations of different industries in the table 5. Let's also include into the table the standard deviation of the fixed costs share within industries, and the critical ratio between «choke price» and marginal costs.

Table 5. The share of fixed costs for different industries.

Industry	The share of fixed costs, %				$\bar{\gamma}$
	M	Min	Max	SD	
Metals & Mining	**15**	4	34	8.8	1.7
Construction	**18**	2	30	8.7	1.9
Oil & Gas	**19**	5	54	13.3	1.9
Auto & Trucks	**20**	13	28	4.3	2.0
Electrical Equipment	**25**	12	44	9.9	2.3
Chemicals	**26**	16	47	10.1	2.4
Food Processing	**29**	4	54	15.8	2.6
Computers	**40**	9	88	24.7	3.6
Personal Care	**42**	22	65	13.8	3.8
Medical Supplies	**53**	26	80	17.0	5.4
Electric Utilities	**61**	39	80	12.1	7.1

6 Conclusions and policy implications

The first result proposed in the paper is that socially effective number of firms is smaller than the equilibrium one for the wide spectre of demand and cost functions, and also for different strategies of companies' behavior. This proposition is satisfied for the markets where output of each company decreases when the number of firms increases, and the competition gets stronger.

Thus the entry barriers, constructed by incumbent companies, do not always decrease social welfare. In some cases it's even good for government not to stimulate excessive competition but on the contrary to restrict the entry of new companies to the market.

At the same time there is a considerable danger of the increasing probability of collusion in a situation of number of firms limitation. So we compared the losses caused by duplicated fixed costs, and by possible collusion, and showed that collusion is the lesser of two evils if the gap between «choke price» and marginal costs is smaller than a certain critical value connected with the share of fixed costs. Thus, we should estimate if the real market value gamma is smaller than the obtained critical point. If so then it's possible to regulate market via entry barriers, otherwise – no.

There are also some other threats of entry limitation which should be taken into account. When the entry barriers system is organized through licenses and permissions it's very important to restrict corruption which is highly probable, especially if the licenses distribution is carried out by officials, but not via auction. Also, as the entry barriers lead to social welfare redistribution (consumer surplus reduces with the simultaneous increase of firms' profits) it's very important to design the effective tax mechanisms of withdrawal of rent, gained because of higher level of market power.

The research is supported by the RFBR grant №16-06-00071, and the European Union's Seventh Framework Program FP7/07-13/ under REA grant agreement number 609642.

References

1. J. Tirole, *The Theory of Industrial Organization* (MIT Press, 1988)
2. O. Shy, *Industrial Organiztion: Theory and Applications* (MIT Press, 1995)
3. D. Carlton, J. Perloff, *Modern Industrial Organization* (Addison-Wesley, 2000)
4. P. Belleflamme, M. Peitz, *Industrial Organization. Markets and Strategies* (Camb. Univ. Press, 2015)
5. N. Aizenberg, A. Filatov, *Mathematical models of imperfect competition* (ISU Press, 2012)
6. M. Spence, Rev. Econ. Stud., **43**, 217-236 (1976)
7. A. Dixit, J. Stiglitz, AER, **67**, 297-308 (1977)
8. C. von Weizsacker, Bell J. Econ., **11**, 399-4 (1980)
9. K. Perry, J. Ind. Econ., **32**, 313-330 (1984)
10. G. Mankiw, M. Whinston, RAND J. Econ., **17**, 48-58 (1986)
11. S. Berry, J. Waldfogel, RAND J. Econ., **3**, 397-4 (1999)
12. C. Hsei, E. Moretti, *J. Polit. Econ.*, **111**, 1076-1121 (2003)
13. A. Hortacsu, C. Syverson, QJE, **119**, 403-456 (2004)
14. P. Davis, J. Ind. Econ., **54**, 293-321 (2006)
15. T. Bresnahan, P. Reiss, J. Polit. Econ., **99**, 977-1009 (1991)
16. E. Zhelobodko, S. Kokovin, M. Parenti, J. Thisse, Econometrica, **8**, 2765-2784 (2012)
17. V. Golikova, K. Gonchar, B. Kuznetsov, A. Yakovlev, *Russia industry at at the crossroads: what prevents our firms to become competitive* (HSE, Moscow, 2008)
18. J. Bain, *Barriers to new competition* (Harv. Univ. Press, 1956)
19. A. Jasquemin, M. Slade, *Handbook of Industrial Organization* (1989)

 DOI: 10.1051/shsconf/20173501074

On attracting investment to Russian economy

Elena Oleinik[1,*]

[1]Department of Business Informatics and Economics and Mathematical Methods, Far Eastern Federal University, Vladivostok, Russia

Abstract. Modern conditions of economic systems development have necessitated more precise directions of investment process for the economic growth to accelerate. The precipitous fall of oil prices that occurred in the second half of 2014 weighed heavily on foreign direct investment flows to oil-exporting countries. The problem of attracting investment is one of the key problems of economic development. The aim of this study is to analyze and evaluate the shifts in the structure of investments, the study of the structure of investments to the Russian economy. Structural changes result from differences in the growth rate of the elements forming an economic system. An integral coefficient has been used for evaluation of structural changes. Multiple regression was used to analyze the impact of various factors on the change in structure of investment. The regression coefficients in the model give quantitative assessment of the pace of change in the structure as it depends on a specific factor. The model estimation shows that the variables "share of investments in manufacturing" (positive influence) and "share of investments in the health, physical culture, education" (negative influence) have the largest t-values. Hence, these two factors influence the structural changes in investments most strongly The evaluation and analysis of the structural shifts may lead to conclusions regarding the efficiency of the structure investment and adjustment of the economic policy.

1 Introduction

The intensity of economic growth is largely determined by active investment activity, and the problem of attracting investment is one of the key problems of economic development. Russia possesses all major factors determining investment attractiveness of the area, such as high level of potential demand at the domestic and international markets, significant reserves of natural resources, well-developed industrial enterprises management structure, available energy resources, production and social infrastructure in the North-Western and Siberian territories, available land and sea transportation routes, possibility of all-year-round supply of products. Foreign investments inflow is vital for overcoming current crisis situation and for economic revival. Rating of countries according to the total volume of attracted investments is presented in the table 1. The precipitous fall in oil prices that occurred in the second half of 2014 weighed heavily on foreign direct investment (FDI) flows to oil-exporting countries. Countries use different concepts of national security, allowing them to take into account key economic interests in the investment screening process. Governments' space for applying national security regulations needs to be balanced with investors' need for transparent and predictable procedures. However, this growth did not translate into an equivalent expansion in productive capacity in all countries.

Table 1. Rating of countries according to the total volume of attracted investments.

Total volume of attracted investments (Millions of dollars)	Years (Millions of dollars)			Cumulative volume of attracted investment	Average growth rate
	2013	2014	2015		
USA	231	92	380	703	2,03
PRC	124	129	136	389	1,03
Canada	71	59	49	174	0,95
Russian Federation	69	21	10	100	0,69

Source: Calculated by the author based on data [1]

The Russian Federation in 2015 recorded FDI flows of $9.8 billion, a 66 per cent contraction from the previous year. FDI flows were mainly in the form of reinvested earnings, as new FDI flows almost dried up. Falling oil prices and geopolitical tensions continued to damage economic growth prospects and erode investor interest in the country

The crisis in the Russian economy affects the changes in investment levels and there is a trend of investment reduction in joint ventures. Foreign investors

* Corresponding author: oleynik.eb@dvfu.ru

are not keen to invest in the development of products with high added value, thus creating competitive products and jobs on the territory of our country. At the same time, Russian economic interests do not always match interests of foreign investors. Investment attraction is vital for integration of Russia into the world economy. However, the question arises: how to evaluate the result of investment policy and investment efficiency?

In the process of implementation of investment projects aimed at economic development of the country and the regions, special transformations take place i.e. the change of economic structure: changes are observed in the ratio of shares and proportions as well as in the direction and strength of the links between its elements. Modern science continues generating unabated interest in the study of problems and identifying patterns of structural shift, their focus, the interdependence between the level of development of the economy and its structure.

2 Literature Review

Select the main factors of factors influencing FDI activity [2-4]

– Macroeconomic and policy factors: regional agreements such as TPP, RCEP, TTIP; changes in tax regimes; commodity prices, including oil; changes in global financial regulations; exchange rate volatility; geopolitical uncertainty; debt concerns in emerging markets.

– Corporate and external factors: technological change, including digital economy; global urbanization; offshore outsourcing of corporate operations; energy security; food security; climate change; migration; cyber threats and data security.

Can be divided countries on two group: investors and recipients. The host country determinants for FDI consist of (a) the general policy framework for foreign investment, including economic, political and social stability, the gislation affecting foreign investment and any other policies affecting FDI locational decisions; (b) economic determinants, such as the market size, cost of resources and other inputs or the availability of natural resources; and (c) business facilitation, including investment promotion. All three groups of determinants interact, enhancing or reducing the attractiveness of countries to foreign investment [5].

Scientific literature provides various approaches to the definition of the category of 'structural shifts'.

-a structural shift is a qualitative change in the relationships between comparable elements of an economic system caused by uneven dynamics of the ratio between their quantitative characteristics. Comparable elements are referred to as elements which belong to the same level or segment of an economic system. A structural shift in the economic system of a higher level cannot be reduced to a simple arithmetic sum of its components [6];

- a structural shift is the change of the internal structure of the economic system, the relationship between its elements, which are accompanied by the transformation of the main (integrated) system properties [7-8].

A structural shift means a change of proportions of the economic system, which occurs under the influence of the structure-forming factors. Structural changes are associated with the intensification of production [9-10].This definition implies that any change in the structure can be seen as a structural shift, which is regarded as the accumulation of imbalances. Changes in the structure of the economy represent a complex system of interrelated changes in the proportions taking place under the influence of the existing technical basis, social mechanisms of production, distribution and exchange, in accordance with social needs, available resources and the achieved level of productivity [11].Structural changes are a certain difference between the new and the old structures, which reflects not only the difference in the organization of production, but also the development of the entire system of economic relations.

Despite the fact that the definition of 'a structural shift' was highlighted in a great number of studies done by Russian and foreign economists, a single definition has not been formulated and subordination between the notions of 'structural shift' and 'structural change' has not been elaborated. The classification of indicators of structural changes is proposed in [12] and completed in [13]. Structural changes of economy under the influence of various factors are investigated and explained in [14–17].

A sufficient number of various indicators used to evaluate the structural shift describes in the literature, but only some of them contain the scale of values and recommendations for using them.

– in [18]: chain and basic index is used the higher the value of the chain index, the more intensively structural shifts occur, and vice versa. The basic index serves as an indication of the extent of the tendency which underlies the structural shifts, and to what extent they are merely the result of irregular fluctuations.

– in [19]: the integrated coefficient that considers the intensity of the changes by individual group and the proportion of groups in comparable structures. L. Kazinets15 Absolute and relative growth rate and the shift intensity. The coefficients indicate the intensity of shifts, but do not give a qualitative characteristic of the processes taking place.

– in [20]: integrated coefficient of structural differences does not depend on the number of gradations of aggregate structure. The author suggests an evaluation scale for measuring importance of structural differences.

– in [21]: the intensity of structural shifts (taking values from 0 to 1) is evaluated through the change in the angular distance between the vectors of economic structure in the initial and final periods.

– in [22]: the importance of the structural transformation evaluates the index, λ the evaluation is considered to be significant if the value of the index is significant λ(a change of 15% in the index λ over 5 years)

The article [21-22] presents a comparative analysis approaches to determining the investment activity and to evaluation of the role of investments in achieving a steady economic advance.

But are these changes in the structure of the positive? Does GDP depending on the integrated shifts?

3 Research Methods

The considered existing evaluation indicators do not allow for a comprehensive study of structural shifts and measuring the proportions between the structural elements. In our opinion, complex evaluation of shifts in the structure of investments requires a combination of some of the methods. Table 2 presents basic approaches, areas of research, types and methods of analysis of structural changes in investment flows.

We offer a comprehensive, step-by-step methodology for the analysis and evaluation of structural shifts.

1. The choice of the study period and the definition of the input parameters. Input parameters: f_j – the relative weight of j-element of the structure in the current (l) and previous (l-1) periods; n- the number of structural elements.

2. Calculation of linear structural shifts (formula 1), that show absolute changes in the relative weight of each element of the structure, but do not give the qualitative characteristics of the processes.

$$S_j^{abs} = f_j^l - f_j^{l-1} = \Delta f_j \quad (1)$$

3. Calculation of the quadratic structural shifts (formula 2), that are calculated based on linear structural shifts. Sabs coefficient allows for getting a summary assessment of the total speed of branch shifts, and Srelat for each j-element of the structure. It is coefficient of uniformity of structure (formula 3). Treated dynamically, these coefficients characterize the trend of change in the industrial structure

$$S_{abs} = \sqrt{\frac{\sum_{j=1}^{n}(f_j^l - f_j^{l-1})^2}{n}} \quad (2)$$

$$S_j^{relat} = \frac{f_j^l - f_j^{l-1}}{f_j^{l-1}} * 100\% \quad (3)$$

4. Calculation of intensity of structural shifts of a branch and branch structure as a whole. Intensity coefficient characterizes the importance of structural shifts [9, 10]. The intensity of structural shifts (taking values from 0 to 1) is evaluated through the change in the angular distance between the vectors of economic structure in the initial and final periods. Sudden changes in intensity coefficients over time indicate instability and uneven development of structural elements.

$$K_{int} = \frac{f_j^l}{f_j^{l-1}} * 100\% \quad (4)$$

5. Calculation of integral structural shifts. Integrated coefficient allows for evaluating the structural differences: the closer the I to 1, the more significant the structural changes are and, therefore, the differences in the structure.

$$I_l = \sqrt[n]{\Pi_{j=1}^{n} f_j^l} \quad (5)$$

6. Evaluation of the significance of structural changes

$$m_l = \sqrt{\frac{1-I_l^2}{n-1}} \quad \text{and} \quad t_l = \frac{I_l}{m_l} \quad (6)$$

If structural changes are significant (ti>ttab; ttab== 2,13), there is a structural shift

Output parameters are a set of integrated assessments of the state and dynamics of the structure.

The proposed method allows a comprehensive analysis of the nature of structural shifts, their interaction and orientation, and thus enables us to evaluate the presence of distortions as well as quantitative and qualitative structural changes.

In the modern economic science, a rather universal approach to construction of econometric models of the economic growth is described, and the following equation is formed in general terms [25]

$$GR = a_0 + \sum a_i Ek_i + \sum b_j CVar_j + \sum c_l Dum_l + \varepsilon_k \quad (7)$$

Here: GR – growth rate of the main indices of the national product in the country; Ek_i – economic variables; $CVar_j$ – additional control variables; Dum_l – dummy variables reflecting the regional effect; a0 – constant; a_i, b_j, c_l – the respective regression coefficients; εk – random error. The econometric approach is based on adding a set of variables describing the economy's structural parameters into the right part of the equation (7). We are following this approach too when studying the influence of investments in the fixed capital on the economic growth rates.

4 Results

The model suggested was tested out on statistical data provided by the Russian Federal service of the state statistics for 1994-2016. The results of calculations in accordance with the suggested methodology are presented in Table 2. Significant integrated shifts marked in bold.

Table 2. Evaluation of structural shifts.

Year	GDP, in percent to the previous year (at constant prices)	Structural shifts, share		Significance of integrated shift
		quadratic	integrated	
1996	145.8	0.646	0.441	1.538
1997	115.1	0.430	0.366	1.317
1998	118.6	1.873	1.081	**5.429**
1999	172.5	2.010	1.324	**6.302**
2000	110	2.013	0.897	**5.117**
2001	81.7	0.492	0.775	**3.211**
2002	104.7	1.074	0.640	**2.310**

2003	107.3	0.522	0.740	**2.923**
2004	107.2	0.620	0.423	1.483
2005	106.4	0.758	0.416	1.462
2006	107.7	0.552	0.640	**2.310**
2007	108.1	0.606	0.291	1.110
2008	105.2	0.269	0.394	1.395
2009	92.2	0.406	0.303	1.141
2010	104.5	0.512	0.376	1.345
2011	104.3	0.173	0.132	0.674
2012	103.4	0.480	0.228	0.941
2013	101.3	0.638	0.322	1.195
2014	95.5	0.644	0.667	**2.452**
2015	101.2	1.310	0.364	1.310

High coefficients indicate uneven structural changes. Their speed and importance allow us to assess quadratic and integrated coefficients. Maximum speed of restructuring and significant differences in the structure of investments was observed in 2000 and 2015. The structure of investments was stable in 2000-2014 (fig.1), the regression coefficient approximately equal to zero (0.004).

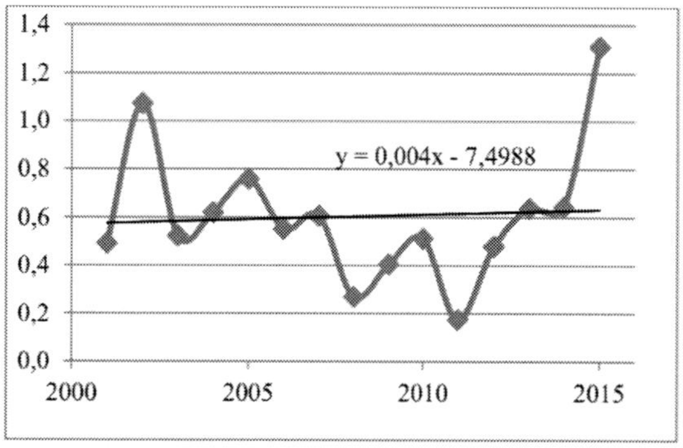

Fig. 1 Regression of an integral structural shift.

Additional calculations are required for a consolidated and integrated assessment of structure dynamics. The correlation coefficient R_{YX2} =0.49, therefore there is a relatively strong linear dependence. The results regression modeling are presented in Table 3. Y is the endogenous variable, GDP growth.

Table 3. Results of regression modeling.

Regression statistics			
Multiple R	0.481907		
R^2	0.232235		
Standardized R^2	0.189581		
Standard error	17.43214		
Observations	20		
Dispersion analysis			
	df	*F*	
Regression	1	5,4446	
Remainder	18		
Total	19		
	Regression coefficients	*t-statistics*	*P value*
Y-crossing	92.72401	11.397	1.15E-09
Integrated shift	30,80066	2,333	0,0314

The model is valid according to Fisher's test. Regression coefficient is valid according to Student's test and positive, therefore, their increase enhances the economic growth. The regression model can be used to predict.

The following designations are used: for share of investments: Z1 – in manufacturing; Z2 – in extractive industries; Z3 – in scientific research and development; Z4 – in the health, physical culture, education. Multiple regression was used to analyze the impact of various factors on the change in structure of investment (Table 4). Y is the endogenous variable, integral shift/

Table 4. Results of multiple regression modeling.

Regression statistics			
Multiple R	0.800771		
R^2	0.641235		
Standardized R^2	0.545564		
Standard error	0.204238		
Observations	20		
Dispersion analysis			
	df	*F*	
Regression	4	6.702523	
Remainder	15		
Total	19		
	Regression coefficients	*t-statistics*	*P value*
Y-crossing	0.2723	0.2618	0.7971
Z1	0.1823	4.5158	0.0004
Z2	-0.0494	-1.4582	0.1654
Z3	-0.5734	-1.9161	0.0746
Z4	-0.3485	-2.2577	0.0393

The regression coefficients in the model give quantitative assessment of the pace of change in the structure as it depends on a specific factor. The model estimation shows that the variables "Z1 –share of investments in manufacturing " (positive influence) and "Z4 – share of investments in the health, physical culture, education " (negative influence) have the largest t-values. Hence, these two factors influence the structural changes in investments most strongly.

Regression coefficients for Z2–Z4 is not valid according to Student's test. The regression model cannot be used to predict. But we can to use regression coefficients for calculation of elasticity coefficients. The following designations are used: S – elasticity coefficient; Y – integral structural shifts of investment. The elasticity coefficients (as a percentage) are:

$$S_{YZ1} = +5,06;\quad S_{YZ2} = +1,38;$$

$$S_{YZ3} = -0,65;\quad SY_{Z4} = -2,55.$$

Additional investments in the manufacturing industry increase the integral structural shift by 5,06 percent, which provides growth of GDP.

5 Discussion

Based on analysis of detected problems and situation evaluation, we consider it necessary to take the following steps in order to attract investments into the Russian economy:

First: improvement of investment climate. Attraction investors is important to create favorable investment climate, that is environment where investment processes take place. To improve the investment climate, the following measures are required to take: creation of insurance investment funds with government participation; stimulation of partner agreements, according to which foreign investor supplies machinery, equipment, technologies, complete plants in exchange of counter deliveries of raw materials and semi-finished goods produced at the supplied equipment; attraction of foreign financial resources in the form of compensation agreements, that is credits which return basis is supply of products made by created enterprise; establishing of favorable tax regime, lowering of electrical energy and transportation tariffs; organization of presentations for priority investment projects; providing by the enterprises of reliable and transparent information, as well as government control over schedule times for priority investment projects; ensuring property protection and investor personal safety.

Second: improvement of economic, as well as regulatory and legal framework for public-private partnership (PPP). The government, entering into alliance with business, obtains not only its financial resources and easing of budget burden, but also more flexible system of project management. Business is also interested in utilization of government resources, guarantees and preferences to solve its tasks. Public-private partnership is a tool of active government interference in the economy with the purpose of elimination and prevention of structural disproportions

Third: development of leasing scheme of investment attraction: establishing of pledge funds to ensure bank investments in leasing with the use of state property; improvement of legal and regulatory framework to protect the interests of leasing deals participants; establishing of state guarantee fund for export when making international leasing of domestically produced machinery and equipment.

The conducted analysis reveals that, in spite of the fact that the volume of investments into the economy of Russia for the period from 1996 to 2015 increased more than threefold, imbalances remain in the structure of investments.

Investors do not invest in the development of the regions of Russia. Short-term investments are aimed at fast and risk-free return on investment. The regions are mainly focused on the production and export of raw materials, rather than processing them, which reduces the level of its economic development. It follows that there are serious problems of attracting investments.

6 Conclusion

Our offers per main areas and methods of investing are shown at the table 5. All these measures will help to attract investments, improve the structure efficiency and economic growth of Russia.

Table 5. Main areas, sources and methods of investment activity.

Area	Sources	Methods
Development of new technologies in the processing industry	Targeted budget financing Extrabudgetary funds	Venture financing Government subsidies
Reconstruction of enterprises	Enterprises' own financial means Bank credits	Leasing Project financing Public-private partnership in the form of top priority investment projects Concession agreements Additional emission of shares Compensation agreements
Creation of enterprises with high added value: vertically integrated structures and raw materials deep processing enterprises	Share capital Targeted bank credits and loans Investment credits	Public-private partnership in the form of top priority investment projects Corporatization of enterprises Bonds issue Compensation agreements
Development of social infrastructure of cross-industry complexes and road networks	Targeted budget financing Bank credits Enterprises' own financial means Mortgage loans Concessionary loan	Outsourcing Public-private partnership in the form of top priority investment projects Implementation of federal targeted programs of territories development

The proposals and approaches to improve the efficiency of the investment policy in the Russia, as well as the method for analysis of investment structure can serve as a basis for further theoretical and applied research in the field of regional investment policy.

References

1. J.X. Zhan, R. Bolwijn, K. Byun, B. Casella, *World Investment Report: Investor nationality: policy challenges* (New York, 2016). http://unctad.org/en/PublicationsLibrary/wir2016_en.pdf
2. P. Nunnenkamp, J. Spatz, Transnational Corporations, **11**, 1-34 (2002)
3. A.A. Sultygova, L.G. Popov, Actual problems of humanitarian and natural Sciences, **3**, 85-88 (2016)

4. E. Aisbett, *Bilateral investment treaties and foreign direct investment: Correlation versus causation* (University of California, Berkeley, 2007). http://repositories.cdlib.org/are.ucb/1032.
5. The Role of International Investment Agreements in Attracting Foreign Direct Investment to Developing Countries materials of conference United Nations Conference on Trade and Development New York and Geneva (2009). http://unctad.org/en/Docs/diaeia20095_en.pdf
6. A. Szirmai, Industrialization as an Engine of Growth in Developing Countries (2009). http://www.merit.unu.edu/publications/wppdf/2009/wp2009-010.pdf.
7. J. Meckl, Journal of Economics, **77(3)**, 241-266 (2002)
8. J. Laitner, Review of Economic Studies, **67**, 545-5619 (2000)
9. C. Greenhalg, M. Gregory, Oxford Bulletin of Economics and Statistics, **63**, 629-646 (2001)
10. F. Caselli, I.I. Coleman, John Wilbur, A Reinterpretation. Journal of Political Economy, **109(3)**, 584-616 (2001)
11. E.M. Urkunchiev, *Structural shifts in the economy of Kazakhstan: theory, methodology, movement* (Turkestan, 2010)
12. L.S. Kazinets, *Growth and structural change in the economy (Performance planning and review process)* (Economy, Moscow, 1981)
13. N.P. Perstenyova, Basic Research, **2**, 478-82 (2012)
14. A. Greenspan, Vital Speeches of the Day, **5(67)**, 130-4 (2000)
15. B. Los, *The impact of research and development on economic growth and structural change* (Universiteit Twente, 2000)
16. I.K. Shevchenko, Y.V. Razvadovskaya, World Applied Sciences Journal, **28(6)**, 860-5 (2013)
17. Y.D. Weissman, V.O. Boos, Regional economy, **4**, 208-19 (2013)
18. V.A. Bessonov, HSE Economics Journal, **2**, 184-219 (2000)
19. K. Gatev, *Statistical evaluation of the differences between structures. Theoretical and Methodological Problems of Statistics* (Statistics, Moscow, 1979)
20. V.M. Ryabtsev, G.I. Chudilin, Regional statistics, (Moscow, Finance and Statistics, 2001)
21. O.V. Spasskaya, MAX Press, 20-39 (2003)
22. L.A. Dedov, V.F. Kapustin, Bulletin of St. Petersburg University, 95-103 (1995)
23. E. Oleinik, A. Zakharova, International Journal of Economics and Financial Issues. Special Issue for Fundamental and Applied Research in Economics and Management: New Perspectives, **6(S8)**, 340-345 (2016)
24. E. Oleinik, A. Zakharova, Indian Journal of Science and Technology, 9(36), 1-7 (2016). DOI: 10.17485/ijst/2016/v9i36/102025
25. R.J. Barro, X. Sala-i-Martin, *Economic Growth* (The MIT Press, Cambridge, 2003)

Comparative study on economic security of enterprises depending on implemented business activities quantity

Tatyana Shkarina[1,*], *Olga* Chudnova[1], and *Olga* Mokhova[1]

[1]Far Eastern Federal University (FEFU), Engineering school, 690018 Russky Island, Russia

Abstract. The results of a comparative study on economic security of enterprises depending on the quantity of business activities are published in the article. The sampling for analysis was conducted based on statistic data of Primorsky Region of the Russian Federation. The control points are the years of the most thorough data collection on the business activities of one-field and diversified enterprises: 2005, 2009, 2013, 2015.

Each enterprise is a dynamically developing economic system. Economic system is characterized with structural diversity of elements and connections both within the system itself and with external parties. In order for this system to function productively, it is necessary to maintain its economic security. Security, in this case, is defined as consistency of system pathology or continuance of its status sequences over time during its metamorphosis.

Economic security of enterprise, much alike commercial activity itself, – is a complex concept; the first characterizes business activities with regard to external influences and can be considered secure if under equal external and internal shifts it is more robust to changes comparing to original status. This said, economic condition of an enterprise may vary from extremely volatile, under which it is borderline bankrupt, to relatively secure.

Security – is an external manifestation of object structure. In order to increase its stability against various factors, primarily, it is necessary to improve the object itself.

Security of business object is an aggregate of independent, singular and private balances. It is higher when the aggregate of stable balances of business object exceeds the quantity of unstable ones.

The goal of given study – is determination of the most economically secure type of enterprise when organizing one-field or diversified production.

One-field enterprises – are the enterprises that produce one type of goods, have one type of sphere of activity, one specialization.

Diversified enterprises – "are enterprises with profile activity concentrated in various businesses, operating in various markets, however, administration and functional supply are most frequently concentrated on one platform".

Aggregation of study sampling was determined based on development of one-field and diversified enterprises conducting business in Primorsky Region.

Timeliness of the study is defined with economic development of Primorsky Region due to creation of zone of porto-franco free port Vladivostok, which has obtained its status since October 12, 2015. The main goal of creating such zone is formation of secure regional development, which is impossible without achievement of economic security of separate enterprises. Capital investments in development of cross-border trade, transport infrastructure and network of logistic centers with specific storage conditions will allow enterprises to establish comfortable external connections. However, achievement of internal security has the role that is not less important and requires clear understanding which business has the highest economic security: the one that focuses on one type of activities or the one that has several types of activities [1-3].

During the conduct of research activity profile of enterprise was not considered. Division of enterprises was based on the only criteria – whether the enterprise has more than one profile. The volumes of statistic data was taken from the official reports of Regional Bureau of Statistics. The control points are the years of the most thorough data collection on business activities of one-field and diversified enterprises: 2005, 2009, 2013, 2015. The aggregates of the control points formed based on enterprise activity and quantity of types of implemented activities.

The research method is the comparative analysis method.

For the purposes of research, 2800 enterprises of medium and large business, which implement activities in Primorsky Region of the Russian Federation, were selected [4].

Considered data is represented in statistic tables by separate aggregations (Table 1 – Aggregation of one-field enterprises, Table 2 – Aggregation of diversified

* Corresponding author: shkarina.tyu@dvfu.ru

enterprises). The aggregations of one-field enterprises are labeled in accordance with years: 2005-a, 2009-в, 2013-c 2015-d. The aggregations of diversified enterprises are labeled in accordance with years: 2005-x, 2009-y, 2013-z, 2015-g.

The type of development dynamics of enterprises is labeled as follows: 0 – enterprise did not conduct activities during the researched timeframe; 1 – enterprise operated as one-field organization during the researched period; M – enterprises functioned as diversified enterprises during the researched timeframe [4-8].

Table 1. Aggregation of one-field enterprises.

Type of development dynamics	2005		2009		2013		2015	
1-1-1-0	a1	151	b1	151	c1	151	D1	0
1-1-1-M		151		151		151		101
1-1-1-1		151		151		151		50
1-0-0-0	a2	345	b2	0	c2	0	D2	0
1-0-0-1		345		0		0		0
1-0-0-M		345		0		0		0
1-1-0-0	a3	100	b3	100	c3	0	D3	0
1-1-0-1		100		100		0		0
1-1-0-M		100		100		0		
…								
M-0-1-0	a19	0	b19	0	c19	2	D19	0
M-0-1-1		0		0		2		2
M-0-1-M		0		0		2		0
Total of enterprises in aggregate	702		811		1014		1116	

Evolution of aggregate changes is shown in Figures 2 and Figure 3. In this case evolution is defined as "expansion" of researched organizations, and, in particular, determination of frequencies of creation, reorganization and saving [9].

Looking at the obtained analysis data on one-field and diversified organizations, it is notable that this is matrix presentation of planar net graphs, table 1 and table 2.

Eye charts are built for the analysis of matrix structure of planar net graphs evolution; as a result, the types of graphs for all researched organizations are the following (Figures 1a and Figure 1b) [10-11].

Table 2. Aggregation of diversified enterprises.

Type of development dynamics	2005		2009		2013		2015	
M-M-M-0	x1	213	y1	213	z1	213	G1	0
M-M-M-1		213		213		213		0
M-M-M-M		213		213		213		213
M-0-0-0	x2	363	y2	0	z2	0	G2	0
M-0-0-1		363		0		0		0
M-0-0-M		363		0		0		0
M-M-0-0	x3	91	y3	91	z3	0	G3	0
M-M-0-1		91		91		0		0
M-M-0-M		91		91		0		0
…								
0-0-M-0	x16	0	y16	0	z16	151	G16	2
0-0-M-1		0		0		151		10
0-0-M-M		0		0		151		121
0-1-M-0	x17	0	y17	0	z17	35	G17	1
0-1-M-1		0		0		35		3
0-1-M-M		0		0		35		35
Total of enterprises in aggregate	704		516		533		550	

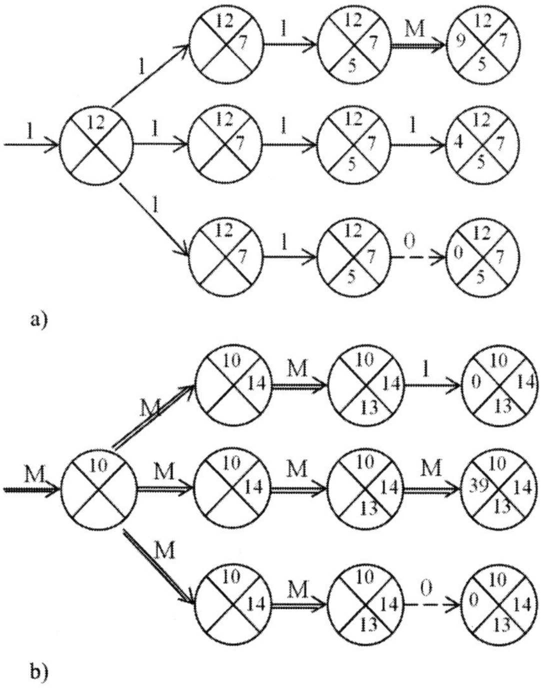

Fig. 1. Planar net graphs of one-field (a) and diversified (b) organizations.

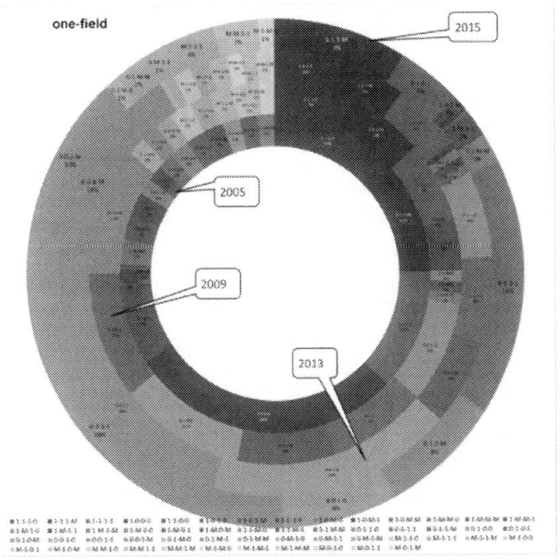

Fig. 2. Evolution of aggregate changes in one-field organizations.

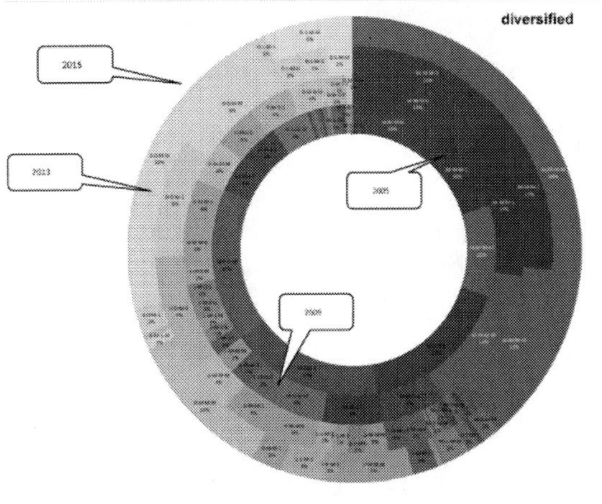

Fig. 3. Evolution of aggregate changes in diversified organizations.

Peculiarity of presented model is the presence of two types of graphs: "organization profile" (0, 2 and M) and segment. Based on graph structure evolution chains of development of researched organizations build up effortlessly, Figures 4 and 5.

We are interested in organizations with highest security. Security is defined as the ability to administrate and maintain the set, yet adjustable operation mode with particular precision and ability to recover it in case of malfunction. For the purposes of security analysis, eye charts were built based on changes in aggregates of organization profiles [12-14].

Analysis of changes in aggregate of one-field and diversified organizations gives and opportunity to form evolution chains of one-field organizations. As a result of analysis, evolution chains were formed as follows, Figures 4 and 5:

1) Organizations that are capable to sustain the set adjustable operation mode with specific precision;

2) Organizations that are capable to regulate the set operation mode with specific precision;

3) Organizations that are capable to recover the set adjustable operation mode with specific precision.

Sustainability		Adjustability		Recovery	
1-1-1-1	12→7→5→4	1-1-1-M	12→7→5→9	0-1-0-M	0→11→0→8
0-1-1-1	0→9→6→16	1-1-M-M	2→1→1→2	0-0-1-M	0→0→18→10
0-0-1-1	0→0→18→28	1-M-1-1	2→1→1→2	0-1-M-M	0→6→9→2
		1-1-M-1	2→1→1→2	0-1-M-1	0→2→1→1
				1-M-1-1	2→1→1→2

Fig. 4. Evolution chains of development of one-field organizations

Sustainability		Adjustability		Recovery	
M-M-M-M	10→14→13→39	M-M-M-1	10→14→13→1	0-1-M-M	0→1→2→6
0-M-M-M	0→4→4→10	M-1-M-1	1→0→1→1	M-M-0-M	1→0→1→1
0-0-M-M	0→0→9→22	M-1-M-M	1→1→1→2	M-0-0-M	17→0→0→0
				M-0-M-1	2→0→1→1
				M-0-1-M	1→0→0→0
				0-M-1-M	1→1→2→1

Fig. 5. Evolution chains of development of diversified organizations.

For the considered organizations, it is necessary to sustain maximum feasibility for as long as possible, meaning to stay in market for a prolonged period of time, and this is nothing less then the concept of critical path in the theory of net graphs [15-16]. Surely, lately this given concept is mainly used for evaluation of duration of implementation of innovative projects and optimization of deadlines. Nevertheless, in our case it is necessary to remain in operating mode – "in project". Thus, we need to evaluate the attained evolutional chains-branches of the graph [17-20].

Each organization profile (atop the graph) is associated with number T_i, called the early occurrence of event i. It is the minimum value of implementation filled in segment (weight) of event i (process with a start point in this extremum must begin not earlier the moment T_i) [21-25]:

$$T = \left| \begin{array}{l} T_0 \\ \max\left(T_i + T_j\right), j - \textit{value of filled in segment(weight)} \\ \{i, j\} \end{array} \right.$$

Calculations of Tj – critical path of evolution chains-graphs in case of «1-1-1-1» are the following:

$$\left. \begin{array}{c} T_0 \\ T_1 = T_0 + "1" = 0 + 12 = 12 \\ T_{1\text{-}1} = T_1 + "1" = 12 + 7 = 19 \\ T_{1\text{-}1\text{-}1} = T_{1\text{-}1} + "1" = 19 + 5 = 24 \\ T_{1\text{-}1\text{-}1\text{-}1} = T_{1\text{-}1\text{-}1} + "1" = 24 + 4 = 28 \end{array} \right\} \Rightarrow 28$$

Thus, upon calculations of all chains-graphs we attain the following (Figures 6 a and b).

 DOI: 10.1051/shsconf/20173501075

The advantage of such analysis is that it allows determining interaction effect and evaluating security of considered organizations.

The analysis shows that sustainability and adjustability is higher in diversified organizations, whereas recovery indicators are the highest in one-field organizations.

Organizations	Critical paths of evolutional chains-graphs						max rang		
	Sustainability		Ajustability		Recovery		Sustainability	Ajustability	Recovery
One-field	1-1-1-1	28	1-1-1-M	33	0-1-0-M	19	46	33	28
	0-1-1-1	31	1-1-M-M	5	0-0-1-M	28			
	0-0-1-1	46	1-M-1-1	5	0-1-M-M	17			
			1-1-M-1	5	0-1-M-1	4			
					1-M-1-1	6			
Diversified	M-M-M-M	76	M-M-M-1	51	0-1-M-M	9	76	51	17
	0-M-M-M	18	M-1-M-1	3	M-M-0-M	3			
	0-0-M-M	31	M-1-M-M	5	M-0-0-M	17			
					M-0-M-1	4			
					M-0-1-M	1			
					0-M-1-M	5			

a)

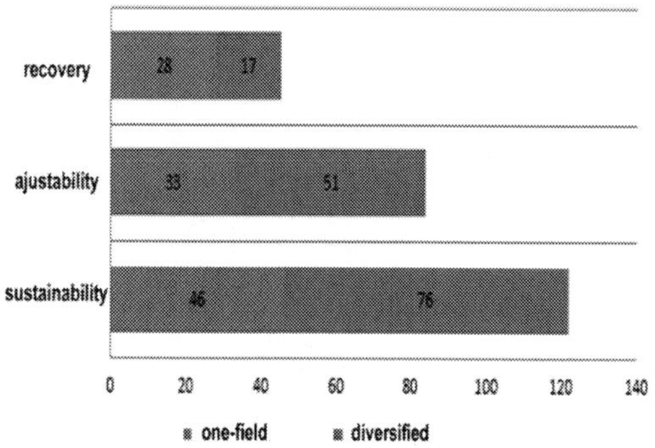

b)

Fig. 6. Max rang of chains-graphs.

These results may be determined by instabilities in requirements for each type of activity under conditions of constantly changing legislature on one hand and opportunities for formation of more secure business at the expense of diversification in accordance with changing conditions on the other.

References

1. About the free port of Vladivostok: the federal law of the Russian Federation of 2015. http://www.consultant.ru/document/cons_doc_LAW_182596/
2. About the territories of the advancing social and economic development in the Russian Federation: the federal law of the Russian Federation of 2014, no. 473-FL, Russian newspaper, **6571** (2014)
3. The territories of the advancing social and economic development. http://minvostokrazvitia.ru/
4. The territorial body of the Federal State Statistics Service for Primorsky Krai. http://primstat.gks.ru/
5. T.Yu. Shkarina, O.A. Chudnova, *Quality Management* (University textbook Far Eastern Federal District, 2015)
6. E.V. Yarotskaya, Scientific-methodical electronic journal "Концепт", **20**, 4391 (2014). http://e-koncept.ru/2014/ 55143.htm
7. T.Yu. Shkarina, S.A. Gruzdeva, Economics and administration: scientific journal, **1**, 102 (2010)
8. ISO 9000:2015 Quality management systems, Fundamentals and vocabular
9. Academic Dictionaries and Encyclopedias. http://dic.academic.ru
10. L.R. Batukova, Journal of Vestnik of SibGAU, **2(15)**, 110 (2007). http://elibrary.ru/ contents.asp?issueid=456573
11. L.R. Batukova, G.S. Mikhalev, Journal of Vestnik of SibGAU, **2(35)**, 212 (2011)
12. K. Berg, *Theory of graphs and its applications* (1962)
13. R. Diestel, *Graph Theory, Electronic Edition* (Springer-Verlag, NY, 2005)
14. A.S. Geida, SPIIRAS Proceedings, **2(17)**, 234 (2011)
15. V.V. Yudin, O.A. Chudnova, Bulletin Of The Russian Academy Of Sciences: Physics, **10**, 1405 (2001)
16. V.V. Yudin, O.A. Chudnova, Journal of Crystallography Reports, **47(2)**, 189 (2002)
17. V.L. Sharstnyov, E.Yu. Vardomatskaya, Management of economic systems, **05** (2006)
18. A.V. Zemtsov, Methodical journal: Bank crediting, **6** (2008).http://www.reglament.net/bank/credit/2008_6.htm
19. GOST R 56716-2015 Project management. Project network techniques. Descriptionsandconcepts. http://docs.cntd.ru/ document/1200127270
20. GOST R 21500-2014 Guidance on project management.http://docs.cntd.ru/document/1200118020
21. E. Börger, R. Staerk, *Abstract State Machines: A Method for High - Level System Design and Analysis* (Springer - Verlag, Berlin, 2003)
22. I. G. Abramova, *Project management based on network models* (Samar. State. Aerospace. Un-t, Comp, 2007)
23. V.N. Burkov, D.A. Novikov, *Elements of graph theory*. http://www.mtas.ru/about/
24. M.A. Korotkov, *Methods of network planning* (2003)
25. R.V. Dopira, R.Yu. Kordyukov, Product and system program, **2(106)**, 22 (2014)

 DOI: 10.1051/shsconf/20173501076

Prospects of using new non-standard employment forms at Russian enterprises

Inga Tsygankova[1*], *Natalia* Ivanova[2]

[1]St. Petersburg State Economics University, 191023, Saint-Petersburg, Russia
[2]Peter the Great St. Petersburg Polytechnic University, 195251, Saint-Petersburg, Russia

Abstract. The article presents the theoretical analysis of non-standard employment new forms, the reasons of their origin, the advantages and disadvantages for the employee. The authors identified the most promising forms of non-standard employment, which give advantages to wage workers and improve their position in the labor sphere. Based on the results of the study of new non-standard employment forms at Russian enterprises the dependence between a particular form employment and the kind of economic activity, in which the enterprise operates, as well as the category (or qualification group) of employees has been identified. The spheres, in which there is a potential for the development of non-standard employment new forms, are shown.

1 Introduction

Non-standard employment has become a constant phenomenon, both on the Russian labor market and on the labor markets of foreign countries. The scale of its spread increases every year. In recent decades, new forms and models of non-standard employment have appeared, which there were not before (outsourcing, freelancing, tele-labor, etc.).

A number of publications by domestic and foreign authors are devoted to the problems of non-standard employment, but it remains an unexplored phenomenon, requiring further research and development, because, on the one hand, various forms of employment have been studied only fragmentarily. On the other hand, they significantly influence on the socio-economic position of the worker in the labor sphere, improving it or worsening. It ultimately affects the development of the country whole economic system.

In the domestic literature there is no single interpretation of the term "non-standard employment". More often non-standard employment includes all kinds of employment, which differ from the classical model of capitalist wage labor [1, p. 29].

Many foreign researchers (S. Flüter-Hoffman, J. Solbrig, K. Linnenkohl, R. Winniger) connect with standard employment such kind of employment as the full-time employment on the basis of an indefinite or long-term employment contract under leadership of a manager or employer [2, p. 6; 3, p. 13, 4, p. 9]. Thus, nonstandard forms by "default" include all the others, that are not related to this definition.

A number of Russian researchers, considering various forms of employment, do not use the term "non-standard employment" [5, p. 77]. Many foreign authors also do not use this term (A. Büssing, A. Drodofsky, N. Warkentin, U. Hellert, H. Klein-Schneider) [6-9].

Some authors use the term "forms of labor relations", rather than "forms of employment", examining so-called sourcing forms (outsourcing, insourcing, crowdsourcing) [10]. But we believe that not all these forms of the relationship between the employee and the employer can be considered as forms of employment or forms of labor relations. According to the Labor Code of the Russian Federation, labor relations are relations, based on an agreement between an employee and an employer on personal performance of an employee's labor function for a fee [11]. So, crowdsourcing is the involvement of a people wide range for using their creative abilities, knowledge and experience to solve any tasks or problems of the enterprise with help of telecommunication and Internet technologies. Often it does not include working at the enterprise and receiving labor reward. In our opinion, crowdsourcing is more a technology to attract new ideas in the organization and search of their implementation opportunities, rather than the form of employment.

Classifications of non-standard employment forms, presented in domestic and foreign sources, are very diverse. For example, according to different authors, non-standard employment includes as temporary, seasonal, home-based employment, employment on calls, part-time, so and informal employment, incomplete, overtime, casual, secondary employment [12], distance employment, borrowed labor [13].

Practically in all publications of German authors forms of non-standard employment are considered from the view of their flexibility in the labor market (N. Warkentin, C. Flüter-Hoffman, J. Solbrig, A. Hoff, C. Hahn, J. Gutman, J. Hüsgen) [8, 2, 14, 15, 16]. A. Hoff

* Corresponding author: icygankova@list.ru

identifies such forms of flexible employment as work on standard schedules with flexible start and end of the working day, the flexible sliding working time, working time on trust with presentation of the final result [15, pp. 5-13]. P. Hanau и A. Veit considers work time accounts as a form of employment and their various models [17]. F. Lorenz и G. Schneider allocates among flexible forms of employment working time on trust, work time accounts and flexible models of part-time work and shift work [18]. The publications of N. Kordey present the advantages and disadvantages of the tele-labor and its various models [19].

2 Formulation of the problem

Among all the variety of the non-standard employment forms, it is possible to highlight out "new", emerging at the beginning of the 21st century, during the period of the formation of the postindustrial economy, and "old", "traditional" forms, which were spread even before the 1990s., both in countries with a market economy, and in the planned socialist economy, that existed in the USSR, or in the economy of the Russia transition period in the 1990s. [20, pp. 72-73]. As a basis, the authors took such classification indicator, as a period of emergence, since the employment forms which are characteristic for the information age were formed precisely in the last decades.

To traditional forms of non-standard employment it is possible to carry: incomplete employment, temporary, non-permanent or accidental employment, self-employment, home-based labor, work on sliding schedules, shift work. The introduction of non-standard employment "old" forms was due, for the most part, to the specifics of the enterprises work (orientation to the internal environment), while the emergence of new forms was caused by the influence of external environment factors.

Thus, the emergence of the majority of "new" forms and models of non-standard employment became possible due to the development of information technologies.

New forms of non-standard employment should be considered in more detail, since they have prospects for development on the Russian labor market in subsequent years. According to the authors, it is necessary to investigate the advantages and disadvantages of non-standard employment forms, since this aspect of the problem influences on the position of the worker in the labor sphere, the economic system as a whole, largely determines the state of the domestic labor market and the prospects of the separate forms development and models of employment in future.

The object of work are selected Russian enterprises, using forms of non-standard employment (practical significance).

The scientific novelty of the research is contained in the author's approach to the classification of non-standard employment new forms, adapted to Russian enterprises conditions, identifying the reasons of their appearance, assessing the impact on the position of the employee and revealing the most promising forms.

3 Specificity of non-standard employment new forms at the enterprise

The main reasons of the non-standard employment new forms emergence, in our opinion, are:

- the aspiration of the employer to reduce labor costs;
- the need to change the organization of labor in order to meet the needs of consumers (the organization of enterprises work or separate departments on weekends and holidays, at night, etc.), that was accompanied by the flexibility of using working time;
- the aspiration of the employee to improve his position in the sphere of labor and employment (increase the chances of finding a job, improve the ratio of free and working time, choose the most suitable days and hours of work).

The non-standard employment new forms can be divided into four groups:

- sourcing forms (outsourcing, outstaffing, staff leasing, insourcing);
- various forms of distance employment (tele-labor, work at a virtual enterprise);
- flexible forms of work (working hours with flexible start and end of the working day, the flexible sliding working time, accounts of working time, etc.);
- freelancing.

Since we evaluate the impact of employment separate forms on the position of the employee at the enterprise, the form of non-standard employment like freelancing, will not be considered, because freelancers are in most cases self-employed. Let us analyze in more detail the first three groups of non-standard employment new forms (table 1).

Sourcing forms were developed in Russia in the second decade of the 21st century. The most common form in this group was outsourcing - transfer by an organization of secondary functions on importance or of non-core functions to another organization for a sufficiently long period (a year and more). Until 2016 on the domestic labor market outsourcing, along with such forms as outstaffing and staff leasing, developed within the framework of borrowed labor [21, 22]. But from January 1, 2016, Russia accepted a law, which significantly limited the use of borrowed labor [23], in order to protect staff, working in these forms. In this regard, the scales of the employment above-mentioned forms using has significantly decreased, but remained on the Russian labor market.

In recent years Russian enterprises have been developing insourcing - the transfer of a "non-profile" project, which is usually performed by a third-party organization, employee or department within the enterprise. Insourcing has appeared on the Russian labor market quite recently and is not connected with borrowed labor. As a form of employment, it differs from part-time employment in that employees, who implement insourcing projects, do not work on them on an constant basis, but from time to time when it becomes necessary. In this case they work permanently in a completely another position.

The appearance of absolutely all sourcing forms was caused by the aspiration of the employer to cut costs. Therefore, these forms give more advantages to the employer than to the employee. Otherwise, he simply avoids of using them.

Adoption of a law, restricting the use of borrowed labor, contributed to improving the position of borrowed workers, at the same time their position on the labor market is worse than ordinary, "standard" workers.

In the last decade distance employment has spread on the world labor market due to the need of the labor organization changing from the side of the employer and in connection with the aspiration of the employee to improve his position in the sphere of labor. Tele-labor is a labor activity, which is organized remotely from the employer and can exist thanks to the development of information and telecommunication systems [24]. Tele-labor can be exclusively remote, or alternative – when an employee comes to the enterprise at least once a week [25, pp. 57-59]. The second model is more spread and gives more advantages (see table 1).

This form of distance employment, like work at a virtual enterprise, appeared on the Russian labor market quite recently. A virtual enterprise is usually understood as an enterprise, consisting of a community of geographically separated economic entities, which cooperate in the production process, using primarily electronic means of communication. In this case, there is generally no main, "parent" enterprise, and employees work at home. They are mobile, use the satellite bureau and communicate among themselves through telecommunications. Virtual enterprises exist in different kinds of economic activity, but in European countries and the United States their number is the highest in information technology and industry.

On the world market examples can serve as Mentor Graphics (design of electronics and electrical engineering, development of software products for industrial design); Katana Software (development of software products); Core Geomatics (geological exploration in the oil and gas and construction sectors); Elbes (viewing and selection of the most successful and interesting projects in the field of culture and their further promotion). In Russia examples of virtual enterprises are NOU "Intuit" – distance learning in the field of information technology, a virtual pharmacies network "Pilulu.ru". At the moment the number of virtual enterprises is small. But on the Russian labor market there are companies, that provide consulting services for virtualizing the customer's business (for example, LLC "Web segment").

Table 1. The non-standard employment new forms at the enterprise and their characteristic *.

Forms of employment	Reasons of appearance	Advantages for an employee	Disadvantages for an employee
1. Sourcing forms (outsourcing, outstaffing, staff leasing, insourcing)	The aspiration of the employer to reduce labor costs	Opportunity to save / receive a working place and / or a certain income. Enrichment of labor and gaining experience of work in an adjacent or even completely different sphere (for insourcing)	Decrease of the employment stability and income from labor activity. Limited opportunities for career growth. Decrease of labor productivity and satisfaction with labor. The emergence of labor conflicts
2. Forms of distance employment (tele-labor, work at a virtual enterprise)	The aspiration of the employer to change the organization of labor in order to meet the needs of the consumer. The aspiration of the employer to cut costs of: - the creation and content of certain working places at the enterprise; - the payment of workers in their region in the event that he can use cheaper, but no less qualitative labor of workers from other regions / countries	The opportunity to allocate working time at own discretion. Minimum control by the employer. The opportunity of a combination of professional life and training, education of children and training. Increasing the duration of free time in the case of effective work. Getting the opportunity of employment for people with limited opportunities and workers from remote or depressed regions	Risk of a decline in the work quality. Risk of partial loss of competence. Decrease of opportunities for career growth and position promotion. The danger of social isolation. The risk of a decline of labor productivity and an increase in the duration of working time in the working place at home
3. Flexible forms of work (working time with flexible start and end of the working day, the flexible sliding working time, accounts of working time, etc.)	The aspiration of the employee to improve his position in the sphere of labor and employment. The aspiration of the employer to change the organization of labor in order to satisfy the needs of the consumer	The opportunity of own working time planning. Higher motivation for labor and satisfaction with labor. The opportunity to combine work and learning	The ban on pay increasing as a result of increase in working time (when the working time accounts using)

* The table was developed by the authors

Flexible forms of work are characterized by flexibility in the use of working time, and therefore give a number of advantages to both the employee and the employer [26, pp. 67-70; 27]. The most widespread is a work on a flexible schedule with a free start and end of the working day, flexible sliding working time, working time accounts. In Russia work on a flexible schedule with a free start and end of the working day and work on sliding schedules with a free start and the end of the working day are more often used in trade and services than in industry. These forms of work have found application in the network of supermarkets "Lenta" (trading), in LLC "SPA Interro" (software development, rendering services in the field of computer technologies) [20].

The working time accounts – a non-standard employment form, contributing to the rational use of working time [28, pp. 14-18]. This form can be introduced at the enterprise for all staff, or for a separate category of employees. For the employee a so-called "personal account" is created, to which the difference between normative and actual working time is credited. The working time accounts contain information on temporary accruals and arrears in time, which must be compensated till a certain term, or within a certain time period (for example, a month). If the actual worked time will be more than the normative time worker should have a rest period. In European countries the accounts are widely distributed (used at the enterprises of JSC "German Post", "Glass Workshops of Lower Saxony", "Brewery of North Rhine-Westphalia", JSC "Volkswagen", etc.). Russian legislation limits the possibilities of their application.

4 The results of the non-standard employment new forms application at Russian enterprises

The research of the non-standard employment certain forms application at the Russian enterprises was conducted in 2013 and 2016. Such an analysis does not allow us to assess the degree of the various forms spread in the country as a whole. But it gives a representation of the development trends of individual forms in enterprises of various kinds of economic activity, and also allows us to understand, for which groups of workers they are predominantly used.

The information, obtained in the course of the enterprise survey (see table 2), show, that non-standard employment was applied at all considered enterprises. But at the same time, the proportion of people, using of non-standard employment forms, was small and amounted in 2013 to no more than 20 % of the total number of staff. An exception was the enterprise of the LLC "SPA Interro", where almost half of the employees worked with the use of non-standard employment. The enterprises used mainly traditional forms of employment: incomplete, temporary employment and shift work.

New forms of non-standard employment, which give advantages to employees, were used only by LLC "SPC Kriochrome" and LLC "SPA Interro". It can be concluded that the most widely spread new forms of work in the field of information technology (the range of forms used in the LLC "SPA Interro" is the most extensive), while in construction – on the contrary, the traditional forms are used.

Distance employment and work flexible forms are used mainly for specialists, while temporary employment and outsourcing are – for low-skilled workers. The choice of standard or non-standard employment is not due to the wishes of employees, but in most cases – to the specifics of the enterprise's activities (equipment operation modes and production technology). At two of the four surveyed enterprises the employee, where using non-standard employment, can plan the beginning and end of the working day himself, that is, it has advantages in the distribution of his working time.

According to the enterprises survey results in 2016, the following ones can be selected.

At the enterprises of LLC "SPC Kriochrome" and LLC "SPA Interro", which were included in the sample in 2013, the forms of employment, used for employees, have not changed, but the share of employees, for whom they are used, has increased (by 15 % at LLC "SPC Kriochrome", by 10 % in LLC "SPA Interro"). The opinion is confirmed, that borrowed labor and temporary employment are mainly used for low-skilled workers (LLC "Alternative"). As in 2013, the choice of standard or non-standard employment is mainly due to the specifics of the enterprise's activities. At some enterprises due to the specifics of the activity the use of work flexible forms is not possible (FSUE "State Corporation for ATM").

5 Conclusion

Based on the results of the theoretical and practical research of the non-standard employment new forms application at Russian enterprises, the following conclusions can be made.

1. Now the size of the spread of non-standard employment new forms at Russian enterprises is low, but the number of workers, using them, is gradually increasing.

2. It is advisable to use new forms of employment at industrial enterprises, since they determine the effectiveness of the economic system functioning, are the sphere, where innovations are applied and scientific and technological progress is taking place.

3. The aspiration of employers to reduce labor costs has led to the emergence or increase of the employment forms prevalence, that negatively influence on the position of employees at the enterprise.

4. The change of the regulatory and legislative framework in the sphere of the borrowed labor application in Russia has improved the position of employees. However their position is worse than that of

ordinary workers at the enterprise, and the level of wage is significantly lower. Despite the fact that the scales of the borrowed labor application has declined, it remains enough a spread phenomenon for Russian enterprises.

Table 2. The use of non-standard employment at some Russian enterprises in 2013-2016 *.

Business name	Kind of economic activity	Forms of non-standard employment, used in the enterprise	The proportion of employees, working in the framework of non-standard employment forms (from the total number of personnel),%	Categories or groups of employees, for whom non-standard employment forms are used	The employee has the right to choose the form of employment (yes / no)	The employee has the opportunity to allocate working time at his discretion (yes / no)
LLC "SPC Kriochrome"	Chemical industry (production of chemically pure solvents)	Incomplete employment (for students-trainees), distance employment (tele-labor)	1-5 (2013) 10-20 (2016)	Specialists of the engineering and technical profile, students-trainees	No	Yes
LLC "Vozrozhdenie"	Construction, production of building materials	Temporary employment (for workers), outsourcing (for workers)	10-20 (2013)	Workers	No	No
CJSC "Kokkomyaki Quarry"	Extraction and processing of natural resources	Temporary employment (for workers), shift work	6-10 (2013)	Workers	No	No
LLC "SPA Interro"	Software development, services provision in the field of computer technology	Distance employment, work with flexible start and end of the working day, flexible part-time employment, flexible shift work	31-50 (2013) 41-60 (2016)	Specialists on the cable networks installation, specialists on video surveillance systems, WEB-developers	Yes, in some cases (depending on position)	Yes
LLC "Alternative"	Laying of networks and pipes	Borrowed labor (outstaffing, outsourcing) (for workers of low qualification, for non-core specialists), temporary, inconstant employment	6-10 (2016)	Workers at the facility, much less often - non-profile specialists	No	No
FSUE "State Corporation for ATM"	Aircraft Management	Shift work without flexible schedule	11-30 (2016)	Specialists, workers	No	No

* The table was developed by the authors

5. The greatest chances for success have those enterprises, in which the interests and benefits of employees and employers are balanced. Such forms of employment as distance work (tele-labor and employment at a virtual enterprise), flexible forms of work contribute to the employees position improvement in the labor sphere.

6. Distance employment has a good development potential in the coming years, as it gives a number of advantages to both and employers.

7. The opportunities of the employment certain forms application are determined by the kind of economic activity and the enterprise work specifics. The enterprises, operating in the field of information technologies, are the most susceptible to new forms of employment. Flexible forms of work and distance employment are used primarily for specialists and other highly skilled employees, borrowed labor is used mainly for workers, engaged in low-skilled labor.

References

1. I.D. Kotlyarov, Issues of economics regulation, **4**, **6**, 29 (2015)
2. C. Flüter-Hoffman, J. Solbrig, *Arbeitszeitflexibilisierung: Erfolgskonzept auch für kleine und mittlere Unternehmen* (Köln, 2003)

3. K. Linnenkohl, *Arbeitszeitflexibilisierung. Die Unternehmen und ihre Modelle* (Heidelberg, 2014)
4. R. Winiger, *Praxishandbuch flexible Arbeitszeitmodelle: Methoden, Konzepte und Vorgehensweisen für die Einführung und Optimierung flexibler Arbeitszeitmodelle* (Zurich, Praxium-Verlag, 2014)
5. I.M. Aliev, N.A. Gorelov, L.O. Ilyina, *The Economy of Labor* (Moscow, Publishing House Yurayt, 2016)
6. U. Hellert, *Arbeitszeitmodelle der Zukunft* (Haufe-Lexware, Freiburg, 2008)
7. A. Büssing, A. Drodofsky, K. Hegendörfer, *Telearbeit und Qualität des Arbeitslebens* (Hogrefe-Verlag, Göttingen, Bern, Toronto, Seattle, 2003)
8. N. Warkenting, Flexible Arbeitszeiten: Eine Übersicht der verschiedenen Modelle. http://karrierebibel. de/flexible-arbeitszeiten-uebersicht-der-modelle/.
9. H. Klein-Schneider, *Flexilbe Arbeitszeit – Vertrauensarbeitszeit* (Memmingen, Bund-Verlag, 2011)
10. R.A. Dolzhenko, Kadrovik, **6**, 54-64 (2016)
11. The Labor Code of the Russian Federation no. 197-FL of 2001. www.consultant.ru.
12. V.E. Gimpelson, R.I. Kapelyushnikov, *Non-standard employment in the Russian economy* (Economics State Higher School Izd. House, Moscow, 2006)
13. R.P. Kolosova, T.O. Razumova, M.V. Ludanik, *The population employment forms in the innovative economy* (Max Press, Moscow, 2008)
14. C. Hahn, *Flexible Arbeitszeit* (C.H. BeckVerlag, 2014)
15. A. Hoff, *Gestaltung betriebliche Arbeitszeitsysteme* (Wiesbaden, Springer Fachmedien, 2015)
16. J. Gutman, J. Hüsge, *Flexible Arbeitszeit: Wie Sie moderne Konzepte und Modelle nutzen* (Bersgel, Haufe, 2005)
17. P. Hanau, A. Veit, A. Hoff, *Recht und Praxis der Arbeitszeitkonten: Wertguthaben, Altersteilzeit, Flexikonten (Aktuelles Recht für die Praxis)* (C.H. BeckVerlag, 2015)
18. F. Lorenz, G. Schneider, *Vertrauensarbeitszeit, Arbeitszeitkonten, Flexi-Modelle: Konzepte und betreibliche Praxis* (VSA, 2004)
19. N. Kordey, *Telearbeit erfolgreich realisieren: Das Umfassende, Aktuelle Handbuch Für Entscheidungsträger Und Projektverantwortliche* (Braunschweig – Wiesbaden, 2013)
20. I.V. Tsygankova, V.A. Bazzhina, O.Y. Nikishina, Russian Entrepreneurship, **24(270)**, 71-85 (2014)
21. A.V. Kumakova, N.G. Ivanova, *SPbGPU Science Week: materials of the scientific and practical conference with international participation. SPbGPU Humanitarian Education Institute*, 233-236 (2014)
22. A.V. Kumakova, N.G. Ivanova, *SPbGPU Science Week: materials of the scientific and practical conference with international participation. SPbGPU Humanitarian Education Institute*, 231-233 (2014)
23. Federal Law 116-FL of 2014 "On Amending Certain Legislative Acts of the Russian Federation". www.consultant.ru.
24. M. Sieber, P. Recknagel, *Ich bin dann mal im Home Office* (Leobell, 2014)
25. I.V. Tsygankova, Saratov State Social and Economic University Bulletin, **1(20)**, 57-59 (2008)
26. I.V. Tsygankova, Man and labour, **2**, 67-70 (2008)
27. M. Zölich, M. Oertig, V. Calabro, *Flexible Workforce - Fit für die Herausforderungen der modernen Arbeitswelt? Strategien, Modelle, Best Practice* (Bern, Hauht, 2016)
28. I.V. Tsygankova, Novosibirsk State University Bulletin, **4**, 14-18 (2008)

Quality management technology impact on small enterprise production activities

Natalia Ivanova[1,*], *Zoya* Lavrova[2]

[1] Peter the Great St. Petersburg Polytechnic University, 195251, Saint-Petersburg, Russia
[2] St. Petersburg State Economics University, 191023, Saint-Petersburg, Russia

Abstract. A modern quality management technology, the so-called quality management system (QMS), can improve the indicators of production activities of any company. Due to the advantages of a small business such project can be implemented faster and at lesser costs. The conducted research of the QMS development in accordance with ISO 9001-2015 at a particular small St. Petersburg enterprise in 2016 resulted in the growth of production output and labor productivity. The obtained results of the standards introduction in the troublesome divisions of a small firm demonstrate the attractiveness of the given technology for the top management of a corporation.

1 Introduction

Enterprises and their types of business associations (firms, companies, corporations, trusts, syndicates, etc.) are a product of the industrial age. Due to realizing of unifying processes under the rapid development of information technologies in the late twentieth century they have been turned into the international network structures [1]. As a result, created a need in the management of them, which led to the creation of a scientific management theory, the main object of which became a large organization. It is known for such companies was typical a mechanistic / bureaucratic type of management (on vertical). This was facilitated by the emergence of management technologies related to various functional managerial spheres of associations, among them strategic, innovative, financial, industrial, environmental, international management, quality management, human resources, organizational behavior and so on) [2].

Particular importance belongs to technologies of quality management, directly related to safety of manufactured products or services for a person and his environment [3]. Among them there are the control charts Shewhart V., G. Gantt chart, seven quality control tools, Six Sigma, etc. In a greater degree they were directed at the quality control of the production process, and in 1951 A. Feigenbaum embodied them in the concept of total quality control (TQC) [4, pp. 11-17]. The importance of quality management technologies is confirmed by the data: marketing researches show, that the selection this or that production on 75% determines its quality [5].

Gradually mechanistic management style was replaced on a more flexible organic (on horizontal) under the influence information technologies in conditions of uncertainty of factors of external environment post-industrial era [6]. This was made possible in the framework of quality management technology, which called a quality management system (QMS). It allows to work out accurately documented procedures (business processes), included in the production process on base of international standards ISO 9000 and under influence of total quality management concept (TQM).

It is considered in the existing practice of international business the development and implementation of a certified quality management system at the enterprise testifies to the high quality of manufactured products (services), forms the confidence of consumers and strengthens the earned image on occupied market segments [7]. So, a certified quality management system of 87-90% of the firms confirms their high level of competitiveness, while at 73% of companies – ensures their a stable financial indicators [8].

2 Formulation of the problem

The quality problem is actual in any historical period. For example, in the industrial age initially manufacturer himself had to solve this problem both in the conditions of becoming production and growing production volume, that inevitably accompanied by a lot of marriage. The manufacturer was directly interested in the quality, because he exercised a choice of production and defined the technical requirements for it. The result of the unifying processes realization at the enterprise and of implementation of information technologies in the post-industrial era became the variety of goods and services

* Corresponding author: ing4@bk.ru

offered by a huge number of different manufacturers who are forced for the conquest the market take into account the requirements of consumers to quality of products / services in terms of safety for human use [3].

Hence, goals of an international standardization and certification are understandable in the conditions of globalization and aim at the development international trade and scientific and technical exchange between countries, that expressed in the creation of international standards ISO 9000. According to data of the international organization ISO [9], more than one million certificates of conformity to ISO 9001 were given out worldwide.

In the scientific literature extensively discusses the creation of a quality management system (QMS) in various enterprises. International quality management practice considers the approaches related to Integrating Lean Six Sigma with ISO 9001-2015 [10] and ISO based on TQM implementation approach as major management approaches, used in foreign companies [11, 12]. At the same time these enterprises are interested in economic and social effectiveness of certification on the ISO 9000, including external customer satisfaction and internal customer satisfaction, which are assessed there [13, 14].

Russian scientists study and use positive international experience in quality management. In particular, in Russia is known method of the German specialist Stefan Hensch, implemented in the company "Hensch development of quality" on the base of process approach and project approach [15].

A number of Russian experts, noting the impact of the quality management system on the success of the company, points to the formal conducting of this project on its development and implementation [16-18]. Often getting the certificate of conformity, which gives the company a competitive advantages, such as when participating in tenders, receipt of benefits in lending and insurance, etc., is not accompanied by the reorganization of the production activities on quality assurance. This is confirmed by the results of the poll workers about the reasons for the ineffectiveness of the quality management system: 73% – lack of training on QMS, 67% – lack of interest of leadership, 63% – lack of quality department and of qualified specialists in quality management, 45% – lack of interest and of involvement of staff, 43 % – insufficient allocation of funds and the lack of systematic accounting the quality costs [19]. Dominant leadership role in resolving quality issues is obvious, on which falls of wine in the amount of 80-85%, and only 15-20% depends on the direct perpetrators [20].

In the articles in addition to these disadvantages the positive results of the QMS functioning at the enterprise are presented. During the period 2013-2015 at the East-Siberian railway ("RZD" branch) marked increase in indicators: labor productivity – by 4%, passenger turnover – by 0.6%; customer satisfaction – by 0.7% [21].

In general, in Russia the situation with the development and implementation of the QMS is improved, it confirms the change in the index of global competitiveness of the Russian Federation: it moved from 53th to 45th place (8 positions) in 2016 compared to 2015 and 45th at the 43th place (another 2 positions) in 2017 compared to 2016 [22, 23].

This is especially true of small businesses whose leaders are busy with not only the execution of strategic functions, but of operating functions a greater extent than in large companies, where they have the functional deputies.

At the same time to small businesses, who organize QMS certification, there are the following characteristics:

1) small, skilled and cohesive team;

2) a wide range of tasks undertaken by each employee;

3) simple and direct communication between employees;

4) limitation in the decision-making (till one person);

5) flexibility, real customer focus;

6) often lack of documentation;

7) limited financial resources [24].

These characteristics together with the smaller scale of the activity of small enterprises offer great opportunities for the introduction of the QMS and reduce the duration of the project stages for its improvement.

Therefore, by the object of the work is selected the small production enterprise (by the example of LLC "ZETA "CONTACT"). Its director has realized the importance of resolving quality problem using QMS implementation technology (practical significance).

Scientific novelty of the research is contained in the planning for the implementation of QMS in large companies like in a small enterprise, starting with the problematic divisions, gradually expanding the coverage of the QMS of all large processes that describe activities of the company. In this case it's enough only one qualified person who studies the specifics of departments, services during a shorter workweek. Thus, the cost of implementing QMS will be minimized, and the quality management system itself will become more attractive for the top-management of large companies.

3 The evolution in the field of Quality Management

In the scientific theory of quality management most often there are 5 stages of quality evolution, accompanied by the standards improvement process.

1. Focus of manufacturer on finished product quality, which aimed at the defective goods rejection by inspection based on the methodology of statistical quality control (SQC) in the earlier period of production (until the 1950s.).

2. Increased attention of manufacturer to the product designing quality and the manufacturing process quality due to the wide implementation of A. Feigenbaum total quality control concept (TQC) (1950-1960s.).

3. Studying of customer needs and suppliers' products quality in conditions of tightening market competition in the 1960-1970s.

4. Inclusion in quality products concept not only quality of its designing and production process, but also

its after-sales service quality, spreading the "quality assurance system" concept (1980-1990s.).

5. The direction of manufacturer's efforts on the errors prevention as a result of intensive development of total quality management (TQM) a new concept, focused on the consumer. Quality management covers all the enterprise activities (1990-2000s.) [4].

At present in the emerging global society is appropriate to speak about the beginning of the sixth stage of the evolution of quality (since 2000), associated with the increasing role of human knowledge in economics and management. It is recognized, knowledge of people creates ideas, forms a team, which their realizes and helps to transform them into innovation. Already now modern technology are beginning to rule the whole world. In this situation the quality acquires a universal character, embodied in types of a public security [3] on the supranational level of cooperation between different countries. This is reflected in international standardization. The evolution of it in compliance with ISO 9000 is considered below.

In 1987 a series of international standards ISO 9000, based on standards of the British Standard Institution (BSI 5750), was developed. It appeared as a response to the globalization of the economy. At the same time, the first edition of standards has not covered a number of specific types of products, including software and services.

In 1994 the technical committee ISO (ISO / TC 176) revised the first edition of standards based on their application experience in different business areas of different countries and on approved recommendations of the national certification bodies. As a result, three standard models appeared: for quality assurance in designing, development, production, installation and servicing (ISO 9001-1994); for quality assurance in production, installation and servicing (ISO 9002-1994); for quality assurance in finished products control and testing (ISO 9003-1994). This is forced the companies to undergo the certification procedure once again while expansion or contraction of business scope took place. In spite of the specified disadvantage of three standards, here has been declared the process approach and became to use the concept of "preventive action" to prevent a potential non-compliance of requirements.

The impact of total quality management concept (TQM) is reflected in the third edition of ISO 9001-2000 "Quality Management System. Requirements", where a universal model for all organizations, size, industry, product or service regardless, has been established. It is based on the concept of industrial business processes: a lot of attention paid to the functioning and controlling processes according to the changes in consumer satisfaction and strivings to continued improvement.

The universality of QMS, developed in 2000, is confirmed by the fact that the fourth edition of standards ISO 9001-2008 [25] offered no significant changes. They concerned the clear understanding of certain requirements, also the definitions of some terms were clarified and simplified. In both versions the accent was made on managers forces transfer from identifying and eliminating inconsistencies to the preventing their occurrence, eliminating or minimizing the possibility of their occurrence during the process of product creation or services and throughout the whole product lifecycle [26]. This led to the appearance of a well-known rule: "tenfold costs increase on the defect elimination by its detection on the next stage of the lifecycle, not on given one" [27]. Thus, corrective actions become nearly the most important in quality management system (QMS). Without managing costs using quality calculating methods, constant increase of products (works, services) quality requirements may cause the increase of costs on its providing [28, pp. 23-43]. Then it is appropriate to speak about the economic and social efficiency of QMS in the context of conformity of intended results and purpose and degree of its achievement taking into account the costs.

Further development of the information society has caused appearance of the fifth standards edition ISO 9001-2015 [29]. It cardinally reviewed QMS created in company, which means its personal choice independence not only in the term of volume but also in term of the documented procedures content [30]. The benefits of companies with implemented certified quality management system are:

- management efforts concentration on the organization risks and opportunities caused by external and internal organization environment, connected with the appearance of such thing as a "risk-oriented thinking" in the text of ISO 9001-2015;
- creation of documented and undocumented information complex, related to organization knowledge (procedures, instructions, regulations, training techniques and other personnel competence, formed during its time of work in a particular company).

Thereby, the new version of ISO 9001-2015 is much more oriented to the management improvement, depended on senior management and the workers themselves. Only the top-manager, who is endowed with the responsibility for quality, can show by his own example and explain the workers with specific competencies the loyalty to the implemented quality management system of the company.

4 QMS introduction results at the small enterprise

As mentioned, the last revision of the international standard ISO 9001-2015 became the most serious of its revision since 2000. First of all, freedom is impressive, which is received by the companies in choosing the means of standard requirements implementation, that is especially valuable for small business organization. Considering that the company practically lacks specific human resources, it is possible to minimize the documentation of carried out activity if this is not necessary. Standard is focused not on reproduction of paper or electronic documents. It analyzes the possibility to use maximum information about internal and external opportunities and threats, t. e. more attention is paid to the process approach, especially to inputs and outputs (process results). The main purpose of this management by processes consists in a reduction of risks. Formed tracking procedures of risk factors are evaluated, are developed of activities to reduce risks and prevent of their appearance. Small business enterprises especially keenly feel the change in external conditions, and this has a significant impact on their functioning and further development.

The second important feature of the fifth version consists in more pronounced responsibility of management for developing and implementation of quality management system. In previous versions of the standard a representative of the quality manual was present. In other words the general director in his order gave responsible to the official for the realization of all necessary measures for the development and implementation of QMS. A result turned that the top management shifted the responsibility to someone another, if not fully, then at least partially. Now it is excluded.

These circumstances are taken into account during development and implementation of quality management system in Limited Liability Company (LLC) "ZETA "CONTACT", which is a manufacturer and supplier of contact network details for railway electrification. The investigated small enterprise is constantly expanding its range of products, which is manufactured for the energy industry. The success factors of the company on the market of St. Petersburg are named the following:

1) extensive practical work experience (since 2006);
2) understanding the needs of customers;
3) well thought-out organization of work processes;
4) rapid response to customer requests [31].

In company, where general director has personal interest in the quality management system, QMS according to the new version (fifth) of ISO 9001-2015 standard is implemented since February 2016. Here the emphasis is made not so much on the development and introduction of QMS documentation, but rather on the real functioning of processes. Therefore, the enterprise standards are developed taking into account the already existing order of organizing and conducting all kinds of activities, not to break the existing system and to comply with all the requirements of standards or other normative and technical documents for manufactured products. Basically, the content of standard is aimed at ensuring that all processes are carried out in system, without leadership intervention and without the so-called "management in manual mode".

Several departments are usually involved in the realization of any standard. At the same time the responsible executor on standard, as well as process manager organizes the work on the given direction in general. In the framework of the standard implementation they have the authority to give instructions to other managers and specialists, who participate in this project.

Today, one year later, only about 15% of the documentation of the supposed quality management system (3 standards out of 18) was implemented. Let us call these standards.

1. STO "Standards of Organization. Requirements". Implemented in June 2016. Responsible executor – representative of the general director on the quality.

2. STO QMS "Monitoring and measurement of product". Implemented in July 2016. Responsible executor – chief of the technical control department.

3. STO QMS "Keeping of materials and finished products. Loading and unloading operations". Implemented in August 2016. Responsible executor – chief of the warehouse.

Carried out the self-assessment of the company by top-management showed that the activities of technical control department (TCD) and the warehouse is the weakest link in the chain of the product life cycle.

TCD performs quality control of the finished product, including control of operations, acceptance tests, periodic tests and typical tests of product quality. Development of the standard has streamlined all of these types of controls and has revealed their phased implementation, accompanied by the filling of the necessary documents proving the correctness and completeness of control (acts, protocols, test logs, etc.). The introduction of this standard in June 2016 has allowed, according to deputy general director, completely release top-management from the management of this process. Now the management of this process is entrusted to the chief of TCD who deal with it.

Activities of warehouse extends to keeping and loading and unloading operations when sending products to consumers. The main difficulties in the warehouse work before the introduction of the quality management system standard consisted in a rapid and accurate selection of materials and components for the production of finished products. Very often, this simple procedure had being dragged on an indefinite period due to the fact that none of the warehouse workers did not know exactly what and where was. So the chief of the warehouse and the storekeepers long enough had to look for the necessary materials and semi-finished products. Introduced in August 2016 standard of organization describes the step by step the whole procedure, which implies a certain technology of placing these materials in the warehouse and eliminates the possibility of their losing, and thus minimizes the search time. In addition, in the standard are laid down certain requirements to conditions of storage and transportation, that preclude damage to materials and finished products during the execution of these processes. All this, according to deputy general director, allowed the 50% reduction of the company's leadership participation time in this process for implement of "management in manual mode".

Thus, we can talk about the positive influence of QMS to productive activities of LLC "ZETA "CONTACT". We will add another indicators, which presented in a table 1. The volume of recycled metal can be selected as an indicator of activities of warehouse and of technical control department. In the warehouse the storekeepers under the supervision of the warehouse chief must bring on charge this volume (on the number and assortment), put it on a warehouse, keep and issue the on requirement to production. The inspectors of TCD during implementation of standard "Monitoring and measurement of product" must subject all manufactured production to acceptance tests. Although they use the selective control, at which try to check out each batch, they perform sampling, test and fill all the necessary documents of the standard (acts, reports, journals, including in electronic form).

Table 1. Main indicators of industrial activity of a small enterprise in 2016*.

Quarter 2016	The volume of recycled metal , t	The average number of staff people, person	Labor productivity, tons / person
1	184.031.6	46	4000.7
2	268.387.2	53.5	5016.6
3	388.815.8	61.5	6322.2
4	380.427.5	61	6236.5

* Authors took figures in LLC "ZETA "CONTACT" and on the basis of them performed necessary calculations

According to table 1, it is obvious 2-fold increase in production volumes in the 3rd and 4th quarters compared to the first, which was accompanied by a set of new staff (15 people.). However, labor productivity still grew every quarter, reaching a maximum in the 3rd quarter, in 4th practically remained at the same level, that exceed by more than 1.5 times magnitude of the 1st quarter.

Since the labor productivity refers to indicators of economic efficiency, to a certain extent as a result of the introduction of the three considered standards, we can say about it rise. At this stage, while has not yet implemented all of the standards, and period of validity of embedded standard is short (less half a year), it is impossible to fully assess the effectiveness of implemented QMS.

Continued implementation of standards has been suspended due to a sharp increase in orders and a large booting of staff. The general director, who was the initiator of the creation of a quality management system in the company, has postponed the introduction, because he did not want that such activities gained formal. In fact after the introduction of the standard each time it is necessary to conducting training with the employees, obtaining feedback from them, the organization of checks the implementation of the standard.

At present STO "Input quality control organization" and related with it STO "Order of the formulation on permit of derogation (the deviation)" and STO "Management of nonconforming product" are prepared for the introduction. Implementation of the quality management system in LLC "ZETA "CONTACT" continues.

5 Conclusion

Carried out analysis of quality management technologies in the industrial epoch and post-industrial era in the context of their influence on the production activities of a particular small enterprise allows make the following conclusions.

1. The unifying processes, which were implemented at the micro, meso and macro levels in the first half of the twentieth century, lead to a general enlargement of production, to forming of associations of enterprises and provided consumers industrial products in large volumes. These processes, certainly had a negative influence on their quality, that caused the appearance of marriage. As a result for reworking defective items or simply recycling them have been required significant additional costs. In these circumstances, naturally, appearance of the concept of total quality control (TQC), which enriched the theory of quality management by technologies of the production process quality control (statistical control of process / 7 instruments of quality control). It is still used by modern companies.

2. In the second half of the twentieth century the further development of created international diversified corporations, offered a variety of products and services, contributed to the increase attention to the human factor. It lies at the basis of the total quality management (TQM) concept, directed to the constant improvement of the quality, which depends on the top management of the company, his responsibility and interest in it, the involvement of staff in this process, his encouragement and motivation. It allowed the development the technology of quality management – the quality management system, which is certified for compliance with international standards ISO 9000. It created a competitive advantages to large companies on a market segments for satisfying demands of consumers. Thus, initially bearing a recommendatory character ISO 9000 in the world markets have become the main measurer or guarantee of ensuring of consumers by products / services of high quality.

3. Often when implementing QMS associations of enterprises are facing with serious financial problems and the need in additional staff for the development of business processes, so for them became cheaper to buy a certificate of conformity and does not to carry out the necessary reorganization in the company. In this situation the image / brand, formed in companies, together with information and communication technologies helps them to survive long enough period of time. About quality it does not have to speak.

4. According to the authors, existing is a problem with the development and implementation of quality management system in large companies includes the low responsibility of top management (often its absent) and the lack of understanding of the QMS influence on organizational performance. In fact in the bureaucratic structure the general director got used to delegate to his deputies its powers. This also is associated with a lack at the top-management of knowledge in this area and highly qualified specialists on QMS. The described practice of step by step implementation it on a real small enterprise is proof of that.

5. Investigation of influence QMS on the production activities of enterprise is carried on the example of a small firm – LLC "ZETA "CONTACT", which effectively works in St. Petersburg. A fundamental

condition of the success the project of implementation ISO 9000 in the company is considered the interest and personal involvement of general director, who continually increases his qualifications, now is learning the MBA program. He found a qualified specialist on QMS, who develops the business processes in the desired sequence for the enterprise, constantly conducts training of staff the quality management technology, patiently explaining of its benefits.

After the introduction of standards in the TCD and in a warehouse the general director could free himself from the operative production problems, and he delegated authorities to his deputy and heads of the relevant departments. So he focused on strategic tasks, that resulted in a 2-fold increase of production volume and growth of labor productivity by 1.5 times during 2016. The growth in orders accompanied by an increase in staff, t. e. capabilities for enlargement in the future of this business are created. It further will be built on scientific basis of quality management, by using it advanced technology.

References

1. N.G. Ivanova, Herald SPbGUTD. Economic, humanitarian and social sciences, **4**, 3-10 (2016)
2. N.G. Ivanova, St. Petersburg branch of the Bobkov Russian Customs Academy. Scientists notes, **2(58)**, 19-29 (2016)
3. N.G. Ivanova, *The new economic reality, cluster initiatives and development of the industry (INPROM-2016)*, 37-43 (2016)
4. O.P. Gludkin, N.M. Gorbunov, A.I. Gurov, Y.V. Zorin, *Total Quality Management: a textbook for high schools* (Hotline-Telecom, Moscow, 2001)
5. I.V. Badiaev, A.Y. Podoinitsyna, Algorithm of the enterprise quality management system implementation, Engineering – From Theory to Practice, **57**, 124-128 (2016)
6. N.G. Ivanova, *Economy and management in the conditions of the global competition: problems and prospects*, 191-196 (2016)
7. Z.I. Lavrova, *Quality management: training manual* (SPbGEU Publishing house, St. Petersburg, 2015)
8. V.V. Okrepilov, Management Consulting, **6(90)**, 40-49 (2016)
9. Briefly about ISO. http:// www.iso.org/iso/ru/home/news_index/mediakit.htm.
10. P.A. Marques, P.M. Meyrelles, P.M. Saraiva, F.J. Frazão-Guerreiro, *IEEE International Conference on Industrial Engineering and Engineering Management*, 894-898 (2016)
11. C.-K. Chen, K. Anchecta, Y.-D. Lee, J.J. Dahlgaard, A stepwise ISO-Based TQM implementation approach using ISO 9001:2015, **7**, **4**, 65-75 (2016)
12. A. Masseli, *IEEE International Conference on Industrial Engineering and Engineering Management*, 194-198 (2016)
13. M. Galetto, F. Franceschini, L. Mastrogiacomo, International Journal of Quality and Reliability Management, **34**, **2**, 231-250 (2017)
14. S. Sumaedi, T. Rakhmawati, International Journal of Systems Assurance Engineering and Management, **8**, 219-229 (2017)
15. S. Hensch, Meat Market, **39(48)**, 4 (2017)
16. Z.I. Lavrova, *Collection of St. Petersburg International Economic Congress materials (SPEC-2015)*, 555-561 (2015)
17. D.A. Komarov, Problems of Modern Economics, **17**, 193-196 (2014)
18. N.I. Usik, A.E. Belorukov, A.V. Vasilenok, Economics and Environmental Management, **4**, 70-77 (2016)
19. E.V. Fedotov, I.I. Sorokolet, *Problems of enterprise development: theory and practice: proceedings of the 15th international scientific-practical conference* (2016)
20. Step ten – quality and management: proceedings of the joint project BIG team and Expert magazine "7 notes of management". http//www.big.spb.ru.
21. M.M. Sharkova, *Student: science, profession, life: materials of III Russian student conference with international part. Omsk railways state university*, 216-219 (2016)
22. The World Economic Forum: Global Competitiveness Ranking 2015-2016. http://gtmarket.ru/news/2015/09/30/7246
23. GCI Global Competitiveness Index 2016-2017. http://reports.weforum.org/global-competitiveness-index/competitiveness-rankings, 2016.
24. M.P. Baryshnikov, E.R. Ladygina, Modeling and development of metal forming processes, **18**, 301-304 (2012)
25. *GOST R ISO 9000-2008. Quality management systems. Fundamentals and vocabulary* (Standartinform, Moscow, 2009).
26. Z.I. Lavrova, Science. Future: technologies and projects: materials of the international scientific-practical conference of young scientists and specialists, **2**, 216-219 (2012)
27. V.I. Gissin, *Quality management* (MarT Publishing Center, Rostov-upon-Don, 2003)
28. V.V. Okrepilov, *Quality Management: textbook* (Polytechnic University Publishing house, St. Petersburg, 2013)
29. *GOST R ISO 9001-2015. Quality management systems. Requirements* (Standartinform, Moscow, 2015)
30. Z.I. Lavrova, *Problems of organizations management improving. Russia and St. Petersburg: economy and education in the XXI century* (SPbGEU Publishing house, St. Petersburg, 2016)
31. www.zetakontakt.ru/about/

Features of additive technologies implementation in aerospace equipment production

Yuri Morgunov[1,*], *Boris* Saushkin[1]

[1]Moscow Polytechnic University, Moscow 107023, Russia

Abstract. Additive technologies (AT) are effective when processing relatively small batches of products, when high costs of materials is compensated by lowering fixed costs which are related to the use of traditional technologies. Aerospace equipment production is characterized, as a rule, within single and serial (small and medium series) types of production, which is why additive technologies will be in high demand in this field of engineering production. There is a necessary condition, regarding the widespread implementation and spreading of new technologies. It is its technical and economic efficiency. This article provides specific information on the comparative characteristics of classic and additive technologies of the common purpose and some pros and cons of these technologies. The main disadvantage of layer-by-layer synthesis of complexly shaped products is a relatively long duration of machine processing, resulting in high production costs, depreciation of equipment, and calculation of net costs. It is also noted that the feasibility study of AT use leads to reducing the number of workers and costs for maintaining on-site life safety. Also the usage of AT will lead to the reduction of shop and factory overhead costs due to cheaper logistics. In the field of aerospace engineering the major and promising direction of work in the field of additive manufacturing is to improve the technologies and equipment for the synthesis of products of ceramic and composite materials.

1 Introduction

Adoption of additive technologies and additive manufacturing at mechanical engineering enterprises is a widely discussed subject of scientific and engineering literature [1–5]. There is a notable demand for such technology in the military-industrial complex [6], particularly in aerospace engineering [7–12].

When making an aircraft, the following advantages of additive technology are of particular importance:

1. *Structural complexity of a product can reduce its mass.* Truly, it costs 12 to 25 thousand dollars to deliver one kilogram of mass to the Earth's orbit; reducing the mass of a long-haul aircraft by 100 kg saves $2.5 million and reduced CO2 emissions by 1.3 million tons over its entire lifetime.

2. *A significant increase in the materials utilization factor (MUF).* In conventional manufacturing of the essential aerospace engineering parts, the MUF is 0.05 to 0.2. Use of AT increases that factor to 0.7 – 0.9, leading to a significant reduction in the cost of expensive materials.

3. *Reducing cycle time and the pre-production costs of new products are of utmost importance, especially in pilot manufacturing.* AT developments enable pilot-plan organizations not only to carry out rapid full-scale modeling of the composite parts of a pilot product, but also to speed up both the manufacturing of complex parts and adjustments in their design in the process of fine-tuning and testing. Reportedly, use of AT in manufacturing the five essential parts of a liquid-jet engine makes the manufacturing cycle five times shorter on average.

4. *Replacing an assembly unit with a single part item makes the composite parts of a product more reliable while also shortening their manufacturing cycle.* For instance, there is only one to three part items in the additive manufacturing of an injector, cf. 138 items in conventional manufacturing.

5. *The critical parts of engine hot sections made of materials of low cutting machinability become less labor-intensive and costly to manufacture.*

6. *Additive shaping technologies are more flexible than any conventional technology because a changeover to another part does not require new workpieces and tooling.* That is why AT are efficient for machining relatively small batches, where the high material costs are offset by cutting the fixed costs associated with conventional technology. Aerospace engineering is generally about single, small-batch and medium-batch manufacturing, which is why additive technologies will enjoy particular demand in this industry.

Calculations performed by one of the leading enterprises in the industry show that additive manufacturing of a number of products, on average, increase the MUF to 0.7–0.9, make the manufacturing

* Corresponding author: morgunov56@mail.ru

cycle for five products five times shorter, whereas such manufacturing requires several times less pre-production time and becomes 5–6 times less costly. Table 1 presents specific calculation results for one such product.

Note that as of 2013, the share of aerospace industry in the world additive technology market is about 12% (2013) and tends to increase. Some technologies that are widely used and are sometimes critical in this industry are essentially additive technologies; however, they are barely discussed in additive technology papers due to their specific nature.

Table 1. Comparison of injector manufacturing technologies.

Options compared	Number of parts	Number of operations per technology	Labor intensity, n/h	Manufacturing cycle, days
Subtractive technology+Assembly	138	12	1025	56
Additive technology+Assembly	3 (1)	2	77	7

This applies to technology related to the winding of carbon, glass, or polymer fiber products on the winding machines where such products are binder-impregnated and then cured [13].

2 Organizational and economic issues; discussion thereof

Techno-economic efficiency of a new technology is a pre-requisite for its wide adoption and spread. The papers on AM economics are not numerous [14, 15], which causes certain difficulties for professionals involved in the implementation and adoption of such technology. From the formal-technological standpoint, adopting an additive technology makes sense for a particular enterprise only if it cuts the per-unit manufacturing costs (the total specific resource costs) ceteris paribus. Therefore, to study the techno-economic feasibility of adopting an additive technology at an enterprise, one first has to calculate the shop costs of making a part or an assembly unit, and the difference in such costs for two technological options:

- the existing manufacturing technology using the mastered methods of workpiece making, machining, and assembly (MT1);
- a manufacturing technology based on additive shaping operations (MT2).

Given the above, the structure of manufacturing and assembly costs for the first option can be written as follows

$$C_1 = \sum_{i=1}^{m} (C_{13i} + C_{1Mi}) + C_{c6} \quad (1)$$

for the second option, as follows:

$$C_2 = C_{pr} + C_{ad} + C_{ps} \quad (2)$$

where C_1, C_2 are the total final product manufacturing costs for the manufacturing options under comparison, C_{13i} is the cost of making a workpiece for the *i*th part of an assembly unit, C_{1Mi} the cost of machining the *i*th part, *m* is the number of assembly unit parts manufactured, C_{c6} is the cost of assembling a composite part with due account of the costs of other components per the assembly unit specifications, $C_{пр}$ is the cost of preliminary operations, $C_{ал}$ is the cost of additive shaping operations, $C_{пс}$ is the cost of subsequent finishing operations.

Note that additive technologies of layer-by-layer powder-based manufacturing do not eliminate residual porosity, which in some cases necessitates further steps like hot isostatic pressing, infiltration, and heat treatment. Reportedly, only 19% of all additive manufacturing products are used as final products.

The general approach to solving this problem is presented in [15]. Let us discuss the effect of this technological option on the most important costs that it would change.

The difference in the costs of materials ΔC_M for the options under comparison is calculated based on the wholesale prices per unit mass of materials and waste, on the material utilization rate, and the mass of waste per product unit [16]. Table 2 presents data on the prices of some powdered materials in the non-Russian market as well as based on commercial offers of some companies in the Russian market [1]. The bottom row contains the weighted average prices of rolled materials (30 to 100 mm in diameter); the prices were obtained by analyzing commercial offers found on the Internet. The data presented show that the price of powder for SLM is an order of magnitude higher than the prices of rolled materials used for workpiece manufacturing.

Table 2. Prices of some powdered materials and corresponding rolled materials (Eu/kg).

Origins of the price	Material				
	Ti-6AL-4V	*Co-Cr*	*AlSi12*	*Spesial steel*	*Inconel 625*
Wholesale price in non-Russian markets	200–300	150	70–90	70–80	75
SLM Solutions	520	210	60	130	160
EOS	485	320	122	235	-
Rolled rods in Russia (RUB)	17–25 (1300–1900)	-	3–5 (230–350)	3–4 (200–270)	-

In case of layer-by-layer sintering (melting) with a laser beam or an electron beam, some portion of the powder in contact with the contour of the solid being formed will be partially sintered or melted and removed in subsequent screening. Supports are used for relatively bulky and complexly shaped parts. This results in waste and the materials utilization factor amounting to 0.7–0.9. For the critical complexly shaped aerospace equipment parts manufactured per MT_1, this factor equals 0.05 to 0.2.

These data allow estimating the ratio of material costs of additive shaping (C_{M2}) to the ratio of material

DOI: 10.1051/shsconf/20173501078

costs in machining ($C_{м1}$); it equals 0.5 to 5. So it is seen that despite the high price of powdered materials, the difference $\Delta C_{м}$ can be positive or negative. Note that powders tend to become cheaper, with forecasts being made their prices will decline by 75%.

Direct worker wages for each manufacturing option are calculated by summing per the number of operations in either MT [16]. The comparative calculations in this case are peculiar because of the following.

First, additive shaping requires considerably less operations to manufacture a part. Second, the per-unit time spent on additive shaping operations usually amounts to dozes of hours. As modern equipment can operate automatically for long periods of time, it makes sense to apply multiple-machine servicing and multi-tasking to operate it. As a result, the machine and labor intensity of the operation differ significantly.

Analysis reveals that the labor intensity of additive shaping operations is relatively low. According to ZAO Globatek CG, only an hour of direct labor is required to manufacture a single part type bush, whereas the machine time amounts to 20–40 hours. As there are no direct labor standards, wages for direct workers involved in additive shaping operations shall be determined experimentally when adopting and mastering new equipment.

When making calculations, one needs to estimate the per-unit time spent on additive shaping operations. Using the classical standard time structure is difficult in this case due to the lack of standards that would help determine its components. At a factory, one has to use statistical methods to determine those components by means of timekeeping and working day photographs. Calculating the machine time is difficult due to the lack of reference formulas and the way these formulas depend on the method of additive shaping. That is why an expression for calculating the machine or operating time is obtained for each additive shaping method based on the existing methodology of valuating the operations with due account of the specifics of the physico-chemical processes occurring in producing bulk solid forms in each case. For example, the following expression it recommended for calculating the product buildup time $t_{о}$ when layering a 3D form, for instance by selective laser sintering.

$$t_0 = t_M + t_{V1} + t_{V2} = \sum_{i=1}^{m}\left(\frac{F_i}{k_p d_p^2 \phi_{Vl}} + t_i^{PS}\right) + t_{V2} \qquad (3)$$

where $t_{м}$ is the machine time determined by moving the light contact spot on the layer contour, $t_{в1}$ is a portion of additional time needed to prepare the shaping of selected layers, $t_{в2}$ are the remaining portions of additional time (unloading, possible technological delays, cycle initiation), m is the number of selected layers, $d_{п}^{эф}$ is the effective diameter of the laser-powder contact spot, $k_{п}$ is the overlapping coefficient, F_i is the area of the ith layer, δ_i is the distance of the route the contact spot center point has to travel to trace that area, $v_{л}$ is the mean contact spot speed when moving along this route, $t_i^{пс}$ is the time to prepare to walk around the jth layer.

For a more accurate calculation of the machine time, one has to take into account the light spot tracing strategy for the contour of area F_i and to calculate the distance of the route for δ_i each layer. This case

$$t_m = \sum_{i=1}^{m} \frac{\delta_i}{v_L} \qquad (4)$$

From the equations above, we conclude the operating time depends on the number of layers, the method of division into layers, the chosen tracing strategy, and the mean contact spot movement speed. As one can see, a significant portion of operating time is spent to prepare the layer for sintering ($t_{в1}$). SLM Solutions discarded the conventional way to form the powder layer by feeding the powder from below with a piston; instead, it has developed and patented a method of feeding the powder 'from above' which requires less strokes of the evener, reduces $t_{в1}$ and improves performance.

The effect of these factors results in a significant different in buildup time for machines made by different companies. Machine time can be estimated less accurately where the 3D performance of the machine in specified modes ($П_v$) and the calculated volume (V_p) of the part are known; such estimation shall take into account possible additional builds.

The main drawback of layer-by-layer synthesis of spatio-complex products is that it requires relatively long machine time. This results in a greater proportion of Ca equipment depreciation costs in cost calculations. As such equipment is very expensive (0.3...1.5 million Euro depending on the manufacturer, the standard size, the options, the delivery terms, etc.), this item of costs can sometimes be crucial. Reportedly, machine depreciation costs can amount in some cases to 70% of the cost of the product. To reduce the costs per this item, it is recommended to maximize operations and operate such equipment without downtime, preferably in two or three shifts. Due to such equipment quickly becoming obsolescent, the payback period shall not exceed five years.

An important advantage of the second manufacturing option is a sharp decrease in tool and tooling specific costs, as a rule $\Delta C_{и} > 0$ and $\Delta C_{пр} > 0$. It is not difficult to calculate the remaining components of costs.

To improve the economic indicators of additive technology adoption, it is recommended to equip the additive shaping shop section with multiple machines having various capacities, such as SLM 50 (100 Watts, 12 μm spot) + SLM 300 (400 W, 50 μm):

When using an additive technology, the enterprise needs less workers and spends less on safety measures; shot and factory overhead costs are reduced thanks to lower logistics costs. These advantages may also be taken into account when studying the techno-economic feasibility of such technologies.

Conclusions

Developing new methods different from layer-by-layer synthesis is an effective way to improve additive shaping performance. We would like to note the brand new additive shaping method based on 3D synthesis from a liquid interface. This method, referred to as continuous liquid interface production (CLIP), boasts significantly higher shaping rate that is 25 to 1000 times, and experts believe could be 1000 times greater than that of layer-by-layer synthesis. It also features structure and property isotropy of manufactures objects, low surface roughness, and high printing resolution. It is expected to enable real-time manufacture of spatiocomplex objects with functional properties matching the finished product specifications.

We would like to stress the trend towards the development of combined technology and equipment, to emphasize how promising this area is for mechanical engineering. Such technologies will further reduce the manufacturing cycle and will feature the pros of both subtractive and additive methods. Solidica has developed Ultrasonic Additive Manufacturing (UAM) technology; Fabrisonic uses this technology in its hybrid SonicLayer R200 machines that are capable of additive-subtractive machining of products made of ductile metals and composite materials with a metal or a polymer matrix. Mitsubishi LUMEX has begun delivering its integrated Avance–25 Hermle complex that combined SLS and 5–axle milling. Hermle has announced it is going to start manufacturing is MPA–40 complex that combined 5–axle milling and 3D-gas-dynamic spraying. Use of such technologies and corresponding machine systems will also necessitate a techno-economic feasibility study with due account of the specifics above.

For aerospace engineering, it is important and promising to enhance the technology and equipment for the synthesis of products of ceramic and composite materials [18].

References

1. M.A. Zlenko, A.A. Popovich, I.M. Mutyilina, *Additive technologies in mechanical engineering* (Politechnical University, St. Petersburg, 2013)
2. I.V. Shishkovskiy, *Fundamentals of high-resolution additive technologies* (Piter, St. Petersburg, 2015)
3. Ya. Gibson, D. Rozen, B. Staker, *Additive Production Technologies* (Tehnosfera, Moscow, 2016)
4. *Public analytical report on development of new production technologies.* Skolkovo Institute of Science and Technology (2014)
5. W.E. Frazier, A Review, J. Mater, Metal Additive Manufacturing Eng. Performance, **23(6)**, 1917-1928 (2014)
6 Yu.M. Mihaylov, *Prospects for the use of additive technologies in the DIC* (Moscow, 2015)
7. O.S. Sirotkin, Aviation industry, **2**, 22-25 (2015)
8. D. M. Chumakov, Works MAI, **78**. http://www.mai.ru/science/trudy/
9. V.A. Isachenko, Yu.P. Astahov, B.P. Saushkin, Technology of mechanical engineering, **1**, 10-14 (2016)
10. Yu.P. Astahov, S.A. Kochergin, Yu.A. Morgunov, B.P. Saushkin, Technology of mechanical engineering, **5**, 14-18 (2013)
11. Yu.A. Morgunov, B.P. Saushkin, Additive technologies, **1**, 30-38 (2016)
12. Yu.A. Morgunov, B.P. Saushkin, Rhythm of Mechanical Engineering, **9**, 26-31 (2016)
13. D.V. Panov, B.P. Saushkin, A.N. Korotkov, Rhythm **7**, 32-36 (2014)
14. E. Atzeni, A. Salmi, Int. J. Adv. Manuf. Tech., **62(9)**, 1147-1155 (2012)
15. Yu.A. Morgunov, B.P. Saushkin, High technology in engineering, **7**, 28-35 (2016)
16. K. M. Velikanova, *Calculations of economic efficiency of new technology* (Mashinostroenie, Leningrad, 1990)
17. J.R. Tumbleston, Science, **347**, **1349** (2015)
18. N. Travitzky, A. Bonet, B. Dermeik, Adv. Eng. Mater., **16(6)**, 729-753 (2014)

«Green» innovations in transport branch: restrictions, ways of implementation

Olga Podolyak[1], *Anastasia* Akulova[1,*], and *Tatiana* Mineeva[1]

[1]Mechanics and Machine Building Institute, Ural Federal University, Russia

Abstract. The course of "sustainable development" is relatively new to the Russian economy and its segments, however even now, experts believe that in the near future, the vast majority of innovations will be "green". The environmental and frugality problems are especially acute for the transport system of the country. The comparison with countries, having the most promising automotive industries, has shown that for developed automotive transport is characteristic developed infrastructure, accompanying vehicle at all stages of the life cycle. The authors consider the organization of the utilization market in the transport system of the country as one of the most promising directions in the transition to resource-saving technologies. The article describes the market mechanisms regulating the utilization market in Europe, reveals the barriers to organization of recycling in the Russian Federation. To overcome these barriers possible market mechanisms are suggested, i.e. the incentive of members of the transport system to create an organization chart of the utilization market. At this stage, a cluster scheme may be offered under which one the control and the responsibility belong to the automaker. This would allow reducing the cost of creating the infrastructure for the recycling system. Were formulated the possible positive effects for each of the participants of transport system. Were formulated the challenges for further research.

1 Introduction

Historical development of the cities, regions and countries is closely connected with the transport development. Addressing to a historical retrospective, it is possible to allocate some epoch throughout which the transport influence on the countries and regions development is most noticeable. The first stage can be considered the Middle Ages and Renaissance when the leading positions in the world belonged to the states having the best fleet. Thanks to the sea transport modern outlines of the world were open, the main trade ways were made, the first higher engineering schools were founded as a naval. The second stage is connected with development of the railroad. The railroad most influenced on the industry growth of the end of the 19th century. It is considered to be that the industrial revolution began with Great Britain where the first railroad was constructed, it is difficult to present a further way of industrialization without the railroad. The following stage is connected with emergence of an internal combustion engine and the invention of brothers Wright, for today it is possible to see that the countries with the developed automotive and aviation industry (Germany, the USA, Japan, etc.) have the steadiest and strong economy.

The modern civilization can't live without transport infrastructure, but overproduction results in need to solve a problem of outdated vehicles processing.

The growth of the intractable waste is the natural tendency caused by economic development. The correct recycling not only protects the environment from pollution, but also is a source of raw materials which can be used again. It should be noted, that in the motor-car industry this problem is the most actual, for example, only during 2013 year the vehicle fleet in Russia grew more than by 5 million units [1]. The vehicle fleet growth steadily leads to:

- the ecological situation deterioration,
- the roads capacity reduction,
- the decrease in the level of the transportations safety.

Since the end of the 20th century the policy and researchers from different countries speak about the economy "sustainable development", directing the main efforts to resource-saving and minimization of negative consequences for environment. According to such approach to economic activity, the majority of innovations will be anyway directed on restoration and rational use of resources, improvement of an ecological situation and decrease in production sector influence on the ecological situation [2, 3].

One of the "steady" development concept approaches is the "return" logistics. The "return" logistics considers the material, financial and information streams movement from the final stage of a product life cycle to a new one (new life cycle). In many European countries the waste processing (recycling) systems successfully function, including the motor-car industry, which

* Corresponding author: aa.akulova@urfu.ru

becomes a potential source of secondary resources and energy. For example, in the USA more than 20 million tons of cars are annually processed.

The first attempts on the cars utilization system creation are conducted now in Russia. Researchers and other interested persons are faced by some important tasks:

- development of an effective system of utilization state program with concrete factors and indicators;
- development of market and state mechanisms for stimulation the cars utilization;
- study of standards and legal questions of the utilizing companies activity
- etc.

The greatest interest is in creation of the economical and organizational scheme of the cars utilization market.

2 The foreign experience in the market and state regulation of the recycling

For many years in the European countries the recycling system for the consumption and production waste was formed, the first regulations (legislative act) appeared in Germany in the 17th century. The term recycling appeared in Germany by 60th years of the XX century. Recycling is the product processing, for the reusing. At the same time processing industry formation begins. Today, Germany has one of the most successful processing industries, therefore for forecasting of the car recycling market development in Russia it is useful to look through the experience of other countries. Feature of recycling systems of these countries is that they are part of the transport system, providing production, service and transport companies with secondary materials and energy resources.

The analysis results of the tools and measures for the utilization market regulation are presented in the Table 1 [4-9]. This tools and measures are widely use in different countries.Maintaining the Integrity of the Specifications.

Experience of foreign countries shows that at the first stages of the vehicles utilization market formation strong intervention of the state is necessary, because at the first period it is necessary to: stimulate producers:

- to use secondary raw materials;
- to create the provisions regulating activity of each participant of the utilization market and their responsibility.

In countries with well-developed automotive industry established recycling systems of vehicles that can recover about 80% of the weight of the car to secondary resources. Principles of construction, management and operation of systems are based on the existing economic structures, especially insurance and environmental legislation. Other conclusion which can be made according to the table is that the main market mechanisms are connected with development of encouragement and penalties, change of tax base and preferential crediting.

Table 1. The main measures used for the utilization market regulation in different countries.

The country	The regulation measures	
	The economical	The legislative
USA	- private recycling facilities, bred to self-sufficiency; - the high payment for the waste disposal; - the cost of processing is lower, than the cost of waste storage; - low tariff rates on transportation of the secondary raw materials; - existence of a large sales market of spare parts from the cars which have to be utilize; - the self-sufficient, private utilization enterprises	- the maintenance minimum of the secondary raw materials in production is established; - waste disposal, made in other states is forbidden; - norms to quality of secondary raw materials are entered; - research works on elimination of technological and economic barriers for the secondary raw materials using in production are stimulated;
EU	- financing of industry due to assignments of car makers (importers), - state grants; - the raised tax rates for the producers who are not applying secondary resources; - preferential crediting and the taxation of the utilizing enterprises; - receiving considerable part of the income from sale of spare parts; - the low tariff rates on transportation of the secondary raw materials; - the pricing regulation in the market of worn-out tires;	- fixing the responsibility for utilization to the manufacturing enterprise; - uniform obligatory marking of details is entered; - the instruction about need to provide the enterprises for utilization with technological documentation on dismantling and a possible reuse of auto components is approved; - a ban on placement of certain auto components on the dumps; - indicators of efficiency and planned indicators of work for the utilizing industry are approved;
Japan	- use of secondary resources in various industries; - granting workplaces at the utilizing enterprises to disabled people; - the payment for utilization of Freon in conditioners, pyros in the airbags, and also operational liquids is assigned to the cars owners;	- legislative stimulation of producers to use the secondary raw materials; - standards of the industrial wastes processing are approved; - the instruction to choose easily utilized materials for production, and also to provide information about the utilization ways;

3 The utilization market organizational scheme creation in Russia

In the Russian Federation the issue of vehicles recycling doesn't have a long history and the starting point is considered the Federal Law "On Production and Consumption Waste" (from 24.06.1998 g). It should be noted that the law regulates the disposal in general, without specific reference to the automotive industry. The main activities described in the bill relate to environmental, while resource conservation is not considered at all.

 DOI: 10.1051/shsconf/20173501079

Since 2012 in Russia utilization charges are exist. These charges are paid by producers and importers of the cars, trucks and buses, also it is necessary to mention that the approved rates are one of the highest in the world practice. [8,10] Thus, in two years of the law action in the budget considerable funds from the utilizing tax are saved, however they can't be used because in the country there is no infrastructure, standard and organizational support of the utilization market.

Today's situation can be presented in the following scheme (Figure 1). From this scheme it is visible that the logistic chain is not closed. So we can say that the principles of the return logistics are not carried out. Today enterprises engaged in recycling of vehicles and their waste in Russia, are not highly specialized and don't have common objectives, performance criteria and technological support. Most of the companies involved in the recycling of vehicles are engaged in the processing of ferrous and nonferrous metals, tires and batteries, and a significant proportion of non-metallic material is sent for disposal. Therefore, despite the presence in the market of individual companies providing utilization services vehicles, talk about recycling system inside the transport complex does not seem possible.

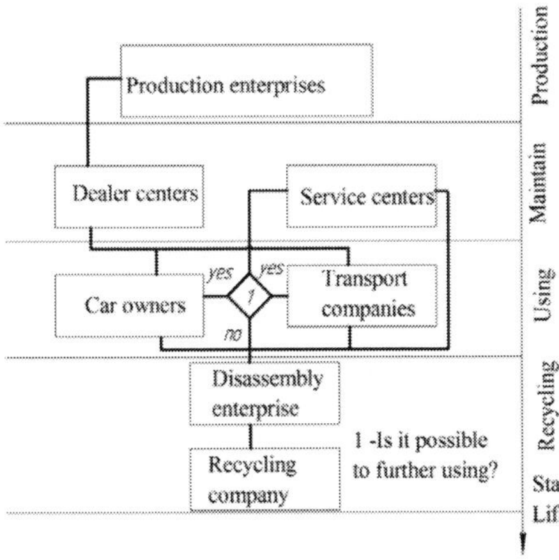

Fig. 1. The material stream movement scheme in automobile industry today.

The main objective of this organizational scheme consists in stimulating updating of vehicle fleet and the car sale made in the Russian territory. Such approach will not allow to gain the considerable ecological effect and to create the profitable industry of recycling. In the conditions of market economy in the country, only the utilization market can be successful, therefore, it has to possess three obligatory signs:

1. The private ownership,
2. The voluntary economic interaction of the independent subjects
3. The competition between the market players.

Functioning of the market is provided with various market mechanisms which main essence is to create the benefits for each participants. In that case when the market does not reach balance only by market mechanisms, state regulation is necessary. So, for the existing scheme it is necessary to provide market and state regulation mechanisms which would stimulate creation of the return material stream, thereby closing a logistic chain (Figure 2).

For today, it may be offered a cluster scheme under which one the control and the responsibility belong to the automakers. The cluster system is characterized by a close relationship between the recycling enterprises and the particular car manufacturer, which guarantees compliance of the recovery technological process, the return of valuable secondary materials automakers, allows to use of existing infrastructure service and dealer centers for receiving vehicles from their owners to shorten the path suitable for further use details. For the successful operation of the proposed scheme is necessary to conduct a series of events.

1. The private ownership,
2. The voluntary economic interaction of the independent subjects
3. The competition between the market players.

Functioning of the market is provided with various market mechanisms which main essence is to create the benefits for each participants. In that case when the market does not reach balance only by market mechanisms, state regulation is necessary. So, for the existing scheme it is necessary to provide market and state regulation mechanisms which would stimulate creation of the return material stream, thereby closing a logistic chain (Figure 2).

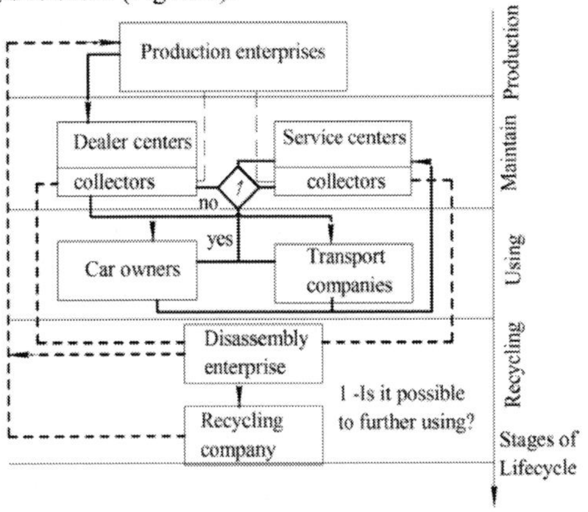

Fig. 2. The offered closed scheme of the material stream movement.

For today, it may be offered a cluster scheme under which one the control and the responsibility belong to the automakers. The cluster system is characterized by a close relationship between the recycling enterprises and the particular car manufacturer, which guarantees compliance of the recovery technological process, the return of valuable secondary materials automakers, allows to use of existing infrastructure service and dealer centers for receiving vehicles from their owners to shorten the path suitable for further use details. For the successful operation of the proposed scheme is necessary to conduct a series of events.

First, it is offered to create the place points for end lifecycle vehicle (EVL) collecting with the dealer and service centers which received accreditation of the car producer. This activity will allow to create additional loading and workplaces for these companies.

Secondly, it is offered to enter various stimulation factors for using the secondary raw materials in the production, such as:

- development of the standards on quality and processing technology of the secondary materials;
- development of price policy for the secondary materials market;
- legislatively to fix the minimum percentage of secondary materials in production by industries.

Thirdly, the base for legitimate utilization activity creation is necessary. We have to take into consideration the secondary raw materials and services in utilization markets development. The utilization activity legislative act have to include establishment of the carrying out licensing process for the utilizing companies, development of the financial stimulation programs for the utilizing companies.

Taking into consideration the scale of the tasks arising when the utilization market is forming, the separate, specially created structure which is coordinating various subjects involved in formation process, performing analytical work and carrying out monitoring process has to be responsible for the result.

Formation of the car recycling market and waste utilization market in general is a necessary measure upon transition to "steady" development which demands the Practice shows that competitiveness of branch depends on a level of epy infrastructure development and ability to use potentials of still unused assets in the long term. Therefore for the Russian conditions it is necessary to develop and submit the functioning scheme of the vehicles utilization market which organically would fit into the existing transport complex and promoted its development.
Nowadays in Russia the first steps in organizational and economic regulation mechanisms development are taken, however, without the corresponding standard and legal base as well as without the infrastructure the recycling market cannot become independent and profitable.

References

1. State report "On the state and Environmental Protection of the Russian Federation in 2013". http://www.mnr.gov.ru/upload/iblock/6c7/ gosdokladeco.pdf
2. A state program of the Russian Federation "Energy saving and increase of power efficiency for the period till 2020". http://government .ru/gov/results/13912/
3. GOST R ISO 14040-2010. Ecological management. Assessment of life cycle. Principles and structure. http://gostrf .com/norma_data/58/58831/index.htm
4. B.B. Bobovich, *Vehicle scrappage and autocomponents* (MGIU, Moscow, 2010)
5. T. Mikkonen, The Recycling of waste works, **6**, 20-21 (2009)
6. I.S. Petrov, The Magazine of automobile engineers, 18-20 (2011)
7. R.L. Petrov, Magazine of automobile engineers, **1**, 44-49 (2014)
8. R.L. Petrov, The Magazine of automobile engineers, **5**, 52-57 (2012)
9. The directive of the European Parliament and Council 2000/53/EU 2000, belonging to the vehicles which served the term
10. Rules of collection, calculation, payment and collecting utilization collecting concerning wheel vehicles and the chassis, and also return and offset of excessively paid or excessively collected sums of this collecting. http://www .consultant.ru/document/cons_doc_LAW_161435/? Frame=1

Structural capacity assessment of machine-building enterprises and associations

Elena Kuznetsova[1,*], *Anastasia* Akulova[1], and *Marina* Zhuravskaya[2]

[1]Mechanics and Machine Building Institute, Ural Federal University, Russia
[2]Ural State University of Railway Transport, Yekaterinburg, Russia

Abstract. The use of "green" logistics technologies for the Russian transport system is designed to reduce the damage to the environment, it becomes a competitiveness factor, and gives the service "quality criterion". Among many areas of "green" logistics, the authors considered the development of piggyback transport and the organization of utilization system as the most promising and priority for the Russian transport market. The advantages and key challenges in organizing piggyback in Russia are revealed. To overcome the current limitations, it is suggested to enable the conditions, that should be taken into account when calculating the tariffs for transportation and take into account the impact of the mode of transport on the ecology. The authors discuss the direct and indirect positive effects of as part of reverse logistics for the transport market of the Russian Federation. The actual task is to form the utilization market as an independent branch of activity. The article analyzes the domestic and foreign experience in addressing these issues, formulates the problems and challenges for further research.

1 Introduction

Economic globalization and scientific and technological progress in the XX century led to the formation of powerful transport networks, to increasing of the types of rolling-stock and to strengthening the role of transport in the world economy. Transportation, as one of the main components of industrial and social infrastructure, allows to ensure freight traffic between the objects of economic activities and spatial mobility of the population. For example, following the results of 2013 passenger traffic grew by 12.8%, freight turnover by 7% compared with 2010, reflecting a significant increase of the industry in recent years [1].

Despite it, at the moment before the transport complex of the country there is a number of strategically important tasks. One of them is the integration of domestic transport in the world transport system that contributes to the development of national economy and achieving the main goal: increasing the competitiveness of domestic goods and services in the global market. Achieving this goal is impossible as long as the Russian transport and logistics companies won't be guided by the global trend of "sustainable development".

According to researches «Towards Sustainable Logistics» [2] about 59% of the surveyed firms believe that "green" technologies of transportation their products will be the deciding factor when choosing a partner. According to a study QSHE Global (5400 survey respondents) [3], we can conclude about the relation to introduction of environmental technologies in the business processes of modern companies:

- 61% - cares about the environment;
- 67% - choose partners, oriented on "green logistics";
- 82% - use the principles of "green" logistics to improve the image;
- -30% - are willing to pay extra for environmental technologies.

Therefore, improving the competitiveness of the transport system of the country and the quality of transport services is not possible without the introduction of the principles of "green" logistics.

At such statement of a question there is a need to review and analyze the various "green" technologies in transport and transport-technological systems; development of various methods and approaches to measure the effectiveness of these technologies; determining the area of responsibility of the state in creating and ensuring market mechanisms contributing introduction of environmentally friendly technologies; formation of matrix of "green" technologies to provide various spheres of transport. These questions, and many others, require additional researches and discussions. Among a set of the directions of "green logistics", the authors considered the most promising for transport of the Ural region the development of piggyback transport and the organization of utilization system.

2 "Green Logistics" as a way of development

In general sense, logistics is considered as the science of planning, management and control of the movement of

* Corresponding author: eltcha@mail.ru

material, informational, and financial resources. Therefore, we can say that modern global problems associated with deterioration of environment, resource depletion, social problems are connected with imperfect approach to the organization of logistical processes, for this reason there is need a methodological framework which would allow eliminating harmful ecological consequences of logistic operations.

In 1987 for the first time were told words about the need to go by means of "sustainable development", meaning that, support and meet the needs of the present without compromising the ability of future generations to meet their own needs [4].

In foreign sources in recent years there are many works devoted to "green logistics" as part of "sustainable development", there are several points of view on its place in the economic activity, are offered a variety of approaches to supply chain management at the junction with the environment. For example, De Burgos and Lorente [5] are talking about improving the environment as an operational objective and the supply chain is secondary. Baumann et al. [6] give special attention to the development of green products in a narrow framework of the logistics chain. Zisdisin and Siferd [7] are conducted a review of environmental procurements. Abukhader and Jönson [8] analyzed the interaction and interrelationship of the environment and logistics Kleindorfer et al. [9] conducted a review and assessment of the publications in the field of "management of sustainable operations", but they consider issues related to supply chain. Srivastava [10] describes Reverse logistics. Elkington [11] argues that the integration of heavily interrelated economic, environmental and social aspects of sustainability gives "triple bottom line."

Russian authors [12] propose to understand the term "green logistics" as "... the scientific and practical activities aimed at optimization and efficient management of direct and reverse material streams and related to them (information, financial, waste streams, hazardous emissions, of different natural resources and energy) in order to minimize the negative impact on the environment".

In the context of the concept of "green logistics" is offered not only to reduce the burden on the environment by reducing emissions and install treatment plants, various bans and increased taxation, is assumed a new approach to the organization of production, through the introduction of innovative technologies, the use of the latest energy-saving efficient equipment, the transition to a resource efficient approaches exploitation of technics and organization of production, rational use of resources and the withdrawal of outdated machines and equipment out of service, using the principles of reverse logistics. The role of the last one in reducing the impact on the environment is that it "closes" the logistic chain, and allows you to manage product life cycle from raw material extraction to disposal. However, given the rising cost of raw materials and energy, the transition to energy-efficient production principle promises to be cost-effective, not only due to the effect of the public good, but also from improving the competitiveness of the product, owing to decreasing of production costs and marketing efforts.

To date, the most striking example of the introduction of "green" technologies belongs to transport and logistics company DHL, which started the service Go Green. Under this program, the customer pays on 3% higher of the tariff, as compensation for the damage caused to the ecology, the money is invested in environmental protection measures around the world.

German Deutsche Bahn for the first time in the European rail branch offered a way of transportation goods, completely eliminating carbon dioxide emissions. Within the project EcoPlus the company receives electricity for their electric locomotives from renewable energy sources by attracting additional funds from the companies whose goods are transported, for example, concern AUDI [13].

Among domestic transport companies adhere to the sustainable development strategy of JSC "Russian Railways", which implements a number of innovative projects accompanied by resource-saving and environmentally friendly technologies: is replaced the railroad tracks, put into operation new diesel locomotives and electric locomotives with increased energy efficiency.

Thus, speaking about road transport, it is possible to formulate two major problems whose solution involves the use of environmental technologies: 1) reduction of the share of harmful emissions into the atmosphere; 2) minimizing resource and energy consumption.

3 Piggyback as an element of eco-oriented logistics

At present, within the framework of the environmental program of JSC "Russian Railways" conducts marketing research attractiveness of piggyback technology in the Russian market, which are more environmentally friendly combined transports "car-rail".

Piggyback first were organized in Central Europe, combining the advantages of road and rail transport, and it is three-stage combined shipping. Initially, goods are transported by road train to railway station, then the vehicle (trailer/semi-trailer) are loaded on a flat-car and transported to the terminal station, from this station the descended from the platform truck delivers the goods to a final consumer. A distinctive feature of these transportations is exactly transportation of road train on a flat-car the main part of the route.

Road transport is one of the most powerful air pollutants, and, unfortunately, in recent years this trend has changed slightly. In Table 1 presents data on air pollution by different sources, from this table you can see, that the road transport emissions account for about 40% of the total, while the share of road transport is only 4.9% of the total freight turnover of the country. Rail transport for comparison, with a market share of 43.2% in freight turnover, has a small proportion of emissions [14].

Table 1. Percentage of emissions by types of source in 2008-2013 (data for Russian Federation).

Year	2008	2009	2010	2011	2012	2013
Total emissions, 1000*tonnes / year, of which:	33691	32560	32220	32488	32309	31871
-from stationary sources, 1000*tons / year	20103	19021	19116	19162	19630	18447
% from stationary sources	59.67	58.42	59.33	58.98	60.76	57.88
-from road transport, 1000*tons / year	13588	13539	13105	13325	12679	13424
% of road transport	40.33	41.58	40.67	41.02	39.24	42.12
- from railway transport, 1000*tonnes / year	n/d	n/d	n/d	140	159.5	192.5
% of rail transport	-	-	-	0.004	0.005	0.006

Currently in Europe, successfully are implemented piggyback both within the country and international transportations, mostly they are intermodal technology. Organizers of the Piggyback route "Gallarate (Italy) - Maze (Belgium)" talk about reducing the transportation costs by 15-17%, decreasing of emissions by 20% and of fuel consumption by 2.5 mln. liters. Company COMBIDEM organizing piggyback in France, talk about reducing transportation costs by 10% and reducing emissions by 70-75%.

Analysis of opinions of researchers and scientists in the field piggyback [15 -19], allowed to form a table with the advantages of this type of transport organization and barriers to its development (Table 2). The main difficulties in the organization of piggyback in Russia are connected with the need to create the appropriate infrastructure and regulatory framework.

Table 2. Advantages and limitations of the organization piggybank in Russia.

Advantages	Limitations
- reducing of harmful emissions; - economic efficiency of transportation for distances over 1000 km; - unloading of highways of megalopolises; - decrease in load of a roadbed; - the accelerated passing of customs points.	- need of creation of a specialized rolling stock; - absence of the working models for these type of transportations organized on "long" routes (in Russia an average distance is 3-4 thousand km, in Europe 700-800 km); - absence of standard and legal base for maintenance of transportations; - absence of tariff base for piggyback

To overcome the current limitations it is necessary to create the stimulating conditions in the market of transport services. The system of fines and encouragement at calculation of tariffs for types of transport is offered [20]. This method takes into account the impact of the mode of transport on the environment in the form of increasing or decreasing coefficients for the fixed and variable costs. Using of these coefficients allows to displace the point of change of scenario of transport services (CP-change point) [21], which is calculated as the ratio of the difference of fixed costs by mode of transport to the difference variable costs by mode of transport, closer to the origin of coordinates, which reduces the area of the effective operation of road transport (value F2, V2) and increases it for the railway transport (F1, V1) (Figure 1). Such a scheme of tariffs directly contributes to the development of piggyback.

However, despite the significant shortcomings of road transport (high share of emissions, low security) refuse completely from road transport it won't turn out as only they allow you to organize the delivery of "door to door", and for some areas it is still the only available type of transport. Therefore, it is necessary to develop an organizational chart of eco-friendly solutions for road transport at every stage of its life cycle.

4 Recycling as part of resource-saving tecnologies

The second direction of development of "green" logistic, obvious for development of the Russian transport market, is solving the problems of utilization as part of "reverse logistics". For a transport complex of Russia, and especially for its automobile constituent, management of a stage of life cycle "recession" is especially actual as more than a half of vehicle fleet of the country has the obsolescence and wear. The balanced development of branch is possible only at continuous improvement of structure of its fixed assets therefore their growth and updating is the most important condition of providing "sustainable development" of branch.

Today the analysis of age structure of fleet of vehicles (Table 3) shows that the considerable part of transport units is operated outside standard service life, another, also considerable, part comes nearer to this term. As a result, ecological indicators, indicators of safety and economic efficiency of work of transport are significantly worsening.

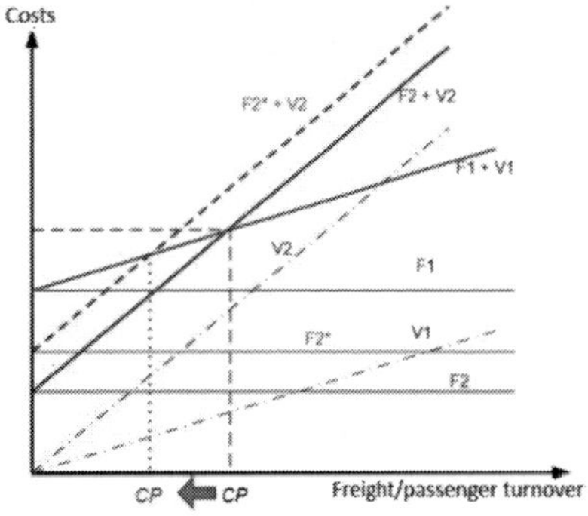

a)

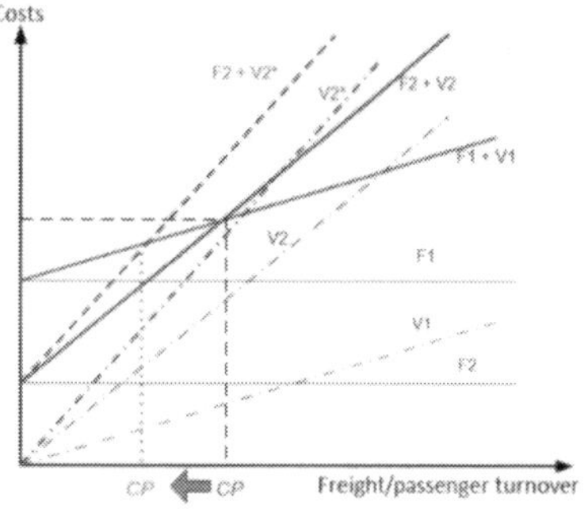

b)

Fig. 1. Displacement of point of change scenario of transport services (CP) with the introduction of green fines for a) fixed costs b) variable costs of vehicles.

Table 3. The age structure of the fleet of vehicles at the end of 2013 [14].

The age structure of the fleet of vehicles, %			
Type of vehicle	to 5 years	5-10 years	More than 10 years
Trucks	21.8	20.1	58.1
Buses	27.0	27.7	45.3
Cars	30.7	24.6	44.7

Many researchers in the framework of sustainable development, have engaged by issues of "reverse logistics" which describes the movement of material streams from a final point of consumption of a product prior to new life cycle.

Creating a system of utilization of vehicles will have an integrated beneficial effect. First, decrease in harmful emissions and costs, increase of safety of transportations and energy efficiency, thanks to stimulation of updating of vehicle fleet. The first experiment conducted in Russia in 2010-2012 on utilization of cars gave a double budgetary effect, and growth of a value added was about 150 billion rub at expenses in 30 billion rubles Second, resource conservation by recycling materials, so, according to Western scholars, recycling of medium car weighing about 1 ton saves 3,300 kg of material resources, reduces energy consumption per 56,000 MJ, reduces emissions on 1945 kg. Third, the reduction of environmental pollution by obsolete cars as well as reducing space required for their storage.

In European countries, the utilization market has long refers to an independent branch of activity. According to the European Automobile Manufacturers Association, the profit from recycling one car in the EU is around 340 euros, and with deduction of payment of land, premises, equipment depreciation - 300. In Germany, European leader in the number of cars (43.4 million units. 81 million inhabitants), is utilized an annual average of 450 th. cars, and in some years, when is started recycling programs, similar to our 2010-2011., 2 million old cars. Another example of the successful formation of the utilization branch is the Netherlands. Coefficient of utilization of vehicles in the Netherlands to 2011 amounted to 96.2% and it is the highest in the world, and most importantly, that the financing of this system every year requires fewer additional resources. Here, from the buyer is taken recycling tax of 45 euros, which is included in the price of a new car.

Unlike foreign countries, the domestic branch of utilization is in a rudimentary stage therefore it is necessary to develop the market mechanisms stimulating participation in questions of utilization not only the private sector and car makers, but also other subjects of the transport complex of Russian Federation.

The directions of "green" logistics considered in article aren't only options on the way to a sustainable development of transport system of the country. The separate attention is deserved by a question of formation of algorithm of a choice of these or those "green" approaches depending on appointment and a stage of life cycle of transport unit.

The authors see the need to further examine in more detail the methods of management the final stage of the life cycle, as every year the relevance of this issue is only growing. In particular, it is planned to create an organization chart of utilization market and its optimization in transport-technological system.

References

1. State report "On the state and Environmental Protection of the Russian Federation in 2013". http://www.mnr.gov.ru/upload/iblock/6c7/gosdoklad eco.pdf
2. Dr. Christof, E. Ehrhart, Delivering Tomorrow: Towards Sustainable Logistics (2012). http:// www.delivering-tomorrow.com
3. K. Alexandrova, The RZD Partner International, **3(39)**, 38-39 (2014)
4. Report of the World Commission on Environment and Development: Our Common Future.
5. De Burgos, J. Jiménez, Céspedes, J.J. Lorente, International Journal of Operations & Production Management, **21(12)**, 1553-1572 (2001)
6. H. Baumann, F. Boons, A. Bragd, Journal of Cleaner Production, **10(5)**, 409-425 (2002)
7. G.A. Zsidisin, S.P. Siferd, European Journal of Purchasing & Supply Management, **7(1)**, 61-73 (2001)
8. S.M. Abukhader, G. Jönson, International Journal of Logistics: Research and Applications, **7(2)**, 137-149 (2004)
9. P.R. Kleindorfer, K. Singhal, L.N. Van Wassenhove, Production and Operations Management, **14(4)**, 482-492 (2005)
10. S.K. Srivastava, International Journal of Management Reviews, **9(1)**, 53-80 (2007)
11. J. Elkington, *Cannibals with Forks: The Triple Bottom Line of 21th Century Business* (Capstone, Oxford, 1997)
12. A. Tambovtsev, T. Tambovtseva, Management and sustainable development, 2/2011 (29)

13. A.A. Kizim, D.A. Kabertaj, Bulletin of the Southern Federal District, **8** (2012)
14. Transport of Russia" Informational and statistical Bulletin.http://www.mintrans.ru/upload/iblock/5cd/stat_2014.pdf
15. M.A. Zhuravskaja, I.O. Макarenko, Transport of the Urals, **3(38)**, 74-79 (2013)
16. V.V. Baginova, D.V. Kuzmin, Modern problems of the transport complex of Russia, **4** (2013)
17. K.V. Holopov, Russian foreign economic journal, **9** (2011)
18. A.G. Kirillova, Rail transport, **2** (2011)
19. D. Kyzmin, A. Belyh, Modern problems of the transport complex of Russia, **4** (2013)
20. M.A. Zhuravskaja, Journal of USURT, **1(25)**, 38-48 (2015)
21. M.A. Zhuravskaja, A.A. Kazakov, P.A. Parsurova, Transport of the Urals, **4**, 50-53 (2012)

Restrictions of process machine retooling at machine-building enterprises

Elena Kuznetsova[1,*], *Ludmila* Tipner[1], and *Alexey* Ershov[1]

[1]Mechanics and Machine Building Institute, Ural Federal University, Russia

Abstract. The competitiveness of the national economy depends on the technological level of the machine-building enterprises production equipment. Today in Russia there are objective and subjective restrictions for the optimum policy formation of the manufacturing equipment renewal. The analysis of the manufacturing equipment age structure dynamics in the Russian machine-building complex indicates the negative tendencies intensification: increase in the equipment service life, reduction in the share of up-to-date equipment, and drop in its use efficiency. The article investigates and classifies the main restrictions of the manufacturing equipment renewal process, such as regulatory and legislative, financial, organizational, competency-based. The economic consequences of the revealed restrictions influence on the machine-building enterprises activity are shown.

1 Introduction

Productions with a high intellectual component in the capital structure, with the elements of new technological ways, form an economic core of any country. The machine-building industry as the multiplier of the economic system development for any territory belongs to such productions. The availability and effective use of mechanical engineering enterprises production capacity is a key to the national economy competitiveness, and the strategy of its use defines the growth path of the economy, innovative prospects, and the possibilities of its development on a modern hi-tech basis.

In the economically developed countries from 30 to 55% of the total amount of industrial output release falls on the share of the machine-building productions (in Germany – 53,6%, Japan – 51,5%, England – 39,6%, Italy – 36,4%, China – 35,2%). In the developed countries profit from the machine-building field provides full modernization of the industry every 8-10 years [1].

The modernization of the Russian industrial enterprises is one of the main problems now: some enterprises need to restore their lost productions, for some the task is innovative development on the basis of complex modernization, for others effective cooperation interactions. For the machine-building enterprise as the minimal structural unit of the economy, independently conducting its production, economic and reproduction activity [2], the recovery function is one of the most important, i.e. the independent distribution of the income from sales, ensuring the spent material, labor, financial and other resources reproduction. The recovery function means the constant need for the product development, which requires modernization and update of the production capacity. Therefore the problem of the manufacturing equipment renewal is relevant for the enterprises which update the production potential, restore it or create cooperation communications to provide the most effective equipment utilization.

The age structure of the equipment used in the mechanical engineering of the Russian Federation is presented in Fig. The average age of the equipment during the last five-years is more than 20 years while the life cycle of the modern hi-tech equipment is 7-8 years.

According to the Ministry of Economic Development of the Russian Federation, in general, until 2025 the domestic economy will require about 2.5 trillion dollars of investments. By the estimates of independent experts, not less the fifth part from this sum should be spend on renewal of the basic productive assets.

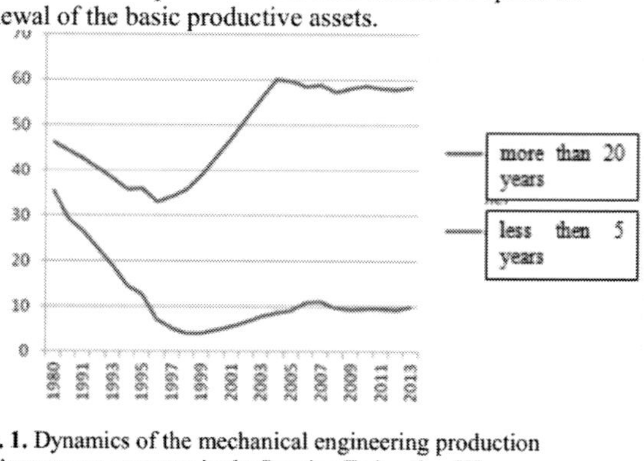

Fig. 1. Dynamics of the mechanical engineering production equipment age structure in the Russian Federation [3].

The average percent of the basic assets renewal on the start of the market relations development in the Russian Federation had been about 10%, in the 90s it decreased to 1-2%. After 2005 the coefficient of funds renovation increased to 5% and practically stabilized.

* Corresponding author: elena.bstm@gmail.com

Now the question of the production assets modernization took on strategic importance.

It should be noted that the problem of the production assets modernization is not new to the Russian economy.

In the post-war years, in the 1950-60s, this problem was studied and discussed by the Soviet scientists Bunich P.G. – problems of basic assets depreciation [4,5], Ivanov E.A. - problems of reproduction and use [6], Strumilin S.G. - questions of functional depreciation of the equipment [7].

In the 70-80s the specific features of the Russian (Soviet) mechanical engineering were the following: it made machines and equipment practically for all industries, acting as the initiator of the fixed capital renovation. And the discussion of the modification questions by scientists and experts was absolutely natural: Palterovich D. M., Zakharov V. G., Baranov D. A. , Khachaturov T. S. [8,9, 10,11] dealt with this issue in the seventies; V. P. Krasovsky, V. M. Rimer, M. M. Goldin in the eighties [12,13,14]; N. M. Alekseenko, V. Ivanchenko in the nineties [15,16].

It is important to note the Ural school of sciences, which was completely devoted to the analysis, studying and forecasting of problems, connected with intra productive reserves and, in particular, with the problems of capacities proportionality and equipment use efficiency, that is reflected in works [17,18]. Already in the mid-nineties the problem of modernization and technical renovation began to be considered not as a task, but as a pressing problem at the level of the country: I. Ivanov, S. Rogov [19]

Obviously, today the Russian machine-building enterprises need to mobilize all available resources (both internal and external) considering the starting condition level of the manufacturing equipment. Many enterprises will be able to count on full support from the state. The role of the state as the main agent of modernization is specific characteristic for all modernization strategies of the end of 19th and 20th century. The governments of the different countries including Russia use a set of economic mechanisms to support the productive assets modernization of the enterprises: funding and direct financing of projects, administration of credit guarantees, tax preferences and support on the legislative level.

The difficult process of the production equipment renewal has the main restrictions: regulatory and legislative, financial, organizational, competence-based.

2 Regulatory and legislative restrictions of the technical equipment renewal process

Some laws can be classified as the regulatory and legislative restrictions. As the Federal law of 18.07.2011 No. 223-FZ "On Procurement of Goods, Works and Services by Certain Legal Entities" [20], and the Federal law of 05.04.2013 No. 44-FZ "On the System of Public Procurement Contracts for Products, Works, Services for State and Municipal Needs" [21] assume realization of the following obligatory procedures:

- carrying out public purchases;
- drawing up and approving the Provision on procurement by each enterprise;
- publication of plans of innovative and hi-tech products procurement for 5-7 years. It should be noted that the business community suggests canceling this restriction as impracticable and risky for the companies. The establishment of the requirement to the publication of short-term plans of innovative and hi-tech products procurement is possible;
- monitoring, audit of the market;
- choice of procurement way at different cost;
- creation of contract service (the contract managing director), which raises expenses of the enterprise;
- creation of the commission on acceptance of the performed works results.

It involves the following consequences:

- expenses increase of the organizations- customers (due to the need of drawing up documentation about purchases, implementation of procurement procedures, etc.)
- increase in transactions terms;
- fall in efficiency;
- expenses increase of the organizations- customers

Public procurement in Russia is an expensive arrangement. Trying to reduce corruption in this sphere, the authorities preferred to complicate technical procedures and document flows for state customers and suppliers. But expenses of the customer during the procurement in the Russian Federation are much higher, than in the EU. The formalized procedure of the public procurement is inefficient for simple inexpensive contracts.

Fight "for honest procurement" led to the adoption of two quite similar laws within two years. The later law [21] along with the introduction of the wider range of procurement ways imposed many additional requirements on customers, including the detailed planning of procurement, justification of initial prices, efficiency monitoring and audit of the procurement. In this regard the question of procurement expenses gained sharp relevance.

During the last years the efforts of officials, businessmen and economists were directed on making the system of public procurement in the country less corrupted and more effective and transparent. Thus in numerous discussions about public procurement the primary attention was paid to the competition increase and prices reduction at the auction. Costs which are incurred by customers, suppliers and regulators (FAS - Federal Antimonopoly Service, industry departments, regional departments of procurement), as a rule, were not considered.

O. Balayeva, A. Yakovlev, L. Korobeynikova and E. Chupandina [22] analyzed the costs of public procurement in Russia and compared the efficiency of various procurement procedures and types: auctions, quotations, purchases from the single supplier (calculations were made on the example of specific organization). The most expensive way of the procurement implementation was an electronic auction – 174 man-hours or 35,5 thousand rubles. The signing of the direct contract with the supplier on average took 62 man-hours or 12,7 thousand rubles. However, the relative expenses of holding the auction was only 1,4% of the auction cost against 8% for quotations and 5,7% for purchases from the single supplier.

The state institution in Russia is obliged to carry out competitive procedures if the volume of purchase exceeds 100 thousand rubles. However the Russian customers cost prefer to carry out formal procedures for smaller contracts too to avoid questions upon audit and possible charges of corruption. As a result of expenses for small purchases make about 20% of their cost. For averages purchases, by Russian standards, (100-500 thousand rubles) this indicator decreases to 6,2%. And only for large procurements (over 500 thousand rubles) it reaches 1,1% of cost. For comparison: by the legislation of the European Union competitive procedures are required for contracts over 125 thousand euros worth. The authors explain so strong divergence in the level of relative expenses of procurements in Europe and in Russia by those distinctions in obligatory procedures.

The conclusion is that the application of the formalized procurement procedures are economically justified for large purchases and it is not necessary to regulate small procurements in details. It is necessary to trust more to the customers – controlling resulting effects of their activity instead of a detailed regulation of their current administrative decisions.

3 Financial restrictions

Limited resource of own funds. Own funds that the enterprises can spend on modernization is a profit after tax payment and the sum of accumulated depreciation. In relation to profit shareholders should make a decision (if the enterprise is a joint stock company) about its distribution. The owners often have a temptation to distribute the received result of the enterprise activity only on dividends payment and on their needs.

The empirical analysis of the Sverdlovsk region enterprises has showed that the limited number of owners who can agree about priorities (or there is a single owner) are the most successful in respect of production capacity creation of the enterprise, because they rigidly divided spheres of responsibility among themselves and observe these arrangements. In this case the decision is made in favor of technical modernization because the owner understands the technology of a production activity and accurately understands that funds of the enterprise, first of all the equipment – are prone to physical deterioration and obsolescence, and demand updating. It is worth reminding that in the Soviet Union the production development funds creation have been fixed on the legislative level since 1965, this funds were used strictly for renovation.

The depreciation deductions are more flexible and available resource, but at the existing deterioration level of fixed capital stock do not provide even simple reproduction, not to mention expanded renovation and modernizations [23].

Borrowed funds are rather expensive. At the present time the average percent of the bank credit is 20-25%. As the profitability of the majority of the processing enterprises does not exceed 10%, it is clear why the credits are inaccessible for them. The enterprises carry out the simple renovations at the expense of own funds, and investment programs are compelled to curtail due to the lack of long-term resources.

Banks act like usurious offices, and give the credits to natural person under the rate from 20 to 70% per annum; the Russian banks have no means for the long-term credits, and the credits for the modernization and technical renovations are, undoubtedly, the long-term credits which are unprofitable for our banks. The main direction of the banking sector for the industrial enterprises is the crediting of current assets and current activity of the enterprises which does not bring updates and increase of production profitability. In this situation the bank credit is a means necessary to ensure the operability of the enterprise. But in the view of financial resources freezing due to the growing debt of the enterprises for the outstanding credits, the banking system needs replenishment of its capital so it resorts to the external loans. The crisis situation of 2008-2009 showed the paradox when the state allocated the funds for competitiveness maintenance to banks, but this money did not reach the enterprises. Moreover, in the crisis of 2008/09 the enterprises which used borrowed funds in the activity, suffered the most. Banks did not restructure the debts of the enterprises, and, on the contrary, toughened the "game rules", and demanded to return the credits urgently, or reconsidered terms of credits towards increase in the credit load for the enterprises.

The operational difficulties regarding finance which can impose restrictions on the renovation process realization are as follows. At the beginning of a year it is objectively more difficult to obtain the credit, than in the middle or the end of the year, even if the enterprise has positive credit history. The reason is the obligatory requirement of the bank to provide the annual account. If the enterprise is planning to receive the credit in the beginning of the year, it is necessary to minimize the time of accounting reports preparation as it is the obligatory requirement of the bank.

Sufficiency of means for a co-financing. Banks, as a rule, do not approve applications on the projects financed only by loan resources: the co-financing is welcomed. Own funds should cover the start of the project and project services, for example, the most obvious payments on the start of the project are:

- advance payments to the supplier of the equipment;
- covering of the production space preparation, personnel and permission documents costs;
- interest repayment until the project reaches a profitable zone. As a rule, the bank grants a delay in the primary loan repayment, but the debt service (payment of interest) begins in the month following the loan receipt.

Also the correct costs planning on modernization belongs to the financial restrictions. There are explicit costs which can be planned rather accurately. They are:

- equipment cost,
- equipment delivery,

• customs registration,
• installation;
• dismantle of the old equipment;
• preparation and service of production spaces;
• retraining of the personnel to work with new equipment.

It is hard to plan the hidden expenses while preparing the technical modernization project, but in certain cases they can make a considerable sum. As the hidden expenses it is possible to consider the following:

• costs of the project and permission documents connected with the new equipment and production volume,
• commercial costs connected with the need of increase in sales,
• costs of support personnel retraining, requalification or reduction;
• costs of new employees hiring;
• costs of the infrastructure modernization which is directly connected with the new equipment (electricity, gas supply, ventilation, etc.);
• expenses on new types of raw materials (if the new equipment assumes changes in technology);
• costs of development and experimental samples production;
• expenses on production downtime (if there is no opportunity to mount the new equipment, without stopping the old).

In general, the restriction of financial character is the long term of return expectation from investment resources.

4 Organizational restictions

The global organizational restriction is the need of complex, system account of technical modernization.

The modernization of the enterprises includes the replacement of the equipment, along with operating conditions changes in the infrastructure of the enterprise. The optimization of the logistic scheme has a key value during the technical modernization: interaction with partners, suppliers, consumers, many of them can be new to the enterprise, or new requirements are imposed to the old partners. It is possible to achieve 15% increase in productivity by the correct labor organization and logistics without additional investments into the technical equipment replacement. The enterprise can get especially unsatisfactory effect, when it passed through the technical renovations, but did not receive worthy results because of the invalid interaction with partners. In absence of the advanced organizational technologies, the purchase of new equipment can appear inefficient waste of the invested funds. The solutions of infrastructure problems, adjustment of a mutually advantageous data exchange between the enterprises are required.

Situational, but not less important organizational restrictions are as follows.

Taking into account balance capacities on different stages. The replacement of the equipment can be carried out at the separate stages of production or cover all of them. Before choosing full or stage-by-stage modernization it is necessary to analyze the productivity of each phase to find the reserve capacities and bottlenecks.

The reserve capacities are necessary for providing the production program if in the result of the technical modernization, the productivity of the equipment, providing the leading process, increased.

The bottlenecks slow down the production, and as the result of modernization, the bottlenecks can be displaced. To avoid this, the design group, working on the organizational segment of the modernization project, should include competent engineers.

If the separate stages of the production works have incomparable productivity (power, capacity), it makes sense to carry out stage-by-stage modernization to increase capacities of the low-productive stages. This approach can lead to the situation, when the enterprise will be in the condition of the constant modernization, but on other hand — it demands less lump investments and allows mastering them gradually, as receiving positive results.

It must be kept in mind that any proposition on increase in equipment work intensity can be considered only at the account of relevant information on the equipment utilization with maintenance and repair work, and about permissible load on the personnel.

While calculation the payback of the investment project, if only part of the industrial complex of the enterprise is exposed to the technical modernization (the most often case in practice), the calculation should be performed, based on the smallest power.

Environmental standards account. Also it is necessary to pay the most thorough attention to the compliance of the planned capacities to environmental standards and requirements of the supervisory authorities.

Extra time account. Strangely enough, technical services are inclined to underestimate the duration of the installation period and trial operation of the updated equipment.

At calculation of the delivery time of the equipment it is necessary to consider the term of transportation, customs passing and other specific factors like features of the national calendars (Christmas vacation, day-offs, holiday seasons, etc.).

The special place in the "organizational" restrictions of the technical modernization should be allocated for the work of marketing service

If before the modernization project the enterprise was in the market with the un-filled capacity (pent-up demand), the enterprise is in an advantageous market position: the modernization will increase the productivity and outputs; the quality of a product, as expected, will improve.

The situation demands marketing services activity when, as the result of the modernization, the productivity and outputs increase, and, so the enterprise plans to expand its presence in the already tapped markets or take on a new one. The task of the market analysis service is to find out the existing and potential demand. In case of need, the service of realization has to increase the production sales volume. If the enterprise does not meet the solvent demand in the market, most likely, the schedule of the credit payments will be broken.

The developed tendency in the market of industrial production also should be kept in mind: the mass markets are inaccessible to the Russian industrial enterprises, and the batch production assumes the formation of the range dynamics based on careful market analysis and anticipatory demand management of the consumer.

5 Competence-based restrictions

It is necessary to mark the competences, necessary during the design and start of the project. The competences of the technical modernization process designer are internal which should be looked for in the internal environment of the enterprise. Practice shows, that this is the task for the economist and engineer, who is an expert both in "old" and "new" technological process and in the corresponding equipment. The most important aspect of the effective modernization realization is a qualitative preparation and drawing of the requirements specification which is defining for the development of pre-project and project documentation on the reconstruction and modernization [24]. Is something is missed at this stage it is difficult to fill it during the subsequent works,

The modern modernization assumes replacement of the universal equipment by specialized which demands absolutely different competences from the workers and service personnel, therefore it the work on search and attraction of "external competences" becomes important. The modern equipment mainly consists of machines with PNC. The operating modes and shift system are written by the technologist-programmer. The machine operator does not interfere with the process; he only needs to look after the work of the equipment. The control, repair, service of the machine is the competence of the service personnel. The modern hi-tech equipment demands special services for maintenance of its exact capabilities: it is necessary to clean, wipe, carry out it regularly autonomous service, to observe culture of the production. At the different enterprises the process is organized differently, but all these competences should be provided. The search of the personnel with the corresponding qualification is a project task: there is a shortage of the competent workers, capable to operate the modern hi-tech equipment in the labor market. The alternative is the preparation of the personnel by the enterprise «for the modernization". The solution of this task demands time which should be considered.

5 Conclusion

In the long term outlook the modernization is a guarantee of the enterprise competitiveness, growth of innovation, labor productivity and production quality.

The dynamics of the technological innovations, movement indicators and condition of the basic productive assets give the grounds to claim that the revealed restrictions do not allow solving the problem of the complex modernization fully.

References

1. V.N. Polovinkin, A.B. Fomichev, Current state and problems of a domestic machine-building, Gasweek, weekly review, 4.07.2013.
2. B. Kleyner, Bulletin of the Samara gosudastvenny university, **123**, 26-35 (2015)
3. O.V. Karasuntseva, *Formation and realization of strategy of increase of level of use of production capacity of the enterprises of mechanical engineering* (Samara, 2014)
4. P.G. Bunich, *Depreciation of fixed assets in the industry* (Gosfinizdat, Moscow, 1957)
5. P.G. Bunich, *Efficiency of use of fixed assets* (Finance, Moscow, 1966)
6. E.A. Ivanov, *Reproduction and use of fixed assets* (Economy, Moscow, 1968)
7. S.G. Strumilin, Economy Questions, **8**, 2-16 (1956)
8. D.M. Palterovich, *Fixed assets: intensification of use and updating* (Knowledge, Moscow, 1974)
9. V.G. Zakharov, *Scientific and technical revolution and updating of fixed assets* (Lenizdat, Leningrad, 1973)
10. D.A. Rams, *Terms of depreciation and updating of the fixed business assets questions of the theory and methodology* (Science, Moscow, 1977)
11. T.S. Khachaturov, Economy questions, **3** (1973)
12. V.P. Krasovsky, *Compensation of fixed assets in the conditions of an intensification* (Economy, Moscow, 1986)
13. M.I. Rimera, *Management of use of fixed assets in the industry* (Economy, Moscow, 1985)
14. M.M. Goldin, *An obsolescence of fixed assets of the enterprise in the conditions of a scientific and technological revolution* (Economy, Moscow, 1986)
15. N.M. Alekseenko, *Reproduction of fixed assets of the industry and efficiency of their use* (Lvovizdat, Lviv, 1990)
16. V. Ivanchenko, The economist, **11**, 15-21 (1994)
17. K.F. Oyner, *Improvement of management of fixed assets and capacities of the enterprises of mechanical engineering* (Publishing house Urals, 1986)
18. V.N. Smagin, *Economy assessment of technical progress and updating of cars* (Chelyabinsk, 1987)
19. I. Ivanov, S. Rogov, The Russian economic magazine, **7**, 20-31 (1995)
20. The federal law of 18.07.2011 No. 223-FZ About purchases of goods, works, services by separate types of legal entities. http://base .garant.ru/12188083/
21. The federal law of 05.04.2013 No. 44-FZ About contract system in the sphere of purchases of goods,

works, services for ensuring the state and municipal needs. http://base .garant.ru/70353464/

22. O.N. Balayeva, L.S. Korobeynikov, E.E. Chupangding, A.A. Yakovlev, University management, **2** (2014)
23. I.V. Yershov, E.V. Cherepanov, *Formation of effective depreciation policy of the industrial enterprise, Messenger of mechanical engineering* (Scientific and technical publishing house Mechanical engineering, Moscow, 2011)
24. E.Yu. Kuznetsova, E.A. Chopovda, Messenger URFA. Economy series and management, **2**, 27-35 (2013)

Modification of the method of construction of value chain industrial enterprises with regard to financial involvement of staff of logistics

A.G. Butrin[1*], and *I.A.* Baev[1]

[1]South Ural State University, 454080 Chelyabinsk, Russia

Abstract. The article presents the author's method of building an efficient value chain in an industrial plant, based on a well-known concept of the value of M. Porter, and other members of the Harvard Business School. Its novelty lies in the fact that, first, the operation captured still poorly known financial stream; second, extended value chain border due to the inclusion of all members of the value within the supply chain management concept. This will speed up capital turnover and reduce costs in terms of integration of all participants in the value chain. Cash flow still remains little known in the concepts of financial management and financial logistics. However, the adaptation of key provisions in the crisis lie is the effective use of financial logistics methods.

1 Introduction

In the current economic downturn in Russia, flowing both integration and disintegration processes, high competition, significant risks, the reorientation of trade flows on import substitution significant reserve to ensure the survival of industrial enterprises, integrated economic formations and their subsequent access to the path of sustainable development is the development and implementation in practice crisis management, aimed at effective adaptation to the environment, reduce costs, maintain economic relations. In these circumstances, it becomes relevant search and improving the methodological basis on which is possible to build a crisis management subjects integrated control system in an industrial complex. Such a sound basis, in our opinion, may be the concept of logistics and supply chain management.

2 The relevance and degree of scientific problem elaboration

One of the basic approaches is well-known concept of the organization of the value chain, set forth in the works of M. Porter [6,7] and his colleagues at the Harvard Business School [1,2,3,4,5,8,9,10]. Theoretical and methodical bases of functioning of the integrated business structures of Russian scientists are reflected in the works of SB Avdasheva [11,17], V.S. Katkalo [16,17], G. Kleynera [19], M. Yu. Shereshevoy [12], A. V. Yudaeva [21] and others.; in logistics concept - M. Kristofera, D. Uotersa [20], V. I. Sergeeva, V. N. Dybskoy, D. A. Ivanova, V. S. Lukinskogo. However, the methodology for the value chain of integrated industry and assess its effectiveness requires further development. The scientific importance of further research in this direction due to the following aspects. Firstly, the classical approach limits the value chain processes of the enterprise and does not fully take into account the significant integration with the chain partners. Secondly, in the classical approach creating value directed only on the interests of the consumer, which may lead to an increase in producer costs, reduce its financial performance and lead to the conflict between the parties. Thirdly, the basic concept of the value chain is formed only on the material flow operations for the promotion of physical goods from point of origin to final consumption, which does not include the current problem of financial resources constraints. Fourthly, the basic concept seeks to reduce losses from the unprofitable operations, which do not lead to the creation of value. However, these actions may require increase traditional costs in the larger extent than losses on the economy. Fifth, evaluation the efficiency of the chain is based only on time approach (as the ratio of the duration of useful operations to the duration of all logistic cycle). However, it is not clear, that these actions lead to achieve the maximum value of the relationship to maximize the profitability of the enterprise in the classical interpretation. The relevance and practical importance of these issues determined the choice of research topic, research problem, object and subject.

Thus, recognizing the fundamental nature of M. Porter is classic concept, we believe it is necessary to put and to solve the problem - to modify the classic approach to building value chains for effective use in the Russian practice. The object of study - value chain of an industrial enterprise, integrated with resource providers and consumers. Subject of research is organizational and

* Corresponding author: butrin_ag@mail.ru

economic relations in the delivery process in the supply and marketing departments.

3 Theoretical part

Proposed modifications to the direction of the methodology and procedure are presented in table 1.

Table 1. Comparative analysis of approaches to the organization of the value chain.

№	Criterion of comparison	The traditional approach	The proposed approach
1	Methodological basis	Systems approach	Intersystem approach (modified systematic approach by expanding its borders)
2	Object of study	A separate company	Supply chain "supplier-producer-customer"
3	Subject of study	Organizational and economic relations in the process of value creation within a single enterprise	Organizational-economic relations in the integrated supply chain participants of create value
4	Logistics concept	Logistics concept	Supply chain management concept
5	The emphasis in the construction of the chain	On the consumer	The balance of interests chain participants (time and cost criteria)
6	Considerations kinds of flows	Material flow	The material, financial and information flows in the trinity
7	Model of chain	The push model	The push model, pull model, hybrid model
8	The approach to assessing the effectiveness of	Transitional approach (as a productive time to total time)	Synthesis of temporal approach, the classical economic (as the ratio of profit to capital) and the target (such as the ratio of the actual value to the target value of the indicator)

Thus, we distinguish two main directions of the modification of the classical method. Firstly, the expansion of the chain due to the inclusion of the financial flow (receivables, payables) and determine its optimal parameters with the assistance of financial infrastructure (bank, factoring company). For example, the company delivers the goods with deferred payment (Eject model). In this case, the obvious conflict of interest is not considered in the classical concept: the value to the consumer - is the immediate receipt of goods, and for the producer - obtaining money with a time lag. But the latter may not happen, possible deferral, delay in payment, etc. In the second method of implementation - prepayment (the pull model) conflict is also evident, but the values are swapped.

Secondly, the formation of value chain, depending on the mode of organization of the supply chain ("pull", "push", and combinations thereof), and the use of the apparatus of financial Logistics (resource regulation parameters of the financial flow in the material parameters, flow-cheap "injection" in the form of bill credit, factoring to speed up payments in the sales and slowdown payment in the supply of the material resources, securitization through structuring financial flows via bill credit and warehouse warrant, etc..) to eliminate "bottlenecks" links in the supply chain [15].

4 The practical significance of the results of experimental studies

Relevance in practice is that the proposed method allows more fully identify and use reserves to reduce operating cycle of industrial enterprise and the corresponding costs, which closely connected with the parameters of suppliers and customers in the supply chain. In a pilot study of the supply chain in one of the largest metallurgical enterprises of the Chelyabinsk region was developed a model of hybrid value chain that combines elements of "pull" and "push" models: producer receives from the customer an advance payment of 50%, and delay in the production of products (the pull part). Producer after the shipment provides to the customer delay in the second half of the money for a period equal to the period of manufacture (push part). The combination of both parts in a single chain provides to achieve the balance between producer and consumer on the term, the financial flows and risks.

For example, the consumer makes prepayment of 50%. (Step 1). After agreeing on the terms of delivery, producer payment for the raw materials (2) and these materials delivery to the manufacturer (3). The technological cycle of the plant includes storing the resulting raw material (4), Production (5) and storage of finished products (6). After shipment to the buyer (7) he fully pay the order after 30 days (8). Value chain is shown in Figure 1, concomitant cost chain - in Figure 2.

Simulation is made in three-dimensional space and allowed us to obtain the best value of the payment time for material resources (the length of the value chain "supplier-producer"), sales (chain length of "producer-consumer"), total expenses and losses (costs associated with creation of chain value), and parameters involved in financial logistics - size prepaid 50%, the value of the factoring discount rate of 10% and a discount of 3% (Figure 3).

Legend: 1 - prepayment of products in the amount of 50%, 2 - payment for raw materials, 3 - transport of raw materials, 4 - storage of raw materials, 5 - the production of finished products, 6 - storage of finished products, 7 - transport of finished products, 8 - Payment for products with delay respect to the shipment.

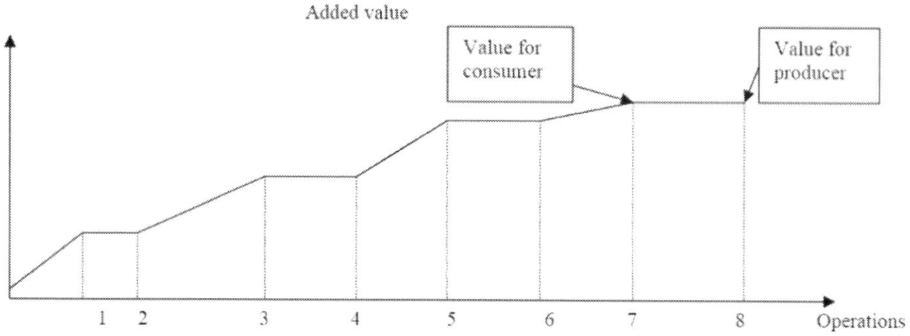

Fig. 1. The value chain as part of the hybrid supply chain.

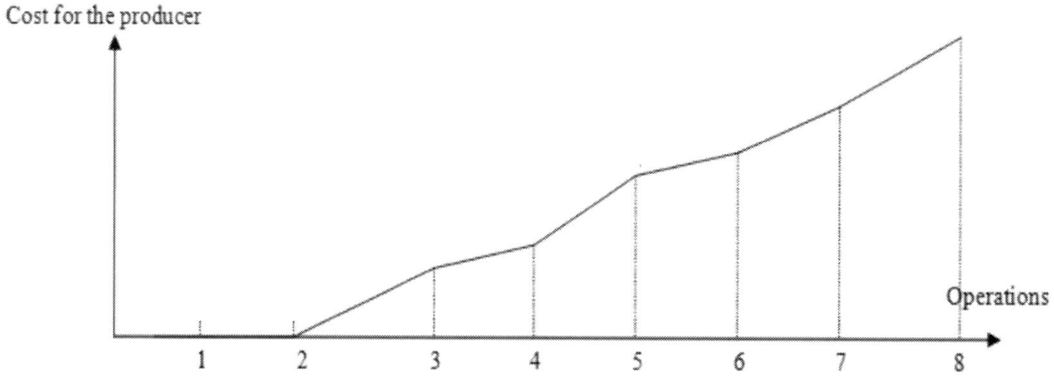

Fig. 2. The chain costs within the supply chain of the hybrid.

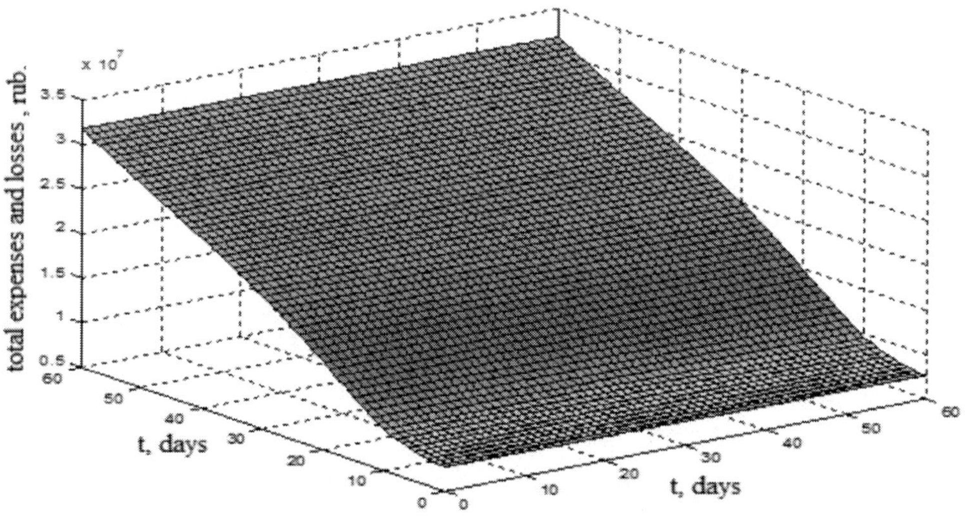

Fig. 3. The value chain "producer-supplier" and "producer-consumer" (two-axis in the horizontal plane) on the criterion of integrated cost.

5 Conclusion

Thus, the growth of knowledge in the field of economy and management of an industrial enterprise is to expand the concept of M. Porter to include in the object of analysis of the key financial flows and offer concrete tools to achieve a balance of interests of participants of the chain. Methodical developments have practical significance and can be used by managers and specialists of economic and technical units of industrial enterprises. Validity and reliability of scientific statements, conclusions and recommendations are confirmed: the conformity of the theoretical results of the fundamental provisions of the theory of systems, logistics theory, efficiency theory; correct application of economic-mathematical modeling methods; obtaining the certificate of state registration of the software product, made in the framework of a set of routines to MatLAB language.

The work was supported by Act 211 Government of the Russian Federation, contract № 02.A03.21.0011.

References

1. K.R. Andrews, *The Concept of Corporate Strategy* (Homewood, IL, Dow Jones-Irwin, 1971)
2. W. McPhee, D. Wheeler, Strategy & Leadership, **34**, **4**, 39-46 (2006)
3. G. Hamel, C.K. Prahalad, *Competing for the Future: Breakthrough Strategies for Seizing Control of Your Industry and Creating Markets of Tomorrow* (Harvard Business School Press, Boston, 1994)
4. A.D. Chandler, *Jr. Strategy and Structure:Chapters in the History of American Enterprise* (MIT Press, Cambridge, MA, 1962)
5. A.D. Chandler, *The Visible Hand:The Managerial Revolution in American Business* (Belknap press, Cambridge, MA, 1977)
6. M.E. Porter, *Competitive Strategy: Techniquesfor Analyzing Industries and Competitors* (Free press, N.Y., 1980)
7. M. Porter, *International competition* (International Relations, Moscow, 1993)
8. R.K. Oliver, M.D. Webber, *Supply chain management: Logistics catches up with strategy* (Logistics, The Strategic Issues, 1992)
9. D.A. Pitta, M.V. Laric, Journal of Consumer Marketing, **21(7)**, 451-464 (2004)
10. D. Walters, G. Lancaster, Management Decision, **38(3)**, 160-178 (2000)
11. C. Avdasheva, V. Dementyev, Russian economic journal, **1**, **13** (2000)
12. R. Badzho, M.Y. Sheresheva, Bulletin of Moscow University. Episode 6: The Economy, **2**, 3-21 (2014)
13. D. Uoters, *Supply chain management* (UNITY-DANA, Moscow, 2003)
14. A.G. Butrin, Z.S. Gelmanova, D.L. Yaruchin, Metallurgist, **59**, **1-2**, 99-103 (2015)
15. A.G. Butrin, *Logistics for the CFO: a tutorial* (Publishing House Abris-Print, Chelyabinsk, 2009)
16. A.V. Buhvalov, V.S. Katkalo, Russian Management Journal, **3**, **1**, 75-84 (2005)
17. V.S. Katkalo, *Evolution of Strategic Management Theory: monograph* (St. Petersburg State. Univ, Executive. wk. management, St. Petersburg, 2008)
18. S.V. Golovanova, S.B. Avdasheva, S.M. Kadochnikov, Russian Management Journal, **8**, **1**, 41-66 (2010)
19. G. Kleyner. R. Kachalov, E. Sushko, Questions economic, **9**, 67-86 (2005)
20. A.G. Butrin, Economy of Region, **2**, 130-140 (2014)
21. A.V. Yudaev, Management of economic systems: electronic scientific journal, **29(5)**, 36 (2011)

The prediction of the economic effect of the energy-saving system implementation: corporate and regional aspects

T.A. Shindina[1,*], and *L.B.* Kotova[1]

[1]South Ural state University, 454080 Chelyabinsk, Russia

Abstract. The solution of the energy-saving problem is connected with two main aspects: technical progress allowing the use of modern equipment and technologies with great efficiency; the development of an economic system which allows achieving development objectives and minimization of resource costs. In our article we are going to talk about the possibilities of formation and results evaluation when implementing energy-saving measures and we are going to present the results of the research on the interdependence of managerial impact and economic results. In this article we study an assessment method of activities at the enterprise according to the results of energy-saving measures, and we are going to present the results of the study of economic factors of their development, we are going to suggest a method for assessing the prognosis based on entropy methods. The development of a group of enterprises that produce rational results as the results of energy-saving measures of some companies taking into account synergy will allow us comment on the overall economic efficiency of the introduction of the energy-saving system in Russia as an element of social development.

1 Introduction

The energy sector of the economy of a country is complex, involving receiving, transferring, transformation and use of various types of energy and energy resources. It is where technological, economic and social components of social development meet and it is the control factor in ecological and economic spheres. The state of the complex and individual enterprises reflects, on the one hand, the environment state; on the other hand it reflects the level of economic development and quality of human reasoning (Fig. 1).

Fig. 1 shows the interconnection of the basic economic categories of the energy sector of the economy of a country taking into account interconnection and diffusion of concepts. Let us study them.

Nowadays useful application of energy primarily is associated with energy-saving. Energy-saving should be understood as a complex of measures dealing with implementing legal, organizational, scientific, production, technical and economic measures aimed at efficient (rational) use (and economical consumption) of fuel and energy resources (FER) with the beneficial effect of their use and for involvement of renewable energy sources in economic circulation.

In fact, energy-saving is the decline in energy consumption, which is possible on the basis of:

- development of the technology of forced reductions in consumption on the basis of technical means, for example, of lower power technology or forced break switches in resources use , etc.
- development of technologies for replenishing consumed energy based on renewable sources use;
- development of society, understanding the importance of lean behavior and creation of a culture of thrift, namely, self-restraint when one deals with energy use and finding ways of replacing energy needs.

In the review of potential areas of energy saving, as well as of policies and programs designed to achieve energy saving, an integrated approach to energy efficiency is very important. The efficiency measured at the macro level (the energy intensity of the economy of a country) is a key indicator characterizing the sustainability of energy sector development and the state in general. In fact, under energy efficiency one can understand the ratio of the results of the use of energy and the cost of their replenishment, which is possible on the basis of:

- management of national income associated with the development of society and global economies, aimed at cost savings in energy consumption;
- reserves management of natural resources, reasonable spending of various kinds of resources, the use of a system of optimization and the principles of careful attitude to nature aimed at reducing costs for energy resources;
- development of engineering and technology with the maximum possible coefficient of efficiency aimed at reducing costs, consumption and energy transfer.

* Corresponding author: shindina@74.ru

DOI: 10.1051/shsconf/20173501083

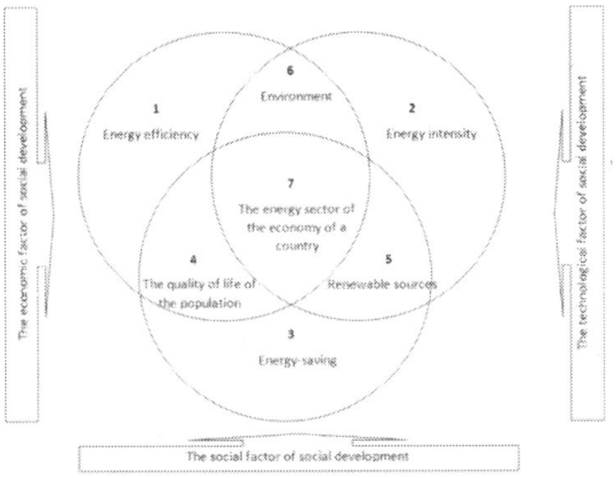

Fig. 1. The interconnection of the basic categories of the energy resources management.

The matters of cost reduction on energy connected with capacitive characteristics, which should be studied in the context of the concept of "energy intensity". Numerical expression of the intensity of the system is the indicator, which is the ratio of energy consumed by the system to the value characterizing the result of the operation of the system. In fact, under energy intensity one can understand power characteristics, reflecting the rate of speed change in continuous processes of consumption and accumulation of energy. The energy intensity shows certain resource endowment, which is possible on the basis of:

- increase in production volumes and reproduction in the dynamics;
- balance in the use of fuel substitutes (oil, gas, etc.);
- increase in the speed of resource replenishment at controlled consumption

In modern conditions of management energy-saving becomes a state priority, as it allows with the help of relatively simple measures of state regulation significantly reducing the load on budgets of all levels, restraining the growth of energy tariffs, increasing competitiveness of economy and increasing the supply in the labour market.

2 Objectives, methodology and stages of the research

The economic target of energy-saving is to minimize equipment costs and engineering services at regional enterprises that entail the effects of regional significance based on the principle of synergy.

The study of the economic results and the short term forecast is the research question which is relevant from the standpoint of ensuring social development of regions. There are two stages when conducting the study:

1. The forecast of economic efficiency of activities concerning energy saving at the enterprise.

2. The forecast of the economic effect in the region as the total of activities of enterprises and organizations that implement energy saving.

We are going to use a correlation and regression analysis, evaluation of plans reliability based on entropy and negentropy, analysis and synthesis of information as a methodological apparatus in order make predictions.

The main role in increasing the efficiency of energy use belongs to modern energy-saving technologies. Enterprises usually implement technologies that provide a significant energy-saving effect. These technologies associated with the energy use (motors with variable speed, heat exchangers, compressed air, lighting, steam, cooling, drying); more efficient energy production, including a modern boiler plants, cogeneration (heat and power), trigeneration (heat, cold, electricity); the replacement of the old industrial equipment with new, more efficient one; alternative sources of energy. In table 1 the main types of energy-saving technologies that are currently used are presented.

The whole complex of energy saving measures presented in the table has an engineering basis, but the essential roots of efficiency are economic in nature. Thus, the main research question which interests us from the point of view of energy saving is the direct economic impact on the enterprise and an indirect effect of regional importance.

3 Discussion on the research outcomes

In the article we study the impact of the above mentioned activities on enterprise costs.

In table 2 and 3 there are the data on direct costs of production of thermal energy and technological heat supply process from 2011 till 2015, and the data on generated energy. As we can see there are quite rhythmic fluctuations of costs from one period to another. The greatest increase in costs occurs in 2013, and further there is a gradual decline that is associated with the implementation of energy saving measures.

The most "vivid" picture is represented by two main indicators: natural gas and equipment repair (Fig.2). The economic indicator which reflects energy saving will be presented as the ratio of direct costs per 1 unit of produced energy, in fact it shows the dynamics of change of costs in ratio to the previous period.

However, the process of inflation and fluctuations in the financial system does not allow limiting in economic research on relative performance changes in costs. As we can see from the data in table 2 despite the fact that in 2014 there is no significant reduction in costs, at some point they increase. But it is not affected by the ultimate level of direct costs per 1 unit of produced energy, because in general there is an increase in development.

What will be with the enterprise along with the full implementation of the energy-saving program?

Predicted changes in the trend line on the indicators tab. 2 show the "conditional reduction" of costs for the next three years (2016-2018). Physical proof is the increase in the growth of energy production while reducing its cost.

The greatest effect from the introduction of the energy saving program can be seen in the reduction of costs for resources, repairs and maintenance of equipment. However, an important factor in the assessment is the uncertainty. For this purpose it is necessary to study the degree of uncertainty and disorder of costs by items of expenditure in relation to the benchmark presented earlier.

This analysis will allow us to assess the regularity of expenditure for the whole period of implementation of the energy-saving program, and it will provide an opportunity to highlight a priority for further detailed consideration of this problem.

Table 1. Styles Classification of energy-saving technologies.

Economic sector **Loss**	**Construction**	**Industry**	**Agriculture**	**Transport**	**Tertiary sector**	**Housing and utility sector**
Decrease in spending amounts	Automation of processes of consumption (forced disconnect, modes change)	Use of low power devices	Technologies of fossil resources, environmental safety	Reduction of losses technologies (risks of theft, loss of transmission, loss of storage, etc.)	Technologie s of resources saving (heat, light, etc.)	Use of metering devices (multirate system)
Reducing of needs	Increase in efficiency coefficients of devices	Change of tariff policy	Biotechnology replacing the energy demand	Tracking device of unauthorized connections	Renewable energy technologies	Smart lighting technology
Reducing energy costs	Generating capacity technologies (the use of minerals, materials substitutes)	Technologie s of resource saving (water saving, fuel, gas, solar energy)	The principle of non-waste production	The energy efficiency classes	Technologie s of resources procurement (making or buying)	Technoligies of energy performance certificates

Table 2. The cost of production of thermal energy and of technological heat supply process, thousands of rubles.

№	**Description of costs**	**2011 year**	**2012 year**	**2013 year**	**2014 year**	**2015 year**
1	Salary	2 078.3	2 259.1	2 455.5	2 780.0	2 931,2
2	Fringe benefit expenses	605.3	657.9	715.1	839.6	888,6
3	Natural gas	23 073.7	23 885.8	25 560.0	29 865.5	30 164,2
4	Maintenance of gas pipeline	926.6	985.7	1 048.6	213.8	352,3
5	Electrical energy	3 868.1	3 792.3	3 830.6	4 088.6	4 598,0
6	Water supply	1 620.1	1 588.3	1 604.3	1 872.2	1 970,5
7	Raw products and materials (reagent)	96.7	95.7	97.7	265.2	173,7
8	Depreciation	238.8	238.8	238.8	238.9	253,3
9	Repair and maintenance of equipment	11 535.6	12 538.7	13 629.0	8 655.3	6 299,5
Total direct costs		**44 043,2**	**46 042.3**	**49 179.6**	**48 819.1**	**47 631.3**

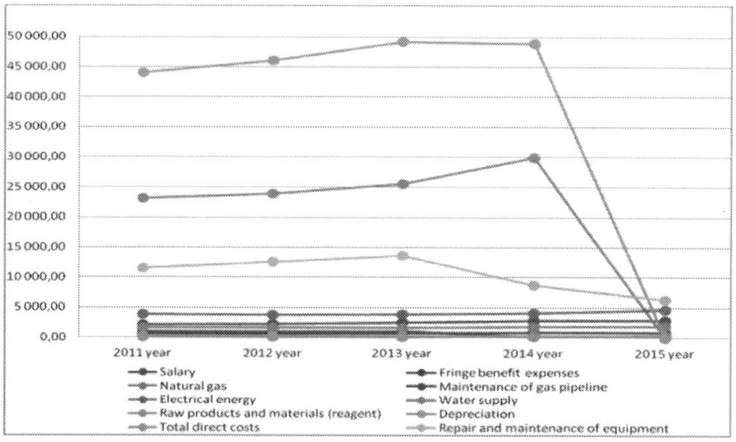

Fig. 2. Dynamics of changes in natural gas costs and repairs and maintenace of equipment, thousands of rubles.

Table 3. Data on energy production of a boiler-house OJSC Electro machine

Indicator	Year				
	2011	*2012*	*2013*	*2014*	*2015*
Thermal energy, Gcal.	39 654.7	42 154.2	45 387.6	49 241.8	45 977.4
Technological heat supply process (steam), Gcal.	10 060.1	9 465.3	9 146.7	8 979.0	8 876.0
Energy production, Gcal.	49 714.8	51 619.5	54 534.3	58 220.8	54 853.4
Direct costs per 1 Gcal, rubles	885.9	892.0	901.8	838.5	794.5
Growth rate, %		0.7	1.1	-7.0	-5.2

In order to account for the level of uncertainty we use the methodology of calculation of the entropy, which allows excluding the influence of random processes. Let us perform a calculation of all studied economic indicators.

The results of calculation are presented graphically (Fig.3)

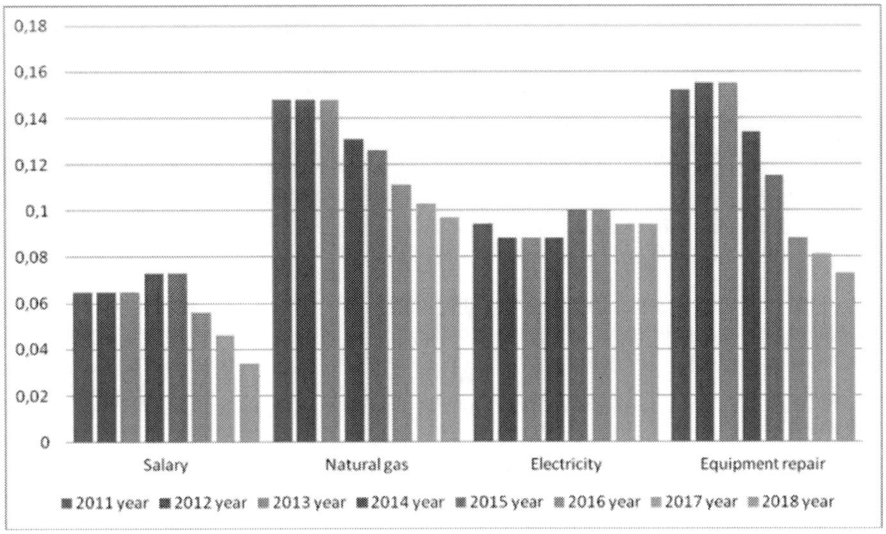

Fig. 3. Entropy of costs at the enterprise.

The calculation will be done according to the formula:

$$H = -\sum_{i=1}^{n}\left(E_i \cdot \ln(E_i)\right) \qquad (1)$$

where H – the level of costs entropy, – the share of costs according to i-kind costs in the total direct costs.

As we can see on the chart the costs on natural gas and repairs and maintenance of equipment have the highest entropy, which shows that these costs are the most difficult to manage and they are more vulnerable to the effects of the produced energy. The payroll costs have the lowest entropy which shows that they are conditionally constant and are not dependent on volume of services.

As a rule managing the cost of natural gas purchasing can only be achieved by upgrading equipment, this is usually a one-time event, which is effective only in a certain period of time and only in natural units (e.g. in monetary terms, regardless of equipment upgrades, there will be an increase associated with the inflation of prices for the supply for this resource). Thus, it is the most "favorable material" for carrying out the research of influence of various factors concerning the reduction of costs and resource saving for repairs and maintenance of equipment.

Thus having considered the dynamics of changes in costs since the implementation of the energy-saving program until its end we can conclude that in spite of significant capital investments in the conditions of market instability, the company will have additional profit increase by reducing costs (at constant income).

4 Conclusion

Building an energy-efficient society in Russia includes forming the system of energy management with an optimally minimized cost, and ensuring the contribution of the energy factor in the country's economic development and in improving living standards.

Energy efficiency from the standpoint of enterprises complex gets a new content - regional and in a regional context it becomes a wide concept, accumulating in its semantic content minimization of specific energy consumption for successful social and economic functioning and development, as well as optimization of the contribution of the energy sector in the GDP, in the social improvement of society and economic well-being of the population and in protection of the environment.

To assess the development of Russian regions in the sphere of energy-saving and energy efficiency it is necessary to carry out a comprehensive analysis of the regional use of certain resources and the benefits from their use.

The spectrum of indicators of technological and economic nature allowed us to identify 10 key indicators that influence the energy efficiency and the economy of the region. In order to form the most correct data, taking into account the result of the pair correlation of values obtained during the process of the analysis, we will rely on some key indicators, namely on an average annual temperature in the region, production and distribution of energy resources (management system of which we have considered above), as well as the emissions of pollutants into the air. The gross regional product will be presented as a dependent value (Fig. 4).

In the results of the correlation analysis we see that when evaluating the development of regions in the sphere of energy-saving and energy efficiency the least impact on the gross regional product value is made by an annual air temperature. The model taking into account the obtained weight coefficients can be represented by the formula:

$$Y = 5{,}85x1 + 441{,}23x2 + 18566{,}82x3 \qquad (2)$$

where x1- average annual air temperature, C°; x2- emissions of pollutants into the air, thousands of tons; x3-production and distribution of energy resources, millions of rubles.

The least change in the production of energy towards the increase or decrease will mostly be displayed in the development of the region as a whole. Emissions also improve the economic performance of the region, but they significantly reduce environmental performance.

So, Russia has vast underutilized potential for energy saving, which is able to solve the problem of ensuring economic growth of the country that is comparable to the growth of production of all primary energy resources

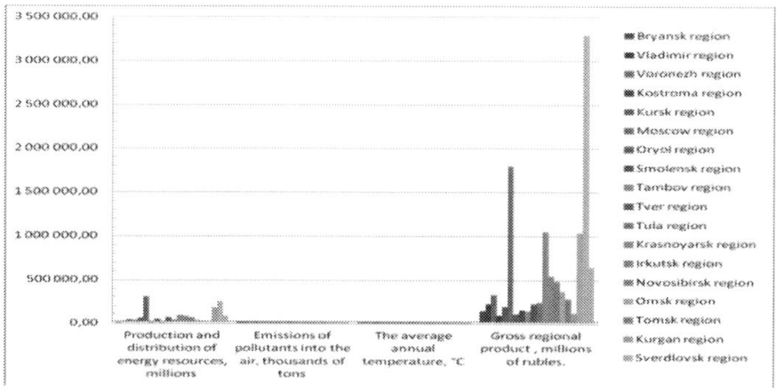

Fig. 4. Main indicators of activities in Russian regions.

The work was supported by Act 211 Government of the Russian Federation, contract № 02.A03.21.0011.

References

1. D.F. Pires, C.H. Antunes, A.G. Martins, *NSGA-II with local search for a multi-objective reactive power compensation problem* (2012)
2. M. Doostizadeh, H. Ghasemi, *A day-ahead electricity pricing model based on smart metering and demand-side management* (2012)
3. L. Yang, C. Dong, C.L.J. Wan, C.T. Ng, *Electricity time-of-use tariff with consumer behavior consideration* (2013)
4. M.O. Abdullah, L.P. Yii, E. Junaidi, G. Tambi, M.A. Mustapha, *Electricity cost saving comparison due to tariff change and ice thermal storage (ITS) usage based on a hybrid centrifugal-ITS system for buildings: A university district cooling perspective* (2013)
5. I. Dincer, M.A. Rosen, *Thermal energy storage: systems and applications* (2010)
6. S.M. Hasnain, *Review on sustainable thermal storage technologies, part I: heat storage materials and techniques* (1998)
7. M. Ostry, P. Charvat, *Materials for advanced heat storage in buildings* (2013)
8. A.M. Khudhair, M.M. Farid, *A review on energy conservation in building applications with thermal storage by latent heat using phase change materials* (2004)
9. N. Zhu, Z. Ma, S. Wang, *Dynamic characteristics and energy performance of buildings using phase change materials: A review* (2009)
10. R. Baetens, B.P. Jelle, A. Gustavsen, *Phase change materials for building applications: A state-of-the-art review* (2010)
11. F. Agyenim, N. Hewitt, P. Eames, M. Smyth, *A review of materials, heat transfer and phase change problem formulation for latent heat thermal energy storage systems (LHTESS)* (2010)
12. B. Zalba, J.M. Marin, L.F. Cabeza, H. Mehling, *Review on thermal energy storage with phase change: materials, heat transfer analysis, and applications* (2003)
13. M. Santamouris, G. Mihalakakou, C.A. Balaras, J.O. Lewis, M. Vallindras, A. Argiriou, *Energy conservation n greenhouses with buried pipes* (1996)
14. C.-L. Su, M.-C. Lin, C.-H. Liao, *A method for evaluating energy efficiency to justify power factor correction in ship power systems* (2013)
15. X.-Y. Guo, *Analysis of energy-saving real option investment based on normal clouds: Nonlinear system and control* (2016)
16. W.-L. Cao, Y.-B. Zhang, H.-Y. Dong, Z.-Y. Zhou, J.-W. Zhang, *Research and application on aseismic energy-saving structural system for rural buildings* (2015)
17. M. Faccio, M. Gamberi, *Energy saving in case of intermittent production by retrofitting service plant systems through inverter technology: A feasibility study* (2014)
18. J.-G. Dong, *Production organization strategy research and simulation based on energy-saving requirements of bulk cargo handling system*
19. S. Siwamogsatham, P. Rattanawan, M. Kitjaroen, P. Sontung, P. Pongpailbool, K. Navanugraha, *Smartly saving energy with a zero power consumption standby system*
20. A. Yoshida, T. Sato, Y. Amano, K. Ito, *Impact of electric battery degradation on cost- and energy-saving characteristics of a residential photovoltaic system*

Motivation of the personnel of hi-tech productions

N.V. Trifonova[1,*], and *O.A.* Shvetsova[1,*]

[1]Saint-Petersburg State University of Economics, Saint-Petersburg, Russia

Abstract. In work the actual problem of motivation of representative group of highly qualified specialists of the domestic knowledge-intensive enterprises is considered. Approaches to research of motivators of innovative processes for problems of modernization of a domestic production are studied. Practice of use of innovative instruments of motivation at the Russian industrial enterprises is studied. The perspective of an integrated approach to human resource management of hi-tech productions, including competence-based approach to preparation innovative the managing director of employees is investigated. Authors have offered new approaches to formation of the steady motivational mechanism of key personnel of the domestic industrial enterprises.

1 Introduction

Periodic changes of market conditions, extensive national modernization of hi-tech sector on the basis of indicative management and development of professional competences of the experts demanded for transformation and a producing knowledge, technologies, innovations are the main factorial reasons for structural shifts in national economy at a present historical stage. Development of technological and not technological field of the Russian economy has cyclic character. Among the reasons of structural shifts of sectors of national economy there are exogenous and endogenous factors which are shown on micro and the macrolevel of national economic system.

Start of accelerating process of structural shifts in innova-tive and hi-tech sectors of economy it is usually connected with functional business cycles (D. Kitchena, S. Kuznetsa), and cycles of development of N. Kondratyev. It is considered to be that the ascending and descending wave of a macroeco-nomic cycle influences intensity of structural shifts in sectors of economy and is followed by change of a priority of the leading innovative technologies. In the descending trajectory of a macroeconomic cycle elements of structure of hi-tech sector of economy are not so steady, are characterized by reduction of innovative and investment appeal of a manufacturing sector, transformation of production demand in the resource markets. Elements of institutional structure of the market of a manpower in various sectors of economy connected with technologies in the field of innovations are subject to influence of macroeconomic cycles. It is known that economic crises do essential harm to structure of production, to the economic relations, a shape of employment of key personnel is got out of stability. The hi-tech sector of economy as a key element of national economy passes a stage of transformational shift as a result of crisis of the relations with foreign investors in Russia, decrease in production of high technologies, changes of forms of the competition of involvement of employees now.

Expansion of use of industrial and system innovations in sectors of economy involves creation of specialized workplaces with higher labor productivity level, strengthening of horizontal integration of structure of production and the regional markets. Result of formation of hi-tech sector of economy in regions of the country are strong industrial clusters which as points of economic growth become the generator of innovative demand, a source of development of the professional market of labor and models of social partnership of territories. Interaction of economic factors of economic growth creates unstable dynamics, cyclic dynamics of structural transformations of sectors of economy and the market of labor, causing change of popular vocational groups of employees that demands new effective forms of motivation of personnel.

2 Approaches to personnel' motivation

In the knowledge-intensive branches of domestic economy it is possible to consider two approaches to motivation of a key human resource. The classical approach based on the administrative technologies corresponding to an era of "industrial economy" and realized within industrial technological way. Innovative (or knowledge-based) the approach focused on the enterprises functioning in more modern technological way (the economy based on knowledge). Features of innovative administrative technologies, and also advantages and restrictions of their introduction in activity of the modern innovative enterprises are defined by heads of data of the organizations.

In M. A. Benediktov's works, A.A. Bureeva, [1,2] is offered author's interpretation of concept and essence of innovative processes as key competence of the knowledge-intensive productions. As innovative process these authors consider process of transformation of

* Corresponding author: shv_olga@rambler.ru

scientific knowledge in an innovation which can be described as a consecutive chain of events in the course of which the innovation "grows" from scientific ideas in concrete new products, technologies or services and extends at applied use.

3 Specifics of human resource management of the hi-tech enterprises

It is possible to allocate the following signs of hi-tech productions:

- Complex high-tech processes.
- Knowledge-intensive products.
- Essential share of costs of research and development.
- Highly qualified specialists with specific key competences.
- Constant staff shortage of qualified personnel.
- High percent of export production in the general volume of the made production.
- Stable foreign communications and technological cooperation.
- The centralized corporate training centers of preparation of personnel.
- Qualitative investment appeal and high dynamism of production [3, 4].

Among branches of hi-tech sector of the Russian economy there are following cluster associations: nuclear energy industry, communication, telecommunications, chemistry and petrochemistry, pharmaceutical industry, power, electronics, aircraft industry, rocket production, defensive industry [3].

In modern economic and political conditions the knowledge-intensive enterprises more than ever demand the latest domestic and foreign the practician in management of a human resource (Table 1).

Table 1. Types of management of human resources in high-tech industries of economy.

The management focused on result - delegation of zones of responsibility - modeling of situations of cooperation of employees	Management on bases of collective participation in the general process - universal quality control of production - interest of the worker and his development in innovative activity, including through educational process and process of mentoring
Management by means of motivational technologies - rational model of remuneration - motivational model of self-realization - motivational model of participation in results of the organization in general	Frame management with active involvement of all employees - encouragement of an initiative of employees - development of communicative skills of employees
Management on the basis of business - the competitive atmosphere in group of employees - implementation of joint projects by workers	

Formation and development of professional competences of staff of the hi-tech organizations can be presented in the form of three "stages":

"The growing experts".

"The frozen experts".

"The experts moving down" [5].

The assessment of requirements and planning of professional competences of personnel of the hi-tech enterprises allows to determine problems of strategic personnel capacity of the enterprise in general (Fig. 1).

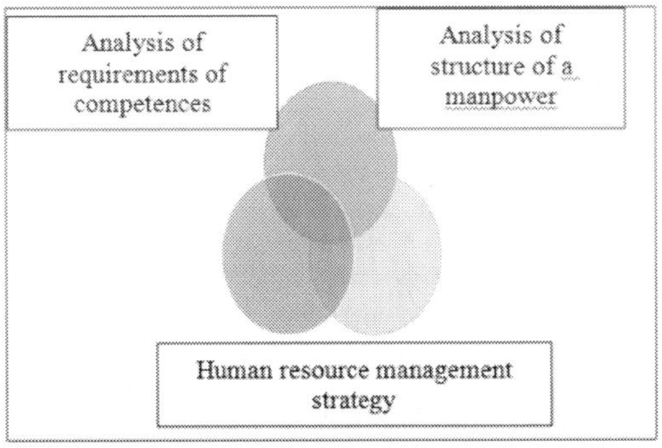

Fig. 1. Determination of professional competences of key staff of the hi-tech enterprises.

On the basis of the analysis of activity of a number of the domestic enterprises it is possible to reveal restrictions of development of professional competences of staff of the companies of high-tech industries of economy:

- Weak level of development of knowledge and skills of linear heads
- Idle ways of development and adoption of administrative decisions
- Insufficiently developed level of creative skills of employees and team building
- Absence of motivation of employees to use of innovations
- Rivalry of employees and conflict behavior in group, high degree of individualism
- Inefficient informal relations of employees, lack of effective system of communication.

It is possible to give practice of use of the concept of the Japanese management in management of the key staff of the enterprises of nuclear energy industry as an example of work with restrictions of development of professional competences (Rosatom State corporation, JSC TVEL, JSC SHK) [6]:

- Flexible management of experts, regulation of procedures
- Adaptive mechanisms of material remuneration
- Reduction of a share of administrative resources in the total number of employees
- Orientation to design and target groups and team building
- Realization of continuous training in system of corporate training

- Universal participation in quality of production and division of responsibility for quality of the received results.

In modern difficult economic conditions many domestic production companies actively financially support problems of production. However isn't enough for formation of fund of material remuneration of personnel of means. This factor increases often the level of a demotivating of employees, and it creates a certain threat: level of loyalty of key personnel to the company decreases, manifestation of deviant behavior, and also increases of coefficient of fluidity of personnel is possible. The company will lose an opportunity to develop and form competitive strategy. The employees however demotivated, but built in innovative process represent strategic (often implicit) threat: at the crisis moments such personnel won't show loyalty to the company, don't create risk - the focused model of innovative search or at all to leave firm in search of new opportunities (obviously in the competing systems).

Often, the companies with effective personnel management try to replace material motivation with non-material instruments of influence, thereby showing the interest in satisfaction of requirements and interests of employees, and the demotivating of the employees who are innovatively focused and professionally taken place is destructive.

In personnel management of the hi-tech enterprises treats non-material ways of impact on the worker:

1. Paternalism – concerning the worker is shown some guardianship, he feels care and attention from the management. The employee is involved in various actions for management of the organization, give loyal opportunities for personal and informal contact, provide broad delegation of powers and division of zones of responsibility, system of side benefits for elimination of the conflicts and intensity in the atmosphere of collective.

2. Patriotism – formation of mission of the enterprise which is congenial to workers it is actual and adequate to modern realities, is achievable.

The motivation of patriotism can be realized as means in the internal environment of the organization, and through the existing strategy and policy of regional representatives, branches and the states in general.

3. Organizational incentives – stimulation by the importance, conditions and the organization of work.

The way of impact on motives of employees allowing to have satisfaction from work performance process follows from the content of the work: it is necessary to organize work of personnel so that it coincided with their natural incentive incentives. Some experts call this process a form of organizational ergonomics.

4. The participation and participation in management – this form of motivation offers involvement of the employee to management, use of open type of information culture, that is granting large volume of administrative information in open access through corporate databases, a vote in decision-making, individual share in profit and the capital of the organization, a possibility of receiving significant percent from order cost when attracting him in the organization.

4 Model of personnel' motivation of the knowledge-intensive productions

In the field of economy and sociology of management and practical experience of the innovative companies 4 stages of process of use of innovative technologies of stimulation of employees in the organizations allow to present research of works.

1 stage. Formation and submission of the specialized program of motivation of personnel with the indication of feature of motivational mechanisms for each group of personnel.

2 stage. Definition of zones of responsibility of each linear head for this or that mechanism of motivation in the divisions.

3 stage. Creation of the working group of representatives of department of personnel and specialists of production divisions for the purpose of performance of control over results of works.

4 stage. Analysis of instruments of motivation, their assessment, definition strong and weaknesses of the mechanism of development and realization of innovations, introduction of measures for their further development.

Authors offer innovative tools of system of motivation and stimulation of key employees of hi-tech productions (Fig. 2).

Motivation of personnel by means of introduction of technologies of a goal-setting	Motivation of personnel by means of modern compensation systems	Motivation of personnel by means of introduction of technology of team building
Motivation of personnel through integration into corporate culture	Innovative technologies of motivation of personnel in the hi-tech companies	Motivation of personnel through delegation of powers
Motivation of personnel by means of introduction of technologies of an assessment of employees	Motivation of personnel through introduction of technologies of preparation and development of employees	Motivation of personnel through use of technologies of management of career

Fig. 2. Model of system of motivation of personnel in the hi-tech companies: innovative approach [6].

The offered model - not just arithmetic sum of components of administrative technologies:

1. Mechanical addition (decrease) any of elements of this system isn't capable to lead to adequate change to the consolidated motivational potential. Motivational potential is a result of synergetic interaction of these tools.

2. Identification of requirements and strengthening of motivational potential. Integrity of the specified model

promotes development of negative consequences, namely the known loss by elements of system of sovereignty. Their work within system paints elements the new motivating quality, releases also increases their primary motivational potential.

3. The "initial" motivational resource of elements of model which is originally present at them as an opportunity thanks to the processes arising in system is transformed to reality.

Participation in administrative processes can get various formats, in a varying degree, proving the efficiency – from rendering preliminary consultations with representatives of employees before acceptance of administrative signals, partnership in the solution of the arising problem signals by means of creation of circles of quality, independent groups before execution by representative bodies of employees (councils, committees, the commissions) of concrete administrative functions.

For management efficiency increase, it is necessary to meet the following conditions:

- the operating employee and employees-performers have to have enough time for performance of this function (in crisis situations similar practice is inexpedient);
- possible benefits from participation in management have to cover cost (for example, the time period which is taken away on participation in management shouldn't exceed time allowed for the main working operations);
- workers show incentives in participation in management;
- competences and qualification of employees have to allow them to take part in management;
- participants of working procedures have to possess competences to interaction within primary informal structures that gives them an opportunity of primary approbation of ideas and uncommon decisions;
- the parties shouldn't treat participation in management as to threat (for example, workers are capable to perceive indications of the management as perhaps dangerous or the head can consider that participation in management doesn't correspond to his status);
- application of practice of participation in management has to be presented as a freedom of choice of working tasks of group (each division has no right to make the decision breaking policy of the company).

Participation of collective in management and having profits by the enterprise as a strategic form of development is accepted by an increasing number business of structures. One of the reasons of a demand of the specified method is the economic aspect of his use: to raise funds "internal resources" (having sold shares of the organization to workers) more favorably and less risky, than from outside.

This participation can be realized at several levels. At the level of a concrete workplace or the working group administrative decisions or improvement suggestions belong to specifications of labor process, communication with colleagues, to labor conditions, daily questions of human resource management; at the level of the organization – this participation of personnel in the solution of questions of a strategic reference point (a production goal-setting, investments, distribution of the income, organizational structure). In addition, these questions concern the concrete labor relations in a situation "the worker – the employer" (compensation, holidays, working hours, social guarantees). At the following level the personnel participate in relationship between the enterprise and labor union which have branch and interindustry character.

Formation of potential of key personnel – planning of a career ladder, work with a personnel reserve, corporate training, mentoring, mentoring, coaching, tutoring, rotation, increase and other.

The specified way of motivation is based on needs for self-expression and on aspiration of each employee to develop, increasing greatly knowledge, skills and experience. When using beneficial influence not only on the level of satisfaction with work of personnel, but also on enhancement of a human resource of own enterprise through formation of competences appears the management of the organization of this type of stimulation. Transformation of administrative processes under the influence of the market relations means prompt dynamics of change of requirements to qualitative parameters of a labor resource. Sharks of business (especially industrial productions) increase investments into preparation of personnel as external market conditions and system of personnel training correspond to requirements of hi-tech productions less. In this situation many enterprises go for increase in investments in development of personnel and consider this action the main competitive factor.

Formation of key competences of staff of the company forms loyalty of personnel which, in turn, possesses steadier abilities and reasonable motivation to performance of tasks. It is natural that it leads to growth of productivity, and respectively, to increase in value of human resources of the organization.

In the social environment of the organization psychological influence on employees can be used, for example, it can be carried out as in a positive form - information transfer about the expert's merits, an assessment of the head and colleagues, recognition of merits; and in negative – censure, punishment, work loss threat.

The specified way of motivation significantly depends on the identity of the head who exercises influence and control.

It is possible to define the most widespread methods of psychological influence on personnel of the organization:

- belief – the impact on other person or a group of persons proved with application of the argument aiming at correction of their judgment, the relation, intention or the decision, for example, at statement of the purposes;

- suggestion – reasonable, but without application of the argument the impact on the person or a group of persons aiming at correction of their state, understanding something and tendency to certain working acts, for example, in situations when the manager carries out actions, unpopular among the employees: reduction of number of personnel, cancellation of additional guarantees, changes in process of training and development;
- self-advance – representation of the principles, the purposes, guarantees and presentation of confirmations of the competence and experience for obtaining personal benefits; for example, the middle manager for increase of the personal authority at negotiation with new interesting clients for obtaining advantages on elections can undertake such behavior, at appointment to the post;
- infection – broadcast of the state or relation to other worker or a group of persons who feel the transferred state or the relation; the state is capable to be broadcast as involuntarily, and it is operated, to get accustomed when forming lines of corporate culture, for example, of public actions, sports rest, introduction of corporate style in clothes;
- awakening of an impulse to imitation – skill to render aspiration to resemble itself; for example, this method can be applicable the charismatic leader with use of an image of the successful company;
- the request – the address to the colleague with a request to meet requirements or desires of the initiator of influence, for example: the aspiration of the head to make subordinated by the participant of a situation leads to formation at the subordinated feeling of the importance; often heads of the manufacturing and knowledge-intensive enterprises put forward requests to satisfy a certain project, to look for resources;
- coercion (in the conditions of rigid temporary deficiency within design decisions and dynamics) – threat of use of the directive abilities by the initiator with the purpose to achieve from the addressee of desirable behavior; for example, the most often specified way is based on understanding of the importance of safety, the head specifies by the subordinate that in his forces to deprive of them all rights; all this exerts strong impact on personnel;
- manipulation – the hidden management of the addressee in participation in experiences, decision-making, performance of the actions necessary for the personal purposes of the initiator, that is head, for example: the desire of the worker direct difficult production site can stimulate a long time him on the organization of works of other experts without additional material incentives.

5 Conclusion

In the innovative and technological environment demand for the creative and creative competences of qualified personnel considering interpersonal communicative abilities and an active exchange of key information amplifies and becomes complicated on the structure. Analytical, search and research competences of the expert of hi-tech production become the most important element of development of professional competences by the purpose of management of corporate knowledge and innovations. Information, knowledge, skills, talents of the hired worker are transformed to the leading factor of formation of competitive advantages in labor market and the labor relations, increasing greatly the value of applied technical skills and innovative culture. The highly qualified specialist as the carrier of knowledge and innovative culture, socially oriented values becomes a source of development of the innovative technologies and new organizational forms of the knowledge-intensive production capable to stimulate introduction of various technologies of adaptation of the market of labor to competence-based requirements of real sector of economy. Professional and corporate values of the expert reflect a level of development of culture, cognitive abilities, talent and socially oriented requirements, are institutionalized by the hired expert and the employer during the period of joint activity that leads to change of structure of social and economic development of production territories, formation new highly specialized, the innovative of the grocery markets and services.

Thus, in modern economic conditions the professional specialist of the enterprises of the knowledge-intensive branches of economy is considered not only as the performer of production tasks, he becomes the active and interested participant of innovative activity of the organization. Participation in management and creation of an innovative product is an element of the system of participation of staff of the enterprise in general including also participation in profits and participation in property. At the same time, participation in profits and property belongs to system of material remuneration, and participation in management – to external (creation of the corresponding conditions from the management) and internal motivation of the non-material stimulating character (development in key personnel of feeling of self-determination and competence).

References

1. S.E. Brewers, L.S. Tarasevich, A.V. Butukhanov, A.I. Mayzel, R. R. Manukyan, I.S. Brewers, A.I. Pogorletsky, N.V. Trifonova, T.G. Tumarov, M.Z. Epstein, *Inter-national management* (St. Petersburg, 2008)

2. T.A. Khudyakova, A.V. Shmidt, *The 26th International Business Information Management Association Conference*, 1626-1634 (2015)
3. V.M. Kutuzov, S.O. Shaposhnikov, O.A. Shvetsova, Planning and Ensuring Training for an Industrial and Economic Complex of the Region, 125-136 (2015)
4. V.V. Olkhovsky, *Methods and sources of formation of intangible assets of the organization, New in economy and management* (MAX. Press, Moscow, 2006)
5. N.V. Trifonova, Questions of new economy, **2**, 71-73 (2009)
6. O.A. Shvetsova, News of "LETI", **3**, 75-85 (2015)
7. O.A. Shvetsova, *The 26th International Business Information Management Association Conference*, 1606-1618 (2015)
8. *GOST R 56407-2015, Lean production: main methods and instruments.*
9. T.A. Khudyakova, A.V. Shmidt, *The 26th International Business Information Management Association Conference*, 1607-1611 (2015)
10. E. Mach, *Erkenntniss und Irrtum* (German, Reidel Publ., 1976)
11. R. Mayer, M. Painter, C. Menzel, B. Perakath, P. deWitte, T. Blinn, Knowledge Based Systems Publ., (1995)
12. *National Institute of Standards and Technology, Integration definition for function modeling (IDEF0)* (1993)
13. Ching-Hong Tsai, How-Jen Luo, Feng-Jian Wang, *11th IEEE International Workshop on Future Trends of Distributed Computing Systems* (2007). DOI: 10.1109/FTDCS.2007.17.
14. R. Braun, *IEEE 17th Conference on Business Informatics (CBI)* (2015). DOI: 10.1109/CBI.2015.41.
15. S. Yamasathien, *Fourth International Conference on Digital Information and Communication Technology and it's Applications (DICTAP)* (2014) DOI: 10.1109/DICTAP.2014.6821684.
16. M. Chinosi, A. Trombetta, *IEEE Conference on Commerce and Enterprise Computing* (2009) DOI: 10.1109/CEC.2009.48.
17. A.H. Maslow, Psychological Review, **50(4)**, 370-396 (1943)
18. A.H. Maslow, *Motivation and personality* (Harper & Row Publ., 1954)
19. J.O. Long, *ITIL 2011 At a Glance* (Springer New York Publ., 2012). DOI: 10.1007/978-1-4614-3897-7.
20. *A Guide to the Project Management Body of Knowledge (PMBOK Guide)* (Project Management Institute Publ., 2013)
21. http://www.metodolog.ru.

Synergetic openness as a social and technical system competitiveness parameter

A.N. Danilov[1,*], and V.Yu. Stolbov[1]

[1]Perm National Research Polytechnic University, Perm, Russia

Abstract. The article introduces the general system competitiveness function in its business processes. It is assumed that in a small time period competitiveness can be viewed as a scalar function of process parameters that can be represented by system order parameters including its synergetic openness. The general problem of synergetic approach-based system openness management is set, part of which is defining optimum order parameters that improve the system competitiveness with restricted resources available. The article introduces a step-by-step management problem algorithm. Finally, a working example is made of an enterprise planning system openness degree management.

1 Introduction

Let us consider a social and technical systems (STS) class, in which the social component prevails over the technical one and which are managed by various internal and external consumer groups [1,2]. Examples include production, information, and educational systems in which management the human factor is crucial. Defining consumer groups allows aggregation of their interests in solving management problems, e.g. in the form of collective proposals or fuzzy preferences [1, 3].

In the market economy environment, the key STS efficiency criterion is its competitiveness in the commodity and services market, defined by the relation between the quality and the price. Let us assume that a system competitiveness is a function depending on its order parameters. It should be noted that each system is characterized by its set of order parameters in a certain time period. Each order parameter is an aggregating value that defines the desirable system condition or its new system attribute.

We will focus on the order parameters which, in our opinion, are characteristic of any social and technical system. One of those is guaranteed quality, broadly construed within the TQM concept [4] intended to satisfy commodity and service consumers' requirements satisfaction. Another order parameter is the synergetic openness of a system, understood as the system's ability to quickly respond to external effects (needs of the society, state, and the market) via innovation organizing and self-organizing mechanisms. Similarly to work [5], we can introduce the system's synergetic openness degree and denote it by α, which defines the system's openness level from 0 to 1. It is to be understood that α effects the system competitiveness non-linearly. In transition to a new social development framework, this non-linearity may be very significant. It seems that the higher is the guaranteed quality and the synergetic openness of a system, the higher is its competitiveness. However, it's far from always so. For example, as noted in [5, 6], in transition from closed to open system, organization, self-organization, and disorganization processes occur simultaneously. Therefore, if the system openness increases, disorganization process can become predominant, which will result in stability loss and system destruction (pos. 1 in Fig. 1). Alternately, if the synergetic openness grows slowly, the system competitiveness can be dramatically increased by successful structural modernization and adjustment to external effects (pos. 2 in Fig. 1.). In this case, the system acquires new attributes and transits to a new equilibrium state.

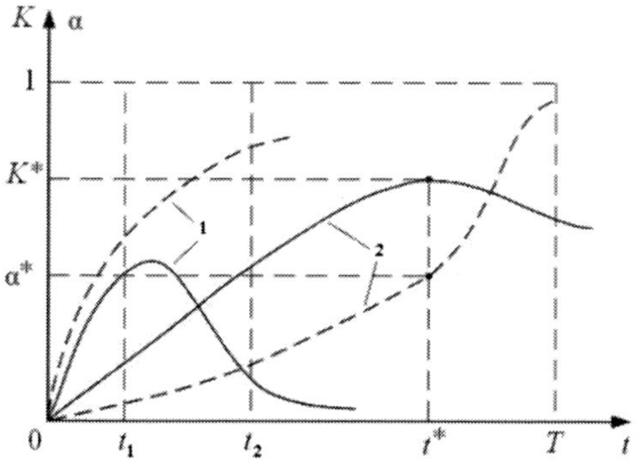

Fig. 1. The dependence of system competitiveness (K) and synergetic openness (α) on time (———— – K; — — — – α).

Hence, at a certain stage of a system development, a need arises to set and solve the problem of synergetic openness management.

* Corresponding author: dan@pstu.ru

2 Latest research and publications overview

Recently, in complex semistructured systems study, the synergetic approach is increasingly widely used, which, unlike the cybernetic approach [7,8], deals with fewer variables known as order parameters. The issue of synergetic approach development is raised in works of I. Prigozhin, I. Stengers, H. Haken, G. Malinetsky, D. Trubetskoy, and other authors [9-12]. In these works, entropy is suggested as a system order strength on the macroscopic modeling level [10]. It is noted that the most challenging issue in using the entropy approach in any STS business process study is the unassessability of a system entropy at different timepoints. However, as shown in V. Shapovalov's works [5], in open social and technical systems one can measure not the entropy itself, but only a parameter defining the system's synergetic openness and demonstrating the integrated effect of the environment (entrostat) on the system. Besides, it is shown that at a certain synergetic openness increase rate the system entropy decreases through ordering and self-organization. In other words, it is stated that in a certain probable state range of the system under research its synergetic openness degree can serve as the negentropy measure [13, 14], i.e. integrated effect of the environment (including that of information) is useful in terms of the system performance and results in entropy reduction via predominance of self-organization over disorganization processes. Therefore, at certain STS development stages it is reasonable to chose the synergetic openness degree of the system as its order parameter and control its increase rate under reduced system resources allocated to restructuring, reorganization, and the system destruction risks [5,6]. Methods and examples of order parameters management for major production systems with account of human factor are set out in works of Yu. Gorsky [14], issues of education systems mega- and macro-levels modeling are addressed in works of Ye. Solodova, V. Kharitonova et al. [15, 16]. However, these works, from our point of view, are not focused enough on systems openness degree management and the effect of transparency on STS competitiveness in transition to a new social development framework. In view of the aforesaid, this research issue is relevant and requires new efficient algorithms to be developed for solving management problems in social and technical systems in the market economy environment.

3 Management task setting

Let us consider the general task setting for a managing synergetic transparency of a system. The system competitiveness is denoted by *K*. Then, at the timepoint $\tau = t$, the system competitiveness value within the core activity $K(t)$ can be defined as function *F* of processes $\overline{\Pi(\tau)}$ as:

$$K(t) = F\,[t; \overline{\Pi(\tau)},\ t_0 \le \tau \le t], \tag{1}$$

where $\overline{\Pi(\tau)}$ is a business process defining the activity and including the range of key processes (planning, organization, production, marketing, etc.), running in the system during the period concerned.

Constructing certain types of function *F* for various system types is a complex problem of organizational and technical systems management theory. Therefore, in practice this problem is being simplified in different ways. For example, instead of function (1) functional relation are established between complex factors of competitiveness and key processes parameters.

It is to be noted that in the simplest case, indicators vector values $\overline{K(t)}$ at timepoint t is just an analytic vector function of $\overline{\Pi(t)}$ process parameters at the same timepoint. Then, it is:

$$\overline{K(t)} = \overline{f}\ (\overline{\Pi(t)}). \tag{2}$$

Let us introduce order parameter vector $x = (x_1(t), \ldots, x_n(t))$, which in the general case is n-dimensional and describes all processes from the system standpoint during some time period [0, *T*]. Let us consider that on the given time interval the competitiveness of the system concerned $K(t)$ is a scalar function as follows:

$$K(t) = f(x(t), \dot{x}(t)), \text{ where } \dot{x}(t) = \frac{dx}{dt}. \tag{3}$$

It is evident that each system tends to increase its competitiveness, i. e. send the function introduced to maximum.

However, for a system competitiveness increase resources *R* are required (material, financial, etc.), which are restricted. Hence, on a certain time period [0, *T*], each system has to solve the relevant optimization problem with the following mathematical setting:

Find such piecewise continuous functions $x^*(\cdot) \in KC([0,T]; R^n)$ affording a maximum to competitiveness function

$$K = K(x(\cdot)) \to \max \tag{4}$$

with restricted resources:

$$R_j(x(\cdot)) \le R_j^*,\ j = 1, \ldots, m \tag{5}$$

and restricted management parameters:

$$x(t) \in X,\ t \in [0, T], \tag{6}$$

where R_j^* is predefined resource constraints, *X* is a set of admissible management parameters values.

It should be noted that the STS production management problem (4)–(6) currently cannot be solved in the general case. It is related to the fact that, first, the general analytic form of function *K* is not known, second, the required functions may have discontinuities on section [0, *T*] (especially during development framework change and transition to innovation technology) and, third, the dimension of this problem in the general case may be very large, which hinders its numerical calculation. Therefore, when using the synergetic modeling methodology for STS development strategy description [15], we distinguish four modeling levels (Fig. 2). The mega-level sets strategic targets for development as a set of system order parameters that define the system development vector according to the global trends in science, education, economy, and the society as a whole under the postnonclassical framework [15].

The system competitiveness may be viewed as some order parameters function. But currently it's an unsolvable problem to describe this function in an explicit form. At this modeling level, we only can provide a qualitative description to the system progress trends using soft models [17].

For example, such model can be represented by an inequality:

$$\frac{dK}{dt}>0,\ t\in[0,T]. \tag{7}$$

This inequality only sets the general trend for the system development. In order to answer the question of how to increase the competitiveness of the system being studied, we need to know the *K(x) dependence*.

This can be done at the qualitative level by introducing additional hypotheses. Such hypotheses may be as follows:

The given STS order parameters may be introduced on a small time interval [0, *T*] as scalar continuously differentiable functions of time, which are linearly independent.

1. The system competitiveness may be represented by a function of the form:

$$K=F\left(x(\cdot)\right)=\int_0^T f\left(t,x(t),\dot{x}(t)\right)dt,$$

where $x(\cdot)\in C^1\left([0,T],R^n\right)$ is an order parameter vector function; f is the prescribed function of 3 arguments.

We can separately study the dependence of the system competitiveness on each order parameter with fixed values of other parameters, i.e.

$$K\left(\alpha(\cdot)\right)=F_1\left(\alpha(\cdot),\dot{\alpha}(\cdot),\hat{x}_2(\cdot),\dots,\hat{x}_n(\cdot),\hat{\dot{x}}_n(\cdot)\right),$$

where F_1 is a function defining the system competitiveness dependence on the synergetic transparency degree α with fixed values of other order parameters $\hat{x}_i$, i=2, …, n.

2. There is such a positive increment of a system's synergetic openness parameter with a restricted increase rate $\dot{\alpha}_{крит}$, with which the system competitiveness increases, i.e.

$$\exists\Delta\alpha=\alpha(t_2)-\alpha(t_1)>0,\ \dot{\alpha}(t)\le\dot{\alpha}_{крит}\ \ t\in[t_1,t_2],$$

that $\Delta K=F_1\left(\alpha(t_2)\right)-F_1\left(\alpha(t_1)\right)>0$.

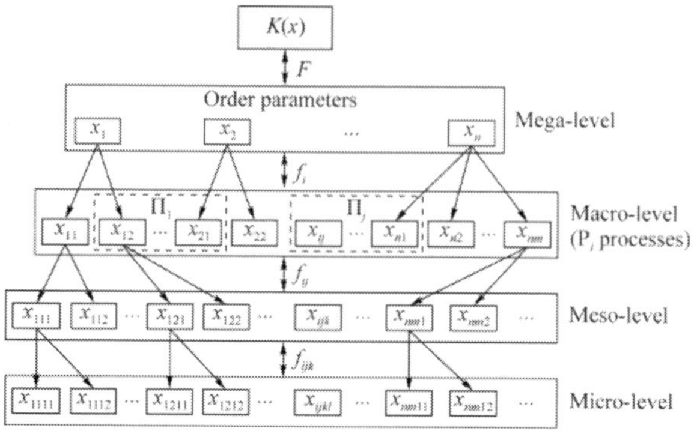

Fig. 2. Hierarchical levels of social and technical systems modeling.

Using the above hypotheses, we can transit from the soft model (7) to another soft mega-level model as the following inequality set:

$$\begin{cases}\dfrac{d\alpha}{dt}>0,\\ \dfrac{dx_i}{dt}>0,\ i=\overline{2,n},\ \ t\in[0,T].\end{cases} \tag{8}$$

The first inequality in the system (8) defines the requirement to the system's synergetic openness degree increase; other inequalities set the requirements to the non-decrease of the rest system order parameters. The System megamodel in the form of the soft model (8) is the conceptual model of the system progress, that shows its current condition and development trend as strategic goals for a certain time period. The systems management problem solving requires further model detalization and the transition from soft to rigid models at the subsequent modeling levels (Fig. 2).

A social and technical system macro-level model (the level of an industrial enterprise, a major academic institution, a large information system) is to describe order parameters change depending on macroparameters that define the system processes quality. Such parameters may be represented by various aggregated resources (human, material, financial, and others) as well as by different risks (economic, environmental, information, etc.). The key social and technical systems processes may include planning, manufacture, promotion, control, etc.

In setting the macro-level management task, the target functions include order parameters established at the mega-level f modeling, and the management parameters are some macroparameters mainly defining the target functions values. Other macroparameters of the system act as the state parameters or define management tasks.

If, for example, the target function is represented by the system's synergetic openness degree α, the macro-level management task may be as follows:

Find such optimum process $\xi^* = (u^*(\cdot), T^*) \in C^1([0,T], R^m) \times R$, in which the function extremum

$$J(\xi) = \int_0^T (\alpha(u(t)) - \bar{\alpha})^2 dt \to \min \text{ is attained}$$

under constraints:

$$\dot{\alpha}(t) = f(\alpha(t), u(t)),\ t \in [0,T]$$

$$\dot{\alpha}(t) \le \dot{\alpha}_{крит},\ t \in [0,T] \quad (9)$$

$$u(t) \in U,\ t \in [0,T]$$

where $u(\cdot)$ are the management parameters represented by macroparameters x_{1j}, $j = \overline{1,m}$; $\bar{\alpha}$ is the preset synergetic openness degree value; f – is the preset function; U – is the preset range of management parameters values.

The meso-level of an STS behavior modeling considers separate subsystems and processes models with a higher level of state and management parameter detalization. For example, in manufacturing systems such models can be represented by operation planning models, for educational systems – separate program planning and implementation models. The number of state and management parameters in such systems can be hundreds and thousands, which requires development of robust computational methods of such problems solving.

The micro-level of STS modeling considers processes at the level of separate system elements. For example, production planning to the level of setting shift and daily targets to individual worker and loading each piece of equipment, building a model of each subject mastering in course of academic institution activities, etc. it is evident that the number of parameters at this level can reach tens and hundreds million, which makes this problem challenging. Therefore, the synergetic approach normally deals only with macro- and meso-level modeling.

4 Management problem algorithm

As noted above, the main challenge in the proposed management methodology implementation is that the competitiveness function in the system's synergetic openness is not known explicitly. Therefore, we suggest to decompose the general management problem (4)-(6), i.e. split it into specific optimization tasks for each order parameter at a certain time interval, which are interconnected. In other words, a solution to each of the specific problems should not deteriorate the state parameters of the entire STS at timepoints considered. For example, let us consider the problem of a system's openness increase at a certain time period. It is obvious that the system's transparency degree increase at the initial timepoint will result in competitiveness increase. However, if the transparency degree increase rate is high enough and the necessary organization and self-organization processes fail to proceed within the system, all product lifecycle processes quality will decrease which will result in the system competitiveness reduction (Fig. 3). Therefore, in case of a system's transparency degree increase, it is crucial to be able to assess not only its performance indicators, but also evaluate the system quality through the level of key processes organization. As the competitiveness dependence on its transparency degree is not known, we suggest a step-by-step management algorithm. At each step, a certain preset system transparency level is established and the time for reaching this level is defined with account of available enterprise resources and system destruction risks.

Now the general production management process can be converted to the following optimization tasks sequence.

Find such optimum time t^*, $t^* \in [0,T]$, with which

$$t^* \to \min \quad (10)$$

and the constraints are complied with:

$$\dot{\alpha}(t) \le \dot{\alpha}_{крит},\ t \in [0,T] \quad (11)$$

$$R_j(\alpha(.)) \le R_j^*,\ j = 1,\dots,m \quad (12)$$

$$\alpha(0) = \alpha_0; \quad \alpha(t^*) = \alpha_1\,. \quad (13)$$

In this case $\dot{\alpha}_{èðèò}$ is the predefined maximum system transparency increase rate; R_j^* is predefined resources; α_1 is the predefined system transparency level.

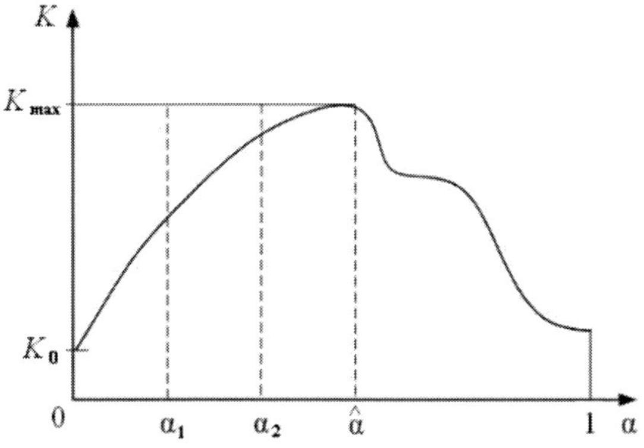

Fig. 3. The dependence of system competitiveness on its synergetic openness degree.

The above general management task decomposition (4)–(6) allows us to reduce its solution to a sequence of separate tasks (10)–(13) according to the following algorithm:

- choose order parameters $x(t)$ of the system considered and assess them at a certain timepoint which is taken as a start point;
- construct a function describing the system's synergetic openness degree dependence α on process parameters;
- define necessary resources R_j^* and synergetic transparency degree increase rate constraint $\dot{\alpha}_{xpum}$, which allows changes in the system with through organization and self-organization without the system destruction;
- solve problem (10)-(13) and define time t^*, with which the system's synergetic openness increased to the preset level;
- assess the system order parameters at t^*.

If other order parameters have not worsened with the new system state defined by the preset degree of its synergetic openness, further transparency increase is possible by solving problem (10)-(13) with a higher predefined α value.

If at least one order parameter has decreased with the new system state defined by the preset degree of its synergetic openness, we need to solve problem (10)–(13) once again with a lower α value or with increased resources for the system upgrade.

5 Working example

Let us consider an example of a problem of synergetic openness management of production planning system of a large-scale manufacturing enterprise working in rapidly changing market conditions and committed to increase its competitiveness.

It is known [18-20] that production planning is a crucial part of production management, in which targets are set and resources are allocated both for the entire production system and for each of its levels (business planning, strategic and tactic planning, operations management). All production planning levels must be consistent with each other. Each level is characterized by a planning horizon $\tilde{A}_i$ and replanning period γ_i (i= 1, ...4).

Let us introduce a parameter that defines the planning flexibility at each structural level of the production planning system (PPS) [16]:

$$\beta_i = \frac{\gamma_i}{\tilde{A}_i}, i = 1,...,4 . \qquad (14)$$

A production system functioning quality is deemed to be defined by parameters β_i, i = 1,...,4 allocation, which characterize the PPS external effects (consumer requirements) adaptability degree at each hierarchical level.

Note that parameter β_i defines the possible ith production schedule change introduction rate. It is evident that $\beta_i \in [0, 1]$. If $\beta_i = 1$, there are no schedule changes, but if $\beta_i = 0$, the schedule changes nearly immediately. I.e. parameters β_i are to be about equal and consistent with each other (for example, $\beta_1 \geq \beta_2 \geq \beta_3 \geq \beta_4$); this helps prevent production costs increase as any reduction of parameters β_i results in extra production expenditures.

Any planning system modernization aimed at the system openness increase requires an initial study of the permissible rate of implementation of the process in accordance with available enterprise resources and capabilities of their structural reorganization and self-organization. Therefore, each specific production requires justification of the planning system openness degree by selecting the optimum value α which depends on a large number of parameters that characterize production: the range and scale of production, the type of equipment used, the available production capacity, technology, etc.

To assess the degree of openness α that defines the state of the production system at a given timepoint, the following formula may be suggested:

$$\alpha = \beta_1 \cdot (1 - \max_i \beta_i) \ , i = 2, 3, 4, \qquad (15)$$

Which shows that the system transparency degree depends on calendar scheduling flexibility in its "narrowest" level. Introduction of factor β_1 for business planning level allows taking into account the character of production in assessment of its planning system openness degree.

Given the following initial data defining the studied production system state at a timepoint taken as the start point:

$\tilde{A}_1 = 3$ years, $\gamma_1 = 1$ year, $\tilde{A}_2 = 1$ year, $\gamma_2 = 3$ months, $\tilde{A}_3 = 3$ months, $\gamma_3 = 1$ week, $\tilde{A}_4 = 1$ week, $\gamma_4 = 3$ days.

Using formula (14), we obtain: β_1 = 1/3; β_2 = 1/4; β_3 = 1/12; β_4 = 3/7.

Using formula (15), we can assess this PPS openness degree at the timepoint considered: α = 1/3(1–3/7) = 0.19.

It is seen that this PPS, from the system approach standpoint:

1) does not feature parameter consistency ($\beta_4 > \beta_3$, $\beta_4 > \beta_2$), which will inevitably lead to a contradiction between the planning level and its actual implementation, i.e. between the scheduled lead time and the production capacity;

2) the transparency degree of this system is defined by the operations management system flexibility, which

in this case is the bottle-neck of the production management system;

3) the tactic planning system is too flexible, which results in unreasonable material costs.

In order to increase the studied PPS transparency degree to allow a more "flexible" response to customer wishes, let us consider the following problem which is a special case of problem (10)-(13).

Find such optimum planning process $(\beta^*(\cdot),\ T^*) \in (C^1[0,T],\ R^4)\times R$, $T^* \in [0,T]$, in which

$$T^* = \int_0^{T^*} dt \to \min \qquad (16)$$

and constraints are complied with:

as equality:

$$\alpha(t) = \beta_1(t)\cdot(1-\max_i \beta_i(t)),\ i = 2, 3, 4,\ t \in [0,T] \qquad (17)$$

and inequality:

to the synergetic openness degree increase rate

$$\dot\alpha(t) \le \dot\alpha_{\text{крит}},\ t \in [0,T], \qquad (18)$$

to resources

$$R_j(\alpha(t)) \le R_j^*,\ j = 1,...,m,\ t \in [0,T] \qquad (19)$$

and boundary conditions

$$\alpha(0) = \alpha_0; \quad \alpha(T^*) = \alpha_1$$

At the start point given some production system characterized by the following PPS parameters: $\beta_{10} = 1/3$; $\beta_{20} = 1/4$; $\beta_{30} = 1/12$; $\beta_{40} = 3/7$; $\alpha_0 = 0.19$.

We need to find the minimum time *T*, sufficient to increase this system transparency degree by 50% with β_i, *i*=1,2,3,4 parameters change rate constraints, i.e.:

$$\left|\dot\beta_i(t)\right| \le \dot\beta_{i\text{крит}},\quad t \in [0,T],\ i=1,2,3,4. \qquad (20)$$

Such constraints are related to the enterprise resources, production modernization capacity, etc. Besides, as was noted above, it is desirable that at *T* parameters β_i, *i* = 1, 2, 3, 4 were roughly equal.

Assume that $\beta_1(t) = const$ (in PPS modernizations targets are not changed). Suppose that the following critical rates of the planning system parameters change are given:

$$\dot\beta_{2\text{крит}} = 0.028;\ \dot\beta_{3\text{крит}} = 0.014;\ \dot\beta_{4\text{крит}} = 0.095,$$

which comply with the enterprise's capacity in management system modernization by increase of production planning flexibility. In this case, the guaranteed product quality is maintained.

Based on relation (15), the following can be written:

$$\dot\alpha(t) = -\beta_1 \max_i(\left|\dot\beta_i(t)\right| sign(\dot\beta_i(t)),\ t \in [0,T],\ i = 2,3,4 \qquad (21)$$

Supposing the time monotonicity of functions β_i, *i* = 2, 3, 4 and taking into consideration constraints (20), from (21) it follows that

$$\dot\alpha(t) \le \beta_1 \max_i \dot\beta_{i\text{крит}},\ i = 2, 3, 4.$$

The initial data analysis shows that the maximum value $\dot\beta_{i\text{êðèò}}$ is reached with *i* = 4, i.e. the maximum system transparency degree change rate is equal to $\dot\alpha = \beta_1\dot\beta_{4\text{êðèò}}$. Using the initial data, we obtain its value: $\dot\alpha = 0.031$.

Now, entering this value to the expression for the required system transparency parameter value at timepoint *T* ($\alpha_T = \alpha(T)$ = 0.19 × 1.5 = 0.285) and using relation $\alpha_T = \alpha_0 + \dot\alpha \times T$, we obtain the minimum time *T* sufficient to increase the system transparency degree to the required value α_T:

$$T = (0.285 - 0.19)/0.031 = 3 \text{ (years)}.$$

Figure 4 shows the obtained solution to the initial problem. It is seen that in order to obtain the required level of the system transparency parameters β_2 and β_4 should be reduced and parameter β_3 should be increased. Note that with the obtained system openness change rate no destructive processes will occur (owing to compliance with the enterprise resource constraints). In this case, at timepont *T*, the PPS parameters will be as follows: β_2 = 1/6; β_3 = 1/8; β_4 = 1/7, which are close to those approved (Fig. 4).

To reach those parameters, it is enough to reduce γ_2 to 2 months, value γ_4 to 1 working day and decrease the planning horizon at tactic level ($\tilde{A}_3$) to 2 months.

Note that in this example, upon PPS modernization, the planning "bottle-neck" which previously was at the operations level, is now at the strategic level which defines customer interaction.

In other words, the enterprise is able to fulfil obligations to be included into the strategic plan through flexibility of replanning at tactic and operations levels of production management. This is essential from production management standpoint, since it shows that production is ready to quickly respond to business environment changes, and no plan changes will result in stoppage in production or product deterioration.

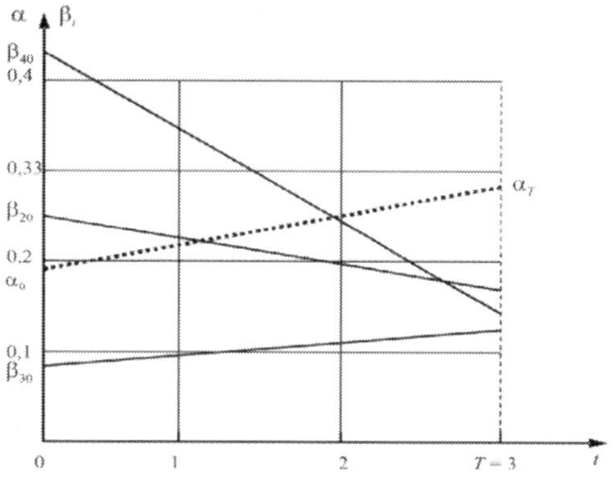

Fig. 4. A production system transparency degree increase via planning system parameters change.

It should also be noted that the reduction of γ_4 value to 1 business day will require a major restructuring of preproduction and operations management processes. Ensuring planning and production within one working day requires automation of shift-and-day tasks preparation and availability of a modern machine pool with skilled operators. This certainly requires major resource investment in production. However, it ensures dramatic increase in enterprise competitiveness through a more rapid response to the external demands, and the investments made pay off in a relatively short period.

6 Conclusion

Application of the synergetic approach to STS modeling in transition to market relations and rapidly changing external interactions was justified. It was noted that at critical moments of any system it is crucial to be able to predict such situations. This is enabled by synergetics that has given rise to various complex systems non-linear modeling and projection methods. Synergetic modeling is distinguished by the fact that, unlike cybernetic and mathematical models that allow accounting more parameters and effects, synergetic models use one or more so-called system "order parameters" that define its behavior and development. Besides, the idea of model integrity prevails over the idea of its completeness, which simplifies modeling and management of complex systems under consideration.

It was demonstrated that one of the robust STS management mechanisms is increasing its openness through introduction of new external interaction tools, that results in the system's entropy decrease and its competitiveness improvement. The introduced concept of a system's synergetic openness allows us to reduce the general open systems management task to managing a specific order parameter; it is demonstrated that increasing this parameter at a certain stage of the system development results in the system ordering, if the increase rate does not exceed a certain critical value that depends on the system's properties and the nature of its interaction with the environment.

A working example is made to demonstrate synergetic openness management of production planning system of a large-scale manufacturing enterprise with constraints to resources allocated for production modernization.

References

1. M.B. Gitman, V.Yu. Stolbov, R.L. Gilyazov, *Management of social and technical systems with fuzzy preferences* (LENAND, Moscow, 2011)
2. K.S. Pustovoit, M.B Gitman., V.Yu. Stolbov, J. Actual Problems of Economics, **135**, **9**, 457-466 (2012)
3. M.V. Gubko, *Mathematical models of optimization of hierarchical structures* (LENAND, Moscow, 2006)
4. S.A. Fedoseyev, M.B. Gitman, V.Yu. Stolbov, A.V. Vozhakov, *Quality management in the modern industrial enterprises* (The Publishing House of Perm National Research Polytechnic University, Perm, 2011)
5. V.I. Shapovalov, *Basics of order and self-organization theory* (Ispo-servis, Moscow, 2005)
6. A.N. Danilov, V.Yu. Stolbov, Mechatronics, Automation, Control, **16**, **6**, 387-395 (2015)
7. D.A. Novikov, *Cybernetics: Navigator. The history of cybernetics, the current state and prospects for development* (LENAND, Moscow, 2016)
8. S. Beer, *Cybernetics and Management* (The English University Press, London, 1959)
9. I. Prigozhin, I. Stengers, *Time quantum chaos* (Librokom Printing House, Moscow, 2009)
10. H. Haken, *Information and Self-Organization: A Macroscopic Approach to Complex Systems* (KomKniga, Moscow, 2005)
11. G.G. Malinetsky, *Mathematical Foundations of Synergetics* (Librokom Printing House, Moscow, 2009)
12. D.I. Trubetskov, *Introduction to synergetics* (Moscow, Editorial URSS, 2004)
13. D.S. Chernavsky, *Synergetics and information (dynamic information theory* (KomKniga, Moscow, 2004)
14. Yu.M. Gorsky, *Systematic analysis and information management processes* (Nauka. Siberian Branch, Novosibirsk, 1988)
15. Ye.A. Solodova, *New models in the education system: a synergistic approach* (Librokom Printing House, Moscow, 2012)
16. V.A. Kharitonova, I.V. Menshikov, O.V. Sannikova, *Methodology and modeling of synergetic development of education* (Regular and Chaotic Dynamics, Izhevsk, 2001)
17. V.I. Arnold, *Hard and Soft Mathematical models* (MCCME, Moscow, 2000)
18. D.A. Gavrilov, *Production management on the basis of standard MRP-II* (SBR, St. Petersburg, 2003)
19. K.S. Pustovoit, M.B. Gitman, V.Yu. Stolbov, J. Actual Problems of Economics, **8**, 440-451 (2012)
20. S.N. Yevstratov, A.V. Vozhakov, V.Yu. Stolbov, J. Automation and Remote Control, **75**, **7**, 1323-1329 (2014)

Innovation project value assessment methodology for high-tech enterprises

M.A. Kossukhina[1,*], and *V.I.* Fomin[1]

[1]St. Petersburg State Electrotechnical University "LETI", St. Petersburg, Russia

Abstract. In the current economic conditions, the innovation project value assessment is a driving force of innovation system development of high-tech enterprises. Innovation projects value assessment is a crucial determinant of technologically advanced industrial enterprises competitiveness in the world market. Therefore it is important to identify the factors that influence on selection of innovation project value assessment methodology for high-tech enterprises. The aim of this paper is to select the innovation project value assessment methodology for high-tech enterprises, that would satisfy their features and to offer the approaches to innovation project value assessment in case of unsuitability of reviewed methodologies. The analytic hierarchy process and fuzzy inference method is used as methods of decision-making in this paper.

1 Introduction

The founder of classical theory of innovation J. Schumpeter pointed out that innovations are not only the production and distribution of new products and services but also application of new production technologies, control of new sales markets as well as creating new organization solutions in economy [1,4,5,6,8]. Innovation activities include all scientific, technological, organisational, financial and commercial steps, which actually lead, or are intended to lead, to the implementation of innovations. According to [2, 10-17] it can be stated that every high- tech enterprise is an innovative company; however, not every innovative company is a high-tech one.

Considering the differences between the high-tech enterprises and innovative companies, it is necessary to outline whether the high-tech enterprises are producers/creators of advanced technologies or rather a users, commonly applying such technologies. From our point of view , a "fusion" approach should be used in this case, as high- tech enterprises usually produces advanced technologies, so that it could be used in their products and (or) services. Normally high- tech enterprises have their own R&D units that produce advanced technologies. In this article, we offer to call them "innovation projects".

Nowadays, innovation projects value assessment is a crucial determinant of technologically advanced industrial enterprises competitiveness in the world market. Therefore, it is important to identify the factors that influence on selection of innovation project value assessment methodology for high-tech enterprises.

2 The review of innovation project value assessment approaches

As innovation projects are being considered as business units of high-tech enterprises, their key features should be outlined. The most important attributes of innovation projects could be divided into four groups as following:

- financial features: total size of the financial expenses for R&D (capital expenses and investment risk);
- team features: flexibility of organization structures, close scientific and technical partnership relations, team work potential, ability to share, use and collect, knowledge;
- means features: fast diffusion of technological innovations, the volume and diversity of collected and processed information, knowledge and unique competences of personnel;
- Idea features. High creativity entrepreneurship, science demand and agility, fast devaluation of the prepared and applied technologies, communication supported with modern IT.

The innovative project, as well as other business functions, requires special tools: the rules, scope, and budget constraints. Let us assume that approaches, which are used to measure business projects value, are common to innovation projects value assessment. Therefore, we offer to consider them taking into account the innovation project features.

Income approach - a set of methods for assessing the valuation of the object, based on the determination of the expected revenue from the object use. There are several variations of the income approach: direct capitalization method, the method of discounted cash flow, and gross

* Corresponding author: masha.kossukhina@mail.ru

income multiplier method [5]. Using the income approach has the following disadvantages: this approach requires a subjective allocation of cash flows, the relevant information that is necessary for the calculations, is not always available from internal reporting.

Comparative approach is a set of methods to assess the valuation of the object, based on a comparison of the object with the objects of evaluation - evaluation of analogues of the object, for which there is information on prices. The main disadvantage of this approach is impossibility of application in case of inadequate data availability.

The cost approach - a set of methods for assessing the valuation of the object, based on the determination of costs necessary for the reproduction or replacement of object of evaluation, taking into account wear and tear and obsolescence. The cost approach views the business as a set of assets and liabilities that are used as building blocks to construct the picture of business value. This approach does not capture the cash flow generating ability of the assets or business and cost and value may have large discrepancy.

All of approaches, that we have considered, except comparative, are based on quantitative methods of evaluation, so they are not taking into consideration qualitative characteristics.

It should be noted that the method of estimating the value of the high-tech enterprise innovation project for more accurate assessment should be based on two groups of scales: qualitative and quantitative. Some researchers, however, considering only quantitative metric [18, 19], because of its simplicity to analyze using statistical methods. Other researchers assert the importance of the qualitative approach in measuring the value of innovation projects [11, 14, 15], as it allows measuring the innovative project value considering its qualitative characteristics.

From our point of view, there is a need to mix quantitative and qualitative approaches to fully assess the value of innovation project.

This article discusses the qualitative modeling method, which allows to transform quantitative values into qualitative, combine them with the values of quality scales method and fuzzy linguistic modeling. Fuzzy modeling suggests the development of qualitative model based on linguistic descriptions. Quantitative data processing can also be performed using fuzzy modeling.

3 Innovation project value assessment methodology

In this paper, we are using a special expansion pack Fuzzy Logic Tools in MatLab environment to simulate the process of innovative projects valuation. It is necessary to build a model, which automatically determines the innovation performance value in the output in interactive mode. It is also required to take into account qualitative variables according to the algorithm, with the usage of the graphic editing tools and visualization.

The substantive interpretation involves the selection and specification of the input and output variables corresponding the fuzzy inference system. In this case, the model is supposed to use six input variables and one output. The type of defined system will be outputted on the Mamdani algorithm, so we are going to use min function as activation function. As logical connective "And" is used in all rules in the knowledgebase, then operation min-conjunction is used as the aggregation method. Center of gravity is used as a defuzzification method.

The fuzzy inference system is shown in Figure 1. It is based on the following variables:

- the novelty of innovation projects;
- the quality of the innovation environment;
- high-tech enterprises income from researches;
- the number of license obtained;
- Skolkovo grants
- the number of IPO;
- the innovation project performance.

The function plot is done for every variable of the fuzzy interference model using the editor of membership functions. Membership functions plots of linguistic variables and their descriptions are shown in Figures 2-4. For all of these terms we will use the triangular function trimf. This function uses three parameters a, b, c, take arbitrary real values and ordered $a \leq b \leq c$ attitude.

In the fuzzy inference systems, linguistic variables have different terms for each variable. As a term of the plurality of input variable «the quality of the innovation environment» will use the set T1 = {"low", "medium", "high"}, to low term we define the parameters of [-1 -0.6 0.2] to medium term we define the parameters [0.2 0.4 0.6] to high term we define the parameters [0.6 0.8 1].

As the term sets of the output variable "Innovation project performance», we will use a set T7 = {"mf1", "mf2", "mf3"}. Fig. 4 shows the membership function of the linguistic variable terms ("low", "medium", "high").

The GUI of rules editor after setting the rule base is shown in Figure 5.

The evaluation of mutual values of linguistic variables in the results of the fuzzy inference system shows for example that combination of average values of: high-tech enterprises income from researches, license obtained, Skolkovo grants and IPO with the low value of innovation environment quality, lead to an average value of innovation project performance.

In contrast to investment projects, innovative projects of high tech enterprises is characterized by specific risks and uncertainties arising from technical and market innovation novelty. Therefore, there is a need for new approaches to assess the effectiveness of such projects. Innovation project value assessment methodology based on fuzzy logic approach, which is offered in this paper, could be used to evaluate innovation project effectiveness as a first-step methodology because of its simplicity.

DOI: 10.1051/shsconf/20173501086

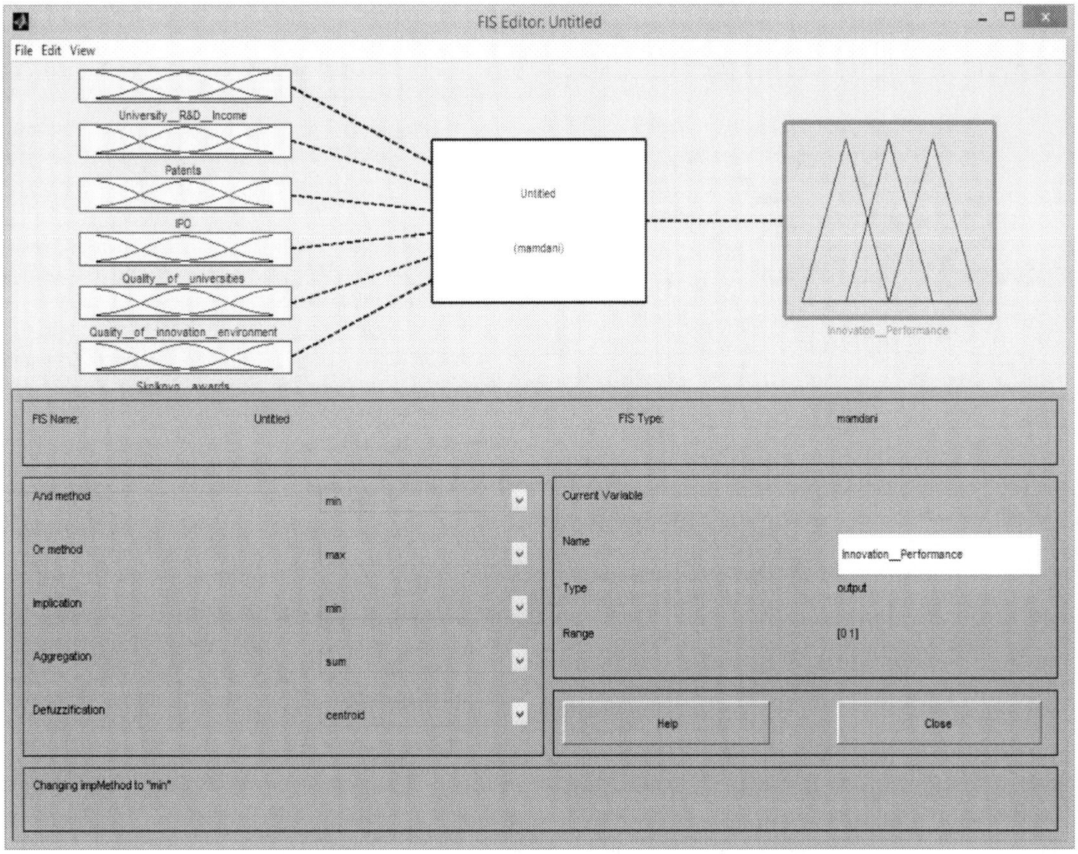

Fig. 1. Graphical user interface of FIS editor after determining input and output variables of the fuzzy inference system.

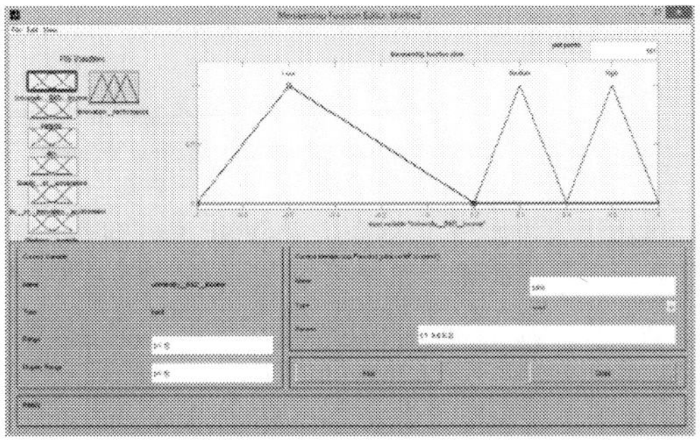

Fig. 2. The GUI of membership functions editor for the input variable " high-tech enterprises income from researches".

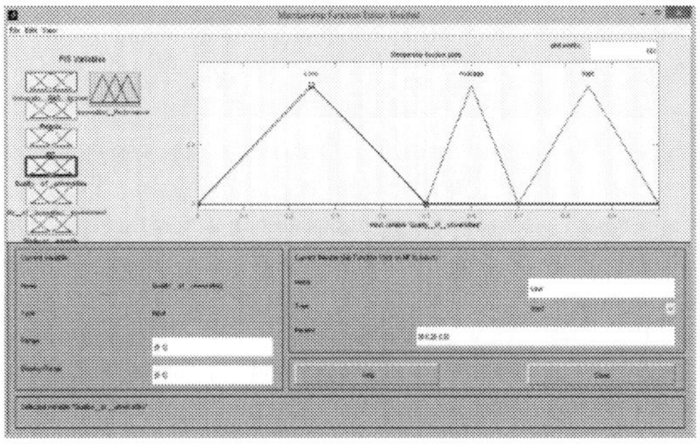

Fig. 3. The GUI of membership functions editor for the input variable "the quality of the innovation environment.

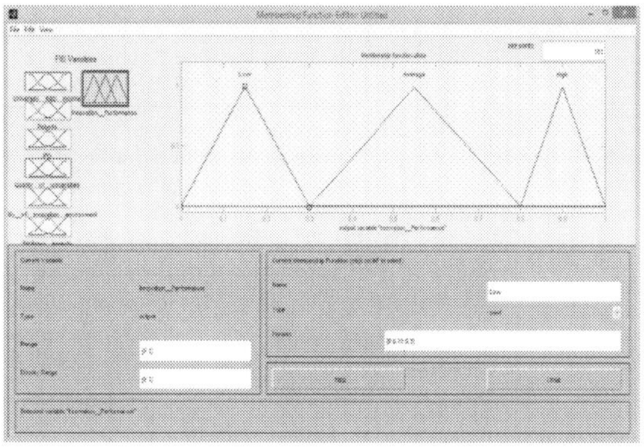

Fig. 4. The GUI of membership functions editor for the output variable "Innovation project perfprmance".

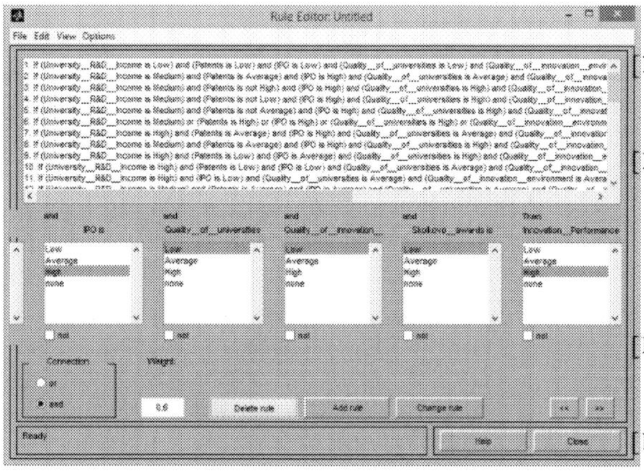

Fig. 5. The GUI of Rules Editor after setting the rule base.

4 Conclusion

The innovation project value measurement is very important for technologically advanced industrial enterprises development, because the innovations are the basic drive for country economic growth. However, there are many difficulties in measuring innovation project value. This paper suggests the model based on fuzzy logic, which helps to avoid some of these difficulties such as the usage qualitative and quantitative indicators separately in measuring innovation project value.

References

[1] V.I. Fomin, *Social technologies in today's society*, **2** (2012)

[2] *Oslo Manual: Guidelines for Collecting and Interpreting Innovation Data* (2005)

[3] Abby Joseph Cohen, *Innovation and Economic Growth, The Goldman Sachs Group* (2011)

[4] I.A. Brusakova, M.A. Kosukhina, *Proc. conference on modern problems of applied informatics*, 6-9 (2012)

[5] I.A. Brusakova, M.A. Kosukhina, *Proc. Conference Proceeding of SCM on* (2015)

[6] I.A. Brusakova, *2015 IV Forum Strategic Partnership of Universities and Enterprises of Hi-Tech Branches*, 77-79 (2015). DOI: 10.1109/IVForum.2015.7388260.

[7] *A Guide to the Project Management Body of Knowledge (PMBOK Guide)* (Project Management Institute Publ., 2013)

[8] I.A. Brusakova, M.A. Kossukhina, *Proc. of the 10th Conference of Open Innovations Association FRUCT and the 2nd Finnish-Russian Mobile Linux Summit*, 24-27 (2011)

[9] M.F. Kossukhina, H. Aminov, *Proc. of the 12th Conference of Open Innovations Association FRUCT and e-Tourism seminar on*, 37-40 (2012)

[10] M.A. Kossukhina, *Proc. of the 12th Conference of Open Innovations Association FRUCT and e-Tourism seminar on*, 41-46 (2012)

[11] I.A. Brusakova, V.I. Fomin, M.A. Kossukhina, S.N. Panin, *Research of development trends of information management in modern conditions* (Publishing house of Saint-Petersburg University. Univ of Management and Economics, St. Petersburg, 2014)

[12] H.J. Zimmermann, *Fuzzy Set Theory and Its Applications* (Kluwer Academic Publishers, Boston, MA, 1996)

[13] A. Kaufmann, M.M. Gupta, *Introduction to Fuzzy Arithmetic Theory and Application* (Van Nostrand Reinhold, New York, 1991)

[14] T.A. Khudyakova, A.V. Shmidt, *Proc. of The 26th International Business Information Management Association Conference*, 1607-1611 (2015)

[15] T.A. Khudyakova, *Proc. of the International Congress on Interdisciplinary Behavior and Social Science*, 292-294 (2015)

[16] P. Romer, 8:1 Journal of Economic Perspectives, 3-22 (1994)

[17] J. Fagerberg, Journal of Economic Literature, **32**, 1147-1175 (1994)

[18] C. Freeman, *The Economics of Industrial Innovation* (Penguin, London, 1974)

[19] C. Freeman, Cambridge Journal of Economics, **20**, **1**, 463 (1994)

[20] K. Beck, *Extreme Programming Explained: Embrace Change* (Addison-Wesley Professional Publ., 1999)

[21] D. Katalevsky, *The Basics of Simulations and System Analysis in Management* (Delo Publ., Moscow, 2015)

The multipurpose industrial company's competitiveness evaluation method for instable environmental conditions

E.V. Kuchina[1,*], and *D.A.* Baeva[1]

[1]South Ural State University, Chelyabinsk, Russia

Abstract. The competitiveness evaluation method for industrial companies was created by authors and described in the article. This method allows considering specific features of contemporary economic conditions. There are two main features: Russian economy is under the external pressure, it instable and needs structural changes. The main advantage of the method is following: it can evaluate both the current and the perspective competitiveness of company considering the aim of increasing competitiveness of Russian economy in general. The evaluation of competitiveness includes five main blocks of company's particular characteristics and accumulates them into several generalizing competitiveness indicators. Such multifactorial model can be successfully used for managing the competitiveness, at the same time the limited number of generalizing indicators makes the final evaluation of competitiveness quite convenient.

1 Introduction

The increasing of industrial companies' competitiveness is one of the most important factors of sustainable development of national economy. The leading experts of strategic management confirm the great importance of this factor [1], [2]. However market participants today are under the pressure of changeable environmental conditions both at the micro-level and at the level of national economy in general. The external sanctions pressure influences the majority of economic processes in Russia and creates the specific conditions for Russian companies.

In contemporary market conditions an industrial company's management has to use specific tools and techniques to provide the decision-making processes which can decrease the negative consequences of external influence. That is why it is really important to develop the existent competitiveness evaluation tools and to create some techniques to increase the competitiveness potential of industrial companies. Moreover, the specific relevance of industrial companies' competitiveness investigation is provided with government policy on import substitution.

All this factors lead us to investigate the problem of evaluation the competitiveness level of industrial companies in dynamic environmental conditions. Also the development of evaluation tools both for current and perspective competitiveness is very important, because companies need to protect and enhance their competitive advantages in the future.

2 The problem to investigate

2.1 The main disadvantages of existing evaluation methods

There are many foreign and Russian scientists who investigated the different sides of competitiveness evaluation problem. We can name such famous researchers as E. Chamberlin, J. Schumpeter, A. Cournot, A. Marshall, D. Ricardo, J. Robinson, P. Samuelson, A. Smith, F. Hayek, M. Porter, I. Ansoff, G. Azoev, L. Balabanova, A. Voronov, P. Zavialov, P. Zabelin, V. Krivorotov, G. Mazhinsky, I. Maksimova, A. Sivachenko, L. Sokolova, A. Temchenko, A. Tkachenko, H. Faskhiev, R. Fatkhutdinov, A. Tsvetkov etc.

According to our analysis there are two main groups of competitiveness evaluation models and methods: graphical and analytical ones.

1) Analytical methods of companies' competitiveness evaluation

- The ratings [3], [4].
- The methods, which use a market share to evaluate competitiveness [5], [6].
- The methods, which use a competitiveness of product to evaluate competitiveness of company [7], [8], [9].
- The methods based on the effective competition theory [10].

* Corresponding author: kuchinaev@susu.ac.ru

- The methods based on the SWOT-analysis model [11].
- The integrative evaluation methods. These models are the most popular and various [12], [13], [14], [15], [16].

2) Graphical methods of companies' competitiveness evaluation

- The methods based on comparative competitive analysis.
- The matrix methods.
- The market attractiveness models.
- The Porter's model [17].

All of these methods have their own advantages and can be used in practice, but all of them have significant disadvantages which ones restrict their usefulness and sometimes make them useless at all.

The main disadvantages are following:

- the methods use complicated calculations or factors which are hard to be evaluated and inaccessible information;
- the majority of these methods are based only on internal factors and don't take into consideration the environmental ones;
- they also don't take into consideration the characteristics of exports and imports, technologies and equipment condition, human resources, effectiveness of management, basic marketing characteristics, which are needed to create a new strategy to promote and sell products in the domestic and foreign markets.

2.2 The way to solve the problem

A macroeconomic volatility influence on industrial companies can be estimated only by using specific companies' competitiveness evaluation methods, not by using the common tools [18]. Therefore the main aim of our investigation was to create the multipurpose company's competitiveness evaluation method for instable environmental conditions, which can be used to increase the long-term competitive capacity.

3 The multipurpose industrial company's competitiveness evaluation method

3.1 The specific factors of industrial company's competitiveness

The diversity of factors which influence industrial company can be classified into two groups: the objective ones (they can be managed by the company) and the subjective ones (they can't be managed by the company and can be also named as environmental factors).

There is the list of objective factors:

- a size of a company;
- a branch of industry;
- characteristics of products, which are produced by company;
- the level of interplant cooperation;
- the direction and volume of company's export or import.

The subjective factors are the following ones:

- the level of scientific and technical development;
- customs and tariff legislation;
- the competitive activity;
- the paying capacity of the main customers;
- the technical legislation;
- internal and external official decisions, which are made according to specific political situation.

3.2 The particular indicators of company's effectiveness

After identification of specific factors of industrial company's competitiveness they should be combined into particular indicators which are connected with the main characteristics of company and its effectiveness. We propose to combine these indicators into five groups, which characterize the different spheres of company's activity: marketing, organizational and human resources, financial position, technical and technological level, characteristics of products.

For estimation of every group the different particular indicators can be chosen and calculated. The majority of them (with equations) can be found in [19]. In addition to the common indicators the share of imported materials and components should be included into financial indicators group, because it is the high risk factor in contemporary economic conditions.

All indicators must be compared with some basic level – for different indicators the basic level could be either the standard or the average value, also it may be the industry leader's level of the indicator. After the comparison the specific deviations must be identified.

3.3 The generalising indicators of company's effectiveness

Only after evaluation of all particular indicators we can start to estimate the generalizing ones, which must be used as a basis for competitiveness criteria, which would be the final result of calculation.

We suggest using the following generalizing indicators:

- the consumer price of the product;
- bringing the product to the consumer index;
- equipment and labor productivity;

- the bankruptcy probability.

The bankruptcy probability can be predicted with using different methods, one of the most common is the Altman's discriminant factor model [20].

If we compare the level of each generalizing indicator of the industrial company with the basic level, we'll be able to make a conclusion about strengths and weaknesses of this company.

3.4 The criteria for company's competitiveness evaluation

We suggest estimating the following criteria of company's competitiveness:

- quality of its product;
- its domestic market share;
- the added value, which is generated by the company.

These criteria are the initial base for comparing the effectiveness and competitiveness of industrial companies between each other and with the baseline. In our opinion these criteria completely characterize the industrial company's competitiveness in contemporary economic conditions in Russia. We chose them according to several important features of such conditions and the main current aims of national economic development.

These criteria are connected with:

- required structural changes in Russian economy;
- required increasing of the processing level of domestic products;
- the increasing of global competitiveness of national products aimed by government;
- the company's success in internal competition during the nation economy is under the external economic pressure and the perspective global competitiveness of domestic industrial companies.

For different purposes of estimation or investigation companies can be ranked according to every criterion separately or according to only one of them which is evaluated as the most significant one for this specific investigation.

If we need the comprehensive evaluation of company's competitiveness and we want to compare easily the current position of investigated company with its competitors' positions, it will be convenient to use the final integrative indicator of competitiveness (1). This integrative indicator can be calculated in this way:

$$C = Q^{V1} \times S^{V2} \times AV^{V3} \quad (1)$$

C – industrial company's competitiveness; Q – quality of the product; S – domestic market share; AV – the added value, which is generated by the company; V1,V2,V3 – the weighting factor of the corresponding criterion.

It is to be said that the multiplicative type of model is the most suitable one to calculate the integrative indicator of competitiveness, because all criteria must be taken into account in total. Even if one of the criteria becomes a zero, the company's competitiveness also becomes a zero.

The following situations may be analyzed as the examples:

- If the industrial company don't have any production which was produced by itself (Q=0), it will not be able to compete in the industrial sphere of economy, it can take a part in competition only in the commercial sphere.
- If the industrial company is not represented in the domestic market at all (S=0), it will be equivalent to total absence of its current competitiveness in contemporary situation when the international trading is vastly limited.
- If the industrial company don't generate any added value (AV=0), it can't be named as the industrial one at all. Moreover, this company obviously doesn't have the aim to increase the processing level of its products, that is why it doesn't have any competitiveness as the industrial company in contemporary environmental conditions.

These three facts lead us strictly to the multiplicative model of the integrative indicator of competitiveness. The next problem to investigate and question to answer is the values of weighting factors because not in every situation the significance of identified competitiveness criteria are equal.

The process of choosing and setting the specific values of weighting factors is one of important parts of the assessment process, because they can significantly influence the balance of power between competitors in the result. In our opinion, for every specific industry and for every specific market situation the specific values of weighting factors must be set and used because the significance of these factors will be different for every industry and situation.

The values of weighting factors must be set according to two important characteristics:

- the level of external economic pressure (Tab. I);
- the necessity of structural changes in the specific industry (Tab. II).

As we can see from Tab. I the significance of two criteria rises according to increasing of external economic pressure. These two criteria are:

- the domestic market share;
- the added value, which is generated by the company.

Table 1. Significance of the Competitiveness Criteria According to External Economic Conditions.

The Criterion	External Economic Conditions		
	Adverse	*Neutral*	*Favorable*
Quality of the product	High	High	High
Domestic market share	High	Medium	Low
The added value, which is generated by the company	High	Medium	Medium

Table 2. Significance of the Competitiveness Criteria According to How the Situation in Industry is Corresponding to the Aims of Structural Changes in National Economy.

The Criterion	The Necessity of Structural Changes in the Industry		
	High	*Medium*	*Low*
Quality of the product	High	High	High
Domestic market share	Medium	Medium	Medium
The added value, which is generated by the company	High	Medium	Low

It can be easily explained. If the international economic relations are limited (when the economy is under the external pressure) the current competitive positions of company can be evaluated only in connection with its domestic market share and independence from import will become the important competitive advantage, vice versa if the international relations are stimulated this factors will become less important.

According to increasing of structural changes necessity in the industry the significance of one criterion (the added value, which is generated by the company) rises. At the same time the criterion of product's quality remains highly significant regardless of necessity of structural changes in the specific industry.

If we combine Tab. I and Tab. II in one, we will create the matrix which let us define the recommended ranges of weighting factors values for all competitiveness criteria according to different combinations of external economic conditions and of necessity of structural changes in economy. These recommended ranges are presented in Tab. III. To choose the specific values from presented ranges correctly we must remember that anyway the sum of all weighting factors in every situation must be equal to unity.

The lower right quadrant of the matrix (combination of favorable external economic conditions and low level of structural changes necessity) describes the most perspective and preferable situation. As we can see the criterion of product's quality becomes the most significant and important in this situation. The upper left quadrant of the matrix (combination of adverse external economic conditions and high level of structural changes necessity) describes the most unfavorable situation which characterizes the current statement of the vast majority of domestic industrial companies. In this case the weighting factors of all competitiveness criteria become approximately equal.

4 Conclusions

The developed company's competitiveness evaluation method can be easily used for detecting both the problems of the particular company and its hidden reserves. This method can be characterized as multipurpose and easy to use in practice; it is based on common statistical database and doesn't use the complicated calculations. The method can be successfully used as a tool for strategic and current management, for continuous monitoring of company's activity to detect the problems and weaknesses, and for making the informed decisions to manage the competitiveness of company. Also one of advantages of this method is its applicability to evaluate not only the current competitiveness of industrial company, but its potential willingness to compete in international dimension too, so it can be useful for increasing the national competitiveness in general.

Table 3. The Recommended Ranges of Weighting Factors Values.

External Economic Conditions	The Necessity of Structural Changes in the Industry		
	High	*Medium*	*Low*
Adverse	V1 [0,2–0,4] V2 [0,2–0,4] V3 [0,2–0,4]	V1 [0,2–0,4] V2 [0,2–0,4] V3 [0,2–0,4]	V1 [0,2–0,4] V2 [0,2–0,4] V3 [0,2–0,4]
Neutral	V1 [0,3–0,5] V2 [0,2–0,4] V3 [0,3–0,5]	V1 [0,4–0,6] V2 [0,2–0,4] V3 [0,3–0,5]	V1 [0,4–0,6] V2 [0,2–0,4] V3 [0,2–0,4]
Favorable	V1 [0,4–0,6] V2 [0,0–0,2] V3 [0,4–0,6]	V1 [0,4–0,6] V2 [0,0–0,2] V3 [0,2–0,4]	V1 [0,6–0,8] V2 [0,0–0,2] V3 [0,2–0,4]

The work was supported by Act 211 Government of the Russian Federation, contract № 02.A03.21.0011.

References

1. Contemporary competition, **3**, 4-37 (2010)
2. Yu.B. Rubin, Contemporary competition, **3**, 38-67 (2010)
3. R.E. Mansurov, The perspectives of sience, 2(4), 125-130 (2010)
4. A.S. Schastlivov, Contemporary tendencies of economics and management: new vision, **1-2**, 210-216 (2010)
5. V.L. Belousov, Foreign Marketing, **5** (2001)
6. L.V. Tselikova, Bulletin of Moscow University, Economics, **2**, 57-67 (2000)
7. A.A. Voronov, An Engineer, **3**, 32-37 (2000)

8. R.A. Fathutdinov, *Strategic marketing* (Business school Intel-Sintez, Moscow, 2008)
9. N.K. Moiseeva, *International marketing* (Peter, St. Petersburg, 2001)
10. A.S. Shalminova, *Innovative business-planning of development in foreground sphere: synopsis* (Publishing house of KFEI, Kazan, 2000)
11. E.A. Ivanova, *The company's competitiveness assessment* (Phenix, Rostov-on-Don, 2008)
12. V.B. Artemenko, E.V. Leontyeva, Invest Region, **2**, 20-23 (2009)
13. K.V. Elin, Baikal research journal, **5**, 16 (2013)
14. A.A. Voronov, The Engineer, **3** (2003)
15. I.M. Kostin, Kh.A. Faskhiev, Foreign and Domestic Marketing, **4** (2003)
16. V.V. Krivorotov, Bulletin of Orenburg State University, **8**, 112-117 (2006)
17. L.Ya. Avrashkov, G.F. Grafova, A.V. Grafov, *Managing of companies' competitiveness: theory and practice: monograph* (Voronezh, Publishing hous of IOTUR, 2009)
18. D.S. Voronov, V.V. Krivorotov, Bulletin of USU, **14**, **5**, 717-740 (2015)
19. Z.V. Khmelnitskaya, E.V. Kuchina, *The increase of productivity and competitiveness of industrial companies: monograph* (Publishing house of USEU, Ekaterinburg, 2008)
20. E.I. Altmanm, The Journal of Finance, 589-609 (1968)

Modeling the formation of an industrial enterprise operating income

T.N. Myznikova[1,*], *E.V.* Bunova[1], and *E.N.* Yaroslavova[1]

[1]South Ural State University, Chelyabinsk, Russia

Abstract. The developed economic and mathematical model of the industrial enterprise's operating income formation allows to reveal cause-and-effect relations and to form a mechanism of influence of initial (base) indicators on the studied index (operating income of a business entity). The model is based on the regression analysis. The base indicators forming the economic and mathematical model are grouped into two blocks: production assets and labour resources influencing the enterprise's financial standing, namely, its operating income. This grouping allows to evaluate efficiency of the influence of production assets and labour resources on the enterprise's potential at taking of justified management decisions. The mechanism of the enterprise's industrial potential management modeling includes analysis of base indicators (factors) of each of the considered blocks. With the help of the developed economic and mathematical model, by setting base values of factors, we can consider their influence on the resulting index – operating income of any business entity; and vice versa, by setting the resulting index we can analyze the influence of base indices for efficient management of the industrial enterprise's operating income.

1 Introduction

Efficiency of a business entity is predetermined by the condition and quantitative composition of its production resources, their quality use and timely renovation. In this regard, almost all industrial enterprises lack floating assets not only for renovation and updating of their production assets, but also for carrying out their current activities. According to the authors [1], most heads of Russian enterprises use this factor for explanation of financial instability of their productions. At the same time, they do not account for duplication of functionalities in different divisions, which leads to disbalance of cash flows and deficiency of resources for re-equipment [2]. It is rather difficult to take a management decision on composition, amount, terms of replacement and sources of financing of production assets renovation. In this regard, the enterprise's fixed assets are subject to moral wear, apart from physical wear, which affects competitive ability of production. Consequently, in order to increase financial stability of enterprises and their competitive power, enterprises need a mechanism of their production potential management. Modeling of the industrial enterprise's production potential management process has become particularly important at the present time characterized by negative factors affecting socioeconomic development of Russia in conditions of the unstable global political and economic situation.

2 Urgency and scholarly Importance

Scholarly importance of the considered issue is confirmed by an extensive discussion in the economic circles devoted to types, forms and principles of modeling the enterprise's potential management processes, including the innovative one.

Issues of sources of financing and financial planning at the enterprise are closely linked with studying of lines of expenditure, which often requires specification of the object of study. It becomes particularly important at modeling of the financial management process, which is confirmed by the authors of the research [3].

N.A. Saltykov [4] considers theoretical aspects of "modeling" category and identifies elements, which, in his opinion, must form this model.

Classification and evaluation of the industrial enterprise's economic potential is presented in the work [5, 6]. The author analyzes potential threats to functioning of the industrial enterprise at taking of erroneous decisions in the course of the industrial enterprise's economic potential management.

In this regard, several scientists [5, 7, 8] are inclined to consider the enterprise's potential from the perspective of innovative development, which is certainly expedient. However, innovations – in the wording of Azgaldov G.G. and Kostin A.V. [9] are defined as a process of introducing novelties accompanied by certain conditions: availability of documents of title for the novelty; production of relevant goods protected with these documents; the goods must be

* Corresponding author: myznikovatn@susu.ru

compatible at the global level, as well as ensure a higher effect than the previously manufactured good.

Based on this assumption, we come to the conclusion that the authors, upon consideration of issues on modeling of the enterprise's innovative potential management, do not describe the innovative but the production potential. At the same time, there are not enough mathematical models able to solve online problems on resource sufficiency formation for the industrial enterprise, or the models are too complicated for practical application.

In the context of the considered problem, the work [10] is worth noticing. However, the author considers the entire set of business processes in one information collection, while local tasks – ensuring of the production resource renovation – are more important for enterprises.

A more detailed evaluation in the course of the enterprise's potential modeling is presented in the work [7]. The author analyzes in detail consequences of resource potential use for the enterprise but considers only one segment – labour resources, which does not reflect the entire scope of the enterprise's production resources.

In the work [11] the author sets a mathematical dependency between the enterprise's financial flows and remuneration of labour. However, in the context of the presented model it is impossible to distinguish connection between the labour potential and the financial result.

Production potential is deeper studied and classified in the work [12]. The author discloses the mechanism for evaluation of the enterprise's resource potential use, formulates its dynamic normal to determine composition and efficiency of use and sets the enterprise's operating efficiency depending on each of the elements. However, limitation of the author's model lies in the fact that in the research the focus is made on sectorial specificity, which reduces the range of application of the models.

In the context of the studied problem be believe it is interesting to overview the enterprise's resource potential modeling process in tasks of project management by means of stochastic dynamic programming. [13]. In this case, the author considers potential development of the enterprise simultaneously by means of extensive (quantitative) and intensive (qualitative) factors.

As it is shown in the works [14, 15], intensive factors are able to increase efficiency of enterprises.

Results of the conducted research have shown that currently there is no integrated practice applied for analysis of the industrial enterprise's operating income.

Thus, urgency of this research is preconditioned, firstly, by the necessity to supplement methods of scientific cognition of the industrial enterprise's operating income formation and management in the strategic aspect. Secondly, it is preconditioned by the necessity to create a practical set tools for calculation and analysis of the industrial enterprise's operating income, which would ensure stable financial standing of the enterprise and its competitive power both in the short- and long-run.

3 Problem statement

It is impossible to build an efficient system of production company management in ever-changing conditions of its functioning based only on intuition. An efficient management system must be based on building and application of the economic and mathematical model allowing to outline interconnections between the disturbing influence and the response of the economic system thereto. As it is described in the work [16], it is necessary to develop an adequate model with outlining of interconnections between the disturbing influence and the response of the economic system thereto and then, if such situation occurs in practice, use such management action, which leads to the set management objective.

This model is aimed at setting of cause-and-effect decreases of production effectiveness and use of production assets and finding of an optimal solution for elimination thereof.

4 Building of the industrial enterprise's operating income formation model

Let us consider stages of building the economic and mathematical model of the industrial enterprise's operating income analysis:

Stage 1. Determination of the indicator, which reflects the enterprise's strategic development. As such indicator let us choose indicator "operating income" defined as income earned as a result of the company's core activity, i.e. reflecting income from production activity and ignoring the enterprise's financial activity. This indicator is a base strategic indicator of production efficiency, which positive value allows the enterprise not only to have financial stability but also to perform planning and implementation of innovative projects at the enterprise.

Thus, let us take indicator "operating income" as the resulting indicator of the economic and mathematical model.

Stage 2. Selection of generalized (concerning processing industry) indicators influencing the enterprise's income.

As the analysis of industrial enterprises' operating efficiency has shown, the key problem of most business entities is decline in productivity of production assets and low level of renovation of production assets. Consequently, let us take the first group of indicators for analysis – performance indicators reflecting productivity of fixed assets. This group will include basic key indicators, such as index of production, coefficient of fixed assets' renovation, degree of fixed assets' depreciation, availability of fixed assets of gross book value as of the end of the year.

Indicators of the second group – indices of labour efficiency of the industrial enterprise's employees, namely: index of labour productivity, index of capital/labour ratio change, average number of people employed in the economy, also influence the resulting operating income indicator.

In order to describe activity of enterprises from the processing industry we have selected indicators: index of labour productivity (ILP), indices of capital/labour ratio change (ICLRC), indices of production (IP), coefficients of fixed assets' renovation (CFAR), degree of fixed assets' depreciation as of the end of the year in percentage (DFAD), average annual number of people employed in the economy (AAN), availability of fixed assets of gross book value as of the end of the year (FA), operating income (OI). The background data are taken from electronic resource http://www.gks.ru and presented in "Tab. 1" and in "Fig. 1".

Table 1. Dynamics of the studied indicators of enterprises from the processing industry (data on the Russian Federation).

Year	Indicator name						
	ILP	*ICLRC,%*	*IP, % versus the previous year*	*CFAR*	*DFAD, %*	*AAN, mln. people*	*FA, mlrd. rub.*
2005	103.7	103.7	106.0	5.4	47.1	11506	1440
2006	101.9	101.9	108.5	5.8	46.8	11359	1575
2007	97.5	97.5	108.4	6.4	46	11368	1963
2008	102.1	102.1	102.6	6.9	45.6	11191	2260
2009	96.3	96.3	95.9	6.2	45.7	10401	2567
2010	103.0	103.0	105.2	5.9	46.1	10292	2860
2011	99.8	99.8	105.6	6.4	46.7	10281	3127
2012	100.8	100.8	103.1	6.5	46.8	10170	3335
2013	99.2	99.2	105.5	6.9	46.6	10173	3672
2014	100.7	100.7	104.3	6.9	46.9	10176	3910

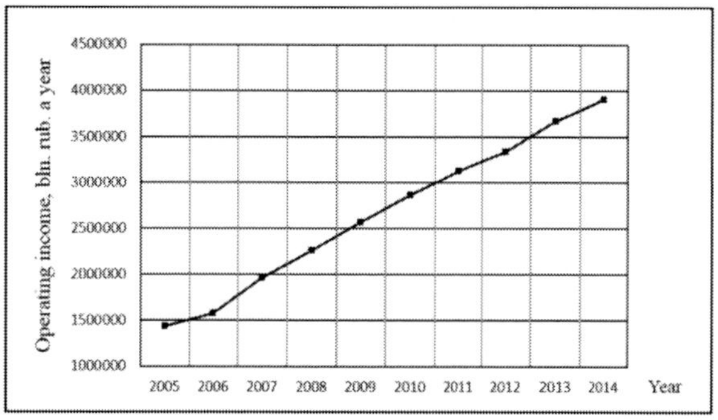

Fig. 1. Operating income of enterprises the processing industry caption.

Stage 3. Building of the economic and mathematical model reflecting general tendencies of functioning of enterprises from the processing industry.

Currently for building of economic and mathematical models the method of principal components [17, 18] and the linear regression analysis [19-21] have become widespread. Eigen value problems can be solved by the linear regression analysis, which allows to analyze indicators by means of the cause-and-effect mechanism of initial (base) indicators' influence on the index under study. Besides, the researches conducted by the authors in the work [20] have shown that regression models can be used for building of adequate economic and mathematical models: "in 70% of cases regression analysis models correctly classify the enterprise's overall financial standing, in 85% of cases – liquidity and business activity, in 80% - financial stability, in 100% – return on investments".

Therefore, at building of the economic and mathematical model of the industrial enterprise's operating income analysis let us use the linear regression analysis.

"Tab. 2" contains parameters (coefficients) of regression by the seven studied factors (indicators) as a result of application of the linear regression analysis based on the data presented in "Tab. 1".

Table 2. Estimated values of coefficients.

Factor (variable)	Coefficient		
	Estimated value	*Symbol*	*Indicator name*
Y result	-5757,25		
X 1	47,6641	ILP	Index of labour productivity
X 2	220,679	ICLRC	Indices of capital/labour ratio change, (in %)
X 3	-0,0024	FA	Availability of fixed assets of gross book value as of the end of the year (mln. rub.)
X 4	-3,43	AAN	Average annual number of people employed in the economy (ths. people)
X 5	158,1	IP	Indices of production (in % versus the previous year)
X 6	1548,34	CFAR	Coefficients of fixed assets' renovation
X 7	-79,94	DFAD	Degree of fixed assets' depreciation as of the end of the year (in %)

So, let us make a generalized economic and mathematical model of the processing enterprise's operating income analysis (1):

$$OI = -5757,25 + 47,66ILP + 220,7ICLRC - 0,0024FA - 3,43AAN + 158,1IP + 1548,34\,CFAR - 79,94\,DFAD \quad (1)$$

The determination coefficient of this model comprises 0,78, consequently, the model built by means of the linear regression analysis is adequate. Graphical proof of adequacy of the built model is presented in "Fig. 2".

As it is clear from "Fig. 2", over the periods, when the income goes downs, estimated and actual values approach to the maximum extent. It has confirmed adequacy of the model and expediency of its application in conditions of lacking resource potential and financial complications.

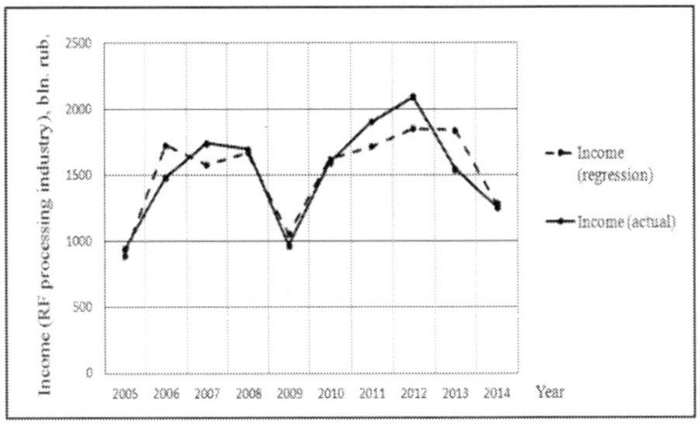

Fig. 2. Dynamics of regression and actual income.

As a result of building the model we can make the following conclusions:

- The coefficient of fixed assets' renovation has the strongest influence on the value of the operating income.
- The index of labour productivity, the index of capital/labour ratio change, the index of production and the coefficient of fixed assets' renovation have a positive influence of the value of the operating income.
- The degree of fixed assets' depreciation, availability of fixed assets of gross book value and the average annual number of people employed in the economy have a negative influence of the value of the operating income.

Stage 4. For building of the operating income management model by means of the industrial enterprise's factor system let us form a system of initial factors influencing the main analytical factors ("Tab. 3").

Table 3. System of Initial Factors Influencing the Main Analytical Factors.

Analytical factor	***Initial factors***
Production factors	
Availability of fixed assets of gross book value, mln. rub.	Book value of fixed assets as of the end of the year, mln. rub. (I)
Indices of production, in % versus the previous year	Amount of equipment, units (Q); coefficient of equipment loading (K); coefficient of defective products manufactured at the available equipment (B).
Coefficients of fixed assets' renovation	Cost of newly commissioned new equipment, mln. rub. (J)
Degree of fixed assets' depreciation as of the end of the year, %	Actual service life of the fixed asset item versus the regulated value (C)
Labour factors	
Average annual number of people employed in the economy, ths. people	Number of production workers employed at the enterprise (L)
Index of labour productivity	Labour productivity (P); growth rate of labour productivity (T); correlation of qualification of works and workers (G); growth rate of personnel qualification (L)
Indices of capital/labour ratio change, %	Capital/labour ratio (F)

Thus, we have formed a group of initial factors, which can be controlled online by the enterprise management, which must allow the business entity to control the condition of its production assets and to form resources for their renovation and/or updating.

At exercising control we propose to use the processing enterprise's operating income model based on general tendencies of processing enterprise functioning with application of initial factors for each business entity of this field (2):

$$OI = -5757{,}25 + 47{,}66 f\begin{pmatrix} P & G \\ T & L \end{pmatrix} + 220{,}7FI - \\ -0{,}024I - 3{,}43L + 158{,}1 f\begin{pmatrix} Q \\ K \\ B \end{pmatrix} + 1548{,}34J - \\ -79{,}94C \tag{2}$$

Resting on the developed model we can take management decisions aimed at stimulation of priority causes influencing the final result.

Thus, the developed model of operating income formation allows to establish a strategy of the industrial enterprise's asset development.

5 Conclusions

Monitoring of scholarly literature corresponding to the research subject allowed to form the most important, according to the authors, set of factors forming the model of the industrial enterprise's operating income management process. In the course of the research the authors have shown that economic and mathematical modeling of the operating income management processes by means of production assets and labour resources is able to increase competitive power, to ensure growth of labour productivity and to achieve the planned financial results. The offered model is a practical set of tools for analysis of the industrial enterprise's operating income.

References

1. A.L. Shestakov, Standards and Quality, 5, 62-68 (2011)
2. T.B. Krasnenko, T.N. Myznikova, *Proc. Modern Society: Scholarly View of the Young*, 86-90 (2015)
3. A.V. Pleshakova, T.N. Myznikova, *Proc. Modern Society: Scholarly View of the Young*, 96-99 (2015)
4. N.A. Saltykov, Bulletin of the Udmurt University: Economics and Law, **2**, 64-69 (2012)
5. T.O. Tolstykh, Bulletin of Kursk State Agricultural Academy, **1**, 61-64 (2011)
6. B. Manganelli, P. Morano, F. Tajani, Wseas Transactions on Business and Economics, **11**, 683-691 (2014)

7. N.M. Salun, Business Inform, **5**, 178-181 (2012)
8. A.G. Khananaev, A.V. Kovalenko, L.N. Zaikina, Scientific journal of the Kuban State Agrarian University, **70(06)**, 1-14 (2011)
9. G.G. Azgaldov, A.V. Kostin, Economic Strategies, **2**, 54-60 (2008)
10. P.P. Pereverzev, Proc. Education and Science in Modern Conditions, 208-211 (2015)
11. E.A. Korobkova, Advanced High Technologies, **3**, 79-80 (2005)
12. A.V. Goncharuk, *Economic Justification of Evaluation of the Mining Enterprise's Resource Potential Use* (Ph.D. Dissertation). http://www.dissercat.com /content/ ekonomicheskoe-obosnovanie-otsenki-ispolzovaniya-resursnogo-potentsiala-gornogo-predpriyatiy#ixzz3zwUcqVxx.
13. A.V. Dokuchaev, *Mathematical Modeling of Resource Distribution in the Task of Project Management by Means of Stochastic Dynamic Programming* (Ph.D. Dissertation, Saratov, 2011)
14. O.G. Turovsts, Organizer of Production, **1**, 34-40 (2006)
15. E.O. Plekhova, Economic Analysis: Theory and Practice, **3**, 64-67 (2008)
16. T.A. Khudyakova, A.V. Shmidt, Modern Management Technologies, **8(44)**, 1-6 (2014)
17. V.V. Mokeev, E.V. Bunova, N.A. Krepak, Bulletin of the South Ural State University. Series: Computer Technologies, Management and Radio Electronics, **4**, 73-81 (2014)
18. O.S. Buslaeva, Problems of Modern Economy, **3**, 6-24 (2014)
19. V.V. Mokeev, Economics and Mathematical Methods, **4**, 82-90 (2010)
20. I.Ju. Glukhikh, Management Consulting, **3**, 185-195 (2011)

The formation of the development model of the innovative industrial cluster and methods for evaluating its economic effectiveness

L.A. Suvorova[1,*], and *L.L.* Zaushitsyna[1,*]

[1]Economics department, Vyatka State University, Kirov, Russia

Abstract. The cluster approach is considered by the authors as the tool to ensure the accelerated development of the country's industrial complex. In the article the authors examine the problem of forming the model of the cluster development in high-tech sectors of industry and the methods for evaluating its economic effectiveness. Unlike traditional approaches, the authors of the article identify a cluster unit as the main structural element of the development model of the innovative industrial cluster. They think that a cluster unit is one member of the cluster (i.e.one enterprise). This point of view is differed from modern scientists` opinion, who view a cluster unit as the complex of enterprises operating within cluster units. The purpose of the study was the development and the economic evaluation of the model of the cluster development. In this research the authors examined the cluster of industrial biotechnologies. They have developed and proposed the model of the development of the cluster of industrial biotechnologies: the Non-commercial partnership (NP) "The biotechnology cluster of the Kirov region". This model takes into account the peculiarities of the innovative production. The authors have calculated the absolute and relative effect from clustering taking into account the effectiveness and profitability indicators of the cluster units activities within the cluster in question and the evaluation of the project activity of the cluster. Thus the authors have proved the economic effectiveness of the proposed model of the cluster development. The received research results allow us to conclude that the designed model of the development of the NP "The biotechnology cluster of the Kirov region" provides a steady growth trend of positive economic effect from the corporate activities of the enterprises within the cluster and increase in the region's competitiveness in the production of high-tech industrial products.

1 Introduction

Nowadays the Russian economy aims at ensuring the economic security and independence of the country. To achieve this goal is possible by solving two main problems. They are: the transition to the innovative way of development and the acceleration of introducing the Import Substitution Programme. These facts form the strategic task for the industrial complex. This task is to increase the production efficiency in the industrial sector of economy.

The defined aims and tasks are characterized by the following main problems. Firstly, the resources are limited. Secondly, the relationships between business entities, partners and consumers of industrial products are complex. Thirdly, there is the lack of efficiency of the market regulation of high-tech industrial products. We think that these problems can be solved through using the cluster approach for accelerating the rate of developing the country's industrial complex [1].

During the strategic development of the economy clusters become one of the elements in the complex mechanism of the new industrial policy of the state. The strategic development of innovative industrial clusters increases the chances of achieving stable positive rates of the country`s economic growth. If we consider the question at the regional level, the innovative industrial clusters and their strategic development are an effective tool to solve many problems connected with the socio-economic development of the region [2].

Despite the considerable interest to the examined topic, one of the obstacles for the wider use of the cluster approach in the industrial complex of Russia is the poor study of both theoretical and practical aspects connected with the development model of clusters in high-tech industries and methods for evaluating its economic efficiency.

The complex of the described circumstances highlights the scientific relevance of the research topic.

2 Defining the task of the research

Despite the large number of scientific articles devoted to the formation and management of cluster development [3, 4, 5], there are no studies in which the authors fully focus on designing the model of development of the innovative industrial cluster and methods for evaluating

* Corresponding author: larsuvorova@mail.ru

its economic effectiveness, that are adopted to the conditions of the Russian regions' economy.

At the same time the analysis of the existing works about the development of industrial clusters in foreign countries (Finland, Sweden, Denmark, Germany, Norway, Netherlands) [6, 7, 8] and in Russian regions [9, 10, 11, 12] led to the following conclusions. They are: 1) the level of clustering in Russian industry is quite low, it significantly falls behind the analogous indicator of the European countries; 2) the national characteristics of industrial policy largely determine the direction of cluster initiatives; 3) it is impossible to create an effective development model of the cluster without the financial and economic, technical and technological, information-communication and other interactions between the participating enterprises; 4) to assess the economic effectiveness of the cluster you must take into account the peculiarities and the main indicators of the development of individual enterprises participating in the cluster [13, 14].

3 The developing of the model of development of the indastrial cluster and the evaluation of its efficiency: methodical bases

According to the authors, an innovative industrial cluster is a quasi-integrated structure consisting of cluster of units, connected by subordinated links.

The model of the development of the innovative industrial cluster is a process including the sequence of actions of cluster units and which allows you to obtain positive economic results from cluster activities. The sequence of actions of cluster units is strictly regulated and it consists of three levels: the innovation designing, production and commercialization of innovations and scaling of innovations.

Unlike traditional approaches, the authors of the article identify a cluster unit as the main structural element of the development model of the innovative industrial cluster. They think that a cluster unit is one member of the cluster (i.e.one enterprise). This point of view is differed from modern scientists` opinion, who view a cluster unit as the complex of enterprises operating within cluster units [15].

While developing the model of development of the innovative industrial cluster the authors of the article paid special attention to the following facts:

1) Firstly, when you form an innovative industrial cluster it is neccessary to take into account the positions of economic theory (A. Marshall [16], M. Porter [17], M. Enright [18], O. Solvell [19]). These positions point out that specialized industries tend to be concentrated geographically and to form clusters, i.e. the community of enterprises of closely related industries, that contribute to the growth of competitiveness of each other at national and international markets.

2) Secondly, it`s necessary to take into account the Resolution position of the Russian Federation Government № 779 "About industrial clusters and specialized organizations of industrial clusters". According to this position the purpose of creating an industrial cluster is to create the complex of stakeholders acting in one field of industry, and that are related to the particular area due to the territorial nearness and the functional dependence, and that are situated on the territory of one subject of the Russian Federation or on the territory of several subjects of the Russian Federation, producing industrial products [20].

3) Thirdly, it`s necessary to consider the innovative industrial cluster as a complex open system, the subjects of which are the manufacturing and service sectors, market infrastructure, educational, scientific and financial spheres; partners, relevant government structures and consumers of the industrial complex products. These subjects are located in a particular area (a country, a state, a county, a district, municipality, etc.) and in addition to the territorial relationship, they also have financial-economic and technical-technological relationship [21].

And, fourthly, it is necessary to consider the management mechanism of developing the innovative industrial clusters as the use of legal, organizational, informational, administrative and economic tools for the purposeful influence on forming the stable long-term and mutually beneficial relationships between the manufacturing and service organizations of the industrial sphere, market infrastructure, education, scientific and financial institutions, other partners and competitors, government and municipal authorities, industrial products customers. This position ensures their effective functioning and development [22].

Creating the model of development of the innovative industrial cluster the authors of the article also took into account the following: a method of forming and the type of the cluster; the specifics of the relationship and the interests of its members; the availability of administrative and territorial boundaries; the level of interaction between participants; possibility, necessity and functionality of referring various socio-economic, socio-political, administrative and legal entities to the certain type and the level of the industrial cluster.

The authors propose a new content of the category "a cluster unit", which is considered as a structural and functional elementary unit in the innovative industrial cluster that performs certain tasks and contributes to achieving the goal of the cluster [21]. At the same time the authors classify cluster units into five types according to their functional purpose (Table I).

All cluster units of the innovative industrial cluster are in certain relations to one another, and this dependence has a character of subordination. The total synergetic effect from developing the innovative industrial cluster is formed by the following way: the innovation cluster units develop an innovative product (technology, etc.), then the production cluster units produce it, and the marketing cluster units sell this product. Technological and institutional innovative cluster units of the innovation industrial cluster has a supporting role as they are engaged in the supply of raw materials, the provision of transport services, advertising, as well as financial aid and insurance.

The distinctive features of the proposed structure of the innovative industrial cluster are the following. The first feature is the division of the cluster into five sub-levels, which are located on the principle of inclusion: each subsequent sublayer includes the previous one, thus increasing the quantitative and qualitative characteristics of the cluster (see Table I).

Table 1. The Structure of the Innovative Industrial Cluster.

Level	The name of the sublevel	The cluster unit type	The type of the organizations of the particular level
Core	Innovation infrastructure	Innovation cluster unit	Organizations developing technologies (research institutes, universities)
Centre	Industrial infrastructure	Industrial cluster unit	Enterprises producing innovative products, created on the basis of the developed technology
	Marketing infrastructure	Marketing cluster unit	Enterprises engaged in selling the products developed on the basis of new technology
Periphery	Technological infrastructure	Technological cluster unit	Technologically depended enterprises
	Institutional infrastructure	The institutional cluster unit	Enterprises, organizations and institutions of the institutional structure

Secondly, in the proposed structure of the industrial cluster the authors have identified the core, which influence the cluster activity as a whole. As a result, the traditionally used approach to the structure of the cluster ("center-periphery") is transformed into the model "core-center-periphery". This "core" consists of two types of subjects: entities that generate innovations and entities that develop innovations.

The authors also offer to create the model of development of the innovative industrial cluster on the principle of inclusion. The innovative industrial cluster is one part of the industrial sector of the country, therefore, the directions of its development should be developed in the same way as the priorities of the development of the national industry and the industrial complex of the region. Practically, this means that the development of the cluster unit is the primary component of the model of the development of the industrial cluster as a whole, which is one of the tools of the region's industrial complex development.

The implementation of the proposed approach to creating the model of development of the innovative industrial cluster can be realized through four stages (Table II).

The authors consider two groups of index of effectiveness to be the basis of the methodology for evaluating the economic effectiveness of the model of the innovative industrial cluster development.

The first group of the index of effectiveness consists of capital productivity, turnover ratio of current assets and labour productiveness. The second group includes different indicators of profitability: return on sales, return on fixed capital assets and return on current capital (there is the same number of evaluation indicators in each group). The proposed method of evaluating the model of development of the innovative industrial cluster is multifunctional. It allows to evaluate all the components of the innovation cluster and to obtain the objective economic results of its operations.

Table 2. Stages of Creating the Model of Development of the Innovative Industrial Cluster.

Stage	Substage	Characteristics of the stage
Formation	The formation of the model of the industrial cluster development	Setting the general purpose of the industrial cluster development as a whole and of its cluster units. Defining the target indicators of the cluster development as a whole and of its cluster units.
	Defining directions, spheres and activities of the biotechnology cluster	Evaluating the effectiveness of each direction of the cluster activities. The financial evaluation of the potential value of the whole cluster and of its cluster units in particular. Determining predictive values of the market share, i.e. the sales volume and the volume of investments to achieve it.
	Defining the composition of the biotechnology cluster, the definition of development alternatives	Making up alternative business portfolios based on the methods of financial modeling. The development of financial indicators characterizing the effectiveness of alternatives. The determination of the optimal composition of the cluster areas, in terms of the optimum ratio of return and risks.
	Defining the role of the cluster "core", the degree of centralization of management functions and the degree of independence of cluster units.	The determination of the status of the industrial cluster "core", the degree of centralization of management functions, the level of control.
	The formalization of the developed model of the cluster development.	The development of the system of internal regulations. The development of internal regulations of cluster management. Creating the relevant positions, regulations and documents.
Realization	The implementation of the common goal of the biotechnological cluster.	The implementation of the cluster development model. Creating the relationship of the model of the cluster development with the current operational management.
Control	Monitoring the implementation of the model and the analysis of deviations.	Organization of the internal control. The financial analysis. Correcting target solutions, update plans, specification of forecasts.
Evaluation	Evaluating the effectiveness of the created model of development.	Evaluating the effectiveness of the created model of the cluster development through targeted economic indicators.

4 The Development Model of the Cluster of Industrial Biotechnologies

The concept of the development of the cluster of industrial biotechnologies in the Kirov region, e.g. the Non-Commercial Partnership "The biotechnology cluster of the Kirov region is realized on the basis of the proposed approach to creating the model of development of the innovative industrial cluster for the period 2015΄2020. The Non-Commercial Partnership "The biotechnology cluster of the Kirov region" (NP "BTC") has been operating in the Kirov region since 2009. This is a classic example of the cluster formed "from below", i.e. it is the union of small businesses that have already experienced the "market testing", produced different products and united to maximize the synergetic effect. Members of the biotechnology finance and control their projects themselves. The NP "BTC" was developed with the support of the Foundation for Assistance to Small Innovative Enterprises in the scientific and technical sphere (Bortnik Fund) and with the assistance of subsidies allocated for the development of small businesses by the regional budget [23]. The priorities of the NP "BTC" are industrial biotechnologies in the field of bioenergy, agriculture, forestry and ecology.

The NP "BTC" is a quasi-integrated structure consisting of cluster units. Each cluster unit passes through certain stages of its life cycle: the initial stage, periods of growth, maturity and decline. In the proposed model of the cluster development the authors offer a set of specific actions for each cluster unit of the NP "BTC" (Table III).

Table 3. Types of Specific Actions for Cluster Units of the NP "BTC".

Stages of the life cycle	Type of the development tactics	Types of financial tactics	Tactics destinations
The initial stage of development or the stage of rapid growth of the cluster unit	***The tactics of rapid growth of:*** - vertical clustering; - horizontal or conglomerate diversification	***The support tactics of the accelerated growth***	The accelerated growth of the potential of forming financial resources
The stage of maturity of the cluster unit	***The tactics of intensive growth:*** - strengthening the competitive position; - expansion of the market; - improving products	***The tactics of financial ensuring of the sustainable growth***	Ensuring the efficient allocation and use of financial resources
The recession or crisis stage of the cluster unit	***The tactics of reducing:*** - structures; - costs; - transition to other organizational forms; - changes in activity area of the company	***The anti-crisis strategy,*** aimed at high mobility of using the released capital for the financial stabilization of the cluster	Ensuring the sufficient level of security

Evaluating the economic effectiveness of the proposed development model of the NP "BTC" was carried out in four stages. During the first stage the total activity results of the cluster units were evaluated. During the second stage the authors analyzed the activity results of the cluster in the context of its infrastructures, as well as in the context of its levels ("core", "center", "periphery") according to two groups of indicators used in the first stage (capital productivity, turnover ratio of current assets, labour productiveness; return on sales, return on fixed capital assets, return on current capital). Further, the authors have evaluated the project activities of the NP "BTC", as the effectiveness of the cluster as a whole depends on the effectiveness of the project activities of the cluster. At the same time, all the innovative cluster projects were divided into three categories: projects in progress, projects in the stage of beginning, projects in the waiting stage. According to project categories and the defined structure of project activities, the authors have evaluated the effectiveness of the projects in progress, taking into account the coefficient of cluster unit's participation in projects. During the fourth stage the authors carried out the calculation of absolute and relative effects from clustering. The positive absolute effect from clustering is expressed in the growth of value added, in absolute growth of the effectiveness indicators (capital productivity, turnover ratio and labour productiveness). And the positive relative effect is expressed in relative growth of the effectiveness indicators (capital productivity, turnover ratio and labour productiveness), and also in the positive dynamics of profitability indicators. The received results of evaluating the economic effectiveness of the proposed development model of the NP "The biotechnology cluster of the Kirov region" are presented in Table IV.

The realization of the proposed development model of the NP "The biotechnology cluster of the Kirov Region" will allow yield benefits in the form of value added on all cluster infrastructures. The results of implementing the given model will be: 1) the inclusion of all cluster units into innovation and project activity; 2) obtaining positive effects in the form of value-added per a cluster unit from a single innovative project; 3) the implementation of the development model will provide a steady growth trend of value added, i.e. the total synergetic effect resulting from the cluster activities. All of this will ensure the achievement of the development purposes of the industrial biotechnology industry in the Kirov region in terms of increasing the share of industrial biotechnology products in the structure of Gross Regional Product (GRP) to 1% by 2020.

5 Conclusion

Summing up, we can conclude that the developed by the authors method for evaluating the economic effectiveness of the development model of the innovative industrial cluster is multifunctional. This method allows to evaluate all the components of the innovation cluster, to obtain the objective economic results of its activity. The implementation of the proposed development model of innovative industrial clusters will allow to yield positive effects in the form of value added on all cluster infrastructure and the steady growth trend of the total synergetic effect according to the results of the cluster activities as a whole.

Table 4. Evaluating the Economic Efficiency of the Development Model of the NP "BTC" for the Period 2015-2020.

Indicators	Before clustering 2001-2009	After clustering, without the develop-ment model	After clustering, with the develop-ment model	The absolute effect from clustering	The absolute effect from clustering based on the development model,+/-	The relative effect from clustering,%	The relative effect from clustering based on the development model,%
Revenue, in thousand rubles	776667	1215520	1383620	438853	606953	156.50	178.15
Expenses, in thousand rubles	719165	754148	764148	34983	44983	104.86	106.25
The financial result, in thousand rubles	57502	461372	619472	403870	561970	802.36	107.31
Fixed capital, in thousand rubles	1324589	1401786	1401796	77197	77207	105.83	105.83
Current capital, in thousand rubles	86822	93113	93113	6291	6291	107.25	107.25
The average number of manual workers on the strength, men	3619	4844	4844	1225	1225	133.85	133.85
Capital productivity, in rubles	0.59	0.87	0.99	0.28	0.4	147.46	167.80
Turnover ratio, per year	8.95	13.05	14.86	4.1	5.91	145.81	166.03
Labour productiveness, in thousand rubles per a man	214.61	250.96	285.66	36.35	71.05	116.94	133.11
Profitability of sales, coefficient	7.4	37.96	44.77	30.56	37.37	512.97	-
Profitability of fixed capital, coefficient	4.34	32.91	44.19	28.57	39.85	758.29	-
Profitability of current capital, coefficient	66.23	495.5	665.29	429.27	599.06	748.15	-

References

1. O.V. Nesmachnyh, V.V. Litovchenko, Basic Research, **9(P.1)**, 162-165 (2014)
2. L.L. Zaushitsyna, L.A. Suvorova, Scientific Review, **6**, 255-260 (2014)
3. T.V. Tsikhan, Theory and practice of management, **5**, 74-81 (2003)
4. G.D. Boush, Economic science of modern Russia, 1, 40-49 (2011)
5. T.V. Mirolyubova, T.V. Larina, Kovalev, *Patterns and factors of the formation and development of regional clusters: a monograph* (Perm State Scientific and Research University, Perm, 2013)
6. Regional Clusters in Europe: European Commission Report, Observatory of European SMES. Enterprise Publications, 3 (2002)
7. PRO INNO Europe paper. The concept of clusters and cluster policies and their role for competitiveness and innovation: main statistical results and lessons learned, **9** (2008). http://www.clusterobservatory.eu.
8. The High Level Advisory Group On Clusters: The European Cluster Memorandum. http://www.innoclusters.ru/uploaded/docs/european_cluster_memorandum.pdf.
9. V.L. Abashkin, M.Yu. Goland, L.M. Hochberg, *Pilot innovative territorial clusters in the Russian Federation* (National Research University Higher School of Economics, Moscow, 2013).
10. I.V. Pilipenko, Proc. Annual Economic Report in 2008 of All-Russian public organizations “Business Russia”.http://www.rsr-online.ru/doc/norm/222.pdf.
11. E. Kutsenko, D. Tyumentseva, Problems of economy, **9**, 93-107 (2011)
12. V. Abashkin, A. Boyarov, E. Kutsenko, Foresight, **3**, **6**, 16-27 (2012)
13. T.A. Khudyakova, *Proc. International Congress on Interdisciplinary Behavior and Social Science 2015*, 295-297 (2015)
14. T.A. Khudyakova, A.V. Shmidt, *Proc. of The 26th International Business Information Management Association Conference*, 1612-1616 (2015)
15. E.A. Udovik, N.A. Ovcharenko, S.B. Chetyz, Problems of Economics and Law, **1**, 134-137 (2012)
16. A. Marshall, M. Marshall, *The Economics of Industry* (Macmillan, London, 1879)
17. M. Porter, Economic Development Quarterly, **1**, **14**, 15-34 (2000)
18. M. Enrghit, U. Staber, N. Shaefer, B. Sharma, Business Networks: Prospects for Regional Development, 190-213 (1996)
19. O. Solvell, *Clusters-Balancing Evolutionary and Constructive Forces* (Ivory Tower Publisher, Stockholm, 2009)
20. http://minpromtorg.gov.ru/common/upload/files/docs/Postanovlenie.pdf.
21. L.L. Zaushitsyna, *The Innovative biotechnological cluster as the basis for achieving the country's regional potential (using the materials of the Kirov region): a monograph* (the Vyatka State University, Kirov, 2015)
22. R.A. Schinova, L.A. Suvorova, L.L. Zaushitsyna, The Bulletin of the Ural Federal University, the section of Economics and Management, **14**, **3**, 457-474 (2015). DOI: 10.15826/vestnik.2015.14.3.027.
23. L.L. Zaushitsyna, L.A. Suvorova, T.V. Baibakova, *Analysis of the conditions of formation, effective functioning and management of innovative technological clusters of the Kirov region* (Kirov, Vyatka State University, 2015)

The process of forming the synergetic effect in the industrial cluster and its economic evaluation

L.A. Suvorova[1,*], *T. V.* Baibakova[1], and *I.V.* Pestova[1]

[1]Economics department, Vyatka State University, Kirov, Russia

Abstract. The synergetic effect of an industrial cluster is considered by the authors as a quasi-effect that has a complex non-linear nature and consists of a number of diverse effects. The authors of the article investigate the formation of synergetic effect in the industrial cluster, as well as the indicators of its economic evaluation. In contrast to the proposed approaches to describing the process of forming the synergetic effect in the industrial cluster, which are based on determining the effects of various kinds (economic, social, environmental, tax, etc.), the authors propose the approach based on the concept "value added". For its economic assessment the researchers consider the following indicator: a synergetic effect of multiplicative-accelerative clustering (SeMAC). The results of the synergetic effect evaluation of the leading clusters of the Russian Federation in the direction of "Biopharmaceuticals, biotechnology and biomedicine" allow us to conclude that the developed scheme of forming the synergetic effect of the industrial cluster. Also the developed indicators of its economic evaluation are common and allow to get the objective data.

1 Introduction

Nowadays Russia implements the Conception of the long-term social and economic development of the Russian Federation for the period up to 2020 (the Conception further).

The strategic goal of the Conception is "to achieve the level of economic and social development according to the status of Russia as the leading world power of the XXI century, which has an attractive way of life and leading positions in the global economic competition and which ensures the national security and the implementation of its citizens' constitutional rights [1]".

To achieve this goal Russia needs to undergo the transition of the economy to the innovative type of development. This is possible if we can form new centers of economic growth. These centers allow to combine the scientific, educational and industrial potential and ensure the rising competitiveness of enterprises, research organizations and educational institutions that have a positive effect on economic development, encourage investment and create new jobs. These centers are innovative industrial clusters [2].

Cluster policy as the basis for development of the country is an alternative to traditional industrial policy. This is because the cluster policy aims at creating and distributing innovative products that are considered to be the basic tools for increasing the competitiveness of enterprises, regions and the whole country. Besides, in the process of clustering branches the synergetic effect appears, and it allows to increase the efficiency of using branches through resource interchange.

The synergetic effect as a quasi-effect increases the speed of circulation of information, and therefore it acts as a catalyst in the development of production, sales, financial, scientific and technical spheres of society.

One of the obstacles for clustering the Russian industrial sector is the lack of examining of theoretical and practical aspects connected with the understanding of the formation of the synergetic effect and its economic evaluation.

2 Defining the task of the research

Today there are a lot of publications concerning this problem [3-16].

Despite this fact, there are no studies that examine the process of forming synergetic effect of the industrial cluster in terms of the formation of value added by combining the cost optimization and accelerating the pace of the industrial cluster development. In addition, there are no studies that assess the synergetic effect in terms of multiplicative-accelerative approach.

3 The developing of the process of forming the synergetic effect of the indastrial cluster

The synergistic effect of the industrial cluster has a complex non-linear nature, consisting of a number of mechanisms for forming different effects.

Pechatkin V., Perfilov V. [18] distinguish four types of effects, forming the synergetic effect. They are: economic

* Corresponding author: larsuvorova@mail.ru

effect, social effect, budgetary effect and ecological effect (Table I).

Table1. The Description of the Clustering Effects that Make the Synergetic effect.

Effect	The Result of the effect
Economic effect	The increase in the gross regional product The increase in investment in fixed capital. The increase in the innovative products output.
Social effect	Creating new jobs and reducing the unemployment level. The increase in real wages of workers engaged in the cluster.
Budgetary effect	The growth of the tax revenue coming to budgets of different levels. Increasing the rate of autonomy of the regional economy.
Ecological effect	Increased investment in environmental protection, increased trapping of pollutants from the atmosphere, reducing the emissions of pollutants into the atmosphere.

Skiba A. [19] defines the following forces forming the synergetic effect of the industrial cluster: strengthening the coherence and the organizational and functional interaction between the elements of the system; the synchrony of the cycles, which characterize the operation of these elements; the consistency of the parameters of external control pulses with the parameters of the internal cycle system.

Patrusheva E. [20] states that: "Provided the non-linearity of the synergetic effect, we can say that the synergetic effect can be obtained as a result of the effect of the partnership, the effect of the flexibility of management decisions... Under the effect of the partnership we mean the economic effect that is manifested in inclusion of individual organizations in one cluster and when the organizations get the advantages of high-quality development of their economic activities on this basis. This effect is reflected while assessing the net current value of the investment project. While assessing one should take into account the expected cash income from the product production, cost optimization in cooperation with the project partners, training, finding new sales markets, etc."

According to the authors, the process of forming the synergetic effect of the industrial cluster is associated with the process of the accelerated formation of value added as a result of realizing the innovative projects of the industrial cluster.

The process of forming value added in the industrial cluster is shown as the scheme on Figure 1.

The formation of value added is realized through combining the costs of optimization and accelerating the growth rate of the cluster.

If we consider the beginning of forming the additional costs as the basis of the synergetic effect, the center is an innovation. Realizing through the industrial cluster potential an innovation turns into a social product, which has the total value-added.

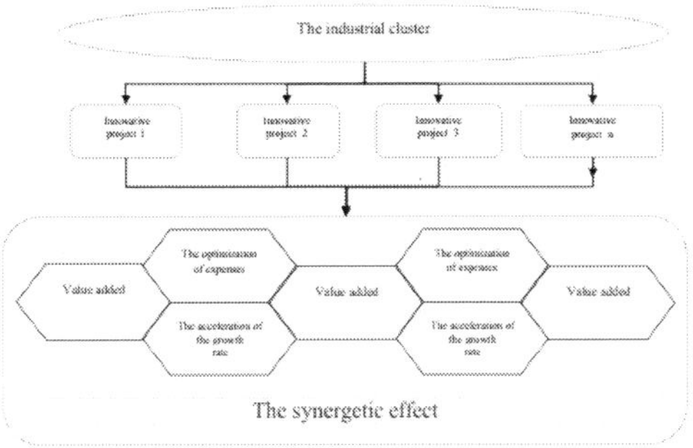

Fig. 1. The process of forming value added in the indastrial cluster.

As a result, the process of forming the synergetic effect of the industrial cluster can be represented in the form of Pascal's triangle (Figure 2).

Figure 2 shows that in order to transform one conventional unit of the innovation (I) into a conventional product (P) you need one conventional unit of financial sources (FS) and jobs (J). Further, financial sources and the product create a new product (P), which in turn provides new sources of finance (FS), and the combination of the product (P) and jobs (J) provides its financial sources (FS) and by implication new jobs (J) and so on. As a result, one conventional unit of the innovation yielded 8 conventional units of financial sources, 6 conventional units of the product and 4 conventional units of jobs.

The presented process can be correlated with the multiplicative and accelerative mechanism [21-23]. In the economy this mechanism acts through the multiplier (income increment) and the accelerator (investment increment), which interact with each other and form resonance. The final result of the accumulated resonance is different effects that all together form the synergetic effect.

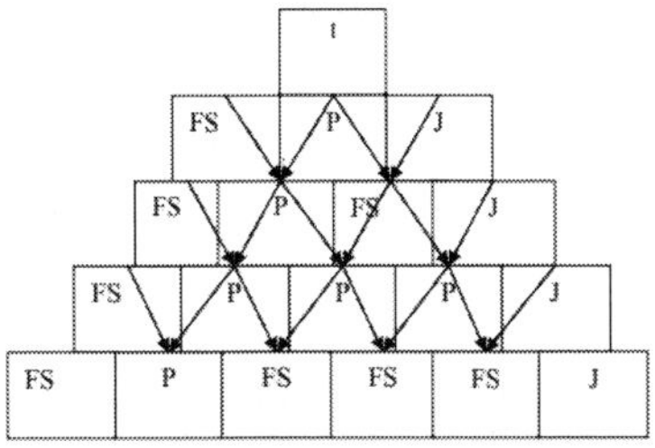

I – innovation, FS – financial sources, P – product, J – jobs

Fig. 2. Forming the Synergetic Effect of the Indastrial Cluster on the Model of Pascal's Triangle.

Thus, obtaining the synergetic effect in the industrial cluster primarily depends on the innovation, potential of the initial dimension of the cluster, as well as on the direction of this potential development. In addition, the synergetic effect of the industrial cluster also depends on the combination of multiplicative and accelerative influence of the innovation.

DOI: 10.1051/shsconf/20173501090

4 Developing the Indicators of Economic Evaluation of the Synergetic Effect of the Indactrial Cluster

The purpose of the economic evaluation of the synergetic effect is to determine the aggregate economic and social benefits received from the operation of the quasi-integrated structure. The defined purpose forms a number of tasks [24]:

1) To evaluate the level of the synergetic effect as a whole.

2) To evaluate the level of the synergetic effect components.

3) To evaluate the impact of the components on the value of the synergetic effect.

The object of the economic evaluation of the synergetic effect is the industrial cluster formed by different kinds and types of organizations [25, 26].

The subject of the economic evaluation of the synergetic effect of the industrial cluster activities is the following: the activities and functions of the organizations within the industrial cluster; property that belongs to the organizations within the cluster; separate management problems of the cluster.

The evaluation of the synergetic effect of the industrial cluster should be based on two structural groups of principles:

1) The group of methodological principles, i.e. the group of general principles relating to the concept of synergy in the industry clustering.

2) The group of methodic principles, i.e. principles that are directly related to the operation of a particular cluster, its specific nature, its economic and financial attractiveness.

The represented groups of principles for evaluating the synergetic effect of the industrial cluster define a number of effects belonging to the synergetic effect of the cluster [26]. Nowadays modern scientists identify eight indicators: the economy of scale, the effect of the accelerator, the practical effect, the investment effect, the multiplier effect, the budgetary effect, the level of clustering and the synergetic effect (Table II).

Table 2. Modern Indicators of Calculating the Synergetic Effect.

The effect name	Indicators
Economy of scale	The consolidated revenues of the enterprises interacting in the industry, in billion rubles
The effect of the accelerator	The share of expenditures for producing innovative products, in percent
The practical effect	The level of the cost reduction, which is realized through optimizing them, in billion rubles
The investment effect	The volume of investments in the production of innovative products, in billion rubles
The multiplier effect	The speed of receiving the consolidated revenues of enterprises cooperating in the industry, in the number of turns per year
The budgetary effect	The amount of taxes received from producing and selling investment products, in billion rubles
The level of clustering	The rate of revenue growth received from selling innovative products, the coefficient
The synergetic effect	The integral effect obtained as a result of industry clustering

According to the authors, some of the indicators proposed by scientists can be correlated with each other. For example, the economy of scale is related to the multiplier effect, the former shows the absolute value, and the latter shows the relative amount, but both indicators display the growth of the output aggregate. The investment effect is related to the effect of the accelerator, the first indicator shows the absolute value of the investment growth, and the second indicator displays the relative amount of the investment growth. In addition, the first consequence of the effects of the multiplier and the accelerator is the practical effect, and the second consequence is the budgetary effect.

Due to this fact, the authors of the article stress the necessity of making up a new concept. It is the synergetic effect of the multiplicative and accelerative synergy (SeMAC). This effect has a complex chain character on the basis of cause-and-effect relationship. this relationship is realized when every subsequent part of the chain generates a particular number of new parts, thereby forming private effects of the multiplicative and accelerative synergies, which together form the total effect of the multiplicative and accelerative synergy. If we consider the beginning of SeMAC effect, according to the authors, it is the innovation that is realized through the industrial cluster potential. The potential of each organization within the industrial cluster is the ratio of potential results to the resources of the element. Since the industrial cluster combines organizations of various activities, the potential of the cluster will be presented by the algebraic sum of the potentials that are created by each element of the cluster. Thus, obtaining the synergetic effect of the industrial cluster depends on the innovations and the initial value of the potential of each industrial cluster organization.

That`s why, the authors of the article propose the universal indicator of the synergetic effect of the cluster. This universal indicator is the multiplicative three-factor model consisting of the synergetic effect multiplier, the synergetic effect and the clustering level – SeMAC (Table III).

Table 3. The Mathematical Explanation of the Model of the Synergetic Effect Evaluation of the Indastrial Cluster.

A unit	Abbreviation	Formula	Explanation
The synergetic effect	Se	Se = MAC	B_c – the cluster consolidated revenues in the current period; B_{c-1} – the cluster consolidated revenues in the previous period; периоде; I_{reew} – Expenses on research, experimental and engineering work (REEW); I_c – the investments of the cluster in the current period
The synergetic effect multiplier	M	$M = B_c / I_c$	
The synergetic effect accelerator	A	$A = I_{reew} / I_c$	
The clustering level	C	$C = B_c / B_{c-1}$	

In order to test the proposed indicators, the authors have evaluated the synergetic effect of Russia's leading clusters in such spheres as "Biopharmaceuticals, biotechnology and biomedicine" (Table IV).

Table 4. Evaluating the Synergetic Effect of Russia`s Leading Clusters in the Spheres 'Biopharmaceutical, Biotechnology and Biomedicine'.

Indicators		2010	2011	2012	2013	2014
The clusters earnings, in mln. rubles		49.96	52.21	59.78	74.42	100.10
The clusters investments, in mln. rubles		18.25	19.07	19.93	20.83	21.76
The share of expenses on research, experimental and engineering work (REEW), in mln. rubles		9.4	9.82	10.27	10.73	11.21
The synergetic effect	Se	0.52	1.47	1.77	2.29	3.19
The synergetic effect multiplier	M	2.74	2.74	3.00	3.57	4.60
The synergetic effect accelerator	A	0.19	0.52	0.52	0.52	0.52
The clustering level	C	1	1.05	1.15	1.25	1.35

The authors found out that if the rate of these clusters development increased by 47% in 2011, then in 2014 it increased more than three times. In other words, the positive policy in the field of innovation and investment activities of the clusters allowed to obtain the significant synergetic effect for a short period of time.

5 Conclusion

Summing up, we can make a conclusion that the process of forming the synergetic effect of the industrial cluster is the complex nonlinear system consisting of the mechanisms of forming value added. The speed of this process and the direction of its realization influence the results of the industrial cluster activities, the level of the industry development and also the dynamics of our country`s economic development.

References

1. The complex program for the development of biotechnologies in the Russian Federation for the period up to 2020, approved by the Government of the Russian Federation (2012)
2. A.S. Huhrin, A.A. Primak, I.A. Semaeva, N.I. Popova, Economy of agricultural and processing enterprises, **12**, 32-37 (2008)
3. E.N. Knyazeva, S.P. Kurdymov, *The basis of synergy. The synergistic vision of the world* (Komkniga, Moscow, 2005)
4. R.A. Schinova, L.A. Suvorova, L.L. Zaushitsyna, The Bulletin of the Ural Federal University, the section of Economics and Management, **14**, **3**, 457-474 (2015). DOI: 10.15826/vestnik.2015.14.3.027.
5. S.V. Smirnov, G.A. Potasheva, InVestRegion, **4**, 26-28 (2007)
6. G.A. Krasnov, V.V. Vinogradov, A.A. Krasnov, Economic sciences. The Bulletin of the Nizhny Novgorod University named after N.I. Lobachevsky, **4**, 219-222 (2009)
7. V.A. Bellavin, S.P. Kurdymov, E.N. Knyazeva, The regimes with difficulties and laws of co-evolution of complex systems. http://spkurdyumov.ru/evolutionism/zakony-koevolyucii-slozhnyx-sistem/.
8. E.G. Lysenko, K.V. Kopach, A.S. Huhrin, *The sustainable development of a private subsidiary farm: the conceptual foundations of strategic management* (Moscow, Rosselhozakademia, 2006)
9. L.L. Zaushitsyna, L.A. Suvorova, Scientific Review, **6**, 255-261 (2014)
10. T.A. Khudyakova, A.V. Shmidt, *Proc. of The 26th International Business Information Management Association Conference*, 1612-1616 (2015)
11. T.A. Khudyakova, *Proc. of the International Congress on Interdisciplinary Behavior and Social Science*, 295-297 (2015)
12. T.M. Khudyakova, O.B. Grekova, V.V. Kravchenko, News of the Voronezh State University of Pedagogics, Russia, **3(268)** (2015)
13. R. Matthews, Economic Strategies, **33-34**, 07-08 (2005)
14. O. Solvell, G. Lindqvist, C. Ketels, *The Cluster Initiative Greenbook* (Ivory Tower, Stockholm, 2003)
15. M. Delgado, M. Porter, S. Stern, Journal of Economic Geography, **10**, **4** (2010)
16. C. Ketels, O. Solvell, *Innovation Clusters in the 10 New Member States of the European Union* (Publications Office of the European Union, Luxembourg, 2006)
17. C. Ketels, O. Memedovic, International Journal of Technological Learning, Innovation and Development, **1**, **3** (2008)
18. V.V. Pechatkin, V.A. Perfilov, The Instrument for evaluating the influence of implementing cluster projects on the sustainable development of Russian regions. http://www.online-science.ru
19. A.N. Skiba, The scientific articles of the System Analysis Institue of the Russian Academy of Sciences, **61** (2011)
20. E.G. Patrusheva, The Regional Economy, Russia, **4** (2015). http://uecs.ru.
21. S.U. Ermakova, *The theory of the multiplier and multiplier effects in the economy* (Ph.D. Dissertation, Ulan-Ude, Russia, 2006)
22. T.G. Kalyugina, V.N. Parahina, InVestRegion, **3** (2012)
23. V.M. Avramchikov, A.N. Antamoshkin, The Bulletin of the Siberian State Aerospace University, **2(54)**, 154-158 (2014)
24. L.A. Suvorova, Proc. Biokirov-2015, 99-103 (2015)
25. L.A. Suvorova, *The synergetic effect of the industry clustering: the analysis, evaluation, forecast: a monograph* (The Vyatka State University, Kirov, 2015)
26. I.V. Pestova, *Proc. Biokirov-2015*, 89-93 (2015)
27. S.U. Ermakova, *Proc. State and society: philosophy, economics, culture*, 94-95 (2005)

DOI: 10.1051/shsconf/20173501091

Analysis of investment appeal of the industrial enterprise by eigenstate method

O.S. Buslaeva[1,*], *E.V.* Bunova[1], and *C.N.* Malovechko[1]

[1]South Ural State University, Chelyabinsk, Russia

Abstract. An analysis of enterprise performance is considered. The problem is solved based on the analysis of the base indicators of functioning of the enterprise in terms of improving economic stability, and also development of autoregulation mechanism of economical stability of enterprise. An eigenstate method is proposed for the analysis of the basic indicators of the enterprise as it allows to construct an economomical stability model of enterprise. Methodology of economic stability analysis of enterprise on the basis of eigenstate method is described. The formulas for calculating the complex indicator of economic stability are given. The effectiveness of the methodology is demonstrated on the example of economic stability analysis of the large trading company.

1 Introduction

The unstable situation in the world economy affected industrial plants the most severely. According to data presented in the article [1], the highest number of enterprises bankruptcies during the economic crisis takes place in the manufacturing industry and it is 35% of all bankruptcy registrations. In that context, the priority development of the industrial enterprise is technical re-equipment of production, allowing the company to obtain competitive advantages and to proceed with a new high-tech and innovative products. Today, businesses that implement infrastructure investments show remarkable growth and have good prospects. To intensify the development of industrial enterprises requires new investments from the outside, i.e. various investments should be involved to upgrade the production, purchase equipment, reduce production costs, release new products, increase sales and labour productivity.

To attract investment company's management need to monitor the investment activity and analyze the investment attractiveness of the enterprise. Currently, the investment attractiveness of many Russian companies is at an insufficient level, as investors consider investment in the enterprise as a high-risk operation. This is due, firstly, to the underdevelopment of Russian stock market, restraining credit policy of banks in relation to enterprises, specific features and priorities of investment policy at the regional level. Secondly, the backwardness of company's management, as currently the management usually focuses mainly on short-term and narrow goal, rather than on the strategic objectives of the business development when it comes to investment attractiveness of the company to investors. In addition, the instability of the economic situation in the country and the geopolitical situation in the world affects the investment attractiveness of enterprises negatively. In this regard, adequate methods of analysis and investment attractiveness evaluation of enterprises is extremely important for investors since it makes it possible not only to minimize their risks, but also to get the greatest benefit from the investment of its assets in one or another company.

There is still no single definition of investment prospects of businesses in the world literature. However, all authors confirm the importance of this concept. For example, the following authors [2] state, «investment grade is important in economic terms because of the attractiveness of new corporate investment, generating new businesses, thus leading to complete economic development." Therefore, we can say that the investment attractiveness of the enterprise - is not only the financial performance of its activities, but also a certain model that consists of qualitative and quantitative indicators characterizing the enterprise functioning in a certain environment. These indicators include assessment of external environment (political, legal, economic, social) and internal positioning of the enterprise in the external environment, the qualitative assessment of the financial and technological potential of the enterprise.

The urgency of an issue of the definition of industrial enterprises investment attractiveness and development of investment attractiveness evaluation and analysis mechanism has led to the certain fact. Today in the economic literature, along with the works of foreign authors, there is a sufficient number of domestic works devoted to the description of construction of methods of industrial enterprises investment attractiveness analysis [3-7]. All variety of

* Corresponding author: buslaevaos@susu.ru

methods can be divided into two groups. The first group includes methods that are based only on the results of the financial activities of the enterprise. The second group is methods that take into account both internal factors (financial performance) and external (legal, economic, social indicators that characterize the country, region or industry).

The first group of the investment attractiveness evaluation methods use only indicators of the enterprise economic activity. As a rule, there is an analysis of the company in the following areas:

1. Asset turnover analysis, taking into account the indicators: asset turnover ratio, current assets turnover ratio, turnover assets duration, current assets turnover duration.
2. Return on capital analysis, taking into account the indicators: the profitability of assets, profitability of current assets, key assets profitability, profitability of products realization, return on equity, profitability index).
3. Enterprise financial sustainability analysis, taking into account the indicators: equity to total assets ratio, financial leverage ratio, fixed indebtedness ratio.
4. Assets liquidity analysis, taking into account the indicators: absolute liquidity ratio, bank-liquidity ratio, debts receivable turnover period, receivable turnover ratio.

The technique described in the Beaver et al. [8] is relating to this group. Beaver et al. [8] find that the financial strength of the firm can be assessed by return on assets (EBIT to total assets), the ability of cash flow from operations pre-interest and pre-taxes to service the principal and interest payments, EBITDA to total liabilities, and leverage (total liabilities to total assets) allowing to predict up to five years prior to the critical event.

One of the methods described in [9] in the analysis of investment attractiveness is to assess the probability of bankruptcy or the company's solvency. Private investors, shareholders and banks are trying to reduce the risk of loss of their investment in the event of default.

In addition, it is offered to evaluate the investment attractiveness of companies using not only the internal measure of the enterprise activity, but also the environment indicators (the second group of methods). At the same time the concept of "investment climate" is introduced. Investment climate is defined in the article [3] as "a set of economic, political, social and other conditions that create a certain level of investments attractiveness." Only taking into account the original assessment of the investment climate investors can proceed to the evaluation of the investment attractiveness of a particular industrial enterprises.

Currently, dominant component analysis [10-13]and linear regression analysis [14-16] are widely used for the construction of models. However, typically dominant component analysis is limited by decrease of data dimension. Eigenstates method is a modification of the dominant component method, but in contrast to it, the main goal of eigenstates method is to build a reference model. To construct the reference model it is necessary to formulate criteria for the enterprise operation as an ideal social and economic system from a perspective of investment attractiveness. These requirements may limit the change in some (basic) indicators of its activity. The reference activity is an activity in which the change is fully consistent with the stated requirements. Application of the eigenstates method allows to show the functional condition of the company at any time as a weighted combination of eigenstates. Each eigenstate is a single-factor model, characterizing a certain process or tendency in the functioning of the industrial enterprise. Under this model, indicators of industrial enterprises activity vary in proportion to the weights of eigenstates and their values depend on a single factor.

This article describes a method of constructing the model of investment attractiveness analysis for the industrial enterprise on the basis of a method of eigenstates. The developed method is a tool for effective analysis of an enterprise operating rates. It also helps to control enterprise assets, as well as manage its costs successfully.

2 The procedure of constructing the model of evaluation and investment attractiveness analysis

To build the model of the investment-attractive industrial enterprises functioning we need to examine a methodology that allows to identify the structure of relationships and to build a generalized indicator of the company investment attractiveness along with reducing the number of analyzed basic indicators of the enterprise economic activity.

The construction of the investment model of the object can be divided into five stages.

The first stage is the formation of an indicators set describing the company's investment attractiveness. For in-depth study of investment attractiveness of industrial enterprises selected indicators should describe not only financial processes but also industrial or ecological ones. All figures should be normalized. Normalization involves error elimination, missing data, and the alignment of indicators change ranges (ie parameters should be changed in one range of numbers, for example from 0 to 1000).

In the second stage the requirements for the industrial enterprises investment attractiveness, which are either a restriction on a change of certain indicators or a set of investment attractiveness indicators with statutory meaning, are formulated. Most often asset turnover analysis, return on capital analysis and assets liquidity analysis are included in the investment attractiveness indicators analysis.

For example, analyzing the investment attractiveness of the region we can say that an increase in investment into the region entails an increase in the gross regional product. The constructed model is an idealized representation of an object behavior.

The third stage involves the calculation of eigenstates of the industrial enterprise. In order to determine the weighting factors of eigenstates it is necessary to solve the problems of eigenvalue matrix. The covariance matrix or the initial second moments matrix can be used as eigenvalue matrix. Each eigenvector has the same dimension as the economic system state vector, and the eigenvector coefficients are interpreted as the weights of eigenstates indicators. Thus, each eigenvector describes enterprise eigenstate, which characterizes a certain tendency of enterprise functioning and development. If to calculate the eigenstates coefficient we use covariance matrix, the vector of mean values is the main component of the investment attractive company model. In addition, if the means do not satisfy the requirements of the enterprise for investments attraction, the model cannot be built. Therefore, to calculate the weight of eigenstates we offer to use the coefficients of the initial second moments matrix. All distinguished eigenstates are independent (the first property of eigenstates [17, 18], ie the development of a trend (eigenstate) does not affect the development of other trends (eigenstates). Thus, the removal of certain eigenstates does not change others eigenstates. Within each trend (eigenstate) values vary in proportion to the weights of eigenstates (the second property of eigenstates [17, 18].

In the fourth stage, the construction of the investment attractiveness model (reference model) of eigenstates obtained in the previous step. The main requirement for the model is that the activities described by the reference model, must comply with the requirements of enterprise investment attractiveness. If the requirements are presented in the form of restrictions on a certain indicators changing, we check the match between initial index modifications of reference activities to the requirements of the enterprise investment attractiveness using the second property of eigenstates. The process of building the reference model is the search of such eigenstates, which combination describes the activities satisfying the requirements of the investment attractiveness of the enterprise. If requirements of the enterprise investment attractiveness are rather strict restriction to the indices changes, there may be no eigenstates combination to describe the indices changes that satisfy these requirements. The number of possible combinations can be substantially reduced by using only eigenstates, which correspond to the requirements of the enterprise investment attractiveness.

If the investment attractiveness requirements are presented in the form of restrictions to indicators change, then according to the second property of eigenstates, base values in eigenstate vary in proportion to weights of eigenstates. It allows to assess the conformity of eigenstates of the enterprise investment attractiveness requirements.

If the investment attractiveness requirements are presented in the form of investment attractiveness indicators and their normative values, values indicators are calculated for each eigenstate. Let the stability indicator is the ratio of x_i and x_j. As part of hth eigenstate, these parameters values are determined by the equation:

$$x^h_{ki} = V_{hi} z_{kh} \quad x^h_{kj} = V_{hj} z_{kh} \tag{1}$$

where V_{hi} – the ratio of hth eigenstate of ith index, z_{kh} – the main component of hth eigenstate for k^{th} observation.

Selected eigenstates are used to generate the enterprise investment attractiveness model (the reference model). The resulting model is an idealization of real activity and serves as a standard for the investigated companies in terms of investment attractiveness.

We differ models with a constant and a variable number of eigenstates. The model with a fixed number of eigenstates uses the same number of eigenstates to calculate indicators:

$$x^{et}_{ki} = \sum_{f=1}^{p} V_{hi} z_{kh} \tag{2}$$

where $h = l(f)$, $l(...)$ – a list of eigenstates numbers, p – the number of eigenstates used to build the model.

The model with a variable number of eigenstates changes the number of eigenstates, depending on the principal components sign. In this case there are three groups of eigenstates: 1) the group of permanent eigenstates; 2) a group of eigenstates with positive main components; 3) a group of eigenstates with negative main components. The eigenstates of the first group are always used, the eigenstates of the second group are used only if the corresponding principal components are positive, and finally, the eigenstates of the third group are used only if the principal components are negative:

$$x^{et}_{ki} = \sum_{f=1}^{p_1} V_{h_1 i} z_{kh_1} + \sum_{f=1}^{p_2} \begin{cases} V_{h_2 i} z_{kh_2}, & if\ z_{kh_2} \geq 0 \\ 0 & ,if\ z_{kh_2} \leq 0 \end{cases} + \sum_{f}^{p_3} \begin{cases} V_{h_3 i} z_{kh_3}, & if\ z_{kh_3} \leq 0 \\ 0 & ,if\ z_{kh_3} \geq 0 \end{cases} \tag{3}$$

Here, $h_1 = l_1(f)$, $h_2 = l_2(f)$, $h_3 = l_3(f)$ where $l_1(...)$, $l_2(...)$, $l_3(...)$, are lists of eigenstates of the first, second and third groups, p_1, p_2, p_3 - the number of the principal components of the first, second and third groups.

At the last stage, analysis of investment attractiveness of the object is carried out. This analysis is done by comparing the actual and reference values of the enterprise performance by using the penalty function method.

Comparing actual performance with reference shows how the actual state of the enterprise corresponds to the reference one. The reference values are calculated on the basis of the standard model of investment attractiveness. Large deviations (eg, deviations of more than 10%) from the reference values can be interpreted as an attendant instabilities object development, and as a consequence, a decrease in its investment attractiveness. Thus, the main objective of the object investment attractiveness management in our view lies in the analysis of the main indicators of the socio-economic object activity, in the early detection of abnormalities that may lead to the destabilization of the development and detection of possible risks of investment loss.

Thus, the main objective of effective enterprise management is the timely detection of deviations, destabilizing activities, identification of possible threats, exploration of the sources and their causes, their solutions in order to maintain the investment attractiveness of the enterprise.

Comprehensive investment potential indicator can be obtained by the equation:

$$I_y = 1 - \overline{f_k} \quad (4)$$

where $\overline{f_{kj}}$ – the mean square root of the penalty function indicators, which is defined by the equation:

$$\overline{f_k} = \sqrt{\frac{1}{r}\sum_{j=1}^{r} f_{kj}^2} \quad (5)$$

Here, f_{kj} represents the penalty function value for the j -th index of kth observation, r – the number of indicators used for the construction of an integrated indicator of investment attractiveness.

3 Industrial enterprise investment attractiveness research

As an example we consider the investment attractiveness of the enterprises of metallurgical industry. To select the parameters that characterize the investment attractiveness of the enterprise it is reasonable to use its financial statements for the 5 years from 2010 to 2014 (external factors are not taken into account, since the investment attractiveness of the country, the region and the industry are considered to be positive). The main indicators of the enterprise economic activity three groups of indicators are taken:

1. Indicators of the company main activities (production factors): revenue, prime cost, finished output volume, profit.
2. The Company's assets: fixed assets, stores, investments, etc.
3. Company Liabilities: undistributed profits, debt, current engagements, borrowed funds.

To build models we accept the criteria specific to the classical methods of enterprises investment attractiveness evaluation. We believe that the investor is willing to invest their capital with minimal risk to the company if: net profit margin is greater than 0.3; current assets turnover is more than 0.25; the ratio of receivables and payables is greater than 1; the ratio of debt capital and the company's assets and return on assets is greater than 0.1; working capital ratio is more than two.

Based on of the second moment's matrix calculation, eigenstates for the selected factors are counted. Fragment of the eigenstates weight numbers calculating are presented in "Tab. 1".

Penalty functions values depend on the principle of comparison of reference and actual values of the enterprise performance. It is proposed to use a cost and result-based method, under which the penalty function value depends on the types of indicators (cost and result).

If the value of the cost index exceeds the reference and is below value of this index at a predetermined level, then the excess is considered a sign of inefficiency and measured by the penalty function.

In order to choose eigenstates that meet the criteria laid down it is necessary to calculate the required indicators for each of eigenstates (fragment calculations are presented in Table 2.

Analysis of eigenstates showed that only one eigenstate is suited to the requirements.

Table 1. Eigenstates weight numbers.

Eigenstates (bln. rub.)	indicator name	1	2	3	4	5
fixed assets	X1	0.096	0.202	0.290	0.758	0.018
stores	X2	0.044	0.127	-0.215	-0.141	-0.116
accounts receivable	X3	0.108	0.050	-0.029	-0.019	0.026
investments	X4	0.275	-0.446	0.223	-0.321	0.451
Working assets	X5	5.062	5.043	15.140	14.169	16.878
owned capital	X6	0.467	-0.185	0.035	0.177	-0.105
long-term liabilities	X7	0.004	0.003	-0.025	0.065	0.003
borrowing costs	X8	0.072	0.122	-0.125	0.058	-0.047
credit indebtedness,.	X9	0.046	-0.057	-0.024	-0.233	-0.240
short-term debt, bln. rub.	X10	0.124	0.095	-0.223	0.036	-0.342
revenue	X11	0.328	0.398	-0.010	-0.176	-0.096
before-tax profit (loss)	X12	0.130	0.356	0.129	-0.123	0.193
net income	X13	0.104	0.287	0.103	-0.118	0.158
net assets	X_{14}	0.466	-0.191	0.052	0.128	-0.088
finished output	X_{15}	0.002	0.001	0.010	0.001	-0.016

Table 2. Indicators of eigenstates investment attractiveness.

Eigenstates number	1	2	3	4	5
net margin	0.31	0.72	-10.32	0.67	-1.65
working capital turnover	0.744	-2.139	0.057	-1.497	-0.914
working capital ratio	3.557	-1.947	0.787	3.276	-0.306
short-term debt and debt receivable ratio	1.141	1.897	7.7633	-1.867	-13.3
the ratio of receivables and payables	2.349	-0.885	1.188	0.082	-0.106
proportion of credit debt in borrowed capital	0.362	-0.576	0.097	-2.315	0.709
leverage	0.240	-0.665	0.266	0.485	-0.882
return on assets	0.196	-1.936	-0.110	-0.570	0.412

On the first of eigenstate we build a reference model of the investment attractiveness of the company. The dependence of the reference and the actual indicators of economic and financial activity of the enterprise are shown in "Fig. 1-2".

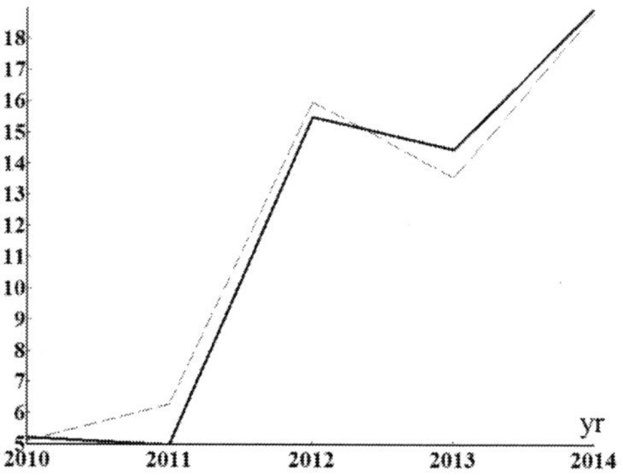

Fig. 1. Owned capital (hereinafter solid line is a model, dashed line is the original values).

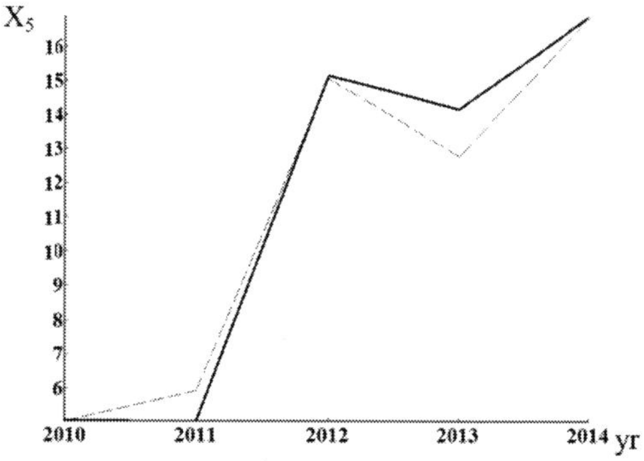

Fig. 2. Working assets.

As can be seen from the graphs, the enterprise operation model that is constructed with the use of the eigenstates method describe the source data rather well.

To match the ongoing by the company's management investment policy with reference complex equation indicator is calculate (6). When calculating the integrated indicator penalty functions are impose on all the factors that have deviations from the reference values more than 10%. The values of the integrated indicator are presented in "Tab. 3".

Table 3. The values of the integrated indicator.

Year	2010	2011	2012	2013	2014
The values of the integrated indicator	0.71	0.72	0.89	0.81	0.82

As can be seen from Table 3, the company's management is conducting inefficient investment policy (indicator value less than 0.9). Although in recent years, since 2012, investment policy of the enterprise has been revised (the indicator began to rise) and policy approaches to the reference.

Similar rates of investment attractiveness of the investigated companies can be obtained using classical methods, for example, Tobin's method [19-20] or the method of calculating the EVA model system indicators [5, 21, 22].

According to the method proposed in [4] for evaluation of investment attractiveness analytical Tobin's Q ratio is calculated. This method relates to the structural and index method of the enterprise investment activity evaluation. Tobin's Q ratio is calculated as the ratio of market capitalization to its net assets at market valuation. If Tobin's Q ratio is less than one, it indicates that the company is undervalued at the market. If the company has an investment activity, the Tobin's Q ratio is growing. If the index is greater than one, it indicates the market interest in the enterprise (market overestimates the enterprise attractiveness).

Tobin's Q ratio describes the efficiency of the net assets to form market value and estimate the enterprise investment attractiveness, but it does not allow to appraise the intensity of investment activity. Calculation of Tobin's Q ratio for the enterprise is presented in "Tab. 4".

As it is seen from Table 4, only since 2014, the market value exceeds the company balance sheet assets (Tobin's Q ratio is greater than 1), and therefore it stimulates investors to invest in venture their own funds.

Similar results are obtained by EVA method calculation [5, 21-23]. EVA model system of indicators (Economic Value Added) is economic value added (EVA) which is a venture profit from ordinary activities after tax, reduced by the amount of fees for the entire capital invested in the company [23].

Table 4. Tobin's Q ratio calculation.

figures	2011	2012	2013	2014
net asset worth, thous. rub.	1171936	11656829	10371044	12113035
net income, thous. rub.	2802860	4315767	1099401	4519763
cap rate	0,4739653	0,61007977	0,2319099	0,2673245
Tobin's Q ratio	0,7227507	0,6068634	0,4571034	1,3958026

EVA index refers to the structural and index method of the enterprise investment activity evaluation. This indicator can be calculated by the following equation:

$$EVA= (P-T)- IC\times WACC = NP- IC\times WACC = (NP/IC - WACC)\times IC = = (ROIC- WACC)\times IC \quad (6)$$

where P – profit from ordinary activities; T – taxation and other compulsory payments; IC – invested capital; $WACC$ – average weighted price of capital; NP – net income; $ROIC$ – net income return on equity.

The advantage of the EVA system is the use of the economic profit principle based on the capital appreciation analysis to assess the enterprise activity. But, it should be noted that the EVA evaluation system provides indirect investment activity of the enterprise.

The concept of economic value added is based on the use of the principle of economic profit founded on cost accounting of debt and equity attraction. The difference between ROIC and WACC is called profitability spread and gives an indication of the relative efficiency of capital use.

For analyzing enterprise we calculated the EVA indicator for the period 2011-2014 year on the initial data basis. The calculation results can be seen in Table 5.

Table 5. EVA indicator calculation.

	2011	2012	2013	2014
net income, thous. rub.	2802860	4315767	1099401	4519763
Invested capital, thous. rub.	-26473	11220720	-765470	5445069
average weighted price of capital	0.14588	0.153729	0.1523213	0.205472
return on equity	0.564777	0.2786174	0.0760470	0.2381555
EVA	0.004005	2.0578E-08	2.0753E-06	1.147E-07

During the reporting periods, EVA index is almost equal to 0, which means the exploring enterprise traces its investment policy rather inefficiently. In this case, the investor should not expect high returns on investments in the company. However, since the EVA is greater than 0, it means that the company earns more than its average weighted cost of capital, and therefore, there is value creation. This means an increase in market value of the company over the carrying value of net assets, which encourages owners to further invest in the enterprise.

Comparison of the calculation results of the industrial enterprises investment attractiveness indices obtained with the help of classical methods such as for example, Tobin's method or EVA method showed the adequacy of the mathematical model of industrial enterprise investment appeal which is constructed using the method of eigenstates. However, in contrast with classical methods, the method of constructing an industrial enterprises investment appeal model using the method of eigenstates allows to take into account the hidden relationship between the studied parameters that influence the investment attractiveness.

4 Conclusions

Thus, in the article the method of the company investment attractiveness analysis, which is based on the investment attractiveness model formation through an analysis of eigenstates is proposed. In the process of constructing a check of compliance of eigenstates with the enterprise economic stability requirements is implemented. As a numerical evaluation of the investment attractiveness of the company it is proposed to use the enterprises investment attractiveness indicator. It is reasonable to use the value-for-cost principle while constructing the indicator. The research showed that the method of eigenstates can be successfully used for the analysis of investment appeal of industrial enterprise. This method can be used not only by investors, but also by company's management, as results obtained in the model show what indicators reduce the investment attractiveness of the enterprise and, therefore, what set of measures it is necessary to implement to improve the investment attractiveness.

The work was supported by Act 211 Government of the Russian Federation, contract № 02.A03.21.0011.

References

1. Z. Topaloğlu, International Journal of Finance and Accounting, **1(3)**, 28-37 (2012)
2. S.C. Guth, M.E. Camargo, António Jorge Fernandes, Maria Elizabeth Pereira Teixeira Rocha, The Scandinavian Journal of Economics, **12**, 249-260 (2015)
3. G.P. Podshivalenko, Financial analytics: problems and solutions, **15(39)**, 7-10 (2010)
4. L.V. Zagorodnova, N.I. Novikov, Bulletin of Kemerovo State University, **2(58)**, 236-240 (2014)
5. O.J. Tolkachenko, Russian Entrepreneurship, **12-2**, 49-53 (2008)
6. T.V. Fedorovich, The economic analysis: theory and practice, **12(117)**, 29-35 (2008)

7. D.V. Doroshin, Directory economist, **8(26)**, 122-128 (2005)
8. W.H. Beaver, M. Correia, M.F. McNichols, Founding and Trends in Accounting, **5**, 99-173 (2010)
9. F.M. Rossi, A. Allini, R. Macchioni, J. Ronen, The Scandinavian Journal of Economics, **12**, 198-210 (2015)
10. G. Stojić, The Scandinavian Journal of Economics, **3**, 293-310 (2012)
11. V.V. Mokeev, V.G. Pluzhnikov, Bulletin of South Ural State University of Economics and Management, **41(258)**, 149-154 (2011)
12. V.V. Mokeev, E.V. Bunova, N.A. Krepak, Bulletin of South Ural State University. A series: Computer technology, management, electronics, **4**, 73-81 (2014)
13. S. Buslaeva, The problems of the modern economy, **3**, 6-24 (2014)
14. I.Y. Deaf, Management Consulting, **3**, 185-195 (2011)
15. V.V. Mokeev, C.L. Solomakho, Investment management, **4**, 26-33 (2011)
16. V.V. Mokeev, Economics and Mathematical Methods, **4**, 82-90 (2010)
17. V.V. Mokeev, D.A. Vorobyev, Bulletin of South Ural State University. Series: Computer technology, management, electronics, **2**, 31-40 (2014)
18. V.V. Mokeev, Scientific and technical information. Information processes and systems, **9**, 3-14 (2014)
19. I.G. Ershova, I.V. Androsov, Vector Science TSU, **4(30)**, 108-112 (2014)
20. S.V. Baguzin, Strategic management, **4(28)**, 242-256 (2014)
21. V. Burešová, L. Dvořáková, Wseas Transactions on Business and Economics, **11**, 525-536 (2014)
22. I.V. Ilin, A.B. Anisiforov, Wseas Transactions on Business and Economics, **11**, 757-764 (2014)
23. O.N. Shcherbakova, Financial Management, **3**, 27-35 (2003)

Intermodal supply chains methodology and economic efficiency study of its implementation (using the example of an Iron and Steel Company)

O.A. Pytaleva[1], *N.V.* Dyorina[2], and *O.V.* Fridrikhson[1,*]

[1]Industrial Transport Department, Nosov Magnitogorsk State Technical University, Magnitogorsk, Russia
[2]Foreign languages in Engineering Department, Nosov Magnitogorsk State Technical University, Magnitogorsk, Russia

Abstract. Currently the global market for intermodal transport (containers in particular) is evolving at a significant pace. The Russian market development of intermodal transport occurs with low speed. It should be noted that low competitiveness of container transport prevents the development of intermodal delivery method within Russia. In addition, this factor stops implementing national transit potential of transport corridors passing through the territory of the country. To date the delivery chains formation is perceived as a secondary process, which provides the scope for production sector and delivers the finished product to the consumer. However, it is the intensive development of intermodal technologies that remains an economic security guarantee and a reserve for increasing the product competitiveness opposed to the rate of finished products production. As a result, the method development of intermodal supply chains formation at large industrial enterprises appears very important. The paper deals with a model and algorithm of forming a system of intermodal cargo delivery chains of the metallurgical enterprises. We performed calculation of the potential economic benefits from their introduction.

1 Introduction

The main trend in the development of Russian and world transport appears an active growth of container traffic volumes, as the basis of intermodal technologies [6, 9, 23]. Experts estimated the ratio of cargo containerization in international transportation that reaches 63%. It is important to note that the coefficient annual growth rate over the last 5 years amounted to 1-2%, and the limit value, according to experts' opinion, is 70%. However, despite the proven high efficiency of container traffic, at present this type of transportation is not attractive to domestic shippers.

Many scientific works of Russian and foreign scientists are dedicated to the improvement of the container transport efficiency. The authors pay great attention to solving the tasks of managing the cargo movement in containers, of organizing various transport modes interaction, of optimization and container infrastructure development. However, the developed methodological tools, as a rule, are designed either to solve specialized tasks, or they contain significant restrictions and simplification.

Objectives and contributions of the study

In order to increase the attractiveness of intermodal delivery technologies, to form highly efficient logistics supply chains it is of great importance to study methodically the issues of forming a container system for the delivery of the products, taking into account economic, infrastructure, technical and technological conditions of the transport market. The formation mechanism should be adaptive to changes in the quantitative and qualitative characteristics of participants in the delivery process and should allow justifying the decisions, both of operational and of strategic levels.

2 Theoretical background

A. Russian Authors' Works Review

Numerous works of Russian scientists are devoted to the issues of improving the efficiency of construction and operation of intermodal supply chains and transport systems. The significant contribution to the solution of these problems has been made in the works of A. T. Deribas, L. A. Kogan, P. A. Kozlov, Y. T. Kozlov, K. M. Lazarev, L. B. Mirotin, S. M. Rezer, A. A. Smekhov, M. F. Trikhunkov, B. A. Lyovin, V. M. Say, V. M. Samuylov et al. An integrated description of the results of the above-mentioned authors was performed in the following works [4, 16].

The works of Russian scientists deal with the following aspects of the intermodal system organization of goods delivery [13, 17]: the formation and development of logistics systems (federal and regional), the principles of rail container traffic are

* Corresponding author: fridrikhsonov@yandex.ru

formulated. The authors reflect methodological foundations of rational distribution of transport infrastructure elements on the territory of Russia, give proposals to increase the transport competitiveness, design streaming concept of material flow promotion, develop proposals of Russia's integration with the international transport space and of the development of regional transport and logistics systems.

B. Foreign Authors' Works Review

The analysis of foreign information sources allowed having summarized the main results obtained by foreign scientists:

• some studies consider the development process of an outline classification of relevant dimensions where the concepts of supply chain and supply network are compared and their distinctive features are highlighted [7].

• the study provides implications for large system change and action steps for change agents in organizations, arguing that by understanding change initiatives through the lenses of complexity and wicked problems, change agents are likely to be more effective [25].

• the following work considers economic behaviour under different economic-political regimes demonstrating continuities and changes during a fundamental social-economic reorientation of an important regional economy, through close observation at the micro and meso-level of, respectively, the workplace, organizations and industry, outlining theoretical, practical and social implications [27].

• the famous author classifies research on distribution network design according to the methodologies adopted and themes tackled. Second, it discusses the main implications for practitioners. Finally, it proposes a few promising directions for future research [20].

• some of the authors investigate the microfoundations of customer knowledge acquisition during logistics innovation development. Specifically, the authors explore the activities and behaviors of employees with customer contact (i.e. boundary-spanning employees) to deepen and broaden their knowledge about customers for the development of innovations [1].

• other scientists facilitate interaction between the domains by increasing the level of joint understanding of the principles used in each domain, and to look at the potential frictions and challenges [5].

• the researchers emphasize that the sorting rules applied in the product recovery process are crucial for the performance in the activity chain from disposer to end-user [10].

• some of the studies give descriptions of benefits from information sharing in supply chains [11].

• the following study analyses behaviour patterns and determined that the segmentation of customers on the basis of both sales and return patterns can facilitate a differentiated service delivery approach [12].

• key findings are that both innovative and non-innovative supply chain operators attribute supply chain synchronization as only a minor indicator of strategic supply chain capability, contrary to the literature; and they also indicate strategic supply chain capability has a minor influence in achieving beneficial outcomes from utilizing industry-led innovation [14].

• it is commonly recognized that the development of a framework for route selection in multimodal transportation that includes a six-phase framework to select an optimal multimodal transportation route. The first phase is to collect the data of each route and select the origin and destination [2].

• formalizing these processes and related activities becomes the differentiating factor in reverse logistic program development and implementation. In addition, providing structure to the reverse logistic effort helps companies to strategically control the related value-added [22].

• weak points of the suppliers as well as dimensions that drive satisfaction were identified [28].

• statement that it is simply not enough to have flexible manufacturing, distribution and procurement systems to achieve supply chain agility. Flexibility in managing demand is also needed. Furthermore, it is the premise of this paper that demand and supply integration inside the firm is critical to achieving supply chain agility [8].

• identifying and explaining the challenges with joint operational decision-making in this context and investigate the precise role of information technology therein [18].

• upgrading from supply chain process modelling towards supply chain business model management; from information to knowledge transfer and from the maturity of supply chain to dynamic capabilities [19].

• research describes the behaviors and processes that impede successful supply chain alliances [21].

• strategic supplier partnership fully mediates the relationship between a lean supply chain strategy and supply chain responsiveness, and that postponement partially mediates the relationship between an agile supply chain strategy and supply chain responsiveness [24].

C. Definitions (conceptual and terminological base)

In order to solve the work tasks we formed terminological apparatus that would unify the basic terms and definitions. The following concepts were used in the study.

Logistics approach means the integration of individual links of material production into a single system, ordered along material flows and it is able to respond to the external environment perturbation adequately.

Logistics supply chains denote many of the elements of the logistics system, linearly ordered by the material (or another one) thread with the purpose of the analysis or design of a certain logistical functions set and (or) costs.

Material flow is the product (goods) considered in the process of using various logistics operations and attributed to a specific time interval.

The flow of services is a complex of specialized tasks performed by each element of logistic system to improve the logistic system effectiveness and achieve its objectives.

Information flow is a term of a message flow in the speech, bond (paper and electronic) and other forms, associated to the material and service flows in the considered logistics system and this flow is intended mainly for the implementation of the control actions.

Financial flow is the directional movement of funds circulating in the logistics system, and between the logistics system and the environment that are necessary for the efficient movement of a particular commodity stream.

Intermodal transport means a goods delivery system under a single transport document and the transfer of cargo from one mode to another in a single loading unit. The basis of modern intermodal transportation is the international standard ISO containers.

3 Methodology

A. Container Transport Market Review

Container Transport Market Review The global market for container transport is developing at a significant rate. In 2015, total worldwide shipments exceeded 190 million. TEU (Fig.1), the total volume of container handling in ports closes to 700 million. TEU. The share of Russian container traffic is 3.02 million. TEU (Fig.2).

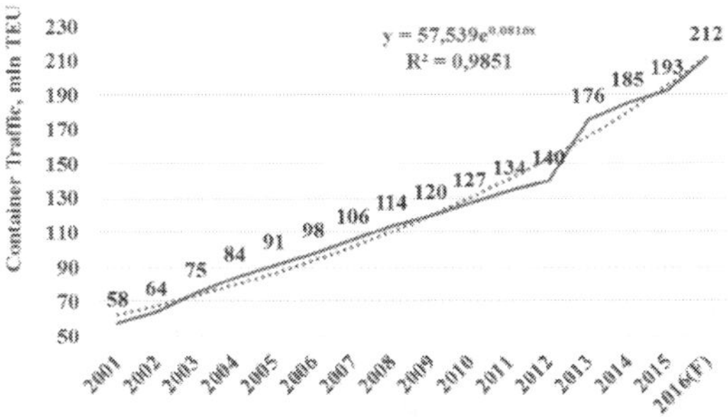

Fig.1. The World's Container Traffic Dynamics.

It should be noted that the main reasons hampering the development of container traffic in Russia are considered to be low speed trailers, considerable time of rolling stock turn-around, lack of integration with railway transport, significant infrastructure constraints, changing the payment system for the use of the owners of the private tracks private cars and containers, lack of studied methodological unit the interaction control of the transport market participants in Russia [15, 26].

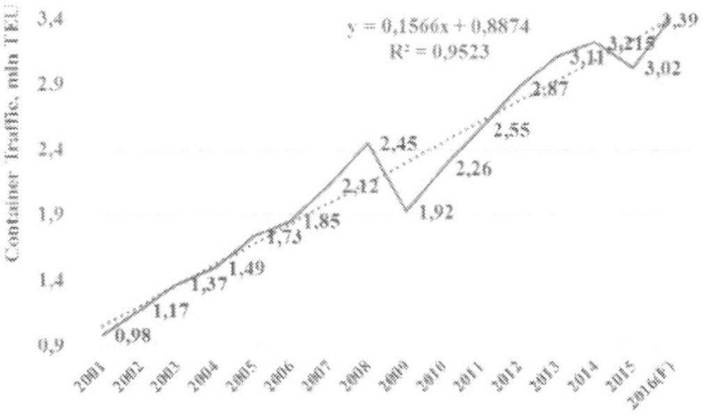

Fig. 2. Russian Container Traffic Dynamics.

B. Methodological tools development

The system of container flow processing for metallurgical enterprise conditions has been examined as the object of the study.

Currently, from 30% to 70% of commercial production of metallurgical enterprises remains suitable for transportation in universal 20 (TEU) and 40-foot containers (FEU), however, in practice, freight containers carried cargoes in the test mode only [3]. Nevertheless, satisfactory results of trial shipments of finished goods in containers have failed to lead to intensive development of containerization.

For the intermodal supply chain formation, it is proposed to consider the containerization process as a system of interrelated flows: the material (the flow of services is combined with the material flow promotion), information, and financial (Fig.3). Allocated flows circulate between the system elements – input, storage, processing, transport, and weekends' ones. For the conditions of metallurgical enterprise functioning, the authors considered three main schemes of product delivery in containers to consumers: direct delivery to a vessel at the port, through regional distribution centers, using port distribution terminals. Promotion and recycling of container flow between participants must be accompanied by financial and informational interaction between participants of the system.

Below there is an algorithm for the formation of the transport and logistics container system on the basis of the developed model (Fig.4).

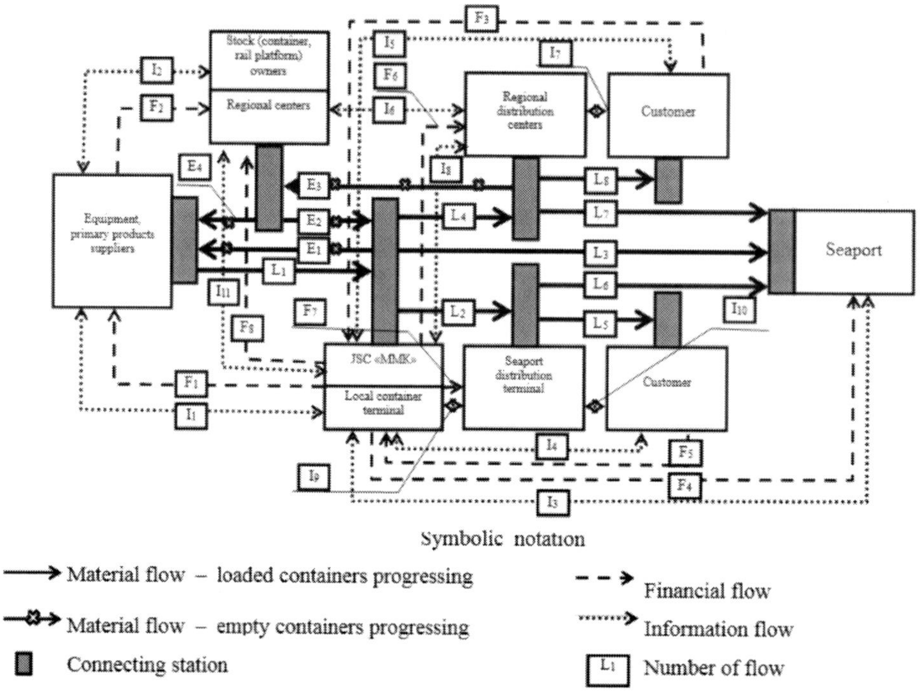

Fig. 3. Product Intermodal Transport Model of a Metallurgical Enterprise.

In order to solve the problem of profit maximization in the intermodal supply chains organization the authors have developed economic and mathematical optimization model, the objective function of which is as follows (1)

$$F(P)=\sum_{i=1}^{n} P_i=\sum_{i=1}^{I} P_1+\sum_{j=1}^{J} P_2 \rightarrow max, \quad (1)$$

where P_i - profits from products sales, rub;
P_1 - profits from products sales, delivered to the consignee on traditional (existing) delivery schemes, rub, as in

$$\sum P_1=\sum n_i \cdot \frac{365}{T_i} \quad (2)$$

P_2 - profits from products sales, delivered to the consignee in containers, rub, as in

$$\sum P_2=\sum n_j \cdot \frac{365}{T_j} \quad (3)$$

n_i – profits from sales of i-mix, delivered to the consignee on traditional (existing) delivery schemes, rub;
n_j – profits from container products sale of j-mix, delivered to the consignee in containers rub;
T_i – the operating cycle time of i-mix products processed, delivered to the consignee on traditional (existing) delivery schemes, days;
T_j – the operating cycle time of container products processing of j-mix, delivered to the consignee in containers, days;
365 – number of days a year.

The following restrictions (4) are imposed on the objective function of economic-mathematical model of the container stream acceleration

$$\begin{cases} Q_1+Q_2=Q; \\ W<\Delta P \end{cases} \quad (4)$$

where W – container flow acceleration costs, rub;
ΔP – profit change as a result of the container stream acceleration according to the delivery scheme, rub;
Q_1– flow of cargo, processed on existing delivery schemes, tons;
Q_2 – flow of cargo, processed by the proposed container schemes of delivery, tons;
Q – annual plan for the finished products shipment, tons.
The calculation results are shown in Table 1.

Table 1. An economic effect estimate.

Delive ry time, days	Total acceler ation costs, mln rub	Cur rent assets turnov er time, days	Num ber of capital turno ver, times	Extra profit, mln. rub	Econo mic effect, mln. rub
6	-	9	36.6		
5	11.92	8	40.6	45.78	33.86
4	23.84	7	45.7	57.99	34.15

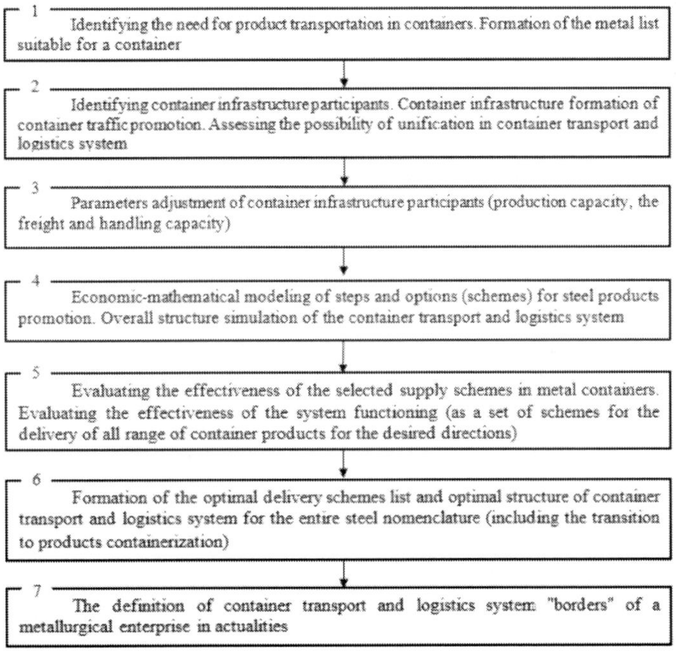

Fig. 4. Container Infrastructure Formation Algorithm.

The following results were obtained in the result of the experiment on the economic and mathematical optimization model (Fig.5).

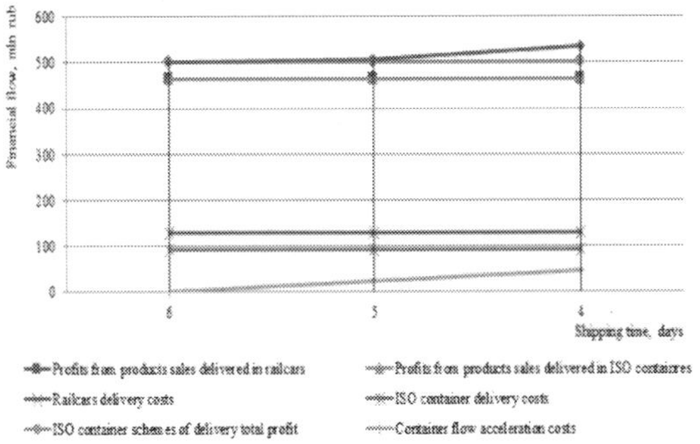

Fig. 5. Payment Iteration of Container Traffic Acceleration Model.

The calculations have shown that using intermodal products supply chains (container delivery scheme) due to a lower rate (price list №10-01) are economically viable for steel transportation in this direction. In addition. accelerating the container stream for 2 days is possible to obtain an additional annual economic impact of 34.1 mln. rub.

4 Conclusions and implications

Currently. despite the high efficiency of container transport. Russian steel companies prefer to deliver products in a traditional way in railway rolling stocks. The formation of the delivery schemes is situational. usually without considering the potential additional costs for products delivery, which are compensated by profit.

The authors developed model and method of forming the transport and logistics products container delivery system of metallurgical enterprises, which allow proving the feasibility of intermodal schemes of delivery of cargoes in containers, to determine the optimal structure of delivery patterns, to assess the potential costs for finished products delivery to the consumer.

The economic-mathematical optimization model has been performed and it enables to maximize profit of participants on the basis of the generated logistic container delivery system of products. The model is based on delivery term reduction to the consumer. The expected economic effect from implementation of the proposed activities is 34.1 mln. rub.

References

1. A. da Mota Pedrosa, V. Blazevic, C. Jasmand, International Journal of Physical Distribution & Logistics Management, **45**, **4**, 313-332 (2015)
2. A. Kengpol, S. Tuammee, M. Tuominen, The International Journal of Logistics Management, **25**, **3**, 581-610 (2014)
3. A.N. Rakhmangulov, A.D. Kolga, N.A. Osintsev, I.N. Stolpovskikh, A.V. Sladkowski, Transport Problems, **9**, **3**, 125-132 (2014)
4. A.N. Rakhmangulov, O.A. Kopylova, Economy of region, **2**, 254-263 (2014)
5. A. Norrman, O. Henkow, International Journal of Physical Distribution & Logistics Management, **44**, **10**, 744-767 (2014)

6. A. Sładkowski, T. Pamuła, Springer. Studies in Systems Decision and Control, **32**, 303 (2016)
7. C. Braziotis, M. Bourlakis, H. Rogers, J. Tannock, Supply Chain Management: An International Journal, **18**, **6**, 644-652 (2013)
8. D.M. Gligor, Supply Chain Management: An International Journal, **19**, **5/6**, 577-591 (2014)
9. H. Clancy, D. Hoppin, J. Moses, American shipper, 69-85 (2008)
10. I. Insanic, Lars-Erik Gadde, International Journal of Physical Distribution & Logistics Management, **44**, **4**, 260-282 (2014).
11. J. Kembro, D. Näslund, International Journal of Physical Distribution & Logistics Management, **44**, **3**, 179-200 (2014)
12. K. Hjort, B. Lantz, D. Ericsson, J. Gattorna, International Journal of Physical Distribution & Logistics Management, **43**, **10**, 852-865 (2013).
13. L.A. Kogan, Y.T. Kozlov, *Container transport system, Study guide* (Academiya, Moscow, 2013)
14. M. Storer, P. Hyland, M. Ferrer, R. Santa, A. Griffiths, The International Journal of Logistics Management, **25**, **3**, 487-521 (2014)
15. M.. Gryaznov, R.A. Franyuk, O.V. Fridrikhson, Bulletin of transport information, **1**, 14-17 (2014)
16. O.V. Fridrikhson, *Formation of transport and logistics container system of metallurgical enterprise products delivery* (Ekaterinburg, 2012)
17. O.V. Moskvichyov, Y.S. Nikonov, Urals transport, **23**, 22-24 (2009)
18. P. Buijs, J.C. ("Hans") Wortmann, Supply Chain Management: An International Journal, **19**, **2**, 200-210 (2014)
19. P. Trkman, M. Budler, A. Groznik, Supply Chain Management: An International Journal, **20**, **6**, 587-602 (2015)
20. R. Mangiaracina, G. Song, A. Pereg, International Journal of Physical Distribution & Logistics Management, **45**, **5**, 506-531 (2015)
21. S.E. Fawcett, M.W. McCarter, A.M. Fawcett, G. Scott Web, G.M. Magnan, Supply Chain Management: An International Journal, **20**, **6**, 648-663 (2015)
22. S.E. Genchev, R. Glenn Richey, C.B. Gabler, The International Journal of Logistics Management, **22**, **2**, 242-263 (2011)
23. S.N. Kornilov, A.N. Rakhmangulov, O.V. Fridrikhson, Proc. Modern problems and ways of their solution in science. transport. production and education, 4-7 (2010)
24. S. Qrunfleh, M. Tarafdar, Supply Chain Management: An International Journal, **18**, **6**, 571-582 (2013)
25. S. Waddock, G.M. Meszoely, S. Waddell, D. Dentoni, Journal of Organizational Change Management, **28**, **6**, 993-1012 (2013).
26. S.A. Tebenkov, Transport business of Russia, **9**, 25-27 (2009)
27. Y. Altman, C. Morrison, Journal of Organizational Change Management, **28**, **5**, 749-769 (2015)
28. Y. Politis, A. Giovanis, S. Binioris, Journal of Modelling in Management, **9**, **2**, 215-237 (2014)

Model of price-dependent management of an industrial enterprise energy consumption

I.A. Solovieva[1], and *A.P.* Dzyuba[1,*]

[1]South Ural State University, Chelyabinsk, Russia

Abstract. The article is devoted to the problems of price-dependent management of energy consumed by industrial enterprises which are involved in purchasing energy in the wholesale and retail energy markets. The article provides the analysis of forming the obligation value to pay for each component of cost of energy. There have been revealed ways and time intervals of energy consumption demand value management in order to minimize the obligation value to pay for cost of energy. The authors have developed a model aimed at managing energy consumption demand which takes into consideration the specific character of forming the obligation to pay in terms of each component of cost bearing in mind to the factors restricting possibilities to manage demand in the time intervals required. The article also covers some description of the parameters developed for information support of the model offered. The practical approval conducted by the authors, permits to state that using the mechanism of price-dependent energy consumption gives industrial enterprises substantial reserves to reduce energy consumption costs. The research results are of practical importance and can be applied in everyday activity connected with energy purchasing by industrial enterprises.

1 Introduction

One of the most vital ways to improve industrial enterprises production efficiency is to reduce the share and value of costs incurred due to electric energy and power consumption which is dominated by electric energy. All the electric energy within the framework of the Russian economy, as well as in most countries of the world, is sold and bought in the energy markets which are characterized by the fact that buyers can regulate the price of the energy bought. One of the underexamined ways in this area is price-dependent energy consumption management.

The domain of price-dependent energy consumption management is researched mostly by American and European scientists [1–6]. A great part of the research data deal with the concept and policy elaboration aimed at encouraging final users to cut down demand schedules of their own. Among the national researchers it's Professor Gitelman L.D. whose contribution in development of energy consumption management is the greatest [7–9]. His works are devoted to the problems of adjustment of foreign experience in consumption management mechanisms to Russian energy markets.

However, among both national and foreign researches one fails to find a work concerned with development of models to manage energy consumption demand schedule of end energy consumers, adjusted to contemporary energy markets.

2 Methodology of research

Price-dependent energy consumption is a model to flexibly manage energy consumption demand of your own based on being responsive to price signals of the energy market to minimize energy consumption costs.

The peculiar feature of price-dependent energy consumption management at industrial enterprises is the possibility to reduce energy consumption costs without cutting down the production plan numbers. Economic effect is achieved due to economically viable rescheduling curves of operational processes and electrical energy demand connected with them [10].

Cost of energy for industrial enterprises of Russia is made up of three basic structural components: electric energy, electric power and electric power transmission.

Figure 1 shows an aggregate structure of electric energy cost for industrial enterprises of Russia.

The value of electric energy and electric power is formed by competitive pricing mechanism while electric energy transmission value depends on regulated price performance [11, 12].

In case industrial enterprises have little influence on price in the energy market, their own values of energy demand schedules can be managed directly through regulating regimes of operational processes and scheduling working hours for equipment consuming much energy.

From our standpoint, all the above components of electric energy cost are subject to management using price-

* Corresponding author: dzyuba-a@yandex.ru

dependent mechanisms making it possible for the consumers to lower costs when buying energy.

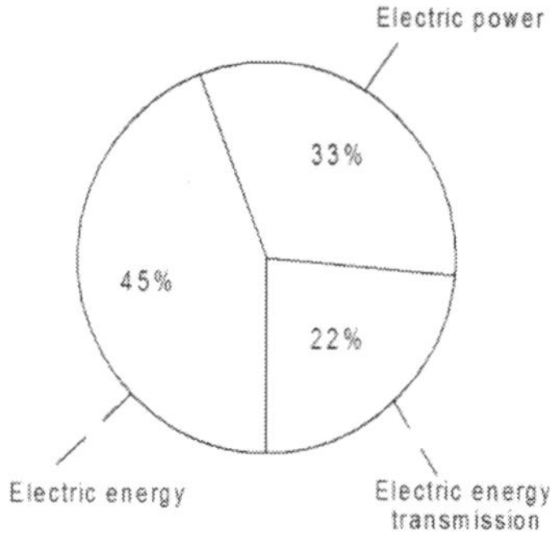

Fig. 1. Structure of electric energy cost for industrial enterprises of Russia.

3 Component "electric energy"

The obligation to pay for electric energy, electric power and electric energy transmission services for each enterprise is calculated individually based on the hourly schedule of energy consumption during the billing month. At that the methods used to calculate the obligation value to pay for each component have principles, rules and peculiarities of their own. The obligation value to pay for electric energy is calculated by multiplying tomorrow market hourly prices by corresponding hourly energy consumption volumes. Tomorrow market hourly prices are exemplified in Figure 2. As can be seen from the figure, tomorrow market prices which are formed for every hour are characterized by substantial diurnal volatility.

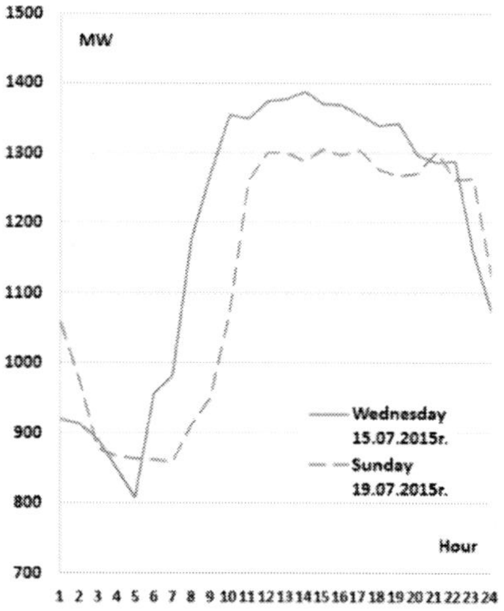

Fig. 2. An example of diurnal hourly curves of tomorrow market prices for a work day and a day off in July, 2016 [13].

Prices at peak hour of working day may be twice as big as post-meridiem prices.

Tomorrow market prices may differ depending on whether it's a work day or a weekend, amounting up to twofold differences.

We consider that price-dependent management of hourly demand schedule must mean channeling energy consumption demand from the time intervals characterized by the highest energy consumption price to economically beneficial periods. It'll allow industrial enterprises to reduce costs of buying electric energy up to 50%.

The periods for price-dependent management of hourly electric energy demand are necessary to be identified with the help of forecasting of the parameters of tomorrow market prices [14, 15].

Component "electric energy" costs are calculated using formula (1) and effect of price-dependent energy consumption management is the difference between the price of electric energy before and after managing energy demand schedule (formula 2).

$$SW_m = \sum_m \left(W^t \times P^t_{PCB}\right) + \sum_m P_{БР} \quad (1)$$

SW_m – value of electric energy bought by the industrial enterprise in m month,

W^t – industrial enterprise electric energy consumption per hour t,

P^t_{PCB} – tomorrow market price (PCB) per hour t,

$\sum_m P_{БР}$ – value of buying electric energy in the balancing sector of the wholesale energy market.

$\sum_m P_{БР}$ – is considered provided electric energy is bought within the wholesale energy market or when buying electric energy in the retail market 5 and 6 price categories are chosen. The authors devoted the series of works [14–16] to researching into electric energy consumption management in $\sum_m Ц_{БР}$ sector.

$$\Delta SW_m = SW_m - SW'_m. \quad (2)$$

where ΔSW_m – cost saving in terms of electric energy payment resulting from price-dependent energy consumption management,

SW_m – costs incurred due to electric energy payment before applying price-dependent energy consumption management,

SW'_m – costs of electric energy payment after managing price-dependent energy consumption.

4 Component "electric power"

The obligation to pay for electric power is formed for every calendar month using hourly electric energy demand schedule for every participant. It is calculated as the mean value of power consumed by the industrial enterprise during the hours which coincide with peak hours of working day of the regional power system during working days of the billing month. The on-peak hours of the regional power system are limited by the peak load planned. Figure 3 shows an example of forming the obligation value to pay for electric power for an industrial enterprise.

Figure 3 demonstrates that the factor upon which the obligation value to pay for buying electric power depends is rather the peak hour of working day of the regional power system than daily electric power consumption maximum of the electric energy end user. The hour of the power system daily maximum is determined only for the working days of the billing month and always happens to coincide with one of the planned hours of peak ratings (Fig. 4).

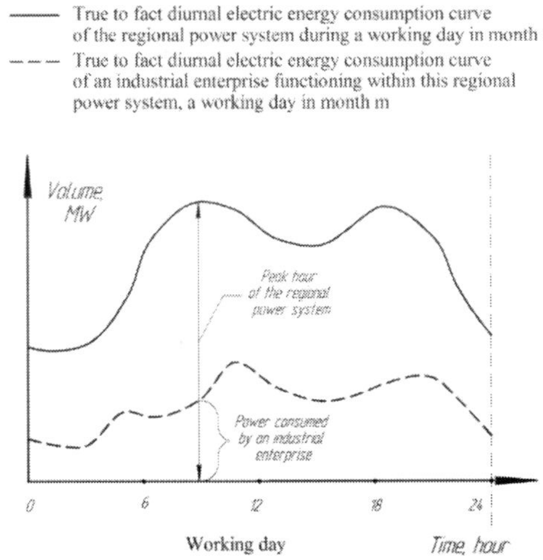

Fig. 3. An example of forming the obligation value to pay for buying electric power by an industrial enterprise [17].

Thus, to reduce the obligation to pay for electric power an industrial enterprise needs to shift peaks of energy demand of their own to the hours which coincide neither with planned peak-load time nor with peak hour of working day of the regional power system.

Forecasting of peak hour of working day of the regional power system can be based on its retrospective data for the last several years, the degree of probability being high enough [19–22]. The approach offered provides industrial enterprises with the tool to lower costs incurred due to purchasing electric power by 10 % - 80 %. Costs in terms of component "electric power" can be calculated using formula (3), while the effect which results from price-dependent management mechanism implementation is determined with the help of formula (5).

$$SP_m = VP_m \times TP_m \tag{3}$$

SP_m – value of power bought by the industrial enterprise in month m,
TP_m – price of power bought by the industrial enterprise in month m,
VP_m – industrial enterprise obligation value to buy power in month m.

$$VP_m = \sum\nolimits_{work,m} W^t_{t\_max\_reion} \Big/ n_{work,m} \tag{4}$$

where $t = t\_max\_region$.
W^t – mount of electric energy consumed by the industrial enterprise during hour t,
t_max_region – hour of coincidental peaking at peak for the Russian Federation entity where the industrial enterprise buys electric energy at hour t during a working day of month m,
$n_{work,m}$ – number of working days in month m.
$t\_max\_region \in T\_peak\_SOT\_nuk\_SO$ – intervals of planned peak-load time, approved by UES System Operator of Russia.

$$\Delta SP_m = SP_m - SP'_m \tag{5}$$

where ΔSP_m – cost saving in terms of payment for electric power value as a result of price-dependent management of electric energy consumption,
SP_m – costs incurred due to paying for electric power value prior employing price-dependent management of electric energy consumption,
SP'_m – costs incurred due to paying for electric power value after employing price-dependent management of electric energy consumption.

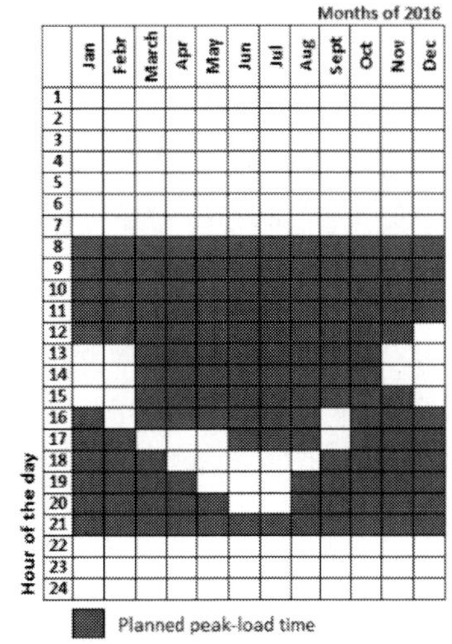

Fig. 4. Chart of peak-load time by months, 2016 [18].

5 Component "transmission services"

The amount of obligation to pay for transmission services is formed for every calendar month. The value of electric energy transmission services can be managed only in case the industrial enterprise chooses the two-part tariffs to transmit electric energy, which consists of two components: the value of electric grids utilization and that of technological power requirements [23]. The value of technological power requirements doesn't depend on the shape of hourly load curves and its share in the total value of services on transmission does not exceed 20%.

The value of electric grids utilization is determined based on the hourly graph of electric energy consumption for the billing month for every participant as the mean value of hourly electric energy maximum consumed by the enterprise during planned peak hours of working day (Fig. 4).

Figure 5 depicts an example of forming the industrial enterprise obligation to pay for electric grids utilization. As it is seen from the figure the management range can be limited to "the blue zone" hours of working days.

Price-dependent management of electric energy hourly demand schedule must involve energy consumption flattering during the intervals of planned peak-load hours. It is worth mentioning that consumers do know these planned peak-load hours a year ahead. Price-dependent management will let industrial enterprises smooth their own energy peak loads and reduce the amount of obligations to pay for electric energy transmission services. The component "transmission services" costs can be calculated using formulas (6–9), while the effect produced due to implementation of price-dependent management is determined with the help of formula (10).

$$SП2_m = SП2_m^{Util} + SП2_m^{Loss} \tag{6}$$

$SП2_m$ – transmission services value for an industrial enterprise using the two-part tariff in month m,

$SП2_m^{Util}$ – transmission services value according to the two-part tariff taking into account electric grids utilization value in month m, (kW×month),

$SП2_m^{Loss}$ – transmission services value according to the two-part tariff considering process losses in electric grids in month m, (kW ×hour) (7).

$$SП2_m^{Util} = T_m^{Util} \times VП2_m \tag{7}$$

T_m^{Util} – rate of tariff for electric grids utilization in month m,

$VП2_m$ – value upon which to calculate obligation to pay for electric grids utilization in month m (8).

$$VП2_m = \frac{\sum_{work,m} max\left(W^t_{T\_peak\_SO}\right)}{n_{work,m}} \tag{8}$$

$max\left(W^t_{T\_peak\_SO}\right)$ – maximum value of energy consumption during the intervals of planned peak-load hours T_peak_SO, which areapproved by UES System Operator, for working day of the month m (Fig. 4) (9).

$$SП2_m^{Loss} = T_m^{Loss} \times \sum_m W \tag{9}$$

T_m^{Loss} – rate of tariff to pay for process losses in electric grids in month m,

$\sum_m W$ – total volume of monthly electric energy consumption for billing calendar month (10).

$$\Delta SП2_m = SП2_m - SП2_m^{'} \tag{10}$$

where – $SП2_m$ costs saving in terms of electric energy transmission services payment due to price-dependent energy consumption management,

$SП2_m$ – costs incurred by paying for electric energy transmission services before applying price-dependent energy consumption management,

$SП2_m^{'}$ – costs incurred by paying for electric energy transmission services after price-dependent energy consumption management was exercised.

Thus, it was revealed that all the three components of electric energy value are dependent on the characteristic of the diurnal hourly curve of energy consumption and subject to price-dependent management (11).

$$S_m = \left(SP_m + SW_m + SП_m\right) = f\left(W_t\right) \tag{11}$$

However, shift in one component cost, as a rule, entails changes in the other two components. This fact should be considered while optimizing the hourly schedule of electric energy loads and making managerial decisions.

6 Actual use experience

Based on the developed approach to managing costs incurred due to purchasing electric energy, as far as each component is concerned, the model of price-dependent electric energy consumption management at an industrial enterprise was worked out (Fig. 6).

From our viewpoint, the parameters of price-dependent electric energy consumption management should be calculated based not on electric energy consumption parameters and market signals alone, limitations caused by technological and economic possibility of production processes management having an impact on daily hourly load curve shape should also be taken into account. The developed model of price-dependent enterprise electric energy consumption management should be employed when a sizable body of data is available. It leads to tough requirements to providing information for the approach offered. One of the key requirements to the information is its high discreteness. In Table 1there are given requirements, worked out by the authors, to volume and discreteness of information necessary for the model to be made use of.

Table 1. Information support data for the model of price-dependent electric energy consumption management.

№	Parameters	Data discreteness, analysis depth
1	Actual electric energy consumption	Hourly discreteness. On-line analysis
2	Planned electric energy consumption	Hourly discreteness. Values for 1-30 days ahead
3	Planned tomorrow market price values	Hourly discreteness. Values for 1-30 days ahead
4	Planned peak hours of working day of the regional power system	1 hour per 1 day. Values for 1-30 days ahead
5	Real peak hours of working day of the regional power system	1 hour per 1 day.
6	Planned peak load time intervals	Interval for every day for 30 days ahead

7	Real schedules of production processes	Hourly discreteness. On-line analysis
8	Planned schedules of production processes	Hourly discreteness. Values for 1-30 days

		ahead

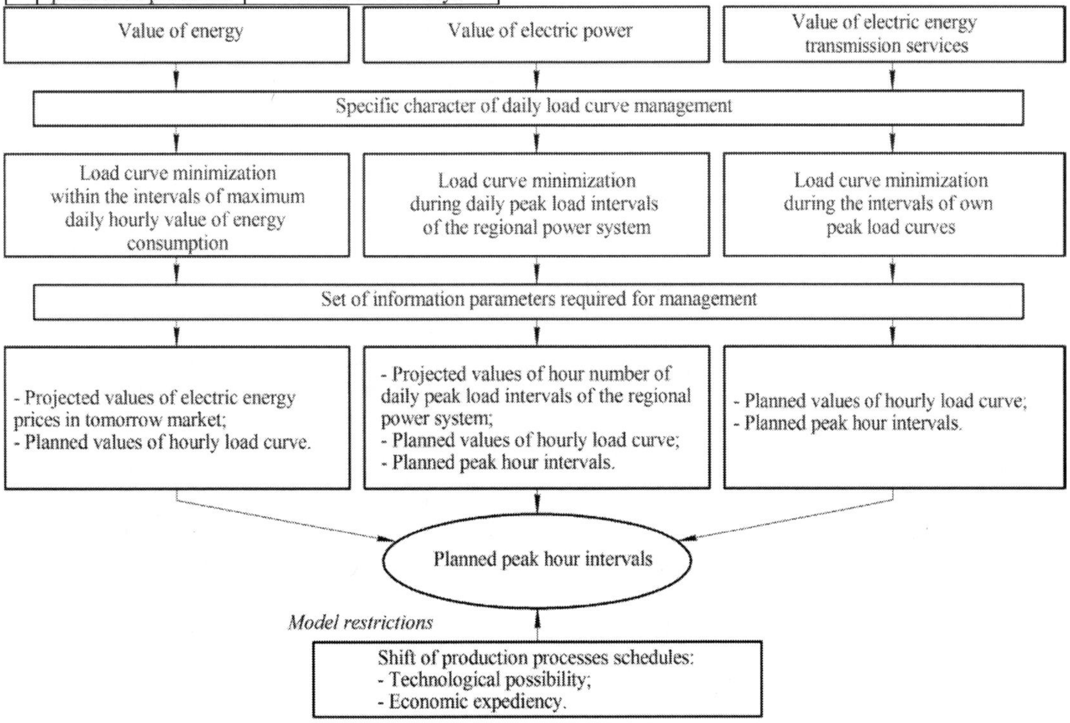

Fig. 5. Model of price-dependent management of industrial enterprise electric energy consumption.

Practical significance and timeliness of the offered method of electric energy consumption cost management is stressed by its practical approval carried out by the authors at iron and steel companies OJSC "Ogneupory" (town of Bogdanovich, the Sverdlovsk region) and CJSC "Volgogradskiy metallurgicheskiy zavod "Krasniy Oktyabr" (city of Volglgrad).

Practical approval revealed that cost saving in terms of electric energy component can reach 20% to 50% of the value before management effect. As for costs of electric power component its cost saving can range from 7% to 50% of

electric power cost [19]. Costs of transmission services component can be cut down by 5–20 %. Thus, practical approval clearly demonstrates that it's possible to lower costs in terms of all its elements, keeping in mind their shares in the structure of costs, by 10% - 40% of aggregate value of electric energy costs; real terms may be as big as several million rubles per annum.

Conclusion

The research conducted leads to a number of important conclusions. Present-day wholesale and retail energy markets make it possible for industrial enterprises through introducing price-dependent energy consumption management to cut down by themselves their costs incurred by buying electric energy. Application of the developed model of price-dependent energy consumption management allows complex management of all existing energy consumption cost components, resulting in substantial economic enterprise-wide effect. The model of price-dependent management should be introduced based on substantial volume of information and take into consideration specific features of production facilities and processes. The scale of the economic effect caused by the implementation of the model offered emphasizes its scientific and practical significance.

The work was supported by Act 211 Government of the Russian Federation, contract № 02.A03.21.0011

References

1. The Law on the US national energy saving policy. http://www.encyclopedia.com/doc/1G2-3407400214.html
2. Energy Conservation and Demand Management Program. Queensland Government. Queensland Government, n.d. Web, 2010. http://www.climatechange.qld.gov.au/pdf/factsheets/1energy-b1.pdf
3. H. Haeri, L. Gage, T. Jayaweera, Assessment of Long-Term, System-Wide Potential for Demand-Side and Other Supplemental Resources.

http://www.pacificorp.com/content/dam/pacificorp/doc/Energy_Sources/Demand_Side_Management/Demand_Side_Management.pdf

4. International Energy Agency Demand Side Management Energy Efficiency. http://www.ieadsm.org/
5. Pacificorp demand-side resource potential assessment for 2015-2034, PacifiCorp, vol. 1-5. http://www.pacificorp.com/es/dsm.html.
6. J. Torriti, Peak Energy Demand and Demand Side Response, 2016. https://www.book2look.com/embed/9781317704447.
7. L.D. Gitelman, B.E. Ratnikov, M.B. Kozhevnikov, Yu.P. Shevelev, *Energy demand side management. Unique innovation for Russian power industry: monograph* (Yekateinburg, 2013)
8. L.D. Gitelman, B.E. Ratnikov, M.V. Kozhevnikov, "Demand-side management: adaptation of foreign experience to Russian conditions," Effective Crisis Management, 2013. http://info.e-c-m.ru/magazine/76/eau_76_207.htm.
9. L.D. Gitelman, B.E. Ratnikov, *Economics and Business in the Electrical Energy Industry: cross-disciplinary manual* (Economics, Moscow, 2013)
10. I.A. Baev, I.A. Solovyova, A.P. Dzyuba, Siberia Science Bulletin, **49(19)** (2015) http://sjs.tpu.ru/journal/article/view/1406/968.
11. RF Government Decree no. 1172 of 27 December, 2010. www.government.consultant.ru.
12. RF Government Decree no. 442 of 04 May, 2012. www.government.consultant.ru/.
13. Official Website of OJSC. http://www.atsenergo.ru/.
14. I.A. Solovyova, A.P. Dzyuba, Bulletin of the South Ural state university. Series: Economics and management, **23**, **30(289)**, 116-123 (2012)
15. I.A. Solovyova, A.P. Dzyuba, Bulletin of the South Ural state university. Series: Economics and management, **7**, **3**, 47-57 (2013)
16. I.A. Baev, A.P. Dzyuba, Regional Economy, **4**, 109-116 (2012)
17. Official Website of Non-Commercial Partnership. www.np-sr.ru.
18. Official Website of OJSC. http://so-ups.ru/.
19. I.A. Baev, I.A. Solovyova, A.P. Dzyuba, Economics, management and investment, **2(4)** (2014). http://euii-journal.ru/24-43.
20. A.P. Dzyuba, *Proc. Energy saving, information technologies and sustainable development: International scientific and practical online conference*, 35-42 (2014)
21. I.A. Solovyova, A.P. Dzyuba, *Proc. Problems of ensuring safe development of contemporary society*, 47-57 (2014)
22. A.P. Dzyuba, I.A. Baev, I.A. Solovyova, L.D. Gitelman, Energy: Challenges and Solutions - WIT Transactions on Ecology and the Environment, **212**, 105-114 (2015)
23. Official Website of Public Joint Stock Company. http://www.rosseti.ru/.

The problem of reproduction of basic assets in industry as the Barrier for neoindustrial transformation of regional economy

T.V. Mirolubova[1,*], and *E.N.* Voronchikhina[1]

[1]Department of the World and Regional Economy, Economic Theory, Perm State National Research University, Perm, Russia

Abstract. Theoretical principles of neoindustrial transformation and basic assets reproduction have been studied in the article. It has been reviewed historical approaches to the disclosure of the examined conceptions for the purpose of defining the extent of every presented concept's reflection of the most important aspects in the analyzed categories. The authors considered one of the internal factors being an obstacle in the new way of Russian development – reproductive process of basic assets in industry. There have been determined the rates of investment growth in capital asset, the growth rates of capital renewals, and the growth rates of basic assets depreciation according to the types of industrial economic activity. There have been revealed general problems of regional industry development connected with reproduction of basic assets. The authors of the article have made a conclusion about the essential improvement of reproductive processes in the regional industry by means of parallel processes' stimulation: elimination of worn assets from the economic turnover and increase of investment in capital assets aimed at renewal of conventional production and conferring high-tech character to the fields identifying basic competencies of industrial region.

1 Introduction

Economic recession experienced by Russia today has revealed the problem of searching new directions of country's development as the most pressing one. The increase of the resource economy's share followed by deindustrialization resulted in decline of competitive strength of the native manufacturing industry. Thereupon, the task of renewal of industrial capital assets on basis of high-tech modernization is becoming particularly urgent. Manifestations of crisis in Russian economy are undoubtedly connected with changes in world oil market, with devaluation of national currency, with sanctions from Western countries, and etc. However, in the years of high oil rent nominally high growth rates of GDP were connected not with diversification of economy and conversion to high-tech and science intensive production but they were principally conditioned by hydrocarbons' stiff prices.

Therefore, economic situation today is connected not so much with external factors as with internal ones, namely with profound reasons of long-term character.

Such scientists as A.R.Belousov [1,2], S.Yu.Glazyev [3] and some others take the view of the fact that difficulties experienced by Russian economy have been caused by the internal reasons. This point of view found the reproduction in the annual President's message on the 4th of December 2014 [4].

At the present time the general strategic imperative of public policy consists in achievement of national economy's qualitative growth assuming its cardinal neoindustrial transformation. Theory principles of forming new industrialization as independent type of society transformation were hypothecated in the work of J.Helbreit about new industrial society [5], then they found continuation and development in works of D.Bell about post industrial society [6], in works of E.Toffler about super-industrial society [7, 8] and M. Castells on the information society [9] (Table I).

Neoindustrial paradigm as vector of national economy development has been considerably substantiated by S.S.Gubanov [10, 11]. The formula for development of Russia has been scientifically identified: neoindustrialization plus vertical integration. Neoindustrial reconstruction implies higher level of technological renewal of Russian productive forces, automation and computerization of all branches, and, first of all, of key, manufacturing industries. Conducting new industrialization as objectively ripen process is explained by the fact that it is exactly industry which provides high growth rates of labour production and makes positive multiplicative influence on the other sectors of economy. "The world of the XXI century continues to be the world of renewed but industrial society" [12, 13].

* Corresponding author: mirolubov@list.ru

Table 1. Theoretical principles of neoindustrial transformation.

Scientists, main works	General ideas
Historical review	
Helbreit J.K. "New industrial society" (1967)	Modern capitalism (second half of the XX century) as "new industrial society" the principle element of which – industrial system – is that part of economy which is characterized by the existence of large corporations.
Bell D. "Forthcoming post industrial society" (1973)	Substantiation of pre-industrial, industrial and post industrial society's existence. Development of the last on the base of technological determinism, i.e. historical process of society's development depends on the level of industrial development.
Toffler E. "Shock of the future"(1970), "The third wave" (1980)	Futuristic concept of new society formation – "super industrial" society. Special attention is given to the predictable trends of future and to the problems of adaptation to those trends.
Kastels M. "Informational era", (monograph, three volumes, 1996-1998)	The concept of "informational society" in the base of which is new information-technological paradigm.
Modern studies	
Gubanov S.S. "Sovereign break. Neo-industrialization of Russia and vertical integration" (2012)	Formula of new Russian development direction: neo-industrialization plus vertical integration.
Afontsev S.A., Galushka A.S., Golubovich A.D., Dmitriev M.G., Idrisov A.B., Osipov A., Pototsky I.V., Titov B.Yu., Shpigel V.V., Yuryev E.L. (All-Russian public organization named "Business Russia") "New industrialization. Concept 25X25" Plan" (2011)	CPI: creation of 25 million new modern high productive workplaces in competitive sector at the expense of priority development of non-resource processing industries of the national economy by 2025 – just "new industrialization".
Bodrunov S.D., Lopatin V.N., Okrepilov V.V. (Institute of New Industrial Development) "Re-industrialization of Russia: improvement of public administration, legal and technological regulation" (2013)	Algorithm of activity aimed at consolidation of public administration mechanisms and scientific-technological progress and innovative development of Russia in the process o formation and realization of public policy.
Coombs R. Miles I. (University of Manchester) "Innovations, measurement and services: modern range of problems" (2000) Barras R. (University of Manchester) "Theory of innovative activity in service industries" (1986)	Problems of service industries in the frames of modern industrial model of development. Product innovations in service industries have "inverse life-cycle". It begins with the use of new technologies aimed at improvement of productive efficiency and finishes by creation of new services opposite to the cycle being the characteristic feature of material production.
Tregenna F. (University of Johannesburg) "Productivity, de-industrialization and re-industrialization" (2011), "De-industrialization: analysis of employment and turnout changes in processing industry" (2009)	Analysis of de-industrialization problem in world economy in the context of employment level changes in processing industries. De-industrialization is considered to be steady decrease in the number of processing industry workers and lowering part of processing industry in GDP is mentioned. The importance of creating industrial policy concept in the transition period from de-industrialization to re-industrialization has been shown.
Sussan A. (International Business at Bethune – Cookman University in Florida) "Strategy of re-industrialization in world economy" (2012)	In the frames of retail trade it is carried out the search of ability to improve social-economic development. Study of modification of competiveness substance and role of technologies.
Aryeetey-Attoh S., Lindquist P., Muraco W., Reid N. (Cambridge University) "Northwestern Ohio: re-industrialization and reduction of harmful substances emission" (2003)	Search of the ways of the greenhouse effect reduction due to technological and structural changes.
Etzioni A. (George Washington University) "Re-industrialization of America (2005)	Social and psychological aspects of re-industrialization. It has been selected seven components of high pace of economic development and living standard - transport, communication, power, innovations, labour resources, finance capital and means of production.
Westkämper E. (Universität Stuttgart) "Re-industrialization of Europe: concept of processing industry development up to 2030" (2014)	General trends of social-economic development, problems of industry development. It has been selected 4 key directions in the European industrial strategy: industrial environmental pollution, ecologically conscious production, cost-making chain, information awareness communication. R&D has been paid special attention.

One of the key problems in the process of realizing neoindustrialized transformation of Russia is the problem of basis assets' reproduction.

Basic assets correspond stocks of materials and capital equipment created in the process of production and used there on repeated occasions or permanently in unvaried natural-material form over a long period of time, persistently shifting their value to products and services [14].

Let's consider the approaches to the subject matter of industrial basic assets reproduction in order to determine the presented concepts' ability to reflect the most important aspects of the analyzed category (Table II).

Table 2. Theoretical principles of industrial basic assets reproduction.

School of economic thought/researchers	General ideas
	Historical review
Mercantilism (Man T., Monkrentyen A.)	The main source of capital accumulation defining the factor of economic development is foreign trade. The principle form of wealth is money in the form of gold and silver; the idea of capital accumulation is in the basis of the concept of reproduction
Physiocraty (Kene F., Tyurgo R.)	Power of nature in all spheres of activity. Reproductive processes of capital were considered primarily in the field of agricultural production where additional product was created. There appeared such notions as "rent", "net effective income", regulations of demand and supply interaction.
Classical political economy (Smith A., Ricardo D.)	Smith A. [15] selected basic and current capital in the structure of production capital. Capital accumulation is the main condition for society's riches growth and the main source of reproduction. Ricardo D. analyzed the means of production. He thought that "the part of the country's wealth which is used in production and consists food, cloths, instruments, raw materials, machines, etc., is indispensable for work" but due to inequality of profit on invested capital the last "is displaced from one activity to the other" [16].
Marxism (Marx K., Engels F.)	Capital accumulation occurs in the process of production and the process of production "… considered in stable connection and in continuous flow of its renewal is at the same time the process of reproduction" [17]. Simple reproduction "…is the simple recurrence of production process in constant scale" [17]. Extended reproduction presupposes systematic addition of the definite part of surplus value to the functioning capital. K.Marx and F.Engels studied the problems of capital replacement through the practice of specially assigned funds' formation, in particular, fund of renewal [18].
Marginalism (Valras L.)	Mathematical scheme of reproduction process description in which casual economic relations are completely changed by the methods of quantitative functional analysis. In methodological aspect L.Valras's theory inspired the efforts to present the process of social reproduction as organic market mechanism of definite products' reproduction.
	Modern research
Arakelyan A.A., Baranov D.A., Kalinin V.P., Kantor E.L., Kovalyov A.F., Korkin S.K., Kuzmenko S.M., Kushlin V.I., Lyubimtsev V.I., Malygin A.A., Pavlov P.M., Rimer M.I., Shulman S.I.	Problems of the enterprise's basic assets revolution, particularly, the problem of separation of the original stage of reproduction process. D.A.Baranov [19], S.K.Korkin [20] think that the original stage of funds motion in the process of basic assets reproduction is the productive one; Yu.I.Lyubimtsev [21] and others mention monetary stage. In the process of production stage the product and its cost are created and formed while in the process of circulation there is only the change of the forms of value (use value, cash value).
Abalkin L.I., Arakelyan A.A., Vilensky M.A., Ivanov Yu.M., Kovalyov A.F., Kurenkov Yu.V., Malygin A.A., Poltorygin V.K., Khoruzhyi L.A.	The definition of the concept of extended, simple and contracted reproduction of basic assets.
Abramov S.I., Amosov A.I., Andrianov V.D., Vodyanov A., Goloviznina O.A., Gorbunov A.A., Gubanov S.S., Zhdanov S.A., Tunin G.A.	Problems of basic assets reproduction at industrial enterprises in modern conditions: limit of enterprise's internal funds, low investment appeal of the most industries, insufficient support of the government. Strategic and tactical directions of those problems solution are considered.

It is necessary to underline that exactly the reproduction of basic assets of industry providing their full renewal can provide neoindustrial transformation of economics.

We put forward a hypothesis that in modern conditions growth rate of investment in basic capital at industry is insufficient for capital assets renewal. Perm Krai, the economy of which has the pronounced industrial specialization can be the example for testing this hypothesis.

In the process of testing there were issues the challenge to define the growth rates of investment in the basic capital of industry, growth rates of capital renewals, growth rates of Perm Krai industrial basic assets depreciation. The following results were obtained.

The necessity of analyzing dynamics and structure of investments in the basic capital is conditioned by the fact that mass renewal of basic assets of industry is, first of all, the result of investments, which in its turn becomes the main factor of economic modernization and development. Let's calculate the investment growth rates according to the types of economic activity in Perm Krai industry. To eliminate the inflation component calculation is made on the base of stable prices of 2005 (Table III).

Table 3. Growth rates of investment in basic capital according to the types of economic activity in Perm Krai industry, prices of 2005, %.

KEA	2009/ 2005	2013/ 2005	2013/ 2009	2014/ 2005	2014/ 2013
Investment in basic capital	113.7	139.3	122.5	141.3	101.4
Mining operations, including:	134.5	199.9	148.6	211.9	106.0
Fuel and energy	138.8	199.9	131.0	217.6	127.8
Apart from fuel and energy	99.2	199.1	116.2	166.2	116.9
Processing industries, including:	121.8	171.7	141.1	171.6	99.9
Production of food including drinks and tobaccos	88.5	97.1	109.7	101.9	104.9
Textile and clothing manufacture	132.9	194.7	146.5	263.1	135.1
Woodworking and wood articles production	86.0	79.1	92.0	73.7	93.1
Pulp-and-paper production; publishing activity and printing industry	83.0	97.4	117.3	75.4	77.4
Coke and oil industry	213.5	560.2	262.3	574.2	102.5
Chemical production	118.6	189.2	159.6	190.0	100.4
Rubber and plastic products manufacture	103.4	99.3	96.0	83.5	84.1
Production of non-metal mineral products	148.9	140.5	94.4	112.1	79.8
Metallurgical production, final metal products	110.8	102.7	92.7	92.7	90.2
Machinery and equipment production	111.1	204.2	183.8	218.5	107.0
Electric, electronic and optic equipment production	111.2	158.4	142.4	148.6	93.8
Production of transport vehicles and equipment	170.7	280.6	164.4	327.5	116.7
Other productions	87.2	129.6	148.7	158.8	122.6

Table III shows that dynamics of investment in basic capital by the types of activity in the period of 2005-2014 is uneven in whole.

During the analyzed decade (2004-1014) growth rates of investment in basic capital amounted to 141,3%, including: in mining operations – 211,9%, in processing industries – 171,6%. If the mining sector demonstrated growth in all KEA (Kinds of Economic Activity) being the parts of it, the achievement of higher growth rates in processing industries was restrained by such branches as woodworking and wood articles production (73,7%); pulp-and-paper production, publishing activity and printing industry (75,4%); rubber and plastic products manufacture (83,5%); metallurgical production and final metal products manufacture (92,7%). These results confirm the concept of slow development of processing industries comparing to mining ones.

The situation emerged in mining operations is more favourable than in processing industries in whole. First of all it is expressed in the fact that growth rates of investment in basic capital of mining sector during the period of crisis exceeded the indicator of processing sector (134,5% against 121,8%). Restoration of mining industry occurred much faster: growth in 2013 with respect to 2009 – 148,6%, while the processing industries showed the growth amounted to 141,1%. Moreover, in 2014 mining sector showed positive increase of 6,0% with respect to 2013 while the processing sector presented some negative growth of 0,1%.

A low growth rate of investment in basic capital in 2014 with respect to 2013 was observed in key industries of Perm Krai. In 2014 chemical production showed the growth of 100,4%; growth in coke and oil industry amounted only to 102,5%.

Relatively positive dynamics was shown by fuel and energy industries, textile and clothing manufacture where growth rates of investment in basic capital in 2014 in comparison with 2013 amounted to 127,8% and 135,1% correspondingly.

Economic potential and character of basic assets renewal at industry are indicated by the factor of renovation and wear of basic assets (Table IV-V).

Table 4. Growth rate of the factor of renewal of Perm Krai industry's basic capital, %.

KEA	2009/ 2005	2013/ 2005	2013/ 2009	2014/ 2005	2014/ 2013
In all KEA, including:	113.3	135.6	119.6	131.1	96.7
Mining operations	136.1	135.3	99.4	134.5	99.4
Processing industries	93.7	98.2	104.8	102.7	104.6

[a] calculated by the authors according to the information of Federal agency of state statistics, Perm Krai [23]

Analyzing the factor of renewal which characterizes the state of basic assets of Perm Krai industrial enterprises we see that negative tendencies of the Russian economy development have the farthest influence on processing industries: in 2009 growth rate of renewal coefficient amounted to 93,7%. And on the contrary, they influenced positively on mining operations (136,1%) in the context of this indicator analysis.

During 2005-2014 it was observed low growth rate of renewal coefficient in processing industry amounted to 102,7%, which is less than the average regional indicator (131,1%). On the contrary, mining sector has the indicator which is higher than the average regional level – 134,5

Low rates of attracting basic capital define the process of basic assets ageing. Estimation of the basic assets deterioration in Perm Krai is given in Table V.

Table 5. Growth rate of basic capital deterioration in Perm Krai industry, %.

KEA	2009/ 2005	2013/ 2005	2013/ 2009	2014/ 2005	2014/ 2013
In all KEA, including:	111.2	117.6	105.7	118.9	101.1
Mining operations	94.1	94.5	100.4	85.4	90.4
Processing industries	107.5	118.1	109.8	116.2	98.4

[b] calculated by the authors according to the information of Federal agency of state statistics, Perm Krai department [23]

Analysis of the calculation results testifies sufficiently high growth rates of basic assets deterioration in the regional industry in whole, i.e. it shows the conservation of negative wear accumulation tendencies.

In the considered period of time more favourable tendency by the indicator of basic assets deterioration has been emerged in mining industry than in processing one. The growth rate of the basic capital wear in these branches has been lower for the decade than on the average in the region - 118,9%. There are better indicators in the mining sector: growth rate of basic capital deterioration reduced by 16,6%, and at the same time the processing sector demonstrated the increase of wear by 16,2%.

Considering the indicators of 2014 in comparison with 2013 it is necessary to mention positive tendencies in mining and processing industries where the growth rate of deterioration level reduced by 9,6% and 1,6% correspondingly. The growth rate of deterioration coefficient in mining and processing industries in 2014 with respect to the previous year (90,4% and 98,4% correspondingly) takes on a value lower than the regional average (101,1%), and at that it is the characteristic feature of the indicators of 2009 in comparison with 2005.

When the share of worn basic assets is considerable economy incurs sufficient loss: first, ageing of buildings, constructions and equipment demands greater placement of funds in capital repairs; second, old facilities decrease products and services competitiveness, and technological backwardness involves production unprofitability.

In comparison of the diagrams presented in Fig. 1 and illustrating growth rates of key indicators of basic assets reproduction in Perm Krai industry it can be mentioned that growth rates of investment in basic capital is insufficient for mass renewal of basic assets. It is especially pronounced in processing industry. So, the investigation carried out by the authors confirmed their hypothesis.

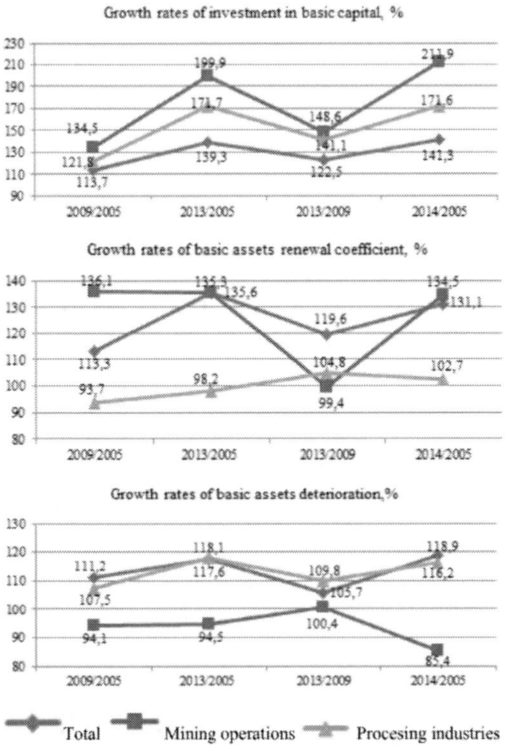

Fig. 1. Growth rates of key indicators of basic assets reproduction in Perm Krai industry.

Results of the research show that in Perm Krai the growth rate of investment in basic capital reduces while the level of deterioration of basic assets becomes higher. At that the situation in mining and processing industries is different: in mining industry it is more favourable while in processing one it is less favourable. At the same time the processes of reproduction in processing industries, and primarily in engineering are the key factor for neoindustrial transformation.

Insufficient growth rate of investment in basic capital being an obstacle for the efficient renewal of industrial basic assets forms the stable tendency of basic assets' deterioration increase in processing industry of the region, and this is the serious barrier for realizing neoindustrial transformation of economy.

The results of this study have practical value consisting in the fact that the analysis of the processes of reproduction which occur in the regional industry showed the necessity of enlarged investing of basic assets in processing industry in the nearest 2-3 years in order to provide their mass renewal. Relying on these results the following practical recommendations to the local bodies can be given:

• it is required from authorities to maintain investment activity of processing industries, to organize purposeful work of favourable investment climate formation in the region;

• it is necessary to promote investment in modern start-ups, to create favourable conditions for them.

On the assumption of this it is thought to be reasonable the realization of public promotion of parallel processes: firstly, removal of worn facilities from economic operation and secondly, increase of investment in basic assets aimed at renewal of conventional industrial production and conferring high-tech character to the fields identifying basic competences of industrial region corresponding to the neoindustrial model of economy.

Thus, the idea of neoindustrialization should transform to the reasonable industrial policy, to the roadmap of movement towards principally new industrial base consisting of complex and highly efficient technological renewal of enterprises' basic assets. And this demands the realization of managerial efforts aimed at support of investment activity of business.

References

1. A.R. Belousov, *Limits of self-development*. http://old.russ.ru/antolog/inoe/belous_o.htm#top.
2. A.R. Belousov, *Evolution of the Russian economy's reproduction system: from crisis to development* (MAKS Press, Moscow, 2006)
3. S.Yu. Glazyev, *Crisis is of artificial character*. http://communitarian.ru/novosti/ekonomika/sergey_glazev_krizis_nosit_rukotvornyy_harakter_27112014/.
4. Key statements of RF President in the message to the parliament. http://tass.ru/poslanie-prezidenta-federalnomu-sobraniyu.
5. J. Helbreit, *Modern industrial society* (AST, Moscow, 2004)
6. D. Bell, *Forthcoming post-industrial society. Experience of social prediction* (Academia, Moscow, 2004)
7. E. Toffler, *The third wave* (AST, Moscow, 2010)
8. E. Toffler, *Shock of the future* (AST Ltd., Moscow, 2003)
9. M. Kastels, *Informational epoch. Economics, society and culture* (SE HSE, Moscow, 2000)
10. S.S. Gubanov, *Sovereign break. Neo-industrialization of Russia and vertical integration* (Knizhny Mir, Moscow, 2012)
11. S.S. Gubanov, Economist. Russia, **9**, 3-27 (2008)
12. V.L. Inozemtsev, Russia in global policy, **6**, 85-98 (2011)
13. O.A. Romanova, Region economics, **2**, 70-80 (2012)
14. Account dictionary. Basic assets. http://www.slovar.plib.ru/dictionary/d30/7785.html.
15. A. Smith, *Study of peoples' nature and wealth* (Eksmo, Moscow, 2007)
16. D. Ricardo, *Fundamentals of political economy and taxation* (Gospolitizdat, Moscow, 1955)
17. K. Marx, *Capital. Criticism of political economy* (Politizdat, Moscow, 1967)
18. K Marx, F. Engels, *Collected works*. http://www.informaxinc.ru/lib/marx/31.html.
19. D.A. Baranov, *Theory of amortization and technological progress* (Gosfinizdat, Moscow, 1965)
20. S.K. Korkin, *Revolution of basic assets in socialist reproduction* (Kazan Univ., Kazan, 1990)
21. Yu.I. Lyubimtsev, *The cycle of reproduction and amortization of basic assets* (Economica, Moscow, 1973)
22. United interagency information-statistics system. http://www.fedstat.ru.
23. Perm Krai federal service of state statistics. http://permstat.ru.

SHS Web of Conferences 35, 01095 (2017) DOI: 10.1051/shsconf/20173501095
ICIE-2017

Improvement of methods of evaluation of investment projects in the context of import substitution

M.S. Kuvshinov[1,*], and N.S. Komarova[1]

[1]South Ural State University, Chelyabinsk, Russia

Abstract. Due to the imposition of sanctions, based more on geopolitical interests, there were additional essential restrictions on sources of external credits, use of import raw materials and technologies, possibilities of attraction to work of the qualified employees which do not have the Russian nationality, participation in the international technology and scientific projects. At similar restrictions of different resources investment decisions are made extremely carefully, and especially careful and reasonable efficiency evaluation of the investments is required. Traditionally used techniques of efficiency evaluation of investment projects not fully consider strategic interests of owners of industrial enterprises and investors. Decrease in risks requires higher quality of project evaluation. In this work the method of multi-level consistently complex integral assessment of investment projects on an indicator of the integral indicator of their efficiency implementing consecutive selection of investment projects from a portfolio on essential indicators of commercial effectiveness, efficiency of production business activities, a financial state and productivity of activity is provided. The developed method of a complex assessment of investment projects is implemented for practical application by the software product allowing to create on the basis of economic-mathematical modeling specific recommendations in economic practice of industrial enterprises during the developing and implementation of investment strategies. The conclusion that for the Russian economy in the conditions of import substitution accounting of all set of interests of investors and owners of the enterprises provides development and implementation of reasonable investment strategies is drawn.

1 Introduction

Paramount tasks of industrial enterprises always there is a development and modernization of production for ensuring competitiveness for the realization account of implementation and expansion of production volumes of new types of products [1, 2]. In modern market conditions on search of sources of investment additional factors in the form of sanctions from one states are imposed on others. Such sanctions are based not only on economic interests, but essential weight is purchased also by geopolitical interests of participants of the market. As a result of it there are essential restrictions on sources of external credits, use of import raw materials and technologies, possibilities of attraction to work of the qualified employees which do not have the Russian nationality, participation in the international technology and scientific projects [3 - 8].

Relevance of research

Under similar conditions of restrictions of financial, material and human resources investment decisions are made extremely carefully, and especially careful and reasonable efficiency evaluation of the investments is required. Traditionally used techniques of efficiency evaluation of investment projects developed for a long time [9 - 11] not fully consider strategic interests of owners of industrial enterprises by opportunities of use of the available or again purchased resources for implementation of the planned investment projects during the periods after their end. To reduce risks, including unpredictable completions investment projects require high quality assessments of their effectiveness.

The operating techniques of an assessment of investment projects are based on procedure of discounting and calculation of mainly base indicators of productivity of investments (the net present profit (NPV), an internal return rate (IRR), an index of profitability (PI), a payback period (PP)). These techniques [11] not always allow to make a reasonable choice for the most effective project. In the projects having high rates of productivity of investments, the investor can be not satisfied since he can be interested not only in profitability and an investment payback, but also other indicators such as reflecting strategic objectives of development of the enterprise regarding efficiency of production business activities, a financial state, social development, etc. which for each enterprise and the investment project can have the structure and the acceptable values.

* Corresponding author: msk1954@mail.ru

3 Problem statement

At accomplishment of a choice of one investment project from a portfolio of projects in the conditions of sanctions restrictions there are questions of the approved application of the main indicators of productivity of project data (NPV, IRR, PI, PP): in projects with the different volume of investment, identical NPV values can be received; occurs in the projects having high NPV value situations when IRR value is much lower, than in projects with lower NPV value; in projects with the different volume of investment and NPV, almost identical PI values can be received; the payback period does not estimate cash flows outside a point of an investment payback and is used only in addition to other criteria of efficiency.

In such situation there is a problem of improvement of quality of examination of efficiency of investment projects on the basis of the complex assessment reflecting, both interests of investors, and interests of the enterprises implementing project portfolios, and based not only on comparison of traditional indicators (NPV, IRR, PI, PP), but also considering all set of the criteria interesting the investors united in a uniform integral indicator of efficiency evaluation of the considered projects. Expansion of completeness of examination increases definiteness of a choice and reduces risks of justification of a choice.

Improvement of techniques of an assessment in this sphere allows to avoid more than probable essential economic losses during the developing and implementation of investment projects and defines relevance of work.

Works of many foreign [2, 12] and Russian authors [13 - 16], including works of a historical retrospective are devoted to problems of efficiency evaluation of the investment projects aimed at the production development [9].

The purpose of work is improvement of methodical tools of a complex assessment of investment projects on the industrial enterprises working in the conditions of sanctions restriction of resources.

4 Development of evaluation methods

Flexible and effective management of investments taking into account action of sanctions factors requires an assessment of their efficiency, as for all term of implementation of the investment project, and on each specific current step during all stages of project lifecycle, including an end stage.

The variety of options and the parties of implementation of investment projects, and their results, creates variety of options of an assessment of their efficiency.

Performance indicators of projects are calculated on the basis of forecast cash flows and depend on regulation of the income which, is defined subjectively and influences their size. Each of the indicators has its own purpose and reflects its incorporation of side, but they have a different dimension and do not always have a direct interdependence of the patterns. Therefore each separate indicator cannot be used as the only criterion of efficiency.

Flexible and effective management of investment processes taking into account changes of the external and internal environment is reached by improvement of quality of efficiency evaluation at the expense of the approved choice of the measure values satisfying sets of requirements of investors and used in efficiency evaluation of projects, definiteness and a correctness of use of the main criteria of efficiency, accounting of specific features of activity of industrial enterprises when carrying out efficiency evaluation of projects from a portfolio and practical recommendations for a choice of the best of alternative options.

More complete examination of investment projects has to be provided due to use of the integral estimates considering both traditional, and other indicators representing essential interest for the investor during all stages of lifecycle including project completion.

Necessary completeness of examination is provided on the basis of an integral quantitative assessment which forms when using of the developed method of multi-level complex sequential estimation of efficiency of investment projects. The structure, stages and functions of such method are defined proceeding from accounting of modern specific features of implementation of investment projects in the conditions of action of the sanctions connected with modernization and technology updating of production, performance improvement of work and quality of products, implementation of innovations, decrease in labor input and costs [17].

A key element of its evaluation of the effectiveness of the investment project is risk monitoring their unpredictable conclusion at all stages of implementation. Incorrectly certain project scope, limitation of determination and risk management, an incorrectness of determination of main assumptions, a formalism or insufficient competence of procedure of carrying out efficiency evaluation of projects are basic reasons of risks of unsatisfactory end. Thus the factors influencing on not predicted project completion, influence the size of cash flows on them on which, in turn, performance indicators directly depend [18].

The variety, heterogeneity and divergence of influence of the initial indicators creating efficiency of investment projects, taking into account risks of the estimates using traditional performance indicators, and unanticipated risks completions when you post all the essential interests of an investor determine the tools implement multilevel method integrated coherent assessment. As such tools acceptability indicators for investors of the used initial indicators, efficiency indicators in the directions of an assessment and the integral indicator of efficiency of the investment project in general act. Thus unacceptability of the current results of an assessment at any stage of work of the used method has to have impact on the level of requirements of the investor to the considered indicators. In connection therewith are defined structure of stages of complex efficiency evaluation of investment projects where consistently from a portfolio constantly decreasing sets of the most acceptable projects are selected.

At the first stage efficiency evaluation of projects in general on traditional indicators of productivity of investments (NPV> 0, IRR> E, PI> 1, PP <T, where E – a discount rate, T – project term is carried out). Accomplishment of these conditions in different degree for several projects from a portfolio creates a problem of a choice of a set of the NPV, IRR, PI and PP values satisfying all set of essential requirements of investors.

To ensure that the integrated impact of investment performance indicators (NPV, IRR, PI, PP) on the effectiveness of the project on their basis is determined by a quantitative indicator of commercial efficiency (EI_{ce}).All calculations were carried out based on two real industrial enterprises when using the group of experts from the interested specialists. Using a method of analysis of hierarchies defined the composition of the most relevant indicators, coefficients of the weight and importance of each indicator used in the formation of EI_{ce}, for the analyzed enterprises. Were their part: the net present profit, an internal return rate, an index of profitability and a payback period with weight coefficients respectively 0,338; 0,406; 0,160 and 0,096. Comparability of indicators is reached by reduction of indicators to uniform basis [19]. As a result, expression of the indicator of the generalized commercial effectiveness of EI_{ce} has an appearance:

$$EI_{ce} = U_{i=1}^{n} IA_i \Big/ \sqrt{\sum_{i=1}^{k} v_i \left(1 - I_i\right)^2 + \sum_{i=k+1}^{n} v_i I_i^2}, \qquad (1)$$

where IA_i – indicators of the acceptability of the considered indicators; I_i – the given values of the considered indicators on the current step of calculation; v_i – weight coefficients; n – total quantity of the considered indicators; k – the quantity of the considered indicators which increase leads to increase in EI_{ce}, and leads increase in the others to reduction of EI_{ce}..

In turn, the given indicators are defined with production scale as:

$$I_{ij} = p_{ij} / \max\left(p_{ij}\right), \qquad (2)$$

where p_{ij} – i values of an indicator for j of the project or option of its implementation; $max(p_{ij})$ – the maximum i value of an indicator from all number of the compared projects or options of their implementation.

The structure of the considered indicators is open system, considers essential interests of specific owners and investors of the specific enterprises and can be changed in any party. As a part of the increasing indicators for the considered enterprises the net present profit, an internal return rate and an index of profitability, and as a part of the reducing indicators – a payback period.

Indicators of the acceptability of indicators of IA_i accept "acceptable" (1) or "unacceptable" (0) values according to the acceptability for the investor. If, at least we do not accept one indicator of the acceptability, EI_{ce} = 0 and it is necessary to refuse implementation of the investment project at this level of requirements of the investor, or to approve with it the level of requirements. If, all indicators of the acceptability conforms to requirements, the stage is necessary complete.

The generalized indicator of commercial effectiveness *of* EI_{ce} allows to resolve issues of approval of a choice of the measure values used in efficiency evaluation of investment projects with set of interests of investors, and also a choice of the main indicator on which it is necessary to draw a conclusion on efficiency of the project as at its determination all standard indicators of commercial effectiveness in total, taking into account their weight importance and an orientation of influence were considered.

In the conditions of action of sanctions modernization and technology updating of production, performance improvement of work and quality of products, implementation of innovations, decrease in labor input and costs are a necessary condition of effective functioning of industrial enterprises on achievement of cumulative strategic objectives of development, including the post-project periods. Therefore even in the projects having high rates of commercial effectiveness of investments for increase of completeness of examination and decrease in risks it is necessary to consider the indicators reflecting efficiency of the production business activities (*PBA*), financial state, etc. which for each enterprise and each investment project can have the structure and the acceptable values.

Indicators of commercial effectiveness have no direct interdependence with *PBA* indicators. In connection therewith, the second obligatory stage of examination of investment projects is impact assessment of its implementation on production business activities of the enterprise on the basis of calculation of the quantitative indicator of the generalized efficiency of production business activities of EI_{pba}.

The generalized indicator of efficiency of influence of EI_{pba} allows to judge influence of implementation of investment process on market provision of the enterprise, efficiency of use of fixed assets, resources and, at the expense of it, on a profit gain. The EI_{pba} indicator is based on components of indicators, taking into account their relevance coefficients v_i.

The significance of the values of the coefficients in the Group of indicators on the impact of the investment process in the enterprise economical activity are defined similarly to the previous case. For the analyzed enterprises coefficients of updating of fixed assets, capital productivity, the impact of the salary, profitability of sales, a market share, scope change of sales, profitability of products, investment profitability, a return materials and power return with weight coefficients respectively were their part: 0,16; 0,204; 0,120; 0,164; 0,098; 0,055; 0,071; 0,057; 0,041 and 0,028.

The indicator of efficiency of influence of implementation of investment process on production business activities of the EI_{pba} enterprise is defined with weight coefficients, an orientation of influence and heterogeneity of the indicators taken into account on expression:

$$EI_{pba} = U_{i=1}^{m} IA_i / \sqrt{\sum_{i=1}^{m} v_i (1-I_i)^2}, \quad (3)$$

where m – quantity of the considered indicators which increase leads to increase in EI_{pba}. Other designations correspond to a formula (1) for coefficients of updating of fixed assets, capital productivity, a impact of salary, profitability of sales, a market share, scope change of sales, profitability of products, investment profitability, a impact of materials and power return

The assessment of financial productivity on the basis of calculation of the generalized indicator of financial productivity of EI_{fp} allows to consider financial return on the invested capital, cash flows, a property gain as a result of implementation of the investment project, change of liquidity and financial stability of the enterprise, and is the third obligatory stage of examination of projects.

Weight coefficients for the considered indicators creating EI_{fp} are calculated by method of the analysis of hierarchies to similarly previous cases. Coefficients of a gain of revenue, gain of profit, a gain of property, overall liquidity and financial stability with weight coefficients respectively were their part for the analyzed enterprises: 0,105; 0,427; 0,099; 0,184 and 0,184. Comparability of indicators is reached by reduction of indicators to uniform basis to similarly previous stages.

The generalized indicator of influence of implementation of investment process on a financial state and productivity of activity of the EI_{fp} enterprise is determined by expression:

$$EI_{fp} = U_{i=1}^{l} IA_i / \sqrt{\sum_{i=1}^{l} v_i (1-I_i)^2}, \quad (4)$$

where l – quantity of the considered indicators which increase leads to increase in EI_{fp}. Other designations correspond to a formula (1) for coefficients, a gain of revenue, a gain of profit, a gain of property, overall liquidity and financial stability.

Improvement of quality of an assessment of investment projects at the expense of the most complete examination of their efficiency on the basis of use of the integral estimates considering indicators of productivity of investments (NPV, IRR, PI, PP), and also other performance indicators of production business activities, a financial state and productivity of the enterprise which are of interest to the investor allows to reduce risks of efficiency evaluation of projects, including risks of their unpredictable completions. Determination of the integral indicator of efficiency of the investment projects IEI is the fourth obligatory stage of examination.

Examination is carried out in the comparative analysis of several investment projects from a portfolio on a rating of integral indicators of efficiency. In case of an assessment of one investment project the comparative analysis is made with the virtual project analog to which on expert evaluation the project analog with the average market indicators accepted in the current sales terms is accepted. In the presence of several options of implementation of one investment project they are accepted to a portfolio. Thus in all options of calculations at each intermediate settlement stage monitoring of probability of not predicted end is carried out. For the chosen investment project, in the presence of opportunity, adjustment of sizes of the most significant parameters influencing increase in an assessment of the integral indicator of efficiency can be carried out.

The selected investment project which most meets expectations of investors on all set of the accepted performance indicators is result of work of the developed method of a complex assessment.

Actually integral efficiency evaluation of projects forms on the basis of the private generalized indicators: indicator of commercial effectiveness of EI_{ce}, indicator of efficiency of influence of implementation of the investment project on production business activities of the EI_{pba} enterprise and indicator of influence of implementation of the investment project on a financial state and productivity of activity of the EI_{fp} enterprise.

The integral indicator of efficiency of the investment project implements the second level of a method, is a complex assessment, is defined according to expression:

$$IEI = 1 / \sqrt{(1-EI_{ce})^2 + (1-EI_{pba})^2 + (1-EI_{ce})^2}, \quad (5)$$

In case of comparison of the estimated project with the project analog and strategic expediency of its implementation the scale of levels of the acceptability for the integral IF_{IEI} indicator is recommended: 0 … 0,66 – risky; 0,67 … 0,75 – satisfactory; 0,76 … 0,9 – acceptable; 0,91 … 1,0 – high where IF_{IEI}^{max} = 1 corresponds to a maximum level of IF_{IEI} of ap of the project analog.

Action of the offered method is implemented by consecutive use of settlement stages when monitoring the acceptability of values of the used parameters according to requirements of interests of investors.

Project evaluation from a portfolio in general on the commonly accepted indicators of productivity of investments carries out allocation of the projects necessary for the enterprise for business development in the conditions of action of sanctions with use of instruments of strategic management. After making decision on compliance of the investment project to strategic objectives of the enterprise transition to project evaluation in general on the commonly accepted indicators of productivity of investments (NPV, IRR, PI, PP) is carried out.

The project scope is defined proceeding from the level of profitability, capacity and tendencies of the market, and also with the analysis of activity of main competitors.

At a choice of main assumptions (according to the decision of developers): in long-term projects the annual rate of discounting should be calculated according to a forecast rate of inflation of every year; the discount rate for the different markets needs to be calculated according to different risk level for the different markets; in real economy as the minimum regulation of the income it is necessary to accept industry profitability of the capital; when using loan financing of projects as the minimum regulation of the income it is necessary to use the size of an annual interest rate on the credit; when using of the mixed financing of projects it is necessary to apply a method of the weighted average capital cost (WACC) to determination of a discount rate; when carrying out sensitivity analysis of the project it is necessary to define sensitivity of the project to different discount rates for identification of limits of stability of the project and acceptable regulation of the income of the project. It is necessary to consider transition of Central Bank of the Russian Federation to a key rate.

For duration of a settlement period the term of return of investments, or the period during which the investment project will make required profit is assumed as a basis. However, thus the increase in lifecycle or term of implementation of such project reduces reliability of calculation and an assessment of performance indicators. It is necessary to consider possibility of flexible decrease (increase) of the production volume and implementation according to demand in subsequent periods reflecting dynamics of probable emergence and removal of sanctions.

In the course of an assessment of long-term projects, owing to irregularity and non-equivalence in time of cash flows, for increase of reliability of efficiency evaluation it is necessary to use more rhythmic step of calculation equal to half-year, quarter or month.

In the presence of a positive integral assessment the decision on expediency of project implementation or on a choice of one of the compared versions of projects of an assessment by criterion of *max* $\{IEI_i\}$ is made. In the presence of possibility of adjustment of the most significant parameters influencing increase in *IEI* taking into account interests of investors their adjustment at the sixth stage with the subsequent transition to implementation of the selected investment project is carried out.

The project chosen on the basis of the offered method to implementation most fully considers all set of interests of investors.

For correction of the most significant the indicators having the greatest weight as a part of the generalized indicators of efficiency and *IEI* which are most influencing the integral indicator of efficiency are selected. Such approach allows to develop actions for management of the factors influencing efficiency of the investment project according to the forecast of change of its integral indicator of efficiency and to create practical recommendations about achievement of target values according to interests of investors.

For determination or approach to target values of private indicators at performance management of projects the method of interval estimates is used. The private indicator of p_{ij} has to is in "a trust interval" which is the solution of an inequality:

$$\overline{p_{ij}} - t\sigma / \sqrt{n} < p_{ij}^{a} < \overline{p_{ij}} + t\sigma / \sqrt{n} , \qquad (6)$$

where p_{ij}^{a} – acceptable i value of an indicator for j of the project or option of its implementation; $\overline{p_{ij}}$ – mean value of i an indicator for j of the project or option of its implementation; σ – an average quadratic deviation of i an indicator for j of the project or option of its implementation from mean value of i an indicator for j of the project or option of its implementation; t – value of argument of function of Laplace; n – amount of forecast i values of an indicator for j of the project or option of its implementation (selection volume).

5 The practical significance and results of testing

The received results of calculations of confidential intervals allow to reveal admissible fluctuations of private indicators of efficiency and to make the decision on implementation of investment projects. When you run a select one project from a portfolio of preferred options that match *max* $\{IEI_i\}$. interval estimation and allows you to set the boundaries of the ranges of allowable fluctuations of the best composed performance indicators. The calculations which are carried out as a result of economic-mathematical modeling on materials of experimental approbation of the developed technique will be approved with the commonly accepted provisions and confirm them taking into account use of set of the indicators essential to investors in a complex.

Developed by authors a method of a multi-level sequential integral estimation of investment projects it is used when developing the software product based on the tabular MS Excel [20] processor by means of which the assessment of investment projects of industrial enterprise LTD "Uralspetstrans" was carried out (Miass) where experimental approbation of the developed method was carried out. The complex evaluation method allowed to make a reasonable choice for benefit of the most effective project (table I).

It is established that using the commonly accepted evaluation methods of efficiency of projects on a set of ratios of NPV> 0, IRR> E, PI> 1, PP <T without integral approach can be made the wrong decision on a choice of one project from a portfolio. Multi-level complex an evaluation method is more exact and allows to make a reasonable choice for benefit of the effective project.

Table 1. Approbation of results of research.

Indicator	**Investments, thousand rubles.**	**Net profit in a year, thousand rubles.**	**NPV, thousand rubles.**	**IRR, %**	**ID**	**PP, month.**	**IEI**
Production of automobile tanks	2140	721	1602	61	1.7	21	2.1
Production of hoses of automobile cooling systems	1558	901	895	51	1.6	18	2.4

6 Conclusion

Thus, to reduce risks, including unpredictable completions investment projects you want to improve the quality of assessments of their effectiveness on the basis of their expertise more fully through the use of integrated assessments that take into account both traditional and other indicators that represent a significant concern for investors. The offered multi-level method of a complex assessment of investment projects of industrial enterprises allows to estimate efficiency more fully by means of the uniform integral indicator considering set of interests of investors, providing thereby development and implementation of reasonable investment strategies.

The work was supported by Act 211 Government of the Russian Federation, contract № 02.A03.21.0011.

References

1. Yu.V. Babanova, V.V. Tomasheva, Finance and credit, **46**, 27-32 (2012)
2. Yu. Brigkhem, M. Erkhardt, *Financial management* (Transfer with English SPb, St. Petersburg, 2007)
3. R. Asankin, *Europe and Turkey will close the markets for the Russian metallurgists*. http://www.rbc.ru/business/28/01/2016/56aa053f9a7947cee59701a6.
4. M.S. Ivanova, World of Science Internet magazine, **1** (2015). http://mir-nauki .com/PDF/26EMN115.pdf.
5. V.S. Luganov, *Economic effects for Russia from economic sanctions*. http://itar-tass .com/opinions/1597242.
6. V. Tkachyov, *Investments in the conditions of sanctions*. http://www.tn.ru/about/press/news/ investicii v uslovijah sankcij.
7. N.V. Klimova, S.Yu. Murashkina, *Influence of sanctions on the investment climate and industrial policy of Russia*. http://cyberleninka.ru/article/n/vliyanie-sanktsiy-na-investitsionnyy-klimat-i-promyshlennuyu-politiku-rossii.
8. The federal law of May 19, 2010 No. 86-FZ "About modification of the Federal law of July 25, 2002 No. 115-FZ. http://pravo.gov.ru/ proxy/ips/?docbody=&nd=102138471.
9. A technique of determination of cost efficiency of implementation of new equipment, mechanization and automation of production processes in the industry.http://www.libussr.ru/doc_ussr/usr_5750.htm.
10. The federal law of February 25, 1999 No. 39-FZ. http://pravo.gov.ru/proxy/ips/?docbody=&nd=102058332.
11. Methodical recommendations about efficiency evaluation of investment projects no. VK-477 of 1999.http://www.consultant.ru/document/cons_doc_LAW_28224.
12. 15 Causes of Project Failure by M. Symonds. http://www.projectsmart.co.uk/pdf/15-causes-of-project-failure.pdf.
13. A.Yu. Ruban, *Forming of the controling mechanism investment projects* (Ufa, 2010)
14. Electronic consulting. Factor analysis. http://www.dist-cons.ru/modules/DuPont/section2.html.
15. S.L. Glekov, *The mechanism of adoption of investment decisions taking into account alternative" efficiency – reliability* (Moscow, 2011)
16. M.G. Litke, *Economic evaluation and management of the innovation projects of small and medium business at the mesolevel* (Chelyabinsk, 2012)
17. N.V. Matveev, *Methods of a complex assessment of investment projects* (St. Petersburg, 2007)
18. M.I. Rimer, *Economic evaluation of investments: The textbook for higher education institutions* (Peter, St. Petersburg, 2014)
19. M.S. Kuvshinov, Economic analysis: theory and practice, **7**, 30-38 (2014)
20. M.S. Kuvshinov, N.S. Komarova, *Calculation of cost efficiency of investment projects of industrial enterprises: program* (RU state registration program, 50201450091)

Modeling prices of wholesale market of electric energy and power by the example of the UPS of the Ural

V.G. Mokhov[1,*], and *T.S.* Demyanenko[1]

[1]South Ural State University, Chelyabinsk, Russia

Abstract. The article oversees forecasting model for deviations of the balancing market index and day-ahead market index according to the maximum similarity sample for different levels of approximation in the context of positive and negative time-series value. The model was being tested on the factual data of the Integrated Power system of the Ural, Wholesale market for electricity and power of Russian Federation. Describes the price formation on the day-ahead market and the balancing market index. The necessity to use accurate forecasting methods consumption and prices of electrical energy and power to reduce penalties when the electric power industry entities on the energy exchange. The testing of mathematical models to predict the balancing market index deviations and day-ahead market based on a sample of maximum similarity with certain approximation equations for positive and negative values gave the prediction error of 3.3%.

1 Introduction

Electric power industry is currently one of the leading sectors of the Russian economy. According to the results of electric power production in the year 2013 Russia took the 3d place. Almost 10% of country's GDP falls to the share of electric power.

The obvious question is whether the electricity is a commodity? Or does it refer to a different object of civil rights? Currently, in the Russian Federation, according to the Civil Code of the Russian Federation, the contract power - a kind of contract of sale [1].

Such a wide distribution of electric energy, and its specificity as a commodity has led to the need for accurate regulation of tariffs.

The market, which allows developing competitive relationship inside the industry, is acknowledged to be the most effective for building relations in electric energy. All the stated above is confirmed by the before-reform state of the electric power market in Russia, when the monopoly energy production conditions and state tariff regulation has led to unsolvable economic contradictions, among of which are the following ones:

- non-payments crisis;
- imbalance between tariffs and production/sales cost;
- quicker growth of the prices on materials and fuel, than on sold electric power;
- cross-subsidization.

Ultimate retail consumer purchases electric energy on competitive electric energy market, alongside with the product itself he is compelled to compensate the associated costs as well [2]. These costs exist because electrical power supply is not possible without supplementary services. These services with time has been formed into the separate markets:

- capacity market;
- derivatives market;
- ancillary services market;
- financial transmission rights market [3].

In today's existing view WMEP operates from September 1, 2006, when the Government of Russian Federation introduced new rules of functioning of the wholesale market. At present, the Russian wholesale electricity market is the youngest in the world and least explored.

In the electricity market, the planned consumption volumes are formed on the basis of bilateral agreements. In reality, the actual amount of electricity is almost always different from the routine. Deviations from targets are sold on the balancing market (BM), and the System Operator of the Unified Energy System of Russia (SO) regularly conducts a competitive selection of applications providers, based on the current forecast electricity consumption.

The causes of deviations are different and qualified own and external initiatives. Own initiative arises because a market participant actions (customer or supplier), external is a result of the command with or accident that led to the forced change in the mode of production and consumption of electricity [4].

Despite the fact that since the beginning of the 50s of the last century is not disproved the concept of information efficiency of stock market [5], a practice encourages researchers to find the direction of the most variable models to predict stock market prices. In Russian conditions, the task of forecasting of the basic parameters of the electricity market is one of the

* Corresponding author: mokhov50@mail.ru

most urgent and practical importance of the fundamental challenges of financial planning subjects of electric power and to optimize their activity on the wholesale market.

Today, developed a large number of time series forecasting models, accumulated data base of real values. Ongoing work on the development of new models and improve the computing platforms and systems. At the same time requirements for the accuracy of forecasting and economic governance are becoming more stringent, so the task of forecasting time series is not only improved, but also more complicated with each passing day [6].

The urgency posed obvious problems, because due to the lack of precision of forecasting models energy consumers are losing their competitive edge, which leads to a deterioration in financial and economic performance of electricity market participants and even crises.

In this paper will propose a mathematical model for predicting the deviation index balancing market (BMI), the market rate on the day-ahead market (DAM) on a sample of maximum similarity with different equations approximations for positive and negative values.

2 The aim and objectives of the research

As mentioned above the aim of this study is to develop a mathematical model predicting deviations from BMI to DAM.

Research objectives:

• research method of formation of the market rate on the day ahead;

• research method of formation of tariffs for the balancing market;

• development of mathematical prediction models;

• testing model, confirmation or refutation of its effectiveness.

In this paper we will consider the formation of tariffs in the current competitive market of electric power industry of the Russian Federation.

3 Formation of prices on the day-ahead market

On the wholesale market of an organized system of agreements between the subjects of the wholesale market. In the electricity market, there are several types of commercial transactions:

• central planning regimes, in which prices are determined based on the result of the auction bids submitted by suppliers and buyers, as this type is called "day-ahead" market or spot market (the DAM market);

• bilateral agreement for the supply between generating companies and buyers of electricity wholesale market entities are free to choose the counterparties to such contracts, but for such a contract, suppliers are required to provide, and buyers purchase amount stated in the contract;

• combined commercial transactions in which generating companies may enter into a tripartite agreement and to participate in centralized planning, submitting price bids.

Wholesale electricity market should include providing sufficient power to cover the electricity demand with regard to the necessary provision of electric power using the most effective resource-saving technologies [7]. Thus, providers are required to maintain power generation equipment in a state of readiness for electric energy production. Monitoring the timely and proper implementation of the investment programs of generating companies formed as a result of the power of trade is carried out by the system operator. Selection of power also provides by the system operator.

The "day-ahead" market is organized in form of bidding auction. The main goal of this auction is to establish prices and volumes of electric energy with maximum mutual advantage. Each pricing application contains 24 of the planned appointment of electricity and 24 values of the price at which the participant is willing to buy/sell a specified amount.

Price bids submitted to the auction organizer for the day. Sellers indicate in the application the minimum price at which they are willing to sell a specified amount of electricity, and the buyers, respectively, the maximum price at which they are willing to buy.

In addition to the application with the specified price, spot price acceptance bids exist, they mean that the buyer is ready to purchase a specified amount of power at any price. Similar applications exist for suppliers, where they are willing to sell a specified amount at any price, even zero.

After the adoption of all applications from the buyers of the auction organizer ranks them in descending order. This means that the buyer is always ready to buy a specified amount at a price lower than he indicated in the application. Thus, a demand curve, which has a stepped form and includes milestones applications. Application vendors are sorted in ascending order of price, thereby forming the supply curve. And the demand curve and the supply curve will undergo spot price acceptance in the first place.

The intersection of demand and supply curves gives the equilibrium price and quantity [8]. Participants of the auction price in applications which coincided with the equilibrium, called pricing. Applications buyers who have the right of the equilibrium price can not buy anything at a cheaper price, as well as suppliers have nothing to sell at a higher price. Scheme of the equilibrium price is shown in Fig. 1.

This auction is closed, as neither buyers nor sellers are not aware of the content of the price submitted by other parties, and can not predict prices and volumes in the past the market. Pricing in this auction is a margin that is the most effective way of organizing the bidding in a competitive environment.

From the foregoing it follows the importance of the accuracy of prediction of consumption and production of electricity. It is clear that errors in forecasting the volume values occur regularly, so in addition to the day-ahead market there is a balancing market, which represents a market of deviations of actual hourly production and consumption of electricity from the planned trading schedule [9]. The volumes on the balancing market are regulated by the System Operator by reducing or increasing the production of electricity. Market participants who have made mistakes in planning can buy or sell electricity on the balancing market, but the price it is very disadvantageous. Fig. 2 shows the deviation of the market price of the day ahead and balancing market index on the example in December 2014, the data are taken from the official website of the wholesale electricity and capacity market (http://br.so-ups.ru/).

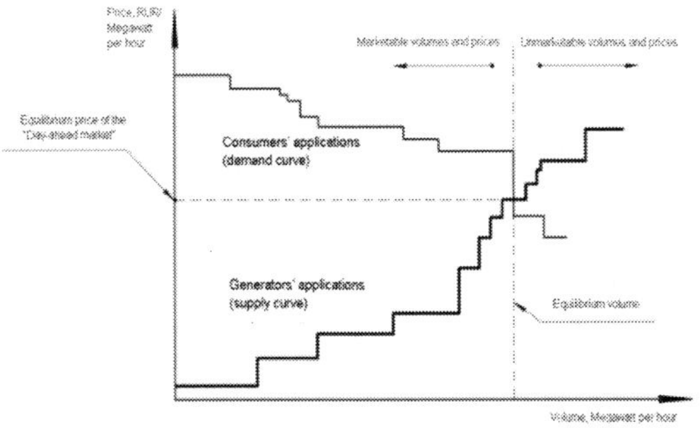

Fig. 1. Forming the equilibrium price of the day-ahead market.

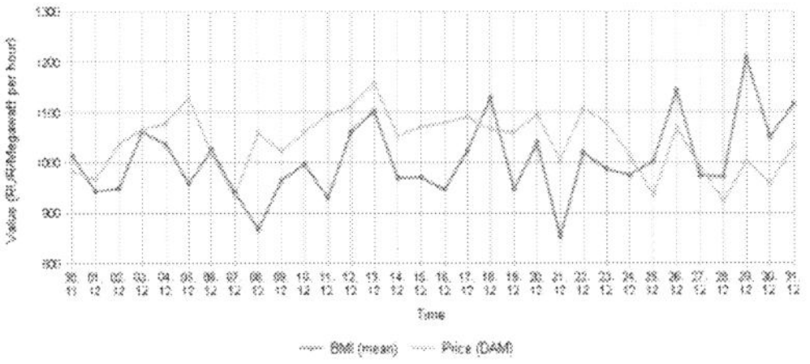

Fig. 2. Price deviation of BMI (Balancing Market Index) and DAM (Day-ahead market) from 30.11.2014 till 31.12.2014 by the example of the UPS (Unified Power System) of the Ural.

4 Formation of prices on the balancing market

Balancing market prices are formed on the basis of market supply and demand curves, in the event of changes in demand, if you need more or less electricity. If the actual power consumption of the power grid has exceeded the plan, the system operator turns considers applications of generators, starting with the immediate right of the equilibrium price. Application generators are joining is not satisfied buyers need electricity (Fig. 3). Price of the balancing market in this case is equal to the point of intersection of the required volume of customers with the supply curve.

If there is a reverse situation and the actual power consumption of the power system was below the plan, the system operator in turn reduces the load power plants, starting with the immediate left of the equilibrium price. Price of the balancing market, in this case will be determined by the price of an application generator, whose stage was at the intersection with the required volume of customers (Fig. 4).

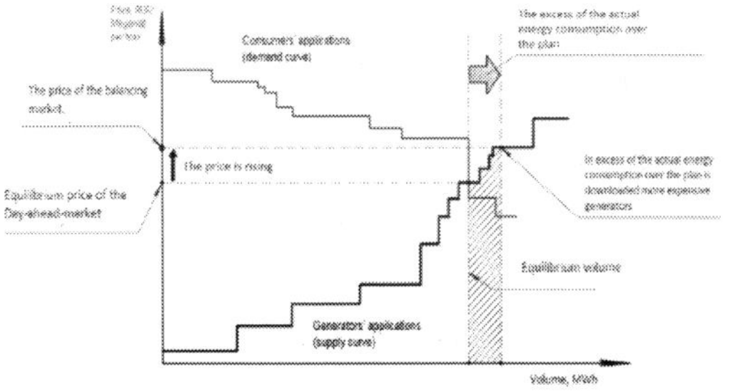

Fig. 3. Pricing on the balancing market in excess of the actual consumption of the power grid on the plan.

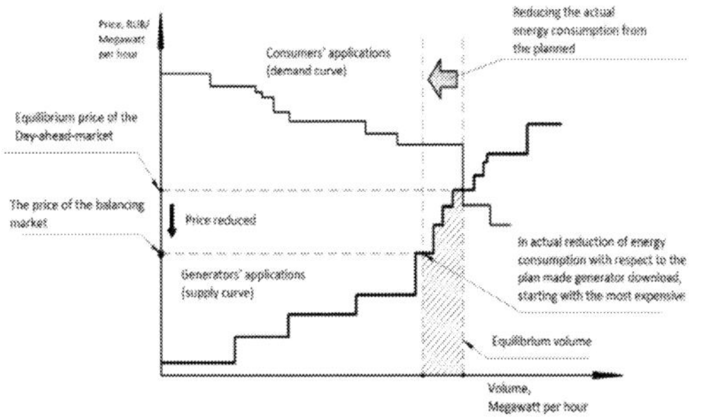

Fig. 4. Pricing of the balancing market with a decrease of actual consumption from the power grid plan.

If any deviations from the target values of energy price becomes unfavorable, compared to the price on the day-ahead market. That is, when errors in planning for consumers, they buy electricity at a higher price and return situation when mistakes in the planning of the suppliers when they sell electricity at a lower price [10]. Thus, the importance of the accuracy of the prediction comes to the fore, as the plan with the smallest errors lead to a reduction in costs and, consequently, to increase profits. Below is a mathematical model for predicting deviations from BMI to DAM sample of maximum similarity.

A large number of studies in the field of forecasting tariffs directly WMEP, but the foregoing should be the importance of forecasting not only tariffs DAM, but deviations BMI from the DAM. With a negative deviation, that is, the BMI is less than the DAM, there is a discharge of expensive generators will work cheaper, so if the electricity supplier wants to participate in the auction, then, knowing the predicted deviation, he applies for SO with valuable lower than forecast DAM. Consumers with a negative deviation to guarantee the participation in the auction may submit the application on a target price below the DAM on the value not exceeding the forecast deviations [11]. With a positive deviation occurs loading more expensive generators to participate in the auction generators may apply to the price above the value of a target DAM forecast deviations.

Thus, the prediction of the BMI deviations from the DAM in tandem with the projected values of DAM rate and volume of consumption will increase economic benefits and efficiency of both producers and consumers.

Thus, the prediction of the BMI deviations from the DAM in tandem with the projected values of DAM rate and volume of consumption will increase economic benefits and efficiency of both producers and consumers [12].

Studies to evaluate the effectiveness of existing prediction models have shown that the most adequate model to show maximum similarity the sample, which was taken as a basis for the proposed model below [13].

5 Mathematical model

Finally, complete content and organizational editing before formatting. Please take note of the following items when proofreading spelling and grammar:

The result of the study was the mathematical model for predicting time series deviations based on a sample of maximum similarity with different equations approximations for positive and negative values, which is presented below [14].

Small scale integration is often used for prognozing the time series [15], that is to say predicting not the series itself or the process, but its change or accession. In other words we get the series $\tilde{Z}(t)=Z(t)-Z(t-1)$, where $Z(t)$ is an original time series, and $\tilde{Z}(t)$ is a series of increase or difference of the adjacent indexes [16]. In this model it is offered to modify first order integration method and use the deviations of BMI and DAM time series (Fig. 2). This method would help to predict the predict the possible electric power cross-pickings and, therefore, to enhance the accuracy of the already existing methods of forecast.

Now we get to the formal description of the model. Let's introduce all the necessary nomenclature:

$Z(t)=Z(1),Z(2),...,Z(T)$ is a time series that represents the volume of consumed energy in the moment of time T.

$Z_t^M=Z(t),...,Z(t=M-1)$ is a piece of the time series Z(t), where M is the length of the sample, considering that $M\in\{1,2,...,T\}$.

The length of the sample is defined as the maximum value from the stable range MAE for prediction value of each of $M\in[P,\frac{0,3Z_t}{P}]$, where P is the length of prognosis. The prognosis is built upon the model, described below. MAE (mean absolute error) is calculated through the following formula:

$$MAE=\frac{1}{n}\sum_{i=1}^{n}|y(i)-\hat{y}(i)|,$$

where t is a moment of time, a datum point in the sample, besides $t\in\{1,2,...,T-M+1\}$

Modeling of the time series with the help of samples is based on the supposition that a time series is a sequence of samples [17]. Grounding on this supposition, we assume 2 samples of the equal length and belonging to one and the same time series with difference in datum points of sequences k [18].

$$Z_t^M=Z(t),...,Z(t+M-1)$$
$$Z_{t-k}^M=Z(t-k),...,Z(t-k+M-1)$$
$$k\in\{1,2,...,t-1\}$$

Let's consider the prognosis algorithm.

Defining the new history sample:

Z_{T-M+1}^M is a sample of a time series, which values are preceding the moment of prognosis T.

Defining the maximum similarity sample:

For each value of time delay from diapason $k\in\{1,2,...,T-M-1\}$ the task of approximation of the sample given is solved $Z_{T-M+1-k}^M$, in other words the sample is being sequentially compared to all other values of time sequence with shift to unitary vector.

Calculating the approximation through classic linear model would be improper, because the values of time series can be either positive or negative. The acquired sample is divided into 2 rows: $Z_{T-M+1-k}^+\geq 0$ - for positive values of the sample and $Z_{T-M+1-k}^-<0$ for the negative.

Than we calculate the approximate value of the sample for positive values of the sample and separately for negative using the following formula:

$$Z_t^M=\alpha_1 Z_{t-k}^M+\alpha_0 I^M,$$

where α_1 and α_0 are coefficients, I^M is a unitary vector.

It is necessary to define the values of coefficients in order to:

$$\sigma^2=\sum_{t=0}^{M-1}(Z(t+i)-\hat{Z}(t+i))^2\rightarrow min$$

in other words in order for the square of deviations of model values from the real ones to be minimal [8].

Function of approximate error looks like this:

$$S_k^M(\alpha_1,\alpha_0)=\sum_{t=0}^{M-1}\sigma_i^2=\sum_{t=0}^{M-1}(Z(t+i)-\alpha_1 Z(t-k+i)-\alpha_0)^2$$

Its value can be calculated with least-squares method [19].

$$Z_X \cdot A = Z_Y \text{, where } A = \begin{bmatrix} \alpha_1 \\ \alpha_0 \end{bmatrix}$$

$$Z_X = \begin{bmatrix} \sum\limits_{i=0}^{M-1} Z^2(k+i) & \sum\limits_{i=0}^{M-1} Z(k+i) \\ \sum\limits_{i=0}^{M-1} Z(k+i) & M \end{bmatrix}$$

$$Z_Y = \begin{bmatrix} \sum\limits_{i=0}^{M-1} Z(k+i)\cdot Z(T-M+1+i) \\ \sum\limits_{i=0}^{M-1} Z(T-M+1\_i) \end{bmatrix}$$

On the basis of the set task we get 2 sets of matrices A:

$$A^{+} = \begin{bmatrix} \alpha_1 \\ \alpha_0 \end{bmatrix} \quad A^{-} = \begin{bmatrix} \alpha_3 \\ \alpha_2 \end{bmatrix}$$

Then coefficients are defined and then approximate values of the sample. After this the values of the correlation module is calculated with the formula:

$$\rho_k^M = \left|\rho(\hat{Z}_t^M, Z_t^M)\right| = \frac{\left|\sum\limits_{i=1}^{M}(\hat{Z}(t+i)-\bar{Z})(Z(t+i)-\bar{Z})\right|}{\sqrt{\sum\limits_{i=1}^{M}(\hat{Z}(t+i)-\bar{Z})^2\sum\limits_{i=1}^{M}(Z(t+i)-\bar{Z})^2}} \in [0,1]$$

From the gathered values the highest possible is picked and its coherence to the time delay is found [20].

With consideration of the maximum value of the time delay, coefficients and the sample length, 2 equations of approximation can be written, which are applied according to positive and negative values of the sample:

$\hat{Z}_t^{+M} = \alpha_1 Z_{t-k}^M + \alpha_0 I^M$ - for positive values

$\hat{Z}_t^{-M} = \alpha_3 Z_{t-k}^M + \alpha_2 I^M$ - for negative values

For the researched series the parametre M=142.

Sample from the new history and sample of maximum similarity from history is represented in the Fig.5, where Sample means sample that precedes the prognosis, SML – sample of maximum similarity (likehood).

The offered model is based on the sample of maximum similarity of the daily digressions from DAM from the history data of 2009-2014 that was acquired for official site of the wholesale electric power market (http://br.so-ups.ru/).

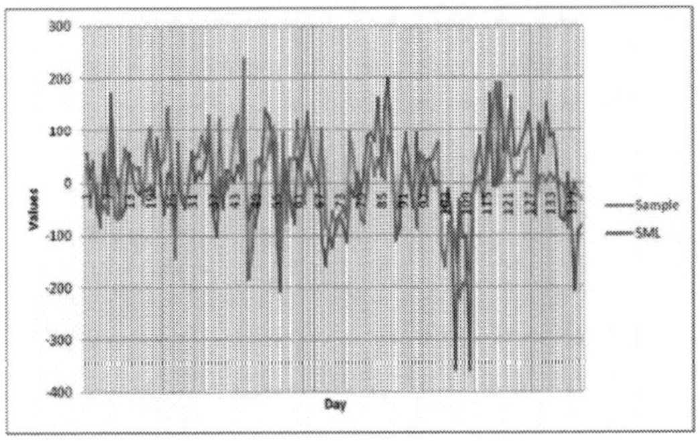

Fig. 5. Comparing the sample of maximum similarity from history and sample of the new history.

The results of prognosis are represented on the Fig. 6, where History means the sample that follows the sample of maximum similarity, values of which approximate while composing the prognosis, Forecast – predicted values, Control – real values.

The forecast error for the aquired results is 3, 3%

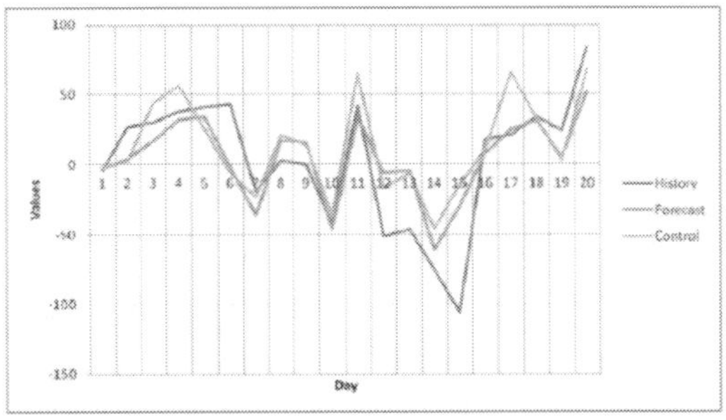

Fig. 6. Prediction 20 deviation values IDB and RSV sample maximum similarity with certain approximation equations for positive and negative values.

6 Conclusion

In the wholesale market, buyers and sellers are generating companies, export/import electricity operators, sales organizations, distribution companies (in terms of electricity purchase for covering transmission losses), large consumers. The subjects of the wholesale market may act as, both sellers and buyers of electricity and capacity. For the wholesale market participant status an organization must satisfy the requirements set out in the approved resolution of the Government of the Russian Federation of December 27, 2010 № 1172 Rules of the wholesale electricity (capacity) market and in the Treaty of Accession to the trading system of the wholesale market. All subjects of the wholesale market are included in Non-Commercial Partnership "Market Council" [21].

In terms of trade, energy is a commodity, and the energy supply contract - a kind of contract of sale.

In a competitive electricity market, the retail price is formed taking into account the cost of additional capacity services markets, derivative financial instruments, financial transmission rights, ancillary services. Commercial transactions in the DAM market are:

- central planning regimes,
- bilateral agreements,
- combined trade deals.

To determine the value of the deviations for different types of initiatives used formulas (cutting), calculated for each hour of the day for each node of the computational model. Cutting is defined as the maximum (minimum) value of the indicator BR and market prices for day-ahead market (DAM), thereby stimulating a more accurate execution of the planned consumption and production of electricity. On the basis of the cost of deviations defined prerequisites and preliminary commitments BR, the difference between which generates imbalance BR. Negative imbalance is distributed among the participants in proportion to their own initiatives. Positive imbalance is distributed between suppliers, proportional to the execution of external initiatives, and consumers, as accurately as possible adhering to the planned consumption. Thus, the BM "fine" participants of the market, allowing the greatest deviations of actual consumption and production of the planned, on its own initiative, and "rewarded" the participants adhering to the planned consumption and as accurately execute commands from [12]. Consequently, the equilibrium market price for a day is formed at the intersection of the curve and the supply curve stprosa received by the auction filed by suppliers and customers applications peresorty electricity sold on the balancing market, but the price formed on it is disadvantageous to both sides of the bargain. There is a need for methods of predicting which give the least prediction error.

In the current formulation, the problem has already become the subject of a large-scale study. Written by a large number of theses and monographs on the electricity market forecasting, both foreign and domestic authors. Works compatriots in this area can be divided into three stages: the Soviet period, the transition period in Russia and the modern period. Among the most active explorers of the modern period of relevant work Sedov AV, Nadtoka II, Chuchueva IA As part of the research Chuchueva IA performed an in-depth analysis of time series forecasting model based on a sample of maximum similarity to approximate WMEP parameters [22].

Currently, the company carried out the accumulation of historical values of economic and physical parameters in the database, which significantly increases the volume of input to solve the problems of forecasting. Consequently, the widespread methods of maximal similarity samples from history that in its different variations show a high prediction accuracy.

The article was discussed in detail the model of forecasting of the balancing market index deviations from the market rate on the day-ahead sample maximum likelihood using a variety of approximation equations for positive and negative values of the time series. prediction error was 3.3%, indicating a high efficiency model. As it has been the model of forecasting of electricity consumption of the sample maximum similarity three time series (source obtained through the integration of first and second order). The highest efficiency demonstrated a method of forecasting time series with the integration of the first order.

From the above we can conclude about the effectiveness of the use of not only individual methods of forecasting in electricity consumption, rates of DAM and the BMI deviations from DAM, but the use of them in combination. Since knowing the predicted values of these three parameters can be adjusted by prices and volumes specified in the application, thereby reducing their risks in participating in the auction.

Subjects with established rules of the electricity power market works economically interested in improving the working methods and improving the accuracy of prediction of its basic parameters.

The work was supported by Act 211 Government of the Russian Federation, contract № 02.A03.21.0011.

References

1. P.F. Drucker, *Management challenges for the 21st century* (HarperBusiness, New York, 1999)
2. I.A. Gheyas, L.S. Smith, *Proc. of the World Congress on Engineering*, **2**, 1292-1296 (2009)
3. R.P. Pradhan, R. Kumar, Journal of Mathematics Research, **1372**, **4**, 111-117 (2010)
4. D. Salvatore, *Microeconomics: theory and applications* (Oxford University Press, Inc, 2003)
5. N. Morariu, E. Iancu, S. Vlad, Romanian Journal of Economic Forecasting, **4**, 213-223 (2009)
6. S.G. Sidorov, A.V. Nikologorskaya, Bulletin ISPU, **3**, 81-83 (2010)
7. S. Singh, Cybernetics and Systems-AnInternational Journal, **31**, **1**, 49-65 (2000)
8. V.L. Zhirnov, *The basics of price determination in electric power industry: lectures summary* (NNSTU publishing, Novosibirsk, 1999)
9. Dj. Box, G.M. Djenkins, *Time series analysis, forecasting and management* (Mir, Moscow, 1974)
10. M. Norizan, A. Maizah Hura, I. Zuhaimy, Proc. Regional Conference on Statistical Sciences, Malaysia, Kelantan, 57-73 (2010)
11. E.E. Tihonov, *Predicting the market conditions* (Nevinnomisk, 2006)
12. N. Draper, H. Smith, *Applied regression analysis* (Wiley, New York, 1981)
13. F.J. Nogales, A.J. Conejo, Journal of the Operational Research Society, **57**, **4**, 350-356 (2006)
14. H.K. Alfares, M. Nazeeruddin, International Journal of Systems Science, **33**, 23-34 (2002)
15. J.W. Taylor, P.E. McSharry, IEEE Transactions on Power Systems, **22**, 2213-2219 (2008)
16. V.G. Mokhov, T.S. Demyanenko, SUSU Herald: Economics and management, **8**, **2**, 86-92 (2014)
17. S. Singh, Cybernetics and Systems-AnInternational Journal, **31**, **1**, 49-65 (2000)
18. I.A. Chuchueva, Information Technology, **12**, 43-47 (2010)
19. I.A. Chuchueva, U.N. Pavlov, Energo-Info, **4**, 46-49 (2009)
20. C. Reinaldo, A. Garcia, *GARCH Forecasting Model to Predict Day-Ahead Electricity Prices. Workshop of Applied Infrastructure* (Berlin, Germany, 2003)
21. R.C. Garcia, IEEE Transactions on Power Systems, **20**, **2**, 867-874 (2005)
22. H. Zareipour, K. Bhattacharya, C.A. Canizares, *IEEE Power Engineering Society General Meeting* (2006)

Research of the elasticity of electric energy tariff demand on competitive market of unified energy system of the Ural

V.G. Mokhov[1,*], and T.S. Demyanenko[1]

[1]South Ural State University, Chelyabinsk, Russia

Abstract. The aim of the study is to analyze the youngest in the world competitive market of electric energy and power of Russia. The hypothesis that there is a significant time lag between the launch date regulated competitive energy market and the actual state of its transition to a stationary operation. To test the hypothesis covered the electric energy consumer's reaction on tariffs changes in the UES of the Ural in the period of time after restructuring RJSC "UES of Russian Federation". Factual data of UES of Russia's System operator shows how indicator values of electric energy consumption elasticity changes: consuming turns from elastic to inelastic according to price. Proved that in 2014 the energy demand matches to a competitive market. As a methodical research tool used 5 types of regression analysis equations for time series of electricity consumption, in addiction by the rate of the day-ahead market for 2009-2014. The research found that the transition to a competitive market of electricity production in Russia in fact was didn't carried out in 2008, but was carried in 2014. Research proved that the electricity market regression analysis applied in predicting short periods of time (day-ahead market), because by increasing the time interval of forecasting accuracy of the forecast is decreases. The research results have great practical significance for the power industry subjects, especially those with high energy intensity of production, as the increase in the accuracy of forecasting reduces fines and total costs for electricity.

1 Introduction

Electric energy is one of the most important products of intrabrach consumption and constitutes the considerable share of the expenditures practically in all branches of economy. As the electrical energy industry is the leading infrastructure branch, it is noncontestantly defines the limitation of economy's development. This calls for outrunning growth of generating capacities as well as deep changes in qualitative characteristics. Electric energy deficiency in some regions (and even in the whole country) inevitably leads to restriction of economic growth [1].

In today's existing view the Wholesale Power Market (WPM) is functioning from September 1, 2006, when the decree of the Government of the Russian Federation introduced new rules for the functioning of the wholesale market. At present, the Russian wholesale electricity market is the youngest and least explored in the world.

During operation the Wholesale Power Market accumulate historical values of economical and physical parameters in database and this drastically increases the volumes of input information for forecasting tasks of electricity. Also hardware and software development facilitated the creation of powerful computing platforms with which the implementation of the difficult forecasting algorithms is possible. In the meantime modern approaches to economic and technical management lodge tough requirements to the accuracy of predictions [2]. As the result, the task of forecasting the time sequences becomes harder with development of informational technologies.

The aim of this research is to determine the type of correlation between the amount of electric energy consumption and the rate of the day-ahead market and the definition of the elasticity of demand for electricity by the rate on the basis of the analysis which we can be seen on the construction of a competitive market.

Based on the goal implies following the research objectives:

• construction 5 types of regression equations (linear, hyperbolic, power, exponential and reverse) to forecast the consumption of electric energy;

• evaluation of forecast errors in forecasting consumption by derived regression equation;

• determination the coefficient of elasticity of demand by the rate for obtained each regression model.

Relevance assigned tasks is obvious, because by the lack of precision forecasting model energy consumers are losing their competitive edge, what leads to deterioration of financial and economic indicators of the subjects of the electricity market and even crises. Electricity market officially became competitive in 2006, but the transition from a monopoly to a competitive pricing can actually display the elasticity coefficient, which shows whether the increase or decrease in tariffs affect the demand [3].

* Corresponding author: mokhov50@mail.ru

2 Theoretical Aspects

Time sequence is a sequence of time ordered index numbers, which characterize the state and level of changing of the event in question [4].

The source elaborates the etymology of time sequence and how statistics of Russia is unaccustomed to this term unlike statistic series and statistic research of dynamics terms [5].

Now there are a lot of methods for time-sequences prediction. They can be divided into intuitive and formalized [6]. Formalized methods are especially interesting because of high accuracy rate.

Quite often in literature two close groups of methods are classified as one group, as it is described in the source [7]. Formalized methods of prediction are methods, which use mathematic description of revealed trends in objects development for acquiring a prognosis. Factographic methods use realities (or fait accompli) already accomplished fact, which are recorded one any storage medium with the help of quantitative or qualitative characteristics as the source of information. In reality both of these groups use one and the same methods of prediction and that's why they are equated so often and classified in one group [8].

Formalized methods are divided into statistic prediction methods, imitational forecasting, historic-logical analysis, analogies methods and advancing methods [9].

Statistic methods define certain equation, which was acquired on the basis of previously researched time-predicted value of the time-series. This equation reflects the correlation between future sequence values and its history and/or external factors [10]. The last three methods are sometimes grouped under the name of "artificial intelligence methods": they imitate the process of human's reasoning in the studying of history or predicting future sequence values.

Methods of statistic analysis, which are good for defining functional dependences in electric energy industry are quite various in their nature. This research will bring light to the method of correlational regressional analysis.

3 Input data and research methods

The essence of method of correlational regressional analysis is in the following:

1) In building and analysis of statistic model as a regression equation (equation of correlation connection) that approximately expresses the dependence of resultative attribute (y) from one or more factors (x).

2) In evaluation of tightness and defining the type of correlation between the resultative attribute and factors exerting influence on it [11].

In our case the task is to compose an equation of correlation (regression equation), which would define the correlation between the electrical energy tariffs (x) and the level of consumption (y), as well as in analysis of tightness between these factors [12].

Let's build the field of set-up attributes correlation. Input data is taken from the official site of System operator of the Unified Power System of Russia (www.http://so-ups.ru/). For calculations the following data were gathered: consuming data and market prices for the 24 hours ahead (Day Ahead Market) on Unified Power System (UPS) of the Ural from 01.01.3013 till 31.12.2013.

The type of correlation connection cannot be defined according to the field of correlation (Fig. 1). Moreover, it can be assumed that the correlation connection between the attributes is virtually nonexistent. However, the reduction of the viewed interval (i.e. till January of 2013 (Fig. 2)) extracts the obvious linear correlation. Therefore, before assuming the existence of the correlation the elasticity coefficient for different types of models (linear, power, exponential, hyperbolical and negative) must be calculated and the level of connection of factor (independent variable) with the result (dependent variable) must be determined.

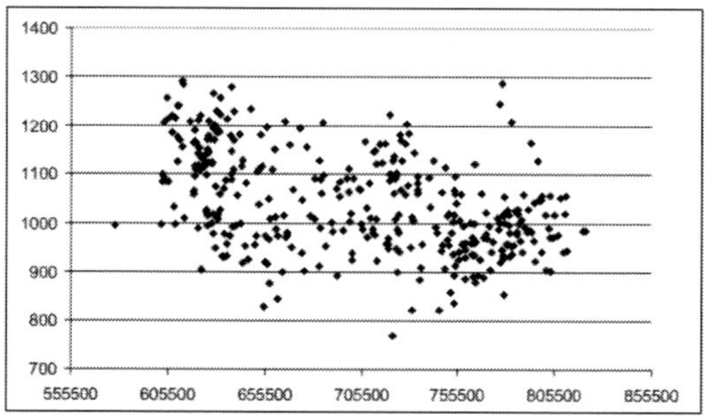

Fig. 1. The field of electric power consumption correlation and DAM tariff for January – December period of 2013 year.

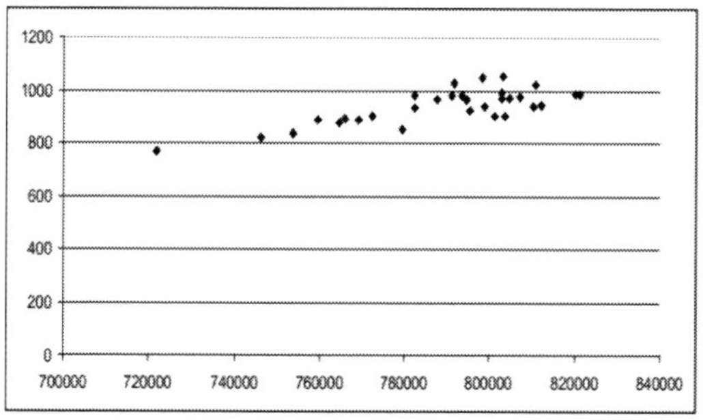

Fig. 2. The field of electric energy consumption and DAM tariff correlation in January 2013.

The second task of this research was to study the dependence between DAM prices and the amount of electric energy consumption. The task of defining the marked neighbouring relationship is in the following: to measure with the help of special values how far correlation connection approaches the actual, according to its force. One of these parameters can be elasticity coefficient [13]. It represents the index of constraint force of factor x with the result y and shows the percents of increase of y when the value of factor changes by 1 It will be appropriate to define elasticity coefficient for each type of regression.

If elasticity coefficient is less than 1 in absolute value then we can say that the variable y along the x is *inelastic*. If elasticity coefficient is more than 1 then we can say that y is *elastic* along the x, because each percent of changing of factor leads to even greater changes of y [14]. If coefficient of elasticity equals 1 it is called *unitary elasticity*. In the limiting case when coefficient of elasticity equals eternity it is called *absolute elasticity*. And, accordingly, when the coefficient of elasticity is zero we call it *absolute inelasticity* [15].

Below are the mathematical models determining the coefficients of regression equations, the coefficient of elasticity of demand at the rate of error detection and time series forecasting.

4 Mathematical models

A. Construction of correlation-regression analysis equations

An equation of linear regression (LR). The resultative attribute is this correlation is changed under the influence of factor attribute, it can be written as:

$$y_x = ax + b,$$

where a is a coefficient of regression.

To find the theoretical equation of correlation in this case is to define the parameters of a right line. Those parameters can be found with the least-squares method which as the result gives the following system of normal equation for finding the parametres of a right line:

$$\begin{cases} a\sum x + nb = \sum y \\ a\sum x^2 + b\sum x = \sum xy \end{cases}$$

Solving this equation systems results in finding the desired parameters [16].

$$\begin{cases} a = \dfrac{n\sum xy - \sum x * \sum y}{n\sum x^2 - (\sum x)^2} \\ b = \dfrac{\sum x^2 * \sum y - \sum x * \sum xy}{n\sum x^2 - (\sum x)^2} \end{cases},$$

where a, b are the parameters of the regression equation; y is a resultative attribute; x is a factor attribute; n is a total number.

Coefficients of the regression equation can be deduced with the following formulas: %.

$$a = \frac{\overline{xy} - \overline{x} \cdot \overline{y}}{\overline{x^2} - \left(\overline{x}\right)^2}, \quad b = \overline{y} - b\,\overline{x},$$

where

$$\overline{x} = \frac{\sum x_i}{n}, \quad \overline{y} = \frac{\sum y_i}{n}, \quad \overline{xy} = \frac{\sum x_i y_i}{n},$$

$$\overline{x^2} = \frac{\sum x_i^2}{n}, \quad n = 365$$

The equation of power regression (PR) looks like:

$$y = b_0 x^{b_1}.$$

Let's linearize the model for more convenient way of finding the coefficients of those models and introduce new variables:

$$v = ln\, y, \quad u = ln\, x.$$

As the result we get the linear regression model:

$$v = a_0 + a_1 u, \text{ where } a_0 = ln\, b_0, \quad a_1 = b_1.$$

Coefficients of this equation are deduced from these formulas:

$$a_1 = \frac{\overline{uv} - \overline{u} \cdot \overline{v}}{\overline{u^2} - \left(\overline{u}\right)^2}, \quad a_0 = \overline{v} - a_1 \overline{u}.$$

Exponential regression (ER) looks like: $y = b_0 + b_1 e^x$.
New variables are introduces to linearize the model:

$$v = y, \quad u = e^x.$$

To make calculations simpler we introduce a multiplier for x, for example 0,001. So the regression equation looks like:
$y = b_0 + b_1 e^{0,001x}$, let's introduce the variables for linearization:

$$v = y, \quad u = e^{0,001x}$$

As the result we get linear regression model:

$$v = a_0 + a_1 u, \text{ where } a_0 = b_0, \quad a_1 = b_1.$$

Hyperbolical regression (HR) looks like:

$$y = b_0 + b_1 \frac{1}{x}.$$

Linearize the given model. To do this we introduce the following variables:

$$v = y, \quad u = \frac{1}{x}.$$

As the result we get the linear regression model:

$$v = a_0 + a_1 u, \text{ where } a_0 = b_0, \quad a_1 = b_1.$$

Reverse regression (RR) looks like: $y = \dfrac{1}{b_0 + b_1 x}$.

Let's linearize this model. In order to do this we should introduce new variables:

$$v = \frac{1}{y}, \quad u = x.$$

As the result we get linear regression model:

$$v = a_0 + a_1 u, \text{ where } a_0 = b_0, \quad a_1 = b_1.$$

B. Determination of the coefficient of elasticity of demand by the rate

Elasticity of demand (W) according to tariff (P) is [17]:

$$E_W = \frac{\partial W / W}{\partial P / P}$$

C. Determination of the forecast error

The accuracy of time sequences forecasting can be estimated with the help of mean absolute percentage error (MAPE) [18]:

$$MAPE = \frac{1}{n}\sum_{i=1}^{n}\frac{|y(i) - \hat{y}(i)|}{y(i)} \cdot 100\%,$$

where $y(i)$ - actual value, $\hat{y}(i)$ - predicted value.

5 Results

The regression equations, based on the data of 2009-2014 years, were calculated in order to get fuller information about the possibility of applying regression models for predicting the electric energy consumption.

The forecasted values were calculated for each of January next year (forecast for 31 days) and on January, 1 of each of the following year (forecast for 1 day). In addition to the construction of forecast values of electricity consumption was researched elasticity coefficient for the 2009 – 2014 years. Summary results are presented in Table 1.

For visualization of the results let plot a graph of actual and forecast values of electricity consumption in January 2014.

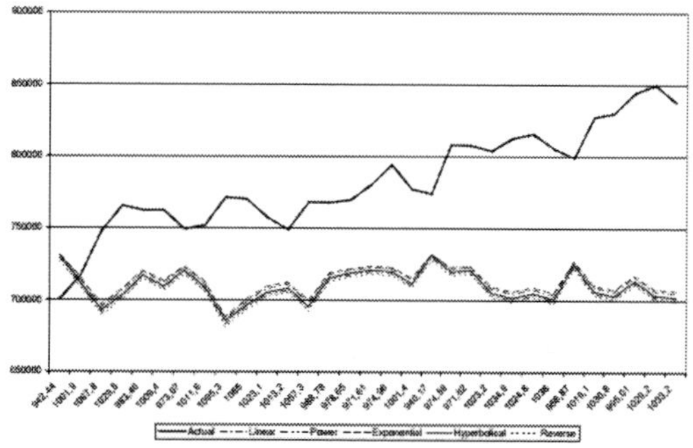

Fig. 3. Actual and predicted values of consumption for regressions built for January 2014.

This diagrams show:
• predicted values, which are obtained by the already found regression equations, are close;
• dynamics of the predicted values is the same;
• with time the amount of forecasting errors increases.

The mean error of the forecast for this period for the regression models in forecasting the first month of each year amounted to 11.06%, and in the forecast of the first day of every year - 4.64%.

On this stage of research the conclusion can be drawn that the most appropriate regression forecasting model of all is the model of exponential regression, because on the average it gives the smallest percent of forecasting error. Its appliance is possible through time sequence prediction.

Its application is possible when forecasting time series at short intervals of time, that is, to predict the parameters of day-ahead market, as the average error for months - 10.44%, and for one day - 4.14%, which is quite satisfactory for the current economic situation.

The main disadvantage of this approach for predicting the amount of electric energy consumption is the fact that the DAM prices values is considered to be fixed. It will never be possible in reality, because in order to predict the amount of consumption you must predict the DAM prices first, using other methods. All this in practice gives even greater percent of forecasting error. Forecasting error for tariffs will add up to the error for amount of consumption [19].

Elasticity coefficient dynamics for each regression model is represented on Fig. 4. From dynamics it is clearly visible that the equations of hyperbolic regression give the highest elasticity coefficient.

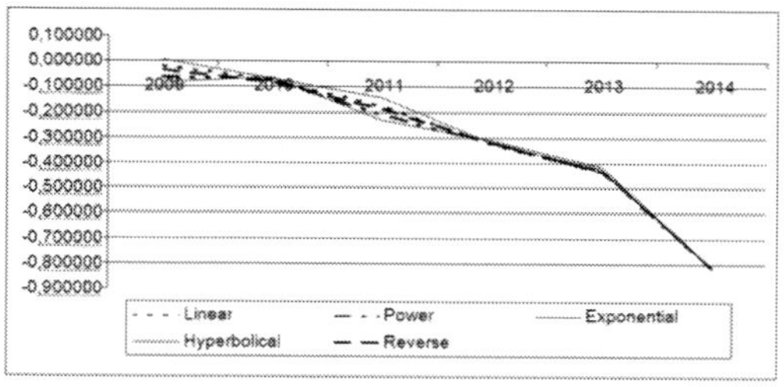

Fig. 4. Elasticity coefficients dynamics for regression models in 2009-2014.

Analysis of the acquired results allows us to draw the following conclusions:

1) The reaction of consumers on changing tariffs for all the analyzed years for all the models is practically the same: with growth of tariff pressure the electric energy consumption becomes smaller, and it is described by the sign «-» in the indexes of the price demand elasticity.

2) 2009-2011 time period is marked with absolute price inelasticity of demand. It is connected with the beginning of reform of price determination in electrical energy industry.

3) In the period of 2012-2014 the elasticity coefficient gradually increases, but in 2012-2013 its indexes are still a lot lower than 1, which is the obvious characteristic of inelasticity.

4) In 2014 coefficient of inelasticity reaches the level, when we can talk about tariff elasticity of demand, because if the tariff growth rate is 10% then the electrical energy demand falls more than on 8%.

Table 1. Calculation of prediction errors for each type of regression equation when forecasting for 1 day and 1 month for 2009-2014 years.

Year	Type	Regression equation	Day forecasting error, %	Month forecasting error, %	Elasticity coefficient
2009	LR	$\hat{y} = -19,822x + 656683,435$	14,88	7,50	-0,02
	PR	$\hat{y} = 952642,266x^{-0,062}$	16,29	8,04	-0,061
	ER	$\hat{y} = 637890,070 + 3203,832e^{0,001x}$	14,10	7,54	0,006
	HR	$\hat{y} = 588503,779 + 34695770,567 \cdot \frac{1}{x}$	16,14	7,60	-0,084
	RR	$\hat{y} = \frac{10^9}{1518,1432 + 0,081859x}$	16,22	8,53	-0,033
2010	LR	$\hat{y} = -57,919x + 719514,4345$	13,83	6,16	-0,073
	PR	$\hat{y} = 1106550,962x^{-0,075}$	14,31	6,72	-0,075
	ER	$\hat{y} = 726350,995 - 23828,193e^{0,001x}$	13,80	6,12	-0,07
2010	HR	$\hat{y} = 628992,782 + 34712848,248 \cdot \frac{1}{x}$	13,72	6,23	-0,062
	RR	$\hat{y} = \frac{10^9}{1382,3267 + 0,148132x}$	14,84	7,17	-0,082
2011	LR	$\hat{y} = -130,020x + 815853,0481$	8,61	4,70	-0,178
	PR	$\hat{y} = 2962106,548x^{-0,213}$	8,69	6,17	-0,212
	ER	$\hat{y} = 797903,412 - 40664,163e^{0,001x}$	8,91	3,41	-0,143
	HR	$\hat{y} = 530694,006 + 152397360,944 \cdot \frac{1}{x}$	8,29	8,17	-0,233
	RR	$\hat{y} = \frac{10^9}{1183,706 + 0,288532x}$	9,21	4,25	-0,186
2012	LR	$\hat{y} = -232,887x + 926583,7874$	10,37	3,57	-0,327
	PR	$\hat{y} = 6405542,995x^{-0,323}$	10,91	3,89	-0,321
	ER	$\hat{y} = 930311,412 - 86451,264e^{0,001x}$	10,24	3,05	-0,323
	HR	$\hat{y} = 479021,242 + 211964999,329 \cdot \frac{1}{x}$	10,69	4,56	-0,31
	RR	$\hat{y} = \frac{10^9}{974,2196 + 0,481210x}$	11,14	3,07	-0,323
2013	LR	$\hat{y} = -289,966x + 1004844,169$	9,11	4,46	-0,431
	PR	$\hat{y} = 14033988x^{-0,432}$	9,58	4,13	-0,429
	ER	$\hat{y} = 990942,326 - 101156,731e^{0,001x}$	8,98	4,43	-0,427

	HR	$\hat{y} = 408526,442 + 303681877,581 \cdot \frac{1}{x}$	9,41	4,34	-0,414
	RR	$\hat{y} = \frac{10^9}{803,241 + 0,606418x}$	9,77	3,86	-0,437
2014	LR	$\hat{y} = -526,296x + 1293032,026$	7,23	0,43	-0,82
	PR	$\hat{y} = 231484376,2x^{-0,826}$	8,73	0,38	-0,821
	ER	$\hat{y} = 1231325,538 - 171564,275e^{0,001x}$	6,60	0,27	-0,809
	HR	$\hat{y} = 129780,315 + 639040284,463 \cdot \frac{1}{x}$	8,78	0,56	-0,813
	RR	$\hat{y} = \frac{10^9}{242,5855 + 1,063704x}$	8,61	0,14	-0,822

Averaged elasticity coefficient for 2014 was around -0,82, so, we can talk about tariff elasticity amount consumption. It can be seen in Fig. 5. When the price increases for 10%, the amount of consumption decreases for 8,2 %.

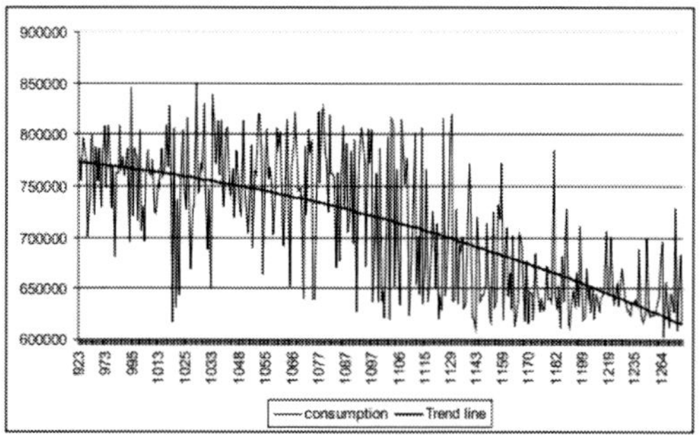

Fig. 5. Dependence of the consumption amounts in the electric energy tariffs in 2014.

Experience of many years in the field of evaluation of electric energy demand functions allowed G. Kouris [20] to draw conclusions even in 1981, which are still solid in our days:

- evaluation of elasticity coefficients crucially depends on the existence of data, model and structure of economy specifications at each moment of time;
- the term "absolute elasticity" is more of an illusion, than reality;
- elasticity coefficients change with time under cause and effect relations influence;
- electric energy demand predictions should be based on statistically evaluated elasticities and not on predicted ones [21].

6 Conclusions

1) The research was based on the data of 2009-2014 years on prices and amounts of electric Energy consumption in CPS of the Ural region, acquired from official cite of System Operator of United Power System of Russia.

2) 5 types of regression models were built (linear, power, exponential, hyperbolical, reverse) for each year under consideration.

3) On the basis of these models predicted values of consumption amounts were calculated for all January and only 1 January of the year, following the researched one. The average predicting error of applying regression models amounted in 11% and 4,64% respectively. The lowest forecast error showed exponential regression model in predicting short time intervals - 4.14%.

4) Restructuring of RJSC "UES of Russian Federation was introduced in 2008, but the transition to competitive market of electric energy manufacturing didn't happened. The electric energy demand gets closer to tariff elasticity only in 2014, and it is depicted in elasticity coefficient 0,82.

5) In 2014 the trend appeared that portrayed the dependence of consumption volume of electricity from the electric energy tariff.

6) It is proved that the electricity market for the regression analysis applied in predicting short periods of time, or for the day-ahead market, as by increasing the time interval of forecasting accuracy of the forecast is reduced.

7) The Market of electric energy and power in Russia of monopoly, when the manufacturer has determined the tariffs, in 2014 became competitive.

The practical significance of the results of the study is that in 2014 electricity producers can no longer dictate the tariffs, and must build up its economic activities to the market behavior of energy consumers.

The work was supported by Act 211 Government of the Russian Federation, contract № 02.A03.21.0011.

References

1. P.F. Drucker, *Management challenges for the 21st century* (HarperBusiness, New York, 1999)
2. D. Salvatore, *Microeconomics: theory and applications* (Oxford University Press, Inc, 2003)

3. F.J. Nogales, A.J. Conejo, Journal of the Operational Research Society, **57**, **4**, 350-356 (2006)
4. V.N. Afanasyev, M.M. Yuzbashev, *Time sequences analysis and forecasting* (Finance and statistics; INFRA-M, Moscow, 2012)
5. E.E. Tikhonov, *Forecasting in conditions of the market* (Nevinnimyssk, 2006)
6. M. Norizan, A. Maizah Hura, I. Zuhaimy, *Proc. Regional Conference on Statistical Sciences*, 57-73 (2010)
7. M.M. Butakova, *Economics and forecasting. Methods and ways of practical calculations* (KNORUS, Moscow, 2008)
8. M. Beenstock, A. Dalziel, Energy Economics, **4**, 90-98 (1986)
9. R. Haas, L. Shipper, Energy Economics, **20**, 421-442 (1998)
10. H.K. Alfares, M. Nazeeruddin, International Journal of Systems Science, **33**, 23-34 (2002)
11. L. Hunt, G. Judge, Y. Ninomiya, Energy Economics, **25**, 93-118 (2003)
12. Dj. Box, G.M. Djenkins, *Time series analysis, forecasting and management* (Mir, Moscow, 1974)
13. R. Nesbakken, Energy Economics, **21**, 493-515 (1999)
14. I.A. Bashmakov, *Evaluation of price elasticity of electric energy demand parameters for different groups of consumers and territorial entities of Russian Federation* (TSENEF, Moscow, 2007)
15. S. Singh, Cybernetics and Systems-AnInternational Journal, **31**, 49-65 (2009).
16. V.G. Mokhov, Proc. Science SUSU, **2**, 83-86 (2009).
17. V.L. Jirnov, *Fundamentals of energy pricing: Lectures* (Publishing center NSTU, Novosibirsk, 1999)
18. P.A. Sinutin, *Formation of system of state regulation of the energy supply in the Russian Federation* (Economic, Moscow, 2002)
19. V.G. Mokov, T.S. Demyanenko, Science SUSU. Section of Economics and management, **2**, 83-86 (2009)
20. G. Kouris, Energy Economics, **5**, 89-99 (1983)
21. J.W. Taylor, P.E. McSharry, IEEE Transactions on Power Systems, **22**, 2213-2219 (2008)

Use of statistical indicators of profit accounting for formation of a product portfolio

L.G. Nesterova[1,*]

[1]South Ural State University, Chelyabinsk, Russia

Abstract. Article is related to perfection of product portfolio policy of the enterprise by formation of a product portfolio by means of using of statistics methods and mathematical criteria of the games theory. Such approach allows to decide many tasks of marketing at the enterprise, including to use results of marketing researches and demand and profit forecasting. As a result of the offered system of calculations of statistical indicators it is possible to receive a matrix of predicted risks and market growth for a concrete product portfolio, and also to estimate potential and market prospects of concrete commodity groups. It will form an optimal product portfolio for the specific market conditions.

1 Introduction

Planning of development of a product portfolio always is an actual task of product portfolio policy of any enterprise. There are various approaches for definition of strategy of such development. This article offers to consider the approach with use of a combination of statistical techniques of the accounting profit and two mathematical criteria of the games theory.

The offered approach of a choice of development strategies of the product portfolio which initial element is the set of possible values of profit, synthesizes into itself elements of several stages of marketing management. As marketing tasks [1-7] within the limits of use of our approach the following can be offered:

• Increase of efficiency of company activity;

• Increase of company stability (decrease of activity risks) in the conditions of uncertainty of market.

As the marketing purposes [1-7] with reference to formation and development of a product portfolio can be following:

• Diversification of company activity;

• Statement of planning system and forecasting in the conditions of variability of market;

• Statement of system of marketing audit (control of development efficiency of existing economic units).

Marketing research [4-6, 8-11] within the limits of our approach has double character:

• It is one of situation analysis methods (really, we define more exact distribution of possible market situations);

• One of controlling methods (we subject to control aprioristic estimations of possible conditions of market).

• Marketing research [4-6, 8-11] executes two functions:

• Definition of possible conditions of market conditions (including, allocation of the data due to strategic segments), that is transition from a condition of full uncertainty to the partial condition; marketing research - the situation analysis tool;

• Correction of already existing expert estimations toward a market situation; marketing research - the tool of control of marketing.

2 Use of statistical indicators of profit accounting for formation of a product portfolio

A. *Matrix of predicted risks and market growth*

Such approach allows to present prospects of a product portfolio within the limits of model General Electric/McKinsey [1-3, 5, 12, 13]. Really, we can obtain the estimated data about predicted rates of increase of the markets of strategic units, as well as to calculate relative predicted risk of investment. For each of the given indicators we can break the markets, the goods, SEU (STRATEGIC ECONOMIC UNIT) into three groups:

• Within the limits of predicted rate of increase:

- Fast growth (more than set top bound of stability criterion);

- Relative stability (within the limits of bounds of stability criterion);

- Falling (less than the lower bound of set stability criterion);

• Within the limits of relative investments risk:

- High risk;

- Average risk;

- Low risk.

Under criterion of stability are understood such bounds of increase rates which are estimated by the person, making the decision, as normal (averages) for the considered market as a whole. As a result of breaking the following matrix to which we recommend to apply

* Corresponding author: nesterovalg@susu.ru

strategies of adaptive growth [1, 7, 12, 14-16]. Name the received model as a matrix of predicted risks and market growth (table I).

Table 1. Matrix of predicted risks and market growth.

		Relative predicted growth		
		Fast growth	**Stability**	**Falling**
Relative predicted risk of investment	Low	Approach	Approach	Deduction
	Average	Approach	Deduction	Deviation
	High	Deduction	Deviation	Deviation

B. Use of statistical indicators

Use of criterion of Laplace [17-19]

$$L = \max_{i}\left(\frac{1}{n} * \sum_{k=1}^{n} PF_{ik}\right) \quad (1)$$

The criterion shows the maximum value of a mathematical expectation of profit receipts for various strategic units. The criterion bases on the precondition that all values of a consumer demand are equiprobable.

Within the limits of our model the person making decision, will receive total ranged in places a row of strategic units according to decrease of row of average values of expected profit. Present row will specify to average values of profit expected by company (caused due to averaged among expected conjuncture) for each of reviewed groups. Actually in the top part of a row (with the highest places) the goods with the greatest average expected potential of release will be located. Name the present row as a row of Laplace.

Finding method:

Originally for each strategic unit we find the size of total profits under all possible conditions of market. Then we choose the strategy which provides to the businessman the maximum possible receipts of profit.

Use of criterion of Bayes-Laplace [17; 18; 19] (for an estimation of efficiency of marketing research)

$$BL = \max_{i}\left(\sum_{k=1}^{n}\left(PF_{ik} * p\left(DP_{k}\right)\right)\right) \quad (2)$$

This criterion shows the maximum value of a mathematical expectation of profit receipts under various strategies of company taking into account the data about the demand levels received on the basis of some (on the basis of marketing research) expert distribution of probabilities of consumer potential values - p(DPk).

Within the limits of our model the person making decision will receive total ranged in places a row of strategic units according to decrease of the weighted-average of expected profit receipts. This row will specify to weighted taking into account the probabilities expected by company, values of profit (due to weighted likelihood average among expected by a conjuncture) for each of considered groups. Actually in the top part of a row (with the least places) the goods with the greatest average probable potential of release will be located. Name the present row as a row of Bayes-Laplace.

Finding method:

At the first stage the businessman sets values of distribution probabilities of possible demand levels (for example, on the basis of the previous experience, or after carrying out of the predicted analysis of the market development). Then, taking into account likelihood conditions a demand determinant, for each strategic unit we find the size of total profit receipts. And, at last, we choose the strategy which provides to the businessman the maximum possible receipts of profit. It will be optimum by the given criterion.

For each strategy of product portfolio development it is possible to present set of profit values in a kind ranged a variation row. It allows us to calculate for each strategy average value of profit, and taking into account distribution of probabilities of market conditions and the weighted-average, as well as an average quadratic deviation and variation factor. Notice that the set of average and weighted-average represents rows of Laplace and Bayes-Laplace accordingly.

Let's designate average value of profit receipts under strategy **i** as AGRi (average gross receipt – average receipt of profit), and probable-average (weighted) receipt of profit as APGRi (average probable gross receipt – average probable receipt of profit). Profit receipts under strategic unit **i** at a conjuncture **k** are designated as PFik.

Average quadratic deviation (AQD) we will count under following formulas [17-20]:

For a row of Laplace

$$\sigma_{i} = \sqrt{\frac{\sum_{k=1}^{n}\left(PF_{ik} - AGR_{i}\right)^{2}}{n}} \quad (3)$$

For a row of Bayes-Laplace

$$\sigma_{BL_{i}} = \sqrt{\sum_{k=1}^{n}\left(p(DP_{k}) * \left(PF_{ik} - APGR_{i}\right)^{2}\right)} \quad (4)$$

Variation factor we will count under formulas [17-20]:

For a row of Laplace

$$v_{i} = \frac{\sigma_{i}}{AGR_{i}} \quad (5)$$

For a row of Bayes-Laplace

$$v_{BL_i} = \frac{\sigma_{BL_i}}{APGR_i} \quad (6)$$

In our approach the variation factor is used for definition of relative risk of investment in an alternative matrix of predicted risks and market growth. For construction of the given matrix it is necessary to divide all goods, SEU (STRATEGIC ECONOMIC UNIT), the markets into three groups due to investment risk levels. For this purpose we suggest to break a row of variation factors into three equal intervals. Length of an interval we will define under the following formula [17-20]:

For a row of Laplace

$$\Delta v = \frac{\max_i v_i - \min_i v_i}{3} \quad (7)$$

For a row of Bayes-Laplace

$$\Delta v_{BL} = \frac{\max_i v_{BL_i} - \min_i v_{BL_i}}{3} \quad (8)$$

Thus, the relative risk of investments for each strategy i can get to one of three groups (an example for a row of Laplace):

High risk maxv - Δv<v_i ≤ maxv;

Average risk minv + Δv<v_i ≤ maxv-Δv;

Low risk minv ≤ v_i ≤ minv +Δv.

Following indicator which is necessary for construction of a matrix of predicted risks and market growth, is average predicted rate of increase. We will count it as follows [17-20]:

For a row of Laplace

$$T_i = \frac{AGR_i - PF_{i1}}{PF_{i1}} \quad (9)$$

For a row of Bayes-Laplace

$$T_{BL_i} = \frac{AGPR_i - PF_{i1}}{PF_{i1}} \quad (10)$$

Here *PFi1* – the size have arrived to an initial stage.

Then it is necessary to divide the goods, SEU (STRATEGIC ECONOMIC UNIT), the markets into three groups of growth, for this purpose it is necessary to set criterion of stability. The criterion of stability will characterize the top and lower bounds of rates of increase of the investigated markets which the person making decision considers expedient to define. On the basis of the entered bounds each strategic unit **i** will get to one of three groups (an example for a row of Laplace):

Fast growth: Ti ≥ the top bound of stability criterion;

Stability: Ti lies within bounds of stability criterion;

Falling: Ti ≤ the lower bound of stability criterion.

Average rate of increase we will define as follows:

For a row of Laplace

$$\overline{T} = \frac{1}{m} * \sum_{i=1}^{m} T_i \quad (11)$$

For a row of Bayes-Laplace

$$\overline{T}_{BL} = \frac{1}{m} * \sum_{i=1}^{m} T_{BL_i} \quad (12)$$

For matrix General Electric/McKinsey it is necessary for us to break all strategic units into three groups according to their potential which is expressed in average profit receipts for a row of Laplace and possible average receipts of profit for a row of Bayes-Laplace. For this purpose we suggest to break a row of necessary average values into three equal intervals. Length of an interval we will define under the following formula [17-20]:

For a row of Laplace

$$\Delta AGR = \frac{\max_i AGR_i - \min_i AGR_i}{3} \quad (13)$$

For a row of Bayes-Laplace

$$\Delta APGR = \frac{\max_i APGR_i - \min_i APGR_i}{3} \quad (14)$$

Thus, **i** within the limits of matrix GE it is possible to define a relative position of: strategic unit as follows (an example for a row of Laplace):

Successful: maxAGR - ΔAGR <AGRi ≤ maxAGR;

Average: minAGR + ΔAGR <AGRi ≤ ≤ maxAGR - ΔAGR;

Unsuccessful: minAGR ≤ AGRi ≤ minAGR + ΔAGR.

3 Conclusion

Table 2. Example of a summary matrix of predicted risks and market growth for commodity categories A,B,C.

		Relative predicted growth		
		Fast growth	***Stability***	***Falling***
Relative risk of investment	Low	A	A	C
	Average	A	C	B
	High	C	B	B

We can formulate the following conclusions:

Use of this approach allows to generate as the general strategy of development of a product portfolio, and to define potential and prospects of its separate units.

This approach will form an optimal product portfolio for the specific market conditions.

The work was supported by Act 211 Government of the Russian Federation, contract № 02.A03.21.0011.

References

1. I. Ansoff, *Strategic management* (Economy, Moscow, 1993)
2. I. Ansoff, *New corporate strategy* (Piterkom, St. Petersburg, 1999)
3. G. Assel, *Marketing: principles and strategy* (Infra Th, Moscow, 2003)
4. R.D. Bazzel, D.F. Koks, R.V. Brown, *Information and risk in marketing* (Finstatinform, Moscow, 2007)
5. F. Kotler, *Principles of marketing. Rostinter* (Moscow, 2005)
6. J. Traut, *New positioning* (Peter, St. Petersburg, 2003)
7. D. Etkinson, *Strategic marketing: situations, examples: Studies. The grant for high schools* (Uniti - Dana, Moscow, 2011)
8. E.O. Putmen, *Marketing of your fields of activity* (New York, the USA, 1998)
9. A. Hayem, *Marketing for teapots* (Dialectics, Kiev, 2010)
10. G.A. Churchill, *Marketing researches* (Peter, St. Petersburg, 2005)
11. R.A. Shmidt, H.A. Right, *Financial aspects of marketing: Studies. The grant for high schools of economic specialities* (Uniti, Moscow, 2012)
12. A.I. Veselov, Marketing in Russia and abroad, **1**, 103-114 (2009)
13. N.U. Chernik, *The commodity policy of the enterprise: Studies* (Publishing house BGEU, Minsk, 2004)
14. R. Watermen, *Faktor of management by results* (Progress, Moscow, 1998)
15. R. Watermen, *Faktor of updating: how the best companies keeps competitiveness* (Progress, Moscow, 2008)
16. R. Foster, *Modification of manufacture: the attacking win* (Progress, Moscow, 2006)
17. E.S. Pozhidaeva, *Statistika: the textbook for high schools* (Academy, Moscow, 2015)
18. S.D. Beshelev, F.G. Gurvich, *Mathematical - statistical methods of expert estimations* (Statistics, Moscow, 1995)
19. P.V. Konjuhovsky, A.S. Malov, *Theory of games: the textbook for high schools of economic directions and specialities* (Urayt, Moscow, 2015)
20. K. Heddervik, *Financial and economic analysis of enterprises activity* (Finance and statistics, Moscow, 2006)

Economic efficiency evaluation of merges and acquisitions in the sector of industry based on nonlinear model of synergistical growth of an industrial corporation value

A.E. Ivanov[1,*], and *A.A.* Vasilchenko[1]

[1]South Ural State University, Chelyabinsk, Russia

Abstract. Numerous research works of the end of the 20th and of the beginning of the 21st century prove that the synergistic effect often declared as the main goal of merges and acquisitions is not generated in fact. This is due to imperfection of the available methodology of its economic evaluation that does not take into account a nonlinear nature of the pooled corporation development. The article suggests a methodology of economic efficiency evaluation of merges and acquisitions in the sector of industry based on identification of synergistically successful acquisition order parameters. These are synergistic effects that with minimal investments in their achievement lead to a disproportionate increase in the value of an industrial corporation. A mathematical model has been created simulating the influence of these investments on the value of an industrial corporation. The model allows one to increase the degree of the decisions validity in merges and acquisitions in the sector of industry.

1 Introduction

In 1980s-1990s on the tide of rapid globalization processes, economists increasingly focused their attention on the economical efficiency of mergers and acquisitions (M&A). Numerous research works of that period reveal that a significant part of mergers and acquisitions do not give the expected results. M. Bradley, A. Desai and E.H. Kim have found that only 7 % of mergers and acquisitions lead to increment in value of the pooled companies [1]. J.W. Hunt in his study draws a conclusion that about 50 % of all M&A do not reach synergies [2]. M.L. Sirower shows that 70 % of the acquisitions yield profit that is less than the premium paid for an acquired firm [3]. According to A. Agarwal and J. Jaffe's research, during the first years after M&A the revenue of pooled companies reduces or remains unchanged [4].

An acquiring company often pays the price for a target firm which exceeds its net asset value. This overpaid amount is considered in the acquirer's financial statements as goodwill. It reflects the acquirer's expectations concerning the potential synergistic effects which may be generated as a result of merger or acquisition. At the beginning of XXI century a mandatory goodwill impairment testing was introduced in many International Financial Reporting Standards as well as many National Financial Reporting Standards. This made it possible to study the validity of M&A initiators' expectations on the basis of information on the amounts recognized as a goodwill impairment. Recent studies show that a significant part of the purchased goodwill is impaired relatively soon [5, 6, 7, 8, 9].

These results point out that the expectations of M&A initiators with respect to the expected synergistic effect of integration are overestimated. This has to do with imperfection of the economical efficiency evaluation methodology of goodwill. Most of the synergistic effect valuating methods applied in planning M&A are based on income approach to business valuation. Value of an enterprise according to income approach is considered as the present value of its future profits. The most common in business practice is the Excess Earnings Method [10]. The methods that are based on Edwards-Bell-Ohlson model and are supposed to use the combination of income approach and cost approach methodologies for business valuation are also often used [11]. A common disadvantage of all these methods is the linear logic that is the basis of future earnings forecast which expects a stable capitalization growth of pooled companies. However, an enterprise is an open system that interacts with the environment. The processes occurring in open systems are characterized by nonlinearity, no equilibrium and equifinality [12]. That is why the nonlinear nature of pooled companies development progress must be taken into account while forecasting future earnings growth rate.

2 Methodology and results

To determine the synergistic effect of an integration it is necessary to evaluate existing values of enterprises

* Corresponding author: ivanov.chel@list.ru

engaged in M&A, and compare this amount with the present value of the industrial corporation formed as a result of the merger or the acquisition (we are going to use further the term «acquisition», considering merger as its particular case). We consider an industrial corporation as a sustainable combination of industrial and infrastructural objects that have a single owner and a common management body. Synergistic effect generating in open systems is caused by their emergence, i.e. appearance of some properties that are absent in isolated elements of the system. Synergistic effect for an industrial corporation results in its revenues increase or in its expenses reduction compared to the revenues and expenses of the acquiring company and the target company operating independently. An acquisition can be regarded as synergistically successful when integration of the acquiring and target companies generates a synergistic effect. We consider by synergistic effect the excess of the industrial corporation value over the sum total of values of isolated companies included in it. This article looks at acquisition implementation feasibility only from the point of view of synergistic effect generating. An acquisition can be economically effective, but fails synergistically if an industrial corporation value is equal to the sum total of values of isolated companies included in it. We have developed a methodology for economic evaluation of acquisitions feasibility in the sector of industry, which includes five stages.

1. Market value estimation of all acquisition participants as independently operating companies.

Discounted cash flow model is used for independently operating companies evaluation [13]. Equation (1) defines the present value of a company (PC) within this model.

$$PC = \sum_{t=1}^{T} \frac{CF}{(1+r)^{t-1}} + \frac{TV}{(1+r)^{T}}, \tag{1}$$

where CF – nominal incomes sum of a company; t – a period of calculation (a year); T – the forecast horizon; r – the interest rate within the forecast period; TV – the terminal value of a company in the post-forecast period.

2. Identification of potential synergy sources and determination of synergistically successful acquisition order parameters.

Open systems, which are far from equilibrium, depend on many external and internal environment factors and have unstable behaviour Their behaviour is determined by order parameters – leading variables that affect systems dynamics. Order parameters stimulate self-organization of a system, while the system's behaviour can vary greatly with insignificant alternation in the amount of the order parameters. Changes of other factors will have a much lesser impact on the behavior of a system, so their influence is negligible. We consider that the expected synergistic effect, which marks off the order parameters of synergistically successful acquisition are those that with minimal investment in their generating lead to a disproportionate increase in the value of an industrial corporation. For example, an acquisition is considered to generate two types of synergistic effects. The first one is the reduction of labor expenses by eliminating duplicate functions of administrative managers. The second is the sales revenues increase through implementation of high-performance incentive structure for sales managers, which is used at the target firm. Investing of $100 to achieve the synergistic effect in the first case leads to $ 1,000 expenses reduction. And a $ 100 investment to generate the synergistic effect in the second case leads to $ 110 revenues increase. In this example, it is the first synergistic effect that is the order parameter of synergistically successful acquisition.

In order to identify potential synergy sources and to determine order parameters of synergistically successful acquisition, a set of measures developed by A.E. Ivanov is suggested for use [14]. This toolkit is based on T.L. Saaty's analytic hierarchy process [15] and involves the use of expert assessment. Top managers of the company, who are planning an acquisition, may act out as experts for this purpose together with consultants, who are called for due diligence. The process of decision justification for an acquisition is represented as a hierarchical model. The goal of an industrial corporation in this model is the sustainable development, which can be reached by means of alternative development options. The acquisition of a target firm is one of these options. The criteria for goals achievement evaluation are the types of synergistic effects that can be generated in operating, financial and investing activities of an industrial corporation. At lower levels of this hierarchical model, experts make paired comparisons of the influence of merged companies' external and internal environment factors, on generating synergistic effects by Saaty's scale of relative importance. Thus, the experts can identify priority factors. At high levels of the hierarchical model, order parameters of synergistically successful acquisition are identified by evaluating their priorities vectors. Levelling of opinions in experts' groups is reached by calculating a geometric mean of priorities values. At the same time, an expert competence weighing system can be introduced as well as adjustments for relative compatibility of expert's estimation matrix.

3. Assessment of an industrial corporation value taking the synergistic effect into account.

Open systems develop nonlinearly. There are bifurcation points, where the performance of systems is unstable. An insignificant effect on the order parameter at bifurcation point can lead to the choice of further development of the system and to the phase transition. The logistic equation is one of the best-studied maps of the open systems development dynamics (2).

$$X_{n+1} = C * X_n(1 - X_n) \tag{2}$$

Logistic equation has an important property. If values of the constant C are small, the argument X_{n+1}

increases exponentially. Large values of the constant C cause the cascade of bifurcations and the system drops down into the state of dynamic chaos. Thus, the system exponentially develops before the point of bifurcation and the dynamics of its development can be described by the Malthusian equation (3).

$$\frac{dx}{dt} = \alpha x, \quad (3)$$

where $\frac{dx}{dt}$ – the change of argument x within a time period t; α – the growth coefficient.

Function (4) is the solution of the Malthusian equation.

$$x = Ce^{\alpha t}, \quad (4)$$

where C – the constant.

An industrial corporation is an open system, so the dynamics of its value changes in the course of time can be described by means of logistic equation. At the same time, we believe that the forecast horizon for industrial corporation evaluation should be limited by the period between bifurcations. We've introduced this restriction, because it is difficult to reliably estimate the changes in the value of corporation at bifurcation points (e.g., during another acquisition process). Therefore, it leads to an unacceptably high risk taking into investor account evaluation while planning an acquisition [16]. Nonlinear equations are very sensitive to any initial conditions changing, so a qualitative forecast is only possible for a short period. A number of empirical studies conducted by McKinsey consulting group proves that synergistic effects are generated during several initial years after a merger or an acquisition [17, 18], so that the main increase in industrial corporation value by synergistic gains falls on this period. Thus, the industrial corporation value within the forecast horizon in the case of achieving synergies will be exponentially growing. Inserting the cash flows sum of separately operating companies in the solution of the Malthusian equation as the constant C, we'll obtain the expression (5) to determine the value of an industrial corporation.

$$SPC_T = \sum_i PC_i * e^{\alpha} = \sum_i (\sum_{t=1}^{T} \frac{CF}{(1+r)^{t-1}}) * e^{\alpha}), \quad (5)$$

where SPC_T – the value of an industrial corporation, taking the synergistic effect into account and considering the cash flows within forecasting horizon; PC_i – the cash flows sum total of an independently operating companies; i – the number of pooling companies; e^{α} – the nonlinear function, which shows the ability of an industrial corporation to generate synergistic effect; α – the synergistic growth coefficient, which represents sensitivity of an industrial corporation value to investment in generating synergistic effect.

In this case the Malthusian equation describes only the dynamics of the industrial corporation changes in value within a framework of the forecast horizon. That is why the cash flow size of an industrial corporation within post-forecast horizon for the purpose of estimating the present value of an industrial corporation (SPC) is a separate item (6).

$$SPC = \sum_i (\sum_{t=1}^{T} \frac{CF}{(1+r)^{t-1}}) * e^{\alpha} + \frac{TV}{(1+r)^{T}}) \quad (6)$$

Synergistic growth coefficient is determined from (7).

$$\alpha = (\frac{dSE}{Idt} - 1) * (\frac{dCF}{SEdt} - 1) * \lambda, \quad (7)$$

where SE – the revenues increase (expenses reduction) of an industrial corporation as a result of investing in synergistic effect generating; I – the volume of investment in synergistic effect generating; λ – the synergistic effect attainability.

The first multiplier (7) shows synergistic effect sensitivity to investment into its achieving. The second multiplier (7) represents the sensitivity of an industrial corporation value to the amount of synergistic effect. To define multiplier we should compute the indicators for each of synergistic effects, which are the order parameters of integration. Then, the geometric mean of these results is calculated. The product of the first and the second multipliers (7) is the growth rate of an industrial corporation value under incrimination of investment in synergistic effects generating.

Possibility of adjusting achievement in synergistic effect is made by means of the third multiplier (7). We use the fuzzy sets theory described in L. Zadeh's works to estimate the possibility of synergistic effects generating. In recent years, methods based on fuzzy sets theory are often used for assessment of investment projects efficiency. In comparison with the classical probabilistic approaches, these methods allow one to take greater account of all the possible diversity of uncertainty types in forecasting development of large systems.

In fuzzy sets theory the possibility distribution is used instead of the probability distribution. The possibility distribution is described by compatibility function of a fuzzy number. The fuzzy number membership function is defined by means of linguistic variable whose values are not numbers but words (8).

$$x = \langle x, T, U, G, M \rangle, \quad (8)$$

where x – the name of the variable; T – the term-set of variable value, i.e. its linguistic value part; U – the universe of discourse; G – syntactic rule which defines the terms of the set T; M – semantic compatibility rule when each linguistic value x is assigned its meaning according to this rule.

We have developed a linguistic variable "Attainability of integration synergistic effect". Its structure is shown in the Table 1.

Table 1. Structure of the linguistic variable "Attainability of integration synergistic effect".

Element	Name	Content
Variable name	x	Attainability of integration synergistic effect
Syntactic rule	G	Very, higher, lower
Set terms	T	High, medium, low
Semantic rule (the triangular form of a fuzzy number is used for describing compatibility function)	M	$M(x)=\begin{cases}1-\frac{b-x}{b-a}, & a\le x\le b;\\ 1-\frac{x-c}{c-b}, & b\le x\le c;\\ 0, x\prec a;\\ 0, x\succ c.\end{cases}$
Universe of discourse	U	[0%;100%]

It makes sense to involve the experts, who previously helped to determine the order parameters, for estimating of synergistic effect attainability that are the order parameters of synergistically successful acquisition. The experts evaluate attainability of every single synergistic effect according to syntactic rule, e.g. "Attainability of labor expenses reduction by eliminating duplicate functions of managers is very high". These evaluations are shown as the fuzzy number a_i. As a result, we have n fuzzy sets evaluations for n order parameters. The total evaluation of the most expected synergistic effect attainability interval is the convex combination of the fuzzy sets $a_1,\dots,a_n$, i.e. the fuzzy set A_i with the compatibility function (9).

$$M_{A_i}(u)=M_{a_1}(u)+\dots+M_{a_n}(u) \quad (9)$$

The average opinion of the expert group is obtained by simple arithmetic averaging of compatibility functions values. The obtained limit defines the factor λ (7).

The synergistic growth coefficient α (7) determines the growth rate of an industrial corporation value as a result of an acquisition.

If α<0 an investment in synergistic effect generating leads to the growth rate decrease of the industrial corporation value, i.e. the acquisition is not feasible in terms of synergistic effect generating.

If α=0 there is no synergistic effect because of one of the three possible reasons:

1) the growth rate of investments in generating synergistic effect and the revenues growth rate (expenses reduction rate) resulting from these investments are equal, that is typical for a linear developing system.

2) the growth rate of the industrial corporation value and the revenue growth rate (expenses reduction rate) caused by these investments in generating synergistic effect are equal, which is also typical for a linear developing system.

3) synergistic effect is unattainable either because it's impossible to generate it or because the order parameters of synergistically successful acquisition have been incorrectly defined.

If α>0 an investment in synergistic effect generating leads to the growth rate increase of industrial corporation value, i.e. the acquisition is advisable in terms of synergistic effect generating. Thus, the values range of the corporation can be achieved with a high degree of confidence which is defined due to factor λ (7) correction.

4. Attainable synergistic effect calculation

Attainable synergistic effect (SE) is determined from the formula (10).

$$SE = SPC - \sum_i PC_i \quad (10)$$

As the value of an industrial corporation is determined by range of values, so the attainable synergistic effect is determined by the range of values as well.

5. Acquisition expenses evaluation

If synergistic effect is attainable, it does not mean that the acquisition is advisable. One may say that it is advisable only if this synergistic effect exceeds acquisition expenses. Inequality (11) is the criterion of advisability of synergistically successful acquisition.

$$SE>AE, \quad (11)$$

where *AE* – the acquisition expenses.

Acquisition expenses include:

• transaction costs, i.e. bargaining for acquisition costs, target firm evaluation costs, costs of financial audit of • the target firm, consultants fees etc.;

• severance payments for the target firm employees who have been laid off;

• expenses for the target firm managerial system restructuring;

• investments in synergistic effect generating.

These expenses can be extended in time, that's why we suggest evaluating them by means of discounted cash flow model. Thus, the difference between attainable synergistic effect and integration expenses will determine the values range of maximum allowable goodwill in the price structure of the target company. It makes no economic sense to pay to a target firm the amount that exceeds upper interval limit.

3 Practical relevance of research and conclusion

Our methodology of economic efficiency evaluation of acquisitions in the sector of industry allows to enhance the validity of decisions in such transactions. Using our methodology, acquiring companies' decision makers will be able to compare attainable synergistic effect with the expenses of an acquisition. Limiting the number of synergistically successful acquisition order parameters provides an opportunity to target resources on achieving the most important synergistic effects. We realize that our methodology is limited by the period between bifurcations followed by limited forecast horizon. That is why studying the behavior of an industrial corporation value close to bifurcation points is an important area of our future researches.

The work was supported by Act 211 Government of the Russian Federation, contract № 02.A03.21.0011.

References

1. M. Bradley, A. Desai, E. Han Kim, Journal of financial economics, **21**, **1**, 3-40 (1988)
2. J.W. Hunt, Strategic management journal, **11**, **1**, 69-77 (1990)
3. M.L. Sirower, *The synergy trap: how companies lose the acquisition game* (The Free Press, USA, 1997)
4. A. Agrawal, J. Jaffe, Advances in mergers and acquisitions, **1**, 7-41 (2000)
5. A.E. Ivanov, Economic analysis: theory and practice, **41**, 35-59 (2015)
6. F. Gu, B. Lev, The accounting review, **86**, **6**, 1995-2022 (2011)
7. R. Camodeca, A. Almici, M. Bernardi, Problems and perspectives in management, **11**, **3**, 17-23 (2013)
8. W. Chen, P.K. Shroff, I. Zhang, *Fair value accounting: consequences of booking market-driven goodwill impairment* (Working paper, SSRN, 2014)
9. Ch.E. Jordan, S.J. Clark, International journal of economics and finance, 7, 9, 159-168 (2015)
10. R.J. Alerding, *The Excess Earnings Method, A BVR Webinar Handbook* (2013)
11. J.A. Ohlson, Contemporary accounting research, **11**, **2**, 661-687 (1995)
12. H. Haken, *Synergetics: Introduction and Advanced Topics* (Springer, Germany, 2011)
13. A. Damodaran, *Investment Valuation* (John Wiley & Sons, Inc., USA, 1996)
14. A.E. Ivanov, Economic analysis: theory and practice, **41**, 60-70 (2013)
15. T.L. Saaty, *The Analytic Hierarchy Process* (McGraw-Hill, USA, 1980)
16. A.E. Ivanov, **42**, 33-42 (2012)
17. M.M. Bekier, A.J. Bogardus, T. Oldham, The McKinsey quarterly, **4**, 6-9 (2001)
18. S.A. Christofferson, R.S. McNish, D.L. Sias, McKinsey quarterly, **2**, 92-99 (2004)
19. L.A. Zadeh, Information and control, **8**, **3**, 338-353 (1965)
20. L.A. Zadeh, Information sciences, **1**, 119-249 (1975)

Efficiency increase of complex production and transport systems management

S. Kornilov[1,*], V. Lukjanov[1], and A. Antonov[1]

[1]Mining Engineering and Transport Institute, Nosov Magnitogorsk State Technical University, Magnitogorsk, Russia

Abstract. This article deals with the problem of the reduced efficiency of management in complex production - transport systems due to the lack of co-ordination in the operation of industrial enterprises and transport carrying out their maintenance. The existing transport service schedules for auxiliary departments do not take into account possible changes in operating conditions, the probability of malfunctions and the amount of reserves, which leads to an increase in general production costs. To solve this problem, we propose to use the interval regulation of production and transport processes in all departments of the complex production and transport systems. Also, such regulation involves the determination of traffic service priority. This will allow passing on from the regulated control of production and transport processes to the situational one, adapted to specific conditions, and reducing losses from untimely transport servicing, which will lead to a stores reduction and efficiency increase of the enterprise circulating facilities use. Testing the effectiveness of interval regulation was performed on the system and dynamics simulation model of liquid iron transportation in the oxygen converter shop of the metallurgical enterprise. It was established that the use of interval regulation processes in iron production and its transportation will allow decreasing non-productive downtime by 21% and the amount of the liquid iron in anticipation of recasting in the oxygen converter shop – by 33%. Economical effect of reducing the liquid iron downtime during transportation to the oxygen converter shop will be about 30 million rubles per year.

1 Introduction

Due to the market development in the country, there is a need to strengthen the links between suppliers, manufacturers and consumers. Transport plays a very important role in this process. Timely organization of transport maintenance allows coordinating the effective work of industrial enterprises in such a complex production and transportation system, as "supplier - manufacturer - consumer" (SMC). By efficiency we mean transport maintenance at certain times due to the SMC process constituents, while stocks of raw materials and finished products should be minimized to speed up the turnover of enterprises working capital.

2 Urgency of the management efficiency problem in the SMC system

One of the factors limiting the efficiency of the enterprises included in the SMC system is the lack of interaction coordination between enterprises, suppliers, companies-producers and enterprises-consumers, as well as transport, carrying out their service and their intercommunication [1, 2, 17]. This factor is stipulated by the lack of centralized control system for all the above constituent parts of the SMC. Only during the interaction of the main and industrial railway transport enterprises in the metallurgical industry as a result of uncoordinated actions, on average, about 1.5 billion rubles a year for a single enterprise is lost by increasing inventory and slowing down its working capital turnover [3], and because of inconsistency in the transport and industrial production subsystems the costs are increased, on average, by 30% [4].

The complexity of the problems in the SMC management efficiency is caused by interdepartmental dissociation, the difference in the duration of the technological cycles and the individual operations of technological processes, the difference between the applied ways and operational methods of regulating the production and transport processes [5- 8, 15].

3 Problem and theory of the SMC management efficiency increase

To improve the effectiveness of the management it is proposed to consider SMC as a single three-level system of interrelated elements with the appropriate levels of management. The first level (system-wide) implements the management of industrial enterprises belonging to the SMC system, taking into account their interaction and influence of the environment, by coordinating their production programs with main transportation plans. The second level (external) stipulates interaction of each industrial enterprise with the main transport connecting these businesses into a system. The third level (internal) implements

* Corresponding author: kornilov_sn@mail.ru

interaction of production subsystems with the industrial transport within each industrial company.

Taking into account such determination of the SMC, the management technique that allows to co-ordinate the technological cycles on all these levels has been developed. This technique is based on the agreed regulation of the production and transport operation duration. For example, let's consider the control at the internal level on the organization of production subsystems interaction at the metallurgical enterprises (the manufacturer in the SMC system) and its industrial vehicles.

As part of this method the integrated approach to the management of the production and transport processes should be applied, according to which all operations should be considered in connection with each other, i.e., the duration of some operations affects the duration of the others. For example, the turnover of the rolling stock in the transportation of liquid iron to the steelmaking department affects the interval between the pig iron tapping from the blast furnace. In accordance with this, the concept of interval regulation, which is defined as the maximum length of time for which the operation can be performed taking into account the fact that adjacent operation will be accomplished in compliance with the production and transport technology. Therefore, adjustment operations of intervals duration allow the modification of data values in a narrower range than these may vary in practice.

Thus, the interconnection of the related production units by a single technology involves a single mutually-coordinated duration of individual operations taking into consideration the requirements for each production subsystem and transport [9, 16]. Let's consider the interaction of the blast furnace and the oxygen-converter shop (OCS), as well as industrial transport under the existing steel plant conditions (Table. 1). The tabulated data were obtained in 2015 by chronometric observations.

Table 1. Definition of Regulation Interval of the Main Production and Transport Operations in the Conveyance of Liquid Iron in the OCS.

Name of Operation	Duration of the Operation, min			Regulation Interval, min	
	Minimum	*Maximum*	*Minimum meaning over the enterprise*	*Minimum meaning*	*Maximum meaning*
The smelting of pig iron in blast furnaces	30	150	60	45	150
Steel production in converters	30	65	40	30	65*
Transportation of liquid iron to the OCS	20	65	43	20	65
Transportation of hot-metal mixer train	20	60	40	20	60

* the maximum value of the adjustment range is determined issuing from the minimization of production costs, however, in terms of actual production, in case of emergency, it can be increased to a greater value.

In determining the duration change of production and transport operations it is necessary to proceed from the fact that there are certain technological ways permitting to reduce or extend the periods of operations. For example, the intensity of the iron-production process can be changed by using the following methods:

a) changing the oxygen and the gas flow rate in the blast;

b) pressure changing under the blast furnace top;

c) changes in the content of iron and coke in the charge;

g) changes in the load mode of the blast furnace.

The above described methods of intensity regulation for producing pig iron can be used both separately and in total. As a result of the methods application the interval between the pig iron tapping from the blast furnace of the volume of 1500 - 2000 m^3 may range from 30 to 150 minutes (Table 1.).

The duration change of operations is mainly due to the dispatching control. For example, the priority is given to a particular carriage at the expense of other conveyance delays. The change of movement speed and the execution of transport operations are rather strictly specified in the Rules for technical operation, in the Regulations for the movement of trains and shunting operations in rail transport, as well as in the local safety instructions.

The limits of regulation will mainly depend on the specific character of a particular company. Consequently, one of the most important tasks is the accurate determination of the values of intervals, which may change the duration of production and transport operations for each enterprise.

The currently applied schedules of production and transport operations which must ensure reliable operation of the enterprise subsystems contain average values of their duration, which does not always ensure the co-ordination of transport and main production subsystems. Furthermore, they do not consider the possibility of malfunction in production and transport subsystems, the probability of which is high enough in real conditions. In particular, the monitoring carried out for the transportation of liquid iron at the existing steel plant in the March - May 2015 led to the conclusion that the weighted average of the probabilities of failures occurrence is 0.51. Thus, the currently used schedules do not reflect the actual production process and, consequently, lose their practical organizational role.

In this regard, it is proposed to replace rigidly fixed time expenditure for production and transport operations in the existing schedules for the time intervals to perform the necessary operations taking into consideration the relationships between them. Intervals introduction will allow the production and transport subsystems to change the rhythm of the work within the controlled intervals and to pass over from strictly normalized to adaptive (situational) management.

It is advisable to make similar calculations for technological cooperation of enterprises-suppliers, producers and consumers, as well as the main and industrial vehicles. In these cases, it is necessary to adjust the volume of traffic

DOI: 10.1051/shsconf/20173501100

between the SMC companies with regard to their interaction with each other and the availability of the stocks of raw materials and finished products. Since the weighting rules for incoming and outgoing goods are strictly regulated, the traffic regulation will be carried out due to interval changes in its arrival and departure [10, 18, 19]. In general, this can be displayed using the following inequality.

1. Determination of the raw materials supply volume and export of finished products from the company-manufacturer can be fulfilled in the following way:

$$\sum_{m=1}^{M} u_c(t)/t_{nc} - \sum_{i=1}^{I} V_{ic}/T_{ic} \le U_c(t) \le \le u_c(t)/t_{nc} + V_{ic}/T_{ic} \tag{1}$$

$$\sum_{n=1}^{N} u_{ug}(t) | t_{ug} - \sum_{k=1}^{K} V_{iug} | N_{iug} \le U_{ug}(t) \le \le \sum_{n=1}^{N} u_{ug}(t) | t_{ug} + \sum_{k=1}^{K} V_{iug} | N_{iug} \tag{2}$$

where $u_c(t)$ – the amount of raw materials processing during the technological cycle, t;

t_{nc} – the duration of raw materials processing in a single technological unit, hour;

V_{ic} – raw materials storage capacity, t;

V_{izn} – finished goods warehouse capacity, t;

i – the number of raw materials warehouses;

κ – the number of finished products warehouses;

n – the number of units for the production of finished products;

m – the number of units for processing of raw materials;

T_{ic} – standard period of raw materials storage, hour;

T_{izn} – standard shelf life of the finished products, hour;

$U_c(t)$ – the volume of raw materials supply to the plant per unit of time, t;

t_{zn} – the duration of the finished products output, hour;

$u_{zn}(t)$ – the volume of finished products output for the production cycle, t;

$U_{zn}(t)$ – the volume of the finished product removal from the enterprise per unit of time, t.

2. Volume of finished products delivery to the enterprise-consumer can be defined as follows:

$$\sum_{m=1}^{M} u_n(t)/t_{nn} - \sum_{i=1}^{I} V_{in}/T_{in} \le U_n(t) \le \le \sum_{m=1}^{M} u_n(t)/t_{nn} + \sum_{i=1}^{I} V_{in}/T_{in} \tag{3}$$

where $u_n(t)$ – the volume of processed products for the production cycle, t;

t_{nn} – the duration of the products processing in a single processing unit, hour;

V_{in} – storage capacity for the arriving products, t;

i – the number of warehouses for the arriving products;

m – the number of processing units for the arriving products processing;

T_{in} – standard shelf life of the arrived product, hour;

$U_n(t)$ – the volume of products delivery to the enterprise - consumer per time unit, t.

Inequalities (1), (2) and (3), (2) are considered together to determine the volumes and delivery intervals in the operational mode, taking into consideration the availability of raw materials and finished products at the enterprises.

Supply regulation throughout the whole way of the route is a very difficult task in practice, besides during the delivery time to the recipient company the need for the goods may be changed.

In this connection, it is proposed to carry out supply control at the adjacent highways to the company, which is designated as the area of attraction to the industrial enterprise (AAIE). AAIE dimensions should be selected in accordance with the time limits - 3-8 hours prior to the goods delivery to the enterprise, depending on the transport infrastructure. The lower control limit is determined by the necessity of accurate information pass about the goods arrival, and the top one – by the average duration of a work shift in the main production unit. With an average speed of goods delivery up to 450 km / day, the AAIE border can be taken within a radius from 50 to 150 kilometers from the industrial enterprise.

When controlling the raw materials supply and removal of finished products within the AAIE the goods arrival and departure intervals obtained from (1), (2), (3) may have close values. In this case, the problem of determining priority of the arrival and departure of goods arises [11, 20, 21]. In this connection, it is proposed to adjust the intervals of arrival and departure separately for each kind of goods and its destination according to the urgency factor which is given by the formulas:

- for the goods arriving at the enterprise

$$K_{ср.пр} = Q_{тек.скл} / Q_{max} \tag{4}$$

- for the finished product ready for departure

$$K_{ср.от} = 1 - (Q_{тек.скл} / Q_{max}), \tag{5}$$

where $Q_{тек.скл.}$ – is the presence of the goods in stock at the moment, t;

Q_{max} – is the maximum storage capacity for the given type of goods, t.

The values of the urgency coefficients may vary from zero to one. At the same time, the highest priority will correspond to a value of zero.

4 Approval of Interval Control Procedures and Determination of Their Effectiveness

To determine the effectiveness of the proposed interval regulation methods the system-dynamic simulation model of the liquid iron transportation from the blast furnace plant to the OCS of the existing steel plant has been constructed. The computer program AnyLogic was used as a simulation tool [12, 13, 14]. Indicators, in particular, the time loss during the liquid iron and transportation operations in the area of the blast furnace shop - OCS at the existing metallurgical enterprise, in March - May, 2015, were compared with the results of simulations when using the developed techniques of the interval control (Table 2.).

On the basis of a simulation modeling it was determined that the weighted average value of all the downtime was reduced by 21%, and the one in the rolling stock with hot iron resulted by 33%. The economic effect only from the reduction of liquid iron temperature losses by limiting the downtime during its transportation to the OCS will be about 30 million rubles per year.

Table 2. The Time Lost During the Liquid Iron Transportation in the OCS.

Name of Inter-operation Downtime	Probability of Downtime		Downtime Average Value, min	
	In real conditions	*On the simulation model*	*In real conditions*	*On the simulation model*
Waiting for free rail ways while driving the locomotive under the blast furnace	0.18	0.15	8.0	6.5
Waiting for driving the rolling stock away from under the blast furnaces for technological reasons	0.74	0.68	12.2	10.8
Waiting for free rail ways when driving the rolling stock away from under the blast furnaces	0.21	0.22	8.1	7.6
Waiting for orders for the rolling stock movement in OCS	0.14	0.11	5.8	5.2
Feed delays to the liquid iron transfer department	0.68	0.42	24.8	10.6
Delays in the OCS and the expectation of the rolling stock removal	0.81	0.55	22.6	17.8
Waiting for orders for the rolling stock movement under the blast furnaces	0.13	0.12	5.2	5.0

5 Conclusion

Thus, the suggested interval regulation in the system "Supplier - Transport Services - Manufacturer - Transport Services - Consumer" means the determination of interval control and urgent factors at every level of their interaction and reciprocal linking, which allows coordinating the work rhythms of production and transport facilities.

The application of this approach will enhance the management effectiveness at all levels of the SMC system, thereby, will reduce inventories by 20 – 30% (depending on the technology of work), will increase the use efficiency of the enterprises circulating assets.

References

1. A.N. Rakhmangulov, V.V. Baginova, N.A. Osintsev, The world of transport, **3**, 108-113 (2010)
2. S.N. Kornilov, A.N. Rakhmangulov, A.D. Kolga, Vestnik of Nosov Magnitogorsk State Technical University, **1**, 115-121 (2014)
3. A.N. Rakhmangulov, *Methodological fundamentals of the railway and industrial transport technological systems functioning* (Moscow, 2013)
4. O.V. Fridrihson, S.N. Kornilov, Modern problems of Russian transport complex, **2**, 85-92 (2012).
5. A.N. Rakhmangulov, N.A. Osintsev, Vestnik of Nosov Magnitogorsk State Technical University, **1**, 16-20 (2013)
6. A.N. Antonov, S.N. Kornilov, Moden problems of Russian transport complex, **1**, 95-98 (2011)
7. A. Sładkowski, T. Pamuła, Studies in Systems Decision and Control, **32** (2016)
8. A.N. Rakhmangulov, O.A. Kopylova, Economy of region, **2**, 254-263 (2014)
9. S.N. Kornilov, V.A. Lukjanov, *Proc. Development of mineral deposits*, 194-197 (2004)
10. S.N. Kornilov, V.A. Lukjanov, *Proc. News of the advanced science*, **30**, 79-85 (2014)
11. V.A. Lukjanov, Innovative science, **7**, 43-45 (2014).
12. A.N. Rakhmangulov, A.D. Kolga, N.A. Osintsev, I.N. Stolpovskikh, and A.V. Sladkowski, Transport Problems, **9**, **3** 125-132 (2014)
13. P.N. Michkurov, A.N. Rakhmangulov, Modern problems of Russian transport complex, **2**, 143-152 (2012)
14. A. Rakhmangulov, A. Kolga, N. Osintsev, I. Stolpovskikh, A. Sladkowski, Transport Problems, **10**, **2**, 109-116 (2015)
15. P.A. Kozlov, Transport of Russian Federation, **10**, 38-41 (2007)
16. S.N. Kornilov, K.M. Varzhina, Modern Problems of Russian Transport Complex, **5**, **1**, 12-16 (2015)
17. K.P. Shenfeld, E.A. Sotnikov, V.A. Ivnitsky, Vestnik of the Railway Research Institute, **3**, 3-7 (2012)
18. A. Fugenschuh, H. Homfeld, A. Huck, A. Martin, Z. Yuan, Transportation Science, **42**, 1-14
19. P. Pellegrini, G. Marli`ere, J. Rodriguez, G. Marliere, Transportation Research Part B, **59**, 58-80
20. T.G. Crainic, G. Laporte, European Journal of Operational Research, **3**, 409-438 (1997)
21. A.F. Borodin, E.A. Sotnikov, Railway Transport, **3**, 8-19 (2011)

The improvement in management of material and production reserves of the enterprise in conditions of uncertainty using the fuzzy sets

M.I. Raskatova[1,*], and *E.B.* Golovanov[1]

[1]South Ural State University, Chelyabinsk, Russia

Abstract. In the article a method of management of material and production reserves of industrial enterprises in conditions of uncertainty has been suggested. The method proposes that a part of a benchmark in economic and mathematical models of determining the volume and completion time for orders for raw products and materials be represented as fuzzy numbers. The application of fuzzy-set theory allows taking into account the uncertainty of the external environment without using the theory of probability. The use of the latter is often difficult. The criterion for choosing the optimal management strategy is the minimization of the objective function of the total cost connected with reserves management. It is suggested to obtain the benchmarks that are expressed by fuzzy numbers with a method of expert evaluation. Using the results obtained with the suggested approach the decision-maker will be able to form a concrete strategy for reserves management in a constantly changing market situation. The suggested technique is universal; its application is possible in various industries.

1 Introduction

Nowadays in the conditions of market economy at industrial enterprises increasingly more attention is paid to improving the system of supply and, in particular, management of material and production reserves, which have a significant share in the cost of finished products production. One would think that much has already been said on the topic of reserves management. There is a large number of methods and models of reserves management. However, determinate models are good for studying this question in a theoretical aspect. In reality enterprises operate in a market environment which is uncertain. In these cases, the management quality can and should be improved just through the use of models that take into account uncertainties. Today the first and the most common method of taking into account the uncertainty is the theory of probability. However, the application of methods of probability is difficult and not always possible due to the high complexity, and often due to the impossibility of collecting all the necessary information on nomenclature of thousands of materials that are used at the enterprise. In addition, when collecting data in a representative volume, conditions of statistical stability and permanence of the situation are necessary, which is impossible, because all enterprises operate in a constantly changing external environment.

Since methods of the theory of probability have such limits, then the question arises: is there any method which allows avoiding these limits?

We suggest moving from the classical formulation of the reserves management problem in which all the parameters are expressed in terms of familiar to us distinct numbers to the same crucial task, but which is formulated and solved using tools of fuzzy sets (fuzzy logic).

2 Main part

Scientists have been preoccupied with problems that concern reserves management for quite a long time. In the twentieth century a large number of methods and reserves management models in determinate and non-determinate conditions were developed [1-8].

In the case when all parameters are set exactly (determinate case), it has the following formulation. Enterprises created a certain amount of material and production reserves which contributes to their non-stop operation. From time to time it is required to replenish reserves. It is necessary to determine the frequency of orders and the size of orders of raw products and materials. A large number of reserves leads to high costs for their storage and freezing of assets, ordering small quantities leads to higher transportation costs. The lack of material and production reserves at the enterprise can lead to a stoppage in production. It is required to determine the optimal reserves management strategy (volumes of orders and frequency of orders), ensuring minimization of the total cost of the establishment and replenishment of a reserve during the planning period.

We suggest using methods of dynamic programming as instruments in order to solve the problem where the decision-making process is divided

* Corresponding author: mirru@yandex.ru

DOI: 10.1051/shsconf/20173501101

into separate stages (steps) [9,10], and simulation modeling [10,11] that allows to simulate various options and to determine the range of acceptable strategies of management of material and production reserves of industrial enterprises that satisfy the limit system. A choice criterion of the optimal strategy is the minimum of the objective function of the total costs connected with reserves management.

$$F^r = \sum_{j=1}^{N} Z_j^r \rightarrow min, \; j = 1,...,N,$$

where Z_j^r – total cost, that appears during j- amount of time in the case of a strategy implementation r.

Total costs consist of five kinds of costs that have the greatest influence on the formation of the delivery plan. These include resource acquisition costs, transportation costs, reserves storage, ordering and costs, caused by the binding of circulating assets (lost profit).

For the purpose of making effective decisions in reserves management of the enterprise in conditions of instability and uncertainty in the external environment in this mathematical model one cluster is added that allows to work with fuzzy benchmarks using the fuzzy-set theory [12-14]. Since the mid-twentieth century the widespread use of models has been started that are based on fuzzy logic in industries from home appliances and to management of complex technological systems and dynamic objects (robots, airplanes, cars, etc.) [15].

The use of the fuzzy logic in solving economic problems is not yet sufficiently widespread. Russian scientists are considering the application of fuzzy-set theory in solving problems of assessing the financial condition of the company, assessment of efficiency of the investment project, risk assessment of the securities portfolio of investment [17-20].

In its application in the management of enterprises a triangular type of a fuzzy number is often used as predictive numbers of benchmarks, in which the most expected value of the parameter (the top of the triangle) and the range of variation of this value (the base of the triangle) are set. The corresponding triangular number X is being written in the following form: X= (x_l, $\bar{x}$, x_r), where $\bar{x}$ is the expected value of a fuzzy number (a real number) , x_l and x_r are respectively the left and right borders of the fuzzy number (real numbers) [14]. In economics the method of extraction of three significant points of benchmarks is often used. Often the pessimistic, most probable and optimistic scenarios of the solution are considered to be these points. However, in this case subjective probabilities are used, the actual value of which can neither be defined nor assigned [20]. In this regard, we suggest using the fuzzy-sets theory, which allows replacing the notion of randomness with the notions of possibility and expectation.

A method of optimal reserves management of raw products and materials

Our suggested method of optimal reserves management of raw products and materials, mediated in the developed economic-mathematical model consists of the following steps:

1. Benchmarks collection.

The benchmark in the model is presented in two forms: distinct and fuzzy. Distinct benchmarks are: the value of safety stock, maximum storage capacity of a warehouse, initial and final reserves of material in a warehouse, cost of storage unit, the capacity of one transport unit, the minimum and maximum delivery amounts.

The benchmarks expressed by fuzzy numbers are the expected demand for materials in each time interval, the expected price of the material and the expected cost of transportation by one vehicle. These data were obtained with the method of expert evaluation.

2. Preparing the total of benchmarks obtained on the basis of an experts survey.

3. Carrying out calculations according to the model of reserves management that is discussed above.

The result of the calculation is the optimal strategy for reserves management, i.e., amounts of orders and time for order replenishment. The criterion for choosing the optimal strategy is the minimization of the total cost over the entire planning period.

4. Since some benchmarks of the studied model of reserves management are suggested to be presented in the fuzzy form, the results of the calculations are fuzzy as well. There are a lot of ways of converting fuzzy numbers into distinct ones [21,22]. We suggest providing a solution in the form of a fuzzy number, thus, it helps to avoid simplifications and the loss of some information during the transition to a clearly determinate solution. On the basis of the findings and recommendations, the decision-maker can use his experience, knowledge, information, that are not considered in the model and he can form a specific strategy for reserves management.

However, in order to ensure that the employee, who does not possess the knowledge of fuzzy sets, could make decisions on the basis of the suggested model, the program has a defuzzification stage, i.e. conversion of fuzzy numbers into a familiar distinct form. In the practice of the application of the suggested method conversion of fuzzy values into distinct order amounts is the most appropriate on the basis of total vehicle load, which will allow reducing the cost of transportation.

3 Results and discussion

Approbation of our developed model has been carried out on one group of furniture enterprises of Chelyabinsk, and in particular on CJSC Factory of special joinery "Cabinetmaker".

Table 1. The expected requirements of a MDF pack 10 mm.

Time intervals	Expert 1			Expert 2			Expert 3		
	Left border	*Average number*	*Right border*	*Left border*	*Average number*	*Right border*	*Left border*	*Average number*	*Right border*
1	31	33	36	31	34	36	30	33	36
2	30	33	36	31	34	37	30	33	36
3	31	34	35	29	33	36	30	33	35
4	30	32	35	30	33	37	29	33	36
5	28	29	31	27	31	34	30	33	36
6	31	33	36	29	33	36	31	34	37
7	31	34	37	32	34	37	31	34	37
8	30	33	34	30	33	37	31	34	37

For this purpose, at the enterprise a group of experts was established, including the head of procurement department, the head of production and warehouse. They provided with the data on the expected material requirements in each time interval, on the expected price of the material and on the expected cost of transportation by one vehicle that areexpressed by fuzzy numbers, i.e. experts set the less expected, the most probable and the most expected value of the indicator.

On the basis of the provided by experts data with the help of additional information regarding the value of safety stock, the maximum storage capacity of a warehouse, the initial and final stocks of material in the warehouse, the cost of a storage unit, the capacity of one transportation unit, the minimum and maximum delivery amounts, the optimal strategy of reserves management has been determined, i.e. delivery amounts and time of replenishment has been found. As an example there are survey data belonging to three experts of CJSC Factory of special joinery "Cabinetmaker" regarding the expected requirements of MDF 10 mm (table 1). The amount of the considered time interval is one week.

In the same way the experts expressed their opinions on two other indicators which are presented in the model in the fuzzy form: the expected price of a MDF pack 10 mm and the expected cost of transportation by one vehicle. Then using the algorithm for calculation of the generalized expert evaluations [23,24] the total of benchmarks has been prepared (table 2). The results of calculations are listed with the help of the suggested model of reserves management in table 3.

It is clear that the most expected value of the minimum cost of the purchased MDF material 10 mm will be 15 665 070 rubles. In the conditions of uncertainty of the benchmarks the variation of the total cost will be from 14 832 432 rubles up to 16 525 086 rubles. CJSC "Cabinetmaker" orders 34 MDF packs every week. This will cost 16 382 435 rubles for 2 months (8 time intervals).

Table 2. The expected MDF 10 mm, packs.

Time intervals	Requirements, number of packs			The price of a MDF pack, rubles			Transportation cost, rubles		
	Left border	*Average number*	*Right border*	*Left border*	*Average number*	*Right border*	*Left border*	*Average number*	*Right border*
1	31	34	36	42842	45277	47260	56700	59128	61544
2	31	35	36	43072	45677	46980	56875	59560	62226
3	30	34	35	42672	45140	47630	57720	60317	62856
4	30	33	36	42842	45473	47365	56780	59060	61340
5	28	34	35	42334	44944	46348	56802	59435	62382
6	30	33	36	43372	44344	47513	56436	59724	61582
7	31	34	37	43405	45507	45346	57284	59350	62383
8	31	33	37	43338	44643	46750	57648	60340	62765

Table 3. The delivery plan of mdf sheets 10 мм, packs.

Time interval	Left border	Average number	Right border
1	31	33	36
2	31	33	35
3	60	65	67
4	0	0	0
5	68	72	73
6	0	0	0
7	64	67	69
8	0	0	0
General costs, rubles	14 832 432	15 665 070	16 525 086

The total cost that corresponds to the delivery plan, calculated with the help of the model for the same time period will be 15 665 070 rubles. Thus, the suggested strategy will reduce the cost to 717365 rubles, or by 4.6%. When considering the annual interval the cost saving will be about 4300 thousand rubles in the absence of major changes in production amounts, prices, etc.

4 Conclusion

The suggested method can calculate storage requirements in other materials that are necessary for the manufacturing process in the factory. The suggested model and methodology are universal; their application is possible at industrial enterprises of other industries.

Thus, the activities of industrial enterprises in a market economy are strongly influenced by the instability and uncertainty of the environment. Therefore, one of the most urgent problems of improving reserves management of raw products and materials at enterprises is finding solutions that take into account the conditions of uncertainty using fuzzy-set approach.

Application of the developed method allows reducing the costs of reserves management which will ultimately lead to an improvement in the overall efficiency of the enterprise.

The implementation of a new methodological approach to the reserves management of raw products and materials will contribute to the growth of efficiency in industry, as an instrument of tactical and strategic enterprise management.

The work was supported by Act 211 Government of the Russian Federation, contract № 02.A03.21.0011.

References

1. A.M. Zevakov, *Logistics of material reserves and financial assets* (Piter, Saint-Petersburg, 2005)
2. K. Lysons, M. Gillingham, *Purchasing and supply chain management* (Infra-M, Moscow, 2005)
3. Michael R. Leenders, Harold E. Fearon, *Purchasing and supply management* (Victoria plus Ltd, Saint-Petersburg, 2002)
4. V.A. Lototsky, A.S. Mandel, *Models and methods of supply management* (Science, Moscow, 1991)
5. S.R. Mikityants, N.N. Goldobina, *Application of mathematical methods in supply management* (LSEI, Leningrad, 1982)
6. G.B. Roubalsky, *Probabilistic and computational methods of optimal supply management* (Knowledge, Moscow, 1987)
7. Oliver W. Wight, *Production and inventory management* (Progress, Moscow, 1978)
8. G. Hadley, Thomson M. Whitin, *Analysys of inventory systems* (Science - Physmath, Mocsow, 1969)
9. N.Sh. Kremer, I.M. Trishin, M.N. Fridman, *Operations research in economics: teaching aid for institutes of higher education* (Banks and Exchanges, UNITY, Moscow, 1997)
10. I.L. Kalihman, M.A. Voitenko, *Dynamic programming* (Higher school, Moscow, 1979)
11. Robert E. Shannon, *Systems simulation modelling - arts and science* (Mir, Moscow, 1978)
12. A. Kofman, Alukha Kh. Khil, *Introduction to the theory of fuzzy sets in the management of enterprises* (Vusheyshaya shkola, Minsk, 1992)
13. A.O. Nedosekin, *Methodological basis of financial activities modelling applying fuzzy-multiple descriptions* (Saint-Petersburg, 2003)
14. L.A. Zadeh, Fuzzy Sets and Systems, **1**, **1** (1978)
15. M.I. Raskatova, Bulletin of Chelyabinsk state university. Series: Economics, **5**, 119-126 (2007)
16. M. Hudec, Proc. Information, Communication and Automation Technologies (ICAT), 1-6 (2015)
17. R.J. Almeida, N. Basturk, U. Kaymak, *Proc. Computational Intelligence for Financial Engineering & Economics (CIFEr)*, 497-504 (2014)
18. V.P. Meshalkin, A.Y. Belozersky, Transport business in Russia, **2**, 189-191 (2011)
19. J. Grable, R.H. Lytton, Financial Services Rewiew, 8, 163-181 (1999)
20. A.O. Nedosekin, Audit and finfncial analysys, **2**, 138-161 (2000).
21. S.V. Matsievsky, *Fuzzy set: a teaching aid* (KSU, Kaliningrad, 2004)
22. A.N. Borisov, O.A. Kroumberg, I.P. Fedorov, *Decision-making on the basis of fuzzy models: examples of application* (Zinatie, Riga, 1990)
23. P. Jackson, *Introduction to expert systems* (Publishing house "Wiliams", Moscow, 2001)
24. L.G. Evlanov, V.A. Koutouzov, *Expert judgements in management* (Econoimics, Moscow, 1978)

Evaluating of innovative projects' effectiveness at industrial enterprises

A. Puryaev[1,*]

[1]Production Management, Kazan Federal University, Naberezhnye Chelny Institute, Kazan, Russia

Abstract. The problem of assessing of innovative projects' effectiveness in industry from the perspective of integrated and compromise approach is actualized. Taking into consideration main directions of the Russian economy development the tooling of efficiency estimation of innovative projects is offered. Evaluating of the effectiveness is presented as solution of multi-criteria optimization problem (using the theory of fuzzy sets and theory of aggregation). The complex of efficiency measurement parameters is developed, which consists of group of critical parameters and economic parameters. Also we offer the mathematical apparatus of innovative projects' efficiency evaluation. The technique, the sequence in the process of evaluating of innovative projects' effectiveness in the industrial enterprises are described in present work.

1 Introduction

Evaluating of investment projects' effectiveness which are related to the introduction of innovations (substitutions) in the industry, i.e., evaluation of innovative projects in the industry from our point of view can not be based on traditional economic (cost) effectiveness evaluation. It would be even a mistake. The innovation allows to obtain the effect in non-economic systems, in most cases impossible being adequately assessed from the economical angle, especially with the help of the ruble and using officially accepted valuation methods (the ratio of benefits to cost, the calculation of cash flows, Internal rate of return – *IRR*, Net present value – *NPV* and etc.). Implementation of innovative projects can significantly improve the competitiveness of products, reduce production costs and supply the commercial products with new consumer properties. And these properties will give effect in other related sub-systems of activity also, especially in the perspective.

It is obvious that any innovation should be new and practically implemented, bring economic or other benefits. For carrying out the process of innovation requires the effective project is essential. Innovation project is a rationale for the introduction of innovation (substitution in particular) and the implementation process itself innovate by the means of investing [1, 2]. The innovative project is a form of construction of the innovative activity of the enterprise with a fixed term, steps and resources aimed at specific results. The advisability of investing for the innovation development is necessary to substantiate so the exact innovative project to satisfy not just economic indicators, but a whole range of settings, which are summarized in the only synthesis criterion. Evaluating of innovative projects' effectiveness is a separate area of study in the field of design solutions, and its methodology goes beyond traditional approaches to the analysis of innovation.

2 Method and research problem

The objective of the study is to suggest a method (approach) for evaluating of innovative projects' effectiveness in industries, based on a single criterion, which summarizes the different estimated parameters of physical nature.

Research methods and solutions of the problem: analysis of the evaluating parameters used in the industry (literature review); application of the theory of fuzzy sets and aggregation theory in the development of the method (approach) of evaluating of the innovative projects' effectiveness.

Results: The methodology (method) of evaluating the effectiveness of (optimality) innovative projects; complex of partial parameter estimation of innovative industrial settings' efficiency; mathematical tooling that allows you to roll quantitative and qualitative parameters out, different in their physical nature (clear and fuzzy sets).

3 Research

To solve the problem must first it is necessary to determine the private parameters of projects' efficiency evaluation. This was achieved by examining existing possible parameters of evaluating in various areas: environmental [3, 4], social [5, 6],

* Corresponding author: aidarp@mail.ru

technological [7, 8], economic [9], national security field [10, 11] and the scientific-technical field [1, 12]. A special role in the development of estimation complex of parameters was played by the program of priority development of science and technology and the list of critical technologies approved on the 7th of July 2011 [13, 14].

In the modern theory of the effectiveness particular relevance is paid to the account in the evaluation process of both quantitative and qualitative factors and parameters different itself and physical nature, which are affecting the efficient operation of production systems in terms of non-linearity (the presence of unforeseen circumstances, "tears", "jumps" or breakthroughs; falls or "explosions", "catastrophes") and uncertainty (under condition of mixed asymmetric information).

The study in a whole has allowed to establish two groups of private valuation parameters of various physical essence, both quantitative and qualitative. Quantitative are encouraged to be submitted in the form of fuzzy sets and fuzzy sets (linguistic variables) with their own term set. The theory of fuzzy sets allows that element can belong to the set, being at the same time with a positive degree of truth, and with a different value of positive degree of truth it cannot belong to the set [15, 16]. If linguistic evaluations are subjective and approximate (which is often can be observed in our case), then to describe them triangular membership functions suit in this case (the simplest in the calculations). Qualitative are encouraged to be submitted in the form of scoring on a specific scale [8]: four-point scale with five gradations.

For the integration process or generalization of all grades in a single criterion (selection) it is proposed to apply the method of desirability function by Harrington, which is considered to be versatile, accurate and appropriate to the task [17-20] using the method of generalization (aggregation).

4 Results

Thus, as a result of the undertaken research of question which is connected with evaluating of innovative projects' effectiveness at the industrial enterprises:

The complex of private evaluation parameters, composed of two groups is worked out:

- a group of critical parameters of efficiency evaluation of innovative projects in industrial enterprises (consists of a sub-group of quality parameters and quantitative parameters). See TABLE I;
- a group of economic parameters of efficiency evaluation of innovative projects in industrial enterprises (officially adopted indicators for assessing the effectiveness of investment projects). See TABLE II.

Table 1. The critical parameters of efficiency evaluation.

№	Group of critical parameters of efficiency evaluation	The essence and rate setting
1.1	Subgroup of qualitative parameters	
1	Compliance with the priority areas of science and technology development (8 directions) and to the list of critical technologies of the Russian Federation (27 technologies).	It matches one or more item (is received and passes the next selection, evaluation) / does not match any (rejected, excluded as inefficient). The list of priority areas: 1. Industry of nanosystems. 2. Information and telecommunication systems. 3. Life sciences. 4. Environmental management. 5. Transportation and Space Systems. 6. Energy efficiency, energy conserving, nuclear energy. 7. Promising kinds of weapons, military and special equipment. 8. Security and counter- action to terrorism. The list of critical technologies is presented in reference [14].
2	The fact of import and (or) the uniqueness	If there is the fact of import substitution during the implementation of an innovative project, the project is accepted (Yes), and passes the next selection, assessment. If there is no fact- the project is rejected, excluded both inefficient (No).
3	The parameter of uncertainty and risk during the implementation of the project (UR, points)	Quality parameter of estimation. The following figures, which will be evaluated and weighed are recommended to include: • macroeconomic risks (inflation, increasing of the suppliers cost); • uncertainty of strategic resources supply; • uncertainty with the sales network (distribution network, distribution channels); • uncertainty in transaction costs of the project (explicit and implicit, for example, the costs of the third parties impact); • political risks (the change of government, the President, the socio-economic structure);

№	Group of critical parameters of efficiency evaluation	The essence and rate setting
		• uncertainty in the macro-sphere
4	Parameter of quality performance of the project's functions (QPF, points)	Quality parameter estimation. Under the quality of performance functions (QPF) we mean quality of the object that performs or should perform the function (group of functions). By subject the project is meant. For this functions, requirements should be stated and weighting coefficients (importance) of each function in the overall composition should be fixed [7, 8].
1.2	The subgroup of quantitative parameters	
5	Parameter of national security (resource security)	Share index of own strategic resources in the value (as of the beginning of the project prices) of the total volume of strategic resources used in the project. "Resource security" option should be provided in the form of a linguistic variable (LV RS) "share of their own strategic resources from the total volume used in the project strategic resources." The term set is recommended T (RS) = {*inadmissible, satisfactory, good, excellent*}.
6	Parameter of reducing the risk of industrial catastrophes and man-made impacts	Greenhouse gas emissions in carbon dioxide equivalent (Emission Reduction Units). The indicator on greenhouse gas emissions in carbon dioxide equivalent to the term set T (ERU) = {*inadmissible, satisfactory, good, excellent*}. Quantitative value, measured in $tonnCO_{2e}$. Assessment on the given strict restrictions is imposed by the supervisory authority (Decision-maker). It is set as a clear set or fuzzy sets (linguistic variable).
7	Parameter of accelerating the growth in gross domestic product (GDP)	The indicator on the average annual growth rate of the real potential added value (in constant prices of the base year), %. For the base year the year of the beginning of the project should be taken in the case of common framework adopted absence. For LV TΔAV term set T (TΔAV) = {*inadmissible, satisfactory, good, excellent*} is recommended [7].

Table 2. The economic parameters of efficiency evaluation.

№	Group of economic parameters of efficiency evaluation	The essence and rate setting
8	Net present value (*NPV*), rub.	Quantitative indicator of "*Cashflow*" methodology. It should be compared with the strict restriction or desirable level set by the investor and (or) the decision-maker.
9	Internal rate of return (*IRR*), %	Quantitative indicator of "*Cashflow*" methodology. It should be compared with the strict restriction or desirable level set by the investor and (or) the decision-maker.
10	Investment payback period (discounted), T_{PP}, periods of Project.	Quantitative indicator of "*Cashflow*" methodology. It should be compared with the strict restriction or desirable level set by the investor and (or) the decision-maker.
11	Capital investments in the project (*CI*), rub.	Quantitative indicator of "*Cashflow*" methodology. It should be compared with the strict restriction or desirable level set by the investor and (or) the decision-maker.

The given complex of parameters estimation allows to implement comprehensive complex approach to evaluating of the innovative projects' effectiveness. Evaluating the effectiveness of innovative projects becomes the task of multifactor optimization, not only for obtaining the cost economic benefit.

1. The mathematical apparatus that allows to roll quantitative parameters out different in the physical essence (in the form of clear and fuzzy sets) in a single optimization criterion.

To solve this optimization problem the desirability function method by Harrington [17,19], which is:

$$d_{ij} = e^{-e^{-y'_{ij}}} \quad (1)$$

$$y'_{ij} = \frac{(y_{max} - y_{ij})}{y_{max}} \quad (2)$$

$$y'_{ij} = \frac{(y_{ij} - y_{min})}{y_{min}}, \quad (3)$$

where d_{ij} – private desirability function with one-side constraint for the *i*-parameter of *j*-innovation project; y_{max}, y_{min} – upper and lower limits of unilateral restrictions on the *i*-parameter private; y'_{ij} – coded (normalized) value of *i*-private parameter of *j*-innovation project, translated to the desirability scale.

Generalized desirability function by Harrington (optimization criterion) of *j*-innovation project (D_j) is defined as the geometric average of ratio by the formula:

$$D_j = \sqrt[n]{d_{1j} \cdot d_{2j} \cdot d_{3j} \cdot \ldots \cdot d_{ij} \cdot \ldots \cdot d_{nj}} \quad (4)$$

To transfer the parameters of evaluation, presented in the form of linguistic variables (fuzzy sets) in the scale of desirability we have develop our value transfer mechanism of the membership function in the desirability function values [18]. On Figure 1 is a schematic example of the graphic transfer mechanism for parameter estimation "Parameter of national security (resource security)."

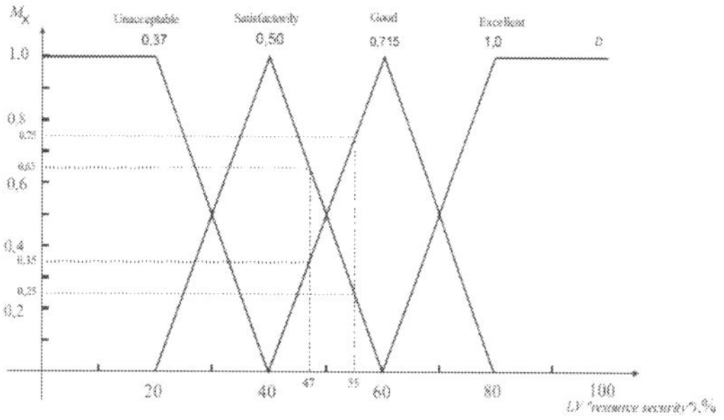

Fig. 1. The functions belonging terms of linguistic variables "Parameter of national security" (LV "resource security").

Below is the method itself (method of approach) of evaluation of innovative projects' effectiveness in the industry.

2. The technique (method, approach) for evaluating the effectiveness of innovative projects at the industrial enterprises has the following strict sequence.

a) Establishing of assessed innovative projects compliance to the presence of private valuation parameters. The possibility of assessment of all the submitted projects of the private parameters would be discovered (see. TABLE I, TABLE II). If all the innovative projects have values to estimated parameters of evaluation, they are comparable (identical), and they can be compared with each other using the results of evaluation. In the case when one project is evaluated so this evaluation step in the procedure is omitted.

b) Setting of limits from the part of the investor (or decision-maker), the supervisory authorities from the values of evaluation parameters in both groups (TABLE I, TABLE II) and their status (min, max, desirable, strict).

c) Transferring of values of critical parameters group (TABLE I) in the value of the desirability function. According to the parameters 1 and 2 subgroup of quality settings is necessary to set private functions of desirability by the following rule. Parameter 1 "*matches*" – $d=0,37$; "*does not match*" – $d=0$ (innovative project is rejected). Parameter 2: "*Yes*" – $d=0,37$; "*No*" – $d=0$ (innovative project is rejected). For the rest of the parameters (qualitative and quantitative) of this group it is necessary to translate values into the meanings of desirability by Harrington according to formulas (1), (2), (3).

d) Calculating of the desirability of a generalized desirability function on the group of critical parameters according to the formula (3). As a result the synthesis intermediate criterion $D_{1\text{-}7j}$ will be received, which allows to reject the options of innovative projects with strict non-compliance with the limits specified by decision-makers (investors, the supervisory authority) according to a group of critical parameter estimates (7 parameters), i.e. when $D_{1\text{-}7j} = 0$.

e) Transferring of values of parameter group of economic parameters (TABLE II) to the desirability function values using the formula (1), (2) and (3).

f) Calculating of the desirability generalized function (optimization criterion) D_j and thus making an assessment of an innovative project effectiveness or setting of the optimal variant of the innovation project (if it is necessary to choose among the existing alternatives). The formula for the calculation is as follows:

$$D_j = \sqrt[n]{D_{1-7j} \cdot d_{8j} \cdot d_{9j} \cdot d_{10j} \cdot d_{11j}} \qquad (5)$$

where $D_{1\text{-}7j}$ - generalized desirability (parameter) of j-innovative project for a group of critical parameter estimates (7 parameters);

d_{8i}, d_{9i}, d_{10i}, d_{11i} - ratios of desirability group of economic parameters (numbered 8,9,10 and 11 of TABLE II, respectively).

5 Conclusion

Evaluating the effectiveness of innovative projects at the industrial enterprise is a multi-criteria task. The effectiveness of innovative projects should be assessed not only by economic indicators, but with a whole parameter complex of evaluating different physical essence. It is necessary to assess the impact of innovation on the quality of life in general. The developed method (methodology) allows to take into account some critical factors (the most important) in the first place. The author proposes a specific list of qualitative and quantitative parameters of this group, which can be supplemented and modified. Projects that do not comply with the limits for critical parameters of assessment must be rejected as ineffective, which do not allow altering the path of industrial development of the enterprise. The remaining projects that were selected by a group of critical parameters should be evaluated according to traditional economic indicators of efficiency, as well as the final selection of determining the optimal project. An innovative project aimed at qualitative change of the development trajectory of industry and society in general will be appropriate, that is, effective.

The proposed approach (methods, methodology) of assessment the effectiveness of innovative projects includes assessment tools: the conceptual apparatus of evaluation, the complex of the parameters estimation consisting of the two groups, the mathematical apparatus of the assessment and allows in the whole to assess innovative projects for effectiveness (optimality). This method serves as an alternative to the traditional economic evaluation of the effectiveness of innovative projects to officially adopted methodology. This technique has the ability to evolve over time and to take into account future development trends of industry and society as a whole.

A further object of the research is the development phase, bringing to the practical application of software products, which allows to facilitate, modernize and instantly respond to emerging changes for improving the methodology for assessing and thus meet the new time challenges in terms of innovative development of the country's economy.

References

1. A.S. Puryaev, E.A. Rybkina, Innovation, **5**, 48-50 (2007)
2. A.S. Puryaev, E.A. Rybkina, Machine builder, **4**, 14-18 (2006)
3. E.V. Hlobystov, *Assessment and modeling of ecological safety of industrial production: regional aspect* (2014) http://www.icfcst.kiev.ua/forform/khlobystov.htm.
4. I.I. Sidorov, *Logistic concept of enterprise management* (Znanie, St. Petersburg, 2001)
5. B.M. Genkin, M.I. Kozlova, Bulletin ENGECON. Economy Series, **4(5)**, 3-9 (2004)
6. S.S. Tkachenko, A.S. Doktorovich, Foundry, **8**, 29-32 (1990)
7. A.S. Puryaev, Modern Applied Science, **9(11)**, 263-275 (2015)
8. G.B. Kac, A.P. Kovalev, *Technical and economic analysis and optimization of machines* (Moscow, Mashinostroenie, 1984).
9. V.V. Kosov, V.N. Livshic, A.G. Shahnazarova, *Methodical recommendations according to efficiency of investment projects* (Jekonomika, Moscow, 2000)
10. V. Mau, Economy questions, **2**, 4-25 (2007)
11. *The Kyoto Protocol to the framework convention of the United Nations on change of climate* (2014). www.74rif.ru.
12. L.I. Abalkin, Economy questions, **12**, 4-19 (2006)
13. L.Je. Mindeli, S.I. Chernyh, Institute for the Study of Science of RAS (2016). http://www.issras.ru/papers/Prior2011_Mindeli.php.
14. *The list of critical technologies of the Russian Federation* (Kremlin, 2016). http://kremlin.ru/supplement/988.
15. L.A. Zade, *Concept of a linguistic variable and its application to adoption of problem decisions* (Mir, Moscow, 1976)
16. K. Asai, *Fuzzy systems for management* (IOS Press, Amsterdam, 1995)
17. E.C. Harrington, Industrial Quality Control, **21**, **10**, 494-498 (1965)
18. Aidar S. Puryaev, International Business Management, **9**, 856-861 (2015)
19. *Chapter 5. The method of desirability function* (2016).http://ecograde.bio.msu.ru/files/library/books/1/glava5.pdf.
20. A.V. Ahaev, *Algorithm of estimation of functional filling of software products on the basis of an indistinct logical conclusion* (2016). http://www.tusur.ru/filearchive/reports-magazine/2013-28-2/169.pdf.

Region innovation and investment development: conceptual theoretical approach and business solutions

D.M. Zozulya[1,*], and *Yu.V.* Smolyar[1]

[1]Don State Technical University, Rostov-on-Don, Russia

Abstract. The article describes essential problems of the region business innovation and investment development under current conditions, issues of crisis restrictions negotiation and innovation-driven economy formation. The relevance of the research is defined by the need of effective tools creation for business innovation and investment development and support, which can be applied, first, to increase efficiency of the region industrial activity, then improve production competitiveness on the innovative basis, overcome existing problems and provide sustainable innovation development in the region. The results of conducted research are represented in the article including region innovation and investment development concept model made up by the authors on the basis of system theoretical approach. The tools of the region innovation development defined in the concept model are briefly reviewed in the article. The most important of them include engineering marketing (marketing of scientific and technical innovations), strategic planning, benchmarking, place marketing and business process modeling.

1 Introduction

According to the Strategy of innovation-driven development of the Russian Federation, the support of innovative business growth at the regional level is considered as one of state policy priorities that determines the need of innovative business development strategies, techniques, models and programs creation for all country regions. The relevance of the research conducted by the authors is defined by the need of effective tools creation for business innovation-investment development and support, which can be applied, first, to increase efficiency of the region industrial activity, then improve production competitiveness on the innovative basis, overcome the existing problems and provide the sustainable innovation development in the region [1]. Under current economic conditions taking into account complicated political and economic environment, the pressure of the Western countries sanctions on the Russian economy [2, 3], it becomes obvious that there is the need and possibility of innovative business problems solution to accelerate the national economy growth as a whole.

2 Problem status and research scientific novelty

The scientific challenge of the conducted research concludes in the need of conceptual approaches, applied models and techniques formulation for the region innovation entrepreneurship development. By now, a number of articles, scientific conference works and theses are published concerning the issues of business innovation and investment development including the regional aspect. The authors have studied the papers of A.A. Dmitriev, V.A. Moiseev, A.K. Klyuev, V.V. Kozyirev, M.V. Charaeva, T.P. Lyubanova, A.Y. Polezhaeva, D.P. Zimin, A.G. Michaylov, E.M. Kozlova, A.G. Ivasenko, Y.I. Nikonova, E.E. Sklyarova, O.N. Kolomyits, V. Blagoev, K. Grenroos, P. Drucker, P. Kotler, J. Culliton, T. Levitt, P. Mazur, J. Scott, J. A. Traut, H. Hershtein, R. Chapman, E.P. Golubkov, Z.K. Okeanova, Y.V. Pashkus, B.A. Soloviev, B.V. Srebnik, O.A. Tretiyak and other scientists in the field of marketing, innovation development, region innovation development, marketing mix, business modeling, innovation infrastructure and strategic planning.

At the same time, it is worth mentioning that specified researches fell short of some theoretical and practical aspects of marketing in the context of innovation business development enabling to evolve competitive strength management functions on the innovative basis for sustainable innovative and investment development of the region in the market environment.

The scientific novelty of the conducted research concludes in the region innovation entrepreneurship development concept model formulation with the use of system and complex approaches and involvement of strategic planning, engineering marketing and place marketing tools.

The practical scholarly importance of the research is expressed in the possibility of concept model implementation in business environment for

* Corresponding author: daria_zozulya@hotmail.com

achievement of sustainable innovation and investment development of region business.

3 Region innovation and investment development concept model

It is commonly known that innovation activity is one of the major factors promoting development of many other activity areas. Fast development of the next-generation technologies justifies the trends of innovation-based economy formation in the developed countries of the world as well as in Russia [4]. However, as we know, a number of barriers and difficulties appear on the way of innovation business development in Russian regions, which in the opinion of the authors include the following:

a) innovation and investment infrastructure problems, including industrial, organizational, personnel, financial, information and international infrastructure [5];

b) the shortage of professional, properly technically, economically and intellectually trained personnel for innovative technological business dealing in conditions of human capital role increase [6];

c) problems of marketing orientation and commercialization of the developed technical products, operations, technologies [7, 8, 9].

Formation of the innovation-driven economy seems impossible without solution of specified problems. On the ground of the research conducted by the authors, study of the available scientific backup and synthesis of a number of hypotheses, the concept model of the region innovation and investment development was made. The model is shown in the figure 1.

Concept model demonstrates the following basic elements of the region innovation development, which are capable to promote the transition to innovation-driven economy upon condition of their cumulative activity:

a) legislative support of the innovation activity not only in the field of quality regulatory support of copyright protection, but also legal framework development for innovation business;

b) innovation activity public awareness campaign in the region with the use of mass media, exhibitions, scientific shows and so on;

c) development of the innovation-oriented culture in the region;

d) operation of the region innovation and investment infrastructure;

e) use of the corresponding tools and innovation development technologies by the innovation enterprises;

f) achievement of the results of efficient innovation development including initiation of new productions, region schools of thought development, support of science, education and business effective interaction, foundation of the region integrated innovation cluster, innovations commercialization support, etc.

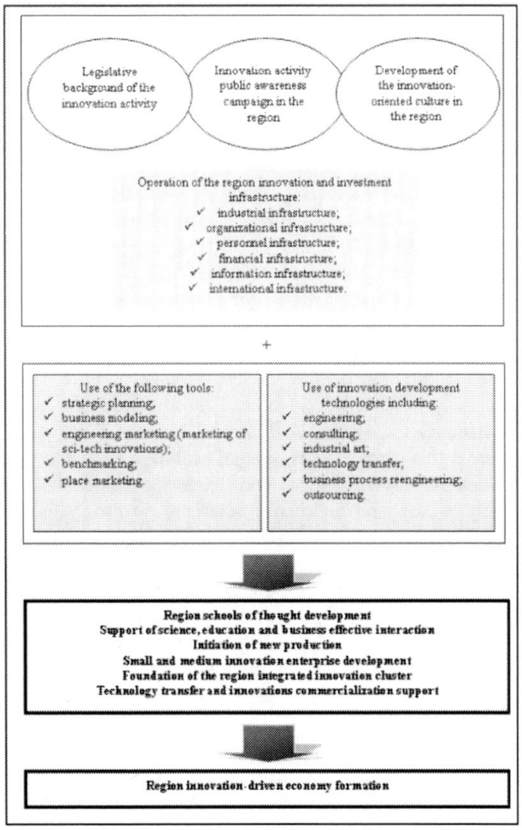

Fig. 1. Region innovationd investment development concept model.

The first block of the region innovation and investment development concept model is presented by mainly state level activity. It should be emphasized that creation and development of innovative and investment infrastructure including all of its parts plays an important role in model possible realization. The industrial infrastructure assumes providing branches and enterprises with technological capabilities, equipment, production capacities, platforms, communications, transport communications, etc. Organizational infrastructure component consists of a number of organizations rendering various services to small and medium innovation business in the field of consulting, communications, premises and equipment rent, initial funding, transport. In general such organizations include business incubators, science and technology parks, technopolises, engineering companies, consulting agencies and so forth. The financial infrastructure is presented by funding investment opportunities, which can be provided to innovation business.

As one of the main problems of innovation and investment development is the lack of funding, credits and investments, it seems necessary to develop specific forms of innovative projects financial support, for example, venture financing, foundation of securities market specially for startup small and medium innovation business with favorable conditions of financial resources raising, organization of independent financial/investment pools for innovation business and so forth. The clusterisation of innovative activity is also one of options of a number of organizational and financial problems solution by means of small and medium innovation enterprises integration in innovation clusters, which represent goal-oriented group of companies, operating on the basis of the scientific knowledge production and highly qualified specialists training centers [10]. At the moment, the activity of twenty five Russian clusters covers six primary industries, including nuclear and radiation technologies, shipbuilding, flying and space vehicles production, pharmaceuticals, biological technologies and medical industry, new materials, chemistry and petroleum chemistry, information technologies and electronics.

The critical problem of personnel infrastructure in the innovation field consists in the shortage of professional, properly technically, economically and intellectually trained personnel for innovation business dealing. This problem shows the need of the system well-balanced education when the student can gain knowledge and master skills not only in basic and engineering sciences, but also in the field of economy, management, marketing (engineering marketing), project and risk management, self-introduction and psychology. One of the most important components of innovative business infrastructure is the information part, the development of which plays an important role in creation, promotion and commercialization of innovative products, operations and services. Information infrastructure part has great marketing significance especially for innovation business suffering from promotion and commercialization difficulties and struggling for competitive advantages. The international component of innovation infrastructure is intended to promote international researches, studies and projects, public awareness campaign of innovation business on a global basis, cross-country technology transfer and so on.

The second block of the region innovation and investment development concept model contains the set of tools and innovation development technologies, which are recommended to be used by innovation enterprises to promote their goals achievement. It is important to note that the analysis of enterprises activity of the Southern Federal District of Russia approved the deficiency of the strategic planning viable mechanism for their development. At the same time, planning is necessary for any enterprise even in crisis conditions because it represents the dynamic process allowing to expect and consider external environment changes, and to adjust enterprise production capabilities and limitations for further development and growth.

The procedure of strategic planning is considered to include not only a simple sequence and interrelation of separate stages, but also maintenance of feedback between them and external environment of the enterprise. Every stage has its logic and promotes optimal decision-making in planning field. The first stage of strategic planning process includes strategic situation analysis and predetermines selection of possible strategy depending on the external and internal environment, taking into account main factors influencing enterprise economy. Key success factors are defined on the basis of enterprise strengths and weaknesses analysis. Strategic management areas qualifying enterprise industry are determined at the second planning stage, as well as strategic management centers, which are selected from enterprise production departments focused on one or several strategic management areas. The third stage is dedicated to the assessment of strategic position with the use of market models. The fourth stage is characterized by enterprise business portfolio selection on the ground of strategic management areas analysis. The fifth planning stage includes development of financial plan, balance of income and expenditure making-out, break-even point calculation, profit forecast. Efficiency of financial policy is defined by use of cash flow model. The sixth stage consists in function planning (technical and economic planning). The seventh stage contains definition of key figures of strategic plan as benchmarks of strategic goals achievement. The structure of key figures can vary as well as the planning period. Key figures of strategic plan can include sales volume, market share, fixed assets value, staffing level, payroll budget and so on. The eighth stage assumes control of all other planning stages depending on market changes, results monitoring and plan updating. Thus, enterprise strategic planning process is focused on definition of the global goal and strategy proved by methods and models of strategic analysis, business portfolio selection and implementation with application of technical and economic planning, and on definition of key figures of the strategic plan [11].

Modern enterprise management should be absolutely market-oriented to gain competitive strength and achieve its goals [12], [13]. Therefore, it seems appropriate to use the tools of place marketing and benchmarking as the comparative analysis study of enterprise activity efficiency and indexes of more successful companies [14], for competitive advantages gaining, for survival and adjustment to external conditions, etc.

Engineering marketing (marketing of engineering activity, marketing of scientific and technical innovations) is supposed to be one of the tools promoting enterprises innovation activity increase, achievement of competitive strength and commercialization of innovations [15]. It is the authors' opinion that scientific and technical innovation means the eventual result of innovation activity (of fundamental studies, exploratory research, applied studies, development, experimental works), which is ready for commercialization. According to T. Levitt, a product is not something engineer talks about, but it is something that consumer needs [16]. Engineering marketing is considered to be a new marketing trend developed to increase engineering activity efficiency by means of engineering solutions market orientation provision. Engineering marketing is the innovative and perspective approach developed by Tatiana P. Lyubanova and Daria M. Zozulya [17]. This new approach is based on the theory that under current conditions all employees of enterprises have to make decisions from the standpoint of market philosophy being the generators of scientific-technological progress able to synthesize enterprise engineering capabilities and consumers' needs successfully.

Such tools as enterprise business model and strategy development on the ground of engineering marketing [18], innovation marketing model drawing up [19], planning with the use of engineering marketing mix, psychoheuristic programming for alternative engineering solutions evaluation and others are used for engineering marketing practical realization in engineers' professional life.

Specific character of engineering marketing mix 9P (people, product, production, price, place, promotion, public relations, provider, processing) in comparison with other concepts assumes that its tools include engineering capacity and are intended for use by engineers (designers, technicians, developers, researchers and so forth) for the purpose of engineering solutions market orientation maintenance. For instance, the first essential tool "people" (meaning consumer needs satisfaction by engineering solutions) includes such way of impact as customization. Customized product is made for specific consumer. Only engineer is capable to develop and produce customized unique product. In addition, he has to synthesize his technical competence, enterprise capability and results of innovations market research and communication with customers.

Examination of engineering marketing mix tools in their interaction is of great importance. Enterprise resources planning with consideration of consumer needs by means of corresponding software use based on specific needs integration into production planning process may serve as a good example. Assortment and production schedule planning is carried out on the basis of consumers requests placed in enterprise data system of suppliers. As a result, product competitiveness can be increased [20].

Except the described above tools of innovation activity growth the concept model of the region innovation and investment development also contains a number of innovation development technologies including engineering, consulting, technology transfer, outsourcing, business process reengineering, industrial art, which can be efficiently practiced by experts for specific problem solution.

Considered blocks of the concept model should function in efficient mode to achieve the goals of innovation development, which, in turn, provide background for the region innovation-driven economy formation.

4 Conclusion

As part of the region innovative and investment development challenges and opportunities research the concept model of the region innovation and investment development was made by the authors. The model is shown and briefly described in the article. Practical realization of the model will promote achievement of innovation development goals, solution of a number of technological business problems and formation of the region innovation-driven economy.

It should be noted that validity of authors' conclusions and research results is confirmed by the application of the following methods and tools: comparison, analysis, synthesis, formalization, statistical methods, economic methods, methods and tools of marketing, engineering marketing, place marketing, Delphi method, business process modeling.

Authors believe that the main task of the follow-up study of this subject consists in the improvement and creation of new tools of innovation business development.

References

1. D.M. Zozulya, V.G. Lisitsin, P.B. Golivkina, V.S. Milovanova, A.A. Mozgaleva, M.E. Povarova, Koncept, **13**, 1856-1860 (2015)
2. Y.S. Toktamysheva, *Parameters system used for evaluation of goals achievement of the region social and economic development strategy* (Aeterna publishing center, Ufa, 2015)
3. Y.S. Toktamysheva, Vestnik Bashkirskogo universiteta, **20**, 142-147 (2015)
4. N.M. Ovanesyan, *Problems of current marketing in Russian economy: theory and practice* Don state technical university publishing center, (Rostov-on-Don, 2014)
5. D.M. Zozulya, Koncept, **4**, 1101-1105 (2013)
6. V.P. Kolesov, *Knowledge-driven economy* (Infra-M publishing center, Moscow, 2008)
7. T.P. Lyubanova, D.M. Zozulya, Vestnik Donskogo gosudarstvennogo technicheskogo universiteta, **10**, **4(47)**, 596-603 (2010)
8. B.A. Soloviev, *Marketing* (Infra-M publishing center, Moscow, 2007)
9. O.A. Tretiyak, *Marketing: new guides of management model* (Infra-M publishing center, Moscow, 2009)
10. B. Twiss, *Scientific and technical innovations management* (Economika publishing center, Moscow, 1989)
11. T.P. Lyubanova, D.M. Zozulya, A.E. Safronov, Modern concepts of scientific studies, **6(15)**, 50-53 (2015)
12. P. Drucker, *Effective management. Business problems and optimal solutions* (Fair-Press publishing center, Moscow, 1998)
13. T. Peters, R. Waterman, *In search of effective management (experience of the best companies)* (Progress publishing center, Moscow, 1986)
14. B.V. Srebnik, *Marketing* (Vyishaya shkola publishing center, Moscow, 2005)
15. T.P. Lyubanova, D.M. Zozulya, *Engineering marketing as a tool of industrial enterprises' development efficiency increase* (LAP LAMBERT, Saarbrucken, 2012)
16. T. Levitt, *The Marketing Mode. Pathways to Corporate Growth* (Irwin, New-York, 1969)
17. D.M. Zozulya, Koncept, **20**, 2106-2110 (2014)
18. D.M. Zozulya, Y.N. Denisenko, Science and Education: Agriculture and economics; predprinimatel'stvo; law and governance, **6(49)**, 43-49 (2014)
19. S.M. Hashin, D.M. Zozulya, A.E. Safronov, *Innovation project management* (Don state technical university publishing center, Rostov-on-Don, 2013)
20. D.M. Zozulya, V.P. Gacenko, High Tech, **7**, 44-49 (2014

Development of a complex system of the enterprise's consumer and partner interaction

Y.V. Astashova[1,*], and *A.I.* Demchenko[1]

[1]South Ural State University, Chelyabinsk, Russia

Abstract. The article considers a methodological approach to formation of the industrial enterprise's consumer and partner interaction system. It shows that the purpose of interaction is a long-term mutually beneficial cooperation or loyalty of consumers and business partners. The offered approach is complex and based on application of tools from different functional spheres of corporate activities. It identifies stages of the interaction process, as well as factors influencing formation and maintaining of loyalty. For building of the interaction system we have used the process approach allowing to ensure management continuity, as well as development of unified indicators of efficiency evaluation. The article presents a model of the interaction system based on the process approach. It develops criteria of the consumer and partner interaction system efficiency based formation of material flow value and cost.

1 Introduction

One of strategic missions of a modern organization is achievement of a dynamic balance with the environment, which key elements are consumers and partners (including suppliers). Availability of stable mutually beneficial links with suppliers and partners, as well as a stable base of loyal customers increases business stability, as well as lays the foundation for its growth and development.

Nevertheless, there is still a widespread approach in business practice, when spheres of the industrial enterprise's consumer and partner interaction are not considered as elements of a uniform system and refer to different functional management loops, which excludes application of a complex approach and, in our opinion, reduces efficiency of interaction processes.

Development of scientific approaches to management of functional business spheres is often focused on solution of tasks within the framework of one discipline sphere, which also restricts applications of the complex approach to consumer and partner interaction management.

2 Urgency of the Issue. Review of Literature

The issue of efficient consumer and partner interaction management is cross-disciplinary. Its individual elements have been developed in detail in different spheres of enterprise management.

Issues of consumer interaction have been traditionally considered in the sphere of marketing in order to form and support consumer loyalty. Currently there is a concept of marketing interaction developing under the auspices of Professor G.L. Bagiev (St. Petersburg). This concept is focused on long-term customer relationship, as well as achievement of objectives of market entities taking part in market processes [1], i.e. it also considers business partners. In this regard, the main focus is made on marketing activities, tasks of business partner interaction management are secondary [2].

The most widespread marketing concept is "relationship marketing" offered by American expert in the sphere of service marketing L. Berry in 1983 as an approach focused on long-term customer relationship [3]. This concept has many interpretations drawing attention to different lines of the company's relationship with its environment. According to Ph. Kotler, relationship marketing is a practice of building long-term mutually beneficial relations with key partners interacting in the market: consumers, suppliers, distributers in order to establish long-term privileged relationship [4]. This definition most fully answers the purposes of this research.

Supplier and partner relationship is the subject of logistics researches. The logistic concept of Supply Chain Management (SCM) is a process of planning, implementation and control of operations of a supply chain in order to satisfy customers' demands as efficiently as possible [5]. Within the framework of the SCM concept a focus is made on integration of the enterprise and its partners (and their business processes) into a supply chain to gain mutual benefits and to satisfy the customer's demands, i.e. provision of a proper product in a proper place and at proper time with minimum expenses.

Within the framework of the SCM concept eight business processes are identified, which essence is

* Corresponding author: julia_asta@mail.ru

considered in the article of Professor O.D. Protsenko [6], including:

• "Consumer relationship management" business process, aimed at customer relationship, development of new products and services;

• "Supplier relationship management" business process aimed at development of reliable links with most important suppliers.

Let us consider two groups of enterprise processes. Firstly, these are business processes focused on consumer and partner interaction, to which marketing techniques and tools can be successfully applied. Secondly, these are business processes ensuring this interaction (determining quality of goods and its compliance with the customers' demands, delivery terms, continuity of the production process, development of new products, etc.).

Thus, we can speak about presence of tendencies for joint use of marketing and logistics techniques at individual sections of the supply chain and in other functional management spheres.

Over the last decade there have appeared publications in the scientific environment devoted to different aspects of marketing and logistics interaction: general issues of interaction between the said spheres and functions [6-10], problems of application of logistics and marketing tools in individual fields and spheres of activity [11-15]. The said publications note similarity of objectives, as well as evaluate potential increase of management performance at combination of the marketing and logistics approach.

In the context of the research problematics it is worth noticing developments in the sphere of management of interaction between business, suppliers, consumers and other related parties. Interaction processes are most widely considered within the framework of SCM, as well as formation of value chains [16-20]. In these researches the logistics approach to relationship management prevails. They analyze forms of interaction between counterparties in the chain and offer principles and mechanisms of interaction in the course of value adding.

Common approaches uniting consumers and suppliers within a uniform interaction system are developed in the stakeholder theory offered by Professor R. Freeman in 1984 [21]. Broadly defined, stakeholders are considered as groups of influence (of related parties), on keeping the balance of which interests the company must be focused. In this regard, respecting of the suppliers' and consumers' interests is considered on a par with the demands of the society, state, investors, credit organizations, etc. [22].

Urgency of development of a complex interaction system is predetermined by economic expediency of keeping permanent relationship both with consumers and partners and suppliers.

Thus, the issue of the research is cross-disciplinary. Implementation of the in-house interaction system requires methodological adjustment of the available concepts in the sphere of consumer and partner interaction and development of a complex approach to interaction management.

3 Statement of Research Objectives

The research objective is development and justification of a complex approach to consumer and partner interaction within the framework of in-house systems.

The interaction system is aimed at long-tern cooperation with the enterprise's consumers and partners, as well as formation of their loyalty. The complex approach is based on identification of factors influencing formation and maintaining of consumer and partner loyalty and contains criteria for evaluation of the interaction system efficiency.

4 Basic Research Findings

Let us consider justification of expediency of building the enterprise's uniform consumer and partner interaction system.

The backbone factor is a common interaction objective. Irrespective of applied approaches (logistics or marketing) and interaction entities (consumers or partners), interaction pursues a common objective – long-term mutually beneficial cooperation and loyalty.

Let us exemplify the requirement of the interaction system unity by fig. 1.

As it is shown in fig. 1, two interaction management subsystems can be outlined for a company: subsystem of company's partners and subsystem of company's consumers. The said subsystems have a common area, which allows to speak about their interaction and to apply the system approach to their management.

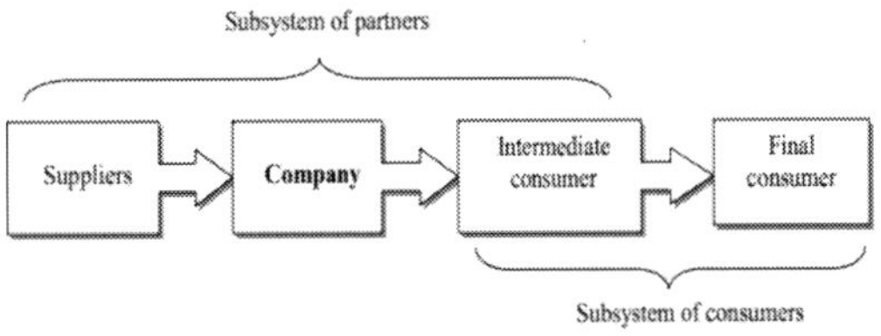

Fig. 1. Structure of the partner and consumer interaction system.

Another factor, which allows to consider consumer and partner interaction as a uniform system, is their interconnection at the level of management tasks.

The company's suppliers and partners form a chain or a network of enterprises jointly working to satisfy the consumers. In this regard, not only final product consumers but also enterprises located below in the chain shall be considered as consumers. Consequently, a chain of enterprises, each of which is a consumer of the enterprise locate above in the chain, is formed through the value chain from the initial supplier to the final consumer. Each enterprise is simultaneously focused on meeting the demands of the partner located below in the chain. Thus, the final consumer's demands are transmitted through the system up to the initial supplier and ensure unity of the chain's efficiency criteria along its entire length. Any failure to meet the demands at any section of the chain reduces the system's ability to satisfy the demands of the final consumer. The most important task of efficient interaction between the system's participants is exclusion of breaks in the demands communication flow, reduction of "hindrances" and corruptions. We can speak about "a stretching principle" of system performance, when the ability to meet the demands of the final consumer and other system participants is a criterion for selection and evaluation of performance of enterprises being the chain participants. Such chain participants interact with each other to achieve their common objective.

Let us consider expediency of application of the complex approach for consumer and partner interaction system management consolidating marketing, logistics and other management approaches and concepts.

Issues of interaction in order to form and maintain customer loyalty fall under the sphere of marketing. For achievement of these objectives marketing applies a marketing complex for segments B2C and B2B. Thus, from the perspective of the value chain the marketing complex can be focused both on consumers and partners.

Use of marketing with respect of suppliers and partners, as well as consumers is based on their satisfaction with a series of interaction factors. Such factors include product quality, cost level, delivery terms and conditions, quality and scope of furnished information, management quality, etc. The influence of marketing on the on the above factors is restricted.

Thus, marketing is not able to ensure high consumer and partner loyalty without application of tools from other functional spheres of the enterprise.

Complexity of the offered approach to interaction assumes application of multi-disciplinary tools, as well as identification of stages within the framework of the interaction process shown in fig. 2.

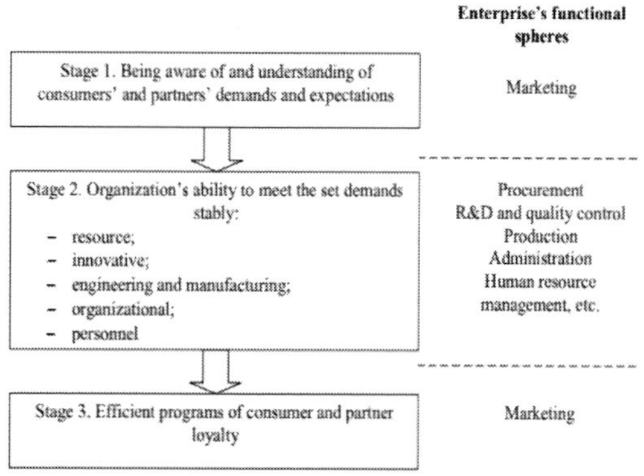

Fig. 2. Stages of the interaction process.

Fragmentary use of the stages shown in fig. 2 or violation of their sequence in practice leads to low efficiency of marketing loyalty programs. Analysis of errors omitted at building of customer loyalty programs shows that unsuccessful programs are most often focused on granting of discounts, bonuses and other motivators of behavior. In this case, a product or service is offered to the consumer, which does not meet his/her expectations, or no competitive price level is provided. As a result, the loyalty program has an adverse effect – loss of customers.

Special attention to product and service quality in interaction processes is also preconditioned by specificity of recently established markets, including the Russian market. Many Russian brands have not yet formed the necessary level of trust among consumers and business partners.

Interconnection of parameters influencing consumer and partner loyalty and determining efficiency of the interaction process is shown in fig. 3.

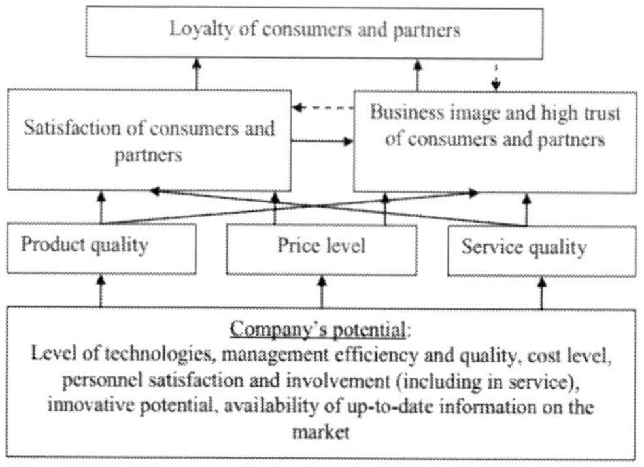

Fig. 3. Model of interconnection of factors influening formation and maintaining of consumer and partner loyalty.

Thus, complexity of the approach to partner and consumer interaction management is based on the following:

• partner interaction and consumer interaction are elements of a uniform system;

• partner and consumer interaction management is based on application of cross-disciplinary tools (marketing, logistics, quality management, human resource management, etc.), and requires involvement of different company divisions in the process;

• interaction management includes all stages of consumer and partner loyalty formation, such as: monitoring and analysis of demands; ensuring consumers' and partners' satisfaction by meeting of these demands; formation and maintaining of loyalty.

Implementation of the complex approach requires development of a methodological approach to building of the interaction system. In our opinion, the process approach to management having the following advantages most fully corresponds to the content of the offered complex approach:

• the process approach ensures consumer orientation by coordination of demands along the entire process chain;

• due to management continuity at the junction of separate processes the process approach allows to integrate different functional spheres of the enterprise;

• use of the process approach allows to establish intercompany systems ensuring management continuity and integrity;

• the process approach enables use of uniform indicators of interaction efficiency evaluation both at the level of the enterprise's processes and within intercompany systems.

The model of the interaction system based on the process approach is shown in fig. 4.

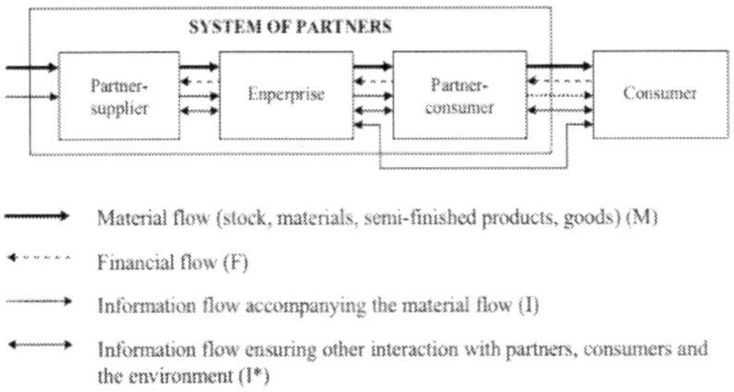

Fig. 4. Proces model of partner and consumer interaction.

As it is shown in fig. 4 partner and consumer interaction can be presented by means of three flows: material, financial and information, which allow to describe interaction in view of the previously identified stages (table 1).

Table 1. Participation of flows in stages of the consumer and partner interaction process.

Stage of interaction	Type of supporting flow
Collection and analysis of demands	I*
Satisfaction of needs	M, I, F
Formation of loyalty	I*

The material flow forms the basis of interaction, therefore, it is suggested to consider its characteristics at development of interaction efficiency indicators.

Let us consider the following characteristics of the material flow: value (v) and cost (c). Value of the material flow is a multi-dimensional characteristics and includes quality of the supplied product, its compliance with the demands and expectations, as well as quality of the accompanying service. Thus,

$$v=f(Q_M, Q_C), \quad (1)$$

where Q_M – quality of material objects (goods, materials, components, etc.);
Q_C – quality of the company's service accompanying the supplied material objects (discipline and completeness of delivery, completeness of information, speed of claims settlement, convenience for the buyer).

The cost of material flow (c) is equal to the value of the expected reverse financial flow. It is determined based on the expenses connected with value adding and level of the desired process profit.

Consumer's (partner's) satisfaction (CS) depends on the flow value v and cost c and can be presented as follows:

$$CS = v/c \quad (2)$$

In this regard, the flow value for the consumer must exceed or at least be equal to its cost, otherwise, consumer's satisfaction cannot be achieved. We can outline two tasks on satisfaction increase: increase of the flow value in the eyes of potential consumers and/or decrease of the flow cost as shown in the expression:

$$CS\uparrow = v\uparrow/c\downarrow \quad (3)$$

It should be noted that the above tasks can be solved both by changes of the material flow parameters (due to improvement of technologies, product and service quality increase) and by means of the informational influence on the consumer, for example, by marketing tools (influencing her/her perception of goods and company in general, forming consumer's expectations, etc.). Development of loyalty programs granting different discounts and bonuses to consumers actually decreases the actual product cost for the consumer.

Let us consider the mechanism of value and flow cost formation implemented within the framework of each process. For positive influence on consumer's satisfaction the flow value growth in the scale of the entire system must exceed its cost growth as shown in fig. 5.

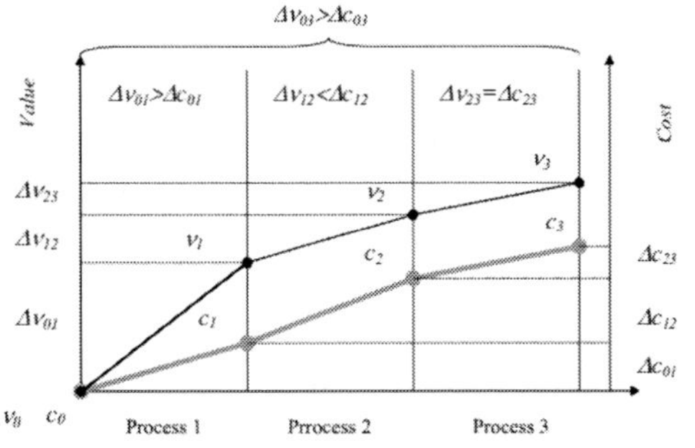

Fig. 1. Material flow value and cost adding in the course of its reformation within the framework of processes.

Let us introduce efficiency criteria for assessment of the consumer and partner interaction system. In general, efficiency can be considered as:

1) ability of the system to achieve the set objectives, which reflects its performance. Whereas the objective of interaction is formation and maintaining of long-term customer and partner relationship, it is expedient to believe that systems or processes, which at the output ensure excess of the flow value above its cost, are efficient.

2) efficiency as ability of the system to achieve objectives in view of the resources spent. It is proposed to believe that interaction systems, in which growth of the flow value exceeds growth of its cost, i.e. $\Delta V - \Delta C > 0$, are efficient.

The correlation of the flow value and cost growth can be assessed not only in the scale of the entire system but within the framework of each process, which allows to make conclusions on contribution of each process to the overall result. Within the framework of the system shown in fig. 5 only process 1 can be acknowledged efficient, whereas its contribution to consumers' satisfaction is determining for the entire system.

5 Conclusions

The offered complex approach uses tools of different functional spheres of activity and is aimed at formation of the industrial enterprise's consumer and partner loyalty. For integration of different elements of the value chain it is proposed to use the process approach. The process approach allows to ensure focusing of the enterprise's activity on the consumer, as well as to form uniform indicators of consumer and partner interaction efficiency.

The work was supported by Act 211 Government of the Russian Federation, contract № 02.A03.21.0011.

References

1. G.L. Bagiev, News of Saint-Petersburg State University of Economics, **4** (2010)
2. I.I. Okolnishnikova, O.U. Juldasheva, Economics & Management Research Journal of Eurasia, **1(1)**, 48-61 (2013)
3. L.L. Berry, G.L. Shostak, G. Upah, Emerging Perspectives on Services Marketing, 25-38 (1983)
4. F. Kotler, *Marketing Management* (Piter Com, St. Petersburg, 1998)
5. R.K. Oliver, M.D. Webber, Logistics: The Strategic Issues, ed. M Christopher, Chapman Hall, 63-75 (1992)
6. O.D. Protsenko, Logistics, **4**, 30-31 (2013)
7. M.A. Vlasova, Economic and Human Sciences, **11**, 77-82 (2011)
8. E.R. Schislyaeva, Practical Marketing, **5**, 9-11 (2004)
9. A.I. Novikov, M.M. Vagramyan, Economics and Management: New Challenges and Prospects, **1**, 304-305 (2010)
10. B.M. Yusupova, A. Porsukova, Issues on Structuration of Economics, **2**, **3**, 19-20 (2008).
11. O.V. Rykalina, Bulletin of G.V. Plekhanov Russian University of Economics, **2**, 87-94 (2010)
12. E.A. Kandrashina, Russian Business, **4**, 55-58, **5**, 80-83 (2005)
13. V.A. Korchagin, Ju.N. Rizaeva, Bulletin of Transport Information, 24-27 (2010).
14. I.S. Shilko, Prospects of Science and Education, **2**, 164-171 (2013)
15. E.Yu. Sarana, European Journal of Natural History, **1**, 44-45 (2010)
16. E.R. Abramova, Academic Bulletin, **2**, 6-9 (2011)
17. S.V. Krasnov, S.V. Sorokin, Bulletin of Volzhsky University named after V.N. Tatishchev, 52-60 (2012)
18. N.P. Karpova, I.V. Fedechkina, Bulletin of Samara State University of Economics, **102**, 39-43 (2012)
19. O.S. Bortnikova, Society: Policy, Economics and Law, **1**, 101-105 (2013)
20. I.V. Gudova, Bulletin of the Siberian State Automobile and Highway Academy, **14**, 61-65 (2009)
21. R.E. Freeman, *Strategic Management: A stakeholder approach* (Pitman, Boston, 1984)
22. M.S. Agafonova, Journal of Economic Theory, **4**, 161-163 (2007)

Graphical-analytic model to form investment policy of industrial enterprise

V.G. Pluzhnikov[1,*], *S.I.* Kukharenko[1], and *S.A.* Shikina[1]

[1]South Ural State University, Chelyabinsk, Russia

Abstract. The article is devoted to the research of graphical-analytic method to form investment policy of industrial enterprise in the capacity of a specific tool to achieve the strategic aim at various stages of life-cycle (LCS). The necessary to formalize the system of estimations is caused by lack of methodological basis of defining quantitative characteristics relying on qualified criteria. To designing graphical-analytic pattern in the capacity of a tool forming investment policy of organisation within the stated development strategy. The comparative analysis of factors that are influencing the investment policy and the stage of organisation life-cycle has been carried out. Theoretical-methodological basis concepts of the development have been considered. All criteria for quantitative estimation of the stage of organisation life-cycle and the performance of investment policy have been proved in the course of the research. The classific characteristics of criteria to estimate the stage of the life-cycle and concepts of investment policy have been formalized. The choice of criteria and interval boundary and its characteristics has been made. The estimation method of the stage of the organisation life-cycle has been introduced. The research, that has been carried out, proves the possibility to develop practical application of graphical-analytic pattern forming investment policy and estimating the stage of life-cycle and type of the investment policy of an analyzed organisation.

1 Introduction

Forming effective investment policy in the frame of the stated development strategy of an organisation is a crucial element involved into the process of strategic management. The concepts of the policy are defined due to subjective and objective factors which influence the functioning and development of an undertaking. One of these factors is the realization of the development strategy of the undertaking should correspond to its stage of the life-cycle (LCS). The scholarly interest in problems of the organisation development at various stages of its life-cycle has increased for the last years (Shirokova 2007). The factors are tightly connected with the fact that each economic system is dependent on subjective assumptions to take some administrative measures. One of the lines of the research is to make an accurate choise of the stated development strategy in the frame of the investment policy. Therefore, a crucial issue of forming the investment policy is to account and control dynamic characteristics of the external environment and investment possibilities of the undertaking to correct its investment policy according to the stated development strategy.

The notion of the investment policy is debatable in scholarly literature. The main definitions of the investment policy. Table 1.

Table 1. Characteristics of the investment policy.

Authors	The concept of the notion
Blank I.A.	A component of the general financial strategy that consists in choosing and realization of the most effective financial investments aimed at high rates of the development and permanent growth of the marketing cost of a company.
Yendovitsky D.A	Factual realization of investment policy (Yendovitsky, 1998)
Charayeva M.V.	A system of arrangements providing a beneficial investment and quick payback period to guarantee the financial stability, financial solvency, increased competitive ability and providing conditions for the future development (Charayeva, 2010).

In the frame of the study the investment policy is considered to be a system of arrangements (business processes) providing a beneficial investment and quick payback period aimed at the realization of strategic objectives for the organisation development. Ultimately, the organisation development, standing on a particular type of the development strategy, may involve the following types of the investment policy. Table 2 (Blank, 2012).

* Corresponding author: Pvg777@bk.ru

Table 2. The main types of the investment policy.

The type	The concept of the notion
Concervative	aimed at minimization of the investment risk as a priority objective
Moderate	aimed at choosing investment objects which obtain mid-market level of the current profitability, rates of capital expansion and risk level.
Aggressive	aimed at maximization of the current profit from the investing of capital the nearest period

The type of the investment policy is defined by the Coefficient of investment activity (Cia).

2 The impact of the development strategy on the investment policy

According to the theory and practice of organisation management there are basic classifications and ways of the investment policy which correspond to a particular type of strategies. The main types and ways of strategies of the investment policy. They present in table 3.

In spite of the traditional point of view new classific characteristics are included in the table: strategies of traditional and specific development. This classification criterion is defined by the correlation of the type of investment policy and the life-cycle stage (LCS):

• strategies of "natural development" are considered as realization of eventual strategic changes in the organisation development using standard strategies and ways of investment policy in accordance with the life-cycle stage (LCS) of the undertaking (Table 3);

• strategies of "specific development" reflect the formation of the business renovation. This process is based on the usage of strategies aimed at diversification of business activities (Table 3).

Thus investment policy is a tool for accomplishing strategic objectives at various life-cycle stages (LCS) of undertaking development. It's characterized by different intensity of the investment policy.

Table 3. The main strategies and ways of the investment policy of natural development at different stages of the life-cycle.

Stages	Types of strategies	Ways of the investment policy
Strategies of the "natural development"		
Birth	The strategy aimed at the exploration of new business types	assimilation, formation new business types, new types of products, technology, materials and new markets; development of new manufacture; recruitment of investors for development of venture business and etc.
Growth	Concentrated growth strategy and/or integrative growth strategy	increase production volumes; capital-raising for acquisition of high-technology equipment and expansion of the circulating capital; Improve results of activity by means of application of new technological systems or new technology; Development of economy management subject by means of expansion with the help of integrative processes within the industrial chain (amalgamation, takeover, consolidation) increase fixed assets based on the engaging of long-term financial sources; optimization of financial sources by means of bringing long-term liabilities, strengthening of financial stability, increase of creditworthiness and investment attraction of an economic entity;
Stabilization	Strategy of economizing costs or integrative growth strategy	preservation of competitive position at the market; providing effective usage of resource organisation possibilities; insurance and acquisition of pertinent guarantees against non-commercial risks and etc. development of the economic entity by means of expansion of integrative processes within industrial chain (amalgamation, takeover, consolidation);
Stagna-tion	Strategy of economizing costs	reduction and restructuring of assets; providing effective usage of resource organisation possibilities; warning of bankruptcy based on winding up of unprofitable enterprises, downsizing, selling a part of the assets and etc.
Strategies of "specific development"		
Stabilization	Strategy of diversification	assimilation, formation of new business types, new types of productes, technology, materials and new markets. development of new manufacture; bringing of long-term capital repurposing, reconstruction, modernization and renovation of separate types of equipment; modernization and implementation of new equipment;
Stagnation	Strategy of diversification	anti-recessionary policy aimed at financial stabilization of the undertaking in the process of crisis recovery; warning of bankruptcy based on winding up of unprofitable enterprises, selling a part of the assets and etc.

3 Impact of the life-cycle stage (LCS) on the business investment policy

There is no a traditional method of the LCS impact on business characteristics in the economic science (Ivashkovskaya and Yangel, 2007). In different surveys (Shirokova, 2007; Kushelevich, 2004; Morgunov, 2006; Glazl, 2000) researchers mention several models/patterns describing organisation development and LCS.

Of special importance can be mentioned: theoretical and empirical models of the LCS organisation of the following authors: Greiner, 2002; Miller, 1984; Lester et al., 2003; Ivashkovskaya, 2007; Shirokova, 2006 and etc. (Lester et al., 2003), (Ivashkovskaya, 2007), (Shirokova, 2006) and etc. The authors argue that the life-cycle is targeting that is made by means of analyzing economic and financial characteristics. However, there is no general understanding of the organisation transition from one stage to another and quantitative estimate methods of qualitative LCS characteristics (Shirokova and Serov, 2006). Although the researchers agree upon the fact that each LCS stage obtains a set of unique qualitative characteristics; sequence of LCS stages; and every stage is a result of the previous one; models have a broad range of organisational contextual characteristics (Shirokova, 2006).

In this work the Greiner's model has been adapted (Greiner, 2002) due to the result of empirical research devoted to the number of stages, choosing of LCS characteristics and transition mechanism from one stage to another. The 5-phased model is included in the LCS theory and constitutes 5 stages of organisation development: birth, growth, stabilization, stagnation and decay (bankruptcy). In this work only the first four stages are considered: birth, growth, stabilization and stagnation. The decay stage is out of the list because the organisation management is subject to specific legal norms (Federal Law, NO 127 - FZ 17.09.09 On Insolvency (Bankruptcy)). So that the 5-point estimate scale of LCS within the description of each point is used. Table 4. (Pluzhnikov and Shikina, 2015).

Table 4. Estimate scale LCS.

No. Name of the stage	Scale meaning	Note
1 Birth	from 0 to 1	The company age is less than 10 years, has informal structure, owner is a head of management. Characteristic meaning can be negative
2 Growth	from 1 to 2	Growth: formalization of management solutions, economic indicators, formalization of organisational structure and investment policy
3 Stabilization	from 2 to 3	Sales gain reduces, deterioration of economic indicators, bureaucratization of the organisational structure and company policy
4 Stagnation	from 3 to 4	Production volume is limited, financial indicators are falling down
5 Decay	from 4 to 5	Production volumes are limited, financial indicators are falling down, deficit of financial resources

The chosen model has shown results of the economic indicators in the process of organisation development. The realization changes in terms of LCS in various aspects: liquidity indicators; possibility to create positive cash flows (Ivashkovskaya, 2007); either for the undertaking itself or for separate business-processes (Pluzhnikov and Shikina, 2014); accounting of aggregates for business growth (Ivashkovskaya, 2007); dependence of investment activity of industrial undertaking (Smagin and Shikina, 2015).

The solution, that is necessary to find out a development stage, is crucial for developing investment policy of the pertinent strategy. Thus, the process for forming particular investment policy should take under consideration the level of business investment activity in correlation with its LCS.

4 Estimation of LCS

Qualitative estimation of LCS is a debatable issue that requires the usage of many theoretical approaches reflecting various market participants' interactions and characteristics. The problem is the lack of traditional theories forming markets and undertakings from the concept of the life-cycle. Contemporary theoretical models cannot choose correctly testing empirical data, so that available characteristics of industrial undertakings are used in order to do research of illustrating variables: organisation age (A), number of employees (N), total assets (TA), total revenue (V), rate of assets increase (I) (Pluzhnikov and Shikina, 2015), Table 5.

Table 5. Criteria of integrated index LCS.

#	Index name	Meaning	Shifts of index			
			Birth	**Growth**	**Stabilization**	**Stagnation**
1	Index of increments of assets	I	≥ 4,0	2,0 ÷4,0	1,0÷2,0	< 1,0
2	Number of people (thous. of people)	N	≤ 0,1	0,1÷1,0	1,0÷10,0	reduce
3	Total Assets (RUB bn)	TA	≤ 0,02	0,02÷1,5	1,5÷15,0	reduce
4	Sales (RUB bn)	V	≤ 0,06	0,06÷1,5	1,5÷25,0	reduce
5	Enterprise age (years)	A	≤ 3,0	3,0÷10,0	10,0÷20,0	20,0÷50,0

The next stage is formalization of interval frame for identification LCS according to the previous stated criteria (Table 5). One part of changing interval frame characteristics of LCS (Table 5) corresponds to Federal Law NO 209 - FZ 24.07.07. On developing small and medium scale entrepreneurship in the Russian Federation; with the Amendments and Additions of 22. 07. 08. № 556 On limits of revenue from sales (works, service) for each category of small and medium business. The other part is carried out by means of expert opinions.

Choosing of numerical expert score of criteria LCS is based on expert opinions of qualitative characteristics. We use the method of quantitative estimation of qualitative characteristics - numerical expert score of Kharrington scale. Due to the chosen method we get quantitative estimation x_i of qualitative characteristics LCS with the usage of formulae 1 – 5:

$$X_1 = \begin{cases} 0,20 & I \geq 4,0 \\ 0,37 & 2,0 \leq I < 4,0 \\ 0,64 & 1,0 \leq I < 2,0 \\ 0,80 & 0 \leq I < 1,0 \\ 1,00 & I < 0 \end{cases} \quad (1)$$

$$X_2 = \begin{cases} 0,20 & N \leq 0,1 \\ 0,37 & 0,1 < N < 1,0 \\ 0,64 & 1,0 \leq N < 10,0 \\ 0,80 & 10,0 \leq N < 20,0 \\ 1,00 & N_t < N_{t-1} \end{cases} \quad (2)$$

$$X_3 = \begin{cases} 0,20 & TA \leq 0,02 \\ 0,37 & 0,02 < TA < 1,5 \\ 0,64 & 1,5 \leq TA < 15,0 \\ 0,80 & 15,0 \leq TA < 50,0 \\ 1,00 & TA_t < TA_{t-1} \end{cases} \quad (3)$$

$$X_4 = \begin{cases} 0,20 & V \leq 0,06 \\ 0,37 & 0,06 < V < 1,5 \\ 0,64 & 1,5 \leq V < 25,0 \\ 0,80 & V \geq 25,0 \\ 1,00 & V_t < V_{t-1} \end{cases} \quad (4)$$

$$X_5 = \begin{cases} 0,20 & A \leq 3,0 \\ 0,37 & 3,0 < A < 10,0 \\ 0,64 & 10,0 \leq A < 20,0 \\ 0,80 & 20,0 \leq A < 50,0 \\ 1,00 & A \geq 50,0 \end{cases} \quad (5)$$

At the next stage we carry out result consolidation x_i with the formula 6 and define the value (estimation) of the integrated index LCS:

$$C_{LC} = \sum_{i=1}^{5} X_i, \quad (6)$$

where C_{LC} - are integrated criterion of LCS; x_i - numerical expert score of qualitative criteria that have been gathered by formulae 1 – 5.

According to the concept of traditional organisation development the investment policy is characterized by a definite level of investment activity and dependent on the LCS (Pluzhnikov et al., 2015).

5 Criterion of investment activity (Cia)

The investment activity is considered as intensity of the company investment work characterizing raise capital dynamics that are aimed at the accomplishment of strategic objectives (Damodaran, 2006; Walsh, 2000; Tikhonov, 2003; Pluzhnikov *et al.*, 2015). Some authors argue that it's connected with the amount of revenues to the rate of capital investment. They approve the positive line between revenue increase and capital investment of a company (Gordon, 1996; Young, 2006; Ashirov, 2000; Yendovizhky and Isayenko, 2007). Therefore the index of investment activity characterizes dynamics of real capital raises as a specific vector index for accomplishing strategic development objectives.

To solve all necessary issues the quantitative criterion of investment activity (Cia) should be used. It is defined through the correlation of increase value of fixed and current assets for the financial reference period (ΔNcA+ΔCA), as a result of general investment (real investments) to the rate of internal development sources (NP+Am) (Pluzhnikov *et al.*, 2015):

$$C_{ia} = \frac{(\Delta NcA + \Delta CA)}{(NP + Am)}, \quad (7)$$

where: Cia – index of investment activity; (ΔNcA+ΔCA) – increase of fixed and current assets value; NP – net profit for the financial reference period; Am – rate of amortization expenses for the reference period.

The criterion of the investment activity Cia is based on the results of three types of activity (investment, operational, financial) in which increase value of aggregated assets characterizes the results of the investment

 DOI: 10.1051/shsconf/20173501105

activity. The rate of internal sources characterizes results of the operational activity. The financial activity is aimed at the balance of cash flows(inflow/outflow) connected with the realization of the investment and operational activities (Плужников *et al.*, 2015).

The criterion of investment activity may estimate not only the dynamic vector of the investment activity but reflect risks of the investment activity in the realization of the stated development strategy. The organisation develops at various life-cycle stages so that different aspects and issues at diverse stages presuppose choosing several criteria to form investment policy.

6 Graphical-analytic model to estimate investment policy

It is extensively used in the economic theory as a tool for analyzing portfolio strategies to define reasonable diversification of undertaking activity. The most popular matrixes are following: BKG (The Boston Consulting Group), GE/McKinsey, ADL/LC (Osel and Wright, 1980), Shell/DPM, SPACE, Hofer/Shendel, Thompson, Strickland, strategic charts (BSC) (Lawrence, 1992) and etc. However, the data of the model doesn't open up opportunities to develop investment policy in accordance with its stated development strategy.

As a tool for developing investment policy in accordance with its stated development strategy the authors suggest graphical-analytic (diagnostic) model to form investment policy (Smaginand Shikina, 2012). The model (4x4) is based on qualitative characteristics reflecting criterion of investment activity Cia (Y-axis) and integrated criterion for estimating life-cycle stage C_{LC} (X-axis) fig. 1).

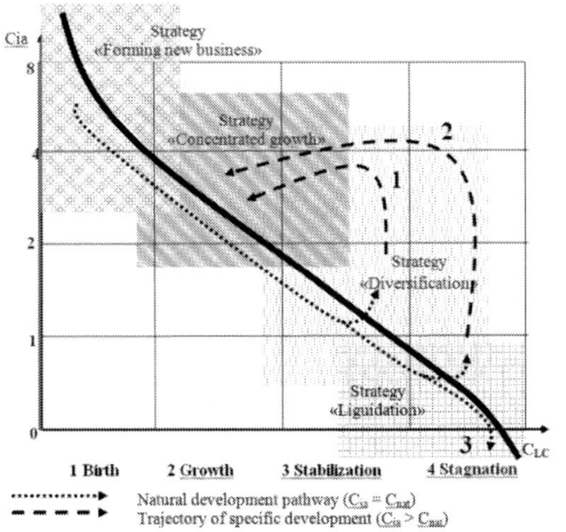

Fig. 1. Diagnostic matrix of strategies (natural, specific development).

Possible trajectories of Cia index transition in the diagnostic model (fig.1) in the development process of the organisation at different life-cycle stages:

1 – trajectory of natural level of investment activity ($C_{ia} = C_{nat}$) is characterized by eventual strategic changes in the process of realization reference strategies of natural development and ways of investment policy in accordance with business life-cycle. Trajectory 3, fig. 1;

2 – trajectory of the specific level of investment activity ($C_{ia} > C_{nat}$) is characterized by intermittent nature of strategic changes in the process of redical business renovation If strategies "diversification" and correlated investment policies are conducted correctly, the trajectory of Cia is directed to the transmission of Cia from the stabilization or stagnation to the growth. Trajectory 1, 2. fig. 1.

Thus, investment policies directed to the realization of strategies which provide Cia rate to the rate that is similar to the natural level of investment activity are rational investment policies (Cnat) (Smagin and Shikina, 2012).

The diagnostic model based on the analysis and estimating vector of Cia defines the type of realized investment policy in accordance with the stated development strategy (Table 6). Defining the type of realized investment policy is based on the comparison of Cia to the rate of the natural level of investment policy (Cnat) with its correlated LCS.

Table 6. Correlation of the type of investment policy and Cia.

The type of investment policy	Criterion of intensity of investment activity (Cia)
Concervative	$K_{ИА}$ < Cnat , (passive)
Moderate	$K_{ИА}$ = Cnat (moderate)
Aggressive	$K_{ИА}$ > Cnat (active, high)

On the basis of research results on data from the table 7 (Smaginand Shikina, 2012), it can be possible to mark LCS and Cia rates which are corresponded to its natural level of investment activity (Cnat).

Table 7. Rate of Cnat on LCS.

№	The type	Rate of Cnat
1	The stage birth - area of strategies "forming new business"	Cia > 4,0
2	The stage growth - area of strategies "growth"	2,0 < Cia < 4,0
3	The stage stabilization - area of strategies "cost saving"	1,0 Cia 2,0
4	The stage "stagnation" - area of strategies "liquidation"	Cia 1,0

Range of Cia rates for different branches in accordance with industry specialization may vary but functional dependence Cnat on LCS has a unique character. fig. 2, table 7.

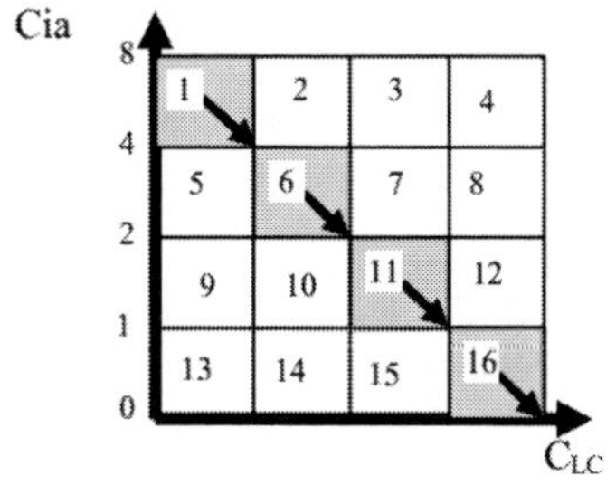

Fig. 2. Area of natural business development.

Diagnostic matrix includes 4 areas introduced in the table 8.

Table 8. Areas of choosing investment policies.

The area	Concept
Mainstream business development (natural)	Corresponds to sequent life-cycle stages of business development (matrix cells or positions 1; 6; 11; 16, fig. 2). On the diagonal rates of investment activity (Cia=Cnat, table 7) for corresponding life-cycle stage (fig. 2);
Eventual development (higher risk)	The rate of coefficient of investment activity (Cia>Cnat, table 7). In result of realizing development strategy another capital raising leads to higher financing business and risk growth (matrix cells or positions 2; 3; 4; 7; 8; 12 fig. 3). Investment policies should be aimed at putting Cia to its natural level (positions 1; 6; 11; 16, fig. 3 a, b, c);
Eventual development (low risk)	In the process of natural development the organisation needs to cary out more active policy and putting Cia = Cnat (fig. 4, a, b, c). If an undertaking has no opportunities to raise necessary volume of investments to provide business development, it should liquidate its business;
Cash generation	The rate of Cia is lower 1.0 that is lower natural level Cia<Cnat (matrix cell or positions 15, fig. 4 c).

To each area of forming investment policy belong a particular rate of assets change and amount of investments. Corresponding ways of investment policy for each matrix position obtain the following economic meaning:

• Position 1. Forming new business (fig. 2) - at this stage internal investment sources (NP+Am) are very small because of inability to generate necessary volume of cash flow so that relative Cia rate of investment activity is large (Cia> 4,0);

• Position 6. Concentrated integrated growth (fig. 2) - characterized by sales growth that is described by significant increase assets. So that external development sources are not necessary. It is reflected in the relative coefficient reduction (2,0 ≤ Cnat < 4,0);

• Position 11. Cost saving (fig. 2). Production volumes and increase of assets are slow so that it leads to the further coefficient decrease (1, 0 ≤ Cnat ≤ 2, 0). According to the further natural development this position is unpromising because it leads to the position 16 (fig. 2). To renovate business a new type of development is necessary that is realization strategies "specific development". Investment policy is directed to putting Cia at the growth stage (trajectory 1 fig. 1);

• Position 16. Liquidation (fig. 2). The assets growth decreases because the production volume falls. Business stagnates, the undertaking business results in "death" without radical changes of natural development strategies (trajectory 3, fig. 1; 2) (Cnat < 1, 0). For prolongation of life undertakings need radical business renovation and realization of specific development strategy that is putting Cia at the growth stage (trajectory 2, fig. 1).

Positions of the diagnostic model (positions 1; 6; 11; 16) points out the area of higher investment activity (fig. 4). Corresponding ways of investment policy for each matrix position obtain the following economic meaning:

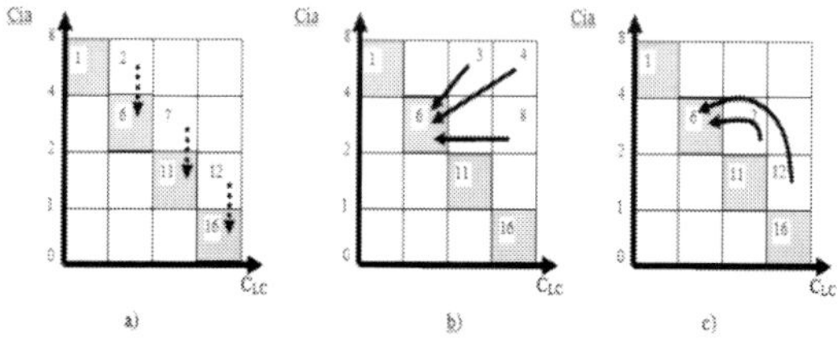

Fig. 3. Area of extra encouraging financial sources of business development.

Corresponding ways of investment policy for each matrix position obtain the following economic meaning:

• Positions 2, 7, 12. The rate of the coefficient of investment activity Cia>Cnat that is it's higher than natural level. This situation can be considered as realization of concentrated (integrated) growth and/or diversification. Assets grow due to encouraging extra financial sources that lead to overaged risk growth. In the frame of natural business development (fig. 3a) the business has overaged assets growth and realization of investment policy, correlated cost saving strategy. It may lead coefficient of investment activity to its natural Cia=Cnat. Or it can be considered as diversification strategy. In this case the rate of Cia>Cnat is higher than the natural level and aimed at the development of new business types.

• Positions 3, 4, 8 should not be considered from the position of natural business development. As a rule they are coefficients of realizing diversification strategies (fig. 3 b).

Positions of the diagnostic model are placed lower than the diagonal (of natural level) and point out the area of low investment activity of a company (fig. 4).

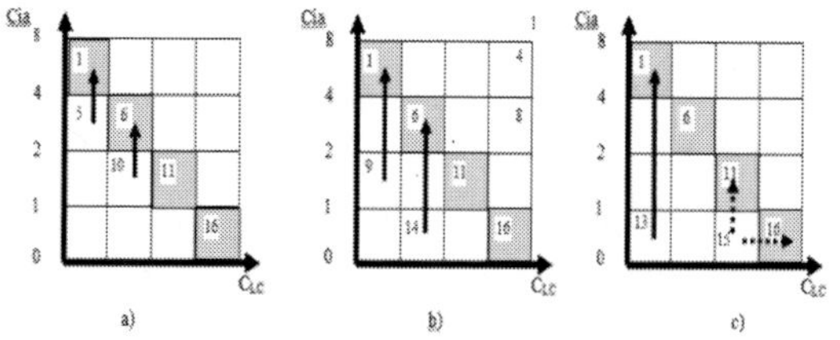

Fig. 4. Insufficient financing of business development.

Corresponded ways of investment policy for each position of the matrix (fig. 4a, 4b, 4c) obtains the following economic sense:

• The level of investment activity is not enough for the positions 5 and 10. The investment policy should be aimed at Cia growth and putting Cia to the natural level (Cnat) so that to natural business development (positions 1; 6; fig. 4a)

• Positions 9, 13, 14. The rate of investment activity is critical, the undertaking can be estimated as a potential bankrupt (fig. 4b, 4c).

• The position 15 of cash generation that is Cia<1,0. Business stagnates. The undertaking has 2 options for development. In the frame of natural development it's necessary to carry out investment policy and correlated cost saving strategies. As a result, Cia transmits from the position 15 to the position 16 (fig. 4c). In the frame of specific development the investment policy is directed to diversification of activity and putting Cia from the position 15 to the position 11 (fig. 4c).

7 Analysis of investment policy

For approbation of the given method the analysis of investment policy of 2 leading pipe and tube production companies in the Russian Federation should be conducted. The research is conducted at several stages. At the first stage the analysis and estimate of the life-cycle stage is carried out.

Undertakings OAO Sinarsky Pipe Manufacturer (OAO SinPM) and OAO Vyksunsky Metallurgical Plant (OAO VMZ) are leading plants in the branch. The main financial criteria characterize their stable statement. The structure of real assets is relatively stable during the whole analyzing period that is pertinent to developmental manufacturing industries with the static technology when the business is satisfied with its position. The life-cycle imitates production life-cycles and applied technologies life-cycles that are characterized for the correlated industrial branches. According to the experience the length of the organisation life-cycle for different types of economic activities may vary but the character of the trajectory will be the same (Shirokova 2006).

According to the data from the financial statements of OAO "SinPM" and OAO "VMZ" on 01.01.2014 we can calculate the quantitative estimate C_{LC}. The calculation results are introduced in the table 9.

Table 9. Results of estimate C_{LC} OAO "SinPM", OAO "VMZ" on 01.01.2014.

#	The criterion	The company OAO "SinPM"		The company OAO "VMZ"	
		The virtual meaning	**Estimate on Kharrington scale**	**The virtual meaning**	**Estimate on Kharrington scale**
1	Index of increments of assets	1.10	0.64	1.05	0.64
2	Number of people (thous. of people)	10 000	0.80	12 000	0.80
3	Total Assets (RUB bn)	21 000	0.80	120 000	0.80
4	Sales (RUB bn)	23 000	0.64	85 000	0.80
5	Enterprise age (years)	more than 50	1.00	more than 50	1.00
	Total		3,88		4,04

Integrated criterion is derivative reflecting the correlation of all estimate criteria LCS. Expert estimate LCS composed 3,88 for OAO "SinPM" and 4,04 for OAO "VMZ" that are corresponded to LCS stagnation. At the stagnation the basic tendency is demand decrease, reduction of competitors and narrowing of product proposal.

At the second stage the criterion calculation of investment activity is carried out using formula 7. The calculation results are introduced in the table 10. Graphical reflection of Cia of the company OAO "SinPM" and OAO "VMZ" are introduced in the fig. 5.

Table 10. Cia dynamics of investment activity.

The company	Period						Average value Cia
	2008	2009	2010	2011	2012	2013	
"SinPM"	0.64	–5.91	–0.08	–0.11	–0.23	0.85	0.31
"VMZ"	0.36	1.00	2.02	0.14	0.79	0.14	0.16

In the period from 2006 to 2014 the Cia vector is placed in the area of the natural level of investment activity characterizing its stability and moderation of investment policy in the frame of natural development strategy (see fig. 5).

At the third stage the analysis and estimate of investment policy according to the Cia vector are conducted. In the period from 2006 to 2014 the Cia vector of the company OAO "SinPM" is placed lower that the natural level of investment activity C_{NAT}, that corresponds to the passive investment policy (see fig. 5, table 10).

In the frame of realizing cost saving the company OAO "SinPM" needs actions on decrease of expanses to the income amounts and restructuring indebtedness. These actions will balance the structure of real assets and their sources. The realization of investment policy may lead the Cia to its natural rate C_{NAT} and come over the position 16 "stagnation"(fig. 2).

In the period from 2008 to 2013 the Cia vector of the company OAO "VMZ" is placed higher than the natural level of investment activity C_{NAT}. According to the rate it corresponds to moderate investment policy (see fig. 5, table 10) in the frame of diversification strategy.

To protect the position at the stagnation stage in the frame of natural development realizing cost saving strategy will lead to the recovery of business activity.

8 Discussion

Comparison of estimate results of the investment policy based on the analyzing Cia vector (analytical model) of the industrial companies introduced on the fig. 5. and financial analysts' opinions introduced in the table 11 may prove the authenticity of the given method.

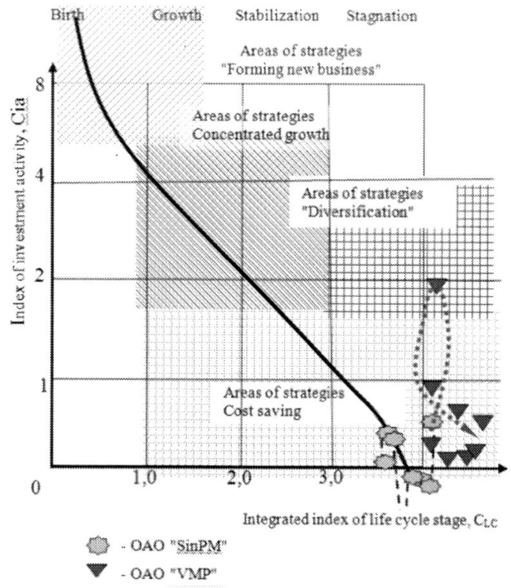

Fig. 5. The Cia vector of the companies OAO "SinPM" and OAO "VMZ" from 2008 to 2013.

Table 11. Comparison of estimate results of investment policy.

The company	External estimate on financial analysts data	Estimate base on the analytical model ($Cia - C_{LC}$)
"SinPM"	**passive** investment policy	**passive** investment policy
"VMZ"	**moderate** investment policy	**moderate** investment policy

The conclusions are proved by analysts financial review introduced on financial web-sites of the company OAO "VMZ" (Fundamental analytics, OAO Vyksunsky Metallurgic Plant) http://www.omk.ru/vmz/smi/5688/?back_url_list=/vmz) and financial analysts opinions OAO "SinPM" (Finanalytics, OAO "Synarsky Pipe Manufacturer" https://sintz.tmk-group.ru).

9 Conclusion

While the research and approbation of the graphical-analytic model to form investment policy in accordance with the stated strategy and its development we can conclude that

• graphical-analytic model may diagnose on the Cia vector the type of realizing investment policy and its correlation to the stated development strategy;

• graphical-analytic model to form investment policy may realize control, adjustment and diagnostics of intensity of investment activity;

• quantitative estimate of the model position may forecast the results of realizing investment policy and expect either entrepreneurial or financial risks.

Calculation of Cia and LCS is conducted in accordance with the financial statement that complicates the research of impact on Cia realizing investment policy in separate business processes.

The work was supported by Act 211 Government of the Russian Federation, contract № 02.A03.21.0011.

References

1. S.O. Ashirov, K.V. Rzayev, *Estimate investment activity of industrial undertakings* (GUU, Moscow, 2000)
2. I.A. Blank, *Investment policy of organisation: eight main stages.* http://www.elitarium.ru/2012/11/21/investicionnaja_politika_predprijatija.html
3. A. Damodaran, *Investment valuation: Tools and techniques for the valuation of any assets* (Alpina Business books, Moscow, 2006)
4. D.A. Yendovitsky, A.N. Isayenko, *Financial organisation reserves: analysis and control* (KNORUS, Moscow, 2007)
5. D.A. Yendovitsky, *Analysis and estimating effective business investment policy: Methodology and methods* (Publishing house of Voronezh state university, Voronezh, 1998)
6. F. Glazl, B. Livekhud, *Dynamic development of entrepreneurship: How pioneer-entreprises and the bureaucracy can become effective* (Dukhovnoe poznanie, Kaluga, 2000)
7. L.A. Gordon, R.J. Iyengar, Journal of Accounting and Public Policy, **15**, 305-325 (1996)
8. L.E. Greiner, News St. Petersburg . Univ . Ser. Management, **4**, 76-94 (2002)
9. I.V. Ivashkovskaya, D.O. Yangel, Corporate Finance, **4**, 97-110 (2007)
10. E.I. Kushelevich, S.R. Filonovich, *Management: XX century – XXI century* (Ekonomist, Moscow, 2004)

11. D.L. Lester, J.A. Parnell, S. Carraher, The International Journal of Organizational Analysis, **11**, **4**, 339-354 (2003)
12. Lawrence S. Maisel, Journal of Cost Management, Summer (1992)
13. D. Miller, P.A. Friesen, Management Science, **30**, 11161-1180 (1984)
14. E.B. Morgunov, *Organizational behaviour: tutorial* (Izd-vo MGUPI, Moscow, 2006)
15. Roger R. Osel, V.L. Wright Robert, *Allocating resources: How to Do It in Multi-Industry Corporations. Handbook of Business Problem Solving* (McGrow-Hill, New York, 1980)
16. V.G. Pluzhnikov, S.A. Shikina, Management of the economic system: the Electronic Journal of Research, **10** (2014). http://uecs.ru/uecs70-702014/item/3087-2014-10-18-08-24-48.
17. V.G. Pluzhnikov, V.N. Smagin, S.A. Shikina, Economic analysis: research and practice, **2(401)**, 2-10 (2015)
18. V.G. Pluzhnikov, S.A. Shikina, Research and practice analytical journal: Economic analysis theory and practice, **44(443)**, 53-64 (2015)
19. V.N. Smagin, S.A. Shikina, Vestnik South Ural State University. Series: Economic and management, **9(268)**, 88-90 (2012)
20. G.V. Shirokova, O.Yu. Serov, Vestnik SPbSU, **8**, **1**, 3-27 (2006)
21. G.V. Shirokova, Editorial commission of Vestnik SPbGU, **8**, **2**, 25-42 (2006)
22. G.V. Shirokova, Russian Management Journal, **5**, **3**, 85-90 (2007)
23. D.A. Tikhonov, *Modeling funding impact on investment activity* (St. Petersburg, 2003)
24. M.V. Charayeva, Financial management, **2**, 97-104 (2010)
25. K. Walsh, *Key indicators of management: How to analyze, compare and control to regulate the data that defines the value of the company* (Case, Moscow, 2000)
26. D. Young, European Management Journal, **15**, **4**, 335-344 (2006)

Identification of business processes of the enterprise by intrinsic states method

V.G. Pluzhnikov[1,*], *S.I.* Kukharenko[1], and *S.A.* Shikina[1]

[1]South Ural State university, Chelyabinsk, Russia

Abstract. The identification of business processes of the industrial enterprise in the course of its operation process is considered in this article. These problems are solved by developing a model of the enterprise by intrinsic states method. Studying of parameters, which characterize the business processes, performs by principal component analysis. The proposed approach makes it possible to verify identity of links between the business processes parameters of the enterprise. The effectiveness of this approach demonstrates in terms of analysis and evaluation of business processes of the particular industrial enterprise.

1 Introduction

Today the role and the value of economic analysis of business processes indicators of industrial enterprises increase in modern economy. The existence of a set of functional strategy with contradictiry purposes is characteristic for the industrial enterprises (further the enterprise) in the course of its development. In this case need arises for the tool allowing to reach compromise as between purposes of an enterprise in general, as the purposes of separate subsystems in business processes (Mokeev, 2014).

In economics for the analysis of the effectiveness of an enterprise processes and the simulation of sustainable development the most common methods are used: financial analysis (Kovalyov, 2004), economic analysis (Watshem and Parramou, 1999), regression analysis (Uotsh, 1999; Pluzhnikov, 2015), simulation (Platonov, 2010), systems analysis using fuzzy logic (Nedosekin, 2003), data envelopment analysis (DEA - Data envelopment analysis) and principal component analysis (PCA) (Mokeev and Vorobyev, 2014).

However, the use of classical analysis and forecasting methods does not meet the quality requirements for results of business process analysis. Therefore it is necessary to talk about the need of development and use of new tools of analysis and forecasting of business processes parameters in the course of its development. This need has led to the creation and development of new methodological approaches in the context of the analysis of the efficiency of an enterprise functioning in general and the individual business processes.

The concept of the business process is currently no single definition (Andersen 2003). Specialists of organizational design represent some different definitions of business processes:

• first, they are treated as a set of different activities in which "input" is used one or more types of resources, as a result of this activity the "output" creates a product that represents value to the consumer (Hammer and Champy, 1997);
• secondly, a set of interrelated or interacting activities which transforms inputs into outputs, representing the value to the customer (Andersen 2003);
• third, as a stable, purposeful set of interrelated activities that transforms "inputs" into "outputs" representing the value to the end-consumer (International Standard ISO 9000: 2000).

Thus, the business process is understood as a set of procedures implementing the specific objectives of the subsystem in the sustainable development of the socio-economic system.

As part of the common goals of development of industrial enterprise are singled out as a research subject the following business processes:
• operating activity - aimed at the formation and support of the current activity;
• investment - associated with the process of investing in competition basis;
• increasing the efficiency of resource use (assets) - aims to increase the returns of resource use.

Research of business processes parameters and the identification of cause-and-effect relationships between them in the course of sustainable operation of business in the rapidly changing market conditions, is carried out by intrinsic states. It is the one of the directions of the principal component analysis method (PCA) Select the main component in the intrinsic method is carried out on the basis of the requirements for the development of socio-economic system. The origins of principal component analysis can be found in the following works: (Pearson, 1901), (Hotelling, 1933), (Frisch, 1901), (Rao, 1964), (Gunst, 1983),

* Corresponding author: Pvg777@bk.ru

(Bendoly, 2009) (Mokeev and Vorobiev, 2014) (Bunova and Krepak 2014).

The intrinsic states method allows to decompose the object state, presenting it as a set of noncollinear components. Because the principal components are independent with each other, their correlation coefficient tends to zero. This allows to pass on to a system of independent vectors (non-collinear) describing the (predictive) multicollinear dynamic system parameters.

Summary of intrinsic states method comes down to the calculation of eigenvectors and eigenvalues of the covariance matrix of the original data. The number of dimensions of the eigenvectors has the same number of dimensions as the original vectors of economic entity condition (and Mokeev, 2014).

Suppose the enterprise is described by a set of parameters x^0_{ki}, where i – parameter number (i=1, 2, 3,...,n), k – point number of timeframe (k = 1, 2, 3, ... ,m). The values of each parameter in different time points (k) form a temporal series, which is denoted by the vector x^0_i. Thus, the space parameter SPP can be represented as a matrix of the original vectors:

$$X^0 = \begin{bmatrix} x^0_{11} & x^0_{12} & \cdots & x^0_{1i} & \cdots & x^0_{1n} \\ x^0_{21} & x^0_{22} & \cdots & x^0_{2i} & \cdots & x^0_{2n} \\ \vdots & \vdots & \ddots & \vdots & \vdots & \vdots \\ x^0_{i1} & x^0_{i2} & \cdots & x^0_{ii} & \cdots & x^0_{in} \\ \vdots & \vdots & \cdots & \vdots & \cdots & \vdots \\ x^0_{m1} & x^0_{m2} & \cdots & x^0_{mi} & \cdots & x^0_{mn} \end{bmatrix}. \quad (1)$$

In this case, the normalizing of the matrix about the arithmetic values is carried out in such a manner:

$$x_{ji} = x^0_{ji} - \bar{x}_i. \quad (2)$$

On the basis of the obtained matrix X the covariance matrix is rated:

$$A = (1/k)X^T X, \quad (3)$$

where X – parent matrix; X^T – transpose of parent matrix.

In the next stage, the calculation of the eigenvalues and eigenvectors of the matrix A is carried out, i.e. , its decomposition:

$$A = VLV^T, \quad (4)$$

where L - scalar matrix, which diagonal sorted by descending eigenvalues, V - orthogonal matrices matrix of eigenvectors.

Eigenvector has property of orthogonality and is rated within the accuracy of factor that allows to get a quantitative estimation of baseline factors intercommunication (Mokeev 2014). This vector has the number of dimension as the original vectors of economic entity condition. Thus, the state of an enterprise can be described by a set of principal components, where each principal component reflects not separate original parameter, and shows the quantitative estimation of the impact of the original factors in each business process group:

$$z_{kj} = \sum v_{kj}(x_{ki} - \bar{x}_i) \quad (5)$$

where Zkj – the value of principal component of j-th new factor at k-th time point, Vkj– coefficient of j-th intrinsic state, corresponding to k – time point and i – base value.

Thus, the state of socio-economic system at any given time can be assumed as a weighted combination of intrinsic states.

The important point is the opportunity to research the original enterprise parameters in the system of independent (orthogonal) eigenvectors reflecting the behavior of the original (multicollinear) indicators. In addition, each main component is based on a group of input parameters defined by the requirements of the selected business process models (Pluzhnikov and Shikina, 2014).

Parameter analysis of business processes of social economic system in terms of an industrial enterprise. So as to analyze the parameters of an industrial enterprise in the framework of the above problem, select the following indicators: non-current assets (NcA); current assets (CA); equity capital (EC); non-current liabilities (NL); current liabilities (CL); cost price (Zcp); net profit (NP); the value of investments (Inv). The value of investments is rated by increment of real total assets value (Shikina and Pluzhnikov, 2010).

As the object of research the industrial enterprise public limited company (OAO "VMZ") is chosen (Pluzhnikov 2015; "VMZ" 2015). The results of activity of OAO "VMZ" over 2006 to 2014 period are presented in Table I. Dynamics of the basic indicators of the enterprise activity is presented in Figure I.

Table 1. Parameters of the basic indicators, RUB bn.

Date	Parameters							
	NcA	CA	EC	NL	CL	Zcp	Inv	NP
01.01.2006	11.9	17.4	14.0	5.3	10.1	26.0	5.7	2.8
01.01.2007	14.6	29.9	20.6	7.7	16.1	43.5	16.4	6.7
01.01.2008	19.0	31.1	29.5	8.3	12.3	48.8	7.1	10.9
01.01.2009	22.0	55.5	35.4	12.0	30.0	51.9	29.1	5.9
01.01.2010	26.9	58.9	40.2	13.1	32.6	59.5	10.2	7.8
01.01.2011	44.2	54.2	57.6	18.1	22.7	67.9	14.5	19.4
01.01.2012	55.5	51.9	61.9	16.0	29.6	72.2	14.3	8.8
01.01.2013	60.8	53.3	64.7	15.3	34.1	62.2	12.2	5.1
01.01.2014	63.1	57.2	68.2	24.6	27.5	61.8	11.8	4.8

The analysis of the dynamics of the base values of OAO "VMZ " indicates their irregular growth, which is caused by factors of external and internal environment.

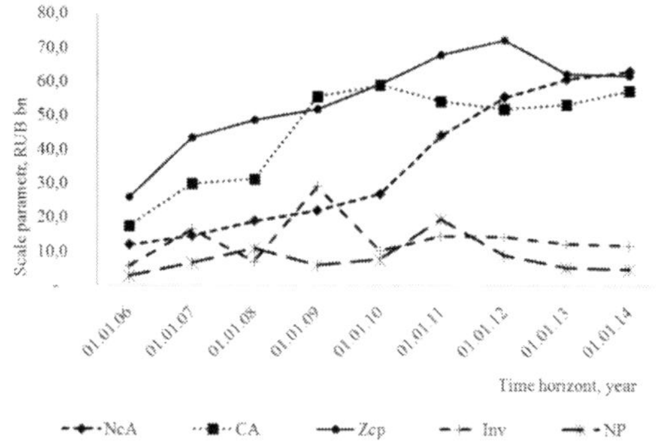

Fig. 1. Dynamics of the basic indicators of the OAO "VMZ" activity over the 2006 to 2014 period.

Evaluation of economic performances is based on analysis of the value and / or the rate of change in the analysis of the effectiveness of operation of the business. But we need not evaluate individual performance, but the total sum of parameter changes. So as to do this, we place demands on the economic-mathematical model that can help assess not only the dynamics of all parameters and trends in their change relative to each other (dynamic business process characterization).

We perform this assessment using the proposed approach. We carry out the selection of considerations for the selection of intrinsic states in accordance with the applied requirements to the specified business processes.

As part of operating activities - increase in production (directly determined by the cost of production Zcp), usually accompanied by the increase: the value of non-current (NcA) and current (CA) assets; the value of property (EC) and non-current (NL) of short-term (CL) liabilities sources; net profit (NP).

As part of the investment activity the growth of investments (Inv) is usually accompanied by the increase of (Kovalev, 2004):

- production (Zcp), more efficient use of non-current assets (NcA);
- current assets (CA) due to increased production (Zcp);
- debt financing sources (NL and CL);
- efficiency of production and net profit (NP),

In the framework of resource efficiency use increasing in production leads to:

- the reduction in requirement for working capital (CA) and non-current (NcA) resources;
- the reduction in requirement for borrowed resources (NL and CL);
- the increase the net profit (NP).

We determine the ratios of their intrinsic states, in accordance with the research (Makeev, 2014). The result is eight coefficients of intrinsic states shown in Table II. The coefficients of the eigenvectors have different values in various states. This means that the parameters growth with a positive value within the limits of its intrinsic state, leads to an increasing in the other and vice versa. Such parametric variation characterizes trends for the intrinsic state. To form the economic model will analyze the intrinsic states in accordance with the requirements for the selected business processes.

Table 2. Coefficients of intrinsic states.

Name title	Eigen-vectors							
	1	2	3	4	5	6	7	8
NcA	0.56	-0.50	-0.24	0.14	0.12	0.04	0.37	-0.45
CA	0.38	0.61	-0.12	-0.30	-0.41	0.02	-0.12	-0.45
EC	0.58	-0.20	0.05	0.08	-0.15	0.23	-0.59	0.45
NL	0.16	-0.03	-0.02	0.06	-0.57	-0.47	0.47	0.45
CL	0.21	0.33	-0.40	-0.29	0.43	0.30	0.36	0.45
Zcp	0.37	0.23	0.60	0.00	0.48	-0.46	0.03	0.00
Inv	0.03	0.41	-0.22	0.88	0.07	0.02	-0.01	0.00
NP	0.03	0.07	0.60	0.12	-0.21	0.65	0.39	0.00

The first intrinsic state characterizes the basic trend of enterprise development (83.59% variance of all parameters), in which the increase in production (Zcp) is accompanied by the growth of all parameters (NcA, CA; EC; NL, CL, Inv, NP). These parameters reflect the business operating.

The second intrinsic state characterizes the trend of enterprise development (9.83% variation of parameters), in which the increase in production (Zcp) lead to the increase in working capital (CA) with the decrease in non-current (NcA) assets that may have caused the increase in efficiency of use of the latter. The process is accompanied by the increase in short-term obligations (CL) and investments (Inv). These parameters characterize the business processes of investment activity.

The third intrinsic state reflects the trend of enterprise development (3.63% variation of parameters), in which the increase in production (Zcp) lead to the increase in net profit (the NP), to the reduction in requirement for borrowed capital (NL, KO) and the reduction in requirement for non-current and current (NcA, CA) assets. These parameters correspond to the business processes aimed at the growth of returns, effective asset application.

All other values of intrinsic states are insignificant (less than 2.95% of variation of all parameters) we leave them outside the research.

Thus, the first intrinsic state reflects the business operating, the second intrinsic state reflects the business processes of investment activity, the third intrinsic state reflects the business processes to improve resource efficiency.

2 Approvement

Calculate the values of model parameters for the selected principal components. Calculation of model parameters is performed according to the method presented in the paper (Mokeev and Vorobyev, 2014). In our case, we choose to test the identified first, second and third principal components (Table II).

We perform analysis of the obtained comparative evaluation by the method of closeness estimate of the actual values to the model ones [11]. Calculate the average of the absolute percentage error (MAPE) by the formula:

$$E_1 = \frac{1}{m} \cdot \Sigma \left| \frac{x_{ki} - \hat{x}_{ki}}{x_{ki}} \right|^* \cdot 100\% , \quad (6)$$

where m – number of observations, – fitted value of the i-th factor in the k-th period, x_{ki} – actual value of the i-th factor in the k-th period.

Obtained results are shown in Table III.

Table 3. Value of average relative approximation error.

Name title	Value description							
	NcA	**CA**	**EC**	**NL**	**CL**	**Zcp**	**Inv**	**NP**
E, %	+2.98	+4.12	+2.27	+9.67	+6.49	+2.80	+23.84	+17.31

The average approximation error stays within the range from 2.27% to 23.84%, which leads to the conclusion that the development of the industrial enterprise corresponds to the model of effective trend. The values of the mean relative approximation error of the values Inv and NP contain higher values, they are listed in the Table III. These deviations can be explained by the influence of unaccounted factors (such as changes in demand, changes in the conditions of attraction of investments, lending rates, etc.). Nevertheless, the results allows for the conclusion about the degree of accuracy of coincidence of the results within the selected three intrinsic states.

3 Conclusion

The developed model is a promising tool for the comprehensive analysis and evaluation of the effectiveness of the business processes of the industrial enterprise.

Analysis of the enterprise behavior results in the form of business processes allows to develop an economic-mathematical model of the real activity of the enterprise and to forecast its activity.

The work was supported by Act 211 Government of the Russian Federation, contract № 02.A03.21.0011.

References

1. B. Andersen, *Business-processes. Tools of the improvement* (RIA Standards and Quality, Moscow, 2003)
2. E. Bendoly, E. Rosenzweig, J. Stratman, A data envelopment analysis Journal of Operations Management, **27**, 310-323 (2009)
3. R. Frisch, J. Nordic Statist, **8**, 36-102 (1929)
4. M. Girshick, Ann. Math. Statist, **10**, 203-224 (1939)
5. R. Gunst, Theor. Meth., **12**, 2217-2260 (1983)
6. M. Hammer, *Reinzhinring's Champi of corporation: The manifesto of revolution in business* (Petersburg university, 1997)
7. H. Hotelling, J. Educ. Psychol., **24**, 417-441, 498-520 (1933)
8. V. Kovalyov, *Financial account and analysis: conceptual bases* (Finance and statistics, Moscow, 2004)
9. V.V. Mokeev, *Method principal component analysis and a method of intrinsic in tasks of the analysis and forecasting* (Prod. of the South Ural State University, Chelyabinsk, 2014)
10. V. Mokeev, E. Bunova, N. Krepak, Bulletin of the South Ural State University, Series: Computer Technologies, Management, Radio Electronics, **14**, **4**, 73-81 (2014)
11. V. Mokeev, D. Vorobyov, Bulletin of the South Ural State University, Series: Computer Technologies, Management, Radio Electronics, **14**, **2**, 31-40 (2014)
12. A. Nedosekin, Audit & Finance Analysis (Russian & English editions), **2** (2000)
13. K. Pearson, Phil. Mag., **2(6)**, 559-572 (1901)
14. A. Platonov, *Development of imitating systems for risk analysis at manufacturing enterprise* (Chelyabinsk, 2010)
15. V. Pluzhnikov, S. Shikina, Economic Analysis of the theory and practice, **44(443)**, 53-64 (2015)
16. V. Pluzhnikov, V. Smagin, S. Shikina, J. The economic analysis: theory and practice, **2(401)**, 2-10 (2015)
17. V. Pluzhnikov, S. Shikina, S. Kuharenko, Bulletin of the South Ural State University, Series: Economy and management, **9**, **4**, 57-62 (2015)
18. V. Pluzhnikov, S. Shikina, J. Management of the economic system: the Electronic Journal of Research, **10** (2014). http://uecs.ru/uecs70-702014/item/3087-2014-10-18-08-24-48.
19. C. Rao, Sankhya A, **26**, 329-358 (1964)
20. T. Uotsh, K. Parramou, *Quantitative methods in finance: Studies. a grant for higher education institutions* (Finance. UNITY, Moscow, 1999)

The practical implementation of enterprise balanced scorecard analysis system

G.A. Pollack[1,*], and *O.V.* Ufimtseva[1]

[1]South Ural State University, Chelyabinsk, Russia

Abstract. The use of balanced scorecard system requires to use some measurable key performance indicators in every part of an enterprise activity, not only for the enterprise itself and its subdivisions, but for every employee. These actions need the introduction/application of a corporate information system. In this work we suggest the most commonly used Russian software "1C: Enterprise Platform" (Accounting system of enterprise). We present the practical implementation of the system of real-time monitoring, analysis and control for the implementation of actions concerned with financial component of the enterprise-balanced scorecard. The formed module (expansion) can be applied in pattern "1C: ERP" (Enterprise Recourse Planning). The conclusion is made about the possibility of the above information system expansion for the implementation of all components of the enterprise balanced scorecard system.

1 Introduction

Modern conception of an enterprise strategic management directs production activity toward customer's needs responds flexibly and makes timely changes in reply to outer challenge. The goal of proper management is to gain competitive preference that can give the enterprise an opportunity to survive in long-term outlook [4, 5, 8, 12, 16].

During the development of administrative resolutions it is advisable to use the system of key indicators which are:
• very important for the rating of company operation;
• estimated in a quantitative sense;
• measured and tracked easily.

Modern technology of strategic management is based on the conception of balanced scorecard (BSC) [1, 2, 9], that gives a company a chance not only to work out an advantageous strategy, but to implement it successfully. The balanced scorecard takes into account not only financial but non-financial indicators which show four important parts of the enterprise activity:
• financial;
• customer relationship;
• inner business-processes;
• staff situation and development of interior infrastructure.

"It is impossible to manage something that can't be measured". One of the main ideas in the basis of balanced scorecard is the idea of measurability. All the factors that are important for the management of the enterprise should be measured and presented in the form of interconnected and balanced key performance indicators – KPI [5].

All the factors that are important for the management of the enterprise should be measured and presented in the form of interconnected and balanced key performance indicators [3].

Performance indicators can be fixed both for the enterprise itself and for each subdivision and employee.

2 Implementation of balanced scorecard analysis system

To use efficiently the system of balanced scorecard at the enterprise, it is necessary to work out and introduce the information system, which can ensure fixation and analysis of key performance indicators for both each employee and the whole enterprise [1, 2, 13, 15, 16].

For most enterprises, financial direction provides the interrelation of all four components of scorecard system [7, 8, 17]. The following goals can be stated for a financial component:
• expenses cut;
• rise in receipts;
• rise in cash;
• profit increase.

The following indicators are necessary to estimate the attainability of every goal.

Goal 1: Expenses cut.

Indicator: share of expenses in receipts.

The indicator is calculated as the ratio of all expenses to receipt volume.

Goal 2: Receipt increase.

Indicators: Receipt amount in value terms (absolute indicator), increase of receipts, %.

* Corresponding author: pollakga@susu.ac.ru

Rise in receipts, % = ((receipts in reporting period – receipts in prior period) × 100% /receipts in reporting period.

Goal 3: Rise in cash.
Indicators: Cash in basic activity, free cash.
Goal 4: Profit increase.
Indicators: Gross profit, acceleration of gross profit, %

Acceleration of gross profit, % = (Gross profit in reporting period – gross profit in prior period) × 100%.

Goal 5: Return on total assets (ROTA).
Indicator: Indicator of total assets profitability

ROTA = EBIT / Total net assets.

In order to attain these goals special measures can be developed. As to a financial component, these are the following measures:

1. Introduction of Budget System for the goal "Rise in cash".
2. Advertising for the goal "Rise in receipts".
3. Formation / use of exterior capital for the goal "Rise in receipts".
4. Outlet expansion for the goal "Profit increase".
5. Range expansion for the goal "Profit increase".
6. Decline in value of additional expenditure for the goal "Expenses decline".

Each measure is characterized by its value and effectiveness of its effect on the goal.

The most popular for working out business resolutions at Russian enterprises are "1C:Enterprise 8.3" [10, 11]. We are going to show based on 1C Platform how to implement the information system for the analysis in real-time efficiency indicators of financial component.

For valuable analysis of balanced indicator system it is necessary to have the possibility of storing in database the information about the goals of enterprise strategic development, the key indicators of succeeding, the measures to be taken, etc. Moreover, in order to plan the strategy and analysis of enterprise development it is necessary to store the history of indicators, to compare assembled information using reports and diagrams.

It required introducing into the pattern some additional catalogs and registers as well as necessary reporting.

The following information is referred to supplemental:

- The enterprise strategic development goals ("Strategic goals" catalog).
- The goals of succeeding ("Indicators" catalog).
- Measures taken to attain strategic goals ("Strategic measures" catalog).

For example, the catalog "Strategic goals" has the following structure (Fig. 1).

Every goal has a list of indicators that influence on its implementation. As some indicators influence on the attainment rate unequally – the characteristic "Indicator significance» is introduced to the table "Indicators". This characteristic takes into account the rate of the above indicator influence on succeeding.

The attribute "Goal value" is some previously calculated amount (planned value). Its actual attainment indicates total goal attainment. The goal actual amount is also used; it is a design quantity and is presented in reports.

It is a sum of indicator significance products multiplied by indicator actual value and divided by indicator planned value:

$$Goal\_value = \sum_{i=1}^{n} \frac{actual\_value_i}{planned\_value_i} \times \times indicator\_significance_i$$

To control the implementation rate of the assigned goal, we use the allowable variation of the assigned goal from the plane value. The history of planned and actual indicator values is stored in the information register (Fig.2).

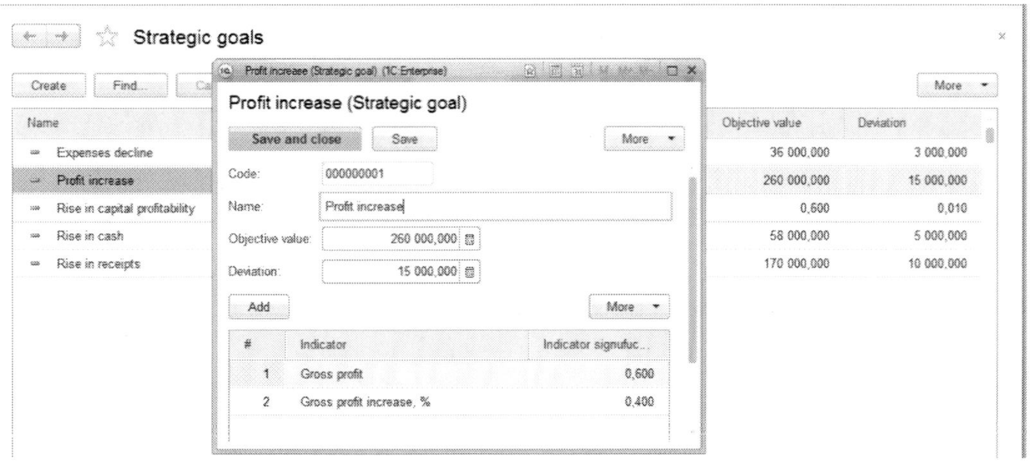

Fig. 1. Structure of catalog "Strategic goals".

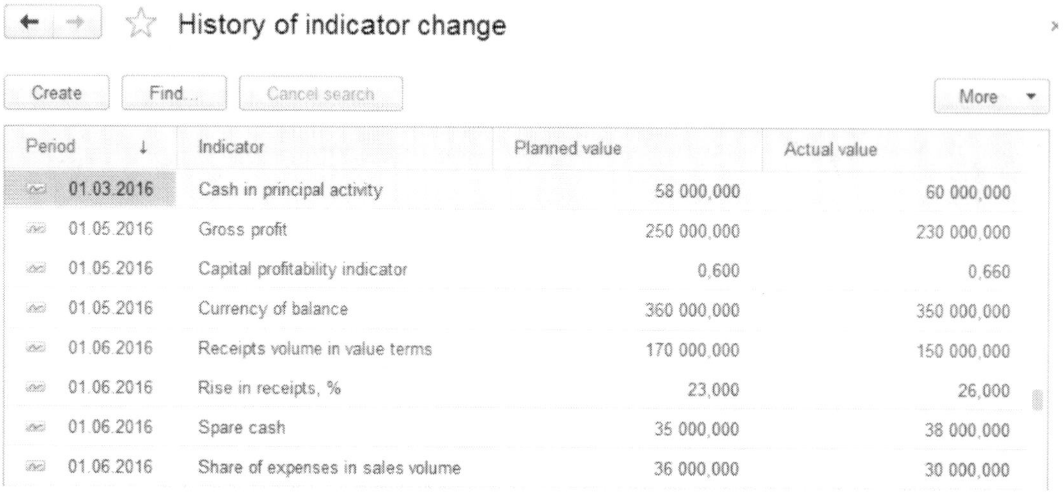

History of indicator change

Create | Find... | Cancel search | More

Period ↓	Indicator	Planned value	Actual value
01.03.2016	Cash in principal activity	58 000,000	60 000,000
01.05.2016	Gross profit	250 000,000	230 000,000
01.05.2016	Capital profitability indicator	0,600	0,660
01.05.2016	Currency of balance	360 000,000	350 000,000
01.06.2016	Receipts volume in value terms	170 000,000	150 000,000
01.06.2016	Rise in receipts, %	23,000	26,000
01.06.2016	Spare cash	35 000,000	38 000,000
01.06.2016	Share of expenses in sales volume	36 000,000	30 000,000

Fig. 2. Register structure "History of indicator changes".

The following reports are suggested to be used for getting some analytical information containing resulting aggregated or detailed information, which is formed according to certain standards.

1. Report "BSC-monitoring in terms of key performance indicators" is a monthly general report where all indicators are presented in the light of the enterprise strategic goals (Fig.3).

2. "BSC-monitoring in terms of implementation goals" report in terms of performance goals is a monthly general report where all the indicators are presented in terms of strategic goals of the enterprise as well as their planned and actual value, indicator significance and color indicator of attainment. The color indicator is green when the planned value has been attained, it is yellow when the value has not changed, it is red when the indicator value has gone down (Fig.4).

3. In the report "The attainment rate of strategic goals", the information about the enterprise goals values is presented. It is necessary for a manager to analyze the information. The results of the report analysis will allow paying attention to some neglect at the enterprise.

4. In the report "The effectiveness of measure influence on the indicator improvement", the list of the above measures is presented. For each measure are presented the following: date of commencement and expiry date, cost and rate of influence on certain indicators. The list of measures can be reviewed over any period.

We used "1C:Enterprise Platform" (Accounting system of enterprise) to implement the analysis information system for the financial component of balanced scorecard. The above module can be introduced in "1C:ERP" (Enterprise Recourse Planning).

The system can be completed with some modules which will analyze indicators which characterize customer relations, interior business-processes, staff situation and development of the enterprise interior infrastructure.

Taking into account that today a lot of Russian enterprises use this software, further work in this direction looks perspective.

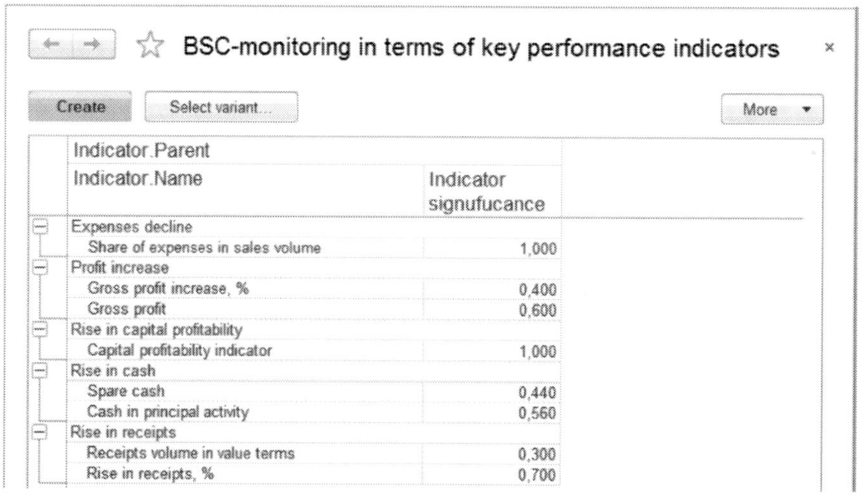

BSC-monitoring in terms of key performance indicators

Create | Select variant... | More

Indicator.Parent / Indicator.Name	Indicator signufucance
Expenses decline	
Share of expenses in sales volume	1,000
Profit increase	
Gross profit increase, %	0,400
Gross profit	0,600
Rise in capital profitability	
Capital profitability indicator	1,000
Rise in cash	
Spare cash	0,440
Cash in principal activity	0,560
Rise in receipts	
Receipts volume in value terms	0,300
Rise in receipts, %	0,700

Fig. 3. Report BSC-monitoring in terms of key performance indicators.

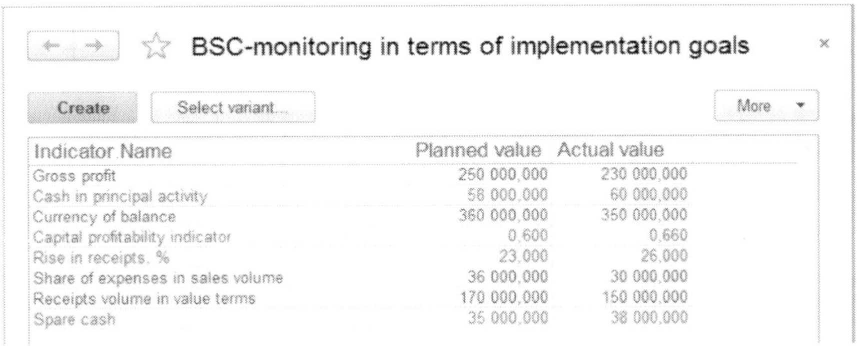

Indicator Name	Planned value	Actual value
Gross profit	250 000,000	230 000,000
Cash in principal activity	58 000,000	60 000,000
Currency of balance	360 000,000	350 000,000
Capital profitability indicator	0,600	0,660
Rise in receipts, %	23,000	26,000
Share of expenses in sales volume	36 000,000	30 000,000
Receipts volume in value terms	170 000,000	150 000,000
Spare cash	35 000,000	38 000,000

Fig. 4. Report BSC-monitoring in terms of implementation goals.

3 Conclusion

The system of the analysis of financial factors has a modular approach that makes it possible to introduce it in the present corporate information system of the enterprise based on "1C:Enterprise Platform" (Accounting system of enterprise). It allows solving the following tasks:

1. To control real-time the attainment of indicator's efficiency both for every employee and for every subdivision and the whole enterprise as well.
2. To analyze detailed and resulting information in terms of assigned goals.
3. To analyze the enterprise key performance indicators (KPI) and estimate them in the view of the attainment of the enterprise strategic goals.
4. To make operative decisions and correct the system of measures those are directed to the attainment of the enterprise strategic goals.

The corporate information system based on "1C:Enterprise Platform" can be completed with modules that will analyze the indicators which characterize the customer relation, interior business processes, staff situation and the development the enterprise interior infrastructure.

So, based on one decision it is possible to implement in practice the analysis of all the components of the balanced scorecard. Taking into account that today a lot of Russian enterprises use this software, further work in this direction looks perspective.

The work was supported by Act 211 Government of the Russian Federation, contract № 02.A03.21.0011.

References

1. R.S. Kaplan, D.P. Norton, *The balanced scorecard: translating strategy into action* (Olimp-Biznes, Moscow, 2006)
2. R.S. Kaplan, S.R. Anderson, *Time-Driven Activity-Based Costing: a simpler and more power-ful path to higher profits* (Harvard Business School Publishing Corporation, Boston, 2007)
3. M. Gavrikov, Corporate Financial Management, **01(13)**, 37 (2006)
4. V.G. Kandalintsev, *Innovation business: use of balanced scorecard* (Delo ANH, Moscow, 2010)
5. V.G. Kandalintsev, *Balanced management of enterprise* (Knorus, Moscow, 2006)
6. S.I. Krylov, **21(420)**, 2-14 (2015)
7. T.V. Logvinova, Financial management, **1**, 20-24 (2007)
8. L. Mullagildina, in Proc. Information technologies in science, management, social sphere and medicine, 273-276 (2015)
9. O. Nils-Göran, J. Roy, M. Wetter, *Performance Drivers. A Practical Guide to Using the Balanced Scorecard* (Vil'yams, Moscow, 2006)
10. M.G. Radchenko, E.Yu. Khrustaleva, *1C:Enterpise 8.2. Developer's practical manual. Examples and standard techniques* (1C-Pablishing, Moscow, 2009)
11. O.A. Romanenko, Economics and enterprise, **113(52-3)**, 772-776 (2014)
12. N.O. Rukhlyada, Theoretical and applied aspects of modern science, 137-142 (2015)
13. R. Rumelt, *Good Strategy. Bad Strategy. The difference and Why it Matters* (London, 2012)
14. N.A. Ryazantseva, D.N. Ryazantsev, *1C:Enterprise 8.0. Management of manufacturing enterprise* (BHV-Peterburg, St. Petersburg, 2006)
15. P. Khorvart, Theoretical and Practical Aspects of Management, **4**, 13-22 (2000)
16. Horvath & Partners, *Balanced Scorecard umsetzen* (Al'pina Pablisher, Moscow, 2008)
17. P. KHorvart, Theoretical and Practical Aspects of Management, 13-22 (2000)
18. A.A. Tsupko, A Series Economy, Finance and Production Management, **2**, 61-66 (2011)
19. V.A. Shchelkova, S.G. Shchelkova, Proc. of the III International Scientific Conference, 223-225 (2014)
20. P.R. Niven, *Balanced Scorecard Step-by-Step: Maximizing Performance and Maintaining Results* (Hoboken, 2006)
21. V. Biryukov, *Using the Balanced Scorecard in Controlling by the Example of the Polygraph Industry* (Moscow, 2010)
22. L.R. Batukova, G.Y. Belyakova, Fundamental research, **2-21**, 4705-4709 (2015)
23. A.V. Krasnikova, Ekonominfo, **23**, 43-47 (2015)

The reliability and efficiency of power companies

I.G. Akhmetova[1,*], and N.D. Chichirova[2]

[1]Economics and organization department, Kazan State Power Engineering University, Kazan, Russia
[2]Thermal power stations department, Kazan State Power Engineering University, Kazan, Russia

Abstract. One of the key indicators of heat supply organizations is the reliability of energy facilities. The index of reliability of heat supply organization is of interest to potential investors, to assess the risks of investing projects. We propose a method of determining the reliability of heat supply organizations with the use of SAW (simple additive weighting method). We are developing a versatile program and the analysis of heat supply organizations on the example of the Republic of Tatarstan. This criterion is of significant interest for potential investors to assess the risks of investing projects. All values for the calculation of the reliability of heat supply organization founded and are publicly available. The method of determining the reliability of heat supply organizations with the use of SAW. Developed a versatile program allows you to analyze the performance reliability of any heat supply organizations.

1 Introduction

The current state of the economy and standard of living are largely determined by the presence of stockpiles of fuel and energy resources and their efficiency. For countries whose economies are based mainly on the import of energy resources, namely efficiency is one of the determining factors of competitive products.

In many countries, legal frameworks allow to provide legitimacy for basic energy efficiency measures. This is particularly important in the cases where special tax incentives or subsidies have been established [17-20].

Among fifteen "old" Member States of the European Union (EU) in ten of them there are state subsidies for energy-saving measures and the conditions for obtaining grants established by law, as recorded in the judicial acts.

In some countries, the law on tax benefits (incentives) is legislated. If some business unit is using energy-saving equipment, or performs services relating to energy conservation, then a tax credit is provided for it. This Act applies in respect of individuals and small groups.

Some programs offer an option for companies: to pay tax on energy consumption or to make investments in activities that improve energy efficiency. Details of tax benefits have been judicially regulated [18].

2 Research relevance

The heat supply infrastructure in Russian Federation requires a systemic modernization. The loss in heat networks of the country reaches 20-30%, which is almost four times higher than in Europe. In order to solve all this problems require significant resources are needed. It is estimated that underinvestment in the industry is almost three trillion rubles. The only way for the development of industry today is to attract private investment. This requires a transition to a modern market model, free of regulation and developing transparency for investors. In the case of successful attracting investments, fixed assets of companies operating in the energy market, as a matter of fact, undergo a substantial renovation and change.

Obtaining private investments and budgets is only possible on the basis of organizations investment-attractiveness and some potential reliability, which guarantees a return on investment. We have developed different methods for determining the reliability of the potential of foreign companies [17-20]. The analysis showed that at the present time, a single and internationally recognized procedure does not exist. In each country there is a specificity of the energy sector, which significantly affects the difference in determining the potential of reliability. At the same time the state of scientific developments in Russian Federation offers new mechanisms, criteria and methods in organizing of rating in the different sectors of economic activity [21].

Among other devices one of the most important tools of managing is assessment and monitoring of the performance of heat supplying companies. Currently, the problems of creating production and investment programs are of significant importance for planning of the industrial power complex organizing practices.

The fuel component in the tariff for electric power up to 80 %, indicating the fact that the scale of possible savings of resources and the absence of the possibility of self-financing investment projects in the Russian Federation's economy (Fig. 1).

The research intentions in this field has repeatedly been made in [2, 3], but a fully formed system of

* Corresponding author: irina_akhmetova@mail.ru

indicators characterizing the state of heat, is currently lacking.

Indicators of reliability established by federal law in Russia actually have been reduced to a single numerical value, which depends on the number of terminations of heating in connection with violations of the heating systems and the volume of the recovery of the resource for the relative year. This particular indicator is rather subjective and can vary over a wide range several years' periods. Using only the above-described indicator for an objective assessment of insufficient heat supply system reliability, but these figures may be used as an integrated component of the reliability index.

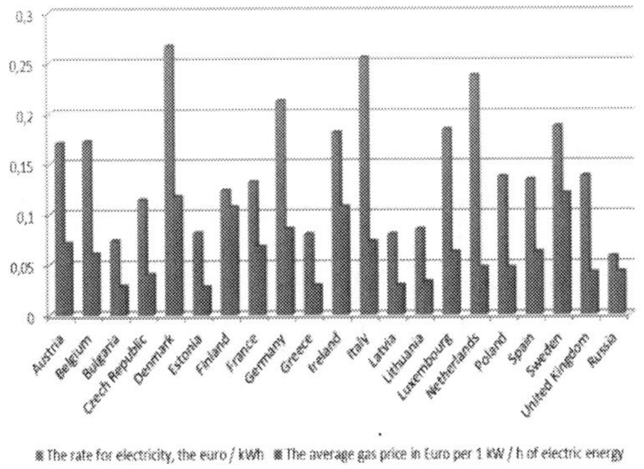

Fig. 1. The fuel component in the tariff for electric energy in different countries.

Methods of determining the integrated index should satisfy the following conditions:
• the full and objectively reflection of the reliability level of the heat supply system which is calculated;
• the information used to calculate the reliability index should be readily available - posted on the Internet in accordance with the disclosure standards.

3 A new method for evaluating the reliability of the power company

To assess the reliability of heat supply organizations are invited to the next set of indicators (Table 1).

During selection of these indicators the results of methodological developments in the field of formation of indicators system, which conduct independent analytical centers and individual researchers, was taken into account. The evaluation system is based on the construction of the comparative ratings of heat supply organizations. Rating is an assessment of the reliability, it is characterized by a numerical value and allows you to compare the organizations involved in one activity together.

In this paper we applied SAW [7] (the method of a simple additive weighting) - a method for determining the reliability of heat supply organizations.

SAW method can be considered reliable only when the consistency of expert opinion is sufficient. Statistically examining the information received from experts, should assess the coherence their opinions and establish the cause of inadequate information. When applying the method of paired comparisons consistency of expert opinion is not checked, it is proposed to use the expert method [15] and a method for checking the consistency of expert opinions proposed by L. Evlanov [16]. These methods are integrated in a single algorithm. Scheme SAW improved method is shown in Fig. 2.

Table 1. Indicators reliability of the heat supply organization.

Indicator	Formula	Note
1. Depreciation of equipment of thermal energy sources	Total wear thermal energy sources for the period / initial cost of fixed assets	Determined to further decrease the degree of serviceability of heat sources
2. Deterioration of heat networks	Total wear of heating networks for the period / initial cost of fixed assets	Determined to further decrease the degree of serviceability of heat networks
3. Number of stopping delivery of thermal energy, heat carrier as a result of technological breakdowns in the heating networks in the 1 km	Method of calculation - the decision № 452	Describes the actual level of damage to the thermal networks
4. Number of stopping delivery of thermal energy, heat carrier as a result of technological breakdowns on sources of heat for 1 GCal / h installed capacity	Method of calculation - the decision № 452	Describes the actual level of damage to the thermal energy source
5. Ratio costs thermal energy cost directed on resource recovery, renewal of fixed assets	Costs for modernization, reconstruction, repair of fixed assets / Cost of thermal energy	Allows us to estimate resources devoted to increase reliability
6. Coefficient of renewal of fixed assets	Cost of fixed assets entered in the period / value of fixed assets	Shows which part of the existing at the end of the reporting period of fixed assets comprise new fixed assets
7. Coefficient of disposal of fixed assets	Cost of retired fixed assets for the period / value of fixed assets	Shows the proportion of core resources available to the beginning of the reporting period, dropped out during the reporting period due to dilapidation and wear

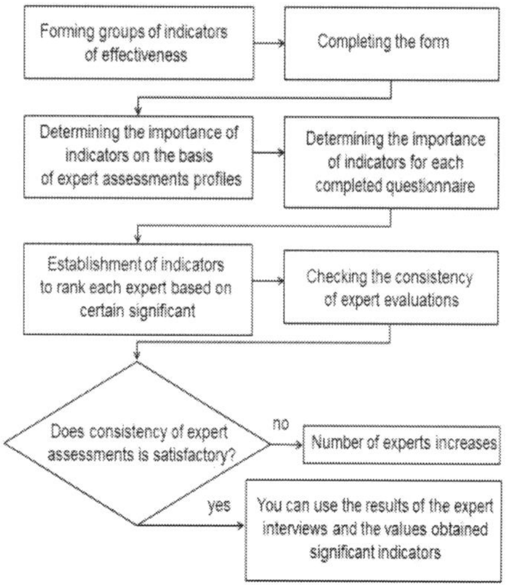

Fig. 2. The flowchart determining the significance.reliability indexes paired comparison method.

Then formed the ranks of indicators, and obtained t_{jk} sets of values (Table 2).

Table 2. Determination of the reliability of heat supply organization.

Expert	Reliability Index			
E_1	t_{11}		…	
E_2	t_{12}		…	
…	…	…	…	…
E_k	t_{1r}		…	
The sum of ranks	$\sum_{i=1}^{k} t_{1k}$	$\sum_{i=1}^{k} t_{2k}$	…	$\sum_{i=1}^{k} t_{nk}$
Average grade	$\bar{t}_1$	$\bar{t}_2$	…	$\bar{t}_n$

A comparative analysis of indicators of reliability of the seven energy companies.

In accordance with the proposed method we investigated the rank of reliability organizations was determined (Fig. 3).

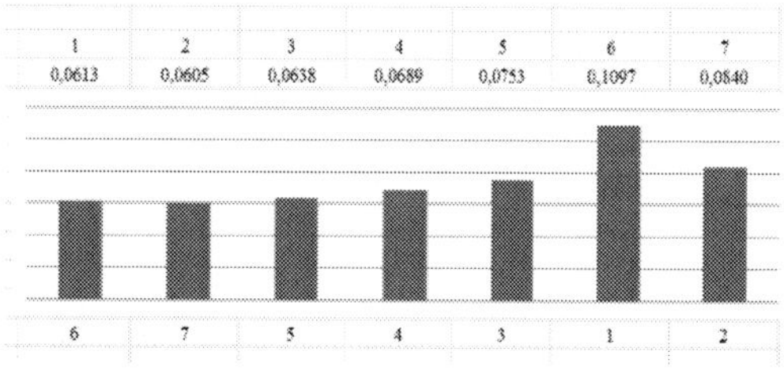

Fig. 3. Determination of the reliability of heat supply organizations rank.

Generalized values of heat supply organizations reliability indicators are shown in Fig. 4, which shows the place of each organization in comparison with the best analogues.

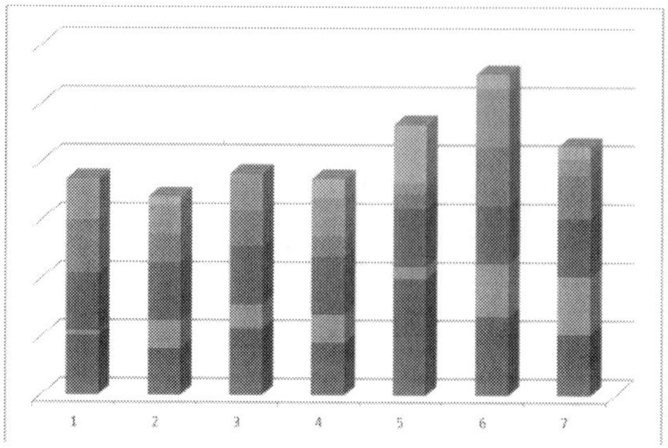

Fig. 4. Generalized values of reliability indicators heat supply organizations: Coefficient of disposal of fixed assets; Coefficient of renewal of fixed assets; Ratio costs thermal energy cost directed on resource recovery, renewal of fixed assets; Number of stopping delivery of thermal energy, heat carrier as a result of technological breakdowns on sources of heat for 1 Gcal / h installed capacity; Number of stopping delivery of thermal energy, heat carrier as a result of technological breakdowns in the heating networks in the 1 km; Deterioration of heat networks1. Depreciation of equipment of thermal energy sources; Depreciation of equipment of thermal energy sources.

4 Conclusion

Thus, we have clarified the definition of the reliability indicator. This criterion is of significant interest for potential investors to assess the risks of investing projects. All values for the calculation of the reliability of heat supply organization founded and are publicly available. The method of determining the reliability of heat supply organizations with the use of SAW (simple additive weighting method). Developed a versatile program allows you to analyze the performance reliability of any heat supply organizations.

References

1. E.Y. Sokolov, *District heating and heat networks: the textbook* (Publishing. House MPEI, Moscow, 2006)
2. *The quality of the energy in the conditions of scientific and technological progress* (Proceedings OIEI, Leningrad, 1975)
3. E.P. Kuznetsov, Heat energy efficiency technology, **2** (1996)
4. E.V. Sennova, S.N. Kiryuhin, S.A. Shimanskaya, News heating, **12(160)** (2013)
5. A.I. Yufa, D.R. Nosulko, *Complex optimization of heating* (TECHNIK, Kiev, 1988)
6. A.A. Son of Jonah, *The reliability of heat networks systems* (Stroyizdat, Moscow, 1989)
7. S. Greco, B. Matarazzo, R. Slowinski, Operational Tools in the Management of Financial Risksyu, Dordrecht: Kluwer Academic Publishers, 121-136 (1998)

8. D.H. Safiullin, *Management strategic potential reliability of power grid companies as a business structures* (Moscow, 2013)
9. K.J. Arrow, Econometrica, 213-243 (1949)
10. C.L. Hwang, K. Yoon, *Multiple Attribute Decision Making Methods and Applications, A State of the Art Survey* (Springer Verlag, Berlin, Heidelberg, New York, 1981)
11. E. Zavadskas, A. Kaklauskas, Proc. Integrating Technology & Human Decisions-Global Bridges into the 21st Century, Athens, Greece, 2051-2053 (1999)
12. A. Kaklauskas, N. Kvederyte, E.K. Zavadskas, *Multiple Criteria Analysis of a Buildings Life Cycle* (Vilnius, Technika, 2001)
13. S. Jakucionis, L. Ustinovicius, STATYBA (Civil Engineering), **VI**, **6**, 469-475 (2000)
14. T.A. Saaty, Journal of Mathematical Psychology, **15**, **3**, 234-281 (1977)
15. E.K. Zavadskas, *Comprehensive evaluation and selection of resource-saving solutions in the Banking* (Vilnius, Mokslas, 1987)
16. L. Evlanov, *Theory and practice of decision-making* (Economics, Moscow, 1984)
17. *Combined Heat and Power: A Resource Guide for State Energy Officials*. www. naseo.org/ data/sites/1/ documents/ publications/ CHP-for-State-Energy-Officials.pdf.
18. Fabio Correa Leite, DecioCicone Jr., Luiz Claudio Ribeiro Galvao, Miguel Edgar Morales, Open Journal of Energy Efficiency, **2**, 22-28 (2013)
19. Peng Wang, *Proc. Sch. of Municipal & Environ. Eng., Harbin Inst. of Technol*, 233-236 (2009)
20. Luca Podofillini, Bruno Sudret, Bozidar Stojadinovic, Enrico Zio, Wolfgang Kröger, *Safety and Reliability of Complex Engineered Systems* (ESREL, CRC Press, 2015)
21. E.G. Gachot, M.V. Stepanova, Energy policy, **3**, 59-66 (2015)
22. *The adoption of "Expertchoice" Decision Support System*. http://www.expertchoice.com/.

 DOI: 10.1051/shsconf/20173501109

Accounting of discounts, acquired by an enterprise when purchasing production inventories

A.V. Shishkina[1,*]

[1]South Ural State University, Chelyabinsk, Russia

Abstract. Different types of discounts for the price of a product are quite common in transactions. They are frequently used by industrial enterprises as well: they receive discounts when they purchase production inventories; they offer discounts when they sell the finished product. Industrial enterprise strive to minimize costs when purchasing raw materials necessary for manufacturing and therefore it is interested in acquiring a discount from supplier. Acquiring a discount is a fact of economic life and must be registered in financial accounts of an enterprise, however nowadays the patterns of accounting of such operations are not regulated by any special normative documents and traditionally applied accounting patterns do not contain information on the amounts of received discounts. It is happening because they comply with the current rules of accounting, according to which inventories are accounted only after deduction of the received discount (so the discount is granted in form of verbal agreement). This is a precursor to the lack of data on the received discounts in financial accounting system of an industrial enterprise, to the impossibility of creation the information base for analyzing the effectiveness of the received discounts on its basis, to the violation of the principle of completeness and continuity of registering all the economic life events in accounting. The article offers the already existing variants of financial accounting of the received discounts, analyzes its advantages and disadvantages from the point of view of compliance with the current legislative system and economic contents of discount operation. Proprietary technology of financial accounting of discounts, received by an industrial enterprise when purchasing production stock, is offered. This technology allows solving the stated methodological issues.

1 Introduction

Self-cost of the finished industrial products has a direct correlation with the purchase price of the materials used for manufacturing it. The lower the cost, the wider are the possibilities of an enterprise either to get more profit without lowering the selling price, or to raise its competitive ratio by lowering the price of the finished goods. That is why managers of an industrial enterprise usually face with the task to obtain the discount form the raw materials suppliers. This discount would lower the purchase value and this issue becomes quite crucial under the conditions of an economic crisis.

In the case of acquiring a discount for purchase items an enterprise's accountant needs to record this operation, because a commercial discount influences the financial result of a discount-receiving enterprise [1] by lowering the contract price. Therefore its acquisition is an economic event and, as any other economic event performed by and industrial enterprise should be accounted. But at present times patterns of financial accounting of the operations with commercial discounts are not regulated by any normative document. Those are the reasons why the researchers of corresponding fields and practicing accountants still haven't reach a single opinion on how to record the acquired discount in the accounts without ruining their methodology.

2 Thematic justification and the aim of the research

Nowadays official notion of a discount has still not been captured in any branches of law [2, 3], the handling procedures of accounting documents when conducting transactions with discounts has not been regulated [4-9], and there are no legislative consolidated patterns of accounting the discounts [8, 10, 11], despite the long-term experience of using the commercial discounts in practice. So the accountant of an industrial enterprise documents and accounts the discounts, which they've got from the raw materials supplier, according to the traditional ways, which, unfortunately, don't exclude the methodological errors.

The main methodological issue of financial accounting of the acquired discounts is in the fact that actually happened and documented fact of economic life – the acquisition of a discount by an industrial enterprise for fulfilling the terms of the contract on supplying the materials – is not recorded in financial accounts of the discount-receiving organization. This violates the

* Corresponding author: anna-shi74@rambler.ru

principal of the continuous recording of all the facts of economic life of an enterprise and denies it an opportunity to form an authentic base for analyzing the effectiveness of the acquired discounts. At present times the information, created within the frames of financial accounting of an industrial enterprise, is the source of data for analyzing the gathered discounts. This information is often less credible than the one that is being formed within the frames of financial accounting (the reason behind that is that accounting is not being regulated legally, responsibility of officials is not so high, the principal of continuous recording and method of information collecting can be violated, etc. It can cause the distortion of the reported data). That's why it is more preferable to use the data of the financial accounting as the source, however the traditional pattern of financial accounting of the discounts, which could solve the stated methodological issue and save from the negative consequences mentioned above, is a critical objective.

The aim of this research is in the devising the method for accounting the discounts, acquired by the industrial enterprise when purchasing materials, that would enable documenting such operations in financial accounting records.

3 Analysis of the existing variants of accounting the acquired discounts

Most researchers, for instance [10, 11], believe that there is nothing difficult in accounting the acquired discounts, but they agree on the idea that their accounting depends on the time of acquiring the discount and finally comes down to documenting several simple accounting records.

The 1st variant. If the discount is granted to an enterprise directly while purchasing the goods and materials the following happens: the account of goods and materials is recorded on the debit side, and the account, intended for payables – on the credit side. At that, purchased goods and materials enter the records with already reduced cost by the amount of the discount, as required by the section 11 of the International Accounting Standards (IAS) 2 "Inventories" and section 6 of Russian Accounting Standards 5/01 "Accounting for inventories".

The approaches for evaluating the cost of purchased inventories of international and Russian accounting rules are the same and that's why the variants of their recording in accounting will be reviewed on the example of Russian chart of accounts.

The example (on the basis of this example other following patterns of accounting will be reviewed). Industrial enterprise placed a request on purchasing materials – 100 tons of metal. The supplier agreed to offer one-time discount in the amount of 3% from the usual price of the price list for the big amount of purchase: 48500 rubles for a ton instead of 50 000 rubles for a ton (without the value-added tax). As the result of the deal discount-receiving enterprise had a full set of documents: sale and purchase contract, containing the conditions for receiving a discount, waybill, sales invoice, and invoice for payment with separate line for registering the discount amount. The fact of non-cash settling of the presented bill is confirmed by a payment order marked with the bank as executed. The following recordings were made in financial accounting of an industrial enterprise when placing in purchased metal in stock records:

- The raw materials delivered by supplier has entered the records (metal, 10 tons), Debit 10 – Credit 60, amount sum 485000 rubles
- The value-added tax has entered the records (18%), Debit 19 – Credit 60, amount sum 87300 rubles
- Settlement with metal supplier has been performed, Debit 60 – Credit 51, amount sum 572300 rubles
- The sum of value-added tax is presented to deduction, Debit 68 – Credit 19, amount sum 87300 rubles

In the given example all the stages of applying the discount has been adhered, the discount is duly registered: it is provided by the supply contract terms and marked out with a separate entry in the waybill, sales invoice, and invoice for payment, the fact of its granting is obvious. But according to the current rules of financial accounting (international and Russian) discount-receiving purchaser does not record the received discount in it, and states the cost of raw materials already reduced by the amount of discount. Therefore actually happened and documented fact of economic life – the acquisition of a discount by an industrial enterprise for fulfilling the terms of the contract on supplying the materials (in this case the purchase amount) – is not recorded in financial accounts of the discount-receiving organization.

The 2nd Variant. The discount is granted after accounting raw materials, in other words retro discount is applied. First, the purchased materials and goods are being recorded in the financial accounting books traditionally – by the actual value without a discount. According to the information from the example given above, the following recordings must be made:

- The raw materials delivered by supplier has entered the records (metal, 10 tons), Debit 10 – Credit 60, amount sub 500000
- The value-added tax has entered the records (18%), Debit 19 – Credit 60, amount sum 90000 rubles
- Settlement with metal supplier has been performed, Debit 60 – Credit 51, amount sum 590000 rubles
- The sum of value-added tax is presented to deduction, Debit 68 – Credit 19, amount sum 90000 rubles

Then financial accounts are being corrected after actually receiving the retro discount. However, how this correction should be performed is still a debated issue.

Some researchers [12-15] believe that the sum of received discount should be included in the miscellaneous revenues of an organization. They justify it with the idea that actual value of inventory stocks, in which they are recorded, cannot be changed or corrected (according to section 12 of Russian Accounting Standards 5/01 "Accounting for inventories"), or with the idea that the discount is seen as writing down of debt of a purchaser. In such cases the following accounting record must be made:

The amount of the discount without the value-added tax is recorded among the miscellaneous revenues, Debit 60 – Credit 91/1, amount sum 15000 rubles

Authors cannot agree with the offered approach. Firstly, such way of accounting the retro discounts does not reflect their economic meaning. Acquired discounts are not the revenue of a discount-receiving enterprise, despite the fact that they indirectly facilitate the small positive economic effect because they help an enterprise to save on expenditures connected with purchasing of material values. Inherently, the offered discount should lower the initial cost of the purchased assets, works and services, but in the reviewed case their cost would not be changed. Secondly, in the researches, conducted before, the authors of the article justified the reasons why the discount and writing out of debt are not the same [16]. It makes their business accounting mismatch, too. Thirdly, tax consequences of applying the retro discount that would lower the initial cost of assets, works and services and so-called discount, which is recognized as miscellaneous revenues of an enterprise, would differ significantly [16].

In this case the opinion of other group of researchers [17-19] presents to be more justified. They adhere to different way of correcting the financial accounting of receiving the retro discount by a discount-receiver. As it was fairly pointed out before, the section 12 of Russian Accounting Standards 5/01 "Accounting for inventories" does not allow changing the actual value of inventory stock, in which they are recorded. However, in the same section the following is stated: "except the cases, emphasized by the Russian federation legislation", and such possibility is directly supported by the section 2 of article 424 of the Civil Code of Russian Federation. Besides, the possibility of applying discounts is described in International Financial Reporting Standards IFRS "Contractual revenue" Therefore the following pattern of accounting records is believed to be legally justified:

• The initial cost of metal is lowered by the sum of the offered discount (reversing entry), Debit 10 – Credit 60, amount sum 15000 rubles
• Value-added tax that falls on the debit sum is restored (reversing entry), Debit 19 – Credit 60, amount sum 2700 rubles

It is positive that this means of correcting the financial accounting of the discount-receiving enterprise, due to the acquiring the retro discount, fully correlate with the current rules of financial accounting. But the same as in the first variant the retro discount itself is not presented in the accounts. This is more about the accounting of made correction due to the fact of receiving a discount than about the accounting of the discount itself.

The third Variant. The completely different opinion concerning the accounting of the discounts acquired by an enterprise belongs to those [9, 20], who believe that purchased goods can be accounted straight away according to the purchased cost and with discount. If a discount-receiver fulfills the terms of acquiring the discount, then there will be no corrections in financial accounting. If after the stated period of discount validity it won't be acquired than the additional accounting records will be made concerning the sum of the lost discount. Therefore in such case the following recordings in the accounts must be made firsthand:

• Purchase cost of metal is recognized including the discount, Debit 10 – Credit 60, amount sum 485000 rubles
• Value-added tax on metal acquired at a discount is accounted, Debit 19 – Credit 60, amount sum 87300
• Settlement with metal supplier has been performed, Debit 60 – Credit 51, amount sum 572300 rubles
• The sum of value-added tax is presented to deduction, Debit 68 – Credit 19, amount sum 87300 rubles

Then, if the requirements for acquiring a discount are met and the discount has been granted, the accounting should not need to be corrected. In the case if the discount has not been granted, the corrections must be made:

• Purchase cost of metal is increased by the sum of a lost discount, Debit 10 – Credit 60, amount sum 15000 rubles
• Valued-added tax from the sum of lost discount has entered the records, Debit 19 – Credit 60, amount sum 2700 rubles
• The metal has been paid in the sum of a lost discount, Debit 68 – Credit 19, amount sum 17700 rubles.
• The value-added tax, connected with the lost discount, has entered the books, Debit 68 – Credit 19, amount sum 2700 rubles.

When this variant of recognizing of the acquired discounts for industrial enterprise is used, the situation can appear in which the discount is yet not been documented, the fact of its granting has not yet been accomplished and, moreover is still in ambiguous, but it has already been registered in enterprise's records as a successful economic event. Such history of events does not comply with the Federal Law "On accounting", which strictly define the sequence of actions: firstly the fact of economic life must actually happen, which must be duly documented with primary document at the moment or immediately after its ending (section 3 of article 9 of the law within named), later on the basis of the primary documents this fact of economic life needs to be recognized in the accounting records. At that, the principle of prudence, stated in section 6 of Russian Accounting Standards 1/2008 "Accounting policy of an organization" is also been violated. Besides, the offered pattern of accounting recording of the received discount is designed not for lowering the initial cost of the assets (works, services) as it is demanded by the term "discount", but for raising it in case if discount-receiver won't fulfill the conditions of the terms of transaction.

4 The reasons behind the issues of financial accounting of acquired discounts

The analysis of the offered variants of accounting of the discounts acquired by the industrial enterprise has shown that each of them has its own serious disadvantages, and this fact wears out the widespread opinion, that offered discounts are "the easiest items for accounting and taxation" [11]. Even the traditionally employed patterns of accounting operations, corresponding with the current legislation, which from the first sight seem to be plain and simple, has one serious drawback – the impossibility of collecting, registering and keeping the information on the gathered discounts in the system of financial accounting and therefore the lack of credible data for conducting the analysis of the effectiveness rate of the applied discounts.

The inability of the financing accounting system to form the information bank on the gathered discounts is conditioned not only with the requirements of International Financial Reporting Standard 2 "Inventories" and Russian Accounting Standards 5/01 "Accounting for inventories", instructing to deduct discounts when defining the costs for inventory acquisition, but also with the compliance with the requirements of section 6.5 of the Russian Accounting Standards 10/99 "Organization's expenditures" according to which all the expenditures is defined "with regard to all the discounts granted to an enterprise under the discount agreement". According to this document discounts are granted viva voice and are not mentioned in the records. This provision is acknowledged by the researcher [10], who believes, that bookkeeping of the granted discounts is not imperative in the business accounting.

However, getting a discount is an operation that influences the financial result of economic entity's activities and so, according to section 8 and article 5 of the Federal Law on Accounting", getting a discount is a fact of economic life, and therefore an object of accounting and should be reflected in its records. Besides, accounting of the acquired discounts is necessary for complying with the requirement of completeness, stated in section 6 of Russian Accounting Standards 1/2008 "Accounting policy of an organization". So it is clear that the documents listed above contain the controversial provisions: some dictate to act in a way that ignores the fact of acquiring a discount (Russian Accounting Standards 10/99 "Expenditures of an organization" and 5/10 "Stock accounting", International Financial Reporting Standards 2 "Inventories"), others call for its accounting (Federal law "On Accounting" and Russian Accounting Standard 1/2008 "Accounting policies of an organization").

Some authors, for instance [8] are fair to point out that discounts accounting is nor regulated enough by the plan of bookkeeping account (Russian and international as well) – they lack universal designated accounts, meant for accounting of the discounts within the frames of any other settled civil contracts.

Therefore, all the circumstances mentioned above – controversies in documents that regulate the organization and bookkeeping process, inconsistency of the existing variants of accounting from the point of view of the immediate discount accounting, the absence of the specific accounts in the plan of accounts, which is intended for registering the applied discount – speak of the fact that methodology of the financial accounting is in the arrearages of the modern practice of widespread application of the discounting system in different fields of business activity.

5 Proprietary technology of the acquired discounts

Narrowly-specialized account named "Acquired discounts" must be included in the working chart of accounts for bookkeeping to form the separate information pool on discounts acquired by the industrial enterprise and to record the facts of acquiring discounts in the accounts of financial accounting. As opposed to Russian standards, international standards of financial reporting does not prescribe how the chart of accounts should look like, so, based only on the structure of Russian Chart of accounts for bookkeeping, we assign number 64 to newly opened account. This account is related to section 6 "Calculations" of Russian chart of accounts, because discounts do not exist solely all by themselves. They appear only when conducting some payments and tied closely to them. It is worth mentioning that account 64 can not only reflect accounts, gathered by an industrial enterprise from suppliers when paying for goods and materials – this account is universal, in other words any acquired discounts can be registered in it.

Method of financial accounting of the discounts acquired by the industrial enterprise is in the following: firstly, the purchased assets (materials, goods, permanent assets, works, services) are being taken to accounting without deducing the discount, and then, if the discount is given, the fact of its acquisition is recorded and after that the initial cost of the account s (works, assets) entered the books. Let's not forget that accounting standards allow to do so when the price of the contract is being changed, and that's what's happening in this case.

The sequence of recording the operations on including materials and goods purchased with a discount in accounts, depends on the time of getting a discount. If it is granted immediately when purchasing with a discount, then the sequence of accounting records is the following:

- The raw materials delivered by supplier has entered the records (metal, 10 tons), Debit 10 – Credit 60, amount sum 500000 rubles
- The discount from supplier is acquired in the amount of 3%, Debit 60 – Credit 64, amount sum 15000 rubles
- Value-added tax is recognized (18%), Debit 19 – Credit 60, amount sum 87 300 rubles
- Settlement with metal supplier has been performed, Debit 60 – Credit 51, amount sum 572300 rubles
- The sum of value-added tax is presented to deduction, Debit 68 – Credit 19, amount sum 87300 rubles

If the retro discount is applied, than the sequence if accounting records is the following – firsthand the metal entered the records with the purchased cost without a discount:

- The raw materials delivered by supplier has entered the records (metal, 10 tons), Debit 10 – Credit 60, amount sum 500000 rubles

• Value-added tax is recognized (18%), Debit 19 – Credit 60, amount sum 90000 rubles
• Then, when all the terms of the contract have been fulfilled and retro bonus granted, the following records are added:
• The discount from supplier is acquired in the amount of 3%, Debit 60 – Credit 64, amount sum 15000 rubles
• The cost of the metal is lowered by the sum of the granted discount, Debit 64 – Credit 10, amount sum 15000 rubles
• The value-added tax is corrected reversing entry, Debit 19 – Credit 60, amount sum 2700.

In case if the settlement with the supplier has already been performed and the sum of value-added tax presented to deduction, the corrections must take place.

6 Practical value of the method of accounting the acquired discounts

The devised method allows solving the main methodological issue of financial accounting of the acquired discounts. It records the fact of acquiring the discount for fulfilling the supply contract terms as a fact of economic life of an industrial enterprise in financial accounting and this restores the principle of completeness and continuity of recording all the facts of economic life of an enterprise in accounting. Also, this method has practical value: it gives opportunity to form the credible base for conducting economic analysis of effectiveness of the received discounts on the basis of financial accounting data.

7 Conclusion

This project makes a scientific contribution in the field of development methods for accounting the discounts, received by an enterprise from raw materials suppliers and this allows shifting from verbal agreements, never reflected in the discounts accounting, to registering such operations in the financial accounts, which later will give opportunity to make the right management decisions.

The work was supported by Act 211 Government of the Russian Federation, contract № 02.A03.21.0011.

References

1. A. Shishkina, Investments and Innovations management, 2, 3-7 (2014)
2. V. Semenikhin, *Organization profits tax, revenues and expenses* (ROSBUKH, Moscow, 2012)
3. L. Zuikova, Economic and legal bulletin, **12** (2010)
4. S. Aipkin, Y. Bogdanova, Accountant solutions, **3**, 6-11 (2009)
5. G. Gorina, Financial and accounting advices, 1, 13-19 (2005).
6. T. Ilyinova, Accountant hotline, **17-18** (2009)
7. S. Utkina, *Organization's expenditures. Accounting and taxation* (Moscow financial industrial academy, 2001)
8. G. Chistyakova, Siberian trade and economics journal, **8**, 52-54 (2009)
9. A. Shaposhnikov, I. Kulikova, Altai State Agrarian University Herald, **2**, 76-80 (2009)
10. A. Bryizgalina, *Accounting: complicated questions from practice of tax consulting, Taxes and financial law* (2012)
11. D. Durnovo, Communication services: accounting and taxation, **4** (2010)
12. V. Semenikhin, *Trade, Encyclopedia* (2011)
13. *On-line updatable publication Applicatory Encyclopedia of an accountant* (Garant, 2013)
14. T. Potapova, Klerk.Ru (2011)
15. L. Fomicheva, Konsultant, **11** (2003)
16. A. Chernenko, A. Shishkina, International accounting, **11(305)**, **2-12** (2014)
17. V. Zakharyin, *Trade. Accounting and Taxation, Taxation Newsletter* (2010)
18. N. Sakharova, Taxation Newsletter, **11** (2008)
19. A. Voiko, Financial newspaper. State issue, **40** (2010)
20. A. Teplyakov, *12100 accounting transactions with commentaries* (Gross-Media, ROSBUKH, 2012)

Methodical assessment and analysis tools of national industrial complex capacity in the regional aspect

Anna A. Ter-Grigor'yants[1,*], and *T.A.* Kulagovskaya[1]

[1]Department of Economic Analysis and Audit, Institute of Economics and Management, North-Caucasus Federal University, Stavropol, Russia

Abstract. Current process of financial instability led to the complex of problems, which solution determines the need to develop the effective management mechanism of regional industrial estates stabilizing their development. The following article is an attempt to develop specific methodical tools to evaluate the environment and the functional results of the national industrial sector with consideration for heterogeneity and uneven territorial development of Russian. Suggested by the authors the implementation of assessment and analysis algorithm of national industrial complexes capacity contributes to a clear objectives statement of industrial policy of territorial authorities and the development on this basis the strategies socio-economic development for a long-term perspective in an unstable market environment.

1 Introduction

Existing financial unrest and the decrease of efficiency of traditional responsive actions to the crisis situations determine the objective necessity to create the competitive and efficient industrial complexes. As the result, there is a need for the development and application of improved methods and tools in the process of building and implementation of industrial policy. All this requires the serious transformation of business systems in industrial, reproduction- functional, institutional aspects. However, this regional component becomes the shaping factor to define the determinants of economic development. Sectorial specificity of the Russian industrial system modernization has grown stronger at the regional levels. In the development of new response principles to the growing non-linearity of the market environment, it is important to take into account regional specificities in development of sectorial structures of national industrial complexes, their various sustainable and structural dynamic types. Pursuant to required reasons, methodology and methods problems of capacity assessment of national industrial complexes appear in a y new light and require fundamentally different, new approaches to their resolving. In the context of this study, national industrial complex should be understood as set of economic entities of various industries located or registered in particular region and forming a variety of vertically and horizontally oriented industrial clusters. National industrial complexes use different types of region resources; its infrastructure thus forming regional socio-economic systems united by common material, labor, financial, information and other flows.

2 Urgency and Scientific Significance

In Russia, in recent years, the major indicators of the industry development slumped. Official statistical data [1, 2] reveal that in the country the industrial production index in six months of 2015 declined by 5.9%. The national total industrial production was the same as in 2007. During 2011-2015 years, annual industrial output without any substantial trend changes with a decline since 2014. The presence of optimal territorial and resource factors, the having of large-scale production facilities combined with inefficient competitive landscape suggest that in Russia there is a weakness in the use of national industrial complexes capacity. As the result, there is a need to find and develop the new management approaches, assessment and analysis of national industrial complexes capacity in the country-wide aspect.

Management improvement of economic development in general and the National Industrial Complex, in particular, as well as the rational development of regional industrial policy taking into account the decision of crisis management tasks involved in both domestic and external scientists Dzh. K. Lafta, B. M. Ayyub, M. T. Shafikov, E. V. Lapin, V. N. Nemtsev, N. L. Zaitsev, G. G. Fetisov, A. V. Andreev, M. V. Ivanova, V. V. Kistanov, I. L. Koz'eva,V. V. Kurnyshev, R. J. Carter, J. L. Gattona, D.W. Walters, L.K. Lau, L. I. Ushvitskii [3-18].

In economics, there is no agreement among scientists on the precise meaning of "capacity". For example, in «Russian defining dictionary» had specified capacity as the range of opportunities and tools in any field [19, p. 518]. In the dictionary of S. I. Ozhegov and N. Yu. Shvedova capacity is considered as a power degree in some respect, the range of opportunities and tools [20, p. 571]. As the result, considering the National Industrial Complex

* Corresponding author: ann_ter@mail.ru

 DOI: 10.1051/shsconf/20173501110

Capacity it should take into account the total of its possibilities and internal resources to increase industrial efficiency and to ensure sustainable development.

3 Main Part

A. Theoretical Part

The best way to present the proposed tools of integrated analysis of National Industrial Complex Capacity and the possibilities of its use is to demonstrate a scheme (figure 1). National Industrial Complex Capacity should be viewed in the context of its main features: real opportunities, resources, availability of reserves, the ability to use the capacity, the level and results of its implementation [21].

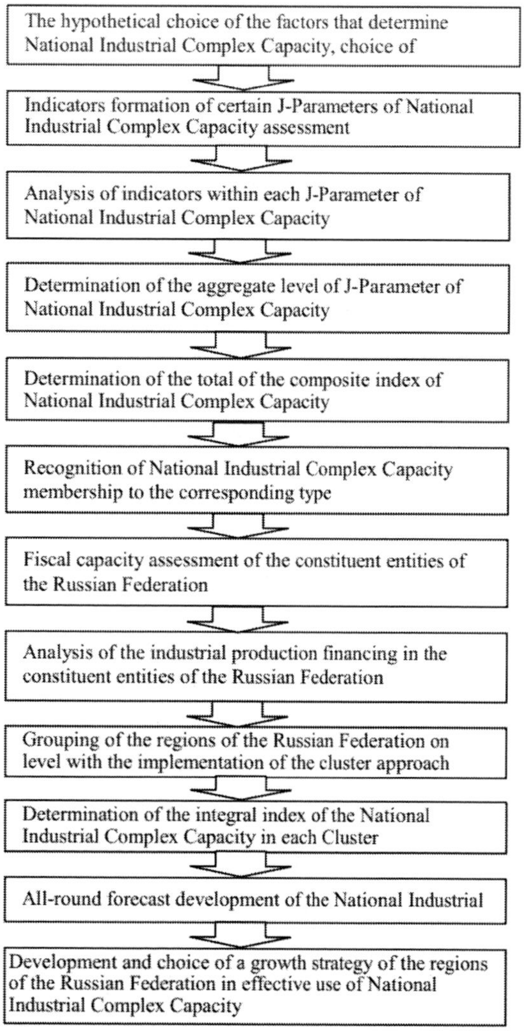

Fig. 1. Block diagram of assessment and analysis of National Industrial Complex Capacity.

B. Setting

Many internal and external, objective and subjective factors have a multidirectional impact on the conditions and results of National Industrial Complex operation, the study of these factors allow identifying the certain parameters of comprehensive evaluation of National Industrial Complex Capacity. Indicators are provided in Table 1.

Table 1. Indicators, Characterizing Parameters Of National Industrial Complex Capacity.

Parameter	Indicators
Resource	The total territory, thou. sq. km.
	The annual average employed in the economy, thou. persons.
	The annual average employed in the industrial production, thou. persons.
	Value of fixed assets, bln. RUB.
	Wear of fixed assets, bln. RUB.
	New fixed assets, bln. RUB.
	Developed and employed advanced production technologies, pch.
Productive	The share of manufacturing in region GDP, %
	The index of production by economic activity «Mining operations», %
	The index of production by economic activity «Manufacturing», %
	The index of production by economic activity «Electricity, gas and water production and distribution», %
	Volume of own-produced and shipped goods, and work and services performed by own efforts, «Mining operations», mln. RUB.
	Volume of own-produced and shipped goods, and work and services performed by own efforts «Manufacturing», mln. RUB.
	Volume of own-produced and shipped goods, and work and services performed by own efforts « Electricity, gas and water production and distribution », mln. RUB.
Market- Infrastructural	Number of businesses and organizations, pch.
	Sales figures (wholesale turnover), mln. RUB.
	Transportation of goods by road of all organizations, million tonnes
	Cargo turnover of motor transport of all organizations, million tonnes
	The density of surfaced public roads, km
	The density of public railways. Railways to 10,000 km2, km
	Dispatching cargo by public railways, density of public railways, million tonnes
Informational	The use of electronic document management in organizations, % of total quantity of interviewed organizations
	Expenditures for Information and Communication technologies, mln. RUB.
	Organizations that had website, % of total quantity of interviewed organizations
	Using special software in organizations, % of total quantity of interviewed organizations
	Using Information and Communication technologies, % of total quantity of interviewed organizations
	Number of personal computers per 100 workers, pch.
	Use of Internet network in organizations, % of total quantity of interviewed organizations
Financial and investment	The institutions with foreign capital, mln. RUB.
	Fixed capital accumulation, mln. RUB.
	Industrial Producer Price Index, %

	Number of privatized property complexes of state and municipal unitary enterprises created during privatization of economic companies, pch.
	Domestic expenditure on Research and Development, mln. RUB.
	Investments in fixed assets, mln. RUB.
	The balanced financial result of performance (gain minus loss), mln. RUB.
Economic	Gross regional product, mln. RUB.
	Profitability of organizations assets on mining, %
	Profitability of organizations assets of manufacturing, %
	Profitability of organizations assets on electricity, gas and water production and distribution, %
	Fixed capital accumulation, mln. RUB.
	Average wade, thou.RUB.
	The specific weight of loss-making organizations,%
Innovative	Organizations performing research and development, pch.
	Expenditures for technological innovations, mln. RUB.
	The volume of innovative products works and services, mln. RUB.
	Staff number, employed in the Research and Development, persons.
	Organizations performing innovations, providing the increase of ecological safety in the production of goods, works, services, pch.
	The innovative activity of organizations, %
	Number of researchers with advanced degrees, persons.

Analysis tools of National Industrial Complex Capacity based on calculating following integrated indices:

- partial estimate of specific J-Parameter of National Industrial Complex Capacity;
- integral index of National Industrial Complex Capacity.

We offer the following formula to calculate the integral index of National Industrial Complex Capacity:

$$P_i = \sum_{j=1}^{4} g_j \cdot d_j , \quad (1)$$

g_j – level of importance J-Parameter of National Industrial Complex Capacity category; d_j – weight of Capacity J-Parameter.

We thought it useful to determine the importance level of J-Parameter of National Industrial Complex Capacity by Pareto approach:

$$g_i = 0{,}8 - 0{,}2 \cdot (j-1) \quad (2)$$

Under this developed method, a differentiated estimation of National Industrial Complexes Capacity of the Russian Federation within a framework of the constituent entities of the Russian Federation from statistical data 2012-2014 г.г had been drawn up.

4 Research Results

The analysis has further compared National Industrial Complex on the basis of complex indices that characterize certain parameters of capacity and identified four typological groups of the constituent entities of the Russian Federation [22, 23] depending on the obtained value of the integral index (Table 2).

Table 2. Grouping of the constituent entities of the Russian Federation on level of National Industrial Complexes Capacity in 2012-2014.

Cluster	The constituent entities of the Russian Federation that integrate the cluster.
Regions-leaders	The Belgorod Region, the Vladimir Region, the Kaluga Region, the Kostroma Region, the Kursk Region, the Lipetsk Region, the Moscow Region, the Tula Region, the Nenets Autonomous District, the Vologda Region, the Kaliningrad Region, the Leningrad Region, the Volgograd Region, the Republic of Bashkortostan, the Republic of Tatarstan, the Perm Territory, the Nizhny Novgorod Region, the Orenburg region, the Samara region, the Sverdlovsk region, the Khanty-Mansi Autonomous District- Yugra, the Chelyabinsk Region, the Omsk Region, the Chukotka Autonomous District
Relatively developed regions	The Ivanovo Region, the Ryazan region, the Smolensk Region, the Tver Region, the Yaroslavl Region, Moscow, the Komi Republic, the Arkhangelsk Region without autonomous region, the Murmansk Region, the Novgorod Region, St. Petersburg, the Republic of Mordovia, the Udmurtian Republic, the Kirov Region, the Ulyanovsk Region, Yamalo-Nenets Autonomous District, the Tyumen Region without autonomous regions, the Republic of Khakassia, the Krasnoyarsk Territory, the Irkutsk Region, the Kemerovo Region, the Magadan Region, the Sakhalin Region
Underdeveloped regions	The Bryansk Region, the Republic of Karelia, the Pskov region, the Astrakhan region, the Rostov Region, the Karachayevo-Cherkessian Republic, the Republic of Marij El, the Chuvash Republic, the Saratov Region, the Kurgan Region, the Altai Territory, the Novosibirsk Region, theTomsk Region, the Republic of Sakha (Yakutia)
Regions-outsiders (depressed region)	The Voronezh Region, the Oryol Region, the Tambov Region, the Republic of Adygeya, the Republic of Kalmykia, the Krasnodar Region, the Republic of Daghestan, the Republic of Daghestan, the Kabardino-Balkarian Republic, the Republic of North Ossetia-Alania, the Chechen Republic, the Stavropol Territory, the Penza Region, the Republic of Altai, the Republic of Buryatia , the Republic of Tuva The Zabaikalye Territory, the Kamchatka Territory, the Primorye Territory, the Khabarovsk Territory, the Amur Region, the Jewish Autonomous Region, Republic of Crimea, Sevastopol

Figure 2 shows quite a negative situation in terms of accumulation and use of the Capacity of National Industrial Complexes: on Regions-outsiders (depressed region) account for 28% of the constituent entities of the Russian Federation, Underdeveloped regions - 17 %, while on the Relatively developed regions account for only 27 %, and Regions-leaders at 28%.

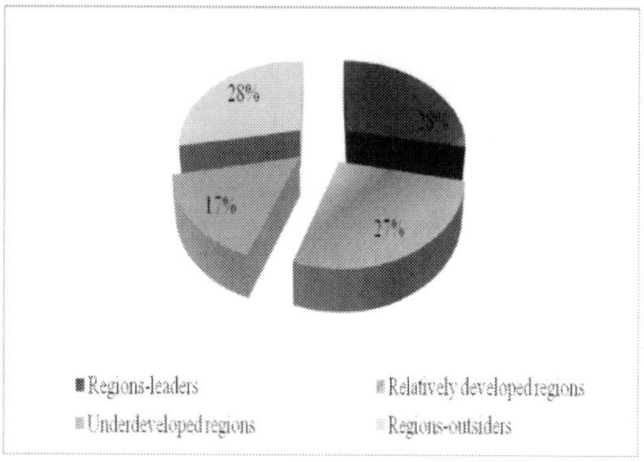

Fig. 2. Typology structure of the constituent entities of the Russian Federation on the level of and use of the Capacity of National Industrial Complexes in 2012-2014, %.

The proposed method assessment and analysis tools of National Industrial Complexes Capacity involves performing of multidimensional various aspects assessment of their formation and functioning in the current competitive environment, which may be useful for some users.

5 Conclusion

A practical point regarding the usefulness of the research is that generalizations and developed methodical approaches to assessment of National Industrial Complexes Capacity allowed selecting the reference regions-leaders and regions-outsiders required special attention and support of the state, suggest optimal basic directions to implement some of their development programs, and in a result, improve the efficiency of domestic industries as a whole.

Thus, the proposed methodological approach to assessment of National Industrial Complexes Capacity contributes to the transformation of understanding of number of processes and tendencies of modern development of Russian industry, modifies the notion about rational delimitation of authoritative powers and regulatory functions as part of the active state industrial policy between the federal, sectorial, regional and local levels of executive power.

References

1. *Russia by the Numbers* (Federal State Statistics Service, Moscow, 2015)
2. *Russia Industry* (Federal State Statistics Service, Moscow, 2015)
3. Dzh.K. Lafta, *Organization Theory* (TK Velbi, Moscow, 2003)
4. B.M. Ayyub, *Risk Analysis in Engineering and Economics* (Chapman and Hall/CRC, 2014)
5. M.T. Shafikov, The socio-humanitarian knowledge, 236-246 (2002)
6. E.V. Lapin, *The economic potential of the enterprise: monograph* (Sumy, ITD Universitetskaya kniga, 2002)
7. V.N. Nemtsev, *Innovative enterprise of ferrous metallurgy: the problems of strategy and risks: a monograph* (Izd-vo Magnitogorsk. gos. tekhn. un-ta im. G.I. Nosova, Magnitogorsk, 2010)
8. N.L. Zaitsev, *The economy of industrial enterprise* (Moscow, INFRA-M, 2008)
9. G.G. Fetisov, V.P. Oreshinm, *Regional economics and management: a textbook* (INFRA-M, Moscow, 2006)
10. A.V. Andreev, *Regional economics* (Piter, St. Petersburg, 2012)
11. M.V. Ivanova, North and Market: Formation of Economic Order, **2(28)**, 146-149 (2011)
12. V.V. Kistanov, *Regional economy Russia* (Finansy i statistika, Moscow, 2011)
13. I.L. Kozeva, E.N. Kubozhev, *Economic geography and regionalism* (KNORUS, Moscow, 2012)
14. V.V. Kurnyshev, V.G. Glushkova, *Regional economy. Fundamentals of the theory and methods of research* (KnoRus, Moscow, 2012)
15. R.J. Carter, *Stores Management and Related Operations* (Macdonald&Evans, UK, 1985)
16. J.L. Gattona, D.W. Walters, *Managing the Supply Chain. A Strategic Perspective* (Macmillan Business. G.B., 1996)
17. L.K. Lau, *Managing business with SAP. Planning, implementation and evaluation* (IGP, 2005)
18. L.I. Ushvitskii, T.A. Kulagovskaya, A.A. Ter-Grigor'yants, I.V. Solovyova, E.S. Mezentseva, International Journal of Economics and Financial Issues, **5(4)**, 1011-1016 (2015)
19. V.A. Lopatsh, L.E. Lopatina, *Russian Dictionary* (Rus. yaz., Moscow, 2001)
20. S.I. Ozhegov, N.Yu. Shvedova, *Explanatory Dictionary of the Russian language: 80000 words and idiomatic expressions* (Azbukovnik, Moscow, 1999)
21. *Regional economics. Natural-resource and environmental foundations* (KnoRus, Moscow, 2012)
22. *Regional economics* (Yuniti-Dana, Moscow, 2013)
23. V.G. Glushkovoi, Yu.A. Simagina, *Federal District of Russia. Regional economy: textbook* (KNORUS, Moscow, 2009)

Management instrumentarium of innovation development of medium-sized industrial enterprises

L. Ismagilova[1,*], *L.* Sitnikova[1], and *N.* Klimova[2]

[1]Ufa State Aviation Technical University, USATU, Ufa, Russian Federation
[2]Institute of social and economic researches of Ufa scientific centre of Russian Academy of Science, ISER USC of RAS, Ufa, Russian Federation

Abstract. The article is devoted to the development management of industrial enterprises related to medium-sized businesses and working in the field of product innovation. In the article there were investigated the methods and management features, the factors, constraining innovation processes in industrial companies. The theoretical background of solving innovation development and maintaining economic stability contradictions were identified. The results of the comparative assessment of small and medium-sized enterprises were unveiled. In paper there were revealed the features and benefits of the transition to the innovative products. A method of positioning matrix and the typology of enterprises in terms of economic sustainability and innovation were proposed. There was worked out a graph-analytical management tool for middle-sized industrial enterprises on the basis of a balance of innovative activity and economic sustainability. For a given industrial enterprise there was formed an effectively balanced portfolio of products at different stages of the life cycle.

1 Introduction

Functioning and development of industrial enterprises in the innovation-oriented environment of the real sector of the production involves in the processing chain businesses of different sizes (large, medium and small forms). The main purpose of the formation of such chains is to maximize the added value at the each stage of the technology cycle, primarily due to the innovative development. However, practice of building such chains has not received proper promotion in the Russian practice, as the evaluation of the role and importance of small and medium enterprises in the Russian economy is ambiguous. With their considerable share in the total number of enterprises, the contribution to GDP remains at 20%. In most statistical studies of small and medium enterprises they fall into one analytical group that does not seem quite correct due to fundamental differences. As noted in [1], adopted criteria for classifying enterprises to groups of small, medium and large are enough conventional and have been introduced in accordance with the established practice of statistical reporting. The features of the medium-sized enterprises in a number of cases are similar to business entities of both large and small size. The analysis of official statistics [2, 3] shows that the majority of medium-sized enterprises (up to 42% of the total) is concentrated in the sectors with high value added. This is the first of all manufacturing industry. At the same time, the average number of medium-sized businesses is closer to the corresponding characteristic of large enterprises.

The increased attention is being paid to the development of small businesses, overshadowed the importance of medium-sized industrial enterprises in modern economy of Russia. The interest in the businesses of this size group may be explained by a number of reasons:

- that is, as a rule, fast-growing companies that could potentially provide an innovative breakthrough in the economy; in [4] it is noted that they are able to generate more than 20% of innovative ideas and become a launching pad for the introduction of new technical and technological developments;
- medium-sized industrial enterprises comparing with the enterprises of small forms have higher capital intensity associated with longer production cycle in the production of added value. Of course, the range and volume of output of midsize enterprises group is significantly lower than the corresponding figures of large enterprises;
- medium-sized enterprises provide a high rate of profitability due to innovation activity. At the same time small businesses achieve high yield in the segments with high margins without adding value (retail, real estate, etc.);
- another important advantages of medium-sized enterprises are the possibilities of their transformation (division or integration), the purpose of which is to choose more effective forms of organizing production. In addition, studies show that large enterprises are more willing to cooperate with medium-sized enterprises. Thus, medium-sized enterprises, on the one hand, are

* Corresponding author: ismagilova_ugatu@mail.ru

able to act as a donor for small businesses, and on the other – are the points of growth and innovative development for large businesses;

• the growth potential of medium-sized enterprises effectiveness is much higher than in small ones. It is comparable to large companies, and even higher in some cases. The share of medium-sized enterprises in providing employment is smaller than their share in the total turnover. This shows quite effective level of using the workforce and a high capacity for innovation development. In some cases, the values of these parameters exceed the industry average levels and general economic level of labor and innovation efficiency.

An important condition for the providing of competitiveness and sustainability of medium-sized businesses is the ability to respond flexibly to the market situation development. These businesses do not have considerable recourses reserves and stability of large enterprises. Their mobility is lower than mobility of small businesses. Therefore, a specific mechanism to ensure stability of this group of companies have to be formed. In modern conditions the basis of this mechanism is implementation of innovation.

2 Relevance

The solution of innovative-active medium-sized enterprises control problem involves the integration of theoretical and methodological approaches in several areas of research. Among them are the competitiveness and innovation management, strategic and economic management, and others.

The analysis showed that the vast majority in developing concepts and tools of strategic management research, including the provision of long-term competitiveness, for example [5-9], is aimed at large enterprises. Much attention is paid to the management of small business development because of its specificity and "the declared" role as a "driver" of economic growth [10], of economic activity of small businesses [11]. The features of medium industrial enterprises management, the choice of effective methods of management, according to the possibilities of innovation growth is not being researched and developed enough nowadays.

Thus, the solution of the problem of insurance medium-sized enterprises competitiveness in the innovation economy requires as target the following objectives:

• systematization of the characteristic features of the industrial medium-sized enterprises, identifying factors of external influence on the efficiency of operations;

• developing the effective management tools for medium-sized enterprises innovation activity as the basis for achieving competitive advantages in integrated criterion of economic stability.

Studies have shown that the implementation of innovations in different periods of medium-sized companies development influence on its resulting characteristics in different ways [12, 13]. In the perspective, innovations contribute to the stability achievement, but the initial moment of their implementation, as a rule, is associated with the violation of financial and economic stability. In this context, the development of methods and management tools for medium-sized enterprises, which would provide coordination of innovative development goals and maintenance an effective production and economic performance is certainly relevant from both the scientific and practical points of view.

3 "Innovative activity" vs "Economic stability"

As noted above innovation changes in technology and organization of production at the initial stage of their implementation are acting in conflict with the main economic interest in any business entity - the preservation of a stable position. One of the main causes of this contradictory situation is high risk and high cost of innovations. As the result there is forming an objective need to find a compromise between the need to improve the innovation activity and the ability to maintain economic stability. For large-scale industrial enterprises, this contradiction is eliminated by attracting sufficient resources and better use of existing capacity. But for medium-sized enterprises, this contradiction can be critical in certain circumstances.

In economic researches there are presented different interpretations of the concepts of "innovation activity" and "economic sustainability" [14-16]. In the context of this study the innovation activity refers to the company's ability to implement innovations faster than the competitors and strengthen the long-term market position. Innovative activity must not only be provided with financial resources. At the same time it must be supported by appropriate financial activity, which is characterized by the speed of funds involving innovative solutions for challenges in a competitive time. This is appropriate to maintain a positive balance of cash flow, which is an indicator of financial stability [17]. To resolve this contradiction of economic and management approach there is proposed the matrix of ranking companies based on multivariate classification and info-graphical model.

The method allows justify management decisions based on changes in the competitive situation of the enterprise status and levels of economic stability. The method is based on multivariate classification process presented in the form of related matrices. Based on the identified characteristics of medium-sized enterprise management it is expedient to classify companies into four competitive statuses:

• Type "C1 - advanced" - is an enterprise with high competitiveness, with strong innovative capacity, high ability to realize the full innovation cycle in a competitive time and financial capacity. The complex of these characteristics determines the high level of competitiveness, which is ensured by the high innovation activity. According to statistics, this type is common to large and medium-sized enterprises - about 8-12%; and less than 1% of small businesses;

DOI: 10.1051/shsconf/20173501111

• Type "C2 - conservative" - the average level of competitiveness with innovative potential, which is insufficient to provide innovative products the market needs. The source of holding competitive position is a temporary position of monopoly, which allows obtain sufficient financial resources for the development of non-innovative movement. According to the estimates this type includes about 20-25% of large and medium-sized enterprises and about 5% of small businesses in Russia;

• Type "C3 - risking" - the average level of competitive enterprises with a significant potential for innovation, and limited financial resources. With these characteristics, they are either involved in innovative projects with a high level of profitability and risk, or middle income projects carried out with a moderate level of risk. Upon successful implementation of projects, enterprises receive financial resources for the next innovation project. In the case of failure the financial stability of the enterprise is disrupted. This type includes about 45% of large and medium, as well as more than 50% of small businesses;

• Type "C4 - surviving" - enterprises with low competitiveness, which are not able to satisfy market demand with innovative products in a competitive time and do not have the financial capacity to deal. These include 10-20% of large enterprises, 45% of medium and 50% of small businesses.

The specification of "economic stability" definition in respect to innovative businesses let us make a conclusion that economic stability performs as the condition of their competitiveness and linked with such key features as innovative and financial activity.

The proposed method of matrix positioning allows, except for the classification of companies on the content and type of competitive status, hold intraspecific classification of enterprises in terms of economic stability. The method presented in the form of a two-step transition from the matrices. This transition defines content management solutions and justifies the strategy of innovative development of the industrial enterprise. Formation of development strategy is based on the sequential selection of the enterprise features in a particular economic and production situation. There were highlighted four levels of positioning companies based on the criterion of economic stability, taking into account the characteristics mentioned above, as well as the possible risks: AES, ES, CES, EI, as shown in Fig. 1. Levels determined by activity performance indicators and active involvement of financial resources.

The combination of high values of these parameters (level AES) provides the coordination of possible contradictions and can achieve a balanced innovation behavior of the average enterprise. Companies with this level of sustainability are highly innovative enterprises, for their effective functioning it is important to maintain as financial stability, and activity in the field of innovation.

Level ES reflects the situation when innovative activity of the enterprise is accompanied by a low level of financial stability or low level of financial activity. In both cases, the economic stability cannot be guaranteed to be secured in the long term.

Level CES reflects the unfavorable situation, characterized by a combination of a high level of innovation activity and low levels of financial stability and activity. It is also considered as unfavorable a situation when insufficient level of financial stability combines with low innovative level. Any manifestation of innovation activity will lead to imbalances and bankruptcy.

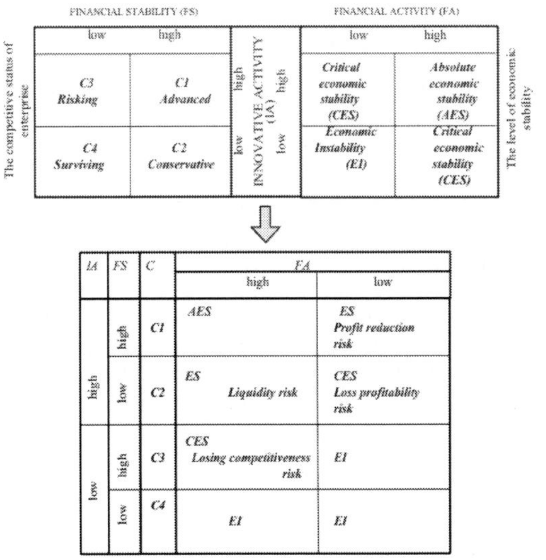

Fig. 1. Enterprise positioning in terms of economic stability.

Various combinations of low levels of indicators (level EI) define crises situations in which business strategy performs irrational innovative behavior, which leads to disparities of resources and strategic perspectives.

The analysis shows that innovative activity and economic stability although conflict, nevertheless working on the long-term competitiveness of medium enterprises [18].

Thus, the specificity and the key problem of organization and management at medium innovative enterprises is the harmonization of administrative decisions in planning and release of innovative products. It is possible to ensure the maintenance of the conflicting interests balance, using the matrix ranking method.

4 Info-graphical tool for innovation-active industrial enterprises management

The elimination of innovative development contradictions and the economic stability of the enterprise is carried out in the development of production plans. The solution lies in the optimization of the product portfolio on the criterion of economic balance of the results obtained from the products at different stages of the life cycle [19-21].

The idea of coordinated management implemented in practice in developed model of infographic (Fig. 2.) and algorithms of analysis of relative indicators: price and output plan, fixed and variable costs, profit (loss).

Info-graphical model as a tool to validate management decisions makes it possible to:

• determine the effective output plan at fixed costs and target value of the relative profits;
• calculate the value of a possible profit at fixed release schedule, costs and prices;
• set the price based on the output plan, costs and targeted level of profitability;
• estimate the required amount of fixed costs for a given output plan, set the variable costs, product prices and the relative profitability;
• identify the effective ratio of fixed and variable costs for specific values of prices, variable costs and profitability;
• set the possible correlation of economic indicators of info-graphical model according to target level of profitability.

Info-graphical model as a tool of management decisions validation, allows us to implement contingency management synchronous in several ways due to the wide field of information. In addition, this model ensures the validity of the decision-making under conditions of conflicting goals and visual coverage of managerial situations, determined by a combination of economic parameters.

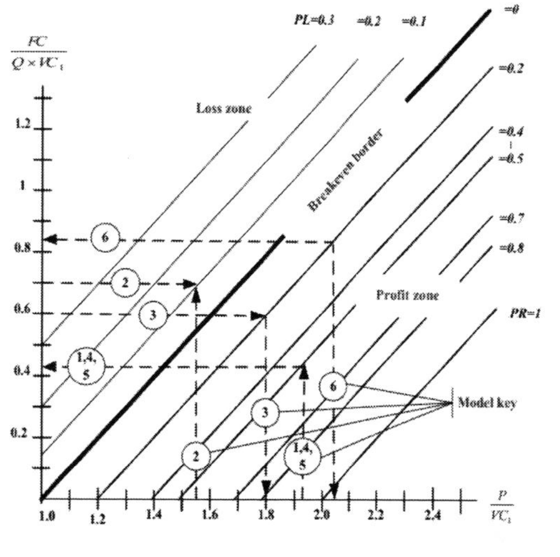

Fig. 2. Info- graphic model of economic indicators planning for the criterion of profitability (FC - fixed costs, Q – output, VC_1 - variable costs per unit, P - price, RP (RL) - Relative profit / loss; 1, 2, 3, 4, 5, 6 - types of management decisions).

5 Practical application

The model above was tested on the one of medium-sized enterprises of "Machinery and equipment" economic activity of the Bashkortostan Republic.

During 2014, the company was unable to provide a profit on new products. For a number of not-innovative competitive products which are at the saturation point, the economic indicators have reached an acceptable level, but not enough to ensure the flow of financial resources allocated for the maintenance of innovative products. The established balance of economic indicators corresponds with the situation "C2-ES" by the method of positioning the matrix (Fig. 1).

One of such products is the directly fuel kiln. According to the info-graphic model (Fig. 3), with the initial values of economic parameters (the price of the product is 2447.37 money units (m.u.), variable costs are 1786.4 m.u., fixed costs are 1031110 m.u., volume of output is 1480 units) the level of relative loss of this product is:

$$RL^{2014} = 1 - \frac{P}{VC_1} \times \frac{1}{(1 + \frac{FC}{QVC_1})} = \tag{1}$$

$$= 1 - 1.37 \times \frac{1}{1 + 0.39} = 0.014$$

The development of strategic decisions about the enterprise transfer into a more favorable situation realize by means of the proposed instruments. As the target criteria there was set a level of relative profitability of the product is not lower than 0.11. Target profitability value (at the same cost) is achievable by varying the price and release plan. Using the info-graphic model it was justified the sequence of administrative decisions formation. With an increase in prices to 2536,69 m.u. on the info-graphical model (Fig. 3) there was determined the appropriate balance of costs:

$$\frac{P}{VC_1} = 1.42. \tag{2}$$

During solving the problem in the graphical-analytical model (Fig. 3) there was defined the ratio:

$$\frac{FC}{Q \times VC_1} = 0.28 \tag{3}$$

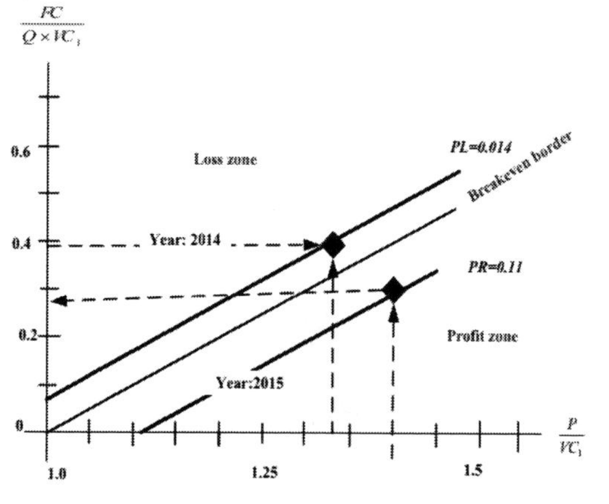

Fig. 3. Modeling of targeted economic parameters.

This ratio corresponds to the release of 2060 units, and profit on the position "directly fuel kiln" amounted to 518220 m.u.

Similar calculations for all product items allowed to determine portfolio structure of products with different positioning on the life cycle stages (Table. 1).

Table 1. Product portfolio.

Position of portfolio	Year	Profit on realization	Changes of profit
Directly fuel kiln (on gas and liquid fuels)	2014	-	+518220
	2015	+518220	
Dosing system	2014	346406	- 262533
	2015	83873	
System of gaz distribution in gazlift well	2014	47093	+ 79700
	2015	126793	
Electro-fulfilment regulating device	2014	68086	+ 259933
	2015	328019	
Oil metering unit	2014	58908	+ 30164
	2015	89072	
Liquid fuel preparation block	2012	31794	+ 31794
Total			+657278

The resulting balanced portfolio provides the smoothing of contradictions of innovative behavior and economic sustainability.

6 Conclusion

As a result of the comparative evaluation there were revealed the features and benefits of medium-sized business entities in the manufacturing of innovative products. There was proposed the classification of medium-sized enterprises according to the criterion of competitiveness, which differs from the existing classification by complex measurement the of innovation activity parameters and financial stability. The problem of resolving economic and managerial contradictions arising in industrial enterprises of the middle class was examinated. there was set and solved the problem of elimination of contradictions between the need of innovation production and financial constraints.

In the paper there was proposed a method of matrix positioning and info-graphic tool to validate management decisions to ensure the competitiveness of innovative enterprises. For a balanced portfolio of products is formed through coordinated control at different stages of the life cycle and thus providing financial support for innovation. The approbation of the developed instruments proved its high importance for the effective management of medium-sized industrial enterprises.

The study was sponsored by the RFH and the Academy of Sciences of Bashkortostan, the project number 15-12-02020.

References

1. V.V. Kotov, The Economic Journal, **23**, 90-101 (2011)
2. *Small and medium business in Russia. 2014: statistical publication* (Rosstat, Moscow, 2014)
3. *Russia by the Numbers* (Rosstat, Moscow, 2015)
4. V.V. Andreev, M.P. Galimova, Intelligence, innovation, investment, **1**, 21-25 (2013)
5. G.B. Kleiner, V.L. Tambovtsev, R.M. Katchalov, *Enterprise in an unstable economic environment: risks, strategy, security* (Economics, Moscow, 1997)
6. D. Aaker, *Strategic market management* (Peter, St. Petersburg, 2007)
7. I. Ansoff, *Strategic Management: classic edition* (Piter, St. Petersburg, 2009)
8. R.M. Grant, *Modern strategic analysis* (Peter, St. Petersburg, 2008)
9. O.N. Melnikov, V.N. Shuvalov, Russian Entrepreneurship, **9(69)**, 100-104 (2005)
10. F.A. Babayev, Problems of Forecasting, **4**, 136-142 (2005)
11. I.A. Baev, M.V. Podshivalova, The economy of the region, **1**, 189-198 (2013)
12. V.V. Andreev, M.P. Galimova, L.V. Sitnikova, Journal of Economic Integration, **7**, 16-22 (2013)
13. L.A. Ismagilova, T.A. Gileva, L.V. Sitnikova, Theoretical and practical aspects of management, **9**, 62-71 (2015)
14. A.D. Kanchaveli, A.A. Kolobov, I.N. Omelchenko, *Strategic management of organizational-economic stability of the company* (MSTU. Bauman, Moscow, 2001)
15. L.T. Gilyarovskaya, A.A. Vehoreva, *Analysis and assessment of financial stability of business* (Piter, St. Petersburg, 2012)
16. S.N. Anokhin, *Technique of modeling of industrial enterprises economic stability in modern conditions* (Saratov State Technical University, Saratov, 2006)
17. L.A. Ismagilova, M.P. Galimova, Herald USAES. Science, education, economy. Series: Economy, **2**, 78-81 (2012).
18. V.A. Dines, V.M. Larin, R.Y. Loskutov, N.S. Yashin, *Competitiveness and economic stability of the enterprise* (Publishing Center SSSEU, Saratov, 1999)
19. O. Bastrikova, L. Sitnikova, Management in Russia and abroad journal, **4**, 77-87 (2009)
20. L.A. Ismagilova, E.V. Orlova, *Proc. of the 8h Int. Workshop on Computer Science and Information Technologies CSIT*, 453-457 (2006)
21. L.A. Ismagilova, T.A. Gileva, M.P. Galimova, Innovation Activity, **1(32)**, 63-73 (2015)

Cost management at the tube-rolling enterprises

A.E. Shevelev[1], E.A. Sheveleva[1], and M.I. Bazhanova[1,*]

[1]South-Ural State University, Chelyabinsk, Russia

Abstract. The article is devoted to the problem of constructing an effective system of cost management of tube rolling enterprises. According to numerous studies for over the past few years tube-rolling consumption has increased inside domestic market. Despite of the positive trends, these enterprises still can't solve the problem of effective cost control, significantly affected their final financial results and, consequently, their competitiveness as a whole. The authors examined the legal aspects of the formation and recognition of enterprise's costs. Main approaches to cost classification of the main members are investigated. Main features of tube-rolling enterprises are analyzed and requirements to their cost management system structure are defined. A cost management model of tube-rolling industry as a logframe of consistent implementation stages, including corrections and feedback, is performed.

1 Introduction

At the moment enterprises of the metallurgical complex and their activity play significant role in the Russian economy. According to numerous modern studies over 28 000 metallurgical enterprises operate in the domestic economy now. In general terms, this industry can be divided into nonferrous and ferrous metallurgy, last is of more value.

An important segment of ferrous metallurgy is tube-rolling industry. Such tube-rolling enterprises, as Nizhnytagilskiy tube-rolling plant "Metallinvest", Severskiy tube-rolling plant, Kamenskuralskiy tube-rolling plant, Pervouralskiy novotrubniy plant, Uralskiy tube-rolling plant, Chelyabinskiy tube-rolling plant, Pervouralskiy novotrubniy plant operate in Ural Federal District.

Currently, the demand for tube-rolling production is growing steadily. For example, the consumption of tubes grew by 6.8% at domestic market, considering that import decreased on average by 20% [1].

2 Relevance

Despite of positive trends tube-rolling enterprises have significant problems with effective cost management system forming. It is known that one of the sustainable economic growth factors is the ability of an enterprise to reduce costs, directly related to production and sale of finished products. Besides, complex production process in this sector, significantly determines the diversity and complexity costs and causes to their impact on the profitability of the activities undertaken. In this regard, maintenance of operative profitability through effective and continuous cost management process becomes of great value, it is undoubtedly of great relevance considering modern trends of financial and economic development.

3 Mane problem

Before turning to the problem of cost management effective system forming, we should identify the peculiarities of tube-rolling enterprises activity and main types of costs. There is no unified conceptual approach to the interpretation of the term "costs" in the economic literature. Often one can come across with the interpretation of this term, its rationale and composition in the connection with such categories as "spendings", "costs price" and "expenses". Consider the content of these terms in details.

International financial reporting standards name "costs" as consumed in production activity resources, which have not been recognized yet as expenses and recorded in the balance sheet at the end of the year as unfinished production, finished products, goods shipped [2].

In the Russian management theory and practice term "costs" is often identified with the term "expenses". For example, in the Tax Code of the Russian Federation [3] is not highlighted significant differences in the interpretation of these two categories. However, detailed study of this problem indicates a serious difference of these terms. PBU 10/99 "expenses of organizations" provides the following definition of the term "expenses": "economic benefits decreasing as a result of the disposal of assets (monetary funds, other property) and (or) occurrence of obligations leading to reduction of capital of the organization, except for reduction of contributions upon decision of participants

* Corresponding author: mbazhanova@mail.ru

(proprietors of property)" [4]. Expenses include such items as production costs of sold products (works, services), wages of administrative staff, depreciation, rental cost, and losses [5]. Based on this we can conclude that "expenses" represent the costs of the defined time period, documental and economical justified, fully undergone their value on sold during this period production [6].

In addition to the interpretation of the term "expenses" according to PBU 10/99 (i. 9) the mechanism of transaction enterprise's expenses to the cost of the produced products (works, services). The cost price of goods manufactured (works, services) is formed on the basis of the following expenses of main types of activity:

• expenses recognized in the reporting and previous periods;
• expenses related to incomes of the following reporting periods, i.e. passing.

In general, cost price is total costs on production and sales of a particular enterprise expressed in monetary terms, which compensation is required for the simple reproduction [7]. Based on this, we can conclude that a significant difference of terms "costs" and "cost price" is manifested in the form of completion of the production and circulation processes. Costs are related to these processes, regardless of their completeness, as well as the nature of relationships with production and sale of products. The cost price implies the completeness of these processes.

General indicator of the number of considered terms is the term "spendings". It's interpretation is often based on the fact of limited resources and possibility of their alternative use. Spendings represent a real or hypothetical monetary measure of the aggregate financial resources used for certain purpose. On this basis, costs can be defined as the aggregate of enterprise's spendings at the moment of purchasing of any services, material values, etc. These costs cause to reducing of economic resources, or increasing accounts payable of an enterprise. In this regard, costs may be attributed either to the organization's assets (in the case of income in the future) or to liabilities (in the case of non-receipt of income in the future and, as a consequence, decrease of retained earnings of the company during the reporting period) [7].

Taking this into account, we may add to the concept of enterprise the following items [7]:

• costs are determined by using labor, material, financial resources, etc., reflecting their amount spent in the production and sales of products (works, services);
• the amount of resources used is expressed in natural and monetary units
• costs are related to the defined objectives of the enterprise;
• consumption of different types of resources must be economic justified by the conditions of the production process and aims of creating values in the production of goods (works, services);
• costs, finally, are transformed to the cost price of production (works, services).For a deeper study of the nature of costs and ways to manage them, it is necessary to make up their classification. Currently there are numerous of ways to classify costs in the economic literature. Consider the most popular of them.

Thus, Drury K [8], Karpova T. [9], propose the "aggregative" group, whereby all of the costs are divided into the following groups:

• costs, information of which is used for the purposes of regulation and control;
• costs used for decision-making and planning process;
• costs, information of which is used for calculation and evaluation of finished products.
• Paliy V. [10] offers the following classification according to:
• the degree of influence the volume of production: fixed and variable;
• the process for assigning to the cost price: direct and overhead;
• the frequency of investments: current and non-recurring;
• to the relation to the manufacturing process: production and non-production;
• the time calculation: planned and actual;
• the degree of aggregation: per unit of product and total.

Lastovetskiy V.E. [11] as the classifications allocates production factors and appointment of certain types of expenses in relation to the nature of resource consumption. These classification features can provide the basis for cost management and calculation of production costs.

Sokolov Y.V. [7] proposed the following grouping of costs according to:

• the type of economic life factors: the cost of activities and events;
• the recognition: accrued and paid;
• the relation to money: monetary and non-monetary;
• the relation to the balance sheet: capitalized and non- capitalized;
• the expectation: ordinary and extraordinary;
• the evaluation: planning and regulatory.

The most correct costs classification in terms of their management were offered by Vakhrushina M.A., Gomonko E.A., Tarasova T.F. Consider these approaches in more detail.

Vakhrushina M.A. [12, 13] proposes to classify all the costs according to the purposes and objectives of management accounting (tab. I).

Professor Mizikovskiy E.A. [14] as the basis of classification of costs took an assumption about their depending upon types of activity of an enterprise (main and other). Generally all costs he proposes to group by economic elements and calculation items.

Efremova A. A. [15] proposed to classify the enterprise's costs according to a period, method of including to the cost price, economic structure, relation to production amount and efficiency.

Table 1. Costs classification according to Vakhrushina.

Objectives of management accounting	Costs classification
calculation of cost price of goods manufactured, cost estimation of reserves and profits	Incoming and pas Direct and indirect Basic and overhead Production and periodic Single and complex Ongoing and one-time
decision making and planning	Conditionally-constant and conditionally-variables Accept and not accept Irrecoverable costs Imputed (lost profits) Marginal and incremental Planned and unplanned

Gomonko E.A. and Tarasova T.V. [7] as a basis of costs classification take the following characteristics: the method of including to cost price of production, composition (homogeneity), the elements, the degree of management, participation in profits and losses valuation, position of a particular enterprise, he coverage by the plan.

These approaches according to the specifics of the enterprise may be supplemented by the following classification:

• depending on the management functions: commercial, administrative and production costs;
• depending on the type of resources consumed: cost of labor, working capital and fixed assets);
• depending on the industry sector of the enterprise: costs of trade enterprises, catering companies, industrial enterprises, etc.;
• depending on the specifics of money circulation: costs of production, the sphere of circulation, etc.

In general terms, costs are the total amount of all available resources of the enterprise in monetary form, used for it's economic activity for a certain period. We can highlight the following feature of the costs, as a management object. Costs are in constant dynamics due to price changes of purchased resources, and the revision of the normative values of the expenditure coefficients.

The formation of an effective cost management system on any industrial enterprise is difficult without understanding the specifics of products, as well as the essence of production and sales processes.

Most tube-rolling enterprises produce the following range of products [16, 17, 18, 19]:

• welded straight-seam tubes (mainly tubes of large diameter);
• seamless tubes of OCTG;
• seamless industrial tubes etc.

Large diameter tubes used for oil and gas pipelines to transfer the hydrocarbons over long distances, and the transportation of crude oil and natural gas over shorter distances from wells to processing plants, storage tanks and loading and distribution of raw materials. OCTG used in oil and gas wells. They are divided into:

• casing pipes used in the drilling and operation of wells;
• pump-compressor pipe for oil and gas production, which are used for withdrawal of gas and oil to the surface;
• oil and gas transportation pipes that are used to transport oil and natural gas, etc.

Seamless industrial pipes are used in various industries. Their use is associated with the transportation of steam, water, gas and air under high pressure. These tubes are divided into hot-rolled, cold-formed, boilers, stainless, bearing, etc.

The described range of products is largely responsible for the specificity of activity of the tube-rolling enterprises. Analysis of most famous enterprises allows to allocate a number of features with respect to different spheres and aspects of their activities.

As to the process of production tube-rolling enterprises have the following features:

• the exact division of the production process to the stages;
• limiting the possibility of establishing inventories of unfinished products at any stage of the process;
• unbreakable connection between the separate parts of the process;
• limited opportunities to interrupt the process;
• relatively long production cycle;
• within a single production plant a large number of nomenclature units is produced;
• presence of downtime due to the transition of production to another size ("handling");
• high demands on compliance with the quality of the finished product with GOST and TU;
• the need for a large number of units and the nomenclature of raw materials needed for production, and others.

In the field of economic and financial planning and accounting enterprises have the following peculiarities:

• raw materials which have been released to production are put off in accordance with standards established by the technical department, but not the requirements of GOST and TU;
• need of the planned and actual costing of finished goods;
• need to address the relationship of different structural units and departments forming a single production chain;
• when using the client's own blanks and blank ("tolling metal") planning process is carried out in different ways;
• distribution of general and ccommercial expenses directly proportional to the volumes of production;
• use of a large number of costing nomenclatures items;
• redistributive cost accounting (often not a separate department acts as a redistribution, and its production sites);
• when metal flow planning not only the return and irrecoverable loss, but also ingots (forgings, billets)

required to adjust the mill and others should be taken into account.

In the management field often the strategy of outsourcing in relation to auxiliary and new types activities is used.

High need for the use of the process approach when making management decisions, as well as a rigid and rather high demands on the selection and maintenance of management personnel are characterized by. In addition, tube-rolling enterprises have often horizontal integration.

For example, currently on the territory of the Russian Federation such largest tube-rolling holdings, as "United metallurgical company", "Pipe metallurgical company", enterprises of group "Chelyabinsk tube-rolling plant", etc. operate.

Supply and marketing sphere also has a number of specific features:

• tube-rolling enterprises, as a rule, have its own sales network, with offices throughout Russia;

• significant fluctuations of the demand for manufactured products under the influence of macroeconomic factors;

• implementation of sales activities mainly in the business market;

• dependence of the pricing from the cost price of products;

• need for monthly purchase of large quantities of various raw materials, etc.

Based on this, and on monitoring of practical sphere, we may name the following specific features, which play a crucial role in the formation of costs at the tube rolling enterprises:

• tube rolling enterprises have complex and organizational structure involving large number of units, in "head" company and in those, which operate according to principles of the strategy of outsourcing

• the existence of forced technological downtime due to the transition of production to another size, a fairly long production cycle, as well as limiting the possibility of creating inventories of work in progress at any technological stage;

• no valuation costs from the perspective of the requirements of GOST and TU, and the lack of any standards when dealing with "tolling" metal;

• developed system application of costing items due to the use of a large amount of inventory items raw materials and other inventory items in the tube-rolling production;

• implementation of management cost accounting mainly by redistribution method;

• presence of both revocable and irrevocable losses, and ingots (forgings, billets) required for mill setup.

4 Solution

The basis of cost management system forming should include [20, 21]:

• need of a comprehensive approach to the management accounting system formation, affecting absolutely all structural units of the enterprise (holding);

• need of using resource approach, to use all available enterprise's resources in a proper way;

• need to use only quantitative indicators, which calculation is based on available and reliable information, to determine the current status and the future of the cost management system;

• formation of developed system of monitoring and internal control at various stages of cost management;

• orientation to the various factors controlling impact, as well as external signals.

Considering the information obtained earlier, the following model of cost management system of the enterprise of tube-rolling industry is offered (Fig. 1).

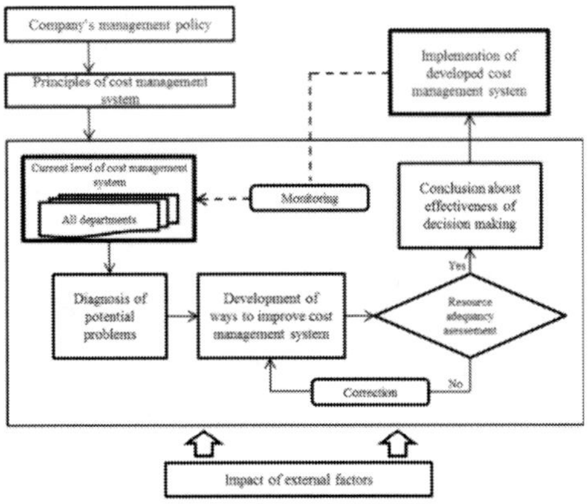

Fig. 1. Cost management model of tube-rolling enterprises.

This model of cost management system of tube-rolling enterprise reflects the sequence of stages of its formation, taking into account the feedback that allows to adjust earlier decisions.

It allows to satisfy all the demands made earlier to the concept of cost management system of tube-rolling enterprise. It should be noted that the implementation of each phase of the proposed model is the subject of a separate research.

Thus, the proposed model, which is a log frame consistent implementation stages, taking into account adjustments and feedback allows making correct management decisions about cost management of an enterprise and quantify the various stages of the implementation of the chosen strategy. This will allow to tube-rolling enterprises operate more efficiently than its competitors.

The work was supported by Act 211 Government of the Russian Federation, contract № 02.A03.21.0011

References

1. http://utmagazine.ru/posts/10561-ekonomika-rossii-cifry-i-fakty-chast-8-metallurgiya.
2. *International financial reporting standards* (SPS Consultant-plus)

3. *Tax Code of Russian Federation – part two* (2000, no. 117-FZ. SPS Garant)
4. *PBU "Expenses of organizations"* (no. 10/99 SPS Garant)
5. M.A. Vakhrushina, *Management accounting: Textbook* (Омega-L, Moscow, 2006)
6. P.P. Taburchaka, *Enterprise economy: textbook for high schools* (Fenix, Rostov-on-Don, 2002)
7. E.A. Gomonко, *Cost management at the enterprise: textbook* (KNORUS, Moscow, 2009)
8. K. Druri, *Introduction to management and cost accountingi* (Audi, Moscow, 1994)
9. T.P. Karpova, *Management accounting: textbook for high schools* (UNITI, Moscow, 2003)
10. V.F. Paliy, *Management accounting of costs and incomes: with elements of financial accounting* (INFRA-M, Moscow, 2009)
11. V.E. Lastovetskiy, *Cost accounting for factors of production and centers of responsibility* (Finance and Statistics, Moscow, 1988)
12. M.A. Vakhrushina, *Management analysis: textbook for high schools* (Омega-L, Moscow, 2008)
13. M.A. Vakhrushina, *Management accounting: from theory to contemporary practice: textbook* (Rossiyskaya gazeta, Moscow, 2005)
14. E.A. Mizikovskiy, *Productive accounting* (INFRA-M, Moscow, 2010)
15. A.A. Efremova, *Cost price: from management accounting to cost accounting cost accounting* (Vershina, Moscow, 2006)
16. http://kutzmet.ru.
17. https://stz.tmk-group.ru.
18. https://sintz.tmk-group.ru.
19. http://www.chelpipe.ru/for_clients/steel_pipes.
20. V.V. Alekseev, *Metallurgy of the Urals from ancient times to our days* (Nauka, Moscow, 2008)
21. V.V. Zapariy, *The history of ferrous metallurgy of the Urals: the 90 years of the 20th century* (Nauka, Moscow, 2003)

 DOI: 10.1051/shsconf/20173501113

Innovation capital in the appraisal system of intrinsic industrial enterprise value

M.K. Akhtyamov[1,*], and *E.A.* Gonchar[1]

[1]South Ural State University, Chelyabinsk, Russia

Abstract. The paper studies innovation capital value of in formation of intrinsic industrial enterprise value. The authors specify a structure of intrinsic industrial enterprise value, establish an innovation capital position therein and describe interrelation between all cost components. Concept and essence of enterprise innovation capital are defined on the basis of a scholarly opinion with concentration on its difference from innovation potential. The scheme of innovation capital formation with involved cost components is described during clarification of nature of innovation capital and its importance in intrinsic industrial enterprise value management. In the paper there is indicated an estimation procedure, in which description differential features of such components appraisal as value of innovation projects in the developmental stage and value of innovation projects in the implementation phase are emphasized based on the author's clarification of essence of innovation capital and identification of its place in a modified model of intrinsic industrial enterprise value. The proposed procedure of innovation capital estimation is approved by an example of two innovation projects, which are in the different stages, but have the same earning capacity and equal financing terms. The calculation data showed that enterprise innovation capital value, which innovation project is under development, is significantly lower than the same indicator of a company, which innovation project is underway. The recommendations on investment policy improvement, made by the authors, which are aimed at innovation capital compounding and intrinsic industrial enterprise value increase, form the result of the paper.

1 Introduction

In the current circumstances intrinsic enterprise value management is an exigency of insufficient investment attractiveness of the most of domestic industrial companies. One of the main company investment attractiveness indicators is overrun of its intrinsic value over the market one. This indicates that enterprise value is undervalued by the market at a particular time and that there is growth potential for it, which provides investment inflow. One of the significant intrinsic value elements for industrial enterprises is innovation capital.

Innovation capital compounding appears as one of the most effective ways to manage intrinsic value and investment attractiveness of industrial enterprises.

The basis of investment inflow in industrial enterprises innovation projects is to be set up that provides terms for their economic recovery and strategic prospects implementation.

2 Timeliness. Scholarly importance

Innovation capital essence and appraisal methodology for it should be considered when it is necessary to emphasize intrinsic industrial enterprise value management, provided by innovation capital compounding. A selected innovation capital appraisal method has influence on objectivity of intrinsic industrial enterprise value appraisal, its investment opportunity as well as its ability to stimulate economic growth. The approach, which should be set as a basis of a selection procedure of an innovation capital appraisal method, is heavily tied to this component positioning in the element system of enterprise intrinsic value.

Modern academic specialists generally give similar concepts of enterprise intrinsic value. M. A. Fedotova defines intrinsic value as appraisal of internal asset capacity based on a forecast of its market price in the future, made by an analyst [14]. In international appraisal and investment practice the terms "intrinsic value" and "fundamental value" are often considered as equal. These two concepts are identical. Lawrence J. Gitman and Michael D. Joehnk consider intrinsic value as hidden or essential stock value calculated in the course of a fundamental analysis [12].

However on paper there are also concepts which differ from the traditional term of intrinsic value. William F. Sharpe, Gordon J. Alexander and Jeffery V. Bailey investigate the terms "investment value" and "intrinsic value" as synonyms, when considering security as an investment project. In their opinion, investment value is the current security price, accounted by well-informed and capable analysts, considering prospect

* Corresponding author: a-mavlit@yandex.ru

evaluation of its demand price level and income from it in the future. In this case it can be considered as fair value of a security [19]. The authors of this paper do not share this opinion. It seems more logical, when investment value is understood as difference between market and intrinsic value due to the fact that it shows growth potential of the first term over the long term and thus provides guidance to an investor on his opportunity to invest in this company. With this in mind, if market value is lower than intrinsic enterprise value, it means that there is investment potential. Otherwise, an enterprise is not investment-attractive.

When the conceptions of intrinsic enterprise value are close except for some of the above-mentioned points, the insight into intrinsic value structure varies significantly. While on subject of market capitalisation and its structure Leif Edvinsson [10] consider, however, basic elements: intellectual capital and financial one. He [10] highlights human and structural capital as a part of intellectual one and divides it into customer and organization one, where the last named consists of innovation, cultural and process capital. Properly speaking, Leif Edvinsson, whose point of view is contrary to the standards of market capitalisation appraisal, listed all major components of intrinsic enterprise value, in our opinion, with exception of such elements as resource capital and commercial goodwill. The key point is intellectual capital structure and innovation capital place therein. Interpretation of innovation capital essence and appraisal procedure for it depend on innovation capital positioning in this system. Therefore, the aim of this paper is to study essence of industrial enterprise innovation capital and to develop an appraisal methodology for it.

3 Thesis statement

The following objectives are set by authors in order to achieve the stated aim within the framework of this paper.

Firstly, it is necessary to clarify a structure of intrinsic industrial enterprise value and to identify the place of innovation capital therein. This problem solving is necessary for understanding decomposition of intrinsic industrial enterprise value and interrelation between its elements that allow us to draw the line of interpretation of elements essence in general and innovation capital in particular.

Secondly, in connection with modified decomposition of intrinsic industrial enterprise value it is necessary to introduce clarity into this concept on the basis of the first problem solving result and through investigation of modern approaches to the interpretation of innovation capital essence.

Thirdly, an appropriate appraisal methodology is to be elaborated by virtue of clarified essence of innovation industrial enterprise capital.

Fourthly, applicative calculations of industrial enterprise innovation capital appraisal are to be executed in order to confirm practical relevance of the author's technique.

4 Theoretical part

Improving the structure of intrinsic industrial enterprise value, we analyze the estimates of market capitalisation, according to Leif Edvinsson, as a basic model and presume that it measures intrinsic value better, although incompletely. As noted above, firstly, there are no such important components as commercial goodwill and resource capital in this model, and, secondly, innovation capital is not positioned entirely correct.

Figure 1 illustrates the author's view on the model of intrinsic industrial enterprise value, taking into account the foregoing points.

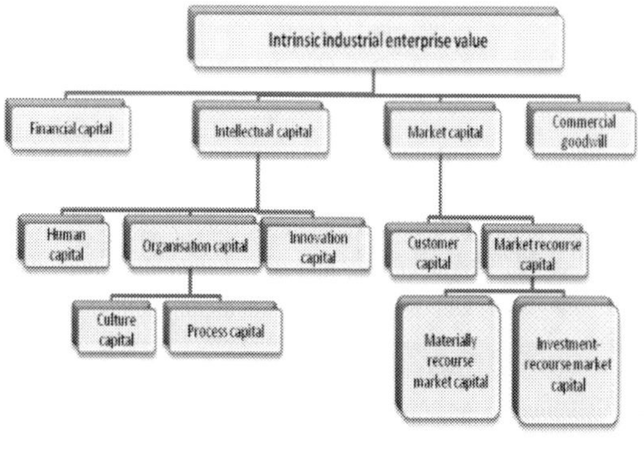

Fig. 1. Decomposition of intrinsic industrial enterprise value.

The signature features of our model of intrinsic industrial enterprise value from the model of market capitalisation, offered by Leif Edvinsson, are the following.

Firstly, in addition to financial and intellectual capital, we have added market one and goodwill in the first hierarchy level, assuming that these components also make their strong contribution to intrinsic industrial enterprise value.

Here we should make a reservation that there is market capital in the Leif Edvinsson's model, but it represents a part of structural capital, included in intellectual capital, and is called there "customer capital". According to the authors, there are some inaccuracies as follows:

1. Market capital is superior to customer one. Customer capital is formed owing to customer base (customers and clients). Then resource (logistic) capital, related with partnership with suppliers, contractors and investors, is equally important. We divide resource capital into the materially resource and investment resource and associate both these elements with market advantages of an industrial enterprise.
2. It is obvious that market capital should not be considered as a part of intellectual capital, taking into account the special nature of the last one. We interpret it as a separate component, which contributes tried and tested market opportunities of an enterprise to intrinsic industrial enterprise value.

Secondly, innovation capital is got by us as an equal component of human and organization capital to the second hierarchy level. According to the Leif Edvinsson's market value model, innovation capital is a component of organization one on a par with cultural and process capital. In the authors' opinion, such an approach does not seem logical, because of the fact that innovation capital should be considered as a component, mainly characterized by already developed innovation projects or those projects, which are under development, where an organizational point is important, but do not have the single role, and has independent significance in formation of intrinsic industrial enterprise value.

We will explain delimitation of innovation capital from human and organization one further during addressing the following issue of innovation capital essence clarification.

Interpretation of innovation capital essence in modern science has received attention of only few authors, while much more attention is paid to innovation potential essence.

So, for example, I. S. Vinnikova gives the following definition of innovation capital: "Innovation capital is a comprehensive system of investment sources formation and their transformation into improved means of production, new economically viable methods of production organization and personnel development in an enterprise for the purpose of profit maximisation" [9]. According to the authors, this definition requires clarification and specification on the following points.

Firstly, we hold that identification of capital with a system is an arguable point. The system is a broader term, which requires with regard to this matter a set of mechanisms of accumulation of different capital types and their transformation into new or improved means of production and technologies. In addition to investment capital industrial enterprise innovation system accumulates all the components, which were previously discussed and are shown in the Figure 1 of intrinsic enterprise value. Moreover, let us turn to the opinion expressed by M. A. Palienko, who deems that formation of innovation capital in industrial enterprises intends a special focus on innovation policy, that is kind of a program for establishing innovation priority in production, which is justified by preliminary calculations, depending on available resources and assigned tasks [15]. Summarizing this discourse, let us clarify the concept of innovation capital. Innovation capital of an enterprise is to be understood as value terms of those innovation projects, which are under development or already represent the intellectual property. At the same time, enterprise innovation capital amount, which, in our opinion, characterize its value, consists of various cost elements of other capital types, involved in the implementation of innovation projects. This includes financial, human, organization, market capital and goodwill which transfer their value to innovation capital in the process of enterprise innovation activity. Formulated differently, a company invests financial and human resources, organizational health and the like in innovation projects, reducing the financial capital value in the current period in exchange for its addition in the future while increasing innovation capital value. We should note that we do not mean decrease of human, organization, market capital and goodwill value at the same time, believing that these capital types are for our purposes inexhaustible and can be supplied faster, if such factors as staff turnover, commercial goodwill decline and others are not taken into account.

Now we should pay attention to difference between the concepts of innovation capital and innovation potential, where the last one is used more frequently.

The term "potential" derives from the Latinism "potentia" that means strength, power, opportunity and ability, which exist in the open form and can manifest under certain conditions [3]. The abstract term of "enterprise innovation potential" is defined as the whole complex of different resource types, including material, financial, intellectual, scientific, sci-tech and other resources required to innovation activity implementation [2,18]. Innovation potential is interpreted as enterprise readiness degree to gain a set innovation objective [1,3,4,5,8,11,13,16,17]. Hence it follows difference between innovation capital and innovation potential. Let us to emphasize the key point: innovation capital is the concept, qualifying the value of resources, invested in innovation projects, which are under development or already represent the intellectual property, taking into account expected profit from them. Innovation potential is enterprise readiness degree to carry out innovation projects, provided that this degree is characterized mainly by financial, human, material and production resources as well as by market opportunities.

Subsequently let us study delimitation of innovation capital from human and organization one and its connection with all components of intrinsic industrial enterprise value.

According to the most common point of view, human capital is characterized by knowledge, abilities and skills of force [6,7,20]. Human capital, involved together with the other forms of capital (organization, financial and market one as well as commercial goodwill) in innovation project implementation, creates innovation capital, which, according to the above given by us arguments, is positioned fairly in intrinsic industrial enterprise value and represents a strategic potential for development. Knowledge and skills of staff, involved in innovation enterprise project implementation are compensated by wage payment that leads to financial capital reduction without the changing human capital value or even with its increase due to new experience and professional training acquisition.

Organization capital, consisting in the value of corporate culture and existing corporate standards and regulations, is also involved in innovation project implementation, ensuring effective communication and creating a new component of intrinsic industrial enterprise value, which is called innovation capital.

Market capital, expressed in value of the long term cooperation with buyers, customers, suppliers, contractors and investors, is also included in the innovation projects implementation, providing easier access to markets of goods and resources and contributing to a new component of intrinsic industrial enterprise value –innovation capital.

Commercial goodwill as a part of intrinsic enterprise value is also very important in the process of innovation capital formation. Goodwill ensures economic benefit from such industrial enterprise characteristics as creditworthiness, tax compliance, investment attractiveness and partner trust.

Thus, all the components, considered by us as a part of intrinsic industrial enterprise value, participate in innovation capital formation and furthermore innovation capital is a separate element requiring the appraisal.

Approaching innovation capital appraisal methodology, we emphasize the following points.

1. The innovation capital consists of two components:

1.1. Value of innovation projects that are under development;

1.2. Value of innovation projects carried out or prepared for implementation.

2. The discounting method is applied under evaluating present value of both innovation capital components. At the same time a discount rate for the first part of innovation capital is knowingly higher than a rate for the second one, as the innovation projects under development are characterized by higher risks.

In accordance with these ideas, innovation capital value can be appraised in terms of modification of a model of total mortgage-investment analysis with following formula:

$$CIC = \left[\sum_{k=1}^{p} \left[\begin{array}{l} \sum_{n=x}^{m} \frac{(NOI - DS)(1 - s_{in})}{(1+i)^n} + \\ + \sum_{n=1}^{m} I_n \frac{(1+j)^n}{(1+i)^n} \cdot \frac{PIC}{PIC + AIC} - BAL \end{array} \right]_k \right] + \left[\sum_{k=1}^{p} \left[\sum_{n=1}^{m} I_n \frac{(1+j)^n}{(1+i)^n} \cdot \frac{AIC}{PIC + AIC} + ML \right]_k \right] \quad (1)$$

where CIC - enterprise innovation capital value, rub.;

NOI - annual net operating income from innovation project, rub.;

DS - annual debenture capital maintenance expenses, raised in an innovation project, rub.;

s_{in} - company take of an innovation project;

I - investments in an innovation project, rub.;

PIC - own investment assets, contributed in an innovation project, rub.;

AIC - sought investment capital, contributed in an innovation project, rub.;

BAL - closing unpaid balance, rub.;

ML - a total sum of debenture capital, contributed in an innovation project, rub.;

j - innovation capital value increment rate per annum;

i - an annual discount rate;

n - an innovation project year;

m - quantity of years for innovation project development and implementation, yr;

x - an expected payback year;

κ - an innovation project number;

p - quantity of enterprise innovation projects.

In the presented model the first part of the formula allows us to calculate innovation capital value, financed by own investments, the second part – innovation capital value, financed by attracted funds (investments of third-parties and credit recourses).

Summarizing the results of the conducted research, we draw special attention to those, which have scientific novelty.

1. The structure of intrinsic industrial enterprise value is clarified as well as the place of innovation capital therein is identified. The outstanding features of the author's model of intrinsic industrial enterprise value from the conventional model of market value of a company, offered by Leif Edvinsson, are as follows:

1.1. there are market capital and commercial goodwill in the first level of the hierarchy in addition to financial and intellectual capital;

1.2. innovation capital is on the second level of the hierarchy as an equal component of human and organization capital;

1.3. innovation capital is deemed to be a component, mainly characterized by already developed innovation projects or those projects, which are under development, where an organizational point is important, but do not have the single role. This component is also self-sufficient in the process of intrinsic industrial enterprise value formation.

2. The scope of the concepts of "innovation capital" and "innovation potential" is defined. Innovation capital is a term, qualifying cost of resources, invested in innovation projects, which are under development or already represent the intellectual property, with account of anticipated gain from them. Innovation potential is a measure of enterprise readiness to implement of innovation projects, provided that this readiness measure is mainly characterized by financial, human, material and production resources, as well as market opportunities.

3. The methodology of innovation capital appraisal is developed. Its distinctive features are the following:

3.1. Innovation capital consists of two components: cost of innovation projects under development and cost of innovation projects, implemented or prepared for implementation.

3.2. The discounting method is applied under appraisal of present value of the both components. At the same time the discount rate for the first part of innovation capital is apparently higher than the rate for the second one, in as much as innovation projects under development are characterized by higher risks.

5 Practical relevance

In order to justify practical relevance of the proposed methodology of appraisal of enterprise innovation capital value we will make practical calculations under two innovation projects, having the same annualized gain and the equal financing conditions (Table 1): 1) under innovation project in the development stage; 2) under innovation project in the implementation stage.

Table 1. The innovation capital value appraisal.

Conditions	The first project	The second project
Annual net operating income from innovation project, mil. rub.	500	500
Annual debenture capital maintenance expenses, raised in an innovation project, mil. rub.	80	80
Company take of an innovation project	0,5	0,5
Investments in an innovation project, mil. rub.	4000	4000
Own investment assets, contributed in an innovation project, mil.rub.	2000	2000
Sought investment capital, contributed in an innovation project, mil. rub.	2000	2000
Closing unpaid balance, mil. rub.	400	400
A total sum of debenture capital, contributed in an innovation project, mil. rub.	800	800
Innovation capital value increment rate per annum	0,15	0,15
An annual discount rate	0,5	0,12
An innovation project year	1-5	1-5
Quantity of years for innovation project development and implementation, yr	5	5
An expected payback year	4	1
An innovation project number	1	1
Quantity of enterprise innovation projects	1	1
Enterprise innovation capital value, mil. rub.	4149	4837

The obtained results of the calculations show that in the context of the same annualized gain and equal financing terms innovation capital value of an enterprise, which innovation project is under development, is 688 mil. rub. lower than innovation capital value of an enterprise, which innovation project is set up. On the basis thereof intrinsic value of the first enterprise is the same amount lower than intrinsic value of the second one under otherwise equal conditions. In the introduced example this is due to different project riskiness level and diverse terms of project start up. Underrun of innovation capital and intrinsic industrial enterprise value can only be compensated through increasing the volume of attracted investment, which will accelerate innovation projects implementation, pull them to financial viability and ensure addition to enterprise financial capital. Employment of this approach requires application of smart investment policy, which implementation will be aimed at increase in investment flow, compounding of innovation capital and financial one as well as at addition of intrinsic value of an industrial enterprise. The presented facts emphasize the importance of innovation capital in the system of management of intrinsic industrial enterprise value and the role of investment in their compounding management.

6 Conclusions

The obtained results affirm possibility of increasing intrinsic industrial enterprise value by innovation capital compounding. Attracted investments, which accelerate innovation projects implementation and their pulling to financial viability that has impact on increase of financial capital and intrinsic industrial enterprise value, are the main regulator of innovation capital value. The following instrumentarium can be applied as the recommendations for capital compounding. It is advisable to reconsider the traditional conditions of outward investments attracting. As follows: venture investors do not really want to finance innovation projects in the stage of their development on the conditions of income acquisition in proportion to the investment in the total capital due to higher riskiness of such projects, as a consequence a planned fraction of the income earned by an investor is to be regulated upward in order to encourage inflow of investments in high-risk projects. When the made recommendation of investment policy improvement is practiced widely, then the efficiency of innovation management could be improved, the processes of innovation capital compounding as well as financial capital increment be accelerated that respectively asserts influence on increasing intrinsic value of an industrial enterprise and its investment attractiveness and creates conditions for a start of some benefits in social and economic development of an industrial enterprise.

The work was supported by Act 211 Government of the Russian Federation, contract № 02.A03.21.0011.

References

1. A.A. Moshkov, S.A. Utkina, Proc. Innovation Economic and Industrial Policy of the Region (2012)
2. A.I. Bazilevich, *Innovation Management of an Enterprise* (UNITI-DANA, Moscow, 2009)
3. A.V. Babkin, A.A. Moshkov, Proceedings of the St. Petersburg State University of Economics, **6(84)**, 46-47 (2013)
4. D.E. Kychkin., A.V., Pautova, V.I. Zolotarev, The transport business of Russia, **8**, 56-58 (2011)

5. D.G. Mikhalenko, E.A. Afonichkina, Scientific and Technical Bulletin of the St. Petersburg Polytechnic University, **5**, **156**, 78-84 (2012)
6. Gary S. Becker, *Economic Analysis and Human Behavior* (State University-Higher School of Economics, Moscow, 2003)
7. Gary S. Becker, *Human Capital: A Theoretical and Empirical Analysis, with Special Reference to Education* (University of Chicago Press, Chicago, 1964)
8. I.F Kefeli, Geopolitics and Security, 1(21), 17-27 (2013)
9. I.S. Vinnikova, *Reserves for Rise of Labor Productivity Based on Innovation Capital Formation* (Nizhni Novgorod, 2005)
10. L. Edvinsson, M. Malone, *Intellectual Capital: Realizing your Company's True Value by Finding Its Hidden Brainpower, New Postindustrial West Wave. Antalogy* (Moscow, 1999)
11. L.V. Glukhova, A.A. Glukhova, Vestnik Volzhskogo universiteta imeni V.N. Tatishcheva, **17**, 107-112 (2009)
12. Lawrence J. Gitman, Michael D. Joehnk, *Investment Activities Basis* (Delo, Moscow, 1997)
13. Le Van Tchan, Z.A. Dikaia, *Analysis and Evaluation of Innovation Potential State of Industrial Enterprises* (Publishing office: St. Petersburg State Politechnical University, St. Petersburg, 2008)
14. M.A. Fedotova, *How Much Costs Business* (Perspektiva, Moscow, 1996)
15. M.A. Palienko, Problems of Improving Organization and Management of Industry Production, **2**, 163 (2014)
16. N.A. Zaglumina, Innovations, **2**, 109-113 (2010)
17. R.Ch. Kushbokova, Z.S. Shamzurzaev, V.S. Misakov, Terra Economicus, **7**, **4-3**, 123-125 (2009)
18. S.N. Shyshkin, Citizen and Law, **5**, 66-69 (2006)
19. U. Sharp, G. Alexander, D. Beiley, *Investments* (INFRA-M, Moscow, 1998)
20. V.I Pivovarov, V.V. Mazur, Proceedings of the Moscow State Technical University MAMI, **1**, **4(18)**, 79-82 (2013)

DOI: 10.1051/shsconf/20173501114

Financial capital appraisal in the system of industrial enterprise development management

M.K. Akhtyamov[1,*], and *E.A.* Gonchar[1]

[1]South Ural State University, Chelyabinsk, Russia

Abstract. The paper studies an objective appraisal of financial capital value in industrial enterprise development management. The authors draw attention to a limited scope of work in the area of financial management under application of a traditional approach to appraisal of enterprise financial capital value and describe a new advanced approach to this problem solving. The paper focuses on reformation of balance sheet statement, in which economic benefit and reserves of financial capital are included by the authors; the author's methodology for appraisal of financial capital value and its application sphere are denoted. In accordance with a case study, we confirm possibility of more accurate appraisal of enterprise investment opportunities by taking into account economic benefit from reserve of liquidity and borrowing capacity, which increases financial capital value. The practical approval results affirm that application of proprietary methodology for appraisal of financial capital and its elements has practical utility in solving problems of determination of required amount of attracted foreign investments and problems of determination of fair value of company take of an external investor, which are aimed at minimizing risks of corporate ownership dissemination and improving managerial efficiency of an industrial enterprise.

1 Introduction

Variability of external and internal environment of an industrial enterprise and its complicated predictability determines need to have a flexible management instrumentarium for well-timed stimulating improvement of positive events and prevention of negative one. Top management of an enterprise should quickly and objectively take decisions about financial capital management in the context of solving of subproblem and general problem of current and strategic management of socio-economic development, increasing intrinsic and market value of a company and improving its competitiveness and investment attractiveness. In the field of financial capital management managers have to constantly solve money management issues through regulation of current activities financing, debt fund raising and investment of temporarily free cash, as well as complete tasks of receivables management and inventory control.

All the listed objectives are aimed at improvement of competitiveness, investment attractiveness and level of socio-economic development of industrial enterprise. Addressing these problems is tightly intertwined among themselves. There are occasions in the management practice, when a positive solution to one problem creates a negative result of another problem solving, so rolling valuation of financial capital, appraisal of baseline and forecast cost and assessment of enterprise economic benefit from possession and disposal of idle financial capital are highly sought in order to avoid unplanned failures in achieving current and strategic objectives in industrial enterprise development management.

2 Timeliness. Scholarly importance

Financial capital value is traditionally determined on basis of sum of balance sheet assets or their sources or on basis of enterprise net asset value. This approach allows us to solve problem of estimating balance sheet value of property, salvage value of property and carrying value of shares of company founders. It also allows to analyze liquidity, solvability, financial stability, economic activity and borrowing power of a company as well as to determine stockout of current assets and to appraise enterprise debt capacity reserve. However scope of financial management tasks is much wider, and it is impossible to solve them by virtue of these measurements. First of all, this is referred to appraisal and management of intrinsic enterprise value, which elements include financial capital. Management of intrinsic industrial enterprise value is directed at its investment attractiveness improvement, stimulating investment in projects, being developed or implemented. Therefore, appraisal of real cost of intrinsic industrial enterprise value components, including financial capital, plays an important role in this matter. Financial capital is directly involved in intellectual capital formation, transforming in human, organization and innovation capital. Formation of market capital dispenses with financial capital because it promotes customer equity and resource capital. Effective use of financial capital

* Corresponding author: a-mavlit@yandex.ru

has also influence on value of commercial goodwill through achieving satisfactory levels of creditworthiness, fiscal discipline and a partnership of trust. All of these factors indicate importance of real financial capital appraisal in the system of industrial enterprise development management.

Issue of appraisal of enterprise financial capital value has not received due cognisance in science yet.

Most scientists cover a wider aspect of cost, considering the appraisal of company or business, real estate (Aniskin J. P. [9] Shcherbakov V.A., Shcherbakova N. A. [19] Dietger Hahn, Hungenberg Harald [11] Kudrina M. W. [13], Esipov V. E. Makhovikova G. A., Mirzazhanov S. K., Terekhova V. V. [20,21], Rutgaizer V. M. [22] Valdaizev S. V. [17], Chernozub O. [14], Jay E. Fishman, Shannon P. Pratt, J. Clifford Griffith, Kate Wilson [10], Cahill G. A., Schneider G. P. [13], Tarasevich E. I. [6,] Black A. [1], Zarev V. V. [23], Volkov D. L. [5], Gribovskii S. V, Ivanova E. N., Lvov D. S., Medvedeva O. E. [16]) and agenda of enterprise value management (including industrial companies): Gershun A., Gorskii M. [2], Nikitushkina I. V. [8] Drogovoz P. A. [15], Beer S. [4].

Nevertheless, in works of these scientists there are presented theories and models of enterprise appraisal, which may be the basis for development of partial models of company value elements appraisal. Among these theories and models are as follows: 1) the theory of economic profit, developed by A. Marshall in the 19th century; 2) the concept of added value, expressed in models of market value added (MVA) and economic value added (EVA), which are designed to evaluate process of enterprise value creating; 3) model of shareholder value added (SVA) (by A. Rappoport shareholder value added is defined as difference between two indicators: share capital value after a operation and the same capital value before that operation [13]); 4) model of options (models of option price formation can be used in order to apprise any asset with features of option) and so-called "formulary" version of discounted cash flow method (Modigliani-Miller formula, which can be used for determination of company cost sources, is one of the "formulaic" versions); 5) Ellwood model, which is based on valuation of property; 6) model of general mortgage and investment analysis; 7) model of innovation industrial enterprise value formation [15] and others. A. N. Sorokin directly associates cost of capital, as a benchmark, with appraisal of enterprise (business) value, arguing that cost of capital is rate of profit, which an investor expects to receive from his investment, taking into account the risks, associated with it [3]. As Cherutova M. I., Trofimova I. G. and Cherutova O.V. stressed, a level of enterprise capital value is the most important measure of a level of its market value [12]. Varfolomeeva V. A., Ivanova N. A. and Arkhipova I.I. understand cost of capital as income, which is to be reached by investments in order to prove their value from an investor's point of view [18].

According to the authors, under appraisal of capital value it is important to take as point of departure interpretation of its essence as economic benefit, which it may bring to an enterprise, but at the same time different nature of cost components is to be taken into account. Concerning financial capital cost, we should proceed from enterprise economic benefit from ownership and disposal of its invested and free parts. This requires paying attention to features of financial capital appraisal, where financial capital is deemed as a specific integral component of intrinsic enterprise value. Meanwhile, taking into account a wider range of investment opportunities of industrial enterprises, it is advisable to consider this issue in a broad perspective. Therefore, we choose development of methodology for appraisal of industrial enterprise financial capital cost in the context of improving system of industrial enterprise development management as a goal of this paper.

3 Thesis statement

For this goal achievement within the context of this paper it is necessary to solve the following problems.
Firstly, it is required to present structure of company balance sheet statement with regard to financial capital, demonstrating its scope and economic benefits.
Secondly, it is obligatory to develop a methodology for assessment of enterprise economic benefit from ownership and disposal of free financial capital and a methodology for appraisal of total financial capital cost and appraisal of its growth.
Thirdly, it is important to define application sphere for this methodology for assessment of economic benefit from enterprise ownership and disposal of free financial capital and the methodology for appraisal of total financial capital cost and appraisal of its growth.
Finally, practical calculations, confirming focus of author's recommendations on improving industrial enterprise development management, are to be carried out.

4 Theoretical part

The suggested by us approach to understanding of financial capital cost as economic benefit from enterprise ownership and disposal of its invested and free parts calls for formation of a separate balance sheet statement in order to reflect structure and changes in financial component of capital (we qualify it as balance sheet statement of financial capital). The requirements, demanded for its drawing up, are as follows.

First of all, balance sheet statement of financial capital is to be made of two parts. The first part should fully comply with the structure of balance sheet in accordance with Russian accounting standards, meaning that it should present reflection of assets of an enterprise and their sources. The second part is to reflect a ratio of economic benefit from ownership of free financial capital, which is at the disposal or is possible to be attracted, (in assets) and reserve, increase of financial capital cost from use of available investment alternatives (in liabilities).

Secondly, economic benefit from ownership and disposal of financial capital is to be divided into two types: 1) economic benefit from liquidity reserve; 2) economic benefit from reserve of borrowing capacity.

Thus, the structure of balance sheet statement of financial capital may be elaborated as follows (Table 1).

Table 1. Balance sheet statement of enterprise financial capital.

Assets (property)	Liabilities (origin of property)
Accounting data	
Fixed assets Current assets	Owner's equity Creditor's equity
Economic benefit and financial capital reserves	
Economic benefit from liquidity reserve Economic benefit from reserve of borrowing capacity	Liquidity reserve Reserve of borrowing capacity Investment growth of financial capital

In the next steps we move to development of methodology for evaluation of the items of the second section of the above given balance sheet statement, specifically economic benefit and reserves of enterprise financial capital.

A. Methodology for appraisal of economic benefit from liquidity reserve

It is commonly known that liquidity should be understood as ability of a company to repay its short-term obligations (current liabilities) duly at the expense of current (working) assets. There is ratio of absolute, quick and current liquidity. Absolute liquidity means that company is able to repay its short-term liabilities timely due to the most liquid part of assets - cash (some experts also add to them even short-term financial investments). Quick liquidity indicates ability of a company to repay its short-term obligations at the cost of the group of current assets, which includes cash, short-term financial investments and receivables. Current liquidity is ability of a company to repay its short-term commitments by all the available current assets, including inventory and input value added tax.

For each type of liquidity there are recommended ratio, the excess of which indicates liquidity reserve. Taking into consideration that such current assets as receivables, inventory and input value added tax are directly related with ordinary business, and it is difficult to analyze them as investment resources to be invested in alternative projects in the short run, we take current and absolute liquidity as a basis for further calculations. We assume with good reason that:

1) difference between actual and statutory value of current liquidity ratio will represent a general reserve of liquidity;

2) all current assets, except for free cash, are assigned to ordinary business at least in the short term.

In addition, we also pay attention to the following points:

1. Reserve of absolute liquidity may be invested in any projects, including not related with enterprise activities.
2. Residual liquidity reserve, formed at the expense of such current assets as receivables, inventory and input value added tax, is to be primarily considered as investments in ordinary business.

Therefore, evaluating economic benefit from availability of liquidity reserve, we emphasize two components in its structure: 1) economic benefit from absolute liquidity reserve; 2) economic benefit from residual liquidity reserve. At the same time economic benefit from absolute liquidity reserve is appraised by us using an average index of alternative investment project profitability; economic benefit from residual liquidity reserve are evaluated using an index of enterprise ordinary business profitability.

Based on the arguments, mentioned above, we represent the model for calculating economic benefit from absolute liquidity reserve (Formula 1):

$$EB_{a.l} = \left[\left(\frac{MR}{CL}\right)_A - \left(\frac{MR}{CL}\right)_R\right] \cdot CL \cdot (1+P_{a.i}) \quad (1)$$

where $EB_{a.l}$ - economic benefit from absolute liquidity reserve, rub.;
MR - cash of an enterprise, rub.;
CL - current liabilities (current assets), rub.;
$P_{a.i}$ - average profitability of investments in alternative projects;
A - actual value;
R - recommended value.

The model of economic benefit from residual liquidity reserve is seen as follows (Formula 2).

$$EB_{p.l} = \left[\left(\frac{CA-MR}{CL}\right)_A - \left(\frac{CA-MR}{CL}\right)_R\right] \cdot CL \cdot \left(1+P_{m.a}\right) \quad (2)$$

where $EB_{p.l}$ - economic benefit from residual liquidity reserve, rub.;
CA - current assets (working assets), rub.;
$P_{m.a}$ - enterprise ordinary business profitability.

Then the general model of evaluation of economic benefit from liquidity reserve may be represented in the following manner (Formula 3).

$$EB_{a.l.r.} = EB_{a.l} + EB_{p.l} \quad (3)$$

where $EB_{a.l.r.}$ - economic benefit from liquidity reserve, rub.

By virtue of the Formula 3, subtracting liquidity reserve, we obtain growth of financial capital, defined by use of liquidity opportunity of a company (Formula 4).

$$GFC_L = \left[\left(\frac{MR}{CL}\right)_A - \left(\frac{MR}{CL}\right)_R\right]\cdot CL\cdot P_{a.i} + \left[\left(\frac{CA-MR}{CL}\right)_A - \left(\frac{CA-MR}{CL}\right)_R\right]\cdot CL\cdot P_{m.a} \quad (4)$$

where GFC_L - growth of financial capital, defined by use of liquidity opportunity of a company, rub.

B. Methodology of assessment of economic benefit from reserve of borrowing capacity

It makes sense to carry out assessment of borrowing capacity of a company by combining financial stability analysis and calculation of degree of financial leverage. On the basis of ratio of debt and own capital, which is an indicator of financial stability of an enterprise, it is possible to calculate borrowing capacity, which means additional possible amount of attracting debt capital without prejudice to financial stability of a company. In this regard it is necessary to compare recommended (statutory) value of the specified ratio with the actual value. However it is to be noted that increment of equity capital to profitability is provided up to a certain limit of debt capital compounding, where after following increase in debt financing leads to decrease in profitability of own funds. This rule is known as the concept of degree of financial leverage. Thus wise, for estimating borrowing capacity reserve of a company it is important to find a balance between the greatest possible approximation to the recommended (statutory) ratio of debt and own capital and ensuring a positive meaning of degree of financial leverage.

As can be seen from above, keeping in mind the recommended (statutory) values of ration of debt and equity capital, which is an indicator of financial stability of an enterprise, borrowing capacity is to be calculated according with this formula:

$$BC = \left[\left(\frac{DC}{OC}\right)_R - \left(\frac{DC}{OC}\right)_A\right]\cdot OC \quad (5)$$

where BC - borrowing capacity of a company, rub.;
DC - debt capital of a company, rub.;
OC - own capital of a company, rub.;
R - recommended value;
A - actual value.

Degree of financial leverage under the recommended (statutory) ratio of debt and own capital may be calculated through the Formula 6, under the actual value – through the Formula 7.

$$EFL_R = \left[\left(1-TR_P\right)\cdot\left(EP-ACIR\right)\right]\cdot\left(\frac{DC}{OC}\right)_R \quad (6)$$

where EFL_R - degree of financial leverage under the recommended (statutory) ratio of debt and equity capital;
TRp - profit tax rate;
EP - economic profitability;
$ACIR$ - average computed interest rate.

$$EFL_A = \left[\left(1-TR_P\right)\cdot\left(EP-ACIR\right)\right]\cdot\left(\frac{DC}{OC}\right)_A \quad (7)$$

where EFL_A - degree of financial leverage under the actual ratio of debt and equity capital.

It is obvious that the basis of taking a decision regarding compounding of debt capital should be fulfilment of the following conditions:

$$\begin{cases} EFL_A > 0 \\ EFL_R > 0 \\ EFL_R > EFL_A \end{cases} \quad (8)$$

The general model of evaluation of economic benefit from borrowing capacity reserve may be presented on the basis of the above calculations (Formula 9).

$$EB_{r.b.c..} = \left[1 + EFL_R - EFL_A\right]\cdot OC \quad (9)$$

where $EB_{r.b.c.}$ - economic benefit from borrowing capacity, rub.

In accordance with the Formula 9, subtracting borrowing capacity reserve, we obtain growth of financial capital due to the use of debt capital of an enterprise (Formula 10).

$$GFC_{D.C} = \left[EFL_R - EFL_A\right]\cdot OC \quad (10)$$

where $GFC_{D.C}$ - growth of financial capital due to the use of debt capital of an enterprise, rub.

Summarizing the above methodology, we introduce the resulting models for assessing economic benefit of an enterprise from ownership and disposal of free financial capital (Formula 11), assessing the total finance capital growth due to the use of liquidity reserve and debt capacity (Formula 12) and assessing the total cost of financial capital (Formula 13).

$$EB_{f.f.c.} = EB_{a.l.r.} + EB_{r.b.c.} \quad (11)$$

where $EB_{f.f.c.}$ - economic benefit of an enterprise from the ownership and disposal of free financial capital, rub.

$$GFC_{sum} = GFC_L + GFC_{D.C} = \left[\begin{array}{l}\left[\left(\frac{MR}{CL}\right)_A - \left(\frac{MR}{CL}\right)_R\right]\cdot CL\cdot P_{a.i} + \\ + \left[\left(\frac{CA-MR}{CL}\right)_A - \left(\frac{CA-MR}{CL}\right)_R\right]\cdot CL\cdot P_{m.a}\end{array}\right] + \left[\begin{array}{l}\left[\left(1-TR_P\right)\cdot\left(EP-ACIR\right)\right]\cdot \\ \left[\left(\frac{DC}{OC}\right)_R - \left(\frac{DC}{OC}\right)_A\right]\cdot OC\end{array}\right] \quad (12)$$

where GFC_{sum} - the total growth of finance capital due to the use of liquidity reserve and debt capacity, rub.

The total cost of financial capital, understood as economic benefit of an enterprise from ownership and disposal of its invested and free parts, is made up of two components:
1) an invested part of financial capital, calculated as net asset value based on the first part of balance sheet statement;
2) a free part of financial capital, which is calculated as the total growth of financial capital due to the use of liquidity reserve and borrowing capacity.

$$CFC = NAV + GFC_{sum} \quad (13)$$

where CFC - total cost of financial capital of an enterprise, rub.;
NAV - net asset value of an enterprise, rub.

Summarizing the undertaken study, we draw attention to those results, which contain scientific novelty.

1. Essence of the notion of "cost of financial capital" is clarified. It is understood as economic benefit of an enterprise from ownership and disposition of its invested and free parts of financial capital. An outstanding feature of such an approach is account of economic benefit from ownership and disposal of its free part in financial capital cost, where this free part constitutes internal investment reserves, which are able to bring extra income to an enterprise.
2. The model of balance sheet statement of financial capital is suggested. It differs from the traditional balance sheet by a section "Economic benefit and reserves of financial capital", where assets include such indicators as economic benefit from liquidity reserve and economic benefit from reserve of borrowing capacity; at the same time liabilities have the figures of liquidity reserve and investment growth of finance capital.
3. The methodology for assessing economic benefit of liquidity reserve is elaborated. It allows considering potential income of an enterprise from possession and disposal of excessive stocks of working capital in value of financial capital.
4. The methodology for assessment of economic benefit from borrowing capacity reserve is developed. It allows taking into account potential income of a company from attracting debt capital in value of financial capital without any prejudice to financial stability of an enterprise and to profitability of own capital consumption.
5. The methodology for appraisal of enterprise financial capital is prepared. Its distinctive feature is a free part of financial capital, included in its body together with net asset value. This part is calculated as its total increment due to the use of liquidity reserve and borrowing capacity.

5 Practical relevance

The proposed methodology of appraisal of enterprise financial capital and its components may be used to solve such problems in management of industrial enterprise as determination of necessary volume of outward investments to be attracted (Formula 14) and determination of fair company take of outbound investor in firm value (Formula 15).

$$I_E = IR - EB_{f.f.c.} = IR - EB_{a.l.r.} - EB_{r.b.c..} \quad (14)$$

where I_E - necessary volume of outward investments to be attracted, rub.;
IR - total demand for investments, rub.

$$S_I = \frac{I_E}{I_E + CFC} \quad (15)$$

where S_I - company take of outbound investor in firm value.

There should be carried out calculations by virtue of an application example, where results of evaluation, made by the proposed methodology, are to be compared with results of evaluation, made by the traditional approach (Table 2, 3).

Table 2. Comparative evaluation of necessary volume of outward investments to be attracted.

Item	Value	
	Author's methodology	***Traditional approach***
Total demand for investments, rub.	4 000 000	4 000 000
Economic benefit from liquidity reserve, rub.	800 000	-
Economic benefit from borrowing capacity reserve, rub.	1 500 000	-
Necessary volume of outward investments to be attracted, rub.	1 700 000	4 000 000

Table 3. Comparative evaluation of company take of outbound investor in firm value.

Item	Value	
	Author's methodology	***Traditional approach***
Net asset value, rub.	3 000 000	3 000 000
Growth of financial capital due to the use of liquidity reserve, rub.	500 000	-
Growth of financial capital due to the use of borrowing capacity of a company, rub.	1 000 000	-
Cost of enterprise financial value, rub.	4 500 000	3 000 000
Necessary volume of outward investments to be attracted, rub. (Table 2)	1 700 000	4 000 000
Company take of outbound investor in firm value	0,27	0,57

Comparing results of the Table 2, we can see that necessary volume of outward investments to be attracted, calculated by the author's methodology, is 2.3 times less than that one, which is calculated by the traditional approach. The difference is due to the use of internal investment opportunities of an enterprise. The comparison results on the basis of the Table 3 show that under the author's technique company take of outbound investor in firm value is reduced by 2.1 times that indicates reducing risk of corporate property dissemination.

6 Conclusions

The results of this study confirm dependence of value of investment potential of industrial enterprises on an applied methodology for assessing cost of financial capital. The traditional approach to financial capital cost appraisal does not include internal investment reserves of an enterprise, such as liquidity and borrowing capacity reserve, leading to overestimation of expected needs in outward investment to be attracted. The author's methodology for appraisal of financial capital value under solving problems of determination of required outward investments volume to be attracted and determination of fair value of company take of outbound investor in firm value promotes to detect and to use internal investment opportunities and decrease dependence on external financing. This contributes to minimize risk of corporate property dissemination without compromising development of industrial enterprises. Timely detection and use of liquidity reserve allows improving efficiency of enterprise working capital management, providing increase in its economic efficiency. Sustainable use of borrowing capacity reserve without compromising profitability of own capital and without possibility of financial stability loss is also aimed at expanding investment opportunities in the best interests of industrial enterprise development.

We are grateful to the sponsors for the opportunity to participate in the conference and to publish the piece. The work was supported by Act 211 Government of the Russian Federation, contract № 02.A03.21.0011.

References

1. A. Black, *Questions of Value* (ZAO "OLIMP-BISNESS", Moscow, 2009)
2. A. Gershun, M. Gorskii, *Balanced Control Technology* (Publishing house ZAO OLIMP-BISNESS, Moscow, 2006)
3. A.N. Sorokin, Problems of improving the organization of production and management of industrial enterprises, **1**, 204-208 (2012)
4. S. Beer, *Brain of The Firm* (Editorial URSS, Moscow, 2005)
5. D.L. Volkov, *The Theory of Value-based Management: Financial and Accounting Aspects* (Publishing house "Vysshaia Shkola Menedzhmenta", St. Petersburg, 2008)
6. E.I. Tarasevich, *Valuation of Real Estate* (SPbGTU, St. Petersburg, 2007)
7. Gerard A. Cahill, George P. Schneider, Problems of management theory and practice, **5**, 92-95 (1998)
8. I.V. Nikitushkina, *Company Cost Management. vol. 1* (Delovaia i Professionalnaia Literatura, Moscow, 2008)
9. J.P. Aniskin. *Financial Activity and Value of The Company: Planning Aspects* (Omega–L, Moscow, 2005)
10. Jay E. Fishman, Shannon P. Pratt, J. Clifford Griffith, Kate Wilson, *Guideline for Business Value Appraisal* (ZAO KVINO-KONSALTING, Moscow, 2000)
11. Hahn Dietger, *Hungenberg Harald. PuK. Wertorientierte Controllingkonzepte* (Finansy i Statisika, Moscow, 2005)
12. M.I. Cherutova, I.G. Trofimova., O.V. Cherutova, Proceedings of the Bratsk State University. Ser.: Economies Andes Management, **1**, 153-160 (2013)
13. M.V. Kudina, *The Theory of Company Value* (Publishing house Forum: INFRA-M, Moscow, 2010)
14. O. Chernozub, *Life after Crisis: Cost-based Approach to Management of The Private Company* (Alpina Publisher, Moscow, 2009)
15. P.A. Drogovoz, *Management of Innovation Industrial Enterprise Value* (Publishing house MSTU n.a. N.E. Bauman, Moscow, 2007)
16. S.V. Gribovskii, E.N. Ivanova, D.S. Lvov, O.E. Medvedeva, *Assessed Value of Real State* (INTERREKLAMA, Moscow, 2003)
17. S.V. Valdaizev, *Business Valuation: 2-d ed., updated and revised* (TK Velbi, Publishing house Prospekt, Moscow, 2006)
18. V.A. Varfolomeeva, N.A. Ivanova, I.I. Arkhipova, Actual Problems of Economics and Management, **3(3)**, 3-11 (2014)
19. V.A. Shcherbakov, N.A. Shcherbakova, *Valuation of Enterprise (Business) Cost* (Omega-L, Moscow, 2006)
20. V.E. Esipov, G.A. Makhovikova, S.K. Mirzazhanov, *Business Appraisal: Complete Practical Guide* (Eksmo, Moscow, 2008)
21. V.E. Esipov, G.A. Makhovikova, V.V. Terekhova, *Business Appraisal* (Piter, St. Petersburg, 2003)
22. V.M. Rutgaizer, *Business Value Appraisal. Teaching medium* (Moscow, Maroseika, 2007).
23. V.V. Zarev, *Assessment of Business Value. Theory and Methodology* (IUNITI-DANA, Moscow, 2007)

 DOI: 10.1051/shsconf/20173501115

Methods of internal control in integrated management accounting system of the enterprise

A.E. Shevelev[1], *E.V.* Sheveleva[1], and *M.Yu.* Gvozdev[1,*]

[1]South-Ural State University, Chelyabinsk, Russia

Abstract. The purpose of internal control procedure of material and information flows is expression of independent opinion about accounting of material and information flows of the organization, development of recommendations about elimination of detected violations and preparing data to implement automated management accounting system, based on integrated system. Internal control procedure of material and information flows in the enterprise can solve complex of problems about identifying deficiencies in accounting, searching for reserves using information resources, identifying opportunities for implementation of automated management accounting system at the enterprise. Internal control was planned and conducted thus to argue that management statements are free of misstatement, and information flows provide required operability of the entire system. Internal control was carried out on a sample basis and included examining, on a test basis evidence supporting the amounts and disclosures in management reports. The disclosure of information about innovative activities, assessment of compliance with the principles and rules of accounting used in the preparation of management reporting, the review of key performance indicators, as well as evaluating the presentation of management reports are presented.

1 Introduction

Implementation of integrated management accounting system using information resources, and it's further use require internal control procedure, which allows to control quality of management accounting and information resource's implementation in all responsibility centers of the enterprise.

Under conditions of crisis economicphenomena's development finding and applying of innovative types of resources - is a key issue for economic activity and further development of the enterprise. The use of information resource in the company's innovative activity and implementation of integrated management accounting system require internal control procedures and it's methods.

2 Relevance

To implement and operate the integrated management accounting system in the company, conducting of internal control of responsibility centres of the organization becomes an urgent problem. It allows to test the quality of management accounting and the functionality of the information resources, and also to check the consistency of accounting documentation flows and information flows between themselves.

3 Mane Problem

In order to effectively manage of the company it is necessary to use the integrated management accounting system. This system is focused on the company's use of information resources, increasing its potential.

The method of internal control procedure in the integrated management accounting system, which provides quality control of accounting and functionality of information resources as well as a consistency of checking records flows and information flows between them, is offered, and the term "information resource" as a set of technical, software and information flows, where the resource is known as an alternative reserve of innovative activity of the company is justified.

At the same time, innovative activity as an independent management object has its own specific characteristics. This is due, primarily, to the length of innovative activities and difficulties in forecasting and evaluation of the results effectiveness, their significant influence to the future development of the company as a whole.

Analysis of scientific publications shows that most of companies have not developed clear methods of innovative resources management and forming complete information about economic activity of a company. Lack of operational management of material and information flows makes it difficult to make management decisions and plan their further work [1, 2, 3].

Most important types of resources:

- human resources;
- material and technical resources;
- financial resources;
- information resource.

* Corresponding author: gvozdevmi@susu.ru

Today in science literature academics have recognized the crucial role of information resource in modern society. However, there are almost no works, giving a complete view of it. Many authors give a narrow interpretation of the term, referring to an information resource only information that is alienated from its creator and is fixed in a tangible medium (Godin V. V., Korneev I. K., Antopolskiy A. B., Kostrov A. V., Gendina N.I., Peshkova O.V.), scientific and technical information (Nizhegorodtsev R. M.) [4, 5, 6, 7] or expand it to an information system, which includes in addition staff (Abraham Block, Timothy D. Wells and Christine Sevilla, Bretagne, Ellen Monk, James O'Brian and George Maracas, Frank Dravis) [8, 9, 10, 11, 12, 13, 14].

In author's opinion, information resource – is a set of installed software that allows planning, accounting, control and administration of the whole available amount of information stored on the media in databases, and technical support, including network equipment, together with the means of collection, transmission, processing and storage of information, and existing information flows between responsibility centers. Unlike most resources that are able to dwindle, information resources tend to be accumulated, be enriched and can be used repeatedly. These properties allow us to speak about the great potential of this type of resources that must be used to solve the strategic management problems.

Analysis of various approaches to the structure of information resource of the company, allows to determine the following types:

• computer equipment and communication equipment;
• licensed software and auto-trade information system;
• the totality of data and documents in libraries, archives, funds, databases, etc., as well as the electronic and print media;
• local area network, the Internet, an information portal.

Typically, companies possess all these types of resources, which together form total information. The use of information resource and forming an integrated management accounting system on its basis, will allow to solve problems of information support and automation of innovative activity management.

Currently problems of taking various automated control systems are widely discussed. Such systems should provide high efficiency of the management to improve the speed and quality of information flows.

Overview of publications on the subject shows a trend of using automated management systems.

Successful western companies which have implemented these technologies can be an example for Russian companies. Many companies develop their own certain fragments of automated systems to solve current problems. Comprehensive implementation of these activities will provide integration of the information resources and will allow creating information infrastructure according to current organizational structure and current flows [15].

Analysis of the publications about automated systems implementation allows us to characterize this process as quiet complex. The main difficulty is to implement and form the management system. In this case, you must adjust this system for existing management processes in the company. It impose a number of restrictions on the automated system and greatly reduces its effectiveness. Sometimes developers have problems that can not be solved, so we have to accept a compromise, or refuse it's further implementation. Such difficulties mainly caused by the lack of opportunities to automate certain management processes [16].

To successful implementation of the integrated management accounting system in the company accuracy measures of internal control of flows existing should be made.

Category of accounting flows, in our view, includes activity indicators, that will reflect the nature of the financial and non-financial activities.

Internal control of accounting and informational flows is a complex event, which can be divided into two components:

• control of account flows;
• control of information flows.

To achieve this goal must be solved following problems:

• to determine compliance with the accounting rules of the selected resources;
• to check the documentation schedule and it's compliance;
• to assess existing information resources of the company;
• to identify accounting deficiencies in various automated accounting systems while their interaction.

Developed methods of internal control in company is presented in the form of a block diagram (Fig. 1).

For successful control of accounting and information flows they should be presented in the form of diagrams. Based on these diagrams, one can define options for interaction and information flows using.

Before using the automated management accounting system it is necessary to understand what management accounting data are receiving at the moment, what information system they are received from, what information system contains the most reliable accounting data and in what form that information is supplied to the managers.

This diagram describes the methods of internal control of accounting information flows in the company. Methods of internal control of each responsibility center of the company includes:

• check of management accountingquality. It includes check of accounting data accuracy, formed by responsibility centers and it's moving between them. At this stage accounting data is assessed for compliance to legislative and internal requirements, appropriateness of data forming, lack of duplication, availability of a responsible person, compliance to document circulation schedule. In case of shortcomings adjustments should be done. Then formation of accounting data flows is compiled;

• check of information resource functionality. It is necessary to evaluate if management accounting system is provided with automated tools.

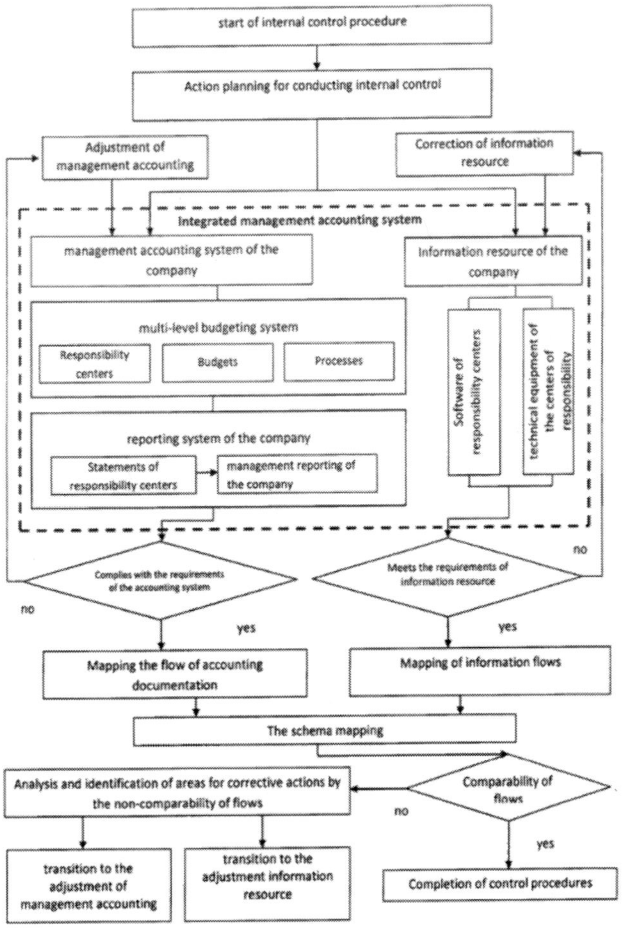

Fig. 1. Methods of internal control in integrated management accounting system of the enterprise.

• check of operability of existing automated systems, their ability of further modernization in accordance with the objectives of management accounting and electronic data transfer method, the presence and movement of data flows. When shortcomings are found, corrective actions are carried out and all defects of information flows formation of using precut software are eliminated.

• flow charts of data and information flows between the centers of responsibility are prepared. It is necessary to evaluate all information, which passed between the responsibility centers and what information systems are most reliable and technically provided, etc.;

• check of consistency of accounting and information flows between themselves. Scheme are compared and flows provision of accounting information are evaluated (possibility to transfer documents in electronic form, data transmission between various kinds of program support, lag or advance of the transmitted data, etc.). In case of flows misalignment some corrective action should be taken. It identifies responsibility centers; causing misalignment inside the center of responsibility(management accounting system and/or information resource), and correction of such sources is performed. Integrated system of management accounting becomes a result of internal control.

Internal control is carried out while planning and implementation of integrated management accounting system, and periodically during its use in cases of changes in management accounting system (changes in legislation, changes in reporting requirements, etc.) and information resource (new software, new devices, etc.).

To assess compliance with the rules of accounting of inventories, you can use the methodical instructions on accounting of inventories (further – Methodical instructions on accounting of inventories) [17].

Legal documents on the organization of information flows in the higher school in the Russian Federation are not developed yet [18]. Therefore, the order of information data transmission needs to be determined by the company. Data transfer is determined by the internal standards and regulations.

There are quite a number of sources normative-legislative regulation on accounting of material flows that need to be followed by relying organization or individual person during the audit. However, information flows control is a fairly recent trend in the Russian practice, therefore, principles of rationality and efficiency should be applied [19].

Under conditions of innovative activity the question of timely information receipt about real situation in the centers of responsibility, as well as about accurate and reliable information of company's activities and liabilities, sharply raises. However, before implementation of integrated management accounting system it is necessary to ensure in correct accounting and information flows organization. For these purposes it is necessary to conduct internal control. This direction is quite new for Russian companies and certain methods have not developed yet.

The joint consideration of flow charts allows to judge about the quality of management accounting and traffic of information flows between centers of responsibility. This procedure is necessary as part of the integration process of convergence of management accounting and information resource, in order to obtain integrated management accounting system. Internal control excludes the replication of the negative factors, accounting factors, accounting informational flows to in the prospective budgeting system and identifies shortcomings of the existing information resource. Thus, follow-up actions for the implementation of elements of management accounting system will be provided with an efficient structure of center of responsibility, and reliable instrument as an information resource of the company.

4 Conclusion

In conclusion, we can draw the following conclusions:
• internal control conducting provides an opportunity to ensure that the accounting and movement of data in the company are conducted in accordance with established rules, which in turn is a prerequisite for the introduction of an integrated managerial accounting system;
• internal control of information flows reveals the presence or absence of required traffic, as well as shortcomings in data transfer organization between centers of responsibility in the company;
• joint consideration of scheme of data and information flows movement allow to determine the causes of the poor functioning of the entire management structure of the company.

Internal control was planned and conducted thus to argue that management statements are free of misstatement, and information flows provide required operability of the entire system. Internal control was carried out on a sample basis and included examining, on a test basis evidence supporting the amounts and disclosures in management reports. The disclosure of information about innovative activities, assessment of compliance with the principles and rules of accounting used in the preparation of management reporting, the review of key performance indicators, as well as evaluating the presentation of management reports are presented. Authors believe that the internal control provides sufficient information about the reliability of management reporting, identifies the features of the topology information and resource of functioning of information flows.

Thus, when information and material control is conducted, complex problems of identifying limitations of accounting, searching of reserves for accounting optimization, definition of automated system configuration of management accounting are solved.
Conclusions and proposals presented in the thesis, have been successfully used in the activities of the office of scientific research, the development of the information portal of the innovation resources of the University in the field of energy and resource saving http://itou.susu.ac.ru/ [20].

The work was supported by Act 211 Government of the Russian Federation, contract № 02.A03.21.0011.

References

1. B. Needles, H. Anderson, D. Goldwell, Accounting, **6**, **9**, **11**, **12** (1991)
2. Y.V. Sokolov, Accounting, **18** (2000)
3. A.D. Sheremet, *Complex economic analysis of enterprise activity (methodology)* (Economic, Moscow, 2012)
4. V.V. Godin. I.K. Korneev, *Information resources management* (UNITI, Moscow, 2000)
5. A.B. Antopolskiy, V.I. Aussem, S.A. Blau, A.I. Gegel, *Research and development of metadata for electronic information resources and services in fundamental science*. www.inforeg.ru.
6. N.I. Gendina, Bibliospere: quarterly scientific journal, **1**, 88-89 (2011)
7. O.V. Peshkova, Series of applied Informatics, **53**, 23-35 (2014)
8. L. Stanton, Proc. of 10-th International Conference of European University Information Systems ENUS, 42-45 (2008)
9. Shtub Avraham, *Enterprise resource planning the dynamics of operations management* (By Kluwer Academie Publihcis, 1999)
10. D. Wells Timothy, Christine Sevilla, *Maximizing the Enterprise Information Assets* (N.W. Corporate Beld, Buch Raton, Florida, 2000)
11. P. Karpova, *Managementaccounting: textbook for high schools* (UNITI, Moscow, 2003)
12. Ellen Monk, *Concepts in Enterprise Resource Planning* (2005)
13. R.Y. Wang, S. Madnick, Journal of Management Information Systems, **5**, **2**, 5-2 (1988)
14. E.E. Lavrisceva, Reproduction: the interrelationship of political-economic and institutional aspects, 56-67 (2010)
15. V.V. Krukov, Proc. Information and communication technologies in education and scientific activity, Novorossiysk, 42-48 (2008)
16. M.V. Ievenko, University, **1**, 96-104 (2008)
17. *The plan of bookkeeping accounts of financial and economic activities of organizations and instructions for its use* (approved by the order of Ministry of Finance no. 94)
18. *Management of information flows in the enterprise*. http://roscomtech.ru/it-services/it-consulting/123.
19. http://ems.ru/1c/?inc=1c-buhgalteriya-8.
20. http://itou.susu.ac.ru.

Methodology of the evolutionary economics in application to the engineering problems

E.B. Kolbachev[1,*]

[1]Platov South-Russian State Polytechnic University (NPI), Novocherkassk, Russia

Abstract. Authors present an idea that the consulting engineering services are a type of knowledge mediation institution. This institution decreases a possibility of the participants' opportunistic behaviour during the purchase and sale deal making. It's actual, first of all, for the knowledge products, such as innovation projects and R&D results, and other search goods. In the article authors present a set of measurement instruments to make a said kind of valuation. Concept of the set is based on one of the evolutionary economics' methodologies: a technological structure method. As a criteria for attributing a project to one or another technological structure are used a level of the real technology's compliance with the extremely efficient, information materialization level, dimension scale of the forming processes and dominant management concept. The approach gives a possibility to minimize the possibility of the existence of the innovation processes participants' opportunistic behaviour, and to increase the efficiency of the innovation activities..

1 Introduction

New industrialization processes of the nowadays may be caused by different reasons in different countries of the world [1]. For example, authors of the well-known report [2] insist that the reasons for development of these processes in the USA are the following: unit labor costs are among the lowest in the industrialized world; energy costs are low thanks to the gas bonanza; the cost of capital is very low given the Fed's ultra-accommodative monetary policy.

The basic reason for the development of new industrialization processes in Russia is a consistent governmental policy of the last years. This policy is a result of understanding both by the government and the society how important the increasing of the country's human capital is for its economic development. The said task becomes vitally important because of a great harm to country's social capital in the beginning of the post-Soviet period due to the destructive Yeltsin-Gaidar government policy and resulted in the elimination of many science intensive sectors of Russian economy [3].

Development of the new industrialization processes result in the stirring up the market of knowledge products, such as innovation projects, R&D results, etc. A complexity of these goods' essence makes the customers (investors) to use the engineering services (particularly, consulting engineering) during making a decision about purchasing one or another product. A demand for this kind of services is now increasing all over the world. But the processes of engineering services markets' development in the world countries go differently [4].

In spite of the said, some general problems exist in the field of consulting engineering and other engineering services. In particular, it's a problem of choosing an acceptable management model. A reason for the situation is a low number of the investigation in the field of determining the economic essence of the engineering services, organizational features of the interactions between consulting firms and customers, evaluation of the knowledge products [5], [6].

One of the significant ways to solve these problems may be paved by investigating an institutional essence of the engineering services, engineering companies' and customers' strategy. Then the results of the investigation, coupled with the evolutionary economy methodology, should be used to work out the evaluation instruments for the innovation projects and R&D results.

2 Institutional essence of the engineering services

Making a decision about acquiring of any good makes necessary a measuring (evaluation) of its characteristic. This measurement requires incurring some costs.

The types of costs that exist during the transactions' profitability analysis or other measurements are defined as measurement costs. The cost for activities, which's goal is to indemnify the participant from losses caused by the measurement errors should also be defined as the measurement costs.

"Measurement costs" as a term first was used by Y. Barzel in his well-known work [7]. But he wasn't a first one who showed the importance of problems that

* Corresponding author: kolbachev@yandex.ru

exist during the measurement processes. In the work [8] by M. Yudkevich it's pointed out that the first one who presented a concept of minimizing the information scattering by market organization was S. Cheung [9]. This concept is a basis for measurement costs' valuation.

In context of the situations we examine, the important role play institutions that decrease a measurement costs' level. The first consideration of this problem was presented in the above mentioned work [7] by Y. Barzel. These institutions also generate the negative incentives that prevent an existence of the adverse selection and opportunistic behavior. Thereby they decrease the negative consequences of the informational asymmetry. It's obvious that the engineering as an institution should be attributed to this kind of institutions.

According to the ideas of institutional theory, engineering may be represented as a transaction institution for the knowledge mediation. It's to be considered as a particular case of the transaction institution, a consulting service, that works as a mechanism of decreasing a measurement costs level [10]. Speaking about measurement costs we have to take in account that the existence of information search costs and measuring of the respective contract attributes should be considered as a factor that hampers the processes of decreasing the informational asymmetry level.

The result of incurring the measurement costs is a partial (not final) decreasing of the price and quantitative ambiguity [11].

According to J. Hirshleifer [12], the existence of a quantitative ambiguity is much more serious problem than price ambiguity. A reason for this assertion is a multidimensionality of the good's or service's qualitative characteristics. The problem is that some aspects of these characteristics may be immeasurable, and in the existence of the subjective appraisal in measuring some good's or service's attributes.

Valuation of the credence qualities often becomes a part of engineering services, because this valuation can't be done by the customer, as opposed to the search or experience qualities. The credence qualities are the ones, which's valuation is a reason for incurring costs, even after the concluding of a contract. Valuation of the credence qualities as a part of engineering services may be hold in combination with the diagnostics. They should be hold before making a final decision about concluding a contract for carrying out a project and may be a reason for essential increasing of the costs level, especially if the customer negotiates with more than one vendor. The above said should be understood as an example of rather often existing situation when the engineering company fills the consuming or the investing company's order to choose one of the forms of the invention for the future acquisition and use.

In this situation the engineering services act as the means of averting both the customer's and the vendor's (or the developer's) opportunistic behavior.

According to the ideas of M. Yudkevich [8], in this situation the customer has a deal with "two markets in the same time": market for goods and services and information market. Rather often the same agent presents the both offers (goods and information) in the both markets. As an example, the engineering company often presents the information about existing inventions and the possible ways of their development by the phase of practical use, and in the same time, makes a design work for this development. The existing situation promotes the opportunistic behavior of engineering company that can be revealed as a part of the engineering company's strategy.

3 Strategies of the engineering companies and their customers

To analyze the possible strategies of the engineering companies and their customers we have to bring the engineering services and the goods that we are going to valuate into the correlation with one of the categories.

In nowadays economy exists a well-known goods and services classification based on ideas by P. Nelson [13], M. R. Darby and E. Karni [14]. P. Nelson marks out two basic groups of the goods: search goods and experience goods, and shows the difference of these groups' quality definition. Search goods' valuation and measurement of their quality should be done before the entering into a contract. Valuation and measurement of the experience goods' quality is being done during their use or exploitation. It's obvious that innovation projects, R&D results and other kinds of knowledge products should be classified as search goods.

Information about the quality level of these products is very important for the potential consumer. Information about quality level may be obtained the same way as about price level. But sometimes obtaining of this information causes so high-level expenses, so the customers look for another ways of getting it.

There are some categories of goods, which's quality can't be determined by the customer himself. The reasons for such situation are the technical complexity of the goods, or some others. In this situation the quality level should be determined by the outside economic agents. Engineering companies' work on project valuation is a kind of outside agents' work. So, the problem of the economic agents' (as of the developers' and intermediates') trustworthiness becomes actual for the consumer who uses their information for the decision-making.

To solve the problem, M. R. Darby and E. Karni [14] suggest to measure the good's general quality level not only by search and experience quality parameters, but use another one, the credence qualities. Valuation of the credence quality level can't be done by the customer and should be handed over to another economic agent.

Under the said conditions, according to the opinion of M. Yudkevich [8], the engineering company as an intermediary may implement one of the following possible strategies:

• "Cooperate" — to "play fairly" in a period of rendering the intermediary services to the customer, without being in collusion with the vendor and taking all measurements to determine the only optimum alternative for the customer's contract;

• "Deviate" — to violate the conditions of the contract with the customer being in collusion with the vendor, or choose goods for the customer basing on an insufficient number of the measurements.

In the same time the customer may implement one of the next two strategies:

• "Stay" — to extend a period of using the intermediary's services;

• "Refuse" — to reject the presented intermediary's services and take a project quality measurements using the own abilities, or use other intermediary's services.

In our reasoning we proceed from the fact that the customer follows a trigger strategy. It means that if the intermediate followed a "Cooperate" strategy in a previous term, then in a current term customer follows a "Stay" strategy and uses this intermediary's services. If the intermediary violated the conditions of the contract in a previous term and followed a "Deviate" strategy, then in a current term and in all further periods customer should follow a "Refuse" strategy. But in the first period the example intermediate follows a "Cooperate" strategy. It's obvious that the thoughtless following this strategy is an impermissible simplification of the situation. In the actual conditions a great number of various factors affect the customers' and the intermediates' behaviour.

Using of the high-grade instruments for evaluation of the inventions makes the opportunistic behaviour (both the developers' and the engineering companies') less possible. These instruments should give a possibility to quantify the qualitative characteristics of one or another invention. During this work the most important becomes an integrated assessment that gives a possibility to compare the inventions with the differing particular parameters. A possible way of such assessment is attributing the invention to one of the technological structures.

In this situation using a natural science or technical science methodology is necessary for solving a management problem of choosing the optimal technological option, methods and instruments.

4 Instruments of the evolutionary economics used for the evaluation of innovation projects and R&D results

During the evaluation of the innovation projects' and R&D results' credence qualities, using the cost parameters looks very difficult because of evaluation's limited possibilities. Another problem is the affecting of the numerous factors that are not in a close proximity with the basic characteristics of the evaluated object, such as the inflation rate, economic situation, etc. It's remarkable that these kinds of cost evaluation method's shortcomings are described in the well-known works by such scientific authorities as M. Friedman [15], L. I. Abalkin [16], G. Odum [17] et al.

Russian academician L. I. Abalkin [16] noted that dissatisfaction with economic science's condition was the first sign of the old paradigm exhaustion and the necessity in a new one. At the first step of its creation the system of analogies can be used — an application of methods of the other sciences such as biology, genetics and thermodynamics that are widely used for the analysis of economic processes' cyclical dynamic.

It seems to be noteworthy the position of M. Friedman [15] suggesting that economic theory can be exact and objective as much as natural sciences. The problems of objectivity and reliability of measurements and estimates in economy were figuratively described by A. Hall [18] who mentioned that a great blemish of the market prices system was the instability of money unit, "...a physicist wouldn't bear a yard-stick made of rubber..."

G. Odum [17] considers a range of logically interconnected problems determining relations between energy, ecology and economic growth. He proves that economic processes exploring only through money circulation is as incomplete as natural phenomena exploring only through mineral cycles. This approach, which is rather constructive, has an essential disadvantage connected with neglecting an entrepreneurial (intellectual) aspect in formation of engineering, organizational and management decisions that define the character of resources transformation into production factors and provide the effectiveness and efficiency of production and other processes in social-economic and technical-economic systems.

It's obvious that carrying out of the engineering services requires the estimation of the invention's technical and economic level. To do this, it's necessary to find a continuum between its technical parameters and parameters of its economic type (but not only the cost parameters).

To solve the said problem it's appropriate to use the evolutionary economics methodology. This methodology was worked out in the post-shumpeterian period of development of the economic science in the well-known researches of such scientists as J. Foster [19], R. Langlois [20] and C. Shubert [21], V. Mayevskiy [22], O. Sukharev [23] at al.

One of the actual lines of nowadays evolutionary economics development is a theory of technological structures (technological orders), devised in the works by S. Glaziev [24], C. Perez [25] and some other authors.

During the integrated assessment of the inventions' credence qualities, which should be hold as one of engineering services, attributing the invention to one of the technological structures is an available way. The procedure of such attribution is described below.

This method is a particular case of a successful using a natural science methodology for the engineering problem solving as the prime transaction institution of knowledge intermediation.

The quantitative characteristics of the technological structure is the materialization level of the production systems' information, which level

 DOI: 10.1051/shsconf/20173501116

increases as far as the previous order is changed by the subsequent. The original approach to this idea is shown by O. M. Yun [26], who examined the technological relations and functions, implemented by the production system (q.v. Tab. I). This approach should be used if we have to analyze the changing informational relations under the different technological structures.

During their evolution, the informational content of the labor processes in the production systems changes greatly. The nature of the proper information medium, which determines the general appearance of the production system of one or another technological structure, also changes. Basing on this statement and the materials by I. Pereyaslova [27], we present Tab. II, in which the informational processes are shown. These processes are being materialized, respectively, on the instrumental, machine and informational steps of the production development.

Another quantitative characteristic of the technological structure is the forming processes' dimension scale. The forming processes should be analyzed for the dominating technology, which technology determines the economic result of the production (q.v. Tab. II).

The dimension scale value decreases as far as the previous structure is changed by the subsequent. From the first to the forth structure the reason for this regularity was based on the increasing of the mechanical engineering products' dimensional accuracy: the dimensional accuracy determined product's performance. The fifth structure appearance is associated with the development of the micro system electronics, which operates the dimensional parameters accuracy within micrometres.

Examining how much the information materialization level and the dimensional scale of the forming processes change, we follow the conceptual proposition that any point on the economic development trajectory is determined by the previous evolution of the production systems in general (I. Pereyaslova [27]).

Table 1. Technological Relations and the Production Functions.

Technological relations type	Function number	Function name
Pragmatic	1	Goal-setting, choosing the product to manufacture
	2	Grounding the product characteristics
	3	Developing the action program for the industrial engineering
Syntactic	4	Choosing the possible technologies
	5	Choosing the technological relations
	6	Grounding the production relations system
Semantic	7	Developing the technological processes system
	8	Fine-tuning of the technological processes
	9	Combination of the manual labor and machinery work
Cognitive and Emotional	10	Developing the system of the instrumental regulators
	11	Means of the instrumental operations' regulating
	12	Regulating of the instrumental process
Material	13	Investment goods reproduction
	14	Product reproduction
	15	Instrumental operating a subject of labor

It's obvious that the sixths technological structure should appear with the next decreasing of the forming processes' dimension scale. This statement corresponds to the result of the actual studies of the natural systems and leaving organisms, and processes in these systems, based on the interaction of the molecular and atomic size objects (size from 0.1 nm to 100 nm).

Table 2. Basic Characteristics of the Technological Structures.

Technological structure	Characteristics of the technological structure				
	Prodution development step	*Basic economic resource*	*Dominant management concept*	*Level of the infor-mation's materiali-zation*	*Dimension scale of the forming processes*
1	2	3	4	5	6
1	Instru-mental	Materi-als (natural stuff)	Basic production management	15–11	1–0.2 mm
2	Machi-ne	Energy		11–10	100–50 µm
3			Production management		50–10 µm
4			Management of the enterprise	9	
			Business management	8–6	10–0.5 µm
5	Infor-matio-nal	Infor-mation	Cost management		
6			Managing the technological efficiency	5–2	100–0.1 nm

A transfer of the forming processes to the nanometer level, which takes place within the sixths technological structure, leads to the conceptual changing of the economic instruments that are used for managing the production systems in general and, specifically, for the engineering services. It looks important that this situation makes possible a practical realization of the extremely efficient technology concept. This concept was worked out in the 1980th by Ju. A. Kalyagin [28], and considerably developed at the present time by I. Pereyaslova [29]. Using this concept, we represent the technology, which assures a maximum possible base product yield (100% selectivity of the process) as the extremely efficient. The level of the real technology's compliance with the extremely efficient (the increasing of this level, first of all, makes the production cost per unit less) can play the role of the efficiency index for the production process.

To make the most out a said methodology to solve the nowadays problems we use the expert appraisal to attribute a project to one or another technological structure. The used measurands are:

• x_1 — level of the real technology's compliance with the extremely efficient;
• x_2 — information materialization level;
• x_3 — dimension scale of the forming processes.

Limits of the parameters variation: [0, 1].

Level of the real technology's compliance with the extremely efficient is being measured in the following way: $x_1 = 0$, if the technology level's deviation from the extremely efficient is more then 50%; $x_1 = 1$, if the technology level complies the extremely efficient; $\in (0, 1)$ in other situations.

Information materialization level is being measured in the following way: $x_2 = 0$, if the materialization level is > 5; $x_2 = 1$, if the materialization level is = 2; $\in (0, 1)$ in other situations.

Dimension scale of the forming processes is being measured in the following way: $x_3 = 0$, if the scale of the forming processes is > 100 nm; $x_3 = 1$, if the scale of the forming processes is ≤ 0.1 nm; $\in (0, 1)$ in other situations.

Level of the experts' opinions consistency is to be determined as a result of variance analyzing of the indexes they rate. A variation coefficient is to be measured as $i = 1, 2, 3$. If the variation coefficient is less then 30% for any index, it becomes possible to estimate an integral index.

The integral index of the project's attribution to the modern technological structure is to be estimated using the Euclidean metric method, according to the weight coefficients:

$$TS = \sqrt{\sum_{i=1}^{3} \alpha_i (1 - x_i)^2}\ , \qquad (1)$$

α_i here are the weight coefficients for the factors that satisfy a condition. $\sum_{i=1}^{3} \alpha_i = 1$ (on the current step of work we determine all the parameters as the equal ranking ones).

In present time we work on the economic and mathematical model of for the direct valuation of the project's attribution to one or another technological structure. It is to be based on indexes of technology's compliance with the extremely efficient (using certain parameters from the stoichiometry and chemical kinetics), information materialization level and dimension scale of the forming processes.

5 Conclusion

The objective of the presented research is determining the economic (institutional) essence of the engineering services, engineering companies' and customers' strategies. Another objective is working out the instruments for the valuation of the innovation projects and R&D results, based on the evolutionary economy methodology.

By the result of research we identified the consulting engineering services as a type of knowledge mediation institution. This institution decreases a possibility of the participants' opportunistic behaviour during the purchase and sale deal making. The said is actual first of all for the knowledge products, such as innovation projects and R&D results. The important part of such transaction should be a valuation of the good's search, experience and credence qualities. As a rule, valuation of these qualities can't be done by the customer (investor) himself and should be handed over to another economic agent (mostly often this agent is an engineering company).

During the interaction between the customer (investor) and engineering company, rendering the intermediary services to the customer should be done on terms of the "fair play", without company's being in collusion with the vendor and taking all measurements to determine the only optimum alternative for the customer's contract.

In the article we present a set of measurement instruments to make a said kind of valuation. Concept of the set is based on one of the evolutionary economics' methodologies: a technological structure method. As a criteria for attributing a project to one or another technological structure are used a level of the real technology's compliance with the extremely efficient, information materialization level, dimension scale of the forming processes and dominant management concept.

A presented approach is giving a possibility to decrease a possibility of the opportunistic behavior and, in other hand, to increase the efficiency of the innovation activities.

It's proved that the ultimate products of the sixths technological structure appear during the implementing of the innovations, which's dimension scale of the forming processes is from 0.1 nm to 100 nm and their information materialization goes in the levels 5–2 (according to the theory by O. Yun).

The results we got show that using of institutional and evolutionary economy methods to manage the engineering services are reasonable and efficient.

References

1. D. North, B. Goutard, Economic Insight, (2014)
2. *The Reindustrialization of the United States. Special Report no. 1187* (Euler Hermes Economic Research Department, Paris, 2013)
3. M. Delyagin, *Russia for the Russians* (EKSMO, Moscow, 2009)
4. K. Fernandez-Stark, P. Bamber, G. Gereffi, *Engineering Services in the Americas* (Duke University, Durham, 2010)
5. D. Ju. Zavadskiy, Fundamental Researches, **7**, (2015)
6. *Engineering Services Trade: the Case of South Africa. Sixth Services Experts Meeting-Domestic Regulation and Trade in Professional Services* (OECD-World Bank, Paris, 2006)
7. Y. Barzel, Journal of Law and Economics, **25(1)**, 27-48 (1982)
8. M.M. Yudkevich, Economic Journal of the Higher School of Economy, **3**, 358-378 (1998)
9. S. Cheung, Journal of Law and Economics, **53**, 53-71 (1974)
10. T. Eggertson, *Economic Behavior and Institutions* (Cambridge, University Press, 1990)
11. G.J. Stigler, Journal of Political Economy, **69**, 213-225 (1961)
12. J. Hirshleifer, American Economic Review, **63**, 31-39 (1973)
13. P. Nelson, Journal of Political Economy, **2(78)**, 311-329 (1970)
14. M.R. Darby, E. Karni, Economica, **12**, 67-88 (1973)
15. M. Friedman, American Economic Review, **58**, **1**, 1-17 (1968)
16. L.I. Abalkin, Issues of the Economics, **5**, 4-20 (1994)
17. G.T. Odum, *Environment, Power and Society* (New York, 1971)
18. A. Hall, *Methodology for System Engineering* (D. van Nostrand Company, 1962)
19. J. Foster, *Biology and Economics, The Elgar Companion to Institutional and Evolutionary Economics* (Aldershot, Edward Elgar, 1994)
20. R. Langlois, *What is Evolutionary Economics? Evolutionary and Neo-Schumpeterian Approach to Economics* (Kluwer Acad. Publ., Boston, 1993)
21. C. Shubert, *Darvinism in economics and the evolutionary theory of policy-making* (Jena, Max Plank Institute of Economics, 2009)
22. V.I. Mayevskyi, Belorussian Journal of the Economy, **4**, (2000)
23. O.S. Sukharev, *Evolutionary Economics* (Finansy i Statistika, Moscow, 2012)
24. S.Ju. Glaziev, International Economy, **5**, 5-27 (2010)
25. C. Perez, *Technological Revolutions and Financial Capital* (Cambridge: Cambridge University, 2002)
26. O.M. Yun, *Production and logics: The informational basis of the development* (Noviy vek, Moscow, 2001)
27. I.G. Pereyaslova, E.B. Kolbachev, The Issues of South-Russian State Technical University (NPI), **5**, 50-55 (2012)
28. Ju.A. Kalyagin, E.B. Tsyrkin, Using the mathematical methods and computer technologies for design and development of the processes for the petrochemistry, 167-172 (1982)
29. I.G. Pereyaslova, V.V. Shmatkov, The Issues of South-Russian State Technical University (NPI), **4**, 6-11 (2013)

Training of innovative engineers for problems of modernization of a domestic production

I. Brusakova[1], *A.* Zvontsov[1], and *O.* Shvetsova[1,*]

[1]Faculty of economics and management, Saint-Petersburg State Electrotechnical University (LETI), Saint-Petersburg, Russia DP Sciences

Abstract. The actual problem of training of innovative managers for problems of modernization of a domestic production is considered in research. Practice of application of innovative engineering in the sphere of modernization of industrial productions is studied. The perspective of an integrated approach to management of innovative processes is investigated, including competence-based approach to training of innovative managers and modern instruments of management of finance innovative business - processes. Authors offered new approaches to formation of innovative capacity of the domestic industrial enterprises. Different ways of innovation creation are discussing.

1 Introduction

At the present stage of social and economic development of Russia it is necessary to realize modernization of domestic knowledge-intensive production. Preparation of engineering shots by the leading technical colleges of Russia is carried out with use of new federal state educational standards of the third generation (FGES 3+). Orientation to professional engineering competences which are defined by the employer – distinctive feature of the materials FGES3+. Existence of the international educational standards of engineering education (CDIO) allows to consider need of creation of infrastructure of engineering education as a condition of efficiency of formation of educational services for engineers.

2 Concept of innovative engineering

University of "LETI" as the innovative higher education institution participating in the TOP program 100 improves educational infrastructure [1]. Application of innovative forms of conducting educational activity (continuous, remote formation, use of innovative educational technologies on the platform e-learning, Learning Space, science and technology park, the small innovative enterprises, the research centers in structure of LETI, the organization of close interrelation in a framework of strategic partnership University of "LETI" with the leading manufacturing enterprises of SPb – allows to position University of "LETI" as the innovative higher education institution capable to organize preparation of engineering shots for problems of domestic economy.

Current trends of innovative development of the country are caused by need of introduction of ideas of innovative engineering for the organization of production of innovative production. The term "innovative engineering" is entered by O. L. Figovsky [2]. The author connects contents of the term with need of creation of an innovative product with use of innovative infrastructure of production. Any product, new to the market, is made with use of the R&D technology, i.e. is result of use of a successful combination of all resources, necessary for his production, – human, logistic, material, production, financial, information, monetary, etc. Transition to production of new innovative products in the "digital" enterprise, use of the CLOUD, SMART, GREEN, GRID technologies – communications – an important component of the modern infocommunication technologies allowing to provide "new properties" to the made product.

The innovative engineer, the engineer innovator has to be trained for realization of all stages of life cycle of innovative processes and projects in new technological way – from market researches of a commodity market and services before design and operation of innovative goods or service, carrying out patenting, commercialization and advance on the market. Skills of the innovative engineer are directly used for the organization of effective activity of the digital enterprise [3].

3 Model of competences of the innovative manager

Important role in innovative engineering in business increase of effective management of a human resource is taken away.

Rational management of innovative processes is possible in the presence of the developed motivational system which is created on purpose:

* Corresponding author: Shv_olga@rambler.ru

• initiations of innovative activity of employees;
• adequate assessment of results of innovative activity of managers and experts of various level and their subsequent encouragement;
• strengthenings of communication communications between divisions of the company;
• integration of innovative activity of staff of various divisions into market success.

The effective motivational innovative system has to cover all levels of management: top-management, average management, experts, working and technicians.

The leading role in formation of innovative climate at the enterprise belongs to top-management. "Ideas then seize the masses when they seize the heads of the first heads" [4].

Key to success – in the head of the innovative project which is appointed not so much for ability to the solution of scientific and technical tasks how many and it is more important, for the qualities of the head and businessman.

Practice has developed a number of requirements to the head of the innovative project. It is necessary for the head for success:
• to possess abilities to think in a complex;
• to be able to operate people and to be able to operate itself;
• to be able to interact with the management;
• to be provident;
• to be capable to maintain stresses, to be an optimist;
• to have perspective business thinking;
• to be able to transfer information;
• to operate effectively time.

The model of the effective innovative manager assumes presence at him of cognitive and functional competence [5, 6].

Cognitive competence.

The innovative manager has to know
• concepts, methods of innovative management;
• content of innovative strategy and policy, instruments of their formation and realization at the enterprise;
• methods of an assessment of level of innovative development of the enterprise;
• bases of the organization of innovative infrastructure;
• methods of search of new ideas and technical decisions;
• standards and methods of ensuring quality and competitiveness of innovations;
• technologies of protection of intellectual property at the enterprise.

Functional competence.

These are skills and abilities of development and realization of innovative strategy and policy, technologies and methods of work in functional spheres of innovative management taking into account conditions and specifics of the concrete organization.

The innovative manager has to be able:
• to develop innovative strategy and policy at the enterprise;
• to organize innovative process at the enterprise according to methodology of system approach to management of innovations;
• to form creative group of developers of an innovation, to diagnose a condition of group, to stimulate creative activity;
• to apply methods of search of new ideas and technical solutions;
• to be able to make the business plan of the project;
• to be able to estimate market appeal of an innovation and its competitiveness;
• to own methods of commercialization of innovations;
• to own methods of planning of innovative process;
• to own methods of an assessment and protection of intellectual property;
• to possess skills of search and work with information on achievements in the special innovative sphere.

In a portfolio of innovative projects of the company there have to be projects of various level:
• projects on development of essentially new products and technologies providing innovative leadership;
• several projects maintaining competitiveness of the existing commodity policy: on expansion of a grocery ruler, modernization, improvement of quality, etc.;
• programs and projects on support of innovative activity of employees [6].

Conveniently and effectively to operate realization of innovative strategy of the enterprise within system of management of projects with use of modern instruments of management.

Tactical aspects of innovative development are realized in the form of involvement of workers in innovative activity: promotions of ideas, improvement suggestions on improvement of technical processes, improvement of quality of production, prime cost reduction, etc.

It is possible to present universal model of competences of the innovative manager:

To specify within what program or a grant work or sponsors was performed.

In case of lack of the similar reference – to remove this text field

1. Universal (it is general administrative) competences

Basic administrative competences:
• ability to analyze a market situation, to reveal, formulate and range problems;
• skills of statement are more whole, preparation and adoption of administrative decisions;
• ability to concrete system of the purposes, to formulate tasks and to prove the choice of ways and methods of their decision.

2. Interpersonal competences:
• ability to form team and is effective in it to work;
• existence and development of leadership skills;
• skills of the presentation and negotiating.

3. Tool competences:
• linguistic – ability to use language means for the solution of professional tasks;
• information – ability to prove the choice and to use modern information products and technologies for the solution of objectives.

4. Subject and specialized competences:
• ability to carry out the marketing and strategic analysis of the organization, to make the ranged list of prospects of development and risks;
• ability to prove the priority directions of innovative development and to form innovative strategy, programs and projects of the enterprise (organization);

• ability to carry out complex diagnostics of the organization, the analysis of her financial and economic state and to develop strategies of financing of innovative development;
• ability to estimate the economic potential of an innovation, to carry out the feasibility study on innovative projects and programs;
• ability to find (to choose) optimal solutions during creation of new knowledge-intensive production taking into account requirements of quality, cost, dates of performance, competitiveness and economic efficiency;
• ability to put and solve organizational tasks, to carry out scheduling of implementation of innovative programs and projects.

The sphere of innovative activity of employees, her activization has high potential of innovative development of the company. Formation at the enterprise of complete organizational and motivational system of mobilization of creative potential of workers on the solution of tasks of increase of competitiveness of business is for this purpose expedient.

Within motivational innovative system the mechanism of an assessment of efficiency of intra company innovations for each division (research and development, production, sale, logistics, marketing, management, finance, shots, etc.) with definition of criteria, indicators of methods of an assessment, forms and methods of encouragement has to be created.

4 Financial management of innovative processes

The rational management of finance of the enterprise, automation of the administrative reporting, opportunity to quickly monitor the movement and target use of money allow to plan and predict development of the company, to make reasonable administrative decisions. Management of finance of the enterprises is a management in the sphere of the monetary relations concerning formation, distributions and use of the monetary income, accumulation and financial resources for the purpose of their most effective use and increase of solvency of the enterprise. The main objective of financial management on mobilization and use of monetary funds of the enterprises consists in maximizing their real liabilities and assets [8]. At the same time the following questions of financial management are essential:
• planning and forecasting of financial activity. Financial services of the enterprises develop financial plans which represent forecasts of output and product sales, development of scientific and technical progress, introduction of new administrative decisions and providing them with financial resources;
• adoption of the most expedient decisions in process of investment of means;
• coordination of financial activity with all services of the enterprises. For example, decisions in the field of marketing influence realization volume that, in the turn, makes impact on the size of profit and investments;
• carrying out operations in the financial market on mobilization of the additional capitals, implementation of own stocks and bonds.

Planning takes the central place in management of finance. This results from the fact that when planning the condition of finance of this or that subject of managing is comprehensively estimated, possibilities of increase in financial resources, the directions of their more effective use come to light. The relevant administrative decisions in the course of planning are accepted on the basis of the analysis of financial information which has to be rather full and reliable. The financial plan has to be full financial display of all planned activity of the enterprise. It defines an integrated approach to development of the system of financial planning: she has to cover all basic structural elements of the enterprise and all essential stages of the processes going to them.

Operational management represents a complex of the actions developed on the basis of the operational analysis of the developing current financial situation; their purpose — receiving the maximum effect of redistribution of financial resources at a minimum of expenses. The main content of operational management consists in maneuvering by financial resources for the sake of timely performance of necessary actions, and also elimination of possible difficulties on separate sites of financial and economic activity of the enterprises. Reserve funds, superplanned financial resources, the credits of banks, etc. can be sources of such maneuvering.

Control as an element of management is carried out both at a planning stage, and at a stage of operational management. He is urged to provide full and timely receipt of money in the centralized and decentralized funds of financial resources, finishing them to the recipient, economical and target use.

At diagnostics of the existing control system of finance at the enterprise, first of all, it is estimated how administrative reporting reflects an economic condition of the company how existing technique of management corresponds to tasks of the enterprise, and are how automated management processes. As a result of the analysis of the existing control system of finance, it is necessary to develop the outline sketch which includes ideal model of work of the enterprise. It is obvious that it is almost impossible to implement the similar project therefore it is necessary to finish it taking into account realities of work of the enterprise. Quite often improvement of management of finance can "bury" in equipment of the company necessary equipment or lack of the software, lack of money on introduction of the ideal project, and also in the low level of preparation of personnel of the company.

Proceeding from realities of work of the company, more "grounded" document including the list of organizational actions which have to be realized at the enterprise is developed. Besides, necessary normative documents are developed, business processes are described, and also modification of the existing automated system is made. Automation of finance – the most up-to-date tendency in management, both the enterprises in general, and finance in particular. The automated system is the main step to effective management and development of business [9]. Automation of business is necessary first of all for dynamically developing organizations which seek to take the leading positions in the market. Automated control systems for finance of the enterprise serve as the tool for obtaining competitive advantage before other organizations.

The automated financial systems at the enterprises will allow heads of an average and the top management to obtain operational information about financial and economic activity of the enterprise, to control the income and expenses, it is essential to raise the profit, to increase sales, to successfully realize investment projects. Owing to what competitiveness and investment appeal of the enterprise will increase, the probability of monetary frauds at the enterprises decreases.

Management of process of functioning and organizational structure within which this process is carried out have significant effect on productivity of production and commercial activity of any enterprise. Correctly chosen structure allows to provide to the enterprise or the company a full employment of personnel, flexibility in use of resources and compliance to requirements of the market. Thus, we inevitably come to need of bigger or smaller reorganization of a control system of innovative activity of the enterprise. For adoption of reasonable and operational decisions the management needs often to count and compare several versions of plans in short terms. Thus, innovations are not only use of high technologies, but also innovations in the sphere of administrative decisions.

Reorganization of system of business management on the basis of new administrative technologies is necessary for realization of the majority of the tasks facing the enterprise. The enterprise intending to work flexibly inevitably comes to need of use of innovative approaches. For a significant amount of the large industrial enterprises, improvement of a control system of finance means, in that number, financial reforming and restructuring. Structural transformations are based on economically reasonable strategy of reforming. According to her the system of financial management by the enterprise as system of planning, the account and control of the enterprise in the course of his financial reforming and restructuring is also developed.

It is necessary to develop system of management of innovative activity which basis - splitting difficult process into simple components on the basis of design approach and forming of peculiar administrative "conveyor" for effective realization of the tasks facing the enterprise. It gives a number of effects: work from process turns into the purposeful movement, labor productivity sharply increases, the quantity of mistakes decreases. It is necessary to develop effective organizational structure, a control system of finance for realization of such approach, to create service of marketing, to provide document flow, i.e. in details to formulate rules of work of the enterprise, in the main the general for all companies of an innovative orientation [5, 6].

5 Conclusion

It is possible to draw the following conclusion: it is necessary to project innovative development of the enterprise in the same way, as well as any system. It is necessary not only define the tasks facing the enterprise but also to create the system providing transition of the enterprise to this state, to define necessary resources, to develop business processes of innovative activity. Only this way it is possible to count on success.

Substantial basis of the system of financial management of the industrial enterprise which is adequately considering the level of influence and ways to avoid financial risks, is budgeting. The management of cash flows of difficult economic complexes what the industrial enterprises are, can't but use schemes of behavioral modeling of possible results of the movement of monetary resources. Development of financial models of an assessment and management of financial risks is a part of the general system of budgeting at the enterprise. They are rather reliable tool of an assessment and management of all system of financial risks at the enterprise as allow to weigh different levels of risks in connection with various factors of impact on them and controllability levels them.

In general, the control system of finance of the enterprise has to provide for the management transparency of a financial state and the movement of financial resources, formation of the most favorable versions of financial plans, the operational accounting of the actual information, control of a deviation from the plan. In this regard, a basis of a control system of finance at the enterprise are: the coordinated analysis of material streams and financial streams, as on separate commodity groups and divisions, and on the enterprise in general; a combination of planning and control of implementation of plans on the basis of coordination of planned documents with system of accounting and the reporting. It gives the chance of formation of the coordinated financial plans both for the enterprise in general, and on his divisions and activities. The created control system of finance of the enterprise has to give the chance to solve the following problems of management of finance:

- assessment of a financial condition of the enterprise;
- a financial assessment of the main directions of activity (expenses, profit, profitability and other indicators on commodity groups, divisions, contractors);
- the choice of the most profitable option of assortment policy (structure of the range and sales volumes on his positions);
- formation of budgets of divisions and control of their performance;
- planning and control of financial streams;

• check of financial feasibility of sales plans and plans of purchases;
• the forecast and prevention of bankruptcy and a swagger - major situations;
• determination of terms, volumes and expediency of capture of the credits;
• analysis, planning and control of profits and losses;
• analysis, planning and control of change of structure of balance (structure of the capital, etc.)

As a result of improvement of a control system of finance at the enterprise, management of manufacturing enterprise has to have more opportunities for adoption of business solutions, the reporting of the enterprise has to become more transparent for investors and shareholders that will allow to increase efficiency of work and investment appeal of the innovative company.

The integrated approach to formation of actual model of management of innovative engineering allows to reconsider modern practice of training of experts, including taking into account the international standards and to develop logically proceeding business processes in the field of financial and information management that allows to transform productions of the domestic enterprises according to actual economic and political realities and to carry out modernization of key resources.

References

1. A.V. Zvontsov, *Proc. Modern education: contents, technologies, quality*, 62-63 (2003)
2. V.M. Kutuzov, S.O. Shaposhnikov, O.A. Shvetsova, *Proc. Planning and Ensuring Training for an Industrial and Economic Complex of the Region*, 125-136 (2015)
3. O.A. Shvetsova, *Innovative aspects of human resource management in business* (St. Petersburg, 2012)
4. I. Brusakova, Proc. of XIX Int. Conf. Systems Analysis in Desier and Management, **2**, 16-22 (2015)
5. T.A. Khudyakova, *Proceedings of the International Congress on Interdisciplinary Behavior and Social Science*, 295-297 (2015)
6. T.A. Khudyakova, A.V. Shmidt, *Proc. of The 26th International Business Information Management Association Conference*, 1617-1625 (2015)
7. O.A. Shvetsova, *Proc. Innovation Management and Sustainable Economic Competitive Advantage: From Regional Development to Global Growth: Proceedings of The 26th International Business Information Management Association Conference*, 1619-1628 (2015)
8. *GOST R 56407-2015*, Lean production: main methods and instruments
9. E. Mach, *Erkenntniss und Irrtum* (Reidel Publ., 1976)
10. R. Mayer, M. Painter, C. Menzel, B. Perakath, P. deWitte, T. Blinn, Knowledge Based Systems Publ. (1995)
11. National Institute of Standards and Technology *Integration definition for function modeling (IDEF0)*, (1993)
12. Ching-Hong Tsai, How-Jen Luo, Feng-Jian Wang, *Proc. 11th IEEE International Workshop on Future Trends of Distributed Computing Systems*, (2007). DOI: 10.1109/FTDCS.2007.17.
13. R. Braun, *Proc. 2015 IEEE 17th Conference on Business Informatics (CBI)*, (2015). DOI: 10.1109/CBI.2015.41.
14. S. Yamasathien, *Proc. 2014 Fourth International Conference on Digital Information and Communication Technology and it's Applications (DICTAP)*, (2014). DOI: 10.1109/DICTAP.2014.6821684
15. M. Chinosi, A. Trombetta, IEEE Conference on Commerce and Enterprise Computing, IEEE Publ., (2009). DOI: 10.1109/CEC.2009.48
16. A.H. Maslow, Psychological Review, **50(4)**, 370-396 (1943)
17. A.H. Maslow, *Motivation and personality* (Harper & Row Publ., 1954)
18. J.O. Long, *ITIL 2011 At a Glance* (Springer, New York, 2012). DOI: 10.1007/978-1-4614-3897-7.
19. *A Guide to the Project Management Body of Knowledge (PMBOK Guide)* (Project Management Institute Publ)
20. http://eltech.ru.
21. http://www.metodolog.ru.

Transaction costs of innovative enterprise

E. Antonenko[1,*], and I. Baev[1]

[1]South Ural State University, Chelyabinsk, Russia

Abstract. The article deals with essence and classification of transaction costs, their distinctive features in case of innovative production. Transaction costs classification by necessity criterion is presented, components of information searching costs and negotiation costs as the most relevant groups of innovative production transaction costs are considered. The mathematical model of negotiations is suggested. This model defines the optimal number of negotiation rounds, determines exact total value of the contract with respect to transaction costs, and estimates the amount of transaction costs for each round. This model has practical significance and it can be used by innovative firm at the stage of commercialization. We conduct a computer experiment for finding the optimal number of rounds by minimal contract value criteria. Our results can be directly extended to real-life negotiations.

1 Introduction

Examination of transaction costs essence and their estimation draw great attention of scientists and experts. It can be explained by two reasons. Firstly, these costs and losses arise from interaction between business partners under free market economy. Secondly, transaction costs are a great part of total costs both at country level [1,2,3,4] and at micro level [5,6]. Moreover, transaction costs are much harder to estimate and difficult to manage since they depend on environmental parameters. Management decisions affect this type of costs not so obviously as on operating costs. It is clear that difficulties in definition and estimation of transaction costs increase in case of innovative enterprise. This article provides variety of definitions and classifications of transaction costs. We explore distinctive features of transaction costs of innovative enterprise and propose a model to determine the optimal contract value to increase commercialization efficiency.

2 Significance, scholarly importance and literature review

A. Transaction costs definition

The problem of transaction costs identification and estimation is one of the key challenges of modern economic science. The acknowledgement of this type of costs marks transformation from perfect market economy to enterprise activity consideration under real-life conditions. Transaction costs theory is significant aspect of current new institutional economy. Transaction costs economics as an independent concept finalized in the late 1970s thanks to Oliver Williamson's paper [7], but the background of the establishment of this theory appeared in a few decades earlier. It is assumed that origin of this conception occurred after Ronald Coase published his paper "The nature of the firm", despite the fact that the term "transaction costs" in this article has not been used [8]. At the same time, John Hicks compared the cost of transferring assets with friction [9]. In 1969, Kenneth Arrow, also came to conclusion that, contrary to theoretical assumptions, the market operation costs cannot be zero [10], so transaction costs are the costs of "running the economic system". Douglass North defined transaction costs as "the costs of measuring the valuable attributes of what is being exchanged and the costs of protecting rights and policing and enforcing agreements" [11]. Oliver Williamson suggested the next definition: "comparative costs of planning adapting and monitoring task completion under alternative governance structures" [12]. Douglas Allen associated transaction costs presence with creation of property rights and retention of title [13]. Steven Cheung [14], Jensen and Meckling [15], Barzel [16], Goldberg [17], Alchian and Woodward [18] are the followers of the definition of transaction costs from the perspective of property rights approach too. Niehans [19] defined transaction costs as «property rights transferring costs». Stavins [20] gave the similar definition: "...transaction costs are ubiquitous in market economies and can arise from the transfer of any property right...".

Great variety of transaction costs definitions give evidence of ambiguousness of their nature. Every scientist applies the definition according to the field of research and commitment to scholarly traditions. However, all definitions have a number of common features. Firstly, this type of costs arises from

* Corresponding author: Elizaveta.Antonenko@susu.ru

interaction of market participants. Secondly, they are connected with the imperfection of the transmitted information. Therefore, from our point of view the most relevant definition is the following: "Transaction costs are the costs originating as a result of market interaction between enterprise and external environment under passing the barriers arising from information asymmetry, administrative barriers, as well as real and opportunity costs carried by firm through this interaction" [24].

B. Transaction costs classification

Transaction costs classification depends on scientist's point of view as their definition does. According to the most accepted classification, transaction costs are divided into seven groups:

1) Costs of administrative barriers overcoming;
2) Information search costs;
3) Negotiation costs;
4) Costs of measurement;
5) Costs of specification and property rights protection;
6) Costs of opportunism;
7) Lobbying costs.

Besides the basic classification, transaction costs can be divided by necessity criterion. Thus, we have two groups: necessary costs and transaction losses. Necessary costs are the costs which enterprise bears in any case for maintaining its activity in process. Transaction losses are a type of transaction costs, which contains losses of time and resources and arises from wrong management decisions. In the context of negotiation costs, necessary costs are the costs of preparing the information for negotiations, representational expenses, business trip expenses, costs of a lawyer, notary officer services and the costs of expert support. Transaction costs in this case are losses from unfavourable terms of bargain, losses due to loss of a partner, goodwill decrease.

3 Transaction costs in innovative production

C. General descriptions of information search costs

Innovative products manufacturing has some specific features, the main one is high sales uncertainty. "Push" strategy application is too risky for innovative enterprises, their marketing focuses primarily on the identification of a potential buyer's needs. Relationship between the company and environment is unstable due to the lack of established links and high uncertainty. As concerns the value and structure of transaction costs, one mentions some specific features about innovative product manufacturing. An enterprise inevitably faces with the costs of licensing, registration and certification while launching a new product. At the same time, in addition to the ordinary types of costs (registration fees, consultancy fees, etc.) costs of specification and protection of property rights (cost of restoration of violated rights, cost of protecting intellectual property, and loss from bad specification) are added. Nevertheless, key components of transaction costs for innovative enterprises are, in our opinion, information search costs and negotiation costs. Information search costs consist of the costs of information search for market situation, information about potential counterparts and losses associated with the incompleteness of received information.

Innovative enterprise activity start with market monitoring for finding a free niche. In this aspect, the information quality is critically important as it rates the reasonability of production startup and direction of enterprise's activities. Losses caused by incomplete information are inversely related with the information quality level whereas in contrast searching costs rise with the growth of information quality requirements. The natural problem of information management is to achieve a balance between the value of information and its utility [25]. The utility of information is not equal to its quality, since generally information of the highest quality is excessive for the company. Components of information search transaction costs are presented in Table 1.

Table 1. Components of information search transaction costs.

Costs	Components
Costs of informatio n search for potential business partner	Costs: of access to database, of obtaining communication services, of attendance professional expositions and events. Losses: of obtaining excessive information.
Costs of informatio n search for market situation	Costs: of access to database, of obtaining communication services, of communication services, of ordering/conducting market research. Losses: of the missing the participation in the free messages (e-mailing, etc.), of obtaining excessive information.
Losses associated with incomplete -ness of received informatio n	All losses arising from: choice of business partner with unfavorable conditions at the market (price, service), choice of an unreliable business partner (delays, incomplete deliveries, etc.), forecasting error due to inaccurate description of the market situation (overproduction, underproduction, incorrectly exposed prices, etc.).

Information search costs management determines the management of costs in general. Using high quality and correct information an enterprise can lower the probability of disputes with business partners and reduce their opportunism.

D. Negotiation costs

After finishing the process of searching information for potential business partners, company starts negotiations. Table 2 shows components of negotiation costs.

Table 2. Components of negotiation transaction costs.

Costs	Components
Costs assotiated with organisation and conducting of negotiations	Costs arising from: representational activities, business trips, using lawyer services, notary officer services, expert support, salary payments to relevant staff. Losses arising from: unreasonably high level of representational expenses, too frequent business trips (compared with the accepted rate in the industry or economic effect of the trip), breakdown of negotiations, overpayment to lawyers (lawyers' fee in comparison with similar qualifications).
Losses associated with failure of negotiations	All losses arising from: loss of a business partner (this follows with losses of discounts; increase of information search costs for a new business partner, etc.); unfavorable terms of bargain (the difference between the desired and the real contract price, level of service, etc.); drawing a loan under unfavorable terms; surplus of stored goods (also related to forecast errors);bargain's failure due to problems with the law, intentional and unintentional errors in the contract on the part of both contracting parties.
Legal registration costs	Costs: of consumables, of coordination with authorities. Losses: from improper execution of documents (amendment of a document, re-signature, and repetitive visits to the authorities).

Let us define the negotiation process. Two players negotiate about a contract of some issues (price, deadline, prepayment, etc.). Negotiations consist of multiple rounds. Players make offers in order, one by one. In every time period the "active" player can do the following:

1) Accept the previous offer;
2) Reject the previous offer and propose a new one;
3) Quit the negotiations.

Players maximize their utilities and rationality is a common knowledge.

A specific feature of innovative enterprise as a side of negotiation process is its "weakness". Under weakness, we consider weak ability to influence on the outcome of negotiations. The objective of any negotiations is to achieve a result in the best possible conditions. For the seller, those will be the highest possible price, acceptable delivery time and the maximum prepayment. For the buyer, on the other hand, it will be the lowest possible price, the shortest delivery time and the minimum value of prepayment. Everything in the contract except time can be specified in terms of monetary value. It becomes possible to define a utility function to compare different contracts. Reservation utility of a player is the minimal acceptable outcome of the negotiation. Therefore, player will refuse to take part in negotiations if their outcome is less then reservation utility.

Let us determine the so-called zone of agreement. Zone of agreement is a set of all feasible price values, which are acceptable to both players [21]. Player's strength affects the part of the zone of agreement where the contract will be signed. It is reasonable to assume that in case of negotiations between a weak buyer and a strong seller contract price will tend to shift to the upper point of the zone of agreement. Successful negotiations directly affect the commercialization of goods.

There are two main classes of strategies: time dependent strategies and behavioral strategies. To evaluate the round in which negotiations are expected to finish we have used a behavior-type strategy [21], specifically tit-for-tat strategy. However, implementation of this type of strategy for evaluation of intermediate offers of players is not acceptable, though it can be used in case both players have equal "strength". In order to consider different values of player's "strength" we have modelled the negotiation process using time-dependent strategy (as was mentioned we got the estimation of the expected duration of negotiations).

It is well known that three general types of strategies exist in time-dependent class of strategies: aggressive, flexible and linear [22]. Aggressive means that the player maintains his initial offer until he reaches the deadline, then he concedes up to his reservation utility. Flexible means that the player quickly concedes and maintains his reservation utility. However, these are too extreme to be used in real negotiations. In real life players tend to use less extreme strategies even if they are not optimal (it is proved that above three dominate the rest strategies). Outcome of negotiations will be determined in accordance with the theorem: "At the deadline, rational agents will offer their reservation utility" [23]. That is the way, how players come to agreement.

We have developed a mathematical model of negotiations, which allows determining the number of rounds of negotiations, total value of the contract, and value of transaction costs for each round. In addition, it is capable to estimate the optimum total contract value, taking into account transaction costs. We have written a computer program using MatLab environment, where total price of the contract is determined according to "strength coefficients" of players (these coefficients are estimated using the method of expert evaluation of players on a number of parameters).

So, to determine the expected number of rounds behavioral strategy is used, in other words, player's behavior is based on a symmetric behavior of the other. In other words, this strategy tries to match the behavior of the counterparty. If the side makes a large concession, the response will be with large concession too, opposite is true – small concession of the one side will be followed by a small concession of the other.

The program estimates the expected number of negotiation rounds. Then using this estimation the program switch to a time-based strategy and compute intermediate players' offers with the help of trigonometric functions.

As discussed, the distinguishing feature of the innovative enterprise is its weakness in the negotiation process. We set up the model experiments in which the strength of the innovative enterprise is fixed at a low level (20% of the possible maximum) and find the expected number of rounds, depending on the strength of the other side.

Figure 1 shows which round is considered as the best in relation to the price of the goods. Consider a point with a maximum strength (100) of the counterparty. Ending negotiations in the fifth round will give players maximum satisfaction.

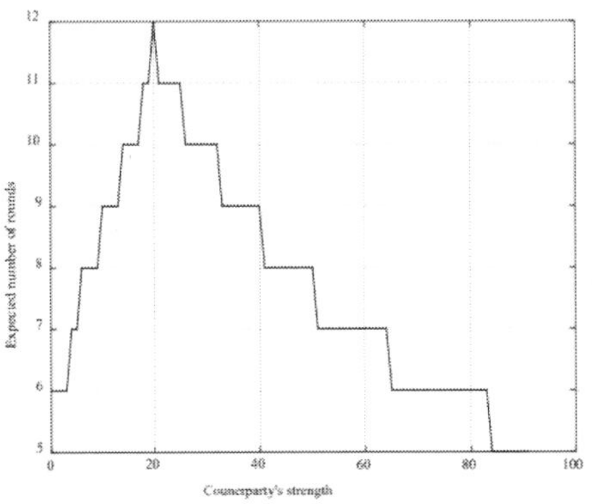

Fig. 1. Expected number of rounds vs counterparty's strenghth.

However, if we look at Figure 2 we see that the minimum value of the contract is achieved in the third round, not in the fifth one. This happens due to the presence of transaction costs. Thus, it makes sense to agree with unfavorable conditions in the third round and do not continue negotiations, as the benefit derived from achieving the desired price in the fifth round, will be neutralized because of the transaction costs.

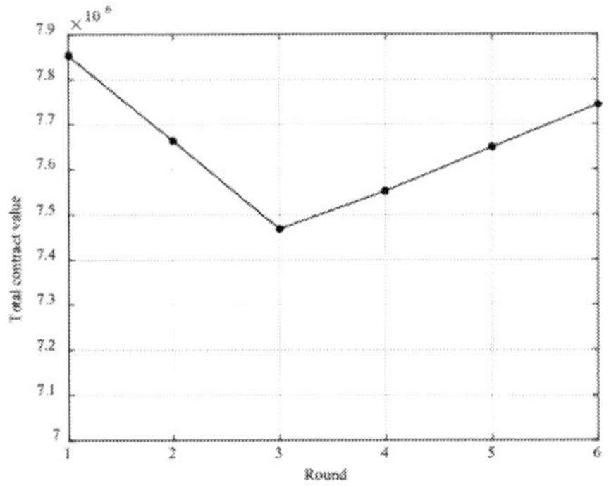

Fig. 2. Total contract value vs number of rounds.

This model has important practical significance. Firstly, it allows innovative enterprises to forecast the expected number of rounds of negotiations and to perform their approximate calculation. Secondly, this program allows user to define the round when negotiations should be stopped, to reach the minimum contract value. The same is true for the case where an innovative company is the seller. In this situation, the model is able to find the round with maximum total contract value. Thus, the system can promote the successful commercialization of an innovative product.

4 Conclusion

Transaction costs identification and estimation is an important part of an enterprise management, the ability to enhance competitive advantages. Greater focus should be placed on management of information search costs in innovative enterprises, as the quality of received information affects the all further enterprise activity. Another important element - costs of negotiating and contracting. Suggested mathematical model allows the estimation of this type of costs and increases efficiency of negotiations, thereby increasing the probability of a successful launch of an innovative product. The computer program based on this mathematical model allows estimating the expected number of rounds of negotiation process and estimating the total contract value with respect to transaction costs. This program recommends finishing the negotiation process in specific round given the optimal outcome of negotiations and can be a part of decision-making systems for support innovative industrial enterprises. Further stage of our study will be devoted to the creation of a mathematical model for management of information search transaction costs.

The work was supported by Act 211 Government of the Russian Federation, contract № 02.A03.21.0011.

References

1. F. Da, The Journal of Quantitative & Technical Economics, **8**, 123-134 (2009)
2. *Dagnino-Pastore, Transaction Costs in Argentina* (ISNIE, 1999)
3. Dollery, Australian Economic History Review, **38(3)**, 207-231 (1998)
4. J.J. Wallis, Long-term factors in American economic growth, (1986)
5. S.S Hosseini, J. Agr. Sci. Tech., **14**, 243-256 (2012)
6. M.M. Polski, *Measuring Transaction Costs and Institutional Change in the U.S. Commercial Banking Industry*. http://www.spea.indiana.edu/ids/pdfholder/ISSN-01-3.pdf.

7. O. Williamson, Journal of Law and Economics, **22(2)**, 233-261 (1979)
8. R. Coase, Economica, **4(16)**, 336-405 (1937)
9. J.R. Hicks, Economica, **2**, 1-19 (1935)
10. K.J. Arrow, The Analysis and Evaluation of Public Expenditures: the PBB System, Joint Economic Committee Compendium, 91st Congress, 1st Section, **1**, (1969)
11. O.E. Williamson, American Journal of Sociology, **87**, 548-577 (1981)
12. D.C. North, *Institutions, Institutional Change, and Economic Performance* (Cambridge University Press, New York, 1990)
13. D. Allen, *Transaction costs. In Encyclopedia of law and economics, volume I: the history and methodology of law and economics* (Cheltenham, Edward Elgar, 1999)
14. S.N.S. Cheung, Economic Inquiry, **36**, 514-521 (1998)
15. M. Jensen, W. Meckling, Journal of Financial Economics, **3**, (1976)
16. Y. Barzel, Journal of Institutional and Theoretical Economics, **141**, 4-16 (1985)
17. V.P. Goldberg, Goldberg, Readings in the Economics of Contract Law, 395-402 (1989)
18. A.A. Alchian, S. Woodward, 26 Journal of Economic Literature, 65-79 (1988)
19. J. Niehans, Journal of Money, Credit, and Banking, **1**, 706-726 (1969)
20. R.N. Stavins, Journal of Environmental Economics and Management, **29**, 133-148 (1995)
21. G. Lai, C. Li, K. Sycara, J. Giampapa, *Literature review on multiattribute negotiations* (Technical Report CMU-RI-TR-04-66, 2004)
22. S. Fatima, M. Wooldridge, N. Jennings, Artificial intelligence, **27**, 381-417 (2006)
23. S. Fatima, M. Wooldridge, N. Jennings, Artificial intelligence, **152**, 1-45 (2004)
24. E.V. Antonenko, Bulletin of the South Ural State University. Ser. Economics and Management, **9**, **1**, 74-80 (2015)
25. E.V. Popov, *Institutes of Minieconomics* ("Economics" Publishing, Moscow, 2005)

Approaches to the evaluation of the effectiveness of innovative development of the machine-building enterprise strategy

T. Gerashchenkova[1,*]

[1]Bryansk State Technical University, Bryansk, Russian Federation

Abstract. For the effectiveness assessment in the strategy of enterprise development it is expedient to carry out the regular monitoring of activities. For that there are must be estimated moment indices, rates of indices increase. A standard sequence is under formation characterizing rates of growth of some indices regarding others. Quantitative indices in sequences formed are established reasoning from the necessity to ensure the innovation development of an enterprise. In this case the degree of a correspondence of actual indices structure preliminarily formed by a standard one witnesses testifies management effectiveness which is quantitatively characterized by rank correlation narrowness. Specifying the priority sequence of indices one can design and establish in programs and plans of enterprise strategic development a model of successful economic activities.

1 Introduction

The purpose of the assessment of the effectiveness of innovative development of the machine-building enterprise strategy is to identify how the characteristics of the chosen model of economic entity in the long term, meet the requirements of efficiency and competitiveness. In this regard, is it necessary to study management and business activities in order to establish a rational choice of approach to achieving the goals, the efficiency of the reaction to changes in the external environment and on-farm situation, revealing the evidence of the enterprise increasing stability, entailing the improvement of organization level of the economic entity as a system.

In order to monitor the effectiveness of the company development in the implementation of innovative development strategies it is advisable to study a set of composite indexes that have a hierarchical structure, using the concept of economic growth proportionality, and indexes characterized by a certain normative structure, adapted and refined in accordance with the essence of the phenomenon under investigation.

2 Rationale

An effective development strategy must ensure a positive trend of promising production and economic parameters of an economic entity, be based on efficient production processes, sustainable competitiveness of the enterprise. In connection with this the analysis of the achievement of targets (settings) and performing mandated activities [1-8] seem to be unreasonable. These approaches are inherently static. Under the strategic approach, involving lengthiness of the process in time, it is important to monitor the processes occurring in their dynamics. Thus, in this case it is necessary to monitor [9, 10] , study growth rates and growth rates of some indexes relative to other.

3 Theoretical part

To assess the effectiveness and management quality of production and economic activity, I.M. Syroezhin [11] proposed to use the method of the investigation of "economic growth structure", which was later developed in the research [12-15] for the definition of "economic growth proportionality"

The study of economic growth proportionality seems to be useful for innovative development strategy effectiveness. However, objects and features of the content of this concept required to adapt the aims presented above.

In fact, the forming index structure, being the subject of evaluation may take the form of any combination of indexes contained in the regulatory framework and reflecting the existing realities and priorities of the management system in the implementation of development strategies. The presence of deviations in the structure shows a decline in the level of quality strategy during the study period. The magnitude of the deviations can be judged on the content and nature of the activities in the area of concern, as well as the mistakes made in the design and implementation of development strategies.

The number of indexes in generated sequences is set on the basis of the purpose of assessment activities, the availability of information and the accuracy of the desired results. In order to assess the company development strategy as a whole a group of system-wide indexes should be considered. The assessment of the innovative development strategy involves the use of a specific detailed structure of indexes.

* Corresponding author: gerash-tatyana@yandex.ru

DOI: 10.1051/shsconf/20173501119

Due to the importance of products manufactured at the machine-building enterprises, which characterizes the level of technological development of the society, the priority in the structure of regulatory indexes is given to the satisfaction of the needs of consumers of products, their desire to purchase products of a given enterprise. The condition for achieving this development parameter is the growth of production, or a cheaper way of production or the execution of the first and second conditions simultaneously. In case of customer satisfaction with innovative products or use of innovative technologies for the production, an appropriate investment is required, so investments will be the last in this chain of indexes.

Thus, the regulatory framework of the indexes should include the following groups of indexes: the objective function - the importance of innovative development results for the consumer, as well as the results of using the potential of the enterprise and the amount of on-going expenses.

We assume the following general conditions under which economic interests of the innovator, the investor and the consumer coincide: the increase in a customer value of the goods must be greater than the increase in the efficiency of the innovator in the processes of the strategy implementation, which, in turn, must be higher than the efficiency gains of investing in the implementation of the activities of the given strategic alternative, ie,

$$\Delta\{P^{kc}\} > \Delta\{E^{inn}\} > \Delta\{E^{ivn}\}, \qquad (1)$$

where: $\Delta\{P^{kc}\}$ – increase of customer value of the product; $\Delta\{E^{inn}\}$ – increase the efficiency of the innovator in the implementation of innovation and investment strategy; $\Delta\{E^{ivn}\}$ – investment efficiency gains in the implementation of the activities of the strategic alternatives.

It is proposed to introduce the indexes, calculated with the help of the index method, into the above mentioned groups, since this approach, reflecting a change in the dynamics of indicators, allows to achieve a single dimension. [16]

It is also necessary to take into account the factor of time money value using a discount rate, on the basis of the source data in terms of value.

Lets consider the figures included in the formula 1 group.

The usefulness of the goods (P^{kc}) - is an assessment of the degree of satisfaction as a result of its use. Usefulness is determined by the characteristics of goods, as well as economic factors (price, transaction costs during its use, etc.). It can be defined as an increment value of production of the enterprises on the basis of the calculation of the product competitiveness index. [17]

This method is based on a comparison of quality and price indexes. Buyers taking the decision to purchase, either prefer quality, losing in the price, or being satisfied with the lower price, are dissatisfied with the quality of purchase. Thus, it seems advisable to get Competitiveness Index (I_{kc}) by multiplying the price index (I_c) and quality (I_k^i).

1. It is necessary to calculate the individual indexes of competitiveness in quality, characterizing the deviation of meaningful quality indicators (by I_k^i) of the products being analyzerd from the basic product indicatorsn [18, 19]. The base products are the products manufactured prior to the implementation of innovative investment strategy.

$$I_k^i = \frac{k^i}{k^b}, \qquad (2)$$

$$I_k^i = \frac{k^b}{k^i}, \qquad (3)$$

where: I_k^i – Index of appropriate quality indicator; k_b, k_i – Basic and investigational value of quality indicators..

Formula 2 should be used if an increase in the dynamics of index values indicates the quality of growth. Formula 3 is intended for cases where an increase in the test speaks about a decrease in the quality of products in general. The definition of quality indexes of both the test, and the sample must be carried out using the same methods. [19]

2. Single parametric indexes (I_k^i) are combined in a group index - ΣI_k. The calculation of the index requires the determination of the weight coefficient of the examined characteristics of evaluated products, the sum constituting a unit. For food products with a large variety of approaches to the subjective assessment of the significance of various quality parameters, it is impossible to give preference to any of them, because if any of the indexes do not comply with the quality level, products become unfit for consumption, regardless of the subjective evaluation. In this regard, the weight coefficients can be represented as an equal share for all tested parameters

$$\Sigma I_k = \sum_{i=1}^{n} \frac{\frac{1}{i} \times I_k^i}{i}, \qquad (4)$$

3. The index of competitiveness for the price is calculated:

$$I_c = \frac{c \times \frac{1}{(1+r)^t}}{c^b \times \frac{1}{(1+r)^{t-1}}} \qquad (5)$$

where: c –retail price of the product being analyzed; c^b – retail price of the basic product .

Since agribusiness is a manufacturer of products not subject to or slightly exposed to obsolescence, the price of the base product can be considered the price of similar products in the previous year, especially since the introduction of new technology or the use of breeding material that is the most common type of innovation in the agricultural sector and the manufacture of food products, all manufactured products in the next period of time will characterize the impact of innovation and investment process.

It should noted that the prerequisite is the similarity of pricing methodology for previously manufactured and a new (or produced by a new technology or using new materials) products.

Total competitiveness index is calculated as the sum of two previously calculated indices:

$$I_{kc} = I_k + I_c \qquad (6)$$

This is followed by the identification of indexes that most adequately reflect the efficiency of innovation and investment activity, taking into account the heterogeneity of performance criteria, from the standpoint of innovators and investors.

From the standpoint of innovation active enterprises the following conditions must be identified:
1) cost growth in the sales of innovative products in total sales (I_Q)

$$I_Q = \frac{Q \times \frac{1}{(1+r)^t}}{Q^b \times \frac{1}{(1+r)^{t-1}}} \tag{7}$$

where: Q – sales of innovative products and services (and products produced by innovative technologies) in the subsequent period; Q^b - the same indicator of the previous period
2) the reduction of resource consumption when using a new technology, in comparison with the previous one (I_R)

$$I_R = \frac{R^b \times \frac{1}{(1+r)^{t-1}}}{R \times \frac{1}{(1+r)^t}}, \tag{8}$$

where: R – resource consumption when using a new technology, or in each subsequent year; R^b – resource consumption before the implementation of the new technology, or in each previous year.
3) growth of enterprise value, estimated by the value of assets before and after the launch of the innovative-investment strategy (may be due to both the growth of value of fixed assets and intangible assets) (I_A)

$$I_A = \frac{A \times \frac{1}{(1+r)^t}}{A^b \times \frac{1}{(1+r)^{t-1}}}, \tag{9}$$

where: A – value of the assets after the launch of the innovative-investment strategy, or in each subsequent year; A^b – value of the assets prior to the implementation of innovative investment strategy, or in each previous year
4) increase in wholesale prices for products compared to previously manufactured analogue (I_c^{inn})

$$I_c^{inn} = \frac{c^{inn} \times \frac{1}{(1+\kappa)^t}}{c^b \times \frac{1}{(1+r)^{t-1}}} \tag{10}$$

where: c^{inn} – the wholesale price of an innovative product, or a product produced by the innovative technology; c^b – the wholesale price of previously manufactured analog of new products, or the price in the previous period.

From the point of view of the investors effectiveness confirmation also requires compliance with several conditions:
1) growth of investment in innovation (I_V^{inv})

$$I_V^{inv} = \frac{V^{inv} \times \frac{1}{(1+r)^t}}{V^{inv\,b} \times \frac{1}{(1+r)^{t-1}}}, \tag{11}$$

where: v^{inv} – the volume of investments in the new portfolio, and in the subsequent period of time; v^{invb} – the volume of investmnents of the previous period.
2) revenue growth of investment in innovation in the total amount of income from investments (I_{np}^{inv})

$$I_{np}^{inv} = \frac{np^{inv} \times \frac{1}{(1+r)^t}}{np^{invb} \times \frac{1}{(1+r)^{t-1}}}, \tag{12}$$

where: np^{inv} – profits under the new investment portfolio; np^{invb} – the amount of profit of the previous period.

The greatest difficulty in the next step is intra-ranking figures because of the diversity of implemented corporate and functional strategies with their corresponding plurality of combinations of priority indicators. In this regard, the logic of construction of a ranked number of indexes should be based on preferences for specific conditions under study. [20]

Based on the practice of the factor analysis of the efficiency in the analysis of interdependence of the categories' costs - Revenue - Profit " [21, 22], we can specify the indexes, using the above mentioned inequality (Formula 1).

$$I_{kc} > I_Q > I_r > I_A > I_c^{inn} > I_V^{inv} > I_{np}^{inv} \tag{13}$$

So we formed a normative set of indexes agreed with the previously established priorities. It reflects the presence of innovation and investment component in the process of achieving strategic goals. In the framework of the adopted in the analysis of the regulatory framework, criterial values of selected indexes can vary within certain limits, which helps to preserve the consistency of the normative set.

By setting a priority sequence of indexes, its possible to design a model of a successful economic activity, and secure it in the programs and plans.

Management efficiency is proved by the degree of conformity of the actual structure of the indexes with preliminary generated normative structure, which is quantified by the tightness of rank correlation. For its definition Spearman rank correlation coefficient is used in mathematical statistics. The works of I.M. Syroezhin provide the evidence and justification of the possibility of using this characteristic to assess a variety of options of disorder ranks indexes in structures [11].

4 Experiment results

Testing at a number of engineering enterprises have confirmed the usefulness of the proposed approach to the determination of the effectiveness of innovative development strategy. The study covered the period from 2005, which made it possible to evaluate the aftereffects of innovative activities used from 2011.

On the basis of the established priorities of the rate of growth indexes we calculated rank correlation coefficients. According to the coefficients the graphs rreflect the dynamics of the change of the Spearman rank activity coefficient (Fig. 1).

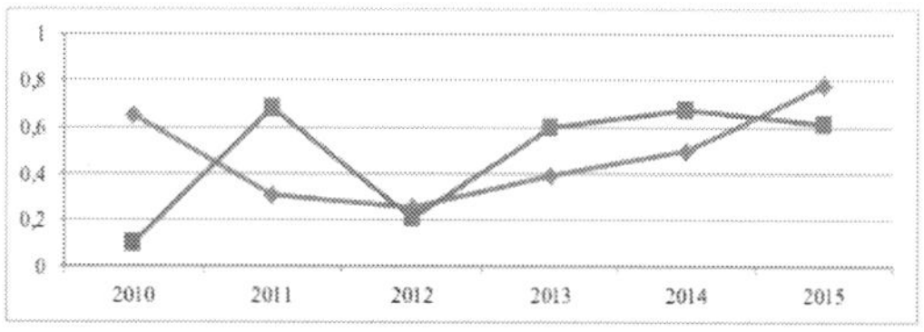

Fig. 1. Spearmen rank coefficient dynamics.

Significant undulation of correlation coefficient, presented in the figure, indicate instability of the enterprise development in the process of the strategy implementation, while the graph having a flat curve, proves the success of the development.

Identification of trends in terms of the effectiveness of the implementation of innovative development strategies sums up the efforts of the company management and serves as a basis for the further conduct of detailed evaluations of the policy carried out by it, a system of interaction with business partners, competitors, financial institutions, infrastructure and complex as well as with the authorities.

It should be noted that this approach is particularly relevant to machine-building enterprises in the conditions of instability of the environment and limited financial resources, as the assessment of the effectiveness of investments in innovation traditional approaches often demonstrates the futility of investing due to a prolonged payback period. Demonstrating positive dynamics of the proposed indexes, corresponding to the predetermined priorities, it is possible to evaluate from a new perspective both efficiency and appropriateness of innovative activities.

5 Conclusion

The analysis and evaluation of innovative development strategies should be carried out with the help of a number of regulatory indexes, comparing the growth rates and simple indexes with those under traditionally implemented scenarios of the economic system development. The effectiveness of this approach is confirmed by the approbation conducted at innovation-oriented machine-building enterprises. Conducted expert surveys confirmed the validity and acceptability of the above approach and widened the understanding of its potential use.

References

1. Y. Brigham, L. Gapensk, *Financial Management: Transl. from English* (1997)
2. I.V. Bryantseva, *The economic stability of the enterprise: essence, evaluation, management* (Publishing house KHGTU, Khabarovsk, 2003)
3. L.T. Gilyarovskaya, D.V. Lysenko, D. Endovitsky, *Comprehensive economic analysis of economic activity* (Prospect, Moscow, 2008)
4. V.N. Lazhentsev, Proceedings of Komi Scientific Center of the Ural Branch of the Russian Academy of Sciences (RAS), **13**, 107-113 (2013)
5. T.B. Leybert, Economics, **9(46)**, (2008)
6. A.I. Solyanik, Control systems and information technology: Scientific and Technical Journal, **1.3(31)**, 410-413 (2008)
7. A.I. Solyanik, O.J. Kravets, Control systems and information technology: Scientific and Technical Journal, **2.1(32)**, 204-207 (2008)
8. N.N. Bondarenko, N.S. Buzygina, L.I. Wasilewska, *Statistics: indexes and methods of analysis* (Modern School, 2005)
9. D.S. Buklagin, Development of information security and innovation: Publications, **143**, 3-12 (2002)
10. B.Y. Serbinovsky, E.V. Rudick, *Monitoring of labour productivity* (South. Federal Univ, Novocherkassk, Lick, 2010)
11. I.M. Syroezhin, *The planned. planning. Plan (theoretical essays)* (Economics, Moscow, 1986)
12. V.A. Vinokourov, G.L. Azoev, *Managing the development of industrial association* (MIM, Moscow, 1989)
13. V.V. Vikharev, Herald of SRSTU (NPI), **3**, 132-135 (2012)
14. V. Novikov, The Economist, **8**, 46-52 (1992)
15. J.S. Merkumov, *The organization and financing of investments* (INFRA-M, Moscow, 2002)
16. T.M. Geraschenkova, Scientific notes of Petrozavodsk State University, **1(138)**, 94-99 (2014)
17. E.V. Tyapushova, Herald of BFU name of I.Kant, **3**, 119-124 (2012)
18. G.L. Bagiev, V.M. Tarasevich, J. Ann, *Marketing* (Economics, Moscow, 2001)
19. N.P. Vashchekin, *Marketing* (Publishing House FBK-PRESS, Moscow, 2004)
20. T. Kaneko, Problems of the theory and practice of management, **3**, 80 (2002)
21. M.A. Ryazanov, Issues of innovative economics, **9(9)**, 3-12 (2011)
22. A.G. Savitskaya, *Analysis of economic activity of the enterprise* (INFRA-M, Moscow, 2009)

Calculation of efficiency of the project of modernization machining shop of the instrument-making enterprise

I. Fomina[1,*], and *V.* Dreshchinsky[1]

[1]Faculty of economics and management, Saint-Petersburg State Electrotechnical University (LETI), Saint-Petersburg, Russia

Abstract. One of the directions of investment activity of the enterprise is investment into fixed assets which include processing equipment. Investment into processing equipment demands presence at the subject of information directorate about expenses and results of operation of the available equipment, knowledge of a current state of development of technology, ability to allocate various versions of projects of modernization, to estimate costs of acquisition, development and operation of the new equipment, abilities to compare projects in the specified parameters, to estimate their expediency and economic efficiency In work modern problems of achievement by the Russian industry of the level of efficiency and competitiveness meeting the requirements of the world market of the knowledge-intensive and hi-tech production, realization of innovative development of the enterprises with ensuring the corresponding growth rates for indicators of economic efficiency are considered. In this article the example of calculation of efficiency of the project of modernization of machining shop of the instrument-making enterprise (St. Petersburg) for participation in the FTP is given. The used numerical values have exclusively demonstration character.

1 Introduction

Achievement by the Russian industry of the level of efficiency and competitiveness meeting the requirements of the world market of the knowledge-intensive and hi-tech production can't be realized out of innovative development of the enterprises with ensuring the corresponding growth rates for indicators of economic efficiency.

One of key actions for increase of technical and economic indicators of fixed assets or their separate parts on the basis of introduction of the advanced equipment and technology, mechanization and automation of production, modernization and replacement of the outdated equipment with more productive new, is modernization of the enterprise [1].

2 Concept of the project

Thanks to the Russian Federal target program "Development of Defense Industry Complex of the Russian Federation for the Period till 2020" [2] many enterprises have had an opportunity to realize projects of modernization and modernization that has allowed them:

To come to qualitatively new level of technological capabilities conforming to the modern international standards of instrument making.

To restore ability to independently update the production capacities at the expense of a sinking fund, lost owing to crisis of national economy and instrument-making branch in 1991-2010.

It is considerable to reduce requirement of the enterprise for unique highly professional personnel which deficiency gains the menacing character [3] recently.

In this article an example of calculation of efficiency of the project of modernization of machining shop of the instrument-making enterprise for participation in the FTP will be reviewed. The used numerical values have exclusively demonstration character [20].

The term of implementation of the project it is expedient to take the term of effective use of the new equipment. According to the experts it is necessary to take term in 15 years.

Key parameters of the project are:

• The amount of financing – 425.6 million rubles.
• The term of performance of an investment stage – 3 years.
• The term of implementation of the project (calculation of indicators of efficiency) – 15 years.
• Profit of the project – 402.7 million rubles.
• Indicators of efficiency of the project.
• In the analysis of efficiency of the project it is necessary to calculate the following indicators:
• Profit of the project (NCF).
Profitability of investments.
• Payback period of the project (PP).
• Internal standard of profitability (IRR).

It should be noted that the discounting method in projects of this sort (so-called "internal" projects) usually isn't applied as in such cases the subject of investment – not the financial investor who is looking for the

* Corresponding author: i-fomina@inbox.ru

maximum return on the capital and carrying out the choice of the investment project in various branches of economy, and the management of the enterprise having alternatives only at the choice of a way of implementation of the project, but not in the choice of the project [4].

General description of the project.

Replacement of the equipment of machining shop of the instrument-making enterprise assumes acquisition within 3 years of 16 processing centers and 2 control and measuring cars:

On this equipment it will be possible to carry out the production program which is carried out today by 155 units of the old equipment, the number of the main and auxiliary workers will be reduced by 59 people, the case spaces occupied today by machining shop (risk analysis is given by tab. 1) will completely be released [5,6].

Table 1. Risk analysis of the investment project

Type of risk	Assessment of probability of emergence of risk	Assessment of extent of influence of risk	Actions for decrease in risk
The risks connected with start and implementation of the project:			
- divergence of expected and actual volumes of material resources	Low	Low	Don't need
- excess of the planned expenses and prime cost growth	Middle	Low	Aren't required (increase in prime cost in the project can go only under the article "depreciation", but her size is more than four times blocked by economy under other articles).
- non-performance of the plan for power, volume of release, productivity	Middle	Middle	Production diversification, creation of an instrument-making cluster on the basis of the enterprise
- a delay of completion of works on start of the project	Middle	Middle	Fixing of responsibility through contracts
- limited availability of the earth, energy, building materials, raw materials, transport	Low	Low	Don't need
- limited availability of working resources, administrative shots, contractors	Low	Low	Aren't required (one of the project purposes – minimization of this risk)
- possible changes of cost of raw materials, materials, equipment, finished goods	Middle	Low	Aren't required (one of the project purposes – minimization of this risk)
Macroeconomic risks:			
- adverse development of a macroeconomic situation in the country	Middle	Middle	Doesn't give in to management
- adverse change of an exchange rate of ruble	Low	Low	Doesn't extent direct impact
- decrease in demand	Low	Low	Production diversification, creation of an instrument-making cluster on the basis of the enterprise
- toughening of procedure of pricing concerning productions/services of society	Middle	Middle	Aren't required (one of the project purposes – minimization of this risk)
Other risks:			
- non receipt of certification, the license in government bodies	Low	Low	Don't need
- bankruptcy of suppliers / contractors	Low	Low	Don't need

3 Calculation of efficiency of the project

Step 1: Investment (tabl. 2)

Table 2. Capital equipment.

№	Price, rubles	Amount	Value, rubles
1	6 780	2	13 559
2	8 051	2	16 102
3	8 898	2	17 797
4	15 254	2	30 508
5	12 712	2	25 424
6	25 424	2	50 847
7	33 898	4	135 593
8	5 085	2	10 169
		18	300 000

Step 2: Retrofitting of the capital equipment

In investment part of the project, except costs of acquisition of the capital production and measuring equipment stated above it is necessary to provide the costs of retrofitting of the acquired equipment connected with it, acquisition auxiliary the equipment and construction works in the case where it is supposed to place the modernized production sites.

Retrofitting of the equipment assumes single investments in a size 33 million 748 thousand rubles [11].

Table 3. Retrofitting of the equipment.

№	Option	Value, rubles
1	Retrofitting	33 748
	Total	33 748

Step 3: Service equipment

The service equipment assumes expenses of 17 million 119 thousand rubles:

Table 4. Service the equipment.

№	Option	Price, rubles	Amount	Value, rubles
1	Thermoins tallation	339	2	847
2	The case is tool	127	16	2 034
3	Compressor	847	2	1 695
4	Conditioner	847	15	12 712
	TOTAL			17 119

Step 4: Installation of the equipment

At determination of cost of construction works it is possible to use given on already carried out re-equipment of part of the case (installation of 5 units of the modern equipment similar to purchased) [7]. Production of construction works assumes investments of 9 million 831 thousand rubles:

Table 5. Investments into constructions work.

№	Option	Price, rubles	Amount	Value, rubles
1	Bases	508	10	5 085
2	Other construction works and materials	1 186	4	4 746
	TOTAL			9 831

Thus, the total need for investments for implementation of the project makes 360 million 697 thousand rubles, and need for their financing (including the VAT) – 425 million 622 thousand rubles.

Step 5: Single income

Except investments the present project assumes also the single income: sale of the released (written-off) equipment at the beginning of the project and sale of the acquired equipment at the end of the term of his operation [12].

The weight of the written-off machines makes 202 tons that at today's market price of scrap metal of 10 thousand rubles for ton (with the VAT) will allow to realize machines for the sum 1 million 708 thousand rubles.

Besides, at the end of operation term the acquired equipment can be realized for 20% of its initial cost that will make 60 million rubles [10].

Table 6. Income from realization.

№	Option	Price, rubles	Amount	Value, rubles
1	Sale of the equipment after operation term	x	x	60 000
2	Sale of the released equipment	8.47	201.525	1 708
	TOTAL			61 708

Thus, investments of the project taking into account the single income will make 298 million 991 thousand rubles.

The current income and expenses on the project.

The current income of the project consists, mostly, of the economy under articles of the current expenses of the enterprise connected with implementation of the project.

Table 7. Distribution of investments in time.

Items of expenditure / income	Year of implementation of the project				
	1	*2*	*3*	...	*15*
Expenses					
Equipment	100 000	100 000	100 000	0	0
Retrofitting	11 249	11 249	11 249	0	0
Service equipment	5 706	5 706	5 706		
Installation	9 831				

Income					
Sale of the released equipment	- 569	- 569	- 569		
Sale of the equipment after operation term					- 60 000
Total investments	126 217	116 386	116 386	0	- 60 000

Such articles are:
• Economy on a salary of the main and auxiliary workers of shop.
• Economy on repair of the equipment which is written off as a result of implementation of the project.
• Economy on the tool spent for work on the written-off equipment.
•Economy on the special equipment which was required for implementation of the production program on the written-off equipment.
• Economy on decrease in marriage [9].

Besides, the project assumes also increase of the income of the enterprise for the following directions:
The additional income from leasing of the floor spaces which are released as a result of implementation of the project.

The additional income (profit) on obtaining orders in cooperation [13].

The main operating costs of the project arise under the following articles:
• Expenses on the tool for work on the new equipment.
• Expenses on special equipment for the new equipment.

Except above-mentioned articles the project of replacement of the equipment mentions also other articles of the income and expenses. But, as the extent of economy, the income and expenses received on these to articles is rather small, we exclude them from consideration [17].

Economy. Salary of the main and auxiliary workers.

The staff list of the modernized shop looks as follows:

Table 8. Staff list.

Group/specialty	**Quantity**	**Salary, rubles**	**Annual salary, thousand rubles**
Management personnel shop	38	22.8	10 407
Main workers, all including:	44	31.6	16 710
operator-serviceman	16	30.0	5 760
foreman	2	50.0	1 200
mechanic	5	27.5	1 650
grinder	2	25.0	600
turner	5	37.5	2 250
milling-machine operator	5	37.5	2 250
workers of a foundry site	7	25.0	2 100
Cutting machine	1	37.5	450
turning and sentry	1	37.5	450
Auxiliary workers, in total	2	25.0	600
including:	2	25.0	600
Total	84	27.5	27 717

Actual data on the number and a salary of employees of machining shop for 2015:

Table 9. Total salary.

Group/specialty	**Quantity**	**Salary, rubles**	**Annual salary, thousand rubles**
Management personnel shop	38	22.8	10 407
Main workers	81	28.4	27 581
Auxiliary workers	24	21.3	6 122
TOTAL	143	25.7	44 110

Thus, the economy on a wages fund makes 16 million 393 thousand rubles a year, and taking into account taxes on a salary – 21 million 967 million a year.

The above-stated data paid off on the basis of 2015. At implementation of the project growth of a salary by 2,81 times in 15 years is considered (1 and 2 – on 10% of growth, from 3 to 5 year – on 20% of growth, from 6 to 15 year – on 3% of growth). Such assumptions are caused by the following reasons:

Continuous growth of deficiency of skilled workers of shots at simultaneous stagnation of system of their preparation is observed that causes higher growth rates of a salary in this segment of labor market.

Indicators of the enterprise for a salary of workers are at the moment lower the middle year.

Strategy of the enterprise has provided the essential growth of a salary in 2016-2018.

Fully the economy on a salary arises for the 3rd year of implementation of the project.

For the term of implementation of the project the economy of a salary and tax charges on a salary will make 684 million 201 thousand rubles.

Economy. Repair of the equipment.

For definition of economy under the article "repair and service of the equipment" data on services of non-productive departments to machining shop and costs of services of the third-party organizations carried on machining shop for the last 6 years have been analysed. The average size of expenses under this article for the written-off equipment has made 4 million 752 thousand rubles a year.

It should be noted the following facts:
• data of the last years are higher than the specified average size;
• eventually expenses on repair and service of the old equipment will grow;
• growth of the production program is predicted.

Owing to these factors in the project growth of costs of repair of the old equipment twice in 15 years is put that in opinions of experts is minimum possible growth.

Fully the economy on repair and service of the equipment arises for the 3rd year of implementation of the project.

For the term of implementation of the project the economy under the article "repair and service of the equipment" will make 130 million 039 thousand rubles.

Economy. Tool and special equipment.

Average costs of purchase and production of the tool and of production and repair of special equipment for the last 6 years have made about 500 thousand rubles and 1 million rubles a year respectively. The specified sizes belong to the written-off equipment.

Fully the economy on the tool and special equipment arises for the 3rd year of implementation of the project. The economy calculates taking into account the planned growth of the production program. [16]

For the term of implementation of the project the economy under the article "expenses on purchase and production of the tool" will make 12 million 440 thousand rubles, and under the article "expenses on production and repair of special equipment" - 24 million 879 thousand rub.

Income. Leasing of the released areas.

As a result of implementation of the project the case spaces occupied today by machining shop which make 3 825 sq.m. will completely be released. At present the area of this case can lease a minimum on 200 rub for 1 sq.m. a month. Respectively, the annual revenue will make at least 7 million 780 thousand rubles.

The income from delivery of the areas for rent arises for the 4th year of implementation of the project as time for write-off and export of the old equipment will be required.

For the term of implementation of the project the income from rent will make 93 million 356 thousand rubles.

Income. The income from obtaining orders in cooperation.

Park of the new equipment from the most modern machines which will be available the enterprises in 3 years after the beginning of the project, will allow to solve many production and technological and many other instrument-making enterprises of the country and city. The solution of a question of development production has only organizational restrictions for today especially as development of such cooperation is provided as the operating strategy of development of the enterprise. [18]

If to assume that the enterprise will be able to allocate about 3% of capacities of the machines which are purchased on the project (4 200 hours of useful work year) and 25% of profitability at production cost of release, then the income of the project under the article "cooperation" (net profit from orders in cooperation) will make 3 million 146 thousand rubles a year. According to the experts this size is minimum possible [8,9].

This type of income is provided, since 4 years of implementation of the project.

For the term of implementation of the project the income under the article "cooperation" will make 37 million 968 thousand rubles.

Expenses. The tool and special equipment for the new equipment.

Expenses on the tool for the new equipment are calculated on the basis of tool expenditure statistics on the operated modern equipment similar purchased, and make 285,7 thousand rubles on 1 machine a year.

Expenses on special equipment paid off proceeding from set service life the specialist of equipment in 5 years and costs of a set, a component of 33 million 748 thousand rubles.

Expenses on the tool and special equipment during the project are taken taking into account growth of the production program twice for the 5th year of implementation of the project.

For the term of implementation of the project expenses on the tool for the new equipment will make 125 million 108 thousand rubles, and on special equipment for the new equipment – 158 million 642 thousand rubles.

4 Conclusion

• Thus, the project has the following indicators of efficiency:

• Profit (NCF) of 402 million 776 thousand rubles.

• Profitability of investments of 135%.

• Payback period of 9 years from the moment of the beginning 6 years from the moment of an exit at full capacity.

Internal standard of profitability of 11,8%.

The provided analysis shows that the project is favorable to the enterprise.

Depreciation of the new equipment.

In the analysis of efficiency of the project it should be noted that except articles of economy and the income given above, in the project there is one more essential plus – the depreciation which is saved up during operation of the new equipment which will allow to update further the equipment on a regular basis without attraction of essential investments from profit and/or from outside.

Depreciation on the economic essence isn't an expense - it is part of prime cost, transfer of already made investment expenses in prime cost small parts for the accounting of these expenses in production price. On an economic essence depreciation is an income, the fund formed by the enterprise for purchase of fixed assets [19].

In the considered project increase in depreciation charges won't affect prime cost and profit of the enterprise as increase in prime cost within 3-4% is compensated by decrease in the current expenses. Therefore, any outflows of money, in connection with increase in depreciation charges in calculation of efficiency it isn't necessary to provide [15].

Everything for the project 312 million 535 thousand rubles will be added to revenues.

If to consider depreciation of the equipment in efficiency calculation, then we will receive the following results:

• Profit (NCF) of 715 million 311 thousand rubles.

• Profitability of investments of 239%.

Payback period of 6 years from the moment of the beginning 3 years from the moment of an exit at full capacity

Internal standard of profitability of 23,4782%.

Payback and internal standard of profitability of the project

The payback period of the project is defined by summation of investments of the project and a flow of the income of the project. Summation is carried out by every year by the accruing result, investments of the project are considered as negative sizes, return of investments – as positive [14].

References

1. *A Guide to the Project Management Body of Knowledge (PMBOK Guide)*, (Project Management Institute Publ., 2013).=
2. Review of Approaches in Modeling Adaptive Business Process, Proc. of the 10th Conference of Open Innovations Association FRUCT and the 2nd Finnish-Russian Mobile Linux Summit, 24 (2011)
3. I.A. Brusakova, *Proc. Modern Problems of Management*, 3 (2015)
4. I.A. Brusakova, M.A. Kossukhina, *Proc. modern problems of applied informatics*, 6 (2012)
5. I.A. Brusakova, M.A. Kossukhina, *Proc. SCM* (2015)
6. D. Katalevsky, *The Basics of Simulations and System Analysis in Management* (Delo Publ., Moscow, 2015)
7. H.J. Zimmermann, *Fuzzy Set Theory and Its Applications* (Kluwer Academic Publishers, Boston, 1996)
8. Avalaible: http://www.metodolog.ru.
9. I.A. Brusakova, *Proc. 2015 IV Forum Strategic Partnership of Universities and Enterprises of Hi-Tech Branches*, 77 (2015). DOI: 10.1109/IVForum.2015.7388260.
10. I.A. Brusakova, V.I. Fomin, M.A. Kossukhina, S.N. Panin, *Research of development trends of information management in modern conditions* (Publishing house of Saint-Petersburg University. Univ of Management and Economics, St. Petersburg, 2014)
11. I.A. Brusakova, Proc. 2015 IV Forum Strategic Partnership of Universities and Enterprises of Hi-Tech Branches, 77 (2015). DOI: 10.1109/IVForum.2015.7388260.
12. K. Beck, *Extreme Programming Explained: Embrace Change* (Addison-Wesley Professional Publ., 1999)
13. T.A. Khudyakova, A.V. Shmidt, Mediterranean Journal of Social Sciences, **6(4)**, 274 (2015)
14. M.A. Kossukhina, *Proc. of the 12th Conference of Open Innovations Association FRUCT and e-Tourism seminar*, **41** (2012)
15. M.A. Kossukhina, *Proc. Modern Problems of Management*, **55** (2015)
16. M.A. Kossukhina, H. Aminov, *Proc. of the 12th Conference of Open Innovations Association FRUCT and e-Tourism seminar*, 37 (2012)
17. S.N. Ostapenko, N.Yu. Fedoseyev, *A methodical grant for manufacturing small and medium-sized enterprises for questions of modernization and modernization* (Moscow, 2010)
18. O.A. Shvetsova, News of LETI, **3**, 75 (2015)
19. O.A. Shvetsova, News FINEC magazine, **4(56)**, 65 (2008)
20. O.A Shvetsova., I.A. Brusakova, A.V. Zvontsov, I.G Fomina, Proc. SGEM, 47 (2015)
21. T.A. Khudyakova, A.V. Shmidt, Mediterranean Journal of Social Sciences, **6(4)**, 274 (2015)

Applied economic model development algorithm for electronics company

I. Mikhailov[1,*]

[1]Russian Scientific Research Institute "Elektronstandart", Saint Petersburg, Russia

Abstract. The purpose of this paper is to report about received experience in the field of creating the actual methods and algorithms that help to simplify development of applied decision support systems. It reports about an algorithm, which is a result of two years research and have more than one-year practical verification. In a case of testing electronic components, the time of the contract conclusion is crucial point to make the greatest managerial mistake. At this stage, it is difficult to achieve a realistic assessment of time-limit and of wage-fund for future work. The creation of estimating model is possible way to solve this problem. In the article is represented an algorithm for creation of those models. The algorithm is based on example of the analytical model development that serves for amount of work estimation. The paper lists the algorithm's stages and explains their meanings with participants' goals. The implementation of the algorithm have made possible twofold acceleration of these models development and fulfilment of management's requirements. The resulting models have made a significant economic effect. A new set of tasks was identified to be further theoretical study.

1 Introduction

Electronic components serve as the basis for electronic systems and in case of the need to create devices for special applications it must pass a series of tests to confirm its compliance with the requirements for this kind of products. Due to the strong growth of modernization projects number in the Russian military–industrial complex and an increase in the number of newly developed devices, as well as tightening their requirements, more and more complicated test conditions, during the last years re-emerged and grew up strong demand for the methods and algorithms those could help to simplify an application development in the field of applied decision support systems. The main goal is to cover the lack of justification and give an evaluation for making administrative solutions more efficiency or at least ease the process of their search.

The probability of making a wrong management decision depends on the degree of conditions uncertainty. The extent of the negative effects rises over time and has continuous increase the longer it is fixed and supported in the course of business. At the intersection of these two critical moments of greatest probability of making a wrong decision and the toughest possible consequences in a case of testing electronic components is the time of the contract conclusion. The most unpredictable and difficult tasks at this stage are future work costs definition, as well as duration justification. To perform the task of determining required time with using native methods, there is an alternative to the following extremes: overstatement of the projected period will lead to the impossibility of finding an agreement with the customer and the loss of a potential client, and when specifying a too short time interval, the danger lies in the inability to carry it out, which can lead to a significant leadership responsibility, especially in the case of a contract in the interest of the state defense ministry.

At the contract conclusion time is difficult to achieve a realistic assessment of time-limit due to the presence of not only the established norms of terms (standard documentation requirements, the pure real time needed to perform machine measurement), but also less predictable components: the duration of the development and establishment of a special measuring equipment, the ability to buy its indispensable equipment, the duration of their delivery, terms of writing special measuring software. It is important to emphasize that these variables must be applied to evaluation amount of time for each incoming product, while in one contract, often indicates the wide range of part types: the former is already partly on the tests and do not require re-preparatory work.

The problem of rational contract price definition is as well not trivial. Cost of the work is formed by a number of terms that depend on two main components: the cost of the components – printed circuit boards, contact devices, as well as the wage bill. Because the contract is concluded and price will be fixed before the actual start of work and receiving of advance payment, by the time of the purchase the most part of foreign components, due to changes in currency exchange rates may rise unpredictable in price, in depend on the political and general economic conditions. Justified estimation of wage bill size is also difficult due to the prevailing structure of work, which largely consists of engineering research, but not of handling and service directly involved measuring and test equipment.

Possible way to solution the above-mentioned complexity lies through creating a model that allows assessing the amount of work for the planned task. That way it is possible to reduce the degree of uncertainty in determining the cost and terms of a projected contract. Therefore, for the construction of such models was

* Corresponding author: igor.mikhailov.petersburg@gmail.com

formulated the sequence of work – in the title of this article mentioned algorithm that was successfully estimated through developing and implementing two modules of information support.

2 Methodological foundation of the algorithm construction

In last years was adopted a series of more than twenty documents, similar to the requirements established by International Electrotechnical Commission [1] in the field of testing of electronic components at the state level standard of the Russian Federation. Meanwhile the current approach to the assessment and valuation amount of labor significantly outdated [2]–[5] and has not been modified since the mid-90s. Since that time, much has changed: not only listed above testing requirements and used for this purpose equipment, but also the whole range of tested products. Because of the transition to a market economy the enterprises management conditions have shifted radically, as well as in connection with a wide spread of information technology was adjusted the principle of collecting, processing and storage of information required to manage the company.

Described in the standards and published in the end of the nineties years XX century approach inherits the features of a planned economy and universalistic attitude to management [6] and tries to establish an interbranch, cross-sectoral, general standards for all kinds of businesses. Shown in this article way of thinking and methodologically it generated algorithm is premised on the dialectical opposites [7] – the existing level of individual business development and calls for the integration of business processes specific test centers and laboratories. It should be noted that this approach to consider "specific cases" is gaining more and more widely accepted in all areas of modern economic and managerial science [8]–[11]. In addition for establishing the algorithm were used the principles laid down in the state standards of the Russian Federation [12], [13], the general methods of scientific research, typical of logical positivism and the heritage of Ernst Mach philosophy of science [14].

3 The algorithm for creating the analytical model amount of work estimation

Step 0: Managerial purposes understanding

This step is not an actual part of the development algorithm, but we felt it necessary to remind that before the start of quite expensive and not always simple works toward model creation it is worth to identify the existing problems and arrange the tasks list that a system must be able to accomplish at the final stage. Only if these purposes cannot be achieved in a simpler way, we may recommend to create the model using this algorithm.

Step 1: Business process scheme drafting

The first step in creating of the analytical model is drafting of "process flow" and "object transition" IDEF3 diagrams [15] for the business processes already existing in the organization. Using the notation IDEF0 [16], as expected, it turned out to be arduous and created difficulty interpreting at complication of the model structure. At this stage along with the business analysts must be joined line managers, who are responsible for reviewed organization divisions.

As a further development of this stage, it seems appropriate to recommend an assessment of the potential benefits of replacing used IDEF3 approach to active developing [17]–[20] BPMN notation, which is due to greater visibility must be well understood not only by business analysts and attracted engineers as experts, but also by all levels of management.

Step 2: Operations lists filling

The second step is reconciliation with engineers-experts, who daily implement instruments on labor intensity. Existing documents and compiled at the first stage in conjunction with line managers business process diagrams must be discussed. It is figured out, that even linear managers which are nearest to real operating processes often ignore or even doesn't know about some important obstacles that could occur during the everyday's work. Therefore, list of operations that was outlined at the first stage together with the managers must be checked to identify unaccounted, unknown or considered by management as minor works. In some cases, these hidden works can be critical in future for the process flow control. It is required to set the execution time for each operation and the conditions under which it is performed, based on the evidence of experts, to have an instrument to branch business process in a different way if needed.

The involvement of staff in this step is justified from a number of reasons: professional with extensive practical experience is an expert in his field and the best source of information. Second, it ensures the absence of opposition from the ordinary performers, no disturbance at the workplaces, the openness and clarity of procedures carried out – which does not allow developing a conflict of blue- and white-collar workers. This assists achieving the involvement of staff – one of the requirements of modern view on efficient management [21], and helps to realize the interviewed experts needs of high estimation and self-actualization [22], [23] – that can serve as an indirect non-material encouragement and recognition on the part of management. The main actors of this stage along with the business analysts are engineers-experts.

Step 3: Gathered data verification

The third step is verification of in the previous stage obtained from the experts' knowledge. This phase aims to perform two tests: the logical and empirical. The first is to eliminate logical and technology contradictions in reconciling the recorded responses through to a re-consultation or referral to an independent third party. Its result is the business process scheme, including the sequence of factually performed operations, consistent and clear to management.

The second, empirical test is needed to clarify by performers' posed time-standards. Obviously, the number of workers will tend to inflate their estimation of the time required to perform discussed task. Such a reconciliation of the data can be carried out through statistical accounting of data for previous periods. Relatively simple marks about needed periods of time for operations implementation, collected as a "black boxes", without any kind of their nature explanation, at this stage can found perhaps the only justifiable practical application. The leading person at the third stage is a business analyst, which if necessary, can engage consultants to address the issues with respect to the logical or temporal mismatch identified in the available data.

Step 4: System complexity determination

The fourth step is to choose a rational number of dimensions involved as multipliers in the system of polynomial functions necessary to calculate amount of work.

The number of dimensions for each work considered separately influences on results reliability and depends on accepted by senior management degree of abstraction. The upper limit of the relationship kinds and business logic system complexity should be consistent with the business informatics experts, whom going to be involved at a later stage for the software implementation of the model. Consequently, fourth stage participants are business analysts, senior management and business informatics experts.

It is important to stress despite the fact that the algorithm realization experience showed that the information obtained from the experts, often in their complexity and number greatly exceeds the amount of details required to implement management's needs to value prediction. However, it should be borne in mind that at this stage a case of special requirements extension on the part of senior management is not excluded at all. In case of their appearance, it would be necessary to get more details of working processes applied to the selected organizational unit, which could lead to a new iteration of the algorithm.

Step 5: Program implementation

The fifth stage contains the program implementation of the compiled information complex, which usually consists of united relationships between gathered process parameters, based common system of polynomial equations. Actual software developing project can be implemented using a variety of approaches [24]–[27], based on the use of general-purpose programming languages, databases, distributed web interfaces. The main participants at this stage are business informatics experts and programmers.

4 The algorithm implementation's results

The algorithm was tested twice to create analytical models for calculating the cost and amount of works carried out by Test Center of Russian Scientific Research Institute "Elektronstandart" (RNIIES) and electronic components assembly department of RNIIES.

The applying of the algorithm have made possible the creation of these models, as well as more than twofold acceleration of their development time, compared with the projected period, calculated by consulting companies for introduction of the DSS and ERP-systems. The result of the integration into business practices was raising the methodological reliability of the labor amount estimation, and therefore, the accuracy of prediction of the organization labor costs, nearly 30% increase in the accuracy of the timing of planned activities. The calculations term for the contract conditions dropped from ten to one day, the influence of the human factor was reduced, arbitrary interpretations in assessing the complexity of the work was excluded.

5 Conclusion

This paper has introduced the five stage algorithm to development of applied economic model that was successfully implemented in electronic company. We assume that the sphere of usage for this algorithm could be expanded towards other branches if for them will remain true the main condition based on using of high skilled workers' experience as company's intellectual capital to improve management efficiency.

Pilot testing of the algorithm has identified a new set of tasks to be further theoretical study. The question was raised feasibility of expanding the algorithm by adding the simulation components [28] that extends the functionality of the analytical model through including of production planning subsystem. Besides, the use of the algorithm provides a basis to initiation of an analysis and a classification of works that must lead to rational choice between analytical, simulation and hybrid models most appropriate for selected kind of investigated working activities. Inspired by the perspectives of further increasing model's functionality, we plan to consider methods to implement neural nets for forecasting of planned works and costs optimization to our algorithm in the future work.

References

1. Series of Russian national standards GOST R IEC 60068; GOST R IEC 61587; GOST R IEC 61508; GOST R IEC 60906
2. Resolution of the Russian Federation Ministry of Labour, **56** (1995)

3. Resolution of the Russian Federation Ministry of Labour, **72** (1994)
4. Resolution of the USSR Ministry of Labour, **69** (1991)
5. Resolution of the USSR Ministry of Labour, **80** (1991)
6. A. Bogdanov, *Essays in Tektology: The General Science of Organization* (Intersystems Publications, Seaside, CA, 1980)
7. K. Hartmann, *Hegels Logik* (De Gruyter Publ., Berlin 1999)
8. E.H. Nathanael, B. Hendradjaya, W. Danar Sunindyo, *ICEEI*, 427-432 (2015). DOI: 10.1109/ICEEI.2015.7352539.
9. I.A. Brusakova, *Proc. IV Forum Strategic Partnership of Universities and Enterprises of Hi-Tech Branches*, 77-79 (2015). DOI: 10.1109/IVForum.2015.7388260.
10. A. Esmaeilzadeh, A.A. Taleizadeh, Journal of Industrial Engineering International, **12**, **1**, 119-135. DOI: 10.1007/s40092-015-0135-5.
11. M.A. Kosuhina, I.A. Brusakova, *Proc. XVIII International Conference on Soft Computing and Measurements (SCM)*, 291-293 (2015). DOI: 10.1109/SCM.2015.7190487.
12. Series of Russian national standards GOST R ISO 9000.
13. Russian national standard GOST R 56407-2015
14. E. Mach, *Knowledge and Error: Sketches on the Psychology of Enquiry* (1976)
15. R. Mayer, M. Painter, C. Menzel, B. Perakath, P. deWitte, T. Blinn, *IDEF3 Process Description Capture Method Report*, (Knowledge Based Systems Publ., 1995)
16. National Institute of Standards and Technology (1993)
17. Ching-Hong Tsai, How-Jen Luo, Feng-Jian Wang, *Proc. 11th IEEE International Workshop on Future Trends of Distributed Computing Systems* (2007). DOI: 10.1109/FTDCS.2007.17.
18. R. Braun, *Proc. IEEE 17th Conference on Business Informatics (CBI)* (2015). DOI: 10.1109/CBI.2015.41.
19. S. Yamasathien, Proc. Fourth International Conference on Digital Information and Communication Technology and it's Applications (DICTAP) (2014). DOI: 10.1109/DICTAP.2014.6821684.
20. M. Chinosi, A. Trombetta, IEEE Conference on Commerce and Enterprise Computing, (2009). DOI: 10.1109/CEC.2009.48.
21. Toyota Code of Conduct, Toyota motor corporation Publ. (2006)
22. A.H. Maslow, Psychological Review, **50(4)**, 370-396 (1943)
23. A.H. Maslow, *Motivation and personality*, (Harper & Row Publ., 1954)
24. J.O. Long, *ITIL 2011 At a Glance*, (Springer New York Publ., 2012). DOI: 10.1007/978-1-4614-3897-7.
25. *A Guide to the Project Management Body of Knowledge (PMBOK Guide)*, (Project Management Institute Publ., 2013)
26. T.A. Khudyakova, *Proc. of the International Congress on Interdisciplinary Behavior and Social Science*, 292-294 (2015)
27. K. Beck, *Extreme Programming Explained: Embrace Change*, (Addison-Wesley Professional Publ., 1999)
28. D. Katalevsky, *The Basics of Simulations and System Analysis in Management* (Delo Publ., Moscow, 2015)

Small industrial enterprises in Russia: survival in conditions of hypercompetition

E.D. Vaisman[1,*], and *M.V.* Podshivalova[1]

[1]South Ural State University, Chelyabinsk, Russia

Abstract. Nowadays, we can already see that Russian industry steady transition into postindustrial economy causes the development of hypercompetition in a number of sectors. The study focuses on the main characteristics of hypercompetition as the consequence of traditional industrial enterprise environment characteristics changing from bad to worse. The authors also provide the results of degree evaluation of hypercompetition resistance of small enterprises in 5 main industries of manufacturing sector. It is also shown that the small firms that produce electrical equipment have a higher probability of success in protecting their competitive advantages. Pulp and paper production as well as printing and publishing industries have the lowest chances of success. To protect enterprises from hypercompetition, the authors offer a model of enterprise's dynamic capability development on the basis of resource-based view. From this point of view, they classify enterprise's resources according to the degree of their influence on the competitive abilities of the latter. More-over, an approach to the evaluation of position of the point of making proactive management decision is formed.

1 Introduction

Over the last years, scientists have actively been discussing the problem of searching fundamentally new industrial enter-prise management techniques due to significant environment changes.

L. Makarov states that "The problems with transformation of Russia's enterprise management are quite common, and thus, the administration of many domestic - owned firms should analyze how the character of strategic management corresponds to the requirements of forming postindustrial and global economic environment" [1].

Postindustrial economy with its Knowledge economy (KE) that nowadays is recognized as the highest stage of development causes a number of changes, among which innovative development and globalization process play a core role. J. Schumpeter wrote long ago that creative destruction is peculiar for innovation decisions [2], and in the conditions of rapid development of Information Technologies (IT) and globalization, wide spread of new information that prompts such destruction takes place.

All these changes led to emerging of such phenomenon as global competition or hypercompetition. The term was introduced by Richard D'Aveni. According to him, the latter is characterized by "constantly increasing competition in form of rapidly appearing product innovations, decreasing time for research and development, aggressive price and competency competition as well as experimenting with new approaches to meet consumer requirements" [3]. According to the experts' evaluations, Russia will face with hypercompetition in the majority of its markets in mid-term perspective, but today some industries are already experiencing its threats. The aim of our research is to assess the resistance potential to hyper-competition threats in the sector of small manufacturing enterprises as well as to develop the directions of management improvement considering the characteristics of hypercompetition.

2 Importance of the issue

The importance of hypercompetition threat proves the traditional analysis of external environment state from a perspective of its four characteristics: interconnectedness of environment factors, its complexity, mobility and uncertainty. Let's characterize within this framework the current changes peculiar to postindustrial economy and reflecting, according to the majority of the specialists, currently emerging characteristics of hypercompetition.

The first characteristic, interconnectedness of environment factors, is connected with reciprocal effect of the latter on one another. As defined (for example, [4]) it is a level of the power with which the change of one factor effects the others. Enhancement of factor interconnectedness turns the environment of today's enterprises into aggressively changing one.

At the moment the growth of interconnectedness of environment factors led to technological convergence. The brightest examples of interdisciplinary research in the past were The Manhattan Project, the USA, and nuclear project in the USSR. Today such research plays a particular role as it is recognized that interpenetration

* Corresponding author: vaisman_elena@mail.ru

will significantly speed up the processes of innovation emergence. Thus, in the EU the Seventh Framework Programme for Research and Technological Development (2007-2013) the idea of convergence formed the basis for support of nanoscience and nanotechnology, further development in IT field and new technologies of information society, biotechnology and ecology. The research of convergence and non natural sciences has been planned, in particular, of enhancement of prognostic activities in social sciences, expansion of research in the fields of ethical implications in computer technologies [5]. The increase of interconnectedness of environment factors led to emergence of competition in some activities, particularly connected with high technologies. It begins to appear simultaneously on different levels: product markets; in resource area; between different entrepreneurial concepts; as a part of enterprise integration, etc. This led to forming such characteristics of hypercompetition as its multi aspect character.

The second characteristic of the environment is complexity which is a number of factors on which the enterprise must react as well as the degree of variability for each of them. Increase of environment complexity is connected both with the development of market institutions meaning the growth of the enterprise contacts, number of transactions, use of more complex, various and more rapidly developing technologies, its expansion on the market, and with virtually constant changes of the "rules" of doing business, that is particularly true for Russian economy. The complicacy of the environment led to such characteristics of hypercompetition as aggressive competitors' behavior, mostly, in price formation [6]. Along with the aggressive price formation a number of goods and technologies imitations are growing; new production processes, brands and competent employees are growing out of control of enterprises.

The third characteristic is mobility that is the speed with which the changes occur in external environment of an enter-prise. The rapid development of informational support is inevitably leading to the growth of environment mobility. It should be noted that the complexity of the decision making in such conditions is also connected with the fact that the speed of changes is different in different industries and even in different subdivisions of the same enterprise. All this causes the change of the structure of enterprises, closing one businesses and opening the other, rapid assortment turnover of enterprise products, reduction of life time of both products and enterprises. Thus, the analysis of the changes in the list of top 100 US enterprises with the biggest capitalization over the last fifteen years has shown that for the first five years their number de-creased on 24 enterprises, for the second - on 26 and for the third - on 41 [7]. All this is the development of such characteristic of hypercompetition as market development dynamics which caused the significant decrease - maximum to 2 years - terms of forecasting activities.

The fourth characteristic of enterprise's external environment is its uncertainty depending on the number of information that an enterprise has about this or that factor on the one hand, and the reliability of this information on the other one. Today's business faces with the increasing degree of un-certainty of the environment among other things because of globalization, and consequently the complexity of effective management decision making is growing. Thus, according to P. Trist, "surviving is becoming fatefully connected with the knowledge level of the organization about its environment [8]. The growth of environment uncertainty appears in the next characteristic of hypercompetition - reduction of life time of the competitive advantage and simultaneous coverage of several fields by competition: expenses, quality, terms, market barriers, "know-how", etc.

It is clear that the characteristics of hypercompetition appear, firstly, to a greater or lesser extent depending on the industry and type of activity. Secondly, this or that characteristic in each industry can be more or less expressed. In any case, hypercompetition significantly complicates functioning of the enterprise and makes it search new methods and management tools as the traditional tools of strategic management in this case are not effective.

3 The assessment of small industrial enterprise hypercompetition resistance potential

One of the issues of enterprise survival in hypercompetition conditions is the business scale. Alvin Toffler in his 1980 book The Third Wave advanced an idea on ten key elements of new economy of postindustrial world, among which, apart from the increasing role of knowledge, innovation, system integration, escalating competition, he noted the significance of small-scale enterprises. As far as postindustrial society is developing, large enterprises, according to Toffler, will be getting out by small ones [9]. Our government, apparently, thinks that small enterprises must become a more sustainable power. Thus, according to the Concept of Long-Term Social and Economic Development of Russian Federation until 2020 to conduct structure changes of production to the benefit of competitive and high-tech sectors, we need to form "the powerful layer of small and medium enterprises playing the significant role on the global markets" [10]. Let's give a closer look at the current state of small manufacturing enterprises from this perspective as the main "powerhouse" of structure changes in industrial field and the growth of economy competitive ability.

The data of the official statistics of Federal State Statistics Service, economic reviews on small enterprises of the RF and its entities, the overall results of sampling analysis and expert estimations as well as the information of international statistics show that domestic small business in manufacturing industry cannot be considered developed. Thus, according to Federal State Statistics Service's estimates, the percentage of small and medium manufacturing enterprises in SME structure since 2008 has been changing insignificantly and accounted for, as a rule, no more than 10%. In comparison with European countries the number of such enterprises in 138-million Russia is in 2 times lower than, for example, in Italy (61 mln. population) [11].

Another big difference of Russian SME sector of manufacturing industry is extremely low percentage of micro-enterprises (20 people maximum) - less than 20%, whereas in European countries their percentage is not lower than 80%. According to Federal State Statistics Service, the percentage of small enterprises with maximum 9 people is in total number of SME in the last 10 years was not lower than 90% [12].

The contribution of SME into manufacturing industry of the RF accounts for 21%, whereas in EU countries this index is in the range of 46%-60%, and 80% in Republic of Korea. The SME contribution to industry sales is minor: Russian SME provides only 16%, whereas EU countries -in 2-3 times more [13].

Regional differentiation that makes Russian small business different as a whole is peculiar for small manufacturing enterprises. Thus, according to Federal State Statistics Service, there are 40% of all small manufacturing enterprises in nine Russian regions - Saint Petersburg, Moscow, Moscow Region, Sverdlovsk Region, Novosibirsk Region, Republic of Tatarstan, Rostov Region, Samara Region, Krasnodar Territory, furthermore, a half of this share is in Moscow, Moscow Region and Saint Petersburg [11].

In total, financial condition of small manufacturing enter-prises can be characterized as unstable. Mid-year sales growth rate in the period of 2010–2013 was gradually decreasing from 13% to 5% considering mid-year inflation rate in this period 6.6% per annum. Taking into account the dynamics of employment and fixed investment, we can say that there is no real expansion of production in the industry for the last 5 years observed. Every fifth small manufacturing enterprise is unprofitable. Low profitability of assets is preserved against the background of considerably high level of financial dependence and comparatively low indices of Current Ratios that on the whole is peculiar for small enterprises due to the financial resources scarcity and their need for "tax optimization" [11].

Among manufacturing sector, small enterprises are more developed in food industry, metallurgical production, pulp and paper, machinery and equipment production as well as electrical equipment production. Together these 5 industries provide more than a half working places in manufacturing sector of small business and give more than a half of the country's sales. The structure was unchanged during 2011–2013. We use Entrepreneur Confidence Index (ECI) of small industrial enterprises for the indirect assessment of environment uncertainty degree. According to Federal State Statistics Service, among the five key industries of manufacturing, optimistic estimates for the last 5 years were observed only in electrical equipment production, benign optimism - in food production. Chronic pessimism was observed in machinery and equipment, pulp and paper and metallurgical production [11].

The difficulty in functioning of small industrial enterprises on the medium term perspective will be worsened not only by threats of increasing hypercompetition but also by the institutional constraints peculiar to national economy. Let's take a closer look at the latter.

1. Administrative barriers. The presence of this type of constraints is recorded by many all-Russian and regional surveys of small and medium enterprise managers [11, 13, 14, 15, 16, 17]. According to OPORA Russia (All-Russian NGO of Small and Medium-sized Businesses), one third of small manufacturing enterprises consider administrative barriers such a heavy burden that it is impossible to adapt to them. Every fifth manager says about the pressure from inspectors that "they don't allow business to keep afloat and develop". About one fourth of the surveyed consider venality to be a usual thing in getting a state order, being inspected, getting the land, getting state aid and joining infrastructure [13].
2. Undeveloped infrastructure including the support infra-structure. According to OPORA Russia survey, every third small manufacturing enterprise notes that the transport infra-structure of the region has either a lot of drawbacks or it doesn't correspond to business requirements at all. Small enterprises also see major obstacles for development in energy infrastructure as for the majority the connection to the electricity grid is not available and the electricity tariffs are very high [13].
3. Taxation system. The problem of high taxes is, probably, the most paradoxical and popular at the same time. By calling it popular we mean its high ratings in top 5 key problems of domestic small business within 20 years of its development. By calling it paradoxical we mean that judging by Index of Economic Freedom tax burden in Russia for the last 20 years is significantly lower than on the average in the world and in many developed countries in particular [18]. This paradox can be explained by the fact that a particular tax payer is interested in the burden not on the macro level, but its assessment in a particular case. For this very reason according to Federal State Statistics Service's survey in 2010–2013, on the average every forth small manufacturing enterprise considered taxation level to be very high [11].
4. Low demand is one of the monumental and deep-rooted problems of Russia's small business, in production sec-tor in particular. The problem of lack in demand was and still is pressing, mainly concerning domestic market as only few small firms as a rule mastered the international market. Therefore, there is a tendency of worsening the situation. Thus, ac-cording to Federal State Statistics Service in 2002–2003, only 38% of small enterprises surveyed considered that the factor had limited the production growth in industry, in 2004–2008 -41% of the surveyed thought the same way and in 2010–2013 - every second manager of a small manufacturing enterprise (in recession year 2009 this index reached 61%) [11]. According to OPORA Russia survey, production distribution is a certain problem for small manufacturing enterprises as the channel of distribution access is limited for many of them [13].

Taking part in the state procurement as a way to form inner demand on the products of domestic SME cannot be named common nowadays. Thus, according to Federal State Statistics Service, among SME entities in 2009–2013 on the average no more than 17% of all orders were in the form of price indication request. In other variants of Federal Acquisitions this index was 2-3 times lower [11].

5. Low investment activity. Chronic lack of demand on the products and services of small industrial enterprises is one of the main reasons of lack of equity funds to expand the output and innovation. From 2006 to 2013 no more than 40% of the surveyed managers noted the lack of financing as a factor limiting the growth of manufacturing production [11]. However, in recent regional surveys of small business owners, the image is more depressing. Thus, for example, in the Republic of Tatarstan 80% of the surveyed announced about the lack of effective interaction between credit- and banking system and small business, artificial and unmotivated obstacles to get bank loan [15].

We can generally say that the perspectives of Russia's small industrial enterprises survival in the conditions of in-creasing hypercompetition and existing institutional constraints will be defined by its investment and innovation activity.

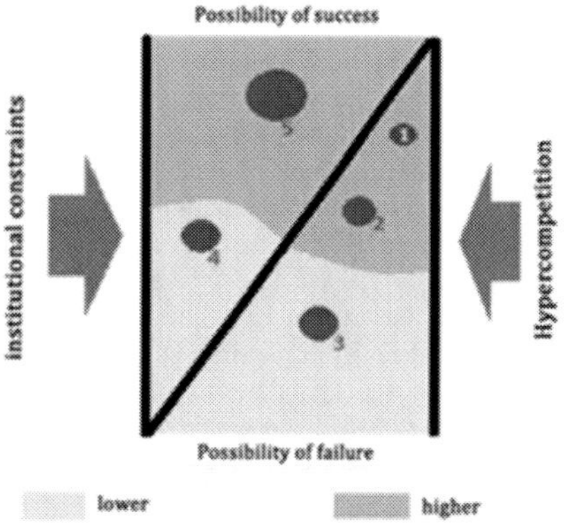

Fig. 1. Small industrial enterprises in the "clutches" of constraints and threats. Note: 1 – pulp and paper production, printing and publishing, 2- metallurgical production, 3 – food production, 4- machinery and equipment production, 5 – electrical equipment production. The diameter is calculated as the ratio of percentage of enterprises introducing technology innovations of the corresponding industry to the similar index of pulp and paper industry.

As for the investment activity of small enterprises, in manufacturing it is focused, mostly, on providing simple reprocessing, and insignificant extensive growth in the sector is not based on innovations. Thus, according to Federal State Statistics Service, 26% of all small enterprises invest in increase of their production capacity, 19% in change of worn-out equipment, and only 5% in introduction of new manufacturing technologies [11]. Analyzing the data of official statistics, we cannot call innovation activity of small manufacturing enter-prises as a whole and in particular in 5 main industries, high. Thus, no more than 5% of small manufacturing enterprises introduce technological innovations, and the percentage of innovation products is insignificant (no more than 2%). Product innovations were lower than industry average level in pulp and paper, food and metallurgical production, higher in machinery and equipment production. And only one industry – electrical equipment production – had innovative activity significantly higher of the average not only among manufacturing but other types of production. There are significant differences in patterns of spending on technology innovations. Thus, in food industry more than 80% of all technology innovations were connected with the purchase of equipment, in pulp and paper – 71.6 %, metallurgical – 52 %. In machinery and equipment production as well as in the production of electrical equipment the spending on research and development of new products (about 39%) prevailed. It is worth noting that this index was also comparatively high in metallurgical production – 20.5% [11]. Thereby, we can summarize that the quality and depth of innovational processes in machinery and electrical equipment production is significantly higher than on the average in the sector analyzed.

Thus, we can talk about the significant potential of successful functioning in the conditions of hypercompetition only considering one industry of small manufacturing enterprises – the production of electronic, optic and electrical equipment. The outsider of manufacturing industry is pulp and paper production, publishing and printing industry. To be fair, we also note that considering the latter such state of business can be also explained by its position on the final stage of its life cycle. Machinery and equipment production, metallurgical production despite the sustainable entrepreneurial pessimism, introduce technological and product innovations approximately on the industry average level. Food production that has every fifth ruble from the whole distribution of manufacturing SME is characterized by benign optimism and innovation activity is below industry average level.

Hence, the probability of success/failure in the "clutches" of hypercompetition and institutional constraints is differentiated in the industries (fig.1). A diagonal and four colored areas on the picture show the degree resistance capacity: high probability of failure (pulp and paper, printing and metallurgy), low probability of failure (food industry), low probability of success (machinery-producing industry) and high probability of success (electrical equipment). It is notable that in Failure zone we included the industries with probability of successful resistance to the threats from hypercompetition and institutional constraints from 0 to 50%. In Success zone – from 50 to 100% respectively.

4 Development of enterprise's dynamic capabilities: resource-based approach

If institutional constraints is a systemic factor, uncontrolled from the side of the business, the survival in the conditions of hypercompetition depends on the efficiency of industrial enterprise management system. As the majority of specialists note, the tools and methods of the strategic management in these conditions become virtually useless and success/failure in a competitive struggle depends only on the speed of assortment renewal, assimilation to innovation products and technologies, decision making how to destroy the competitor's advantages. Such management corresponds to the modern concept of dynamic capabilities.

If to accept that dynamic capability of an industrial enter-prise appears in its capability to form, develop and rearrange resources in accordance with the opportunities and environment threats [for example, 19, 20], then, considering the characteristics of hypercompetition, it is quite logical to classify all business resources into three groups according to the subject of their influence to competitive ability:

a) basic – minimal set of resources which is necessary to enter this or that market and function on it;

b) competitive – resources providing an enterprise with competitive advantages in the current period of time depending on the activity type it can be: access to certain types of resources, unique competences, expertise, unique technologies, etc.;

c) strategic – resources providing an enterprise unique opportunities for development in the future and conquering the future leadership on the market.

It is clear that with the course of time one type of resources can be transformed into another one. Thus, the strategic resources formed in the past period are currently transforming into the competitive ones, replacing or adding to already existing ones competitive advantages of an enterprise. Resource structure can be changed depending on whether the enterprise has time to change its market conduct and accept a proactive decision to provide its activity with strategic resources.

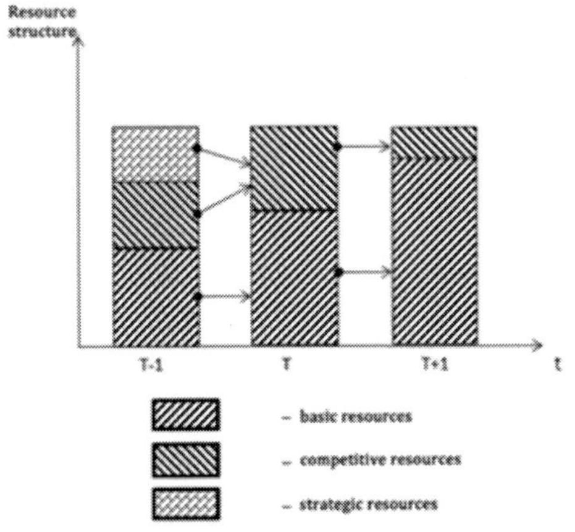

Fig. 2. Resource structure changing in the situation of delay.

If we imagine the situation of delaying when in hope (or, how it usually happens in real practice, ungrounded certainty) on long-term keeping of competitive advantages conquered in past, the enterprise doesn't make proactive decisions to change strategic resources, the structure of resources is changing the way it is shown on fig. 2. The interpretation of the picture is as follows: as we can see, the refusal from a proactive decision in the current period of time led to the resource structure change and as a result the percentage of basic ones increased and the percentage of competitive ones decreased. In the future resource structure can become worse without making proactive decisions up to the complete loss of competitive advantages of the enterprise.

The system can provide an enterprise with competitive advantages when in every preceding period the changes in outer environment are predicted and the decision to change market conduct with attracting all the necessary strategic resources and their transformation into competitive ones is made rapidly.

We consider it reasonable to introduce the term of SP-point (Strategic Prediction Point) that is the strategic point of predicting future changes. Accepting proactive decisions at the moment will allow the enterprise to get competitive advantage in a rapidly changing environment. On the conceptual level we offer to define the time coordinate of the point of making proactive decision on the basis of the following reasoning.

The later the decision about changes will be made the quicker the enterprise will lose its position on the market, and consequently, first, market share will decrease at the cost of a part of consumers choosing the competitor's products instead, then the sales will drop and the profit will decrease. At the same time in such conditions the expenses on changes will grow. Let's put Y1 as the correspondence of the profit loss to the period duration from the moment of making changes.

On the other hand, the earlier the decision about the changes is made the bigger the value of transaction costs will be needed to get ready to these changes. From our point of view, predicting the changes will require in this or that degree virtually all types of transaction costs. Correspondence of transaction cost growth to the changes in the enterprise activities will be marked with Y2.

Then cumulative costs including the total loss because of delay in decision Y1 and transaction costs connected with pro-active decision Y2, will be marked as Y3 (fig. 3).

The correspondence of Y1 and time can be shown as follows:

$$Y_1 = k_1 \cdot k_2^{(k_3 \cdot x + k_4)} + k_5 \qquad (1)$$

then
$$x_1 = \frac{\log_{k_2}\left(\frac{Y_1 - k_5}{k_1}\right)}{k_3} \quad (2)$$

The correspondence of transaction cost value Y2 and time in general will be as follows:

$$Y_2 = \frac{k_6}{(k_7 \cdot x + k_8)} + k_9 \quad (3)$$

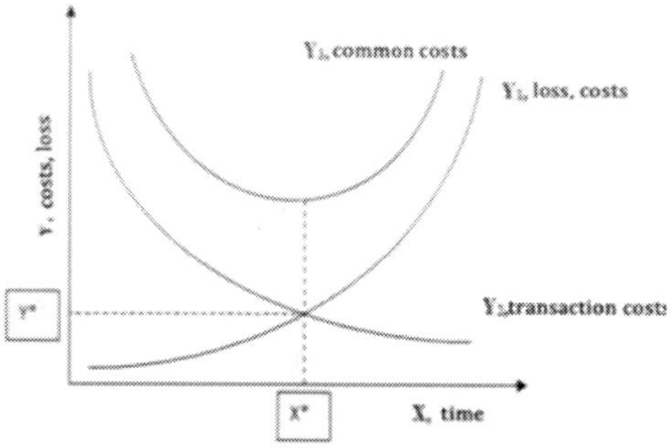

Fig. 3. Defining the moment of proactive decision making.

hence
$$x_2 = \left(\frac{k_6}{Y_2 - k_9} - k_8\right) / k_7 \quad (4)$$

desired point [X*;Y*] if $x_1 = x_2$ r

$$\frac{\log_{k_2}\left(\frac{Y_1 - k_5}{k_1}\right)}{k_3} = \left(\frac{k_6}{Y_2 - k_9} - k_8\right) / k_7 \quad (5)$$

derivative of the formula to define the moment of proactive decision making:

$$Y_3 = k_1 \cdot k_2^{(k_3 \cdot x + k_4)} + k_5 + \frac{k_6}{(k_7 \cdot x + k_8)} + k_9 \quad (6)$$

where k1…k9 – coefficients characterizing transformation spending and losses and transaction costs.

Hypercompetition characteristics mentioned above virtually require from the process of making proactive management decisions constant tracing of the state and predicting the tendencies of changes of different areas: resource market, consumers' and competitors' behavior, product feature, market prices, technology development, etc. The ability to detect coming changes and rapidly transform strategic resources and competitive or, in fact, dynamic capabilities is the key condition of enterprise's survival in today's environment.

5 Conclusion

Hypercompetition is an inevitable threat for today's business and the probability of successful resistance to it, as it was shown, is significantly differentiated not only within the business scale, according to Toffler's ideas, but also the types of activities of industrial enterprises. Nowadays, among Russia's manufacturing small enterprises, electrical equipment production has the greatest chances to protect and develop their competitive advantages. Machinery and equipment production has a little lower chances. As for the observed method of protection from the growing tension of hypercompetition, it is useful for the enterprises of any scale and activity type, but for small enterprises of food production, pulp and paper, metallurgical production its introduction can be considered the key factor of survival.

The work was supported by Act 211 Government of the Russian Federation, contract № 02.A03.21.0011.

References

1. L.M. Makarov, http://www.orekhovasv.ru/Is_PDF/E/E7.pdf.
2. J.A. Schumpeter, *Theorie der wirtschaftlichen Entwicklung: e.Unters, ber Unternehmergewinn, Kapital, Kredit, Zins u. d. Konjunkturzyklus, von Joseph Schumpeter* (Duncker und Humblot, Berlin, 1987)
3. R. D`Aveni, *Hypercompetition: Managing the Dynamics of Strategic Maneuvering* (The Free Press, New York, 1994)
4. http://www.inform.od.ua/articles/examen/vn_sreda.htm.
5. http://protown.ru/information/hide/4455.html.
6. M. Brun, Problems of the theory and practice of management, **3**, 104 (1998)
7. G. Hamel, *Das revolutionäre unternehmen*, (Econ Verlag, München, 2000)
8. http://www.nanonewsnet.ru/nanoweek.
9. A. Toffler, *The Third Wave* (Pan Books Ltd, London, 1981)
10. http://base.consultant.ru/.
11. Small and medium enterprises of Russia (Rosstat, Moscow, 2014)
12. http://www.emitentes.pt/images/media/docs/200_logos_19991209_ForEC_Relatorios_sobre_PMEs.pdf.
13. www.opora.ru.
14. http://www.gsom.spbu.ru/images/cms/data/faculty/gem_2013_final20_all.pdf.
15. V.V. Shlyichkov, Herald of economics, law and sociology, **2**, 97-102 (2015)
16. http://www.trust.ru/sme/indexmsb/.
17. http://rcsme.ru/ru/library/show/8464.
18. http://www.heritage.org/index/.
19. K.M. Eisenhardt, J.A. Martin, Strategic Management Journal, **21(10-11)**, 1105-1121 (2000).
20. D. Dzh. Tis, E. Shuen, Vestnik SPBGU. Ser. Management, **4**, 133-184 (2003)

Communication support in manufacturing enterprise: coordination mechanism and management particularities

T.V. Simonyan[1,*], *N.M.* Ovanesyan[1], and *E.A.* Midler[2]

[1]Don State Technical University Rostov-on-Don, Russia
[2]Southern Federal University Rostov-on-Don, Russia

Abstract. In the environment of market economy the economic system is competitive only if it defines its own communication system more properly, reacts to the market needs and requirements, creates and organizes competitive manufacturing production, takes actions that lower risks, and provides a transformation of in-process interaction. The participants of these relations was initiated by the formation of the elements of communication, together with market actors, is gradually getting involved in the modern process of innovation, developing new strategic decisions and initiatives.. Therefore, manufacturing enterprises have to undertake considerable efforts and radically new approaches in order to make use of communication tools for strategic development. Recent gradually adapting to the current state of the market, setting up its coordination mechanism on the perception of communication support.

1 Introduction

The current stage in the development of Russian enterprises poses entirely new tasks in the area of systemic changes towards enhancement in the managerial field, that causes an increased attention to management process in complex economic systems: especially here the majority of funds is centered; issues regarding economical expenditure of material, labor, financial resources are solved; more advanced new technologies are applied and effective communications are realized. Manufacturing enterprises as the construction initiators of the of branched communication networks with various market entities are motivated to involve them in the innovation process and development of effective strategic decisions. In these circumstances communication tools greatly expand and transform its functions as well as interaction of the producer with the communication range acquires a multilevel and multi-channel character. Communication tools are only effective, when they are adapted to the market condition and coherent with motivations of entities to the perception of communication support. Ease of Use

2 Communication support of management process as structural element of business-system

The analysis of strategic alternatives allows substantiating and defining the basic approach to the formation of enterprise strategy, on the base of which its mission, goals and ways to achieve are formed. At the same time there is a clear understanding of the need and feasibility of a strategic approach implementation, aimed at creating systematic prerequisites for the formation of complex communication support for the manufacturer development.

Depending upon the economic situation in the market, the integrity of complex economic system can be supported, but can also be developed on the base of corporativity, taking into account not only the cooperation, according to activity functions but also to the relations within the system.

A corporate integrated management system represents an open system; which activity depends upon the internal and external environment. The peculiarity of such management system is in the study of the influence of external and internal factors, not only on business processes but also in their own system dependency on communication support [1].

The most important tendency in the development of corporate integrated systems is reduction of the number of decisions, made in the top echelon, increased ability to handle extensive information and respond quickly to current situations. It is achieved through the provision of larger authority at lower levels, i.e. transfer of managerial decision-making centers to the levels at which the problems arise.

Allocation within the corporate integrated system of independent business-units, interconnected by business processes, leads to a constant complication of communication interconnections and structures of implemented socio-economic objectives and, consequentially of these trends, a build-up of internal connections occurs within the system.

Accordingly, the volume and diversity of administrative interactions increase in solving industrial-

* Corresponding author: cimonyan.t@gmail.com

 DOI: 10.1051/shsconf/20173501123

technological and socio-economic tasks that certainly leads to transformation of content and nature of management, maxims, methods and functions of management itself [2].

Information and communication flows, circulating within integrated management system, must comply with the maxims of having a meaningful interpretation of the following: adequacy – elimination of the information deformation; usefulness – absence of unnecessary information within the data, complicating the managerial processes; particularity – inclusion of only relevant data in the management system; authenticity – reflection of management situations through objective information; verification – examination of the veracity of information via analysis of past information [3, 4].

The communication support system of management process is the very component of an overall management system in the corporate integrated business-system, which allows the unification of the corporate information space through delivery of information, necessary to ensure the system stability with the help of monitoring feedbacks. In the frame of the system, the collection, analysis, storage and procurement of data are ensured for managerial decision-making with the aim of supporting management process by the usage of resources, including materials, labor, financial assets and manufacturing facilities.

In the process of the system management, based on communication support, the strategic component is the defining element [5].

Since the development of the strategy represents a dynamic process, the strategy must be reviewed and also consequently updated on a regular basis, it includes planned actions, i.e. implementation of planned strategy and unplanned strategic decisions, which are necessary in the event of unforeseen circumstances.

At the initial stage the strategy development process should be assessed by the totality of external and internal factors which determine the basic approach towards corporate integrated business-system in order to the strategy formation. (Figure 1).

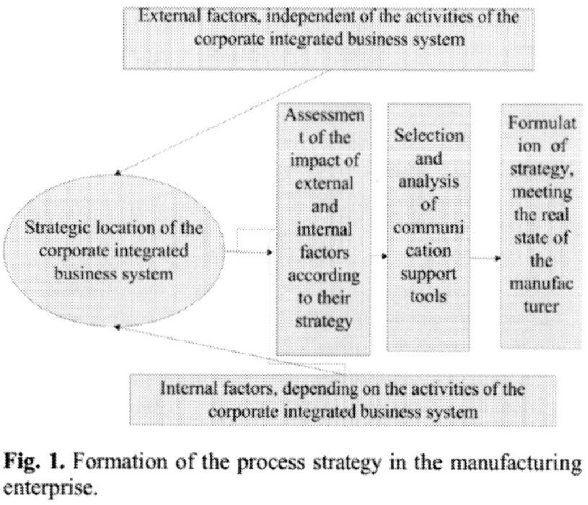

Fig. 1. Formation of the process strategy in the manufacturing enterprise.

The implementation of strategy communication support requires primary decision of managerial tasks: creating organizational prerequisites for successful strategy implementation; budget management for a viable distribution of funds; policy determination of the corporate integrated systems to ensure the strategy implementation; motivation of labor for more effective work results; comparison of remuneration with the achievement of expected results; creating a favorable atmosphere within the corporate system in order to achieve successful performance towards stated goals; usage of advanced experience for continuous improvement, etc. All of this raises the need for complex development of all elements of communication support in the manufacturing enterprise.

One of the main benefits from using integrated communication support is the effect of synergy: coordinated with each other communications represent a much greater influence than each way individually.

The benefits from integration of communications are: increase in sales and profits; achievement of purity within the communication channel as a result of elimination of barriers; cost savings due to the removal of parallel and duplicated use of communications within the framework of a unified marketing program. Moreover, for every company to achieve its mission, it is highly relevant to have their own communications and combinations, by the choice of which it is necessary to take into account main components, criteria and consistency of stages in communication strategy development. The strategy model traditionally represents its constituent components and their dependencies of the company's objectives. At the same time, the links, which are extremely important, are not explained, when different entities (consumers, governing structures, investors, shareholders and others) interact in order to achieve the goals of the company and, of course, those of the end consumer. People are referred to as key figures in a dynamic communication system. However, the lack of a definition of those figures and their behavioral motives gives the model solely a descriptive character, instead of making them a managerial tool of communication support in a particular market sector [6, 7].

In addition, out of sight is the impact of this support on the competitive behavior of consumers. There is no correlation between sets of communications and conditions, in which they realize their potential, namely aspects of the consumer area, the specifics of the manufacturing activity type in the economic industrial sector in any given territory (region), with production-technical and economic condition of the actual manufacturer, and also the character and nature of the relation between manufacturer with business partners (communication agents, infrastructure entities of the business-area, involved in the process of function and development of the manufacturer). The consideration of the changing relations between the producer, the consumer in a wide range of its behavioral decisions, market partners and other entities of the communication area forces the producer to seek out rational solutions in the area of formation of communication toolsets for activating interactions not only with the consumers, but also interactions with other participants. Communication support, understood as a toolset and process component of interaction, does not provide feedback (response) within the analysis of consumer preferences in the context of the assigned vector of the enterprise strategic development.

3 Enterprise communication competence of: features of its formation

Manufacturing enterprise, creating communication support, reasons from the fact, that effectiveness of its activity and competitive potential is determined by quality and quantity of the information which is required in order to take managerial decisions [8]. Managers in the enterprise build their relationships with various market participants, using communication tools, focusing on the effectiveness of its action and rationality in terms of consumption of various resources (financial, labor, material, etc.) [9].

The formation of communication competence is based on the following principles [10]:

a) equation of interests between participants of exchanges;

b) rationality in the use of communicative resources that can alter its form and content, in accordance with the target audience;

c) pragmatism in the behavior of market participants, offering their communication mix;

d) concentration of enterprise efforts in building a targeted impact on one or another consumer type;

e) integral perception of participants' interests in interaction and accounting for vector of motives in determining a single reference point for achieving the desired results.

The implementation of these guidelines requires, on the one hand, sufficiently large and organized information about all interaction participants, on the other hand, streamlining of its accumulation in the enterprise information subsystem in the through the distribution of roles and functional responsibilities of experts, included in the process for its collection and processing in different directions of the enterprise activity.

The communication process of "awareness" and "perception" in this very case is the way of existence and movement of semantic information, which provides a change of forms of its display and their further development. Moreover the actual communication process does not appear as a single act, but as a system of interconnections in a unified communication field, in which information is retrieved into one of its subsystems.

The formation of communication competence involves:

a) differentiation of consumers in types according to their competitive behavior in changing market conditions;

b) collection of analytical data on dynamics of parameters, representing sales volume, consumer potential of the individual segments, the level of communication costs, character and changes in the business climate in specific markets in a denote time interval and etc.;

c) systematization of the information received in those formats that allow us to use effective technical means for its processing and communicating to the stakeholders;

d) carrying out various analysis within the context of formulated managerial tasks and its objectives.

The tasks, that challenge the company, require creating a model of management in the enterprise information subsystem, sufficient in content, managerial and communicative aspects for solving problems of the formation of the database, necessary for the enterprise managers to take rational managerial measures [11, 12].

Communication tools of an enterprise, replenished in the course of the emergence of new technical means and development of network technologies, provides a new quality of relations between them and the market. Thus, targeting (targeted impact) is fixed, interactivity is increased and coordination of entities' interactions is improved in the communication area. The enterprise "gets more motivated to be closer to the customer, and therefore sees more sense in it." [13].

Three components of communication competence of enterprise, understood as a level of readiness and ability of reciprocity in the enterprise, - motivational, pragmatic and cognitive – are interrelated and complementary. Each of them has a different degree (level) of activity and openness. The combination of communication competence components at different levels allows us to construct a typology of communicative conditions of the entities. This typology is based on combinations of different levels of motivation, on balanced interaction between entities, communicative coherence between entities, as well as the enterprise knowledge about the entities. The higher the level of motivation activity of the market entity in relation to other is, the higher the level of its readiness for interaction with other participants of the communication process (consumers, competitors, business partners, communication agents and etc.) is.

Functionality of communications, manifesting itself through a process aspect of building economic and other relations between entities, first of all, involves transfer of information via channels, installed by the enterprise to the corresponding communicants, showing interest and forming motivation towards the acquisition of products, goods (services) or ownership interests in the enterprise as a market entity. The accumulation of the obtained information about competitors, consumers, market communication agents requires a certain consistency and organizational design, which can be represented as an enterprise management subsystem or organizational structures of different scales and subordination. Creating an information subsystem (in the form of an information center), serving as a separate organizational unit of the enterprise communication field can be regarded as a rational solution for the enterprise [14].

4 Method of efficiency assessment in communication support

A key element in the generated information array appears the consumers himself – with his interests, ideas about the external environment and producer of goods, motivation in relation to purchases. As far as the consumer has the role of a processor in decision-making, regarding purchase and consumption, he should know about himself, about the goods, about the producer and their communication agents, that could provide him with at that moment truthful or long term decision about purchase and consumption of the product in need. Manufacturer should as well know and collect information about the consumers, who represent participants of his communication interaction as well as consumers, which may under certain external influences from communication agents engage in the process of purchase and usage of goods. All the information should be accumulated in the enterprise data center, which can include various subsystems, for example, the internal monitoring of business processes and their management; search of information with operational, tactical and strategic character; collection and processing of information on communications; collection and processing of external information; transfer and passage of information; control of information receipt; searching and processing of information about interaction problems between participants; security support of communication field and its evaluation; consumer relation management (CRM); enterprise resource planning (ERP). Such a set of subsystems is required by the enterprise for structuring of received information on the state of external and internal environment (information about the level of use of annual production capacity, the production structure of the main enterprise products, etc.), consumer perceptions (information about the personal consumer information area); the brand initiator of communication impact; the cost of transactions in various kinds of communication funds and so on [15, 16]. Multidimensionality and plurality of information on various aspects of enterprise activities requires the establishment of certain fundamentals, which largely determine the construction of communication networks and choice of means of communication interaction on entities' interactions which differ from each other (whether they are consumers or governmental and municipal authorities, or market communication agents – representatives from various professional fields of activities). Not of less importance is the information on the state and condition in the market of resources, on the perception of communicative messages by consumers, and also the information on communicative activity of the competing enterprises. The variety of information that can be included in the information subsystem, determines the multiplicity of goals of her creation, tasks and methods for their achievement.

The filling of an information array requires the application of specific analysis tools that are oriented on solving strategic tasks for the enterprise. For this we need a separate analysis of the individual potential of the entities, competitor analysis, market analysis, macro-environment analysis as well as an analysis of opportunities and risks.

Accordingly the highest level of communication support efficiency can only be achieved by coordinated interaction of all communication tools for the transmission of a unified message to the target market and the provision of complex impact on the target consumer. The application of an integrated approach to its planning and implementation provides the following benefits:

- unified creative design of a message for the target market;
- maximized sequence of messages;
- efficient use of all communication tools;
- more precise focus of messages to the market;
- improving the efficiency of internal and external communications and their interaction;
- cost savings and efficient use of resources [17].

As part of this study we suggest a methodology to assess the effectiveness of communication support on the base of factorial dependence:

$$Э_t = a + C_{u,n,a_e} + K_{M_t} + M_{K_t} + C_{O_t} + KK_{M_e} M_{K_e} + k_t C_{O_t} \\ + k_3 C_{u,n,a_e} K_{M_t} + k_4 K_{M_t} M_{K_t} + k_5 C_{O_t} M_{K_t} + k_6 M_{K_t} C_{O_t} \quad (1)$$

where in *Эt* – income (RUB./Y.); a – uncontrollable factors of the previous periods; *k1, k2, k3, k4, k5, k6* – coefficients, which reflect the degree of pair bond integration, determined communication support tools; $C_{u,n,a}$ – connection with investors, business-partners and shareholders; K_M – crisis management; M_K – communication management; C_o – public relations;

K_{Mi}, M_{Ki} – influence of crisis management on international communications;

C_{Oi}, K_{Mi} – influence of public relations on crisis management;

M_{Ki}, K_{Mi} – influence of international communications on the realization of crisis management strategy;

$C_{u,n,a,}$ K_{Mi} – influence of connections with investors, business-partners and shareholders on the realization of crisis management strategy;

C_{Oi}, M_K – influence of public relations on international communications;

M_{Ki}, C_{Oi} – influence of international communications on public relations.

Estimated factorial dependence only shows paired connections between communication support tools; although in real conditions, there might be more complex systematic interrelationships of three, four or more elements. Regarding the complexity of given dependences, the question of their mathematical modelling is left open, and it requires special software products and explains the absence of empirical confirmation.

5 Conclusion

Thus, to ensure the functionality of an information subsystem of communication support, in our opinion, requires the allocation of continuous monitoring and control of very significant parameters in consumer behavior – the type of its competitiveness. Understanding of the dynamics of changes, initiated by communications in specific target groups of consumers, along with understanding of dynamic structures of payment capability of these target groups will provide a science-based approach to the organization and development of communication interaction, strengthen its target orientation, purposefulness, informational and economic efficiency.

References

1. V. Zinchenko., V. Zusman, *Intercultural Communication from Systematic Approach to a Synergetic Paradigm* (Flinta, Moscow, 2007)
2. N. Ovanesyan, E. Midler, Bulletin of Don State Technical University, **11**, **9(60)**, 1719-1725 (2011)
3. G. Klejner, V. Tambovtsev, P. Kachalov, *Company in an Unstable Economic Environment: Risks, Strategy, Security* (Jekonomika, Moscow, 1997)
4. N. Ovcharenko, A. Voronov, Practical Marketing, **7** (2010)
5. E. Khandamova, Bulletin of ENGEKONA. Economy Series, **3(22)**, 245-252 (2008)
6. A. Krylov, *Communication Management. Theory and Practice*, (Publishing house of the National Institute of Business, Moscow, 2002)
7. G. Khamel, K. Prokhlad, *Competing for the Future* (Olimp-Bizness, Moscow, 2014)
8. I. Belskikh, Marketing in Russia and abroad, **5(55)** (2006)
9. B. Andersen, *Business Processes. Tools of Perfection* (Standards and quality, Moscow, 2003)
10. T. Simonyan, S. Imamkuliev, Proc. Marketing and Management: Science Practice, 97-109 (2015)
11. R. Gilyarevskii, *Information Management: Management of Information*, (Knowledge, Technology, Moscow, 2009)
12. T. Simonyan, D. Datshinskij, Bulletin of Don State Technical University, **11**, **9(60)**, 1700-1709 (2011)
13. G. Dei, *Organization with Market Orientation: How to Understand, Attract and Hold Valuable Clients*, (Jeksmo, Moscow, 2008)
14. L. Glukhikh, *Theory and Methodology of Development and Realization of Competitive Manufacturing Enterprise Strategies* (KuBGU, Krasnodar, 2012)
15. D. Schults, B. Barns, *Strategic Brand-Communication Companies* (Grebennikov Biznes Buks, Moscow, 2003)
16. V. Eliferov, V. Repin, *Business Processes: Reglamentation and Management* (Infra-M., Moscow, 2005)
17. M. Porter, *Competitive Advantage: How to Reach High Results* (Alpina Biznes Buks, Moscow, 2005)
18. V. Zinchenko, V. Zusman, Z. Kirioze, *Intercultural Communication from a systemic approach to the synergetic paradigm* (Flinta, Moscow, 2007)
19. *Problems of Modern Marketing Russian Economy: Theory and Practice* (DGTU, Rostov-on-Don, 2014)
20. *Problems of Eonomy and Management of Enterprises, Branches, Complexes* (Center for the Development of Scientific Cooperation, Novosibirsk, 2009)

Commercial effectiveness assessment of implementation the energy efficiency raising of the building project due to introduction of automatic heat consumption control

Yu.N. Zvonareva[1,*], *Yu.V.* Vankov[1], and *V.V.* Shlychkov[2]

[1]Department of Industrial power system and systems of heat supply, Kazan State Power Engineering University-KGEY, Kazan, Russian Federation
[2]Institute of Business & Innovative Technologies, Kazan National Research Technical University – KAI, Kazan, Russian Federation

Abstract. Introduction of the automated metering stations and regulation (AUU) located directly in the heated building besides creation of comfortable conditions indoors leads to decrease in consumption of thermal energy. The annual expected effect of realization of the offered actions (installation of metering stations and automatic control) can make up to 22% consumed and that isn't less important, the paid thermal energy. In general, efficiency of implementation of the project on introduction of AUU can be characterized by considerable decrease in heat consumption of the building and, respectively, reduction of a payment for the consumed energy resources. In this paper we evaluated the effectiveness of implementation of increase of energy efficiency of the building investment project (hereinafter SP). We calculated the ratio of expenses and the results considered actions for inhabitants of an apartment house located in Kazan after installation of a weather-dependent regulation. As a result of calculation of the imitating model created on the basis of basic data and the investment project plan the main results of determination of economic efficiency of the Project have been received. For the analysis and increase of reliability of a settlement assessment of efficiency of the investment project calculations at different options of a set of basic data are executed.

1 Introduction

Optimal thermal energy consumption in compliance with the required comfort in rooms of buildings is the main objective of energy saving in the sphere of municipal heat supply to which much attention is paid today.

According to experts, excess heat consumption of the housing stock is about 30-40% [1, 2].

Along with power inefficient use of thermal energy, annually there is a growth of tariffs. For example, during 2010-2015 in Kazan growth of a tariff was 36,1% for thermal energy.

Introduction of the automated thermal points (AITP) so popular for today allows to carry out complex regulation of thermal energy consumption for needs of heating and hot water supply of object according to his actual requirements. It gives the opportunity to save means of thermal energy suppliers and its consumers considerably. The greatest economic effect of introduction of AITP is reached in the conditions of the centralized systems of heat supply in which there are seasonal (autumn-spring) "overheating", according to the operational data reaching 30% of annual settlement consumption [3, 4, 5].

2 Goals and objectives

Earlier we published the results of researches which showed that upon transition from TsTP to ITP the actual decrease of thermal load for heating residential apartment houses, averages 30,4% [6]

For an assessment of effectiveness of implementation the energy efficiency of the building investment project (hereinafter the SP) carried out calculation of expenses and results ratio of average 9-floor residential apartment house located in Kazan [7].

The purpose of this work is providing to investors the information about the investment project sufficient for decision-making of expediency of participation in its realization.

The assessment of commercial effectiveness of SP was carried out taking into account the following requirements [8, 9]:

- the capital investments put in calculations are indexed according to price indexes for capital investments and elements of its technological structure;
- all calculations - in rubles of the Russian Federation;
- all calculations are executed including VAT;
- the cost of installation works is accepted the partner company, equal according to calculations;
- design cost wasn't taken into consideration.

* Corresponding author: Skulinaun@mail.ru

The assessment of indicators of efficiency of investments on the considered object was made according to the "Methodical recommendations about an assessment of investment projects efficiency" approved by the Ministry of Economics of the Russian Federation, the Ministry of Finance of the Russian Federation, the State committee of the Russian Federation on construction, architectural and housing policy (to No. BK 477 of June 21, 1999).

Financial and economic calculations are executed with using the certified computer program "Project Expert 7.55" developed by the Expert Systems firm (Moscow) intended for the creation and the analysis of financial model of a new which yet isn't created or already operating enterprise regardless of its branch accessory and scale that allows to define the main indicators of economic efficiency on the basis of the integrated indicators and expected economic estimates [10].

As capital investments are taken the investments in a volume necessary for implementation of the project.

All types of integrated results and expenses are expressed and compared in the discounted form.

The model of cumulative construction (CCM), as the most adequately reflecting discounting indicators for objects of power branch is applied to calculation of a discount rate. In this method the discount rate is the sum of a risk-free rate of profitability and set of awards for the separate, relating to this project risks estimated in the expert way.

The general discount rate pays off as the sum of all components and by results of calculation is accepted equal 13,75% [11].

In calculations were considered value added taxes, income tax, the property tax.

Taxation rates on each of points are accepted according to the tax code of the Russian Federation.

3 Results and discussion

In the course of work generated the indicators by groups of objects of inflation which characterize changes of receipts cost and payments under the main articles of the project in accordance with expected macroeconomic indicators.

In work was carried out the integrated assessment of energy efficiency increase project of heating system in a house.

Volumes of necessary capital investments for implementation of the project are determined proceeding from the integrated estimation of cost of the main and auxiliary, costs of the installation and construction works (ICW), and also an other expenses [12, 13]. The data provided by producers and suppliers of the equipment, contract organizations, cost indexes of similar objects in the territory of the Russian Federation, and also information and analytical materials on power branch are assumed as a basis.

The total cost of the project makes 430,476 thousand rubles.

Duration of the project is accepted 240 months (20 years) that corresponds to AUU service life declared by the producer. The beginning of the project is set: 01.08.2015 y.

For the planned volume of deliveries was accepted the calculated value of economy of thermal energy after the AUU installation for an assessment of economic efficiency of the considered project it according to the earlier calculations [14].

As a result of calculation of the imitating model created on the basis of original data and the investment project plan the main results of determination of economic efficiency of the Project were received.

Results of calculation are reduced in Table 1.

Table 1. Main Financial perfomanceable.

Indicator	Value
Simple period of payback of initial expenses (RB), m	31
The discounted period of payback of Initial expenses (DPB), m	35
Average standard of profitability (ARR) on the project, %	51,55
The net provided income (NRV), rub	1 055 747,4
Index of profitability (PI)	3,45
Internal standard of profitability (IRR),%	66,6

Ratio of the discounted and simple period of payback of initial expenses are presented in Figure 1.

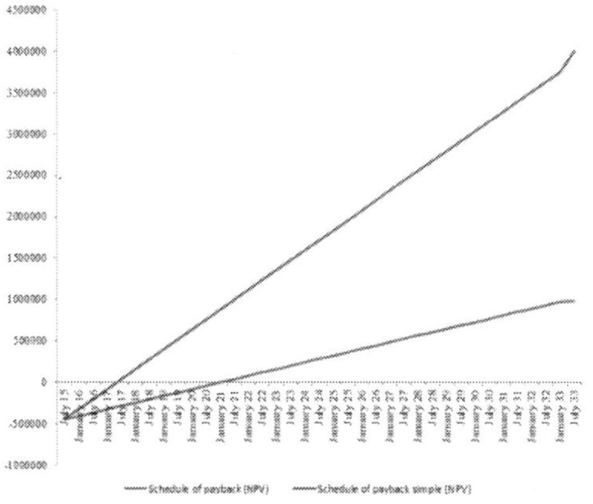

Fig. 1. Schedule of payback (NPV).

Thus, proceeding from the analysis of indicators of economic efficiency [15], the considered versions of the project are economically expedient.

For the analysis and increase of reliability of a settlement assessment of the investment project efficiency are executed the calculations at different options of a set of basic data, such as:

- price of realization of finished goods (heat carrier);
- costs of production (direct charges);
- general investment expenses;
- production sales volume (the thermal charge consumed by the considered costumer).

Range of change of above-mentioned factors from-50% to + 50% with a step of 10%.

Results of calculations for each of the considered options are presented in the form of schedules in Figures 2-4.

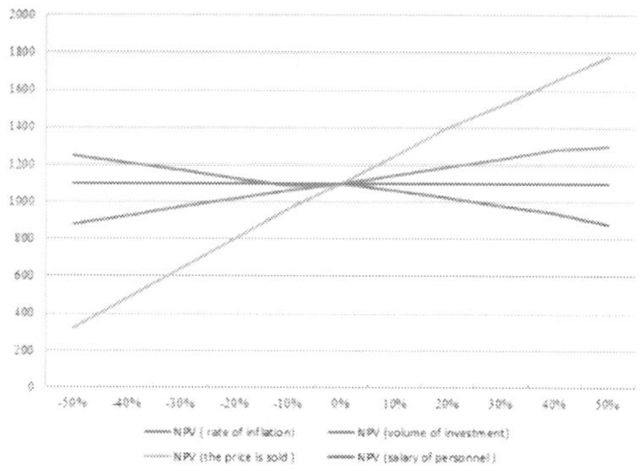

Fig. 2. Analysis of sensitivity (NPV).

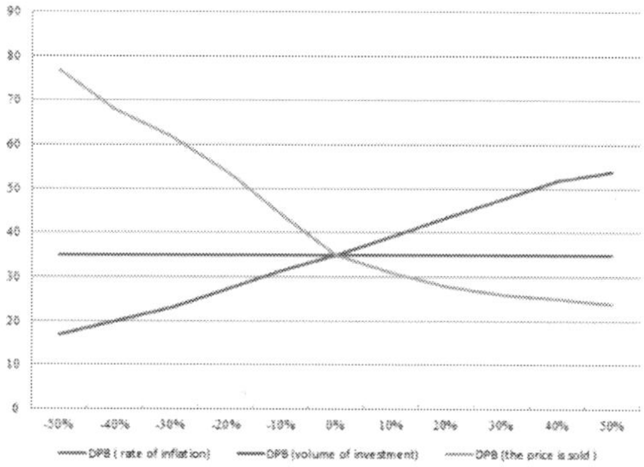

Fig. 3. Analysis of sensitivity (DPB), m.

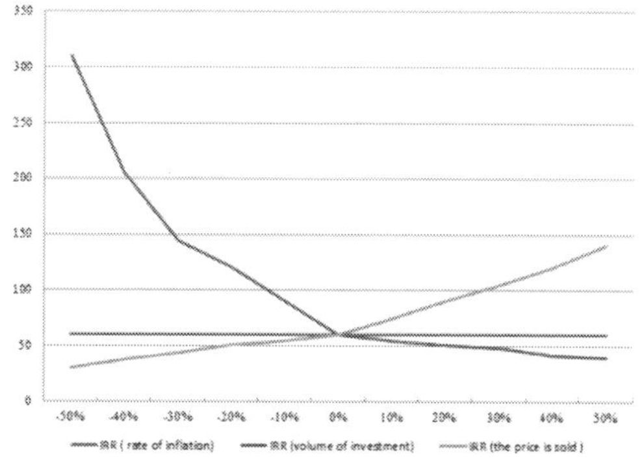

Fig. 4. Analysis of sensitivity (IRR), %.

The analysis of the results received due to calculations showed:

1. At change of tariffs there is a change of NPV coinciding in the direction and the return change of a payback period.
2. The change in price of fuel leads to the unidirectional change of a payback period and back influences the net discounted income and has smaller impact on indicators of payback of the project, than change of tariffs.
3. The increase in cost of the equipment leads to decrease in economic indicators.

It should be noted that at all considered changes the project remained steady.

In the considered range of deviations (+/-50%) the Project remains profitable irrespective of the following factors:

• at increase in cost of raw materials and materials (direct expenses) from planned by every year;
• at increase in the volume of investment.

The given researches of economic risk factors allow to note high stability of the main indicators characterizing efficiency of the project: the net discounted income, the internal standard of profitability, the index of profitability and a payback period. The considered project turned out to be the most sensitive to change of a tariff for thermal energy [16, 17].

Stability of the project is caused by existence of a stable sales market of thermal energy.

The risk on the project can be caused by the following reasons:

• increase in the volume of investment in connection with instability of an economic situation;
• change of the size of tariffs for net output of thermal energy.

4 Conclusion

Capital expenditure from 430,746 thousand rubles is necessary for the realization of actions the energy efficiency increase project of heating system of the considered average house.

The annual expected effect from realization of the offered actions can save up to 30,4% consumed and, that isn't less important, the paid thermal energy [18, 19].

In terms of money, according to a tariff for thermal energy for 2015 y. a decrease in heat consumption will allow homeowners to reduce the cost of the provided services for the sum about 347,52 thousand rubles with VAT in a year. Thus it should be noted that due to annual growth of tariffs for thermal energy, the economy will increase.

The carried-out calculations showed that the internal standard of profitability of the Project is more than chosen discount rate, and it indicates a high degree of the Project reliability and small dependence on change of investment climate.

According to experts, capital expenditure for introduction of AUU at average useful service of the capital equipment of 20 years, pays off for the period from 3 to 3,5 years [20].

For our Project the simple payback period (taking into account growth of a tariff) made 2 years and 7 months, and taking into account discounting - 2 years 11 months.

In general, efficiency of implementation of the Project can be characterized by considerable decrease in heat consumption of the building and, respectively, by reduction of a payment for the consumed energy resources.

References

1. A. Chistovich, The Russian Expert review, **2** (2006). http://www.protown.ru/information/articles/3325.html.
2. B.E. Korenkov, I.A. Smirnov, L.P. Needle, Power system, **11** (1992)
3. B.M. Madorskij, V.A. Shmidt, *Operation of central heat distribution stations, systems of heating and hot water supply* (Moscow, 1971)
4. I.S. Shein, A.V. Izvekov, *Some questions of optimization of functioning of city thermal knot. Radio electronics, electrical equipment and power* (MEI, Moscow, 2010)
5. B.V. Sazanov, *Thermal power plants: Manual for technical schools of heat power specialties* (Energy, Moscow, 1974)
6. Yu.N. Zvonareva, Yu.V. Vankov, Proc. 21 century: fundamental science and technology VII, USA, **2**, 131-133 (2015)
7. Yu.N. Zvonareva, Yu.V. Vankov, L.A. Polenov, L.A. Pavlov, Energy Resource and energy conservation in the Republic of Tatarstan, 77-79 (2015)
8. V.E. Popov, Regional economy, **1**, 59-69 (2009)
9. E.Ya. Sokolov, *Central heating and thermal networks* (MEI publishing house, Moscow, 2001)
10. T.A. Volosatova, Inženernyj vestnik Dona, **4** (2013). ivdon.ru/ru/magazine/archive/n4y2013/2054.
11. I.S Shein, A.V. Izvekov, Reliability and safety of power, **2(17)** (2012)
12. L.O. Dikul, Economy, statistics and informatics, **4**, 30-35 (2011)
13. Yu.N. Zvonareva, *Proc. Engineering and technology* (2014)
14. Yu.N. Zvonareva, Yu.V. Vankov, The Tomsk Polytechnic University, **11** (2015)
15. *Employment of Danfoss automation at heating points of systems of central heat supply of buildings. Benefit. RB.00.H8.50* (OOO Danfoss, Moscow, 2014)
16. M.F. Drigo, Management in Russia and abroad, **1**, 18-24 (2008)
17. V.V. Pyrkov, *Modern thermal points. Automatic equipment and regulation* (Takispravi, 2007)
18. Yu.N. Zvonareva, Proc. Radio electronics, Electrical equipment and Power (2012)
19. Yu.N. Zvonareva, Yu.V. Vankov, S.A. Nazarychev, Engineers Don Gazette, **4** (2015). ivdon.ru/ru/magazine/archive/n4y2015/3315.
20. T.A. Makarenja, S.V. Stash, Inženernyj vestnik Dona, **3** (2013). ivdon.ru/ru/magazine/archive/n3y2013/1839/.

Methods of economic reliability assessment for industrial enterprise in the market economy conditions

R.A. Timofeev[1,*], *V.V.* Shlychkov[2], and *D.R.* Nestulaeva[1]

[1]Kazan State Power Engineering University, KSPEU, Kazan, Russia
[2]Kazan National Research Technical University named after A.N. Tupolev – KAI, KNRTU-KAI, Kazan, Russia

Abstract. The paper dwells upon prospects, topicality and scholarly importance of research into formation and assessment of economic reliability in industrial enterprises as socio-economic systems. The main notions and interconnection of the categories of "stability" and "reliability" of socio-economic systems are presented, based on which the conclusion that these notions are not identical is made. From the standpoint of system-synergetic approach, economic reliability comprises reliability of management system, technical development, conservation activities, human resources management, financial and economic development, marketing activities, and production management. The paper presents main methodological principles of integral estimation of economic reliability of an enterprise in modern conditions. It is suggested to apply E. Harrington's desirability function in order to give a quantitative estimate of economic reliability. In conclusion it is pointed out that assessment of industrial enterprise's economic reliability can become an effective mechanism of assessing the efficiency of harnessing the manufacturer's potential and identifying the reserves and inefficient usage of enterprise resources.

1 Introduction

The problem of economic reliability of Russian industrial enterprises has become an important issue in domestic economic knowledge only in the last decade. The current interest in this problem is accounted for by transition of Russian economy to market relations, the processes of macroeconomic stabilization in late 1990s [1] and the growing influence of the major world commodity market environment on the industrial reproduction processes [2]. Notwithstanding scientific efforts in this area certain substantive and evaluative issues of economic reliability of industrial enterprises in the real sector of Russian economy have not yet received due theoretical attention and remain unsettled [3].

From the standpoint of sanction wars and international limitation of access to investment resources and new technologies for Russia [4], as well as critical state of real sector of Russian economy [5] the topicality of the given research lies in the following:

• low economic reliability caused by earlier unfulfilled economic tasks is one of the main reasons for bankruptcy of the majority of industrial enterprises [6];

• low financial solvency and, as a result, low economic reliability of the enterprise limit its financial potential making further expansion of production almost impossible [7];

• high probability of faulty operation leads to loss in production process reliability, idle periods and erratic work which, in the long run, increase production costs [8];

• low economic reliability of an industrial enterprise has a negative impact not only on competitiveness of end products but on the competitiveness of the whole enterprise [9].

Thus, the notion of "economic reliability" is not only the main component of enterprise's effective development; it is also a guarantee of its further lucrative existence in market conditions.

Taking into account turbulence in real sector of Russian economy and risks for economically active agents, information concerning estimate of enterprise's economic reliability is in-demand for both internal and external users which determines topicality and viability of systematic research into it.

2 Problem definition

Reliability of an industrial enterprise is generally viewed in terms of assessing its technological potential which, in our view, appears unsound. Economically from the standpoint of providing the target competitive power and enterprise efficiency, economic reliability implies not only equipment availability, qualified personnel and potential opportunity to manufacture goods [10] but its stability for political disruptions, management mistakes and business partners' blunders, etc. [11], significantly affecting the process and results of economic activities. An enterprise is considered to be "reliable" or "unreliable" for its business partners depending on its capacity to react adequately and efficiently to external and internal claims. It should be noted that the terms "reliability" and "stability" are not convertible as it is often presumed in the existing literature on the subject.

* Corresponding author: vestnik_eps@mail.ru

In general, stability is defined as an ability of an object to return to equilibrium after termination of the external forces effect. Nevertheless, an equilibrium system is a closed system with constant parameters. Economically stable enterprise operating on the market and trying to enhance its profitability cannot be in equilibrium as it is constantly developing, therefore, its structure, links and functioning are modifying [12]. Realized value of economic stability is formed by means of overcoming external and internal effects. Due to its specificity, the level of economic stability is a short-term indicator; it is not a long-term given. In the long run, economic reliability of an industrial enterprise must be understood as its ability to operate efficiently and flawlessly for a given period under the current conditions and under the action of constantly changing external and internal factors [13]. Based on the results of the research conducted the authors present main notions and interrelation of reliability and stability of economic and social systems (Table 1).

As can be seen from the above, nonidentity of such categories as "stability" and "reliability" is of theoretical and practical importance as it accounts for the existing differences in the methods of economic stability and reliability assessment for industrial enterprise in market conditions.

Table 1. Main concepts and interconnection of categories of "reliability" and "stability" of economic and social systems.

Authors	Definition
Dolyatovskiy V.A., Dolyatovskaya V.N.	The category of stability is an ability to maintain sufficiently small deviations after perturbation effect. Large deviations capable of upsetting system's equilibrium state make the term "stability" meaningless and should be treated in terms of system's reliability [14]
Mishin V., Astashkina I.	Stability is a particular case of system's reliability [15]
Popov V.N.	Reliability as a property of a system manifests itself as an ability to operate efficiently, preserving stability of certain parameters at a given time [16]
Tseitlin V., Stetyukha A.	Economic reliability is an ability of the system to achieve given economic result (possible only provided technological and operational reliability of the system and its compounds) [17]
Kabanov V.N., MIkhailov S.N.	Economic reliability is understood as, on the one hand, an ability of an object to be implemented under certain interaction with the environment and, on the other hand, as a quantitative estimate of an object, explicitly relating probability with time or other parameters of the implementation process [18]
Kolokin A.L., Mikhailov S.N., Marchenko G.N.	Economic reliability is an economic state of the system which optimizes the use of property, economic, financial and professional resources, its stable development and settlement of economic claims and demands of various interest groups in conditions of sensible (tolerable) economic risk [19]

3 Theoretical part

In the process of developing methodology for economic reliability assessment the authors addressed an industrial enterprise in terms of system-synergetic approach that allowed presenting it as an economic and social system comprised by a scope of elements, or subsystems, determining aims and conditions of its functioning [20].

Economic reliability of an industrial enterprise is determined by the following elements: reliability of management system, technical development, conservation activities, human resources management, financial and economic development, marketing activities, and production management. Reliability of each subsystem is defined by significant index chosen with the help of Delphi approach or autonomously.

Furthermore, the choice of significant indices is substantiated by strategic factors determining reliability for industrial enterprise's subsystem. In general, the conducted research has demonstrated that all market factors can be divided into:

1. External and internal. External factors affecting the industrial enterprise are the level of economic stability, the level of inflation and unemployment, living standards, social conditions, advances and technological level, state legislation, custom's rules, etc. The internal factors that are connected with manufacturing activities are resource efficiency, resources for output expansion and change of product portfolio, industrial and management structure of an industrial enterprise providing the effect, corresponding to enterprise's objectives and sensitive to market changes.
2. Controlled and uncontrolled. Among the controlled factors are internal and external ones, necessary information about which is obtainable in full measure and within limited time. Uncontrolled factors are determined by lack of necessary information for an industrial enterprise. Meanwhile, in some cases information may be partial or incomplete but limited by time window, for instance, data about competitors, demand tendencies, political changes, etc. It might be problematic to forecast uncontrolled factors and their exact timelines. Therefore, accounting uncontrolled affects is accompanied by certain risk. On the one hand there may be potential losses of enterprise's resources, but on the other hand, the enterprise can derive a significant benefit.
3. Controllable and uncontrollable factors. Controllable factors can be affected. Uncontrollable factors, which can be controlled and uncontrolled, are the ones that the enterprise cannot affect since there are no means or ways to affect them. External factors related to production sphere are mostly uncontrollable and uncontrolled. The control of consumption and distribution factors can be maintained only on condition of data collection vehicle availability even though effect on these factors is limited. Internal factors, as a rule, are subject to control providing proper organization of business and operations.

Thus, the aforesaid contributes to the analysis of underlying determinants of production efficiency. Herewith it becomes obvious that factors determining the reliability for each subsystem will have their own nature and significance [21].

The reliability of financial and economic development is conditioned by the following factors:

• Factors for stable financial state of an industrial enterprise and its independence from loans. Among them are inventories, accounts receivable, cash assets, short-term liabilities, net worth, etc.

• Factors affecting resilient net worth that leave the opportunity to manoeouvre them, including net worth, investments, etc.

• Factors determining stock and expenses by own sources, as well as factors affecting capital investment, production stock and in-process inventory.

Reliability of marketing activities is determined by the factors affecting market share, sales (revenue), storage, shipment and transportation of goods and expenditures for marketing research.

Reliability of management is conditioned by factors determining development strategy, management system and processes, informatization and computerization, and corporate government.

Reliability of environmental activities is determined by:

• Factors characterizing the degree of exceedance of statutory air contamination (impact area, radius of sanitary protection area).

• Factors characterizing the degree of exceedance of emissions release in comparison with permissible exposure limit.

• Factors characterizing the degree of exceeding the real adverse physical impact (noise, ultrasound, infrasound, electromagnetic emanation) as compared to specified value.

• Factors, characterizing population density in the zone of enterprise's negative effect and, consequently, enterprise's potential danger for population (specified population density, average population density in the zone of enterprise's negative effect).

Reliability of the personnel is determined by trust and insistence on high standards among members of staff, business qualities (education, expertise, skills), culture, communicativeness, self-discipline, psychological and physical state.

Based on the conducted analysis, the authors suggest the following scope of strategic indices of economic reliability of an enterprise:

• current liquidity ratio, financial stability index, receivable turnover;

• profitability of sales, return on equity;

• depreciation coefficient of fixed production assets, proportion of monetary assets in enterprise's incoming receivables;

• staff turnover, average monthly wages of 1 employee;

• automation index of management systems;

• amount of fatal industrial accidents;

• index of normative ecological hazard, proportion of environmental payments in product costs.

The proposed methodology of economic reliability assessment is based on determination of the universal scale (wherever possible) for comparing development levels of the object under study and benchmark results.

In accordance with the given statement of the problem we represent the methodology of calculating the economic reliability indices for an industrial enterprise as an economic and social system:

1. The input data:

X_i - set of reliability indices for subsystems of an industrial enterprise determining its economic reliability; $i = \overline{1,N}$, where n – the number of indices;

2. Average ratio is calculated as follows:

$$\overline{X} = \frac{\sum_{i=1}^{N} X_{ij}}{N} \quad (1)$$

3. Square deviations of indices are calculated in the following way:

$$\sigma_i = \sqrt{\frac{\sum_{i=1}^{N} (X_{ij} - \overline{X_j})^2}{N}} \quad (2)$$

4. Standardized values of the chosen indices are determined as:

$$Z_{ij} = \frac{X_{ij}}{\sigma_j} \quad (3)$$

5. Reference values of the indices are set:

$$X^* = (X_1^*, X_2^*, \ldots, X_N^*) \quad (4)$$

6. Standardized values of the references are calculated as:

$$Z_j^* = \frac{X_j^*}{\sigma_j} \quad (5)$$

7. Weights of attributes in economic reliability function are calculated in the following way:

$$\alpha_j = \frac{Z_j^*}{\sqrt{\sum_{j=1}^{n} (Z_j^*)^2}} \quad (6)$$

8. Economic reliability function is formed as follows:

$$y = \alpha_1 Z_1 + \alpha_2 Z_2 + \ldots + \alpha_n Z_n \quad (7)$$

9. The value of economic reliability function is attributed to indices value at time:

$$y_i = \alpha_1 Z_{i1} + \alpha_2 Z_{i2} + \ldots + \alpha_n Z_{in} \quad (8)$$

10. The value of economic reliability function is calculated at reference values of the indices:

$$y^* = \alpha_1 Z_1^* + \alpha_2 Z_2^* + \ldots + \alpha_n Z_n^* \quad (9)$$

11. The level of economic reliability at time t_i is determined as [22]:

$$S(t_i)=\frac{\sum_{j=1}^{N}\alpha_j\frac{X_{ij}}{\sigma_j}}{\sum_{j=1}^{N}\alpha_j\frac{X_j}{\sigma_j}} \text{ or } S(t_i)=\frac{y_i}{y^*} \quad (10)$$

The suggested integral assessment of economic reliability for an industrial enterprise is of dual character. Firstly, an industrial enterprise is capable of achieving potential positive results in its business and operations. Secondly, effective use of its potential guarantees achieving certain results, in other words, actual outcomes of business and operations are a constituent part of business entity's potential. E. Harrington's desirability function was used to measure the economic reliability of an industrial enterprise [23]. The results are shown in table 2.

Table 2. Economic reliability scale of industrial enterprise.

Values	Qualitative evaluation of economic reliability of an industrial enterprise
[0.76–1]	High level of economic reliability of an industrial enterprise
[0.51–0.75)	Stable level of economic reliability of an industrial enterprise
[0.36–0.50)	Medium level of economic reliability of an industrial enterprise
[0.21–0.35)	Low level of economic reliability of an industrial enterprise
(0–0.2)	Critical level of economic reliability of an industrial enterprise

4 Conclusion

The suggested method of economic reliability assessment for industrial enterprise and economic reliability scale allow us to assess reliability and stability of a given enterprise on the market compared to others and the efficiency of its resource use.

The method of economic reliability assessment for industrial enterprise proved effective in assessing the efficiency of harnessing the manufacturer's potential and identifying the reserves and inefficient usage of enterprise resources. In addition, this method can be applied to assessment of industrial enterprise at a given period as well as in the dynamics, which, in its turn, allows to use moment and tempo indices (i.e. duplication of initial indices) to get generalized estimate of economic reliability, characterizing not only the current state but also prospects of socio-economic development of industrial enterprise in real conditions of Russian economy.

References

1. V.V. Shlychkov, S.M. Kulishch, R.A. Timofeev, *Resource-Based View of Economic Reliability for Industrial Enterprise* (Kazan, 2009)
2. R.A. Timofeev, *Formation of Economically Reliable Strategy of Developing Industrial Territorial Production Complex.* http://test.vak.ed.gov.ru/common/img/uploaded/files/TimofeevRA.doc.
3. N.V. Vedin, N.F. Gazizullin, A.Sh. Hasanova, Problems of modern economics, **3-4**, 39-48 (2003)
4. A.Sh. Hasanova, The Review of Economy, the Law and Sociology, **4**, 101-105 (2014)
5. I.G. Alafuzov, R.A. Timofeev, Vestnik of Samara State University of Economics, **8(70)**, 5-8 (2010)
6. M.N. Smagina, I.G. Alafuzov, The Review of Economy, the Law and Sociology, **3**, 90-93 (2015)
7. E.N. Kadyshev, M.E. Kadyshev, V.V. Smirnov, Audit and financial analysis, **2**, 317-326 (2012)
8. P.A. Bataykin, *Development and realization of a behavioral paradigm of modern Russian Economy Regulation* (Moscow, 2014)
9. V.V. Shlychkov, S.M. Kulish, Economic sciences, **54**, 277-281 (2009)
10. S.M. Kulish, Kazan Pedagogical Journal, **2(109)**, 168-172 (2015)
11. V.V. Shlychkov, D.R. Nestulaeva, Asian Social Science, **19**, 82-89 (2015). DOI:10.5539/ass.v11n19p82.
12. I.K. Kiyamov, *Institutionalization of economic space of the modern Russian state* (Kazan, 2012)
13. S.M. Kulish, The Review of Economy, the Law and Sociology, **6**, 25-32 (2008)
14. V.A Dolyatovskiy, V.N. Dolyatovskaya, *Management Systems Research: Workbook* (2004)
15. V. Mishin, I. Astashkina, *The Stages of Management Systems Research.* http://www.inventech.ru/lib/analis/analis0025/.
16. S.A. Popov, *Strategic Management: Understanding is More Important than Knowledge* (Moscow, 2003)
17. V. Tseitlin, A. Stetyukha, Transport and Telecommunication, **3**, **2**, 61-64 (2002)
18. V.N. Kabanov, S.N. MIkhailov, Management in Russia and Abroad, **6**, 112-114 (2007)
19. A.L. Kolokin, S.N. Mikhailov, G.N. Marchenko, Proc. Modern Problems of Transforming Economy, Cheboksary, 171-175 (2002)
20. V.V. Shlychkov, A.Sh. Khasanova, N.V. Vedin, Indian Journal of Science and Technology, **8(S10)** (2015). DOI: 10.17485/ijst/2015/v8iS10/84883
21. R.A. Timofeev, *Assessment of economic reliability of the enterprise on the basis of management of fuel and energy resources* (Ph.D. Dissertation, Nizhny Novgorod, 2005).
22. N.V. Vedin, V.V. Shlychkov, A.Sh. Khasanova, Indian Journal of Science and Technology, **8(S10)** (2015). DOI: 10.17485/ijst/2015/v8iS10/84877.
23. J. Harrington, Industrial Quality Control, **21(10)**, 494-498 (1965)

Management features of the innovative development of the enterprise taking into account cost-effective use of resources

Yu.V. Babanova[1,*], *V.P.* Gorshenin[1], and *A.Yu.* Dolinskaia[1]

[1]South Ural State University, Chelyabinsk, Russia

Abstract. Modern business is developing in the conditions of innovation-driven economy in which science and technology are the basis of competitive ability and a productive force. Most of the managerial decisions are taken in the context of limited resources and high uncertainty, because they depend on many factors, the dynamics of which is not always possible to estimate with reasonable accuracy. In this article special method is presented, which allows you to formalize the process of management decision, in terms of innovative product development. This method includes matrix analysis and vector calculus. The purpose of this method and model is to develop a set of alternative options for the product development and the selection of the most cost effective projects based on the integration of consumer requirements, innovative product specifications and the manufacturing capability of the organization (initial conditions for the project realization).

1 Introduction

Modern business is developing in the conditions of innovation-driven economy in which science and technology are the basis of competitive ability and a productive force. In contrast to the earlier stages, when the competitive position was determined by the closeness to resources, in today's world the key role is played by creativity and the ability to use opportunities beyond controlling the company's resources at this time, that is called – innovation. The transition from idea to innovation requires significant overhead. The term "resources" in accordance with the interstate standard GOST for resource saving 30166-95 [1] includes the value, stores and income opportunities. In general, the resources are divided into natural and economic (physical, labor, and financial).

An integral part of effective management is the economic, technical, scientific, practical and information activities, the complex of organizational and technical actions and activities that accompany all stages of the product life cycle, which are aimed at the efficient use and economical use of resources. In accordance with the national standard of resource saving GOST R 52104-2003 efficient use of resources defines resource economy or cost-effective use of resources [2].

2 The main body

Requirements for cost-effective use of resources are divided into three groups [2]. The first group includes requirements of resource quality defining perfection of processes, products, and services, such as the material composition and quantity of the used materials, weight, size, volume of products, etc. The second group includes the requirements of resource intensity (manufacturability); determining the possibility of achieving economic resource costs during the manufacture, repair and utilization of products, as well as carrying out various works and the provision of services taking into account the requirements of environmental safety. The third group includes requirements of resource efficient products, and also the ability to determine the economic resource costs in the operation, repair and utilization of products, as well as the performance of work and provision of services.

These groups of requirements are interrelated:

• during product development, work and services planning (set project requirements of resource quality and resource efficient, recommendations on resource intensity);
• during the manufacturing of goods, performance of works and service provision (specified claim (control) requirements for resource intensity (for manufacturability);
• during production operations, performance of works and provision of services (specified claim (control) requirements of resource intensity and resource efficient);
• during utilization of the products (resource intensity and resource efficient requirements are established).

Most of the managerial decisions are taken in the context of limited resources and high uncertainty, because they depend on many factors, the dynamics of which is not always possible to estimate with reasonable accuracy. This is caused by the lack of full information and the variability of the conditions of the organizational environment. Innovative projects are among the most high-risk to investments [3]. Any innovative project implemented by the company, is an open system. It is a part of a whole range of projects, ensuring the achievement of the eventual result. As a consequence there is a necessity not only for justifying the

* Corresponding author: babanovayv@susu.ru

competent management reasons to launch an innovative project, but also for drawing up an effective portfolio of innovative projects to ensure the achievement of strategic business goals, and determine the optimal sequence of start these projects.

After the analysis of the available scientific approaches, innovative tools and techniques of project management it was determined that not enough attention is paid for justifying the selection of the innovative project and its interrelation with other projects (implemented or planned for implementation in the enterprise). The solution of this problem is of scientific and practical interest, because it will simplify the process of selection and at the same time improve the quality of management decisions.

As part of the research work it was revealed that modern methods of justifying management decisions are advisory for the most part. Currently, for modeling economic situations, including an innovative direction, specialists increasingly used physical (mechanical) analogy, in particular, the tensor calculus, graph analysis [4] or the cognitive modeling, vector analysis, which allow us to understand what is happening in areas far removed from the natural sciences. On the basis of the scientific understanding the processes, it is possible to develop possible answers to the new changes. Excluding charges of mechanical philosophy, it may be noted that it is not about transferring the laws of physics and mechanics in economics, but the recognition of the effectiveness of the mathematical description and analysis of the systems under consideration and objects, using the positive experience, proven in physics and mechanics.

The team of authors of the International Faculty of the South Ural State University developed a method of integration and matrix analysis [5], based on international experience [6], which allows you to formalize the process of management decision, in terms of innovative product development. The proposed method combines matrix analysis and vector calculus. Matrix analysis is a tool that allows revealing logical connections between given parameters. Vector calculus allows interpreting the given data from matrix analysis in a graphical representation of the identified interrelations, and their consequences.

Management decision-making about selection of the innovative project can be viewed as a process consisting of ten stages (Table 1). The purpose of this process is to develop a set of alternative options for the product development and the selection of the most cost effective projects based on the integration of consumer requirements (because it is a source of income), innovative product specifications or a combination thereof (competitive advantage of the product), and the manufacturing capability of the organization (initial conditions for the project realization).

The selection of the innovative project is closely linked to the elements of cost-effective use of resources.

The process of management decision-making about choosing the innovative projects includes the following steps:

Step I. Research objective statement

For making the objective statement of the study is necessary to determine the object, which will be under directed attention. Thus, you choose a product which will be introduced innovative changes. This can be done in the following ways:

- based on the needs of the organization;
- on the basis of the actual requirements of the market;
- on the basis of future market needs.

After defining the object of research, it is necessary to give it a general characteristic. In the description, particular attention should be paid to the customer's needs, which would be satisfied the product under study. Analysis of substitute-goods (products that perform the same or similar function as the product in question) would allow more fully describe the consumer properties, ensuring customer satisfaction.

To identify the necessity of innovative changes in the product you should analyze the financial and economic results obtained by the organization from the output, and hypothesize about their positive change as a result of increased sales. Define SMART-goal of increasing sales volume and to define its place in the tree of the organization's goals.

The result of first step (the process of management decision-making on a choice of the innovative project) would be a reasonable objective of the research, which is reflecting the specialty of the product, trends and conditions of innovative changes.

Step II. Consumer requirements analysis

Defining features of the investigational product on the previous step, custom requirements and product properties satisfying these needs, you should use the survey and a tree diagram, prepared on its basis, to identify all consumer demands for which the client is able and willing to pay. You should define the requirements that determine demand. Moreover, you should remember about real and potential size of the demand.

The survey begins with the selection of the audience, which should represent the whole set of potential consumers of the product under research. This is called a representative sample [7]. Then conduct a survey of a selected audience; based on the results to determine what requirements must have a product so these people want to buy it. It is important to draw up a survey questionnaire, which is offered to your potential customers. The questionnaire card should contain questions with closed and opened answers. The respondent has a limited time for responses, only 3-5 minutes, if it would be longer, the qualities of responses will suffer. In addition, the formulation of the issues requires the participation of at least three experts: an engineer for this type of product, a psychologist or a linguist who knows how to ask questions, and an analyst who understands how the data will then be processed.

The survey results are presented in the form of a list of expected benefits that the customer can get from the product. After that it is necessary to make a classification of all expected benefits on the basis of "affinity". For each group of benefits is assigned the common name in the form of consumer demand (CD*i*). The presented set of requirements is ranked according to the selected scale (R*mni*).].

Table 1. Process model of innovation project selection and development.

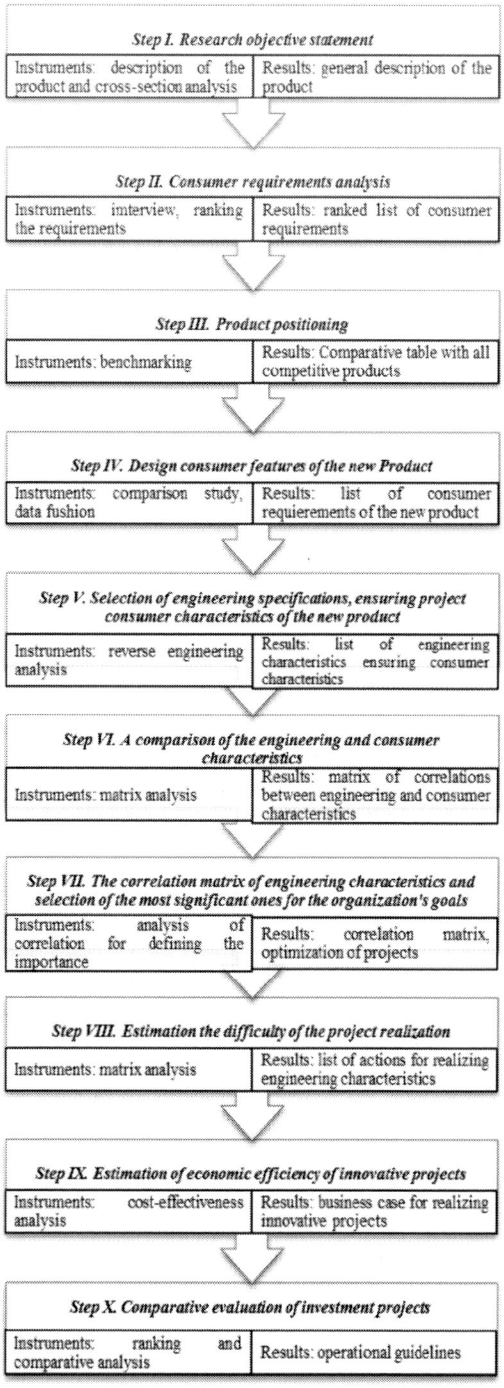

The end result of the second step should be ranked identification of the most important product characteristics, according to the consumer's wishes.

Step III. Product positioning

At this step, the evaluation of satisfaction level of each consumer demand is happened. The evaluation is based on estimation of satisfying demands by similar competitive products or substitute-goods (P*ci*), and own product before innovative changes (P*pi*), if it had previously produced.

The used method is called the product benchmarking, and it enables the opportunity of the rating of consumer requirements for improving the research object parameters in order to increase its value for the customer and increase its competitiveness.

Step IV. Design consumer features of the new Product

Next step is to create a list of objective values in points for each consumer requirements (P_{npi}), which the new product should obtain, to ensure a high level of demand.

Objective values of consumer requirements, without need of changes are assumed to be the base:

$$P_{npi} = P_{pi} \quad (1)$$

Other targets are assumed equal to or higher than its competitors:

$$P_{npi} \geq P_{pi} \quad (2)$$

Using this data, the degree of improvement can be calculated (3):

$$K_{pi} = P_{npi} / P_{pi} \quad (3)$$

Here is also set the weight (V*mni*) of each consumer requirement as the project goal (4):

$$V_{mni} = K_{pi} \times R_{mni} \quad (4)$$

Further, the sum of weights of project goals (ΣV_{mni}) is determined. Also it is determined the weight percentage of each goal in the total amount: ($V_{mni(oe)}$) (5):

$$V_{mni(oe)} = V_{mni} / \Sigma V_{mni} \quad (5)$$

Step V. Selection of engineering specifications, ensuring project consumer characteristics of the new product

At the step V, key engineering (technical) characteristics of the designed product (EC*j*) are identified. These characteristics are not related, but they are enabling to secure previously selected consumer characteristics. In addition, for a deeper analysis of engineering characteristics their quantitative parameters are determined, which are suitable for the basis own product, competitors' products, and to designed your own product.

Step VI. A comparison of the engineering and consumer characteristics

A comparison of engineering and consumer characteristics is by means by a matrix where the customs' requirements (CD_i) are laid vertically, and the engineering characteristics (EC_j) are laid horizontally. At the level crossing, the mutual coupling coefficients (A_{ij}) are indicated. Each coefficient shows: how much each engineering characteristic ensures the implementation of the consumer characteristics of the new product.

The coefficient, which reflects the strength of mutual influence of engineering and consumer characteristics, can vary from zero to one. One ("1") means the maximum (full) interaction between the factors. Zero ("0") means that any connection between factors is missing. Intermediate data values show the tendency of dependence.

Except the coefficient, reflecting the strength of mutual influence of engineering and consumer characteristics, it is necessary to know the orientation of this force. The direction of the force is reflected by sign: if with the growth of engineering characteristics, customer satisfaction increases, so the coefficient is positive, if it falls – the coefficient has negative sign.

The common sense, the theoretical concepts, expert judgment, published data, observation, experience, special planned experiment are commonly used to determine the value of the coefficient.

Step VII. The correlation matrix of engineering characteristics and selection of the most significant ones for the organization's goals

The implementation of some engineering characteristics affects the feasibility of others, it is necessary to reveal how many such characteristics affect each other. Next step is matching the engineering characteristics with each other. The interrelation between characteristics can be reflected through the coefficient K_{ij}. The shaded squares will be situated on diagonal, this area should not be filled, and the rest of the cells need to be filled. At this step for completing the table, experts of engineering services should be involved.

A comparison of data from the previous steps allows analyzing the connection between custom requirements and the interconnected engineering characteristics by using coefficients of interrelation A_{fij} between 2 main coefficients CD_i and EC_j (6):

$$A_{fij} = A_{ij} + \sum_{i=1-n}^{j=1-k} K_{ij} \times A_{ij} \quad (6)$$

where $i(n)$ – number (quantity) of consumer requirements;
$j(k)$ – number (quantity) of engineering characteristics.

This step allows you to create a clarified matrix reflecting the tripartite interrelation between engineering characteristics and custom requirements.

Step VIII. Estimation the difficulty of the project realization

At this step it is necessary to identify activities that are essential to the innovative projects realization. After that you need to match these activities with the previously selected engineering characteristics, and the last ones are taken as the main goals of the project. As a result, the share of costs will be identified, which are required for the implementation of each project. The result of this step is a chosen innovative project, which is optimal for cost-result ratio.

Step IX. Estimation of economic efficiency of innovative projects

It is necessary to determine indicators of economic effectiveness of the project implementation for each engineering characteristics. Economic analysis of each project is held, this analysis allows determining the following parameters: cash flow, net present value, the rate of profit for capital, return on sales.

Step X. Comparative evaluation of investment projects

The purpose of the last step in the selection of the innovative projects is ranking all innovative projects according the feasibility of the projects and the order for their implementation.

The main goal of this process – the selection of the most economically efficient innovative project – is achieved. The scientific novelty of the suggested method (integrated-matrix analysis) consists of combining a number of matrixes that enable to justify an innovative project on basis of analysis of product utility, its competitive position and engineering capabilities of the enterprise. The main difference between suggested method and well-known method of structuring the quality function is the presence of analytical coefficients of interrelation between different engineering and consumer characteristics, which take into account the effect of influence. The score estimation of interrelation between the engineering characteristics is presented in basic model. This score is established by the expert.

Interpretation of the information obtained as a result of integrated-matrix analysis, can be carried through calligraphic display of norm-based scores.

The object of the study has a number of consumer requirements Y_i, and as the result of interrelation it has one main goal Y (total maximum level of consumers' requirements satisfaction). The following equation will be true (7):

$$Y = \Sigma Yi \quad (7)$$

If by every consumer requirement understand the purpose of the project and define through Yi, and under $A_{\Sigma fi}$ – we define the range of sum of engineering characteristics of each (*i-number*) consumer requirement, the consumer demand can be represented as a vector (8):

$$\Sigma Yi = A_{\Sigma fi} \times \cos\varphi i \quad (8)$$

where φi – the angle between the direction of the main goal of the project and the vector of the CDi.

This angle determines how this consumer requirement is consistent with the overall objective of the project and is associated with a rating of R_{mni} of this CDi in relation to the overall goal of the project. For example, the overall goal of the educational institution – the graduation of highly skilled professionals and the basic consumer requirement is a minimum level of cost to the training of a good specialist. Rank of such consumer requirement will have a negative value with respect to the common goal. The degree of influence (or rating) is determined by the angle "φi" between the received direction to the early achievement of the maximum function Yi and accepted direction of the consumer requirements' vector.

There are cases, for example for each factor:

1. $\varphi i = 0^0$: the requirement is directed to the early achievement of the maximum overall function Y;
2. $\varphi i = 90^0$: the requirement does not change the overall function Y;
3. $\varphi i = 180^0$: the requirement resist the achieving of the maximum overall function Y.

The remaining intermediate positions of the vector affect in different way to the achievement of the maximum overall function Y.

The general expression defining the impact of requirements (or vectors) to the final function Y has the form (9):

$$Y = \Sigma(A_{\Sigma fi} \times \cos\varphi i) \quad (9)$$

This is the basic equation of the harmonic connection of general objectives of the project and specific custom requirements.

To investigate the dependence of the required function Yi from variable factors, many of which in the initial state may not be appropriate to apply a numerical expression. To solve this problem it may be possible to use rationing arrangements of the valuation system. The procedure for calculation is described below:

1. Defined the engineering characteristics (ECj), affecting on this consumer requirement.
2. By calculation the amplitude value A_{fij} for each i-number factor by j-number engineer characteristic is determined. After that you use the estimation scale from 0 to 10 points; the sum of all the engineering characteristics for each i-number of custom requirements is displayed.
3. Calculate the factors in relative units.
4. In addition to the amplitude of the requirements, rating is affecting on the achieving of the common goals. It should be defined how the rating of each requirement is consistent with the main goal of the project.

You can accept that scores impact assessment index orientation customer requirements to achieve the common goal ranges from (10) – the direction of the vector is fully consistent with the direction of achieving the common goal; to (–10) – the direction indicator completely counteracts the attainment of a common goal. With 10 points vector is analyzed in accordance with the general direction (angle error Public and Private vectors is equal to zero). At – 10 points angle between the common and the private areas is 180^0. With 0 points is equal to the angle between the vectors 90^0, and this factor, although it has a certain peak value, but its course of action does not affect the rate in the overall result.

Let "q" be the score point of the deviation's range of each requirement from achieving the common goal.

The value of scores is defined by experts:

- qa = 10 – the action of the particular requirement and the total target vector are in full agreement;
- qb = – 10 – the action of the particular requirement fully resist the total target of achieving goal;
- qc = 3 – partially agreed (3 out of 10 points).

5. We make the transferring the deviation of grading system qa (qb, qc) into degrees of the misalignment angle of the total and partial vectors, based on these conditions:

- (+10) points => $\varphi = 0^0$;
- 0 points => $\varphi = 90^0$;
- (–10) points => $\varphi = 180^0$.

6. Determine the angle of alternation «φ» [in angle degrees] of arbitrary vector of consumer requirements from the main purpose direction of taking into account the score mismatch «q» (10):

$$\varphi = q \times Ra + 90 \quad (10)$$

where Ra – is the coefficient of the conversion adopted points of vectors deviation from direction of achieving goals to general function to deviation amounts of the angles at the conditions:

1) $q = -10 \Rightarrow \varphi = 180^0$;
2) $q = 0 \Rightarrow \varphi = 90^0$;
3) $q = 10 \Rightarrow \varphi = 0^0$.

Then, the conversion factor takes the form (11):

$$R_a = \frac{\Delta\varphi}{\Delta q} = \frac{0-90}{10-0} = -9, degree/\ point \quad (11)$$

The function of the transfer of points to degrees takes the form (12):

$$\varphi = (-9)\ [degree/point] \times q\ [point] + 90\ [degree] \quad (12)$$

7. Further, the formula (13) produces the transferring the values of deviation angles "φ" from degrees to radians and next step is calculation the values of cosφ.

$$\varphi\ [deg] = \pi \times \varphi\ [deg] \quad (13)$$

8. According to the obtained values of the amplitudes of the ball distribution vectors $A_{\Sigma fi}$ and directions of their alignment with the overall objective function (the value of $\cos\varphi i$) calculated the total value of the desired function the Yi. This is a common normalized value of the objective function, a certain fixed state requirements

and consumers *Yi* ("a" factor "b" and "c"). Factors, in turn, defined parameters normalized amplitudes $A_{\Sigma fi}$, and angles *qi*.

3 Conclusion

Thus, the proposed method of accounting treatment of interrelation between engineering characteristics, consumer characteristics and vector (harmonic) mapping of studied consumer requirements extend the analytical representation of project changes required to improve the consumer requirements of the object. In addition, these methods are calculated and graphically implemented in the program Microsoft Excel, which simplifies and extends their application to the development of innovative projects on the basis of resource.

The work was supported by Act 211 Government of the Russian Federation, contract № 02.A03.21.0011.

References

1. *GOST 30166-95, Resource. Basic statements, Interstate standards (the Russian Federation)* (2002)
2. *GOST R 52104-2003, Resource. Terms and definitions, Russian National Standard* (2004)
3. S.A. Samovoleva, Technological Business, **4**, (1999)
4. E.K. Kornoushenko, V.E. Maksimov, IPU Research study, **2-3** (1998)
5. Yu.V. Babanova, Economic analysis: theory and practice, **19(274)**, 45-51 (2012)
6. Akao Yoji, *Quality Function Deployment (QFD). Integrating customer requirements into product design* (Productivity Press, Portland, 1990)
7. W.G. Cochran, *Sampling Techniques* (Statistics, Moscow, 1976)
8. H.K. Rampersad, *Managing Personal and Organizational Change: with Total Quality Management* (Olymp-Business, Moscow, 2005)
9. H.K. Rampersad, *TPS-Lean Six Sigma: Linking Human Capital to Lean Six Sigma: A New Blueprint for Creating High Performance Companies* (Standards and quality, Moscow, 2009)
10. H.K. Rampersad, *Total Performance Scorecard: Redefining Management to Achieve Performance with Integrity* (Alpina Publisher, Moscow, 2005)
11. T. Luyster, D. Tapping, *Creating Your Lean Future State: How to Move from Seeing to Doing* (Standards and Quality, Moscow, 2008)
12. A. Farraharov, *Energy and resourse saving in building and manufacturing sphere* (2016)
13. O.N. Kolodina, Modern tendencies in economics and management: new opinion, **11**, (2011)
14. A.V. Kleyman, A.Yu. Shumakov, Chelyabinsk State University Vestnik, **7**, 132-138 (2008)
15. Y. Babanov, V. Gorshenin, Proc. Schumpeterian Readings, 70-76 (2014)
16. V. Gorshenin, S. Alyukov, SAE Technical Paper, **1**, 22-25 (2014)
17. V.A. Barinov, Management in Russia and abroad Journal, **1**, (2007)
18. S.A. Surkov, Yu. Boyko, A.A. Hrupalov, A.M. Lukashev, Management in Russia and abroad Journal, **1**, (2007)
19. S.A. Evseeva, Problems of modern economics, **2(50)**, (2014)
20. R.S. Martynov, *Resource-saving management methods at an industrial enterprise* (2010)

Industrial traumatism management oriented to result

V.S. Parshina[1], *T.B.* Maruschak[1,*], and *A.A.* Gusev[1]

[1]Chair "Management in social and economic systems", Ural State University of Railway Transport (USURT), Yekaterinburg, Russia

Abstract. To improve industrial traumatism management we should direct at the analysis of the causes of the employee's dangerous actions and elaboration of the measures to minimize the identified causes on an employee. The specific aim of managing the system of safe labour is to prevent the reduction of the activity effectiveness that comes from falling of capacity, diseases, traumas and employees death. The condition for the successful realization of the management decision is specifying of goals concerning the certain structural subdivision, its quantitative expression, methods of achievement, the involved functions. On the basis of the study of the modern psychological and normative theories of decision-making proposed is the technology of work with administrative decisions including the elaboration stage, making and acceptance of decisions. This technology is aimed at minimization of risks related to the accident rate and traumatism.

1 Introduction

Management of the industrial traumatism level is aimed to ensuring efficiency of activity through optimization of expenses to ensure safety and gain maximum economic effect from reducing the accident rate. The system of industrial traumatism management that has been existing in this country for long years was oriented to the observance of the formulated requirements and did not raise issues related to the end-point in this sphere. Perfection of the management system in market conditions allows to obtain an additional effect. We will consider the aspects of industrial traumatism management basing on our research in structural subdivisions of railway and time-tested experience of the developed countries.

2 Management of the industrial traumatism

To show the ability of the designated management processes we will refer to the statistical data generalized by Henrikh. The essence is in a stable dependence of severe injuries with everyday violations. The author shows quantitative correlation of fatal, severe and easy cases of industrial traumatism (Fig. 1). The frequency of injuries depends on the amount of dangerous action, as can be seen from the figure. Using this principle we can conclude that industrial traumatism can be controlled.

The identified proportions depend on the nature of production and the principle has not been refuted for the past years. The statistical study of the traumatism on the railway structural subdivisions allows the identification of certain correlations. Presented in [1] calculations based on 18 railways show that there is a middle correlation degree of the general coefficient of production accidents frequency in the workplace and the coefficient of industrial accident with the fatal industrial end. The questioning revealed that not all mortal accidents were subject to investigation, cases resulting in an error. That is, the objective assessment of industrial traumatism state allows to approximate to the proportions derived by Henrikh. The study of the above-mentioned coefficient dynamics on the concrete railway and the experience of the leading structural subdivisions will provide an informative base for administrative decisions.

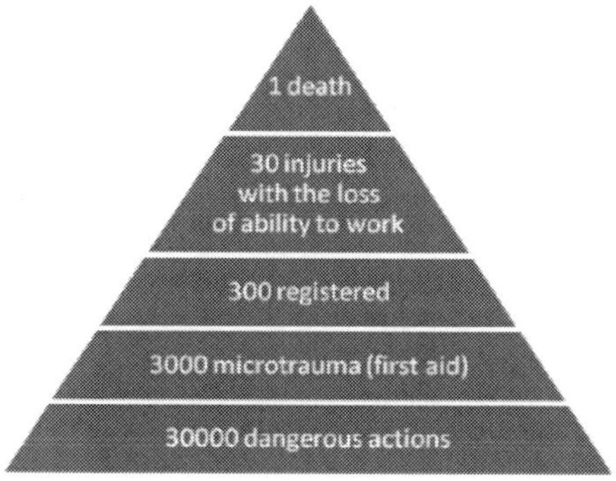

Fig. 1. Henrikh,s pyramid of accidents.

To improve industrial traumatism management we should direct at the analysis of the causes of the employee's dangerous actions and elaboration of the measures to minimize the identified causes on an employee. In the table 1 we reduce the most common reasons of an employee's dangerous actions grouped according to Methodology [2] and suggest appropriate management tools. The same function lie in the basis of management of the industrial traumatism that are in managing of organizational and technical systems defined in the control theory.

* Corresponding author: tbmar@mail.ru

Table 1. The main groups of causes of employees, dangerous actions and the proposed management.

The most common causes of the employee's dangerous actions	Groups of causes in accordance with Mythology	Management tools allowing to minimize the influence the reasons
1.The low level of professional competence. 2. The lack of the practical experience. 3.Inefficiency of training.	**A. Can not** The employee does not possess the necessary knowledge for this work, relevant skills, methods, techniques ways of working	1. Improvement of mentoring. 2. The use of innovation forms of training, instruction, control of knowledge and skills. 3. Improvement of the system of the professional selection of personnel.
1. Predominance of penalty elements in the system of employees' motivation. 2. Problems in the relationship with the supervisor or co-workers. 3. Ineffective (excessive or insufficient) control.	**B. Does not want** The employee is able to perform the work (operation) efficiently and safely, but he has no desire to follow the requirements of safety, i.e. absence of psychological motivation to follow them.	1. Actualization and creative rethinking of the existing system of motivation for the needs of a particular organization needs. 2. Identification and promotion of employees with a high level of motivation. 3 Improvement of the professional personnel selection system. 4. Improvement of the system of organization and payment.
1. Poor health 2. Adverse emotional and psychological state of a person	**C. Is not able** The employee is in such physical or psychological state that despite the ability and desire acts dangerously.	1. Improvement of moral- psychological climate in the team. 2. Elaboration of measures aimed at preventing such negative phenomena among co-workers as distress or professional burnout. 3. Elaboration of the program of working with the families of the violators of labour requirements. 4. Search of the reserves for perfection of labour organization. 5. Perfection of the professional personae selection system.
1. Lack of inadequate quality of personal protective equipment. 2. Imperfection or failure of the technological process. 3. Depreciation of machinery, equipment mechanisms. 4. Lack of or inadequate quality of tools and materials. 5. Adverse working. 6. Lack of time (hurry) 7. Lack of the information about the organization of work and changing conditions of work.	**D. Not provided** The employee does not perform the prescribed action because is not necessary conditions (tools, materials, devices, information, etc.)	1. Search for the reserves to improve the organization of production and labour. 2. Perfection of the system of material and technical support of the enterprise. 3. Involment (survey) of employees in making decisions concerning the choice of personal protection, tools, etc. 4. Perfection of the effectiveness of the system of informational support.

But at forming the effective management influence in the field of industrial safety the peculiarities of the management must be taken into account. The specific aim of managing the system of safe labour is to prevent the reduction of the activity effectiveness that comes from falling of capacity, diseases, traumas and employees death.

The total results from the safety culture are not limited only by reducing injuries. Work in safe conditions improves the employees' moral state that effects productivity.

3 Successful realization of the management decision

Management realization is done through the following functions: accounting, analysis and assessment, planning, organization, coordination and regulation, status monitoring, motivation. The composition of management functions at strictly defined management objectives is known: safety measures, account and analysis. Management system must maintain a certain level of safety preventing its decline. The control system must maintain a certain level of security, preventing its loss.

The condition for the successful realization of the management decision is specifying of goals concerning the certain structural subdivision, its quantitative expression, methods of achievement, the involved functions. To determine the impact of the object in the railway structural subdivisions proposed is the research based on Isikava's chart and reasons of industrial traumatism should be highlighted: human factor, influence of environment, organizational, technical and educational reasons (Fig. 2).

The management organ makes decisions taking into account external and internal information. Management functions determine the activity directions at taking decisions on safety labour management. Functions of accounting, analysis and assessment are directed at forming the necessary information about the object of management for the development of management decisions. Certification materials on working conditions, analysis of traumatism and control result are used. etc.

Our research in the railway structural subdivisions confirmed by results of the analysis [3] allowed revealing the following average statistical data of injury causes; 52% of the total number of injuries are of organizational character, 32% depend on the human factor, 6% are of technical and 4% of training ones. Further is the conclusion that the prevailing part of violations is connected with the human activity (organizational reasons and human factor). Further each of these reasons is subdivided into 5 – 8 concrete directions that are estimated in relative terms. After that meaningful management objects are set.

There are some features of the elaboration technology and making the effective management decisions aimed at reducing, industrial traumatism and accident rate in the process of production. The full minimization of risky management functions realization is possible only in the conditions of the definite certainty. We should note that the human factor is the most difficult forecast condition of the management decision realization. Decisions made in risky conditions are the decisions resulting in non-definition but the every possible result can be defined. The management activity in the above-mentioned sphere must be considered as a risk, and efficiency of management decisions should be determined as a possibility of minimization of risks.

On the basis of the study of the modern psychological and normative theories of decision-making [5–12] proposed is the technology of work with administrative decisions including the elaboration stage, making and acceptance of decisions. This technology is aimed at minimization of risks related to the accident rate and traumatism.

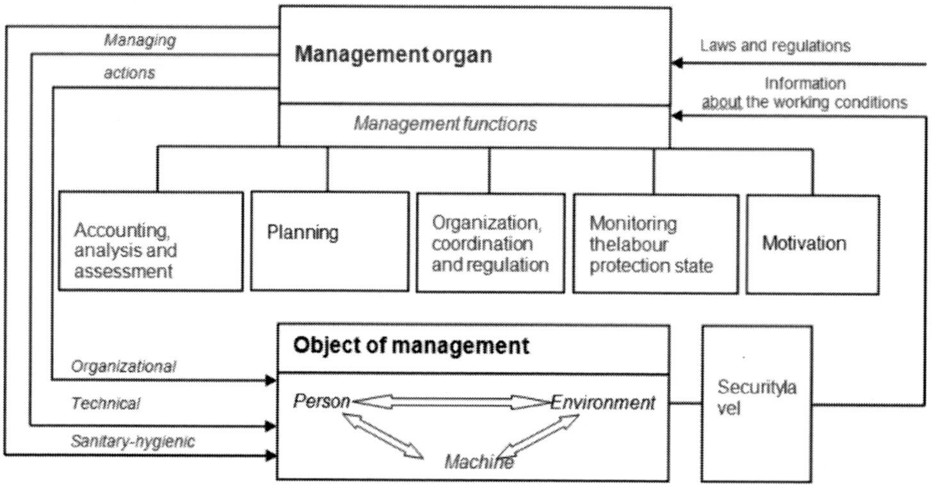

Fig. 2. Functional diagram of safety management.

When making decisions every variant of development of the events should be estimated as a probability. This allows estimating each of the efficiency value and choosing an option with the variant with a maximum sum of the results based on minimal display of subjective factors.

The results of the research into effectiveness of administrative decision realization show that it can be defined as an integral assessment of the effective implementation of all stages of industrial traumatism reduction level including the degree of risk [12–18].

To expert (qualitative) methods are more often used for the risk assessment, they are based on the subjective assessment of the expected performance parameters [19, 20].

We believe that for the more precise evaluation, quantitative methods of the degree of risk assessment can be appropriately used.

If risks related to various performance aspects and negative factors manifestation occur independently of each other (the rule of mathematical adding of risks), the probability of their manifestation (*Po*) is estimated by the rules of the probability theory for the probability sum of independent events. For two factors:

$$Po = p_1 + p_2 + p_1 * p_2, \qquad (1)$$

and the total measure of risk (Mp) is:

$$Mr = \Sigma (p_i * P_i) / \Sigma p_i. \qquad (2)$$

If the risks related to one performance area and negative factor manifestation occur independently of each other (the rule of risk absorption) the probability of their manifestation is assessed by the maximum value:

$$Po = \max \{p_i\}, \qquad (3)$$

where *Po* is the overall risk degree in this area;
p_i is the degree of the particular risks in one area.

The lack of a unified risk management system makes all the particular risks to be considered separately and to use the mathematical rule of their adding.

Implementation of such traumatism management system on an enterprise leads to the possibility of taking them into account in the framework of one type of activity, which means using the risk absorption rule and allows to reduce the overall calculated risk level.

If risks related to various performance aspects, and negative factors are manifested in dependence from one another (the rule of logical adding of risks), the risk is calculated as the sum of the risk of a single event on the other chances:

$$Po = \Sigma (p_i * q_j), \qquad (4)$$

and the total measure of risk (*Mp*) will be:

$$Mp = \max \{Mp_i\}. \qquad (5)$$

In other words, the overall measure of risk will be determined by the risk measure of the most risky phase of the project, which means that the level of security of the whole project depends on the degree of risk of injury at each stage. Stage characterized by the least formed injury management system will determine the overall calculated risk level associated with injuries.

The advanced experience of the developed countries shows the monitoring of the process of industrial safety where *DuPont BBC* program is used. During the research we got an answer of the questioning on the reasons of employees, failures to observe the safety requirements. Then the program allows the typical errors that result in traumas. Thus, the task of BBS is to determine that an employee is in a such condition that will lead to failure and prevent an accident.

The use of the existing works concerning industrial traumatism management allows to effectively dispose the facilities and obtain significant results.

References

1. B.A. Anfilofyev, E.V. Lukenuk, Y.V. Dementyeva, Vector of Science, TSU, **2(24)**, (2013)
2. "The methodology for assessing the impact of human factors on the origin of the case injured by the job and determining the percentage of the involved workers," Approved by the order of the open joint-stock company, (2014)
3. S. Young, *Management System of an Enterprise: translated from English* (Sov. Radio, Moscow, 1972)
4. J. Ober-Krie, *Enterprise Management: translated from French* (Progress, 1973)
5. V.S. Parshina, Railway Economy, **12**, (2008)
6. V.S. Parshina, Railway Economy, **7**, (2014)
7. V.M. Ponomaryev, Transport of the Russian Federation, **1**, (2011)
8. B.M. Rapoport, *Optimization of Management Decisions* (Teis, 2001)
9. V.N. Spitsnadel, *Theory And Practice Of Optimal Decision Making: textbook* (Business Press, St. Petersburg, 2002)
10. Ya.R. Reylyan, *Analytical Basis For Decision Making* (Finance and Statistics, Moscow, 1989)
11. E.V. Shikhin, A.G. Chkhartashvili, *Mathematical Methods And Models In Management* (Delo, Moscow, 2009)
12. V.I. Busov, *Management Decisions* (Yurait, Moscow, 2013)
13. Ye.I. Brazhko, G,V. Serebryakova, V.B. Utkin, *Management Decisions* (RIOR, Moscow, 2008)
14. A.A. Gusev, USURT Vestnic, **2**, 94-103 (2013)

15. V.N. Kodin, S.V. Lityagina, *Work On a Management Decisions* (KnoRus, Moscow, 2010)
16. Ye.V. Lavrova, Ye.V. Stroeva, *Development Of Managerial Decisions* (INFRA-M, Moscow, 2013)
17. V.S. Parshina, A.A. Gusev, Railway Economy, **6**, (2013)
18. V.S. Parshina, A.A. Gusev, Innovative Transportation, **5(6)**, 44-52 (2012)
19. V.A. Trefilov, *Safety Management At Work (Health And Safety Measures): textbook* (Publishing house of the Perm. state. tehn. Univ., Perm, 2009)
20. N.V. Khokhlov, *Risk Management* (Unity-Dana, Moscow, 2001)
21. I.K. Adizes, *Managing change* (Mann, Ivanov and Ferber, Moscow, 2014)
22. A.O. Blinov, *Change management: training manual* (Dashkov and Ko, Moscow, 2014)
23. W. Bridges, *Managing the company through a period of structural* (Williams, Moscow, 2008)
24. I. Gerchikova, *Management: the textbook* (UNITY-DANA, Moscow, 2009)
25. L.D. Gitelman, *Transforming management. The leaders of the reorganization and management consultants: a training manual* (Delo, Moscow, 1999)
26. A.A. Ogarkov, *Management of organization: textbook* (Eksmo, Moscow, 2009)
27. A.G. Porshnev, *Management organization: textbook* (INFRA-M, Moscow, 2008)

Production capacity of metallurgical enterprises in modular structure of management accounting

E.S. Zambrzhitskaia[1,*], *D.G.* Neshporenko[1], and *O.V.* Dolgopolov[1]

[1]Nosov Magnitogorsk State Technical University (NMSTU), Lenin Street, 38, Magnitogorsk city, Chelyabinsk Region, Russia

Abstract. This article is the result of constant research in development of management accounting at modern Russian metallurgical works. The required steps of renovation are presented for the modernization of management accounting system in the case of production capacity. The necessary factors with defined values are systematized for modern enterprises. The main ones are those which are regulate the measure of the payload of production capacity. For the purposes of effective management accounting at metallurgical enterprises it is suggested to use the system of controlling parameters which were formulated for the enterprise for production of rolls. Later in the article the necessary changes to the modular structure of management accounting business are described and expanded with new structural element – "Management of production capacity". The suggested methodical approach will allow the company management to respond quickly to rapidly changing external environment and, as a consequence, to make effective management decisions.

1 Introduction

The economic policy of any state is aimed for such purposes and scales, that the gross domestic product was more optimal.

Gross domestic product is a macroeconomic indicator that reflects the market value of all final goods and services (that is aimed for direct consumption) produced per year in all sectors of the economy in the State for consumption, exports and savings [1].

It is important to note that the solution of any macroeconomic problem is better to start with each individual company, its divisions and jobs. The maximum possible output of the production units of the enterprise production for a given period is determined by the index of industrial output (production capacity) [2].

It should be noted that in modern conditions of managing, the production capacity as the micro-level indicator (PM) have been neglected. Most economists practitioners believe that the figure static,, unchanged over time. In other words, it is assumed that the value of the production capacity is determined at the design stage of the production system and is independent of external and internal factors affecting the continued operation of the enterprise.

Meanwhile, in the scientific and specialized literature, there are well-established view that the efficiency of use of production capacity depends on a number of factors associated with the presence and the real possibilities of maneuver available to the enterprises manufacturing resources - means of labor, objects of labor, human labor, professional management, etc [3, 4].

It is important to note that the calculation of the production capacity of the structural units of the enterprise issue is more technical and technological. At the same time, the question of efficiency of use of production capacities and interfaces of structural units in the same industrial complex has distinct economic aspects [5, 6, 7].

2 Problem Statement

The existing theories of planning the production capacity follow the idea of factors, influencing the value of PC (production capacity), which are [8, 9, 10]:
1) factors for organization of production processes (level of equipment maintenance, regularity and rhythm of production, utilization of production space, the degree of use of the equipment);
2) factors of labor potential (training and skills development, motivation to improve equipment utilization, development of creative potential of employees, progressive forms of work organization);
3) factors of technical and technological issues (the perfection of machinery equipment, technology products, equipment downtime for repairs, advanced technology, cooperation in the production);
4) factors of the organization and management of production (outstaffing, logistic service production, focusing production capacity).

Three-dimensional factor model E.V. Novikov allows you to select over thirty key factors [11]. The variety of conditions (factors), improving production capacity and the extent of its use make the mathematical modeling of the process indicator too sophisticated.

* Corresponding author: jenia-v@yandex.ru

For these purposes, E.V. Novikov suggested to select the components of the direct interaction of (baseline factors) and dependent, derived from the basic factors.

As the basic factors taken following – the demand for products and performance that defines the capabilities of the enterprise to meet the market demand.

The model of economic growth by A.M. Fattakhova describes the production capacity as a dependence on the industrial, economic and financial activity of the economic entity, which in turn provides a stable activity of the company, which is building some good, satisfying the needs of the market and retaining competitive advantage [12, 13].

The presented analysis of the factors affecting the value of the index of production capacity, allows us to conclude that the production capacity is a reflection of "resultant" of all used production and management factors on the particular enterprise.

3 Main Part

3.1. Analysis of the structure of the production system of metal-working enterprises

The issue of management of production capacity is significant for all sectors of the economy, but it acquires a special urgency for metalworking companies where the production process is asset-intensive and multi-layered.

For management purposes, the production capacity is proposed to define the controlled parameters. The value of these parameters should allow to assess the production capacity of the value sensitivity to changes in the internal and external environment of the enterprise.

It is mandatory to be specifically developed a clear and relatively simple methodology for quantitative (numerical) assessment of efficiency of production capacity, which allows for specific managerial situations to determine the ratio of its value at low labor costs (man-hours).

The effect of controlled process of this indicator will depend directly on the degree of objectivity, reliability, progressiveness of the selected parameter.

Let us consider the proposed methodological approach on a concrete example of a metalworking enterprise, which produces the following kinds of production:

1) increased accuracy shaft;
2) spline grooves shaft;
3) whittled grooves shaft.

The entire manufacturing process is carried out at six production sites.

The production site No. 1 is a stamping shop. The stamping is a process of plastic deformation of the material with changes in body shape and size. The presses used for the process of stamping are devices to deform the material by means of mechanical influence.

For giving the product a certain shape is used at the factory shaft stamping.

The shaft stamping is an operation of metal forming, and obtainment of axisymmetric parts of the cylindrical workpiece by the simultaneous action of the radial and axial loads. An axial load of the workpiece is generated due to the movement of the punch, the radial load is generated through the running of its lateral surface in the rollers or shafts. Thus, the shaft stamping is a way for the complex local deformation, in which in a single technological process takes place a combination of one of the basic forging operations, specifically firmware or precipitation with transversal rolling or running.

The shaft stamping improves the physical and mechanical properties of processed metal, ensuring the required arrangement of its fibres, which improves operational properties of the resulting parts.

The production site No. 2 is a turning shop. The turning is a cutting machining of external and internal surfaces of rotation, including a cylindrical and conical, trimming, cutting, chamfering, processing fillets, slotting, cutting internal and external threads on lathes.

The production site No. 3 is a thermochemical treatment. The thermochemical treatment is a heating and holding of materials at high temperatures in reactive environments (solid, liquid, gaseous).

The thermochemical treatment is carried out with the purpose of enrichment of the surface layers of products of certain elements. They are named by specialists "the saturating elements" or "the components of the saturation". As a result of thermochemical treatment is formed the diffusion layer, namely changes the chemical composition, phase composition, structure and properties of surface layers. The change in chemical composition causes changes in the structure and properties of the diffusion layer.

The production site No. 4 is a planer shop. The planing is a machining process of materials by cutting by chip removal, carried out by the relative reciprocating movement of the tool or product. For the metal planing treatment there are used the different machines: transverse- and longitudinal-planing, hem-planing, planing-slotting machines and the others.

The production site No. 5 is a milling shop. The milling is a mechanic machining process in which the cutting tool (mill) performs a rotational movement, and the workpiece makes a forward movement.

The production site No. 6 is a grinding shop. The grinding is a mechanical or a manual operation on the metal processing. It's a variety of the abrasive treatment, which, in turn, is a type of cutting. The mechanical grinding is usually used to a processing of metals to specified size with micron precision, as well as to achieve the smallest surface roughness of the product acceptable valid the GOST (State Standard).

To obtain the increased accuracy shaft it is necessary the workpiece passed the following production areas: punching → heat treatment → turning → planing.

To obtain the spline grooves shaft it is necessary the workpiece passed the following production areas: punching → turning → heat treatment → milling.

To obtain the whittling grooves shaft it is necessary the workpiece passed the following production areas: punching → turning → heat treatment → grinding.

Taking into account the peculiarities of the technological process described above, the company operates a network of the technological interactions, which is presented in a graph form in fig. 1 [14, 15]. The circles mean the kinds of products, the arrow shows the expenditure of some kinds of products on the production of others. The rectangles indicate the links of the production system.

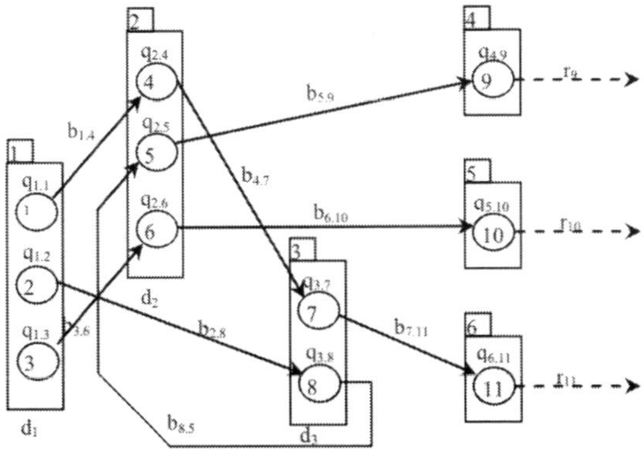

Fig. 1. The structure of the production system (metal working factory). In fig. 1 are the following notation: q – the production capacity of the links on the products; b – the coefficients for direct costs of the products on the products; r – the product range structure of the final production.

Thus, the considered example shows us modern metal working factories present themselves as a comprehensive multi-stage system, an important indicator of which is the amount of the production capacity and the efficiency of their use.

3.2. Determining the system of controlled parameters

In the aim of the organization of management accounting of production capacities at the studied enterprise can be offered as the monitored parameters the following indicators [16 - 20]:

• an availability of integrated information (automated) system of management accounting, monitoring and control of industrial capacity and the level of its use;
• the diagnostics of the organizational and technical level of the main shops at the industrial enterprise;
• an assessment of the manufacturing opportunities regarding the labour expenses, natural and monetary terms on the problems decision of modernization, reconstruction and technical re-equipment at the enterprise;
• analysis of uneven production load jobs, of imbalance of the capacities of the production facilities and productive capacity of the structural subdivisions;
• the identification of free capacities, of reserves and bottlenecks in the production;
• the self-cost per unit of production;
• the indicators of the labour productivity (the indices of labor productivity associated with number of personnel, the actual value of revenue generated);
• the value of the differential costs of the operation production capacities of the enterprise by autonomous business units taking into account the market conditions;
• the assortment structure of finished production and its compliance with the market conditions (the structure of production capacity to market demand); etc.

3.3. The results of the study

After the systematizing the proposed control options it may be possible to allocate the main management capacity element of metal working production of the enterprise. It may be the creation of the automated unit of information automatic base of management accounting, monitoring and control of industrial capacity and of the level of its use.

According to the modular structure of the management accounting proposed by O. Kaverina currently, there are three segment of management: budgeting and monitoring of the execution of budgets, calculation of the cost of production and the special managerial decisions shown in Fig. 2.a [21].

According to the authors, it is reasonable in this model to make an additional segment which is responsible for control of the value production capacity at the business entity presented in Fig. 2.b.

Budgeting and monitoring execution of budgets	**Calculation of the** product **cost** and the processes cost
Special management decisions	

a)

Budgeting and monitoring execution of the budgets	**Calculation of the** product **cost** and the processes cost
Special management decisions	**The production capacity** of an enterprise and structural unit

Fig. 2. The modular structure of management accounting by the method of O. Kaverina before change (a) and after change (b).

4 Conclusion

The lack of management accounting of production capacity and the level of their use suggests that still may be not resolve an independent task by the objective assessment of the indicators of production capacity of industries with the complex technological cycle, in terms of conjugate objects.

In conclusion, it should be noted the proposed methodological solutions for improving the efficiency of the use of production capacity can be used not only on metal working production but also in other sectors of economics.

The main conditions of the effective administrative decisions are:

1) the creation of the module "Management of manufacturing capacity" in the structure of management accounting at the enterprise;
2) detailed study of the system-controlled parameters of the effective use of production capacity, in the

DOI: 10.1051/shsconf/20173501128

conditions of dynamically changing external environment.

References

1. T.Yu. Matveyeva, *An introduction to the Macroeconomics: textbook* (HSE Publishing House, Moscow, 2010)
2. E.M. Korotkov, I.Yu. Soldatova, *The foundation of Management: textbook, Publishing & Trading Corporation* (Dashkov and Ko, Moscow, 2013)
3. E.S. Zambrzhitskaya, A.V. Logacheva, M.V. Logacheva, Young scientist, **10(69)**, 238-242 (2014)
4. M.I. Tertyshnik, I.A. Dmitrieva (Ageeva), Economy of Russia and its regions: issues of efficiency and development strategy, 241-246 (2005)
5. N.I. Gavrilenko, R.P. Kuksin, O.V. Baguzova, Vestnik of IIEL, **4(5)**, 58-60 (2011)
6. K.S. Krivyakin, Journal of VSTU, **2**, 36-39 (2010)
7. S.N. Tsimbalyuk, T.V. Kozlova, Actual problems of modern science, technology and education, **2**, 280-284 (2010)
8. G.I. Shepelenko, *The production planning at the enterprise. Economics and management: textbook for students of economics* (Publishing Centre «March», Rostov-on-Don, 2004)
9. M.V. Dadalova, Journal of the university of cooperation, Economics and Law, **4**, 290-294 (2008)
10. V.S. Tikhonov, M.A. Brajnikov, Higher education, business, enterprise, 1st ed. high school collection of scientific papers, 116-120 (2003)
11. E.V. Novikov, Journal of South-Russian State Technical University. Series of sociology and economics, **1**, 71-74 (2012)
12. A.M. Fattakhov, *Using production capacity and strategic stability of the oil refining enterprises* (Ufa, 2005)
13. G.V. Danilov, I.G. Ryzhova, E.S. Voynova, Economy analysis: the theory and practice, **3(168)**, 34-39 (2010)
14. G.V. Danilov, I.G. Ryzhova, E.S. Voynova, St. Petersburg State Polytechnical University Journal. Economics, **4(102)**, 87-90 (2010)
15. V.N. Kononov, I.G. Ryzhova, E.S. Voynova, Topical issues of economic science, **12-2**, 167-176 (2010)
16. R. Zeman, M. Vochozka, E.S. Zambrzhitskaya, N.E. Ivanova, R. Píchová, M. Popílková, K. Kabourková, *Management accounting: textbook for high schools education* (Magnitogorsk, 2015)
17. V.T. Chaya, N.I. Chupahina, *Management accounting: a textbook and a workshop for the academic bachelor* (Moscow, Publisher House YURAYT, 2016)
18. G.Ya. Ostayev, *Management accounting: a textbook* (Business and Service, Moscow, 2015)
19. O.E. Nikolayeva, T. Shishkova, *Classic management accounting: a textbook* (Krasand, Moscow, 2014)
20. E.S. Zambrzhitskaya, Remote course Management accounting: e-training complex, Chronicles of the combined fund of electronic resources: Science and Education, **11(78)**, (2015)
21. O.D. Kaverina, *Management accounting: theory and practice: textbook for students* (Publisher House YURAYT, Moscow, 2014)

Energy management of industrial enterprise

E.A. Lyaskovskaya[1,*], and *M.R.* Kurbangaliev[1]

[1]South Ural State University, Chelyabinsk, Russia

Abstract. In the intensifying condition of economic situation and increasing competitiveness in domestic and foreign markets, the most important way to develop competitive ability of an industrial company is to reduce energy costs in the production process. Insufficient level of the efficiency of energy resources usage affects an industrial company's performance indicators and its investment attractiveness. A promising way of solving this matter is to develop and implement a strategy of rational energy consumption, which is aimed at the realization of company's potential to optimize the consumption of electric energy by using internal and external resources in order to minimize energy costs. The strategy of rational energy consumption defines how an industrial company acquires electric energy and uses it to sustain the production. While developing and implementing the strategy, one should use a systemic and complex way and consider the following: peculiarities of electric energy and power as products; the structure of electric energy market and the possibilities of its consumers; peculiarities of price-formation on electric energy market; technical and technological, organizational and administrative, social and economic parameters of a company, characteristic features of its resource potential and production processes; the results of company's energy efficiency audit and energy problems; company's reserves that can increase its energy efficiency. An integral strategy of energy consumption includes a strategy for energy preservation and efficiency and a strategy for energy costs management. Both strategies are interrelated and serve for one purpose, which is minimizing the energy costs. This division helps simplify the analysis, search for alternatives and realization of energy management on operative, tactical and strategic levels, considering the regional and industry-specific peculiarities of an industrial company, its financial performance and efficiency.

1 Introduction

Lately there has been a reform in Russian electric-power industry, which resulted in a completely new industry structure, new procedural and institutional form of enterprises in the industry, new open market of electric energy and power. Due to this the prices on electric power boosted, growth rates became threatening high, overcoming ones in the developed countries (USA and EU). The increasing prices on energy affect Russian industrial businesses as major energy consumers , which in the first place shows up in the energy costs and, as follows, in the production costs. The industry competitiveness increased after Russia has become a member of WTO. The expansion of Chinese businesses, economic measures against our country, the negative air in the economic situation that still remains since the crisis of 2008 – all this increase competitiveness even more. An important direction in this view is lowering energy costs, namely the costs of electric energy and power consumption. For an industrial company these costs can be lowered by effective rational management of energy consumption, based on energy saving, increased energy efficiency and management of energy costs.

Many Russian scientists have studied the field, such as I.A. Baev, B.I. Kudrin, L.D. Ghitelman, A.V. Sedov, E.E. Tikhonov, A.I. Tatarkin, V.G. Mokhov, V.V. Karpov, V.I. Gordeev, V.P. Stepanov, J.A. Fokin and others. Despite impressive results achieved by them, there is still no sufficient way to overcome the problems of strategic energy management on Russian market.

2 Problems in reforming electric-power industry in Russian Federation

In 2001-2008 years Russian electric-power industry has been reformed. It was adapted to modern conditions of economic management. During the reformation process dramatic changes took place in the industry (the industry structure was completely changed, a new normative framework and regulation system were formed etc.). These changes affected not only companies in the sphere of energy supply, but also the consumers of energy [1-9]. Present situation in the electric-energy and power industry can be described like this

- Poor technical performance in russian power suppliers.
- Absence of motivation to increase efficiency, energy saving, rational planning of producing and consuming electric energy, frequent cases of non-payments from consumers of energy.

* Corresponding author: elen_lea@mail.ru

• Absence of informational and financial transparency among power suppliers.
• Low profitability of energy sphere.
• Lack of investments.
• Cases of "rolling black-outs", frequent "black-outs" in particular regions.
• Decreasing reliability of power suppliers in many regions of Russia.
• New independent players have no possibility to enter the market.
• Absence of conditions for healthy competitiveness, uncompetitive price-formation.

During the reformation electric energy prices were constantly rising [10], power grid system contributed to this significantly. The origins of the increase in network rates are as follows
• Mistakes in the implementation of rab – rate-regulatory system, which is aimed at attracting investments to power grid system.
• Absence of proper supervision over the spending of the assets of investment projects.
• Dishonest regional power grid organizations.
• Unsolved issue of cross-funding and redistribution of electricity payments among several groups of consumers, when electric energy costs for citizens are partially covered by industrial consumers.

Summing this up, we can draw an unsettling conclusion: the reformation cannot be called complete, as nearly none of the planned goals is ultimately reached. In the producing sector there is still no way to effectively stimulate business initiative and competitiveness that can contribute to energy preservation, lowering production costs and increasing production efficiency. In the trading sector, just like in the producing sector, there are still no means of creating healthy competitiveness. The problem is common for the whole industry – the industry does not have any effective state supervising and proper state regulation.

3 Analysis of market of electric energy and power in the Russian Federation

Russian market of electric energy and power has sophisticated structure due to specificity of electric energy and power as products, vast territory of the Russian Federation, insufficiently developed power distribution structure and others. Due to certain features, determined by physical properties of electric energy, peculiarities of its production and storing processes, electric energy is a specific product. At first, electric energy cannot be stored in amounts, corresponding to the scales of energy system. Only constant balance between the production and the consumption can provide stable functioning of market in this sphere. At second, from consumer's point of view all producers act as one producer, and from producer's point of view – all consumers act as one consumer [11-14]. At third, planning and estimation of production and consumption of electric energy is needed n order to organize normal functioning of energy system. The amounts of production are directly linked to the amounts of consumption here, so it is important to evaluate the amount of consumption, which are not constant. At fourth, timely balancing of the whole energy system is obligatory.

Impossibility to store electric energy and the absence of substitute products determine the necessity of accumulation of backups of generating capacities, power grids and fuel at power plants [12]. It leads to the fact that consumers pay not only for the amount of electric energy they consume, but also for the possibility of consuming. Price for actual consuming is price for electric energy as the main product on the electric energy market. Price for the possibility of consuming is price for power as a specific product, that provides every consumer with a possibility to use backup generating capacities to satisfy consumer's demand in electric energy [11-14].

Presently there is a sophisticated system of price formation on the Russian market of electric energy and power. This system lets industrial consumers choose their rates and therefore change the final price for consumption of electric energy and power. Only for that industrial consumers should develop their energy management strategies, hold technical and organizational activities managing price levels. Implementation of such activities does not presume direct reduction of consumption of electric energy, which can be achieved only by implementing energy efficiency activities [11-14].

4 Problems in energy consumption efficiency among Russian industrial companies

It is historically developed that industrial sector in the Russian Federation, inherited from the Soviet Union, was a rather energy demanding economy sector. It was determined by following factors
• Structure of production, developed during a long period of time.
• High moral and physical exhaustion of major production assets of industrial companies.
• Technological undevelopment of industrial sector.
• Severe environmental conditions.
• Cheap energy resources, including electric energy [15-16].

In the middle of 1990s energy demands of Russian economy was 2 -3 times higher than the same indicator for developed countries. The main reason was absence of interest from owners of industrial companies in the modernization of production and in rational use of energy resources, including electric energy. The last can be explained by following factors
• Electricity rates, artificially lowered by state.
• Low level of accounting, control and regulation of electric energy consumption [17-19].

In the 2000s government developed a number of measures, aimed at reducing energy demand of industrial sector. Federal law "About energy preservation and increasing energy efficiency" became one of the means of realization of energy preservation and increasing energy efficiency [5]. After Russia got membership in WTO the problem of managing energy consumption on industrial companies become more and more vital.

The moving forces for implementation of activities aimed at energy preservation and increasing energy efficiency are

• The increase of prices of electric energy.
• The decrease of costs for electric energy.
• The decrease of production costs.
• The increase of profitability and competitive ability of a company.

Besides, there are also factors that impede the implementation of activities aimed at energy preservation and increasing energy efficiency. The usually are following [17-19,22]

• Absence or lack of financial resources for development, verification, implementation and monitoring of energy preservation activities.
• Absence of motivation among personnel.
• Low level of knowledge in the field of energy preservation and increasing energy efficiency among management.
• Lack of information on the methods of energy preservation and increasing energy efficiency.
• There are usually two major directions of energy preservation and increasing energy efficiency in an industrial company [20-21].
• Extensive energy preservation, based on quantitative decrease of energy consumption.
• Intensive energy preservation that implies qualitative change in the system of power supplying and consumption.

In order to manage energy consumption in an industrial company, energy management systems should be implemented. The notion of energy management is in a way a synonym to energy consumption management and includes structure, resources, processes and procedures necessary for realization of planning, organization, providing the realization, evaluation, analysis and monitoring of energy policy and achieving energy goals of a company I order to increase its competitive ability [17-19].

5 Strategy of rational energy consumption in an industrial company

In our view, strategy of rational energy consumption in an industrial company – is a part of company's general strategy, aimed at the realization of company's potential to optimize the consumption of electric energy by using internal and external resources in order to minimize energy costs. The strategy defines how an industrial company acquires electric energy and uses it to sustain the production. According to its major goals, strategy of rational energy consumption in an industrial company can be divided into two parts

• Strategy for energy preservation and efficiency.
• Strategy for energy costs management.

Each of the parts serves for the general purpose of strategy of rational energy consumption but use different ways of achieving it.

This division helps simplify the analysis, search for alternatives and realization of energy management on operative, tactical and strategic levels, considering the regional and industry-specific peculiarities of an industrial company, its financial performance and efficiency.

The sequence of stages of strategic management of energy consumption in an industrial company can be presented like this (Fig. 1).

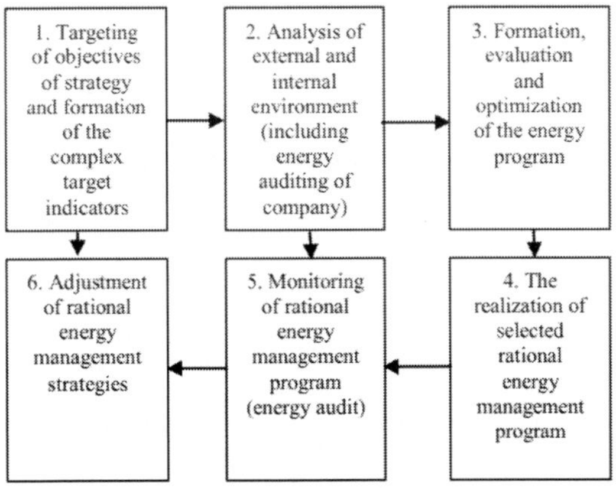

Fig. 1. Stages of development of strategy of rational energy consumption in an industrial company.

Characteristics of strategy of rational energy consumption in an industrial company are presented in table 1.

Table 1. Characteristics of strategy of rational energy consumption in an industrial company

Strategic components	Detalisation of strategic components
goal	Minimize energy costs for a company.
issues	Energy preservation, energy efficiency increase on production. Lowering the costs of acquired electric energy and power.
substrategies	Substrategy for energy preservation and efficiency, aimed at reducing the amounts of electric energy consumption. Substrategy for energy costs management, aimed at reducing the costs of acquired electric energy and power.
factors that affect the costs of energy consumtion and possibilities for energy preservation in a company	Technical factors (quality of materials, equipment, hardware, availability and quality of commercial accounting system etc.). Technological factors (technology, list of equipment, technological working patterns etc.). Organizational factors (company's working patterns and schedules, level of planning, operative production schedule etc.). Social factors (level of qualification and loyalty among employees, personnel motivation, necessity to provide certain working conditions etc.). Ecological factors (air temperature, time of day, meteorological conditions, etc.). Economic factors (accounting, analysis and management of costs, rates for electric energy and power).

Strategic components	Detalisation of strategic components
ways of energy preservation and incresing company's energy efficiency	Realization of constant energy economy. Cuts of energy losses in company's system of energy supply. Implementation of less energy demanding technologies and equipment. Electric energy efficiency increase during production process. Efficiency increase of company's system of energy transmission and distribution.
stages of energy management and increasing company's energy efficiency	Energy consumption examination (energy audit). Development of a set of activities aimed at increasing energy efficiency, it includes • Organizational activities aimed at increasing energy preservation and energy efficiency of production process. • Technical activities aimed at increasing energy efficiency of main and auxiliary production lines. • Development of computer-controlled system for accounting controlling electric energy consumption. • Activities aimed at increasing power supply system efficiency. Technical and economic grounding for suggested activities. • Implementation of previously developed set of activities. • Monitoring of results of implementation of activities aimed at energy preservation, for futher analysis.
stages of company's energy costs management	Operative management of energy consumption can be achieved by immediate reaction on changes in internal and external environment of a company. It can be also achieved by constant interaction with production systems of a company and decrease in amount of discrepancies between planned and fact energy consumption, which leads to minimized fees that company pays for instability on a balancing market. Strategic management of energy consumption can be achieved by development and implementation of organizational and technical activities aimed at lowering energy costs, which, in its turn, can be achieved by optimizing energy consumption patterns (decreasing energy consumption during peak-load hours and increasing it gradually during the last part of day).
ways of reducing costs of acquired electric energy and power	Cooperation with a reliable power supplier, offering the most economically beneficial terms. Choosing the most economically suitable rate for electric energy and power according to company's energy consumption pattern, or optimization of present energy consumption pattern in order to correlate it with the most economically beneficial rate for electric energy. Adaptation of company's structure of power grid equipment and improvement of electric energy and power commercial accounting system. Changing company's energy consumption pattern.
factors that affect company's energy costs management	External factors • Activity of federal and local governments (FRS, FAS, local executive body of power operating in the field rates regulation). • Activity of guaranteeing suppliers and retail energy sales companies. • Activity of commercial infrastructure companies of electric energy and power wholesale market ("TSA", JSC). • Activity of companies of technological infrastructure of the market ("SO UPS", JSC). • Generating companies and power suppliers activity. • Other electric energy consumers. Internal factors • Technological pattern and schedule of production. • Technical specifications of electric equipment used by company • Structure and specifications of external power supply system. • Structure and specifications of electric energy and power commercial accounting system.

It should be noted that the strategy for energy preservation and efficiency is aimed at reducing the amounts of electric energy consumption. Strategy for energy costs management is aimed at reducing the costs of acquired electric energy and power. Both strategies are interrelated and affect each other. While carrying out one strategy, one should take into account the aims of the other one.

The division of strategy of rational energy consumption into parts makes it possible to

• Show the factors of external and internal environment , that affect the formation of energy costs, for their detailed analysis and increasing the quality of evaluation of their impact.
• Simplify the procedure of formation of target values for the evaluation of the efficiency of rational energy consumption management, and also simplify the procedure of planning of these target values.
• Elaborate complex sets of organizational and technical activities aimed at general strategic goals.
• Simplify the procedure of planning and the mechanism of realization for activities prepared for implementation.
• Increase the quality of evaluation (monitoring) of implemented activities.
• Elaborate the necessity for correcting actions, that can improve other activities aimed at rational energy consumption in a company, also define the directions of these actions.
• Organize company's energy management in a proper way.
• Define the directions of strategic and dynamic energy management.

The realization of strategy of rational energy consumption is a powerful mechanism that helps

• Achieve the most suitable energy consumption pattern according to present conditions of electric energy and power market.
• Increase financial and economic stability, efficiency, competitive ability and investment attractiveness of an industrial company.

6 Conclusions

On the basis of the analysis of Russian energy sector reformation, it is considered here as ineffective. It is stated that the prices for electric energy have grown significantly, and due to this appears the problem of increasing energy costs for industrial companies. In order to solve the problem we offer and ground the necessity of implementation of strategy of rational energy consumption.

The analysis of peculiarities of electric energy and power as products and the structure of electric energy and power market has let us state and ground the necessity of regulation of energy consumption patterns of industrial companies taking into account both short-term and long-term condition of electric energy and power market.

The notion of "strategy of rational energy consumption" is specified, its goals, subjects, directions and mechanisms are defined. The characteristic feature of this strategy is its division into two substrategies: substrategy for energy preservation and efficiency and substrategy for energy costs management.

The work was supported by Act 211 Government of the Russian Federation, contract № 02.A03.21.0011.

References

1. *Russian Federation Government Resolution, On the reform of the Russian power* (2001)
2. *Federal Law, On Electric Power Industry* (2003)
3. *Russian Federation Government Resolution, On approval of the wholesale market of electric power and capacity of the Rules and on Amendments to Certain Acts of the Russian Federation on the issues of functioning of the wholesale electricity market and power* (2010)
4. *Russian Federation Government Resolution, On the functioning of retail electricity markets, the full and (or) partial restriction mode electric power consumption* (2012)
5. *The Federal Law, On energy saving and energy efficiency improvements and on Amendments to Certain Legislative Acts of the Russian Federation* (2009)
6. *GOST R ISO 50001-2012, Energy management systems. Requirements with guidance for use*
7. *Russian Government Resolution, On approval of rules of non-discriminatory access to electricity transmission and delivery of these services* (2004)
8. *Regulation of the Russian Government, On the Energy Strategy of Russia for time period untill 2030* (2009).
9. *The Russian Federation Government Resolution, On the mechanism of promoting the use of renewable energy in the wholesale market of electric energy and power* (2013)
10. *Analytical report: analysis of electricity reform and proposals to increase its efficiency*, (Moscow, Russia, 2013)
11. E.Z. Dambaeva, Bulletin of Irkutsk State Economic Academy, **2**, 78-81 (2010)
12. R.B. Naumkin, Bulletin of the Kuzbass State University, **6**, 121-127 (2014)
13. D.V. Balandin, Vestnik St. Petersburg State University, **8**, 167-188 (2005)
14. I.A. Gerasimov, Russian journal of scientific publications, **1**, 58-61 (2012).
15. M.Y. Mishina, O.N. Polyakova, Herald Bryansk State University, **3**, 62-64 (2012)
16. I.A. Baev, I.A. Soloviev, A.P. Dzyuba, Economics, management and investment, **2**, 23-43 (2014)
17. A.I. Dovgyallo, D.A. Dovgyallo, S.O. Nekrasov, A.I. Dovgyallo, *Energy management: electronic edition* (Samara, Russia, 2011)
18. G.F. Kayachev, M.P. Govorushkin, Bulletin of the Krasnoyarsk State Agrarian University, **11**, 20-24 (2014)
19. V.S. Salnikov, A.A. Bryksina, Izvestiya of the Tula State University. Technical science, **12**, 149-155 (2013)
20. I.V. Burenina, D.A. Gamilova, S.V. Alexeyeva, Economy and Management. Scientific and practical journal, **5**, 13-19 (2015)
21. D.C. Eltyshev, N.I. Khoroshev, **11**, 1472-1475 (2014)
22. V.V. Odinochenkov, I.V. Gavreeva, M.Y. Mishina, Herald Bryansk State University, **3**, 98-105 (2013)

Engineering Center as a tool for HEI's scientific and innovation potential integration in the real economy

M.S. Minkin[1,*], *M.A.* Izotov[1], and *O.G.* Ananova[1]

[1]Don State Technical University, Rostov-on-Don, Russia

Abstract. Currently, one of the biggest challenges facing the state HEI is the problem for commercialization of intellectual activity results using technology transfer funds into the real economy. However, if the creating products processes ends in Research and Development stage, everything is quite transparent and clear here, but at the stage of transition to the process of self-production and sales there appears a number of difficult problems solvable. To simplify the interaction between the university and the enterprises of the real economy are invited to create an engineering center as a separate entity, affiliated with the university, whose tasks will include the issues of technology transfer to the real economy, as well as the establishment of an external production site, to address the commercialization of tasks results of intellectual activity without involving in this process the intermediate third-party organizations.

1 Introduction

Innovative activity and small business – two main components of a sustainable development of modern economy. The market relations assume existence of the enterprises of various sizes and forms of ownership. It is necessary to allocate especially small forms of the enterprises for implementation of innovative activity as small enterprises differ in high efficiency in decision-making, fast adaptation to influences of external factors, high turnover of means and small management expenses, and also a susceptibility to innovations in production and administrative activity.

Small business in the innovative sphere is the most important structural element of new economic business models and a nutrient medium for system changes in technology and the social relations. The small innovative enterprises in the person of their most dynamic and successful representatives act for economy, on the one hand, as the test business ground, and with another – the excellent base on which the national and global companies are under construction.

It is very important that the priority directions of formation of innovative infrastructure, mechanisms of support of the small innovative enterprises and investment projects of their development became key contents of regional programs of support of development of small innovative business. A main objective of the small innovative enterprise as commercial organization, receiving profit on the basis of introduction of scientific developments in production and production is. Unfortunately, now not all enterprises are ready to introduction of results of innovative activity because of high cost of cost of the new equipment, materials, increase in costs of a salary to new qualified personnel and other reasons. Many enterprises now, seeking for maximizing profit, go on the way of reduction of product cost due to decrease in her quality, and it often leads to decrease in demand for products.

According to the Federal Law of August 2, 2009 № 217-FL "On making amendments to some legislative acts of the Russian Federation concerning the process of establishment by budget scientific and educational institutions the business entities to practical implementation of intellectual activity results" the budgetary scientific institutions, and also the created scientific institutions, have the right to be founders of economic societies, in particular the small innovative enterprises for the purpose of practical application of results of intellectual activity (programs for electronic computers, databases, inventions, useful models, industrial samples, selection achievements, etc.), exclusive rights to which belong to these scientific institutions.

Creating of the engineering center as a separate legal entity, affiliated with HEI can become a successful mechanism for commercialization of scientific and innovative potential of state HEIs. The proposed organizational and legal form of the commercial organization will be Limited Liability Company. It will be registered in accordance with the Federal Law of the Russian Federation (Federal Law of August 2, 2009 № 217-FZ "On making amendments to some legislative acts of the Russian Federation concerning the process of establishment by budget scientific and educational institutions the business entities to practical implementation of intellectual activity results"), according to which the budgetary scientific institutions and universities, which are budgetary educational institutions were granted the right to create, together with other legal

* Corresponding author: msi_58@mail.ru

entities or individuals economic companies without having the special consent from the federal enforcement authorities, which they are administered by. This law opened up the new opportunities to increase the efficiency of the higher education interaction with high-tech sector to increase competitiveness of the economy.

The business profile of such entities' economic activity is a practical application (implementation) of the intellectual activities results (RIA), which exclusive rights belong to this budget institution. The RIA selection is carried out by institution independently on the database analysis results accounting RID of this budget institution and the results of market needs investigations for the planned areas of practical application (implementation) RID.

The business entity that is planned to create is a subject to state registration in accordance with the Federal Law of 8 August 2001 (Ed. July 21, 2014) № 129-FZ "On the State Registration of Legal Entities and Individual Entrepreneurs".

The main advantage of this business entities created under the Federal Law of August 2, 2009 № 217-FZ is a possibility to apply the simplified taxation system, as well as the minimization of insurance premiums deductions during the wages paymentf. The Federal Law №272-FZ of 16.10.2010 sets for the business entities applying the simplified taxation system the total size of insurance premiums in the amount of 28.2%.

The educational institution, being the participant of economic community, is engaged in activities to support them. These activities may take the concluding the research and development agreements; providing the innovative services infrastructure; attracting the employees to work part-time in business companies; HEI's departments and divisions hardware and software.

2 The analysis of project risks

As an example we should consider the creation of the agricultural engineering center project.

Engineering Centre is created at the junction of two industries: agricultural machinery (technological equipment for the compound feed industry) and industrial biotechnology (production of industrial enzymes and microorganisms). The total project cost is 104 million. rub., at their own expense.

To confirm the effectiveness of the described systems we shall carry out the project risks calculation.

One admits to share the qualitative and quantitative analysis of project risks. One of the most common quantitative analysis methods is to analyze the sensitivity of the project, in other words to study how the project effectiveness will change depending on changes in any parameter. The sensibility analysis of the investment project is to assess the impact of the changes in any single project option to the project performance indicators, provided that the other parameters remain unchanged. The account must be taken that the project is more sensitive to the parameter's changes, the higher the risk and the more significant the parameter will be reflected in the risk reduction activities.

The main factors for the resulting test sensibility analysis are:
- prices of products (sales volume);
- the level of unit costs;
- the size of the investment costs.

These parameters' changes are carried out in the design model at 10, 20 and 30% both in the direction of increasing and decreasing values. At the same time as a final indicator of the project the net present value (NPV) is considered, due to the fact that this very one indicator is the feasibility indicator of the innovative project. The sensibility analysis results are shown in the table and in the graph.

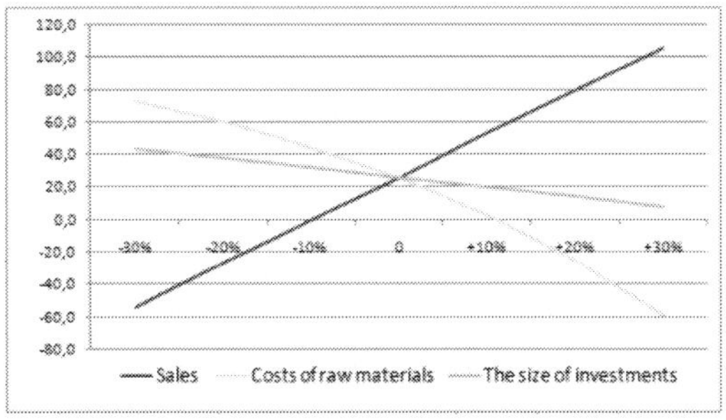

Fig. 1. Net Present Value of the project less the amount of subsidies.

The project is more sensitive to price fluctuations for raw materials: an increase of a factor of 20% leads to the formation of a negative net present value (NPV) of the project in the period under consideration. However, the risks associated with this factor are regulated by the registered character of the engineering center services and the specifics of the raw materials list drawing up (engineering center being a solution developer has the ability to pick up raw materials from the point of view of economic expediency of a specific contract), as well as the fact that the inflationary processes' impact on the budgeting of the materials and supplies costs are provided by this financial model originally.

Table 1. Net Present Value (NPV) of the project.

Parameter	Net Present Value (NPV) of the project, mln. Rub.						
	-30%	-20%	-10%	0	+10%	+20%	+30%
Sales (sales price)	-54.1	-27.4	-0.8	25.8	52.4	79.0	105.6
Costs of raw materials	72.3	60.0	44.7	25.8	2.6	-25.4	-59.1
Investments size	43.7	37.7	31.8	25.8	19.8	13.9	7.9

The maximum sensibility of the project demonstrates the value of sales (sale price). On reducing of this parameter to 10% of the project net present value (NPV) becomes zero in the period of 4 years 3 months. However, the actual risk of such changes is small due to the number of factors: the virtual absence of direct competition, the availability of stakeholders, integrated approach to the services provision, as well as their wide range, combined with the presence of the work plan for the promotion of products in the industry and assessment of prospective sales market in the absence of direct analogs.

In order to assess the degree of the most risky settings impact on the actual Engineering Centre cash flows it is advisable to analyze the sensibility of the project to their vibrations, using as a final value of the engineering center net profit index after reaching its design capacity.

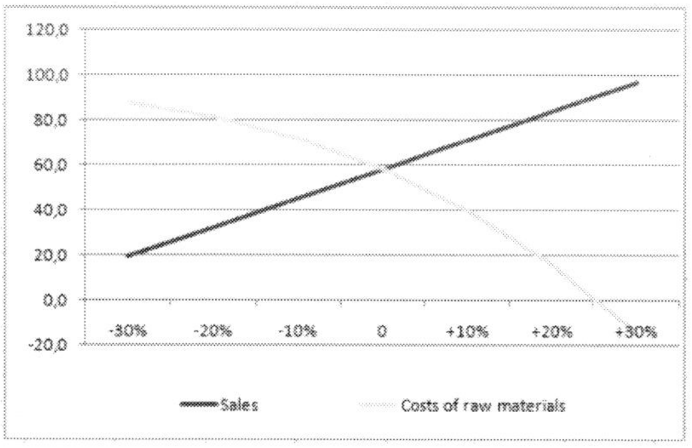

Fig. 2. Net Present Value of the project.

This one-parametric sensibility analysis of the project net profit value to the changes in sales and costs rates and materials shows the high stability of the basic economic indicators (net profit) to the changes in sales and selling prices of the services. With 30% deviation of this factor from the planned value engineering center will remain profitable. Decline in net profit will amount to 67% (from 58.1 mln. Rub. To 19.2 mln. Rub.). This indicates a high degree of the project financial stability.

In order to analyze the project financial stability it is advisable to determine the breakeven point, that is one of the most important landmarks of the project risk assessment. This type of analysis shows the principle feasibility of the project in terms of achieving the set profit value. Breakeven point characterizes the situation when total revenues will be equal to total costs. Calculation of breakeven point has been made in the planning interval after reaching its design capacity.

The table show the breakeven point when you change the volume being provided by the Engineering services center:

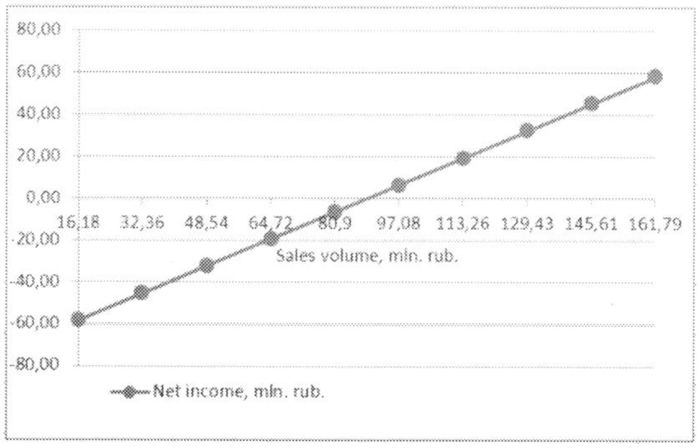

Fig. 3. Net Present Value of the project less the amount of subsidies.

At the rate of breakeven point, it follows that the situation comes on the sales of 89.22 million rubles. per year (55.1%) from the planned value when Engineering Center's revenues are equal to expenses. It follows that even if there is the decrease in the volume of services by more than 40% of the forecast, the Engineering Center will be self-financing.

Thus, the main project risk factor is the size of the raw materials costs. The methods for risk factors control and reduction planned for implementation during the project are:

• Cost accounting is an introduction and continuous development of the cost accounting systems for the provision of services in accordance with a few basic techniques: documentation, inventory, permanent evaluation;

Table 2. The net profit of the project.

Parameter	The net profit of the project, mln. rub.						
	-30%	-20%	-10%	0	+10%	+20%	+30%
Sales (sales price)	19.2	32.2	45.1	58.1	71.0	83.9	96.9
Costs of raw materials	87.6	81.0	71.4	58.1	40.0	16.4	-14.0

Table 3. The breakeven point.

Parameter	**10%**	**20%**	**30%**	**40%**	**50%**
Sales volume, mln. per year	16.18	32.36	48.54	64.72	80.9
Net income, mln. rub.	-58.43	-45.49	-32.54	-19.60	-6.66
Parameter	**60%**	**70%**	**80%**	**90%**	**100%**
Sales volume, mln. per year	97.08	113.26	129.43	145.61	161.79
Net income, mln. rub.	6.29	19.23	32.18	45.12	58.06

• Centralized control of pricing and established planned targets, in terms of the cost of services rendered;
• The use of resource-saving, low-waste and non-waste technologies;
• Strengthening of individual and collective responsibility, and material interest in the sustainable resource management;
• Material costs rationing.

3 Risk assessment out of the formal analysis

The risks that emerge in the implementation stage of capital investments are:
• failure to fulfill obligations by the supplier (low quality, defects in equipment, technology);
• untimely delivery and equipment installation of;
• the incompatibility of domestic equipment imported;
• the failure of the timing of construction;
• the excess of the estimated project cost;
• force majeure, damage to property.

The risk of the equipment suppliers' default from the obligations can be characterized as "low"; the technological scheme of the production planned for the engineering centers, involves the use of the adapted and widespread equipment. The market suppliers of this equipment are quite enough to make a choice of high quality.

The risks of force majeure and property damage also are assessed as "low". This conclusion leads to the stability of the economic and geopolitical situation in Russia, as well as the fact that the placement of the region and the activities of companies are characterized by the absence of any climatic threats.

Risks associated with the operation of the enterprise are:
• not reaching the design capacity (the emergence of technological constraints or raw);
• the products do not sell well in the desired value and settlement periods (the emergence of price caps, the revaluation or underestimation of the market competition in the marketing);
• inflation costs (faster growth of prices for raw materials, energy and other costs of the article.)
• untimely supply of raw materials.

The availability of pre-interest of several potential customers, a wide and diverse range of the services, the virtual absence of equivalent competitors in the market and growing the industry market will reduce the risks associated with the possible absenteeism at full capacity.

Sufficient random of raw material's suppliers for production services as well as the nature of services, most of which is a product of intellectual work, allow us reckon on a high viability of the project without dependence on one particular supplier of raw materials. The sensibility analysis had demonstrated the importance of the raw materials prices that would be taken into account in Engineering Center's activities.

Pricing Policy of the considered project and integrated approach to the service package formation leave the stock "price leverage" to promote a competitive offer products on the market under consideration and control the of the negative impact of inflation.

4 Conclusion

The calculations show a high resistance to changes in project size of investment costs: even with the increase of this factor within the project net present value (NPV) remains positive in a period of 4 years and 3 months. This is due to the relatively short investment period, as well as high levels of economic efficiency of the project.

References

1. E.S. Klimovich, Accounting in budget and non-profit organizations, **21** (2006)
2. N.V. Novozhilov, Bulletin of CSU, **4** (2007)
3. T.M. Bezborodova, K.N. Poladenko, STAGE, **2(20)** (2015)
4. A.V. Palashenkov, Bulletin of Omsk State University. Series. Right, **4** (2011)
5. L.A. Bakhvalov, D.A. Zhuravlev, PROC, **2-1** (2008)
6. Z.I. Abdullaev, Wax, **9(57)** (2013)
7. A.M. Polinsky, SR, **1** (2014)
8. W.J. Shalbolov, M.A. Albanova, L.A. Kazbekov, Physics of fluids. Series: Economics and law, **10(44)** (2014)
9. V.V. Shmat, I.A. Lutkov, Economics and modern management: theory and practice, **38** (2014)
10. I.U. Chekmazov, E.V. Iodine, Socio-economic phenomena and processes, **9**, (2014).
11. V.N. Arslanov, I.S. Gazizov, N.G. Kashapov, R.A. Celbraties, V.V. Khomenko, E.V. Shamsutdinov, Z.G. Shigapov, MAPS, **4** (2014)
12. V.V. Pod'yapolskii, MAPS, **2** (2015)
13. S.A. Tikhonova, Science. Innovation. Education, **9**, (2010)
14. O.A. Saprykin, Innovations, **3** (2008)
15. E.S. Taushev and N.I.Smorodinova, Actual problems of aviation and cosmonautics, **8** (2012)
16. O.V. Puchkova, R.M. Shigapov, Economics and modern management: theory and practice, **4(48-1)** (2015)
17. I.K. Novikova, A.V. Polutornyi, I.A. Ivanov, GORN, **7** (2011)
18. A.A. Kozlov, Management of innovation: theory, methodology, practice, **1** (2012)
19. K.A. Maksimov, M.E. Bryleva, Journal of the science of Science, **3(16)** (2013)
20. S.E. Aznauryan, A.A. Solonenko, Economics and business: theory and practice, **4**, (2015)
21. J.E. Amadi-Echendu, A. John, *Proc. Portland International Conference on Management of Engineering & Technology*, Cape Town, 1491-1496 (2008)
22. S. Su, H.D. Wang, *Proc. 2008 4th International Conference on Wireless Communications, Networking and Mobile Computing*, 1-4 (2008)
23. Y.K. Zhai, Q.Q. Dong, *Proc. 2007 International Conference on Wireless Communications, Networking and Mobile Computing*, Shanghai, 6305-6308 (2007)
24. Z. Ke-ying, L. Ruixia, *Proc. 2009 International Conference on Information Management, Innovation Management and Industrial Engineering*, 141-145 (2009)
25. H. Feng, *Proc. Management and Service Science (MASS)*, Wuhan, 1-3 (2010)

Implementation of information and analysis support of the industrial enterprise's logistical management based on the tools of the fuzzy set theory

G.M. Greyz[1,*]

[1]South Ural State University, Chelyabinsk, Russia

Abstract. Management of industrial enterprises' logistical systems is based on application of rather heterogeneous and not always certain information. Presence of different types of uncertainty in the complex hierarchical system of industrial enterprises' logistical management gives grounds for analysis support of management solutions based on the fuzzy set theory. Use of the fuzzy set theory allows to link together and adequately consider all the necessary heterogeneous information. In this regard, information on functioning of the logistical system must be presented in a specific form as membership functions. It is justified in the article that the tools of the fuzzy set theory can be applied for description of parameters of the industrial enterprises' logistical system and justification of decision-taking in the sphere of logistical management. Within the framework of the system of information and analysis support of industrial enterprises' logistical management it is proposed to use tools of problem "determination of the fuzzy set image" and its variety – "definition of the sub direct fuzzy set image" in order to choose the best variant of combination of key efficiency indicators of logistical management complying with the present complex of criteria. Application of the fuzzy set theory also allows to determine fuzzy values of factors, as a result of which the enterprise's logistical system has obtained the existing or objective set of features. For analysis of factors influencing the key efficiency indicators of the industrial enterprise's logistical management it is proposed to use tools of problem "definition of the fuzzy set pre-image at a fuzzy binary relation".

1 Introduction

Development and taking of management decisions in the logistical management system must be based on obtaining and processing of specific information messages. The author has proved that in order to increase efficiency of the industrial enterprise's logistical management the necessary information must be generated and processed by a special system of information and analysis support of logistical management [2].

According to the author, the system of information and analysis support of logistical management must ensure:

• selection of a system of key parameters-characteristics of functioning of the industrial enterprise's logistical system;

• agreement of strategic and tactical objectives of logistical management with the adopted system of indicators;

• conversion of the selected indicators into a form necessary for their processing and analysis by the system of information and analysis support of logistical management;

• analysis of actual values of parameters-characteristics of functioning of the industrial enterprise's logistical system for their compliance with the system of target criteria;

• identification of a potential connection of deviations of the logistical system's actual parameters and the management decisions taken;

• analytical check (forecasting) of consequences of the planned management decisions in the sphere of logistical management for their compliance with the preset system of target criteria.

According to the author, the system of information and analysis support of logistical management is a key link of interfunctional coordination in the industrial enterprise management system.

2 Theoretical framework and literature review

In order to ensure stable development of the logistical system it must have such system property as homeostasis [9]. The author believes that the system of information and analysis support of logistical management constitutes one of the conditions for achievement of the homeostasis state by the logistical system ensuring its adaptability to external and internal disturbances. Efficient functioning of the industrial enterprise's logistical system is maintained on account of entry and processing of relevant information in its management system.

* Corresponding author: ggreyz09@mail.ru

In the author's concept the system of information and analysis support of the industrial enterprise's logistical management is defined as a system contributing to achievement of homeostasis by the enterprise's logistical system and meant for: measurement of key indicators of the industrial enterprise's logistical system; complex evaluation of logistical management efficiency based on the criteria entered in the system; monitoring of consequences of management decisions in the logistical sphere.

Information and analysis support of the industrial enterprise's logistical management is provided by this system on account of available inverse relationship with the logistical system of the enterprise itself and the logistical system of the next level including the industrial enterprise and its counterparties.

Subject to the author's methodology, the system of information and analysis support is a necessary link of the industrial enterprise's logistical management, which obtains the necessary information on the enterprise's logistics activity, converts it into a specific form necessary for analysis, analyses it in terms of effectiveness and uses the obtained results for evaluation of efficiency of current management decisions in the sphere of logistical management and monitoring of consequences of planned management decisions.

The system of information and analysis support of the industrial enterprise's logistical management consists of: subsystem of formation and evaluation of key indicators of logistical management, matrix model of information and analysis support of logistical management and monitoring of management decisions in the sphere of logistical management.

The author of this research believes that the subsystem of formation and evaluation of key indicators of logistical management must be based on the concept of balanced scorecard. The offered subsystem of formation and evaluation of key indicators of logistical management provides for a transfer from verbally stated objectives to relevant groups of indicators of logistical management and indicators themselves. A set of indicators establishes selection of a certain interpretation of objective statement from a potential range of these statements. The indicators, which reflect a specific objective, are united into relevant groups by objectives.
The analysis conducted by the author of this research has shown that currently there is no ready set of groups of key indicators of logistical management and relevant criteria suitable for application in the author's system of information and analysis support of logistical management.

In the course of the research it has been determined that in order to increase mobility of initial data collection and to simplify their analysis for entry in the subsystem of formation and evaluation of key indicators of the industrial enterprise's logistical management it is necessary to use six groups of key indicators. They, primarily, include such groups of indicators as financial indicators, as well as time, expenses, customer satisfaction and inventories.

It should be also noted that the methodological approach to creation of a complex of information and analysis support of the industrial enterprise's logistical flow process management offered in the research assumes application of the sixth group of key indicators, which contains the author's logistical indicators reflecting specificity of the industrial enterprise's activities. These indicators reflect the level of inefficiency of material flow organization at the considered industrial enterprise.

In several cases, including at carrying out activity in the sphere of logistics, there appears a need to take decisions in conditions of lacking complete and certain (clear) information. For some of such situations classical mathematics assumes application of the probability theory. However, fuzziness of some phenomena does not always have a probabilistic nature. Besides, it is often impossible to apply the probability theory due to lacking test data. Use of the probability theory as applied to uncertain quantities does not give a proper effect, whereas the source of uncertainty at decision-taking is not a contingency but fuzziness [2,4,6,13]. In order to solve this problem it was necessary to introduce a new definition of uncertainty, and one of such definitions was provided by the fuzzy set theory offered by Professor of the University of California Lotfi A. Zadeh [5,20]. His work "Fuzzy Sets" [20], which appeared in 1965, laid the basis of the fuzzy set theory and establishment of intellectual systems able to interact adequately with the human being.

As opposed to the customary set theory, where there are only two potential memberships of the subset element: it can be either a member or not a member of this subset, in the fuzzy set theory fuzziness is formalized by introduction of the definition of the degree of membership of an element in a fuzzy set.

The fuzzy set theory was further developed in the 1980-s, when computer systems using fuzzy control algorithms started to appear. The fuzzy set theory laid the basis for building complex process control systems, as well as found application in home appliances, diagnostic and other expert systems [7,10,14,15,17-19].

The fuzzy set theory allows to convert expert estimates in terms of the natural language into the quantitative form, which allows to simulate economic entities based on opinions of experts observing these or similar entities.

Sufficient heterogeneous information is needed for logistical system management. These are values of different parameters, their potential variation parameters, as well as verbal information obtained from experts in the form of so-called linguistic variables.

Presence of different types of uncertainty in the complex hierarchical system of industrial enterprises' logistical management gives grounds for information and analysis support of management decisions in these systems based on the fuzzy set theory, which allows to take justified decisions in view of different types of uncertainty [1,8,11,16].

Application of the fuzzy set theory allows to link together and adequately consider all the necessary heterogeneous information. At the same time, all the information on functioning of the logistical system and objective functions must be presented in a specific form as membership functions.

The circumstance, that at description of different systems, including logistical systems, effective values of their functioning parameters are mostly fuzzy – there is no actual border between one effective value and other values close to it, counts in favor of application of the fuzzy set theory. An insignificant change of the system parameter usually leads to a small change in efficiency, therefore, application of the fuzzy subset membership function better corresponds to a real situation than application of preset (clear) parameters of logistical system efficiency.

In several cases at complex system management there is no need to take a clear decision for each point of time, whereas expenses on obtaining of information necessary for this decision and change of system parameters can exceed the effect obtained from this management decision. There frequently appears a situation, when not clear values but the most probable estimate of these values, as well as their potential change range can be set for several system parameters. Conditions and specificity of management problems generally permit fuzzy decisions. A significant factor of setting fuzzy conditions and objectives is complexity of multi-level logistical systems of industrial enterprises and presence of numerous links between their subsystems. Setting of clear restrictions and objectives is actually possible only for simple single-level systems.

In view of the above-mentioned, tools of the fuzzy set theory can be applied for description of parameters of the industrial enterprises' logistical system and lay the basis of the industrial enterprise's logistical model and justification of decision-taking in the sphere of logistical management. At creation of this model constituting the nucleus of the author's system of information and analysis support of the industrial enterprise's logistical management [2], the author has used the concept that any enterprise being an open logistical system is in the process of continuous changing. The enterprise's internal processes and their interaction (or, at least, interinfluence) and changing environmental conditions precondition continuous changes in various parameters of the enterprise's activity, including logistical parameters. Measurement of these parameters reflects their current values at a certain point of their changes and is a unique photo recording "floating", not exactly clear values of the dynamic process. The author believes that this uncertainty justifies application of the fuzzy set theory for study and evaluation of logistical systems.

3 Materials and methods

A. Methods Description

Parameters of the logistical system, which values are fuzzy sets, are used for reflection of the economic reality and simulation of the economic entity. Within the framework of the industrial enterprise's logistical model tools of the fuzzy set theory are used for solution of two tasks:

1) Analysis and evaluation of key indicators of the industrial enterprise's logistical management efficiency
2) Evaluation and monitoring of management decisions in the logistical sphere

Let us review, what opportunities are provided by the fuzzy set theory for solution of the set problems. At analysis and description of elements of the fuzzy set theory necessary for conducting of this research the material presented in [12] was taken as a basis.

In several cases for evaluation of efficiency of any activity, including logistical management, the set of criteria and calculation of the integrating efficiency indicator are used. However, calculation of the integrating efficiency indicator does not always solve the problem on evaluation of the logistical system management optimality degree. In several cases calculation of the integrating indicator does not give an answer to the question on optimality of this or that combination of key parameters of the logistical system. One and the same value of the complex logistical management efficiency indicator can be obtained at different combinations of parameters of the developed evaluation system. At the same time, specific features of an individual industrial enterprise result in specificity of combinations of objective evaluation criteria. The problem lies is selection of one best combination complying with the preset complex of criteria out of several combinations of the logistical system parameters with close integral values. For such selection within the framework of the methodology of information and analysis support of the industrial enterprise's logistical management the author proposes to use the tools of problem "determination of the fuzzy set image at fuzzy binary relation" and its variety – "determination of the subdirect fuzzy set image".

Statement of problem "determination of the fuzzy set image" can be presented as follows: there are several economic entities, each of which has a set of specific properties (features), and there is a criterial set of these features, in accordance with which the economic entities are evaluated. After solution of this problem we can find an economic entity, which has at least one feature from the criterial set.

Within the framework of this field of research sets of indicators formed by the subsystem of formation of key indicators of the industrial enterprise's logistical management are used as economic entities [3]. Such sets consist of a set reflecting the current state of the logistical system, indicators of past and future (forecast) periods. The criterial set of features is based on the requirements to be met by the key indicators of logistical management converted into fuzzy sets (ratings). After solution of this problem we can evaluate, which of the considered sets of ratings has a larger degree of membership in the criterial set of requirements.

Let us consider this problem in terms of the fuzzy set theory. It is necessary to find an image of fuzzy set A (**criterial set** of features x) in set Y (economic **entities** y), provided that we know fuzzy binary relation R from set X to set Y (sets of properties (features) of economic entities). After solution of this problem we find membership degree q for each element y in the image of fuzzy set A at the preset fuzzy binary relation (presentation) from set X to set $Y - R$. Values of features x, binary relations r and membership degree q of element y in the image of fuzzy set A subject to the laws of the fuzzy set theory must be within interval [0;1], i.e., the closer value q to one, the larger degree of compliance of the complex of the economic entity's properties with the preset criterial set of features A.

In general terms, fuzzy set B constituting a direct image of set A can be written as $B = A \blacklozenge R$. $\blacklozenge$ – selection symbol (*max – min*). In the matrix form this problem can be solved as follows:

$$(p_1, p_2, \dots, p_n) \blacklozenge \begin{bmatrix} r_{11} & r_{12} & \cdots & r_{1m} \\ r_{21} & r_{22} & \cdots & r_{2m} \\ \cdots & \cdots & \cdots & \cdots \\ r_{n1} & r_{n2} & \cdots & r_{nm} \end{bmatrix} = (q_1, q_2, \dots, q_n) \quad (1)$$

where $p_1, p_2, \dots, p_n$ – degree of membership of features $x_1, x_2, \dots, x_n$ in the fuzzy set (criterial set), which is written as $A = \{(x_1 | p_1); (x_2 | p_2); \dots (x_n | p_n)\}$;

$r_{11} \dots r_{nm}$ – elements of matrix R (fuzzy binary relation from set X to set Y) , which preset fuzzy values of features x for elements y (economic entities) of fuzzy set Y;

$q_1, q_2, \dots, q_m$ – degrees of membership of elements $y_1, y_2, \dots, y_m$ in the image of fuzzy set A, which can be interpreted as a degree of conformity of the complex of economical entities' properties with the preset criterial set of values of these features;

n – number of features x;

m – number of elements y (economic entities);

Solutions of expression (1) in the coordinate form are

$$\begin{aligned} &max(min(p_1; r_{11}); \dots min(p_n; r_{n1})) = q_1 \\ &\dots\dots\dots\dots\dots\dots\dots\dots\dots\dots \\ &max(min(p_1; r_{1m}); \dots min(p_n; r_{nm})) = q_m \end{aligned} \quad (2)$$

Based on the calculation results by formulas (2) we obtain set $B = A \blacklozenge R = \{(y_1 | q_1); (y_2 | q_2); \dots; (y_m | q_m)\}$ on the ground of which we can evaluate, which economic entity y has a larger degree of membership in the image of fuzzy set A. The larger value q (potential range of values [0;1]), the larger the compliance of the economic entity with the preset criterial set of features A by availability and value of properties.

Apart from problem "determination of the fuzzy set image", for analysis and evaluation of key efficiency indicators of the industrial enterprise's logistical management we can use a variety of this problem – "determination of the subdirect fuzzy set image". The difference is that as a result of solution of this problem we determine a set of indicators, which has not one but all the features specified in the criterial set and which best of all corresponds to this set by the complex of its properties.

In general terms, fuzzy set B constituting a subdirect image of set A can be written as $B = A \blacktriangleleft R$. In view of the introduced symbols in the matrix form this problem can be solved as follows:

$$(p_1, p_2, \dots, p_n) \blacktriangleleft \begin{bmatrix} r_{11} & r_{12} & \cdots & r_{1m} \\ r_{21} & r_{22} & \cdots & r_{2m} \\ \cdots & \cdots & \cdots & \cdots \\ r_{n1} & r_{n2} & \cdots & r_{nm} \end{bmatrix} = (q_1, q_2, \dots, q_n) \quad (3)$$

Values of symbols of expression (3) coincide with the values presented in expression (1). The difference lies in introduction of operation $\blacktriangleleft$, which content becomes clear from recording of expression (3) in the coordinate form

$$\begin{aligned} &\min((p_1 \blacksquare r_{11}); \dots; (p_n \blacksquare r_{n1})) = q_1 \\ &\dots\dots\dots\dots\dots\dots\dots\dots\dots\dots \\ &\min((p_1 \blacksquare r_{1m}); \dots; (p_n \blacksquare r_{nm})) = q_m \end{aligned} \quad (4)$$

The so-called cut-off operation ($\blacksquare$) can be presented as follows:

$$p \blacksquare r = \begin{cases} 1 & at\, p \le r \\ r & at\, p > r \end{cases} \quad (5)$$

Based on the calculation results by formulas (4) we obtain set $B = A \blacktriangleleft R = \{(y_1 | q_1); (y_2 | q_2); \dots (y_m | q_m)\}$, on the ground of which we can evaluate, which economic entity y has a larger degree of membership in the subdirect image of fuzzy set A. Analogously to determination of the image, the larger value q, the larger the compliance of the economic entity with the preset criterial set of features A by the complex of properties. Within the framework of the matrix model we can also analyze factors influencing the key efficiency indicators of the industrial enterprise's logistical management. For this purpose, within the framework of the methodology of information and analysis support of the industrial enterprise's logistical management the author proposes to use tools of problem "determination of the fuzzy set pre-image at a fuzzy binary relation".

Statement of problem "determination of the fuzzy set pre-image" can be presented as follows: there is a set of factors, which values are unknown, and an economic entity having a set of specific properties (features), and there are known cause-and-effect relations between the factors of influence on the economic entity and the properties (features) of this entity. Upon solution of this problem we have to restore fuzzy values of factors, as a result of which influence the economic entity has obtained the existing or desirable (target) set of features.

Within the framework of the system of information and analysis support of logistical management problem "determination of the fuzzy set pre-image" is transformed as follows: a) we select a set of factors significant for logistical management (*fuzzy set A*), b) we set in a fuzzy form the degree of influence of the selected factors on the key efficiency indicators of logistical management (cause-and-effect relations between the factors of influence on the economic entity and the properties (features) of this entity), i.e. in this case relations of transfer from fuzzy set A to fuzzy set B, (*fuzzy binary relation R of transfer from set X to set Y*), c) we determine current and preset target values of key efficiency indicators of logistical management in the form of ratings by groups of indicators (set of specific properties (features) y) (*known fuzzy set B*). Solution of this problem results in fuzzy values of factors, under the influence of which the logistical management system has obtained the existing or desirable (target) set of indicators.

Let us consider this problem in terms of the fuzzy set theory. It is necessary to restore fuzzy set A on the basis of known fuzzy set B, if we know fuzzy binary relation R of transfer from set X to set Y , i.e. in this case relations of transfer of fuzzy set A to fuzzy set B. There can exist the following relations between fuzzy sets B and A: $B = A \blacklozenge R$ (determination of ⧫-pre-image), $B = A \blacktriangleleft R$ (determination of ◀-pre-image) and $B = A \blacktriangleright R$ (determination of ▶-pre-image) [6]. Selection of this or that model determines the relevant function of membership of fuzzy set A and its calculation procedure.

Let us consider calculation of ⧫-pre-image. In the matrix form this problem can be solved as follows:

$$(q_1, q_2, \dots, q_m) \blacktriangleright \begin{bmatrix} r_{11} & r_{12} & \dots & r_{1m} \\ r_{21} & r_{22} & \dots & r_{2m} \\ \dots & \dots & \dots & \dots \\ r_{n1} & r_{n2} & \dots & r_{nm} \end{bmatrix} = (p_1^{\blacklozenge}, p_2^{\blacklozenge}, \dots, p_n^{\blacklozenge}) \quad (6)$$

where $p_1^{\blacklozenge}, p_2^{\blacklozenge}, \dots, p_n^{\blacklozenge}$ – fuzzy values of factors, as a result of which influence the logistical management system has obtained or could have obtained key efficiency parameters $q_1, q_2, \dots, q_m$;

$r_{11} \dots r_{nm}$ – elements of matrix R in the fuzzy form preset the degree of influence of the selected factors p of the key efficiency indicators of logistical management q.

$q_1, q_2, \dots, q_m$ – fuzzy values of key efficiency parameters of the logistical management system.

degrees of membership of elements $y_1, y_2, \dots, y_m$ in the image of fuzzy set A, which can be interpreted as a degree of conformity of the complex of the economic entities' features with the preset criterial set of values of these features;

n – number of factors x;

m – number of elements y (key efficiency parameters);

Let us write the content of operation ▶ from expression (6) in the coordinate form on the analogy of expression (4)

$$\begin{array}{c} min((r_{11} \blacksquare q_1);\ (r_{12} \blacksquare q_2);\ \dots (r_{1m} \blacksquare q_m)) = p_1^{\blacklozenge} \\ \dots\dots\dots\dots\dots\dots\dots\dots\dots\dots\dots\dots \\ min((r_{n1} \blacksquare q_1);\ (r_{n2} \blacksquare q_2);\ \dots (r_{nm} \blacksquare q_m)) = p_n^{\blacklozenge} \end{array} \quad (7)$$

The cut-off operation (■), on the analogy with expression (5) can be presented as follows:

$$r \blacksquare q = \begin{cases} 1 & \text{при } r \le q \\ q & \text{при } r > q \end{cases} \quad (8)$$

Based on the calculation results by formulas (7) and (8) we restore the initial set (⧫- pre-image)

$A = B \blacktriangleright R = \{(x_1 \mid p_1^{\blacklozenge});\ (x_2 \mid p_2^{\blacklozenge}); \dots (x_n \mid p_n^{\blacklozenge})\}$ which after application of fuzzy relation R was transferred to fuzzy set $B = \{(y_1 \mid q_1);\ (y_2 \mid q_2); \dots (y_m \mid q_m)\}$.

Let us consider implementation of the second function of the matrix model – evaluation and monitoring of management decisions in the logistical sphere.

For implementation of this function within the framework of information and analysis support of the industrial enterprise's logistical management it is proposed to use the tools of problem "fuzzy inference models". According to the author, these tools of the fuzzy set theory can be efficiently used for monitoring of management decisions in the logistical sphere.

The content of this transfer can be presented as follows. The inference basis is formed by statement r = «if a, then b», which is called implication in classical logics. In the fuzzy set theory statement r is transformed into «if A, then B», where A and B are fuzzy sets interconnected by fuzzy relation R. This fuzzy relation is also called implication and denoted by symbol $A \Rightarrow B$.

After determination of the specific function of implication $A \Rightarrow B$ it becomes possible depending on input fuzzy set A^+ to build output fuzzy set (effect) $B^+ = A^+ * (A \Rightarrow B)$.

In accordance with the author's approach the algorithm of the fuzzy inference model application is adapted to the sphere of information and analysis support of logistical management. Logical premise A+ is a specific set of management actions, each of which has a different degree of "power" (intensity) preset by the fuzzy values within range [0,1].

According to the author, such management actions (X_i) in the sphere of logistical management include:

- operating-calendar planning of finished product output;
- operating process control;
- maintaining of quality standards;
- inventory control and management at all production stages;
- implementation of logistical information systems;
- investments in production automation;
- implementation of advanced cost management systems ("ABC-costing" or "target costing" or "kaizen costing");
- investments in R&D in the sphere of logistics, etc.

In this case the complex of results of these management decisions can be presented as effect B^+.

The following output parameters of the industrial enterprise's logistical system are considered to be results of these management actions (efforts) (Y_i):

- improvement of product quality;
- cutting of the production lead time;
- performance improvement;
- cutting of logistical costs;
- inventory minimization (optimization) at all production stages;
- cost reduction;
- profit mark-up, etc.

Logistical concept or opinion of a group of experts must be used as implication $A \Rightarrow B$ in the form of fuzzy relation R. This approach allows to make forecasts based on the opinion of several groups of experts, and aggregation operation must be used to obtain an overall conclusion.

Let us consider in detail the procedure of finding effect B^+ of the complex of management actions A^+. In order to simplify description of the algorithm let us restrict complex A^+ by five potential management actions, and their result B^+ - also by five output parameters.

Then, the output management decisions will take the form of fuzzy set $A^+ = \{(x_1 | p_1); (x_2 | p_2)(x_3 | p_3)(x_4 | p_4)(x_5 | p_5)\}$. Let us assume that for finding of implication $A \Rightarrow B$ in the form of fuzzy relation R the opinion of two experts (group of experts) is used. Their fuzzy sets have the following form << if A_i, then B_i>>, $i= 1,2$ and can be presented as follows

$$A_1=\{(x_1 | a_1^1); (x_2 | a_2^1); (x_3 | a_3^1); (x_4 | a_4^1); (x_5 | a_5^1)\}.$$
$$B_1=\{(y_1 | b_1^1); (y_2 | b_2^1); (y_3 | b_3^1); (y_4 | b_4^1); (y_5 | b_5^1)\}. \quad (9)$$
$$A_2=\{(x_1 | a_1^2); (x_2 | a_2^2); (x_3 | a_3^2); (x_4 | a_4^2); (x_5 | a_5^2)\}.$$
$$B_2=\{(y_1 | b_1^2); (y_2 | b_2^2); (y_3 | b_3^2); (y_4 | b_4^2); (y_5 | b_5^2)\}.$$

For building of the implication we use function $\theta(a,b) = a \blacksquare b$ and present implication $A_i \Rightarrow B_i = A_i \blacksquare B_i$ in the form of matrixes

$$A_1 \blacksquare B_1 = \begin{pmatrix} r_{11}^1 & \cdots & r_{15}^1 \\ \vdots & \ddots & \vdots \\ r_{51}^1 & \cdots & r_{55}^1 \end{pmatrix} \quad (10)$$

$$A_2 \blacksquare B_2 = \begin{pmatrix} r_{11}^2 & \cdots & r_{15}^2 \\ \vdots & \ddots & \vdots \\ r_{51}^2 & \cdots & r_{55}^2 \end{pmatrix} \quad (11)$$

Matrix elements are as follows:
$r_{11}^1 = a_1^1 \blacksquare b_1^1; r_{12}^1 = a_1^1 \blacksquare b_2^1; \ldots; r_{55}^1 = a_5^1 \blacksquare b_5^1$ for matrix (10) and analogously for matrix (11). The rule of the cut-off operation ($\blacksquare$), for this variant of comparison can be presented as follows:

$$a \blacksquare b = \begin{cases} 1 & at\ a \leq b \\ b & at\ a > b \end{cases} \quad (12)$$

Then, on the basis of matrixes (10), (11) and output management decision A^+ we will build local inferences B_1^+ and B_2^+. As an inference rule let us take $* = \blacklozenge$.

$$(p_1; p_2; \ldots; p_5) \blacklozenge \begin{pmatrix} r_{11}^1 & \cdots & r_{15}^1 \\ \vdots & \ddots & \vdots \\ r_{51}^1 & \cdots & r_{55}^1 \end{pmatrix} = (q_1^1; q_2^1; \ldots; q_5^1) \quad (13)$$

$$(p_1; p_2; \ldots; p_5) \blacklozenge \begin{pmatrix} r_{11}^2 & \cdots & r_{15}^2 \\ \vdots & \ddots & \vdots \\ r_{51}^2 & \cdots & r_{55}^2 \end{pmatrix} = (q_1^2; q_2^2; \ldots; q_5^2) \quad (14)$$

The solutions of the expression (13) in the coordinate form are the following expressions:

$$max(min(p_1; r_{11}^1); \ldots min(p_5; r_{51}^1)) = q_1^1$$
$$\cdots\cdots \quad (15)$$
$$max(min(p_1; r_{15}^1); \ldots min(p_5; r_{55}^1)) = q_5^1$$

Values $q_1^2 \ldots q_5^2$ are determined analogously. Then, we will obtain fuzzy sets, which determine the result of the action of the management decision taken

$$B_{(1)}^+ = \{(y_1 | q_1^{min}); (y_2 | q_2^{min}); \ldots (y_5 | q_5^{min})\}$$
$$\quad (16)$$
$$B_{(2)}^+ = \{(y_1 | q_1^{max}); (y_2 | q_2^{max}); \ldots (y_5 | q_5^{max})\}$$

Values q^{min} and q^{max} are selected by comparison of resulting sets q from formulas (13) and (14) for determination of local inferences.

Fuzzy inference models allow not only to determine the membership function for effect B of fuzzy expression $A \Rightarrow B$, but also to solve a reverse problem: by setting fuzzy set B as objective we can find input set A, as a complex of management actions in the logistical sphere. In this case the problem is realized as finding of the pre-image of fuzzy set B (prime cause of obtaining the base set of key indicators).

4 Conclusions

Thus, it has been justified that within the system of information and analytical support of logistics management of an industrial enterprise the inventory of the fuzzy sets theory can be used for reaching the following tasks:

1. Multicriterial assessment of effectiveness of logistics management according to the key parameters of the logistics system and selection of the best combination of these parameters, that corresponds to the determined set of criteria.
2. On the basis of the determined causal relationships between the impacts of the logistics system and properties (evidences) of this system: reconstruction of fuzzy values of factors whose impacts stipulate the present or desired (aimed) set of properties of the logistics system.
3. On the basis of the accepted logistics concept or the opinion of an expert group: analytical prediction of the results received due to the impact of a certain set of managerial actions on the logistics system.

The work was supported by Act 211 Government of the Russian Federation, contract № 02.A03.21.0011

References

1. A.E. Altunin, M.V. Semukhin, *Models and Algorithms of Decision Taking in Fuzzy Conditions: monograph* (Publishing house of Tyumen State University, Tyumen, 2002)
2. G.M. Greyz, Bulletin of the South Ural State University, Series: Economics and Management, **8**, **3**, 121-127 (2014)
3. G.M. Greyz, V.M. Katochkov, V.V. Volozhanin, Bulletin of the South Ural State University. Series: Economics and Management, **8**, **3**, 128-135 (2014)

4. L.A. Gusev, I.M. Smirnova, Automatics and Telemechanics, **5**, 66-85 (1973)
5. L.A. Zadeh, *Definition of Linguistic Variable and its Application to Making of Approximate Solutions* (Mir, Moscow, 1976)
6. L.A. Zadeh, *Fuzzy Sets and their Application in Image Recognition and Cluster Analysis, Classification and Cluster: collection* (Mir, Moscow, 1980)
7. A. Kaufman, *Introduction to the Theory of Fuzzy Sets* (Radio and Communication, Moscow, 1982)
8. B.L Kuchin, A.E. Altunin, *Gas Supply System Management in Complicated Operating Conditions* (Nedra, Moscow, 1987)
9. L.B. Mirotin, I.E. Tashbaev, *System Analysis in Logistics* (Ekzamen, Moscow, 2002)
10. S.A. Orlovskiy, *Problems of Decision Taking at Fuzzy Initial Information* (Nauka, Moscow, 1980)
11. M.V. Semukhin, *Fuzzy Set Theory. Study guide* (Tyumen State University, Tyumen, 1999)
12. V.I. Ukhobotov, *Selected Chapters of the Fuzzy Set Theory: manual* (Publishing house of Chelyabinsk State University, Chelyabinsk, 2011)
13. R.E. Bellman, M. Gierts, Inform. Sci., **5**, **2**, 149-156 (1973)
14. S.S.L. Chang, Kybernetes, **6**, 203-208 (1977)
15. D. Dubois, *Prade Fuzzy Sets and Systems: Theory and Applications* (Acad. Press, New York, 1980)
16. W.Y. Kickert, Automatica, **12**, **4**, 301-308 (1976)
17. H. Tanaka, K. Asai, IEEE Trans. Syst. Maan and Cybern, **2**, 325-328 (1984)
18. B. Urban, V. Hansel, Proc. Fuzzy inf., Marseille, 313-320 (1983)
19. R.R. Yager, J. of Cybern, **10**, 1-18 (1980)
20. L.A. Zadeh, Information and Control, **8**, **3**, 338-353 (1965)

Cash flow optimization in industrial enterprise

T.N. Myznikova[1,*], and *N.V.* Zhdanova[1]

[1]South Ural State University, Chelyabinsk, Russia

Abstract. Optimization of cash flows of the industrial company provides economic entity necessity and sufficiency of financial resources for sustainable activities. Cash optimization techniques are grouped into two blocks: theoretical - is mainly foreign methods and applied techniques that are mostly used by Russian authors. Models described in the literature are not allowed for the particular industry in the formation of cash. The mathematical models described in the literature do not allow to take into account industry characteristics in the formation of funds. The proposed methodology by authors allows to predict cash amounts based on business company. The balance of cash flows is provided by the budgeting system. The company's the released money can send funds for investment purposes. Effectiveness of confirmed by practical testing methodology on the existing machine-building enterprise.

1 Introduction

The relevance of cash flow optimization for any business entity is acquiring greater significance under conditions of shortage of production resources, including cash funds.

Cash flow can be generated from various types of activity: operating, financing and investing. Along with this, an amount of flow, its vector and dynamics allow us to solve a variety of tasks, among other things in the interests of different participants of business process.

Cash flows from various activities collide with each other, altogether solving the general problem of industrial enterprise - providing ever-increasing economic benefits. For this reason there is a need to create a model of cash flaw optimization for industrial enterprises that is able to meet production, financial and investment goals.

On the one hand, models of cash flow optimization presented in modern scientific literature do not take into account sectoral affiliation of a business entity, which reduces their credibility and practical significance. And on the other hand, they do not help to identify factors that affect total financial result from their use.

It can be assumed that the system of cash funds budgeting can provide an optimal cash flow ratiofor a machine-building enterprise that is sufficient for its effective operating, investing and financing activities.

2 Scientific development of thetopic

Considerable attention to issues of cash flow management has been payed since the second half of 80s. In the beginning of 21st century, the number of scientific and practical researches in this area has grown considerably. The USA leadsin the number of researches (in particular, Ohio State University, New York University, University of California, Los Angeles and other North American universities). A significant part of researches is dedicated to management of the cash flow generated by financial investments. To us, particularfeatures of interest present researches of cash funds management in models of multivariate time seriesthat use past values of indicators as explanatory variables in time series of regression [1].

A number of researchers use the factor model to optimize cash flow. In particular, Martins, P. establishes the relationship between the mainresulting indicator – cash funds and factorsthat form it: workforce and production. Along with this, authors combine them into a single system based on monoproduction (production of one type of goods) without reference to the specifics of an industry, which reduces the usefulness of the presented model [2].

In the paper of Chen, W.-N., Zhang, J., Chung, H.S.-H., Huang, R.-Z., Liu, O the change of cash funds value is taken into account and discounting is introduced. The practical component of the proposed mathematical modeling is a multi-dimensional assessment of analyzed factors. However, even this authors' approach does not take into account the effect on the result of internal factors, such as lack of liquidity, while considering cash flow over time [3].

Works of Russian authors that address the problem of cash flow optimization in enterprises can be classified into two parts. The first one represents works of a theoretical nature. These include works the economic analysisfounders of the Soviet and post-Soviet period (Savitskaya, Kovalev, Sheremet , Lukasiewicz [4-7]), the second one represents works of an applied nature, includingworks that take into account sectorial specificity.They are much rarer. Works of Sanakoev E.V., Novolodsky D.A., Sinkina A.A. are noteworthy [8,9,10]. Along with this, these authors view cash flow without reference to relationship between all activities and industry specifics.

* Corresponding author: miz108410098@mail.ru

Table 1. Factors of changesin net cash flow (NCF) for current operations12.

Factors of changes			
Increase of NCF	Decrease of NCF	Factor model	Comment
Growth of net profit	Decrease in net profit	$\Sigma[VMP_i(P_i\text{-}CP_i)]$	VMP_i– volume of manufactured production, goods, services; P_i – price of realization; CP_i– cost price of manufactured good
Decrease in losses	Increase in losses		
Increase in amount of deprecation	Decrease in amount of deprecation	$\Sigma[DA_i^j \cdot RD_i]$	DA_i^j –depreciable asset; RD_i – rate of depreciation of the asset
Decrease in accounts receivable	Increase in accounts receivable	$\Sigma[\frac{SP \cdot SCAR}{365}]$	SP – shipped products; SCAR– speed of collection of accounts receivable
Decrease in stock	Increase in stock	$\Sigma[Q_i \cdot P_i]$	Q_i –quantity of goods and materials; P_i– price of goods and materials
Increase in short-term financial liabilities	Decrease of short-term financial liabilities	Factor model is compiled on the basis of types of expenses and revenue receipts	
Increase in amount of provisions for future expenses and revenues	Decrease of amount of provisions for future expenses and revenues		

Table 2. Factors of changesin net cash flow for investment operations.

Factors of changes			
Increase of NCF	Decrease of NCF	Factor model	Comment
Increase of revenue from sale of fixed assets	Decrease of revenue from sale of fixed assets	$\Sigma[Q_i \cdot P_i]$	Q_i– quantity of sold fixed assets; S_i – price of realization of fixed assets
Increase of revenue from sale of intangible assets, long-term financial assets6 repurchased shares	Decrease of revenue from sale of intangible assets, long-term financial assets6 repurchased shares	$\Sigma[Q_i \cdot P_i]$	Q_i – quantity of sold assets; P_i – price of realization of fixed assets
Increase of amount of interest and dividends on securities	Decrease of amount of interest and dividends on securities	$\Sigma[TAFI_i \cdot SFI_i \cdot Pr_i]$	TAFI – the total amount of financial investments; SFI_i – structure of financial investments; Pr_i – level of profitability of financial investments
Decrease of investments in purchase of fixed assets	Increase of investments in purchase of fixed assets	$\Sigma[Q_i \cdot P_i]$	Q_i – quantity of purchased fixed assets; P_i – cost of purchased assets

Decrease of investment in capital construction	Increase of investment in capital construction	$\Sigma[Q_i \cdot VPW_i \cdot P_i]$	Q_i – the number of objects under construction; P_i – estimated cost of performed work; VPW_i – volume of performed work on each subject
Decrease of investment in purchase of intangible assets	Increase of investment in purchase of intangible assets	$\Sigma[Q_i \cdot P_i]$	Q_i – quantity of purchased assets; P_i – cost of purchased assets
Decrease of investment in purchase of long-term financial assets	Decrease of investment in purchase of long-term financial assets		
Decrease in the number of repurchased shares of enterprise	Increase in the number of repurchased shares of enterprise	Factor model is compiled on the basis of types of expenses and revenue receipts	
Decrease in amount of other payments	Increase in amount of other payments		
Increase in amount of other revenue receipts	Decrease in amount of other revenue receipts		

3 Dynamics and analysis of cash flow

After the text edit has been completed, the paper is ready for the template. Duplicate the template file by using the Save As command, and use the naming convention prescribed by your conference for the name of your paper. In this newly created file, highlight all of the contents and import your prepared text file. You are now ready to style your paper; use the scroll down window on the left of the MS Word Formatting toolbar.

Inflows and outflows of cash funds associated with the use of external financingare traditionally reflected in financing activity [4, p. 494].

Results of net cash flow calculations for current, investment and financial operations allow to determine the total net cash flow:

$$NCF = NCF_{CO} + NCF_{IO} + NCF_{FO} \quad (4)$$

The advantage of this method is the possibility to identify dynamics of all factors that form net cash flow. However, the use of both direct and indirect methodsin combination to obtain accurate data on the composition and amount of cash funds.

The second stage of the complex analysis of funds is the analysis of cash flowbalance. To assess the degree of uniformity of forming of cash flows and inflows and their synchronicity certain indicators are calculated:

- Standard deviation (Formula 5):

$$\sigma = \frac{\sqrt{\Sigma(x_i - \overline{x})^2}}{n} \quad (5)$$

where x_i is the value of the i-th indicator;

$\overline{x}$ - the average value of index.

A large standard deviation value indicates a large spread in the values of presentedsetwith an average value ofset; a small value, respectively, shows that the values in a set are grouped around the average value.

- The coefficient of variation, to determine which formula 6 is used:

$$V = \frac{\sigma}{\overline{x}} \cdot 100. \quad (6)$$

A collection is considered to be homogeneous if the coefficient of variation does not exceed 33% [11, p. 137].

- The correlation coefficient is calculated as follows:

$$r = \frac{\sum x \cdot y - \frac{\sum x \cdot \sum y}{n}}{\sqrt{(\sum x^2 - \frac{(\sum x)^2}{n}) \cdot (\sum y^2 - \frac{(\sum y)^2}{n})}}, \quad (7)$$

where x is the amount of positive cash flow for the i-th period;

y is the amount of negative cash flow for the i-th period;

The closer the correlation coefficient is to one, the smaller is spread of fluctuations between values of positive and negative cash flow.

The third stage of the cash flow analysis is the ratio analysis or analysis of cash flow intensity and effectiveness.

Table 3. Factors of changesin net cash flow for financial operations.

Factors of changes			
Increase of NCF	Decrease of NCF	Factor model	Comment
Increase in long-term loans and borrowings	Decrease in long-term loans and borrowings	$\Sigma[S_i \cdot T_i \cdot I_i]$	S_i – sum of loans and borrowings; T_i – period (term) of loans and borrowings; I_i –the interest rate of loan
Increase in short-term loans and borrowings	Decrease in short-term loans and borrowings		
Decrease of payments on long-terms loans and borrowings	Increase of payments on long-terms loans and borrowings		
Decrease of payments on short-terms loans and borrowings	Increase of payments on short-terms loans and borrowings		
Increase in amount of received dividends and interest on short-term financial investments	Decrease in amount of received dividends and interest on short-term financial investments	$\Sigma[TAFI \cdot SFI_i \cdot Pr]$	*TAFI* – the total amount of financial investments; SFI_i– structure of financial investments ; Pr_i – level of profitability of financial investments
Decrease in amount of dividends paid to company shareholders	Increase in amount of dividends paid to company shareholders		
Increase in income from share issue	Decrease in income from share issue	$\Sigma[Q_i \cdot P_i]$	Q_i – quantity of issued assets; P_i – cost of issued assets
Increase in amount of target financing of revenue receipts	Decrease in amount of target financing of revenue receipts	Factor model is compiled on the basis of types of expenses and revenue receipts	
Increase in amount of other revenue receipts	Decrease in amount of other revenue receipts		
Decrease in amount of other payments	Increase in amount of other payments		

The intensity of cash flow is characterized by process of its movement, which includes three stages: procurement, production andsaling. The total duration of these stages gives the total length of the operating cycle:

$$OC = CPGM + CPWP + CPFP + CPAR, \quad (8)$$

where *CPGM* is a circulation period of goods and materials;
CPWP is a circulation period of work in progress;
CPFP is a circulation period of the finished product;
CPAR is a circulation period ofaccounts receivable.

Analysis of time cycle of cash funds circulating allows us to establish on which of the steps there is a delay and to develop activities aimed at the more intensive use of cash funds.

Analysis of cash flow effectiveness is carried out with epyuse of number of factors that are shown in Table IV [12, p. 395; 13, p. 179]

Listed in Table IV coefficients should be analyzed in dynamicswith calculatingof deviations

This technique of cash funds analysis allows to find out, measure and analyze cash flow of an enterprise, as well asoptimize cash funds on the basis of obtained data [14-16].

In the analysis of cash flow of enterprises unbalanced cash flow is identified - an imbalance, prevalence of cash outflow over cash inflow.

Table 4. Indicators of cash flow effectiveness.

Coefficient	Calculation formula	Comment
the efficiency ratio of cash flow	$R_{ef}=\frac{NCF}{CF_o}$	*NCF* – net cash flow

the adequacy ratio of cash flow	$R_{ad}=\frac{NCF}{PLB+\Delta S+DP_i}$	PLB_i –payments on loans and borrowings; ΔS –increase in stock balance; DP – dividends paid
the efficiency ratio of cash flow	$R_{liq}=\frac{CF_i}{CF_o}$	CF_o – cash outflow
The profitability ratio of cash flow	$R_{pr\,cash\,inflow}=\frac{NP}{CF_i}$	CF_i – cash inflow
	$R_{pr\,cash\,flow}=\frac{NF}{\overline{CB}}$	$\overline{CB}$ – average cash balance
	$R_{pr\,net\,cash\,flow}=\frac{NP}{NCF}$	NP – net profit of company
	$R_{pr\,cash\,outflow}=\frac{NP}{CF_o}$	CF_o – cash outflow

The results cash flow optimization should be reflected in preparation of enterprise cash funds budget for the timely adjustment of cash flow and ensuringsolvency of enterprise.

3 Practical implementation of the computational model of cash flow optimization

Grouping of theoretical and practical developments, as well as the results of original researches on this subject was tested on the materials of production machine building company ZAO(privately held corporation) "UUMK", which produced 22 heavy autograders (models A-98 and DZ-122B) in 2012. In 2016, it is planned to increase production by 30%. The average retail price is going to be 3.6 million rubles with quarterly growth, taking into account projected inflation within 15-17% (according to the Central Bank of the Russian Federation [17]. Sales budget is presented in Table V

Table 5. Sales budget ZAO "UUMK" for 2016.

Production	I quarter	II quarter	III quarter	IV quarter	Total
Sales volume, in units					
Autograder A-98	3	5	4	4	16
Autograder DZ-122B	4	3	3	2	12
The unit price, in thousands of RUR					
Autograder A-98	4 550	4 778	5 016	5 267	-
Autograder DZ-122B	2 680	3 095	3 250	3 413	-
Revenue, in thousands of RUR					
Autograder A-98	13 650	23 888	20 066	21 069	78 672
Autograder DZ-122B	10 720	9 286	9 751	6 825	36 582
Total	24 370	33 174	29 816	27 894	115 254

Thus, it is planned for 2016 to produce and sale 28 autograders with expected value of 115,3 mln. rubles.

After completing sales budget production budget is needed.

The balance of finished goods at the beginning of 2016 is known and is equal to 2 for A-98 model and 1 for DZ-122B model with basic complexion.

Production budget is shown it Table VI.

Table 6. Production budget of ZAO "UUMK", in units.

Production	I quarter	II quarter	III quarter	IV quarter	Total
Sales					
Autograder A-98	3	5	4	4	16
Autograder DZ-122B	4	3	3	2	12
The balance of finished goods at the beginning of year					
Autograder A-98	1	1	1	1	-
Autograder DZ-122B	1	1	1	1	-
The balance of finished goods at the end of year					
Autograder A-98	2	1	1	1	-
Autograder DZ-122B	1	1	1	1	-
Autograder A-98	2	5	4	4	15
Autograder DZ-122B	4	3	3	2	12

The set level of finished goods stock is a prerequisite for trouble-free shipment and rhythm of product supply.

The data on volumes of production allows to start the development of direct material costs. To produce autograders, ZAO "UUMK" buys imported goods and components. Due to depreciation of the national currency rate, the purchase of foreign materials is expensive and, as one of the measures for optimizing company's financial condition is reducing costs, the recommendation will be to purchase goods produced by domestic enterprises.

Material budget is presented in Table VII

In 2015 the total cost of purchasing goods amounted to 64 mln. rubles, which is by 30% higher than value obtained in preparation of material cost budget.

The budget (estimate) of direct material costs is based on the immutability of consumption rates and prices of materials. In the case of changing these settings during budget period values of estimate must be re-calculated [18, p. 348].

The next stage is drawning up the budget of direct labour cost. In this budget, we take into account only labour cost of production workers. Average wage of a production worker amounts to 25 thousands of rubles.

The budget of direct labour costs is presented in Table VIII.

Table 7. Budget of direct material costs of ZAO "UUMK" in 2016.

Autograder A-98	I quarter	II quarter	III quarter	IV quarter	Total
Quantity of production to manufacture, in units	2	5	4	4	15
Direct material costs per unit, in thousands of RUR	2 050	2 153	2 260	2 373	-
Direct material costs, in thousands of RUR	4 100	10 763	9 041	9 493	33 396
Material stock at the end of period, in thousands of RUR	215	181	190	185	-
Material stock at the beginning of period, in thousands of RUR	1 230	215	181	190	-
The amount of cost for purchasing materials, in thousands of RUR	3 085	10 728	9 050	9 487	32 350
Cost of purchasing for producing 1 unit, in thousands of RUR	1 543	2 146	2 262	2 372	-
Quantity of production to manufacture, in units	4	3	3	2	12
Direct material costs per unit, in thousands of RUR	1 450	1 523	1 599	1 679	-
Direct material costs, in thousands of RUR	5 800	4 568	4 796	3 357	18 520
Material stock at the end of period, in thousands of RUR	91	96	67	218	-
Material stock at the beginning of period, in thousands of RUR	1 450	91	96	67	-
The amount of cost for purchasing materials, in thousands of RUR	4 441	4 572	4 767	3 507	17 288
The amount of cost for purchasing materials, in thousands of RUR	1 110	1 524	1 589	1 754	-
Total amount of costs on purchasing materials, thousand of RUR	7 527	15 300	13 817	12 995	49 638

Table 8. Budget of direct labour costs of ZAO "UUMK" in 2016.

Autograder A-98	I quarter	II quarter	III quarter	IV quarter	Total
Quantity of production to manufacture, in units	2	5	4	4	15
Direct labour costs per unit, in thousands of RUR	420	422	424	426	-
Amount of direct labour costs, in thousands of RUR	840	2 111	1 697	1 705	6 353
Quantity of production to manufactu-re, in units	4	3	3	2	12
Direct labour costs per unit, in thousands of RUR	230	231	232	233	-
Amount of direct labour costs, in thousands of RUR	920	693	697	467	2 777
Total amount of labour costs, in thousands of RUR	1760	2 804	2 394	2 172	9 130

Additionally, to optimize cash flow of ZAO "UUMK", it is necessary to make plans for payments to suppliers for materials and payments of buyers for delivered goods. According to cash inflow plan, the payment for delivered products will be based on advance payment in the amount of 70% with defer of payment (3 months). There also will be repayed receivables of 2015.

Cash inflow plan is shown in Table IX.

Table 9. Cash inflow plan for realized products of ZAO "UUMK" in 2016.

Value	I quarter	II quarter	III quarter	IV quarter	Total
Accounts receivables at the beginning of 2016: 14 635, thousand RUR					
Revenue from sales of products, thousand RUR	24 370	33 174	29 816	27 894	115 254
Payments for products, thousand RUR	18 52 3	32 72 8	33 75 0	31 39 8	116 398
Accounts receivables at the end of 2016, thousand RUR					13 490

According to plan of cash flow, by the end of 2016 accounts receivable can be equal to 13, 5 mln. rubles, which is 9% less than in 2015.

Table 10. Cash inflow plan for realized products of ZAO "UUMK" in 2016.

Value	I quarter	II quarter	III quarter	IV quarter	Total
Accounts payable at the beginning of 2016 , thousand RUR					20 735
Material costs, thousand RUR	7 664	15 437	13 954	13 132	50 186
Payments for products, thousand RUR	5 905	13 128	17 288	19 023	55 344
Accounts payable at the end of 2016 ,.thousand RUR					15 577

Thus, a decrease in accounts payable will amount to 5 mln. rubles, which will significantly affect the liquidity of enterprise.

Cash flow budget is a planning document that includes expected inflow and outflow of cash funds during the planned period. The inflow is formed on the basis of sources of cash funds income and outflow – on the basis of directions of use [19, p. 19].

Cash flow budget of current operations is presented in Table XI.

Table 11. Cash flow budget of ZAO «UUMK», current operations in 2016.

Value	I quarter	II quarter	III quarter	IV quarter	Total
CASH FLOW FROM CURRENT OPERATIONS					
TOTAL INCOME	18 523	32 728	33 750	31 398	116 398
Including sales of products, goods, work and services	18 523	32 728	33 750	31 398	116 398
TOTAL PAYMENTS, including:	14 039	24 219	28 683	30 014	-96 954
payments for goods, work, services	5 905	13 128	17 288	19 023	-55 344
labour payments	2 950	3 998	3 592	3 375	-13 915
payments of interest on loans and borrowings	380	380	380	380	-1 520
payments of taxes and duties	126	2 011	2 698	2 487	-7 322
other payments and transfers	4 678	4 701	4 725	4 749	-18 853
Cash flow balance of current operations	4 483	8 509	5 068	1 384	19 444

The expected net cash flow is positive due to increase in entity's revenue, a decrease of production cost because of use of domestic materials, optimization of customer payments for purchased products, as well as payments to suppliers for materials supplied. Cash flow of financial and investment operations are reflected in cash flow budget as they have a quick turnover, big amounts and short repayment terms [20]. Therefore, the budget will be not on a quarterly, but on a year basis. The cash flow budget for investment and financial operations is shown in Table XII.

Table 12. Cash flow budget of ZAO "UUMK", investment and financial operations in 2016.

Value	Total
CASH FLOW FROM INVESTMENT OPERATIONS	
TOTAL INCOME	4 120
Including interest on financial investment, return of loans	4 120
TOTAL PAYMENTS	3 540
Including payment for purchase of fixed assets(including income bearing investments tangible assets) and intangible assets	-2 300
Payments for financial investments	-1 240
The result of cash flow from investment operations	580

CASH FLOW FROM FINANCIAL OPERATIONS	
TOTAL INCOME	11 200
including loans and borrowings	11 200
TOTAL PAYMENTS	11 028
including repayment of loans and borrowings	-10 822
Payment of dividends	-206
The result of cash flow from financial operations	172

In 2016 ZAO "UUMK" plans to acquire a machine tool that will be able to fo multioperational machining of details with complex configuration. In addition to purchase of this machine, the enterprise plans to carry out a repair of shop premises. For this purpose, it is planned to take an investment loan in the sum of 11 mln. rubles with repayment term of 18 months and an average rate of 15%.

In 2015 the company acquired debt securities that provide income in 2016.

The expected net cash flow from investment and financial operations is positive. Total net cash flow is equal to 20 196 thousand rubles, which will make it possible for organization t carry out further modernization of products at the expense of its own funds without involving expensive external sources of financing.

Conclusion

The monitoring of scientific literature on optimizing cash flow, as well as testing of the hypothesis of research on materials of industrial enterprise allowed to develop a method of cash flow optimization for of ZAO "UUMK" company, which is one of the four largest manufactures of road machines along with ZAO "Bryanskiy arsenal", ZAO "ChSDM", ZAO "Zavod Dormash" and ZAO "PO "Irmash".

Financial analysis showed that the company is financially unstable at the end of 2015 with a high probability of bankruptcy and loss of solvency.

Authors propose a model of cash funds optimization included in budgeting of ZAO "UUMK" that suggests:

• the use of "accelerating-decelerating pay transactions" which provides acceleration of accounts receivable turnover – payment for delivered products will be based on 70% advance payment, defer of payment (3 months) and slowing down payments to suppliers and contractors – 50% will be paid on condition of advance payment with deferred of payment (6 months) in the short term:

• the use of financial instruments for both raising and allocating cash funds, reduction of production cost by reducing costs of enterprise in the long term.

The performed analysis of ZAO "UUMK" targets led to conclusion about the appropriateness of using the optimization model based on obtained data that will provide:

1) positive trend in revenue of 9 mln. rubles due to increase in production of autograders which, as well as increase in machinery prices;
2) decrease of 15 mln. rubles in production cost;
3) the use of "accelerating-decelerating pay transaction system" that will reduce accouts receivable by 1 mln. Rubles, accounts payable – by 5 mln. rubles;
4) Net cash flow and net income suggest that enterprise has its own sources for self-financing of its activities.

Thus, the hypothesis of cash flow optimization through budgeting system was confirmed during the practical experiment in industrial plant.

The work was supported by Act 211 Government of the Russian Federation, contract № 02.A03.21.0011.

References

1. K.S. Lorek, G.L. Willinger, Source of the DocumentAccounting Review, **71(1)**, 81-102 (1996)
2. P. Martins, Lecture Notes in Economics and Mathematical Systems, **682**, 115-121 (2016). http://www.scopus.com/inward/record.url?eid=2-s2.0-84951798889&partnerID=40&md5=1deebebbbcabc6ec4fa7ef9ce98ce7fb.
3. W.-N. Chen, J. Zhang, H.S.-H. Chung, R.-Z. Huang, Liu, IEEE Transactions on Systems, Man and Cybernetics Part C: Applications and Reviews, **40(1)**, 5196736, 64-77 (2010). http://www.scopus.com/inward/record.url?eid=2-s2.0-73049110195&partnerID=40&md5=52b48a883f1a0088f50dc5072c086e3a
4. G.V. Savitskaya, *Analysis of economic activity of enterprise* (New knowledge, Moscow, 2011)
5. V.V. Kovalev, *Financial analysis: methods and procedures* (Finance and Statistics, Moscow, 2012)
6. A.D. Sheremet, *Comprehensive analysis of economic activity: Textbook* (INFRA-M, Moscow, 2011). http://www.books.ru/books/analiz-finansovykh-operatsii-metody-modeli-tekhnika-vychislenii-32980/ © Books.ru.
7. E.V. Sanakoyev, *Cash flows of companies and their optimization 2011* (Vladikavkaz). http://www.dissercat.com/content/denezhnye-potoki-predpriyatii-i-ikh-optimizatsiya#ixzz4415Quda.
8. D.A. Novolodsky, *Improving cash flow management system in enterprise* (Ph.D. Dissertation, Novosibirsk, 2004). http://www.dissercat.com/content/sovershenstvovanie-sistemy-upravleniya-potokami-denezhnykh-sredstv-na-predpriyatii#ixzz4417T96uO.

9. A.A. Sinkina, *Cash Flow Management: A case study of industrial enterprises in the Kemerovo region* (Kemerovo, 2006)
10. http://www.dissercat.com/content/upravlenie-denezhnymi-potokami-na-primere-promyshlennykh-predpriyatii-kemerovskoi-oblasti#ixzz4419KraVv.
11. B. Kolass, *Management of financial activity of enterprise: problems, concepts and methods: manual* (Economics, Moscow, 2013)
12. *Economic Analysis: Textbook for universities* (Unity-Dana, Moscow, 2011)
13. M.A. Vahrushina, *Financial Statement Analysis: Textbook* (University textbook, Moscow, 2011)
14. A.T. Gasparyan, T.N. Myznikova, Proc. Young researcher, 291-295 (2015)
15. L.G. Zaharova, T.N. Myznikova, *Proc. Trends and prospects of financial and economic development of Russia*, 50-52 (2015)
16. T.B. Krasnenko, T.N. Myznikova, Optimization of cash flows in company. In the collection: Modern society: scientific view of young people Collection of articles and abstracts of the XI International scientific and practical conference of students, undergraduates and graduate students, 86-90 (2015)
17. The Central Bank of the Russian Federation. Inflation in the consumer market. http://www.cbr.ru/statistics/?PrtId=macro_sub.
18. E.Y. Dobrovolsky, *Budgeting: step by step* (Peter, Saint-Petersburg, 2013)
19. On Accounting: Federal Law of December 6, 2011 no. 402-FZ, as amended on April 4, 2014, legal reference system ConsultantPlus.
20. S.A. Sterlyadev, Modern scientific researches and innovations, **8**, 19 (2014)

The economic efficiency of investment in the development of reserves of small groups of geographically contiguous gold deposits

S.I. Evdokimov[1,*], *S.V.* Galachieva[1], and *D.V.* Tebloeva[1]

[1]North Caucasian Institute of Mining and Metallurgy (State Technological University), NCIMM (STU), 362021, Republic of North Osetia-Alania, Vladikavkaz, Nikolaev st., 44

Abstract. The object of the research is a group of geographically contiguous low volume gold deposits. The subject of the study is an economic justification for a way to involve economic turnover to get a positive commercial result on a specially formed group of gold deposits, in which individual field development is unprofitable. A small production volume, combined with high capital and operating costs are objective reasons for the reduction in investment attractiveness of the deposits which have reserves of gold of <25 tons in the "supply and demand" chain. The economic result will be fundamentally different if the production capacity of the company is increased by combining geographically contiguous low volume deposits in the group where the total reserves of the mountain holding are represented as one deposit. A criteria was developed and it was determined that the economic and technological constraints for the deposit to enter the group would be outpacing growth of the recoverable value of reserves as compared with a decline in their gold content. Deposits of the group are rationally developed on a rotational basis with a seasonal operating mode of small mining enterprises (with an annual underground mining of 5000-50000 tons up to 100,000 tons, with private ownership> 50%), equipped with a mobile processing complex with deep processing technology on highly liquid commodity products on site. An economic-mathematical model was devised to determine the rational placement of the processing capacity of the group.A simulation was conducted and an economic evaluation was performed on the effectiveness of investments in individual and group mining projects. The simulation results show that the joint exploitation of the reserves of the group of deposits, the internal rate of return on investments exceed the rate of return of funds to the bank deposit, the return on investment is above the level of inflation. The group project complies with the strategic line of small mining companies in terms of cost recovery terms, availability of financial sources to cover expenses, provision of stable means of income and obtaining competitive advantage.

1 Introduction

Gold mining companies have no competition in product sale, since gold is a standardized commodity with almost absolute liquidity. According to the agency Thomson Reuters GFMS, the total volume of global demand for gold is more than 4100 tons. Gold is widely used in electronics and pharmacology, but the traditional and the largest consumer of it (with a share of aggregate demand consistently above 50%) is the jewelry industry. The relationship between its supply and demand on the part of the gold mining industry - the most stable of the world gold market segments - is a key factor in determining the price of gold. The largest consumers of gold are India and China, the undisputed leader among the gold-producing countries which also increases its production volumes on an annual basis. Gold is a popular investment vehicle. Against the background of high risks of inflation in the major developed economies, central banks hold a growing share (11-13%) in the structure of demand[1-4].

2 The relevance of the task to be solved

In Russia there are about 30% of proven gold (with the total balance reserves of over 39 million ounces of metal) in the undistributed fund, concentrated in about 100 small sources of gold (with gold reserves of up to 800 thousand ounces) [5]. The reason lies in the fact that individual exploitation of these georesources is unprofitable [1, 6-9], due to the small production volume, arising from the need to comply with the normal depreciation life of the mining enterprise, which is not sufficient for the payback on the production costs. Considering the problem of growth of the gold production in Russia as a relevant state-significant program, it can be argued that the solution to the problem of increasing investment demand for low volume raw mineral sources of gold by reducing the level of costs and payback period of investment will increase the amount of metal handed over to

* Corresponding author: eva-ser@mail.ru

GOKHRAN in the next 10-20 years to 650-970 thousand. ounces annually.

3 The theoretical part

According to Taylor's formula relating the rational development period T (years) of the deposits with its reserve balance reserves M (t)

$$T = \frac{\sqrt[4]{M}}{5}$$

specific cost value (capital, operational and organizational) are reduced with an increase in the value of reserves ΔM. Consequently, the way to increase the commercial efficiency of operation of small deposits is to form (including to account for licensing and the creation of a group of auction items) a group for joint operation [10-13].

Mixing ores with the same gold content, but with different enrichment might be disadvantageous when separated [14-16].Gold content a_{Σ} (g/t) in the total reserves of the group of deposits MΣ is calculated based on the content of gold α_i (i = 1 ... n) and the reserve amount in different deposit group M_i (t)

$$a_{\Sigma} = \frac{M_1 a_1 + M_2 a_2 + ... + M_n a_n}{M_1 + M_2 + ... + M_n}$$

The deposit may be included into the group for joint exploitation under the condition that upon joining its reserves with that of the group, the criteria shown in equations (1) and (2) are met

$$\frac{M_{\Sigma}}{M} \frac{a_{\Sigma}}{a} \geq \frac{Y}{Y_{\Sigma}} \quad (1)$$

where is the economic efficiency criterion of separation of ore mixture

$$Y = \left[\frac{Q\,(C_{dob} + C_{usr} + C_{ob})}{y} + C_{T1} + C_{T2} + \Sigma C + C_{met}(\beta) \right] \varepsilon_{met}(\beta) \quad (2)$$

here M - available reserve of the basic gold deposit α; *Q*–processed ores; C_{dob}, C_{usr}, C_{ob} - cost of production, averaging and enrichment of 1 ton of ore, respectively; y - concentrate output; C_{T1}, C_{T2} - the cost of transportation of 1 ton of ore to the mill and 1 ton of concentrate to the transport hub; $C_{met}(\beta)$ –the cost of production of metallurgical processing of 1 ton of concentrate from its β content of gold; ΣC–other expenses for 1 ton of ore; ε_{met} (β) –extraction of gold with metallurgical limits depending on the quality of concentrate. The geological and technical parameters of the gold deposit, «Pavlik» (Magadansk region) were used as the base for the economic and mathematic model.

The profitability of extracting gold (R) depends on the sum of expenses on extraction and enrichment of the ore (3), the gold content in the ore (αΣ), gold extraction (εΣ) at the mill, transport costs (bS), the price of 1 g of gold (C_{Au}) and mathematically related equations of the models (3)-(5):

$$R\left(\alpha_{\Sigma}, S, \varepsilon_{\Sigma}\right) = f\left[\left(C_{Au} \times \alpha_{\Sigma} \times \varepsilon_{\Sigma}\right) \times (3 + bS)^{-1}\right] \quad (3)$$

$$\varepsilon_{\Sigma}\left(\alpha_{\Sigma}, R, S\right) = f\left[R \times \left(C_{Au} \times \alpha_{\Sigma}\right)^{-1} \times (3 + bS)\right] \quad (4)$$

$$\alpha_{\Sigma}\left(\varepsilon_{\Sigma}, R, S\right) = f\left[R \times \left(C_{Au} \times \varepsilon_{\Sigma}\right)^{-1} \times (3 + bS)\right] \quad (5)$$

The practical significance of the model (3)-(5) is the ability to determine the optimal placements of the mills within the deposit groups, based on the calculation of the maximum distance for ore transport considering the acceptable economic damage for the subsoil user (given the profitability).

4 The results of economic-mathematical modeling

The reduction of unit cost with an increase in production capacity of the mining company was proven (Fig. 1) with the analysis of the investment plans to develop a series of deposits in the Northern-Eastern regions of the Russian Federation.

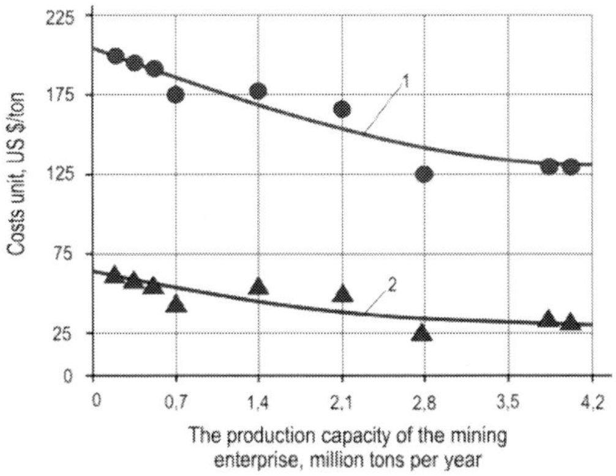

Fig. 1. Unit cost as function of production capacity of the mining enterprise, 1 - Capital expenditures for the annual processing of ores; 2–unit exploitation costs for drill, extraction, enrichment and transport.

Figure showing the unit costs as a function of the production capacity of the mining company

From the simulation calculation results of the models (5)-(7) it is deduced that it is rational to place the mills near the deposits with the most extraction significance in terms of reserves (Fig. 2).

Table 1. Results of the efficiency evaluation of investment in mining projects.

Item no	Name of indicator	Individual processing of georesources from Petinikan-Bastachsky Square							Joint developm entofgrou p
		Oreoccu rrence 1	Oreoccu rrence 2	Oreoccur rence 3	Oreoccurr ence 4	Oreoccurr ence 5	Oreoccurr ence 6	Industrialplac ers	
1	The volume of sales (Q) of gold in thousand ounces	22.645	135.935	65.207	19.309	45.058	163.459	4.516	456.129
2	The average weight selling price of gold P, $ per ounce of gold equivalent	1275	1275	1275	1275	1275	1275	1275	1275
3	Revenues from sales (TR) of gold, $ mln.	28.872	173.317	83.139	24.619	57.449	208.410	5.758	581.564
4	The total cost (AISC) of production of gold, $ mln.	32.9	108.7	52.3	27.8	65.6	130.0	5.8	376.3
5	Operating costs for drilling, extraction, transporta tion and refining, $ mln.	25.9	66.6	32.1	21.8	51.6	79.3	4.2	266.8
6.	Capital costs for extraction and enrichment , $ mln.	7.0	42.	20.2	6.0	14.0	50.7	1.6	109.5
7.	Net present value (NPV), $ mln.	-2.1	34.0	16.2	-1.7	-4.3	41.2	-0.02	107.8
8.	Profitabilit y index PI of discounted investment	-0.61	+1.43	+1.64	-0.60	-0.63	+1.70	-0.04	+3.77
9.	Discounte d payback period (DPP), years	Doesnot payoff	5	5	Doesnotpa yoff	Doesnotpa yoff	6	Doesnotpayoff	5
10.	Thediscou ntrate E%	-	15	15	-	-	15	-	18
11.	Internal rate of return IRR,%	4.3	26.3	28.4	4.2	4.5	29.2	14.7	35.2

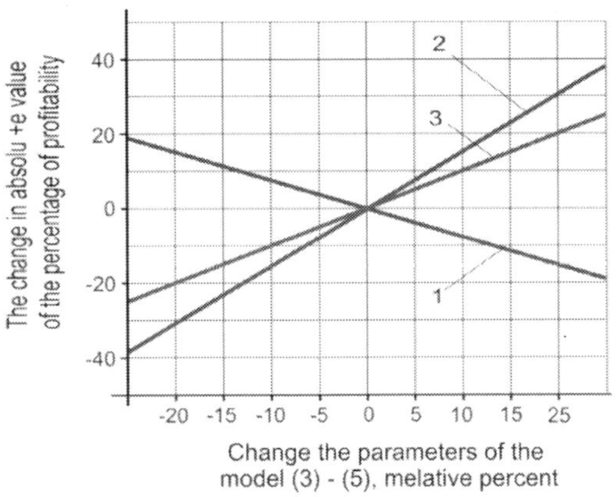

Fig. 2. The affinity of the profitability of the project to changes in the transportation distances of the ores (1), its gold content (2) and its extraction at the mill (3).

The methods described in [1, 5, 17, 18, 19, 20], were used to conduct a simulation of the individual and group exploitation of geographically contiguous low volume gold deposits and the waste from the alluvial gold mining on Petinikan-Bastachskaya square (Yano-Kolyma gold-mining province). The georesources of the group are treated as a single deposit site (single auction item), the reserves of which are developed by small mining enterprises of the holding. Raw material processing facilities are absorbed consequently just as the mining system on a rotational basis with a seasonal mode of operation. Collapsible mobile processing complex (CMPC) is used a processing unit, which reduces capital costs by 25%. The complex is equipped with CMPC technology to obtain the commercial product at the site in the form of fused gold ligatures, which thus reduces the costs of refining.

The simulation results show that for the transition from individual to joint development of deposits, the overall economic effect is significantly positive, the total gold reserves of the groups provide a return on investment within acceptable terms for the investor, the internal rate of return on capital investment exceeds the rate of return from the placement of funds into bank deposit, which in turn favours growth of investment demand for low volume deposits.

5 Conclusion (findings)

A cost-effective way to operate geographically contiguous gold deposits with small reserves is their joint development into a single auction lot. The economic criteria for the selection of the deposit in the group is to anticipate the growth of recoverable value of inventories as compared to the decline in gold content. The optimal location of the processing capacities within the group is determined by the model, linking economic and mining parameters of gold production.

It is rational to develop deposits in the group on a rotational basis with seasonal operation mode of small mining enterprises, equipped with mobile enriching complexes with deep processing technology to obtain a highly liquid commercial product on site.

This article was prepared with the support of the agreement on granting of subsidies № 14.577.21.0142 (RFMEFI57714X0142).

References

1. N.Yu. Samsonov, M.A. Yagolnitser, *Group development of small gold deposits* (Novosibirsk, IEIE SB RAS, 2012)
2. A.A. Petrosov, V.A. Ivanov, Mining information and analytical Bulletin, A separate article (special issue), **12**, 3-13 (2012)
3. B.K. Mikhailov, Mineral resources of Russia: Economics and management, **3**, 26-35 (2012)
4. I.M. Petrov, Mineral resources of Russia: Economics and management, **3**, 85-89 (2011)
5. N.Yu. Samsonov, *Study areas for cost-effective development of small and medium-sized gold deposits* (Institute of Economics and industrial engineering of Siberian branch of RAS, Novosibirsk, 2010)
6. N.Yu. Samsonov, Economy of region, **2**, 139-146 (2010)
7. N.Yu. Samsonov, Regional Economics and sociology, **1**, 306-316 (2010)
8. A.A. Sobolev, Mountain information-analytical Bulletin, **3**, 176 (2013)
9. V.V. Kharitonov, U.N. Kurilchuk, Mining magazine, **9**, 100-106 (2014)
10. V.A. Ivanov, *Organizational and economic justification of public-private partnership in the development of small-scale gold deposits* (Moscow state mining University, Moscow, 2013)
11. V.A. Kabirov, *Assessment of economic efficiency of development of group of geographically contiguous ore (metal) deposits* (Saint Petersburg, National mineral resources University, Saint Petersburg, 2014)
12. A.B. Anisimova, *The Economic rationale for involvement in the development of contiguous iron-ore deposits of the Polar Urals* (Russian state geological prospecting University, S. Ordzhonikidze, Moscow, 2011)
13. A.B. Anisimova, The proceedings of the mining University, **191**, 30-33 (2011)
14. M.A. Peshkov, *Substantiation of requirements to quality of ore and technogenic raw materials for*

the integrated development of copper-pyrite deposits of the Urals (Moscow Institute of comprehensive exploitation of mineral resources Russian Academy of Sciences, 2014)

15. Y. Roslavtseva, *Feasibility of mining operations during incremental development of small deposits in the open way* (Irkutsk state technical University, Moscow, 2014)
16. L.I. Boguslavskaya, *Increase of economic efficiency of the mining company on the basis of control of its capitalization* (St. Petersburg state mining University, St. Petersburg, 2012)
17. M.L. Polumiskova, *Development of economic-mathematical models and tools for assessing the investment attractiveness of gold mining companies* (VNIIPVTI, Moscow, 2010)
18. M.L. Polumiskova, Economic science, **10**, 399-404 (2008)
19. M.L. Polumiskova, Economic science, **3**, 272-275 (2010)
20. V.N. Kalmykov, A.A. Gogolin, A.N. Ivashov, Mining magazine, **12**, 3-7 (2015)

Accounting engineering for sustainable development

A. Sidornya[1,*]

[1]Don State Technical University, Rostov-on-Don, Russia

Abstract. The article deals with the sustainable development of industrial enterprises in Russia, accounting for sustainable industrial growth of the national economy, tools of accounting engineering aimed at creating an information basis of transformation the Russian economic model to knowledge based economy. The proposed mechanism of ownership control of industrial enterprises in the context of long-term planning of the national economy. Theoretical bases of accounting engineering, its tools are defined. A brief review of the literature on the problem of accounting engineering is provided. A practical example of the application of the accounting engineering logic for the industrial enterprise is reviewed. It describes the research results obtained during the last 25 years of Russian scientific school of accounting engineering. Conclusions and recommendations on the use of accounting engineering to sustainable development of the Russian economy are formulated.

1 Introduction

Drivers of modern economy essentially differ from economy of last century. If power of industrially developed countries consisted in the recent past in physical resources, today, significant natural resources make the Russian economy weakness and vulnerable. The fact is that the vast natural resources of Russia for a long time compensated the low economic efficiency, low labor productivity, an unacceptable level of corruption, and low quality of decision-making. Is no more than several of Russian companies, which management has a sufficiently high level (Wimm Bill Dann, Tinkof Bank, Ozone, Magnet, Yandex, ABBYY, Sberbank), but majority of them senior management formed by foreigners. This situation is unacceptable for Russia, in which the higher education level was high enough in the Soviet period. However, the problem of quality of managers' education – is not the only one. There is an urgent problem of the quality of management information support – quality economic, and first of all accounting, information.

Accounting education in Russia has transformed in the last few years. Russia goes to the international financial reporting standards (IFRS). The IFRS course is included into the basic program of a bachelor degree of financial accounting.

Programs of advanced training of professional accountants are also provided by studying of last modifications of IFRS. However, IFRS is not a "cure against all diseases" of the national economy. Using IFRS as the basis for the design of management accounting systems, as suggested by some authors in Russia – is inappropriate, because management accounting cannot be placed in any frame.

Solving the problem of improving the quality information support of decision makers is beyond the scope of IFRS. This problem is in the area of trade secrets, insider information and engineering of company's business processes. This problem cannot be solved by the implementation of any kind, even the best (IFRS), mandatory standards. Each company has its own unique culture, strategy, business model, organizational model, life cycle phase, the production cycle, business process maps, assets portfolio, the relation with consumers, structure of financing, etc. And each company has its own unique characteristic of the needs for management information. That is why in the last 25 years in Russia the scientific school of "accounting engineering", which main concept is the design of management accounting under the specific company based on the "structured working chart of accounts" and tools of "derivative balance statements", was widely adopted.

Despite the fact that the terminology "accounting engineering" carries a "technical" context and the school promotes the widespread use of instruments of exact sciences: graph theory, the adjacency matrix, algorithms, fractals, patterns, etc. The main goal of accounting engineering is add more economy in the accounting.

Throughout the last century accounting has steadily lost its economic nature turning into political science. People who were born in the Soviet Union know the famous saying of Lenin: "Socialism – it is accounting." These people do not need to explain how much accounting can be politicized (enough only to remember the experiment undertaken in the Soviet Russia on transition from the monetary measuring to the «workdays» in accounting). However, the market economy countries accounting is politicized too. So

* Corresponding author: gorodrostov@gmail.com

one of the famous accountants A. Hopwood 30 years ago wrote: Accounting begins to be considered more and more as a social but not simply technical phenomenon. Although slowly, but begin to take into consideration of his origin and that it gives the environment [1]. Accounting researchers have long been interested in the extent to which politics affects accounting standard-setting. Even casual observation of the standard-setting process suggests that politics can have a first order effects on how accounting standards are set. Beginning with Watts [2] and Watts and Zimmerman [3], researchers have sought to develop and test economics-based theories of standard-setting that capture these political forces [4].

What kind of impact the policy has had on accounting?

One of the most striking examples of the impact of policy on accounting is its division into types: financial and tax. This division is purely political and designed to provide the ability to manipulate numbers. Let's assume that the enterprise earned profit, there are 4 participants who agreed to divide it equally on 25%. And so, for one of participants the accountant counted profit in the amount of $100 and paid $25 of dividend, for another he counted profit in the amount of $40 and paid $10 dividends, for the third counted profit in the amount of $10 and paid $2,5 dividends, and for the last the profit made $4 and respectively dividends - $1. All received on ¼ profits. All is the fair. However the first received $100 and the last $1.

Do everything is fair? Whether it is possible to manage the enterprise on the basis of accounting information driven by political expediency?

School of Accounting Engineering, no claim to absolute truth, attempts to reduce the all variety of economic transactions and accounting entries of economic actors to the highest degree of abstraction: property, risks, reserve system, intellectual capital, a margin of synergy safety, a zone of financial risk, etc. on the basis of the most demanded achievements of economic science: theory of the firm, transaction cost theory, agency theory, stakeholder theory, the concept of sustainable development and others.

2 Overview of literature

The theoretical basis of accounting engineering was developed mainly English-speaking researchers J.Briloff [5], M. Schilit [6], R. Parker [7], J. Griffith [8], J. Elliot [9], which circle of scientific interests stretched generally for earning management and so-called "creative accounting".

At the same time, its methodological basis comes from Germany, where engineering mechanisms of accounting control: "zero derivative balance statements" is widely adopted. Which purpose is "a reflection of the relations of property and debt after the closing of accounts" that would "give at the end of the reporting period, a reliable estimate of property status enterprise "[10]. Such famous German scientists as Willenbucher [11], Kayser [12], Simon [13], and Beigel [14], were ideologists of inventory approach and an valuation of assets at current value. In the end, the Higher Court of Germany (RG) has become for these positions noting that "the practice of leaving in the balance sheet at the acquisition price paid by the flawed"[15].

However, it should be noted that the term "accounting engineering" in these papers are not used. The term "accounting engineering" appeared in the late twentieth century in in the works of C. Drury, A. Kure, M. Capron [16]. So Alan Kure identified of financial engineering instruments and devices (balance engineering and engineering of cash flows) [17].

In Russia the first who used the term "accounting engineering" was the famous Russian scientist, the founder of the scientific school of accounting engineering, Viktor Tkach. In 1993 he goes from Rostov State University of Economics, where he as an assistant professor at the Department of Accounting, to Rostov State University of Civil Engineering (RSUCE). Here, in his own words, he "faced with a completely different, engineering, thinking" [18]. Since 1993, Viktor Tkach created a scientific school where adherents of more than 100 are unique scientists today: the authors of dissertations, numerous articles and monographs. V. Tkach is the author qualimetrical textbook "Accounting engineering theory"[19], monographs "Accounting engineering: concept, theory and practice", "The theory, systems and models of accounting engineering network" and more than 200 publications.

Terminology "accounting engineering" found in recent English-language studies. So P.D. Nicoleta claims that accounting engineering and creative accounting is the same, and their only distinction lies for the purpose of these types of accounting [20]. L. Michalczyk argues that «accounting engineering - legally accepted manipulation of the financial result by interested parties - is a serious problem for the economy» [21]. And M. R. Young in general claims that good news is that accounting engineering - is rather difficult in application [22].

It is possible to note that now the term "accounting engineering" reliably entered in Russian-speaking and English-speaking economic literatures. But the negative context that is embedded in the English-speaking authors it is not typical for Russia. Moreover in Russia to use the balance of engineering tools is seen approaching the objects of accounting control to the true economic values.

3 Setting of a problem

The concept of accounting engineering was the response to excessive politicization of financial accounting. Even in 1973, the American professor C.T.Hongren wrote that the accounting standards being burrows limit the social behavior of economic agents, and later they should be accepted and other sectors of society. This acceptance can be forced or voluntary. In a democratic society to achieve the consent is particularly difficult [23]. And in 1991, also American Professor Parker writes, "Accounting is a social network, enclosing a very significant consensus"[24].

However, at the same time, a political consensus is embodied in accounting further alienates him from his economic substance. It must to say that this process is inevitable: the object of accounting is a property, and property is political category. Therefore, the economy and politics are inseparable. Recent events around the sanctions against Russia are the clearest example of that. However, Russia's problems are more related not to the sanctions, it related with finiteness of its development model built on the basis exhaustible natural resources. That it is not possible to ensure the sustainability of its development.

However, the problem of sustainability has several dimensions: political stability, economic stability, demographic stability, environmental sustainability, institutional sustainability. Accounting engineering makes complex impact on all above-named elements of a sustainable development: the economic - through risk control, reserve system and strategy of development; the political - through justice of distribution of value in current generation; the demographic - through ensuring justice of distribution of value between generations; the ecological - through ensuring safety of an ecosystem (the natural capital); the institutional - through ensuring safety of the property rights. But all of this, - is only return to the economic origins of the accounting.

4 Theoretical essays

Problem of excessive politicization of accounting led to the fact that in order to return him to the economic substance is necessary to redesign it. It is the task of accounting engineering. So, Michalczuk writes that a theoretical basis of accounting engineering is the combination:

a) kuhn's post-constructivist anti-cumulativism - to the extent to which it establishes a theoretical basis for observation and refutes the idea that science is merely a system for collecting knowledge,

b) general systems theory - its theory of sciences and the logic of argumentation based on the principles of causality,

c) earnings management - to the extent it combines accounting and company management,

d) agency theory and the theory of transaction costs - as far as management and organizational sciences are concerned,

e) prospect theory - as far as defining key concepts used for describing, classifying and measuring values is concerned, detailed accounting theory - as a basis for analysis within the variant accounting framework, as well as the analysis of earnings and outgoings flow, the notions of accountability and responsibility - to determine who is responsible for the quality of information generated by means of variant accounting [21].

At the Russian school select the following basic concepts, theories and mechanisms of accounting engineering:

a) General Systems Theory (Bertalanffy L., Tkach V., etc.);
b) Theory of the firm (Coase R., Maksimenko A., etc.);
c) The agency theory (Eisenhardt K., Kurseev D., etc.);
d) Stakeholder theory (Freeman E., Pavlenko I., etc.);
e) The concept of sustainable development (Barbier E., Sidornya A.[25], etc.);
f) Situationally - matrix simulations (Sorter G., Kolvakh O., etc.);
g) The fractal theory (Barnsley M., Krohicheva G., etc.);
h) Behavioristic theory (Watson J., Grafova T., etc.);
i) Stochastic analysis (Howard M., Lesniak V., etc.);
j) The theory of network efficiency (Castells M., Arkhipov E, etc.);
k) The theory of constraints (Goltratt E., Aksenova, etc.);
l) The theory of synergism (Geneen H., Sidornya A., etc.);
m) The theory of economic cycles (Schumpeter J., Bogataya I., etc.);
n) The all-inclusive income concept (Hendriksen E., Sidornya A., etc.);
o) The business modeling concept (Ostervalder A., Sidornya N., etc.);
p) The theory of capital structure, dividend theory (Modigliani F. and Miller M. , Evstafieva E., etc.).

The basis of accounting engineering method is the algorithm of several stages. The first stage is defined by the object of engineering controls: it may be a business process, property, costs, risks, reserves, intellectual capital, a synergistic effect, etc. The next step is designed «structured working chart of accounts» for the primary control of engineering data. The third stage is determined by the initial operator: the balance, chart of accounts chapters, a group of related accounts, mega accounts. Since accounting engineering method is based on the balance it is important that the initial operator balanced. In the fourth step the aggregated entry made on the balance sheet and off balance sheet accounting unaccounted factors influencing the control object. In the fifth stage, made «zero balance» by the hypothetical separate sale of all assets and satisfaction of liabilities. The sixth stage is determined by the fair value of an institutional unit. At the seventh stage, the disaggregation of influence controlled and an uncontrolled factor on object of engineering controls is made.

5 Practical essays

Application of accounting engineering is successfully performed at the largest enterprises of the south of Russia various forms of ownership. This tools is effective in both the private (corporate) and the state (universities, hospitals, municipal) property. On the platform of accounting engineering in Russia, received more than 160 patents and certificates of registration of computer programs and databases, including the following «derivatives balance sheets»: immunization, monitoring, socio - economic, adaptive, hedged, outsourcing, gradualist, transactional, situational, integrated risk, strategic, subsidiarity, substation, synergistic, fractal, etc.

The internet resource: http: www.resoursum.com provide some accounting engineering tools can be used online. Suppose the net assets of the company amount to $ 1,000. The share capital of $ 600. Its market value is $ 1,500. After putting on entry unrecorded and off-balance liabilities - net assets amounted to $ 400, and after putting on entry unrecorded assets - net assets amounted to $ 1,300. After the procedure of separately hypothetical sale of assets and satisfactions of liabilities to meet debt-free money amounted to $ 800. The following conclusions can be drawn from the above data: : recognized capital gain ($400 (1000 - 600)) is actually overestimated by $200 (1000 - 800). Real capital gain amounted of $900 (1500 - 600). Including: the impact of recognized factors of $200 (800 - 600), synergetic enterprise security margin of $ 700 (1500 - 800), where the influence of the engineering factors of $300 (1300 - 1000), the impact of uncontrollable factors of $400 (700 - 300).

Also in this example, you can keep track of the «rate of current consumption» of the enterprise. It is evident that the company recognized income of $ 400 (1000 - 600). However, the increase in net assets in the amount of $ 300 upon the registration unaccounted factors, formed as follows: registration with unrecorded liabilities (provision for ecosystem restoration) reduced net assets by $ 600, and registration with unreported assets gave an increase of $900. «Rate of current consumption» is defined as: the fair value of the company minus the initially invested capital, minus the provision for restoration of ecosystems and is $300 ((1500 - 600) - 600)). This rate shall be distributed among the present generation: the owners and society. In other words, the profit of $ 400 only $300 can be directed to the payment of dividends, staff development, social projects and charity. $ 100 must to be sent to the reserve for ensuring the sustainability development to the point until it is $600. Reserve for ensuring the sustainability development should be subject to an annual audit of the need for adjustments.

Depending on what factors are registered in the implementation of engineering transactions by accounting engineering tools can monitor the impact on the sustainability development: risk, reserve, reengineering of business processes, the various risk situations, transaction costs, inflation, changes implemented strategies, changes of property rights, stakeholders, existing restrictions on the use of assets, information and logistics networks, intellectual capital, synergies, cash flow dynamics, government grants and expenses, etc.

6 Conclusion

In conclusion, it must be admitted that the management of the considered enterprise has no control over a significant part of the factors that affect the welfare of the owners. But provides a high level of creating new value is sufficient to enable the implementation of current consumption. At the same time, opportunity of increase of efficiency of management decisions its large (unaccounted factors accounted for $ 400).

However, the lack of practice create reserve for sustainability development, both on the enterprise in this paper, and the vast majority of real-life Russian companies, may result in them becoming a "financial pyramid" in which the amount of current consumption exceeds the rate, and so consumption is made from the capital , and this automatically makes the economic model (business model) – exhaustible. That eliminates the sustainability development in economic and political, demographic, environmental and institutional aspects. It is possible to tell that accounting engineering is urged to evaluate and provide in an explicit form, degree of congruence goals the persons making decisions at all levels of management in private and public sector, to criteria of sustainable development of Russia.

References

1. S. Burchell, C. Clubb, A. Hopwood, Accounting, Organization and Society, **10**, **4**, 5-27 (1985)
2. R.L. Watts, Australian Journal of Management, **2**, 53-75 (1977)
3. R.L. Watts, J.L. Zimmerman, The Accounting Review, **53**, 112-134 (1978)
4. B. Gipper, B. Lombardi, D.J. Skinner, *The Politics of Accounting Standard-Setting: A Review of Empirical Research.* https://www.business.unsw.edu.au/research-site/publications-site/australianjournalofmanagement-site/2013-australian-journal-of-management-symposium-site/Documents/Skinner-ajms-20130612.pdf.
5. A.J. Briloff, *Unaccountable Accounting* (Harper & Row, New York, 1972)
6. H.M. Schilit, *Financial Shenanigans* (McGraw Hill, New York, 2002)
7. R. Parker, C.W. Nobes, *Financial Reporting in the United Kingdom and Australia, Comparative International Accounting* (Prentice Hall, NJ, 1995)
8. I. Griffiths, *Creative Accounting* (Sidgwick & Jackson, London, 1986)
9. B. Elliott, J. Elliott, *Financial Accounting and Reporting* (Prentice Hall, 2011)
10. C. Garreis, O. Fuchsberger, *Das Allgemeine Deutsche Handelsgesetzbuch* (Berlin, 1891)

11. Willenbucher, *Das allgemeine Deutshe Handerlsgesetzbuch, mit Ausschlub des Seerechts, fur die Praxis erlautert* (Berlin, 1891)
12. P. Kayser, *Gesetz, betreffend die Kommanditgesellschaften auf Aktien und die Aktiengesellschaften* (Berlin, 1884)
13. H. Simon, *Die Balanzen der Aktiengesellschaften* (Berlin, 1899)
14. R. Beigel, *Das Buchfuhrungsrecht der Aktiengesellschaft* (Leipzig, Ludwig Huberti, 1900)
15. Dicision of RG on 1887.
16. I. Tkach, *Management accounting of changes* (RSUCE, Rostov-on-Don, 2010)
17. *Enhancing Shareholder Wealth by Better Managing Busoness Risk. Director General International Federation of Accoutants* (New York USA)
18. D. Kurseev, *Tkach - founder of the scientific school "Engineering system of accounting, auditing and management in macroeconomics"* (AzovPechat, 2015)
19. V. Tkach, *Accounting Engineering Theory* (Azov, 2013)
20. P.D. Nicoleta, *Creative accounting – art.* http://www.oeconomica.uab.ro/upload/lucrari/920071/52.pdf.
21. L. Michalczyk, Financial Internet Quarterly "e-Finanse", **8**, (2012).
22. M.R. Young, *Financial reporting after the subprime crisis: new challenges for audit committees.* http://www.willkie.com/publications/2010/month/financial-reporting-after-the-subprime-crisis.
23. J. Rishar, *Accounting: theory and practice* (FiS, 2000)
24. L.D. Parker, Advances in Public Interest Accounting, **4**, (1991)
25. A. Sidornya, *Social accounting: Theory and practice* (2011). http://sites.google.com/site/rostovnauka/zurnal-nauka-i-praktika/no3-2011.

The development of the methodical approach to value business appraisal within the integration policy for industrial enterprises

M. Karelina[1,*]

[1]Nosov Magnitogorsk State Technical University, Magnitogorsk, Russia

Abstract. At present, a positive trend in the development of corporate integration processes in industrial economics of the Russian Federation is frequently accompanied by a negative dynamics in their efficiency indicators. This causes the necessity for development of an integrated approach to integration transaction valuation using mathematical and statistical methods of research in order to increase the competitiveness of the Russian industry under conditions of the economic sanctions. The paper presents the methodical approach developed to value business appraisal of industrial enterprises based on the integration of profitable, cost-based, sales comparative approaches and the method of real options using weights Fishburne. In this paper, we propose a methodical approach that provides an integrated value appraisal of the target company and preliminary substantiates the expediency of mergers and acquisitions. In addition, this methodical approach is the foundation for high quality and reliable evaluation of the synergistic effect.

1 Introduction

Mergers and acquisitions taking place in business are aimed at increasing its scope and market share and becoming important factors for increasing the competitiveness under conditions of the economic sanctions against Russia, volatility of foreign exchange markets, and price reduction of raw materials. A technological breakthrough in the field of telecommunications and the deregulation of national markets created an entirely new environment for business dealing [1-2].

At the same time, the complex of problems related to business valuation at mergers and acquisitions has been insufficiently developed. The International Valuation Standards published by the International Valuation Standards Committee as well as guides to the standards issued in 2007 represented the definitions of different types of costs, valuation types, including at the mergers and acquisitions. However, the range of problems under reviewing is insufficient to obey the standard provisions in practice of industrial enterprises.

The development of the merger and acquisition market requires the more serious development of approaches to the integration transaction evaluation in the theory and additional amendments of existing standards. During the study, it was found that there are a number of problems which often take place in practice at evaluating industrial companies but they are usually referred to neither in educational nor in scientific literature:

• there are companies that are difficult to model and evaluate because of specific problems with their reporting;

• there are companies which balance of the structure is expected with significant changes, for example, as a result of share buybacks;

• there are high cyclic companies at which evaluating it is difficult to determine what level of extrapolation should be selected for long-term extrapolation;

• there are companies with a high proportion of intangible assets, often not involving in the balance sheet, what makes difficult to determine the actual return on capital;

• there are fast-growing and highly profitable companies, the main problem in this case is to determine what rates of drawdown of growth indicators, and profitability *ratios* towards normal levels should be expected.

Thus, there is a need to develop a methodical approach to value business appraisal within the integration policy for industrial enterprises based on the existing Russian and foreign experience with the use of mathematical and statistical methods.

2 A methodical approach to value enterprise appraisal in merger and acquisition transactions

Currently, the dynamics of prices for raw materials in Russia, as well as throughout the world, is a special challenge to the established system of pricing for the production of metallurgical companies. Under these circumstances, the Russian steel company announced its decision to purchase one of the coal companies, which in 2008 was included in a government list of strategic companies in Russia. This should allow the creation of a

* Corresponding author: marjyshka@mail.ru

highly vertically integrated structure and result in an increase in the value of the steel company.

The determination of the value of the coal company by the discounted cash flow method is based on the expectation that the metallurgical company will not be able to pay for this business the sum greater than the present value of future incomes from this business. The mathematical model for value company appraisal by the discounted cash flow method can be presented as [3]:

$$V^{DCF} = \sum_{i=1}^{n} \frac{FCFF_i}{(1+r^{CAPM})^{i-0.5}} + \frac{TV}{(1+r^{CAPM})^n} - D, \quad (1)$$

where $FCFF_i$ is *free cash flow to the firm* of *i*-th year; r^{CAPM} is the discount rate determined on the base of the capital asset pricing model; *TV* is terminal value; *D* is net debt of company.

The capital asset pricing model is based on the analysis of data arrays of the stock market. The discount rate according to the CAPM (Capital Asset Pricing Model) is defined as [4-5]:

$$r^{CAPM} = r_f + \beta(r_m - r_f), \quad (2)$$

where r_f is the rate of earnings on risk-free investments; β is a beta coefficient, which is a measure of systematic or market risk relating to macroeconomic and political processes; r_m is the market average rate of return.

At the appraisal of coal company's value the risk-free rate r_f was determined at the level of profitability of the Russian Federation federal loan bonds (the issuer is the Ministry of Finance of the Russian Federation, the general services agent is the Central Bank of the Russian Federation). Then, based on the average profitability of FLB the rate of earnings on risk-free investments is equal to r_f=10.11%.

The average market rate of return r_m is determined on the base of the revenue position on the security market in the sector to which the valued company belongs. Since among all coal companies of the Russian Federation the shares of Southern Kuzbass Coal Company are longest-listed on the stock exchange (RTS since 2003) ordinary income shares of Southern Kuzbass Coal Company were taken as the average market rate of return r_m. Thus, r_m is 16.72%.

The beta coefficient β is a measure of systematic (market) risk for the coal companies in developing markets (including Russia) and is equal 0.37. Then, based on expression (2) the following was obtained:

$$r^{CAPM} = 10.11 + 0.37(16.72 - 10.11) = 12.56\%.$$

A cash flow to equity model was chosen as a cash flow model in order to evaluate the coal company by the DCF method. At the same time with efficient management of the company, the term of its life tends to infinity. The terminal value of the project is determined to take into account incomes which business can bring beyond the forecasting period [6-9].

In the present study, the Gordon model was chosen to calculate the company's value in post-forecasting period, according to which the determination of business value is carried out by income capitalization of the first post-forecasting year on the capitalization rate, which takes into account the long-term growth rate of cash flows. Then, the terminal value is determined in according to the expression (3):

$$TV = \frac{FCFF_{(n+1)}}{r^{CAPM} - \xi}, \quad (3)$$

where $FCFF_{(n+1)}$ is *free cash flow to the firm in the* post-forecasting period; ξ is long-term (conditionally permanent) growth rates of cash flow in the residual period.

Net cash flow on capital in the post-forecasting period is equal to 194.67 million USD. In accordance with the market conditions of coking coals, experts from Economic Directorate of the Ural Metallurgical Company estimated long-term rates of cash flow growth in the residual period at the level ξ=0,5%. Therefore, terminal value is equal to:

$$TV = \frac{194.67}{0.1256 - 0.005} = 1614.19 mln\ USD.$$

Net debt of the coal company at the time of the integration transaction was D = 526 million USD. At that time the coal company value according to the method of discounted cash flows was:

$$V^{DCF} = 822.86 + (1614.19 * 0.2025) - 526 \approx \approx 623.65 mln\ USD.$$

The cost approach is based on the principle of substitution: the company's cost is no more than charges for replacing of all its integrated parts. The basic formula of the cost approach is as follows [10]:

The company's value = Assets - Liabilities.

The method of net assets under the cost approach is based on a balance adjustment in connection with the fact that the book value of assets and liabilities of the company does not always correspond to their real value. Under business valuation when using the method of net assets (Order of the Ministry of Finance of the Russian Federation and Federal Securities Market Commission of January 29, 2003 On approval of the Procedure of Appraisal of Net Assets of Joint Stock Companies No 10n, 03-6 / pz) the traditional balance of the enterprise is replaced by the adjusted balance, in which all the assets and liabilities are shown on the market or any other suitable current value. Then, the value of the coal company in accordance with the method of net assets is $V^{NA} \approx 227.45 mln\ USD$.

A comparative approach is a combination of methods for value company appraisal, based on the comparison of the valued object with its analogues in respect of which there is information on prices. In Russia, two methods of the comparative approach: capital market and transaction market are commonly used for company valuation. The transaction method is based on a direct comparison of the valued enterprise with analogous ones in respect of

 DOI: 10.1051/shsconf/20173501135

which the price of a controlling interest or the company as a whole is known.

The use of valuation multipliers within the transaction method is based on the assumption that analogue companies have a sufficiently close relationship between the price and the most important financial indicators. Thus, the price of the enterprise can be determined by multiplying finance indicator by the corresponding multiplier.

With a relatively small sample of analogue companies, there is a problem - the error of statistical calculations. At the same time when determining any multiplier according to its sampling values (for individual analogue companies), one can assume that these values are normally distributed [11-14].

Then, when determining the required number of analogue companies n, in order to a multiplier M with a given probability α lies within a certain range $(\mu(M_i)-\beta\mu(M_i),\mu(M_i)+\beta\mu(M_i))$, the problem was solved on finding the centre of the normal distribution with a known dispersion [15-18]. Thus, for relationship between dispersion and mathematical expectation equal to 0.15, and the probability of α equal to 95% entering the range $\mu(M_i)\pm 13\%$, the sample size of analogue companies must be equal to 5.

A key point in searching analogue companies is to find information on transactions with blocks of shares. Since the information on mergers and acquisitions without the participation of the state is enclosed, the data on the sale of large blocks of shares is generally relative to the auctions for privatization held by the state. We present the data on potential analogue companies for the coal company under consideration in Table 1.

Despite the fact that many researchers consider multipliers calculated on profits the best accurate valuation, the author of the study had primarily denied them. This is because under present conditions Russian companies try to overvalue the cost of their production in every possible way in order to minimize income tax.

Table 1. The data on the auctions for the sale of coal company shares.

Name of the company	Block of shares, %	Selling price of a block (mln RUB)	Estimated value of 100% of a block (mln USD)
Southern Kuzbass	26%	55.14	7.07
Kuzbassrazrezugol	26%	166.51	21.35
Krasnoyarsk Coal Company	76%	339.41	14.89
Mezhdurechensk Coal Company	100%	120.13	4.00
Khakasugol	39%	23.43	2.00

The use of the multiplier on the book value of equity in cost calculation of the coal company also seems inappropriate since the book value of Russian companies too often deviates significantly from its market value. Thus, the relationship "company's price - income" is a widely used multiplier. Income indicators are subject to the least misstatement among all the financial indicators of the Russian companies, however, this multiplier should also be used with great caution since at different proportions of the debt to equity capital ratio its use may result in significant misstatements. Since at selection of analogue companies the debt to equity capital ratio was taken as a criterion for the comparability of companies, we selected this multiplier for the study.

Considering that the coal company is a company in the raw material sector, we also used the multipliers based on the characteristics of resource stocks for its valuation:

- company's value - coal production;
- company's value - prospected coal reserves.

The multiplier values calculated are shown in Table 2. In this case, when calculating the multipliers we took financial and natural indicators for the last fiscal year.

Table 2. The multiplier values for value coal company business appraisal in analogue companies.

Name of the company	Multipliers		
	Company's value – income	Company's value – mining	Company's value – prospected coal reserves
Southern Kuzbass	0.87	0.73	4.26
Kuzbassrazrezugol	0.93	0.80	4.81
Krasnoyarsk Coal Company	0.95	0.63	4.78
Mezhdurechensk Coal Company	0.73	0.74	5.55
Khakasugol	0.78	0.80	4.57
Average value:	**0.85**	**0.74**	**4.79**

We calculated the price of the coal company on the base of the multiplier values (see Table 3).

The use of different multipliers gives various options for the company's value. So the average price of the coal company with the use of the multiplier "the company's value - income" amounted to 422.88 million USD; with the use of the multiplier "the company's value - coal production" - 451.92 million USD; with the use of the multiplier "the company's value - prospected coal reserves" - to 511.57 million USD. The choice of the result is based on giving a percentage weight to each cost option. Assuming that all three multipliers are equal in importance the value of the coal company results in $V^T \approx 463.12 million\ USD$.

At present, increased significance of the latest methods of business valuation allows them to be used in practice both for valuation by foreign investors and for making incompany decisions that are more considered and focus on cost management in the long term. Among such methods is a method of real options.

Table 3. Calculation of the coal company's value using multipliers.

Coal company performance indicator, mln USD		Multiplier value		The average sales price of coal company, mln USD
gross incomes	501.03	company's value – income	0.85	425.88
coal mining	610.70	company's value – mining	0.74	451.92
prospected coal reserves	106.80	company's value – prospected coal reserves	4.79	511.57

The most important feature of an optional method of business valuation is its ability to take into account the rapidly changing economic environment in which the company acts. At the same time, a ROV-method has not yet been recognized in full; limits and possibilities of its application are still the point for discussion. Nevertheless, experts in the field of evaluation pay attention to this method in different countries that results in conclusion that its research is an urgent task and for Russia.

The Ohlson model and Black-Scholes model are the best-known models for determining the company's value based on the method of real options. The peculiarity of the Ohlson model is the use of abnormal incomes. The concept of abnormal incomes is based on the following assumptions. Let b_t be the book value of the company at time t and d_t - dividends paid during the year in the period from t-1 to t [19-20]. Then the total income of the shareholder (denoted as x_t) for the period t is the dividend plus the addition of book value of the company:

$$x_t = b_t - b_{t-1} + d. \tag{4}$$

At the beginning of the period under review, that is, at time t-1 the company owner by the end of the time interval from t-1 to t expects to get income equal to rb_t-1. This is a consequence of the determination of the discount rate r. This value, that is, the income based on the discount rate is called average income:

$$x_{норм} = rb_{t-1}. \tag{5}$$

Abnormal income is the discrepancy an investor receives between the actual return on capital and the expected return, that is, the return based on the discount rate:

$$x_t^a = x_t - rb_{t-1}. \tag{6}$$

The most common way to use the value of abnormal incomes for obtaining the company's cost value is the capitalization of abnormal income and the addition of this value to the company's cost resulted from the cost of its assets. This capitalization of abnormal income is carried out using the conventional capitalization rate, which is equal to the difference between the discount rate and the average rate of income growth.

The model described is the Edwards-Bell-Olson model (EBO). In the light of this, the company's cost in the Edwards-Bell-Olson model will be equal to:

$$V^{EBO} = b_t + \frac{x_t}{r-g}. \tag{7}$$

where V^{EBO} is the company's cost in the Edwards-Bell-Olson model; g is the expected average annual growth rate.

In the Olson model, the dynamics of abnormal incomes was proposed in contrast to their invariance condition, as in the Edwards-Bell-Olson model. In accordance with this, the economic meaning of the term "abnormal incomes" has been changed.

According to the Olson model, the company's value at time t is equal to the initial "accounting" value of the company plus the sum of discounted abnormal incomes:

$$V_t^{OL} = b_0 + \sum_{i=1}^{n} \frac{E_t[x_{t+i}]}{(1+r)^i}. \tag{8}$$

where $E_t[x_{t+i}]$ is a predicted value of abnormal incomes.

Thus, in order to find the final cost value it is necessary to set an explicit expression for the predicted abnormal income values. Olson proposed the following time dynamics of the abnormal income behaviour:

$$\begin{aligned} x_{t+1}^a &= \omega x_t^a + v_t + \varepsilon_{1,t+1} \\ v_{t+1} &= \gamma v_t + \varepsilon_{2,t+1} \end{aligned}. \tag{9}$$

where ω is the given empirical value indicating which part of the abnormal income in the previous year remains in the current year; v is the value characterizing the difference between the expected and actual value of abnormal incomes; γ is the given empirical value indicating which part v of the last year is preserved in the current year ($\gamma < 0$); ε is the normally distributed error with zero mean (considered noise, not related to the speed of the abnormal-income decrease).

Expression (9) shows that abnormal incomes with time tend to zero magnitude. This means that, in comparison with other companies the valued company eventually loses its original "advantages" relating to the situation on the market and its income tends to the determined average market rate of return, that is, to a discount rate r.

Substituting the expression for abnormal incomes (9) in the formula for the final cost value (8) we obtain the desired result:

$$V_t^{OL} = b_t + a_1 x_t^a + a_2 v_t, \tag{10}$$

where

$$a_1 = \frac{\omega}{(1+r-\omega)};$$

$$a_2 = \frac{1+r}{(1+r-\omega)(1+r-\gamma)};$$

$$v_t = \gamma(x_t^a - \omega x_{t-1}^a).$$

The Committee for appraisal activities of the Chamber of Commerce and Industry of the Russian Federation recommends taking for ω and γ the values equal to 0.62 and 0.32. Then a_1 is equal 1.226; a_2 is equal 2.764; $v_t = 366821.98$ thousands RUB. Substituting the values obtained in (10) the coal company's cost value results in $V^{OL} \approx 348.80 mln\ USD$.

According to the Black-Scholes model, the company's cost value is represented as the call option value, consisting of the company's assets and liabilities. At the same time, the Black-Scholes model is considered as the "ultimate" case of the binomial model. The initial premise of the model is the normal probability distribution. Based on this premise the basic formula of the model is made (in relation to a European call option on the right to purchase the asset):

$$V^{B-S} = S * N(d_1) - X * e^{-r_f t} * N(d_2), \qquad (11)$$

where V^{B-S} is the value of the appraised company; S is the value of a company's asset; X is the value of liabilities; r_f is the risk-free rate; t is duration of company's debt;
$N(d_1)$ and $N(d_2)$ are cumulative normal probabilities of density distribution function, where:

$$d_1 = \frac{ln(\frac{S}{X}) + (r_f + \frac{\sigma^2}{2})t}{\upsilon\sqrt{t}} \qquad (12)$$

$$d_2 = \frac{ln(\frac{S}{X}) + (r_f - \frac{\sigma^2}{2})t}{\upsilon\sqrt{t}}$$

It should be noted that the company's value calculated on the Black-Scholes model decreases with the increased parameter X and increases with the growth of other parameters. Also $d_1 > d_2$, and as a consequence $N(d_1) > N(d_2)$.

Initial data for the calculation of the coal company's cost in accordance with the Black-Scholes model is presented in Table 4. The value of assets was determined in accordance with the approved order of the Ministry of Finance of the Russian Federation of 05.08.1996, No72 and of the Federal Commission on Securities Market of 05.08.1996, No149, that is, according to the results of the cost assessment method. The nominal value of liabilities was determined in the same way.

Table 4. Initial data for calculation of the coal company's value

Initial data	Value
1. Asset value S, thousand RUB	20146402
2. Nominal value of liabilities X, thousand RUB	12989591
3. Duration of company's debt, year	2.97
4. Risk-free interest rate r_f, %	10.11
5. Variation of company's value, %	43.57

The substitution of the initial data in (11) and (12) results in the value of the coal company equal to $V^{B-S} \approx 379.48 mln\ USD$ in accordance to the Black-Scholes model.

Table 5 shows the calculated value of the coal company in accordance with the income approach (a discounted cash flow method), the cost approach (a net assets method), the comparative approach (a transaction method) and the method of real options (the Ohlson model and the Black-Scholes model).

Table 5 illustrates that the maximum value company appraisal was determined by the discounted cash flow method ($V^{DCF} \approx 623.65 mln\ USD$), and the minimum one was determined by the net assets method ($V^{NA} \approx 227.45 mln\ USD$).

Table 5. The coal company's value based on various methods.

Method	Company's value
1. DCF method	$V^{DCF} \approx 623.65 mln\ USD$
2. Net asset method	$V^{NA} \approx 227.45 mln\ USD$
3. Transaction method	$V^{T} \approx 463.12 mln\ USD$
4. The *Ohlson model*	$V^{OL} \approx 348.80 mln\ USD$
5. The Black–Scholes model	$V^{B-S} \approx 379.48 mln\ USD$

Figure 1 represents a model of integrated value business appraisal in merger and acquisition transactions. At the same time, for the integrated business valuation it is necessary to determine the weighting coefficient for every appraisal value of the valued company. For this, we suggest to place all values in decreasing order of their importance and to determine the significance of the i-th value according to the Fishburne rule [21].

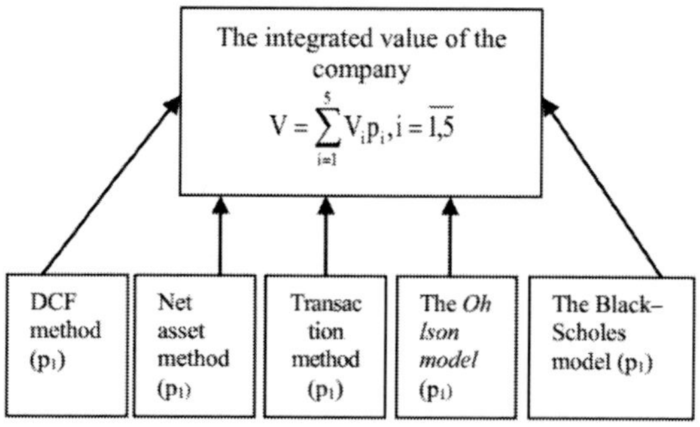

Fig. 1. The model of the integral value company appraisal in merger and acquisition transactions.

Let us consider the order of weights Fishburne construction. The system of weights reducing in accordance to the arithmetic progression rule best meets the system of decreasing preference of n alternatives:

$$p_i = \frac{2(n-i+1)}{(n+1)n}, i = \overline{1,n}. \qquad (13)$$

and a set of equal weights best meets the system of indifferent to each other alternatives:

$$p_i = n^{-1}, i = \overline{1,n}. \qquad (14)$$

To determine a set of Fishburne weights for the mixed system of preferences when, along with preferences, the system includes the relationship of indifference, it is necessary to determine the numerators r_i of rational fractions according to the recursive scheme:

$$r_{i-1} = \begin{cases} r_i, F_{i-1} \approx F_i \\ r_i + 1, F_{i-1} \succ F_i \end{cases}. \; r_n = 1, i = n...2. \qquad (15)$$

Then the sum of the numerators obtained is the common denominator of fractions Fishburne $K = \sum_{i=1}^{n} r_i$ and:

$$p_i = \frac{r_i}{K}. \qquad (16)$$

On the base of considerations of generally accepted appraisal practices and practical experience of business appraisers of International Academy of Appraisal and Consulting and Committee for appraisal activities of the Chamber of Commerce and Industry of the Russian Federation:

$$V^{DCF} \succ V^{T} \succ V^{OL} \approx V^{B-S} \succ V^{NA}.$$

Then the weights Fishburne for the integrated value business appraisal of the coal company takes the following form (see Table 6).

The integrated value appraisal of the coal company was calculated based on Table 6 $V_{инт.} = 463,99 mln\ USD$.

Table 6. Weights fishburne for value appraisal of the coal company.

	V^{NA}	V^{B-S}	V^{OL}	V^{T}	V^{DCF}
p_i	$\frac{1}{12}$	$\frac{1}{6}$	$\frac{1}{6}$	$\frac{1}{4}$	$\frac{1}{3}$

In accordance with the traditional three levels of the company's value model (the level of control, the level of liquid minority blocks of shares and the level of unliquid minority blocks of shares) at the acquisition of the coal company buyout in addition a control premium arises.

In many areas of business appraisal in Russia, the empirical data on the base of which it is possible to calculate the size of the control premium is closed and not available in many cases. Therefore, as an estimate of the control premium we took the data presented by Mergerstat Review. In compliance with that data, the premium for the acquisition of controlling stake shares of public companies in the USA in 2005-2008 on average fluctuated from 33% to 40%. Then the final integrated value appraisal of the coal company on average fluctuates from 617.11 mln USD to 649.59 mln USD.

3 Conclusion

The statistical approach proposed to value business appraisal within the integration policy for industrial enterprises provides an integrated value business appraisal, forms a preliminary justification of expediency of mergers and acquisitions and is a foundation for high-quality and reliable valuation of the integration transaction, which allows one to make a considered investment decision. All these will contribute to the increase in the competitiveness and the stable industry development in Russia.

The work is performed under the grant of the President of the Russian Federation for state support of young Russian scientists — PhDs (MK-5339.2016.6)

References

1. M. Polikarpova, Vestnik of Nosov Magnitogorsk State Technical University, **3**, 95-102 (2011)
2. M. Polikarpova, *Risk analysis and the effectiveness of mergers and acquisitions of Russian companies* (Moscow, NPO MAX Group, 2012)
3. A.G. Gryaznova. M.A. Fedotova, *Business valuation textbook. second edition* (Moscow, 2009)
4. D. Fishmen, Sh. Pratt, K. Griffit, K. Uilson, *Guide to Business Valuations* (Kwinto-Consulting, Moscow, 2008)
5. V.Popkov, E. Yevstafyeva, *Business Valuation. Diagrams and tables: Textbook* (Piter, Saint Petersburg, 2007)
6. A. Damodaran, *Valuation of assets* (Potpourri, Moscow, 2011)
7. N. Chebotarev, *Valuation of Enterprise* (Dashkov and Co., Moscow, 2014)
8. O. Maslenkova, *Valuation of Enterprise (Business)* (KnoRus, Moscow, 2011)

9. F. Evans, D. Bishop, *Assessment of companies in mergers and acquisitions: the creation of value in private companies* (Alpina Pablishar, Moscow, 2004)
10. S. Valyuh, B. Gadzhimet, E. Hours, Energy - XXI century, **3(65)**, 70-79 (2007)
11. S. Aivazian, I. Enyukov, L. Meshalkin, *Applied Statistics. Dependency Research* (Finance and Statistics, Moscow, 1985)
12. S. Aivazian, V. Mkhitaryan, *Applied Statistics and Econometrics fundamentals. Volume 1: Theory of Probability and Statistics, specialty Applied* (UNITY-DANA, Moscow, 2001)
13. I. Eliseeva, *Statistics* (Yurayt Publishing, Moscow, 2012)
14. V. Gmurman, *Theory of Probability and Mathematical Statistics* (Higher School, Moscow, 1977)
15. N. Kremer, *Probability theory and mathematical statistics* (Unity, Moscow, 2003)
16. V. Mkhitarian, *Data analysis* (Yurayt Publishing, Moscow, 2016)
17. Y. Sazhin, N. Rodionov, *Statistical methods for research on the socio-economic processes* (Saransk, Publisher Mordovia University, 1998)
18. E. Khrustalev, O. Khrustalev, The economic analysis: theory and practice, **45**, 11-16 (2011)
19. *Guidelines on the business valuation using the Ohlson model and the Black-Scholes model* (International Academy of Valuation and Consulting, Moscow, 2010)
20. I. Ansoff, *The new corporate strategy* (Piterka, St. Petersburg, 1999)
21. P. Fishburn, *Utility theory for decision* (Science, Moscow, 1978)

A statistical approach to the analysis of merger and acquisition efficiency in the Russian industry

M. Karelina[1,*], and *V.* Mkhitarian[2]

[1]Nosov Magnitogorsk State Technical University, Magnitogorsk, Russia
[2]National Research University Higher School of Economics, Moscow, Russia

Abstract. At present, the success of economic institution transformations, as well as creating an efficient economic system with a fundamental new nature of corporate relationships are impossible without the statistical recording of factors contributing to the efficiency of merger and acquisition transactions in the Russian industry. The paper proposes a method for analyzing the efficiency of merger and acquisition transactions of enterprises in the industrial sector of the Russian economy, based on simulation methods. The methodical approach developed to analyze the efficiency of the integration transactions of Russian industrial companies allows one to consider individual preferences of investors, as well as to give a complex statistical evaluation of the strategic economic benefits from M&A transactions. This method enables to evaluate the probability and stability of the synergistic effect values within the increase of competitiveness of Russian industrial enterprises on the domestic and foreign markets.

1 Introduction

The current economic environment creates quite strict operating conditions for enterprises in the industrial sector in the economy of the Russian Federation. Only those companies can maintain and strengthen their competitive positions whose management is always concerned with the qualitative and quantitative growth; that develops effective business strategies for realization of which there may not be enough internal capacity. When internal reserves of increasing efficiency in activity have been exhausted, external development that is realized by means of mergers and acquisitions (M&A) becomes an indispensable condition to ensure the continuity of the enterprise life cycle [1,2].

Issues of analysis of investment project efficiency with varying degrees of detail are considered in the works of Russian and foreign specialists. Vladimirova I.G., Gvardii S.V., Nikonova I.A., Rudyk N.B., Semenkova Ye.V., Chekun I.N, Breg C., Bishop D., Galpin T., Gokhan P., Depamfilis D, Koller T, Kouplend T, Lazhu A., Murrin Dzh., Porter M., Rid S., Khendon M., Evans F. analyzed the possible economic effects of mergers and acquisitions aimed to forecast their results and efficiency increasing.

Employees of such international consulting and auditing companies as Deloitte, Ernst and Young, KPMG, McKinsey, PricewaterhouseCoopers significantly contributed to the study of the merger and acquisition practice, to the investigation of their effects, the causes of failure and the development of methods and techniques for analysis of the expected results of the integration transactions.

Despite the presence of a large number of works devoted to the analysis of individual issues relating to merger and acquisition efficiency, a study of investigations in the field of M&A showed that the greatest attention has been paid to the integration processes taking place in the banking sector. Transactions that are carried out in the industrial sector of the economy and have serious specificity remain insufficiently studied.

Mergers and acquisitions are a high-risk, expensive, and, at the same time, highly profitable kind of investment project whose potential can be realized only by transaction careful preparation, planning, analysis, and its efficiency justification [3]. All this causes the necessity for the *development of a statistical approach to the analysis of merger and acquisition efficiency in the Russian industry.*

2 A systematic approach to the analysis of integration transaction efficiency of enterprises in the industrial sector in the economy of the russian federation

In this work we determined the synergistic effect as a key parameter that affects the decision making process on integration within the framework of the justification of the merger and acquisition efficiency [4].

There are a number of approaches to the formation of an information base for analysis of the integration activity efficiency; for example, you can use the accumulated statistical information on the industrial enterprises, as well as promising data. For the solution of

* Corresponding author: marjyshka@mail.ru

this issue, we took into account the world practice of analysis of the integration project efficiency based on discounted cash flows generated by the companies - participants of the integration [5].

When identifying the origin of the synergistic effect one should take into account only those opportunities that were not available for autonomous companies and resulted from mergers. Comparing and contrasting the plan of the strategic development of the merged company with the development plans of independent companies involved in M&A transactions may help to solve this problem.

After the data on the supposed sources of the synergistic effect of mergers and acquisitions has been collected and the costs associated with their implementation have been described, you can proceed to the compilation of the forecasting bookkeeping balance sheet and profit and loss report. When forecasting the individual articles of these reports you should take into account the influence of auxiliary facilities that give rise to the synergistic effect [6].

At the same time, financial indicators, which are the basis of decision making on the expediency of the integration project are based on the point valuation for the most plausible data. This valuation is also point and does not provide complete information about the synergistic effect in the M&A transaction. The transition from the point valuation to probability distributions of indefinite variables and the use of simulation technique allows you to get more exact information about the synergy as a foreseeable result of integration [7].

The need for simulation in studying the integration activity of industrial structures is caused by differences between the Russian market and the global total market. Simulation results supplemented by statistical estimation can provide the head of the integration project with more exact information about the extent of the impact of key factors on the expected results and possible scenarios.

The peculiarity of the use of the Monte Carlo method in the analysis of the synergistic effect in M&A transactions is related to the ability to work with a model, as in case of conventional investment calculations and risk management of integration transactions derives from multiple calculation of the model. This method is most fully characterizes the whole range of uncertainties which a real integration project may face and by setting restrictions allows one to take into account all information available to the project analyst.

Let us consider the algorithm proposed for estimating the synergistic effect in mergers and acquisitions with the use of simulation technologies:

1) the choice of activity options of integrated companies directly affecting the synergistic effect and included in the calculation model;

2) the construction of cash flow model of the merged company;

3) setting the type and characteristics of probability distributions for the parameters of the merged company;

4) carrying out statistical tests, obtaining random realizations of parameter values;

5) a statistical analysis of simulation results and interpretation of the results obtained.

In this article, we examined the purchase of one of the coal companies by a Russian steel company, incorporated in 2008 in a government list of strategic companies in Russia, as the example of the merger and acquisition transaction. This should allow the creation of a highly efficient vertically integrated structure possible and result in increased cost of the metallurgical company.

Forecasting and carrying out calculations resulted in a value that evaluates the synergistic effect based on generated incomes of the two business segments of the mergered company ($SE = 348{,}80$ million U.S. dollars). Since the forecast was created in the form of an algorithm, it is possible to re-count the results of the forecast for the different set of input data values. This makes possible to implement the Monte Carlo method for statistical tests [8-11].

Sensitivity analysis found that the most critical parameters in evaluation of the synergistic effect are

• cost value of a ton of metal production (each 1% decrease under otherwise equal conditions results in increased SE by 16.75%);

• cost value of a ton of coke concentrate (each 1% decrease under otherwise equal conditions results in increased SE by 8.22%);

• coke concentrate production (each 1% decrease under otherwise equal conditions results in increased SE by 3.68%).

In practice, normal and uniform distribution laws are more common in the generation of the input parameters of the projects. In this study, we used the normal distribution law to generate the input parameters. Table 1 presents the values of parameters required for generation of random numbers for the most sensitive factors for synergy valuation.

Table 1. The values of parameters required for generation of random numbers according to the normal distribution law.

No	Indicator	Expected value μ	Standard deviation σ
1	Cost value of metal production, million USD	16915.37	6121.02
2	Cost value of 1ton of coke concentrate, USD	117.87	16.82
3	Coke concentrate production, kt	4284.07	544.25

The most effective methods for determining the number of tests are methods for reducing dispersion [12-14], such as:

• use of Chebyshev's Theorem;

• use of Spalding formulas;

• estimation of set standard deviation as a measure of dispersion of sample values.

Chebyshev's inequality says that at a given number k and a random sample $x_1, x_2, \ldots, x_n$ of size n at least $1-1/k^2$ measurements are close to the average value of μ at a distance of no more than k of standard deviations. This inequality is valid for any distributions of sets, and you can use Chebyshev's inequality to determine sample size, which has the form

$$\mu_D(w,i) = \min_{\omega\in W}(1, 1-\mu_A(v)+\mu_B(i)) \qquad (1)$$

If we want to estimate the parameter of a simulation model into the interval $A \in D, B \in D, v \in A, i \in B$, with a probability of 0.95, then n is 320. If we want to estimate the parameter of a simulation model into the interval with a probability of 0.05, then n is 720.

Table 2 shows the sample size obtained by using Spalding formulas.

Table 2. The sample size corresponding to the desired precision in calculations and significance value.

Desired precision	Significance value		
	0.10	0.05	0.01
0.10	149	185	266
0.05	596	740	1063
0.04	913	1156	1655
0.03	1650	2053	2950
0.02	3721	6084	6643
0.01	14884	18496	26569
0.001	1488400	1849600	2656900

Among *dispersion reduction techniques* for finding the sample size n evaluating set fluctuation is the most effective. The problem of estimating dispersion of a set can be put as a task to find such s^2 evaluation that

$$P\left\{(1-d)\sigma^2 \le s^2 \le (1+d)\sigma^2\right\} = 1-\alpha, \qquad (2)$$

where $0 \le d \le 1$ is a number characterizing the similarity degree of estimation s2 to true dispersion σ2 [15-16].

Maze and Cox demonstrated that it is more convenient to use in this equation χ2 – the statistic (n-1)s2/ σ2 with (n-1) degrees of freedom, which allows you to find the confidence probability independent of σ2 [17-18]. If n is large enough then χ^2 distribution can be approximated by a normal distribution, then

$$i_j \in C_\alpha, \qquad (3)$$

where $F(C) = \frac{1}{\alpha_{max}} \int_0^{\alpha_{max}} M(C_\alpha)d\alpha$ is the standard normal statistic for the original probability.

If we want s^2 to be different from σ^2 no more than in 5% with the probability of 0.95 then n ≈ 3075; if we want s^2 to be different from σ^2 no more than in 10% with a probability of 0.95 then n ≈ 770.

The next stage of the simulation was the stage of statistical tests implemented by using the Monte Carlo method [19-20]. The values of the input parameters were generated according to the normal distribution, the sample size was 3075. Simulation experiment was carried out in the AP Excel environment with the help of a random number generator.

3075 experiments at a significance level of 95% were generated and a sample of estimates for the synergistic effect was obtained based on the normal distribution; all these resulted in the calculated statistical characteristics of distribution:

- μ(SE)=313,67 million USD;
- σ(SE)=117,59 million USD;
- υ(SE)=37,49% (coefficient of variation);
- min(SE)=75,15 million USD;
- max(SE)=519,13 million USD;
- Median (SE)=337,87 million USD.

Simulation allows you to determine not only the average and the most probable estimation of the synergistic effect, but also to estimate the probability of obtaining a certain annual cash flow in different segments of the merged company that is necessary for the compilation of production and financial programs of a vertically integrated holding company.

Within the possible deviations of the synergistic effect from the forecasted, it is possible to argue that the synergistic effect will fall below the value of 361,460,000 million USD with the probability of 44%. Since the range space of the synergistic effect resulted from statistical simulation the dispersion of values characterizes the uncertainty of the estimated SE, measured by the standard deviation σ(SE) is 117,590,000 million USD. Then the coefficient of variation is equal to 37.49%.

At the same time the probability that the value $\mu_{S_5}(x) = 1-x, x \in I$ is greater than μ(SE)+σ=431,260,000 million USD is equal to 11.84%. The hit probability of the value of the coal company in the range of (μ(SE)-σ, μ(SE)) is equal to 34.13%.

Table 3 presents the scale of the level for compliance risk assessment of evaluation of the synergistic effect in the merger and acquisition transaction to the coefficient of variation developed by the authors, according to which an investor considers both the final price for the target company and the possibility of implementing an integration project under the existing conditions.

Table 3. The compliance scale of the level for risk assessment of the synergistic effect to the coefficient of variation.

Risk level	The coefficient of variation
Light risk	<0.1
Moderate risk	0.1-0.2
Strong risk	0.2-0.5
Critical risk	>0.5

In this integration project the level of risk assessment of the synergistic effect is estimated as strong (37.49%). But since the metallurgical company gives the priority to this M&A- project, the implementation of the integration project in order to create a vertically integrated holding company in the industry of Russia is expedient at a cost of no more than $V_{unm.} \approx 463,99$ million USD.

3 Conclusion

Thus, the statistical approach proposed to the analysis of the efficiency of merger and acquisition transactions in the industry of the Russian Federation makes it possible to predict the synergistic effect in the integration transactions. The values obtained allow systematically exploring the factors of the change in the performance of the integrated company in the real sector of the economy; to estimate risk assessment of the synergistic effect; to give grounds for management decisions relating to a particular integration project which should be accepted or rejected within the framework of improving the competitiveness of the industry of the Russian Federation.

The work is performed under the grant of the President of the Russian Federation for state support of young Russian scientists — PhDs (MK-5339.2016.6)

References

1. E. Campbell, K. Summers, *Lachs strategic synergies* (Peter, St. Petersburg, 2004)
2. M. Polikarpova, *Risk analysis and the effectiveness of mergers and acquisitions of Russian companies* (NPO MAX Group, Moscow, 2012)
3. V. Mkhitarian, M. Polikarpova, Problems of the theory and practice of management, **2**, 114-123 (2013)
4. P. Lapshin, A. Khachaturov, Management in Russia and Abroad, **2**, 21-30 (2005)
5. E. Khrustalev, O. Khrustalev, The economic analysis: theory and practice, **45**, 11-16 (2011).
6. T. Rubinshtein, Russian Journal of Non-Ferrous Metals, **6**, 516-518 (2011)
7. R. Shannon, *Systems simulation. The art and science* (Mir, Moscow, 1978)
8. R. Barlow, F. Proschan, *Mathematical theory of reliability* (SIAM, 1996)
9. K. Brownlee, *Statistical theory and methodology in science and engineering* (Wiley, 1965)
10. N. Kremer, *Probability theory and mathematical statistics* (Unity, Moscow, 2003)
11. N. Kremer, *Mathematical statistics* (Economic Education, Moscow, 1992)
12. I. Eliseeva, *Statistics* (Yurayt Publishing, Moscow, 2012)
13. V. Mkhitarian, *Data analysis* (Yurayt Publishing, Moscow, 2016)
14. Y. Sazhin, N. Rodionov, *Statistical methods for research on the socio-economic processes* (Publisher Mordovia University, Saransk, 1998)
15. S. Aivazian, V. Mkhitaryan, *Applied Statistics and Econometrics fundamentals. Volume 1: Theory of Probability and Statistics, specialty Applied* (UNITY-DANA, Moscow, 2001)
16. S. Aivazian, I. Enyukov, L. Meshalkin, *Applied Statistics. Dependency Research* (Finance and Statistics, Moscow, 1985)
17. E. Wentzel, L. Ovcharov, *The theory of stochastic processes and its engineering applications* (Nauka, Moscow, 1991)
18. E. Wentzel, L. Ovcharov, *Theory of Probability and its engineering applications* (Nauka, Moscow, 1988)
19. E. Chetyrkin, I. Kalikhman, *Probability and Stick styles* (Finance and Statistics, Moscow, 1982)
20. V. Gmurman, *Theory of Probability and Mathematical Statistics* (Higher School, Moscow, 1977)

Lean production: experience among Russian organizations

Yu.S. Bakhracheva[1,*], and *Ya.Ya.* Kayl[2]

[1]Volgograd State University, Prosp. Universitetsky, 100, Volgograd, Russian Federation
[2]Volgograd State Socio-Pedagogic University, Prosp. Lenina, 27, Volgograd, Russian Federation

Abstract. The article gives analysis of the implementation of lean production, one of the basic models of resource management, in the context of the current unstable economic situation in Russia, occurring due to the geopolitical situation in the world. The concept of lean production is defined. The basis for resource management: effective use of available resources, minimization of overall costs, increase profits, maximize the capitalization of the company. Strategic enterprise management based on the optimal allocation of internal resources is the key factor of success. Lean production is widespread in Russia since 2001. However, there are difficulties in the introduction, one reason for which is the "human factor" and the lack of a unified methodology of practical application.

1 Introduction

The beginning of 2015 was marked for Russia by immersion in a crisis economic condition, informed the aggravated geopolitical situation in the world. In the current environment, it is necessary the rational use of domestic resources. Economic diversification is a complex process that can be implemented through state regulation. However, the role of enterprises is the main and lies in the effective management of private economic resources.

In Russia at the enterprises begins to show interest in possibilities to improve competitiveness and modernization of production through Lean Production. The highest activity is demonstrated by large enterprises employing more than a thousand people located in the Ural or Volga Federal districts relating to the engineering industries, ferrous or non-ferrous metallurgy. This is due to the greater availability of information about the experience of implementing the tools of lean production and the presence of these businesses in the global market: they have to compete with foreign producers, to meet international standards of product quality [1].

2 The introduction of Lean Production in the Russian economy

Resource management is a field of scientific knowledge, based on the efficiency of enterprises. The basis for resource management: effective use of available resources, minimization of overall costs, increase profits, maximize the capitalization of the company. One of the main models of resource management is Lean Production, also known as the Toyota Production System. The concept is built on the simultaneous achievement of the following internal purposes, providing the company competitive advantage [2]:

• constant reduction of production costs and sales;
• continuous improvement of product quality;
• optimization of the flow of material according to the logic of Just In Time.

The basic idea of Lean Production is to improve the efficiency of enterprise activity from within existing resources through continuous improvement of all processes. In Russia Lean Production appeared relatively recently, but already managed to get support at the highest level. Dmitry Medvedev at meeting of Commission for modernisation in 2011 noted that to enhance the competitiveness of domestic enterprises and their effectiveness it is essential to develop production systems. In this sense he's very interested in the concept of "Lean Production and Lean Thinking" [1].

In 2001, he was first recorded working example of introducing the concept of Lean Production in domestic practice at the plant of the Ford Motor Company in Vsevolozhsk, Leningrad region. Currently, this model of resource management applied in all the sectors of the Russian economy.

3 The results of the implementation of Lean Production in industry

There is an assumption that the concept of lean production – Lean Production is more applicable for industrial companies [3; 4; 5]. The tools of Lean Production implementing the enterprise's fundamental, existing from the Soviet era (for example, OJSC KAMAZ, Engels instrument-making association "Signal", GAZ group, etc.). Doing this foreign enterprise, which came to the Russian market, is already actively using this model (e.g., Ford Motor Company,

* Corresponding author: bakhracheva@yandex.ru

Toyota Motor Manufacturing, group of companies NCC, Volkswagen group, etc.). Thus, there is no uniform method of introducing the concept in Russian companies, which is one of the reasons for unsuccessful attempts of development of the model. However, the implementation of lean production is very high. For example, in the sector of oil production the effectiveness of implementation of Lean Production:

• cost reduction by 30% (annual savings of 11.5 million rubles);

• reduced work in progress by 50% (elimination of downtime of wells – additional income of 130 million rubles per year);

• the release of the production area by 30%;

• the release of labour by 25% (reduced labor cost – more than 1.3 million rubles a year).

In the automotive industry (OJSC KAMAZ, GAZ group) resulted in reduction of costs by 30% (saving 65 million rubles per 1 week).

4 Analysis of experience of implementation of Lean Production in the service sector

Typical examples applying the concept of Lean Production are a giant organization covering a network of its branches throughout Russia, such as JSC "Sberbank", JSC "Russian Railways", "Russian Post". At the moment they are at different stages of implementation of lean production principles. Quality restart patterns, covering the whole country, is a difficult and complex process [6].

Sberbank since 2008 quite successfully it takes place. Within the first transformation after his appointment to the post of Chairman of the Board German Gref announced the creation of Sberbank's production system, which is based on the principles of lean manufacturing. Production system of Sberbank is focused on increasing customer satisfaction, improving operational efficiency, increasing motivation and increase employee satisfaction [7]. The starting action towards becoming system involved the formation of working groups and invitation specialist, with experience of working in a company applying the tools of Lean Production. The whole existing system of the Bank was dismantled to detail as part of the implementation of Lean Production, strategy of development of Bank till 2014 it is Noteworthy that in the initial stages, productivity growth was about 25% and subsequently increased. The principles of production system of Sberbank was successfully set in the key areas of the Bank, improving efficiency by 30-50% [8].

Achieving high performance was the result of a number of projects [8]:

• transformation of the retail network. Developed and implemented a new model of working implemented flexible schedules, developed a new incentive system tying the employees ' income to sales results and service quality;

• optimization of work with small businesses. By freeing up time were introduced the role of the client managers, who are tasked with quality customer service;

• standardization and optimization of work of accounting departments. Standards have been created for each operation, implemented a mechanism to manage the workload of employees;

• development and implementation of standards and their continuous improvement;

• creation of the methodology of "Do-it-yourself step by step".

The main objective of the Sberbank's development strategy until 2018 as part of the production of the system is its widespread and successful application.

JSC "Russian Railways" and "Russian Post" has not yet achieved such considerable success as a savings Bank, but are actively implementing Lean Production. One of the steps to becoming Lean Production on the railway was the project "Improving the organization of technology development". At year end, the economic effect of innovations in monetary terms amounted to more than 7.5 million rubles. [9, 10, 14].

Achieved the following technical results:

• reduced time for manufacturing operations;

• minimization of the total idle time of the transit of the carriage on the tracks of stations;

• reduction of irrational use of locomotives as well as downtime of trains awaiting processing.

"Russian Post" began to develop the concept of lean production since the end of 2013. The guidance was adopted a number of measures:

• daily morning conference calls with affiliates;

• create normal working conditions for customs officials;

• optimization of mainline traffic;

• the decline in the proportion of agents in favor of the conclusion of direct contracts;

• creation of integrated logistics unit (appeared obselescence direction of transport and sorting, as well as new patterns – unit planning and the unit for work with claims);

• create a new sorting center in Vnukovo (to unload Moscow), purchased Italian equipment – automated sorting.

Implemented activities provoked an increase in efficiency, bringing the company to the next level of development [11]:

• the volume of international mail, compared to 2012 increased by 4 times;

• over half the share of agents has declined from 75.7% to 15%, while rates decreased by about 15% at the expense of margin, had the agents trucking: 2013 – 9 contractors, the rate 40,86 RUB/km, 2014 – 32 contractor, 27,28 tariff RUB./km;

• in 2014, the savings on gas logistics amounted to 900 million rubles – about 11% of the budget mainline traffic (including repairs);

• all international mail is processed by the Moscow center at the Vnukovo airport, in Moscow arrives about 70% of mail (up to 95%), while the remaining 30% come to Kazan, Yekaterinburg and Vladivostok;

• in 2014, the volume of parcels delivered has increased 1.6 times compared to 2012, average delivery time within two years decreased from 90 to 14 days, the monthly number of written complaints decreased from 250 000 to 20 000-40 000.

5 Problems of implementation of Lean Production at the Russian enterprises

According to experts, the main problem in the implementation of Lean Production is "the human factor" [12], significantly affect the realization of the ideology of lean manufacturing.

Let's consider the main reasons for the rejection of innovations:

- fear of worsening working conditions, pay, load increases, etc.;
- lack of conviction that change from confusion over the objectives of the change;
- the surprise prepared by the operations manual, important changes;
- fear of failure to adapt to changes, increase efficiency of work in the new situation;
- violation of the established order, habits and relationships;
- lack of respect and trust to the person conducting the change.

It is also possible to allocate three basic principles of the procedure:

- fiercer all resist change are exactly those who most need them;
- resistance to change - the result of incorrect techniques of their implementation;
- passive opposition to change is often associated with past experiences: people who have already survived a lot of reorganizations that had not benefited, are particularly suspicious.

For successful introduction of system "Lean Production" it is necessary to use principles of participation, in other words, worker participation in enterprise management. Participate is the participation of the employees in the management of the company along with its owners, employers, professional stewards (managers and others).

Most effectively these principles can be implemented through: inclusion of labour collective of the enterprise in the decision-making processes; employee participation in the profits (income) of enterprises; the availability of equity ownership among employees through equity (or shares) of the company.

In the leading foreign countries of Japan, Germany, Sweden and others, all of these principles has been successfully applied.

In Russia today the practice of implementing the principles of participation at the enterprises is still lacking. This is evidenced by open sources of information. Usually workers of the enterprise (organization) work according to clearly defined rates and pay systems, stimulating their personal performance. Ordinary employees usually do not get anything from the profits of the enterprise.

Today in industrially developed countries (Norway, Sweden, France, Japan, Denmark, USA, etc.) in enterprises has a program of employee participation in stock ownership [13]. The advantage of this practice is that it does not require additional costs from an employee, but can provide income in the form of dividends. In many Western countries it is recommended to concentrate in the hands of managers and external investors is not more than 30% stake in the company.

Experience shows that the introduction of such a system within two years leads to productivity growth on the average on 10 - 15% [8]. So in Japan, productivity in companies that implement plans of employee participation in the share capital, 7% higher than those where workers do not own shares in [13].

Of course, to introduce the basic principles of participarii it is necessary to regulate at the state level and the development of corresponding normative-legal acts. To achieve a good economic effect and improve the performance of employees, you need to:

1) the adoption of Lean Production at the government level and the state regulation of activities of enterprises applying this model of resource management. In the Republic of Tatarstan, the first among the subjects of the Russian Federation, was adopted at the national level and is functioning successfully long-term target program "Implementing methods of "Lean production" for 2012-2013". In 2013 economic effect of the introduction of this model of resource management at national enterprises amounted to more than RUB 6 billion in the first nine months of 2014, 3 billion rubles;

2) the involvement of foreign consultants for implementation of Lean Production and training of domestic consultants;

3) the implementation of advocacy activities with objectives, tasks and principles of Lean Production to all levels of organisational structures of enterprises.

The paragraphs is far from complete, but the basic complex of measures on transition of Russian enterprises to modern Japanese organizational technology.

6 Conclusion

In conclusion, it is worth noting that you should carefully and thoughtfully apply to the use of lean manufacturing at the enterprise, to carry out the mandatory preliminary assessment of the feasibility of implementation. Moreover, currently there is active development of information automation solutions to lean manufacturing, the use of which, together with traditional management systems can greatly help to optimize the implementation of lean production. Summing up the results of the systematization of the results of lean implementation in various sectors of the Russian economy; we can distinguish the following main conclusions and results:

1. The future of national economic sectors is directly dependent on the actions made in the current crisis. The strategic management of enterprises based on the optimal allocation of domestic resources are the main factor of success.
2. Lean Production is a effective concept to achieve growth and competitiveness in the market in the conditions of unstable economic development.
3. There are successful examples of the development of the concept both in industry and in services.

4. Lean production is widespread in Russia since 2001, However, there are difficulties in the introduction, one reason for which is the "human factor" and the lack of a unified methodology of practical application.

This work was supported by the Russian Foundation for Humanities (project No №16-12-34014).

References

1. O.S. Vikhansky, Expert-Ural, **7(270)**, 35 (2007)
2. E.S. Balashova, V.N. Yuriev, *Economic mechanism and tools of the resource management for industrial enterprise* (Publisher Polytechnic University, Saint Petersburg, 2014)
3. Imai M. Kaizen, *The key to Japan's competitive success* (Alpina Publisher, Moscow, 2014)
4. J. Liker, *14 management principles from the world's greatest manufacturer* (Alpina Publisher, Moscow, 2013)
5. T. Ohno, *Toyota production system. Beyond large-scale production* (Institute for strategic studies, Moscow, 2013)
6. Digest Mc Kinsey, *New trends in management* (Alpina Business Books, Moscow, 2010)
7. *Sberbank Production System.* http://www.leanforum.ru/content/files/library/PSS.pdf.
8. *Sberbank's Development Strategy until year 2018.* http://www.sberbank.ru/common/img/uploaded/files/S.
9. Ya.Ya. Kayl, Yu.S. Bakhracheva, Science Journal of Volgograd State University. Technology and Innovations, **3(18)**, 19-25 (2015). DOI: http://dx.doi.org/10.15688/jvolsu10.2015.3.2.
10. R.S. Medvedev, Yu.S. Bakhracheva, Science Journal of Volgograd State University. Technology and Innovations, **3(12)**, 93-97 (2014). DOI: 10.15688/jvolsu10.2014.3.11.
11. *How did the new management change the "Post Office of Russia".* http://www.leanforum.ru/library/r1/2386.html.
12. *Modern concepts of production management: Lean Production.*http://www.remmag.ru/admin/upload_data/remmag/07- 5/RoundTable.pdf.
13. I. Jac, J. Sedlar, A.A. Zaytsev, A.V. Zaytsev, E & M: Ekonomie a Management, **16**, **3**, 75-84 (2013)
14. Ju.S. Bahracheva, E.V. Akatova, Bulletin of Volgograd State University. Series 10: Innovation activities, **5**, 6-11 (2014). DOI: http://dx.doi.org/10.15688/jvolsu10.2014.5.1.
15. Ju.V. Gaponenko, Bulletin of Volgograd State University. Series 10: Innovation activities, **1**, 6-10 (2014)
16. *The age of lean production - Lean manufacturing, Livshits V. national server of modern prose Prose.ru.* http://proza.ru/2007/03/25-282.
17. L.M. Semenova, V.B. Khlebnikov, Yu.S. Bakhracheva, Science Journal of Volgograd State University. Technology and Innovations, **2**, 49-56 (2013). DOI: 10.15688/ jvolsu10.2013.2.7.
18. L.M. Semenova, V.B. Khlebnikov, Yu.S. Bakhracheva, S.V. Semenov, Science Journal of Volgograd State University. Technology and Innovations, **7**, 120-126 (2012)
19. *Russian Statistical Yearbook 2015: Statistical Compilation* (Rosstat Publ., Moscow, 2015)
20. *The Strategy of Railway Transport Development in the Russian Federation Till 2030* (2008)

 DOI: 10.1051/shsconf/20173501138

Improved methods of complex estimation of efficiency of activity of industrial organizations

E.H. Taipova[1], *Y.V.* Lysenko[1], and *T.N.* Myznikova[1,*]

[1]South Ural State University, Chelyabinsk, Russia

Abstract. The article substantiates the need for combined technological, technical, financial and economic indicators that characterize the efficiency of production and economic activity of an industrial enterprise. Stages of performance evaluation: from single to integrated assessment criteria. It named the main drawback of the existing methods of constructing a system of indicators, which leads to subjectivity, to a lack of real understanding of the financial and economic state of the industrial organization. It is shown that multivariate statistical methods of classification can not only minimize their number, but also to detect correlations for each cluster, as well as to build a multi-factor regression models, which are the basis of perfection of a technique of an integrated approach to assessing the results of functioning of the organization. Thus, the algorithm for determining the integral index of evaluating the performance of the company contributes to an objective vision of the ways to improve operational and financial performance of industrial organizations.

1 Introduction

The relevance of this article lies in the fact that the increase of efficiency of functioning of industrial organizations - one of the most important economic problems, the solution of which depends on the level of development of the industrial sector, and hence the rise of living standards of the population. To solve this problem is impossible without substantial investment and development of innovative technologies. Primarily, it refers to a precocious, high technology and high-tech industry, as the industry.

To support the industry and its transition to an innovative path of development it is important to ensure rational use of available resources and internal reserves to increase production, improve the efficiency and competitiveness of the industry's products. An integrated approach to solving a wide range of issues relating to the justification of the priorities of reconstruction and further development of the industry. This approach is based on an objective and comprehensive assessment of the processes, the aim of which is to determine the state of the organization and the potential for their further development.

Modern economics offers a variety of methodology for assessing the effectiveness of using a significant amount of production and financial performance. But, as practice shows, such orientation adversely affect the adoption of the strategic and operational management decisions leaders, which is difficult to navigate in a variety of disparate indicators and their derivatives financial ratios.

Thus, the issue of improving the assessment of efficiency of functioning of industrial organizations in modern conditions becomes relevant.

Hence the purpose of this article is to develop an algorithm for evaluating the performance of industry organizations.

2 Approaches to the assessment of performance industry organizations

The theoretical analysis of the effectiveness of methods for assessing the activities of industry organizations has identified the following stages: an initial assessment carried out by a single criteria, then was used a system of indicators, at the present stage apply integrated assessment, obtained by the methods of multivariate analysis (Figure 1). [1, 2, 15].

In the first stage we have been allocated such individual factors as profit margins. But these figures do not allow to fully and objectively evaluate the effectiveness of the functioning of institutions, as their selection is carried out at random, based on the subjective views and preferences of analysts on the importance of certain criteria. Accordingly, the selected criteria do not reflect the real state of the organization, do not allow to build a reasonable forecast of its development.

The next step is evaluation of the functioning of organizations has become based on the use of indicators. However, at present there is no universal system of common indicators to obtain a unique solution. This is due to the fact that all organizations

* Corresponding author: miz108410098@mail.ru

DOI: 10.1051/shsconf/20173501138

have their particular industry. Therefore, evaluation of the effectiveness of activities designed only for a defined range of administrative tasks (identification of credit, drawing up the strategic development plan, receiving the auditor's report, etc.), carried out on the basis of a particular situation.

In addition, the main drawback of currently existing methods of constructing a system of indicators is that the selection criteria carried out by experts. Hence, the subjectivity of the results, the lack of real understanding of the situation in the organizations, which complicates the work of the analyst. At the same time the desire of some of the authors included in the measurement of assessing the effectiveness of as many indicators are not always justified, as a number of indicators often duplicate each other, and some of them do not carry the semantic load.

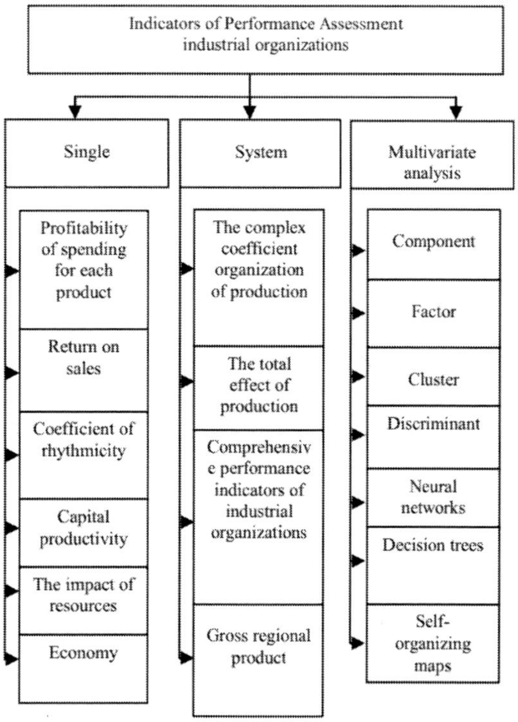

Fig. 1. Classification of methodological approaches to the assessment of the effectiveness of the industry organizations.

Our studies show that at the present stage to remove all the deficiencies noted: bias, subjectivity and one-sidedness - the procedure allows the calculation of the integral index, the methodological basis for the formation of which is multidimensional analysis. Methods cluster, discriminant, factor, component, and other types of analysis allow you to search and find out unlimited, heterogeneous and dispersed information unobvious and complex patterns, to acquire new knowledge. Thus, the integral value obtained by the methods of multivariate analysis, makes it possible, in contrast to the prior art, not only to assess the effectiveness of the organization, but also to suggest ways of its development, see the directions of optimization of management and production structures to build a competent marketing and investment politicians.

3 Cluster analysis

The main method to categorize multivariate observations is cluster analysis, with which you can break the set of monitored parameters of the set of industrial organizations into homogeneous subsets [3, 13, 11].

Data territorial body of the Federal Service on the performance of industrial companies within five years of the state statistics of the Chelyabinsk region is used as a source of information. In order to avoid distortion of the results of cluster analysis of the influence of baseline measurement units last were normalized in terms of "revenue from sales of products." Dendrogram conducted multivariate classification is shown in Fig. 2, where it is possible to distinguish three typological groups of industrial organizations of Chelyabinsk region [4, 5, 9].

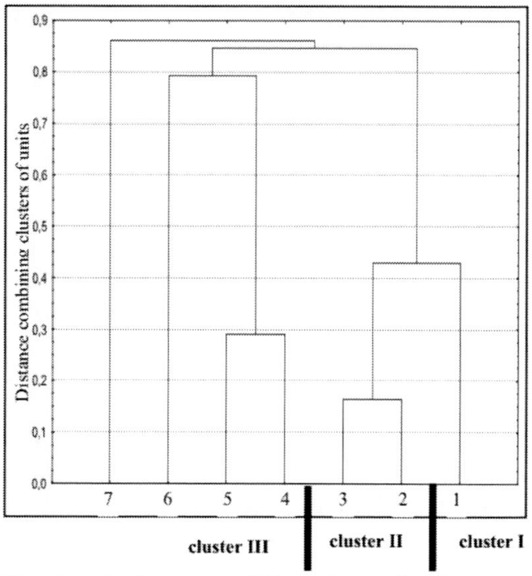

Fig. 2. Dendrogram multidimensional classification of industrial organizations: 1 - OJSC "Chelyabinsk Tube Rolling Plant"; 2 - JSC "Trubodetal"; 3 - OJSC "Chelyabinsk Metallurgical Plant"; 4 - OJSC "Chelyabinsk Mechanical Plant"; 5 - JSC "Chelyabinsk Compressor Plant"; 6 - LLC "Chelyabinsk Tractor Plant - URALTRAC"; 7 - OJSC "Chelyabinsk Forging Plant".

Cluster I comprises a single organization, which is one of the major industrial producers in the area: "Chelyabinsk Tube Rolling Plant". This cluster is characterized by high levels of efficiency of its functioning.

Cluster II combines objects traditionally produced in terms of the effectiveness of secondary evaluation and having an average competitiveness. It includes OAO "Chelyabinsk Metallurgical Plant" and OJSC "Chelyabinsk Mechanical Plant".

Cluster III concentrates a remaining industrial organizations. This group is characterized by low efficiency of activity: the simple process of reproduction it is not provided.

4 The discriminant analysis of multiple

The resulting group typical industrial organizations were subjected to further study using the statistical analysis of multiple discriminant, the essence of which is that between the selected variable characteristics and grouping correlation set. Thus, for each cluster we identified five relevant variables (Table 1) which determine the efficiency product [6, 12, 20].

Table 1. Indicators that determine the efficiency of the industrial organizations on the results of the discriminant analysis.

Indicator	Cipher indicator	Cluster		
		I	II	III
The volume of gross output, cond. u	X_1	+	+	+
The volume of commodity products, cond. u	X_2	+	+	+
Production costs, rub. / Conventional units.	X_3	+	+	+
The complexity of production, man-hours / conventional units.	X_4	+	+	+
The volume of shipped products, cond. u	X_5	+	+	+
Time between unit product, min	X_6	–	–	–
Unit Weight, t	X_7			
The production cost of one standard unit of production, rbl.	X_8	–	–	–
The average number of employees, people.	X_9	–	–	–
The volume of sales, cond. u	X_{10}	–	–	–
The average selling price, rub. / Conventional units.	X_{11}	–	–	–
Cost of sales, rub. / Conventional units.	X_{12}	–	–	–

Checking the closeness of the connection between the derived variables (multicollinearity) based on the analysis of the matrix of paired correlation coefficients allowed to exclude from further calculation of a number of indicators: the gross output (X1) and the volume of commodity output (X2), as their simple correlation coefficients were more than 0.8 .

To show the impact of the remaining indicators: production cost (X3), the complexity of production (X4), the volume of products shipped (X5) - the financial results of activity of the organizations of each cluster (YI, II, III) were constructed multivariate regression models of industrial organizations focused for-profit (Table 2).

Table 2. Regression models of industrial companies focused on making a profit.

Cluster number	The regression equation	Coefficient determination (R^2)
I	$Y_I = -0{,}42 - 0{,}38X_3 + 1{,}30X_4 + 0{,}74X_5$ $2215 < X_3 < 4933;\ 3 < X_4 < 10;\ 291 < X_5 < 421$	0.953
II	$Y_{II} = 2{,}49 + 0{,}33X_3 - 0{,}63X_4 - 2{,}28X_5$ $4272 < X_3 < 5306;\ 6 < X_4 < 13;\ 360 < X_5 < 419$	0.875
III	$Y_{III} = -1{,}25 - 0{,}65X_3 + 1{,}44X_4 - 0{,}21X_5$ $9716 < X_3 < 19167;\ 75 < X_4 < 118;\ 88 < X_5 < 226$	0.726

According to Table. 2 You can make the following conclusions.

In a cluster I profit organizations, achieved through a combination of intensive and extensive factors of production. This is evidenced by the fact that the financial result is determined by a decrease in the cost per unit of production and increasing the complexity.

In cluster II the growth of a productive indicator opposite situation cluster I. Algebraic signs with phrases X3 and X5 indicate the extensive nature of production and, moreover, inefficient use of resources.

A characteristic feature of cluster III is a surplus of labor force, as reflected in the resulting regression model. Use in this segment, as compared to other, larger proportion of human labor in conjunction with the inefficient use of resources makes manufacture unprofitable.

In general it can be noted that the industrial organization of the Chelyabinsk region at the moment, except for a few large ones, do not focus on the application of innovative technologies for the reduction of material, energy and labor resources in the production process [7, 14, 19].

5 Algorithm for Integral indicators of organizational performance

The developed algorithm for estimating the efficiency of organizations in industry involves the following steps (Fig. 3) [8, 16, 18].

The developed algorithm is adapted to the industrial Chelyabinsk area institutions. Values levels intervals determined according to step 7 shown in Table 3.

Table 3. Qualifier detection level range of performance indicators.

Indicator	Level range				
	Very low	Low	Average	Tall	Very tall
X_1	$x_1<8300$	$8300<x_1<16600$	$16600<x_1<24900$	$24900<x_1<33200$	$33200<x_1$
X_2	$x_2<60$	$60<x_2<120$	$120<x_2<180$	$180<x_2<240$	$240<x_2$
Weights levels	0.9	0.7	0.5	0.3	0.1
X_3	$x_3<102$	$102< x_3<204$	$204<x_3<306$	$306< x_3<408$	$408<x_3$
Weights levels	0.1	0.3	0.5	0.7	0.9

The results of evaluating the performance of industrial enterprises of the Chelyabinsk region are presented in Table 4.

Table 4. Determination of the effectiveness of the type of Industrial Organizations.

Name Industrial Organization	Existing indicators			The proposed integral index	Rank efficiency
	Cost margin,%	Return on sales,%	Return on equity,%		
"Chelyabinsk Tube Rolling Plant"	34	40	39	0.825	5
JSC "Trubodetal"	3	3	1,3	0.759	4
OJSC "Chelyabinsk Metallurgical Plant"	26	21	2	0.761	4
"Chelyabinsk Mechanical Plant"	–1,2	–0,7	0	0.235	2
CJSC "Chelyabinsk Compressor Plant"	–63	–168	–4	0.099	1
LLC "Chelyabinsk Tractor Plant - URALTRAC"	–19	–24	0	0.693	4
OJSC "Chelyabinsk Forge-and-Press Plant"	–74	–285	–1	0.231	2

According to the traditional method of evaluating the performance of only three organizations are positive margins.

In developed algorithm, one industrial organization has a high level of efficiency of functioning - OJSC "Chelyabinsk Tube Rolling Plant". Above average level of efficiency of functioning "Trubodetal", OJSC "Chelyabinsk Metallurgical Plant" and JSC "Chelyabinsk Tractor Plant - URALTRAC". Stragglers - by a combined indicator of efficiency of operation of less than 0.3 - are the three companies within the cluster III: "Chelyabinsk Mechanical Plant", JSC "Chelyabinsk Compressor Plant", OJSC "Chelyabinsk Forging Plant".

This algorithm for determining the integral index of functioning efficiency evaluation reveals the real potential of the organization. The special significance of the proposed algorithm is that it can help managers can see ways of improving operational and financial performance, and potential investors receive a tool to see the investment attractiveness of the interest of their organizations [10, 17].

6 Conclusion

1. The tendency to assess the effectiveness of industrial enterprises shows the transition from the simple use of disparate criteria that describe the technological, technical, economic, financial and other state organizations, to the integral index, obtained by the methods of multivariate analysis, formed taking into account all the characteristics of the variety, typical of large economic systems, under which can be understood as a branch, and individual industry organizations.
2. Conduct a cluster analysis allowed to divide the totality of these objects into three groups, characterized by homogeneity, the presence of common problems and similar disabilities affecting the results of operations.

Using discriminant analysis, combined with a procedure to eliminate the multicollinearity between the indices has made it possible for each cluster to identify those that best meet the objective of increase of productivity of functioning of industrial organization through identified indicators: production cost, labor intensity of production, the volume of products shipped.

3. The use of multivariate analysis has allowed a high degree of reliability of approximation to construct multivariate regression models. The analysis identified the characteristics of combinations made it possible to establish their optimal values that maximize the financial results for each cluster.

An algorithm for evaluating the performance of industrial enterprises on the basis of a multi-dimensional approach to recording industry characteristics, which allows you to get the value of the integral index and determine the type of organizational effectiveness. In addition, the resulting criterion allows to reflect the real potential of the existing organization and enables public authorities and municipal governments, as well as prospective investors to make informed management decisions to improve the efficiency of the business entity.

1. Definition of indicators reflecting the efficiency of the industrial organizations: X_i, $i = 1,2, ..., n$
X_i – rate;
n – index number sequence

2. Scaling performance (bringing disparate indicators in a comparable form): $X_i^{н} = \frac{X_i}{X_{i\,max}}$,
where X_i – the actual value of the indicator;
X_{max} – maximum value of the i-th index

3. Conduct discriminant analysis – the main means of solving problems of pattern recognition, which is used for deciding which variables are separated (ie, "discriminate") arising from the data sets:

$$Э = f(X_1^{н}; X_2^{н};..., X_n^{н})$$

4. Elimination of multicollinearity (when the pair correlation coefficient greater than 0.8): $Э = f(X_3^{н}; X_4^{н}, X_5^{н})$

5. Determination of the level of significance r:

$$r_{ij} = \frac{1}{n_{ij}},$$

where r_{ij} – significance level i-th index in the j-th interval;
n_{ij} – sequence number of i-th index in the j-th interval.

6. Formation interval level classifier performance:

Indicator	Level range				
	1	2	3	4	5
$X_i^{н}$	Very low	Low	Average	Tall	Very tall

7. Determination of the length of the interval I of each indicator:

$$I_i = \frac{x_{i\max}^{н} - x_{i\min}^{н}}{5}$$

8. Calculation of the integral efficiency index K:

$$K = \sum_{i=1}^{n}\sum_{j=1}^{m} g_{ij} \cdot r_{ij} \cdot \lambda_{ij},$$

where λ_{ij} – quality value of i-th index in the j-th interval,,
$\lambda \in (0;1)$;
g_{ij} – weighting factor i-th index in the range of j-th, $g = 0–1$.

9. Determination of the type of organizational effectiveness:

Range of values integral index	Rank	Type effectiveness
0.0–0.2	1	low limit
0.2–0.4	2	Low
0.4–0.6	3	Average
0.6–0.8	4	Above average
0.8–1.0	5	High

Fig. 3. The algorithm for evaluating the performance of industrial organizations.

The work was supported by Act 211 Government of the Russian Federation, contract № 02.A03.21.0011.

References

1. V.V. Joffe, *Estimation of economic stability of industrial enterprise* (Irkutsk, 2005)
2. V.M. Simchera, *Methods of multivariate statistical data analysis* (Finance and Statistics, Moscow, 2008)
3. M.V. Bushmanova, T.A. Dubrova, N.A. Mochalkin, *Cluster analysis. Conducting multivariate classification methods of observation of cluster analysis in "Statistica" package* (Bauman, Magnitogorsk, 2002)
4. *Chelyabinsk region: Statistical Yearbook, the Federal State Statistics Service, the territorial body of the Federal State Statistics Service of the Chelyabinsk region* (2008)
5. *Scenario conditions for the functioning of the Russian economy, the basic parameters of the forecast of social and economic development of the Russian Federation for 2016 and the planning period of 2017 and 2018*
6. S.A. Nedosekin, Audits and financial analysis, **2**, (2000)
7. P. Fishburne, *Utility theory for decision-making* (Science, Moscow, 1978)
8. S.A. Nedosekin, O.B. Maksimov, *The new comprehensive index evaluation of the enterprise financial condition.* http://sedok.narod.ru/sc_group.html.
9. V.I. Guard, *Analysis of economic activity in the industry* (High School, Minsk, 2005)
10. I. Prokopenko, K. Nort, *Performance management and quality modular program, in 2 hours* (Case, Moscow, 2001)
11. A.M. Dubrov, V.S. Mkhitaryan, L.I. Troshin, *Multivariate statistical methods* (Moscow, Finance and Statistics, 1998).
12. A. Khalafyan, *Industrial Statistics. Quality control, process analysis, design of experiments in the package STATISTICA* (LIBROKOM, Moscow, 2013)
13. R. Kaplan, D. Norton, *The Balanced Scorecard. From strategy to action* (Business Olympus, Moscow, 2008)
14. A.M. Pearls, M.K. Pearls, Russian Entrepreneurship, **6**, **2**, (2010)
15. I.V. Matyas, Math. ASU, **2-1(66)**, 253-262 (2010)
16. A. Vasiliev, *Integral indicator of the efficiency of economic activity of the enterprise.* http // ekonomika.ru.
17. E.M. Panevina, Russian Entrepreneurship, **11**, **2(147)**, 56-62 (2009)
18. T. Copeland, T. Koller, D. Murrin, *The cost of companies. Assessment and Management* (Business Olympus, Moscow, 2008)
19. EVA & Strategy. Stern Stewart & Co. Research. The Americas, **4**, (2000).
20. Rappaport, *Creating shareholder value: a guide for managers and investors* (Free Press, New York, 1998)

Effect of high-quality technical capacity characteristics to the impact of industrial production

Y.V. Lysenko[1,*], *M.V.* Lysenko[1], and *G.A.* Zelenskaya[1]

[1]South Ural State University, Chelyabinsk, Russia

Abstract. The article deals with problems of the qualitative assessment of the technical capacity of the economic industrial organizations, in which the development of methods for assessment and management of heterogeneous quality of resources requires establishing their resultant value at any ratio of the structural elements (resources) and the definition (increase or decrease) the effectiveness of their quantitative characteristics. The extent of the qualitative assessment of the adequacy of technical capacity of industry organizations has a direct impact on the efficiency of its formation. The article made specific calculations to identify the functional relationship between the productivity of industrial production and the main characteristics of the technical capacity of industry organizations Chelyabinsk and Sverdlovsk regions. It was concluded that the negative correlation between the quantitative security technology and efficiency of industrial production - the last influence energy power technical capacity. The equations of regression, depending on the factor variable, made their forecast.

1 Introduction

The relevance of research topic of this article is associated with the necessary knowledge of qualitative and quantitative characteristics of the technical capacity of industry organizations, cost-effective conditions of its formation, which allows managers of industrial companies to assess the effectiveness of its use and make fundamental strategic decisions.

Currently, the production process the main task performs formation of resource potential: optimizing resources and increasing the volume of industrial production. This primarily concerns the technical capacity, in which the results of scientific and technological progress have a direct mapping.

The purpose of the study – to establish the relationship between the productivity of industrial production and the main characteristics of the technical capacity of industry organizations. PURPOSE: To determine the presence and type of relationship between indicators of equipment manufacturing resources and the final results of production activities of industrial organizations Chelyabinsk and Sverdlovsk regions, in absolute, relative, specific values; determine the extent of the impact of technical equipment on the production volume, revenues and earnings; determine the functional relationships between quantitative and qualitative characteristics of the technical resources and results of operations. The subject of study is the technical potential of industrial production, the subject – industry organizations Chelyabinsk and Sverdlovsk regions. The sample includes those industrial organizations that stably throughout the study period 2010-2014.

2 Analysis of technical equipment

Investigation of the effect of high-quality technical capacity at the performance characteristics of industrial production includes several stages [1, 12].

In the first stage of research revealed the technical equipment of the impact on the final results of operations, to this end, consider the following indicators:

X_1 – income (loss) from sales, mln rub.; X_2 – the revenue from sales, mln rub.; X_3 – production area, m^2; X_4 – production capacity at constant prices, mln rub.; X_5 – output in comparable prices, thousand rub.; X_6 – the average selling price per 1 unit, thousand rub.; X_7 – the cost of commodity products, mln rub.; X_8 – plasma installation "Agate", ed.; X_9 – plasma installation "Pella", ed.; X_{10} – average number of employees, pers.

With a package of applied statistical programs based correlation matrix of the studied indicators of industrial organizations Chelyabinsk and Sverdlovsk regions in real terms. An example is shown in Table 1.

The outcome of the industrial enterprises in 2014 revealed no direct relationship between security technology and industrial production efficiency. The correlation between these two indicators is not only weak, but also negative.

* Corresponding author: lysenkoyulia@mail.ru

Table 1. Correlation analysis of the main economic indicators of industrial organizations (end of 2014).

Productive resources	Profit (loss) from sales, mln rub. (Y_1)	Revenues from sales, mln rub. (Y_2)	Production capacity at constant prices, mln rub. (Y_3)
Chelyabinsk region			
Production area, m^2 (X_3)	-0.16	0.56	0.87
The plasma unit "Agate" units. (X_8)	-0.15	0.32	0.62
The plasma unit "Pella", ed.; (X_9)	-0.15	0.39	0.68
The average number of employees, pers. (X_{10})	-0.18	0.45	0.74
Sverdlovsk region			
Production area, m^2 (X_3)	0.55	0.68	0.88
The plasma unit "Agate" units. (X_8)	0.60	0.43	0.74
The plasma unit "Pella", ed.; (X_9)	0.57	0.51	0.86
The average number of employees, pers. (X_{10})	0.40	0.27	0.59

The outcome of the industrial enterprises in 2014 revealed no direct relationship between security technology and industrial production efficiency. The correlation between these two indicators is not only weak, but also negative.

A significant degree of correlation observed in 2014 between the production capacity of comparable prices and security of industrial enterprises manufacturing facilities, human resources, plasma "Agate" units (the correlation coefficient ranges in the Chelyabinsk region – 0.62 ... 0.87, the Sverdlovsk region – 0.59 ... 0.88). Between the proceeds from the sale of products and technical equipment, as well as between profit (loss) from production and realization of communication resources noticeable: the correlation coefficient is equal to the Chelyabinsk region – 0.15 ... 0.56, the Sverdlovsk region – 0.27 ... 0.68. Communication on the profit (loss) from sales of products is lower than the production capacity at constant prices, and revenue from sales – due to the fact that the profit (loss) from the sale of products to a greater extent influenced by external factors: selling price, conditions and market size, the debt owed to creditors, payments to the budget and off-budget funds, and so on.

Thus, the functional dependence of the production capacity at constant prices (Y) of productive resources is as follows:

Chelyabinsk region

$$Y_3 = -5339.639 + 5.473X_3 - 412.561X_8 + 719.393X_9 + 344.091X_{10} \quad (R^2 = 0.88),$$

Sverdlovsk region

$$Y_3 = -11202.833 + 1.331X_3 + 83.404X_9, \; (R^2 = 0.89)$$

In all cases, "Agate" standard error for installations Plasma deviation was significant, in this connection, the figure in the regression models was not reflected. This suggests that the impact of the amount of industrial production of plasma systems "Agat" has no direct impact.

At the second stage of the study was a hypothesis about the possible presence of a closer link between performance and specific industrial production values and an attempt was made to process than the absolute values of the indicators and their change, that is, the growth rate [2, 8, 11, 20]. But in this case, it was confirmed by the lack of communication between the studied parameters. So, using specific parameters in the calculations, it was found that in most cases the specific outcomes of economic activity, the higher the smaller the size of the production area and following security appliances.

3 Analysis of quality performance technical building

The contradiction between the results obtained in the absolute and specific indicators require analysis of the impact of the qualitative characteristics of technical resources on the final economic indicators. As such, it considered the security of industrial production energy capacity (energy saturation characteristic of the process) and the degree of technological renovation of the park [3, 6, 14, 19].

Divide data set on energy capacities of the main technology available in industrial organizations in the three intervals is shown in Fig. 1, 2.

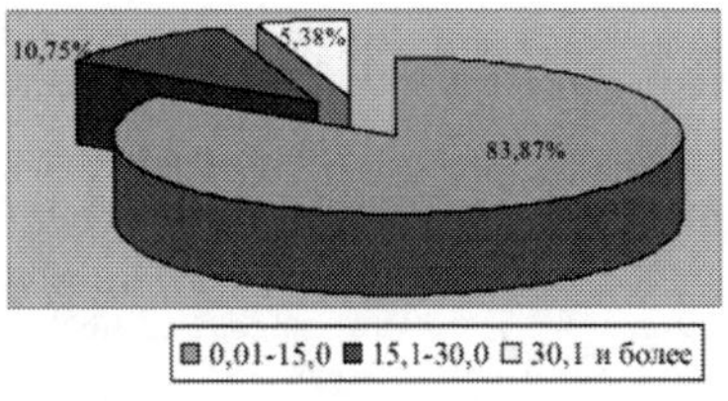

Fig. 1. Provision of industrial enterprises of the Chelyabinsk region power capacities, 2014, thousand. Hp.

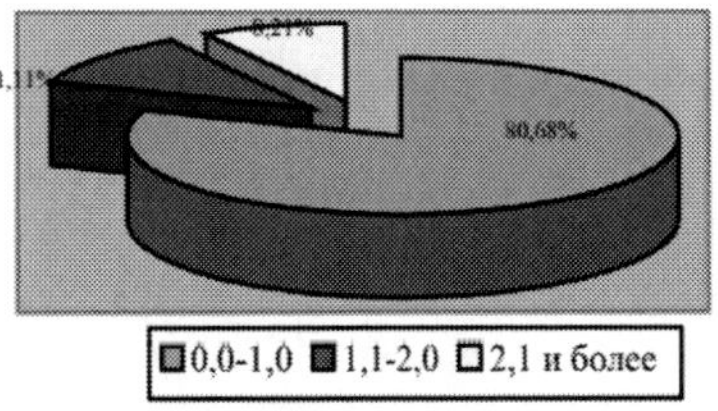

Fig. 2. Provision of industrial enterprises of the Chelyabinsk region energy capacities, in 2014, ths. KWh / person.

Power supply of industrial enterprises in terms of cardinality from 0.01 to 15 thousand. Hp Conditionally separating industrial organizations into three groups, it was found that 83.87% of the number of all investigated companies of the Chelyabinsk region provided with equipment with a total power capacity of 0.01 ... 15.0 thousand hp, 10.75% – Power appliances 15.01 ... 30.0 thousand hp., 5.38% – machinery, power over 30.01 thousand hp.

On industrial organizations of Chelyabinsk Region energy facilities in th. KW / person. The situation is about the same. Only 8.21% of the industrial companies of the Chelyabinsk region to ensure energy capacity more than 2.1 ths. KWh / person.
Divide data set on energy capacities of the main technology available in the Sverdlovsk region, in the three intervals is shown in Fig. 3, 4.

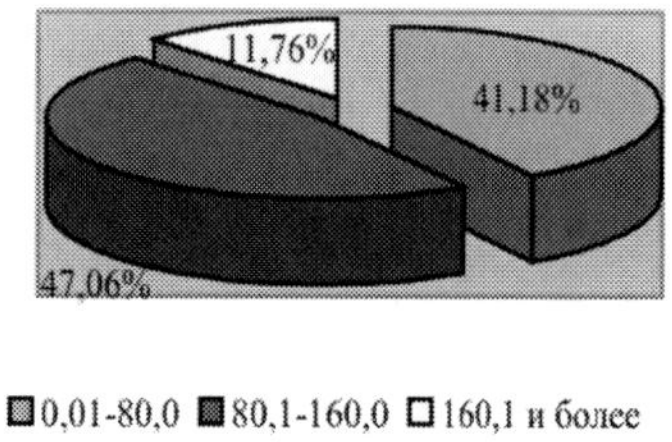

Fig. 3. Availability of industrial enterprises of the Sverdlovsk region energy capacity, 2014 hp.

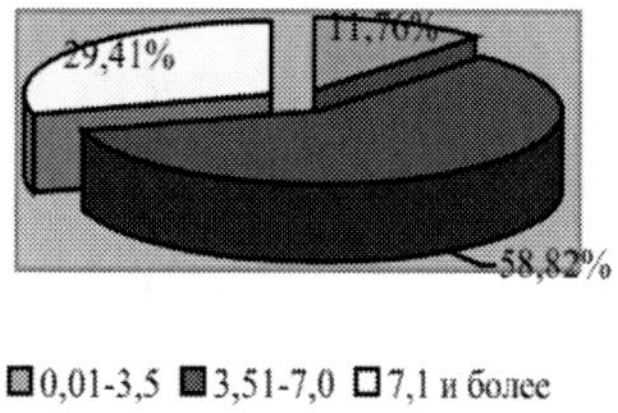

Fig. 4. Provision of industrial enterprises of the Sverdlovsk region energy capacities, in 2014, ths. KWh / person.

Power supply industrial enterprises of Sverdlovsk region is in terms of cardinality of 80.1 to 160.0 hp Conditionally separating industrial organizations into three groups, it was found that 47.06% of the number of all investigated areas of the Sverdlovsk region provided with equipment with a total power capacity of 80.1 ... 160.0 hp, 41.18% – 0.01 ... technology power 80.0 hp, 11.76% – appliances, power over 160.1 hp

On industrial organizations of the Sverdlovsk region about the same installed power situation. Only 29.41% of the industrial companies of the Chelyabinsk region to ensure energy capacity more than 7.1 ths. KWh / person.

Correlation analysis of the link between technical resources in industrial companies of the Chelyabinsk region and the production area, as well as indicators of achievement (production capacity, revenues and profits from sales of products) is presented in Table. 2.

Table 2. Correlation between technical resources, production area and efficiency of industrial enterprises, 2014.

Correlation	Power capacity of agricultural machinery	
	hp	kWh / person
Chelyabinsk region		
Production area, m^2	0.69	0.44
Production capacity at constant prices, mln rub.	0.65	0.50
Revenues from sales, mln rub.	0.29	0.16
Profit (loss) from sales, mln rub.	0.02	-0.04
Sverdlovsk region		
Production area, m^2	0.57	0.57
Production capacity at constant prices, mln rub.	0.42	0.50
Revenues from sales, mln rub.	0.44	0.22
Profit (loss) from sales, mln rub.	0.29	0.03

The analysis showed that industrial activity results are determined by the energy capacity of the technical capacity [4, 13, 15, 21]. Indicators for which the coefficient of the absolute value of the correlation is greater than 0.80 are multicollinearity. Thus, in 2014 has not been a close relationship between technical resources, production area and the impact of the industrial organizations.

According to the results of factor analysis using principal component analysis [5, 10, 16, 18] were identified two factors that contribute most to the total dispersion of –60.8% (Table 3).

Table 3. Factor loadings of initial data.

Variable	Factor 1	Factor 2
Chelyabinsk region		
Production capacity at constant prices	**–0.876**	–0.304
Revenue from product sales	**–0.887**	–0.304
Profit (loss) from sales of products	0.305	0.295
Load one plasma unit "Agat"	–0.252	**0.735**
Load one plasma unit "Pella"	–0.410	0.560
Load on one machine with numerical control	–0.281	**0.752**
The total variance	1.957	1.691
The proportion of the total variance	0.326	0.282
Sverdlovsk region		
Production capacity at constant prices	**–0.941**	–0.101
Revenue from product sales	**–0.754**	–0.298
Profit (loss) from sales of products	–0.657	–0.308
Load one plasma unit "Agat"	–0.683	0.114
Load one plasma unit "Pella"	**–0.878**	0.036
Load on one machine with numerical control	–0.490	**0.843**
The total variance	3.362	0.918
The proportion of the total variance	0.560	0.153

The calculations use absolute data on industrial organizations, describing their interest in obscheoblastnyh terms of capacity, revenue, profits from sales of products, the load on the plasma unit "Agat" plasma unit "Pella" and the machine with numerical control (CNC).

The bold color highlights those values of the studied parameters, which interprets the main components. For example, in the Chelyabinsk region, the first component is closely related to production capacity at constant prices (in the Chelyabinsk region R^2 = –0.876, in the Sverdlovsk region R^2 = –0.941), and revenue from sales of products (in the Chelyabinsk region R^2 = –0.887, in the Sverdlovsk region R^2 = –0.754), and the second component includes a load on a plasma unit "Agat" (Chelyabinsk region R^2 = 0.735) and load on a CNC machine (in the Chelyabinsk region R^2 = 0.752, in the Sverdlovsk region R^2 = 0.843). The first factor can be interpreted as the level of security of the Ural federal district organizations the main economic results of production activities, the second – as a provision of the technical potential. Two factors scattering diagram is displayed in Fig. 5 and 6, which show the load for each variable.

Fig. 5, 6 clearly shows that industrial activity results are not correlated with other variables.

4 Building a regression model of dependence from factor variables

The next step is an attempt to build a regression model was made according to the factor variables [7, 9, 17], which is selected as the load structure technology, in three scoring criteria: capacity at constant prices, million rubles. (Y1); revenue from sales, mln. (Y2); profit from sales, mln. (Y3) (tab. 4).

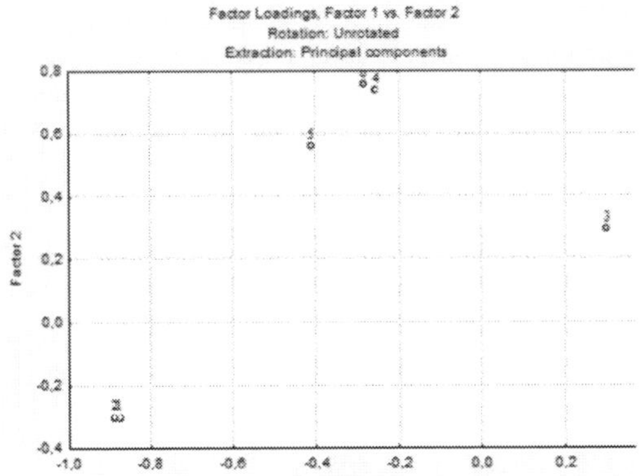

Fig. 5. Factor loadings of variables using principal component analysis for the Chelyabinsk region (1 – capacity at constant prices; 2 – Revenue from sales of products; 3 – Profit (loss) from sales of products; 4 – load on one plasma unit "Agat"; 5 – load on one plasma unit "Pella"; 6 – the load on the CNC.).

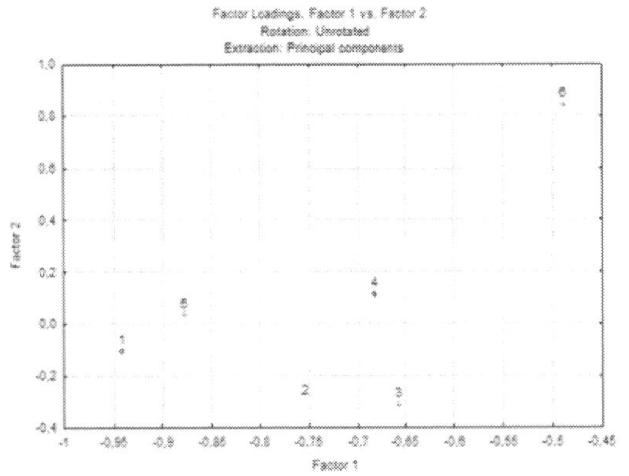

Fig. 6. Factor loadings of variables using principal component analysis for the Sverdlovsk region (1 - capacity at constant prices; 2 - Revenue from sales of products; 3 - Profit (loss) from sales of products; 4 - load on one plasma unit "Agat"; 5 - load on one plasma unit "Pella"; 6 - load on one CNC).

Table 4. Significant regression coefficients of equations.

The load per unit of agricultural land technology	Y_1	Y_2	Y_3
Chelyabinsk region			
The plasma unit "Agat"	–0.632	–0.628	–0.632
The plasma unit "Pella"	0.604	0.599	0.603
Machine with CNC	–0.104	–0.096	–0.103
Sverdlovsk region			
The plasma unit "Agat"	0.207	0.039	0.312
The plasma unit "Pella"	0.645	0.777	0.331
Machine with CNC	0.063	–0.211	–0.116

According to the above Table. 4 According to the plasma installation "Agate" and CNC have a negative impact on the effectiveness of the signs. The increase in their number leads to lower production and financial results of industrial organizations.

The relationship between the profit per standard unit sales and the main types of industrial machinery will be as follows:

$$Z = f(X;\ Y) = 42.542 - 0.007X + 0.66Y$$

with constraints $X \in (12.86; 6700), Y \in (10; 2700)$;

$$Z = f(X;\ Y) = -131.086 + 0.635X + 0.738Y$$

with constraints $X \in (178;\ 670)$, $Y \in (36;\ 366)$, where Z – profit per standard unit of sales, mln rub.;
X – load per plasma unit "Agat", that is;
Y – load per plasma unit "Pella", tons

Graphically, the relationship between the profit per standard unit of sales and technical resources is shown in Fig. 7. The coefficient of determination, given in brackets, showing a significant relationship between the study variables.

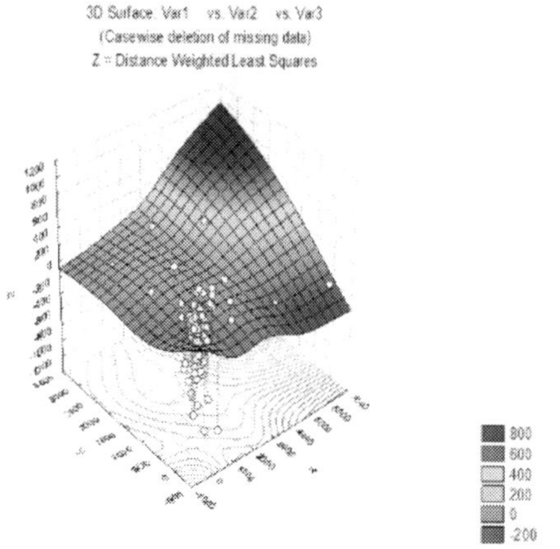

Chelyabinsk region

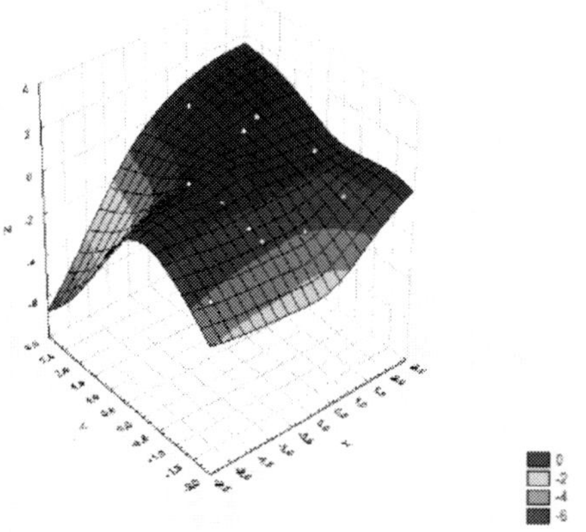

Sverdlovsk region

Fig. 7. Linear regression technical resources and profit per standard unit of sales (*Note*: *Z* – profit per standard unit of sales mln rub.; *X* – load on one plasma unit "Agat", that is; *Y* – load on one plasma unit "Pella", tons).

In Fig. 7 clearly shows that the presence of the load on the plasma unit "Agat" or plasma unit "Pella" in the park the main types of equipment for industrial companies leads to an increase in the results of their productive activity.

High ($R^2 \geq 90.5\%$) value of the coefficient of correlation between the final performance indicators (revenue, profit, production capacity) and the load on one of the main types of equipment allows to forecast specific independent variables. scatter plot for one variable – load on one plasma unit "Agat", where the dependent variable production capacity stands at comparable prices. Similar charts describe the dependence of revenues, profits from the sale of products on a single plasma unit "Agat" and one plasma unit "Pella".

Table. 5 shows the single-factor regression equation for scattering of variables with a confidence level of 90.5%.

Table 5. Regression Equations (forecast) of individual variables.

Regression equations	The correlation coefficient (R^2)
Chelyabinsk region	
$Y_1 = 0.058 - 0.05X_1$	0.905
$Y_1 = 0.658 + 0.164X_2$	0.955
$Y_2 = 0.037 + 0.074X_1$	0.807
$Y_2 = 0.043 + 0.204X_2$	0.895
$Y_3 = -0.013 + 0.037X_1$	0.734
$Y_3 = -0.013 + 0.045X_2$	0.741
Sverdlovsk region	
$Y_1 = 0.374 + 0.533X_1$	0.933
$Y_1 = 0.39 + 0.771X_2$	0.971
$Y_2 = 0.415 + 0.493X_1$	0.893
$Y_2 = 0.43 + 0.682X_2$	0.882
$Y_3 = 0.179 + 0.433X_1$	0.733
$Y_3 = 0.159 + 0.43X_2$	0.730

This confirms the close relationship between the analyzed variables.

5 Conclusion

The analysis suggests the following conclusions.

1. The amount of plasma systems "Agat" has no direct impact on the efficiency of production of industrial organizations. Thus, the detection of the functional dependence of the production capacity at constant prices by the amount of productive resources plasma systems "Agat" in the regression equation is not reflected.
2. The hypothesis of the existence of a strong connection between the specific indicators as well as between the relative values of these parameters (growth rate) was not confirmed.
3. It was found that the results of production activities affect the technical capacity of power generation capacity.
4. There is a significant relationship between the type of the number of purchased units of plasma "Pella" and economic indicators of the efficiency of industrial production – revenue and profit from the sale of products. According to the plasma unit "Agat" such a connection has not been revealed.

The work was supported by Act 211 Government of the Russian Federation, contract № 02.A03.21.0011

References

1. N.A. Vasiliev, The messenger of the Saratov State Socio-Economic University, **4**, 78-82 (2012)
2. V.V. Syroizhko, A.A. Andreev, A.A. Solyannikova, Gazette Saratov State Socio-Economic University, **9-10**, 323-329 (2012)
3. I.V. Gontareva, Vestnik of SSU: a series of "Economy", **4**, 133-139 (2011)
4. V. Denisyuk, The Economist, **5**, 76-81 (2004)

5. S. Koverga, A.S. Peredreeva, S.P. Kuzmenko, Marketing and Management of Innovations, **4**, 2, 205-209 (2011)
6. A. Korenkov, *Managing business growth potential: a monograph* (GROTTO, 2004)
7. B.M. Mizyuk, Economics and Management, **3**, 31-39 (2001)
8. A.N. Polozova, L.V. Bryantseva, D.N. Hrohordin, Audit and Financial Analysis, **4**, 360-369 (2008)
9. T.N. Prizhigalinskaya, D.S. Ternovskiy, The messenger of the Belgorod University of Cooperation, the housekeeper and the right, **1**, 30-35 (2012)
10. N. Serebryakov, N.V. Grishchenko, PES Finance. Economy. Strategy, **3**, 55-59 (2012)
11. V.V. Syroizhko, M.A. Voronov, I.I. Kulneva, Audit and Financial Analysis, **1**, 142-145 (2014)
12. *Electronic textbook according to the statistics StatSoft*. www.StatSoft.ru/ home/ textbook/.
13. Zhukovskaya Century Meters, I.B. Muchnik, *Factor analysis in socio – economic research* (Statistics, Moscow, 1976)
14. K. Iberia, *Factor analysis* (Statistics, Moscow, 1980)
15. D. Lawley, A. Maxwell, *Factor analysis as a statistical method* (Mir, Moscow, 1967)
16. Ya Perch, *Factor analysis* (Statistics, Moscow, 1974)
17. A. Byul, P. Cepel, *SPSS: the art of information processing. The analysis of statistical data and restore hidden patterns* (Diasoftyup, St. Petersburg, 2001)
18. B. Markin, *Analiz qualitative features and structures* (Statistics, Moscow, 1980)
19. A. Kolemaev, N. Kalinina, *Theory of probability and mathematical statistics: a Textbook* (UNITY-DANA, Moscow, 2003)
20. L.A. Soshnikova, Tamashevich *Century, the Website of the Year, M. Schaefer, Multivariate statistical analysis in Economics: Textbook* (UNITY- DANA, Moscow, 1999)
21. E.V. Popov, I.B. Fominykh, E.B. Kisel, *Statistical and dynamic expert systems, the PPM Chapot* (Finance and statistics, 1996)

Enterprises' communication system diagnostics

I. Penkova[1,*], *A.* Bodnar[2], and *M.* Rebezov[3]

[1]Institute of Economy and Management Crimean Federal University by V.I. Vernadsky, Simferopol, Russia
[2]Automobile Road Institute, Donetsk National Technical University, Gorlovka, Ukraine
[3]South Ural State University, Chelyabinsk, Russia

Abstract. The article is devoted to the issues assessing the development of information and communication technologies. There have been analyzed the level of innovative technologies application in enterprises, as well as their impact on improving the efficiency of organizations' productivity and management. The analysis allowed us to estimate the level of use of innovative technologies in enterprises in developed, developing and transition economies. The use of information and communication technologies is at a low level. This is evidenced by the interest obtained in the paper and numerical characteristics on such indicators as the average level of use of fixed telephone lines, the average number of Internet users, and the use of broadband Internet lines and mobile phones. The article shows the effect of the level of use of information and communication technologies to improve the efficiency of production and management of enterprises. There was a relationship between the number of companies using different types of information and communication technologies on their size. The analysis of indicators for micro and small enterprises. The dependence between the amount of use in its activities by domestic enterprises Web site and e-mail on their size. A parallel with similar enterprises in developing countries , using in the various activities of information and communication technology and having an overall yield 5.1% higher as opposed to those who do not use them.

1 Introduction

In order to identify the advantages of enterprises' communication system operating it is advisable to examine the main trends in the development of information and communication technologies in the Russian Federation, to assess the value and the dynamics of its application on different types of enterprises, to identify the main operations results and the level of spending on implementation of information and communication technology, and to analyze the dynamics of every components of the communications system within the country's businesses not specified by kind as well.

2 Theoretical and practical framework. results of investigation

At the end of 2014 year there were registered 486 000 enterprises and organizations of various types of ownership in the Russian Federation of which 332700 enterprises were of state and municipal property, 4212200 companies were private and 34300 had mixed Russian ownership type (Fig. 1) [19].

Aiming at the fining out the main point of management advantages on the enterprises dealing with the different economic activity we will analyze in progress every informational and communicational component of the communication system such as namely technical, informative, organizational, administrative and social one.

The technical component of the communications system is one of the most quickly and dynamically developing, and significantly influencing the speed of the communication system and the enterprise as a whole. It is necessary to involve into this component the facilities applied by enterprises namely PCs, local and global networks, software, and personal sites.

Investigating the results of information and communication technologies use by the end of 2014 it should be noted that the company preferred to employ PCs in their activity, that is 93.8%, as opposed to the other kind of computer techniques - 26.6%. This evolution can be traced in the whole range of research.

To function and transfer the information 89.8% of companies draw on in their work the global information network, this greatly increases the speed of information transmission and accelerates its swapping, both within the enterprise and with the external environment (Fig.1).

The number of enterprises using e-mail in their work as a tool to support communication with customers 24 hours daily have recently increased. This figure has risen from 56% in 2005 to 84.2% in 2014. E-mail is also mean for internal communication between departments and divisions of the enterprise and it's needed in special

* Corresponding author: panacea_inessa@mail.ru

DOI: 10.1051/shsconf/20173501140

protection from breaking due to the large volume of confidential information.

At the same time every fourth company using computers has Web-site functioning in the Internet and Wi-Fi access to its internal computer network [4].

A significant impact on the of the enterprise communications system operation has software installed that can perform various functions from transferring information to being helpful in decision-making.

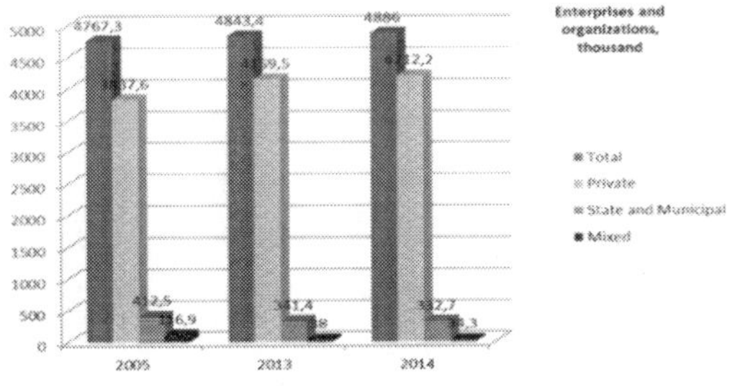

Fig. 1. The dynamics of the total number of enterprises and organizations in the Russian Federation (end of the year).

At the end of 2014 more than 86% of enterprises used various kinds of specialized software tools ranging from email to a system that allowed access to databases for solving managerial and economic tasks, sales and purchases monitoring, which indicated an increase in the influence of the technical component of the communication system on the effectiveness of business management and decision-making rate (Fig.2).

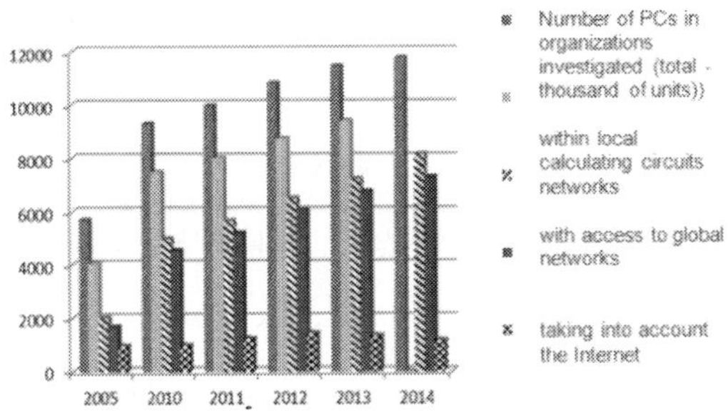

Fig. 2. Personal computers in enterprises.

Number of enterprises using the Internet to interact with suppliers, goods and services consumers increased from 2005 to 2014 by 15.5%. Simultaneously, the number of companies, which desired to obtain information about products, services and works, increased by 29%. The percentage of companies that provided information about their products increased by 22% within the same period of time (Fig. 2) [20].

At the end of 2014 the expenditures on the technical component of communication systems formed 72.8% while purchasing the computer equipment costs did not increase, as far as there were no need to replace personal computers annually, there was enough to make their modernization and repairing (Fig. 3) [9].

Thus, nowadays enterprise actively involve information and communication technologies in the activity, in self-learning how to work with clients through the Internet and personal sites. They do not reduce the technical component costs of the communications system, introducing new information support to improve the quality and the work rate that will eventually affects the enterprises performance as a whole.

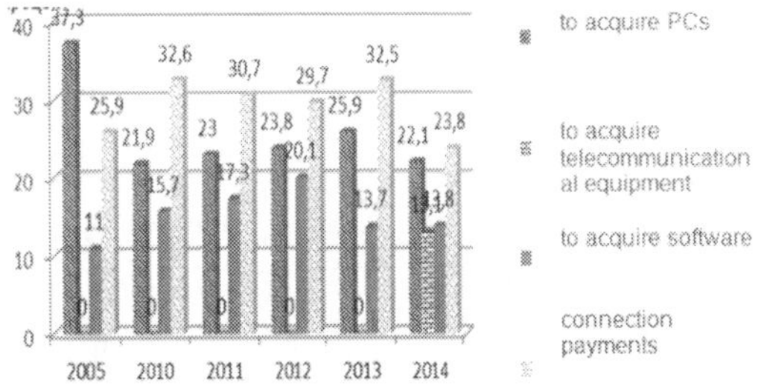

Fig. 3. The enterprises' expenditures on technical component of the communications system (percentage of total).

Information component of enterprise communications systems can be characterized by a degree of incoming information automatic receipt, workflow, and data loss level during its transmission.

More than a half of the enterprises surveyed in various economic activities apply electronic document management system during the period estimated (Table 1.) [16-20].

Table 1. The electronic document management use in enterprises by economic activity (as a percentage of the total number of enterprises surveyed within the corresponding activity).

	Businesses and organizations use an electronic document management system			
	2011	**2012**	**2013**	**2014**
Total	61.9	60.4	61.7	58.9
Mining	58.8	57.4	58.6	62.0
Manufacturing	58.0	55.3	57.0	58.4
Production and distribution of electricity, gas and water	56.2	55.0	57.5	58.2
Building	59.6	53.5	55.4	53.9
Wholesale and retail trade; motor vehicles, motorcycles repairs	57.5	54.8	55.1	57.0
Hotels and restaurants	52.9	51.4	5:20	54.3
Transport and connection	62.9	60.3	61.5	60.1
Finance	71.6	70.7	70.7	65.1
Real estate, renting and business activities	53.0	50.6	51.3	49.8
Public administration and defense; social insurance	72.8	72.7	75.1	68.8
Higher professional education	68.6	69.4	68.5	67.7
Recreation and entertainment, culture and sport	41.9	43.2	45.8	43.8
Other activities	66.2	63.7	66.4	62.3

The leaders in this matter are the enterprises of public administration, military and social insurance. Their automation level has reached up to 68.8%. The lowest

 DOI: 10.1051/shsconf/20173501140

automation rate in the enterprises activities is within recreation, culture and sports (43.8%).

Regardless of the economic activity type more than a half of the companies enlarged the data automatic exchange up to 52.7%. from 2013 to 2014. Having analyzed the data on the automatic exchange of information between internal and external information systems there were found out the incontestable leaders such as the organizations of higher education, that is 64.3% of the information circulated within these organizations in electronic form at the end of 2014 (Fig. 4).

The adoption of electronic document management and automatic data exchange between internal and external information systems requires enterprises' related costs which are shown in Figure 4 [16-20].

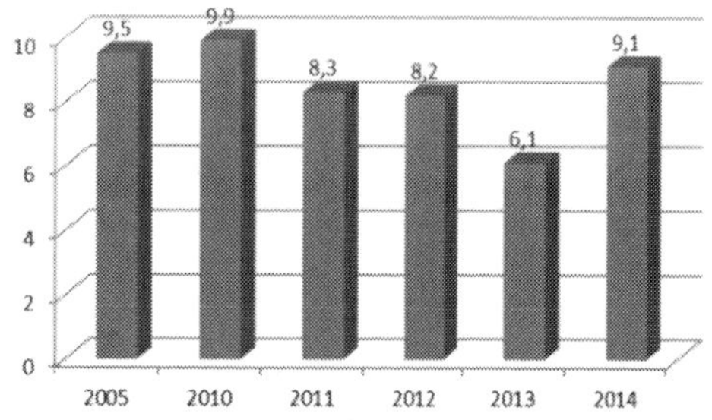

Fig. 4. Electronic documents circuit building-up costs (Percentage of total).

Thereby, the information component of communication system has become increasingly important in providing high-quality and timely information in the business management process.

The employment of personal computers, special software, electronic document circulating imposes on the enterprise the relevant requirements for staff training.

The social component of the communications system is advisable to be estimated with such indicators as the number of employees who need advanced training, the staff turnover, the conflict frequency.

The total labor force in the Russian Federation at the end of 2014 amounted up to 75 428 000 people of which 71 539 000 dealt with the economy and 3 889 000 people were unemployed (not taking into account the information on the Crimean Federal District) [20].

Level of training, namely, the level of education has significant impact on the development of the social component of the communication system.

Analyzing the distribution of population by type of employment by economic activity and the education level we should note that the vast majority of workers employed in the economy have higher education - 32.2%, while in the sphere of research and development corresponding level reaches 73.2%, within financing - 69.4%.

In spite of this, training and skills improvement is an important direction to develop any enterprises implementing in the activities information and communication management tools, as communication management include the use of innovative technologies and data exchange tools, personal computers and its software.

It is also significant to point out a slowdown in labor productivity among enterprise employees by 6.2% during the period estimated. This leads to the idea of the necessity to improve the staff skills despite the enterprises' activity [4-5].

Consequently, the companies, increasing personnel costs, enrich the staff skills, provide training and retraining of staff, thus ensuring an efficient work organization.

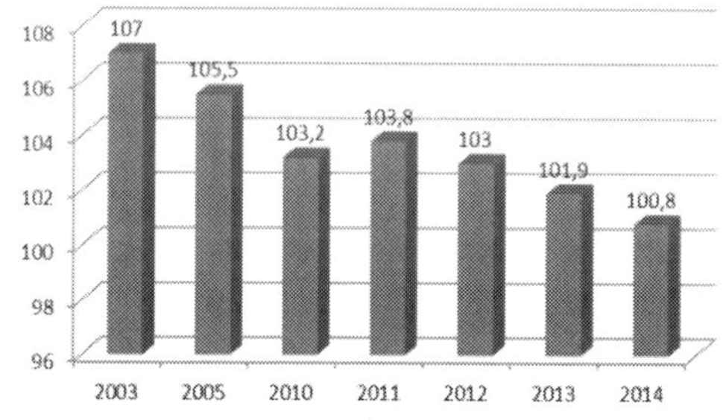

Fig. 5. The productivity growth (decline) rate (comparing to the previous year).

The organizational component of the communications system can be characterized by the feedback occurrence, the standard of the long-term goals achievement and the current assignments fulfillment.

Among the existing organizational structure of enterprise administration communication system is expediently to be managed within a mixed structure, as it is very flexible and one of the most progressive. It is composite because it includes features of organizational structures types which are widespread all over the world:

1) horizontal (planar) covers three management level which perform different functions:
- The 1st level (higher) - Head-office;
- The 2nd level (Intermediate) - Middle-office (main regional establishments, local administrations, independent branches);
- The 3rd level (lower) - Front-office (the departments like A, B, C, D.).

2) The Staff Line-functionally structure includes the Chairman of the Board headquarters - Head-office (HO), and the Directorate of the head-office and branches - Middle-office, and in addition linear functional vertical connections are commonly used.

3) Divisional customer and functionally oriented structure includes:

3.1) profitable business streamline;

3.2) Business managing (consumables) pointing;

3.3) supporting direction.

In addition, the management structures are distinguished as a divisional-product, a divisional-regional, a network an a design matrix.

Management component development status can be measured by the level of communications centralization, the relative degree of details and the number of timely response to environmental changes. However, nowadays in the administrative structure does not often hold the departments and managers directly involved into communication management. As a rule, these types of activity are worked over by the outside experts who are engaged in setting up automated systems, workflow automation [17,11,12].

This approach does not allow communication system to be wholly under control, and it affects only some of its components.

The research done for companies and organizations of different economic activity has correlated costs for information and communication technology with a company turnover as their operational results at the end of 2014. It is concluded that the information and communication technologies costs make up an average of 2% of the enterprises turnover, excepting public administration and military security, but it can influence the enterprise revenue value (Fig. 6) [19,11].

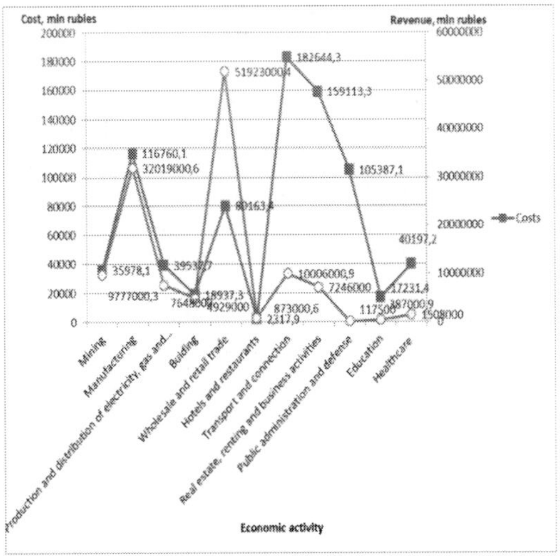

Fig. 6. Comparison of the of information and communication technology costs with the enterprise revenue value of economic activities (million rubles).

Therefore, enterprises currently make aware of the need for communications management. However there is no up-to-date approach or method developed to investigate communications management system as a whole. There is quite uncoordinated management of its separate elements.

Based on an analysis of actual approaches to the formation of criteria for a communication system evaluation, from the standpoint of the qualitative and quantitative assessment regarding information as a communication facility, here is the measuring indicators combination for every component of the communications system. To assess communication system development statues it is proposed the method comprising 15 indicators characterizing the components of the communications system [6,9,10]:

- Information INF (X1 - automated receipt of incoming information, X2 - workflow automation, X3 - data loss).
- Technical TEX (X4 - the software adequacy, X5 - hardware sufficiency, X6 - communication channels loading);
- Social SOC (X7 - the need for training, X8 - turnover, X9 - the level of conflict);
- Organization ORG (X10 - feedback, X11 - long-term goal accomplishment, X12 - current task implementation);
- Management UPR (X13 - communications centralization, X14 - detailed information, X15 - communication dynamics).

The appropriate management decisions are found out for the trade and service companies investigated such as "Aleko-Service", "Alfa-Bank", Gorlovka electromechanical plant. And the following results were obtained, that allows them to identify prior areas to advance (Table 2) [1-3].

Table 2. Acceptance of administrative decisions in the building management system sales and service enterprises "Aleko-service" pjsc "Alfa-bank", Gorlovka electromechanical plant.

Level Communication System Development	**The standard of the communications system components development**	**Events**	**Managerial impact essence**
Alfa-Bank			
SYS_KOM = 0.251 The communication system is not sufficiently developed	INF = 0.749> 0.5	-	-
	TEX = 0.605> 0.5		
	SOC = 0.273 <0.5 The social component is not sufficiently developed	Labour Systems Improving	1. To mprove the labor motivation system. 2. To Introduce feedback channels.
	ORG = 0.749> 0.5	-	-
	UPR = 0.61> 0,5		
Trade and service company Aleko-Service			
SYS_KOM = 0.273 The communication system is not sufficiently developed	INF = 0.626> 0.5	-	-
	TEX = 0.26 <0.5 The technical component is not sufficiently developed	New communication technologies introducing, the technical platform improving	1. To introduce a new information system to respond to client requests by one program. 2. To invest in technical equipment
	SOC = 0.5> 0.5	-	-
	ORG = 0.749> 0.5		
	UPR = 0.66> 0.5		
Gorlovka			

electromechanical plant			
SYS_KOM = 0.259 The communication system is not sufficiently developed	INF = 0.564> 0.5	-	-
	TEX = 0.71> 0.5		
	SOC = 0.749> 0.5		
	ORG = 0.584> 0.5		
	UPR = 0.251 <0.5 The management component of the underdeveloped	Administrative and control system improving	1. To increase the efficiency of the information received.

As a result of the research conducted through epy trade and service companies "Aleko-service" there has been revealed that the communication system is not sufficiently developed, as evidenced by the low rate of the outcome indicator SYS_KOM = 0,251. This is due to the fact that the technical component requires the decision-making to enhance the efficiency of the enterprise as a whole. At the final stage of a functional algorithm there have been formed the means to improve communications systems for the enterprise tested, i.e., a list of the administrative decisions is worked out to improve the technical equipment of the system, to introduce new information technology, to invest in technical equipment [7,8,10].

3 Conclusion

It has been grounded that the communication system includes the technical, informational, social, managerial and organizational components, which provide processes of information exchange within the enterprise and with the external environment to achieve its objectives. Having analyzed the enterprises' communication system dynamics there was estimated the current development level of its elements. Investigated companies increase the expenditures on the technical component of communication systems, personnel, improving the staff skills, staff training and retraining. The management components analysis shows that no mechanism is able to manage any entire information and communication system, and therefore, the necessity of its development has been specified.

The work was supported by Act 211 Government of the Russian Federation, contract № 02.A03.21.0011.

References

1. Yu. Lysenko, P. Egorov, Economics of Ukraine, **1**, 86-87 (1997)
2. G. Lysenko, V.Yu. Zakharchenko, Naukovi pratsi NDFI, **3**, 44-49 (2012)
3. A.V. Leonenkov, *Fuzzy modeling in MATLAB environment and fuzzyTECH* (BHV-Petersburg, St. Petersburg, 2005)
4. M.R. Ryan, M.R. Frater, *Communications and information systems* (Argos Press, P/L, 2002)
5. N.O. Shpak, *Fundamentals of communication management industry* (Publishing Lviv Polytechnic National University, Lviv, 2011)
6. N.O. Shpak, Management and Entrepreneurship in Ukraine: stages of formation and development problems, 145-149 (2010)
7. E.S. Surovtseva, G.Ya. Rubin, Production Organizer: theoretical and scientific journal, **2**, 54-56 (2008)
8. E.S. Surovtseva, Proc. Management of economic systems: theory, methodology, practice, 61-64 (2007)
9. A.A. Shubin, I.V. Simenko, Science and studio Premysl, **3(15)**, 83-92 (2009)
10. N.O. Shpak, A.M. Ulyanov, Management and Entrepreneurship in Ukraine: etipy formation and development problems. Bulletin of the National University Lviv Polytechnic, **575**, 39-45 (2006)
11. *Information economy report 2014 ICTs, economics of cloud technologies and developing countries* (United nations conference on trade and development UNCTAD, New York and Geneva, 2015)
12. *Information economy report 2012 ICTs, production software and developing countries* (United nations conference on trade and development UNCTAD, New York and Geneva, 2012)
13. *Information economy report 2011 ICTs as an Enabler for Private Sector Development* (United nations conference on trade and development UNCTAD. New York and Geneva, 2011)
14. *Information economy report 2010 ICTs, Enterprises and Poverty Alleviation* (United nations conference on trade and development UNCTAD, New York and Geneva, 2010)
15. *Information economy report 2009 Trends and Outlook in Turbulent Times* (United nations conference on trade and development UNCTAD, New York and Geneva, 2009)
16. *Russian Statistical Yearbook* (2011). http://www.gks.ru/bgd/regl/b11_13/Main.htm.
17. *Russian Statistical Yearbook* (2012). http://www.gks.ru/bgd/regl/b12_13/Main.htm.
18. *Russian Statistical Yearbook* (2013). http://www.gks.ru/bgd/regl/b13_13/Main.htm.
19. *Russian Statistical Yearbook* (2014). http://www.gks.ru/bgd/regl/b14_13/Main.htm.
20. *Statistical Yearbook* (2015). http://www.gks.ru/bgd/regl/b15_13/Main.htm.

Increasing economic stability of the industrial enterprise as a result of its market value modeling

M.G. Efimova[1,*]

[1]South Ural State University, Chelyabinsk, Russia

Abstract. The article deals with an urgent problem of the Russian and global economic stability. The issues of management of the modern industrial enterprise in the conditions of variable economy are considered in this research. Also the article highlights the importance of evaluation of industrial enterprises in Russia nowadays. The management model of the industrial enterprise cost based on sustainability criteria is constructed. The author focuses on new approaches to mathematic modeling of business evaluation. The article contains graphs and charts related to performance of enterprise system and different business strategies based on economic sustainability. Relevant problems of management by means of business assessment for enterprises owners and top managers are defined. The aim of the article is to suggest measures for the Russian industry to manage the cost of enterprises and consequently, to make national economy management more effective.

1 Introduction

The situation in world economy, developed by 2016, highlights the existence of strategic, fundamental goal of enterprise development as the most important factor of competitiveness. The growth of market value of the enterprise can become such a purpose in most cases. The Russian and foreign practice shows that the tendency of increasing enterprise cost is the index of stability of functioning and developing the enterprise, and also promotes public welfare increase and improvements in social and economic sphere of the state.

First of all, competitiveness and stability of the enterprise development surely depend on conditions of internal environment of the enterprise. However, the biggest danger is represented by factors of external environment of the enterprise which are not influenced by management and control from the management of the enterprise executives [4, 11].

So, the increase of speed changes in market conditions, and continuous fluctuations of external environment have a rather strong impact on steady performance and development of modern enterprises. It is reflected in crisis phenomena everywhere in economy, frequent geopolitical changes, terrorist threats, sanctions aimed at restriction of opportunities of the Russian economic entities participation in international trade, investment relationships and projects of technological cooperation; fluctuation of exchange rates and energy costs, etc.

For example, economic sanctions imposed on Russia in the context of political crisis development in Ukraine caused a severe damage to positions of the Russian business; and decrease of the world oil prices in the second half of 2015 created significant threats for revenues of the state budget and stability of an exchange rate. As a result, by the end of the year the economy entered the stagnation phase and faced real risks of transition to economic recession at the beginning of 2016.

The enterprises need to provide appropriate and timely responses to disturbing influences of external environment to preserve and strengthen economic and social stability in the state, and also to achieve its global strategic objectives. It will provide both economy in general, and the enterprise in particular with the necessary adaptive potential which being combined with other capacities of the enterprise (such as production, finance, labor, investment, etc.) will guarantee a sustainable development and successful achievement of the demanded results [3, 20].

Thus, there is a need to consider external environment and identify the factors which have an impact on the cost of the enterprises to further anticipate, prevent or minimize their negative influence.

2 Theoretical framework and literature review

The works of such foreign scientists as G.S. Harrison, J.R. Hicks, K. Walley, G. Teyl, R. Rayleigh, Sh.P. Pratt, R. Link, G.M. Desmond are devoted to theoretical development of the problem of cost and its assessment techniques.

The works of such modern Russian scientists who studied the problems of business evaluation as S.A. Tabakova, L.D. Revutsky, I.I. Kolesnikova, A.G. Gryaznova, M.A. Fedotova, A.M. Sambursky,

* Corresponding author: efimovamg@susu.ac.ru

M.Y. Hodorovsky, E.A. Raush, A.V. Razmakhova, S.A. Kazantseva, V.V. Balakirev, V.V. Grigoriev, A. Volkov, M. Kulikov, V.A. Scherbakov are worth noticing.

The significant contribution to research of issues of a sustainable development and effective management of the industrial enterprise was made by scientists of the Ural economic school of sciences: A.I. Tatarkin, I.A. Bayev, L.A. Bayev, A.V. Schmidt, T.A. Khudyakova, O.A. Nikolayevskaya and others.

Examination and analysis of publications of scholars on the existing problem, and also on experience of economic activity of the industrial enterprises in modern conditions allowed to reveal practical difficulties while implementing a sustainable development strategy which can result in improper business management techniques, including the absence of effective assessment and forecasting model of its cost, taking into account the unavoidable uncertainty of external factors.

The object of the research is the cost of the industrial enterprise in conditions of instability of external environment and methods of management of the industrial enterprise in these conditions.

The goal of the article is to develop the model of the industrial enterprise cost in terms of economic stability in conditions of instability of the external environment based on probabilistic and statistical approach.

3 Materials and methods

A. Description of methods

In modern market environment the evaluation of existing industrial enterprise cost should be a part of its work algorithm and one of tools designed to support the executive staff in management decisions. The enterprise is at growth stage, in other words in economically stable position, if the result of estimation the enterprise market value increased in comparison with the assessment for the previous period. Cost decrease shows a disadvantageous situation in the firm and requires acceptance of immediate measures from the executives [17-19].

The main criterion of economic development of the enterprise as an ability of the enterprise to compete in the market now and in long-term periods is the indicator of market value. [1].

Therefore, for the functioning enterprises it is quite urgent to make an assessment of their market value for prediction of negative impact of environment fluctuations [2].

The management needs to use systematic analysis to develop a technique of industrial enterprise management, taking into account its cost in the conditions of variable economy. It is obviously possible for several reasons which refer to application of systematic approach, namely: the enterprise is an open, goal-based, dynamic system [5].

The conception of the industrial enterprise as a controlled and open system focused on a goal, assumes the possibility to develop an algorithm of managing its financial and economic activity which will allow this system to achieve its strategic objective of functioning and developing in case of possible fluctuations of external and internal environment [9].

Such an algorithm should involve a criterion by which the decision will be further made. In other words, it is necessary to determine the parameter, which can help managers to analyze and carry out management feedback. Therefore, if interrelations between input parameters of the enterprise economic system and a resultant indicator, i.e. cost of the enterprise, are previously known, the process of administrative decision-making will become clear and transparent [4, 5].

B. The algorithm of seeking ways to enhance efficiency of the administrative system.

Any executive board needs to create an algorithm of seeking the ways to increase the effectiveness of the administrative system which is based on creation of the industrial enterprise cost model in terms of economic stability. This algorithm can consist of a number of stages. It is rational to accept the following stages as the principal ones:

1) Theoretical research of the system of the industrial enterprise.

This is the stage for goal-setting, identification of needs which will become the basis of administrative actions, determination of an overall strategic objective of the enterprise, identification of tasks to achieve the goal at this stage.

2) Study of general properties of the enterprise system.

To achieve this, it is necessary to set goals and form the model by means of which the influence of external environment factors on a resultant indicator will be studied.

3) Model setting.

The next stage is characterized by development of the model based on probabilistic and statistical approach, as well as quantitative determination of resultant indicator limits in case of a stable position of the system.

4) Model implementation.

Procedures which happen at this stage are listed below: training of the personnel to work with the model, practical application, comparison of results of management with the data of previous periods.

C. Graphic model

It can be possible to present a graphic model of the analysis of the industrial enterprise market value on the basis of statistical data. The cost of the enterprise (Ct) should fall within the specified goal interval range from C0 to C1, where C0 and C1 – respectively the minimum and maximum border of industrial enterprise market value in a stable position of the system. Figure 1 shows this.

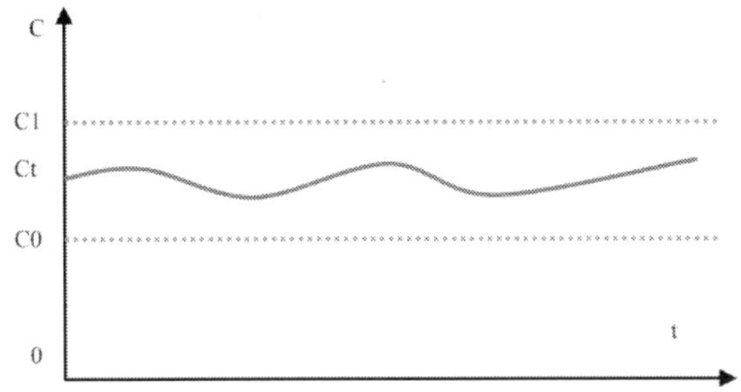

Fig.1. Graphic model of the market value analysis of the industrial enterprise implementing the strategy of stabilization and financial strengthening.

The goal of any enterprise is development and increase of activity efficiency. The graphic model of the analysis of industrial enterprise market value will be quite different at implementation of growth strategy or development. This option can be seen in figure 2.

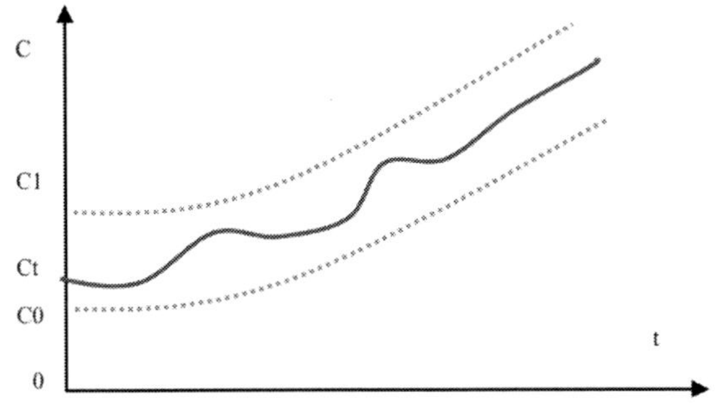

Fig.2. Graphic model of the market value analysis of the industrial enterprise implementing the strategy of growth and development.

Parameters constantly vary depending on disturbing influences of the external and internal environment, and there is a possibility for them to move beyond steady functioning limits.

Therefore, the enterprise needs to develop a system of economic stability control. Also it means to develop tools for integrated management of economic enterprise stability [8, 10, 11].

It should be noted that there is a need to take into account not only the financial stability of the enterprise, which reflects its liquidity and solvency, but also the economic stability that characterizes the probability of achieving the strategic goals of the company [9].

The model allows to identify the most hazardous impact, which can bring the resulting indicator "market value of the industrial enterprises" out of balance and, consequently, the stability of the whole system out of the zone, as well as to carry out the analysis of the sensitivity of the resulting figure to the disturbing influences of the environment [6,7].

D. Mathematical economic model

This method while developing a model, however, cannot provide the executive board and external analysts with the opportunity to confidently describe correlation between the function (an indicator "the market value of the industrial enterprise") and intensity of parameters fluctuations of external and internal environment of the industrial enterprise [12].

Therefore, there is a need to estimate probabilistic dependence of a resultant indicator on fluctuations of parameters which will enter this model.

To solve this problem, it is necessary to divide the entire time interval, which is taken for analysis, into $n \to +\infty$ number of ranges. In each time point it will be possible to determine the probability of a random variable hit, namely the indicator "market value of the industrial enterprise" in the interval [C0; C1], which characterizes a zone of a steady condition of system (1) [14-16].

$$P(C0_t \le \overline{C_t} \le C1_t) = \Phi(\frac{C1_t - \overline{C_t}}{\sigma}) - \Phi(\frac{C0_t - \overline{C_t}}{\sigma}), \quad (1)$$

where P – the probability; $C0_t$ – the minimum border of market value of the enterprise at a steady condition of system; $C1_t$ – the maximum border of market value of the enterprise at a steady condition of system; $\overline{C_t}$ – random variable of market value of the enterprise in a time point named t; Φ – Laplace's function; σ – standard deviation of a random variable S.

The probabilistic and statistical model is used for the description of uncertainty which needs to be considered at the administrative decision-making [13].

4 Conclusions

Summing up, it should be emphasized that changes of market condition are inevitable and can be characterized by various speed and intensity. It also results in short-term or long-term violations in proportionality of market process; deviations from the main objective of development. And, as the market is inherently prone to spontaneity, very flexible in the development, responsive to the many social and economic impact, moreover, it is very dependent on political and psychological effects, and the like, therefore, its parameters are subject to fluctuations as randomly and constantly manifested (cyclical and seasonal).

Therefore, under condition of instability of environment factors, the relevance and practical importance of the development of the economic and mathematical model of enterprise value, which will determine the probability of deviation of actual results from planned ones in terms of the system dynamics and possible effects of the environment, significantly increase.

Thus, the criterion which characterizes the model of industrial enterprise market value in terms of the economic stability based on probabilistic and statistical approach is a random variable probability of hitting "the market value of industrial enterprises" in the area of sustainable functioning of the system [14-16].

In conditions of unsustainable economy in the modern world the main tasks to be solved by senior management of the industrial enterprise are as follows:
• the constant analysis of market situation;
• timely identification of the risks, associated with instability of economy;
• the development of result application methodology of evaluation of the industrial enterprise cost in management;
• timely administrative decision-making on the basis of the evaluation results analysis;
• the choice of optimum measures of management and preventive actions;
• accurate planning, forecasting and obligatory systematic plan revision.

The executive board has to make decisions to change or promote such strategy which allows the industrial enterprise value fall into the desired range of values. It must be done on the basis of modeling made and the analysis of results. According to them the type and a pattern of management are chosen. The actions involving increase of enterprise efficiency and, in turn, increase of its cost, are selected as well.

The work was supported by Act 211 Government of the Russian Federation, contract № 02.A03.21.0011.

References

1. T. Kouplend, *Business value, estimation and management* (Olimp-Bussnes, Moscow, 2007)
2. I.A. Baev, V.I.Shiryaev, E.V. Shiryaev, *Dynamic theory of the firm* (SUSU Publishers, Chelyabinsk, 2001)
3. T.A. Khudyakova, A.V. Shmidt, Proc. of The 26th International Business Information Management Association Conference, 1617-1625 (2015)
4. M.S. Kuvshinov, Economics & Management Research Journal of Eurasia, **1**, 3-24 (2013)
5. T.A. Khudyakova, A.V. Shmidt, *Proc. of The 26th International Business Information Management Association Conference*, 1626-1634 (2015)
6. T.A. Khudyakova, A.V. Shmidt, Mediterranean Journal of Social Sciences, **6**, **4**, 274-279 (2015)
7. O.A. Nikolayevskaya, *Management of development of the industrial enterprises on the basis of cost approach* (Creative economy, Moscow, 2013)
8. A.V. Shmidt, T.A. Khudyakova, *Proc. of The 27th International Business Information Management Association Conference* (2016)
9. A.V. Shmidt, T.A. Khudyakova, *Proc. of The 27th International Business Information Management Association Conference* (2016)
10. T.A. Khudyakova, Proc. Innovative development of modern science: collection of articles, 293-295 (2015)
11. T.A. Khudyakova, Actual problems of economics, **12**, 444-450 (2012)
12. T.A. Khudyakova, *Proc. of the International Congress on Interdisciplinary Behavior and Social Science 2015*, 295-297 (2015)
13. T.A. Khudyakova, *Proc. of the International Congress on Interdisciplinary Behavior and Social Science 2015*, 292-294 (2015)
14. T.A. Khudyakova, A.V. Shmidt, *Proc. of The 26th International Business Information Management Association Conference*, 1607-1611 (2015)
15. T.A. Khudyakova, A.V. Shmidt, *Proc. of The 26th International Business Information Management Association Conference*, 1612-1616 (2015)
16. S.A. Kushtanok, S.G. Spirina, New technologies, **2**, 66-74 (2014)
17. S.G. Spirina, *Modeling of internal and external factors of financial stability of economic subjects* (Krasnodar, 2013).
18. M.G. Efimova, *Proc. Breakthrough economic reforms in the conditions of risk and uncertainty*, 53-55 (2015)
19. V.I. Vagizova, Problems of modern economy, **4** (2009)
20. A.A. Nikolaenko, Proc. Science SUSU, 1517-1525 (2015)

The management level of logistics customer service industrial enterprise

Y.V. Butrina[1,*], *V.N.* Tishina[1], and *P.Y.* Tishin[1]

[1]South Ural State University, Chelyabinsk, Russia

Abstract. This article deals with some aspects of the logistics service of industrial enterprise customers, current approaches to its management, namely, measurement, evaluation and optimization. It describes the effect of the level of customer service in the business results. How the concept of "logistics service users" is formulated. The author's index of evaluating the effectiveness of the logistics customer service – "logistics leverage". Author's universal measurement system of the logistics maintenance level, serving industrial enterprise customers is presented, including 11 indicators criteria values and integral indicator. The optimal values of parameters can be different for different categories of customers who prefer low, medium and high level of logistics services. It substantiates the practical significance of the study through tested at Chelyabinsk Electrode Plant.

1 Introduction

Modern business conditions are characterized by a fairly high level of competition. In this situation, low price (or slightly different from the prices of competing companies), high-quality products and services, a wide range of products, the continuous improvement of products (works, services) are not always the main factors of success in the market. Many consumers must be listed along with quality service.

Companies are turning to the level of customer service as a mean to improve competitiveness. Currently, customers prefer enterprises that can deliver the goods by a certain date, in the required amount, convenient packaging, reasonable lot size, in accordance with the ordered assortment, etc. [1]. In some cases, the level of service is given more importance than the cost of the provision of services [2]. In Russian companies there are problems in the service, namely the "poor quality of service, lack of reliability, ..., lack of proper openness and "transparency"... [3].

Currently, experts point out that the competition between enterprises is enhanced by the level of customer service. Customers are becoming more demanding, expecting more personal service.

2 Relevance: scientific significance of the issue with a brief review of the literature

Logistic management of industrial enterprises in the US and Western Europe have been paying considerable attention to the 70-ies of the last century. One of the first industries to apply the concept of logistics, the automotive industry has become. The importance of logistics customer service and the basic approaches to its management were first reflected in the works of world-renowned scholars and practitioners as R.H. Ballou [4], D.J. Bowersox, D.J. Closs [5], D. Waters [2], M. Christopher [6], J. Gattorna [7, 8], J.S. Johnson, D.F. Wood, D.L. Vordlou, P. Murphy-jr. [9], M.R. Linders, H.E. Fearon [10] and others.

M. McGinnis after order fulfillment analysis revealed that currently the level of service is given more importance than the cost of the provision of services [11].

Since 2000, the number of scientific and practical research in this area has increased significantly. First-class service as the basis of competitive advantage considered J. Tschohl [12]. Some researchers, such as D. Mourtzis, considering the current trend of individualization of logistics services in the production [13]. There was a separate area – "client-oriented service" [14]. Some authors pay attention to the performance of logistic service [15].

Among the Russian scientists, including those who engaged in management consulting business leaders, their work should be noticed, namely V.I. Sergeyev [16], L.B. Mirotin [17], B.A. Anikin [18], Y. Tashbayev [19], A.A. Chebotaev, D.A. Chebotaev [20] and other

In studying the works of the above we had identified about 90 different indicators, in varying degrees, allowing to measure the level of logistics services [21]. The analysis of these parameters allowed to express the following conclusions.

1. Authors are limited only by the concepts of performance without giving definitions.

* Corresponding author: butrinayv@susu.ru

2. There is a weak formalization of logistical service level parameters and their definitions.
3. No classification performance.
4. Some indicators do not directly relate to the logistics activity, for example, partners honesty.
5. Do not set the criteria for measuring the limit performance, service standards.
6. No integral indicator of the level of service.
7. The indicators do not form a system, some indicators contradict each other.
8. There is no indication on the priority of one over the other indices.
9. Not considered data collection mechanism and evaluation of the service level indicators.

In the literature, there are various methods of evaluating the performance of suppliers, developed assessment scale [22], which can be conventionally used to measure the level of customer service of any company. To select the parameters you can use surveys of consumers. After that service standards need to bring to the customer, it is an important tool of competition. Possible market segmentation and providing different levels of service for different market segments.

To manage the logistics activities of industrial enterprises needed to measure the quality of logistics services. For this purpose, defined (criteria) parameters of quality of service and management of these parameters is performed so that the discrepancy between the expected and actual levels of selected criteria (parameters) were minimal. For this there are different methods of assessment: questionnaires to customers and potential customers, expert opinions, statistical methods, etc.

The difficulty lies in the fact that most of the parameters of quality of service can not be measured quantitatively, i.e. receive a formal evaluation. In most cases we have to use qualitative descriptions like: above, below better – worse reliably – is not reliable , less likely – more likely , more affordable – less available , etc.

The following analysis of the literature showed that there are currently no methods of evaluating the level of customer service that fully meet the following requirements: consistency, completeness of information, the use of measurable indicators, a comprehensive system of indicators and criteria, reliability of results, ease of collecting baseline data, use of results, etc.

Therefore, author's method of estimating the level of service is shown in order to improve the level of customer service evaluation developed.

3 Level of logistics customer service

The main element of the service level is the speed (time) services, it is important for the majority of consumers. Customer service is the responsibility of marketing and logistics. Marketing identifies and often creates the need for a particular product (works, services) through advertising. A quality logistics helps to meet this need. Serving users is "Product" of logistics [19]. But marketing and logistics services interpreted in different ways.

Let's introduce the concept of "logistics service users". It is necessary to distinguish the level of service on the level of logistics services, which is its integral part. That is, to the level of performance of logistics services will refer only to those which depend on the level of enterprise logistics. For example, indicators such as honesty, friendliness of staff, etc., in our opinion, does not reflect the efficiency of enterprise logistics for the consumer, but in general are associated with the service and are marketing competence.

One of the known definitions of logistics means the ability to ensure the availability of the right product in the desired quantity and quality specified in the right place at the scheduled time for a particular customer with the best cost. This definition emphasizes the importance of service. Customer service – a set of activities carried out during the execution of orders, to meet their needs and aimed at the creation of the clients idea of the reliability of the relationship with this company [9]. Customer service – the process of creating substantial benefits, containing added tax, subject to maintaining an effective level of costs in the supply chain. This way, the customer service is considered as a process aimed at supply chain management [19]. Logistics service performs an integrating function. Such a function is shown in the interaction of the functional areas of logistics in order to meet consumers. Order fulfillment is at the center of attention service system. Warehousing, inventory management, transportation, order management, etc. – is part of the implementation of the order and customer service.

If we consider the logistics services from the point of view of the industrial enterprise, it is a set of technologies, techniques, methods of enterprise interactions with customers to ensure continuous communication between the time of ordering and completion of its implementation with a view to long-term customer satisfaction.

The level of logistics services – is the degree of customer satisfaction expressed in the proper execution of orders, there is no error, the effective provision of services and constant striving to improve the level of service, as well as under the level of customer service standards, terms of the contract or the usual requirements for quality of service [19].

Assess the level of customer service is only possible when it is measured, ie obtaining quantitative assessment of the degree of customer satisfaction. Measuring the level of customer service in logistics is not possible without the use of specialized tools in the form of a system of performance indicators. It allows you not only to measure the level of achievement of planned results (such as reliability, accuracy of orders, etc.), But also to monitor the dynamics on the integral and particular indicators.

Service level indicators of each company can sufficiently differ from each other due to their sphere of activity, types of products or services, consumer demand, etc., and therefore. Nevertheless, the key factor here is the turnaround time as the period between placing and order fulfillment.

In the literature, there are various methods of evaluating the performance of suppliers, developed assessment scale [21], which can be conventionally used to measure the level of customer service of any company. To select the parameters you can use surveys of consumers. After that service standards need to bring to the customer, it is an important tool of competition. Possible market segmentation and providing different levels of service for different market segments.

According Rodnikov A.N., the level of service is a comprehensive index, calculated on set criteria, the composition of which may vary in different systems. The most important – is the ability of the logistics system to ensure delivery of the right product to the consumer due to the time and space required. Calculated as the probability of the delivery of the ordered goods in the required time in the right place and is not less than 0.95 in real logistic systems [23].

Customer service level should be viewed as an integral component, which includes many options logistics services, such as lead time, a willingness to supply, range, fidelity deliveries, accuracy of the order, the availability (probability of occurrence of a shortage of goods, the saturation rate of demand; the frequency of the completeness of inventories coverage orders for delivery); features (speed of order execution, supply continuity, flexibility, ie non-standard orders execution); reliability (compliance regulations demand stocks); Territorial convenience and others. In the study of a large number of different works on logistics and supply chain management, as well as the results of economic activity of the enterprises we have identified and analyzed about 90 of such indicators.

The company, created its own system of service level indicators, can measure the effectiveness of the level of service, and also to compare the results with those of other enterprises. This allows you to define new targets to improve the effectiveness of its activities. Focused logistics service strategy is a means to improve market efficiency and competitive differentiation for firms employing logistics [24]. "The level of logistics customer service directly affect the market share of the company, its overall logistics costs and, ultimately, profitability, determining not only the loyalty of existing customers, but also the number of its potential customers, which will be the actual" [16].

In many areas the result of quality service are repeat orders from customers. Proven methods of high-quality customer service can be applied to the whole logistics channel [9]. Currently, consumer demand is largely determined by the level of service. Studies in European enterprises show that the reputation of the company has a greater influence on customer orientation. This factor is most important for more than 90% of enterprises [26].

It is necessary to monitor the rate of logistics services in the dynamics. In the process of monitoring identified gaps, and where the company lags behind the demands of consumers, corrective actions can be carried out. It is advisable to analyze the level of service direct competitors for benchmarking procedure. When performance indicators exceed market requirements, there are two possible ways of development. You can either cut the process to a suitable level in order to reduce costs at the same time, or continue to develop them as a criterion for receiving the order.

Tasks of improving logistics services have to be extremely specific. For example, "97% of all orders must be carried out completely and accurately, and the goods shipped within 24 hours after receipt of order" [9].

It is important that in this case used clear and measurable indicators, eg [5]:

1) availability of stocks according to the classification for the ABC analysis: 99% of the goods of group A, 95% for products of Group B, 90% for the products of the group;
2) delivery of 98% of orders within 48 hours after receipt of the order;
3) additional delivery on the understaffed orders must be made within 5 days, and others.

Large network of mass US retailers such as Wal-Mart, Kmart and Target show to suppliers following requirements: at least 98%compliance with an order, lack of orders return, shipping within the prescribed period, marking the universal codes of goods, equipment for specific customers, advance notification of shipment [9].

Currently, buyers dictate their demands. The level of customer tolerance for errors of suppliers is gradually reduced. In the United States ten years ago, the buyer was in a 1% error in carrying out the order, five years ago – 0.5%, and today – only 0.1% error [9]. As for Russian companies, the distribution centers network "Pyaterochka" supermarket, the number of errors in order picking in recent years has decreased from 5% to 0.5% [26].

To improve the level of logistics services customers need to develop measures to improve logistics customer service.

We offer measuring logistics performance index of "logistics leverage" the essence of which lies in the comparison of service levels and costs required to achieve it:

$$LL = level\ of\ service\ /\ logistics\ costs \qquad (1)$$

LL indicates what level of service (percentage) achieved an attachment 1 ruble (1 ths. Rubles to 1 million rubles, etc.).

On the one hand, this index contains two basic criteria logistics.

On the other hand, they are mutually opposite, since a high level of maintenance typically requires significant largest costs. However, they are interdependent. The indicator is a system, because It evaluates the efficiency of the entire logistics of the company.

Unit *LL* – %/ruble (%/thousand rubles,%/mln, etc.). As the need to reduce logistics costs, while seeking to improve customer service levels, the value of the index should strive to the maximum or target value.

Let's proceed to the measurement and evaluation of the level of logistical service of industrial enterprise customers.

Level of performance of logistic service systems can be quite different in different circumstances. However, the key factor here is the lead time as the period between the placement and execution of the order. Indicators of logistics services can be categorized as followed:

1) the measurability: quantitative, qualitative;
2) integrated assessment: complex, simple;
3) with respect to time during the period (floating), at time (static);
4) comparing the degree: absolute; relative;
5) the frequency of observations: the permanent, periodic;
6) by priority (importance) for consumers: significant, less significant, insignificant;
7) on the application: universal, special;
8) scale: general, single;
9) the nature of the evaluation: positive, negative;
10) on units of measurement: cost, in terms of time, in physical units (units of measurement of mass, volume, etc.).
11) by the number of influencing factors: single-factor, multi-factor.

From the set of indicators discussed in the literature and used in practice management companies (90), detected and analyzed by us, as the level of logistics service system of indicators will take the ones that are most objective and complete, in our opinion, reflect the efficiency of logistics for customers, provide measurability and clarity, and which can be adjusted depending on the specific conditions if necessary (see table 1).

Since the term of the order is the main parameter of the level of service, it uses 3 different parameters (serial numbers 1, 2, 3) to measure it in the system.

Further indicators are determined by weight. Determining the weights should be done by skilled team of experts from the number of customers of this company. The group should be a representative sample of the total number of consumers. Matrix allows to set priorities among the indicators. Matrix that is ready to calculate each indicator rating has the form shown in the table II.

Comparison of produced is done in pairs.. For example, with the first is compared with the second, then the third, with fourth, etc.; then the second is compared with the third, fourth, fifth, etc. Most important from the point of view of experts criterion is attributable to a INDICATE-value "1" less important – "0". After this result for each indicator is added, and all amounts are reduced to a common denominator, that is the total number of indicators. That is how we obtain the weight of each indicator.

Table 1. The level of logistic service indicator.

Indicator of level measurement	The meaning of the formula	The criterion of the effective ness of the logistics service level
1. Term of order fulfillment	The period of time between the moment of registration of the application and implementation of supply (provision of services, execution of works)	→ min
2. Compliance with the expectations of the consumers to the actual time of the order	The number of applications, timing of which correspond exactly to the expectations of customers * 100 % / total number of client requests.	→ 100%
3. Precise execution of orders on terms (continuity)	Number of orders, timing of which in fact correspond exactly to the terms of contracts * 100 % / total number of customer orders.	→ 100%
4. Lack of product scarcity (the ability to perform work, services)	The probability of the absence of production deficit.	→ 100%
5. The completeness of the range	Number of SKUs into the * 100 % / number of required SKUs.	→ 100%
6. Accuracy of order fulfillment	The number of applications, the run options that in fact correspond exactly to the terms of contracts * 100 % / Total orders.	→ 100%
7. Willingness to orders	The number of orders that have the same customer requests * 100 % / number of client requests.	→ 100%
8. Flexibility of order fulfillment	The number of changes made to orders being implemented * 100 % / number of customer requests for change orders.	→ 100%
9. Information support	Number of fast and accurate responses to customer requests / total number of client requests.	→ 100%
10. Correspondens to the level of service costs	Striking a balance on two criteria: the level of logistics services and the costs it.	→ 100%

Table 2. Determination of the scale service level indicators.

Indicator	1. Term of order fulfillment	2. Compliance with the expectations of the consumers to the actual time of the order	3. Precise execution of orders on terms (continuity)	4. Lack of product scarcity (the ability to perform work, services)	5. The completeness of the range	6. Accuracy of order fulfillment	7. Willingness to orders	8. Flexibility of order fulfillment	9. Information support	10. Correspondens to the level of service costs
1. Term of order fulfillment										
2. Compliance with the expectations of the consumers to the actual time of the order										
3. Precise execution of orders on terms (continuity)										
4. Lack of product scarcity (the ability to perform work, services)										
5. The completeness of the range										
6. Accuracy of order fulfillment										
7. Willingness to orders										
8. Flexibility of order fulfillment										
9. Information support										
10. Correspondens to the level of service costs										
Points										
Weight of the indicator										

Then the integral indicator of the level of logistics customer service is calculated. It is a weighted average value, ie, the sum of products of values of service parameters and their weights divided by the sum of the weights.

4 Practical implementation of the method of logistics level of customer service at the industrial enterprise

Certain provisions of the presentation-service logistics management system are repeatedly reported and approved at the scientific-practical conferences of different levels, as well as various business forums held in Chelyabinsk.

The results of the consulting activities of authors of this article allow us to judge the effectiveness of the proposed method.

Here are details of the results of measuring the level of logistics services to the industrial plant of "Energoprom – Chelyabinsk Electrode Plant" in March 2016 (see table 3).

Table 3. The level of logistic service indicator "energoprom – chelyabinsk electrode plant".

Name of the indicator	Index of the indicator,%	Weight of the indicator
1. Term of order fulfillment	75	0.15
2. Compliance with the expectations of the	76	0.11

consumers to the actual time of the order		
3. Precise execution of orders on terms (continuity)	87	0.14
4. Lack of product scarcity (the ability to perform work. services)	84	0.09
5. The completeness of the range	74	0.11
6. Accuracy of order fulfillment	82	0.09
7. Willingness to orders	78	0.09
8. Flexibility of order fulfillment	72	0.07
9. Information support	90	0.06
10. Correspondens to the level of service costs	71	0.09

So, the integral service level logistics enterprises in March was 73.68%, which corresponds to the average level. The results do not meet the company's management as well as the planned increase in the number of clients. In the future, Chelyabinsk Electrode Plant will regularly measure and evaluate the level of logistics customer service, and also implement measures to improve service levels.

5 Conclusion

So, guiding logistics strategy for customer service, organizations industrial enterprises can achieve sustainable and long-term competitive advantage Despite the importance of the level of service currently in the theory and practice are not fully worked out the question of organizing a system of indicators, criterion scale evaluation and an integral indicator to measure the level of service.. For reliable response level logistics service system of indicators should be realistically measured, objective, related to the current and past results, comparable with other companies and other time periods, understandable to all interested parties, allow to identify problem areas.In this regard, the authors proposed an approach of measuring and assessing the level of logistics customer service, which should allow industrial companies to become more efficient.

The developed method allows to estimate the level of logistics customer service for individual indicators, as well as get an overall assessment of the level of service by means of an integral indicator of the level of service to individual orders section, categories of consumers and for the whole company, to conduct benchmarking procedure. This technique can also be used to assess suppliers and business partners (eg, contractors).

The work was supported by Act 211 Government of the Russian Federation, contract № 02.A03.21.0011

References

1. A.M. Gadghinski, *Logistics* (Marketing, Moscow, 2013)
2. D. Waters, *Logistics: An Introduction to Supply Chain Management: Palgrave Macmillan* (Palgrave, 2002)
3. Logistics, **2**, 20-22 (2003)
4. R.H. Ballou, *Business Logistics Management* (Prentice-Hall International, Inc., 1993)
5. D.J. Bowersox, D.J. Closs, *Logistical Management* (New York, 1996)
6. M. Christopher, *Logistics and Supply Chain Management: Strategies for reducing cost and improving service* (Prentice-Hall, NJ, 1998)
7. J. Gattorna, Gower Handbook of Supply Chain Management (Taylor & Francis Ltd Gower, 2003)
8. R.P. Kampstra, J. Ashayeri, J.L. Gattorna, The International Journal Of Logistics Management, **17**, **3**, 312-330 (2006)
9. J.S. Johnson, D.F. Wood, D.L. Vordlou, P. Murphy, *Modern logistics* (Williams, Moscow, 2002)
10. M.R. Linders, H.E. Fearon, *Manage supply and inventory. Logistics* (LLC Victoria plus, St. Petersburg, 2002)
11. D. Waters, *Logistics. The supply chain management* (Uniti-DANA, Moscow, 2003)
12. J. Tschohl, *Achieving Excellence Through Customer Service 7th edition best sellers publishing minneapolis* (2013)
13. D. Mourtzis, Challenges and future perspectives for the life cycle of manufacturing networks in the mass customisation era, 1-20
14. Y. Lin, S. Pekkarinen, S. Ma, International Journal Of Logistics Management, **26**, **1**, 195-214 (2015)
15. O. Bouzaabia, van Riel, A.C.R., J. Semeijn, Journal of Service Management, **24**, **2**, 112-129 (2013)
16. V.I. Sergeev, *Corporate logistics. 300 answers to questions of professionals* (INFRA-M, Moscow, 2005)
17. L.B. Mirotin, Y. Tashbaev, O.G. Poroshina, *Efficient logistics* (Publishing House Examination, Moscow, 2003)
18. B.A. Anikin, *Logistics* (INFRA-M, Moscow, 2003)
19. L.B. Mirotin, Y. Tashbaev, A.G. Kasenov, *Logistics: customer Service* (INFRA–M, Moscow, 2002)
20. A.A. Chebotaev, D.A. Chebotaev, *Logistics and distribution management* (Economics, Moscow, 2012)
21. Y.V. Butrina, *The methodology of optimization of the level of customer service and costs in the logistics system business (monograph)* (SUSU, Chelyabinsk, 2012)
22. V.E. Nikolaichuk, *Procurement and production logistics* (St. Petersburg, 2001)
23. A.N. Rodnikov, *Logistics: terminologists. Dictionary* (INFRA–M, Moscow, 2000)
24. E.A. Golikov, *Marketing and logistics* (Pashkov, Moscow, 2001)
25. A. Kranke, Loss image loss of customers, **1**, 10-14 (2004)
26. SCM day – talk about important, **9**, 52-53 (2004)

DOI: 10.1051/shsconf/20173501143

Risk management as a modern sustainable development of the enterprise management tool

K.V. Aikhel[1,*]

[1]South Ural State University, Chelyabinsk, Russia

Abstract. Actively embedding in the international market processes, domestic industrial enterprises improve their management tools to create competitive structures. Spread one of the modern approaches to Business Engineering defines risk management as the starting point of building processes of the company. Adapting the method of analyzing hierarchies for quantitative risk assessment will increase the efficiency for the implementation of the new management approach, taking into account the existing time limits and the initial data, which in turn will ensure the stability of enterprises operations in the face of uncertainty. From the point of view of prospect of development, stability of the modern enterprises is also reflection of their competitiveness.

1 Introduction

Stability and sustainable development of enterprises are the signs of conduct of its proper functioning. Degeneration of the domestic industry after the recognition of disability centralized economy of the Soviet Union was focused on the creation of new organizational structures and operating principles that ensure survival in the marketplace [17, 19]. Therefore, the problem of introduction of modern management approaches to generate real business machines that can stably exist in an open international economy for the long term (Figure 1).

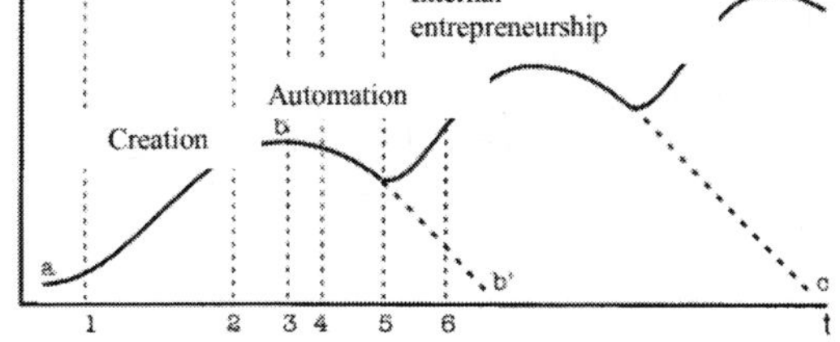

Fig. 1. Organizational Development Cycles.

The turbulence of the markets in the beginning of the XXI century, can undergo a huge number of bifurcations, has formed a new economic paradigm [20, 21]. In view of the fact that complex systems have the properties of high sensitivity with respect to fluctuations, the basis of a new approach to management has become a knowledge economy. The growing pace of globalization coexist with internal policies of the klienting modern enterprises, pulling technology and lean manufacturing philosophy system theory of constraints - the active implementation of the principles of controlling [6, 16]. Blend in with the modern quantitative management school, all these approaches are focused on providing much needed for sustainable development of modern industrial enterprises.

The utilitarian sense resistance characterizes the ability of an object to resist external influences. In practice, most of the economic stability of a company is analyzed only from the perspective of its financial situation [9]. Company Economic sustainability - a much more complex concept, defining the role and place as a subsystem in the system of higher order regional, federal and industry [13, 15].

Opportunities to maintain and enhance the economic sustainability of the company depends on how well its management are able to avoid the possible threats and to eliminate the harmful effects of the risk of external and internal environment, select the appropriate options for the management of actions under uncertainty [1, 18]. The principle of sustainable imbalance is one of the main principles of the modern theory of self-organization - synergy. Effective development of the organization in the innovation economy is determined by a combination of strategies, active use of domestic capacity for purposeful change of the external environment and continuous adaptation to it. The risk, as the competition is the driving force behind the sustained development, he - one of the objective moments to create and innovate.

The dynamics of the overall economic performance of the domestic industrial enterprises 2010-2014, due to differences in the post-crisis recovery. Given the structure of imbalances in the spheres of production, not much volatility in the overall pace of economic development, decreasing from 110% to 102%, clearly

* Corresponding author: aikhelkv@susu.ac.ru

reflects the situation of recession [2]. Weak innovation activity, imbalance in the technical and technological characteristics, inertial development - digitized visual evidence of low competitiveness and unpreparedness of Russian companies to work in a crisis.

According to official statistics, in 2013 the financial sustainability of the country's industrial enterprises is low:

• slightly less than 5000 of unprofitable enterprises in the sector of manufacturing industries (about 30% of the total);

• the total debt of operating companies of manufacturing sector at 18 billion rubles;

• reduction of the current liquidity ratio from 146.1% to 131.2%;

• a significant increase in negative ratio of its own working capital from 8.2% to -35.1% [10, 12].

A survey of three hundred Russian companies showed that only 29% of companies participating in the study have a functioning corporate risk management system. At the same time 90% of the companies of the real sector of the Russian economy do not have a system of financial motivation of managers to better risk management, and, in spite of the awareness of the need to introduce a risk-based approach, 80% of companies do not conduct regular training of personnel in the area of risk management [5].

Recovery of the domestic industry performance by means of the introduction of new management approaches is of particular relevance in a given time period. Adaptation and detail complex available tools of modern business engineering, despite the duration and intensity of fundamental organizational changes will be in the nature of health measures for a number of Russian enterprises.

2 Theoretical framework and literature review

Research problems of economic risk is actively implemented in many countries with developed market economies. In the 1960s, scientists in the United States, among which includes T. Mos, R. Mer B. Hedzhers, developed a comprehensive theory of the risk of a commercial organization. A significant contribution to this subject made by such well-known western writers like William Ting, W. Anderson, G. Magnusson, W. Arsur, C. Stone and OE in the Russian Federation in the sphere of problems of risk management began to be updated relatively recently, probably in the last 20-25 years. Russian researchers have turned their attention to the study of the problem in the post-transition period market economy. Among them, AP Algin, IT Balabanov, PG Hornbeam, JJ Kinev, GB Kleiner, L. Skamay, VD Shapiro et al. Summarizing and supplementing studies, practical implementation issues were actively engaged in the duma N., N. Kondratyev, A. Korezin, Yuri Kuznetsov, A. Petrov, A. Yudanov.

A series of existing standards for risk management of different levels of influence, such as FERMA: 2002, the COSO: 2004, ISO 31000: 2009 AS / NZS 4360: 2004, are quite similar algorithms for its implementation [4]. Based on the development of a common policy, taking into account the individual characteristics and goals of the enterprise, through the stages of identification and risk assessment, passing to make informed management decisions. A significant role for risk monitoring, analysis and optimization of management costs, providing cyclical and continuous monitoring.

Compliance with the development of the organization's management system on the basis of risk management to the mainstream internal openness and accessibility, is the need to create certain conditions and the possibility of involvement of employees in organization management. To ensure that all its employees are motivated for the development of new analytical and management tools and is actively involved in identifying risk factors for the development of his company's business. [11, 22]

Current in the real sector of the economy the concept of Business Intelligence (the BI), implies the existence of a clear pattern, clear enterprise team and allows you to connect the input of business resources to its final result for the interest period (Figure 2).

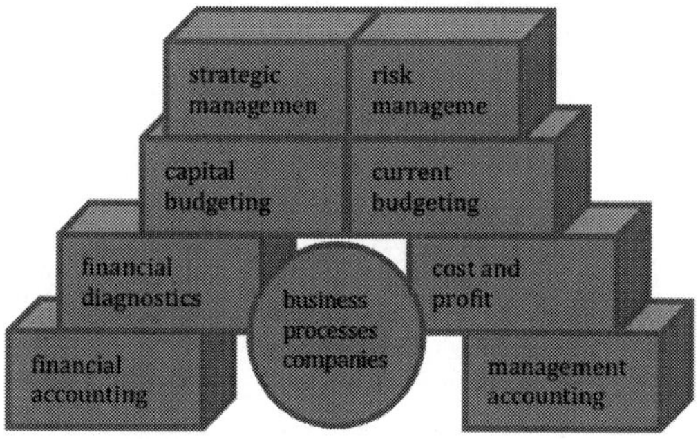

Fig. 2. The concept of Business Intelligence [8].

Popularized the practice of visualization purposes and the formation of a system of key performance indicators - Key Performance Indicator (KPI) - will be of the greatest value in the context of the application of risk-based approach, based on data from well-organized management accounting. In this case reached full integration of business processes, their continuity, preventive, efficiency, simplicity and accessibility, with a reference to the long-term dynamic growth, which in turn would lead to significant sustainability indicators of the national economy as a whole.

3 Materials and methods

A key technical challenge of implementation of risk management in practice, is the choice of tools to assess the significance of risk events on the possible consequences of their occurrence. The presence of high-quality methods of assessment (risk rose, events, trees, method of analogies, expert characteristics) should be combined with appropriate quantitative. The use of the latter is accompanied by a number of difficulties associated with limited access to the original information. Due to poorly organized fixation and low reliability of statistical information, both internal and external to the domestic industrial enterprises, mathematical valuation methods (such as the back-analysis, extrapolation, probability theory tools) is not always possible to obtain reliable results. Accordingly, the importance of acquiring expert assessments and predictive models. If possible, use the most correct scripts let you receive the simulation model, game theory, stress-testing methodology. In the absence of sufficient quantity or diversity of the initial information the general trends of the risks can be estimated using expert opinions.

A. Description of the methodology.

One of affordable, relatively easy to implement and provide the opportunity for structuring of management, is the analytic hierarchy process - methodological basis for solving problems of selection of alternatives through their multi-criteria for rating. hierarchy analysis method created by T. Saaty American scientists at the end of the XX century and has grown now into a vast multidisciplinary branch of science that has rigorous mathematical and psychological studies and numerous applications.

The essence of the method is to decompose the problem into simpler component parts and further processing sequence face judgment, decision-making, for paired comparisons. The problem is structured and presented in a tree or hierarchical structure of the network, as a result of the relative degree of interaction of the elements in the hierarchy can be expressed. Computational aspects of the MAI associated with operations over back-symmetric matrices of paired comparisons (expert judgment). As a result of mathematical operations on each of the matrices can be calculated judgments priorities compared elements objects on the levels of the hierarchy and the degree of consistency of judgment.

In the hierarchy of elements are allocated in two types: elements- "parents" and elements- "descendants". The matrices of paired comparisons are being built for the elements-all "descendants," referring to the corresponding elements - "parent." Each cell of the matrix of pairwise comparisons really does not contain a single number (the result of a direct comparison) and a vector (including all indirect comparisons through comparison with other factors). Accounting for these additional comparisons can significantly improve the reliability of the results.

B. Economic-mathematical model

After placement scores within a matrix of pairwise comparisons, direct processing of data hierarchy implies a valuation of the estimates (Formula 1) and the calculation of the priorities of the vector (Formula 2):

$$n_{ij} = \frac{a_{ij}}{\sum a_{ij}} \qquad (1)$$

where n_{ij} - rationing scoring in each matrix of pairwise comparisons of the cell; a_{ij} - Scoping pairwise comparison of elements of a certain level in the hierarchy; Σa_{ij} - aggregated numerical score within each column of the matrix.

$$w = \frac{\sum n_{ij}}{k} \qquad (2)$$

where w - the vector of priorities defined for each element of a certain level of the hierarchy; Σn_{ij} - the aggregate value of the normalized scores within each row of the matrix; k - the number of elements of the hierarchy level, compared within this matrix [7].

As a result of the direct hierarchy data, obtained totals for each level of each element. Processing data hierarchy in reverse order starting from the 2nd level of the hierarchy and implies taking into account the priorities of the vectors obtained at the higher levels. Calculations carried out by multiplying the corresponding vectors and matrices.

Hierarchy analysis method also involves checking the consistency of the resulting matrix. From linear algebra we know that positive definite back-symmetric matrix having a rank equal to 1, the maximum eigenvalue equal to the dimension of the matrix, that is, n. When compared to the actual situation the calculated maximum eigenvalue λmax will differ from the corresponding own number for the ideal matrix. This difference characterizes the so-called mismatch real matrix. And, accordingly, characterizes the level of confidence in the results. The higher the contrast, the less confidence.

An important element of the settlement of the MAI is to define the so-called coherence index (SI):

$$AI = (\lambda max - n)/(n - 1) \qquad (3)$$

He is the one who gives information about the degree of violation of the numerical (dramatically) and transitive (ordinal) consistency.

There is a need to pay a special attention to the iterative process of using AHP. If found to be insignificant experts any hierarchy of elements or, on the contrary, the need to include additional elements cognitive model, the adjustment can be performed with or models except the respective blocks.

4 Conclusions

Adaptation of Analytic Hierarchy Process in Risk Assessment, ultimately, allows to generate the rating, which determines the priorities of future management decisions. If necessary, the assessment can be carried out in two stages: the relative probability of occurrence of risk events and the possible amount of damage. The data obtained in this case can be constructed matrix of integrated assessment, including the permissible area of critical and catastrophic situations. Convenience method as is t)he availability of a quick update of the results of calculations, when the state of the individual factors of the external and internal environment.

The ability to automate the analytic hierarchy process and its application opportunities not only for stations of the risk assessment, but also for the follow-management decisions, monitoring and prioritization, cause its key role in the implementation of risk management in the enterprise. Availability of relevant structured data source management accounting and statistics will enable to minimize the level of subjectivity in the assessment.

Multivariance risk management involves a combination of knowledge of standard and unique features of financial combinations, flexibility and originality of certain ways of acting in a particular economic situation. Availability of general theoretical guidelines and practical tools available, a reference to the personality development of the enterprises, professional understanding of trends and the current market situation, support real economic activity data management accounting - the basis of the successful implementation of risk-based approach in the context of the modern concept of Business Intelligence (BI).

The work was supported by Act 211 Government of the Russian Federation, contract № 02.A03.21.0011

References

1. M.A. Betilgiriev, Management of economic systems (2012). www.uecs.ru.
2. O.I. Izryadnova, Economy. Taxes. Right, **2**, 69-80 (2015)
3. A.M. Iylshev, A.V. Karavayeva, Problems of modern economy, **4**, (2007). http://www.m-economy.ru.
4. N.V. Kapustina, *Methodology of management of development of the organization on the basis of risk management* (Economics, St. Petersburg, 2015)
5. A.M. Maps, *Risk management in the system to ensure the economic stability of the company* (Moscow, 2014)
6. T.A. Khudyakova, of the International Congress on Interdisciplinary Behavior and Social Science, 292-294 (2015)
7. T. Saaty, Decisions. Analytic Hierarchy Method, (Radio and Communications, Moscow, 1993)
8. V.P. Savchuk, *Business Intellegenc: principles, technologies, training*. www. delovoymir.biz.
9. S.G. Spirina, Russian economic magazine online, **1**, 19 (2014)
10. *Trends in the development of the industrial complex 2015: Report of the Ministry of Economic Development, the Department of sectors of the economy*. www.economy.gov.ru.
11. T.A. Khudyakova, A.V. Shmidt, *Proc. of The 26th International Business Information Management Association Conference*, 1607-1611 (2015)
12. *The Federal Service of State Statistics*. www.gks.ru/bgd/regl/b14_48/Main.htm.
13. T.A. Khudyakova, A.V. Shmidt, *Proc. of The 26th International Business Information Management Association Conference* 1617-1625 (2015)
14. M.G. Efimova, *Proc. Breakthrough economic reforms in the conditions of risk and uncertainty*, 53-55 (2015)
15. I.A. Baev, A.V. Shmidt, Bulletin of Ural Federal University, Series: Economy and Management, **3**, 50-63 (2012)
16. N.Y. Varkova, *Production management* (Publishing center SUSU, Chelyabinsk, 2009)
17. A.V. Shmidt, T.A. Khudyakova, *Proc. of The 27th International Business Information Management Association Conference* (2016)
18. A.A. Nikolaenko, Proc. Science SUSU, 1517-1525 (2015)
19. E.S. Pishchulina, Service: economics, technics, education (2015)
20. V.V. Zhuravlyov, *Proc. International scientific-practical conference*, 37-39 (2015)
21. A.V. Shmidt, T.A. Khudyakova, *Proc. of The 27th International Business Information Management Association Conference* (2016)

SHS Web of Conferences 35, 01144 (2017) DOI: 10.1051/shsconf/20173501144
ICIE-2017

Risks of a town-forming enterprise in the risk system of a company town

V.S. Antonyuk[1,*], I.V. Danilova[1], and G.V. Erlikh[1]

[1]South Ural State University, Chelyabinsk, Russia

Abstract. The article is devoted to review of risks of a town-forming enterprise in the risk system of the company town counterparties (infrastructure, population, small and medium business, local self-government authorities). It contains an algorithm of stepwise study of risks of a town-forming enterprise from the perspective of their interconnection with the risk system of the company town counterparties. Approbation of theoretical and methodical provisions by the example of town-forming enterprise Asha Metallurgical Plant OJSC, Asha, Chelyabinsk region, with the help of the graph theory allowed to analyze interconnections and dependence of risks of the town-forming enterprise and to make a conclusion on its negative role in creation of a crisis situation in the economy of the company town. The article is completed with a complex of lines for neutralization of risks of AMP OJSC: TFE's infrastructural and technological risk management; personnel risk management policy; mitigation of consumer, product and supply risks; competitive struggle risk management and advertising risk management.

1 Introduction

A special form of the settlement and production system is company towns, which town-forming basis is laid by one or few industrial enterprises.

Town-forming enterprises (TFE) are legal entities, in which the number of employees comprises no less than 25 % of the economically active population of a revenant settlement [1].

Currently there are about 319 Russian towns with single-industry production [2], their territory is occupied by 25 % of the urban population, and the aggregate gross regional product comprises 40 % of the GRP of the RF regions [3, p. 141].

Vitality of the research is connected with the fact that a town-forming enterprise by virtue of its monopoly position has a considerable influence on all the counterparties of the monotown: it determines the level of household income and well-being of the population, influences social and technical infrastructure, limits the value of the demand for products of small business and local industry, influences the income basis of the local budget. Besides, domination of one focused enterprise considerably strengthens vulnerability of the monotown and makes it a "hostage" of the global market situation.

2 Overview of economic references

Problems of the general theory of risks (scientific works of Vishnyakov Ya. D, Radaev N.N. [4]), as well the theory of uncertainty and risks of industrial enterprises (works of McCarthy M.P., Flynn T.P [5], Burkov V.N., Kireeva E.A. [6], Kleiner G.B., Tambovtsev V.L., Kachalov R.M. [7]) have been rather widely studied in the economic literature.

Problems on functioning of TFEs of company towns in the part of studying development trends in the crisis conditions, as well as their recovery ability have been studied in the works of such economists as: Trifonov V.A., Lobanov M.M. [8], Kleiner G.B. [9], Ryakhovskaya A.N., Kovan S.E., Kryukova O.G. [10], Turgel I.D. [11], Kokh I.A. [12], Lyubovny V.Ya. [13].

It is also worth noticing methodical approaches to study of TFEs and monotowns, among which most interesting are guidelines on preparation and implementation of complex investment plans of company town development [14].

3 Theory and methodology of analysis of risk dependence of a town-forming enterprise

The task of the article is studying of risks of a town-forming enterprise in the part of their interconnection and dependence on the system of internal and external risks of a company town for development of proposals on their neutralization.

Risk (risqué (French)) is literally defined as a cliff, a rock [15, p. 616]. In the economical literature risk is defined as potential occurrence of events with negative consequences as a result of certain decisions or actions [16, p. 898]. In the theory of risks the notions of factor (cause), type of risks and type of losses (damage) are differentiated [17, p.12].

The algorithm of solving of the set task consists of the following stages:

* Corresponding author: antvs@list.ru

 DOI: 10.1051/shsconf/20173501144

• analysis of risks of individual monotown counterparties (TFE, population, infrastructure, small and medium business, local self-government authorities) with identification of risk factors, probability of occurrence and damage;
• ranging of risks by the monotown components;
• analysis of interdependence of TFE's risks on the risk system of the town counterparties and building of cause-and-effect links with the help of the graph theory;
• development of a complex of measures aimed at neutralization of risks of a town-forming enterprise.

4 Risks of monotown counterparties

This procedure has been approved by the example of analysis of the situation in monotown Asha of Chelyabinsk region in 2010, which forming enterprise is Asha Metallurgical Plant OJSC (AMP OJSC). Initially we conducted expert analysis, which allowed to identify most typical risks of the monotown counterparties:
• risks of AMP OJSC (infrastructural; technological; personnel; consumer; product risks; as well as risks connected with owners);
• risks of population (risks of labor power migration; risks of population's paying capacity; risks of reduction of living standards and human capital assets);
• risks of infrastructure (depreciation; expiration of the life cycle of infrastructural elements; lack of capacity / availability of infrastructure; influence of external natural and man-made factors; lack of financing; title to infrastructure);
• risks of local industry and small business (infrastructural; social and consumer);
• risks of local self-government authorities (political and credit risks; risks of volatility of outgoing payments and volatility of collected taxes; risks of undistributed liabilities).

Then, we **ranged the risks** by the monotown components, when the range of each of them is determined as a product of the average weighted (by the degree of influence) probability of activation of each factor by the average weighted damage assessment, as a result of which we obtained "the rose of risks" by the monotown components (fig. 1).

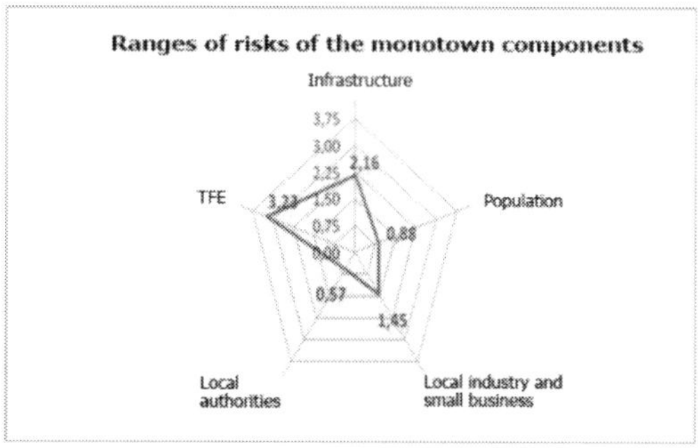

Fig. 1. "Rose of risks" by the monotown components.

Table 1. Risks of AMP OJSC in the risk system of Asha, Chelyabinsk region.

#	Risk	Dependent risks
1	TFE's supply risks	TFE's product risk
2	TFE's competitive struggle risk	TFE's consumer risks, TFE's personnel risks
3	TFE's advertising risk	TFE's competitive struggle risk, TFE's consumer risks
4	Risk connected with the TFE's distribution system	TFE's supply risks, TFE's consumer risks
5	TFE's consumer risks	TFE's personnel risks; risk connected with infrastructural financing
6	TFE's product risks	TFE's consumer risks
7	TFE's personnel risks	TFE's supply risks; TFE's advertising risk; risk connected with the TFE's distribution system; TFE's product risks; risk of the population's paying capacity; risk of reduction of the population's living standards
8	TFE's infrastructural risks	TFE's supply risks; Risk connected with the TFE's distribution system
9	Risks of tax collection volatility	Risk of reduction of the population's living standards; risk connected with infrastructural financing
10	Risk of the population's paying capacity	Risks of tax collection volatility; consumer risks of local industry and small business; risk connected with infrastructural financing
11	Risk of reduction of the population's living standards	Social risks of local industry and small business
12	Infrastructural risks of local industry and small business	Risks of tax collection volatility
13	Consumer risks of local industry and small business	Risks of tax collection volatility; риск, risk connected with infrastructural financing
14	Depreciation / wear of infrastructural elements	Expiration of the life cycle of infrastructural elements; lack of capacity / availability of infrastructure
15	Risk connected with infrastructural financing	Depreciation / wear of infrastructural elements; expiration of the life cycle of infrastructural elements; lack of capacity / availability of infrastructure
16	Expiration of the life cycle of infra-structural elements	TFE's infrastructural risks; risk of reduction of the population's living standards; infrastructural risks of local industry and small business; lack of capacity / availability of infrastructure
17	Lack of capacity / availability of infrastructure	TFE's infrastructural risks; risk of reduction of the population's living standards; Infrastructural risks of local industry and small business

As it is clear from the ranging, risks connected with the activity of the town-forming enterprise (range 3,23) and infrastructural risks (range 2,16) have the maximum negative influence on the activity of Asha.

5 Dependence of risks of a town-forming enterprise

At the next stage we studied interconnections and dependence of TFE's risks on risks on other town counterparties with the help of the graph theory [18, 19, 20]. Table 1 contains a system of risks of AMP OJSC in the monotown risk system and their interconnections.

In figure 2 external, internal and external-internal risks are shown as points, and the system of interconnection and interdependence of risks – as arcs, wherein a high degree of interconnection is marked with a thick line.

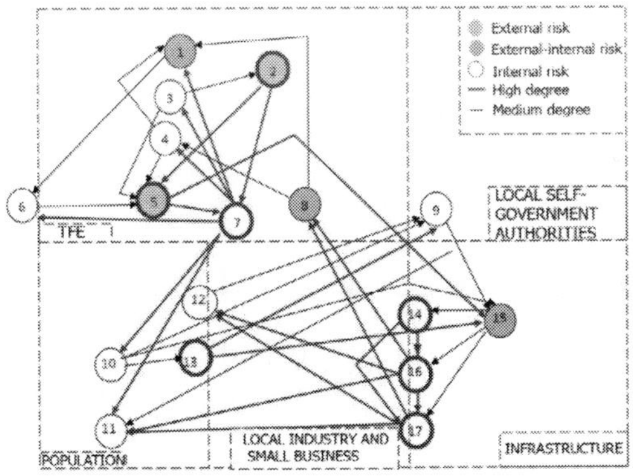

Fig. 2. Dependence of risks of AMP OJSC in the risk system of Asha, Chelyabinsk region.

We can make the following conclusions of the conducted analysis of risk dependence.

Firstly, risks of the town-forming enterprise and infrastructure are most vulnerable for a company town. Thus, out of 17 types of risks 8 are inherent of AMP OJSC and 4 – of the technical infrastructure.

Secondly, the main types of risks for the town-forming enterprise are: competitive struggle risk, consumer risks, personnel risks, and for infrastructure – depreciation, expiration of the life cycle of infrastructural elements, lack of capacity / availability of infrastructure.

Thirdly, AMP OJSC is characterized both by internal (advertising risk; risk connected with the distribution system; product risk; personnel risks) and external-internal and external risks (supply risks; competitive struggle risk; consumer risks and infrastructural risks). Technical infrastructure is basically characterized by risks internal in relation to the town, which is quite explainable, whereas infrastructure is a tertiary sector for the town counterparties.

Fourthly, town-forming enterprises are a rather unstable form of enterprises, whereas they are characterized by a high degree of field specialization and a low level of field diversification, which makes them rather dependable on the global economic situation. In modern conditions this situation is aggravated by the global crisis and metal price collapse tendencies.

6 Conclusion

Thus, practical value of the research is that it allows to identify most typical risks characteristic of Asha Metallurgical Plant OJSC and to develop lines for their neutralization. The most important lines include:
1) TFE's infrastructural risk management by means of:
• elimination of failures in operation of the logistical infrastructure (failures in transport operation, problems with availability of storage areas, etc.);
• search of new partners or specialists able to ensure support service;
2) Technological (production) risk management on account of:
• motivation of operational personnel;
• increase of the product quality control level;
• upgrade of the production management quality;
3) Personnel risk management aimed at non-admission of dismissal of qualified personnel;
4) Consumer risk management, for which purpose we need:
• availability of information on the market condition;
• marketing analysis of the market and stimulation of product demand;
• increase of marketing allocations;
5) Product risk management on account of:
• improvement of consumer and functional product properties, increase of awareness;
• improvement of pricing;
6) Supply risk management aimed at increase of the materials quality control level and improvement of acquisition planning;
7) Competitive struggle risk management by increase of awareness on the market condition, its competitors and new players;
8) Management of risks connected with the distribution system and availability of sales partners;
9) Advertising risk management on account of:
• bringing of the product information to the knowledge of potential buyers and improvement of advertising financing.

Thus, system risk management of a town-forming enterprise built on a periodic basis and in the mode of continuous monitoring of external and internal risks will allow to ensure compensation of its risks, balanced state of the company town and improvement of the population's well-being in conditions of an extremely unstable global market situation.

The authors make conclusions on the research work and give recommendations on application of the obtained results.

The work was supported by Act 211 Government of the Russian Federation, contract № 02.A03.21.0011

References

1. *Federal law no. 127-FZ dated 2002, On insolvency (bankruptcy)* (2002)
2. *Resolution of the RF Government no. 1398-r* (2014)

3. O. Roy, *Basics of State and Municipal Administration: Manual* (Piter, St. Petersburg, 2013)
4. Ya.D. Vishnyakov, N.N. Radaev, *General Theory of Risks* (Akademia Publishing center, Moscow, 2007)
5. M.P. McCarthy, T.P. Flynn, *Risk: Risk Management at the Level of Top Management and Board of Directors* (Alpina Business Books, Moscow, 2005)
6. E.A. Kireeva, V.N. Burkov, *Proc. XII All-Russian Conference on Management Problems (ARCPM)-2014*, 5309-5319 (2014)
7. G.B. Kleiner, V.L. Tambovtsev, R.M. Kachalov, *Enterprise in an Unstable Economic Environment: Risks, Strategy, Security* (Economics, Moscow, 1997)
8. V.A. Trifonov, M.M. Lobanov, News of Tomsk Polytechnic University, **309**, **3**, 198-201
9. G.B. Kleiner, ECO, **6**, 187-192 (2007)
10. A.N. Ryakhovskaya, S.E. Kovan, O.G. Kryukova, *Prevention of Bankruptcy of Town-Forming Organizations of Company Towns: monograph* (Magistr: Infra-M Scientific and Publishing Center, Moscow, 2012)
11. I.D. Turgel, *Russian Company Towns: from Survival to Stable Development* (Publishing house of the Ural State University of Economics, Yekaterinburg, 2010)
12. I.A. Kokh, Issues of Management, **9**, 75-84 (2009)
13. I.A. Lyubovny, *Company Towns in Crisis Conditions: Condition, Problems, Recovery Opportunities* (Dortranspechat CJSC, Moscow, 2009)
14. *Guidelines on Preparation and Implementation of Complex Investment Plans of Company Town Development.* gov.cap.ru/HOME/24/oip/17.doc.
15. L.P. Krysin, *Illustrated definition dictionary of foreign words* (Eskimo, Moscow, 2011)
16. A.N. Azrilian, *Great Dictionary of Economics* (Institute for New Economic Thinking, Moscow, 2002)
17. I.N. Kalashnikova, Economic Analysis: Theory and Practice, **13(220)**, 10-18 (2011)
18. K. Berzh, *Graph Theory and its Applications* (Foreign Literature, Moscow, 1962)
19. T.A. Khudyakova, A.V. Shmidt, *Proc. of The 26th International Business Information Management Association Conference*, 1612-1616 (2015)
20. T.A. Khudyakova, A.V. Shmidt, *Proc. of The 26th International Business Information Management Association Conference*, 1617-1625 (2015)
21. V.N. Burkov, A.Yu. Zalozhnev, D.A. Novikov, *Graph Theory in Organizational System Management* (Sinteg. Organizational System Management, Moscow, 2001)
22. V.N. Burkov, V.N. Zinchenko, S.V. Sochnev, G.S. Khulap, *Exchange Mechanisms in the Economy of the Transition Period* (Institute of Control Sciences of the Russian Academy of Sciences, Moscow, 1999)

Performance management of investment projects based on real option theory and method of analysis of hierarchies

L.A. Baev[1,*], and *O.V.* Egorova[1]

[1]South Ural state University, Chelyabinsk, Russia

Abstract. The article is devoted to the development of the method for evaluation and control of investment projects efficiency in conditions of high uncertainty. There are marked the limitations of the traditional evaluation methods of investment projects efficiency and proposed more relevant approach based on the application of real option theory, which allows evaluating projects with consideration of their multivariate and opportunity of decision making at every stage of their development. The article shows that, despite the high relevance of real option theory, its practical use is very difficult for a number of reasons. The main problem in estimating the value of real options is the need to use a sufficient number of a priori statistics regarding the project, which in practice is difficult. The authors propose a method in which the probabilistic characteristics of the model project are replaced with the correct expert ratings, named the modified ROV – method (MROV). The method is based on the binomial model of real option value assessing, based on the building of a decision tree, at each node of which there is possible the development according to the "optimistic" or "pessimistic" forecast. Since the development of the project under any scenario depends on a number of criteria environments, the task of assessing the value of real options becomes a multicriteria task. One of the most reasonable and practically relevant methods of solving such problems is the method of analytic hierarchy process (AHP), which is used in the basis of the developed MROV-method. The article also details the algorithm of the method application to determine quantitative estimates of the relative probabilities of the development of the project on "optimistic" or "pessimistic" scenario in each node of the binomial tree. Obtained estimates are used as weights in the calculation of the net present value of the optional investment project for a decision on the feasibility of its realization and inclusion in the project evaluated real options.

1 Introduction

Analysis and assessment of investment projects efficiency is one of the most difficult tasks of economic management. The complexity of this task is caused by the high degree of uncertainty of conditions of the projects realization. On the one hand, it is necessary to foresee the future, and on the other hand all indicators of efficiency of projects calculated on the basis of strict script implementation and does not take into account possible changes.

2 The actuality of the research

Topicality of the research is defined by need of development of methods of assessment and management of investment project efficiency as multi-stage processes with a high level of diversity and uncertainty.

In recent years, the approach to the evaluation of investment projects based on application of the theory of real options becomes commonly used. It allows evaluating the project taking into account the various options step-by-step development.

The evaluation methodology of real options is in the stage of rapid growth [1-8,11]. An increasing number of companies use it for the formation of its development strategy. The idea of the approach is that the uncertainty of the project realization conditions contains not only risks but also opportunities that must be considered at the planning stage of projects, putting in the project the necessary conditions of their realization, that is real option [9,10]. Thus, for the decision it is required the valuation of the option [12-15].

Despite the high relevance of the theory of real options and the potential for development on its basis a new direction of a priori adaptive performance management of investment projects, the practical use of the ideas of this theory very difficult for a number of reasons.

3 Existing approaches and solutions

Existing approaches to assessing the value of real options come from the unity of the theoretical and methodological basis of real and financial options [3,7,16,17].

* Corresponding author: baevla74@mail.ru

Financial option – an option to purchase or to sell a financial asset in the event of a predicted changes in its future value.

Real option – the possibility of correction of the cash flows of the project, in case of predicted changes of its realization.

This allows us to conclude that since theoretically financial and real options are similar, technologically, the existing traditional different approaches to the valuation of financial options can be used for the evaluation of the real options [3,12,16]. It is important to understand that even theoretically this approach is possible only for "simple" options, which have only one source of uncertainty and which are based on an underlying asset. In this case, due to the variety of real options, it is required the development of a greater number of special methods and techniques. Discussing this area of research, we need to understand that the development of a methodology of evaluation of financial options, applied to real options does not remove the need for correct quantitative estimation of the probabilistic characteristics of the different trajectories of development of the project, which, as previously noted, is not practicable. Not rejecting the idea of using financial options theory to valuation of real options, let's focus on construction methods, eliminating this restriction.

Fig.1 shows the developed algorithm of selection of the most appropriate methods of financial options assessment, applied to the problem of real options evaluation, taking into account existing constraints and problems associated with their application.

In Fig.1 there are the following symbols:

C_{esp}^{call} - the price of the European call option; S – the price of the underlying asset at the time of purchase of the option; X – the strike price (the price of the underlying asset); r – the continuously compounded risk free rate; e – base of the natural logarithm; T – the expiry time; $N(d1)$ – function of the normal distribution, which can be regarded as the risk-neutral probability that the option will gain; $N(d2)$ – function of the normal distribution, which is seen as the probability that the option will be exercised.

Fig.1 shows that the known transformation methods of assessment do not remove the underlying problem of the need to use a sufficient number of a priori statistics on the project.

In addition, the analysis shows that the Black-Scholes Option Pricing Model and its modifications will be correct for a very narrow class of the optional projects even in the purely theoretical case of known stochastic characteristics of the project variables.

The binomial model provides a much higher flexibility [18-20].

In accordance with the binomial model the entire period of the option is broken into a number of intervals, during each of which the value of the underlying asset may increase with the probability q or decrease with the probability $(1-q)$. The most known method of real options valuation (ROV – real option valuation method) is based on this approach, which, however, requires a stochastic characteristic of decision tree – function of density of distribution of probability of growth or decline of the asset value in the tree nodes.

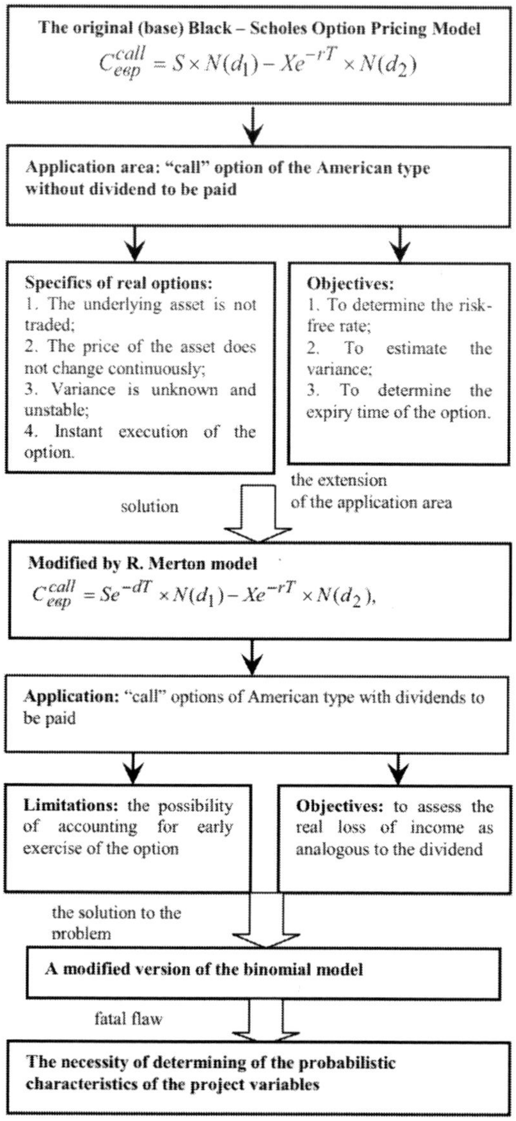

Fig.1. Algorithm of the real options evaluation methods selection.

According to the authors, this fundamental limitation can be eliminated as follows.

1. It must be emphasized as a well-known postulate that the purpose of the analysis and synthesis of investment projects is not a precise calculation of the optimum parameters and maximization of the known criteria of effectiveness, but the choice of the most rational conditions and trajectory of the project.

2. The correct method of expert estimation of the relative probability of positive or negative outcome of the previous

administrative decision in the node of the decision tree must be used.

However, the issue of real option valuation is not only about of how much money we earn, it is also about of the use of the method from positions of real opportunities to receive additional positive net cash flows in principle. To answer on this question it is enough to consider two alternatives: optimistic and pessimistic.

Thus, the managerial task of the real options valuation comes down to the valuation of the relative probability of the outcomes of a binomial option. Therefore, a binomial model of valuation of the real option is taken as the bases of the proposed for development MROV-method (MROV–modified real option valuation method).

The proposed method is essentially a modification of the known ROV method, with correct replacement of "accurate" assessment of probability of branches on the decision tree for the expert assessment of the comparative probabilities of optimistic and pessimistic outcomes of previous management decisions.

The meaning is that according to the binomial model, at each stage of the project (at each node a binomial decision tree) there are two possible ways of its development (Fig.2).
1. Optimistic variant: value of the criterion is higher (better) than baseline value $X_1 > X_o$ with a probability p_1;
2. Pessimistic variant: value of the criterion is lower (worse) than baseline value $X_2 < X_o$ with probability p_2.

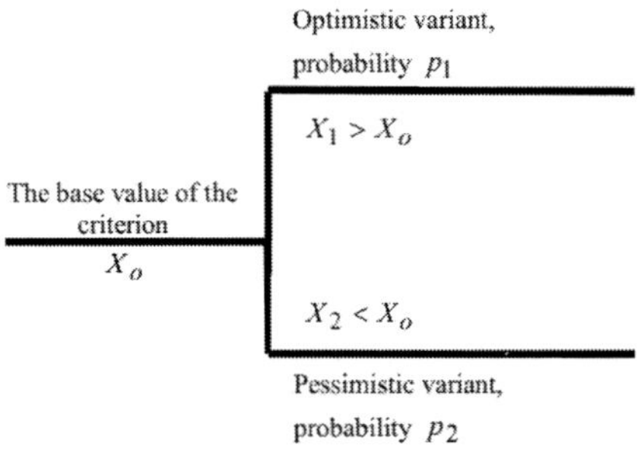

Fig. 2. The ratio of probabilities of alternatives.

It is important to note, that while judging, the sum of the probabilities p_1 and p_2 may not be equal to 1, since other options are possible when the value can be in the range between X_1 and X_2.

It should be noted that the proposed approach is more correct than the classical ROV-method. The fact that the binomial model of the real options valuation is based on constructing of a decision tree, in each node of which there are two types of events: the development of the project in an optimistic or pessimistic forecast. Moreover, the degree of optimism and pessimism is determined by the project managers of the enterprise that realize the project. The absolute probability of each of the specific forecasts will be equal to zero, therefore, the assessment of probabilities can be correct only in the sense of the answer to the question, how much more likely an optimistic situation, relatively pessimistic, or Vice versa. So the unranked absolute probability of the ROV-method is replaced by the estimated comparative probability of the MROV-method.

The development of the project on a particular forecast depends on a number of criteria of the environment in which the project is located. Thus, the problem of estimating of the real options value becomes a multi-criteria problem which can be solved by using one of the methods of expert estimation, specially designed for such tasks. The essence of these methods is to build a hierarchical structure that allows to disaggregate the overall task into particular tasks.

One of the most reasonable and practically relevant methods of solving such problems is the Analytic Hierarchy Process (AHP). This method offers to split the global problem into a number of subtasks, allowing the same type of alternatives and the further processing of decision makers judgments from paired comparisons. Thus obtained judgments become the ratings in the scale of relations and correspond to the so-called hard, quantitative estimates [21,22].

The advantages of the method (AHP) allow to select it as the technological basis for MROV-method, which can be carried out correctly, transparently and comprehended.

Algorithm of MROV-method implementation mean the use of AHR to determine the ratio of the relative probabilities of the project realization within optimistic and pessimistic economic forecast in binomial model of the optional investment project. Meanwhile the option realization is considered to be the optimistic scenario, and the missing of the option realization is considered to be a pessimistic scenario.

Omitting the details of the method applying within the proposed approach, we note that after the processing of the data, obtained from experts, by the procedure of synthesis of multiple judgments, quantitative estimates are determined. These estimates refer to the priorities of the analyzed criteria influence on the outcome of the project realization at any stage, and also the priorities of each of the alternatives relative to the criteria.

Evaluation is conducted at each node of the binomial decision tree, in thus case the hierarchical structure of the criteria may be individual for each node and can have any number of levels and criteria.

On the basis of the obtained results ranking values of the probabilities of the project development under any scenario in each node of the binomial tree are determined. The quantifying model of the relative values of the probabilities of the project development within optimistic or pessimistic scenario is a linear additive model, which is represented by formula (1):

$$p_i = a_1 \cdot x_{i1} + a_2 \cdot x_{i2} + a_3 \cdot x_{i3} + a_4 \cdot x_{i4} + \dots , \qquad (1)$$

where p – ranking value of the probability of the project development on the *i*-th scenario (optimistic or pessimistic);
a_1, a_2, a_3, a_4 - values of the priorities of criteria influence on the outcome of the project realization on the evaluated phase;
$x_{i1}, x_{i2}, x_{i3}, x_{i4}$ – values of the priorities of the *i*-th alternative relative to the relevant criteria.

As a result of the using of AHP, based on the formula (1), we obtain a ranking value of the relative probabilities of project realization according to the optimistic scenario (p_1) and pessimistic scenario (p_2).

At the final stage the decision about the feasibility of the option investment project realization is made. This decision is based on the net present value calculation, which can be obtained by sequential discounting of the weighted values of the project cash flow. The ranked values of the probabilities obtained within AHR are used as the weights there.

Schematic representation of the described algorithm of the MROV-method implementation is presented in Fig.3.

4 Conclusion (findings)

The proposed approach can correctly implement methodological ideas of the optional approach to the development and analysis of investment projects, taking into account not only the uncertainty associated with risks, but also embedded in the project the options of its realization due to the uncertainty of the positive opportunities of the project development.

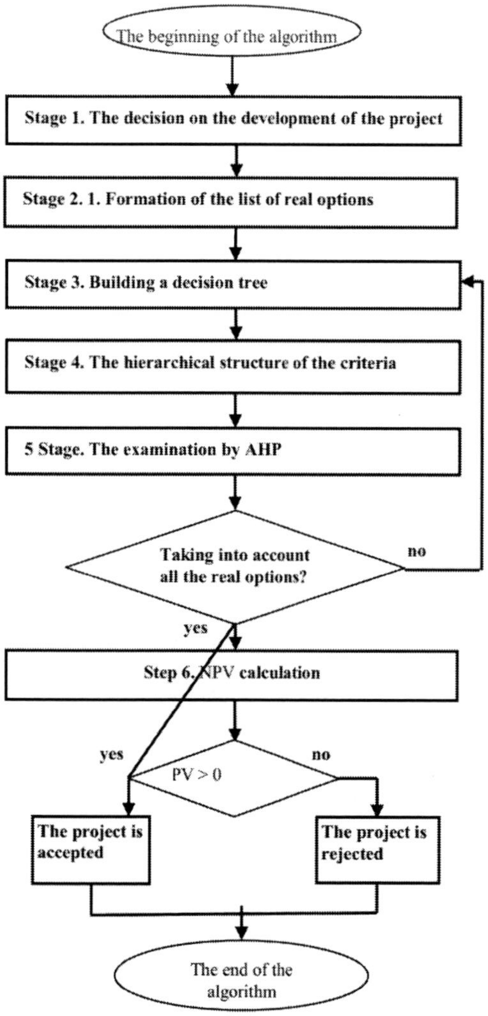

Fig. 3. Algorithm of the project valuation by MROV-method.

References

1. S. Vine, *Options: Full course for professionals* (Alpina Pablisher, Moscow, 2015)
2. V.A. Galanov, *Derivatives* (Infra-M, Moscow, 2014)
3. A. Damodaran, *Investment valuation: Tools and techniques for the evaluation of any assets* (Alpina Pablisher, Moscow, 2015)
4. M.A. Limitovskiy, *Investment projects and real options in developing markets* (Yurayt, Moscow, 2010)
5. I.V. Lipsits, V.V. Kossov, *Economic analysis of real investments* (Magistr, Moscow, 2010)
6. N. Bruslanova, Financial director, **7-8**, 13-20 (2004)
7. A.V. Bukhvalov, Russian management journal, **1**, 3-32 (2004)
8. T. Kouplend, T. Ostrovski, Vestnik McKinsey, **2**, 113-132 (2002)
9. P. Boyle, Journal of Financial and Quantitative Analysis, **23(1)**, 1-12 (1988)
10. T. Copeland, Ph. Keenan, The McKinsey Quarterly, **3**, 128-141 (1998)
11. A. Borison, *Proc. 7th Annual International Conference on Real Options* (2003)
12. R. Geske, Journal of Financial Economics, **7**, 375-380 (1979)
13. B. Kamrad, P. Ritchken, Management Science, **37(12)**, 1640-1652 (1991)
14. R.M. Stulz, Journal of Financial Economics, **10(2)**, 161-185 (1982)
15. L. Trigeorgis, *Real Options - Managerial Flexibility and Strategy in Resource Allocation: The MIT Press* (1996)
16. R. Breyli, S. Mayers, *Principles of corporate Finance* (ZAO "Olympus-Business", Moscow, 2010)
17. T. Copeland, Ph. Keenan, The McKinsey Quarterly, **2**, 38-49 (1998)
18. *A guide to body of knowledge project management (PMBOK guide)* (2010)
19. L.A. Baev, *Framework for the analysis of investment projects* (Chelyabinsk, 2007)
20. G. Birman, S. Shmidt, *Investments. Economic analysis of investment projects* (YuNITI-Dana, Moscow, 2003)
21. T. Saati, *The decision-making. Method of analysis of hierarchies* (Radio and communication, Moscow, 1993)
22. E.V. Shikin, A.G. Chkhartishvili, *Mathematical methods and models in management* (Delo, Moscow, 2002)

Modernisation of production as a factor of influence on economic stability of the industrial enterprise

N.Yu. Varkova[1,*]

[1]South Ural State University, Chelyabinsk, Russia

Abstract. The present article considers key problems of modernising production process at industrial enterprises, also referred to as reengineering, in modern conditions of Russian economics. Having analysed factors that affect operation of business, the author presents the model of interdependence between modernisation of production and sustainable development of the enterprise. Based upon analysis of Russian and foreign sources there were formulated fundamentals of the modernisation process and presented an algorithmic mechanism of enabling modernisation at the industrial enterprise. The said mechanism takes into account main aspects of sustainable development of the industrial enterprise and evaluation parameters as to the modernisation efficiency.

1 Introduction

In modern ever-changing conditions when competition is getting stronger and stronger, and there act environmental factors that bring influence on the enterprise activities, coping with different types of crisis, inhibiting bankruptcy, and also planning measures intended to achieve a steady sustainable development are seen as a recently established business priority of industrial enterprises. In this context it has become urgent to design a system of modernisation of industrial production understood as a process comprising investment, innovation and other constituents related to the sustainable development process.

In economic science the notion of modernisation represents evolution and renovation in technological and social development of the society, both in global scale and at the level of a particular country, branch or business enterprise [1]. This process involves enhancing properties of the production system and its separate components, using investment resources, and selecting criteria of economic efficiency, which serve as a ground for making decisions on production processes improvements. Thus, we can detect interrelations and mutual influences between the production modernisation process and economic stability of the business enterprise.

According to Rosstat (Federal State Statistics Service), as of today the average age of machinery and equipment in Russia is about 11.2 years [2]. This situation is mainly accounted for imports. For example, the share of imports in machine-tool industry is more than 90%, in heavy machine-building – 60–80%, in electronics manufacturing industry – about 80–90%. It implies that the ability to maintain the technological level of Russian economics to a great extent depends on imported services (servicing and repair) and on substitutions from abroad due to moral and physical depreciation of equipment and associate technologies [3]. A number of countries stiffened sanctions against Russia after it joined Crimea in March 2014 and in connection with the political crisis in Ukraine. As a result, domestic industries are facing enormous challenges, which include the risk to lose an access to western technologies and cooperation with foreign investors, and to deal with limitations on purchasing import equipment and complementary parts. The Government of the Russian Federation adopted regulations aimed to reject imported equipment and to introduce the policy of import substitution. Presently there has been composed a list of goods to be purchased only from manufacturers of Russia, Byelorussia and Kazakhstan [3].

In order to cope with these challenges, it is necessary to build up systems designed to facilitate modernisation, project and technological activity of the industrial enterprise, as well as the system aimed to manage scientific and technological reproductive performance, innovational, educational, and scientific processes that take place inside the enterprise and in association with other market participants [4, 5].

2 Theoretical framework and literature review

Issues connected with managing modernisation of production at the industrial enterprise and related interactions have been examined by many Russian academic economists such as T.Yu. Bazarov, V.P. Barancheev, V.V. Glukhov, V.E. Dementiev, S.B. Korobko, V.K. Lozenko, E.V. Popov, A.M. Permyakov, L.P. Kleeva, G.B. Kleine, V.A. Balukova, Yu.V.

* Corresponding author: Varkovan80@mail.ru

Levitsky, I.I. Mazur, I.A. Sadchikov, V.D. Shapiro and other scientists.

Questions of reengineering, production modernisation and a business enterprise are examined in works of such western authors as I. Ansoff, S. Beer, K. Blanchard, T. Vaghorn, P. Drucker, S. Klein, N. Rosenberg, D. Sahal, R. Foster, M. Hammer, J. Champy and others.

Sustainable development and stability of economic systems are subject of theoretic and practical research conducted and presented in scientific papers of such foreign and Russian scholars as J.L. Laft, M. Robson, J. Schumpeter, J. Haywood, D.R. Ageeva, A.V. Bezgodov, O.V. Bratkov, L.V. Volkova, D.S. Kondaurova, A.G. Koryakov, O.V. Mikhalev, V.M. Tumin, A.D. Sheremet, E.M. Korotkov, T. Sukhorukov.

Analysis of prominent papers on the given research issue shows that lately there has been developed a set of approaches to managing the production modernisation at the industrial enterprise, and there has been specified its impact on economic stability and development of production systems without recessions. It should be noted that these matters are still conventionally considered in the frame of the approach based on designing such a system of measures that shall enable an industrial enterprise to surmount the crisis, while modernisation of production must prevent recessions, rather than facilitate overcoming them when they already happen. Therefore it is essential to specify the very notion of modernisation of production and its influence on the economic stability of the industrial enterprise.

The object of the research: reengineering (modernisation) of the industrial enterprise and the extent of its influence on the enterprise.

The goal of the article is to develop the model that will reveal the influence of reengineering production on economic stability of the industrial enterprise.

3 Materials and methods

A. Description of methods

At the present stage of economic evolution it is common to determine if the industrial enterprise has achieved stability or not. It is defined by positive dynamics of quantitative and qualitative transformations of elements and business processes at different stages of its life cycle [8,9]. Such transformations include renovation of production facilities, launching a new product, modernising and reengineering the technological base. Accordingly, the efficiency of sustainable development in many ways is defined by the level of investment and innovation potential of the enterprise, serving as the key factor [10]. Any industrial enterprise is an open dynamic system, since it constantly interacts with external environment, which is susceptible to significant fluctuations. From external environment the industrial enterprise receives resources in the form of raw materials and supplies, fuel and energy, capital and information, labour force, equipment and other assets that are becoming part of its internal environment. Partly resources are processed, converted into products and services, which return to external environment [6,7]. In the course of implementing modernisation projects aimed to maintain financial stability of the industrial enterprise, there arises a necessity to consider its risk profile in order to ensure fast response to external and internal threats.

It is essential that managers of the enterprise apply a certain mechanism designed to facilitate modernisation. The principal element of the said mechanism is to determine the direction of modernisation, which would enable sustainable development. This mechanism represents the process of implementing the production modernisation as a sequence of specific stages. Each stage involves evaluation and analysis of certain economic activity parameters of the industrial enterprise.

The designed mechanism allows to achieve a prescribed pace of sustainable development on the basis of an integrated approach to modernisation of the industrial enterprise.

In view of sustainable development being an integrated complex with a definite structure and also a process influenced by a great number of factors, the author offers a special algorithm that will aid to assess the level of the development stability based on using modernisation methods. The offered algorithm can include a sequence of the following principal stages:

1) Analysis and forecasting stage. It involves identification and analysis of principal technological aspects of the business operation. This stage presents conducting a technological audit and analysis of production facilities. Energy efficiency, ecological parameters, flexibility, automation of production, as well as parameters of launching new products are to be assessed at this stage. Key factors of external environment of the enterprise are subjected to analysis.

2) Estimating and forecasting stage. At this stage indices of sustainable development of the industrial enterprise are built up. Top managers shall estimate efficiency of core activities, analyse the revenue structure and assets of the enterprise, structure of borrowed funds, net profit, and to specify the return on assets.

3) Formation stage. This stage of the mechanism involves identifying such directions of modernisation that will provide for sustainable development of the enterprise.

4) Control and adjusting stage. It implies comparison of planned results against standard indicators, as well as launching feedback mechanism.

B. Graphic model of implementing modernisation

The model of the mechanism aimed to enable modernisation of the industrial enterprise presents a system of main criteria that affect its economic stability. Fundamental provisions of modernisation are as follows. External environment of the industrial enterprise is characterised by a great number of highly volatile parameters, and in such conditions the enterprise has to respond really fast. It implies objective complexity of achieving sustainable development [11].

Analysis of external environment of the enterprise should include investigation of such key factors (factors of direct impact) as politics and economics influences, impact made by competitors, product consumers, suppliers, resources and materials, as well as sociocultural factors. When making an analysis of internal environment of the enterprise, it is essential to distinguish three groups of principal influencing aspects, namely, technical (production and technological parameters, indicators of mastering new products, of automation, ecological properties), economic (investments, ratio of spending resources), organisational (attributes of the enterprise staff, management methods and styles, their efficiency).

The model of mutual influences of modernisation and sustainable development is graphically presented in Fig.1.

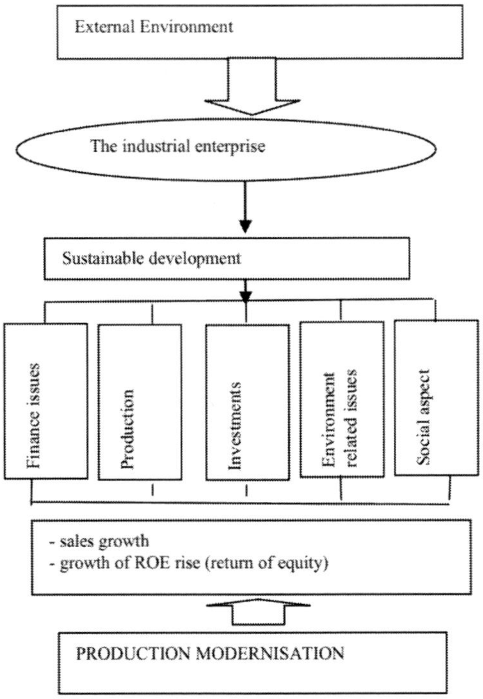

Fig. 1. Graphic Model of mutual influences of sustainable development and production modernisation.

Sustainable development can be considered as a process specified by enhanced economic efficiency indices, technological efficiency and quality of production. To improve indices of economic efficiency and technical excellence is possible by means of preparing a strategic plan of complex modernisation. The continuity of the modernisation process is to be provided by relevant elements of the organisational structure of management implemented at the industrial enterprise.

Below we present the model of the mechanism aimed to enable modernisation at the industrial enterprise. The present model testifies that the functioning of the industrial enterprise depends upon an impact of factors of external and internal environment. When analysing the enterprise operations, it is also necessary to bear in mind the impact made by basic indices of sustainable development and effectiveness criteria applied to production modernisation [12].

The presented mechanism enables top managers of the enterprise to consider all aggressive and most disturbing influences when performing the process of complex modernisation of production, as well as to analyse how these influences are reflected in sustainable development of the industrial enterprise [13,14].

C. Economic and Mathematical Model

One of principal factors of the effective functioning of any enterprise in competitive conditions is how timely and adequately it reacts to changes of external environment that create a situation, in which decisions are made under risk. In this context evaluation of possible risks and risk management becomes a key issue of strategic management, and, therefore, of corporate planning.

Hence, risk management in performing the process of production modernisation is one of core directions of the corporate planning, since in this very aspect there arises a risk of making non-optimal solutions [17,18].

We offer an approach to evaluation of strategic risk based on applying such an integral index as a risk-benefit ratio. It implies that the risk level is measured by correlation of the anticipated profit and anticipated loss when comparing different variants of modernisation strategies (1).

$$Ki = Zi/Ri, \quad (1)$$

where Ki is a risk-benefit ratio of the strategy variant i; Zi is an anticipated profit of the strategy variant i; Ri is an anticipated loss of the strategy variant i. The risk-benefit ratio Ki demonstrates what income falls within one rouble of loss.

4 Conclusions

By this means, within the framework of the presented concept of sustainable development, operation of the industrial enterprise ceases to be an uncontrolled, spontaneous process, and becomes self-organised. The control mechanism has complex nature, it is designed to achieve new strategic goals and to facilitate stability at all life-cycle stages.

The interrelation between the process of production modernisation and sustainable development presented in the article can be also viewed at the level of the indices that characterise the given processes. Indicators of modernisation efficiency include: net profit margin; turnover rate of the total assets; equity capital; productivity of labour; assets value; capital productivity and equipment productivity ratio; cost of production of sold goods [5,19].

Indicators of sustainable development of the enterprise include: the coefficient of stable economic growth; profitability of sales; turnover rate of productive assets; profitability of production assets; the financial lever effect; the equity capital level; profitability of production assets; return on assets against the net profit [5,20].

The article presents the model of correlation between sustainable development of the industrial enterprise and modernisation of production. The author offers the mechanism aimed to enable modernisation of production at the industrial enterprise. Applying the given mechanism, managers of the enterprise can make a management decision on selecting a direction of modernisation, and, accordingly, on selecting the strategy required to implement it and to achieve sustainable development of the industrial enterprise.

The work was supported by Act 211 Government of the Russian Federation, contract № 02.A03.21.0011

References

1. N.A. Dubrovina, Finance and Credits, **3**, 21 (2014)
2. I.V. Krasnopevtseva, Economics and Management in Machine-Building Industry, **6**, 21 (2012)
3. *Rosstat: information and analytical materials. Electronic data.* Moscow (2016). http://www.gks.ru/wps/wcm/connect/rosstat_main/rosstat/ru/statistics/science_and_innovations/science/
4. A.O. Kalabaeva, Ekonominfo, **17**, 9-13 (2012)
5. K.P. Lichko, M.A. Romanyuk, E.I. Krivomaz, Herald of Moscow State Academy of Business Administration. Series: Economics, **2(22)**, 121-132 (2013)
6. R.A. Fatkhutdinov, *Production Engineering* (INFRA-M, Moscow, 2000)
7. R.A. Fatkhutdinov, *Management of the Organisation Competitiveness* (EKSMO, Moscow, 2004)
8. T.A. Khudyakova, A.V. Shmidt, *Proc. of The 26th International Business Information Management Association Conference*, 1617-1625 (2015)
9. T.A. Khudyakova, A.V. Shmidt, *Proc. of The 26th International Business Information Management Association Conference*, 1607-1611 (2015)
10. T.A. Khudyakova, *Proc. of the International Congress on Interdisciplinary Behavior and Social Science*, 292-294 (2015)
11. T.A. Khudyakova, *Proc. of the International Congress on Interdisciplinary Behavior and Social Science*, 295-297 (2015)
12. T.A. Khudyakova, A.V. Shmidt, *Proc. of The 26th International Business Information Management Association Conference*, 1626-1634 (2015)
13. T.A. Khudyakova, Actual problems of economics, **12**, 444-450 (2015)
14. T.A. Khudyakova, A.V. Shmidt, *Proc. of The 26th International Business Information Management Association Conference*, 1612-1616 (2015)
15. D.R. Kondaurova, Proc. Economics, Management, Finance: Materials of the II International Scintific Conference, 130-132 (2012)
16. N.V. Rodionova, *Crisis Management* (Publishing House UNITY, Moscow, 2012)
17. Yu.I. Ryzhikov, *Simulation Modelling. Theory and Technologies* (Publishing House "Korona Print", St.-Petersburg, 2004)
18. V. Zhukov, Finances Business, **8**, 18-21 (1996)
19. A.G. Aganbegyan, Money and Credits, **3**, **4**, (2010)
20. A.S. Pelikh, *Organisation of Entrepreneurial Activity* (Publishing and Scintific Centre MarT, Rostov-on-Don, 2015).

Construction and assembly works planning: optimization or search for solutions

E. Gusev[1,*]

[1]South Ural State University, Chelyabinsk, Russia

Abstract. The article analyzes the instability causes of construction and assembly works (CAW) schedules which leads to the construction works and logistics instability. The "nature" of multi-variant management solutions is considered. The contradiction of requirements to the basic organizational and technological model (OTM) to solve scheduling problems of CAW and construction works is revealed. The basic "an object construction technology model" as "an object techno-logical dependencies models" is considered. A new approach is proposed to solve the problem of CAW scheduling, which includes two-step managerial decisions development at the CAW planning stage and construction works organization stage. At the first stage, the problem is reduced to finding a solution (and not to determining an optimal variant). At the second stage, the problem is solved through optimization.

1 Introduction

The period of 1940–70th saw a strong penetration into the economy and managing the economic-mathematical modeling (EMM) and a new direction in mathematics - optimization problems solving. During this period, mathematicians made a significant progress in modeling various sectors of the economy management, including construction, developed methods for optimizing management decisions, methods for decision making under the uncertainty and incomplete information conditions, etc. Anyway, a scientific research or scientific publication without economic and mathematical modeling and optimization was difficult to imagine [1, 2, 3].

Such optimization techniques as "Calibration" and "Smoothing" are widely used in the construction industry. To re-solve economic and administrative problems, methods of linear and dynamic programming, simulation modelling, multi-criteria optimization and many others are applied. This period can be labelled as an euphoria of universal optimization in economy and economy management.

However, in 1970, in the book "Mathematics puts the experiment" N. Moiseev questioned the optimization method in solving economic and management problems. The quintessence of the book is the phrase emphasizing "the significance" for using optimization methods: "and rolled the "wheel" of optimization along one way, and the economy – along the other one" [4].

Many systems and subsystems of management in construction were developed and a large number of scientific articles and various publications were issued [5, 6, 7, 8, 9, 10, 11, 12, 13, 14]. In the former Soviet Union, the most popular and completed, in the opinion of the author, was the system of "A-Plan", which was developed based on the construction trust "Tallinstroy", headed by L. Golub [15]. This system was awarded with a gold medal of the USSR Economic Achievements Exhibition. However, after about ten years of implementation and operation, the main ideologist of the system admitted that its results were more frustrating than positive.

In any case, various objectives analysis for construction and assembly works (CAW) scheduling and their practical application shows that the optimization plans do not lead to the desired result, which was expected by the developers and users. To evaluate the causes of inefficiency in solving CAW scheduling problems, the scheme (Fig. 1) is to be examined.

2 The basic sequence of a balanced caw schedule development

As a basis for CAW planning (Stage 1), organizational and technological models (OTM) in the form of linear[16], network diagrams, and sequence diagrams are used. In this case, the selected model can be optimized or non-optimized. Therefore, the optimal criterion in the scheme is highlighted with a dotted rectangle.

The second stage of the "Scheduling" module is solved only through optimization.

Management theory considers any process of managing an entity as a uniformity for objectives selection, their achievement programs (or plans) and implementation mechanisms that compensate for all kinds of interference. With regard to a construction company, developing the objectives that determine the stability is to make planning decisions (the system of technical and economic indicators), which would ensure its viability. In fact, the planning decisions are a doctrine that is

* Corresponding author: gusevev@susu.ru

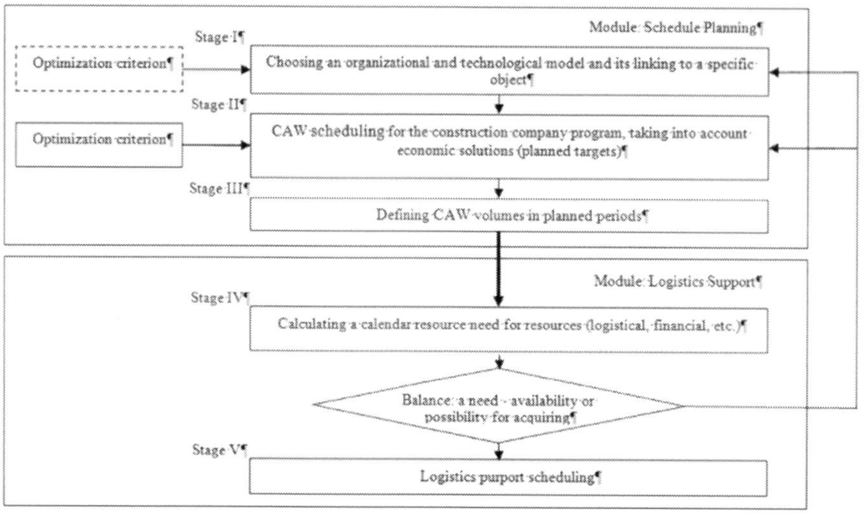

Fig. 1. The basic sequence of a balanced CAW schedule development for a construction enterprise.

determined by external conditions, but it is not external to the construction company itself. Changes of external conditions entail the mandatory change of the doctrine [4].

For developing a CAW schedule, the first and second stages of the "Scheduling" module are assumed to be multi-variable in nature and, therefore, the solution of the problems at these stages requires optimization.

The solution of optimization problems is known to be related to a multiple choice of possible ways to achieve the targets (doctrine), followed by finding a extremum.

For CAW scheduling, such a multiple choice is "provided" in two ways:

1) the top managers of construction companies are considered to have a variety of options for CAW time planning;

2) the number of restrictions, requirements and conditions are purposefully reduced to facilitate the economical and mathematical model and reduce the complexity of problem solving [17,18].

The first way involves obtaining an optimal plan for project (s) construction, that is a construction schedule, and this schedule should "adapt" the entire infrastructure (finance, logistics, human resources, etc.). Operating a well-known industrial installation "Accord" has shown the futility of this approach.

As to the second way, let's consider the diagram shown in Fig. 2. Point A is the level of the current conditions of a construction company (the socio-economic system) and point B is technical and economic parameters (doctrine) to be achieved in a given (fixed) period.

Construction companies operate under a variety of restrictions, terms and conditions. Therefore, the increase in-various limitations in the process of CAW scheduling leads to a decrease in the area of management decision-making. There may be three options (Fig. 2).

Option 1: The combination of various types of restrictions leads to inability to achieve the planned technical and economic indicators (TEI). In this case, the decision-maker (DM) has to make a difficult choice:

a) to adjust TEI to the worst compared to a planned one;

b) to take extra costs to ensure the planned TEI achievement.

Option 2. The author does not exclude the possibility of an area formation for management decisions making during CAW scheduling, that is an area of several ways of achieving the set goals (planned TEI). In this case, this problem is solved using the optimization methods. It is imperative to know these ways instead of creating multiple choice by artificial reduction of input data quantity, purposeful reduction of constraints quantity, as input of "additional" constraints into economical and mathematical model allegedly "leads to the optimization effect partial loss" [17, 18].

Option 3. A construction company is set under various rigid constraints - financial, capacity, logistical, time, technological, organizational, etc. This set of constraints ultimately determines the CAW scheduling specific parameters and distributing the amount of construction works in time. Current practice analysis shows that the task of a construction company manager (management) is not to choose the best option on a certain criterion, but to find a way leading to the planned decisions (assignments, TEI) implementation, to anticipate the factors that may affect the program implementation, and take them into account at the planning stage.

Let's consider another aspect. The scheme in Fig. 1 is essentially the second component of management theory – developing a program (plan) for goals achievement.

In the scheme in Fig. 1 a special attention should be paid to the next important point. Among all the CAW scheduling stages, it is only at the fifth stage in the module "logistics support" when the documents developed have legal force (contracts with different enterprises and organizations) are developed.

The optimal CAW scheduled plan developed and balanced with the logistics provision is a design

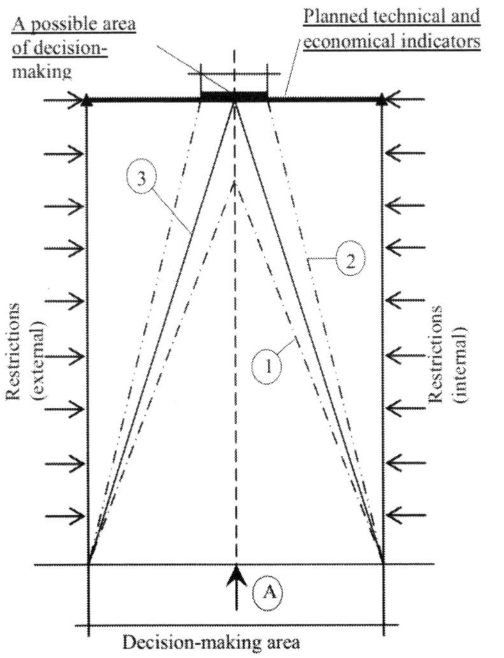

Fig. 2. Schematic diagram of the dependence of the decision-making (multi-variant) area on the number of restrictions considered.

decision. During the plan implementation (the third component of the management theory), as a result of various external and internal factors, it requires a corresponding correction (adaptation). But, the logistics support plan "works" for the initial schedule of construction, and changes in contractual obligations lead to costs rise, increased construction costs and the imbalance of administrative decisions.

3 The structure and interrelation of the schedule elements

The reason for this discrepancy is the following. The planned construction schedule of an entity (complex of entities) is a source of information for solving many logistical problems. Therefore, the requirements to the information must be strictly implemented - it should be stable, unchangeable. At the same time, construction schedules are the organizational and technological models, primarily reflecting organizational solutions and technological stages of inter-related works. As a result, opposite requirements for the schedule are presented, both objectively and dialectically legitimate. On the one hand, the schedules must be resistant to various disturbing influences, stable, objective, since the information obtained from them is used to solve the exact calculation problems of works distribution in time, logistics, the construction company potential balances and planned indicators checking, the subcontractors work, etc.

On the other hand, the models are required to be easily adjusted. This requirement belongs to the general

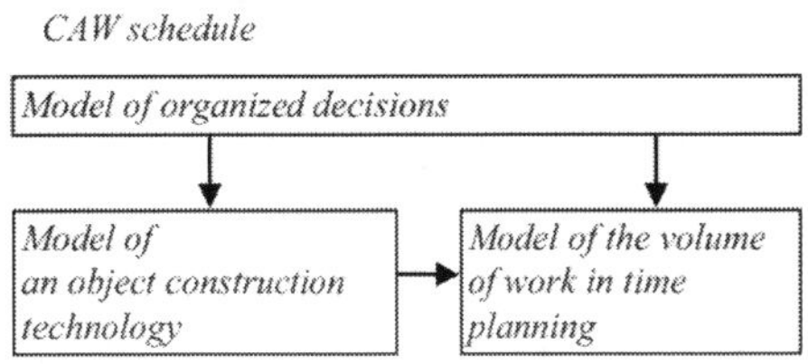

Fig. 3. The structure and interrelation of the schedule elements.

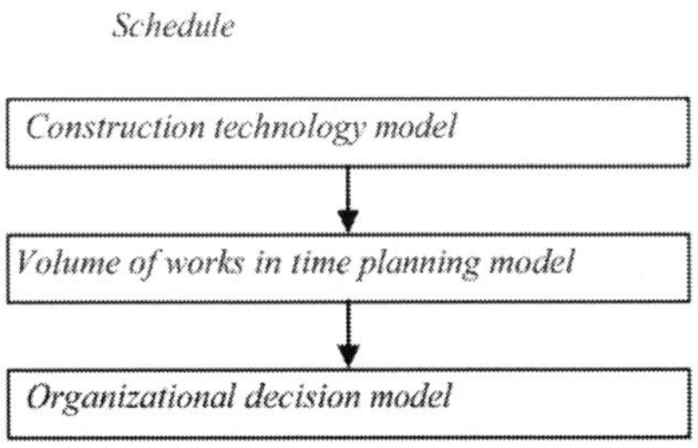

Fig. 4. New structural model interconnection of the BAW scheduling.

management theory, as all the external and internal influencing factors cannot be taken into account in making management decisions.

The revealed objective contradiction of requirements to the schedule allows to rethink its "internal" content. To fully understand the causes of CAW schedules instability the concept of "three models" should be considered [19]. Its essence lies in the fact that the schedule cannot be considered as a homogeneous "monolith", described in time, volume and resources (such as power) parameters. It can be represented as a synthesis of three interrelated models. These models reflect an object construction technology, construction organization and the volume of work in time planning (distribution), which allows to obtain a set of CAW scheduling parameters (Fig. 3).

The decision organization model includes a wide range of activities related to the building production organization. The main objective of the model is to define time parameters.

The work volume planning model generates a decision on the objects of working time distribution. The main goal of this model is to develop solutions that determine what kinds of work, to what extent and on what items should be made in each specific period of time taking into account various types of restrictions and requirements.

Construction technology model is rather weakly expressed in the schedule. The main goal of this model is the buildings and structures designing technology.

According to the scheduling rules for construction and assembly works (BAW), needed for the construction project, the organizational model holds the priority position in this interconnection. After organizational decisions, based on the technological sequence of interrelated works, the volume of work in time is calculated.

With such structural interconnection of "internal" models planning the volume of work is like "a secondary stage" that is after the time parameters modelling we determine those volumes of works for the project that should be done in one or another planning period.

Considering these models in terms of their stability and resistance to various factors, one can draw the following conclusion.

The construction technology is quite limited in choice. Basically, there are two technological schemes of building construction: from the foundation to the roof; from the roof to the foundation. Therefore, the technology model is the most stable (here we assume that the technological stages of interrelated works is a component part in a construction technology) [20].

The organizational model is the most unstable as under the influence of various factors during the construction period, organizational decisions are changing repeatedly. This leads to some changes in the planning model. Organizational decisions instability is objective and this leads to instability in the volume of work in the logistical sup-port. Generated "internal «interconnection in the schedule has leads to a shift in understanding the scheduling. The purpose is not to allocate resources, such as power, for the works and facilities for the construction organization immediately but to determine the volume of work which lead to achieving engineering-and-economical performance taking into account construction organization capacity, various restrictions and technological requirements. For this approach construction technology model should play the major role as a base model for the management decisions development concerning BAW planning and construction organization (see Fig. 4).

Taking into consideration such interconnection, organizational decisions are made only when volumes of work and the conditions at the construction site are known. In this case, the volume parameters are more stable and less affected by various factors. This approach allows to divide scheduling BAW task into two stages. At the first stage the types and volumes of work to fulfill the planned-economic tasks (decisions) for a planning period are determined, taking into account the output, various resources endowments, construction technology requirements, etc. The obtained results are additional restrictions on the second stage, where organizational decisions are specified.

4 A project technological dependencies model

In order to implement a two-stage scheduling BAW, a basic model should be made based on the construction technology model which is the most stable to the influence of various factors.

Currently the term "construction technology" is associated with the manufacturing stages of works in the project and their interconnection. However, there is no quantitative relationship between interrelated works.

The term "construction technology" can be defined as follows: "it is a qualitative and qualitative assessment for the technological links between the works determining the work planning possibility and industrial organization depending on the results of the previous ones". In this case, the point of the construction technology modelling is to establish technological links between various works and to determine a minimum volume of the previous works that give the possibility to plan a technologically interconnected volume of the following stage.

Such model describing the technological links between works stages and their quantitative assessment in the be-ginning and after their completion of work is designed and presented in Fig. 5 (T_{dir} – directory (normative, contractual) duration of the objection struction; t_j – duration of time area; t_{jn}; t_{jo} – technologically possible beginning and completion of work; $t^{n}_{j+1,j}$, $t^{o}_{j+1,j}$ – minimum technological delays during the beginning and completion of the following work (j+1) from the point of the beginning and completion of the previous work (j); t^{kp}_{j} – criticality point after which execution of the volume of work demands maximum intensity and can lead to disruption of work T_{dir}). It is important that there are no organizational decisions when calculating quantitative assessments. This increases the model stability when planning BAW and construction organization. As technological dependences for the initial and final works determine the technological stages of the project works, we will term it as a project technological dependencies model (OTDM).

OTDM calculation is reduced to time assessments for technological dependencies for the initial (not earlier than the initial) and the final (not earlier than the final) works; time area for each stage (as opposed to the duration of the work according to OTM); criticality points of stages. "Not earlier than the initial one" means that follow-up work (j + 1) cannot start technologically if functional minimum volume of work is not done (planned) on the preceding j; "not earlier than the final one" means that follow-up work (j + 1) technologically may not end earlier if the minimum volume of work technologically essential after the previous work (j) is not be done (planned).

The first stage of scheduling BAW is connected to the volume of work time distribution (for the planning period). Based on the task set to the building organization –fulfillment of planned technical and economic tasks (decisions) – work method statement which should be balanced with planned tasks (decisions) and logistic provision plan are to be developed.

The construction company does the projects within the tough limits due to many different restrictions of social, organizational, technological, technical, logistical nature and other requirements [21]. Such variety of restrictions and requirements questions the fact that the construction company leaders have a number of decisions about scheduling BAW. The author in his further reasoning is based on the following hypothesis: "Taking into account sufficiently large number of

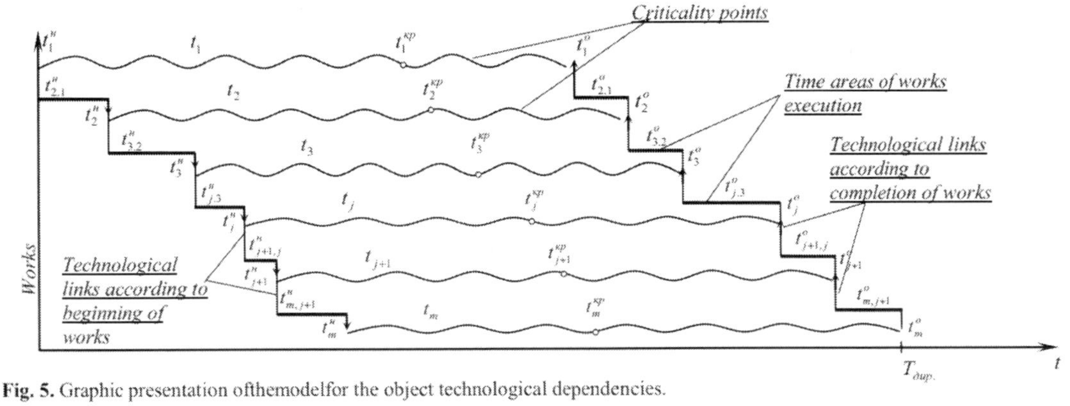

Fig. 5. Graphic presentation ofthemodelfor the object technological dependencies.

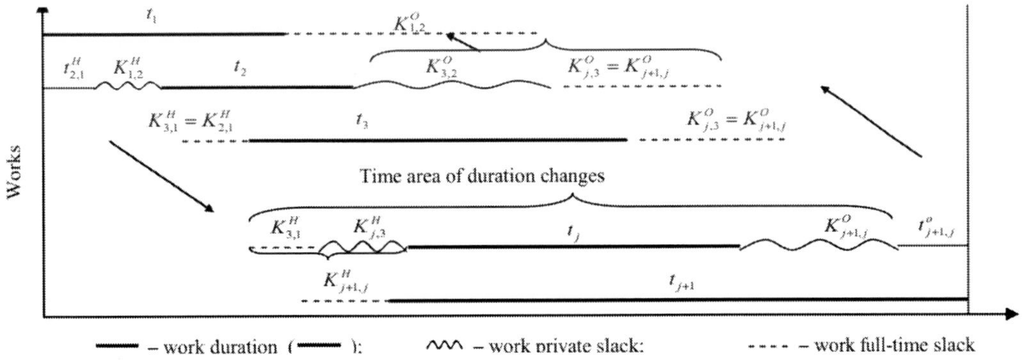

Fig. 6. Line graph with slacks by the beginning and completion of works.

restrictions, the task comes to finding ways to achieve the goal (scheduled TEI), or to the only one alternative". Therefore, it is necessary not to reduce the number of restrictions but gradually, in the course of evaluation to fill with them, thereby the reliability and stability of the results are increased. Therefore, the solution of this problem is aimed not to choose the best option according to the specified criterion but to find the way (or ways) to achieve goals under the given conditions that is the problem content is to find such a set of project works in quantity and financial terms which ensures fulfilling the planned economic tasks (decisions) by a construction company. The solution is to find a set of work volumes which construction company should perform in each planning period of the planning year to ensure their own viability.

The task has a twofold purpose. On the one hand, the task is to make a set of works of the planned technical and economic indicators, on the other hand, the task is to check the validity of these indicators that is whether the construction process can ensure these indicators taking into account various restrictions and requirements. Therefore, the process of solving the problem is closely linked to the work of planning, construction process, logistic and other functional de-apartments in a construction company.

The task of the current (operational) scheduling BAW for the construction company at the first stage which is based on the unconditional fulfillment of the planned tasks (decisions), and technical and economic indicators (set by appropriate planning departments) can be defined as follows: the production program of a construction company is formed from «n» projects. Planning tasks are defined with the breakdown by performers and customers.

It is required to determine types and volume of work for each scheduled period of the year ($V^{ns}_{i,j,\tau}$) which ensures the annual goal achievement (with breakdown by planning periods) by a construction company.

During the time distribution of the works volume in addition to the goals, a number of requirements and conditions related to the construction manufacturing, maintenance by labor, material, technical and financial resources, labor conditions, and so on should be taken into account.

The second stage is connected with the development of organizational decisions. First of all it is the determination of intensity, interchangeability, duration of work, the beginning and the end of each construction process in the time area, combining the technologically-related works, etc. But the basis for all the organizational issues is the labor resources allocation of the project and types of works because only the accepted allocation determines other organizational decisions to a large extent. Organizational decisions development is directly related to optimization.

In scheduling there are numerous criteria, and the criterion system. One of the most complete reviews of the methods and criteria for the construction production organization is A.A. Gusakov's review [20].

When calculating the construction object schedule, further restrictions must be considered: the works time area, certain volumes of works distribution for time periods, the availability of the critical points of the initial stage, quantitative and qualitative parameters of technological dependencies, etc. It is important that the distribution of BAW volumes by time periods may be irregular. And this leads to calculating the time schedule for each planning period. The obtained construction schedule (time parameters) must be within the boundaries of the work time areas, certain OTDM (Fig. 6). A special feature of this calculating a line graph method is to identify private and full-time slacks.

Currently considered approach to the scheduling BAW development is implemented in the form of software product OTDM and problems statement at the first and second stages.

The work was supported by Act 211 Government of the Russian Federation, contract № 02.A03.21.0011.

References

1. K. Aksyonov, E. Bykov, O. Aksyonova, Proc. Intelligent Systems Modelling & Simulation (ISMS), 310-314 (2013)
2. K. Kim, J. Walewski, Y. Cho, J. Manage. Eng., (2015). 10.1061/(ASCE)ME.1943-5479.0000374, 04015038.
3. U. Klanšek, J. Constr. Eng. Manage., (2015). 10.1061/(ASCE)CO.1943-7862.0001074, 04015088.
4. N.N. Moiseev, *Mathematics set up an experiment* (Nauka, Moscow, 1970)
5. G. Ma, A. Wang, N. Li, L. Gu, Q. Ai, Journal of Construction Engineering and Management, **140(12),** 04014055 (2014)
6. J.T. O'Connor, W.J. O'Brien, J.Ouk Choi, Practice Periodical on Structural Design and Construction, 04015014 (2015)
7. E.H. Oh, N. Naderpa-jouh, M. Hastak, S.Gokhale, Journal of Construction Engineering and Management, **142(2)**, 04015067 (2015)
8. M.F.F. Siu, M. Lu, S.AbouRizk, Journal of Construction Engineering and Management, **142(1)**, 04015048 (2015)
9. Y. Gajpal, A. Elazouni, Construction Management and Economics, **33(7)**, 531-553 (2015)
10. A. Francis, Procedia Engineering, **123**, 162-168 (2015)
11. T. Horenburg, W.A.Günthner, Computing in Civil Engineering, (2013)
12. S.M. Chen, P.H. Chen, L.M. Chang, Automation in construction, **21**, 99-113 (2012)
13. O. Salem, S. Mohanty, Journal of Construction Engineering and Management, 134(7), 501-508 (2008)
14. A. Baldwin, D. Bordoli, *Handbook for construction planning and scheduling* (John Wiley & Sons, 2014)
15. L.G. Golub, E.N. Lyashchenko, *MIS of a construction trust* (Stroyizdat, Moscow, 1976)
16. V.I. Varapaev, *Models and methods of scheduling in construction management automated systems* (Stroyizdat, Moscow, 1975)
17. K.A. Antanavichyus, *Multilevel stochastic simulation of industry splanned solutions* (Vilnius, Mokslas, 1977)
18. K.A. Antanavichyus, *Designing and optimization in construction management* (Stroyizdat, Moscow, 1979)
19. E.V. Gusev, *Technological designing and balanced planning of BAW* (RIK "Redaktor", Chelyabinsk, 1990)
20. A.A. Gusakov, *Organizational and technological reliability of construction operations* (Stroyizdat, Moscow, 1974)
21. P. Lorterapong, M. Ussavadilokrit, Journal of Construction Engineering and Management, **139(4)**, 414-422 (2012)

Adaptive inventory management system of an industrial enterprise in the unsteady market environment

N. Dzenzelyuk[1,*]

[1]South Ural State University, Chelyabinsk, Russia

Abstract. The article examines the features and problems of methodology of formation of inventory management in a volatile market environment. Currently, there was a need to change the development paradigm of the scaling operations to the more efficient use of resources and capital. In this regard, the high importance of the newly received inventory control problem as an essential component of enterprise resources. The author sets out the basic methodological principles of inventory management systems. The article from the standpoint of the theory of systems and automatic control systems proposed structure adaptive inventory management system. As a criterion of the effectiveness of the joint action is considered the minimum criterion of total costs and expenses associated with the unmet demand. The simulation results confirm the efficiency of the developed model under unforeseen and significant fluctuations in the parameters of stochastic demand.

1 Introduction

Despite the rather long period of inventory management theory development, the issue of creating effective and practically implemented inventory management systems is still relevant.

The search for new solutions in this area is actively carried out [1-5]. The importance of this area is fairly high due to the high proportion of stocks in the capital of companies as well as low capital productivity of the economy as a whole.

Today's economy has developed an urgent need to move from the paradigm of development through the expansion of operations to the paradigm of the capital productivity and resource efficiency improvement. This is true not only for the Russian economy, but also for the world economy as a whole.

Confirmation of the above is the large number of publications related to the search for a variety of practical solutions in the field of and increasing of assets turnover as a whole and inventory turnover as the most "bound" part of the current assets [6-9].

In this regard, it is sufficient to note that Russian companies have a high current asset to total asset ratio. Amount of inventories is more than 8 percent of economy in Russia, while the return on assets not exceeds 7 percent [10, 11].

Thus, the inventories today is one of the most important components that improve profitability and ensure competitiveness of business simply because they continually involve substantial financial resources that form the profits of an enterprise, and efficient use of which requires continuous monitoring. Therefore, the enterprise inventory management means the implementation of inventory control and market environment control, the analysis of their interaction and influence on the operation of the enterprise as the subject of market economy, decision-making, aimed at saving money and increasing the rate of assets profit.

2 Some aspects of construction of model of inventory management system

The main difficulties of effective current assets management appear, as a rule, in the inventory management system, due to the necessity of sufficiently regular purchase, transportation, storage and sales in an unsteady market with high dynamism and uncertainties relevant to inventory management parameters. This fact makes the focus on the examination of inventory as an unsteady stochastic object of management.

The study of inventory management problems from the proposed positions allowed to formulate the following methodological principles of construction of efficient inventory management systems:

- The principle of the market environment unsteadiness;
- The principle of the management systems adaptivity;
- The principle of adequate complexity and dynamism of management methods and algorithms;
- The principle of the generation of multi-criteria quasi-optimal solutions based on heuristic realtime simulation modelling.

* Corresponding author: dzenzeliukns@susu.ru

Without dwelling on the principles (substantively presented by the author in [12]), we note that such basic approach should be used in the construction of models of practically implemented inventory management system (hereinafter, IMS).

Author's development of the model, which meets the formulated principles, is discussed below.

Interpretability, adequacy of the actual conditions and universality are worth noting among the most important requirements for developing of IMS models. Nowadays the development of inventory management theory is distinguished by increasing use of modern analytical and mathematical apparatus of the theory of automatic control systems theory. The most frequently cited are the principle of maximum, dynamic programming, statistical optimization, identifying algorithm and filtering algorithm, adaptive approach, linear and nonlinear programming and invariant strategies class [13-17]. The effective practical implementability in the general case can only be achieved when the derived IMS models simultaneously have all three properties listed above (interpretability, adequacy and universality). However, studies indicate that the actual nature of their interaction does not meet the specified requirements [18].

For this reason, the determination of the morphology and algorithms of simple, adaptive inventory management systems and the justification for their adequacy become the most important problem. It is known that the most rational, and sometimes the only method of study and decision-making in the conditions of high uncertainty and complexity of the unsteady object of management is a simulation modelling method [19]. Let us discuss some features of IMS development which meet the above mentioned principles.

The basis of the system is formed by IMS structure proposed by A.A. Pervozvanski [20]. The being studied IMS, in this case, is considered as a management system consisting of an object of management (in our system - stock reserve), and controlling system consisting of a regulator (executive body) - a supplier that operates since the formation of the control signal (in our system - purchase order), which, in turn, is formed considering results of demand forecast as an external disturbance, and measurement of the current level of inventory (the condition of the object of management).

This approach is a focused specification of the classical approach to the management systems construction and can be used for the construction of the morphology of the adaptive system. However, the solutions proposed by A.A. Pervozvanski do not meet the actual conditions of stochasticity and unsteadiness of the economic environment. Moreover A.A. Pervozvanski a priori considered the opportunity to separate the optimal regulation of reserve stock and current inventory, without proving the admissibility of such a separation in a real situation.

Therefore, the universal method of simulation modeling should be used .to obtain managerial decisions, in contrast to analytical approaches which provide particular solutions with strong assumptions.

Starting from the fundamentals of applied systems theory and automatic control systems theory and considering the structure proposed by A.A. Pervozvanski, the common morphological structure of IMS on demand can be synthesized (Fig. 1).

Simulation modelling of the proposed structure as an unsteady stochastic system allowed us to estimate the possibility of extension of the results of A.A. Pervozvanski model, to more general and real conditions of the functioning of the IMS as well as to remove a number of fundamentally strong assumptions for known theoretical solutions to the optimal inventory management tasks. When choosing the IMS efficiency criterion, a Du Pont formula was primarily considered for the determination of return on assets, as this formula allows the integral estimation of enterprise efficiency. However, studies by the authors show that, when all types of costs had taken into account, this criterion was reduced to the criterion of aggregate costs minimization, which is used in the subsequent construction and research of the adaptive IMS.

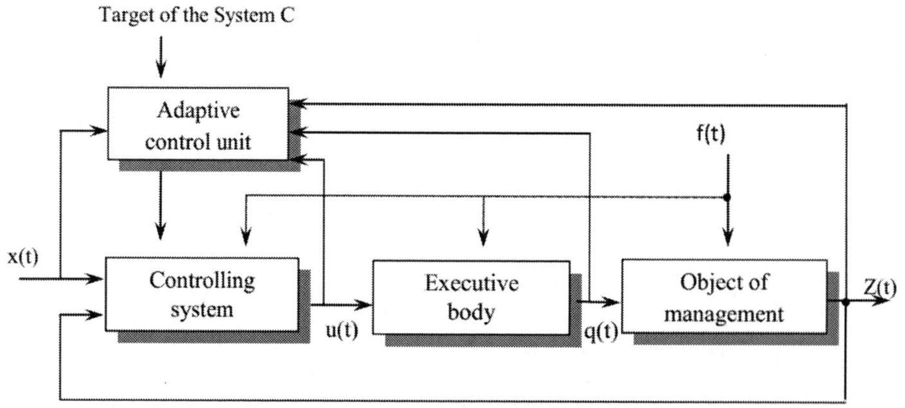

Fig. 1. Common morphological structure of IMS. x(t)- regular environmental disturbance, u(t)- control parameter, f(t)- external environmental action (irregular disturbance), z(t)- current inventory status, q(t)- goods delivery.

Omitting the proof of possibility of substituting unknown parameters of optimal solutions rule by point estimates of such parameters, we should discuss the assumption that environment characteristics are known. Randomized demand x_t is chosen to take account of the influence of the external environment (hereinafter the demand and its determinants - x_t, f). The author prefers adaptive approach as attractive and appropriate for two reasons. First, the principle of universality is performed, namely the management system functioning structure and parameters are invariant to changes in the environment. Second, these models offer the simplicity of the structure used. In this case, unlike their original structure, they have the subsystem which allows to estimate of unknown factors. In our case they are ($\hat{X}_t,\hat{f}$), which recover the correspondence between control parameters $\bar{q},r$ and scores of factors $\hat{X}_t,\hat{f}$, i.e. correction $U(\bar{q},r)$ by its estimate:

$$U(\bar{q},r,\hat{\bar{x}}_t,\hat{f}) \equiv \hat{U}(\bar{q},r) \quad (1)$$

Quasi-optimal (adaptive) management strategy, constructed in accordance with the considered method, is a particular case of two-tier *(R,r)* strategy, symbolically:

$$U(q,r)=\begin{cases} R-z, & z \le r, \quad R=q+r \\ 0, & z>r \end{cases}, \quad (2)$$

where z – current stock situation;
r – reorder point,
control vector $U(q,r)$ is defined by point estimates of mathematical expectation $\bar{x}_t$ and variation coefficient $var\{x_t\}$ of the demand ($\hat{\bar{x}}_t$, $\widehat{var\{x_t\}}$)

Then, operative decisive rule for this strategy is to optimize the size q and reorder point r in next period $t+T$ based on procedures of determination and forecasting of the demand x_t and estimation $\hat{z}_t$ of the current stock situation z_t, made in basic period (t), so

$$q_{t+T}=\begin{cases} q^*_{t+T}, & \hat{z}_t \le r_t+\tau\cdot\hat{\bar{x}}_t, \\ 0, & \hat{z}_t > r_t+\tau\cdot\hat{\bar{x}}_t, \end{cases} \quad \text{when } T>,(3)$$

where τ – period of delivery,
$r^*_{t+T}=\{r\in\Omega_r \,/\, C_r(var\{x_t\})=min\}$,
Ω_r - range of possible values of r,
$C_r(var\{x_t\})$ - aggregate costs of reserve stock management, depending on random component of the demand (variation, presented by indicators estimate (demand variation coefficient $var\{x_t\}$)).

Therefore, the derived structure of optimal order quantity q^*_{t+T} is similar to optimal order structure, proposed by A.A. Pervozvanski for the linear control law, i.e. consists of deterministic $\bar{q}^*_{t+T}$ and random $\tilde{q}^*_{t+T}$ components, which are defined separately:

$$q^*_{t+T}=\bar{q}^*_{t+T}+\tilde{q}^*_{t+T}, \quad (4)$$

The essential difference lies in the procedure of a finding of the components $\bar{q}^*_{t+T}$ and $\tilde{q}^*_{t+T}$. The finding procedure has significant features of the adaptive method, which fundamentally distinguish the developed strategy of those already known.

First, the definition of deterministic component $\bar{q}^*_{t+T}$, which provides accounting of stable (average) component of the demand x by the condition of minimum of storage expenses h and forming A days-on-hand inventory (excluding the stockout), although defined by Wilson formula, is based on use of not the target intensity value of the demand $\bar{x}_t$, but on its preliminary estimate $\hat{\bar{x}}_t$:

$$\bar{q}^*_{t+T}=\sqrt{\frac{2\cdot\hat{\bar{x}}_t\cdot A}{h}} \quad (5)$$

Second component of an order $\tilde{q}^*_{t+T}$, used for compensation of stochastic component oa the demand x_t and necessary to ensure the target level of a satisfied demand, is calculated as the deviation of levels (in next and basic period) of the inventory r, which is used as an operation threshold, symbolically:

$$\tilde{q}^*_{t+t}=\Delta\hat{r}^*_t=\hat{r}^*_t-\hat{r}^*_{t-T} \quad (6)$$

Thus, this formulation of the synthesis problem in the form of adaptive IMS provides the flexible feedback from the two components of the demand (deterministic and random), so the inventory management capabilities are significantly expanding by including a reserve stock r in a control loop, so the reserve stock becomes a functional relation to the variability of the demand x_t.

Random order component $\tilde{q}^*_{t+T}$ forming block appears similar to operating management channel, where deterministic component $\bar{q}^*_{t+T}$ is formed. Obviously, only those parameters of the demand are used which characterize its variation (volatility), such as variation coefficient $var\{x_t\}$ and standard deviation σ_{Xt}., not a stable component (average intensity $\bar{x}_t$) is used.

Returning to the criterion of efficiency of considered IMS, the description of its components should be determined.

For example, optimization condition for the parameter r of the reserve stock storage expenses C_{ir} and expenses of stockout itself $C_{v\eta}$ is written as:

$$r^*=\{r\in\Omega_r \,/\, (C_{ir}+C_{v\eta})=min\}, \quad (7)$$

where $C_{ir}=k_{ir}r$, k_{ir} – coefficient that indicates the alternative investment profitability;
$C_{v\eta}=k_\eta\cdot\bar{\eta}(r)$, k_η – costs per day, related to unsatisfied demand because of stockout;

$\overline{\eta}(r)$ - link function between the average quantity of stockout days $\overline{\eta}$ in considered period and the level of the reserve stock r.

A quantity of stockout days $\overline{\eta}$ in any cycle of operating management with duration T by random component of the demand depends on estimate of the demand variability $var\{x_t\}$, i.e. on particular intensity estimates $\overline{x}_t$ and on the standard deviation σ_{Xt} so as $\overline{\eta}(r) \equiv \overline{\eta}(r;\hat{\overline{x}}_t;\hat{\sigma}_{Xt})$. This parameter ($\overline{\eta}$) is also used during the system synthesis for its performance evaluation. As it is interesting enough to control the "loyalty" of the buyer to an "empty" warehouse, this parameter is also used in the system as a controlling action (with estimation of number of stockout days effecting the system parameters).

Of course, in general case of different baselines (laws of demand distribution density and its parameters) a formula for a function $\overline{\eta}(r;\hat{\overline{x}}_t;\hat{\sigma}_{Xt})$ is also different, and it can easily be derived experimentally during the simulation modelling as a family of curves $\overline{\eta}(r;var\{x_t\})$. Latest can be converted into the corresponding family of curves of the costs $C_{v\eta}(r;var\{x_t\})$ for consideration in the target function $\boldsymbol{C_r}$ during definition of the optimum values $\boldsymbol{r^*}$ and match making between control parameter $\boldsymbol{r^*}$ and demand variation values $\boldsymbol{var\{x_t\}}$.

Accordingly, the optimization condition can be written as:

$$r^* = \{r \in \Omega_r \,/\, (k_{ir} * r + k_\eta * \overline{\eta}(r;var\{x_t\})) = min\}, \quad (8)$$

The structure of adaptive IMS, obtained by the author, reflecting the outlined strategy and relevant to considered inventory management model, is shown in Fig. 2.

The use of such a structure and the proposed strategy can greatly simplify the selection of the optimal control parameters of the IMS.

No doubt, for the estimation of the adaptive solutions effectiveness, i.e. proximity of the adaptive $\hat{U}(q,r)$ and the «optimal» $U^*(q,r)$ rules for IMS solutions, it is necessary to evaluate the convergence, consistency and efficiency of estimates of the unknown parameters of demand. It is also important to evaluate the rate of convergence of calculated estimates.

Evaluation of the convergence rate of the demand parameter estimates $\hat{\overline{x}}_t, \hat{\sigma}_{Xt}$ as well as true values $\overline{x}_t$ and σ_{Xt} in the context of a being solved problem of the IMS construction, is necessary due to limited period of delivery. It is obvious that the formation of the demand parameter estimates should take place within a time not exceeding the period of delivery, taking into account the time of the order T-τ.

Analysis of simulation modeling results indicated that the convergence rate of demand estimates $\hat{\overline{x}}_t, \hat{\sigma}_{Xt}$ to true values $\overline{x}_t, \sigma_{Xt}$ with accuracy better than 0,5% does not exceed the specified values in the range of variation coefficient $var\{x_t\}$ from 0,2 to 2, and with $var\{x_t\}$=5, which is extremely rare in reality, does not exceed the period of delivery.

The research results confirm the constructibility of the considered principles of methodology and inventory management system construction, as well as the performance of the proposed algorithms for the estimation of unknown parameters of demand. Thus, it can be stated that the proposed adaptive strategy and the model of operational inventory management could well be used in harsh environments typical of the unsteady economy, namely in terms of significant and unforeseen fluctuations in the stochastic parameters of demand during delivery period.

3 Conclusion

The conducted studies allow to state a constructibility of the proposed methodological principles at the current stage of inventory management theory development and a possibility of construction of an adequate universal adaptive management model with well interpretable results.

The work was supported by Act 211 Government of the Russian Federation, contract № 02.A03.21.0011

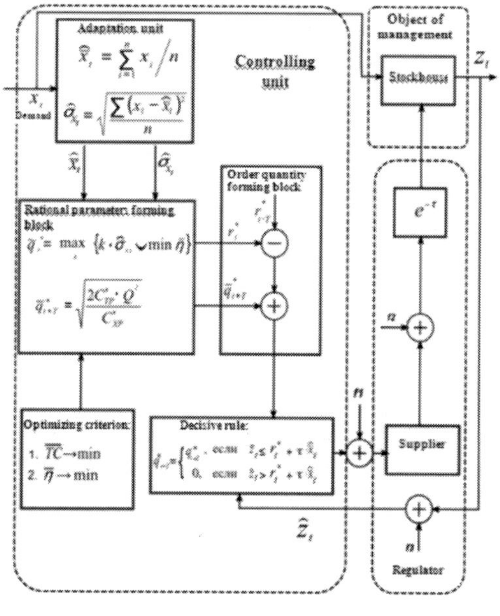

Fig. 2. Adaptive IMS structure.

References

1. J. Buchan, E. Koenigsberg, *Scientific Inventory Management* (Prentice-Hall, 1963)
2. A.G. Butrin, Economy of a region, **2**, 130-140 (2014)

3. N.I. Vorobyova, V.S. Lukinskii, V.V. Lukinskii, Logistics and Supply Chain Management, **3(62)**, (2014)
4. V.S. Lukinskii, *Proc. Logistics: modern trends of development: materials of the XIII International scientific-practical Conference*, 24 (2014)
5. O.A. Sviridova, Organizational development modeling: the round table reports, Scientific Technologies (2014)
6. I.A. Baev, A.V. Shmidt, Bulletin of Ural Federal University. Series Economics and Management, **3**, 50-63 (2012)
7. A.G Butrin, A.I. Kovalyov, Economic analysis: theory and practice, **3** (2011)
8. N.I. Reshetko, Y.I. Bashkatova, Electronic scientific publication of "Naukovedenie", **7**, **1** (2015)
9. E.N Troyanova, N.P. Ivantsova, Social sciences, **1**, **6-1(9)**, 103-107 (2015).
10. *Russian Finances, 2014: statistics digest* (Rosstat, Moscow, 2014)
11. N.S. Dzenzelyuk, Management in Russia and Abroad, **3** (2013)
12. L.A. Baev, N.S. Dzenzelyuk, Y.N. Tarasov, Theoretical and Practical Aspects of Management, **1** (2014)
13. Y.I. Ryzhikov, *The theory of lines and inventory management* (Piter, Moscow, 2001)
14. E.V. Shikin, A.G. Chkhartishvili, *Mathematical methods and models in the management: Educat. Guidance* (Delo, Moscow, 2002)
15. V.I. Smagin, S.V.Smagin, Tomsk State University Journal, **290** (2006)
16. V.I. Smagin, S.V.Smagin, Tomsk State University Journal, **3(4)** (2008)
17. K.P. Goloskokov, K.A. Davydov, Economic systems Management, **2** (2012)
18. Y.N. Tarasov, N.S. Dzenzelyuk, Economic systems Management: Electronic scientific journal of Kislovodsk institute of economy and law
19. O.I. Babina, Simulation. The Theory and Practice (2009)
20. A.A. Pervozvanski, *Mathematical models in production management and inventory management* (Nauka, Moscow, 1975)

Trends in the structures development of the regional machine-building complex

I.V. Ershova[1,*], *O.O.* Podolyak[1], and *M.A.* Prilutskaya[1]

[1]Mechanics and Machine Building Institute, Ural Federal University, Ekaterinburg, Russia

Abstract. In the process of market reforms of the Russian machine-building complex several distinct periods can be revealed. In this article the authors define periods of mass disintegration and spontaneous integration (since the beginning of the reforms until the financial crisis of 1996), post-crisis stabilization, directional specialization (2000-2008) and evolutionary development (since 2010). The economic consequences of the enterprises mergers and divisions are shown on the example of machine-building enterprises of the Middle Urals. The aim of this study is to substantiate the methodical approach to the selection of the optimal organizational structure for the machine-building business. The necessity of taking into account the extent of the personnel diversification and the production volume has been revealed for the optimum organizational structure determination in the machine-building associations. The authors have analyzed sales profitability of the 2745 machine-building enterprises, depending on the production scale and industry sector. The factors affecting the development of cooperative ties and outsourcing have been defined. The authors have made a conclusion that it is necessary to form technological chains as a new kind of business associations.

1 Introduction

The number of registered machine-building enterprises has increased dramatically during the privatization reform: from 6 to 60 thousand. At the same time there was a significant decline in machinery production, reduction of return on assets, increase in number of unused equipment and fixed assets deterioration. Even the global crisis of 1996 or 2009 did not lead to such significant disturbances in the work of machine-building enterprises, as the beginning of the spontaneous disintegration in the industry.

The retrospective analysis of changes in machine-building business structures has revealed some factors that affect the performance of the machine-building enterprises. Taking into account the factors identified in the organizational structure design of the machine-building associations allowed to develop a methodical approach to the creation of new innovative structure of machine-building complex. The relevance of this approach is related to the necessity of development of Russian machine-building industry as a key modernization sector of the economy, especially in the context of economic sanctions.

2 The economic consequences of institutional changes

The reform of the Russian economy has several clearly identified periods. The first phase is characterized by the division of the machine-building associations into independent enterprises and selling formerly state-owned assets to private owners. This phase can be called the stage of mass disintegration. The well-known machine-building association "Uralmash" is the significant example of this process. More than 30 independent companies of different sizes have been created on the basis of this association. According to Russian researchers [1], the number of enterprises during this period increased by 10 times and reached 58.4 thousand in 1997.

At first the integration processes of newly established enterprises occurred spontaneously. Certain orderliness in the creation of associations can be seen after the first Russian financial crisis of 1996-1998. The reform stages and their implications are listed in Table I.

Classification of the newly established associations according to the purpose and way of creating includes 4 main types.

1. *"Buy everything"*. This is particularly true for the period of property redistribution from mid to late 90s.

The associations were formed by the owner from the number of absolutely technologically unrelated companies. The main formation strategy of such associations was conglomerate diversification

2. *Segregations from the mono-enterprise.* The subsidiaries are segregated from the parent company in order to reduce costs and risks.

* Corresponding author: eiv.eopm@e1.ru

Table 1. Reform stages and their implications.

Stage and its characterization	Trends	Number of enterprises	Organizational forms	Indices of change in production of machinery and equipment by the end of the period
Until 1992	Planned economy	Up to 6 thousand	State-owned enterprises	100%
1992-1994. - The first property redistribution. Corporatization and privatization of state-owned enterprises	The emergence of private ownership in the industry, the division of assets	Up to 60 thousand	The emergence of multi-disciplinary regional associations established on the "buy everything" principle, the transfer of non-core divisions into independent companies	31%
1994-1998. - The second property redistribution. The struggle for the controlling stake	Monetary privatization. Bulk purchase of shares from outsiders	Up to 60 thousand	The emergence of private Russian machine-building corporations, the acquisition of Russian companies by foreign owners	84%
1998-2005. - The first Russian financial crisis		Up to 30 thousand	Mass bankruptcy of enterprises	140%
2001-2008. - The third property redistribution	Corporatization of state-owned enterprises	Up to 30 thousand	The emergence of Russian state corporations, mainly in the field of defense industry	125%
Since 2008 - the crisis and post-crisis development	Strengthening of the state support and regulation. Leasing schemes of hardware upgrades	Up to 50 thousand	Bankruptcy of medium sized enterprises, transfer of non-core assets into subsidiaries	84%
Since 2013 – political restraints	Strengthening of the state support for the defense enterprises	Up to 50 thousand	Stabilization of the business structure	96%

3. *Technological unity.* The companies in the association are connected by a single production cycle. The association is developing according to the related diversification strategy, forming the integral complex of enterprises. The owners create a union on the principle of the technological chain unity. [2]
4. *Belonging to the state ownership.* The associations are created to ensure the efficiency of the state-owned enterprises, mainly producing weapons and military equipment.

Currently, more than 1/3 of the machine-building enterprises are part of large associations. Independent companies have sales volume of about one billion rubles.

This study is limited to the analysis of the legal structures activities. Legally registered associations assume a managing company existence, a structure, which will coordinate the work of all enterprises in this association (technology exchange, joint development of new products, co-operation between enterprises), and allocate resources between companies. For such structures there is a need for the clear management system.

Although the authors believe that the term 'business association' can be used both for legally registered structures and unregistered structures. For example, a structure with strong technological and financial constraints can be seen as a business association. As an example of such association we can take a conglomerate consisting of a parent company and subsidiaries segregated from it. This view is consistent with the model of the enterprise, introduced by G. Kleiner [3], since such structures have five necessary sets: property, labor, institutional, informational and business model complexes.

The limitations of the research object only to legally registered structures allows solving questions of the resources concentration scale and the allocation of resources within the structure.

3 The effectiveness of the resources concentration in the machine-building association

The sample of 2745 Russian machine-building enterprises from various industrial branches was taken to analyze the concentration effects. We have discovered the reduction of sales profitability in all enterprises having large sales volume (see Fig. 1).

The revealed tendency is typical both for industries with minimum return on sales volatility and for the sectors with the maximum gap between the performance of separate enterprises. Undoubtedly, the profitability of sales is influenced by a variety of other factors, which explains the "spread" of indicator for individual enterprises (Table 2). It is necessary to mention that return on sales was calculated as the ratio of net profit (from all activities of the enterprise) to sales revenue (only from the main activity).

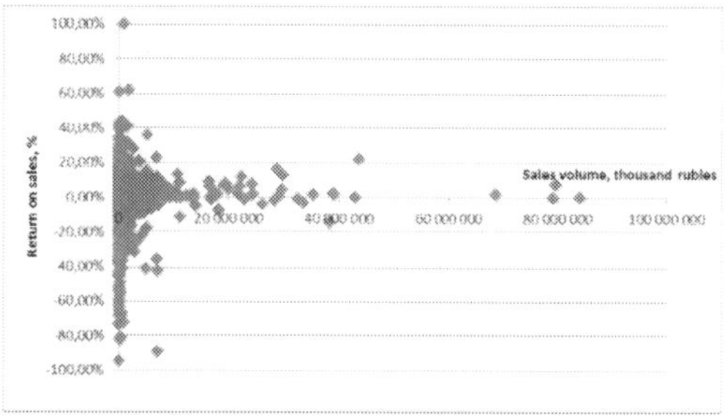

Fig. 1. The distribution of legally registered companies and associations in terms of sales volume and sales profitability in 2013 (calculated by authors according to the electronic resources base FIRA.).

As some enterprises demonstrated income and profit from other types of activities much greater than from their main activity it came out that return on sales was evaluated as 100% and even more (which is impossible for main activity calculations solely).

Because of this approach to calculation the return on sales for some enterprises has reached such catastrophic values as -221.33% and even -977.6% (which means that this company actually sold out their property).

However, the overall trend of the indicator is as follows: increase in the scale of production leads to decrease in sales profitability. Economic-organizing meaning of this trend is that the managing costs of large organizations rise significantly. At the same time optimally structured associations have the opportunity to spread risk by diversifying activities.

However, the overall trend of the indicator is as follows: increase in the scale of production leads to decrease in sales profitability. Economic-organizing meaning of this trend is that the managing costs of large organizations rise significantly. At the same time optimally structured associations have the opportunity to spread risk by diversifying activities.

However, the overall trend of the indicator is as follows: increase in the scale of production leads to decrease in sales profitability. Economic-organizing meaning of this trend is that the managing costs of large organizations rise significantly. At the same time optimally structured associations have the opportunity to spread risk by diversifying activities.

Table 2. Volatility of return on sales according to the machine-building sector.

NCEA code	NCEA code formulation. Production of:	Number of enterprises	Return on sales, %		
			Min.	Weighted average	Maximum
29.1	mechanical equipment	375	-221.3	6.11	100.0
33	medical equipment, measuring devices and optical instruments, watches	237	-18.98	5.27	39.76
29.22	materials handling equipment	205	-68.07	4.85	214.2
31.2	electricity distribution and control equipment	295	-111.9	4.63	62.01
29.23	industrial cooling and ventilation equipment	203	-38.31	3.38	42.65
31.1	electric motors, generators and transformers	150	-36.83	3.13	30.14
35	ships, aircraft and spacecraft and other vehicles	186	-124	1.54	35.10
34	motor vehicles, trailers and semi-trailers	489	-123.6	0.65	61.15
29.5	machinery and equipment for special purposes	418	-68.5	-0.14	43.91
29.3	machinery and equipment for agriculture and forestry	97	-127.7	-0.71	24.27
29.4	machines	87	-977.6	-8.40	16.54

As an example, we give the return on sales dynamics of one machine-building holding – "Sinara - Transport Machines". The specialization of this holding is transport machinery (Figure 2).

The analysis of other machine-building holdings showed that the most successful associations have the centered type of diversification, deliver a full range of services and incorporate research organizations.

4 The methodical approach to the organizational structure formation of a diversified machine-building association

In order to measure the degree of optimal diversification we propose to use the matrix of the product diversification and the degree of administrative resource diversification, since human resources (personnel) is the most difficult, inert, expensive at any change component of the enterprise resource potential.To assess the level of diversification we use Herfindal-Hirshman index proposed by Berry C.H. [4], which was originally designed to measure the degree of industries concentration.

Depending on the ratio of production and personnel diversification degree we propose to form the "diversification matrix" classification (Figure 3), which shows the rankings of diversified businesses. The threshold values of the index are selected according to empirical studies, such classification is used by Strategic Management Journal in its practice [5].

The index values greater than 0.7 points show that the union is highly diversified.

To optimize the structure of such association a range of measures is provided. First of all, we propose to analyze the possibility of outsourcing for products and services with a small share in the production volume. The second measure is the expansion of autonomy in strategic decision-making by individual enterprises. The third measure is strengthening the innovative potential of the association to form a technological chain.

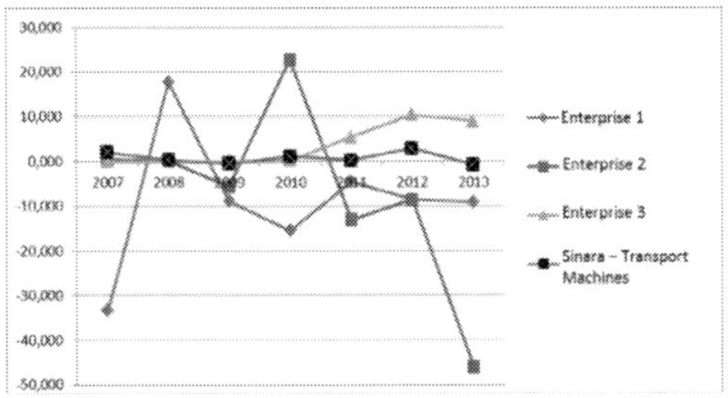

Fig. 2. Return on sales dynamics of individual machine-building enterprises and the holding as a whole, %.

The concept of the "technological chain" is interconnected with the concept of "customer value chain" and allows defining the strategic prospects of the existence of this or that production in terms of meeting the consumers needs. The structure of the technological chain should include the "core" of the chain – an innovative company; manufacturing enterprises, providing the main functions; and buffer zone enterprises, performing secondary functions. The concept of the technological chain is more in line with the contractual business association, rather than their joining in the legally registered form. However, in Russia, the processes of outsourcing and co-operation are underdeveloped, which forces businesses to keep "buffer zone" enterprises within the association.

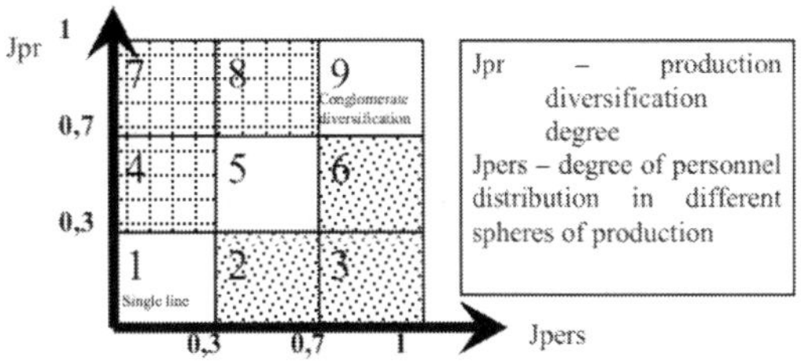

Fig. 3. Conformity assessment matrix of conformity of the association organizational structure to the product diversification.

According to the survey the managers of the machine-building enterprises named the unpreparedness of the legal framework, the terms and supply quality violation by contractors as reasons for the slow spread of the production processes outsourcing in Russia.

5 Conclusion

The phase of spontaneous integration of the machine-building enterprises has been replaced by the ordered diversification phase.

The structure of the regional machine-building complex now on 2/3 consists of independent companies with the sales volume slightly exceeding 1 billion rubles and 1/3 of large enterprises being a part of the private and public associations.

The associations of the machine-building enterprises implement the centered diversification strategy, getting rid of non-core assets, but the implementation of this strategy is complicated because of the legal framework and industrial outsourcing practice absence.

The further development of the enterprises is possible towards the creation of non formalized associations in the form of the technological chain, with the innovative enterprises as the core.

References

1. S.G. Kara-Murza, S.A. Batchikov, S.Yu. Glazyev, *Where there is Russia. White book of reforms* (Algorithm, Moscow, 2008)
2. I.V. Ershova, P.P. Krylatkov, *Economic and structural models of machine-building enterprise complete development* (URFA, Ekaterinburg, 2011)
3. G.B. Kleyner, Bulletin of the Samara state economic university, **1**, 26-35 (2015)
4. C.H. Berry, Journal of law and Economics, **14**, **2**, 371-383 (1971)
5. V. Ramanujam, P. Varadarajan, Strategic Management Journal, **10**, **6**, 523-551 (1989)

6. C. Grimpe, K. Hussinger, *Formal and Informal Technology Transfer from Academia to Industry: Complementarity Effects and Innovation Performance.* http://ftp.zew.de/pub/zew-docs/dp/dp08080.pdf.
7. I.M. Guly, A.P. Dorogovtsev, Economic and Social Changes: Facts, Trends, Forecast, **4(8)**, 22-32 (2009)
8. B.O. Ivanovich, G.I. Vladimirovich, Economy of Region, **1**, 125-132 (2014)
9. S.A. Golovikhin, Regional economy: theory and practice, **3**, 26-37 (2013)
10. I.V. Ershova, P.P. Krylatkov, O.O. Podolyak, Proc. Problems of ensuring safe development of modern society collection of works IV of the international scientific and practical conference, 7-13 (2014)
11. Z.Kh. Kilchukov, Modern aspects of economy, **2(210)**, 75-80 (2015)
12. Federal State Statistics Service: official site. http://www.gks.ru/.
13. M.A. Prilutskaya, E.V. Cherepanova, V.G. Ulitin, *Proc. Problems of ensuring safe development of modern society collection of works IV of the international scientific and practical conference,* Ekaterinburg, 213-216 (2014)
14. V.A. Trapeznikov, *Development of the production cooperation in regional machine-building a kompleks* (Ekaterinburg, 2011)
15. D.V. Syradoyev, *Proc. Prospects of development of modern science collection of articles of the international scientific and practical conference. international center of the innovative researches,* Ufa, 79-84 (2014)
16. O.O. Podolyak, E.Yu. Kuznetsova, P.V. Mezentsev, Bulletin of Ural Federal University. Series Economics and Management, **14**, **1**, 96-114 (2015)
17. T.A. Khudyakova, A.V. Shmidt, Mediterranean Journal of Social Sciences, **6**, **4**, 274-279 (2015)
18. T.A. Khudyakova, Actual problems of economics, **12**, 444-450 (2015)
19. FIRA PRO, First Independent Rating Agency (FIRA)
20. D.A. Ruban, Bulletin of Ural Federal University. Series Economics and Management, **14**, **2**, 78-96 (2015)
21. F.O. Walumbwa, M.A. Maidique, C. Atamanik, Organizational Dynamics, **43**, 284-293 (2014)
22. O.I. Botkin, I.V. Grebyonkin, Regional economy, 1, 125-132 (2014)
23. M.M. Makhmudova, Management Issues, **1(38)**, 143-152 (2016)

Formation of the sustainable development mechanism at enterprises of mining industry as a condition of overcoming crisis

V.V. Zhuravlyov[1,*]

[1]South Ural State University, Chelyabinsk, Russia

Abstract. The present article addresses problems of sustainability at enterprises of gold mining industry of the Russian Federation in the period of recession. The author analyses environmental factors and tendencies of global economic development that have an impact on Russian business. On the basis of studying academic papers of foreign and Russian researchers the author offers a simulated model for operations of the gold mining enterprise, which is considered as a stochastic non-linear dynamic system. The model allows to identify diapasons of stability, to make a quantitative evaluation of the economic status of the enterprise in conditions of crisis phenomena. The author offers a mechanism of sustainable development for enterprises of mining industry that involves applying a complex of methods and techniques aimed to enable the company to perform its goal function.

1 Introduction

The world financial crisis, marked aggravation of political situation all over the planet, and instability of economic ties resulted in the situation when many states are turning their close attention to formation of gold-exchange reserves and increasing the gold reserve stock. Gold mining is one of key branches that guarantee economic security of the country. In recent years we have witnessed a stable growth of the amount of gold mined in the Russian Federation; number of enterprises engaged in this activity is increasing, purchases of precious metals are on the rise, as well as the demand of the Russian state addressed to representatives of the "gold business". Strategic role of the gold-mining industry certainly contributed to its progressive advance. Nevertheless, recessionary phenomena in the world economy together with sanctions imposed by the USA and European Union, a de facto stagnation of the Russian economics caused great difficulties for enterprises of the mining industry. Stability of the industry depends on a number of both controlled and uncontrolled factors of external and internal environment.

Gold-mining has certain number of peculiarities, for example, it is not an infrequent case when prognostic recovery factors prove to be wrong, many of developed deposits are found being unprofitable. The principal deposits developed as far back as in the Soviet times are located in hard-to-reach places in Eastern Siberia and Yakutia. Gold deposits are explored and mined in difficult climatic conditions. Quite often there exist neither reasonable infrastructure nor roads, so it is possible to deliver equipment, materials, instruments, fuel only during a certain time of the year. Other unfavourable external forces include purchasing prices for an ounce of gold, rouble exchange fluctuations, inflationary processes, etc. [1].

The internal environment of the gold-mining enterprise is specified by the state of maintenance, technologies, equipment, qualifications of personnel, a possibility to introduce novelties, the volume of own financial resources [2]. The equipment status depends on climate and the level of excellence of workers, and in its turn it influences the correlation between the operating time and the time spent on repair works; it also determines expenditures connected with equipment handling. Small mining enterprises are facing problems connected with their intention to buy and operate very expensive, but high-quality, user-friendly and productive foreign pieces of equipment. At the same time, larger companies that have been able to renovate their production facilities in time, are now anxious about such matters as, above all, further training for their staff, labour motivation, and quality of works.

Hence, controlled and uncontrolled factors of external and internal environment, which determine the actual situation in gold-mining, call for analysis, intensive study and timely response by means of formation and perfection of a mechanism of sustainable development at enterprises of mining industry in the context of economic recession.

2 Theoretical framework and literature review

Theoretic aspects of development, formation and perfection of the mechanism of sustainable development at industrial enterprises are considered in works of such foreign researchers as R. Daft, J.L.

* Corresponding author: eltcha@mail.ru

Laft, M. Robson, Joseph Alois Schumpeter, J. Heywood and others.

Works of Russian scholars devoted to methodological issues of stability of economic systems also present a great interest, namely works of D.R. Ageeva, A.V. Bezgodov, O.V. Bratkov, L.V. Volkova, D.S. Kondaurova, A.G. Koryakov, O.V. Mikhalev, V.M. Tumin.

Theoretical and practical problems of economic stability at mining and gold-mining enterprises are studied in research papers of R.A. Arend, R.S. Moiseev, T.V. Petrov, A.N. Strekhin, V.M. Shek.

Theoretic research in the field of businesses evaluation is presented in papers of such foreign and Russian scholars as G.M. Desmond, G.S. Harrison, R. Link, R. Reily, C. Wally, V.V. Balakirev, V.V. Grigoriev, A.G. Gryaznova, E.D. Chikina, V.A. Shcherbakov and others.

Problems of modelling parameters and stability conditions for enterprise are presented in scientific papers by T.S. Bulyshev, I.V. Dodonov, A.I. Kibitkin, V.D. Kovalyov, K.A. Miloradov, L.B. Senetskaya.

Reviewing academic papers on the given issue, as well as studying practices of economic activities at industrial enterprises in modern conditions, highlighted practical obstacles in implementing the sustainable development strategy. Practical problems are also evidenced by imperfection of management procedures applied by enterprises. One of such imperfections is that there is no effective model designed to assess and forecast the value of the enterprise, which would take into account an inherent uncertainty of external factors.

The object of the research is sustainability of a Russian industrial gold-mining enterprise in volatile external environment, when exposed to recessionary influences.

The goal of the article is designing and formation of the sustainable development mechanism of an industrial enterprise and introducing this mechanism into practical activities of mining enterprises of the Russian Federation.

3 Materials and methods

A. Methods Description

An industrial enterprise is a dynamic open goal-based nonlinear system [3]. The system dynamics can be presented as a change of properties, parameters, which brings about absolutely new states of the system and thus provides for stability while launching certain qualitative characteristics [4].

Sustainable development is ensured by such a response to external impacts that allows to minimize their negative influence through anticipation and making expert managerial solutions [5,6]. Effectiveness of such solutions accounts for the precision of the enterprise management mechanism, which includes assessment of the economic stability of the enterprise and clearly defined algorithms of crisis management [7]. Examination of works on the theory of systems, unstable state of industrial objects, together with considering specific features of mining enterprises, demonstrated that in gold-mining industry we should view enterprises as stochastic non-linear dynamic systems.

Having reviewed research papers by V.V. Kovalchuk, L.B. Senetskaya, A.V. Shmidt, T.A. Khudyakova, the author offers the simulation model aimed to model business processes based on analysis of financial, economic, technical and other parameters in the state of instability [8,9]. The said model is to serve as an instrument of quantity assessment with regard to stability of gold-mining enterprises.

For managers it is essential to make an adequate consequence of decisions based on assessment of the state of the enterprise, which would facilitate overcoming recessionary phenomena.

An adequate assessment of the current situation and timely managerial decisions enable the enterprise to implement the strategy of sustainable development provided there has been formed a mechanism designed to regulate unstable economic systems [10, 11].

The sustainable development mechanism for enterprises of the gold-mining industry should be defined as a system of consecutive managerial impacts of the managerial subsystem made on the microeconomic system; herewith, these impacts are aimed to ensure that the enterprise performs its functions and has a longer life cycle in the framework of implementing a proper strategy.

The mechanism of sustainable development of gold-mining enterprises includes fundamental action blocks that correspond to the managerial function, a system of monitoring environment, parameters of making decisions, and obtaining a feedback. Altogether it provides for the possibility to anticipate external challenges and to respond to them in a timely and precise manner [12, 13].

Algorythm of search for methods aimed to enhance efficiency of the management system

To launch the mechanism of sustainable development of gold-mining enterprises it is necessary to introduce the simulation and block models of business development, which requires application of a step-by-step approach with the following stages:

The forecasting stage, which implies examination of trends of social and economic development, consideration of statistical data aimed to identify possible scenarios of the enterprise business activity.

The analysis stage, which involves an all-around analysis of different aspects of the enterprise operation. The key criterion here is completeness of information, which allows to estimate how balanced and proportionate the enterprise is, and also to consider its potential for sustainable development.

Organisation and administration stage, which consists in setting up the mechanism of sustainable development of the industrial enterprise, and introducing the model of making assessments and taking decisions.

Procedures and pragmatics stage, which involves designing necessary managerial solutions, recommendations, principles, management methods.

Control and adjustment stage allows to compare results of development management with data from prior periods, and to detect deviations.

B. Graphic model

An existing experience of modelling activities of industrial enterprises demonstrates that it is essential to conduct an intensive study and thoroughly consider all external and internal factors, without which realistic evaluation of the state of the enterprise is totally impossible [14].

The simulation model of the enterprise activity involves estimating the probability of loss occurrence at the end of the simulated period with provision for changes in the environment parameters, identification of diapasons where the state of the enterprise is not stable, and their quantitative assessment necessary to form the sustainable development mechanism (Fig. 1).

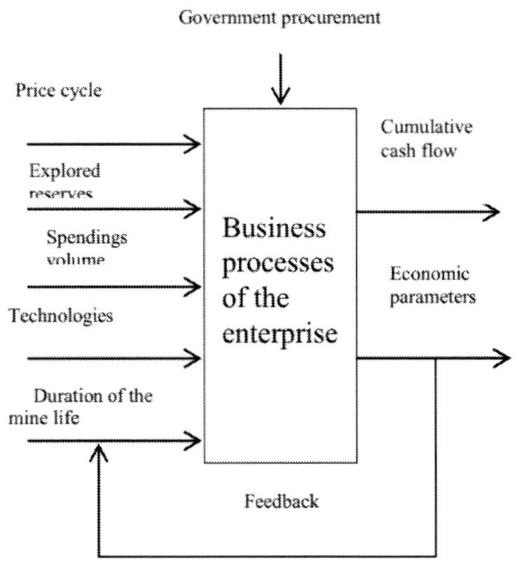

Fig. 1. The simulation model of value analysis for a gold-mining enterprise.

This simulation model serves as the basis for conducting assessment of statistical probability concerning occurrence of recessionary phenomena and prospective losses occurrence at the end of the fixed period of activity of the enterprise, which would take into account changes in environmental parameters characteristic of the given industry sector. The model deals with three diapasons that correspond to possible stability states of the enterprise, where the first is the diapason of return, and the third is the diapason of necessary bankruptcy, which can occur if managerial actions during the transition diapason were unsatisfactory.

Business practice suggests that the gold-mining enterprises most exposed to instability are enterprises that mine deposits with expected reserves of less than 50 tons, with the average duration of the deposit mining (life of mine) up to 7 years during phases in the price cycle when the price for gold is low [15].

Each of the mentioned stability diapasons is specified with characteristic values for dynamic and static parameters. Dynamic parameters consider the daily average rate of changing cumulative cash flows. This rate has a tendency to decrease when the enterprise moves to the transition diapason, and then to the diapason of necessary bankruptcy, when the monetary funds of the enterprise decrease drastically and the enterprise generate losses.

In order to enable a timely response to instable operation patterns at the enterprise, it is necessary to activate the mechanism aimed to control the state of such economic agents that got in a hazard-prone area, in a crisis situation. Core elements of the given mechanism are schematically presented below (Fig. 2).

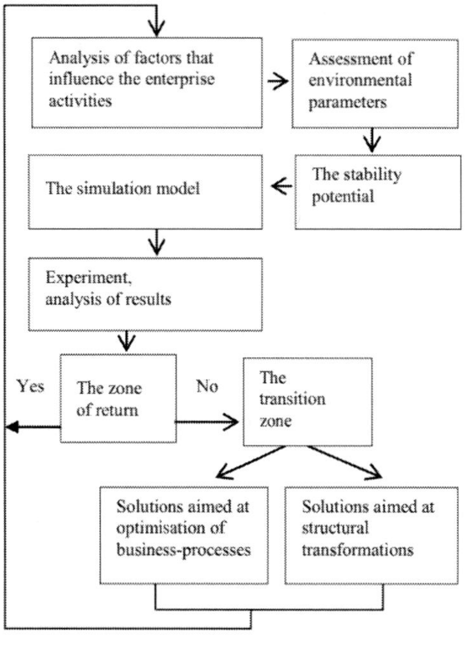

Fig. 2. The sustainable development mechanism for a gold-mining enterprise in a crisis situation.

The present model takes into account the fact that operation of the mining enterprise is subjected to numerous external and internal environment factors. It also provides for monitoring the enterprise stability parameters, developing and applying certain instruments designed to activate the stability potential and perform an integrated management at the gold-mining enterprise [10].

The given model allows to detect environmental challenges and to undertake adequate steps to overcome crisis situations by means of developing appropriate managerial solutions. For example, solutions aimed to improve business processes at gold-mining enterprises imply the following actions: using advanced resource saving technologies; optimisation of processes of developing gold reserves with due account for climatic conditions of Yakutia, Kamchatka, Far North; providing advanced vocational training and career enhancement for staff members; launching new forms and methods in the field of employee engagement. The described complex of measures, as evidenced by practices of the enterprises under analysis, i.e. AO "Serebro Magadana" (Silver of Magadan), OAO "Polymetal", allows managers to increase labour productivity and volumes of extracting precious metals by 5–7% [16].

It is essential to apply a rational approach when replacing import technologies and equipment with domestic ones, which are cheaper and more reliable in conditions of permafrost, since they allow to continue the process of extraction and beneficiation of ores in conditions of great drops in temperatures in wintertime [17, 18].

Managerial solutions connected with structural transformations imply merging, setting up partnerships between enterprises. This is due to the fact that maximum efficiency is achieved by enterprises, which mine deposits with expected reserves from 100 to 300 tons, with the average duration of the deposit mining (life of mine) 15 – 20 years.

Implementation of managerial solutions made in the framework of the sustainable development mechanism results in increased net cash flows, decreased losses risks, a consolidated competitive position of the enterprise [19]. Applying the sustainable development mechanism produces a synergetic effect, accounted for operations improvement, reduction of expenses and increase of income of the mining enterprise [20].

C. *Economic and mathematical model*

The simulation model of business operations in the gold-mining industry sector allows to model activities of economic agents in the situation of uncertainty and expect improvements only under the condition that that there was performed a proper analysis of necessary parameters that depend on the impact intensiveness on the part of internal and external environment factors.

In order to ensure maximum precision of quantitative estimation of parameters of the diapasons that correspond to stability states of the gold-mining enterprise, it is recommended to use the estimate of probability of occurrence of the losses risk P. In this case the time interval under analysis should be divided into diapasons. The offered model is based on assessing the probability of occurrence of a prospective loss at the end of the period subject to modelling.

$$P=A(d,r,g,s)+B(d,r,g,s)K+C(d,r,g,s)K \quad (1)$$

where P .is probability; ABC – coefficients whose values depend on d,r,g,s; d – duration of the deposit mining (life of mine); r,g,s – change in price for gold, rent of the minefield (a lot with the deposit), exploration (search) of the actual reserves; K – a number of units X in the amount of financial assets of the enterprise for the beginning of the period; X– characteristic value, which stands for the means necessary to exploit the deposit.

4 Conclusions

In conclusion we should note that analysis of theoretic sources and conceptual approaches used in studies devoted to how economic systems behave in conditions of crisis, together with summarising practical experience of assessing stability of industrial enterprises suggest that enterprises in the sector of gold-mining should be viewed as stochastic dynamic nonlinear systems.

In order to estimate probability of occurrence of a prospective loss we offer to employ a method of the simulation modelling of activities at the gold-mining enterprise, which involves changing parameters of internal and external environment. The simulation model allows to distinguish three stability diapasons, specify their borders, perform a quantitative estimation of the position of the enterprise in the context of recessionary phenomena, estimate a probability of loss occurrence for the economic agent for the end of the period, which is subject to modelling.

We offer the sustainable development mechanism for enterprises of gold industry that involves analysis of environmental factors, assessment of the sustainable development index (SDI), monitoring external and internal environment, management methods and tools that allow the enterprise to respond crisis phenomena in a timely and successful manner, and to adjust the development strategy.

The work was supported by Act 211 Government of the Russian Federation, contract № 02.A03.21.0011.

References

1. R. Arend, Questions of Economics, 24-36 (2006)
2. A.S. Gorevsky, *Crisis Management* (INFRA-M, Moscow, 2011)
3. T.A. Khudyakova, A.V. Shmidt, *Proc. of The 26th International Business Information Management Association Conference*, 1626-1634 (2015)
4. A.G. Koryakov, Questions of Economy and Law, **4**, 42-49 (2012)
5. T.G. Kasiyanenko, *Business Valuation: textbook for bachelor students* (The Publishing House URIGHT, Moscow, 2014)

6. A.V. Shmidt, T.A. Khudyakova, *Proc. of The 27th International Business Information Management Association Conference* (2016)
7. D.R. Kondaurova, *Proc. Enterprise, Economics, Management, Finance*, 130-132 (2012)
8. I.A. Baev, V.I. Shiryaev, E.V. Shiryaev, *Dynamic theory of the firm, Monograph* (SUSU Publishers, Chelyabinsk, 2001)
9. V.V. Kovalchuk, L.B. Senetskaya, Bulletin of Moscow State Technical University, **13**, **1**, 136-140 (2010)
10. T.A. Khudyakova, *Proc. of the International Congress on Interdisciplinary Behavior and Social Science*, 292-294 (2015)
11. O.V. Mikhalev, *Economic Stability of Household Systems: Methodology and Practice of Research and Application Analysis* (Publishing House of St. Petersburg Academy of Management and Economy, St. Petersburg, 2010)
12. T.A. Khudyakova, A.V. Shmidt, *Proc. of The 26th International Business Information Management Association Conference*, 1612-1616 (2015)
13. T.A. Khudyakova, *Proc. of the International Congress on Interdisciplinary Behavior and Social Science*, 295-297 (2015)
14. N.V. Rodionova, *Crisis Management* (Publishing House UNITY, Moscow, 2012)
15. T.A. Khudyakova, Actual problems of economics, **12**, 444-450 (2015)
16. M.V. Lysenko, Fundamental Research, **4(1)**, 56-61 (2013)
17. T.A. Khudyakova, A.V. Shmidt, *Proc. of The 26th International Business Information Management Association Conference*, 1617-1625 (2015)
18. O.A. Nikolayevskaya, *Management of development of the industrial enterprises on the basis of cost approach, Monograph* (Creative economy, Moscow, 2013)
19. Z.P. Rumyantseva, *Organisation Management: study guide* (INFRA-M, Moscow, 2010)
20. R.S. Golov, *Innovation and Synergetic Development of Industrial Enterprises (Theory and Methodology)* (Dashkov and Co, Moscow, 2012)
21. T.A. Khudyakova, A.V. Shmidt, Mediterranean Journal of Social Sciences, **6**, **4**, 274-279 (2015)

Research and assessment of competitiveness of large engineering complexes

V.V. Krivorotov[1,*], *A.V.* Kalina[1], and *V.D.* Tretyakov[2]

[1]Department of Economics of Industrial and Energy Systems, Ural Federal University named after the first President of Russia B.N. Yeltsin, Ekaterinburg, Russia
[2]Project Office, CJSC "Energomash (Ekaterinburg) – Uralelectrotyazhmash", Ekaterinburg, Russia

Abstract. The urgency of the problem of ensuring the competitiveness of manufacturing and high-tech sectors is shown. Substantiated the decisive role of the large industrial complexes in the formation of the results of the national economy; the author's interpretation of the concept of "industrial complex" with regard to current economic systems. Current approaches to assessing the competitiveness of enterprises and industrial complexes are analyzed; showing their main advantages and disadvantages. Provides scientific-methodological approach to the study and management of competitiveness of a large industrial complex; the description of its main units is provided. As a Central element of the scientific methodology approach proposed the methodology for assessing the competitiveness of a large industrial complex based on the Pattern-method; a modular system of indicators of competitiveness is developed and its adaptation to a large engineering complexes is made. Using the developed methodology the competitiveness of one of the largest engineering complexes of the group of companies Uralelectrotyazhmash, which is the leading enterprises in electrotechnical industry of Russia is assessed. The evaluation identified the main problems and bottlenecks in the development of these enterprises, and their comparison with leading competitors is provided. According to the results of the study the main conclusions and recommendations are formed.

1 Introduction

The modern world is characterized by a high level of economic activity in its different manifestations and various dynamic changes that affect socio-economic development of the society. In such conditions, competition is the main driving force of development of economic entities. As a consequence, the problem of increase of competitiveness of economic entities, from private companies to the largest transnational companies (TNCs) and states' economies as a whole is of high scientific and practical significance in the modern dynamically changing world. Research in this direction is presented by numerous works by leading scholars and practitioners. These studies develop a variety of approaches to providing, enhancing and managing the competitiveness of economic entities, which strongly differ in the basic principles and ideology of the research, and from positions of the methods and models. However, despite the diversity of ongoing research, the vast majority of them relie on a scientific approach involving primary research dealing with problems of competitiveness, and then practical implementation of the scientific principles in the activities of economic entities. Such positions are followed by a group of authors in solving the problems and challenges identified in this article.

2 The urgency of the problem of competitiveness of machinery industries in modern Russia

Dynamics of world economic development shows that in fact since the nineteenth century the basis of all the changes, transformations, became the scientific-technical progress (STP). First of all, STP is associated with the interconnected development of science and technology, using in production advanced achievements in science and invention, leadership in innovation activity, etc. Thanks to the active manifestation of STP many countries not rich in natural and material resources, have achieved tremendous success, becoming world leaders in certain areas and segments of activity and achieving high performance levels and the quality of life for their inhabitants.

As the experience of the development of the leading economies in the world shows, the basis of STP are the so-called high-tech industries which are primarily associated with the manufacturing sector of the economy, but mainly with such segment as machinery production. Namely, in the engineering sector there are the most high-tech and science-intensive products with the highest degree of readiness and the highest demand in the markets of finished products.

Consequently, we can say that the status of engineering industries largely determines the

* Corresponding author: v_krivorotov@mail.ru

competitiveness and development prospects of the state economy as a whole. Therefore, the task of maintaining and enhancing the competitiveness of engineering industries and their individual segments becomes a priority associated with the current and prospective development of the economies of individual countries and regions. The basis for its solution is the development and implementation of competitive strategies of development of machinery enterprises that ensure their work at a level appropriate to leading world analogs from the standpoint of competitiveness. Thus, the formation of competitive strategies should be based on the approach, seamlessly taking into account the different sides and aspects that determine the competitiveness of enterprises in their continuous comparison with the characteristics of the external environment (competitors). Another key aspect of the modern development of the world economy is globalization processes, accompanied by the gradual formation of a common economic space, "erasing" of borders and barriers in foreign relations between states, the formation of TNCs, the scale of which is comparable to the economies of individual countries. As a result, today the central object of activity of economic systems becomes industrial complex (IC) associated with large integrated structures, which are the main drivers of the development of socio-economic systems. In this paper the authors give the following interpretation of the concept of IC: "the industrial complex is an integrated structure consisting of the main enterprises of the production cycle, integrated into a single technological chain of production, from development, production to sales and after-sales service, and auxiliary and service enterprises, collaborating and benefiting from it in the form of a synergistic effect" [1, p. 76].

Experience shows that the use of approaches to assessing and ensuring the competitiveness of individual companies does not allow to consider all features and functions of IC. The problem is that in the industrial complex, the enterprises have close links within the production cycle, from ore extraction and raw materials to its processing, the manufacture of final products and delivery to consumer, and these connections form a certain synergy, which, in fact, increases the competitiveness of these structures compared with private enterprises. Thus, it is necessary to develop new scientific and methodological approach to the assessment and management of competitiveness of major IC, which form the modern economic landscape of the world economy. This approach will allow to assess the current IC's position on the background of leading competitors, identify the main "point of application" of control actions in order to increase competitiveness and to form an optimal portfolio of activities required in the context of resource constraints.

One of the main components of this approach should be a comprehensive method of assessing the competitiveness of the IC, allowing on the one hand to carry out its current rating in comparison with the main competitors and to identify bottlenecks and "strengths" of the considered IC. On the other hand, this technique and the obtained results should be the basis for development of competitive strategies of the IC and its constituent companies.

3 The analysis of modern approaches to evaluation of competitiveness of economic entities

As shown above, competition is the main driving force of development in market models. As a result, the approaches to assessing the competitiveness of various actors of the current socio-economic systems have been widely developed: from small companies to national economies and transnational corporations.

The founders of papers on the study of competition and competitiveness are the representatives of classical political economy – A. Smith and D. Riccardo [2, 3] with their perfect competition models, which were ahead of their time. So, A. Smith identified competition as the main driving force of the optimization of parameters "price-quality" of products in order to capture and retain competitive advantage.

In the future, as the development and complexity of socio-economic systems and processes there were other, more complex and sophisticated models of competition. Primarily, these include:

1. *The structural approach* developed in the work of A. Cournot [4], F. Edgeworth [5], J. Robinson [6], E. Chamberlin [7] and many other famous scientists. In this approach when investigating competition the emphasis is shifting from rivalry firms with each other on the analysis of market structure, that is, those conditions in which firms exist.
2. *The functional approach* proposed by J. Schumpeter in the early 20th century, who considered competition as the rivalry of old and new [8]. Further, this approach has been developed in the works of many known scientists. For instance, F. Hayek interpreted the competition as a discovery procedure, in accordance with which as the main criteria of competitiveness of the company performs the innovative activity [9].
3. *The marketing approach* based on the idea of consumption control, which, in turn, involves the environment and, in particular, consumers manipulation by the producers and distributors. Among the scientists adhering to this approach, highlight J. Galbraith [10], F. Kotler [11] L. L. Radder and Lowe [12] and many others.

Along with the development of approaches to the study of competition the approaches and methods for assessing the competitiveness of managing subjects have been broadly developed. Speaking on the assessment of the competitiveness of enterprises and IC in the first place, should be highlighted the approaches and methods that take into account *the competitiveness of the products*. In most methods based on a specific approach, the evaluation of the goods is proposed to be carried out by the use of economic and qualitative indices of competitiveness of goods, which characterize, respectively, the cost and technical (qualitative) characteristics of the investigated products [13 – 14 and others]. Not less popular are *the methods based on the competitive strategy* of mainstreaming the company's position in the market, attractiveness of the industry, but also based on strategic behavioral recommendations. In this group of methods should allocate the Boston consulting group matrix "relative market share / market growth Rate" [15], Mc Kinsey matrix "Position / attractiveness of the industry" [16], M. Porter's "Strategic goal / Strategic advantage" matrix [17].

In the *methods based on the concept of the value chain* of the company, the latter's activity is divided into primary, associated with the production and distribution of the product, and secondary, related to support activities that provide inputs and infrastructure, which allow to perform primary activities. The competitiveness of enterprise is considered from the perspective of efficient implementation of the above activities relative to its competitors [18].

Widespread in practice became methods based on *the business valuation*. The basic idea of these methods is that the higher the value of the enterprise, the more economic benefit it can bring to its owner and consequently, the higher is its level of competitiveness [19 – 22 and others].

In recent times widespread for the tasks associated with the assessment of the competitiveness of enterprises and IC have become *methods based on game theory*. The essence of this group of methods lies in the choice of optimal strategy of behavior of economic entities aimed at obtaining maximum gains or reduce losses in competition with other players. As the optimized criterion can be the performance of the entity: bank branches' expenditures [23], market positioning [24, 25], the production price [26] and other indicators.

In addition to these methods, in the literature one can find links to other less common groups of methods (graph-analytical methods, potential methods, integrated methods, etc.). However, in most cases they use approaches and techniques that are an integral part of the above mentioned methods.

Summing up the analysis of methods and approaches to assessing the competitiveness of enterprises and IC, it can be noted that in the vast majority of research and evaluation of the competitiveness of enterprises and IC focuses on one of the groups of factors determining competitiveness (product, marketing, management, etc.). On the other hand, there are almost no approaches based on comprehensive consideration of various influencing factors. Existing methods of estimation of competitiveness of business entities do not take into account the complex factors of competitiveness inherent to the big IC. In this regard, the apparent urgency of developing a comprehensive methodological approach to assessing the competitiveness of major IC, comprehensively taking into account various aspects of their activities and organization is obvious.

4 Scientific and methodical approach to the assessment and management of competitiveness of large industrial complexes

In the developed approach IC, as noted earlier, is considered as an integrated production structure united around the technological chain of production, sales and after-sales service of products, which may include a sufficiently large number of enterprises (in some cases several dozen or more).

The proposed approach is a systemic approach, according to which the object is seen as a holistic set of elements in the set of relations and links between them [27]. In this case, there are three major interrelated units aimed at sequential solution of the problem of the assessment and management of competitiveness of the IC:

1. The Section of evaluation of the current level of competitiveness of the IC on various aspects of financial and economic activity.
2. The Section of selection of the optimal portfolio of control actions aimed at improving the competitiveness of the IC.
3. The Section of the formation of predictive strategies for the development of ICs aimed at ensuring the competitiveness of long-term competitive advantage of ICs in the forecast period.

The main tasks in the first section, are complex factorial assessment of competitiveness of the studied ICs with the identification of strengths and bottlenecks in their development and definition (based on evaluation) the main directions of development of the IC in the foreseeable future, as well as "points of application" of control actions.

The second section solves the problem of selecting the optimal portfolio of control actions in order to maximize the increment of the competitiveness of the ICs in terms of existing resource constraints.

Finally, the task of the third section is elaborating and implementing the strategy for the development of ICs in the forecast period (through the various stages of the prospective period), which would allow to ensure the survival and maintenance of sufficient levels of competitiveness of the IC and its constituent business-units in the future.

A central element of the considered scientific and methodical approach, linking and "coordinating" the work of all of the above section is a technique of a complex estimation of competitiveness of the IC.

Methods of assessing the competitiveness of the IC is based on the Pattern-method widely used to solve such problems [28], which has been successfully applied by the authors before [1, 27, etc.]. In accordance with the proposed approach the methodology involves the following main provisions and principles of the assessment of the competitiveness of the IC:

• comparison of indicators of competitiveness of the investigated IC with some standard (base) model with predetermined characteristics;

• integrated indicator of competitiveness as the ultimate goal of the assessment calculations and comparisons;
• the block structure of competitiveness indicators of IC.

As shown above, the ultimate aim of assessing the level of competitiveness of the IC in accordance with the methodology is the determination of the integral competitiveness index that combines various sides and aspects of IC's operations. In this case, the realized and potential competitive capabilities of the IC are estimated. The composite index of competitiveness is calculated by two major characteristics:

$$C_{IC} = \sqrt{C_{cur} \cdot C_{pot}}, \quad (1)$$

where C_{cur} is the level of IC's current (implemented) competitiveness; C_{pot} is the IC's competitive potential.

Among the main advantages of the presented approach can be identified:
• the possibility of quantitative measurement of the competitiveness index that allows you to compare multiple ICs or multiple variants of development of the same IC;
• a comprehensive consideration of various criteria for the development of the IC without giving priority to some chosen criterion;
• the methodology is "open" and allows relatively easily to change the composition of indicators, C_{cur} and C_{pot} depending on your goals, the characteristics of individual ICs and industries, as well as available software;
• in the equation (1) it is possible to combine the results of evaluations of individual aspects of an IC, made on the basis of various methodical approaches and algorithms.

Each of the characteristics presented in the equation (1) is broken into smaller components – blocks that characterize certain aspects of IC's operation (production, marketing, corporate culture, etc.). In turn, each block consists of individual indicators of competitiveness – *Ci* (5 – 7 or more indices in the block), the ultimate characteristics of competitiveness of the investigated ICs in comparison with the main competitors or key characteristics of the competitive environment.

Calculating the *Ci* is the mapping of the relevant indicator of the competitiveness of the IC with the same index of the base model:

$$C_i = \frac{I_i}{I_{basis,i}}, \quad (2)$$

where I_i is the value of indicator i of IC's competitiveness, $I_{basis,i}$ – the basic valueof i competitiveness indicator.

Indicator of current competitiveness,

C_{cur} is tinguishes the following components (blocks):
1) the operating efficiency and market position (C_{IC1})
2) the competitiveness of IC's key products (C_{IC2})
3) the condition and efficiency of functioning of IC's industrial-technological base (C_{IC3})
4) the efficiency of staff and personnel policy (C_{IC4})
5) the quality of IC's organization and management (C_{IC5})
6) investment and innovation activity of IC (C_{IC6})
7) risks associated with the activity of IC (C_{IC7})

Indicator of the IC's competitive potential (C_{pot}) is broken down to such components (blocks):
1) the potential use of manufacturing capacity (P_{mc})
2) the market potential of the IC (P_{market})
3) accordance of IC's personnel qualification to demands of scientific and technical progress (P_{pers})

A key feature of the scorecard methodology is the orientation to available sources of information (enterprises' data reporting, public information data on competitors, raw materials and product markets, statistical services, etc.). This distinguishes the proposed method of assessing the competitiveness of ICs from many of the same techniques where the evaluation rely heavily on the expert assessment procedure and survey data, the implementation of which in many cases is difficult.

5 Assessment of competitiveness of engineering complexes on the example of a group of companies uralelectrotyazhmash

The proposed scientific and methodical approach was implemented in assessing the competitiveness of one of the largest Russian engineering complex, formed by a group of companies JSC "Energomash (Ekaterinburg)-Uralelectrotyazhmash" and LLC "Elmash (UETM)", which are the leading enterprises of electrotechnical industry of Russia. The received results became a basis for formation of competitive strategies of development of these enterprises.

In evaluating competitive environment for the group of companies Uralelectrotyazhmash the world's leading manufacturers of electrical equipment were considered– Siemens (Germany), ABB (Sweden), Alstom (France), Hyundai Heavy Industries (South Korea), Toshiba (Japan), as well as a number of Russian manufacturers working in this area –JSC "Elektrozavod", LLC "Togliatti transformer", "CJSC "SverdlovElectro" Group, "Power machines – Toshiba. High voltage transformers", the company "High-Voltage Union", CJSC "Elektroapparat". In the formation of basic values for individual indicators of competitiveness, which is necessary for carrying out comparative analysis in accordance with the Pattern method, in most cases, the values by corporations Siemens, ABB, Alstom, is the world's leading manufacturers in the world were taken.

Competitiveness assessment was conducted in the period 2012 – 2015, the results of the calculations are presented on Figure 1. The assessment at this stage due to the information restrictions and high uncertainty of the current situation of the world and Russia was carried out only in terms of the level of current (implemented) competitiveness of the IC (C_{cur}) and constituent components, which, in fact formed integral index of competitiveness – C_{IC}. In addition, in the calculations for the same reasons, was not considered an indicator of risks associated with the activity of IC (C_{IC7}).

As shown by the assessment, *the unit 1 "Operational efficiency and market position"* in the group of UETM demonstrated a failure in 2014, while overall long-term trend towards a slight decrease in the level of competitiveness. This is due primarily to lower net income of UETM caused by falling order volumes because of the reduction of investment programs of key customers with a high level of fixed costs. Nevertheless, the indicator of operational efficiency of UETM surpassed the analogue competitor Alstom Grid.

On *the block 2* calculations were carried out in an express assessment on the main tyoes of high voltage equipment: live-tank circuit breakers VGT-UETM 126 kV dead-tank circuit breakers VEB-UETM 245 kV (compared with analogues produced by Siemens), as well as gas-insulated switchgear KRUE-UETM on 126 kV (compared with the analogous ABB for weight and size indicators). Overall, it can be stated that the products manufactured by UETM correspond to the world standards in technical performance and quality indicators.

Block 3 "Condition and efficiency of functioning of industrial-technological base" assessment shows relatively low values of indexes of competitiveness. The level of depreciation of group UETM exceeds the similar indicator of the leading competitors. Approximately similar results shows an energy intinsity indicator.

The most problematic position of the group companies of UETM *is the level of the investment activity (block 6).*

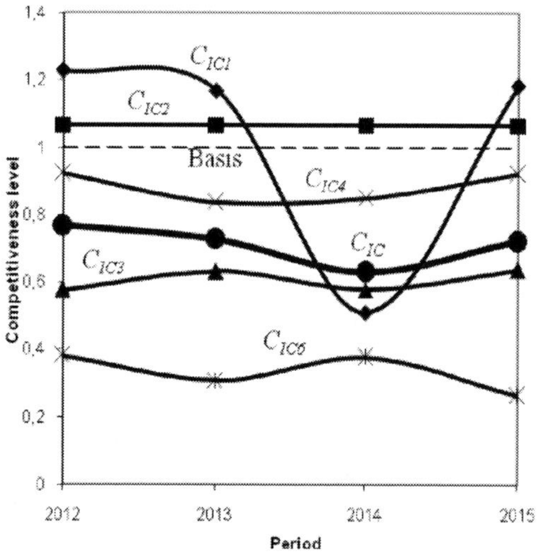

Fig. 1. Assessment of competitiveness of the group of companies Uralelectrotyazhmash compared to leading competitors.

First of all, with high innovation activity and R&D expenses there were practically no investments in updating production facilities.

Overall, the value of *the integral index of competitiveness* of UETM from 2012 to 2014 has been steadily reducing, which is caused primarily by shortage of working capital funds to finance orders and gain profit as well as investing in upgrading the production basis. However, at the end of the period (2015) there has been a fracture of the negative tendency, and the integral index of competitiveness began to return to the state of 2013. The evaluation found that when implementing investment projects aimed at improving the competitiveness of the group UETM, the target is to give preference to those which focus on:

1) development of new market niches and attracting new orders for the purpose of working capital replenishment.
2) renewal of fixed assets due to the adoption of energy efficient technologies as well as technology excluding the residual material suppliers with an effective increment of the proportion of value added by an enterprise in the manufacture of its products;
3) upgrade infrastructure to increase the energy efficiency.

Direct selection of investment projects within identified priority areas should be conducted using the model of optimal portfolio selection of control actions according to the criterion of the maximum increment of the competitiveness in conditions of limited resources. In detail, the algorithm for constructing the model and conducting the selection procedure is considered in [27].

6 Conclusion

The study and the results obtained allow to note the following:

1. Based on the analysis of modern approaches to the study of competition and competitiveness of economic entities a scientific and methodical approach to the assessment and management of competitiveness of large industrial complexes, which are the central elements of economic activity in modern economies is developed. The proposed approach relies on the use of methods of system approach and system analysis and takes into account the relationship between the state of the IC and control actions oriented to change of this state towards meeting a set of requirements from the standpoint of competitiveness.
2. A comprehensive method of assessing the competitiveness of the IC that has the property of universality and is applicable to ICs, which operate in various industries, and based on the method of Pattern is developed. In accordance with the methodology, assessment of competitiveness level is carried out by the comparison of the investigated IC with some basic model that characterizes the similar figures of the leading competitors or the competitive environment in general. The ultimate goal of the evaluation of the developed method is the determination of the combined (integral) index of competitiveness, which is calculated on the base of the aggregate competitiveness indicators describing different aspects of development and functioning of the IC.

3. Practical implementation of the developed methodological tools applied to the problem of evaluation of the competitiveness of IC group UETM, which is the largest engineering complex in Russia is carried out. The results of the assessment revealed strengths and weaknesses of the activities of UETM in comparison with leading competitors, factorial causal analysis is conducted. The obtained results allow to begin closely to develop a roadmap for the development of the IC and to define the main strategic orientations of this development.

This research was supported by the Government of the Russian Federation (Act 211, agreement № 02.A03.21.0006).

References

1. V.V. Krivorotov, A.V. Kalina, V.D. Tretyakov, S.E. Erypalov, The Competitiveness of Socio-Economic Systems: Challenges of the New Era, 75-138 (2014)
2. A. Smith, *Inquiry into the Nature and Causes of the Wealth of Nations* (Petrocom, Petrozavodsk, 1993)
3. I.A. Stolyarov, *Anthology of economic classics* (Econav–Key, Moscow, 1993)
4. A. Cournot, *Recherches sur les principes mathematiques de la theorie des ri-chesses* (Paris, 1838)
5. F.Y. Edgeworth, *Mathematical Physics: An essay on application of mathematics to the moral sciences* (1881)
6. J. Robinson, *The Economics of Imperfect Competition* (Progress, Moscow, 1986)
7. E. Chamberlin, *The Theory of Monopolistic Competition: A Re-orientation of the Theory of Value* (Ekonomika, Moscow, 1996)
8. J. Schumpeter, *Theory of Economic Development* (Progress, Moscow, 1982)
9. F. Hayek, *The Road to Serfdom* (Ekonomika, Moscow, 1992)
10. J. Galbraith, *The new industrial society* (Progress, Moscow, 1969)
11. F. Kotler, *Marketing management. Express course* (Piter, Saint Petersburg, 2003)
12. L. Radder, L. Louw, Long Range Planning, **31**, **4**, 549-559 (1998)
13. L.N. Chaynikova, V.N. Chaynikov, *Competitiveness of the Product* (Tambov State Technical University, Tambov, 2007)
14. R.A. Fatkhutdinov, *Strategic Marketing* (Intel-Sintez Business School, Moscow, 2000)
15. B. Henderson, *The Product Portfolio*. https://www.bcgperspectives.com/content/Classics/strategy_the_product_portfolio/.
16. McKinsey & Company, Enduring Ideas: The GE–McKinsey nine-box matrix (2008). http://www.mckinsey.com/insights/strategy/enduring_ideas_the_ge_and_mckinsey_nine-box_matrix.
17. M.E. Porter, *Competitive Strategy: Techniques for Analyzing Industries and Competitors* (Alpina Business Books, Moscow, 2005)
18. M.E. Porter, *On Competition, updated and expanded edition* (LLC "ID Williams", Moscow, 2010)
19. T. Copeland, T. Koller, J. Murrin, *Valuation: Measuring and Managing the Value of Companies* (Olymp-Business, Moscow, 2005)
20. M. Scott, *Value Drivers. The Manager`s Framework for Identifying the Drivers of Corporate Value Creation* (Olymp-Business, Moscow, 2005)
21. T.A. Khudyakova, A.V. Shmidt, *Proc. of The 26th International Business Information Management Association Conference*, 1607-1611 (2015)
22. T.A. Khudyakova, A.V. Shmidt, Mediterranean Journal of Social Sciences, **6**, **4**, 274-279 (2015)
23. V. Cerasi, B. Chizzolini, M. Ivaldi, Applied Economics, **34**, 2213-2225 (2002)
24. B. Doberman, H. Gatingnon, G. Sargsyan, Using Attraction Models for Competitive Optimization: Pitfalls to avoid and Conditions to Check, INSEAD Working Papers, 2006/27/MKT.
25. V. Kadiyali, K. Sudhir, R. Vithala, Methods in Marketing. International Journal of Research in Marketing, **18**, 161-186 (2001)
26. S. Sriram, V. Kadiyali, Johnson School Research Paper Series, **14-07** (2007)
27. V.V. Krivorotov, A.V. Kalina, T.V. Matveeva, A.Yu Bayranchin, *Improving the Competitiveness of the modern Russian Territorial-Industrial Complexes* (Urfu, Ekaterinburg, 2013)
28. Pattern, from Wikipedia, the free encyclopedia. https://ru.wikipedia.org/wiki/%D0%9F%D0%B0%D1%82%D1%82%D0%B5%D1%80%D0%BD.
29. V.D. Tretyakov, V.V. Krivorotov, A.V. Kalina, Bulletin of Ural Federal University. Series Economics and Management, **5**, 140-149 (2013)

Modelling management process of key drivers for economic sustainability in the modern conditions of economic development

E.S. Pishchulina[1,*]

[1]South Ural State University, Chelyabinsk, Russia

Abstract. The text is about issues concerning the management of driver for manufacturing enterprise economic sustainability and manufacturing enterprise sustainability assessment as the key aspect of the management of enterprise economic sustainability. The given issues become topical as new requirements for the methods of manufacturing enterprise management in the modern conditions of market economy occur. An economic sustainability model that is considered in the article is an integration of enterprise economic growth, economic balance of external and internal environment and economic sustainability. The method of assessment of economic sustainability of a manufacturing enterprise proposed in the study allows to reveal some weaknesses in the enterprise performance, and untapped reserves, which can be further used to improve the economic sustainability and efficiency of the enterprise. The management of manufacturing enterprise economic sustainability is one of the most important factors of business functioning and development in modern market economy. The relevance of this trend is increasing in accordance with the objective requirements of the growing volumes of production and sale, the increasing complexity of economic relations, changing external environment of an enterprise.

1 Introduction

Economic sustainability management is one of the most significant drivers for modern businesses development in modern market economy. The relevance of this trend is increasing in accordance with the conditions of changing external environment.

The current state of the issue of enterprise economic sustainability management is a lack of a clear mechanism for monitoring the prospects of development of both the enterprise and the environment and individuals who are to make sound managerial decisions [14].

As to management methods, the current market tools cannot be considered fully formed. It is especially true for inter-firm relationships, as the interaction has not been worked out while necessary communication links have not been provided. Significant differences in setting goals to complete the tasks of individual enterprises complicate management decision-making aimed to harmonize common interests. Thus, it makes the lack of behavior management capabilities of individual objects of economic activity prominent [3].

Modern Russia's economy is characterized by a strong dependence on the world oil prices. The fall of these prices, which began in 2015 and continues up to this day led to a significant fall of the ruble and, consequently, to a sharp rise in prices and reduction of the purchasing power of the Russian population. In addition, economic sanctions imposed against Russia by many of European countries and the United States, forced to introduce responsive measures against these countries. All these factors have led to a sharp decline in the economic stability of Russian enterprises, and will continue to contribute to the prominent tendency towards bankruptcy and withdrawal from the market of those companies that are not able to adjust to changes in the external environment.

Therefore, under the circumstances, in order to make the company remain economically stable in the market, efficient and profitable it is sufficient to reveal the relationship. Also it is essential to determine the dependence of drivers for economic sustainability and to make adequate managerial decisions aimed to improve it [7].

2 Theoretical framework and literature review

The current state of economic stability of enterprise management problems can be represented as a lack of a clearly defined mechanism to ensure dynamic and effective monitoring of the long-term prospects for the development of both the manufacturing enterprise and the environment and to take adequate methodologically based managerial decisions.

Problems of economic sustainability of economic systems, economic development and growth are

* Corresponding author: pishchulinaes@susu.ac.ru

reflected in the works of such Russian scientists as L. Abalkin, I. Aleshin, A. Schmidt, A. Borodin, V. Mau, A. Spitsyn, S. Gubanov, K. Khubiev, A. Neshitov, T. Khudyakova O. Sukharev, A.Goluba, I. Baev. Strategy and management issues are considered in the works of V. Efremov, N. Kondratyev, P. Pokrytan, V. Shkardun, R. Fatkhutdinov, O. Vikhansky, R. Nureyev, F. Ufimtsev, R. Popov and others. Among the foreign scientists who study the issues of economic growth and strategic management, M. Porter, I. Schumpeter, D. Aaker, I. Ansoff, *Jean Paul Blandinieres*, D. Norton, M. Meskon, J.Keynes and P. Druckershould be called.

The general laws of development of economic systems that are applicable nowadays have been studied in the works of local and foreign economists, but the issues relating to the economic sustainability of enterprises in the conditions of modern Russian economy characterized by a heightened sensitivity to external and internal factors have not been highlighted properly.

The object of the research is the management process of drivers for economic sustainability of manufacturing enterprises in the conditions of changing external environment.

The goal of the article is to develop a model of managing drivers for economic sustainability in the modern state of the economic development.

3 Materials and methods

A. Description of methods

In this study, the basics of calculating economic sustainability of a manufacturing enterprise by integrating the indicators of economic growth, the economic balance of external and internal environment and economic sustainability of the manufacturing enterprise have been developed.

First of all it is required to use financial and economic analysis of the enterprise based on the index method to generate results that will characterize the state of the economic system and the economic growth of the enterprise. Under this analysis the information obtained by the survey of the employees as well as the data gained by questioning the managers is collected and processed.

Then it is required to create a model of the economic sustainability of the manufacturing enterprise. Based on the values obtained for each employee, an individual model of the economic sustainability of the manufacturing enterprise is created. Then, a group model and a general model of the economic sustainability of the manufacturing enterprise are created [17].

Based on these results some conclusions concerning the assessment of the economic sustainability are drawn, the effect of the each individual driver for the economic stability of the manufacturing enterprise is determined. Following these conclusions there is a search and analysis of the problems in the enterprise management. Reserves are determined to improve the economic sustainability of the manufacturing enterprise. Then some recommendations how to improve the economic sustainability of the manufacturing enterprise are prepared [10].

B. Characterisics of the drivers for the enterprise economic sustainability perspective

Economic sustainability is the internal state of an economic entity which is formed under the influence of plenty of factors.

Drivers for economic sustainability can be classified according to various criteria:

1. By the way of influence: stabilizing and destabilizing, i.e. having either a positive or a negative effect on the level of the economic sustainability of the enterprise.
2. By the place of occurrence: internal and external. Internal factors are created by the internal environment as a consequence of the enterprise performance as a whole or its employees' performance. External factors are created by the external environment without any participation and against the will of the company or its employees.
3. By the nature of exposure: direct and indirect. Direct factors have a direct impact on the level of economic sustainability of the enterprise. Indirect ones do not have a direct immediate impact on the level of sustainability, rather an indirect impact.
4. By the degree of dependence: objective and subjective. Objective factors are due to processes that develop independently or due to force majeure (e.g. natural disasters). Subjective factors are created deliberately by certain groups of people based on their interests.

In the most common form the economic sustainability of a manufacturing enterprise is determined by three major drivers:

1) economic growth as the determining factor;
2) economic balance (the state of internal and external environment of a manufacturing enterprise), it is the factor that ensures the proper functioning of all available enterprise subsystems;
3) the level of control that enables the full development of the enterprise in the course of its performance [13].

The perspective aims to maximize the growth of the economy, including the enterprise, with minimal deviation values of the economic system and / or the enterprise from the balance state. It is these two factors that are fundamental in building a model of economic sustainability, which is proposed to be used for the enterprise management and assessment of its performance.

C. Graphic Model

The basis of constructing a model of economic sustainability is a different combination of the above-mentioned factors. These combinations can be divided into 4 groups:

1) if the company is characterized by economic growth, while maintaining the state of balance, it is 'relative economic sustainability';
2) if the company is characterized by economic growth and the imbalance of its internal and external environment, it is 'critical economic sustainability';
3) if the company is not characterized by economic growth and at the same time there is the imbalance of its internal and external environment, then we can talk about the state of 'economic instability';
4) if the company is not characterized by economic growth, but it is in the state of balance of its internal and external environment, then this condition is called 'stagnation'.

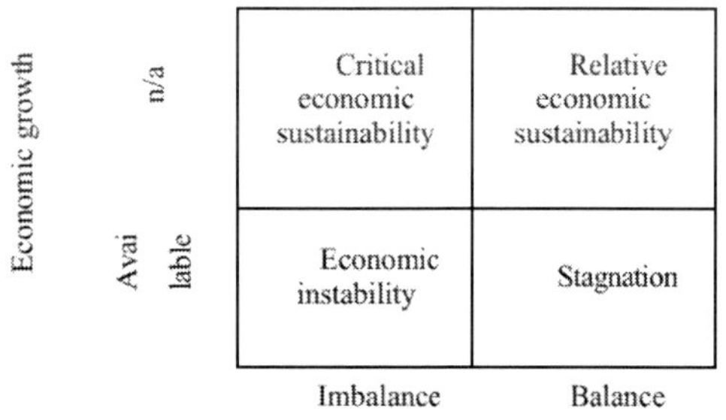

Fig.1. Model of Economic Sustainability.

The proposed model of economic sustainability can be used to improve the efficiency of company management, while rating companies by the indicator of economic sustainability and etc [5].

D. Economic-mathematical model

The assessment of economic sustainability is central to the management of the economic sustainability of an enterprise.

It should be noted that at the present time scientists and economists do not share the same opinion whether any specific indicators of the company should be considered as indicators of its economic sustainability. First of all, this is shown by a great variety of different approaches to the assessment of the economic sustainability of an enterprise under the current scientific methodology.

Today a lot of original approaches to the assessment of the sustainability of enterprises are offered. They are aimed at early detection of symptoms of performance worsening and potential bankruptcy. Diagnostic methods have been developed for more than 70 years, but even today they continue to attract the attention of analysts and remain the subject of scientific debate. The assessment of the economic stability of enterprises was deeply analyzed and a great progress in its methodology and means of implementation was noted especially in the last decade.

The economic-mathematical model presented in this study is based on the integration of the economic growth of the enterprise and its state.

The indicator of the economic growth of an enterprise is determined by the following formula:

$$E_i = \sum_{j=1}^{n} y_j , \quad (1)$$

Where y is a factor influencing the economic growth of an enterprise; n is a number of evaluated factors influencing the economic growth, j = 1 ...n.

The value of the index of economic balance of an enterprise is determined similarly to create a model of economic sustainability (E_i):

$$E_i = \sum_{t=1}^{m} x_t , \quad (2)$$

where x_t is a factor influencing the economic balance of the enterprise; m is a number of evaluated factors affecting the economic balance, t = 1...m.

An individual model of economic sustainability shown as a vector (EM_i) is created on the basis of the obtained values:

$$EM_i = (E_i ; EP_i) \quad (3)$$

The indicator of group economic sustainability is figured out as follows (EM_g):

$$EM_g = \sum_{i=1}^{p} EM_i , \quad (4)$$

where p is a number of values in this group, i = 1...p.

The indicator of general economic sustainability of a whole enterprise (EM_o) is figured out by the formula:

$$EM_o \sum_{g=1}^{k} EM_g \quad (5)$$

where k is a number of groups under economic sustainability assessment, g = 1...k.

4 Conclusions

Thus, the method of assessment of the economic sustainability of a manufacturing enterprise is an integrated approach to the assessment of the enterprise: analysis of all existing kinds of accounting and reporting combined into a single procedure with the study of the views of the employees and managers of the enterprise.

The management mechanism of the economic sustainability of a manufacturing enterprise implies some changes in the system of control and planning at a manufacturing enterprise in accordance with the identified shortcomings and reserves of its performance [16].

The economic policy of a manufacturing enterprise should aim to search for the optimal combination of the economic growth rate and the balance of the factors of external and internal environment of a manufacturing enterprise.

Management of the economic stability of a manufacturing enterprise by adjusting economic policy on the basis of identified shortcomings in its performance and untapped reserves is to achieve such a combination of methods of managerial decisions, which will ensure the effective performance of the manufacturing enterprise [19].

The integrated control mechanism of the economic sustainability of a manufacturing enterprise must include all of the functions of management and be based on the principles of continuity and coherence of actions.

The economic sustainability of an enterprise develops not only under the influence of any specific factors of external and internal environment, but also under the influence of unpredictable and difficult to predict factors of both economic and non-economic nature.

However, despite the availability of a large number of economic sustainability control mechanisms the problem of effective management of the economic sustainability of a manufacturing enterprise remains disputable and requires further improvement of existing and / or development of new effective approaches.

The current economic conditions do not encourage the domestic industry to strengthen; actually they hinder their sustainable development. But even in these difficult circumstances businesses must have the internal capacity and ability to adapt to the environmental influence, or economic sustainability.

The work was supported by Act 211 Government of the Russian Federation, contract № 02.A03.21.0011.

References

1. T.A. Khudyakova, A.V. Shmidt, *Proc. of The 26th International Business Information Management Association Conference*, 1626-1634 (2015)
2. T.A. Khudyakova, A.V. Shmidt, Mediterranean Journal of Social Sciences, **6**, **4**, 274-279 (2015)
3. T.A. Khudyakova, Actual problems of economics, **12**, 444-450 (2015)
4. A.V. Shmidt, T.A. Khudyakova, *Proc. of The 27th International Business Information Management Association Conference* (2016)
5. A.V. Shmidt, T.A. Khudyakova, *Proc. of The 27th International Business Information Management Association Conference* (2016)
6. T.A. Khudyakova, A.V. Shmidt, *Proc. of The 26th International Business Information Management Association Conference*, 1607-1611 (2015)
7. V.P. Meshalkin, A.Y. Belozersky, Russian Entrepreneurship, **9(81)**, 92-95 (2006)
8. N.N. Dmitriev, Pulp, paper and cardboard, **7**, 57-62 (2006)
9. *Economy and ecology: mechanisms of interaction.* http://bibliofond.ru/view.aspx?id=489159.
10. M.S. Abryutina, Financial management, **3**, (2002)
11. E.I. Krylov, *Analysis of financial state and investment attractiveness of an enterprise* (Finance and Statistics, Moscow, 2003)
12. A.V. Grachev, *Analysis and management of the financial stability of an enterprise* (Finpress, Moscow, 2002)
13. N.N. Kreynina, *Financial state of an enterprise: assessment methods* (Business and Service, Moscow, 1997)
14. V.V. Kovalev, *Financial analysis: methods and procedures* (Finance and Statistics, Moscow, 2001)
15. I.A. Baev, V.I. Shiryaev, E.V. Shiryaev, *Dynamical Theory of Firm: monograph* (South Ural State University Publishing House, 2001)
16. R.Y. Loskutov, *Reserves of improvement of economic sustainability of a manufacturing enterprise* (Saratov, 1998)
17. I.V. Piskunova, *Evolution of the interpretation of enterprise sustainability.* http://ecsocman.hse.ru/data/2011/05/20/1266918948/13.pdf.
18. D.M. Shotylo, *The essence and content of the production system.* http://ekportal.ru/page-id-1785.html.
19. A.B. Borisov, *Great Dictionary of Economics* (Moscow, 2003)
20. V.I. Roshchin, *Economic sustainability of enterprises and the implementation of their economic interests* (Cheboksary, 2000). http://www.dissercat.com/ content / ekonomicheskaya-ustoichivost-predpriyatii-i-realizatsiya-ikh-ekonomicheskikh-interesov.
21. A.B. Oleinik, *Enterprise economic sustainability in the modern conditions* (Volgograd, 2002). www.dissercat.com/content/ekonomicheskaya-ustoichivost-predpriyatiya-v-sovremennykh-usloviyakh.
22. T.A. Khudyakova, A.V. Shmidt, *Proc. of The 26th International Business Information Management Association Conference*, 1612-1616 (2015)

Hi-Tech sector's forming and development under the conditions of new industrialisation

O.A. Romanova[1,*]

[1]Federal State Budgetary Institution of Science "Institute of Economics of the Ural Branch of the Russian Academy of Sciences" Ekaterinburg, Russia

Abstract. The article shows that new industrialisation is a key component of the new economic development model, which has been designed in Russia. The following special characteristics of this process are defined: the synchronisation of the processes of the creation of new hi-tech economic sectors and of the modernisation of traditional ones; coordination of changes in technical and economic sphere and in social and institutional sphere; interactive managerial decisions for the realization of these changes. It is shown that new industrialisation is impossible without the creation of indigenous companies in principally new industrial markets. The new trends of global economy development which will define it in the future according to the new technological wave are described. This allows to specify the strategic vector of the new industrialisation of the Russian economy. Convergent technologies are grounded as the intellectual heart of the hi-tech sector of the Russian economy. The importance of technology transfer and the need of transformation of the system of innovations support are stressed. The industrial policy combines the priority guidelines of the development of the state, regions and municipalities taking into account their resource, science and technology and institutional capabilities. Hence, the industrial policy can be an effective mechanism of the development of hi-tech economic sector.

1 Introduction

The creation of an effective model of the Russian economy is one of the most important factors for its modern crisis overcoming. The central component of this model is the new industrialisation aimed at providing the intensive type of extended reproduction, a new quality of human capital, forming a high-tech based economy of Russia as a key factor of its competitiveness and homeland security [1]. These problems solving requires the reorientation of various engineering and technology systems to non-polluting ones. However this concerns not only the hi-tech economical sector, but also the traditional industry whose modernisation is based on so-called "nature-like" technologies and supposes the change of resources involvement with the focus on natural turnover of resources.

2 New industrialisation

The systematization of the consequences of the deindustrialisation of the Russian economy, taking into account new trends of reindustrialisation of the economy of the developed countries, analysing the Russian theoretical and practical research on neo-industrialisation have allowed to formulate the author's view of these problems and to propose the author's concept of new industrialisation. In our opinion, the new industrialisation is "the simultaneous process of the creation ша new hi-tech economic sectors, and also the effective innovative modernisation of traditional ones with agreed quality changes in technical and economic sphere and in social and institutional sphere, provided by means of interactive technological, social, political and managerial changes" [2].

Within this article, a special attention is paid to such an important process of the new industrialisation as hi-tech economical sector development [3-5]. Intensive technological development is the crucial characteristic of the second half of the XX century. Due to this has allowed to the leading countries in this domain to determinate the global geopolitics' transformations. A lack of the clear objectives and priorities for the scientific, technological and innovation policy, uniform organisation of research activity, low share of hi-tech industry based on domestic and not on import technologies have not allowed the Russian industry to be adequate to technological, structural and ecological parameters, to the requirements of the new technological wave.

Recently the idea of the achievement of the global technological parity of Russia and leading technological countries became actual [6, 7]. In such conditions the Russian economy needs to make an attempt to develop a number of domestic '"horizontal" technologies playing a crucial role in the creation of new industries. Such a "technological thrust" can contribute to the formation of a new technological environment and to technological

* Corresponding author: econ@uran.ru

inequality decreasing, that is one of the aims of the industrialisation.

New industrialisation is also impossible without the emergence of indigenous companies in principally new industrial markets which currently do not exist. In the opinion of many analysts, and the author shares it, Russian companies can reserve for themselves a certain niche only in the markets which have not yet been created [8-10].

The formation of the strategic vector of the new industrialisation of the Russian economy should take into account the trends of the global economy development which will define it in the future according to the new technological wave [11, 12]. It is possible to distinguish among the most important trends those, which are connected with *the growth of the markets based on network solutions*, which leads to the profound changes in the added value chain; with the formation of *the network approach to the organisation of research*, when in the framework of large multidisciplinary centres the cooperation of research and technological organisations is developed, which are consolidated into territorial innovation systems by innovation infrastructure - centres of public access, open laboratories, etc.; with the growing role of the companies which find comprehensive and cost-effective solutions of emerging problems by the combination of the best accessible technologies and various forms of demand [13]. Other trends are not less important, among them the trends which demonstrate complicating of ethical, motivational and psychological problems due to the situation of the society which is not ready to receive number of innovations; trends defining the profound changes of competency profile in-demand in the labour market which lead to critical changes in the employment pattern; trends of the creation of competency portfolio based on the assessment of companies anticipated need which determines the creation of a new model of higher education.

3 Convergent technologies are the intellectual heart of the hi-tech economical sector

Convergent technologies traditionally include nanotechnologies, biotechnologies, information technologies, cognitive technologies. Their mutual influence as well as interpenetration is so named "NBIC-convergences". This term was coined in 2002 by M. Roco and W. Bainbridge who prepared the report "Converging Technologies for Improvise Human Performance" in World Technology Evaluation Center. The specificity of the development of these technologies consists in their interdisciplinary character, and they can largely predetermine the development level of the technological base of the domestic economy and its research intensity. However such an important factors of the global development as the creation of new forms of activity, the possibility of social reality design remind ignored. The technologies developed in this domain are called socio-humanitarian technologies. Their development is promising in number of directions, among which the creation of high technologies from the point of view of the convergence of humanities and natural-science knowledge, as well as of the design of the models of active environment for innovation development, aimed at the multiple distributed sources of innovations.

We can speak not only about NBIC-technologies, but also about NBICS-technologies due to their importance. The consideration of the regularities of the processes of the interaction between technological sphere, individual and society along with the development of the methods and means for influencing them appears to complete the particularity of the development of the new economy. The treatment of the seventh wave as a socio-humanistic one appears us to be substantiated [14]. According to the above mentioned, the convergent technologies are the technologies having a great practical potential, defining a principally new technological basis of the economy which is compliant with environment protection requirements, and being critically important for the socio-economic development of the country and for its national security.

The realization of the main statements of NBICS-technologies within the new industrialisation of the Russian economy will allow to elaborate unconventional applications of scientific research results, to actualize new principles of intersectorial technology transfer, and also their transfer from science to the real sector, to activate the methods of the control of technological platforms in the domain of high-technologies. The emergence of principally new products and services based on the technologies convergence and the implementation of the logic and objectives of their development in the scientific and technological policy of Russia, in the system of sectorial and territorial management which is common for the country and in the whole society, will permit to create new industries resulting from the convergence. The specificity of this process is the transition to the utilization of renewable energy sources, and to the energy saving high-technologies with the ecological priority at all the stages of the development and realization of high-technologies.

Despite all positive features of NBICS technologies application, the ever increasing technological threats must not be underestimated. They are due to the consequences of NBICS' high use and to constantly growing distance between the technologies and ethical and civilization values.

4 New institutions for innovation support

The innovation potential of the domestic economy is composed by the import of technologies in form of patents, know-how, license agreements, joint ventures creation. Another significant direction is the involvement of the business in the technological operations of foreign companies. It is also important to utilize the results of the Russian scientific research. The creation of companies' own scientific and technological reserve is one of the most significant ways for increasing the innovation potential of the country. However the Russian innovation potential remains insufficient, which results in a low country's rate in the global rating. At the World Economic Forum in Davos 2015 Russia received the 45th place in the global rating, and by innovation potential it is only 65th. A successful convergence of NBICS technologies appears to be able to have a significant impact on the improvement of the innovation parameters of the Russian economy.

The development of innovation strategy based on the model of the demand not only from buyers, but also from the technology sector, is likely to be efficient due to the existing structure of the Russian economy, the specific character of the scientific and technology system and of the institutional system.

Government institutions play a special role in the framework of the interacting public and private innovation organisations. This is the state policy which determine the configuration of the innovation and institutional profile of the system, or of the operating mode of business environment, motivation system of research activity, degree of the market orientation of the basic research, practical orientation of the higher education sector. However nowadays the fact of unrealized expectations from the innovation policy in Russia is generally recognized [15]. The government has showed different initiatives in the domain of innovations. In particular, there are special economic zones as the central platforms for the innovation business development. Although they are still existing, as well as the technology platforms, the state has almost lost interest to them. The initiative of the government to enforce the state corporations to innovative activity has neither met expectations. As a result, the government raises the question of redesigning the existing development institutions without the necessary analysis of their activity, determination of positive results and evident failures [16, p.45].

The "open government" proposes to turn around the management structure of the development institutions by creating a new state corporation. Nevertheless, this suggestion is arguable. The creation of Project Office affiliated to the Central Office of the Government appears to be better substantiated. The principal aim of this new institution is the coordination of the activity of the innovative organisations and elimination of duplication of their responsibilities. The expediency of the creation of an Agency of technology development is widely discussed in Russia. It's supposed to have the status of a nonprofit organisation focused on technology transfer, it is scheduled to be formed in June, 2016 [18]. The author of this concept is the business community represented by "Business Russia".

In author's opinion, the Agency of technology development can became an effective component of the national innovation system if the functions of assuring the whole chain of an efficient technology transfer will be assigned to it. This includes the collection of data on the existing domestic and foreign promising projects taking into account the specific character of industries and markets; fact-based analysis of various technological solutions for more detailed development of the long term strategy; continuously updating the information about the existing technologies and competences both domestic and foreign; systematic selection of the most effective technologies and their transfer to the private sector after the quality review of projects; legal support of economic entities; assistance for changing the structure of import technologies. It may be noted that nowadays a half of the expenses is concentrated in the engineering which means that Russia generally imports already mature technologies. The share of the purchase of patents, licenses, know-how connected with the new productions, new technologies is only 10%.

It is necessary to coordinate the research of the Agency with the studies conducted within the National technology initiative (NTI). NTI includes system solutions for the determination of key technologies taking into account the global development trends and the need to create new global markets. Three hypothesises have been formulated for the choice of these markets. They include the increasing demands of population, network principle as the basis of new markets in the future, the wishful planning of the studies in the framework of NTI future, from so-called "preferred reality" to the present [10]. The system work within NTI is organized according to the NTI Matrix. It defines 9 basic markets, determines technologies, institutions, infrastructure and resources which are necessary for these markets formation. The road maps created for each of these 9 new markets contain a concrete schedule of the realization of technology initiatives focused at the development of key markets segments. In fact, NTI suggests new institutional basis for resolving the challenges of new industrialisation.

5 Industrial policy as a tool for the hi-tech economical sector development

The new industrialisation has no alternative, and in this context the industrial policy is a universal mechanism of the development of the competitive economy with balanced structure. Thus its forms and methods have significantly changed. The industrial policy acquires a multisubject character when the corporate and public sector (Chamber of Commerce and Industry, Russian Union of Industrialists and Entrepreneurs, Mainstay of Russia and others), civil society institutions along with the state play an important role in its formation. The state plays a key role not only as being present in the economy, but as a SMART-state, or as a state which determines the aims and objectives, priorities of the development of the country. Such a priority is nowadays the creation of hi-tech economic sector, increase of general business and innovation activity. The global and domestic practice shows that this is the industrial policy which plays the crucial role in these processes [19-22].

In Russia the Federal Russian Industrial Policy Act applies since 30th of June 2015. It determines the principles, new instruments of industry support, as well as the powers of public and municipal authorities. A considerable advantage of this Act is the implementation of such novelties as the creation of the Foundation of industry development, special investment contract and others which can become a supplemental factor for the acceleration of the hi-tech production development. The agreed view of priorities of the scientific and technological development at the Federal as well as at the regional level is a necessary condition for it. Regional industrial policy acts should not contradict to the strategic aims of the Federal Russian Industrial Policy Act. They should be focused on both the national and regional tasks solution. Unfortunately, the Federal Act only defines the defence industry as a strategic priority. The legislative support of the development of other hi-tech industries including convergent technologies, should be documented at the federal level as well as at the local level, since industry policy is firstly a sectorial policy which creates preferences for selected branches [23].

6 Conclusion

Consequently, three new institutions have been created for the acceleration of the formation and development of the hi-tech sector in addition to the existing ones. There are the Project Office affiliated to the Central Office of the Government, Agency of technology development and National technology initiative. Their agreed activity, elimination of possible duplication are capable to considerably contribute to the new industrialisation of the Russian economy. The implementation of the supplementations to the Federal Russian Industrial Policy Act and regional industrial policy acts aimed at the stimulation of the hi-tech production and convergent technologies will allow to provide a gradual achievement of the technological parity with the developed countries based on the single legal basis.

The article has been supported by the Russian Foundation for Basic Research, the project N 16-06-00403 "Modelling of motivational potential of multisubject industrial policy under the conditions of new industrialisation."

References

1. S. Gubanov, *Sovereign breakthrough. Neoindustrialization Russia and vertical integration* (Moscow, 2012)
2. O.A. Romanova. Economy of region, **1**, 43-56 (2014)
3. M.A. Bendikov, *The high-tech sector of the industry trend of Russian state mechanisms of innovative development* (Nauka, Moscow, 2007)
4. Yu. Koval'chuk, I. Stepnov, Problems of the theory and practice of management, **4**, 8-17 (2013)
5. V.E. Dementev, Economics and Mathematical Methods, **49**, **4**, 33-46 (2013)
6. N. Volchkova, POISK, **8**, (2015)
7. *Conceptual Framework of the National Technology Initiative project* (RAS, 2015)
8. T. Edovina, *We expect a radical restructuring of major industries* (Merchant, 2015)
9. D. Denisov, Business Journal, (2015)
10. *National Technology Initiative* (Petersburg International Economic Forum, 2015). http://government.ru/media/files/T9Crayp8PsBQU6hdVAl0SsDlu2XvCvYG.pdf.
11. A.I. Tatarkin, O.A. Romanova, Economist, 8, 21-38 (2013)
12. A.I. Tatarkin, O.A. Romanova, N. Yu. Buhvalov, Bulletin UrFU: a series of economics and management, **3**, 13-21 (2014)
13. A. Chulok. http://www.rusventure.ru/ru/press-service/massmedia/detail.php?ID=61386.
14. V.E. Lepskij, *Reflective-active environment of innovation development* (Kogito-Centr, Moscow, 2010)
15. V.M. Polterovich, Economic science of modern Russia, **4(55)**, 17-29
16. A. Mekhanik, Expert, **6(974)**, 44-45 (2016)
17. T. Edovina, Kommersant, **12**, (2016)
18. *Transcript of the meeting of the Presidium of the Presidential Council on economic modernization and innovative development of Russia.* http://i-russia.ru/sessions/reports/.
19. D. Rodrik. http://www:project-syndicate.org/commentary/the_return-of-industrial-policy.
20. P. Bianchi, S. Labory, *From "old" industrial policy to "new" Industrial development policies* (Edward Elgar, USA, 2006)
21. T. Altenburg, *Industrial Policy in Developing Countries* (Bonn, 2011)
22. O.A. Romanova, Modern competition, **3**, 44-57 (2008)
23. D. Rodrik, *Industrial Policy for the Twenty-First Century* (Centre for Economic Policy Research, London, 2004)

Developing integrated performance assessment and forecasting the level of financial and economic enterprise stability

T.A. Khudyakova[1,*], and *A.V.* Shmidt[1]

[1]South Ural State University, Chelyabinsk, Russia

Abstract. The article deals with the problem of assessing and forecasting the level of financial and economic enterprise stability through the integrated indicators development. Currently, many enterprises operate under variable external environment, which imposes a strict requirement to consider this uncertainty. For the evaluation, analysis and prediction of the sustainability of the enterprise in the conditions of crisis we believe it possible and necessary to use the apparatus of probability theory and mathematical statistics. This problem solution will improve quantitative assessing the financial and economic stability level, forecasting possible scenarios of the enterprise development and, therefore, based on the proactive management principles and adaptation processes will greatly increase their effective functioning, as well as reduce bankruptcy probability.

1 Introduction

In order to improve the efficiency of implementing enterprise stability controlling, a set of indicators assessing the financial and economic enterprise stability dynamics, at the exit from the economic system, should be developed.

Currently, much research in forecasting enterprise activity and its stability relies on different growth rates and growth increment indicators defined as the relation of a forecasted indicator to the current indicator value. At the same time, the forecasted indicator is often defined deterministically.

However, an enterprise is known to operate and develop in a variable environment. The external environment of an economic entity bears the sweeping changes by affecting both the enterprise activity as a whole, and its resulting indicators.

Thus, forecasting an enterprise activity indicator is not exactly possible. This fact makes it necessary to review expected indicators in a probabilistic and statistical aspects.

2 Theoretical framework and literature review

In our research, the current approaches to define an integrated indicator for assessing and forecasting the financial and economic enterprise stability level were analyzed (25 research works). A lot of attention was paid to the attempts to assess enterprise stability level. This problem was studied by A. Aroshidze, A. Babich, V. Bor, A. Duraeva, O. Zaitcev, A. Ilicheva, L. Irkhina, E. Kazyuka, A. Kanunnikov, N. Homyachenkova etc. However, only a few authors study the financial and economic enterprise stability in temporary aspect and even less authors study it in the stochastic aspect.

We believe that in unstable economic situation, forecasting stability level is extremely actual. For this purpose, we propose using the probability theory for solving this applied problem. Wetzel was engaged in promoting the probability theory.

These approaches symbiosis combined with the authors' suggestions allow to increase the accuracy for forecasting the enterprise stability level, and therefore, based on adaptation and preventive measures to increase the activity effectiveness of both a separate enterprise, and national economy as a whole.

3 Materials and methods

A. Description of methods

The analysis was performed in the context of the indicators use, i.e., the information that can be obtained as a calculation result: indicators evaluation or prediction indicators. The analysis was also performed in the context of the calculation principles of integral enterprise stability indicator, these principles were formulated after a detailed study of the authors' approaches: using weighting coefficients, calculating arithmetic mean for a number of enterprise activity indicators, calculating geometric average for a number of enterprise activity indicators, probabilistic models. Besides, the type of enterprise stability: only financial, only economic or

* Corresponding author: khudiakovata@susu.ac.ru

both, allowing evaluating the integral indicator, was analyzed.

Based on the analysis of determining the integral evaluation indicators, and financial and economic enterprise stability forecasting indicators, we can draw conclusions on the current approaches to this indicator determination (see Table 1).

The study of the current approaches to determining the integral enterprise stability indicator revealed the following regularities. First, most of the authors follow the concept of stability evaluation in determining the indicator and eliminate the issues of stability level forecasting in a given period. Though some authors refer to the existing indicators dynamics, which should be considered in the space-time continuum, however, the proposed calculation models, based on simulation and mathematical methods, do not allow predicting the indicator level in the future. This approach is observed in almost 88% of the studied works. The authors of such works are: V. Bor, L. Irkhina, E. Kazyuka, E. Slabinskiy, A. Babich, P. Nefedov, D. Yunusova, B. Shchurov, M. Dmitriev, T. Saakyan, and others. Only 12% of the scientific approaches to determining the integral stability coefficient include tools for forcasting financial and economic situation at the enterprise in the context of stability for future periods. Such authors are: Yu. Suleymanova, I. Pavlova, A. Shmidt.

Table 1. The main approaches to determining the integral assessing indicators, and financial and economic enterprise stability forecasting indicators, identified on the basis of research analysis.

Factor used in determination	Number of authors, people
Integral indicator is used only to evaluate the stability level at the output of the system	22
Integral indicator is used to predict the stability level based on the temporal aspect	3
Integral indicator is determined on the basis of weighting coefficients set by an expert	15
Integral indicator is determined by calculating the arithmetic mean for a number of enterprise activity indicators	1
Integral indicator is determined by calculating the geometric average for a number of enterprise activity indicators	3
Integral indicator is determined on the basis of probabilistic models	2
Integral indicator is determined on the basis of other methods	6
Integral indicator is the indicator for only financial enterprise stability	3
Integral indicator is the indicator for only economic enterprise stability	4
Integral indicator is the indicator for financial and economic enterprise stability	18

However, in our view point, in a period of significant fluctuations of the external enterprise environment, due to the crisis, as well as sanctions against the Russian Federation, it is crucial for the managers of different levels and stakeholders to have the tools, which primarily would allow forecasting enterprise development in the dynamic environment conditions as only this instrument type will allow mitigating the negative impact of enterprise environment disturbing factors, based on proactive management. When the purpose of an entity is not just functioning, but developing – achieving new levels, a need of developing forecasting indicators increases significantly.

Scientific literature analysis in the context of the methods, proposed by the authors for calculating integral enterprise stability level indicator, revealed the following main principles of its determination:

• the calculation is based on weighting factors, determined primarily by experts. The approaches, based on the method of weighting coefficients, presented in the majority of works, the authors are: E. Nekrasova, A. Kanunnikov, M. Makova, G. Khusnullina, V. Strokov, V. Kozlov, L. Danchenok, A. Il'icheva, A. Aroshidze, O.Zaytsev, V. Bor, L. Irkhina, E. Kazyuka, Yu. Suleymanova, D. Yunusova, Yu. Perskiy, V. Lepikhin, E. Semenova, I. Pavlova. The share of this group is approximately 56%.

• based on calculating the geometric average for a number of enterprise activity indicators. This approach was used in the works by M. Makovoya, A.Il'icheva, N. N. Khomyachenkova. It accounts for about 11%.

• based on calculating arithmetic mean for a number of enterprise activity indicators. This approach is applied in the research by A. Babich. It accounts for 3.7%;

• other approaches for determining the integral indicator. This group includes works outside the group with the general trend – individual scientific approaches. However, in most of these authors' works the integrated enterprise stability indicator was formally expressed: E. Slabinskiy, A. Duraeva, E. Nekrasova.

• only two works are based on the necessity of applying probabilistic approaches in predicting the enterprise stability level (the authors are: Kozlov V.A., Danchenok L.A., Shmidt A.V.) – about 7%.

The analysis was also performed in the context of stability type, allowing evaluating and/or forcasting the integral stability indicator. We can note that in modern research, the authors tend to divide principles of financial and economic stability. However, still there are works in which enterprise stability means only financial stability, and works in which, despite the fact that in calculation methods both financial and economic indicators are used, authors consider stability in terms of one aspect: either financial stability or economic stability.

Thus, the analysis results show that 12% of scientific approaches to determining the integral enterprise stability indicator are aimed at evaluating only the financial stability (the works of A. Duraeva, T. Saakyan, I. Pavlova), also 16% – only economic stability (the works of V. Strokov, Yu. Suleymanova, B. Shchurov, M. Dmitriev, A. Shmid), the remaining 72% of works consider the enterprise stability in the financial and economic aspect (E. Nekrasova, A. Kanunnikov, M. Makovoy, G. Khusnullina, M. Makovoy, N. Katkova, E. Matushevskya, V. Kozlov, L. Danchenok, A. Il'icheva, A. Aroshidze, O. Zaytsev, Khomyachenkova A., V. Bor, L. Irkhina, E. Kazyuki, E. Slabinskiy, A. Babich, P. Nefedov, Yu. Perskiy, V. Lepikhin, E.Semenova, D. Yunusova).

B. The basic requirements for the simulation model of forecasting of financial and economic stability of the enterprise

Thus, as an analysis result of modern scientific approaches to determining the integral enterprise stability indicator, the following indicator patterns can be observed:

• first, it allows only assessing, not forecasting the entity stability level (with the exception of only 3 works);

• second, the calculation is carried out mainly on the basis of calculating geometric average for growth rates of financial and/or economic indicators;

• third, almost all proposed methods are based on principles of determinism, when input enterprise parameters are set by a certain number. In models, stochasticity is not provided, while in a dynamic environment of entity functioning, determinism is not possible (with the exception of only 2 works).

• fourth, in most cases, it can be used to evaluate the financial and economic stability level. However, in the economic stability analysis, only different sets of technical and economic indicators are taken into account. The approach to determining the stability level in the context of goal-setting is used only in one work (the author is A. Schmidt).

As an analysis result, we formulated the basic principles of constructing integral financial and economic enterprise stability indicator:

• The indicator should be able to be used for assessing and forecasting stability, which is especially important in the period of crisis phenomena in the economy.

• Assessing and forecasting entity stability must be comprehensive, and consider not only financial and economic enterprise activity, but the degree of its strategic and tactical goals achievability, i.e., the integral stability indicator should reflect the financial and economic stability. And economic stability should be considered in the context of goal-setting.

• The integral indicator should be based on probabilistic and statistical approaches, which will improve the calculations accuracy, compared to the majority of authors' methods, based on the expert assessing.

• The integral indicator should allow assessing and forecasting financial and economic enterprise stability, regardless of the input factors in a simulation model, for example, the size of enterprise profits, but not its cash flow, is used as an input parameter for the economic stability analysis, etc.

• The integral financial and economic enterprise stability indicator should consist of both – financial stability and economic stability indicators, to conduct a separate analysis in this context, which will allow correlating generalized evaluation with assessing individual entity stability aspects.

• The indicators, included in the assessment methodology of financial and economic stability, must be quantifiable.

• The use of integral financial and economic stability indicator should give possibility for implementing generalized comparative assessing various mini-economic systems activities and have relativity aspects. The integral stability indicator of various groups should have the same calculation method.

• Methods of assessing and forecasting economic and financial stability should be based on the dynamics of the model input parameters, and the resulting value of financial and economic stability level should be continuous in the whole domain of the determination.

• For interpreting the financial and economic stability level, there must be a rating scale with the characteristics of the established ranges.

Thus, the developed integral indicator for the financial and economic enterprise stability level, as well as financial stability indicator and economic stability indicator taken separately, should be considered as random variables, distributed according to the particular distribution law and other law with corresponding parameters.

C. Mathematical economic model

Analyzing the financial and economic stability level of an enterprise, defined as the product of two independent events, i.e. system's economic and financial stability, limitations, imposed on the stability condition, should be taken into account. Evaluating the economic stability should be based on the interpretation of the term, which implies that the economic entity operates smoothly if its resulting indicator reaches the goal area [20]. We assume the probability of cash flow reaching a certain area as a company's goal, which is consistent with its stability in a certain extent [21]. However, the proposed technique may use other types of indicators as a resulting economic stability indicator, such as product sales profit, gross profit, etc.

Thus, the probability of the desired economic and financial stability level growth (ΔPfs) may be generally used as the integral indicator for the enterprise's economic financial stability level, i.e. the criteria for evaluating stability controlling results. This integral indicator can be presented as follows:

$$\Delta P_{fes} = P^{**}_{fes} - P^{*}_{fes} \qquad (1)$$

where ΔP_{fes} – the probability of achieving a desired economic and financial stability level using enterprise's controlling system; P^{**}_{fes} – the probability of achieving a desired economic and financial stability level without using enterprise's controlling system; P^{*}_{fes} – the probability of achieving the economic and financial stability.

We propose determining the economic financial and stability achievement as a probabilities achievement product either for economic or financial stability, which is possible due to our interpreting the categories of "economic stability" and "financial stability", as in this case these events are independent.

$$P_{fes} = Pes \cdot P_{fs}, \quad (2)$$

where P_{es} – the probability of achieving only the economic stability; P_{fs} – the probability of achieving only the financial stability.

The probability of achieving the company's financial stability, which can be expressed, for example, using the financial stability coefficient, should take into account the limitations imposed on the coefficient. It reflects the company's dependence on the external suppliers and is calculated as a part of the equity base in the balance sheet total. We propose calculating the probability of achieving the economic stability as the probability of the resulting indicator (the company's cash flow) reaching the goal area that is determined by the top managers or stakeholders at the stage of goal setting when taking the strategic and tactical aims.

Thus, the probability of achieving the economic and financial stability can be calculated using the formula:

$$P_{fes} = P_{es}\left(CF_{min} \le CF \le CF_{max}\right) \cdot \cdot P_{fs}\left(k_{fs\,min} \le k_{fs} \le k_{fs\,max}\right), \quad (3)$$

where CF_{min} – the lower boundary of the goal area; CF_{max} – the upper boundary of the goal area; CF – the random variable of the company's cash flows; kfs – the random coefficient variable of the company's financial stability.

Using the rules for estimating the probability of a random variable determined by the normal law in a given interval [23], the probability of the financial and economic stability can be calculated using the Laplace function.

$$P_{efs} = \int_{CF_{min}}^{CF_{max}} f(CF)dCF * \int_{k_{fs\,min}}^{k_{fs\,max}} f(k_{fs})dk_{fs}. \quad (4)$$

In Russia, the financial stability safety margin is ranged from 0.5 to 0.7. When the value is less than 0.5, a company is considered to have the financial instability, and when the value is more than 0.7, it is considered to use its financial resources inefficiently. Thus, the lower limit of the second integral in the formula (4) may be changed to 0.5, and the upper – to 0.7 respectively. Although in practice, these boundaries may vary depending upon the analysis objectives.

$$P_{efs} = P_{es}\left(CF_{min} \le CF \le CF_{max}\right) \cdot \cdot P_{fs}\left(0.5 \le k_{fs} \le 0.7\right) = = \left(F\left(\frac{CF_{max} - \overline{CF}}{\sigma_{CF}}\right) - F\left(\frac{CF_{min} - \overline{CF}}{\sigma_{CF}}\right)\right) \times \times \left(F\left(\frac{0.7 - \overline{k_{fs}}}{\sigma_{k_{fs}}}\right) - F\left(\frac{0.5 - \overline{k_{fs}}}{\sigma_{k_{fs}}}\right)\right) \quad (5)$$

where F – the Laplace function; $\overline{CF}$ – the mathematical expectation of the random cash flows variable; σ_{CF} – the mean squire deviation of the random cash flows variable; $\overline{k_{fs}}$ – the mathematical expectation of the random coefficient variable for financial stability; σ_{kfs}– the mean squire deviation of the random coefficient variable for financial stability.

4 Conclusions

The study helps to solve a number of essential problems.

First, using modern approaches to developing the enterprise's stability integral indicator, the author's approach is formulated, based on the stochasticity principles.

Second, the integral indicator is proposed, which allows to assess both the financial stability level as the total solvency and the economic stability level defined in our approach as the possibility to reach the goal level of the resulting figure (e.g. a cash flow).

Third, the authors propose the model based on probabilistic and statistical approaches. In this model, both input and output parameters are subjected to stochasticity principles.

Thus, the tools produced by the study provide the effective assessment, forecasting and controlling financial and economic stability of the enterprise, which has a great significance during the world global crisis.

This publication was prepared within the framework of a research project №15-32-01051 supported by the Russian Humanitarian Scientific Fund. The work was supported by Act 211 Government of the Russian Federation, contract № 02.A03.21.0011

References

1. A. Aroshidze, Economic Sciences, **6**, **115**, 30-34 (2014)
2. A. Babich, Bulletin of the Samara State University of Economics, **7(45)**, 10-15 (2008)
3. V. Bor, *Improving stability and effectiveness management in small service businesses* (2009)
4. A. Duraeva, *Students' scientific forum*. http://www.scienceforum.ru/2014/570/4201.

5. N. Homyachenkova, Bulletin of the Tver State University. Series: Applied Mathematics, **4(37)**, 81-96 (2010)
6. A. Ilicheva, *Developing the mechanism for evaluating sustainable development of regional industrial complex* (Moscow, 2014)
7. L. Irkhina, *Methodical aspects for assessing the economic stability of industrial enterprises performance* (2007)
8. E. Kazyuka, Bulletin of ISEA, **4(60)**, 89-92 (2008)
9. A. Kanunnikov, Bulletin of OrelGIET, **1(19)**, 121-126 (2012)
10. N. Katkova, E. Matushevskaya, BIZNES INFORM, **1**, 252-257 (2015)
11. V. Kozlov, L. Danchenok, *Problems in managing the stability development and business performance, Monography* (MESI, Moscow, 2012)
12. M. Makova, Bulletin of VEGU, **4(60)**, 53-60 (2012)
13. M. Makova, G. Husnullina, *Improving economic sustainability based on the resource provision optimization* (BashGAU Publishing House, Ufa, 2007)
14. E. Nekrasova, *Developing the effective development system of the enterprise* (Publishing House of the Institute of Economics and Management of the Udmurt State University, Izhevsk, 2004)
15. P. Nefedov, *Formation of the enterprise for sustainable development mechanism* (2010)
16. J. Persky, V. Lepihin, E. Semenova, Bulletin of Perm University. Series: Economics, **1(24)**, 103-110 (2015)
17. E. Slabinsky, National Academy of Sciences of Ukraine.http://dspace.nbuv.gov.ua/bitstream/handle/123456789/4403/st_26_23.pdf?sequence=1.
18. V. Strokov, Engineering magazine: Science nd Innovations, **11**, **11**, 199-211 (2012)
19. Y. Suleymanova, Management Issues, **2(19)**, 224-229 (2012)
20. T.A. Khudyakova, *Proc. of the International Congress on Interdisciplinary Behavior and Social Science*, 292-294 (2015)
21. T.A. Khudyakova, *Proceedings of the International Congress on Interdisciplinary Behavior and Social Science*, 295-297 (2015)
22. B. Schurov, M. Dmitriev, Humanitarian, socio-economic and social sciences, (2015). http://www.online-science.ru/m/products/economi_sciense/gid2918/pg0/
23. E. Wetzel, *Probability Theory* (Nauka, Moscow, 1969)
24. D. Yunusova, *Organizational and economic mechanism to increase the economic stability of an industrial enterprise (based on the Republic Dagestan example)* (Makhachkala, 2009)
25. T.A. Khudyakova, Actual problems of economics, **12**, 444-450 (2015)
26. O. Zaitcev, Papers of Russian Scientific and Practical Congress: Problems of Economics and Statistics at the national and regional scales, (2006)

DOI: 10.1051/shsconf/20173501155

Risk reduction based on controlling technologies on industrial enterprises

T.A. Khudyakova[1,*], and *A.V.* Shmidt[1]

[1]South Ural State University, Chelyabinsk, Russia

Abstract. During the global economic instability, demand in developing toolkit which helps to mitigate the negative impact of environmental factors on efficiency of activity of the enterprise is increasing dramatically. However, in practice implementation of any of the tools causes the growth of additional costs, which, for purposes of effective functioning of mini-economic system must be covered by the profit's increase from the introduction of this toolkit. The task's implementation was achieved by processing a significant amount of scientific and statistical literature and calculation of integral indicators of the effectiveness of controlling system's implementation at the enterprises. The analysis revealed a positive effect of the implementation of controlling system on the quality of decisions, efficiency of management, productivity, investment attractiveness, effectiveness of workflow and financial stability, and calculated the average integral efficiency indicators. The obtained results allow to make an unambiguous conclusion about the effectiveness of the controlling system's implementing, which is especially important during the period of economic globalization and the growth of perturbations in the external environment, the efficiency reserves identified in the operation of the controlling system can and should be a certain buffer to reduce the negative impact of external factors on the sustainability of the enterprise.

1 Introduction

In the times of the crisis caused by the external environment fluctuations the geopolitical and socio-economic area variability increases dramatically. The disturbances that come from the outside have a significant impact on the Russian industrial cluster condition, reducing its financial and economic sustainability. Controlling system is aimed at neutralizing the impact of the disturbances on company activity as the terminal segment of economy. However, changes in the company structure and functionality will result in higher costs. In this regard, before the stage of sustainability controlling system designing the stakeholders should understand the general trends of the company efficiency rising due to the measures taken. The experience of leading Russian and foreign companies that have already been using controlling system in their activity can be useful in this case. Analysis and synthesis of data they have on the rise in the company efficiency, as well as the common trends identification can simplify the prediction phase greatly.

2 Theoretical framework and literature review

Nowadays the methodological issues in the field of controlling systems formation and development as the basis for enhancing business entity sustainability by means of preventive control, aimed at reducing the negative effects of disturbances coming from the enterprise operational environment, are not given much attention to. There are no methodological approaches to the company controlling system construction ensuring the highest possible elimination of the environmental uncertainty, aimed at improving its financial and economic sustainability. The controlling models implemented in some companies perform mostly recording, controlling and planning functions, without predicting the environmental uncertainty impact on the financial and economic performance of the company. This is due to the controlling evolution in the Russian Federation: the first controlling systems introduction was mentioned only after perestroika in Russia. Only at the turn of the XXI century the approaches to the business entity strategic management began to form and dominate the recording concept. That is why there is practically no quantitative assessment of the controlling system introduction impact on the company financial and economic activity in research papers. Only a small group of Russian researchers tried to quantify the impact of the controlling system introduction on the business entity activity efficiency, despite the fact that the positive aspect of the controlling introduction is mentioned almost in all modern scientific studies.

* Corresponding author: khudiakovata@susu.ac.ru

3 Materials and methods

3.1. Analysis of the impact of research on the introduction of the controlling system to the efficiency of the enterprise

In their article "Controlling in Russia" Falko S., Russell K. and L. Levin, considered to be the founders of the Russian controlling, noted that according to experts estimations due to introducing a controlling system in the company "it is possible to ensure a 15–30 % growth of integrated performance indicators" [1]. And the controlling introduction at the enterprises engaged in innovative developments will improve efficiency by 50–75% [1].

In their theses Syuzeva O.V. [2], as well as Yusupova S.Y. [3], referring to the Chamber of Commerce official data, noted improving in the decisions efficiency, which is calculated as the increase in profit to the increase in costs due to the controlling system introduction at industrial enterprises, stating the growth from 11% to 18.9% [2]. At the same time there is an increase in administrative staff working efficiency from 9.6% to 16% [2, 3].

Another Russian research work which deals with the effectiveness of the controlling system implementation at enterprises is the thesis by Necheuhina N.S. [4], where, despite of the fact that controlling system introduction at industrial enterprises is considered only with respect to the recording at basic production, the rise of management efficiency, the efficiency of decision-making, as well as labor productivity increase are specially noted [2]. The growth of these indicators is comparable with the values given in the previous paper.

Yusupova S.Y. in her thesis "Theory and practice of controlling system implementation in information society" draws our attention to the fact that "the use of controlling systems for the global, strategic decisions is most effective. In this case, due to a more careful study the decisions quality and their effectiveness are getting higher, resulting in a significant increase in the management performance" [3] (Table 1).

Table 1. Long-term financial decisions efficiency based on controlling measures[a]

Index	Decision quality rate of growth, %	Management efficiency rate of growth, %
Indicator values range	18.9–43.4	18.9–38.7
Mean value	34.02	30.98
Integrated Index, %	32.46	

[a] Composed by the author

Impact analysis of the controlling system introduction at foreign enterprises on the basis of research done by PricewaterhouseCoopers international company (Table 2).

Controlling system implementation efficiency at foreign industrial enterprises operating in the field of consulting and audit, has shown the trends similar to the Russian ones, although the increase in the resulting indicators of the enterprise financial and economic activity, as well as the increase in the integral index, characterizing overall indicators growth during the controlling system implementation, are less than in Russia. This trend can be explained by the fact that foreign companies have already been using the strategic management and controlling tools for a long time, as for Russia, controlling is a new development, so its first implementation has much greater output.

Table 2. Controlling system implementation efficiency at foreign industrial enterprises[b]

Company	Management efficiency rate of growth	Production process efficiency rate of growth
Sony	25.0	18.4
Samsung	24.2	18.0
LG	18.7	25.9
Akai	11.8	8.4
Kodak	11.7	31.9
Nokia	10.2	25.3
Mazda	9.9	22.9
Siemens	9.3	21.5
BP	9.1	19.6
Shell	8.6	19.0
AIG Brunswick Capital	7.2	14.9
Indicator values range[b]	7.2–25.0	8.4–31.9
Mean value[b]	13.25	20.53
Integrated Index, % [b]	16.49	

[b] Composed by the author

The analysis of the performance indicators changes at foreign industrial companies resulting from the controlling systems introduction (Table III) enables to reveal the following positive trends in their development. Firstly, introduction of controlling enabled companies to increase the investment attractiveness level by 9% on the average (maximum increase showed by IBM Company is 34% on the average [3]. Secondly, documents circulation efficiency increased almost by 10% on the average (maximum increase showed by Mazda Company is 27.6% [3]. Thirdly, it was possible to increase the financial sustainability level by more than 8% on the average (maximum increase showed by Cristall Gross Corporation is 22.5% [3].

Other official sources of information on the controlling system introduction at foreign enterprises are the materials of the RF State Duma Financial Monitoring and Control analytical department [5] (Table 3). On the basis of these documents analysis, we can come to the conclusion that due to the controlling system introduction the investment attractiveness level increases by 8.9% on the average, documents circulation efficiency – by 9.7%, financial sustainability level – by 8.3%.

Thus, the analysis of investment attractiveness, documents circulation efficiency, financial sustainability rates of growth at foreign industrial companies allows us to conclude that controlling introduction helps to improve the situation at the enterprise and raises its efficiency.

Table 3. Dynamics of foreign industrial companies indicators due to the controlling introduction

Index	Investment attractiveness rate of growth, %	Documents circulation efficiency rate of growth, %	Financial sustainability rate of growth, %
British Petroleum	+11.4	+5.2	+6,3
Tesco	+9.3	+11.6	+8.9
METRO GROUP	+7.2	+15.3	+22.1
Cristall Gross	+6.5	+11.3	+22.5
Ford	+1.2	+14.8	+6.3
Mazda	+16.8	+27.6	+9.9
Bayer AG	+3.5	+5.2	+2.2
Schwarzkopf	+6.3	+8.9	+9.3
Oriflame	+3.3	+9.3	+5.7
Motorola	+6.8	+9.3	+11.0
Chivas	+3.9	+1.6	+5.2
Huawei	+16.5	+11.9	+3.5
Ziegler	+0.2	+0.96	+0.45
Singer Corporation	+1.6	+2.9	+3.6
Renault	+13.3	+6.9	+9.2
IBM	+33.9	+11.8	+6.9
Indicator values range c	1.2-33.9	0.96–27.6	0.45–22.5
Mean value c	8.9	9.7	8.3
Integrated Index, % c	8.95		

c. Composed by the author

And, despite the fact that the rates of growth vary in different enterprises, the trend of changes in 16 companies analyzed is positive. Therefore, conclusions are drawn that the controlling system introduction at industrial enterprises can improve business entity sustainability, which is especially important in times of crisis.

In the paper written by Borgardt E.A. and Nosova V.M., dealing with the controlling system development at OJSC "KAMAZ", the following results of the introduction are noted, "the administrative staff productivity has increased by 9.6%, decision-making effectiveness has increased by 11% on the average, the long-term financial decisions effect due to the controlling measures is almost 20%" [6].

One more example of the successful controlling system implementation is the "AT & T Canada" company – the largest mobile operator in Canada that implemented the weighted indicators system in the company (Balanced Scorecard (BSC)). Due to the introduction the following indicators dynamics has been achieved: sales income has increased by 15%, output per one worker has increased by 11%, and the company market value has become 4 times higher [7].

3.2. Basic results

Having analyzed the research on quantitative assessment of the controlling system implementation efficiency at the company, we have come to the following conclusions.

First, all researchers have noted the positive dynamics of the enterprise financial and economic indicators as well as performance indices due to the controlling system implementation, despite the fact that controlling in the majority of papers is considered with limited functionality by performing control and recording functions or just the function of budgeting.

Second, the introduction of controlling gives from 8.7% to 18.9% increase of the decision efficiency, or 13.6% on the average.

Third, there is also an increase in administrative staff working efficiency from 9.6% to 17.2%, or 13.4% on the average.

Fourth, management efficiency rises from 7.2% to 38.7%, which is 18.5% on the average for the group of industrial enterprises.

Fifth, the decisions quality also increases. This increase is 34.0% on the average, ranging from 18.9% to 43.4%. in different companies analyzed.

Sixth, introduction of a controlling system at the industrial plant causes production efficiency increase from 8.4% to 31.9%, 20.5% on the average.

Seventh, the use of the controlling system even with limited functionality has a positive effect on the overall level of the company financial sustainability. An average increase of this type of indicator is 8.3%, ranging from 0.45% to 22.5%.

On the basis of results generalization the integrated index of controlling system introduction efficiency at industrial enterprise has been calculated, which is 16.42%.

4 Conclusion

The results obtained allow us to draw a definite conclusion about controlling system implementation efficiency for the enterprise, which is especially important because of economy globalization and disturbances growth in the environment. Efficiency reserves revealed during the controlling system operation should be used as a specific buffer to reduce the negative influence of environmental factors on the stability of the enterprise.

In addition, we believe that an integrated approach to the controlling system implementation in the context of both financial and economic aspects, the use of extended functionality, combining not only the control functions, recording and budgeting, but also the function of preventive management in a variable environment, based on the evidence-based simulation models will significantly increase the rate of controlling system introduction efficiency at industrial enterprise, and it will also reduce the financial and economic sustainability loss probability of mini-economic system.

The analysis of research containing quantitative assessment of the controlling system implementation effectiveness at the company brings us to the following conclusions. Firstly, all researchers stress the positive dynamics of the enterprise financial and economic indicators owing to the controlling system implementation.

This publication was prepared within the framework of a research project №15-32-01051 supported by the Russian Humanitarian Science Foundation. The work was supported by Act 211 Government of the Russian Federation, contract № 02.A03.21.0011

References

1. S.Ya. Yusupova, *Theory and practice of implementing controlling system in the information society* (Moscow, 2008)
2. N.S. Necheukhina, *Accounting events modeling within the system of controlling in industrial enterprises* (Izhevsk, 2010)
3. S. Falco, K. Russell, L. Levin, http://www.cfin.ru/management/controlling/controlling.shtml.
4. O.V. Suzeva, *Organisational and methodical support of controlling formation and development in industrial organizations* (Moscow, 2010)
5. The information bulletin, **5**, 6-9 (2006)
6. The Pricewaterhouse Cooper's bulletin, **2** (2002)
7. D.A. Yunusova, *Organizational and economic mechanism of increasing economic stability of industrial enterprise (on the Republic Dagestan example)* (2009)
8. Materials of Analytical Department of the State Duma of the Russian Federation for control and financial monitoring, 23-69 (2004)
9. A.N. Asaul, M.K. Starovoytov, R.A. Faltinskiy, *Cost Management in construction* (St. Petersburg, 2009)
10. E.A. Borgardt, V.M. Nosova, Bulletin of VSU. Series: Economics and management, **1**, 108-119 (2014)
11. K.A. Sakhovskaya, N.A. Tsaplina, Modern trends in economics and management, **24**, 54-58 (2014)
12. T. Hubert, S.G. Falco, Controlling, **55**, 3-9 (2015)
13. A. Aroshidze, Economic Sciences, **6**, **115**, 30-34 (2014)
14. A. Babich, Bulletin of the Samara State University of Economics, **7(45)**, 10-15 (2008)
15. V. Bor, *Improving stability and effectiveness management in small service businesses* (2009)
16. http://www.scienceforum.ru/2014/570/4201.
17. N. Homyachenkova, Bulletin of the Tver State University. Series: Applied Mathematics, **4(37)**, 81-96 (2010)
18. A. Ilicheva, *Developing the mechanism for evaluating sustainable development of regional industrial complex* (Moscow, 2014)
19. L. Irkhina, *Methodical aspects for assessing the economic stability of industrial enterprises performance* (2007)
20. E. Kazyuka, Bulletin of ISEA, **4(60)**, 89-92 (2008)
21. A. Kanunnikov, Bulletin of OrelGIET, **1(19)**, 121-126 (2012)
22. N. Katkova, E. Matushevskaya, BIZNES INFORM, **1**, 252-257 (2015)
23. V. Kozlov, L. Danchenok, *Problems in managing the stability development and business performance*, (MESI, Moscow, 2012)
24. M. Makova, Bulletin of VEGU, **4(60)**, 53-60 (2012)
25. M. Makova, G. Husnullina, *Improving economic sustainability based on the resource provision optimization*, (BashGAU Publishing House, Ufa, 2007)
26. E. Nekrasova, *Developing the effective development system of the enterprise* (Izhevsk, 2004)
27. P. Nefedov, *Formation of the enterprise for sustainable development mechanism* (2010)
28. J. Persky, V. Lepihin, E. Semenova, Bulletin of Perm University. Series: Economics, **1(24)**, 103-110 (2015)
29. E. Slabinsky, National Academy of Sciences of Ukraine.http://dspace.nbuv.gov.ua/bitstream/handle/123456789/4403/st_26_23.pdf?sequence=1.
30. V. Strokov, Engineering magazine: Science and Innovations, **11**, **11**, 199-211 (2012)
31. Y. Suleymanova, Management Issues, **2(19)**, 224-229 (2012)
32. T.A. Khudyakova, A.V. Shmidt, *Proc. of The 26th International Business Information Management Association Conference*, 1626-1634 (2015)
33. T.A. Khudyakova, Actual problems of economics, **12**, 444-450 (2015)
34. B. Schurov, M. Dmitriev, Humanitarian, socio-economic and social sciences, (2015). http://www.online-science.ru/m/products/economi_sciense/gid2918/pg0/
35. E. Wetzel, *Probability Theory* (Nauka, Moscow, 1969)
36. D. Yunusova, *Organizational and economic mechanism to increase the economic stability of an industrial enterprise (based on the Republic Dagestan example)* (Makhachkala, 2009)
37. O. Zaitcev, *Proc. Papers of Russian Scientific and Practical Congress: Problems of Economics and Statistics at the national and regional scales* (2006)

Development of industry in Far East of Russia based on use of innovative investment attraction practice

Denis B. Solovev[1,*], and *Alena* E. Merkusheva[2]

[1]Far Eastern Federal University (FEFU), Engineering school, 690018 Russky Island, Russia
[2]Student in Far Eastern Federal University (FEFU), Engineering school, 690018 Russky Island, Russia

Abstract. The article considers the issues relevant at the present stage of economic development of the Far East of the Russian Federation. One of instruments of this economic development of the region is forming the territories of the advancing development which can be considered as one of the possible ways of interaction of industrial enterprises of the subjects of the Far Eastern Federal District with partners from the Asia-Pacific states. The features of the territories of the advancing development, the criteria for classification of the subjects of the Far East as the territories of the advancing development, and the main preferences for the residents of these territories are noted. The algorithms for the estimation of prospects of the activities of the territories of the advancing development are also given in the article. Based on the conducted theoretical researches, the criteria for evaluation of the efficiency of activities of the territories of the advancing development are determined. Based on the analysis of the criteria, a mathematical model for the estimation of efficiency of advancing development territories activities was developed, which clearly demonstrated that in the current economic environment the optimum quantity of the territories of the advancing development in the Far East, securing the maximum performance indicator of their activities, is 8.

1 Ways of innovative economic development of the Far East of Russia

The Asia-Pacific Region (APR) is one of the most dynamically developing regions of the world [1]. The Russian Federation, owing to the geographical location, also treats this region. In connection with the difficult geoeconomic country situation, which developed in 2014, activization and building-up of a cooperation with Asia-Pacific countries is profitable and natural to Russia.

The Far Eastern Federal District, considering its geographical provisions, becomes the leading platform of interaction of the Russian Federation with Asia-Pacific countries. It should be noted the developed unique situation: The Far East takes only the 6th place from 8 possible, proceeding from analysis results of the main socio-economic indexes [2]. In addition to lagging of the region on the main socio-economic indexes from other federal districts of Russia, it has a small indicator of an index of innovative development of the federal district, in comparison with other federal districts [3]. The index of innovative development of territorial subjects of the Russian Federation measures achievements of regions in the field of development, implementations and development of innovations; includes indicators of such primary groups as social and economic conditions of innovative activities, scientific and technical potential, innovative activities, quality of innovative policy [3].

For strengthening of innovative development of the Far East, the Government of the Russian Federation together with the Ministry of the Russian Federation of development of the Far East (the Ministry for the Development of the Russian Far East) with the assistance of scientists was developed and adopted the Federal law "About the Territories of the Advancing Social and Economic Development in the Russian Federation" [4]. This Federal law provides requirements to creation of the Territories of the Advancing Development (TAD), financing of creation of infrastructure, the existence termination, management, features of regulation of the separate relations in the territories of the advancing development [5].

During project development of the Law, the Ministry for the Development of the Russian Far East analysed international experience of creation of TORAHS in Asia-Pacific countries based on which were marked out the main criteria by which the separate territories of territorial subjects of the Russian Federation located in the Far East can be carried to the territories of the advancing development.

The main criterion is demand of the investor for placement of investment projects in the territory of the advancing development, to additional criteria refer:

1) location;

2) technical characteristics of the parcels of land;

* Corresponding Author: solovev.db@dvfu.ru

3) availability (proximity) of transport and engineering infrastructure;

4) characteristics of engineering networks and objects of engineering infrastructure;

5) cost of creation of transport and engineering infrastructure;

6) availability or proximity of a manpower;

7) rights to the parcels of land and restriction of the rights.

In addition, the ministry offered the model of the competitive territory of the advancing development including three spheres and phases of development of TAD, namely the kernel is tens hundreds of hectares, the environment consisting of tens of square kilometers and the ecosystem representing completely developed and the functioning territory of the advancing development.

The structure and the constituting TAD developed model not strongly differ from models of the territories of the advancing development of other countries. The main differences consist in conditions, which provide the countries to residents of the territories of the advancing development. The logic is that TAD in the Russian Federation shall provide to residents the best conditions of implementation of activities, within the borders. To achieve the set result conditions of activities of foreign TAD [6-22] were analysed, the received analysis results allowed to constitute table 1.

Table 1. Conditions of implementation of activities of the foreign territories of the advancing development.

Country	Conditions of implementation of activities of the advancing development			
	Income tax	Wage tax	Receipt of the construction license	Connection to system of power supply
Republic of Korea	10%	8,8%	29 days	28 days
USA	33%	7,7%	27 days	68 days
China	15%	26,7%	67 days	41 days
Japan	26,4%	25,6%	193 days	105 days

Having considered conditions of implementation of activities within TAD of Asia-Pacific countries, it is possible to see: in case of development of preference for residents the ministry considered international experience of creation and implementation of activities by the territories of the advancing development and offered the most attractive conditions which consist in the following [6]:

1) 0% the income tax, on property, on the land for the first 5 years;

2) 0% import and export customs duties;

3) 0% of the VAT for import for conversion;

4) 7,6% of insurance premiums instead of 30% for the investor the first 10 years;

5) receipt of the land and ready infrastructure is free;

6) the accelerated order of a VAT return of the exporter;

7) there are no checks without consent of the Ministry for the Development of the Russian Far East of the Russian Federation;

8) "one window" for the investor;

9) free customs zone;

10) the simplified state control;

11) the accelerated and facilitated ministerial procedures, including receipt of the construction license, passing of customs.

2 The investment offers created in the Far East for foreign investors

The foreign enterprises interested in development of the business outside the states can become interested in the following territories of the advancing development of the Far East provided in table 2.

Table 2. The territories of the advancing development of the Far Eastern Federal District (FEDO).

	Name and subject of TAD	Specialization of TAD
1	TAD "Nadezhdinskaya" Primorsky Krai	Light and food industry, transport and logistic orientation
2	TAD "Komsomolsk" Khabarovsk Krai	Industrial orientation
3	TAD "Khabarovsk" Khabarovsk Krai	Industrial, transport and logistic orientation
4	TAD "Beringovsky" Chukotka Autonomous Area	Mining industry
5	TAD "Kangalassa" Republic of Sakha (Yakutia)	Industrial park
6	TAD "Mikhaylovsky" Primorsky Krai	Livestock production, crop production, production of food
7	TAD "Kamchatka" Kamchatka Krai	Tourist and recreational, port and industrial, agro-industrial orientation
8	TAD "Belogorsk" Amur region	Agro-industrial orientation
9	TAD Priamurye" Amur region	Industrial, transport and logistic orientation

It should be noted that around each of the territories of the advancing development, the area of an economic impact in 1500 km [23] was designated and the quantity the population, entering it is calculated. Population of the FEFD for January 1, 2015 is given in the table 3 [2].

Table 3. Population of the Far Eastern Federal District.

The Far Eastern Federal District	All population, man.
Republic of Sakha (Yakutia)	956896
Kamchatka Krai	317269
Primorsky Krai,	1933308
Khabarovsk Krai	1338305
Amur region	809873
Magadan region	148071
Sakhalin region	488391
Jewish Autonomous Oblast	168368
Chukotka Autonomous Area	50540
Total	**6211021**

Based on it, it is possible to draw a conclusion that products and services TAD, which are located in the territory of Primorsky Krai, Khabarovsk Krai, and the Republic of Sakha (Yakutia) will be in the greatest demand. TAD, being in the territory of these subjects, will not be competitors each other as each of them has the specialization influencing products, which are turning out them or services. Here it is necessary to pay attention to proximity of these territories to foreign countries, such neighborhood increases the possible product demand and services issued TAD.

Having chosen one territory of the advancing development by means of developed preferences for their residents, foreign investors including from the Asian countries, can receive certain benefits. For example, in connection with the developed economic situation in the Russian Federation when it is focused on import substitution, competition level considerably decreases; it is profitable to the companies of the Asian countries created within TAD and gives the chance to position the company as the company of the Far East of the Russian Federation. Conditions and selection process of projects and the companies for the territories of the advancing development it is in details stated in the Federal law "About the Territories of the Advancing Social and Economic Development of the Russian Federation" [4], and also on the official site of the Ministry for the Development of the Russian Far East [24], in appropriate section with the list of necessary documentation for assignment of a resident status. Rules and process of assignment of a resident status of TAD are accurately regulated, where as selection criteria of projects of creation of TAD have more flexible nature, and depending on interest of the state in these projects, can change.

Should not forget also about benefits for the Russian Federation, which bear in themselves the territories of the advancing development. First, this attraction of financial resources on development of the Far East of the Russian Federation; secondly, creation of the territories of the advancing development will provide a large number of workplaces for employment of the population that in turn, will give help in fight against unemployment of the population.

3 Technique of an efficiency evaluation of activities of investment offers in the Far East of the Russian Federation

It is quite difficult to estimate at the moment efficiency of activities of the territories of the advancing development for the reason that only 3 TAD, from planned to 9, are on implementation phase. However, on the basis of the analysis of the projects realized in the Far Eastern Federal District, it is possible to mark out three main criteria by which this assessment is possible:

1) the number of residents of TAD as inflow of financial resources on development and functioning of the Far East depends on the number of the entities which became residents;

2) the proximity of TAD to potential consumers, than is closer this territory to densely populated regions to the developed infrastructure, those the high probability of the product demand and service which are made in the territory of the advancing development;

3) availability of the developed and approved plan of creation and development of the territories of the advancing development, as the main regulating document regulating activities of a certain territory of the advancing development.

Based on the obtained data from the stated above criteria the model of efficiency of activities of the territories of the advancing development provided in the figure 1 was constructed.

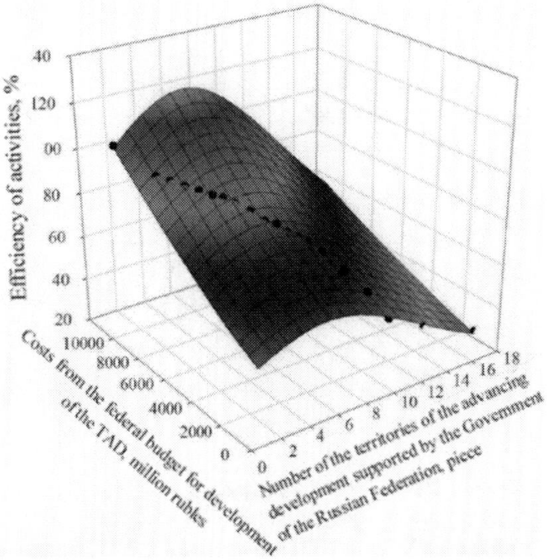

Fig. 1. Model of efficiency of activities of the territory of the advancing development. Three-dimensional regression on Lorenzyan's formula.

As a result of the removed mathematical model of efficiency of the activities of the territories of the advancing development provided in the figure 1, authors of article revealed – optimum quantity of TORAHS in the Far East in case of which the maximum performance indicator of their activities is reached, 8 equals.

The result received during the analysis is connected with the fact that on average for development of one TAD from the federal budget by the state spends about 10 billion rubles. The amount of the allocated funds has under itself the following reasons: many kernels of the territories of the advancing development have the destroyed infrastructure which recovery requires big costs from the state that significantly influences quantity of the opened TAD. Achievement of a high rate of efficiency of activities of TAD requires reducing quantity of costs from the federal budget for their development, by attraction of bigger amount of private financing. However, increase in a share of private investment in TAD is complicated by the fact that there is a lack of the number of residents of the territories of the advancing development. However in case of observance of an optimum ratio of the state and private investments into the territories of the advancing development growth of a performance indicator of their activities to 120% and more is possible.

The analysis of the received dependences shows that to increase in costs from the federal budget for development of the territories of the advancing development there is a reduction of number of the TAD supported by the government. Owing to what, it is possible to determine necessary costs from the federal budget for development of TAD by the received dependences, depending on their quantity, proceeding from receipt of maximum efficiency of activities by them. Values of costs from the federal budget for development of the TAD, depending on quantity supported by the government of TAD are approximated in expressions:

$$E_{4\text{-}6}=\frac{513}{\left[1+\left(\frac{N-4,8}{0,65}\right)^2\right]\cdot\left[1+\left(\frac{C-4043,8}{2316,31}\right)^2\right]},\delta=7\%;$$

$$E_8=\frac{549}{\left[1+\left(\frac{N-122}{25,3}\right)^2\right]\cdot\left[1+\left(\frac{C-5520}{5985,4}\right)^2\right]},\ \delta=6\%;$$

$$E_{10\text{-}13}=\frac{1129}{\left[1+\left(\frac{N-5,822}{0,89}\right)^2\right]\cdot\left[1+\left(\frac{C-11935}{8328}\right)^2\right]},$$

$$\delta=5\%;$$

$$E_{20}=\frac{1575}{\left[1+\left(\frac{N-5,157}{1,13}\right)^2\right]\cdot\left[1+\left(\frac{C-16203,6}{13274,5}\right)^2\right]},\ \delta=1\%.$$

E_i – efficiency of activities i - the territories of the advancing development, %; N – number of the territories of the advancing development supported by the government, piece; C – costs from the federal budget for development of the territories of the advancing development, million rubles; δ - a relative error between the calculated values and dependences received as a result of an experiment on the basis of the analysis of the international experience of implementation of the territories of the advancing development [25-29].

For establishment of statistical certainty of the received regression equations Fischer's criterion was used. Use of this criterion assumes calculation settlement the statistician and their comparison with analytical values. The received regression equations are considered as values if calculated values the statistician of criterion of Fischer is less than corresponding values received in practice.

When checking the equations, in all cases calculated values turned out slightly less analytical. Therefore, these equations can be considered statistically significant, and they can be used for determination of efficiency of activities of the territories of the advancing development.

Besides, for the analysis of expressions relative errors of δ between calculated values, determined with use of the received regression equations and the dependences received in the analytical way were calculated. The maximum value of the calculated relative errors does not exceed 7% that speaks about rather high precision of the received dependences.

Conclusion

1. Results of the conducted research showed that development of the Far East including innovative, goes unevenly, is not present single "engine" which would be aimed at the complex development of the region.

2. One of the ways of complex development of the region initiated by the state is creation of the territories advancing development. In the territories of the advancing development based on the large enterprises and associations interested in receipt of the status of residents, the Government of the Russian Federation sees "engine" of the region.

3. Residents of TAD shall receive the foreign investments necessary for successful implementation of their activities; the developed economic situation in the Pacific Rim will only promote this process.

4. Participants of a working group on forming of strategy of innovative development are sure of prospects of development of the territories of the advancing development, that conditions of implementation of activities on TAD similar to the Russian conditions do not exist.

5. Russia wins first place on the area among the countries of the world, uniform development of regions of which requires fixed costs from the federal budget. The developed mathematical model shows at what moment of costs from the federal budget for development of TAD can be reduced or suspended.

6. For assessment efficiency of activities of the territories of the advancing development, in FEFU the mathematical model that showed was developed that the optimum quantity of the territories of the advancing development equals the 8th, their bigger quantity will reduce efficiency. In addition, it revealed the need for development of other mathematical models directed to reasons for number of the entities necessary for a program implementation of development of the region. The mathematical model of efficiency of activities of TAD was developed for unique conditions of the Russian Federation.

Successful implementation of the concept of creation of the territories of the advancing development consists in carrying out the profound analysis of foreign experience which results could be used during creation of TAD of the Far East, taking into account special conditions of the region.

7. Analyzing the international experience of creation of the territories of the advancing development, it is possible to tell: the model of the territory of the advancing development created in Russia on the basis of the Federal law "About the Territories of the Advancing Social and Economic Development in the Russian Federation" and taking into account the revealed criteria and features will provide competitive conditions of investment that will favorably affect the investment climate of the country, and the preferential mode of activities, will promote effective use of limited resources and rapid economic growth.

References

1. About results of presidency of Russia in APEC. Institute of world economy and international relations of name E.M. Primakova of the Russian Academy of Sciences. http://www.imemo.ru/files/File/ru/conf /2013/30012013_KADtez.pdf
2. Federal State Statistics Service. http://www.gks.ru/
3. G.I. Abdrakhmanova, L.M. Gokhberg, K.A. Ditkovsky, E.A. Islankina, Rating of innovative development of subjects of the Russian Federation, **4** (2016)
4. About the territories of the advancing social and economic development in the Russian Federation: the federal law of the Russian Federation of December 29, 2014, no. 473-FL (2014)
5. Putin: the cooperation with Asia-Pacific countries is priority for Russia. Politikus News agency. http://politikus.ru/events/ 35195-putin-sotrudnichestvo-so-stranami-atr-prioritetno-dlya-rossii.html
6. About results of activities of the Ministry of the Russian Federation of development of the Far East in 2013 and the first half of 2014. The Ministry of the Russian Federation of development of the Far East. http://minvostokrazvitia.ru/upload/kollegiya-DV.pdf
7. G. Dominguez, Deutsche Welle (2015)
8. L. Rubini, M. Di Tommaso, E. Barbieri, International Encyclopedia of the Social & Behavioral Sciences, **2**, 207 (2015)
9. Y. Li, Z. Zhang, Social and Behavioral Sciences, **96**, 182 (2013)
10. Z. Min, M. Lijun, Z. Bo, Y. Zhihong, Journal of Corporate Finance, **36**, 15 (2016)
11. L. Qing, L. Yi, Journal of International Economics, **97**, 392 (2015)
12. T. Fishman, H. Schandl, H. Tanikawa, Ecological Economics, **113**, 76 (2015)
13. W. Yu, Social Science Research, **39(6)**, 1088 (2010)
14. J. DeBacker, B. Heim, A. Tran, Journal of Financial Economics, **117(1)**, 122 (2015)
15. G. Squires, S. Hall, Land Use Policy, **33**, 81 (2013)
16. J. Ham, C. Swenson, Journal of Public Economics, **95**, 779 (2011)
17. L. Reynolds, S. Rohlin, Journal of Urban Economics, **88**, 1 (2015)
18. B. Park, Political Geography, **24(7)**, 850 (2005)
19. H. Park, E. Rene, S. Choi, A. Chiu, Journal of Environmental Management, **87**, 1 (2008)
20. J. Kim, Journal of Asian Economics, **16(6)**, 973 (2005)
21. K. Kim, E. Kim, Habitat International, **29(2)**, 183 (2005)
22. Y. Lim, Journal of Banking & Finance, **35(2)**, 456 (2011)
23. About approval of the Concept of investment development Tomsk till 2025 (with the forecast till 2030). Administration of the region of the Tomsk region.http://www.investintomsk.com/files/doc/rato _ot_29_12_2014_957-ra_koncepciya_invest_razvitiya_ to.pdf
24. The territories of the advancing social and economic development. http://minvostokrazvitia.ru/
25. J. Wang, Journal of Development Economics, **101(2)**, 133 (2013)
26. M. Levien, Development and Change, **44(2)**, 381 (2013)
27. C. Long, X. Zhang, Journal of International Economics, **84(12)**, 112 (2011)
28. M. Levien, Journal of Peasant Studies, **39(3-4)**, 933 (2012)
29. D.B. Solovev, International Journal of Electrical Power and Energy Systems, **73**, 107 (2015). http://dx.doi.org/ 10.1016/j.ijepes.2015.04.011.

AUTHOR INDEX

AUTHOR INDEX

AUTHOR INDEX